AF411677

THE UNITY
OF
EVOLUTIONARY BIOLOGY

THE UNITY OF EVOLUTIONARY BIOLOGY

Proceedings of the
Fourth International Congress
of
Systematic and Evolutionary Biology

VOLUME II

*University of Maryland
College Park, USA
July 1990
Co-hosted by the Smithsonian Institution*

Edited by
Elizabeth C. Dudley

DIOSCORIDES PRESS
Theodore R. Dudley, Ph.D., General Editor
Portland, Oregon

© 1991 by Dioscorides Press
(an imprint of Timber Press, Inc.)
All rights reserved

ISBN 0-931146-19-4 (two-volume set)
Printed in Hong Kong

DIOSCORIDES PRESS
9999 S.W. Wilshire, Suite 124
Portland, Oregon 97225

Library of Congress Cataloging-in-Publication Data

International Congress of Systematic and Evolutionary Biology (4th :
  1990 : University of Maryland)
   The unity of evolutionary biology : proceedings of the Fourth
International Congress of Systematic and Evolutionary Biology,
University of Maryland, College Park, USA, July 1990 / co-hosted by
the Smithsonian Institution ; edited by Elizabeth C. Dudley.
       p.   cm. -- (Ecology, phytogeography & physiology series ; v.
3)
   Includes bibliographical references and index.
   ISBN 0-931146-19-4
   1. Evolution--Congresses. 2. Biological diversity--Congresses.
I. Dudley, Elizabeth Corning. II. Smithsonian Institution.
III. Title. IV. Series.
QH359.I58 1990
575--dc20   90-27811
                                                      CIP

Contents

DISCUSSION GROUP

DEVELOPMENTAL PROCESS AND EVOLUTIONARY CHANGE—SYMPOSIUM

DISCUSSION GROUP

VOLUME II

POPULATION AND COMMUNITY EVOLUTION

EVOLUTION AND ECOLOGY OF SMALL POPULATIONS—SYMPOSIUM

SYMBIOSIS AND COEVOLUTION

GENETIC PROCESSES IN EVOLUTION

CELLULAR AND MOLECULAR LEVELS OF EVOLUTION

PLENARY ADDRESS

ORIGIN AND EVOLUTION OF MITOCHONDRIAL AND PLASTID GENOMES—
SYMPOSIUM

NATURAL SELECTION IN MOLECULAR EVOLUTION—SYMPOSIUM

INDEXES

POPULATION AND COMMUNITY EVOLUTION

The Effects of Bottlenecks on Genetic Variation, Fitness, and Quantitative Traits in the Housefly

Edwin H. Bryant and Lisa M. Meffert

Abstract. The major concern of conservation programs for endangered species has been to minimize the effects of inbreeding and maximize retention of genetic variation. However, little is known about the dynamics of inbreeding depression, genetic variation, and fitness of complex genetic traits in finite populations. We provide empirical evidence on this issue by following changes in genetic variation and fitness in experimental housefly populations subjected to five serial founder-flush cycles. Losses in electrophoretic variation over the five bottlenecks largely followed neutral expectation; however, quantitative genetic variation of morphometric traits increased after the first bottleneck and then declined only to levels of the control by the end of the five founder-flush cycles. Components of fitness for the bottleneck populations initially declined and then rebounded to the level of the control by the end of the experiment, at which point there was little relationship between fitness, genetic variation, and bottleneck size. This was presumably due to sufficient genetic variation remaining in the bottleneck lines for natural selection to restore the loss in fitness (inbreeding depression) evident early in the experiment. The results demonstrate that severely bottlenecked populations may retain the levels of fitness and genetic variation of a large outbred population. Conservation models based solely on neutral traits affected by purely additive genetic variation may be poor estimators of the actual changes in quantitative traits affecting fitness in populations that have undergone constriction in size.

A central goal of conservation biology is to preserve evolutionarily viable populations of species whose numbers are diminishing to apparently critical levels. Normally outbred species that undergo population bottlenecks are expected to lose genetic variation and reproductive capability, presumably due to increased homozygosity of recessive deleterious alleles (i.e., inbreeding depression). To avoid this possibility, management programs have focused on direct manipulation of effective population size in order to minimize loss of genetic variation (e.g., Seal & Foose, 1983; Ralls & Ballou, 1983, 1986; Lande, 1988; Simberloff, 1988). The underlying assumption behind these programs is that retention of genetic variation will increase both the short term survival and the long term evolutionary capability (adaptability) of the managed species (Soulé & Wilcox, 1980; Senner, 1980; Frankel & Soulé, 1981; Foose, 1982; Ralls & Ballou, 1983; Shoenwald-Cox et al., 1983; Soulé et al., 1986; Soulé & Simberloff, 1986).

Although it is well known that severe inbreeding in normally outcrossed species can lower fitness by affecting many reproductive traits (see review by Charlesworth & Charlesworth, 1987), the long-term effects of small population size on fitness are less well known. Frankham (1980) found, for example, that severe bottlenecks reduced short- and long-term responses to artificial selection in *Drosophila*, but less intense inbreeding may

---

Drs. Bryant and Meffert are with the Department of Biology, University of Houston, Houston, TX 77204, USA.

592

create little long term reduction in fitness since selection can eliminate recessive alleles as they become homozygous (Wright, 1977; Lande, 1988). In the long term there may not be a clear relationship between the coefficient of inbreeding, F, calculated from population structure or known pedigrees, and average reproductive capability. Little correspondence between level of inbreeding and reproductive capability was found, for example, in standardbred horses (Cothran et al., 1986) or in an Australian parrot (Daniell & Murray, 1986). The successful establishment of inbred laboratory lines of many organisms is testimony that highly homozygous lines can be fit, although in most cases the number of unsuccessful lines is unknown.

Electrophoresis is commonly used for assessing genetic variation in wild or captive populations (e.g., Lande & Barrowclough, 1987). Electrophoretic heterozygosity has been found to be correlated with a variety of fitness measures in natural populations of many species (e.g., see review by Allendorf & Leary, 1986). However, low electrophoretic variation does not necessarily mean that the population is devoid of genetic variation for adaptively important traits, nor that it is lower in fitness (Hedrick et al., 1986; Lande, 1988). Several examples in which natural populations have persevered despite extremely low levels of electrophoretic variability include elephant seals (Bonnell & Selander, 1974), polar bears (Allendorf et al., 1979), cheetahs (O'Brien et. al., 1987), and sika deer (Feldhammer et al., 1978). While electrophoretic variation may decrease in response to a population bottleneck (McCommas & Bryant, 1990), additive genetic variance for polygenic traits can increase (Bryant et al., 1986; Lopez-Fanjul & Villaverde, 1989). Most adaptively important traits are not only polygenic but may also be influenced by nonadditive as well as additive gene processes; how genetic variation for more complex traits may respond to reduced population size may be quite different than predicted by the simple additive neutral models currently used in managing captive populations (Foose, 1980, 1982; Chesser et al., 1980; Bryant et al., 1986; Goodnight, 1987, 1988). Consequently, it may be premature to base plans for conserving genetic variation primarily on simple genetic models for additive gene action that may not apply directly to adaptively important traits (e.g., Soulé et al., 1986).

As yet there are few experimental studies demonstrating a close relationship between reproductive success and genetic variation, particularly with regard to genetic variation for polygenic traits in bottlenecked populations. During the past several years we have been studying changes in genetic variation and fitness of the housefly, *Musca domestica*, in relation to experimental bottlenecks. We summarize our results here and discuss their implications with respect to conservation biology. Specifically, we will explore the associations of bottleneck severity and recurrence with electrophoretic and polygenic variation and with components of fitness among experimental lines of the housefly. These results provide information as to how genetic variation and fitness may change in response to a bottleneck and may be useful for assessing the effect of current management plans on wild and captive species.

MATERIALS AND METHODS

The details of experimental protocol can be found in Bryant et al. (1986; 1990). McCommas & Bryant (1990), or Bryant & Meffert (1991), and only a summary is provided here. An outbred ancestral population was established by a single mass collection of flies from a refuse site and was maintained as an outbred control throughout the experiment. In the second generation, bottleneck lines were initiated by the offspring from a single pair, from four pairs, or from 16 pairs of flies. When the populations had flushed to normal size (about 1000 flies in 5–7 generations), they were assayed for additive genetic variation by offspring-parental covariances for eight morphometric traits on parents (mid-parental values) and female offspring (averaged over three daughters) reared under standard conditions of 80 eggs per 18 g of CSMA larval medium (Bryant, 1969). All measurements were taken with an ocular micrometer and converted to natural logarithms before analysis; these traits were (see Bryant et al., 1986, for exact descriptions of the traits): wing length, wing width, head width, inner-eye separation, thoracic suture length, scutellum length, scutellum width, and metafemur length.

Estimates of additive genetic variation and heritability were obtained separately for the replicate lines and the control after each bottleneck period using approximately 40 families per line, each family consisting of two parents and three daughters. An additive genetic covariance matrix for a bottleneck size class was obtained by a weighted (by number of families) average of the separate covariance matrices for each replicate line. Our results are summarized here as additive genetic variances and narrow-sense heritabilities averaged over the eight morphometric traits to obtain an overall response of morphometric variation to the bottlenecks. Regressions of these averaged estimates for additive genetic variance and heritability in the control line onto the sequence of repeated samplings through the course of the experiment were not significant, so the six estimates on the control line were pooled (as above). Observed variation of individual estimates around this pooled value were used for significance tests of deviations of bottleneck lines from the control (Bryant & Meffert, 1991).

After the flush phase, 30 flies were sampled from each line for starch-gel electrophoretic assays at four initially polymorphic loci: esterase (EST; EC 3.1.1.1). superoxide dismutase (SOD; EC 1.15.1.1). and two phosphoglucomutases (PGM-1; PGM-2; EC 2.7.5.1). Three measures of fitness, also taken after the flush phase, were as follows: 1) the percentage of egg-to-adult survival under standard conditions of 80 eggs per 18g CSMA larval medium; 2) the percentage of single male-female pairs that produced no offspring; and 3) the percentage of virgin single pairs that initiated mating within two hours. All percentages were arcsine transformed before comparison to parallel control values. These procedures were repeated for five successive founder-flush cycles for two replicate lines of each bottleneck size, designated a and b. The total experiment encompassed approximately 35 generations.

RESULTS

After the first bottleneck, additive genetic variance increased significantly relative to the outbred control for all three bottleneck sizes (Fig. 1a), and heritability increased for bottlenecks of intermediate size (Fig. 1b). Genetic variability for all bottleneck sizes generally remained above the level of the control through the course of the five founder-flush cycles (Bryant & Meffert, 1991), and after the fifth bottleneck additive genetic variance and heritability were still significantly higher than the outbred control in the four-pair lines (Fig. 2). Remarkably, additive genetic variance for the single-pair lines was not significantly lower than the control despite the severe inbreeding of five successive bottlenecks to a single male-female pair. The levels of additive genetic variation in every bottleneck size after five serial founder-flush cycles were not significantly lower than that of the outbred population. Thus, one could not infer the recent history of repeated bottlenecks in these populations by comparing their levels of additive genetic variance to an outbred control.

In contrast, the losses in heterozygosity for the single-pair and 16-pair lines over the course of the experiment, and for the four-pair lines up to the third bottleneck episode, were near to the $1/2N$ per episode expected for neutral alleles (Fig. 3; see also McCommas & Bryant, 1990). In later bottleneck episodes, the four-pair lines appeared to recover in heterozygosity to a level comparable to the control. By the end of the experiment, both

594

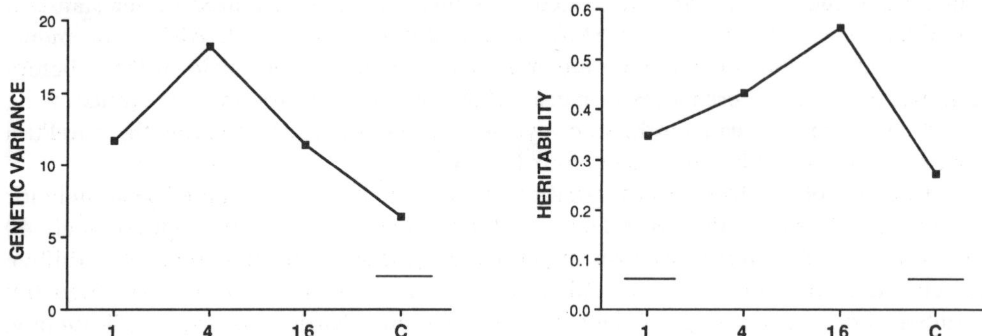

Figure 1. Average additive genetic variances and narrow sense heritabilities for the three bottleneck sizes in relation to the outbred control after the first bottleneck. An underscore indicates no significant difference between the estimate of genetic variability for a bottleneck size and that for the control.

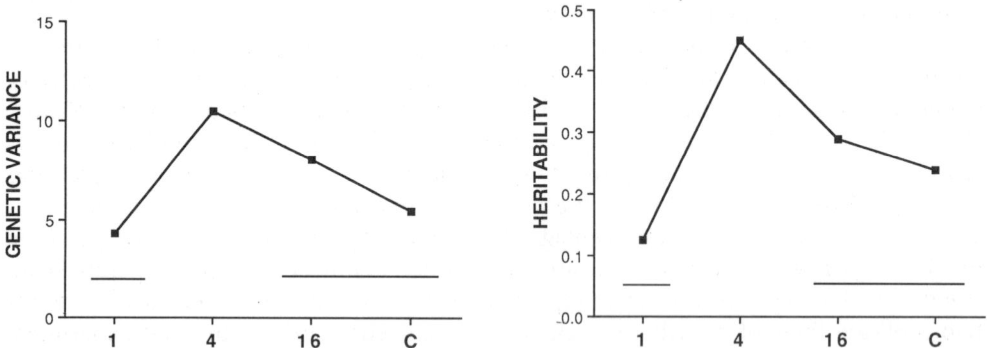

Figure 2. Average additive genetic variances and narrow sense heritabilities for the three bottleneck sizes in relation to the outbred control after the fifth bottleneck. An underscore indicates no significant difference between the estimate of genetic variability for a bottleneck size and that for the control.

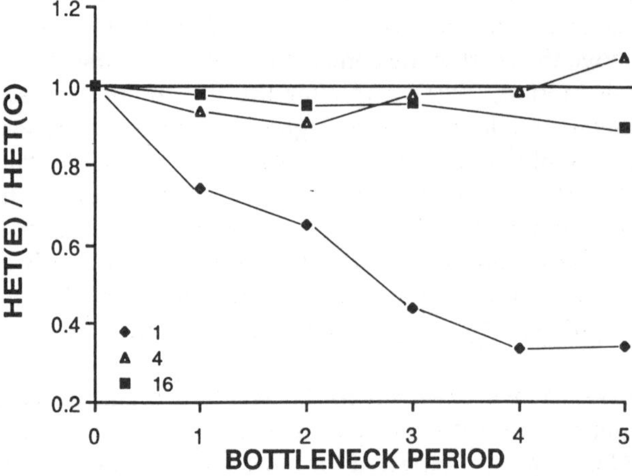

Figure 3. Average electrophoretic heterozygosity for the bottleneck sizes in relation to the control over the five serial bottlenecks.

single-pair lines were fixed for three of the four enzyme loci, whereas within the four-pair and 16-pair bottleneck treatments, one line remained polymorphic for all four loci while the other was fixed at a single locus. No locus that became fixed was later polymorphic, so changes in quantitative genetic variance were not likely to have been caused by inadvertent contamination of lines. These results clearly show that low electrophoretic variation relative to an outbred population does not imply low genetic variation for polygenic traits.

Despite the disparate responses of quantitative and electrophoretic traits to the bottlenecks, there was a moderate but non-significant correlation in level of average genetic variation for electrophoretic and morphometric traits among the bottleneck lines after the first founder-flush cycle (r = 0.57; Figure 4a). Repeated bottlenecks apparently affected average genetic variation among the lines in a concerted way, for the correlation between heterozygosity and average heritability was stronger after the fifth bottleneck episode (Fig. 4b; r = 0.88, significant for one-tailed Student-t). Nevertheless, average heritability for the bottleneck lines was no longer uniformly above the control after the fifth bottleneck. Also, the level of variation was not linearly related to bottleneck size, since the highest heritabilities and heterozygosities occurred in the four-pair rather than the 16-pair lines.

After the first founder-flush cycle all bottleneck lines were lower than the outbred control for the three measures of fitness, showing significance on average for the single-pair and four-pair lines (Fig. 5). The average fitness deviation of individual lines

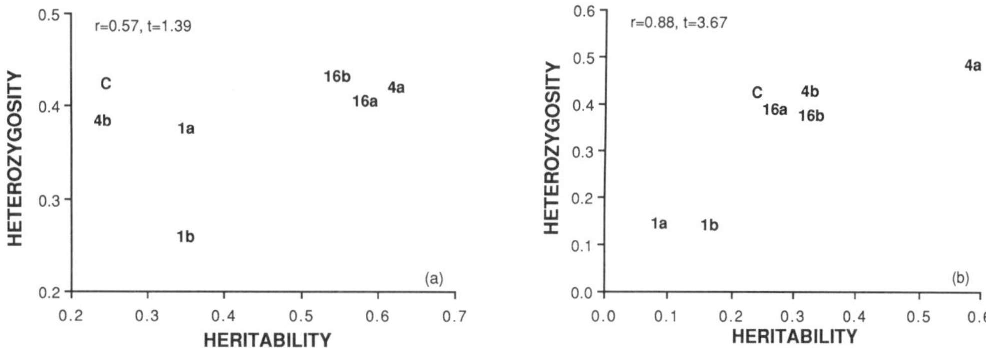

Figure 4. Relationship between electrophoretic heterozygosity and average morphometric heritability traits after the first (a) and fifth (b) bottlenecks. Correlation coefficient calculated among the bottleneck lines (excluding control) is shown along with its Student-t value.

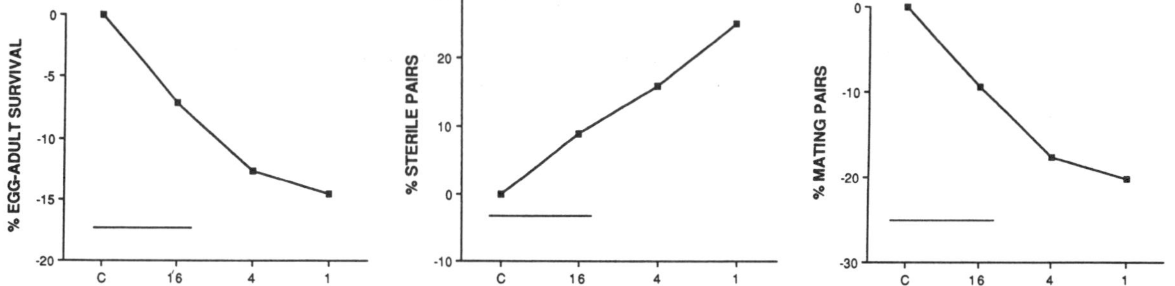

Figure 5. Fitness deviations of bottleneck lines from the outbred control for three components of fitness after the first bottleneck: percentage egg-to-adult survival, percentage of sterile male-female pairs, and percentage mating of single male-female virgin pairs. Underscore indicates a nonsignificant difference from the control.

(arithmetic mean of the three arcsine percentage deviations in fitness) throughout the course of the five bottlenecks are shown in Figure 6. Although all bottleneck lines showed an initial reduction in fitness, nearly all lines showed partial or full recovery from the inbreeding depression by the end of the experiment. The exception was the one single-pair line that continued to decline in fitness throughout all five bottlenecks. The percentage deviation in fitness from the control for the bottleneck lines during each successive bottleneck episode can be related to the expected inbreeding coefficient based on successive bottlenecks of size N, where $F = 1 - (1-1/2N)^t$. Initially there was a strong relationship of fitness to F, but this was ameliorated during the last few bottlenecks (Fig. 7). When the exceedingly low fitness of the one single-pair line is excluded from the analysis the slope of fitness onto F is nearly zero after the fifth bottleneck. Hence, after five consecutive bottleneck episodes, there was little relationship between the inbreeding coefficient calculated from known bottleneck events and the level of inbreeding depression.

There was a significant correlation among bottleneck lines after the first bottleneck between average fitness deviation and genetic variability, measured either by heterozygosity of electrophoretic loci or by average heritability of morphometric traits (Fig. 8); however, at the end of the experiment fitness deviations were uncorrelated with either measure of genetic variation (Fig. 9). The bottleneck populations clearly retained sufficient genetic variation for these fitness traits to be able to adapt to the novel conditions of the laboratory and overcome the initial inbreeding depression they experienced. As a result, there was no relationship of fitness to genetic variation measured either by electrophoresis or by morphometrics.

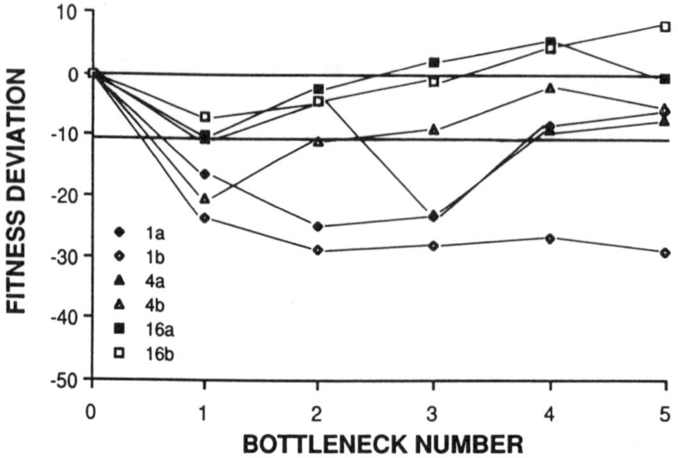

Figure 6. Fitness deviations averaged over the three separate components plotted over the five bottleneck episodes for the six replicate bottleneck lines. Lower 95% confidence bound for significant deviations from the control is shown.

Figure 7. Regression slope of fitness deviation onto inbreeding coefficient, F, for the three bottleneck sizes after each bottleneck episode. The strong inbreeding depression evident after the first few bottlenecks is largely ameliorated by the end of the experiment.

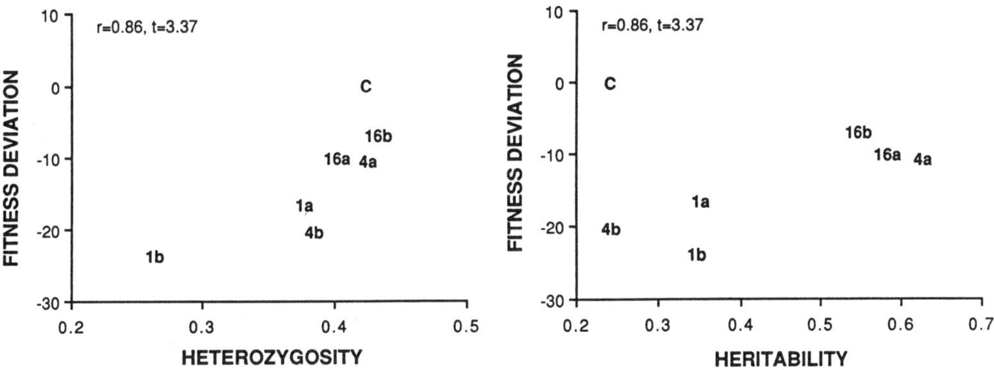

Figure 8. Fitness deviation in relation to electrophoretic heterozygosity (a) and morphometric heritability (b) after the first bottleneck. Correlation coefficient calculated among the bottleneck lines (excluding control) is shown along with the Student-t value.

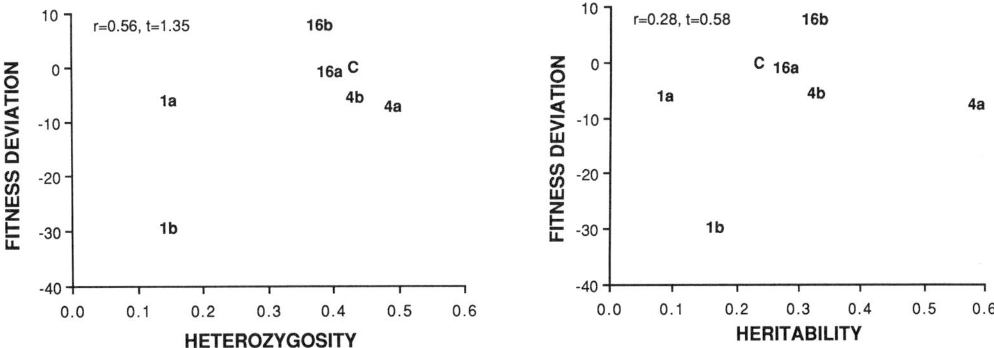

Figure 9. Fitness deviation in relation to electrophoretic heterozygosity (a) and morphometric heritability (b) after the fifth bottleneck. Correlation coefficient calculated among the bottleneck lines (excluding control) is shown along with its Student-t value.

598

DISCUSSION AND CONCLUSIONS

These data clearly do not support the doctrine that bottlenecks will decrease additive genetic variation and cause permanent inbreeding depression. Rather, additive genetic variation for quantitative traits increased in response to a bottleneck (see also Bryant et al., 1986) and remained equal to or greater than that for the outbred population despite repeated severe bottlenecks. As a result, it would not be possible to infer the occurrence or severity of historical bottlenecks by comparing levels of additive genetic variance among our ancestral and derived populations. While the exact mechanisms for the increases in additive genetic variance are unknown, one possibility is that nonadditive genetic variation can be converted to additive genetic variation as a result of inbreeding (Robertson, 1952; Goodnight, 1987, 1988; Tachida & Cockerham, 1989). Since traits associated with fitness are more likely to be affected by such nonadditive genetic processes as dominance and epistasis (Wright, 1931; Robertson, 1955; Fisher, 1958; Mather & Cooke, 1962), genetic variation for adaptively important traits is also likely to respond quite differently to bottlenecks than would neutral or purely additive traits. Our morphometric traits may have responded to the bottlenecks as they did because of their correlation with fitness through general body size (Bryant et al., 1986).

The increases in additive genetic variance for the morphometric traits in our study may not necessarily reflect similar increases in additive genetic variance for adaptively important traits. However, nearly all of our bottleneck lines recovered in fitness to the level of the control during the course of the five serial founder-flush cycles. In addition to the fitness traits reported here, we also investigated levels of mating activity and ambulatory activity and found the bottleneck lines were comparable to or higher than that of the outbred control (Meffert & Bryant, 1991a,b). Sufficient additive genetic variation for these fitness traits must have been retained in the bottleneck lines to allow them to recover from the initial inbreeding depression, even for the most severe of bottlenecks possible (i.e., reduction to a single mating pair). Hence, there appears to be no inevitable long-term effect of bottlenecks on the average fitness of these populations. However, conservation biology is as equally concerned with short-term adaptedness of managed populations as with their long-term adaptability (Foose et al., 1986). It is unknown whether or not the long-term adaptability of our bottleneck lines was compromised. We do know, however, that despite the influence of extreme bottlenecks the lines were able to adapt to the novel environmental conditions of the laboratory. In this sense, the lines retained sufficient adaptability to respond to a novel environmental regime.

Although our populations were assayed for only four loci polymorphic in the outbred control, electrophoretic variation in our experiment, particularly for severely bottlenecked lines, decreased in general accord with neutral expectation over the course of the five bottlenecks. However, low electrophoretic variation did not correspond to low genetic variation for morphometric or fitness traits. Electrophoresis may be a valuable tool for detecting a historical severe bottleneck or for evaluating the level of genetic subdivision of captive populations (Foose & Ballou, 1988), but it is likely to be a poor indicator of the actual genetic variation for complex polygenic traits or of the evolutionary potential of populations. Indeed, bottlenecks may open avenues of multivariate variability that might allow faster evolution to novel environments than would occur in large outbred populations (Bryant & Meffert, 1988).

We would expect to see a correspondence between the level of inbreeding, as calculated from pedigree or population structure, and fitness. As Lande (1988) has pointed out, deleterious recessive alleles may be eliminated from the population when the rate of inbreeding is gradual, such that fitness may be affected only transiently. It is less clear how evolution may proceed in traits that are influenced by epistasis, but the expected increase in additive genetic variation following a bottleneck would likely allow for adaptive evolution as well (Goodnight, 1987, 1988). Hence, the correspondence between the inbreeding coefficient and fitness immediately following a bottleneck will diminish or disappear after the populations have evolved for some time. Thus, as populations evolve there will be a decoupling of fitness from the level of inbreeding experienced by the population. A natural example may be the collared lizard, which has existed as locally adapted inbred populations in the Ozarks for several thousand years (Templeton, 1986).

Obviously the housefly, an organism with a very high reproductive capacity, may have remarkable abilities to recover from the deleterious effects of bottlenecks, thus making it an apparently poor model for less fecund higher organisms, particularly endangered vertebrate species. The potential deleterious effects of bottlenecks depend upon both the growth rate after a bottleneck, such that slow growth may augment genetic loss, and on the size of the bottleneck per se (Nei et al., 1975). While the growth rate of "higher" organisms may be much lower than that of the housefly, the sizes of bottlenecks experienced by most endangered species have not yet been as extreme as those of our experiment. Similarly, the founding numbers of animals for captive propagation programs have been higher than the bottleneck sizes of our experiment. Figure 10 shows that the effective population sizes in our experiment, calculated over a founder-flush cycle for each of the three bottleneck sizes, encompass the average founding population size of 25.8 for the 40 Species Survival Plans (SSP) reported by Foose (1990). In addition, because the expanding population size for these endangered species (which has doubled for most captive species since their founding) is often composed of some of the original founders as well as their descendants, the rate of allelic loss is less for these species than for the discrete generation species simulated by our experimental protocol. Hence, for many endangered species the combination of higher founder size, longer generation time, and slower rate of allelic loss may compensate for the lower growth rate after the bottleneck, making their effective bottleneck sizes within the range of our experimental model.

These experiments suggest that genetic variation for quantitative traits may not be rapidly eroded in small populations and that even very small populations may retain sufficient genetic variation to adapt to new environmental conditions. Nevertheless, captive populations are generally managed to reduce inbreeding and to retain maximal allelic variation for future adaptation (Allendorf et al., 1979; Foose & Ballou, 1988). Most rare alleles in an outbred population are either neutral or deleterious (Kimura, 1983), and it is not clear that salvaging them in a captive breeding program would benefit either current populational health or future evolutionary flexibility, in the expectation that some of the

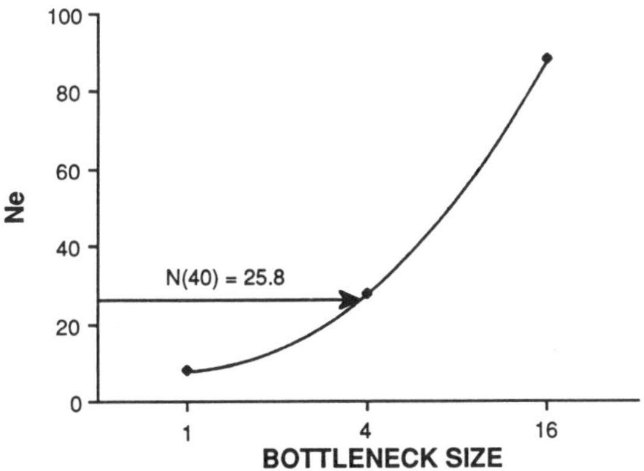

Figure 10. The effective population sizes (harmonic mean of population sizes during a founder-flush cycle) of our three experimental bottleneck sizes having taken into account the rapid flush to a population size of 1000. Average founder numbers for 40 SSP species reported by Foose (1990) is 25.8 and falls within the range of our effective population sizes.

600

alleles may become advantageous later. Nor is it clear that adaptation to current environ-ments per se, whether natural or captive, will reduce future evolutionary flexibility, particularly for quantitative traits whose genetic variability may be maintained in rela-tively small populations (Lande, 1978; Lande & Barrowclough, 1987). Hence, breeding plans that seek to minimize loss of allelic variation, including rare alleles, by equalizing founder contributions may only serve to retard adaptation and short-term population viability without ensuring long-term evolutionary flexibility. Maintenance of genetic variation in captive or wild populations is likely of far less importance to the ultimate survival of the endangered species than is the preservation of vital habitat, even when population sizes have been drastically reduced.

ACKNOWLEDGMENTS

This study was supported by grants from the National Science Foundation (BSR-8198128 and BSR-8800977) and by the University of Houston Coastal Center. We thank Mark Johnston, Stan Mays, and Phil Hedrick for their constructive criticisms of an earlier draft of this paper.

LITERATURE CITED

Allendorf, F. W., Christiansen, F. B., Dobson, T., Eanes, W. F. & O. Frydenberg. 1979. Electrophoretic variation in large mammals. I. The polar bear, *Thalarcros maritimus*. *Hereditas* 91:19–22.

Allendorf, F. W. & R. F. Leary. 1986. Heterozygosity and fitness in natural populations of animals. *In:* M. E. Soulé (ed.). *Conservation Biology: The Science of Scarcity and Diversity.* Sinauer: Sunderland, MA.

Bonnell, M. L. & R. K. Selander. 1974. Elephant seals: Genetic variation and near extinction. *Science* 184:908–909.

Bryant, E. H. 1969. The fates of immatures in mixtures of two housefly strains. *Ecology* 50:1049–1069.

Bryant, E. H., McCommas, S. A. & L. M. Combs. 1986. The effect of an experimental bottleneck on quantitative genetic variation in the housefly. *Genetics* 114:1191–1211.

Bryant, E. H. & L. M. Meffert. 1988. The effect of an experimental bottleneck on morphological integration in the housefly. *Evolution* 42:698–707.

Bryant, E. H. & L. M. Meffert. 1991. The effects of multiple bottlenecks on quantitative genetic varia-tion. Ms. in preparation.

Bryant, E. H., Meffert, L. M. & S. A. McCommas. 1990. Fitness rebound in serially bottlenecked populations of the housefly. *Amer. Natur.,* 136:542–549.

Charlesworth, B. & D. Charlesworth. 1987. Inbreeding depression and its evolutionary conse-quences. *Ann. Rev. Ecol. Syst.* 18:237–268.

Chesser, R. K., Smith, M. H. & I. Lehr Brisbin. Jr. 1980. Management and maintenance of genetic variability in endangered species. *Int. Zoo Yearbook.* 20:146–154.

Cothran, E. G., MacCleur, J. W., Weitkamp, L. R. & S. A. Guttormsen. 1986. Genetic variability, inbreeding, and reproductive performance in standardbred horses. *Zoo Biol.* 5:191–201.

Daniell, A. & N. D. Murray. 1986. Effects of inbreeding in the budgerigar *Melopsirracus undulatis* (Aves:Psittacidae). *Zoo. Biol.* 5:233–238.

Feldhammer, G. A., Chapman, J. A. & R. L. Miller. 1978. Sika deer and white-tailed deer on Maryland's eastern shore. *Wildl. Soc. Bull.* 6:155–157.

Fisher, R. A. 1958. *The Genetical Theory of Natural Selection.* Dover: NY.

Foose, T. J. 1980. Demographic management of endangered species in captivity. *Int. Zoo Yearbook* 20:154–166.

Foose, T. J. 1982. *Species survival handbook.* AAZPA: Wheeling, WV.

Foose, T. J. 1990. *Midyear report—1990.* AAZPA: Wheeling, WV.

Foose, T J. & J. D. Ballou. 1988. Management of small populations. *Int. Zoo Yearbook* 27:26–41.

Foose, T. J., Lande, R., Fleshness, N. R., Rabb, G. & B. Read. 1986. Propagation plans. *Zoo Biol.* 5:139–146.

Frankham, R. 1980. The founder effect and response to artificial selection in *Drosophila. In:* A.

Robertson (ed.), *Selection Experiments in Laboratory and Domestic Animals.* Commonwealth Agriculture Bureau: Slough, UK.

Frankel, O. H. & M. E. Soulé. 1981. *Conservation and Evolution.* Cambridge University Press: Cambridge, UK.

Goodnight, C. 1987. On the effect of founder events on epistatic genetic variance. *Evolution* 41:80–91.

Goodnight, C. 1988. Epistasis and the effect of founder events on the additive genetic variance. *Evolution* 42:441–454.

Hedrick, P. W. Brussard, P. F., Allendorf, F. W., Beardmore, J. A. & S. Orzack. 1986. Protein variation, fitness, and captive propagation. *Zoo Biol.* 5:91–99.

Kimura, M. 1983. *The Neutral Theory of Molecular Evolution.* Cambridge University Press: Cambridge, UK.

Lande, R. 1978. The maintenance of genetic variability by mutation in a polygenic character with linked loci. *Genet. Res.* 26:221–235.

Lande, R. 1988. Genetics and demography in biological conservation. *Science* 241:1455–1460.

Lande, R. & G. F. Barrowclough. 1987. Effective population size, genetic variation, and their use in population management. *In:* M. E. Soulé (ed.), *Viable Populations for Conservation.* Cambridge University Press: Cambridge, UK.

Lopez-Fanjul, C. & A. Villaverde. 1989. Inbreeding increases genetic variance for viability in *Drosophila melanogaster. Evolution* 43:1800–1804.

Mather, K. & P. Cooke. 1962. Differences in competitive ability between genotypes of *Drosophila. Heredity* 17:381–407.

McCommas, S. A. & E. H. Bryant. 1990. Loss of electrophoretic variation in serially bottlenecked populations. *Heredity* 64:315–321.

Meffert, L. M. & E. H. Bryant. 1991a. Mating propensity and courtship behavior in serially bottlenecked lines of the housefly. *Evolution:* 45:293–306.

Meffert, L. M. & E. H. Bryant. 1991b. Divergent ambulatory activity and grooming behavior in serially bottlenecked lines of the housefly. Ms. in preparation.

Nei, M., Maruyama, T. & R. Chakraborty. 1975. The bottleneck effect and genetic variability in populations. *Evolution* 29:1–10.

O'Brien, S. J., Wildt, D. E., Bush, M., Caro, T. M., Fitzgibbon, C., Aggindey, I., & R. E. Leakey. 1987. East African cheetahs: evidence for two population bottlenecks. *Proc. Nat. Acad. Sci.* 84:508–511.

Ralls, K. & J. Ballou. 1983. Extinction: lessons from zoos. *In:* C. M. Shonewald-Cox, S. M. Chambers, B. MacBryde and L. Thomas (eds.), *Genetics and Conservation: A Reference Manual for Managing Wild Animal and Plant Populations.* Benjamin/Cummings: London, UK.

Ralls, K. & J. Ballou, (eds.) 1986. Proceedings of the workshop on genetic management of captive populations. *Zoo Biology.* Vol. 5.

Robertson, A. 1952. The effect of inbreeding on the variation due to recessive genes. *Genetics* 37:189–207.

Robertson, A. 1955. Selection in animals: synthesis. *Cold Spr. Harbr. Symp. Quant. Biol.* 20:287–297.

Shoenwald-Cox, C. M., Chambers, S. M., MacBryde, B. & L. Thomas, eds. 1983. *Genetics and Conservation: A Reference Manual for Managing Wild Animal and Plant Populations.* Benjamin/Cummings: London, UK.

Seal, U. S. & T. J. Foose. 1983. Genetics and demography of the Siberian tigers in North America with evidence of inbreeding depression. *Zoo. Biol.* 2:241–244.

Senner, J. W. 1980. Inbreeding depression and the survival of zoo populations. *In:* M. E. Soulé & B. A. Wilcox (eds.), *Conservation Biology: An Evolutionary-Ecological Perspective.* Sinauer, Sunderland, MA.

Simberloff, D. 1988. The contribution of population and community biology to conservation science. *Ann. Rev. Ecol. Syst.* 19:473–511.

Soulé, M., Gilpin, M., Conway, W. & T. Foose. 1986. The millenium ark: how long a voyage, how many staterooms, how many passengers? *Zoo. Biol.* 5:101–113.

Soulé, M. E. & D. Simberloff. 1986. What do genetics and ecology tell us about the design of nature preserves? *Biol. Conserv.* 35:19–40.

Soulé, M. E. & B. A. Wilcox. eds. 1980. *Conservation Biology: an Evolutionary-Ecological Perspective.* Sinauer: Sunderland, MA.

Tachida, H. & C. C. Cockerham. 1989. A building block model for quantitative traits. *Genetics* 121:839–844.

Templeton, A. R. 1986. Coadaptation and outbreeding depression *In:* M. E. Soulé (ed.). *Conservation Biology: The Science of Scarcity and Diversity.* Sinauer: Sunderland, MA.

Wright, S. 1931. Evolution in Mendelian populations. *Genetics* 16:97–159.

Wright, S. 1977. *Evolution and Genetics of Populations,* Volume 3. *Experimental Results and Evolutionary Deductions.* University of Chicago Press: Chicago, IL.

Management of Genetic Variation in Captive Populations

Jonathan D. Ballou

Abstract. Captive populations can only contribute to species conservation if the populations are genetically and demographically managed under cooperative national and international captive propagation programs. The primary goal of these programs is to establish demographically secure populations with sufficient genetic diversity to retain the population's ability to adaptively respond to selection if and when individuals from the population are used to re-establish or reinforce wild populations. An important aspect of these programs is the strategy used to select which individuals should be bred each generation to optimize the retention of genetic diversity. A computer model is used to evaluate the effects of three mate selection strategies on the retention of genetic diversity in populations with complex pedigrees: maximum avoidance of inbreeding, equalization of founder contribution, and minimizing mean kinship. Managing solely by founder contribution is shown to be ineffective in retaining genetic diversity, while managing by mean kinship appears to be a excellent mate selection strategy.

INTRODUCTION

Zoological parks can contribute to conservation in three significant ways: 1) conservation education to increase public awareness of global conservation issues: 2) captive propagation of endangered and threatened species for long-term preservation; and 3) behavioral, genetic, reproductive and nutritional research on exotic species to increase our basic knowledge of these otherwise often inaccessible animals. This paper focuses on the second contribution: the role of captive propagation in the conservation of species threatened with extinction. In particular, it discusses several issues relating to genetic considerations of managing captive populations for long-term conservation.

CONSERVATION GENETIC GOALS FOR CAPTIVE PROPAGATION PROGRAMS

Goals of captive propagation programs can vary from those designed to domesticate animals solely for exhibit purposes to those formed to save species from extinction (Frankham et al., 1986). The primary objective of captive propagation for conservation purposes is to establish and maintain populations of threatened and endangered species—not to replace wild populations but to reinforce and reestablish wild populations when (and if) the opportunity to do so becomes available (Foose, 1983). The development of captive

Mr. Ballou is with the Department of Zoological Research, National Zoological Park, Smithsonian Institution, Washington, D.C. 20008, USA.

breeding programs designed to accomplish this goal has been a major focus of effort among zoological institutions over the last 15 years, and a number of international, national and regional programs have been developed to assist with this objective.

The one common thread that underlies all conservation-oriented captive propagation programs in zoos is that, without exception, these captive populations are small and can not contribute to the conservation of their species unless they are managed cooperatively and scientifically at a national or international level. Population sizes at individual institutions are too small to constitute viable populations and animal collections at different institutions must be managed as one population to accomplish conservation objectives. Even so, most captive populations, especially of endangered species, have extremely small effective population sizes. Conway (1987) summarized the distribution of number of individuals per species in zoo populations. Of the approximately 1100 mammal species held by zoos, about 70% are populations with fewer than 25 individuals. Since effective population sizes usually range in the area of 25–30% of actual population sizes, most captive populations have effective sizes of fewer than 10 individuals.

Due to their small effective sizes, inbreeding is common (and often results in reduced survival and reproduction; Ralls et al., 1988; Ralls and Ballou, 1983) and genetic diversity is rapidly lost through genetic drift. These populations are also exposed to a much different form of genetic selection then they would experience in the wild. Artificial selection (conscious and unconscious) for traits adapted to a captive environment is probably strong. In sum, many processes occur in the captive environment that change the genetic constitution of these populations. Therefore, if the goal is to eventually use captive populations to reinforce and restore wild populations, then the most effective conservation strategy that can be implemented is to try to minimize change in the genetic constitution of the populations and maintain as much of their genetic diversity as possible while they are in captivity.

BREEDING STRATEGIES FOR CONSERVATION OF GENETIC DIVERSITY

The development of a detailed captive breeding program involves two processes: 1) conducting genetic and demographic analyses of existing population data to evaluate current status and future projections; and 2) formulating specific year-by-year recommendations for each animal in the population (Foose and Ballou, 1988). A critical component of this process is the breeding strategy used to select breeding individuals for the purpose of maintaining genetic diversity. An ideal breeding strategy would dictate which animals to breed and to whom they should breed in order to optimize the retention of genetic diversity in subsequent generations. This is not as straight-forward as it might seem. For example, strategies that minimize immediate loss of genetic diversity do not necessarily minimize loss of genetic diversity over the longer term (Crow and Kimura, 1970).

At a very general level, basic population genetics theory provides broad guidelines for this purpose. Since retention of heterozygosity is a function of the population's effective size, genetic diversity can be maintained by maximizing the population's effective population size. The theory of effective population size has been well developed (Wright, 1931; Crow and Kimura, 1970; Lande and Barrowclough, 1987). Maximizing the population's effective size can be accomplished through maximizing the number of breeders in a population, equalizing their sex ratio, and minimizing their variance in offspring number (ideally to zero). Under this strategy, the effective size can be almost double the actual size of the population (Crow and Kimura, 1970).

Maintaining complete pedigrees on captive populations, however, provides an even more accurate method for developing genetic management recommendations. Complete pedigrees allow exact computation of genetic parameters in a population and therefore

604

provide us with all the information required to identify individuals that need to be bred to assure that the maximum amount of genetic diversity is passed from one generation to the next.

Unfortunately, genetic management is rarely implemented when the captive population is founded. More often than not it is initiated well after the population has been established. At this late stage, the pedigrees are usually extremely complex, often highly inbred and have already lost significant levels of genetic variation. Characteristically, the effective size of the population in the early generations is often very small due to disproportionate breeding of individual founders or their close descendants. The genetic diversity potentially contributed by many of the founders has already been lost. Therefore, genetic management must try to compensate for complications already built into the pedigrees by lack of genetic management in the past. This complicates the process of identifying optimal breeding strategies and clearly necessitates the careful analysis of captive population pedigrees if genetic management is to be successful.

Pedigrees are analyzed primarily with respect to three characteristics: 1) the distribution of each founder's genetic contribution to the population's current gene pool (founder contribution; Foose et al., 1986); 2) the loss of each founder's alleles due to genetic drift imposed by the historical structure of the pedigree (founder gene survival; Thompson, 1986); and 3) the relationships among living animals in the managed population (Thompson, 1986; Ballou, 1983; Boyce, 1983). Each of these characteristics, alone or in combination, has been used to provide the basis for selecting individuals to maintain genetic diversity.

One of the earliest strategies recommended for selecting breeding individuals was Maximum Avoidance of Inbreeding (MAI: Flesness, 1977; Senner, 1980). MAI is a systematic strategy of breeding that utilizes a recursive pattern of breeding each generation (Fig. 1). This strategy minimizes the rate at which inbreeding accumulates in the population but somewhat paradoxically does not guarantee maximum retention of heterozygosity over the long term. This strategy ignores the historical pedigree structure of the population and establishes breeding pairs according to the formula specified for the number of animals in the population. Mates and numbers of offspring are defined by this strategy (Fig. 1).

Another strategy uses founder contribution distributions to identify priority breeders. On the assumption that more genetic diversity is retained if all founders contribute equally to the gene pool, this strategy's goal is to equalize the average founder contribution of all founders to the current population (Foose et al., 1986). Priority breeders are identified as descendants of under-represented founders (Fig. 2). This will increase the contribution of under-represented founders and lower the contribution of over-represented founders. This strategy identifies the animals that should be bred but does not address who should breed with whom or how many offspring are to be bred by each pair. Pairings are determined by evaluating the relationships among the proposed breeders, taking care not to mate animals too closely related. Offspring numbers are determined according to the degree to which the pair's ancestors are under- or over-represented (Ballou and Foose, in prep).

The strategy of equalizing founder contribution ignores founder gene survival. Founders who have lost a large proportion of their alleles due to drift in the population do not have as much genetic diversity to contribute as founders whose genomes are more intact. Therefore, a modification of the above strategy is to take into consideration gene survival by defining the objective of founder contribution management as not being an equal distribution across all founders but rather a distribution where the contributions are weighed by each founder's gene survival (Fig. 3). Other aspects of the strategy remain the same (Ballou and Foose, in prep).

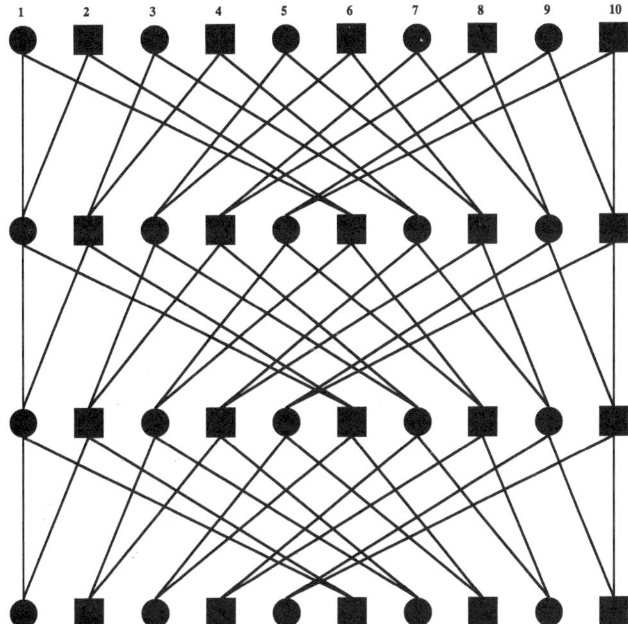

Figure 1. Maximum avoidance of inbreeding mating scheme applied to a population of 10 animals (5 males, 5 females). Animals are labeled from 1 to 10. In general, the recursive mating formula for N animals (with x designating even number animals) is then: x mates with x−1 to produce offspring x/2 and (x+N)/2 in the next generation. Exact adherence to this breeding scheme may require within-family selection to assure proper distribution of sexes among breeders.

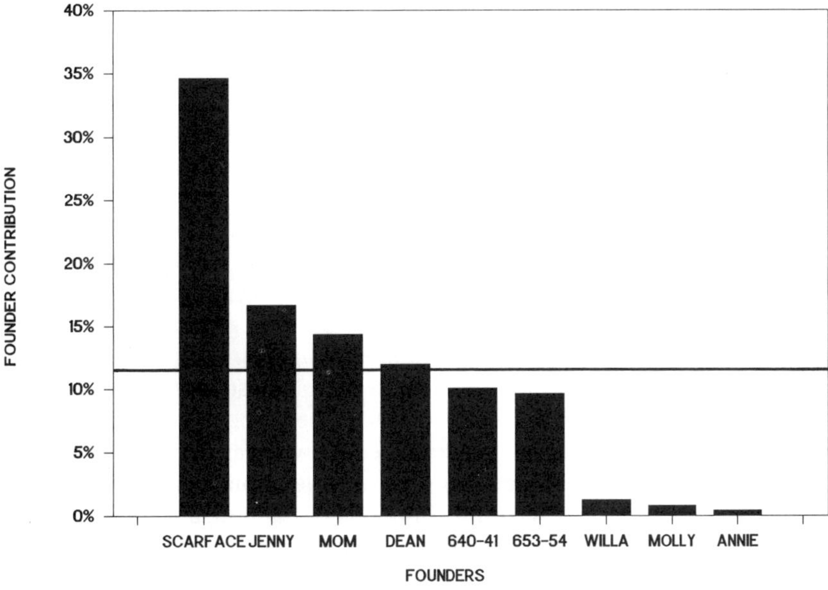

Figure 2. Genetic contribution of each of the 9 founders to the gene pool of the 1989 population of black-footed ferrets. One genetic management strategy for preserving genetic diversity recommends selecting breeding animals to equalize the contribution from all founders at the parity line shown (11.1%).

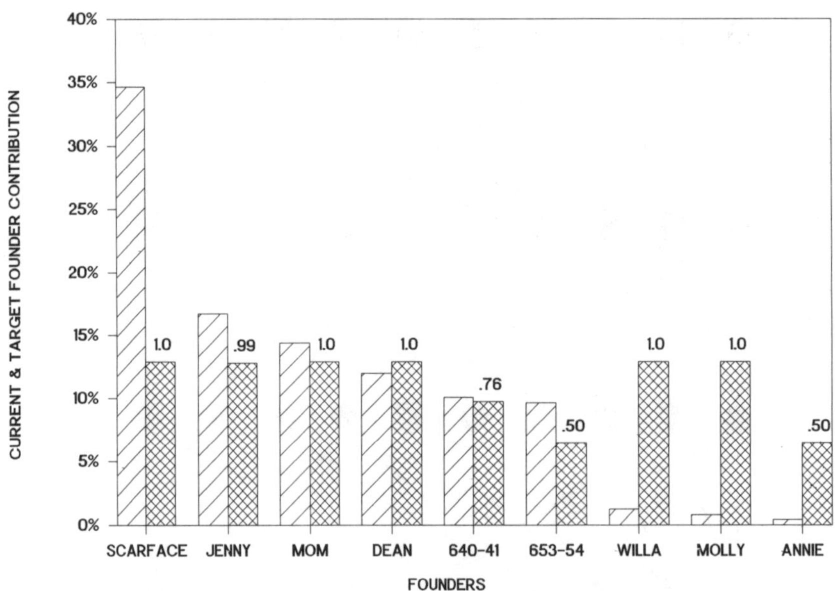

Figure 3. Current (striped) and Target (hatched) founder contributions for the 1989 captive population of black-footed ferrets. The Target founder contributions define the objectives for managing founder contribution in the population. As opposed to equal founder contribution (see Figure 2), target founder contributions reflect the proportion of each founder's genome surviving in the population (number above each bar).

A more recent strategy uses the concept of mean kinship. Mean kinship is the average of the kinship coefficients between an individual and all living individuals (including itself) in the population. Mean kinship values are calculated for each individual in the population. An individual who shares many alleles with the rest of the population (has alleles identical by descent) will have a higher mean kinship than an individual with few relatives in the population. Priority breeders can be identified by ranking individuals according to their mean kinship and selecting those whose mean kinship values are low relative to the rest of the population.

EVALUATING MATE SELECTION STRATEGIES

While other animal breeding schemes are available, those discussed above are the ones primarily considered for captive management programs with conservation objectives. Because each focuses on a different criterion for selecting mating preferences, these strategies can have significantly different effects on the genetic character of the population—particularly when applied to complex pedigrees.

While theory has been developed for how some of these strategies affect genetic diversity in theoretical populations (Falconer, 1981), there has never been a systematic evaluation of how effective these breeding strategies are when applied to existing populations with complex pedigrees resulting from several generations of unmanaged matings.

As a preliminary step towards accomplishing this evaluation, a computer model was developed to simulate mate selection strategies in hypothetical populations and compare the results of different selection strategies on the retention of both heterozygosity and allelic diversity in the population. The model simulates a hypothetical population consisting of 30 sexually reproducing individuals with non-overlapping generations. Each generation, parents and pairings are selected according to the breeding strategy being tested to create 30 offspring for the next generation. This process is repeated to produce 50 generations, each with 30 individuals. Each simulation to 50 generations is repeated 1000 times.

The model compared the effects of four different breeding strategies: maximum avoidance of inbreeding, managing to equalize founder contribution (FC), managing by mean kinship (MK), and what is termed the "Unmanaged" strategy. The "Unmanaged" strategy represents the kind of variance in reproductive success often observed in unmanaged zoo populations and was based on the variance in family size of the captive population of golden lion tamarins between the years 1970 and 1980 (before the population was being managed for maintenance of genetic diversity: Ballou, 1990). Details of the model are provided in Ballou (in prep.).

To evaluate the effect of these different strategies on complex pedigrees, an initial pedigree was created as the "seed" from which to start the modeling. This initial pedigree was founded by 30 unrelated individuals and, although it consisted of only two generations, had already lost genetic representation of 14 of its original 30 founders, had an average inbreeding coefficient of a little over 8%, and had an expected heterozygosity of 89% of the original level. Two unique alleles were assigned to each of the original 30 founders, and at the second generation only 16 of the original 60 alleles remained. Starting with this pedigree, the model tracked expected heterozygosity, number of original alleles remaining, and observed average inbreeding for each of the four mating strategies modeled.

PRELIMINARY EVALUATION RESULTS

Figure 4 shows the decrease in expected heterozygosity over 50 generations for the four mating strategies. As expected, the unmanaged population loses the most heterozygosity because of its small effective size. The loss of heterozygosity when managing by founder contribution alone is initially rapid but levels off at around 40%. Closer evaluation of the results shows that managing by founder contribution alone can lead to forced line breeding in the population. The equilibrium heterozygosity of 40% represents heterozygosity between lines with zero heterozygosity within lines. Both the mean kinship and MAI strategies result in high levels of heterozygosity retention, with mean kinship retaining a slightly, but significantly higher, heterozygosity than MIA ($p < .001$).

Retention of allelic diversity, as measured by the number of original alleles remaining in the population, is shown in Figure 5. Results are similar to those of expected heterozygosity. Both "Unmanaged" and founder contribution strategies prove to be poor strategies for allele retention, and again mean kinship exhibited the highest (and statistically significant) levels of allele retention ($P < .001$).

Change in average inbreeding (1 minus observed heterozygosity) is plotted in Figure 6. Not surprisingly, over the long run, MAI results in minimal levels of inbreeding, with mean kinship a close, but statistically significant second ($p < .001$). Because managing by founder contribution forces line breeding in this situation, average inbreeding is highest for this strategy.

It is worth noting the average inbreeding dynamics within the first 10 generations (Fig. 7). The first two generations are data from the pre-defined complex pedigree. Inbreeding peaks at the third generation (the first generation modeled) for the MAI strategy because historical knowledge of the pedigree is ignored and because some of the matings determined by the recursive MAI formula are between closely related animals. However, as the MAI strategy takes effect, inbreeding is quickly reduced. In both the founder contribution and mean kinship strategies, inbreeding is sharply reduced in the third generation because priority matings are among unrelated animals with uncommon alleles/founder contributions. However, inbreeding rapidly rises with founder contribution management because relationships among animals are ignored.

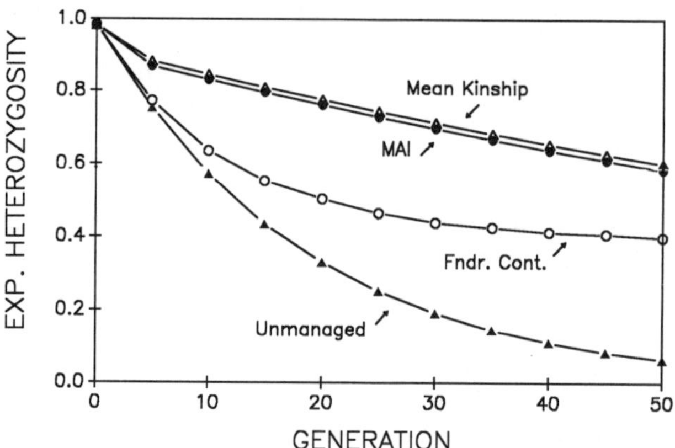

Figure 4. Expected heterozygosity retained in the modeled population over 50 generations for the four different breeding strategies. MAI — Maximum Avoidance of Inbreeding strategy. Fndr. Cont. = managing to equalize founder contribution.

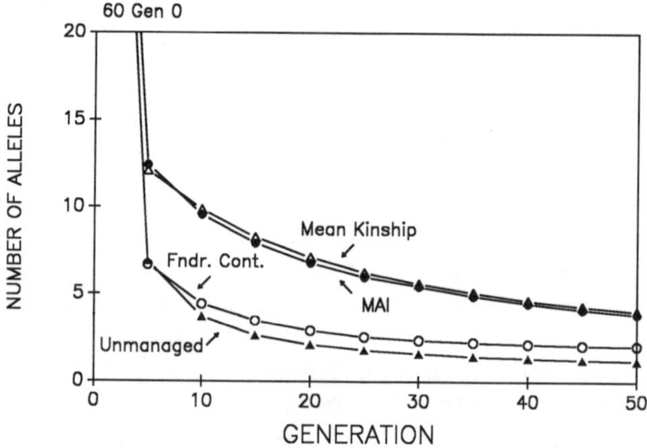

Figure 5. Number of original alleles retained over 50 generations for four different breeding strategies. The population started with 60 alleles (2 assigned to each of the 30 founders). MAI — Maximum Avoidance of Inbreeding Strategy, Fndr. Cont. = managing to equalize founder contribution.

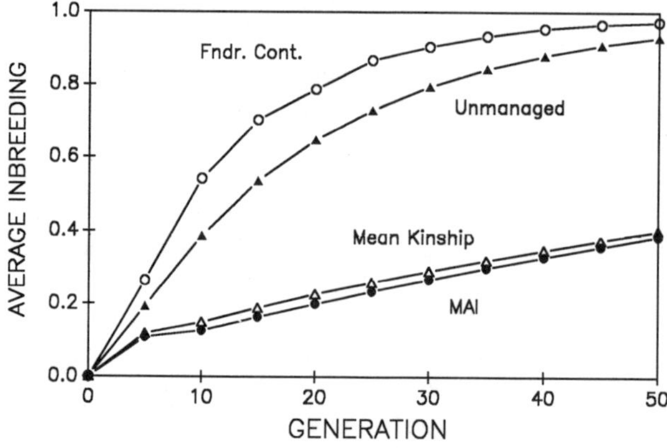

Figure 6. Average level of inbreeding over 50 generations for four different breeding strategies. MAI — Maximum Avoidance of Inbreeding Strategy. Fndr. Cont. = managing to equalize founder contribution.

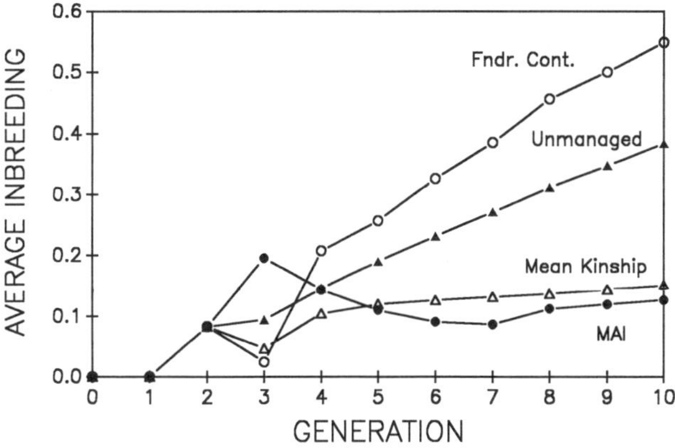

Figure 7. Changes in average level of inbreeding in the first 10 generations for four different breeding strategies. MAI — Maximum Avoidance of Inbreeding Strategy, Fndr. Cont. = managing to equalize founder contribution.

DISCUSSION

These preliminary results indicate that in complex pedigrees (at least this typically complex pedigree), management by founder contribution alone does not provide optimal retention of genetic diversity (both expected heterozygosity and allelic diversity), while mean kinship seems to serve as an excellent strategy for making management decisions.

Although many captive populations use information on founder contribution to assist with developing management strategies (Ballou and Foose, in prep.), other factors are routinely taken into consideration as well—including mean kinship and minimizing inbreeding. The model results, in which blindly determined mating strategies are based solely on founder contribution rankings, therefore do not realistically reflect how these strategies are implemented in captive populations. Nevertheless, these results do suggest that over-reliance on founder contribution information alone is not an effective strategy for maintaining genetic diversity.

The results of the mean kinship and MAI strategies were similar over the long term. The MAI strategy is expected to retain high levels of variation from one generation to the next because the effective size of the population is maximized. The MAI strategy, however, makes no effort to compensate for the pedigree complexities and therefore does not retain as high levels of diversity as the MK strategy. In the early generations, the MK strategy gives breeding preference to individuals with fewer relatives in the population. Variance in relationships among individuals decreases over time and eventually all animals are equally related to each other. At this point, the MK strategy mimics the MAI strategy: all individuals produce the same number of offspring each generation and the effective size is maximized. Any difference between levels of diversity retained by these two strategies is thus determined by how effective the MK strategy is at compensating for historical effects.

As these results are preliminary and are based on modeling the effects of strategies applied to only one complex pedigree, a caveat is in order. Further analyses need to be performed on a wide variety of pedigrees, varying both in structure and depth before these results can be generalized. However, these preliminary results do provide impetus for further analyses along similar lines (Ballou, in prep).

610

CONCLUSIONS

The objective of captive propagation programs designed for long-term conservation of endangered or threatened species is to establish demographically secure populations with sufficient genetic diversity to retain the population's ability to adaptively respond to selection if and when individuals from the population are used to re-establish or reinforce wild populations. This objective requires careful management of the genetic and demographic characteristics of the population. Such management can only be based on detailed genetic and demographic analyses of the historical, current, and future trends of the specific populations being managed. However, general strategies, such as the optimal mating selection strategies discussed here, are needed to guide the formulation of specific management recommendations.

Intensive genetic management for conservation is a relatively new field, and analytical methods as well as general guidelines from related fields such an animal science, population and molecular genetics, evolutionary biology, and mathematical statistics have been critical in developing the theory upon which management guidelines are based. We need to continue to enlist the expertise and experience of others working in these fields to advance the science of conservation management.

ACKNOWLEDGMENTS

I thank Bob Lacy, Kathy Ralls and Kathy Cooper for their useful comments at various stages of this research.

LITERATURE CITED

Ballou, J. D. 1983. Calculating inbreeding coefficients from pedigrees. Pp. 509–520. *In:* C. M. Schonewald-Cox, S. M. Chambers, B. MacBryde, & L. Thomas (eds.), *Genetics and Conservation.* Benjamin/Cummings, Menlo Park, CA.

Ballou, J. D. 1990. *1989 Golden Lion Tamarin International Stud Book.* National Zoological Park: Washington, D.C.

Ballou, J. D. & T. J. Foose. In prep. Demographic and genetic management of captive populations. *In:* S. Lumpkin & D. G. Kleiman (eds.), *Wild Mammals in Captivity.*

Boyce, A. J. 1983. Computation of inbreeding and kinship coefficients on extended pedigrees. *J. of Heredity* 74:400–404.

Conway, W. 1987. Species carrying capacity in the zoo alone. Pp. 20–32. *In: Proceedings 1987 AAZPA Annual Conference.* AAZPA: Wheeling, WV.

Crow, J. F. & M. Kimura. 1970. *An Introduction to Population Genetic Theory.* Harper and Row, New York.

Falconer, D. S. 1981. *Introduction to Quantative Genetics.* Longman, Inc.: New York.

Flesness, N. 1977. Gene pool conservation and computer analysis. *Inter. Zoo. Yearbk.* 17:77–81.

Foose, T. J. 1983. The relevance of captive propagation to the conservation of biotic diversity. Pp. 374–401. *In:* C. M. Schonewald-Cox, S. M. Chambers, B. MacBryde, & L. Thomas (eds.), *Genetics and Conservation.* Benjamin/Cummings: Menlo Park, CA.

Foose, T. J. & J. D. Ballou. 1988. Population management: theory and practice. *Inter. Zoo Yearbk.* 27:26–41.

Foose, T. J., Lande, R., Flesness. N. R., Rabb, G. & B. Read. 1986. Propagation plans. *Zoo Biol.* 5:139–146.

Frankham, R., Hemmer, H., Ryder, O. A., Cothran, E. G., Soulé, M. E., Murray, N. D. & M. Snyder. 1986. Selection in captive populations. *Zoo Biol.* 5:127–138.

Lande, R. & G. Barrowclough. 1987. Effective population size, genetic variation, and their use in population management. Pp. 87–124. In, M. E. Soulé (ed.): *Viable Populations for Conservation.* Cambridge University Press: Cambridge.

Ralls, K. & J. D. Ballou. 1983. Extinction: lessons from zoos. Pp. 164–184. *In:* C. M. Schonewald-Cox, S. M. Chambers, B. MacBryde, & L. Thomas (eds.), *Genetics and Conservation.* Benjamin/Cummings, Menlo Park, CA.

Ralls, K., Ballou, J. D. & A. R. Templeton. 1988. Estimates of lethal equivalents and the cost of inbreeding in mammals. *Cons. Biol.* 2:185–193.

Senner, J. W. 1980. Inbreeding depression and the survival of zoo populations. Pp. 209–224. *In:* M. E. Soulé & B. A. Wilcox (eds.). *Conservation Biology.* Sinauer Associates: Sunderland, MA.

Thompson, E. A. 1986. Ancestry of alleles and extinction of genes in populations with defined pedigrees. *Zoo Biol.* 5:161–170.

Wright, S. 1931. Evolution in Mendelian populations. *Genetics* 16:97–159.

Using DNA Fingerprinting to Assess Kinship and Genetic Structure in Avian Populations

Patricia P. Rabenold, Kerry N. Rabenold, Walter H. Piper, Mark D. Decker, and Joey Haydock

Abstract. DNA fingerprinting (using Jeffreys' multilocus probes) has proven useful in the analysis of reproductive status, breeding system, and genetic structure in three avian species. In cooperatively breeding stripe-backed wrens (*Campylorhynchus nuchalis*) and bicolored wrens (*C. griseus*), reproduction by males previously thought to be nonreproductive was readily detected despite high local genetic homology. In all cases of parentage outside of the dominant pair, paternity was shared among male relatives with a single unrelated breeding female. These results confirm the expected low effective population size resulting from limitation of successful breeding to a few lineages. We found broad overlap among classes of relatedness in proportions of bands shared; however, close relatives are reliably distinguished from unrelated individuals. Polymorphic fragments on the W chromosome of female stripe-backed wrens are useful in tracing matrilines. A population that has gone through a recent sharp decline has an unexpectedly high diversity of matrilines compared to a nearby, more productive population, owing to a higher frequency of establishment of immigrant breeders. Communally roosting black vultures (*Coragyps atratus*) proved monogamous but showed much higher background band-sharing despite their high vagility and fluid behavioral associations. Difficulty in assigning parentage in vultures resulted from lower variability.

INTRODUCTION

DNA fingerprinting is a molecular technique that creates a visual record of individual variation at hypervariable minisatellite loci (Jeffreys et al., 1985a,b). These genomic regions consist of tandem repetitive arrays of a common core sequence in which allelic variation is produced mainly by extreme variability in the number of repeats. The technique is appropriate for assignment of parentage because it screens dozens of loci simultaneously, because those loci are hypervariable, and because they assort in Mendelian fashion (Burke and Bruford, 1987, Burke et al., 1989, Wetton et al., 1987, Gyllensten et al., 1990). Because of the high mutation rates responsible for creating the useful polymorphisms at these loci (Jeffreys et al., 1988), rapid accumulation of new alleles makes the technique less appropriate for population-level applications (Lynch, 1988). Here we summarize our applications of this technique to assign parentage in three avian populations, report a novel set of fragments clearly appropriate for population-level application, and compare results among species to evaluate the usefulness of this analytical tool.

We have been using this technology to test our assumptions of the genetic structure of populations of three bird species for which long term behavioral and demographic data

Drs. Rabenold, Rabenold, Piper, Decker, and Haydock are with the Department of Biological Sciences, Purdue University, West Lafayette, IN 47907, USA.

612

exist. These species exhibit different social organizations and demographies but currently live in small or declining populations. Two species of South American wrens live in naturally fragmented populations due to patchily distributed habitat and limited dispersal, and a temperate population of vultures exists in an area of long-term decline in abundance. We will report results in detail for one wren species and make comparisons with selected aspects of vulture results.

STRIPE-BACKED WRENS

Stripe-backed wrens (*Campylorhynchus nuchalis*) are cooperatively breeding passerines that inhabit scrub woodlands in the llanos of Venezuela (Rabenold, 1990). Groups of two to ten adults live on permanent territories that they defend year-round. Fourteen years of behavioral observations on more than 100 groups of individually marked wrens have suggested that reproduction within these groups is the exclusive domain of a single dominant pair (the "principals"). The other adult members ("auxiliaries") help rear young produced in the group and are usually the offspring of one or both principals. We were interested in testing our assumptions regarding the breeding structure of these groups in order to gauge the importance of the indirect component of inclusive fitness in favoring retention of young in natal groups. Deferred reproduction and helping could be favored if the cost, unexpressed reproductive potential, is more than compensated for by the benefits of increasing the production of nondescendent kin. This benefit depends precisely upon the genetic relatedness between auxiliaries and the young they help to rear.

Groups with two or more auxiliaries produce six times the number of juveniles produced by unaided pairs or trios (Rabenold, 1984). The effect of group size creates a strong skew in reproductive success among lineages, as successful groups tend to continue to be so, while small groups rarely rise above the threshold group size for success. While the absolute population sizes are small (66 birds including 40 putative breeding adults in 20 Samán groups in 1989, 150 birds including 70 putative breeders in 35 Guácimo groups), the strong skew in reproductive success results in an even smaller effective population size.

Dispersal can be accurately characterized for this species because all recruits are color-banded in individual codes within a 10 km^2 area, and censuses are routinely performed throughout the study area and within a 1 km belt surrounding it. Females disperse more often than males and must do so to breed (97% of principal females of known origin have bred away from their natal territories), while males usually ascend to breeding status on their natal territories after the deaths of their fathers and older brothers (68% of breeding males of known origin have bred on their natal territories). Females normally disperse to a breeding position only one or two territories away from their natal territories. Male reproductive options appear to depend mainly on natal group composition (the number of older males and their survival), while female reproduction requires fierce competition for breeding positions when they occur in neighboring groups. Female removal experiments showed that females discriminate among breeding positions that occur on the basis of reproductive potential (determined by group size) and that females from natal groups near breeding openings win contests more often than those from farther away (Zack and Rabenold, 1989).

Assignment of Parentage

The accuracy with which DNA fingerprinting can assign parentage depends on the similarity of banding patterns of unrelated individuals; as variation increases, unrelated individuals will be more dissimilar and it becomes more likely that the actual parents can be identified. Assigning parentage in stripe-backed wrens should be an especially difficult application. Because of the short-distance dispersal of females and the high natal philopatry of males, the potential for local inbreeding (and therefore high levels of band-sharing) is high, and parental assignment requires distinguishing among several first-order males (brothers or fathers and sons), themselves potentially inbred, as potential fathers.

We analyzed parentage for 69 offspring produced by 22 different social groups (Rabenold, et al., 1990). Blood was collected from 260 birds in 1988 and 1989; 110 of these constituted complete membership of 17 social groups that reproduced during 1988 and 1989, and 28 additional samples provided tests of parentage for five other groups that had reproduced earlier. Histories of group membership and behavioral status were available for the previous 5–12 years.

For each group, genomic blots of *Hae*III and separate *Hin*fI digests were hybridized with Jeffreys' probes 33.6 and 33.15. Genomic blots of *Alu*I digests were also hybridized for especially large or complicated families. Five scorers located bands exclusive to each adult on autoradiographs and recorded their presence in offspring lanes. For 53 of 68 offspring, one adult female group member and one adult male group member each provided a minimum of two exclusive bands (range 2–18, mean 5.8 ± 3.4), and the entire offspring banding pattern could be derived only from those two adults. Secondly, two of us tallied the number of bands for each offspring that were unattributable to each adult combination of potential parents (intragroup). The "best fit" assigned parents, by this matrix of unattributable bands, in no case contradicted parents assigned by matching exclusive adult bands. The second technique was important for large families with several closely related potential breeders and allowed assignment of parentage for all but one juvenile.

To define the "background" proportion of bands shared (proportion shared among unrelated birds), we ran gels of unrelated birds, selecting birds related at a level less than cousins, based on long-term pedigree information. By calculating the proportion of bands shared (x) among this panel of "unrelated" birds, we are able to calculate the average allele frequency (q) across the family of loci screened by these probes and from q derive measures of confidence in our ability to exclude uncles as fathers (Table 1). However, the probability of mistakenly assigning the uncle as the father (3.6×10^{-5} in Table 1) pertains only to groups containing two first-order male relatives as potential fathers. The likelihood that an offspring band could be attributed to only one adult male in a group with n first-order male relatives including the true father is the likelihood that the father would not share that band with any of his brothers $(1-s)^{n-1}$ (see Table 1), or 0.05 for a group containing four first-order adult male relatives like our most difficult group. With an expected number of exclusively paternal bands (p) of 21.5 for both probes, only one exclusive band should match any particular male with an offspring in such a group. Matrices of unattributable bands for each combination of potential parents provide much greater resolution, as mother-uncle dyads should produce $p(1-s)$ or 8.2 unattributable bands on average.

Of the 69 young produced in 34 group-years, 62 (90%) were offspring of the principal male and principal female of their groups, and 6 were offspring of the principal female and an auxiliary male (one juvenile of uncertain history was not related to any other birds in its group). In no case was an auxiliary female identified as the parent of a juvenile. The six juveniles fathered by auxiliary males were found in four groups in which the reproductive auxiliary was not the principal female's son (because of her immigration during the lifetime of the auxiliary). Assignment of parentage in three of these four groups was obvious based on matching exclusive adult bands (Fig. 1); the auxiliary males contributed bands not shared by either principal to the offspring. The fourth group was complex, and assignment relied on the matrix of numbers of unattributable bands. In all four groups,

Table 1. Results of analysis of DNA fingerprints produced with enzyme *Hae* III and Jeffreys' probes 33.15 and 33.6 for a panel of 24 unrelated stripe-backed wrens.

Probe:		33.15		33.6	
Sex:		Females	Males	Females	Males
f	(# bands)	22.7±2.9	24.6±3.0	29.9±6.0	31.4±8.4
x	(prop. shared)	0.23 *	0.31	0.22 **	0.30
q	(mean allele freq.)[a]	0.15		0.14	
h	(heterozygosity)[b]	0.92		0.92	
s	(sib band-sharing)[c]	0.62		0.62	
m	(expected # maternal bands)[d]	14.4		18.5	
p	(expected # exclusively paternal bands)[e]	9.3		12.2	
P_u	(probability of misassigning uncle as father)[f]	0.012		0.003	
	using both probes		3.6×10^{-5}		

* $p < 0.10$
**$p < 0.05$
a: where $x = 2q - q^2$ (Jeffreys, 1985a)
b: $h = 2(1 - q)/(2 - q)$ (Georges et al., 1988)
c: $s = (4 + 5q - 6q^2 + q^3)/4(2 - q)$ (Georges et al., 1988)
d: $m = f(1 + q - q^2)/(2 - q)$ (Georges et al., 1988)
e: $p = f[1 - (1 + q - q^2)/(2 - q)]$
f: $P_u = s^p$

paternity was shared with the principal male: 1/2, 2/4, 1/2, and 2/3 juveniles were sired by the auxiliary. Of the groups in which parentage was determined, 16 contained at least one auxiliary male, and that male was assisting a father or brother with an unrelated female in 13. Shared paternity clearly does not always occur when auxiliary males assist principal pairs consisting of first-order male relatives and unrelated females (4 of 13 groups).

That males can sometimes reproduce while in subordinate auxiliary status helps explain the apparent patience of males in these positions. However, the expected reproductive success of auxiliary males (10% of the total juveniles produced during the last four breeding seasons per 212 auxiliary male-yrs = 0.08 young/year) does not compare well against either the reproductive success of males breeding without helpers (0.26 young/yr, n = 91 male-yrs during 1986–1989 breeding seasons) or their indirect contributions to genetic representation by increasing sibling production (0.64 siblings/yr; Rabenold, 1985). The ability of principal females to monopolize maternity helps explain the energy females expend in aggressive competition for principal status in large groups.

Genetic Structure of Populations

The viscous dispersal of stripe-backed wrens should have important effects on gene flow and population differentiation. Because stripe-backed wrens live in naturally fragmented populations owing to patchily distributed habitat, the non-dispersal of males and very short-distance dispersal of females create conditions favoring strong differentiation among populations (Barrowclough, 1983).

The birds sampled for the parentage study were drawn from two populations separated by 1 km: the Samán population since 1977 (now 20 groups) and the Guácimo population since 1985 (now 35 groups). Dispersal between populations is uncommon; only six wrens (four of which reproduced) in 13 years have dispersed from the Samán to the Guácimo population, and only two dispersers from Guácimo to Samán have reproduced since 1985. In contrast, 15 dispersers within Samán and 29 within Guácimo have reproduced between 1985 and 1989, reflecting the spatial limitation of most natal dispersal (85% within 0.5 km; Rabenold, 1990). The Samán population has declined from 180

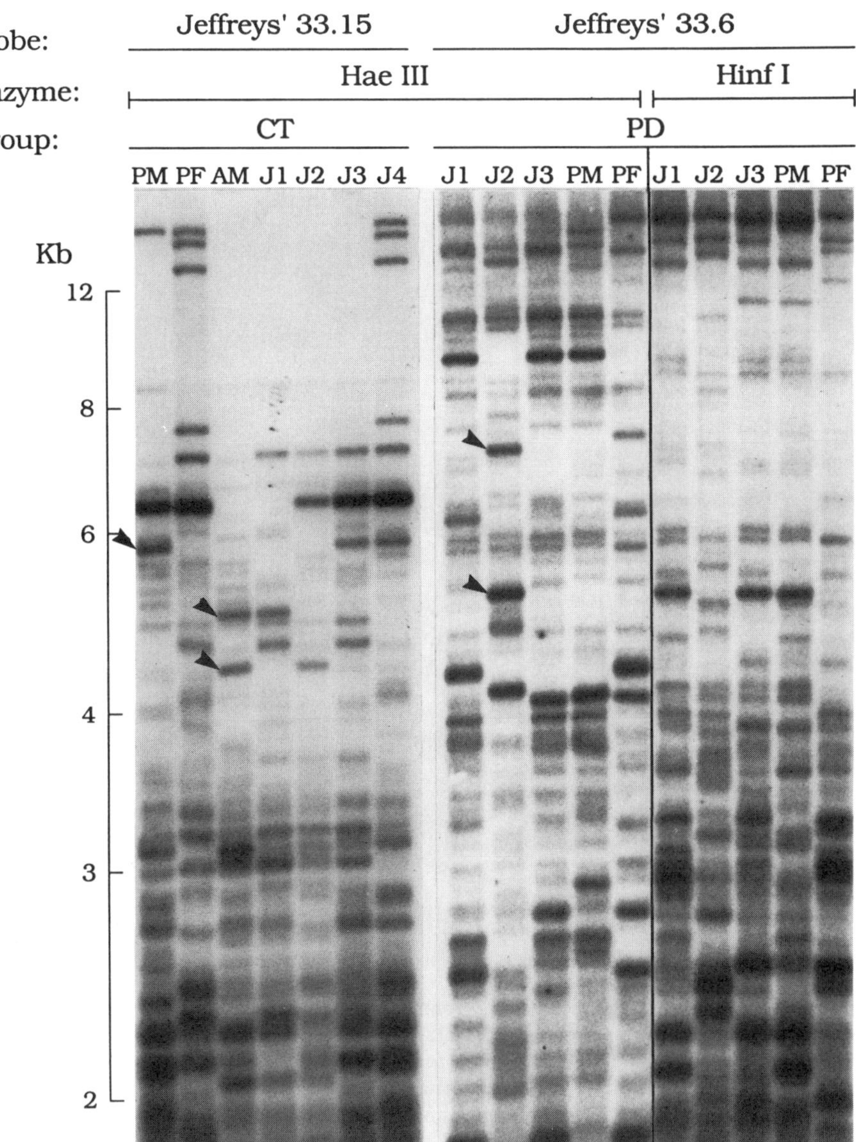

Figure 1. DNA fingerprints from 2 groups of stripe-backed wrens. 1a: Group CT: The principal male (PM) and principal female (PF) were identified as parents of juveniles 3 & 4, while juveniles 1 & 2 were offspring of the auxiliary male (AM) and principal female. Arrows indicate obvious exclusive adult bands confirming this assignment. 1b: Group PD: Juvenile 2 is the only juvenile of 69 examined not closely related to the rest of its group. Arrows indicate obvious nonmatching bands. Assignment of parents for this juvenile has not been possible among adults in the 10 neighboring groups. Reprinted by permission from *Nature* Vol. 348, pp. 538–540. Copyright 1990 Macmillan Magazines Ltd.

birds in 1978 to 66 in 1989 (0.33 birds/ha), while Guácimo has remained large throughout (0.75 birds/ha). At least part of the decline in Samán can be attributed to poor reproduction (0.63 juveniles/group-year 1985–1989, down from 0.92 juveniles/group-year 1977–1984). Reproductive success in Guácimo is 1.06 juveniles/group-year from 1985 to 1989.

We discovered distinctive genomic regions in female samples during analysis of parentage (Rabenold et al., in press a). *Hae*III digests of DNA from female wrens

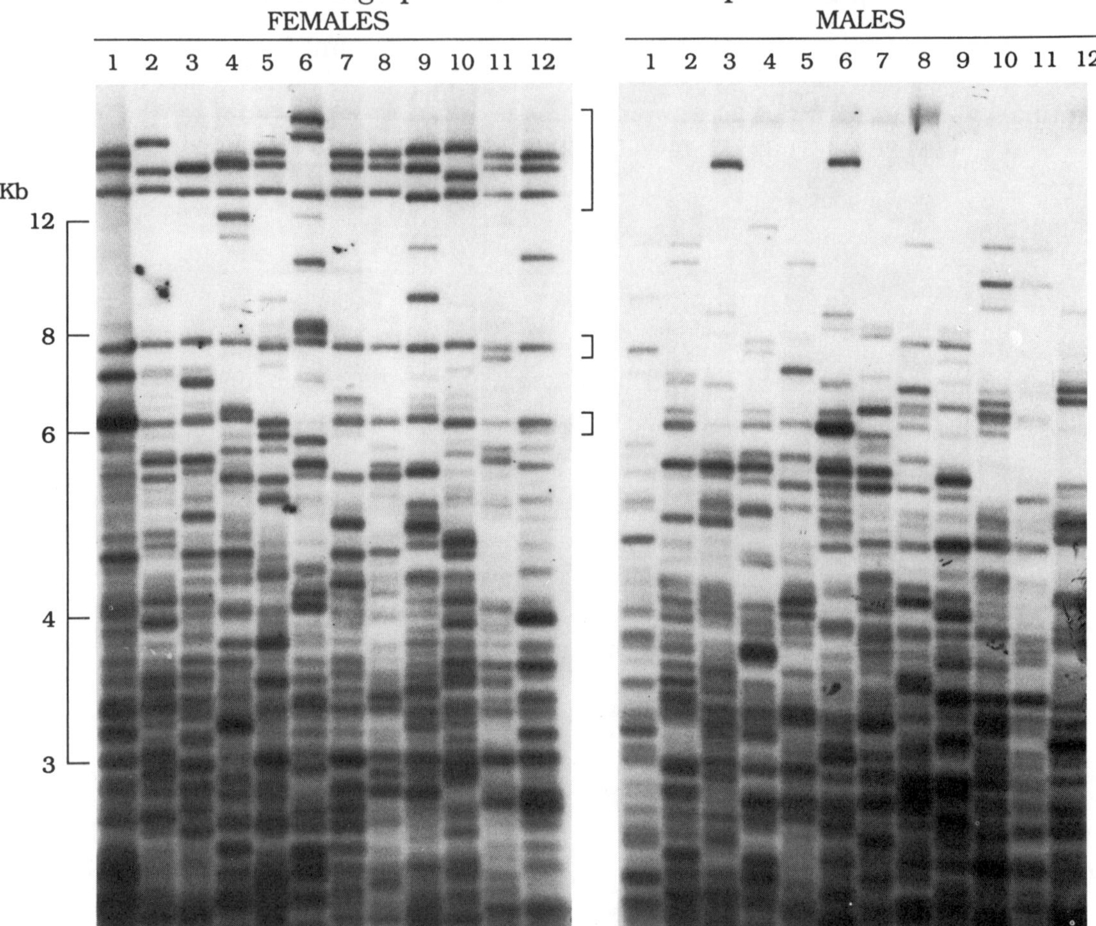

Figure 2. DNA fingerprints of 24 stripe-backed wrens whose sex had been independently determined by behavioral observation or morphology related to breeding. Female-specific fragments occur in the regions indicated by brackets to right of female lanes. Five of the 8 pattern variants are shown: Pattern A, female lanes 1, 5, 7, 8, 11, 12; B, lane 9; C, lane 3; D, lanes 2, 10; E, lane 6. Large fragments in the same region occur in male lanes 3 and 6. Reprinted by permission from *Animal Behavior* (in press). Copyright Academic Press Inc. (London) Ltd.

hybridized with Jeffreys' probe 33.15 produce a pattern of usually 5 large fragments which are distinct from the hypervariable patterns presumably derived from autosomes (Fig. 2). Three of these fragments are larger than 12 kb and two are near 6 and 8 kb; males only rarely have a single fragment larger than 12 kb. Eight pattern variants existed among 84 adult females living in the two populations in 1989. The female-specific fragments are transmitted from mother to daughter (24 pairs in 16 groups) but not transmitted to sons (27 pairs), suggesting that they are derived from regions on the W chromosome not involved in crossing over with the Z chromosome.

The relative frequencies of the eight sex-linked haplotypes differ significantly between populations (Rabenold et al., in press b). Frequencies of types A through H for 35 Samán females are 0.34, 0.23, 0.20, 0.09, 0.09, 0.03, 0.00, 0.03; for 49 Guácimo females frequencies for types A—H are 0.71, 0.14, 0.06, 0.02, 0.02, 0.00, 0.02, 0.02 ($X^2 = 12.3$, 3 df, p = 0.01, collapsing categories D-H; Fig. 3). The Guácimo population (which is large and stable) is strongly dominated by type A, while Samán (the low-density population) is

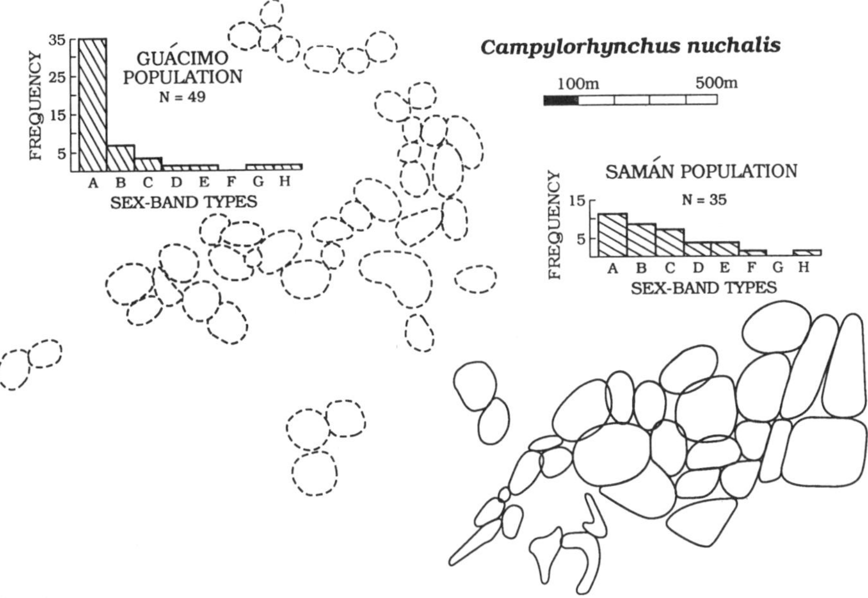

Figure 3. Geographic relationship of the Saman (lower right, solid lines) and Guácimo (upper left, dashed lines) populations of *Campylorhynchus nuchalis*. Histograms represent counts of females sorted by pattern of sex-specific fragments in each population. Not all territories outlined are currently occupied. Other marked groups exist, separated from those shown, to the northeast and southwest.

more diverse (Simpson's reciprocal diversity index, interpreted as the equivalent number of equally represented types, is 1.89 for Guácimo and 4.42 for Samán).

The higher genetic diversity found in the low-density population can be explained in light of demographics. In the Samán area, the mortality of breeding females has not been matched by production of juvenile females in the last five years. During this time, 18 of 35 (51%) vacancies for breeding females have been filled by birds originating outside the Samán population. High genetic diversity is therefore associated with a known high frequency of immigration. In contrast, only 13 of 52 (25%) female breeding positions in the Guácimo population have been filled by birds originating from outside the population.

These results suggest that normal behavioral barriers to dispersal are relaxed at low densities and migration rate increases accordingly. In this case, the mechanism is a strong advantage of natal group proximity in competition for productive breeding positions that can demonstrably be monopolized (Zack and Rabenold, 1989). In the low-density Samán population, few young female contestants reside in the neighborhood of a vacancy, so that females from farther away are more likely to compete successfully than in the relatively dense Guácimo population.

BLACK VULTURES

Black vultures (*Coragyps atratus*) inhabit an enormous range covering most of South and Central America and southeastern North America and occupy a variety of habitats. However, local decline in the southeastern U.S. is common, and local extirpation in parts of their former range has resulted in isolation of some remnant populations. Migration among populations has not been documented, but such mobile animals could easily interbreed over very large areas.

Black vultures aggregate nightly in large communal roosting groups, even through

618

the breeding season when they nest as isolated pairs. Roosting groups vary in size throughout the year, numbering as many as 300 individuals in North Carolina winter roosts (Rabenold, 1987a). Roost membership is fluid (Rabenold, 1987b), so individuals interact with hundreds of others weekly, including during their fertilizable periods. Within roosting groups, members of nuclear families associate preferentially, and different nuclear families of unknown relatedness associate predictably in behavioral alliances (Rabenold, 1986). Evidence that vultures use communal roosts as centers for the transfer of information regarding the current location of food (Rabenold, 1987a), along with behavioral association of nuclear family members, suggests that the evolution of this form of cooperation may be favored by kin association, such that birds sharing information with close kin increase the representation of their genes in future generations because they contribute to the survival and reproduction of close relatives.

Our aims in applying DNA fingerprinting to black vultures were to ask whether mating is monogamous, as suggested by behavior, and whether associated families represent extended genetic families. Our tests of parentage in black vultures have revealed no deviations from monogamy among 40 young produced in 15 families, although, unlike the wrens, there exist alternative adults in the population that cannot be excluded as parents. Parental exclusion was hampered by relatively high levels of band-sharing among unrelated individuals. Dyads of behaviorally allied adults not mated to each other and not known to be close relatives show surprisingly high frequencies of co-occurrence. Many such dyads also show high genetic similarity, but the analysis is only suggestive to date.

Assigning Parentage—Technical Considerations

For stripe-backed wrens, distributions of proportions of bands shared by dyads of individuals for three categories of relatedness overlap, but the extremes are distinct to the extent that any birds sharing 45% or more of their bands are at least half sibs, and any sharing 60% or more are almost certainly first-order relatives (Fig. 4). For black vultures, the two extreme categories overlap much more and the variance about each mean is larger (Fig. 5; related birds here consist of known parent-offspring dyads, while unrelateds consist of mates and dyads of young birds paired with known non-parents). The mean proportion of bands shared among unrelated birds in wrens is 0.25 averaged across two probes with one enzyme, and 0.34 averaged across three probes with one enzyme for the vultures. The higher genetic similarity and greater variance within categories for vultures results in several parent-offspring dyads sharing 85% of their bands. By conventional parentage assignment, any adult in the population that could provide the missing 15% would qualify as the second parent. With $x = 0.34$ and $f = 12.3$, the number of bands shared with offspring in order to be assigned as the second parent is only 1.8, which will occur with a likelihood of $0.34^{1.8}$ or 0.14. The use of several probe-enzyme combinations and reliance on matrices of unattributable bands in combination with band-sharing distributions (the actual second parent should share a far greater proportion of bands than a second parent misassigned as illustrated above) allowed assignment of parentage for 39 of 40 vulture offspring (Decker et al., in prep).

The surprising technical result that the vultures, with their extreme mobility and relatively large populations, have much higher band-sharing than the wrens, with their extremely viscous dispersal and small populations, has two possible explanations. First, the mutation rates of these minisatellites vary across species (Burke and Bruford, 1987, Wetton et al., 1987, Jeffreys et al., 1988). Higher mutation rates in wrens could produce the observed variation within wren families rivaling that found between families in vultures. We have calculated the mutation rate at these loci for wrens as 0.0027 mutations/locus/generation, and a preliminary calculation of vulture minisatellite mutation rate is at least that high (>0.003).

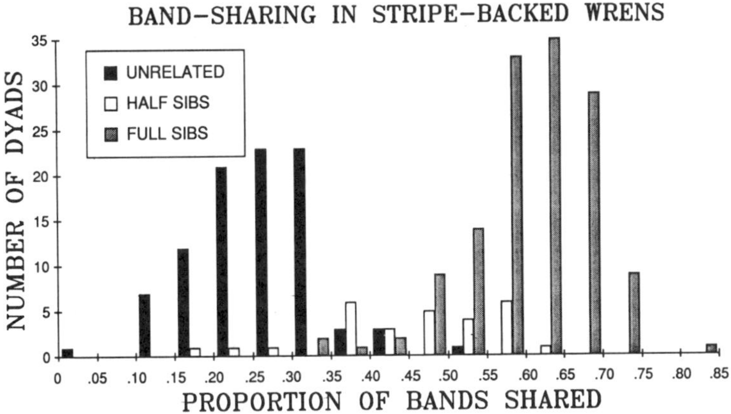

Figure 4. Proportions of bands shared in dyadic comparisons of stripe-backed wrens, *Campylorhynchus nuchalis*, by genetic relatedness.

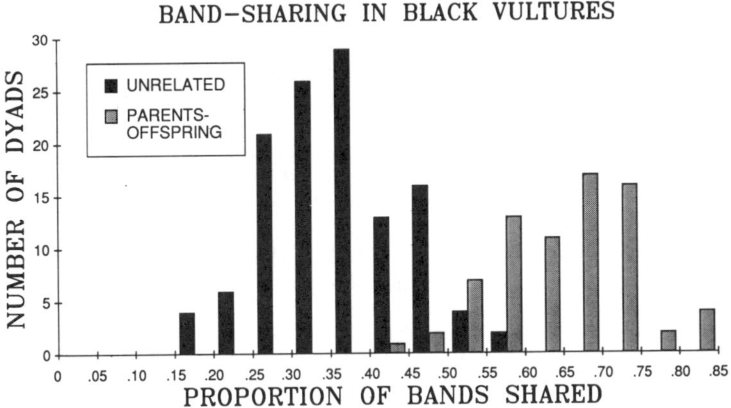

Figure 5. Proportions of bands shared in dyadic comparisons of black vultures, *Coragyps atratus*, by genetic relatedness.

An alternative explanation is that vulture demographics drive the relationship. It is possible that skew in reproductive success among lineages is even more pronounced in vultures than wrens and that natal dispersal is effectively just as viscous in vultures despite their mobility. Our 14-year study of a North Carolina population of black vultures has shown birds marked as nestlings beginning to breed at age eight or more not far from their parents, and reproductive success is highly skewed (among 33 breeding pairs monitored since 1977, one pair is responsible for 18% of the 120 young produced). In general, because black vultures are long-lived and mobile, we cannot quantify their demographics with nearly the certainty possible with the sedentary short-lived wrens.

In summary, this technique has confirmed some of our assumptions about the genetic structure of these populations. In stripe-backed wrens, reproductive success is skewed among lineages, resulting in small effective population sizes because of constraints imposed by social organization. Breeding by auxiliary males that was previously undetected is entirely shared within male lineages. Fierce competition among females for breeding opportunities has a strong role in the genetic structuring of populations. When a declining population fails to produce many females, local resistance to immigration is less effective, migration rate increases, and the expected drop in genetic diversity is compensated.

DNA fingerprinting has illuminated these systems considerably. We can clearly determine breeding systems and the distribution of biological fitness in some populations, and make headway toward understanding the mechanisms of gene flow and differentiation among populations. When the underlying genetics of these loci are better understood, we will know the degree to which we can use DNA fingerprinting to make inferences regarding demographics of previously intractable species.

ACKNOWLEDGMENTS

We are indebted to Tomás Blohm for use of his field station, Hato Masaguaral. Our thinking has been influenced by discussions and comments on earlier manuscripts by Jeff Bennetzen, Jerram Brown, Alec Jeffreys, Jeff Lucas, Dennis Minchella, Sam Skinner, Peter Waser, and David Westneat. Dennis Minchella provided lab facilities, and Steve Zack provided census information on the Guácimo wren population. This work has been supported by the National Science Foundation, most recently by BSR-8818038.

LITERATURE CITED

Barrowclough, G. F. 1983. Biochemical studies of microevolutionary processes. Pp. 223–261. *In:* A. H. Brush and G. A. Clark Jr (eds.), *Perspectives in Ornithology.* Cambridge University Press, London.

Burke, T., & M. W. Bruford. 1987. DNA fingerprinting in birds. *Nature* 327:149–152.

Burke, T., Davies, N. B., Bruford, M. W. & B. J. Hatchwell. 1989. Parental care and mating behaviour of polyandrous dunnocks, *Prunella modularis*, related to paternity by DNA fingerprinting. *Nature* 338:249–251.

Georges, M., Lequarre, A-S., Castelli, M., Hanset, R. & G. Vassart. 1988. DNA fingerprinting in domestic animals using four different minisatellite probes. *Cytogenet. Cell Genet.* 47:127–131.

Gyllensten, U. B., Jakobsson, S. & H. Temrin. 1990. No evidence for illegitimate young in monogamous and polygynous warblers. *Nature* 343:168–170.

Jeffreys, A. J., Royle, N.J., Wilson, V. & Z. Wong. 1988. Spontaneous mutation rates to new length alleles at tandem-repetitive hypervariable loci in human DNA. *Nature* 332:278–281.

Jeffreys, A. J., Wilson, V. & S. L. Thein. 1985a. Hypervariable 'minisatellite' regions in human DNA. *Nature* 314:67–73.

Jeffreys, A. J., Wilson, V. & S. L. Thein. 1985b. Individual-specific 'fingerprints' of human DNA. *Nature* 316:76–79.

Lynch, M. 1988. Estimation of relatedness by DNA fingerprinting. *Mol. Biol. Evol.* 5:584–599.

Rabenold, K. N. 1984. Cooperative enhancement of reproductive success in tropical wren societies. *Ecology* 65:871–885.

Rabenold, K. N. 1985. Cooperation in breeding by nonreproductive wrens: kinship, reciprocity and demography. *Behav. Ecol. Sociobiol.* 17:1–17.

Rabenold, K. N. 1990. *Campylorhynchus* wrens: The ecology of delayed dispersal and cooperation in the Venezuelan savanna. Pp. 157–196. *In:* P. B. Stacey & W. D. Koenig (eds.), *Cooperative Breeding in Birds.* Cambridge University Press, London.

Rabenold, P. P. 1986. Family associations in communally roosting black vultures. *Auk* 103:32–41.

Rabenold, P. P. 1987a. Recruitment to food in black vultures: evidence for following from communal roosts. *Anim. Behav.* 35:1775–1785.

Rabenold, P. P. 1987b. Roost attendance and aggression in black vultures. *Auk* 104:647–653.

Rabenold, P. P., Rabenold, K. N., Piper, W. H., Haydock, J. and S. W. Zack. 1990. Shared paternity revealed by genetic analysis in cooperatively breeding tropical wrens. *Nature* 348:538–540.

Rabenold, P. P., Piper, W. H., Decker, M. D. and D. J. Minchella. In press a. Polymorphic minisatellite amplified on avian W chromosome. *Genome.*

Rabenold, P. P., Rabenold, K. N., Piper, W. H. and D. J. Minchella. In press b. Density-dependent dispersal in social wrens: Genetic analysis using novel matriline markers. *Anim. Behav.*

Wetton, J. H., Carter, R. E., Parkin, D. T. & D. Walters. 1987. Demographic study of a wild house sparrow population by DNA fingerprinting. *Nature* 327:147–149.

Zack, S. W. and K. N. Rabenold. 1989. Assessment, age and proximity in dispersal contests among cooperative wrens: field experiments. *Anim. Behav.* 38:235–247.

The Possible Contribution of Retroviral-like Element Insertion Mutants to Inbreeding Depression

John F. McDonald, Amy K. Csink, and A. Jamie Cuticchia

Abstract. Retroviral-like transposable elements (RLEs) are a large and pervasive class of transposable elements. RLEs are abundantly represented within the genomes of vertebrates, invertebrates, plants, and yeast. Spontaneous or stress-induced insertion of RLEs in or near coding sequences can result in the sudden acquisition of new regulatory phenotypes. The regulatory phenotypes associated with RLE insertion variants can be shielded from natural selection by (host) chromosomal suppressor genes. Thus, RLE mediated regulatory and developmental mutations may be harbored in natural populations in significant numbers. Although this regulatory variation may constitute an important source of long-term evolutionary potential, in the short run it may be expected to be released in inbred populations and contribute to the detrimental effects associated with inbreeding depression.

INTRODUCTION

A fundamental goal of conservation biologists is to reduce the probability of extinction of endangered species by instituting breeding programs which may help lessen the effects of "inbreeding depression", i.e., the tendency of populations to experience significant decreases in viability and/or fertility when inbred. Obviously, the most effective conservation measures to counteract the negative effects associated with inbreeding depression will be those that are based on a complete understanding of the biological factors which underlie the process. However, our current knowledge of the biological basis of inbreeding depression is incomplete.

Although it is axiomatic that inbreeding reduces the level of heterozygosity in populations, the biological consequences of this reduction in heterozygosity are not entirely clear. The classical explanation of inbreeding depression is that it is the result of the unmasking of recessive lethal and sub-lethal alleles that are carried in the founding population and made homozygous by inbreeding. Based on the assumption that this explanation is essentially correct, breeding programs have been designed by conservation biologists to maximize the levels of genetic variation present in endangered species and thereby minimize the probability of extinction. However, recent laboratory and field studies of inbred populations suggest that the genetic changes experienced by inbred populations may extend beyond the simple uncovering of recessive lethal alleles (e.g., in this volume). As more information is accumulated on the molecular and genetic processes

Drs. McDonald, Csink, and Cuticchia are with the Department of Genetics, University of Georgia, Athens, GA 30602, USA.

622

which may have an influence on the phenomenon of inbreeding depression, additional genetic strategies may emerge for helping to insure the survival of endangered species.

For the past several years our laboratory has been engaged in studying the role of retroviral-like transposable elements (RLEs) in regulatory evolution. RLE insertion mutants are frequently associated with novel regulatory phenotypes and recent studies have identified several instances where RLE insertion events have been at the basis of the acquisition of new regulatory and developmental phenotypes over evolutionary time. In addition to being frequently associated with the acquisition of novel regulatory phenotypes, RLE insertion mutations are unique in their ability to be phenotypically suppressed by chromosomal genes. In this report, we present data which indicate that at least one such class of chromosomal suppressor alleles is represented in natural populations at significant frequencies. We believe that as a result of this naturally occurring suppressor variation, RLE-mediated regulatory and developmental mutations may be harbored in natural populations in significant numbers. In the long run, this regulatory and developmental variation may constitute an important source of evolutionary potential. In the short run, however, this regulatory variation may be expected to be released in inbred populations and thereby significantly contribute to the detrimental effects associated with inbreeding depression.

THE *DROSOPHILA* RLE *COPIA* IS VARIABLY EXPRESSED IN NATURAL POPULATIONS, AND AT LEAST SOME OF THIS VARIATION IS ATTRIBUTABLE TO TRANS-REGULATORY CONTROLS

We have recently completed a survey of *copia* expression in flies representing 37 populations world-wide of *Drosophila melanogaster, D. simulans,* and *D. mauritiana* (Csink & McDonald, 1990). The results demonstrate that, although *copia* elements are present in all three species, *copia*-encoded transcripts are detectable only in *D. melanogaster.* Levels of *copia* transcripts vary nearly 100-fold among flies representing geographically diverse populations of *D. melanogaster* and this variation is not correlated with variability in *copia* copy number. Analysis of transcript levels in inter-population hybrids demonstrated that much of this variability may be attributed to the action of trans-acting controls.

TRANSCRIPTION IS A RATE-LIMITING STEP IN THE RLE LIFE CYCLE; RLE EXPRESSION IS SUBJECT TO HOST CHROMOSOMAL CONTROL

Retroviral-like transposable elements replicate in a manner similar to infectious retroviruses (Boeke et al., 1985). The inserted element is transcribed by RNA polymerase II, assisted by a variety of tissue- and stage-specific transcription factors. Some of the transcripts are processed into mRNA while some remain full length. The full-length transcripts are packaged into capsid particles where they are reverse-transcribed into DNA copies of the element. The DNA copies are integrated back into the genome by means of an element-encoded integrase. The activation of RLEs by the cellular transcriptional apparatus is made possible by the presence of eukaryotic promoter and enhancer sequences contained within the element's long terminal repeat (LTR). The presence of these sequences within RLE LTRs usually results in the element being expressed in a tissue- and/or life-stage specific manner (McDonald et al., 1988).

THE MUTANT PHENOTYPES ASSOCIATED WITH RLE INSERTION MUTANTS MAY BE SUPPRESSED BY CHROMOSOMAL GENES; THESE SUPPRESSOR GENES ARE TRANS-REGULATORS OF RLE EXPRESSION

An interesting property of RLE insertion mutants from the evolutionary perspective is the fact that mutant phenotypes are not rigidly determined but may be modified by the action of chromosomally encoded suppressor genes (Kubli, 1986; Parkhurst & Corces, 1985). Analysis of RLE insertion mutants in yeast (Winston, 1988), mice (Sweet, 1983), and *Drosophila* (Parkhurst & Corces, 1985) indicate that the mutant phenotype may differ dramatically or only slightly from the progenitor wild-type allele depending upon the genetic background in which the mutant allele is expressed.

Suppressor genes have been identified and mapped to numerous fixed chromosomal locations in the genomes of yeast (Winston, 1988), *Drosophila* (Kubli, 1986), and mice (Sweet, 1983). It has been shown that suppressor genes act by exerting trans-regulatory effects on RLEs. The RLEs then repress functions responsible for the mutant phenotype. In both yeast and *Drosophila*, the transcriptional activity of RLEs inserted upstream of chromosomal genes has been shown to interfere with the adjacent gene's expression Winston, 1988; Kubli, 1986). Genes which repress RLE expression are therefore able to partially or completely suppress mutant phenotypes associated with this type of RLE insertion mutant. For example, our laboratory has characterized a naturally occurring *Drosophila melanogaster* alcohol dehydrogenase (*Adh*) allele which contains a *copia* RLE inserted 240 bp upstream from the distal (adult) transcriptional start site (Strand & McDonald, 1989). The results of our characterization demonstrate that patterns of *Adh* expression are quantitatively disrupted at life-stages and in tissues where *copia* is actively expressed.

SUPPRESSOR GENE VARIATION IS PRESENT IN NATURAL POPULATIONS OF *DROSOPHILA*

Our finding that natural populations of *Drosophila* contain trans-repressors of *copia* expression implies that these populations also harbor the potential to phenotypically suppress regulatory variants associated with *copia* and perhaps other RLE insertion variants. To test this hypothesis we produced hybrids between flies homozygous for a naturally occurring dominant trans-repressor of *copia* expression (an isofemale line derived from the Loukanga, Congo population) and flies having moderate levels of *copia* expression but homozygous for the *Adh copia* insertion variant described above. *Copia* levels in the F1 progeny were nearly absent, which is consistent with the presence of a dominant repressor of *copia* expression being carried by the Loukanga flies. Interestingly, however, *Adh* expression in the F1 hybrids was found to be restored to wild-type levels, which is consistent with a trans-suppression of the mutant *Adh* phenotype.

Further evidence that natural populations of *Drosophila* contain the potential to suppress RLE insertion variants comes from a series of preliminary studies where females homozygous for X-linked RLE insertion mutants have been crossed to males collected from an Athens, Georgia population. Examination of the F1 males, which are hemizygous for the RLE insertion mutant, indicate that the mutant phenotype has been trans-suppressed in approximately 50% of the progeny. Thus, dominant suppressors of RLE insertion mutants are apparently present in natural populations of *Drosophila* in moderate frequencies.

EVOLUTIONARY SIGNIFICANCE OF SUPPRESSOR GENES IN NATURAL POPULATIONS

Since RLE insertion mutants are often associated with the acquisition of new trans-regulatory controls, it has been postulated that they may contribute significantly to the evolution of new regulatory and developmental networks (McDonald, 1990; Temin, 1982). However, the major changes in phenotype typically associated with RLE insertion variants make the theory that these mutations are responsible for major evolutionary shifts subject to the same criticism leveled against earlier macromutational evolutionary theories (e.g., Goldschmidt, 1940): that variant alleles having major phenotypic consequences should, in most instances, be quickly eliminated from populations by natural selection. However, since the mutant phenotypes associated with RLE insertion variants can be partially or completely suppressed by chromosomal loci (which our results indicate are present in natural populations), RLE insertion mutants may be shielded from natural selection and thereby maintained in populations for extended periods of time (McDonald 1990; Cuticchia & McDonald, in preparation). As a consequence, we believe that RLE insertion variants may have long-term evolutionary potential. In populations being observed over short periods of time, however, RLE insertion variants may appear to be selectively neutral or nearly neutral due to the fact that they are being phenotypically suppressed. Thus, RLE variants present in populations may be mistakenly interpreted as being of little or no long-term evolutionary importance (e.g., Langley et al., 1988).

RLE INSERTION VARIANTS, INBREEDING DEPRESSION, AND CONSERVATION BIOLOGY

Our results suggest that a significant component of the decrease in fitness associated with inbreeding depression may be attributable to the unmasking of normally suppressed RLE insertion variants. Although the number of insertion variants exerting negative selective effects may be large, the number of suppressor loci controlling their expression may be small. Current models of inbreeding depression incorporate the postulate that the detrimental genes uncovered by inbreeding are recessive. According to our model the mutant alleles need not be recessive since dominant RLE insertion variants can be trans-suppressed as efficiently as recessive mutants.

There have been relatively few detailed surveys of the quantity and quality of genetic variation that is released upon the inbreeding of natural populations. The best studies available are analyses of *Drosophila* populations. In these studies, inbreeding of flies collected in the wild resulted in the uncovering of large amounts of cryptic morphological and viability mutants. Interestingly,not all of the variant alleles uncovered by inbreeding were found to be recessive (e.g., Spencer, 1957).

Among the most intensively studied classes of naturally occurring variants are those affecting viability. Many lethal genes have been found to be cryptically maintained in natural populations of *Drosophila*. Although most of these lethal genes are recessive, it has been found that the lethal effects associated with many of these alleles can be "suppressed" by the action of other naturally occurring genes (e.g., both papers by Magalhaes et al., 1965). Whether or not these "suppressible" lethal alleles are RLE insertion variants remains to be determined.

If RLE insertion variants are an important source of naturally occurring variation affecting fitness, it will become important to identify and characterize host genes that are capable of suppressing these mutant alleles. Future conservation strategies may include efforts to insure that such "suppressor genes" are present in populations of species which are few in number and considered to be in danger of extinction.

LITERATURE CITED

Boeke, J., Garfinkel, D., Styles, C. & G.Fink. 1985. *Ty* elements transpose through an RNA intermediate. *Cell* 40:491–500.

Csink, A. K. & J. F. McDonald. 1990. *Copia* expression is variable among natural populations of *Drosophila. Genetics.* 126:375–385.

Goldschmidt, R. B. 1940. *The Material Basis of Evolution.* Yale University Press: New Haven.

Kubli, E. 1986. Molecular mechanisms of suppression in *Drosophila. Trends Genet.* 2:204–208.

Langley, C. H., Montgomery, E., Hudson, R., Kaplan, N. & B. Charlesworth. 1988. On the role of unequal exchange in the containment of transposable element copy number. *Genet. Res.* 52:223–235.

Magalhães, L. E., de Toledo, J. S., and A. B. da Cunha. 1965. The nature of lethals in *Drosophila willistoni. Genetics* 52:599–608.

Magalhães, L. E., da Cunha, A. B., de Toledo, J. S., de Toledo, S. P., Souza, H. L., Targa, H. J., Setzer, V., & C. Pavan. 1965. On lethals and their suppressors in experimental populations of *Drosophila willistoni. Mutation Res.* 2:45–54.

McDonald, J. F. 1990. Macroevolution and retroviral elements. *BioSci.* 40:183–191.

McDonald, J. F., Strand, D. J., Brown, M. R., Paskewitz, S. M., Csink, A. K. & S. H. Voss. 1988. Evidence of host-mediated regulation of retroviral element expression at the post-transcriptional level. Pp. 219–234. *In:* M. E. Lambert, J. F. McDonald & I. B. Weinstein (eds.), *Eukaryotic Transposable Elements as Mutagenic Agents.* Cold Spring Harbor Press: Cold Spring Harbor, NY.

Parkhurst, S. & V. Corces. 1985. Forked, gypsys and suppressors in *Drosophila. Cell* 41:429–437.

Spencer, W. P 1957. Genetic studies on *Drosophila mulleri. Univ. Texas Publ. No 5721*:186–205.

Strand, D. J. & J. F. McDonald. 1989. Insertion of a *copia* element 5' to the *Drosophila melanogaster* alcohol dehydrogenase gene (*Adh*) is associated with altered developmental and tissue-specific patterns of expression. *Genetics* 121:707–794.

Sweet, H. O. 1983. Dilute-suppressor, a new suppressor gene in the house mouse. *J. Hered.* 74:305–307.

Temin, H. H. 1982. Viruses, protoviruses, development and evolution. *J. Cell Biochem.* 19:105–118.

Winston, F. 1988. Genes that affect Ty-mediated gene expression in yeast. Pp. 145–153. *In:* M. E. Lambert, J. F. McDonald, I. B. Weinstein (eds.), *Eukaryotic Transposable Elements as Mutagenic Agents.* Cold Spring Harbor Press: Cold Spring Harbor, NY.

Conservation of Genetic Diversity in Rare and Endangered Plants

Kent E. Holsinger

Abstract. Two decades of research on the distribution and abundance of genetic diversity in plant populations has revealed two important facts. First, broad generalizations about the influence of life-history characteristics and geographic range on the genetic structure of plant populations are possible, but these variables can explain less than 50% of the differences among species. Second, rare species are usually less genetically diverse than their widespread congeners, but they distribute that variation within and among populations in similar ways. Because geographic range is a poor predictor of genetic structure, however, rare species in some genera are more diverse than widespread species in other genera. Most of the data on genetic diversity in plant populations is based on analysis of allelic variation at electrophoretically detectable loci, but conservationists are primarily interested in genetic variation at loci important in ecological and adaptive responses. Unfortunately, electrophoretic loci may not provide an accurate description of genetic variation at other, ecologically significant loci, but they may provide useful markers of changes in a population's status. The protection of existing populations must be the primary means by which genetic diversity is conserved, but it should be supplemented by the collection of propagules for conservation in off-site collections. Although preliminary genetic surveys are essential for the design of a sampling program for crop germplasm, samples for an off-site collection of rare plants can be obtained without prior genetic surveys. An integrated approach to rare plant conservation that focuses on the protection of existing populations and uses off-site collections as an insurance policy against catastrophe has a good chance of success.

INTRODUCTION

In 1988 The Center for Plant Conservation estimated that nearly 700 species of vascular plants might go extinct in the United States before the year 2000. That is roughly equal to the number of bird species known to breed in North America north of Mexico. Thus, the task facing plant conservationists is enormous. We must try to prevent the extinction of a huge number of species, even though we know almost nothing about the biology of many of them. Protection of existing populations will obviously be the cornerstone of any conservation plan, but that alone will not suffice. Detailed analyses of autecology and demography are necessary to identify threats to the viability of natural populations, and conservation management will usually be directed toward the mitigation of direct ecological threats to a population's continued existence (Huenneke et al., 1986; Lande, 1988a, b; Holsinger and Gottlieb, 1991; Menges, 1991). The ultimate success of any conservation effort, however, depends not only on our ability to protect the habitat in which a rare plant is found and our ability to manage its population dynamics, but also on our

Dr. Holsinger is with the Department of Ecology & Evolutionary Biology, U-43, University of Connecticut, Storrs, CT 06269, USA.

ability to conserve the genetic diversity that provides raw material for adaptive responses to environmental change. In this paper I shall describe some of the techniques that can be used to conserve that genetic diversity. In particular, I shall demonstrate that off-site conservation of genetic diversity is a practical reality for plant species and should be used to supplement the protection of natural populations wherever possible.

To do this, I must first describe some of what we know about the genetic structure of rare plant populations. Because much of what we know about the genetic structure of plant populations, whether common or rare, is based on the analysis of variation at electrophoretically detectable loci, the first part of this paper summarizes a bit of what we know about factors determining the distribution and abundance of electrophoretic variation in plant populations. Of course, electrophoretic variation is not usually something that plant conservationists are concerned about saving. Rather, we are interested in what variation at electrophoretic loci can tell us about patterns of variation at other, ecologically significant loci. Thus, the second part of this paper illustrates some of the limitations of electrophoretic data, but also suggests how they can be wisely used. In the final part of this paper I shall discuss some practical recommendations for the conservation of genetic diversity in rare plants that can be implemented today.

PATTERNS OF ELECTROPHORETIC VARIATION

The genetic structure of plant populations is influenced by many things. Hamrick and his colleagues have, in a series of influential studies over the past 10 years (Hamrick et al., 1979; Loveless and Hamrick, 1984; Hamrick and Godt, 1990), described many of them. But there are two characteristics that are particularly important for our purposes: the mating system and geographic range. The mating system is arguably the single most important characteristic determining the genetic structure of plant populations, influencing both the total amount of variation found within a species and the partitioning of that variation within and among populations. The influence of geographic range on the genetic structure of a species will obviously affect the design of strategies to conserve genetic diversity in rare species.

Plant taxa that reproduce predominantly by outcrossing have, on average, almost 40% more loci with electrophoretically detectable variation than those that reproduce predominantly by selfing (Table 1). In addition, the mean heterozygosity is an order of magnitude lower in selfers than in outcrossers (Brown, 1979; Gottlieb, 1981). In plant taxa that reproduce predominantly by selfing over 50% of the total electrophoretically detectable variation is found among populations, while only 12% is found among populations of outcrossers (Hamrick et al., 1979; Loveless and Hamrick, 1984; Hamrick and Godt, 1990). In short, selfers tend to be less genetically diverse than outcrossers, but what genetic diversity they do exhibit is more the result of differentiation between populations than of differences between individuals within a population.

Table 1. Breeding System and Genetic Structure of Plant Populations[1]

	P^2	A^3	G_{st}^4
Selfing	41.8	1.69	0.510
Mixed	56.8	1.93	0.155
Outcrossing	58.1	2.20	0.148

[1]from Hamrick and Godt (1990)
[2]proportion of polymorphic loci
[3]number of alleles per polymorphic locus
[4]proportion of genetic diversity due to allele frequency differences among populations (Nei 1973b)

In contrast to the effect of breeding system, the geographical range occupied by a species influences only the total amount of genetic variation present, not its partitioning within and among populations (Table 2). Geographically widespread species are polymorphic at almost 50% more of their loci than endemics, and the average number of alleles at polymorphic loci in widespread species is 2.29; it is only 1.80 in endemics. In contrast, G_{st} values for endemics are not significantly different from those for widespread species. Widespread species have an average G_{st} of 0.210, and endemics have an average G_{st} of 0.248. In short, the genetic structure of geographically widespread species is similar to that of endemics. The only difference is that geographically widespread species are more genetically diverse than are endemics.

Unfortunately, these broad patterns do not allow precise predictions of either the amount of electrophoretic diversity present or its distribution among populations. In the most extensive analyses to date (Hamrick and Godt, 1990; Hamrick et al., 1991) less than 50% of the species-to-species heterogeneity in genetic structure has been explained by life-history characteristics and geographic range. The inability to make precise predictions about the genetic structure of populations from life-history characteristics is often attributed to the specific ecological and evolutionary history of a species, and this is clearly at least part of the explanation. Karron (1987, 1991) found that widespread species in a genus had both a greater percentage of polymorphic loci and a larger number of alleles at polymorphic loci than geographically restricted members of the same genus, consistent with the general pattern identified by Hamrick and his colleagues. Karron's analysis also shows, however, why geographic range alone is a poor predictor of genetic structure.

Some plant genera are characterized by a relatively high level of genetic diversity, and others by a level that is relatively low. Thus, geographically restricted species in some genera may be more genetically diverse than widespread species in others. For example, geographically restricted species of *Layia* are polymorphic at 67% of their loci, and each polymorphic locus has, on average, 3.0 alleles (Gottlieb et al., 1985; Warwick and Gottlieb, 1985). Geographically widespread species of *Astragalus*, on the other hand, are polymorphic only at 29% of their loci, and each polymorphic locus has only 2.7 alleles, on average (Karron et al., 1988). In short, only direct observation can provide an accurate picture of the genetic structure of any plant species. As a result, preliminary surveys of genetic variation are the rule in plans for the conservation of crop germplasm. They will also be required in plans for genetic conservation of rare plants, if those plans depend on an accurate understanding of the genetic structure of natural populations.

Table 2. Geographic Range and Genetic Diversity in Plant Populations[1]

	P^2	A^3	G_{st}^4
Endemic	40.0	1.80	0.248
Narrow	45.1	1.83	0.242
Regional	52.9	1.94	0.216
Widespread	58.9	2.29	0.210

[1] from Hamrick and Godt (1990)
[2] proportion of polymorphic loci
[3] number of alleles per polymorphic locus
[4] proportion of genetic diversity due to allele frequency differences among populations (Nei 1973b)

SIGNIFICANCE OF ELECTROPHORETIC VARIATION

The patterns that I have so far described are based entirely on the analysis of allelic variation at electrophoretic loci. There are good theoretical reasons to expect that the amount of differentiation at electrophoretic loci is poorly correlated with the amount of genetic differentiation at other loci determining morphological or life-history traits (Lewontin, 1984). On the other hand, both the level of electrophoretic diversity and its partitioning within and among populations are broadly correlated with life-history characteristics that affect the extent of inter-population gene flow in plants (Loveless and Hamrick, 1984; Hamrick and Godt, 1990). Thus, electrophoretic variation could provide a marker for the genetic structure of populations at ecologically significant loci. That is certainly the hope of most plant conservationists that use the technique. How well does it serve this purpose?

Few studies have been done that directly address this question, but those that have been done are not encouraging. Of the 22 studies included in Hamrick's (1989) survey, only 13 showed an association between variation at electrophoretically detectable loci and variation in other genetically controlled characteristics. Electrophoretic differentiation appears to be tightly correlated with differentiation at other loci only in species that are highly selfing. Using Nei's genetic distance (Nei, 1973a) as a measure of genetic differentiation at electrophoretic loci and a multivariate morphological distance as a measure of differentiation at quantitative trait loci, for example, Price et al., (1984) found a high correlation between differentiation at electrophoretic and quantitative trait loci in three highly selfing species, but not in an outcrosser.

Even in species that are highly selfing, however, the pattern of variation at electrophoretic loci may be quite different from that at other loci. For example, Brown et al., (1978) analyzed natural populations of *Hordeum spontaneum* throughout the Middle East for allozyme diversity and for morphological variation in spikelet characters. They found that 17% of the electrophoretic diversity was due to differences in allele frequency between major geographical regions, 32% to differences between populations within regions, and 51% to differences between individuals within populations. In contrast, they found that an average of 38% of the diversity in the spikelet characters was due to regional differences, 38% to differences between populations within regions, but only 24% to differences between individuals within populations. Thus, to the extent that the spikelet characters measured are genetically determined, the distribution of genetic variation among populations for loci involved in determining spikelet morphology is quite different from that at the electrophoretic loci examined. In short, the data currently available suggest that electrophoretic data are not good markers for the distribution of genetic diversity at other loci.

The reason for this discrepancy is rarely obvious, but there is one well-studied case in which the reason seems clear. Emerson (1939) surveyed populations of *Oenothera organensis* for allelic variation at a locus determining the gametophytic self-incompatibility reaction. In a sample from only 135 plants in a population of fewer than 1000 individuals he identified 34 distinct alleles. An electrophoretic survey of these same populations detected variation at only one of fifteen loci surveyed (Levin et al., 1979). It is clear that an individual with a rare self-incompatibility allele has a significant reproductive advantage over common ones because it can participate in more matings. This frequency-dependent advantage can lead to the maintenance of a large number of alleles even in a small population because the rarer any allele becomes the stronger the selective force opposing its elimination (Wright 1939, 1960, 1965).

Although it appears that electrophoretic loci provide a poor guide to differentiation at other loci because they are subject to different forms of selection (e.g. Holsinger and Gottlieb, 1991), electrophoretic analyses may still be useful for detecting changes in the

630

genetic structure of populations. For example, the amount of heritable variation in a number of life-history characteristics increased after population bottlenecks in experimental housefly populations (Bryant et al., 1986), but the number of alleles at electrophoretic loci in these same populations showed a precipitous decline (McCommas and Bryant, 1990). Thus, electrophoretic variation may be a sensitive indicator of bottlenecks in a population's recent history. In a similar vein, Prober et al., (1990) were able to determine from the levels and distribution of electrophoretic variation in *Eucalyptus parviflora* that it was probably once more widespread and that its present restricted distribution is due to habitat fragmentation.

CONSERVING GENETIC DIVERSITY

The evidence I have presented so far suggests that generalizations about how to conserve genetic diversity in rare plants will be impossible. We cannot accurately predict the distribution and abundance of electrophoretic variation from life-history characteristics alone. Even if we could, we would know little about the distribution and abundance of ecologically significant genetic variation. Thus, it would appear that genetic conservation programs must be designed individually for each species of concern. That would be a depressing conclusion, because it requires far more work than we have time for. Fortunately, such a depressing conclusion is not necessary. Generalizations about how to conserve genetic diversity in rare and endangered plants are possible, even in the face of all this uncertainty (e.g., Holsinger and Gottlieb, 1989, 1991).

Management of existing natural populations of rare and endangered plants will usually be focused on ecological manipulations, e.g., exclusion of exotic competitors, outplanting of seedlings, modification of existing disturbance regimes. Populations large enough to buffer against extinction as a result of environmental stochasticity are generally large enough to mitigate any genetic threats to population viability (Lande, 1988a, b; Menges, 1991). In addition, plants tend to mate only with near neighbors and to disperse seeds locally in a small area around the maternal plant. As a result, adaptation to specific microhabitats within the populations is common, and individuals separated by more than a few meters or tens of meters evolve almost independently of one another, resulting in the maintenance of substantial amounts of polygenic variation (Goldstein and Holsinger, unpublished). In short, management efforts in existing natural populations of rare and endangered plants should be directed toward mitigation of ecological threats to survival, except in those cases where population sizes have been so drastically reduced that genetic threats to population viability become important.

Although the primary focus of efforts to protect rare and endangered plants must be the protection of existing natural populations, that is not enough. Floods, fires, or other natural catastrophes may derail our best efforts to protect natural populations. In some cases, establishment of new populations in suitable habitat may be appropriate. In either case, a seed source other than existing natural populations may be required. Off-site collections in botanical gardens or seed banks can serve such a role. Unfortunately, if these collections are maintained as living plants, we cannot hope to mimic natural conditions closely enough to prevent adaptation to the cultivated environment, no matter how carefully a controlled crossing program is designed and how much effort is put into matching garden conditions to those in natural populations. If, on the other hand, off-site collections are maintained as seed banks, much of the genetic diversity present in existing populations can be captured and stored almost indefinitely. Thus, seed banks can provide an insurance policy against catastrophic loss of natural populations and a source of seed for re-establishment, supplementation, and research. The problem is simply one of designing an appropriate sampling program.

The design of appropriate sampling schemes usually depends on an accurate understanding of the genetic structure of a species (Brown, 1989). Particular attention must be paid to geographic differentiation among populations because alleles that are locally common are frequently rare in the species as a whole. Because life-history characteristics provide only a rough guide to the genetic structure of any species, preliminary genetic surveys must be a part of conservation programs for crop germplasm. Similarly, if sampling schemes for the conservation of genetic diversity in rare plants depended on the genetic structure of these populations, preliminary genetic surveys would be essential in the design of genetic conservation programs for rare and endangered plant species. With nearly 700 plant species at risk of extinction in the next decade, such an approach is clearly impractical. Fortunately, it is also unnecessary.

In a survey of rare plants in southeastern Australia Brown and Briggs (1991) noted that 22 of the 47 species occurred in only a single population, 38 in 5 or fewer, and none in more than 10. Similarly, nearly all of the 1000 species in The Center for Plant Conservation's highest priority ranking for conservation in the United States occur in five or fewer populations, and even for those threatened species of long-term concern, about 3100 in all, the vast majority occur in fewer than 20 populations (Holsinger and Gottlieb, 1989, 1991). Thus, only a few populations need be sampled to obtain a representative sample of the genetic diversity in rare plant species. By taking samples from as few as five populations, we will have sampled all populations of the most critically endangered species and a substantial fraction of the populations of species less immediately endangered. Selection of populations that are geographically isolated from one another and that occur in distinctive habitats is likely to capture most of the ecologically significant genetic variation that exists if the sampling within populations is adequate.

As Marshall and Brown (1975) noted over 15 years ago, a sample of between 50 and 100 maternal plants is sufficient to have a 95% chance of capturing all alleles with a frequency of 5% or greater within that population, regardless of breeding system, mode of reproduction, growth habit, or population size. The number of alleles captured in a population sample is approximately a logarithmic function of sample size (Brown, 1989). Thus, we gain as much in terms of allelic diversity from our first sample of 10 individuals as we do from the next 90. As Brown and Briggs (1991) and Holsinger and Gottlieb (1991) have argued, there are many reasons to suspect that alleles in low frequency within a population have received an inordinate amount of attention from conservation biologists. Thus, there is little justification for within population samples to exceed 50 to 100 maternal plants.

In short, rare plants are found in few populations. By applying a little common sense to the selection of populations to be included in a genetic sample, e.g., by selecting geographically isolated populations or populations that occur in distinctive habitats, the chances of excluding adaptively significant genetic variation that is locally common but rare in the species is minimal if we take samples from just few populations of rare and endangered plants. If we take samples from as few as five populations we will have sampled from all existing populations of the most critically endangered species and from a substantial fraction of the populations of those species that are less immediately endangered. Investigations for management or scientific purposes may require a detailed genetic analysis, but a representative sample of the genetic diversity present in a rare species can be obtained without any preliminary survey (Holsinger and Gottlieb, 1989, 1991).

A great deal remains to be learned about the ecology and genetics of rare plants and about techniques for managing them. Fortunately, we need not await the results of these studies to begin conservation efforts. The directions in which we must move are clear. Rare plants generally occur in only a few populations. Thus, detailed preliminary studies to select populations for protection are not needed. We simply need to protect those few that still exist. In addition, the potential for long-term storage of seed provides plant conservationists with an option not readily available to animal conservationists: off-site conservation can conserve most of the genetic diversity present in a rare plant species without a complex breeding program. An integrated approach in which off-site collections provide an insurance policy against catastrophic events in natural populations has a good chance of success, if we act quickly enough.

ACKNOWLEDGMENTS

I am indebted to Don Falk and Les Gottlieb for many long discussions about the issues raised in this paper, but neither of them can be held responsible for what I have written. This work was supported, in part, by a grant from the University of Connecticut Research Foundation.

LITERATURE CITED

Brown, A. H. D. 1979. Enzyme polymorphism in plant populations. *Theor. Popul. Biol.* 15:1–42.

Brown, A. H. D. 1989. The case for core collections. Pp. 136–156. *In:* A. H. D. Brown, O. H. Frankel, D. R. Marshall & J. T. Williams (eds.), *The Use of Plant Genetic Resources.* Cambridge University Press, Cambridge, UK.

Brown, A. H. D. & J. D. Briggs. 1991. Sampling strategies for genetic variation in *ex situ* collections of endangered plant species. *In:* D. A. Falk & K. E. Holsinger. *Conservation of Rare Plants: Biology and Genetics.* Oxford University Press: New York (in press).

Brown, A. H. D., Nevo, E. Zohary, D. & O. Dagan. 1978. Genetic variation in natural populations of wild barley (*Hordeum spontaneum*). *Genetica* 49:97–108.

Bryant, E. H., McCommas, S. A. & L. M. Combs. 1986. The effect of experimental bottlenecks upon quantitative genetic variation in the housefly. *Genetics* 114:1191–1211.

Emerson, S. 1939. A preliminary survey of the *Oenothera organensis* population. *Genetics* 24:524–537.

Gottlieb, L. D. 1981. Electrophoretic evidence and plant populations. Pp. 1–46. *In:* L. Reinhold, J. B. Harborne & T. Swain (eds.), *Progress in Phytochemistry*, Vol. 7. Pergamon Press: Oxford.

Gottlieb, L. D., Warwick, S. I. & V. S. Ford. 1985. Morphological and electrophoretic divergence between *Layia discoidea* and *L. glandulosa. Syst. Bot.* 1:181–187.

Hamrick, J. L. 1989. Isozymes and the analysis of genetic structure in plant populations. Pp. 87–105. *In:* D. E. Soltis & P. S. Soltis (eds.), *Isozymes in Plant Biology.* Dioscorides Press: Portland, OR.

Hamrick, J. L. & M. J. W. Godt. 1990. Allozyme diversity in plant species. *In:* Pp. 43–46. A. H. D. Brown, M. T. Clegg, A. L. Kahler & B. S. Weir (eds.), *Plant Population Genetics, Breeding, and Genetic Resources.* Sinauer Associates: Sunderland, MA.

Hamrick, J. L., Godt, M. J. W., Murawski, D. A. & M. D. Loveless. 1991. Correlations between species traits and allozyme diversity: implications for conservation biology. *In:* D. A. Falk & K. E. Holsinger (eds.), *Conservation of Rare Plants: Biology and Genetics.* Oxford University Press: New York (in press).

Hamrick, J. L., Mitton, J. B. & Y. B. Linhart. 1979. Relationships between life-history characteristics and electrophoretically-detectable genetic variation in plants. *Ann. Rev. Ecol. Syst.* 10:173–200.

Holsinger, K. E. & L. D. Gottlieb. 1989. The conservation of rare and endangered plants. *Trends Ecol. Evol.* 4:193–194.

Holsinger, K. E. & L. D. Gottlieb. 1991. Conservation of rare and endangered plants: principles and prospects. *In:* D. A. Falk & K. E. Holsinger (eds.), *Conservation of Rare Plants: Biology and Genetics.* Oxford University Press: New York (in press).

Huenneke, L. F., Holsinger, K. E. & M. E. Palmer. 1986. Plant population biology and the management of viable plant populations. Pp. 169–183. P. R. *In:* B. A. Wilcox, Brussard, and B. G. Marcot (eds.), *The Management of Viable Populations: Theory, Applications, and Case Studies.* Center for Conservation Biology, Stanford, CA.

Karron, J. D. 1987. A comparison of levels of genetic polymorphism and self-compatibility in geographically restricted and widespread species of *Astragalus* (Fabaceae). *Evol. Ecol.* 1:47–58.

Karron, J. D. 1991. Patterns of genetic variation and breeding systems in rare plant species. *In:* D. A. Falk & K. E. Holsinger (eds.), *Conservation of Rare Plants: Biology and Genetics.* Oxford University Press: New York (in press).

Karron, J. D., Linhart, Y. B., Chaulk, C. A. & C. A. Robertson. 1988. The genetic structure of populations of geographically restricted and widespread species of *Astragalus* (Fabaceae). *Amer. J. Bot.* 75:1114–1119.

Levin, D. A., Ritter, K. & N.C. Ellstrand. 1979. Protein polymorphism in the narrow endemic *Oenothera organensis. Evolution* 33:534–542.

Lande, R. 1988a. Genetics and demography in biological conservation. *Science* 241:1455–1460.

Lande, R. 1988b. Demographic models of the northern spotted owl. *Oecologia* 75:601–607.

Lewontin, R. C. 1984. Detecting population differences in quantitative characters as opposed to gene frequencies. *Amer. Natur.* 123:115–124.

Loveless, M. D. & J. L. Hamrick. 1984. Ecological determinants of genetic structure in plant populations.. *Ann. Rev. Ecol. Syst.* 15:65–95.

Marshall, D. R. & A. H. D. Brown. 1975. Optimum sampling strategies in genetic conservation. Pp. 53–79. *In:* O. H. Frankel & J. G. Hawkes (eds.), *Crop Genetic Resources for Today and Tomorrow.* Cambridge University Press: Cambridge.

McCommas, S. A. & E. H. Bryant. 1990. Loss of electrophoretic variation in serially bottlenecked populations. *Heredity* 64:315–321.

Menges, E. S. 1991. The application of minimum viable population theory to plants. *In:* D. A. Falk and K. E. Holsinger (eds.), *Conservation of Rare Plants: Biology and Genetics.* Oxford University Press: New York (in press).

Nei, M. 1973a. The theory and estimation of genetic distances. Pp. 45–51. *In:* N. E. Morton (ed.), *Genetic Structure of Populations.* University of Hawaii Press: Honolulu.

Nei, M. 1973b. Analysis of gene diversity in subdivided populations. *Proc. Nat. Acad. Sci. USA* 70:3321–3323.

Price, S.C., Schumaker, Kahler, A. L., Allard, R. W. & J. E. Hill. 1984. Estimates of population differentiation obtained from enzyme polymorphisms and quantitative characters. *J. Hered.* 75:141–143.

Prober, S. M., Tompkins, C. Moran, G. F. & J. C. Bell. 1990. The conservation genetics of *Eucalyptus paliformis* Johnson et Blaxell and *E. parviflora* Cambage, two rare species from south-eastern Australia. *Aust. J. Bot.* 38:79–45.

Warwick, S. I. & L. D. Gottlieb. 1985. Genetic divergence and geographic speciation in *Layia* (Compositae). *Evolution* 39:1236–1241.

Wright, S. 1939. The distribution of self-sterility alleles in populations. *Genetics* 24:538–552.

Wright, S. 1960. On the number of self-incompatibility alleles maintained by a given mutation rate in a population of a given size: a reexamination. *Biometrics* 16:61–85.

Wright, S. 1965. The distribution of self-incompatibility alleles in populations. *Evolution* 18:609–619.
Keystrokes in holsing.2: 15472

Quantitative Genetic Changes in Small Populations

Thomas Mitchell-Olds

Abstract. Population bottlenecks may result in loss of genetic variability and reduce the potential for adaptation to future environmental challenges. Changes in quantitative genetic variances following a bottleneck are extremely dependent on details of genetic architecture, such as gene action and allele frequency, that are unknown in natural populations. Different measures of variability (heterozygosity, the number of rare alleles, and components of quantitative genetic variance) change at different rates. If rare alleles provide an important source of genetic variance for future population adaptation then bottlenecks may severely reduce the evolutionary potential of endangered and sensitive species. There is large stochastic variation among different bottleneck lines. Consequently, genetic changes in a particular population or species are difficult to predict, and results from empirical studies are difficult to intepret.

Changes in genetic variability due to population bottlenecks have been discussed for many years (Robertson, 1952, Nei et al., 1975, Maruyama & Fuerst, 1985, Lacy, 1987). A population bottleneck of effective size N results in moderate declines in expected heterozygosity, but large losses of rare alleles. Theoretical and empirical analyses of bottleneck effects on quantitative genetic variation have been presented by Bryant et al. (1986) and Lande and Barrowclough (1987). Barrett and Kohn (1990) reviewed population genetic aspects of bottlenecks, but little information is available on changes in quantitative genetic variance. This paper considers changes in quantitative genetic variance due to population bottlenecks. Although results from simple theory have considered mean changes in genetic variance, it is important to examine the *distribution* of changes in genetic variance, as well as complications that may occur in multilocus models.

METHODS

I performed Monte Carlo simulations of changes in quantitative genetic variance in small populations. Each replicate of a bottleneck population contained N diploid individuals, each with 11 diallelic polymorphic loci. Each bottleneck population was drawn from an infinite parental population (generation 0) with a range of allele frequencies (two loci with allele frequencies at 0.1, 0.3, 0.7, and 0.9 each, and three loci at 0.5). Based on the multilocus genotype, I assigned a genotypic value to each individual. In order to simulate several modes of gene action, the genotypic value was assigned either additively (sum of

Dr. Mitchell-Olds is with the Division of Biological Sciences, University of Montana, Missoula, MT 59812, USA.

allelic effects over all alleles and loci: "additive"), logarithmically (the natural logarithm of summed allelic effects, giving decreasing returns to scale: "nonlinear"), or additively with an epistatic component (additive effects at nine loci, while two loci gave high scores for alternate homozygotes, AABB or aabb, and low scores for other genotypes: "epistasis"). There was no environmental influence on phenotype (heritability = 100%). In order to estimate the heritable variance, V_H, in the bottleneck population (generation 1), I simulated a large (N = 250) post-bottleneck population (generation 2) created by random mating of individuals from generation 1. The total genetic variance after the bottleneck was calculated as the variance of genotypic values in generation 2. Although this model did not incorporate linkage, linkage disequilibrium is expected to develop due to finite population size. Levels of heritable variation were calculated from the regression of offspring (generation 2) on midparent (generation 1). The sampling error of V_H is due to stochastic sampling of loci undergoing Mendelian segregation in the parent and progeny generations. Note that inbreeding is expected in these bottleneck populations, so that parent-offspring regression will include components of additive and nonadditive genetic variance (Cockerham, 1963; Mitchell-Olds & Rutledge, 1986). Thus, the heritable variance, V_H, differs from the additive variance, V_A, due to nonadditive genetic variance, inbreeding, and the effects of linkage disequilibrium. Each bottleneck size was replicated 1,000 times. There was no mutation, migration, or selection. Finally, I calculated the probability that genetic variance may increase as a consequence of genetic drift.

I also calculated the distribution of changes in genetic variance due to a single locus with a bottleneck of size N. This may be computed exactly by evaluating all possible genotypic outcomes during a single generation bottleneck, their probability, and the genetic variance under those circumstances. The expected level of additive or total genetic variance during or after a bottleneck population is

$$E\{Pn_1,n_2,n_3 \cdot Vn_1,n_2,n_3\},$$

where Pn_1,n_2,n_3 is the multinomial probability of observing n_1, n_2, and n_3 individuals of genotypes *aa, Aa,* and *AA*, respectively, $N = n_1 + n_2 + n_3$, and Vn_1,n_2,n_3 is the additive or total genetic variance in a population given this distribution of genotypes.

RESULTS

Mean levels of heritable genetic variance declined in all treatments (Table 1). At the smallest bottleneck size ($N = 2$), final levels of V_H were significantly less than the single locus additive prediction, H, (p < 0.001) in the additive model. The nonlinear model of gene action resulted in slightly higher levels of V_H than the additive model. Proportional losses of V_H were greater with epistatic gene action than in the additive model. In most cases, the total genetic variance exceeded the heritable variance, and both declined following the bottleneck. Despite the declines in mean genetic variance following a bottleneck, some bottleneck populations actually exhibited increased genetic variation due to chance (Table 1). In 20–44% of the replicated bottleneck populations the level of heritable variance increased above its initial level. Large stochastic changes in the level of heritable variance were evident. With $N = 32$, V_H exhibited increases and decreases up to about 50% of the original value (not shown). V_H ranged widely between 0 and 300% of its initial levels when $N = 2$ (Fig. 1). The distribution of V_H differed slightly between additive and epistatic models (mean V_H following the bottleneck equaled 68% vs. 60% of the original V_H, in the additive and epistatic $N = 2$ models, respectively). However, these distributions are so broadly overlapping that very large samples would be required to distinguish between underlying patterns of additive and epistatic gene action in any real experiment.

Table 1. Mean genetic variance after a bottleneck.

	Additive		Nonlinear		Epistatic		H	P_I
N	V_G	V_H	V_G	V_H	V_G	V_H		
2	75	68	77	73	66	60	75	20
8	94	94	95	98	89	88	94	36
32	98	99	99	99	99	99	98	44

The total genetic variance (V_G) or heritable genetic variance (V_H) under several models of gene action after a single generation bottleneck of size N, as a percentage of original V_G or V_H, respectively. Estimates of mean V_H can exceed V_G due to sampling error. H is the expected heterozygosity, which is identical to the additive genetic variance under a single locus, strictly additive model of gene action. Data show the mean of 1,000 Monte Carlo replicates. P_I is the probability that V_H will increase following a population bottleneck (probabilities are multiplied by 100). See text for details.

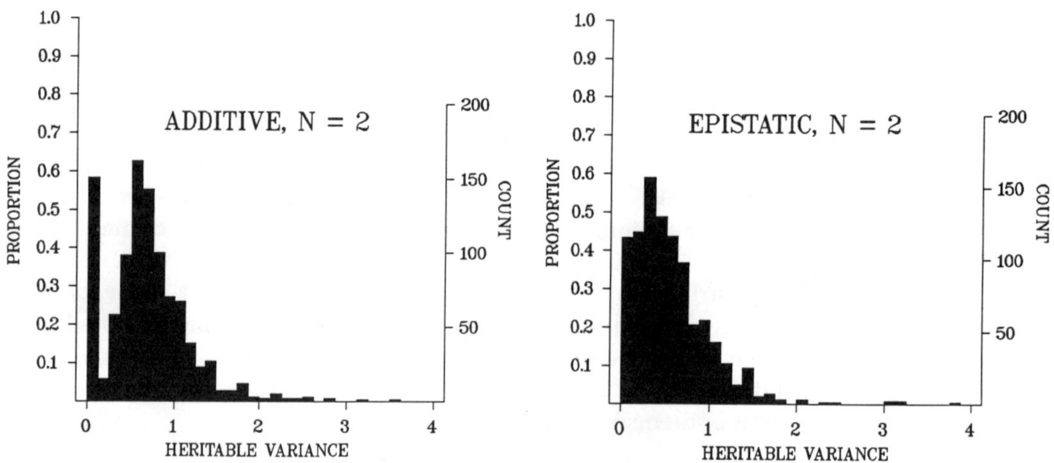

Figure 1: The distribution of estimates of the heritable variance in bottlenecks of N = 2. Fig. 1a: strictly additive gene action. Fig. 1b: epistatic gene action. See text for discussion.

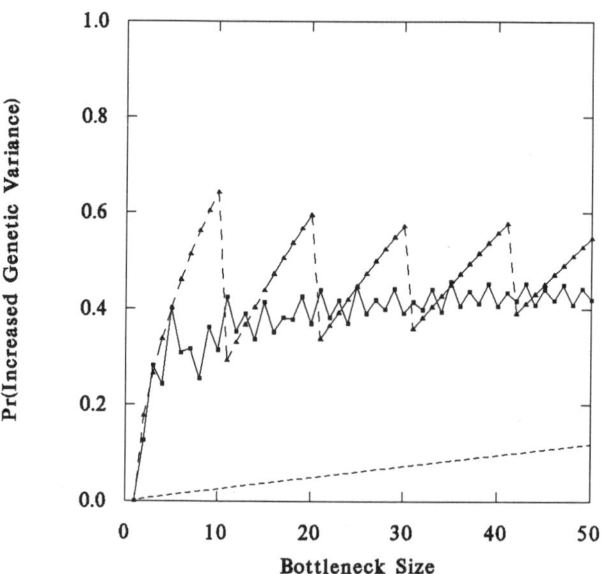

Figure 2: The probability P_I that the genetic variance during a single generation bottleneck will increase above its initial level. Shown are a common (0.5) additive allele (solid line with squares), a rare dominant allele (dashed line with triangles), and a rare recessive allele (dashed line). These are exact results for a single locus model. Table 1 reports results from a multilocus Monte Carlo simulation. See text for discussion.

Calculated changes in genetic variance due to a single locus are exact, since variances and their probabilities were evaluated for all possible outcomes. With additive gene action the genetic variance declined to $1 - (1/N)$ of its original level during the first bottleneck generation, then rebounded to $1 - (1/2N)$ in an infinitely large population following the bottleneck (not shown). With dominance gene action, changes in additive and total genetic variance are extremely dependent on levels of dominance and allele frequency (Robertson, 1952). Genetic variances may decline rapidly (rare dominant allele) or may increase several-fold (rare recessive allele). The probability P_I that genetic variances increase due to drift fluctuated erratically, depending on levels of dominance, allele frequencies, and bottleneck size. For example, Fig. 2 shows P_I during the first bottleneck generation.

DISCUSSION

Changes in quantitative genetic variance in bottleneck populations are unpredictable, being extremely sensitive to allele frequency and levels of dominance (Robertson 1952). Changes during the first bottleneck generation (when there is a limited distribution of genotypes) may differ substantially from large, post-bottleneck populations under Hardy-Weinburg equilibrium, where genetic variances are dependent on gene action and allele frequency. For example, with additive gene action the genetic variance declines to $1 - (1/N)$ of its original level during the first bottleneck generation, then rebounds to $1 - (1/2N)$ in an infinitely large population following the bottleneck. These simulations indicate that genetic analyses of loss of heterozygosity may underestimate the deleterious consequences of population bottlenecks on quantitative genetic variance, which is the raw material for future evolutionary change. Furthermore, if rare alleles provide an important source of genetic variance for future population adaptation (e.g., disease resistance, drought tolerance, etc.), then bottlenecks may severely reduce the evolutionary potential of endangered and sensitive species.

Small populations are expected to lose genetic variation. Although the expected level of genetic variance declines, genetic drift may cause some populations to actually have increased levels of genetic variance (Fig. 2). The probability, P_I, that a particular bottleneck population will have increased V_G due to chance fluctuates erratically, depending on allele frequency and levels of dominance (Fig. 2). Despite the uneven behavior of P_I, it is clear that small populations will, on average, have reduced V_G, but may often exhibit increased V_G due to the chance effects of genetic drift. While the average genetic behavior of small populations is clear, the large stochastic variation in outcomes of population bottlenecks must also be considered.

Goodnight (1987, 1988) has discussed changes in V_A within and among population subdivisions. When small populations undergo genetic drift, then some types of epistatic genetic variance, V_I, may be converted to V_A, and thus be available to support evolutionary change. If V_I contributes a large proportion of the total genetic variance, as may be expected for major components of fitness in equilibrium populations, then V_A may increase in small, subdivided populations (Goodnight 1988). Such increases were not observed in these simulations, although other models of epistatic variance (different allele frequencies and modes of gene action) may reveal mean increases in additive genetic variance. These simulations actually found greater losses of heritable variance when epistasis was present. However, the importance of such changes in natural populations is unknown.

Bryant et al. (1986) reported increased V_A for morphometric characters in bottleneck lines of house flies. Although each bottleneck size, N, was replicated several times, analyses of genetic variance were based on a single pooled estimate from each N. Since

638

quantitative genetic variances have an appreciable probability of increasing or decreasing under several models of gene action (Table 1 and Fig. 2), underlying patterns of gene action cannot be inferred from these experimental results.

The fundamental conclusion from this analysis is that changes in quantitative genetic variances following a bottleneck are extremely dependent on details of genetic architecture that are unknown in natural populations. Populations may lose substantial genetic variation in a bottleneck, especially if small population size persists for a number of generations. Different measures of genetic variability (heterozygosity, the number of rare alleles, V_G, and V_A) change at different rates. There is large stochastic variation among different bottleneck lines. Although population bottlenecks are expected to reduce genetic variability, the genetic changes in a particular population or species are difficult to predict.

LITERATURE CITED

Barrett, S.C. H. & J. R. Kohn. 1990. Genetic and evolutionary consequences of small population size in plants: implications for conservation. *In:* D. A. Falk and K. E. Holsinger (eds.), *Conservation of Rare Plants: Biology and Genetics.* Oxford University Press: Oxford, U.K., in press.

Bryant, E. H., McCommas, S. A. & L. M. Combs. 1986. The effect of an experimental bottleneck on quantitative genetic variation in the housefly. *Genetics* 114:1191–1211.

Cockerham, C. C. 1963. Estimation of genetic variances. Pp. 53–94. *In:* W. D. Hanson and H. F. Robinson (eds.), *Statistical Genetics and Plant Breeding.* NAS-NRC Pub. #982: Washington, DC.

Goodnight, C. J. 1987. On the effect of founder events on epistatic genetic variance. *Evolution* 41:80–91.

Goodnight, C. J. 1988. Epistasis and the effect of founder events on the additive genetic variance. *Evolution* 42:441–454.

Lacey, R. C. 1987. Loss of genetic diversity from managed populations: interacting effects of drift, mutation, immigration, selection, and population subdivision. *Conservation Biology* 1:143–158.

Lande, R. & G. F. Barrowclough. 1987. Effective population size and its use in population management. Pp. 87–123. *In:* M. E. Soulé (ed.), *Viable Populations for Conservation.* Cambridge University Press: New York.

Maruyama, T. & P. A. Fuerst. 1985. Population bottlenecks and non-equilibrium models in population genetics. II. Number of alleles in a small populations derived from a large steady-state population by means of a bottleneck. *Genetics* 111:675–689.

Mitchell-Olds, T. & J. J. Rutledge. 1986. Quantitative genetics in natural plant populations: a review of the theory. *Amer. Natur.* 127:379–402.

Nei, M., Maruyama, T. & R. Chakraborty. 1975. The bottleneck effect and genetic variability in populations. *Evolution* 29:1–10.

Robertson, A. 1952. The effect of inbreeding on the variation due to recessive genes. *Genetics* 37:189–207.

The Channel Island Fox (*Urocyon littoralis*) as a Model of Genetic Change in Small Populations

Robert K. Wayne, Sarah B. George, Dennis Gilbert, and Paul W. Collins

Abstract. Small island populations provide a model for genetic changes expected in small captive or managed populations. However, small island populations have high extinction rates and rarely persist for substantial time periods. The Channel Island fox, *Urocyon littoralis*, is unusual because populations as small as a few hundred individuals have persisted for more than 10,000 years. This dwarf island form is distributed among the six Channel Islands, 30–100 kilometers off the coast of southern California. Since the island populations vary in two important determinants of genetic variability, effective population size and time of founding, they provide a natural experiment for testing ideas about the decline of genetic diversity in small populations. We use several techniques to measure genetic variability of island foxes, including morphometrics, allozyme electrophoresis, mtDNA restriction site analysis, analysis of hypervariable minisatellite DNA (genetic fingerprinting), and comparative karyology. Our results indicate that several island fox populations have reduced levels of morphologic variability and low or nondetectable levels of genetic variation, yet they have persisted for several thousand years. In contrast, other island populations have greater levels of genetic variability than expected. These data indicate the need for more sophisticated genetic models and for studies that examine the relationship between genetic variability of a population and its persistence over time.

INTRODUCTION

Conservation biology is largely a study of animal and plant populations of small size (Frankel & Soulé, 1981). A major concern in small populations is the loss of genetic variability since such losses are thought to negatively affect many aspects of fitness, including fecundity, juvenile mortality, growth rate, and disease resistance (Ralls & Ballou, 1983; Wayne et al., 1986a; Allendorf & Leary, 1986; O'Brien & Evermann, 1988; Ralls et al., 1988). The loss of genetic variability in small populations may be accelerated further due to random demographic factors, which cause dramatic decreases in population size (Gilpin & Soulé, 1986). In such cases, genetic variability may be reduced and inbreeding increased such that the resulting loss of fitness makes further population reductions more probable. Thus, small populations may be caught in a cycle of population decline until

Dr. Wayne is with the Department of Biology, UCLA, Los Angeles, CA 90024, USA. Dr. George is with the Section of Mammalogy, Natural History Museum of Los Angeles County, 900 Exposition Blvd., Los Angeles, CA 90007, USA. Dr. Gilbert is with the Biological Carcinogenesis and Development Program, Program Resources Incorporated, NCI-FCRDC, Frederick, MD 21702, USA. Mr. Collins is with the Santa Barbara Museum of Natural History, 2559 Puesta Del Sol, Santa Barbara, CA 93105, USA. Please address correspondence to Dr. Wayne.

640

they finally go extinct (Gilpin & Soulé, 1986). Moreover, since small populations are often established by just a few individuals, the founding of such populations may result in a loss of genetic heterozygosity if the intrinsic rate of increase is small (Nei et al., 1975). Even given a high rate of increase, allelic diversity inevitably will be reduced (Allendorf, 1986), and the alleles that are lost may reduce fitness or the ability to respond to environmental perturbation.

Several theoretical and computer-modeled studies have been developed to estimate the loss of genetic variability in small populations (e.g., Lande, 1976; Falconer, 1981; Fuerst & Maruyama, 1986; Harris et al., 1986; Lacy, 1987). Similarly, zoo studies have been extremely useful in establishing the specific costs, in lethal equivalents, of reductions in heterozygosity (Ralls et al., 1988). However, the imposed breeding structure of captive populations may differ dramatically from those in the wild. Moreover, little information exists concerning the persistence of small wild populations for long periods of time. Studies of small populations of various sizes and lengths of persistence are needed to provide data on the decline of genetic variability in nature that could be used to improve theoretical models.

In this study, we summarize data on genetic variability of six island populations of the Channel Island fox, *Urocyon littoralis*. These populations vary greatly in size and time of isolation and hence provide a natural experiment for assessing the decline in genetic variability experienced by isolated populations of wild mammals. Several genetic techniques are used, including standard karyology, allozyme electrophoresis, mitochondrial DNA restriction site analysis and genetic fingerprinting. We also assess morphologic variability in 29 cranial characters of island foxes (Collins, 1982; in press, a). The primary goal of this paper is to interpret these results from a conservation genetic perspective rather than to discuss the data in detail as this is done elsewhere (Gilbert et al., 1990, Wayne et al., in press).

The Study Population

The Channel Island fox, *Urocyon littoralis*, is a diminutive island species related to the mainland gray fox, *U. cinereoargenteus*. It is about two-thirds the body weight of the mainland species and morphologically distinct from it (Collins, 1982). The island fox is found on six of the Channel Islands, each with its own recognizable subspecies (Collins, 1982; Fig. 1). The three northern islands are closely situated to one another and contrast with the southern islands that are separated by deep water channels, 50–80 kilometers in width. The islands differ in size, and the effective population size of foxes on each island varies from approximately 150 to 1,000 individuals (Table 1). The subfossil record is fairly complete on most of the islands and suggests that island foxes first appeared on one of the northern Channel Islands, Santa Rosa, approximately 16,500 years ago (Fig. 1). At this time the three northern islands were connected in one single landmass and were separated from the mainland by only a 4–5 kilometer wide channel. Approximately 11,500 years ago Santa Cruz separated from the other northern islands, followed by the separation of Santa Rosa and San Miguel about 9,500 years ago (Johnson, 1983). The first record of foxes on the southern islands dates from about 3,400 years ago, and all the southern islands had been colonized by 800–2,200 years ago (Fig. 1; Collins, in press, b). Foxes may have arrived on the southern islands through transport by Native Americans who first colonized the islands approximately 9,000 to 10,000 years before present (Collins, 1982).

Table 1. Ranks of fox populations by effective population size, and variability in morphology, allozymes, fingerprinting profiles, and mtDNA.

Population	Population[1] Size (N_e)	Morphology[2] (variance log measurements)	Fingerprints (average percent difference)	Allozymes (Nei's heterozygosity)	MtDNA (number of genotypes)
San Miguel	1 (157)	1 (0.012)	2 (4.7)	2 (0.008)	2 (2)
San Nicolas	2 (247)	2 (0.013)	1 (0.0)	1 (0.000)	1 (1)
San Clemente	3 (551)	5 (0.017)	3 (8.5)	3 (0.013)	1 (1)
Santa Rosa	4 (922)	4 (0.016)	5 (23.7)	5 (0.055)	2 (2)
Santa Catalina	5 (979)	5 (0.017)	6 (25.3)	1 (0.000)	3 (3)
Santa Cruz	6 (984)	3 (0.015)	4 (10.0)	4 (0.041)	2 (2)
Mainland	7 (>10,000)	6 (0.022)	ND	6 (0.097)	4 (7)

[1] Effective population size is estimated to be 0.5 that of the actual population size. This conversion is based on the model of Reed et al. (1986). The terms of their model were estimated from the limited field data on island foxes and extrapolation from data on mainland foxes (Fritzell, 1987; Garcelon et al., 1989). Assumptions are: 2 year generation time, equal sex ratio, the number of breeding males and females is 0.5 times actual population size; one young born to each sex each year, and 0.5 probability of survival to reproduction.

[2] This is the average variance for the log of 28 cranial and dental measurements.

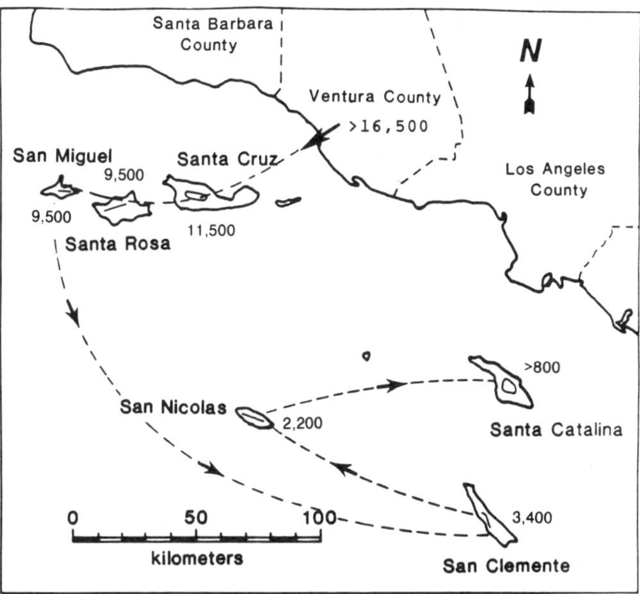

Figure 1. Approximate size and location of the six southern California Channel Islands where island foxes are found. Solid lines on each island indicate transect lines along which fox samples were obtained. Dotted lines indicate hypothesized colonization route of island foxes. Times of colonization are indicated in numbers of years before present (see text).

642

<h1 style="text-align:center">MATERIALS AND METHODS</h1>

Effective Population Size

The estimated population density of foxes on the islands varies from 7.6 to 10.1 per square kilometer (Gilbert et al., 1990). The effective population size, given generous assumptions about mortality and fecundity, is approximately 50% of the total population size (Table 1). Considering that there is less habitat presently available to foxes on the islands because of human disturbance, the harmonic mean of effective population size over time is probably greater than that of the present population. Thus, the present effective population size is used as a minimum estimate to calculate the expected loss in heterozygosity (see text).

Morphologic Measurements

Twenty-nine morphologic measurements were made on crania and dentition of 213 female island foxes from each of the Channel Islands and 76 mainland gray foxes from Santa Barbara, Ventura and Los Angeles Counties (Collins, 1982). Foxes from these counties are likely to be closely related to the ancestral stock that founded the Channel Islands (Fig. 1, Collins, in press, a). Morphologic measurements were log-transformed prior to computation of variances so that the variance could be compared among measurements and fox populations of different overall size (Wayne et al., 1986b). Measurements showing significant non-homogeneity of variance with Bartlett's test were evaluated with non-parametric rank order tests: the Wilcoxon signed rank test and a sign rank test (Wilkinson, 1988). The significance of differences in average ranking could then be evaluated with log-data that is not necessarily normal. Differences in the average overall ranking was assessed using Friedman's two-way ANOVA test (Wilkinson, 1988).

Protein Electrophoresis

Red cells isolated from whole blood of 24–40 foxes from each island were thawed in buffer and subjected to three freeze-thaw cycles as in Wayne and O'Brien (1987). After centrifugation the supernatant was used in protein electrophoresis and staining of 20 soluble blood proteins including ADK-1, ADK-2, CA, EST-D, HB, GPD-1, GPD-2, GPI, LAP, LDH-1, LDH-2, IDH, MDH, NP, 6-PGD, PGM, PEP-B, PEP-C, PEP-D, and SOD (George, 1986; Wayne & O'Brien, 1987). Estimates of genetic polymorphism and heterozygosity from each island were calculated from allozyme data (Nei, 1987). Allele frequencies for each population at each genetic locus were used to calculate Nei's (1978) genetic distance modified for small sample size.

Mitochondrial DNA

Total genomic DNA was isolated from white blood cells separated from whole blood of 20–30 foxes from each Channel Island and 20 foxes from the mainland (see Wayne et al., 1989). Two to three ug of DNA were digested separately with the following 7 restriction endonucleases found to be polymorphic in island and mainland foxes (Wayne et al., in press): *Dra*I, *Hinc*II, *Hin*DIII, *Hha*I, *Msp*I, *Stu*I, and *Sty*I. After electrophoresis through 1% agarose gels, DNA was transferred to Nylon membranes and probed with cloned, radiolabeled domestic dog mtDNA. Fragment identity was assessed on the basis of comigration of fragments separated in the same gel. Because island fox mtDNA genotypes differ by the loss or gain of a single site, restriction maps were readily inferred from the patterns of fragment loss or gain (Lansman et al., 1981; Avise et al., 1987).

Hypervariable Minisatellite DNA Analysis

Genomic DNA from foxes was digested with the restriction endonuclease *Hin*FI and probed with hypervariable minisatellite DNA clone 33.6 as outlined by Gilbert et al. (1990). The difference value (D) was calculated as the number of fragments that differed between two individuals divided by the total number of fragments present in both individuals. The average percent difference (APD) is simply the average of all D values for each island times 100 (Gilbert et al., 1990). The inter-island APD is the mean of the D values between individuals in any two populations.

Standard Karyology

A 1 ml sample of blood from a male fox from each of the Channel Islands except San Miguel was added to 10 ml RPMI media (Gibco) supplemented with 15% fetal calf serum and phytohemaglutten (Difco) as a mitogen. After three days incubation at 37°C, 2 drops of colcemid (10 ug/ml, Gibco) was added and the sample was pelleted by centrifugation and processed as in Wayne et al. (1987).

RESULTS

Patterns of Variability

As expected, most measures of variability increase with estimates of effective population size (Table 1). Morphologic variation is significantly less on the two smallest islands, San Miguel and San Nicolas, and largest in mainland populations (Wilcoxon signed rank test, $p < 0.05$; Wayne et al., in press). Similarly, variability in DNA fingerprint restriction fragment profiles is second lowest on the smallest island, San Miguel, and surprisingly absent among foxes on San Nicolas (Gilbert et al., 1990). For both morphologic variation and hypervariable minisatellite DNA, Santa Cruz foxes have lower than expected levels of variability. The allozyme data indicate relatively low levels of variation in foxes from San Miguel and San Clemente, and no apparent variation in foxes from San Nicolas, in agreement with the fingerprint data. These heterozygosity values are significantly lower than the level of variability found in mainland foxes (t-test with arcsine data, $p < 0.05$; Archie, 1985). However, Santa Catalina Island, which has the second largest effective population size, has no detectable allozyme variation. This unexpected result may reflect the pattern of island colonization (see below).

Mitochondrial DNA variation corresponds most poorly with effective population size. The smallest island, San Miguel, has the same number of genotypes as Santa Cruz, the largest island. San Nicolas foxes, however, have only one genotype as might be expected from their small population size and low levels of variation in other measures. Finally, our survey of karyologic variation of island foxes from all the island populations except San Miguel reveals a uniformity in diploid number and morphology. Island foxes have a karyotype with a diploid number of 66, composed of 62 acrocentric chromosomes, a submetacentric pair and typical mammalian sex chromosomes. This complement is identical to that found in the mainland gray fox (Wayne et al., 1987).

Distance values based on DNA fingerprint restriction fragment profiles or allozyme frequencies suggest considerable differentiation among island populations and between island foxes and the mainland gray fox. For example, Nei's allozyme genetic distance between island and mainland foxes has an average value of 0.115 which is as large as that observed among discrete canid species (Wayne & O'Brien, 1987). Therefore, significant genetic differences may accumulate due to drift and founder effect in as little as 16,500 years (cf. Baker & Moeed, 1987). The extent of drift reflects population size: the largest islands, Santa Cruz and Santa Rosa, have genetic distance values to the mainland fox that are less than half the value of the other populations (Table 2). The genetic distance among foxes on different islands ranges from 0.0 (San Nicolas vs. Santa Catalina) to 0.053 (Santa Cruz vs. San Nicolas). Foxes on the three southern islands and San Miguel clearly form a group of closely related populations with genetic distances near zero.

The degree of inter-island genetic differentiation indicated by the comparison of DNA fingerprints of foxes from different islands is substantial relative to that among foxes on the same island. For example, the APD between foxes on different islands ranges from 43.8% to 84.4% compared to 0.0% to 25.3% among foxes on the same island (Tables 1 and 2). San Miguel and Santa Rosa, the two islands in closest proximity, are the most similar. Foxes on the northern islands appear to be more closely associated with each other (mean APD equals 53.6) than they are to foxes on the southern islands (mean APD equals 71.2). Similarly, the mean APD equals 62.0 among foxes on the southern islands vs. a mean APD of 71.2 between foxes on southern and northern Channel Islands.

Table 2. Average percent difference matrix (upper diagonal) based on shared restriction fragments in the fingerprinting analysis and Nei's genetic distance (lower diagonal) based on allele frequencies in the allozyme analysis.

	Smi	Sro	Scr	Sni	Sca	Scl	Uci
San Miguel (Smi)	—	43.8	64.8	68.6	59.2	74.1	88.1
Santa Rosa (Sro)	0.031	—	52.2	75.1	68.1	84.4	86.4
Santa Cruz (Scr)	0.049	0.005	—	67.8	63.2	80.7	89.6
San Nicolas (Sni)	0.000	0.033	0.053	—	55.6	73.3	93.2
Santa Catalina (Sca)	0.000	0.033	0.053	0.000	—	57.0	84.2
San Clemente (Scl)	0.000	0.029	0.048	0.000	0.000	—	90.4
Mainland gray fox (Uci)	0.140	0.065	0.064	0.134	0.145	0.134	—

DISCUSSION

Patterns of Variability in Island Foxes: A Phylogenetic Framework

Combined data from the fingerprinting, allozyme, mtDNA analyses and the archeological evidence argue for the colonization scheme outlined in Figure 1 (Gilbert et al., 1990; Collins, in press, b). As suggested by the archeological and geological data, the northern islands were likely the first islands colonized, followed by isolation of foxes on Santa Cruz, then Santa Rosa and San Miguel (Fig. 1). The first southern island colonized was likely San Clemente by immigrants from San Miguel, followed by San Nicolas and Santa Catalina. Foxes were probably transported to the southern islands by Native Americans (Collins, 1982, in press, b).

This colonization scheme may account for lower then expected levels of allozyme variation in foxes from Santa Catalina Island. On this island, the absence of allozyme variability among foxes seems to reflect the number of prior founding events. For example, there is a drop in heterozygosity from Santa Rosa (H = 0.055) and Santa Cruz (H = 0.041) to San Miguel (H = 0.008) and San Clemente (H = 0.013), and then to San Nicolas (H = 0.00) and Santa Catalina (H = 0.00) which follows the hypothesized phylogenetic pattern of colonization of the Channel Islands (Table 1, Fig. 1). Conceivably, the immigrant foxes from Santa Catalina were reduced in genetic variability as result of a historical process involving sequential founding events and small population sizes. Because mutation rates of allozyme loci are so low (10^{-7} per year, Nei, 1987) there has not been sufficient time to restore Santa Catalina foxes to higher levels of heterozygosity. In contrast, hypervariable minisatellite DNA loci have a much higher mutation rate (10^{-3} per gamete; Burke, 1989) and variation has been restored to a level reflecting the higher effective population size of foxes on Santa Catalina. Thus, allozyme variability may be strongly biased by the number and severity of past founding events or population bottlenecks.

Likewise, the poor correspondence of mtDNA variability and population size may reflect low mutation rates of the mtDNA sequence relative to hypervariable minisatellite DNA and morphologic loci and infrequent migration events. For example, the high number of genotypes in foxes on Santa Catalina Island may reflect its position as a Native American trade center (Collins, 1982, in press, b). Such gene flow of mtDNA genes in the near absence of nuclear gene flow has been found in a number of other vertebrates (Ferris et al., 1983; Spolsky & Uzzell, 1986; Cronin et al., 1988) and is expected since the nuclear contribution of a single or several immigrants would be rapidly diluted in a large resident population through recombination. The mtDNA genotype does not recombine and will increase in frequency so long as daughters produce more than one female offspring. Moreover, drift more strongly effects changes in mtDNA genotype frequency because the effective population size for the mtDNA genome is ¼ that of a nuclear gene. Thus, mtDNA frequency variation is strongly affected both by infrequent migration events and the smaller effective population size.

In sum, the five techniques used in this study differ qualitatively in the genetic variation that they measure. The variation of morphologic characters and hypervariable minisatellite DNA loci directly reflect differences in effective population size, suggesting they are good indicators of genetic variability in rapidly evolving loci for recently diverged populations. Allozyme variability generally corresponds with effective population size, but allozyme loci evolve slowly and genetic variability may be biased by the phylogenetic history of the founding population. Certainly, variability in morphologic characters, which may be important for the ability of populations to respond to changing conditions, is sometimes not indicated by the levels of allozyme variation (Table 1). Similarly, mitochondrial DNA variation may be affected by occasional immigration events and stochastic changes in genotype frequency, and thus may not be a good index of genetic variability in small isolated populations.

Genetic Variation and Population Persistence

In San Nicolas Island foxes there is an absence of variability in fingerprint restriction fragment profiles, in allozymes, and in mtDNA genotypes, as well as reduced levels of morphologic variation. These results indicate breeding among close relatives is common. Only in colonies of highly eusocial naked mole rats and inbred strains of mice has even a near absence of DNA fingerprint restriction fragment variability been observed (Jeffreys et al., 1987; Reeve et al., 1990). Since restriction fragment sizes are thought to change through unequal exchange at meiosis or mitosis, through replication slippage or deletion/insertion of repeat units, and not through point mutations, the absence of variability likely reflects low recombination rates and inbreeding rather than an absence of point mutations (Jeffreys et al., 1985; Wolff et al., 1989). Therefore, recombination in San Nicolas foxes has apparently not restored variability as rapidly as it has been lost through inbreeding. San Nicolas is the second smallest Channel Island and clearly the most remote as it is separated from the other islands by 80 to 100 km (Fig. 1). Thus, the small population size and high degree of isolation of San Nicolas foxes may have led to their low levels of morphologic and genetic variation.

The consequences of inbreeding in foxes on San Nicolas are not obvious. Although the frequency of skeletal abnormalities in San Nicolas foxes is high (Collins, 1982), reproductive capacity appears unaffected. In 1980, the population is thought to have declined to 110 individuals or fewer, but it rebounded to as many as 500 animals by 1984 (Kovach & Dow, 1985). Given a persistent small population size, inbreeding-tolerant individuals are likely to be at a high selective advantage. Thus, over the 2,200 year history of foxes on San Nicolas, the fox population could have become dominated by inbreeding-tolerant individuals. An analogous scheme was used in the founding of inbred mouse strains and has been applied successfully to the breeding of the Speke's gazelle (Morse, 1978; Templeton and Reed, 1984). The persistence of foxes on San Nicolas and other Channel Islands for several thousand years despite reduced morphologic variability and low or nondetectable levels of genetic variability suggests inbreeding schemes may sometimes be a viable long-term strategy for maintaining small captive populations. However, as in inbred strains of mice, the success rate of the initial founding populations is low and most go extinct due to inbreeding depression (Morse, 1978). Moreover, inbred populations may be more susceptible to disease, parasites and environmental perturbations (O'Brien & Evermann, 1988; Quattro & Vrijenhoek, 1989).

Significantly, in our serum survey of 194 island foxes we found no evidence for the presence of distemper virus, which is a common disease among canids. Distemper is almost always fatal in gray foxes (Nicholson & Hill, 1984). If this disease reached the related populations of island foxes, several island populations might be reduced below levels from which they could recover. The population of San Nicolas foxes would likely be the most threatened because of its small size and lack of genetic variability. In fact, the importance of genetic variation may be more significant as a response to infrequent catastrophic events that occur on a hundred or thousand year time scale. (Thorne & Williams, 1988; May, 1988). Thus, the success of inbreeding schemes for captive management may be short-lived and the concept short-sighted.

Models of the Maintenance of Genetic Variability

Our results indicate that more variability may be retained in some small populations than expected from theoretical considerations. For example, assume the simple model defined by $H_t = H_o (1 - 1/[2N_e])^t$ where t = number of generations, N_e is effective population size; and H_t, H_o are heterozygosity at time t and the initial heterozygosity, respectively (Falconer, 1981). The expected heterozygosity, given our estimated founding time and effective population size and assuming no mutation, selection, or immigration, is near zero for the island populations (Wayne et al., in press). This expectation agrees with the levels of variability found in foxes from the southern Channel Islands, but on Santa Cruz and Santa Rosa, foxes have much higher than expected heterozygosities, ranging from an H of 0.041 to 0.055, or 42% to 57%, respectively, of the H value of the mainland population (Table 1). These results indicate the need for more refined models that can explain the maintenance of heterozygosity in very small populations. Moreover, the persistence of several island populations with very low levels of genetic variability for as long as 9,500 years suggests a more precise understanding of the relationship between genetic variability and population persistence is needed so that effective cost-benefit decisions can be made about preservation of genetic variation in captive populations. For example, the population size required to preserve 95% of the average heterozygosity is twice that necessary to preserve 90% of the average heterozygosity (Foose et al., 1986).

Genetic Variability and Reserve Design

A prolonged debate has developed concerning the merits of a single large or several small reserve (SLOSS) systems for conservation of endangered species (Quinn & Hastings, 1987). The island fox populations fit a several small reserves model without connections for gene flow between reserves. In this system. all populations have apparently persisted for approximately 2,200 to 9,500 years even given small effective population sizes and low genetic diversity (Table 1). In addition, genetic data suggests that migration among islands is low (Wayne et al., in press). Thus significant gene flow is not a prerequisite to population persistence as suggested in some models and may accelerate the spread and persistence of pathogens in island populations (Dobson, 1988).

In fact, the potentially rapid rate of evolution in small isolated populations may have allowed unique adaptations to develop for island-specific parasites or environments. The Channel Islands differ significantly in average precipitation, and hence the plant communities and prey base differ greatly among the islands (Raven, 1965; Von Bloeker, 1965). Similarly, ectoparasites and viral diseases of foxes are unevenly distributed among the islands (Wayne, unpublished data). These different selective conditions on each island increase the likelihood that at least one population will persist given general changes in environmental conditions. Thus, the division of the island fox into small, isolated populations may enhance the survivorship of the species. However, founder effect and genetic drift have also resulted in each population possessing unique morphologic and genetic characteristics and several subspecies have been be defined (Hall, 1981; Collins, 1982, in press, a; Gilbert et al., 1990; Wayne et al., in press). As a result, we may need to be concerned with the preservation of each island population as a separate genetic entity. The apparent paradox is that through division of a species into discrete populations their collective persistence may be enhanced but this may result in the formation of new genetic entities that may be candidates for conservation in their own right.

CONCLUSIONS

1. The five techniques used in this study differ qualitatively in the genetic variation that they measure. Variation in cranial and dental characters and hypervariable minisatellite DNA loci seem best for comparing relative genetic variability of small, recently isolated populations.

2. Small populations of foxes on several Channel Islands apparently have persisted for as long as several thousand years with nondetectable or low levels of genetic variation. This suggests that genetic variation is not essential for long term persistence of small isolated populations of mammals. However, such populations may be extremely vulnerable to infrequent periodic catastrophes such as epizootics.

3. Populations on the two largest northern islands have much higher levels of genetic variation than expected from simple models. This indicates the need for more sophisticated models and more empirical data relating genetic heterozygosity in populations to their long-term persistence.

4. Small isolated populations of foxes have persisted on the Channel Islands for thousands of years, suggesting this is a viable design for reserves of terrestrial vertebrates. However, the partitioning of individuals into small populations increases the likelihood that they will rapidly become morphologically and genetically distinct hence defining new evolutionary units that may require preservation.

ACKNOWLEDGMENTS

Logistical support was provided by: the United States Department of the Navy, Point Mugu and North Island Naval Air Stations; the United States National Park Service, Channel Islands National Park; The Nature Conservancy, the Santa Cruz Island Preserve; and the Santa Catalina Island Conservancy. Financial support was provided by the Nature Conservancy and the Genetic Resources and Conservation Program of the University of California. D. Garcelon, F. Hertel, and B. Van Valkenburgh provided comments that greatly improved the manuscript.

LITERATURE CITED

Allendorf, F. W. 1986. Genetic drift and the loss of alleles versus heterozygosity. *Zoo Biol.* 5:181–190.

Allendorf, F. W. & R. F. Leary. 1986. Heterozygosity and fitness in natural populations of animals. Pp. 57–76. *In:* M. E. Soulé (ed.) *Conservation Biology: the Science of Scarcity and Diversity.* Sinauer Associates Inc.: Sunderland, MA. 584pp.

Archie, J. W. 1985. Statistical analysis of heterozygosity data: independent sample comparisons. *Evolution* 39:623–637.

Avise, J. C., Arnold, J., Ball, R. M., Bermingham, E., Lamb, T., Neigel, J. E., Reeb, C. A. & N.C. Saunders. 1987. Intraspecific phylogeography: the mitochondrial DNA bridge between population genetics and systematics. *Ann. Rev. Ecol. Syst.* 18:489–522.

Baker, A. J. & A. Moeed. 1987. Rapid genetic differentiation and founder effect in colonizing populations of common mynas (*Acridotheres tristis*). *Evolution* 41:525–538.

Burke, T. 1989. DNA fingerprinting and RFLP analysis. *Trends in Ecology and Evolution* 4:136–139.

Collins, P. W. 1982. *Origin and Differentiation of the Island Fox: A Study of Evolution in Insular Populations.* M. A. Thesis, Univ. Cal. Santa Barbara. 303pp.

Collins, P. W. Origin and differentiation of the island fox: a study of evolution in insular populations. *In:* F. G. Hochberg (ed.), *Recent Advances in California Islands Research.* Santa Barbara Natural History Museum: Santa Barbara, California, in press (a).

Collins, P. W. Interaction between Island foxes (*Urocyon littoralis*) and Indians on the islands of the coast of Southern California: I. Morphologic and archaeological evidence of human assisted dispersal. *J. Ethnobiology* in press (b).

Cronin, M. A., Vyse, E. R. & D. G. Cameron. 1988. Genetic relationships between mule deer and white-tailed deer in Montana. *J. Wildl. Managem.* 52: 320–328.

Dobson, A. P. 1988. Restoring island ecosystems: the potential of parasites to control introduced mammals. *Conservation Biology* 2:31–39.

Falconer, D. S. 1981. *Introduction to Quantitative Genetics.* Longman. New York. 340pp.

Ferris, S. D., Sage, R. D., Huang, C. M., Neisen, J. T., Ritte, U. & A. C. Wilson. 1983. Flow of mitochondrial DNA across a species boundary. *Proc. Natl. Acad. Sci. USA* 79:2290–2294.

Foose, T. J., Lande, R., Flesness, N. R., Rabb, G. & B. Read. 1986. Propagation plans. *Zoo Biol.* 5:139–146.

Fritzell, E. K. 1987. Gray fox and island gray fox. Pp. 408–421. *In:* M. Novak, J. A. Baker, M. E. Obbard & B. Malloch. (eds.), *Wild Furbearer Management and Conservation in North America.* Ministry of Natural Resources: Ontario, Canada.

Frankel, O. H. & M. E. Soulé. 1981. *Conservation and Evolution.* Cambridge University Press: New York. 327pp.

Fuerst, P. A. & T. Maruyama. 1986. Considerations on the conservation of alleles and of genetic heterozygosity in small managed populations. *Zoo Biol.* 5:171–180.

Garcelon, D. K., Roemer, G. W. & G. P. Frederick. 1989. Preliminary report on the status and demographics of the island fox on San Clemente and Santa Catalina Islands. Report to the Nongame Section, Wildlife Management Branch, California Department of Fish and Game, Institute for Wildlife Studies, Arcata, California.

George, S. B. 1986. Evolution and historical biogeography of soricine shrews. *Systematic Zoology* 35:153–162.

Gilbert, D. A., Lehman, N., O'Brien, S. J. & R. K. Wayne. 1990. Genetic fingerprinting reflects population differentiation in the Channel Island fox. *Nature* 344:764–767.

Gilpin, M. E. & M. E. Soulé. 1986. Minimum viable populations: process of species extinction. Pp. 19–34. *In:* M. E. Soulé (ed.), *Conservation Biology: The Science of Scarcity and Diversity.* Sinauer Associates Inc.: Sunderland, MA. 584pp.

Harris, R. B., Metzgar, L. H. & C. D. Bevins. 1986. *Generalized Animal Population Projection System* (GAPPS): *a User Manual* Montana Cooperative Wildlife Research Unit. University of Montana: Missoula, MT.

Hall, E. R. 1981. *The Mammals of North America.* John Wiley and Sons: New York.

Jeffreys, A. J., Wilson, V. & S. L. Thein. 1985. Hypervariable minisatellite regions in human DNA. *Nature* 316:76–79.

Jeffreys, A. J., Wilson, V., Kelly, R., Taylor, B. A. & G. Bulfield. 1987. Mouse DNA 'fingerprints': analysis of chromosome localization and germ-line stability of hypervariable loci in recombinant inbred strains. *Nucleic Acids Res.* 15: 2823–2836.

Johnson, D. L. 1983. The California continental borderland: landbridges, watergaps and biotic dispersals. Pp. 481–527. *In:* P. M. Masters & N.C. Flemming (eds.), *Quaternary Coastlines and Marine Archaeology: Towards the Prehistory of Land Bridges and Continental Shelves.* Academic Press: London.

Kovach, S. D. & R. J. Dow. 1985. *Island Fox Research on San Nicolas Island*. 1985. Ann. Report. Department of the Navy. Pacific Missile Test Center: Pt. Mugu, CA.

Lacy, R. C. 1987. Loss of genetic diversity from managed populations: interacting effects of drift, mutation, immigration, selection and population subdivision. *Conservation Biology* 2:143–158.

Lande, R. 1976. The maintenance of genetic variability by mutation in a polygenetic character with linked loci. *Genet. Res. Cambridge* 26:221–235.

Lansman, R. A., Shade, R. O., Shapira, J. F. &. J. D. Avise. 1981. The use of restriction endonuclease analysis to measure mitochondrial DNA sequence relatedness in natural populations. III. Techniques and potential applications. *J. Mol. Evol.* 17:214–226.

May, R. M. 1988. Conservation and disease. *Conservation Biology* 2:28–30.

Morse, H. C. 1978. *Origins of Inbred Mice*. Academic Press: New York. 719pp.

Nei, M., Maruyama, T. & R. Chakraborty. 1975. The bottleneck effect and genetic variability in populations. *Evolution* 29:1–10.

Nei, M. 1978. Estimation of average heterozygosity and genetic distance from a small number of individuals. *Genetics.* 89:583–590.

Nei, M. 1987. *Molecular Evolutionary Genetics*. Columbia University Press: New York. 521pp.

Nicholson, W. S. & E. P. Hill. 1984. Mortality in gray foxes from East-Central Alabama. *J. Wildl. Managem.* 48:1429–1432.

O'Brien, S. J. & J. F. Evermann. 1988. Interactive influence of infectious disease and genetic diversity of natural populations. *Trends in Ecology and Evolution* 3:254–259.

Quattro, J. M., & R. C. Vrijenhoek. 1989. Fitness differences among remnant populations of the endangered Sonoran topminnow. *Science* 245:976–978.

Quinn, J. F. & A. Hastings. 1987. Extinction in subdivided habitats. *Conservation Biology* 1:198–208.

Ralls, K. & J. Ballou. 1983. Extinctions: lessons from zoos. Pp. 164–184. *In:* C. M. Schonewald-Cox, S. M. Chambers, B. MacBryde & L. Thomas (eds.), *Genetics and Conservation: a Reference Manual for Managing Wild Animal and Plant Populations*. Bejamin Cummings: Menlo Park, CA.

Ralls, K., Ballou, J. D. & A. Templeton. 1988. Estimates of lethal equivalents and the costs of inbreeding in mammals. *Conservation Biology* 2:185–193.

Raven, P. 1965. The floristics of the California Islands. Pp. 57–68. *In:* R. N. Philbrick (ed.), *Proceedings of the Symposium on the Biology of the California Islands*. Santa Barbara Botanical Garden: Santa Barbara, CA. 363pp.

Reed, J. M., Doerr, P. D. & J. R. Walters. 1986. Determining minimum population sizes for birds and mammals. *Wildlife Society Bull.* 14:255–261.

Reeve, H. K., Westneat, D. F., Noon, W. A., Sherman, P. W. & C. F. Aquadro. 1990. DNA "fingerprinting" reveals high levels of inbreeding in colonies of the eusocial naked mole-rat. *Proc. Natl. Acad. Sci. USA* 87:2496–2500.

Spolsky, C. & T. Uzzell. 1986. Evolutionary history of the hybridogenetic frog *Rana esculenta* as deduced from mtDNA analysis. *Mol. Biol. Evol.* 3:44–56.

Templeton, A. R. & B. Reed. 1984. Factors eliminating inbreeding depression in a captive herd of Speke's gazelle (*Gazella spekei*). *Zoo Biol.* 3:177–200.

Thorne, E. T. & E. S. Williams. 1988. Disease and endangered species: the black-footed ferret as recent example. *Conservation Biology* 2:66–74.

Von Bloeker, J. C., Jr. 1965. The land mammals of the Southern California Islands. Pp. 245–264. In: R. N. Philbrick (ed.), *Proceedings of the Symposium on the Biology of the California Islands*. Santa Barbara Botanical Garden: Santa Barbara, CA. 363pp.

Wayne, R. K., Forman, L., Neuman, A. K., Simonson, J. M. & S. J. O'Brien. 1986a. Genetic monitors of zoo populations: morphological and electrophoretic assays. *Zoo Biol.* 5:215–232.

Wayne, R. K., Modi, W. S. & S. J. O'Brien. 1986b. Morphological variability and asymmetry in the cheetah (*Acinonyx jubatus*), a genetically uniform species. *Evolution* 40:78–85.

Wayne, R. K. & S. J. O'Brien. 1987. Allozyme divergence within the Canidae. *Systematic Zoology* 36:339–355.

Wayne, R. K., Nash, W. G. & S. J. O'Brien. 1987. Chromosomal evolution of the Canidae: I. Species with high diploid numbers. *Cytogenetics and Cell Genetics* 44:123–133.

Wayne, R. K., Kat, P. W., Fuller, T. K., Van Valkenburgh B. & S. J. O'Brien. 1989. Genetic and morphologic divergences among sympatric canids (Mammalia: Carnivora). *J. Heredity* 80:447–454.

Wayne, R. K., George, S., Gilbert, D., Collins, P., Kovach, S., Girman, D. & N. Lehman. Genetic change in small, isolated populations: evolution of the Channel island fox, *Urocyon littoralis*. *Evolution*, in press.

Wilkinson, L. 1988. SYSTAT: *The System for Statistics*. SYSTAT Inc.: Evanston, Illinois. 821pp.

Wolff, R. K., Plaetke, R., Jeffreys, A. J. & R. White. 1989. Unequal crossing over between homologous chromosomes is not the major mechanism involved in the generation of new alleles at VNTR loci. *Genomics* 5:382–384.

Estimation of Effective Population Size of Grizzly Bears by Computer Simulation

Fred W. Allendorf, Richard B. Harris, and Lee H. Metzgar

Abstract. We present a new method to predict the rate at which small populations will lose genetic variation and apply it to the grizzly bear (*Ursus arctos*). The simulation model is a discrete-time, stochastic computer program that follows the life-history and kinship of each individual. We determined heterozygosity by calculating the inbreeding coefficient of individuals, pedigree F, which is the probability that two genes at a locus in an individual are identical by descent from an allele in the foundation population. Thus, F provides an exact measure of heterozygosity, $H=(1-F)$, at selectively neutral loci over the entire genome.

We estimate the genetically effective population size to be approximately 25% of total population size under a wide variety of demographic conditions. However, the introduction of even a few bears per generation from other populations greatly increased the effective population size. We conclude that many extant populations of grizzly bears can only be maintained by intensive management that includes movement of bears among currently isolated populations.

INTRODUCTION

The fragmentation and isolation of populations is of increasing concern in management of endangered species. Loss of genetic variation in isolated populations of large mammals is especially serious because of their low population densities and high spatial requirements. Thus, even the largest protected reserves may be too small to maintain genetically viable populations of large mammals (Soulé et al., 1986; Belovsky, 1987).

The most useful concept to estimate the expected rate of loss of genetic variation is effective population size (N_e). Knowledge of effective population size allows prediction of the expected time when reduced genetic variation is likely to threaten continued existence of an isolated population. In spite of universal agreement about the importance of effective population size for making management decisions (Soulé, 1980), considerable confusion persists about its estimation in natural populations. This is especially true for large mammals because of their complex demographics and numerous departures from the genetically "ideal" population.

In this paper, we introduce a new method for estimating N_e by computer simulation in a consideration of grizzly bear (*Ursus arctos*) populations. An estimation of the rate of loss of genetic variation in grizzly bear subpopulations is needed in order to determine population sizes required to maintain genetically viable subpopulations. Moreover, it is also important to determine what management actions can be taken to reduce the rate of loss of genetic variation in the remaining subpopulations.

Drs. Allendorf and Metzgar (Division of Biological Sciences) and Dr. Harris (Cooperative Wildlife Research Unit) are with the University of Montana, Missoula, MT 59812, USA.

The United States Endangered Species Act of 1975 declared the grizzly bear to be a threatened species. The number of grizzly bears in the contiguous 48 states has declined from an estimated 100,000 in 1800 to less than 1,000 at present (Servheen, 1985). Similarly, the range of the species within this area is now less than 1% of its historic range. The current verified range of the grizzly bear is approximately five million hectares in six separate subpopulations in four states (Servheen, 1985).

The reduction in range has isolated subpopulations as continuous habitat was divided and movement corridors disappeared. Population decline accelerated because these isolated subpopulations were small and subject to stochastic demographic influences. Current recovery goals for the remaining subpopulations are based upon estimates of the minimum viable population (MVP) size that has a 95% probability of survival for 100 years (Allendorf & Servheen, 1986).

Current estimates of MVP for the grizzly bear are based upon a comprehensive series of computer simulations of demographic structure (Shaffer & Sampson, 1985; Shaffer, 1983). Recovery targets for four of the six subpopulations are 70–90 individuals (U.S. Fish & Wildlife Service, 1982). The genetically effective population size (N_e) of such subpopulations is far short of the recommended effective number of approximately 500 that is necessary to maintain evolutionarily significant quantities of genetic variation (Lande & Barrowclough, 1987). These N_e are also likely to be lower than that of the generally accepted minimum number of 50 necessary to avoid serious loss of genetic variation in the short term (Soulé, 1980).

Well known studies with domestic animals have shown that loss of genetic variation has a variety of harmful effects on development, reproduction, survival, and growth rate. Studies with a variety of species in zoos indicate that similar effects probably occur in wild populations of animals (Ralls et al., 1986; Ralls et al., 1988; Ralls & Ballou, 1986). For example, natural populations of lions (*Panthera leo*) that have lost genetic variation through recent population bottlenecks have more developmentally abnormal sperm and lower testosterone concentrations than adjacent populations that have not lost genetic variation through a bottleneck (Wildt et al., 1987). Thus, subpopulations within the recovery targets of 70–90 individuals will lose genetic variation at a rate likely to decrease their expected longevity.

THE MODEL

The rate of loss of genetic variation generally has been measured by change in average heterozygosity per individual per locus (h). Heterozygosity is expected to be lost at an approximate rate of $(1/2N)$ per generation in the theoretical "idealized" population of $(N/2)$ males and $(N/2)$ females that are all equally likely to contribute a sperm or egg to the next generation (Wright, 1969). However, a wild population of N individuals will lose heterozygosity much faster than the rate of $(1/2N)$ expected in the ideal population. For example, unequal sex ratios, fluctuations in population size, and non-random reproductive success of individuals will all increase the rate of loss of heterozygosity.

Sewall Wright (1969, p. 211) defined effective population size (N_e) as whatever must be substituted in the formula $(1/2N)$ to describe the actual loss in heterozygosity. A variety of methods provide estimates of N_e under different violations of the assumptions of the ideal population (Wright, 1969). Several problems restrict application of these estimations to wild populations. First, these formulas cannot be combined to estimate rate of loss of genetic variation in a wild population in which all of the assumptions are not likely to hold. Second, many of the parameters needed to estimate N_e with these formulas are virtually impossible to estimate in wild populations. In addition, most populations do not consist of a single random mating group. Existing formulas for estimating N_e have not been

652

designed to incorporate effects of gene flow between geographically separated local populations. The simulation model was a discrete-time, stochastic computer program that followed the history and kinship of each individual. Values of parameters used in the simulations were taken from studies of grizzly bear populations in Montana, Wyoming, and British Columbia (Harris & Allendorf, 1989; Harris & Metzgar, 1987a,b). The simulation model is described in a paper in which we evaluated published techniques to estimate effective population size (Harris & Allendorf, 1989).

Four events occurred during a simulation year: breeding, natural mortality, birth, and weaning of juveniles. At each event, an individual's fate (e.g., dying) was determined by comparing its sex and age-specific probability with a random number from a uniform (0,1) probability distribution (Harris & Metzgar, 1987a,b). We performed 50 replicates for each combination of parameter values. Each replicate lasted 100 years; the first 24 years were omitted from the regression analysis to provide a delay of 2–3 generations in which loss of heterozygosity became asymptotically linear.

We determined heterozygosity with the algorithm of Boyce (1983) to calculate the inbreeding coefficient of individuals, pedigree F, which is the probability that two genes at a locus in an individual are identical by descent from an allele in the foundation population (Wright, 1969). Thus, F provides an exact measure of heterozygosity, $H=(1-F)$, at selectively neutral loci over the entire genome. We estimated the rate of loss of heterozygosity by regressing the natural logarithm of H in newborns on time measured in generations. Effective population size was estimated by solving for N_e using the rate of loss of heterozygosity per generation of $1/2N_e$ ($N_e = 1 / (-2\,e^m + 2)$, where m is the slope of the natural log of mean heterozygosity of the newborn cohort on time, measured in generations). The mean inbreeding coefficient of newborns was calculated at five year intervals, rather than each year, to save computing time.

RESULTS AND DISCUSSION

A critical, but poorly known, component for estimating N_e is the distribution of the reproductive contribution among males. We simulated three male mating systems with the intent of embracing the range of actual values: (1) random, in which adult males were picked at random (with replacement) to be sires; (2) our best estimate, in which adult males of a prime age group (8–18 years) had greater probabilities of being sires; and (3) extreme, in which there was an extreme advantage to prime age males. We detected surprisingly little effect of changing the distribution of male reproductive success on rate of loss of genetic variation (Table 1).

The rate of loss of genetic variation through time depends upon both N_e and mean generation interval because $1/(2N_e)$ is the expected rate of loss per generation. Therefore, results are presented (Table 1) by two measures: the ratio of effective to census population size (N), and the amount of variation expected to be lost in 100 years. The former measure varies more because the amount of variation lost depends upon both the N_e:N ratio and the mean generation time. In general, conditions that lower N_e:N tend to lengthen mean generation intervals (Table 1).

Table 1. Results of simulations showing mean generation interval, N_e/N ratio, and total percent loss of heterozygosity (H) in 100 years in grizzly bear populations of mean size of N=100.

Male mating	Generation (years)	N_e / N	Loss of H (%)
Random	10.0	0.311	15.6
Estimate	10.8	0.280	15.8
Extreme	11.1	0.260	16.4

We first tested for effects of population size on the ratio between effective and census population size using three different mean census sizes: 50, 100, and 200. We found no indication of an effect of population size on the population size ratio (Harris & Allendorf, 1989). We selected a population size of 100 to carry out the main body of simulations.

Our results suggest that the effective population size of grizzly bears is approximately 25–30% of census size (Table 1). This estimate may be high because of other factors in real populations that our simulations did not consider. A dominance hierarchy among breeding males is one such factor. All males of a given age had equal probability of reproductive success in our simulations. Dominance hierarchies within males of similar age may further reduce effective population size.

Thus, even fairly large isolated subpopulations, such as the 200 or so bears in Yellowstone National Park, United States, are vulnerable to harmful effects of loss of genetic variation. A moderate decrease in genetic variation in this population may decrease reproductive rates, further reducing population size and, in turn, accelerating the rate of loss of genetic variation. Artificial movement of bears among naturally isolated subpopulations is required to decrease the rate of loss of genetic variation. We therefore extended our simulations to determine the amount of gene flow needed to reduce the rate of loss of genetic variation in subpopulations to an acceptable level.

Gene flow was incorporated by introducing three year-old males or females that were assumed to be unrelated to all other animals in the population. Once introduced into a population, an immigrant had the same sex and age-specific life-history probabilities as other bears in the population; thus, not all introduced bears reproduced. These simulations were done with the random male mating system.

The introduction of a few bears greatly reduced the rate of loss of genetic variation (Table 2). For example, introduction of one male bear per year (approximately 10 per generation) increased mean N_e from 31 to 194 (Table 2). This agrees with analytic results that have shown that one migrant per generation is expected to limit genetic divergence among subpopulations (Wright, 1969). Those results have also indicated that the effectiveness of gene flow is determined by the number, not the proportion, of individuals exchanged among subpopulations. Accordingly, our results apply to subpopulations of other sizes.

Introduction of males caused a greater increase in N_e than females. This occurred because of the greater variability in male mating success. A few exceptionally successful introduced males had a pervasive effect on the pedigree of relationships. However, introduced males in nature may be less likely to be incorporated successfully into a population than females because of the greater movement of young males. Thus, these results may overstate the benefits of introducing males rather than females.

Our results support the notion that even large and protected reserves are too small to maintain viable populations of large mammals if they are isolated (Soulé et al., 1986; Belovsky, 1987). Genetically viable populations can only be maintained in such reserves by artificial exchange among reserves. However, even if all available isolated preserves are genetically connected, there is insufficient habitat available for many species (Ralls & Ballou, 1986). A combination of protected natural habitat preserves and ex situ preservation in zoos will become necessary for many species. Zoos will allow an increase in total

Table 2. Mean N_e with introduction of unrelated females or males in simulated grizzly bear populations of mean size N=100.

	Immigrants per generation				
	0	1	2	5	10
Females	31.1	36.6	40.7	67.2	123.6
Males	31.1	42.5	49.6	113.2	194.2

number of animals to be maintained and can also serve as sources for individuals to be used in gene exchange programs.

ACKNOWLEDGMENTS

This work was supported by the U.S. Fish and Wildlife Service, through the office of Grizzly Bear Coordinator, Chris Servheen. The simulation language was written by Collin D. Bevins; A. J. Boyce allowed us to adopt his pedigree analysis algorithm into this program. We thank Nils Ryman for his suggestions and Kathy Knudsen for her comments on this manuscript. This manuscript was completed while FWA was an employee of the National Science Foundation.

LITERATURE CITED

Allendorf, F. W. & C. Servheen. 1986. Conservation genetics of grizzly bears. *Trends in Ecology and Evolution* 1:88–89.

Belovsky, G. E. 1987. Extinction models and mammalian persistence. Pp. 35–57. *In:* M. Soulé (ed.), *Viable Populations for Conservation.* Cambridge University Press: Cambridge.

Boyce, A. J. 1983. Computation of inbreeding and kinship coefficients on extended pedigrees. *J. Heredity* 74:400–404.

Harris, R. B. & F. W. Allendorf. 1989. Genetic effective population size of large mammals: assessment of estimators. *Conservation Biology* 3:181–191.

Harris, R. B. & L. H. Metzgar. 1987a. Estimating harvest rates of bears from sex ratio changes. *J. Wildlife Management* 51:802–811.

Harris, R. B. & L. H. Metzgar. 1987b. Harvest age structures as indicators of decline in small populations of grizzly bears. *Int. Conf. Bear Research and Management* 7:109–116.

Lande, R. & G. F. Barrowclough. 1987. Effective population size, genetic variation, and their use in population management. Pp. 87–123. *In:* M. E. Soulé (ed.), *Viable Populations for Conservation.* Cambridge University Press: Cambridge.

Ralls, K., Ballou, J. & A. Templeton. 1988. Estimates of lethal equivalents and the cost of inbreeding in mammals. *Conservation Biol.* 2:185–193.

Ralls, K., Harvey, P. H. & A. M. Lyles. 1986. Inbreeding in natural populations of birds and mammals. Pp. 35–56. *In:* M. E. Soulé (ed.), *Conservation Biology: The Science of Scarcity and Diversity.* Sinauer: Sunderland, MA.

Ralls, K. & J. Ballou. 1986. Captive breeding programs for populations with a small number of founders. *Trends Ecol. Evol.* 1:19–22.

Servheen, C. 1985. The grizzly bear. Pp. 400–415. *In:* R. L. Di Silvestro (ed.), *Audubon Wildlife Report.*

Shaffer, M. L. 1983. Determining minimum viable population sizes for the grizzly bear. Pp. 133–139. *In:* E. C. Meslow (ed.), *Bears—Their Biology and Management.* International Association for Bear Research and Management.

Shaffer, M. L. & F. B. Sampson. 1985. Population size and extinction: A note on determining critical population sizes. *Am. Nat.* 125:144–152.

Soulé, M. E. 1980. Thresholds for survival: maintaining fitness and evolutionary potential. Pp. 151–169. *In:* M. E. Soulé & B. A. Wilcox (eds.), *Conservation Biology. An Evolutionary-Ecological Approach.* Sinauer: Sunderland, MA.

Soulé, M., Gilpin, M., Conway, W. & T. Foose. 1986. The millennium ark: How long a voyage, how many staterooms, how many passengers? *Zoo Biology* 5:101–114.

U.S. Fish and Wildlife Service. 1982. *Grizzly Bear Recovery Plan.*

Wildt, D. E., Bush, M., Goodrowe, K. L., Pacer, C., Pusey, A. E., Brown, J. L., Joslin, P. & S. J. O'Brien. 1987. Reproductive and genetic consequences of founding isolated lion populations. *Nature* 329:328–331.

Wright, S. 1969. *Evolution and the Genetics of Populations. Vol. 2. The Theory of Gene Frequencies.* University of Chicago Press: Chicago. 511 pp.

Energy and Community Evolution

Virginia C. Maiorana and Leigh M. Van Valen

The biotic world is, among other things (but basically) a system of energy flow. For any community or larger biota we can consider this flow itself and ask such questions as how it is partitioned, what causes and regulates the flow and its partitions, how these change over time at various scales, and what processes cause these changes. One does not need to be a systems ecologist in order to appreciate and study such questions. The phenomena are real, and they deserve real attention from evolutionary biologists.

The role of energy flow in shaping the structure of communities through time was the focus of a roundtable discussion which we organized. Two additional discussants were John Damuth (University of California. Santa Barbara) and Richard K. Bambach (Virginia Polytechnic Institute and State University). James H. Brown (University of New Mexico) added considerably to a lively exchange, with other participants including Rolf Sattler (McGill University), Janet A. Sherman (Pennsylvania College of Technology), Blaire Van Valkenburgh (University of California, Los Angeles). Charles W. Thayer (University of Pennsylvania) and J. Frederick Grassle (Rutgers University). Because the "community physiologists" and systems ecologists who had been invited could not attend, the discussion leaned heavily toward aspects of community structure. Each presentation was followed by a discussion, with a more general discussion at the end. This report summarizes the presentations and discussions with some relevant comments added during the preparation of this report.

Van Valen opened with comments on his ongoing work with the early radiation of placental mammals and how this work exemplifies an aspect of evolution that is neglected and even denied—that of biotas. We usually think of evolution as something which happens to phyletic lineages: they change, branch, and maybe even merge. A somewhat broader view encompasses also the evolution of clades and taxa. These can all be called *organismal evolution.* Orthogonal to this is *trait evolution,* where the focus is on specific traits like eyes or cryptobiosis or repetitive DNA. Organismal evolution and trait evolution partition nearly the same phenomena in different ways.

What they miss is the evolution of biotas themselves. Such *biotal evolution* is not just a theater for the evolutionary play, nor is it just an epiphenomenon of organismal evolution. The ecological interactions among organisms evolve as do the organisms themselves. Which is primary is a matter of our perspective, not of the biological processes. Despite their fuzzy boundaries, biotas have emergent properties that evolve also. By looking at the evolution of energy flow and its causes we can study natural selection within communities and larger biotas in a natural and causal way. Biotas are not superorganisms, but they are organized with diverse sorts of processes and regularities. Biotal evolution

Drs. Maiorana and Van Valen are with the Department of Ecology and Evolution, University of Chicago, 915 East 57th Street, Chicago, Illinois 60637, USA.

deserves as much attention as other aspects of evolution. All four discussants centered on this aspect of evolution in their presentations.

Van Valen next presented an outline of biotal evolution derived from his work on the evolution of mammals in the Paleocene of North America. For mammals (or any other group in fact) there are problems at the moment because of difficulties in describing communities or biotas adequately, particularly fossil ones. Because the nature of the biases in the preservation or collection of fossils are inadequately known, one cannot really remove them to get back the original community that formed the assemblage of fossils one has to study. Instead of attempting the impossible Van Valen simply used species as surrogate units for the flow of energy, a procedure that ignores a lot of information which cannot be adequately retrieved at present. Given diet, inferred from teeth, one can weight the energy flow of a species by its trophic level based on what is known in modern mammalian communities.

The time period considered, the Paleocene and the immediately adjacent latest Cretaceous and basal Eocene, can be divided into a little over 20 distinguishable intervals. The information is based on his own compilations and the data base is perhaps as reliable as possible at present. This period of time marks the beginning of the Cenozoic radiation of the placentals. Van Valen compared intervals at the beginning and later in this early stage in the placental radiation with respect to the amount of selection, a measure independent of the duration of a time interval. The amount of selection is defined as the variance in the absolute value of the realized fitness of whatever unit one is considering. The absolute realized fitness is here measured by the ratio of ancestral to descendant species numbers. This definition assumes that all changes are due to selection, one way or another; if not the case, the amount of selection is somewhat reduced. Some of his calculations weighted the species by energy flow, but this did not make a great deal of difference.

Some relevant and fairly striking patterns emerged from the analysis: the amount of selection is one to two orders of magnitude greater in the earlier interval, at the beginning of the placental radiation, than in the later and longer interval nearer biotic evolutionary equilibrium. This is true of all taxonomic categories, from species to order, in one manner of partitioning the biota, as well as for each trophic group and among trophic groups in another manner of partitioning. Even the multituberculates, a group of mammals which began a slow extinction during this period of time, shows less selection later in their extinction than earlier. The herbivorous mammals are one to two orders of magnitude more selected than insectivores in the basal Paleocene but not later in this epoch.

One can also look at the cause of selection in a formal sort of way by extinction, branching, persistence, and phyletic evolution. Not surprisingly, persistence is greater in the later interval than the earlier even though the later one is several times longer than the earlier. Extinction is also more important in the later interval and in fact none was detectable at the very beginning of the placental radiation. The greater importance of extinction later occurs despite the fact that the marsupials became almost extinct and the multituberculates greatly declined in abundance in the early interval. Not surprisingly, the branching of lineages is more important earlier, as is phyletic evolution. This sort of analysis can be extended in various ways.

Using the analogy of a logistically growing population, Brown raised the problem of how to compare fitness in an empty biospace, as occurs during the initial radiation of a taxon, with that in a filled biospace. Entities that leave more descendants in an unfilled biospace may be adaptively lousy compared to those leaving fewer in a crowded biospace, Van Valen agreed that his measure of fitness (the ratio of descendants to ancestors) is not a general definition but works with the type of data he used. He defined fitness as the propensity, in a spatiotemporal set (or manifold) of environments, to acquire trophic energy for growth and reproduction. Moreover, the analogy Brown used is not quite applicable to the situation of the radiating placentals because the amount of biospace used is not known to have differed over the interval examined. Some forms were ecologically more generalized earlier and subdivided later. Diversity does not measure abundance. Expansion of a group over evolutionary time usually relates to what it can take from others.

Changing the focus a bit. Sattler asked whether the evolution of Gaia is a case of biotal evolution. Van Valen argued against Gaia; because there is only one Gaia there is nothing to select against to cause her evolution. One can't even consider Gaia to have an ontogeny because an ontogeny is a program directed to some end point and that is not appropriate here. But the biota of the world changes and evolves, e.g., the number of species, the probability of extinction, and the like. These aspects are important to study.

Brown emphasized the importance of knowing the rates of change in primary and secondary productivity during the fluctuations in the Earth's biota. Is there more than intuition to enable us to know how productivity fluctuates? Van Valen noted the lack of foolproof methods at getting at this directly and Bambach mentioned he will be touching on this question in a crude way later.

Damuth next discussed problems of ecological scale and the partitioning of energy among species of different body sizes. Before getting to this topic he raised a more general issue which is exemplified in his specific analysis. Damuth thinks that the next revolution in evolutionary thinking will come with the unification of evolutionary biology with systems ecology by means of what he hesitantly calls systems thinking. Since the system of energy exchange that constitutes an ecosystem is both a cause of evolutionary change and is something that is changed by evolution, a unification of these two perspectives will transform both sets of theory (evolutionary and systems) and the way we actually think in these fields. At present the theory is embodied in two separate bodies of thought that are not easy to merge without the danger of losing important elements of one or the other. We need to create some kind of translation between the more biological adaptive evolutionary processes and the more mechanical or physical processes of energy exchange, a goal that will not easily be achieved. We can start, however, by identifying the important relationships between the processes of evolutionary biology and those of energy flow, with the hope of finding some correspondence and thus a beginning of a translation. Some of the particularly vexing problems that arise in making a translation are illustrated by the interspecific relationship between ecological density and body size.

In previously published work Damuth has shown a significant relationship between the ecological population densities of mammalian primary consumers and body mass, with a slope of -0.73. This relationship enables us to answer a question of adaptive potential—are species of different sizes equally able to evolve to extract energy from the environments they inhabit or is there a bias along the body size gradient? First, however, one needs to know how much energy individuals can extract as a function of body size. Since metabolic rate increases with body size with a positive slope of about 0.75, population energy use is essentially independent of body size on the average. Body size per se does not influence how much energy a species can extract from its habitat.

With a body range of six orders of magnitude for mammals and 11 from soil mites to elephants, it is hard to believe that this regular relationship is maintained by physical attributes of the environment or fortuitous combinations of factors impinging on each species. More likely, biotic interactions are maintaining this relationship over evolutionary time. Diffuse interactions may maintain the upper bound of energy use. Any species expanding too much gets its toes stepped on by an increasingly large number of other species while creating increasing selective pressure on them to respond to its advance; after a while it becomes impossible to outrun all potential competitors and predators and one's expansion is stopped. Species with very low rates of population energy flow, on the other hand, are susceptible to random-like fluctuations and thus suffer higher extinction.

This analysis seems to be heading in the right direction—it is a pattern that involves the current adaptive success of species and their population energy flow and that seems to be explained by ecological interactions over evolutionary time. However, the analysis doesn't quite bridge the gap between the two frameworks because it sits totally within the adaptive selective framework by its use of ecological densities, densities which consider only habitat suitable for a species. Factored out of the analysis was the distribution of habitat patches and their size; the flow of energy of a species is unknown unless we have information on the availability of habitat patches suited to it.

Crude density adds in the geographic component but in a way not easily separable from local, ecological density. What happens to the relationship if crude densities are used? For this analysis Damuth used the arthropods collected from fogging several tropical trees. The difference in the relationship was the addition of small, rare species in the lower left corner of the plot. For these arthropods, the trees are arbitrarily chosen areas to sample. For some species you are presumably collecting only chance individuals that do not regularly use that area as a habitat. One tree over and these rare species may be quite abundant. The chance-caught small species will be well below its normal density, but a large one less so. Using crude densities for mammals gives roughly the same results as ecological ones. Why? Perhaps, being large mammals, we often choose areas of habitat which are roughly similar to areas of habitats that mammals inhabit. In sampling for tiny creatures we tend to take areas that are quite large relative to their actual habitat use.

An interesting question that arises from this is, what area would need to be sampled to produce a density plot for mammals resembling that for arthropods from a tropical tree? Damuth attempted this to do this for a large area in Colorado and adjacent states using vegetation maps and habitat-specific densities for about six mammals. The results of arbitrarily chosen plots began to show some species with low densities because their habitats were sparsely distributed. But in order to get comparable results for arthropods and mammals he had to go from a tropical tree to an area of several thousand square kilometers. A very real problem, then, is how to compare very large things with very small. It is essentially impossible to sample mammals in areas large enough to obtain comparably crude densities because undisturbed habitats are no longer large enough, if they ever were. This leaves us with scaling up the results of the small, but in order to do this we need to understand enough about how they perceive and use their habitats to obtain comparable values for ecological densities. At the moment we cannot do this adequately. Until we can deal with the problem of how vastly different biologies divide up space in vastly different ways we will be unable to put the adaptive-selective world into some kind of energy-exchange context. There being severe problems with empirical approaches, theoretical ones might lead to an understanding of what densities at the different parts of the scale mean relative to each other. Meaningful theory must be empirically based and at this point we adequately understand the habitat use of few species. At that point Damuth stopped for discussion.

Brown questioned whether even ecological densities throughout will give a negative ¾ slope for the entire size range of organisms on Earth. A long dialogue between Damuth and Brown generated by this statement boiled down to the following points:

In comparing densities as a function of size one must restrict the analysis to trophically meaningful groups, and not necessarily to the trophic level in the sense of Eltonian pyramids of energy flow. These groups of species ought to be using the same basic source of energy in about the same way. (One cannot really compare the densities of the arthropod fauna in a decaying log with those in nearby leaf litter because the density of the energy base itself is quite different. If it were only feasible, an energetic unit rather than a spatial one should be used to compare numbers of individuals of different species.)

Are there truly rare species, as Brown claims, or are they artifacts of how we obtain densities, as Damuth thinks? Comparing densities in a meaningful way, with respect to the amounts of energy controlled by the species, means knowing the geographic dispersal of the habitats in which species live. However, as Damuth emphasized in his presentation, we rarely know this information and thus are at a bit of a standstill at present. The ecological densities that Damuth used in his plots cannot be used to calculate the energy control of a species because the appropriate habitats are not spread evenly over a species' range. But this is precisely the information that Brown finds meaningful. For example, he observed that the most abundant species in local habitats also tend to have the widest geographic ranges. (Subsequently, Maiorana found that this does not seem to hold for large African mammals.) Combining geographic range with density, being careful to exclude areas where a species is absent, does not equalize species' control of energy. Some have a lot and some have little. This tells him more about how ecosystems are organized with relation to energy than do ecological densities. Damuth, however, is not as interested as Brown in such broadscale questions, so he doesn't care about the restricted information his plots give about total energy control of a species. Damuth is concerned more about the adaptive potential of species and what that might mean to competitive success or failure over evolutionary time.

Bambach wondered whether crude densities might be more equivalent to the information yielded by the fossil record because of the time-averaging integration that goes into the formation of fossil assemblages. Damuth answered yes and no. Sometimes fossil assemblages are not such a mix. However, to address many paleontological questions one has to work at the same broad scale as Brown does, and thus one has to be concerned with how abundance interacts with geographic range. One can perhaps ignore the lower-level processes reflected in ecological densities at these broader scales, but that doesn't mean that they are not important to organisms.

Brown concluded this part of the discussion by bringing up a repeated pattern in breeding bird census data: that for any species more censuses find one individual than any other number, the frequency declining in a hollow curve toward larger densities. Because the data represent breeding densities they are not quite comparable to the densities Damuth collected for mammals. To Damuth they do support his view that species tend to spread out from habitats where they are relatively abundant into others where they may not even be able to maintain a population without immigration—what Van Valen has called the unfavorable region.

Maiorana started by showing some puzzling data noticed during the pursuit of other projects. Simply summarized, carnivores, defined as animals that eat primarily other members of their own broad adaptive radiation, are disproportionately represented as species in faunas or communities relative to their biomass or abundance as individuals. The individual or biomass representation is about what is expected for the energy flow between trophic levels, but the fewer carnivorous individuals divide themselves more finely into species than do their herbivorous relatives. For example, in the arthropod community of decaying logs, carnivorous species contained only 5% of the individuals but about 50% of the species; on living collard plants, carnivores were only 2% of the individuals but 54% of the species; in a beech canopy, they reached 15% of the individuals but 35% of the species. Terrestrial-mammal communities show the same pattern. For example, within the Serengeti bushland habitat, large carnivores represented about 1% of the individuals or biomass of large mammals but 27% of the species. In a similar type of habitat in Venezuela with many fewer large mammalian species, carnivores were 1.3% of the individuals—less than 1% of the biomass—but 50% of the species. Fossil assemblages show the same pattern. In an Oligocene large-mammal assemblage and in a Cretaceous and a Jurassic dinosaur assemblage, carnivores were 7.8% of the individuals, 1.6% of the biomass, but 17.24% of the species. Even among Permian reptiles, for which the amount of energy flowing between trophic levels has been thought to be greater than in dinosaur or mammal communities, and where carnivores did represent a higher proportion of individuals (21%), they were an even higher proportion of the species (47%).

Carnivores are rare because of the large loss of energy between trophic levels, but why should they subdivide this small amount of available energy so greatly among themselves, such that the average carnivorous species has much less energy control than the average herbivorous one? This question arises from the theory or point of view that each species attempts to maximize its control of trophic energy in evolutionary time. So if you look at a radiating clade such as mammals, which began as insectivores and then gave rise to herbivores and later carnivores, why should so many carnivores have evolved when they seem to have so much less control of the trophic energy than most herbivores? Why not simply be an herbivore with greater population density, more energy control, and thus presumably a lower chance of going extinct, given that density does relate to a species' probability of extinction? Maiorana presented two factors that she believes partially answer this question. They both boil down to the fact that carnivores as species may be as successful in their control of trophic energy as are many herbivores.

The most important factor is the apparent dominance of the herbivorous trophic level by only a few of the many herbivore species that compete for the living photosynthetic tissues of plants. If a species can't be a dominant herbivore it might be as well off eating that species rather than competing with it. Support for this notion comes from two studies: that of the invertebrate fauna on collard plants from a three-year study by Richard Root, and that of large mammals of the Serengeti ecosystem from data gathered by H. Hendrichs. To estimate dominance Maiorana used trophic-energy flow of the species, measured as body mass to the 0.75 power times the number of individuals.

For the 94 herbivores collected from the collards, the average trophic, energy flow was 106 units, of which 91.4% was accounted for by only three species. This left an average of 9.5 units for the other 91 species. If 10% to 24% of the energy present in the herbivore trophic level is available to predators, the 180 predator species had an average energy flow of 6 to 13 units, about the same as the average subdominant species in the system. For the 39 regular species (present in each year of the study), the results were similar. A lack of information on body masses of the carnivores made it impossible to compare the distribution of energy control of the species directly.

The pattern for large mammals in the Serengeti is similar. The four dominant herbivore species control 82% of the trophic-energy flow, while the other 18 have an average of only 50 units. If 1% to 3% of this energy intake is available to higher trophic levels, the eight carnivores average but 6 to 19 units of energy. However, whether comparing the frequency distributions of their numbers or of trophic-energy flow directly, they overlap with the lowest half of the herbivores; an unknown proportion of their food is from smaller species

Considering that carnivores do as well as the less successful half of the herbivores, their way of life is not as disadvantageous as it first appeared. Because so few species of herbivores hog the available resource, most herbivores are as rare as carnivores. When a second major factor, that of geographic range, is also considered carnivores gain more than herbivores in their energy control as species. Both arthropod and mammalian carnivores have larger geographic ranges than related herbivore species. Some quantitative evidence for African ungulates and carnivores indicate the latter having approximately three times the geographic range of herbivores. Because the dispersion of suitable habitats may differ over their range for the two trophic groups, it is unclear at this time whether this difference may not be even greater, herbivores being more tied to kinds of vegetation than carnivores.

Both the strong dominance and smaller geographic ranges of herbivores can be explained by the same factor—the chemical and structural warfare that terrestrial plants engage in with their predators. When one taxonomic group comes up with a successful strategy it may well dominate the particular plant resource to the virtual exclusion of most other existing or potentially evolving herbivores. The more these successful plant eaters hog the available productivity, the greater the advantage of becoming a predator on them. Low dominance of herbivores over enough evolutionary time is predicted to be associated with fewer carnivorous species, which average the same control of trophic energy as most of the herbivores. If too many carnivores happen to evolve, so that they do have lower energy control than most herbivores, one might expect to see reversals in food habits if these become possible. Among mammals omnivory is rampant and some carnivores have given rise to herbivorous descendants (even "hypercarnivorous" viverrids having partly done so: the guts of omnivorous viverrids deserve study). Thus, there can be flexibility in changing diet when it becomes a way to increase one's flow of energy. This is also true for many arthropods (e.g., beetles, ants, mites, or bugs, but not spiders). A similar relationship may be predicted for plankton and, for detritus-based communities like the soil or the benthos, detritivores replace herbivores in the prediction.

Brown immediately asked whether mammalian carnivores can really go back, with Van Valkenburgh pointing out that cats are so good at what they do and their skull is so modified in structure, particularly the teeth, that it may be impossible for them to switch to herbivory. Van Valen noted that it is not just the skull but also the gut of cats that may inhibit them from a radical change in diet. Mammalian herbivores have a highly specialized gut that is not easily lost or even regained. He suggested in another context that the gut of humans suggests a long period of heavy reliance on meat because of a simpler structure relative to other primates. Maiorana pointed out that meat is a much richer and concentrated source of food and is faster to digest and that the energetic advantages of this might have to be worked in as well as geographic range. On the negative side may be greater susceptibility to obtaining parasites from one's prey. It will be difficult to come up with exact knowledge of the energy control of different species but the rough approach taken here may be adequate for the types of questions she finds interesting.

Sherman wondered where the insectivores fit in, considering that placental mammals initially pursued insects. Maiorana thinks of them as indirect herbivores, getting at the plants through the arthropods that feed on them directly. Since as much as 24% of the energy intake of herbivorous arthropods may be available to their predators, insectivores may do better with insects than with the more easily available plants directly, particularly if they are small. Mammalian insectivores are more common in forest habitats where arthropods are more dominant than foliage-feeding mammals, in contrast to open, grassland habitats, where large mammalian grazers and their predators have greater control of the community energy flow. As Sherman then pointed out, the attempt to separate trophic components of the mammalian radiation in order to study patterns in the evolution of energy flow is further complicated by the fact that many mammalian herbivores rely on arthropods to obtain essential amino acids. She noted that when brought into zoos, many herbivorous species do not thrive because they do not get this small but essential component of their diet.

Damuth found the distribution of body size among meat specialists and omnivores within the mammalian order Carnivora intriguing. The middle-sized and very largest are the most omnivorous. In response to a question by Sherman, Maiorana felt her analysis can prompt looking at the fossil record in a new way. From the time the amphibians emerged on land there have been numerous taxonomic turnovers in the currently dominant group, each of which has been associated with a radiation of large herbivores and their predators—lots of material to investigate the ideas presented here. Van Valkenburgh would like to study what makes the dominant herbivores and carnivores so special and see if these traits show up during these times of turnover. Maiorana noted that strong ecological dominance seems to be reflected in the classification of a group. A special trait enabling the animal to dominate its food in one community often gives it superiority elsewhere, which is then often associated with greater speciation and then overlap in the same area. Some preliminary studies suggest that this taxonomic dominance seems more common among herbivorous taxa; some carnivores also evolve special traits that give them ecological dominance, as Van Valkenburgh noted, and these groups (such as *Felis*) often show taxonomic dominance as well.

Brown brought up the problem of comparing the energy control of birds and mammals, with the former being much smaller on average no matter what you add in (geographic range, metabolic rate, longevity, etc.). Sherman objected somewhat to Maiorana's suggestion that flight enabled birds to subdivide the habitat more finely than mammals by pointing out that lots of mammals do live in trees and seem to have no trouble traversing the space between, and thus might be expected to subdivide the habitat more finely like birds. No one brought up in this context the bats, which are unusually species rich for mammals, particularly in the tropics, where their food of insects and fruit is especially abundant and available all year. Maiorana conjectured that the big difference between birds and mammals may involve a difference in what leads to speciation. Birds are diurnal, with visual cues that lead to reproductive isolation. The bird-mammal difference may be analogous to the grasshopper-cricket one Dan Otte discussed earlier in the Congress. Birds may speciate more and thus be rarer on the average than mammalian species. It may be inappropriate to compare them at the species level; perhaps the generic level is more meaningful in terms of evolutionary adaptive units—one then avoids species that are no more than geographic or habitat isolates. Unfortunately, obtaining the sort of data used here is difficult for species and, at this time, almost impossible for genera, but methods might be devised. This sort of analysis would seem important to the types of questions Brown is asking. It is unclear, though, why one would expect mammal species to have the same average energy flow as bird species.

Bambach provided a very long perspective on seafood through time, switching attention away from the terrestrial mammalian systems which had so far dominated the discussion. His very descriptive overview documented without a doubt significant changes in energy use and energy flow in the marine ecosystem. Bambach began with a slide showing his compilation, digested from Jack Sepkoski's compilation, of the number of marine families during the Phanerozoic. Although it is somewhat dangerous, Bambach used diversity as a proxy for abundance and biomass. Studies have shown strong correlations (0.8–0.9) of family diversity with specific and generic diversity. Whether we look at generic or specific diversity, at trace-fossil abundance, or at species within habitats over time, the resulting pattern is consistent and coherent. We are not going to change our overall picture of the history of life with future discoveries despite the incompleteness of the fossil record.

This record documents a change in diversity with time, episodic but increasing. The increase in diversity has been accompanied by a change in the types of taxa that dominate the system and by changes in the use of ecospace, including expansion of the more energetic modes of life. Three major faunas are replaced by one another through time: the Cambrian by the Paleozoic after the major extinction in the Ordovician, and the Paleozoic by the modern after the Permian extinction. These faunas differ not only in the taxa that dominate each but in the way they use the available marine ecospace. In a diagram he developed several years ago Bambach divides ecospace into a variety of trophic/habitat groups. The Cambrian fauna filled most of the ecospace but not equally so. Carnivory existed, but was not a popular mode of life. Few swimming and infaunal types existed, although there was some shallow burrowing. In the middle to upper Paleozoic fauna, in which diversity stayed constant for a long period of time, the epifaunal mode of life became extremely crowded; it was the major focus of faunal evolution at that time. There was an increase in the more energetic pelagic modes of life but the infaunal realm did not fill appreciably; there was little energy devoted yet to deep burrowing. In the emergence of the modern fauna in the Mesozoic and Cenozoic, the rest of the pelagic space and the infaunal space filled up. Bambach contends that these modes of life require greater expenditure of energy basically because they involve more movement than the relatively passive epifaunal modes. Carnivory increased dramatically as the ecosystem became more densely populated.

The pattern of an increase in diversity and increase in use of ecospace is reflected in the diversification patterns of the individually radiating groups involved. For example, the cephalopods have not changed their mode of life very much through time and they have maintained the same diversity as well. In contrast, the bivalves have been expanding exponentially since the Permian extinction and some have changed their mode of life by invading infaunal habitats. The gastropods also expanded but it is only the predatory groups that have done so; the herbivorous groups have not diversified greatly over time. So both bivalve and gastropod expansions were mostly the result of adding new modes of life rather than expanding old ones. Within the malacostracan arthropods, the larger, more muscular, more active predatory forms such as crabs and shrimps expanded enormously relative to the more passive filter-feeders in this group. These arthropods not only increased in diversity but have also been abundant, suggesting an increase in the amount of energy flowing through the system as well. Within echinoderms, the sedentary, stalked, suspension-feeding crinoids were replaced in diversity by the mobile, jaw-bearing echinoids. The weak lanterns of Paleozoic echinoids limited the types of herbivory they could engage in. The modern fauna has added forms with stronger lanterns and more varied modes of herbivory, as well as infaunal groups. Finally, the Paleozoic chordates with little biomass and a more passive way of life gave way to the active, highly muscular, and mobile chordates that expanded enormously after the Permian. In group after group, the large increases in diversity during the evolution of the modern fauna was caused by addition of new modes of life rather than expansion of old. Moreover, these new modes of life were of the more energetic styles. It seems that the changes in energy use during evolution have occurred at the time of origin of new groups when they become committed to the adaptive innovations or morphological breakthroughs that let them invade new ecospace. It becomes important to look at the way these key innovations have evolved. Bambach believes there may be parallels with terrestrial systems such as the one Van Valen discussed. He believes the early radiation is extremely rapid and then the group settles down for a long period of stasis during which little of significance happens to its energetic role in the ecosystem, even though continued taxonomic turnover may occur.

A third line of evidence, a strong correlation between the number of species and the number of guilds in a fossil community, emphasizes the point that the increase in marine diversity through time involved addition of new modes of life rather than finer division of existing modes. For example, during the long stretch of the Paleozoic where total diversity was constant, so were the number of guilds; there was effectively no change or reorganization in the system despite enormous taxonomic turnover. But in the Neogene, a doubling in the number of species is associated with nearly a doubling in the number of guilds. At the local level there was a reorganization of the Neogene system; the number of species per guild is roughly the same for the two time periods. The diversity within habitats is not increased by greater species packing but by the addition of new modes of life. The efficiency of energy use has not altered; instead, new sources have been added. This is a very important point which raises the question of where the energy to support these new modes has come from. We really don't know, and our attention in our studies of energy flow through ecosystems through time needs to focus on the source of change at the base of the food chain to support a larger number of more energetic types of organisms. The larger biomasses as individuals means they were being fed well, so what happened? Diatoms and coccolithophores appeared in the Mesozoic and these may have been an important additional source of primary productivity, but we still know little about why the Mesozoic would begin to support absolutely more biomass than the Paleozoic.

Bambach concluded by emphasizing the parallels that seem to exist between marine and terrestrial ecosystems.

Where does the energy come from to support new modes of life; from whom is it taken? This initial query by Damuth led to a variety of ideas being tossed about. Damuth wondered whether one can pick up whether it came from guilds that were not using it very efficiently. Bambach thinks you really have to look at the base of the food chain because there are only a few cases of a new group competitively ousting an old. Mostly what happens is that the old ones hang on. That's what happened with the brachiopods when the bivalve mollusks replaced them in their old haunts after their decimation in the Permian extinction. The more energetic mollusks may have been competitively superior to the brachiopods but, if so, where did they get the additional energy to sustain their more energetic mode of life that enabled them to oust the brachiopods? It apparently wasn't there in the Paleozoic, since all the major ways of life had appeared by the Devonian and merely hung around doing little until the Mesozoic. He believes that although the bivalves had the energetic potential to invade new modes of life, they did not yet have the energy resources that supported these modes.

Maiorana suggested that an increase in the metabolic rate of the primary producers could have fixed carbon at a faster rate and thus provided a base to support a larger biomass of consumers. There is evidence that metabolic rate has increased over time in all sorts of organisms at all levels of the food chain. Angiosperms have higher rates than gymnosperms, and this may have been important in terrestrial ecosystems since the metabolic rates of the important primary consumers (dinosaurs and, later, mammals) also seem to have increased at this time. Is it possible that the same thing happened in the marine system? Bambach thinks it is likely and is an important thing to look at, although he has had trouble getting information from the botanists about metabolic rates. Maiorana considered later that an increased metabolic rate of producers was not essential to increase overall productivity; an increase in the metabolic rate of consumers alone can lead to faster harvesting and turnover of nutrients that may increase the production levels in the sea in a way analogous to grazing on land. Van Valen noted that the microbial loop, and other consumption by inadequately preserved organisms, is not controlled and, because it is trophically very nonrandom, could have decreased to provide some or most of the energy needed. Perhaps isotopic fractionation may help a bit in looking at this awkward possibility.

Van Valen pointed out that you can get an increase in the number of guilds even with a stochastically constant flow of energy through time, at the expense of individual density. Bambach replied that the density of individuals did increase dramatically through time. This is suggested by trace fossils and especially by the much higher density of shells in Neogene deposits than in Paleozoic ones of the same depositional setting. Grassle then wondered if there were instances where biomass did not increase with an increase in species diversity. Thayer suggested that the trace-fossil abundance record may be revealing. There is not much reworking of Cambrian-Ordovician sedimentary rocks by infaunal organisms—in contrast to Cenozoic rocks, which are enormously more worked. Although controversial, the evidence suggests that the difference was the result of a greater abundance of activity. Bambach also pointed out that the depths of the burrows have increased dramatically in offshore habitats, where all food filters down from the open ocean. Bambach believes that when there was food there, the organisms went after it, dismissing Grassle's objection to lack of oxygen, since many of the organisms pump their own oxygen down their burrows.

Brown switched from this focus on primary productivity to ask whether predators may have had something to do with the change in marine diversity. In terrestrial systems they can cause major restructuring of the flow of energy without any change in level of primary productivity. Maybe we get an increase in the number of shelly fossils because they started to get away from predators more successfully. Bambach agrees that there is an interplay there and went on to describe what the record tells about the origin of top predators and the defense response of their prey. Although shell forms altered and predation may have encouraged more infaunal and epifaunal dwelling, it caused no major change in the marine system aside from its addition as a major way of life.

Going back to productivity, Brown wondered how much increased angiosperm productivity may have fed the seas; there now seems to be a lot more onshore production than once thought, and perhaps the angiosperm revolution on land in the Mesozoic helped raise the productivity of the seas and led to the upsurge in consumer diversity. Although Bambach thinks it may have been part of the picture, indications of increased productivity (appearance of diatoms, coccoliths, and symbiotic forams) occurred before the angiosperm radiation.

Brown also wondered about the apparent failure of the modern fauna to level off in diversity like that of the Paleozoic. It seems to be a long time after the key innovations came in for the system to be increasing, especially since Bambach emphasized how rapidly things change in the beginning. Bambach agreed that the increase in diversity has been rather continuous, and suggested that the extreme extent to which the marine realm was devastated by the Permian extinctions may have been partly responsible. Only 5 to 10% of the fauna may have survived. There's a lot of empty ocean to be refilled and that takes time. On top of that is the interactive nature of the diversification. One cannot get crabs diversifying until there are enough things for them to eat. While the innovations may have appeared early, there may have been a long lag before their potential could be realized. We just don't know, but the answers are probably in the fossil record and can eventually be teased out.

Van Valen noted that the possibility of an increase in primary productivity raises the sticky point of what regulates primary productivity in the sea. Nutrient regulation enters in here and some of the organisms, such as nitrogen fixers, are not preserved, which makes rather a mess of sorting out what is an enormously complex problem. That led to some discussion of problems in trying to detect what was happening to nutrients at different times in the past. The fossil record can get at some aspects, but as we do not understand the current marine system where we have everything still there, the prospects of getting complete answers is not very good. Grassle made the point that when one puts in the microbial food web, the entire food web of the ocean becomes vastly more complex and that for most of the organisms we sample in the fossil record, the amount of primary productivity that gets to the bottom is more important than the oceanic productivity. Bambach took that as a good point to conclude discussion of his presentation. He notes that the amount of organic carbon preserved in fuel resources can tell how much primary productivity is being lost from the system. During the 40-million-year interval in which half of the world's oil reserves were preserved, the average daily loss of organic carbon was the amount contained in 50 barrels of oil. The amount is trivial on a world-wide basis, which indicates how difficult it is to detect changes in the world's productivity.

If nothing else, the more energetic modes of life appearing in the modern marine fauna have had a great deal more edible meat in them.

SYMBIOSIS AND COEVOLUTION

Detection of Chemosymbiosis in the Fossil Record: The Use of Stable Isotopy on the Organic Matrix of Lucinid Bivalves

E. A. CoBabe

Abstract. All members of the bivalve family Lucinidae appear to house chemolithoautotrophic bacterial symbionts in their gills. The symbionts may provide these bivalves with a considerable portion of their nutrition. This additional, independent food source may be evolutionarily advantageous to lucinids. While the evolutionary history of chemosymbiosis can be evaluated by traditional methods of paleontology (e.g., variations in biogeography and survivorship across extinction boundaries), this requires the assumption that taxa that have chemosymbionts today have always possessed them. Isotopic geochemistry may allow the detection of chemosymbiosis in both modern and fossil shells, providing an independent means for confirming chemosymbiosis in the fossil record.

It has been well documented that the carbon stable isotopic values from the tissues of chemosymbiotic organisms are significantly more negative than those values from non-symbiotic organisms. Carbon isotopic signals from the organic matrix of modern bivalves indicate that the organic matrix material also reflects this isotopic difference between symbiotic and non-symbiotic bivalves. This signal will allow exceptionally well-preserved shell material to be evaluated for chemosymbiosis. From this work, questions of the evolutionary history of chemosymbiosis can be addressed

INTRODUCTION

Since the discovery in the early 1980s that the ecosystem at the hydrothermal vents is supported by chemosynthesis and that several of the more unusual organisms there rely on chemosynthetic bacterial symbionts (Cavanaugh, 1980; Cavanaugh et al., 1981; Cavanaugh, 1983), this trophic strategy has been documented in a wide variety of organisms and from an increasing number of different kinds of environments, including mangrove swamps, mud flats, grass beds, and deep sea basins (Felbeck et al., 1983b; Felbeck, 1983; Dando et al., 1985; Schweimanns & Felbeck, 1985). While it is certainly true that chemosymbiosis is an unusual way to make a living, questions remain as to its long-term evolutionary advantage. Are there any quantitative differences in the evolutionary histories of organisms that possess chemosynthetic symbionts and those that do not? To try to answer this question, I am looking at the evolutionary history of bivalves in the family Lucinidae, a group of bivalves in which all members appear to possess chemo-

Dr. CoBabe is with the Museum of Comparative Zoology, Harvard University, Cambridge, MA 02138, USA.

symbionts today. Lucinids inhabit a wide range of environments, often with non-symbiotic infaunal bivalves. These environments include mangrove swamps and eelgrass beds—environments that are relatively easy to make collections in. Most importantly, lucinids are well-represented in the fossil record, making them an ideal group to evaluate chemosymbiosis over geologic time.

WHY SHOULD CHEMOSYMBIOSIS BE ADVANTAGEOUS?

Lucinid bivalves house thiotrophic symbionts in specialized cells in their gills (Southward, 1986). The symbionts convert CO_2 into carbohydrates in the same way plants fix CO_2 during photosynthesis. However, instead of using sunlight as the driving energy source, the bacteria utilize the energy caught up in the hydrogen-bonds of hydrogen sulfide (Felbeck, 1983; Felbeck et al., 1983a; Felbeck et al., 1981). The symbionts are used as a primary food source by the bivalve, though lucinids do maintain a functional, albeit reduced, digestive tract (Allen, 1958). In essence, the chemosymbionts provide the bivalve with an independent, internal food source, relying only on the continuous supply of CO_2, O_2, and HS^- in the environment. This internal food source may allow lucinids to expand into environments, regardless of the biological productivity of the environment. Lucinids, by possessing this food source and expanding into a wider range of environments, may be able to weather some kinds of ecologic crises more readily than other infaunal bivalves.

Evolutionary advantage in lucinids encompasses two kinds of questions: 1) How far back in time have lucinids been chemosymbiotic? Once possessing the symbiosis, do lucinids maintain it for a long period of geologic time or does this relationship appear and disappear with regularity through time? 2) Do lucinid species survive ecologic crises more readily than non-symbiotic bivalves? Do they have longer species durations? Do their species cross diastems more often, surviving local and mass extinctions more readily?

The uniformitarian assumption that lucinids have always been symbiotic can be tested using both sedimentology and stable isotopy. Erle Kauffmann has suggested that the occurrence of bivalves in disaerobic facies is evidence for chemosymbiosis on the grounds that disaerobic sediments will have high levels of sulfide and low levels of oxygen (Kauffman, 1988). These conditions are ideal for a chemosynthetic system, particularly when the presence of any bivalve exclusively in these sediments indicates conditions would not have supported bivalves using a photosynthetically-based food system. Lucinids occur in disaerobic sediments throughout the Gulf Coastal Plain. This suggests that lucinids have been chemosymbiotic for some time. However, the bacterial symbionts may provide a biogeochemical signal that can be detected in the shells of their bivalve host, confirming chemosymbiosis in these organisms.

It has been documented by several workers that the carbon isotopic values from the tissues of bivalves with chemosymbionts is significantly lighter than the carbon isotopic values of non-symbiotic ecologic counterparts (Spiro et al., 1986; Conway et al., 1989; Southward et al., 1981; Southward, 1986, 1987). These values are a result of the fractionation of carbon by the bacteria during carbon fixation. Carbon isotope ratios may be on the order of 7 to 15 ppt more negative for chemosymbiotic organisms than for non-symbiotic organisms (Table 1).

In order to use this offset in carbon isotopic ratios to detect chemosymbiosis in the fossil record, it must be demonstrated that the isotopic values of a bivalve's tissues is the same as the isotopic values of the shell's organic matrix. In other words, is the organic matrix a good proxy for the organism itself? Secondly, it must be shown that a signal for chemosymbiosis can be consistently detected in both recent and fossil shell material. To

Table 1. Carbon isotopic values indicating a signal for chemosymbiosis in modern tissues (from Southward, 1986).

Location and Species	Symbionts Present	Depth (m)	Tissue	$\delta^{13}C\%$
England				
Lucinoma borealis	+	0	gill	−28.1
			body	−25.3
Dosinia lupinus	−	0	gill	−17.6
			body	−16.8
Norway				
Myrtea spinifera	+	33	gill	−24.2
			body	−23.4
Thyasira flexuosa	+	55	gill	−29.3
Thyasira sarsi	+	90	gill	−31.0
			body	−29.2
Arctica islandica	−	33	gill	−18.8
			body	−18.9
Bahamas				
Codakia orbicularis	+	0	gill	−23.9
			body	−23.8
Lucina nassula	+	0	whole body	−23.0
California				
Solemya reidi	+	100	whole body	−30.0

address these questions, a suite of modern lucinids were collected from southern Florida, along with non-symbiotic infaunal bivalves living in the same environment. Fossil lucinids were collected from a number of extraordinarily preserved faunas in the Gulf Coastal Plain. With fossil material, there is the expectation that if the original mineral phase of the shell is still intact, then the possibility exists that there will be indigenous organic matrix molecules preserved as well (Robbins & Brew, 1990; Hoering, 1980).

MATERIAL AND METHODS

Shell material (both modern and fossil) was washed with distilled water. No additional chemical cleaning was performed. Shells were ground under liquid nitrogen. This was done to limit the amount of heat produced during grinding—heat that could potentially cleave organic molecules. Two different decalcification protocols were used (Fig. 1). One using 12M HCl, is harsh and aggressive, but takes only 3 to 12 hours. The second, EDTA, is much more gentle, but decalcification takes anywhere from 7 to 10 days. Clearly, HCl is time efficient and cheaper than EDTA. EDTA is better at removing humic acids. However, besides being more time consuming and expensive, removal of EDTA from the sample (both the soluble and insoluble components) is difficult. This presents a problem for stable isotope work, since EDTA contains both carbon and nitrogen and will potentially overwhelm any biological signal. Samples were suspended in dialysis tubing and rinsed for several days in distilled water. Each shell sample was separated into soluble and insoluble components using centrifugation. Soluble and insoluble components of each sample was lyophilized. Lyophilized samples were prepared for isotopic analysis using sealed tube combustion procedures. Samples were combusted for three hours at 850°C. Isotopic analyses were performed using a VG Prism and a VG Micromass 602E mass spectrometer.

672

Decalcification Protocols

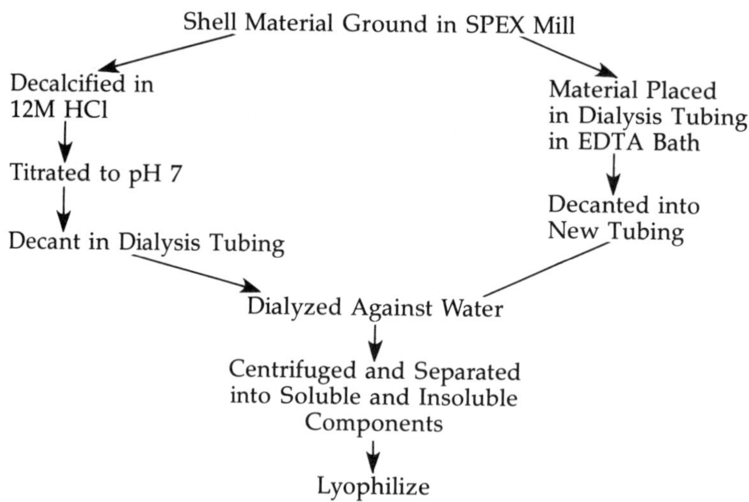

Figure 1. Decalcification protocols used to extract organic matrix molecules.

RESULTS

Initial carbon isotopic data from modern lucinid bivalves and ecologic counterparts collected from the southern Florida Keys are presented in Table 2. It appears that the organic matrix of the bivalve shell is a good proxy for the isotopic values of the organism itself. Further, a detectable signal for chemosymbiosis is seen in the modern shells. Note that, generally speaking, the carbon isotopic signal of the bivalve tissue is within 1 ppt of the soluble and insoluble components of the shell organic matrix. Further, there is about a 7 ppt offset of the isotopic value in the symbiotic bivalves *Lucina pensylvanica* and *Codakia orbicularis* when compared to the non-symbiotic *Arca*. Detection of chemosymbiosis in the fossil record is a possibility. Over the next year, this signal will be sought in a series of extraordinarily preserved fossils from the United States. If successful, the isotopic chemistry may allow an occasional look at the trophic strategy of lucinid bivalves in geologic time.

Table 2. Carbon isotopic values from tissue and shell organic matrix components of some modern bivalves.

	$\delta^{13}C\%$
Lucina pensylvanica	
soluble, shell	−23.979
foot	−23.384
Codakia orbicularis	
soluble, shell	−23.401
insoluble, shell	−22.140
foot	−24.818
Arca sp.	
insoluble, shell	−13.980
foot	−15.403
Sediment organics	
soluble	−15.508
insoluble	−12.969

ACKNOWLEDGMENTS

This work would not be possible without the support of Dr. Noreen Tuross of the Smithsonian Institution and Bruno Marino of Harvard University. Isotopic work was completed at Harvard University in the Atmospheric Chemistry Lab of Dr. M. McElroy. Thanks to Paul Morris for helpful comments on this manuscript.

LITERATURE CITED

Allen, J. M. 1958. On the basic form and adaptations to habitat in the Lucinacea. *Phil. Trans. Ser. B* 241:421–484.

Cavanaugh, C. M. 1980. Symbiosis of chemoautotrophic bacteria and marine invertebrates. *Biol. Bull.* 159:457.

Cavanaugh, C. M. 1983. Chemoautotrophic bacteria in marine invertebrates from sulfide-rich habitats: a new symbiosis. Pp. 609–708. *In:* W. Schwemmler & H. E. A. Schenk (eds.), *Endocytobiology,* vol. II, Walter de Gruyton and Co.: Berlin-New York.

Cavanaugh, C. M., Gardiner, S. L., Jones, M., Jannasch, H. W. & J. B. Waterbury. 1981. Prokaryotic cells in the hydrothermal vent tube worm *Riftia pachyptila* Jones: possible chemoautotrophic symbionts. *Science* 213:340–342.

Conway, N., Capuzzo, J. & B. Fry. 1989. Role of endosymbiotic bacteria in the nutrition of *Solemya velum:* evidence from a stable isotope analysis of endosymbionts and hosts: *Limnol. Oceanogr.* 34:249–255.

Dando, P. R., Southward, A., Southward, E., Terwilliger, N. & R. Terwilliger. 1985. Sulphur oxidizing bacteria and hemoglobin in the gills of the bivalve mollusc *Myrtea spinifera. Mar. Ecol. Prog. Ser.* 23:85–98.

Felbeck, H. 1983. Sulfide oxidation and carbon fixation by the gutless clam *Solemya reidi:* an animal-bacteria symbiosis. *J. Comp. Phys.* 152:3–11.

Felbeck, H., Childress, J. & G. Somero. 1981. Calvin-Benson cycle and sulfide oxidation enzymes in animals from sulfide-rich habitats. *Nature* 298:291–293.

Felbeck, H., Liebezeit, G., Dawson, R. & O. Giere. 1983a. CO_2 fixation in tissues of marine oligochaetes (*Phallodrillus leukodermatus* and *P. planus*) containing symbiotic chemoautotrophic bacteria. *Mar. Biol.* 75:187–191.

Felbeck, H., Childress, J., & G. Somero. 1983b. Biochemical interactions between molluscs and their algal and bacterial symbionts. Pp. 331–358. *In:* P. W. Hochachka (ed.) *The Mollusca.,* Vol. 2: *Environmental Biochemistry.* Academic Press: New York.

Hoering, T. C. 1980. The organic constituents of fossil mollusc shells. Pp. 193–201. *In:* P. E. Hare et al. (eds.) *Bioeochemistry of Amino Acids.* J. Wiley and Sons: New York.

Kauffman, E. 1988. The case of the missing community: low-oxygen adapted Paleozoic and Mesozoic bivalves and bacterial symbiosis in typical Phanerozoic seas. *Geological Society of America Annual Meeting Abstracts.* p. A48.

Robbins, L. & K. Brew. 1990. Proteins from the organic matrix of core-top and fossil planktonic foraminifera. *Geochimica Cosmochimica Acta* 54:2285–2292.

Schweimanns, M. & H. Felbeck. 1985. Significance of occurrence of chemoautotrophic bacterial endosymbionts of lucinid clams of Bermuda. *Mar. Ecol. Prog. Ser.* 24:113–130.

Southward, A., Southward, E., Dando, P., Rau, G., Felbeck, H., & H. Flugel. 1981. Bacterial symbionts and low $^{13}C/^{12}C$ ratios in tissues of Pogonophora indicate unusual nutrition and metabolism. *Nature* 293:616–620.

Southward, E. 1986. Gill symbionts in thyasirids and other bivalve molluscs. *J. Mar. Bio. Ass. U.K.* 66:889–914.

Southward, E. 1987. Contributions of symbiotic chemoautotrophs to the nutrition of benthic invertebrates. Pp. 83–118. *In:* M. A. Sleigh (ed.) *Microbes in the Sea.* Ellis Horwood, Ltd.: England. Spiro, B., Greenwood, P., Southward, A. & P. Dando. 1986. $^{13}C/^{12}C$ ratios in marine invertebrates from reducing sediments: confirmation of nutritional importance of chemoautotrophic endosymbiotic bacteria. *Mar. Ecol. Prog. Ser.* 28:233–240.

Symbiosis in Evolution

Mary Beth Saffo

Abstract. Symbiosis is an important biological phenomenon in need of serious attention by evolutionary biologists. Topics of discussion included the biogeographical and taxonomic distribution of endosymbioses, the impact of symbiotic interactions in evolutionary processes (including speciation, large-scale taxonomic radiations, and evolutionary rates), lateral gene transfer, the consequences of modes of symbiont transmission on symbiont-host evolution, and the dynamics of symbiotic mutualisms.

> *The history of ideas is paved by constraints of language.*
>
> Leo W. Buss, 1987.

Symbiosis is a common biological phenomenon of increasing experimental and ecological interest. Neverthless, symbiosis, especially mutualistic endosymbiosis, has received only the minimal attention of evolutionary biologists.

To redress this imbalance, a roundtable was organized to address two major questions. 1. What impact does symbiosis have on the rate and direction of evolutionary processes? Does it serve as a source of evolutionary innovation? 2. How can symbiosis contribute to evolutionary theory, especially to theoretical studies of the evolutionary dynamics of mutualism? Participants included Mary Beth Saffo (UC Santa Cruz), Lynda Goff (UC Santa Cruz), John Thompson (Washington State University, Pullman), Robert May (Oxford) and John Maynard Smith (Sussex), as well as members of the audience (named below). As befits an emerging field, the discussion elicited a number of difficult questions, interesting hypotheses, and few definitive answers.

THE IMPACT OF SYMBIOSIS ON EVOLUTION

A number of observations suggest that symbiosis should have a significant impact on evolutionary processes.

First, symbioses pervade today's ecosystems. Endo- and ecto-parasites are ubiquitous associates of animal, plant, fungal, and protoctistan hosts. Many of those endosymbioses traditionally considered to be mutualisms—reef-building corals and other algal-invertebrate associations; termites, ruminants, and other herbivorous animals with intestinal symbionts; symbioses between chemosynthetic bacteria and pogonophoran, bivalve, and oligochaete taxa in hydrothermal vents and other sulfide-rich habitats; lichens; mycorrhizae; nitrogen-fixing bacteria in several plant and animal taxa—are key members of marine and terrestrial communities. Other endosymbioses—symbioses of

---

Dr. Saffo is with the Institute of Marine Sciences, University of California, Santa Cruz, CA 95064, USA.

insects with intracellular bacteria, molgulid ascidian tunicates with bacterial and protistan symbionts (Saffo, 1990), luminescent bacteria with fish and invertebrate taxa, fungal "endophytes" in grasses and other land plants (Clay, 1988)—are less well known but nevertheless very widespread. If symbiosis is ecologically important today, should it not have played some role in earlier evolutionary events, and should it not be of evolutionary importance today?

Several studies have postulated that autotrophic-animal symbioses have existed at least since the Cambrian. The dependence of modern reef-building corals on photoauto-trophic endosymbionts has led several paleobiologists to propose that earlier corals (Cowen, 1988), non-cnidarian reef-builders such as the bivalve rudists (Kaufmann & Johnson, 1988) and the spongelike archeocyaths (Rowland & Gangloff, 1988), along with several calcareous reef-dwellers (benthic foraminifera: Strathearn, 1986) also harbored endosymbiotic autotrophs. Similarly, the ubiquity of chemoautotrophic bacteria in living lucinid bivalves has suggested to several paleobiologists that extinct lucinids were also symbiotic (Campbell, 1989; CoBabe, this volume). The distinctive stable carbon isotope signals of both photoautotrophic and chemoautotrophic organisms provides the potential to test some of these hypotheses on fossil taxa (Strathearn, 1986; CoBabe, this volume).

Symbiosis is implicated in parts of pre-Cambrian evolutionary history as well. Seilacher (1989) and others (McMenamin, 1986) have proposed that the Ediacaran "Vendozoa" harbored chemoautotrophic or photoautotrophic symbionts. It is now widely accepted that mitochondria, chloroplasts, and possibly other eukaryotic organelles are of symbiotic origin (Margulis, 1981; Fisher, 1989).

Second, the ubiquity of endosymbionts throughout several higher-order taxa suggests that endosymbiosis has played a role in the radiation of a number of animal families (molgulid ascidians: Saffo, 1991; leignathid fish: Mcfall-Ngai, 1991) and even phyla (Pogonophora: Vetter, 1991). Malloch et al. (1980) have argued that mycorrhizal associations played a pivotal role in the evolution of land plants.

Symbiosis can play a role in speciation. Commenting that biologists rarely discuss how symbiosis might drive speciation in hosts, Thompson noted that several cases of symbiont-induced speciation have nevertheless been discovered in arthropods. In a number of insect, isopod and mite species harboring mycoplasma, spiroplasma, and rickettsia endosymbionts, maternally-inherited reproductive incompatibility can be "cured" by use of antibiotics. Symbiont-driven reproductive isolation has been demonstrated so far in fewer than 20 species (Nardon & Grenier, 1991; Thompson, 1987, 1989; Somerson et al., 1984). But Thompson raised the possibility that the phenomenon might be more common; suggesting that symbiosis should be considered as a possible cause for "failure" of some hybridization experiments involving arthropod species.

Thompson posed two possible mechanisms by which symbiosis could induce reproductive isolation. The first (Thompson, 1987, 1988, 1989) proposes that differential coevolution of symbionts and hosts in different environments could lead to faster reproductive isolation than traditional mechanisms of gene isolation. The second notes that symbiosis, such as in the association between ovule-parasite species of *Greya* (Lepidoptera) and its plant hosts, can lead to extreme levels of host specificity.

Do all of these observations suggest that symbiosis has played a signficant role in major evolutionary innovation? Maynard Smith concluded that while symbiosis explains some major evolutionary events, such as the origin of eukaryotic cells, one should not expect symbiosis to play a role in every evoluionary event; in a similar vein, Law (1989) suggested that symbiosis has only rarely played a role in major evolutionary innovation, but that when it does happen, it is important.

There was considerable discussion of the effects of symbiosis on rates of evolution. Some lines of fossil evidence suggest that mutualistic endosymbiosis might be correlated both with rapid species radiation and long species duration. For instance, E. CoBabe

(Harvard) noted that lucinid bivalves, whose living representatives contain chemoautotrophic symbionts, show a fossil record of rapid early evolution (rapid reduction in the digestive system, and progression into different environments), with subsequent species durations that are significantly longer than those of non-symbiotic infaunal bivalves. The closely related thyasirids, with species that also contain chemoautotrophic symbionts (Southward, 1986) share a broadly similar history, marked by rapid evolution in the early Cretaceous, with little subsequent changes up to Recent times (Kauffman, 1967). L. Stathoplos (Smithsonian) commented that the fossil record of benthic foraminifera (whose living species contain symbiotic algae) shows rapid shell thinning (providing greater light transmittance to photosynthetic symbionts) compared to other morphological changes.

Other considerations suggest that symbiosis might have a constraining effect on evolution. Cowen (1988) has argued that the slow recovery of reef ecosystems from extinction events in the fossil record is best explained by the dependence of reef-builders on symbiotic algae: because the symbiosis takes time to evolve, Cowen proposed, "there is a lag between the time when reefs could be successful and the time that they re-evolved after a crisis." L. Van Valen (Chicago) noted that (putatively photoautotroph-containing) fusulinids undergo rapid extinction in the fossil record.

There was no consensus regarding the possible genetic mechanisms for symbiosis-induced (or symbiosis-correlated) evolutionary change, other than the importance of such mechanisms and the need for data. Nevertheless, a number of points were raised.

A. Coleman (Brown) questioned whether the multiple-genome systems of symbiotic associations might be more difficult to change than a single genome. Thompson argued that the additional genetic material acquired by symbiosis could provide instead richer potential for evolution.

John Maynard Smith asked whether obligate symbiosis is ever reversible evolutionarily, noting that "it's clear we're not going to reverse [our acquisition of] mitochondria." Coleman and Goff stated that *Paramecium bursaria* can be induced to lose zoochlorellae only with great experimental difficulty, but that non-photosynthetic parasitic flowering plants, while retaining (non-photosynthesizing) plastids, do lose photosynthetic genes over time. On the other hand, Saffo noted that several chronic symbioses are at least temporarily reversible: cnidarians exocytose algal symbionts at low temperatures or other suboptimal conditions, mycorrhizal symbioses dissociate at high soil levels of phosphorus, and nitrogen-fixing symbioses dissociate in high nitrate levels. Van Valen (Chicago) summarized the paradox: "Corals dump their symbionts; a plant can't do this [with chloroplasts]." A key difference between the cited reversible and irreversible examples may be their mode of transmission: organelles and the zoochlorellae of *P. bursaria* are transmitted vertically (direct transmission), while mycorrhizae and nitrogen-fixing symbionts are transmitted horizontally (indirect transmission), an issue emphasized by Maynard Smith.

Maynard Smith argued for the importance of a population genetics perspective on endosymbiosis, posing as a sample question: "Supposing there were a gene mutation in a symbiont making it less likely to be swallowed or ingested or taken hold of by the host, would the gene spread or not?" The central issue for the solution of such problems is the mode of transmission of the endosymbiont. He noted that the elegant model of Anderson and May (May & Anderson, 1983), of the evolution of the myxoma virus in rabbits in Australia, "hinges crucially on the pattern of transmission." In considering a number of non-pathogenic endosymbioses, Maynard Smith distinguished between direct transmission (as in eukaryotic organelles, and in intracellular bacteria in aphids, weevils and other insects), where symbionts are transmitted with the host each generation, and indirect transmission, where symbionts must re-find the host each generation. He discussed the large number of such endosymbioses where symbionts are released to re-find hosts: those in which symbionts rejoin a host not necessarily genetically related to their previous host (luminous bacteria in leignathid fishes; sulfur-oxidizing bacteria in the vestimentiferan pogonophorans; mycorrhizae; and most lichens) and those in which symbionts rejoin a host with some genetic relation to the previous host (as in termites).

A number of symbioses employ both methods of transmission: the protistan endosymbiont *Nephromyces* is transmitted indirectly to molgulid ascidians, but the intracellular bacteria of *Nephromyces* itself are transmitted directly with their protistan host (Saffo, 1990). In algal associations with clonal invertebrates, symbionts can be transmitted either indirectly to genetically related or unrelated offspring, or directly to asexually reproduced offspring.

Maynard Smith's perspective suggests the importance of still other distinctions informed by population biology. In considering the role of symbiosis in particular evolutionary processes, it is also important to distinguish—aside from the impact of particular symbionts on host fitness—between symbioses where, typically, only a small percentage of the host population is infected by symbionts, and "chronic" symbioses, where 100% of a host population is regularly infected by symbionts (Saffo, 1991).

Lynda Goff recounted the several instances of lateral gene transfer in nature between prokaryotes, and between prokaryotes and eukaryotes (bacteria and yeasts, and transfer of T-DNA from *Agrobacterium* to the genome of host plants). The relative rates of horizontal gene transfer among prokaryotes were a matter of discussion. Maynard Smith questioned whether (given the correspondences of cytochrome-c-based and rRNA-based phylogenies in the taxonomy of purple bacteria) horizontal gene transfer among prokaryotes might be less common than previously thought. A. Coleman (Brown) noted that a significant rate of horizontal gene transfer would greatly complicate studies of molecular biology.

Although lateral gene transfer between eukaryote species has not yet been demonstrated experimentally (Goff, 1991), it seems probable that such transfer may well occur in the several symbioses where a eukaryotic symbiont genome inhabits the cytoplasm of a eukaryotic host. Goff outlined several such cases. For instance, *Paramecium bursaria* contains cells of *Chlorella*, which themselves contain three (nuclear, mitochondrial and chloroplast) genomes. Several marine opisthobranch molluscs, such as *Elysia viridis*, maintain functional chloroplasts, derived from siphonaceous green algae ingested by the mollusc, in the cytoplasm of some cell types. An example of heterokaryotic symbiosis includes *Peridinium balticum*, which contains a dinoflagellate nucleus and a chrysophyte nucleus in the same cytoplasm; during sexual reproduction, the two nuclear types undergo synchronous meiosis (Chesnick & Cox, 1989). A second example is provided by Goff's own studies (1991) of the cellular interactions between parasitic red algae and red algal hosts: in these symbioses, host cells are at least transiently heterokaryotic, and in several striking cases apparently transformed by parasitic nuclei.

SYMBIOSIS AND MUTUALISM

As the second major topic, May, Saffo, and Thompson all discussed the evolutionary dynamics of mutualistic symbioses as a particular subcategory of mutualism. May raised the general question: "Why do textbooks [while devoting entire chapters to competition and predator-prey interactions] still only genuflect toward mutualism? It is still a problem." Among a number of factors (see May 1982, 1984; Keddy 1990), May discussed one possible answer: that "the simplest models of predator-prey and competition tell you something useful, but simple models of mutualism tell you nonsense." Perhaps the situation could be improved if mutualism models could be drawn from a more realistic biological base.

678

Saffo (1991) outlined some ways in which chronic (usually considered mutualistic) endosymbioses point up the need for revision of the most simplistic notions of mutualism. While mutualisms have been considered in the past to be rare or ecologically unimportant phenomena, chronic endosymbioses are in fact ubiquitous; many of them, as outlined earlier, have major ecological impact. Though the notion persists that mutualisms are limited largely to the tropics, a number of chronic endosymbioses (Saffo, 1991) are in fact more prevalent in higher latitudes than in the tropics. Endosymbioses such as lichens and ectomycorrhizae are conspicuous, and sometimes even dominant, members of non-tropical communities. In the Antarctic, for instance, lichens account for 90% of the photosynthetic flora.

Saffo, May, and Thompson all stressed the diversity and dynamic complexity of mutualistic interactions. May pointed out the need for "people to think more formally about the range of mutualisms"—from interactions between two free-living organisms to those between the blended partners of a "meta-organism," and from facultative to obligate mutualisms—and the need to include the interplay of life history, ecology, and mode of transmission in theoretical studies of mutualism. There is a theorem, said May, of the dynamics of host-parasite interactions: there is no general answer about direction of evolution of host-parasite interactions. There are examples among human diseases to prove any point. "Whether you end up with cooperation or conflict" said May "—evolution towards mutualism or towards increasing virulence—is dependent on transmission, pathogencity and host resistance. And there are no simple answers."

Thompson and K. Clay (Indiana) described symbiotic and non-symbiotic interactions which are simultaneously mutualistic and antagonistic. A species of *Greya* moth, for instance, pollinates its plant host only when it oviposits in the plant ovule; thus, it is really only a mutualist when it's a (parasitic) antagonist. Also, many (symbiotic and non-symbiotic) mutualisms are based, in some sense, on parasitism: that is, in providing enhanced defense against disease, they exist, as Clay noted, "only in the face of parasite or pest attack."

Using several chronic endosymbioses (vesicular-arbuscular mycorrhizae, algal-*Hydra* symbioses and molgulid ascidians) as examples, Saffo pointed out that for many such symbioses, definition of the association as mutualism is in itself problematic, since benefits of the symbiosis to both partners have not always been demonstrated (Douglas & Smith, 1989; Saffo, 1991). More importantly, several experimental studies, particularly on mycorrhizae (Saffo, 1991), have shown that environmental conditions can affect the outcome of endosymbiotic interactions, which in the same symbiosis can range from mutualism to parasitism to dissolution of the symbiosis—in *physiological* as well as evolutionary time. This environment-dependent view of interaction outcomes in endosymbiosis was echoed by Thompson's perspective on plant-insect (and other) mutualisms; once you get past "mitochondria and all that," said Thompson, mutualisms, in their dependence for existence on antagonists or specific nutrient conditions, are "all about distributions of outcomes." Again, distinction between modes of transmittance becomes important: the notion of environment-dependent outcomes seems to apply more easily to horizontally (indirectly) transmitted symbiotic and non-symbiotic mutualisms than to vertically (directly) transmitted symbioses.

Saffo stressed the need to *apply* such thinking to theoretical studies and textbook explanations of mutualism, and especially to experimental tests of mutualism. For this purpose, many cyclically reestablished (indirectly transmitted) endosymbioses are quite tractable, experimental material, and provide a promising, useful tool with which to test theoretical models of mutualism (Saffo, 1991).

Does mutualism reduce diversity, or enhance it? May suggested that, at the community level, "there is no doubt that mutualism leads to diversity," but the question remains: is the diversity thus promoted a more fragile one? From another point of view,

Maynard Smith cited Law (1988), who proposed that endoparasites of plants reproduce sexually, while primarily mutualistic endosymbionts of plants reproduce asexually, at the expense of species diversity. At the population level, May recounted the current view that host-parasite interactions are a powerful force for creating intraspecific genetic polymorphisms, since rarer host genotypes will be at an advantage and more common ones at an disadvantage. May hypothesized that mutualism, in contrast, might be a powerful force for destroying intraspecific genetic variability. Thompson noted the need to distinguish between mutualisms "driven by parasitism" and those "driven by nutrient-poor environments," suggesting that May's hypothesis applied to the latter, rather than the former, sort of mutualism.

As is habitual in symbiosis research, the roundtable tended to take a "host bias" in its consideration of symbiosis in evolution, discussing the impact of symbiosis on several host taxa, but virtually ignoring the impact of symbiosis on the evolution of microbial endosymbionts. An occasional reversal of perspective might be instructive. For instance, as outlined earlier, several genetically divergent organisms have served, successively, as dominant reef-builders during evolutionary time, and it has been inferred that all of these harbored photoautotrophic endosymbionts. Have the *endosymbionts* also differed significantly in each of these host taxa, or have they provided a consistent genetic thread throughout the evolutionary history of reefs? How, in general, have microbial endosymbionts changed through evolutionary time?

The roundtable was ultimately constrained by the general limits of perspective, of biological background, and of language. As Maynard Smith said in a later session of the Congress (understating the breadth of his outlook), "My trouble is that I was brought up on birds." The general trouble is that evolutionary models are biased by the systems we know best; in the particular case of symbiosis, the biologists studying cellular aspects of endosymbiosis have rarely interacted with evolutionary theorists interested in interspecies interactions. In frustration with some of the effects of this intellectual apartheid, P. Chabora (City University of New York) complained that the roundtable tended to leap from very specific descriptions of particular genetic or cellular interactions to overly generalized inferences about "coevolution," "parasites," "mutualists," "hosts," and "symbionts," mixing and matching symbiotic interactions as convenience dictated. In response, Thompson underlined the need for a new sort of biologist ("the ultimate dilettante") equipped to deal with the genetics of coevolution in a *single system*. The production of biologists equipped to connect the experimental and conceptual language of evolutionary biology with that of cellular and molecular biology would be at least a good start. As Maynard Smith said (in urging students of symbiosis to exploit the extraordinary rich analogous examples of bacterial-phage interactions [Bouma & Lenski, 1988; Lenski, 1988]): "It would be a pity if these branches of science don't take a look at what the other one is doing."

ACKNOWLEDGMENTS

I thank the Graduate Division and Institute of Marine Sciences, UCSC, P. Martinez, N. A. Whitehorn, and the organizers of the 4th International Congress of Systematic and Evolutionary Biology, for support. I also thank L. LaPorte, C. Hickman, and C. Lee for bibliographic assistance.

680

LITERATURE CITED

Bouma, J. E. & R. E. Lenski. 1988. Evolution of a bacteria/plasmid association. *Nature* 335:351–352.

Buss, L. W. 1987. *The Evolution of Individuality*. Princeton University Press: Princeton, NJ. 202 pp.

Campbell, K. A. 1989. A Mio-pliocene methane seep fauna and associated authigenic carbonates in shelf sediments of the Quinault formation, SW Washington. *Geol. Soc. Amer. Abstracts.* 21(6):A 290.

Chesnick, J. M. & E. R. Cox. 1989. Fertilization and zygote development in the binucleate dinoflagelleate *Peridinium balticum* (Pyrrophyta). *Amer. J. Bot.* 76:1060–1072.

Clay, K. 1988. Fungal endophytes of grasses: a defensive mutualism between plants and fungi. *Ecology* 69:10–16.

Cowen, R. 1988. The role of algal symbiosis in reefs through time. *Palaios* 3:221–227.

Douglas, A. E. & D. C. Smith. 1989. Are endosymbioses mutualistic? *Trends Ecol. Evol.* 4:350–352.

Fisher, A. 1989. The wheels within wheels in the superkingdom Eucaryotae. *Mosaic* 20:3–13.

Goff, L. J. 1991. Symbiosis, interspecific gene transfer and the evolution of new species: a case study in the parasitic red algae. *In:* L. Margulis & R. Fester (eds.), *Symbiosis as a Source of Evolutionary Innovation: Morphogenesis and Speciation.* MIT Press. (in press).

Kauffman, E. G. 1967. Cretaceous Thyasira from the western interior of North America. *Smithsonian Misc. Coll.* 152:1–159.

Kauffman, E. G. & C. C. Johnson. 1988. Morphological and ecological evolution of Middle and Upper Cretacous reef-building rudistids. *Palaios* 3:194–216.

Keddy, P. 1990. Is mutualism really irrelevant to ecology? *Bull. Ecol. Soc. Amer.* 71:101–102.

Law, R. 1988. Some ecological properties of intimate mutualisms involving plants. Pp. 315–341. *In:* A. J. Davy, M. J. Hutchings & A. R.Watkinson (eds.), *Plant Population Ecology.* Blackwell: London.

Law, R. 1989. New phenotypes from symbiosis. Trends Ecol. Evol. 4:333–334.

Lenski, R. E. 1988. Dynamics of interactions between bacteria and virulent bacteriophage. Adv. Microb. Ecol. 10:1–44.

Malloch, D. W., Pirozynski, K. A. & P. H. Raven. Ecological and evolutionary significance of mycorrhizal symbioses in vascular plants (a review). *Proc. Nat. Acad. Sci., USA* 77:2113–2118.

Margulis, L. 1981. *Symbiosis in Cell Evolution*. W. H. Freeman: San Francisco, CA. 419 pp.

May, R. M. 1982. Mutualistic interactions among species. *Nature* 296:803–804.

May, R. M. 1984. A test of ideas about mutualism. *Nature* 307:410–411.

May, R. M. & R. M. Anderson. 1983. Parasite-host coevolution. pp. 186–206. *In:* D. J. Futuyma & M. Slatkin (eds.), *Coevolution.* Sinauer: Sunderland, MA.

McFall-Ngai, M. 1991. Luminous bacterial symbioses in fish evolution: adaptive radiation among the leiognathid fishes. *In:* L. Margulis & R. Fester (eds.), *Symbiosis as a Source of Evolutionary Innovation: Morphogenesis and Speciation.* MIT Press. In press.

McMenamin, M. A. S. 1986. The Garden of Ediacara. *Palaios* 1:178–182.

Nardon, P. & A. M. Grenier. 1989. Symbiosis in weevil evolution. *In:* L. Margulis & R. Fester (eds.), *Symbiosis as a Source of Evolutionary Innovation: Morphogenesis and Speciation.* MIT Press. In press.

Rowland, S. M. & R. A. Gangloff. 1988. Structure and paleoecology of lower Cambrian reefs. *Palaios* 3:111–135.

Saffo, M. B. 1990. Symbiosis within a symbiosis: intracellular bacteria within the endosymbiotic protist *Nephromyces*. *Mar. Biol.* 107:291–296.

Saffo, M. B. 1991. Symbiogenesis and the evolution of mutualism: lessons from the *Nephromyces*-bacterial endosymbiosis in molgulid tunicates. *In:* L. Margulis & R. Fester (eds.), *Symbiosis as a Source of Evolutionary Innovation: Morphogenesis and Speciation.* MIT Press. In press.

Seilacher, A. 1989. Vendozoa: organismic construction in the Proterozoic biosphere. *Lethaia* 22:229–239.

Somerson, N. L., Ehrman, L., Kocka, J. P. & F. J. Gottlieb. 1984. Streptococcal L-forms isolated from *Drosophila paulistorum* semispecies cause sterility in male progeny. *Proc. Nat. Acad. Sci. USA* 81:282–285.

Southward, E. C. 1986. Gill symbionts in thyasirids and other bivalve molluscs. *J. Mar. Biol. Assoc. U. K.* 66:889–914.

Strathearn, G. E. 1986. *Homotrema rubrum:* symbiosis identified by chemical and isotopic analyses. *Palaios* 1:48–54.

Thompson, J. N. 1987. Symbiont-induced speciation. *Biol. J. Linn. Soc.* 32:385–393.

Thompson, J. N. 1988. Variation in interspecific interactions. *Ann. Rev. Ecol. Syst.* 19:65–87.

Thompson, J. N. 1989. Concepts of coevolution. *Trends Ecol. Evol.* 4:179–183.

Vetter, R. 1991. Symbiosis and the evolution of novel trophic strategies: thiotrophic organisms at hydrothermal vents. *In:* L. Margulis & R. Fester (eds.), *Symbiosis as a Source of Evolutionary Innovation: Morphogenesis and Speciation.* MIT Press. In press.

Coevolution: Insects—Parasitoids

Peter C. Chabora and H. Roberta Koepfer

INTRODUCTION

The following statement and questions were distributed to discussants prior to the Congress. In the broadest sense, coevolution is a process by which two or more species influence the rate or direction of each other's evolution. The characteristics of one species are evolving in response to the trait(s) of another species, which itself is responding to the characteristics of the first species. For insect-parasitoid systems, the theoretical expectation would be a cyclic pattern of escalating defenses and counter-defenses that are highly specific traits evolving at similar rates, possibly at the same time. In a more practical sense, less stringent requirements reduce the adherence to such precise definitions.

Insect parasitoid-host interactions, by their very nature, fit the theoretical and empirical requirements for explorations into the mechanisms and process of coevolution. The following questions were posed to guide and focus this discussion.

1. What is the basis for invoking concepts of coevolution in explaining the interactions between parasitoids and their host species? Why are parasitoid-host interactions envisioned as theoretically satisfying and empirically proper for the development of coevolutionary thought?

2. What is the strength and nature of the evidence supporting the theory of coevolution? Has the development of coevolutionary theory rested on interpretations of certain biological phenomena which appeared so appropriate that to seek verification appears almost heretical?

3. Is it valid to ascribe special coevolutionary mechanisms to interactions between parasitoids and their hosts? Might these putative coevolutionary processes simply represent the action of natural selection on each of the interacting species?

4. Consider the broad aspects of the biology of endo- and ectoparasitoid-host interaction and speculate on the character of the coevolutionary patterns that could be predicted for each parasitoid type. Since ectoparasitoids can escape host immune responses, can we postulate differences between endo- and ectoparasitoids in their patterns of coevolutionary adaptation? Is there any evidence to substantiate speculation that coevolutionary processes have yielded parallel patterns within each type of parasitoid? Would we expect stronger coevolutionary responses between endo- or ectoparasitoids and their hosts?

5. Host immune responses that result in the encapsulation of parasitoid egg and larval stages, and thereby inhibit parasitoid success, have been considered an important illustration of parasitoid-host coevolution. Are the available data sufficient to differentiate between encapsulation as a coevolutionary response and encapsulation as a response to natural selection on the individual species? Can modification of an immune response by a third species (e.g., virus in calyx fluid) be interpreted as a coevolutionary response?

682

6. Theoretical speculation of coevolutionary processes concentrates on interactions between a single parasitoid species and a given host species. However, natural systems usually consist of several parasitoids (a guild) that are not necessarily host species specific, and that attack a variety of host species which may or may not be related (diffuse coevolution). Can we also define such a multifaceted complex of exploiter and victim species as a coevolving system? Is there evidence that evolutionary pressures from the interaction of one pair of parasitoid-host species can induce parallel and concurrent responses when in combination with another species? Could one speculate coevolutionary adaptation at the generic level? At the deme or family level? Or are such interspecific interactions taxonomically neutral?

7. From a theoretical perspective, can we speculate that the differences in reproductive biologies of endo- and ectoparasitoids (schedule of offspring production, offspring sex ratio, development time, etc.) may be interpreted as an integral component of coevolutionary responses? Is there any evidence to support this line of inquiry?

8. Sex ratios, particularly among parasitoids, have been the focus of numerous theoretical and empirical investigations. Mechanisms invoking natural selection at the single species level (the parasitoid) have been used to explain extraordinary sex ratios due to female bias among the offspring. Aspects of host biology, other than size, have usually been neglected in these investigations. Can we infer coevolutionary aspects from sex-ratio phenomena, and is it parsimonious to do so? Are secondary level causes for sex ratio modification, such as bacteria and viruses, also attributable to coevolutionary mechanisms?

9. Both similarities and differences in numerous behavior patterns have been demonstrated within and between related parasitoid groups. Is there evidence that behaviors such as host habitat finding, host searching, and oviposition are coevolutionary responses to the host species? Is there evidence that genetic and/or behavioral changes on the part of either parasitoid or host have altered the level of interspecific interaction?

10. Population dynamical aspects of parasitoid-host systems have usually focused on features of parasitoid reproductive success (host finding behavior, net reproductive rates and schedules, longevity) and their effect on (evolving?) host populations. Empirical evidence suggests that host populations show markedly higher rates of "evolution" than do parasitoid populations. In theory, this would result in lowered parasitoid reproductive success. Other lines of investigation, including electrophoretic and molecular techniques, suggest that parasitoids exhibit relatively low levels of genetic variability. Paradoxically, parasitoids often show high species diversity. However, it is our contention that in spite of high species diversity, parasitoids are genetically and evolutionarily conservative. This conservatism may be reflected in the highly stereotyped parasitic life-style, which is reinforced through the system of arrhenotoky (diploid females, haploid males); complex courtship and mating behaviors involving chemical, visual, auditory, and tactile components; and the prevalence of parthenogenetic species. Thus, from a coevolutionary point of view, parasitoid-host systems consist of one partner (the host) that exhibits a set of evolutionarily labile characteristics, while the other is evolutionarily stable and conservative. Given that the generation times of interacting species are similar, is this apparent dichotomy in evolutionary rates paradoxical in view of coevolutionary explanations?

DISCUSSION

Eleven years ago on this date, Dan Janzen's paper entitled "When is it coevolution?" was received by *Evolution*. The paper was important then and is equally pertinent to today's discussion, for we must first precisely define coevolution as it is to be applied when discussing parasitoid-host systems. Evolutionary biologists and ecologists have demonstrated a considerable capacity for creating hypotheses and models designed to examine complex phenomena which from anecdotal and formal empirical observations appear to be especially appropriate for studies of coevolution. Interacting populations, whether they be plant-herbivore or pollinator, predator-prey, and pathogen or parasite-host, afford a seemingly endless array of opportunities to explore aspects of coevolutionary adaptation.

Following the fusion of evolutionary and ecological thought, there emerged a fundamental concept, "balance in nature," that guided early investigations into interspecific interactions. Stability was considered to be the ultimate result of long term evolutionary and community processes. It was thought clearly necessary for certain interacting systems to evolve defense and counterdefense mechanisms, thereby thwarting potentially destabilizing influences that would lead to extinction. The groundwork for hypothesis development was clear, and parasitoid-host models were ideal subjects for exploration. Many of the pioneering investigations, by George Salt for example, described magnificently intricate interspecific relationships and helped to entrench concepts of coevolving defense and counter-defense mechanisms. Subsequent inquiries into the evolution and dynamics of interacting populations by Pimentel and his colleagues presented empirical support of genetic feedback mechanisms of population regulation.

In order to establish a consistent starting point, the discussants (Table 1) bandied about definitions of coevolution, particularly with respect to insects and their parasitoids. Thompson stressed that we must remember that two or more distinct gene pools influence one another, and deemed this to be a critical difference between coevolution and general evolution. There was a clear consensus on that point: coevolution *must* involve reciprocal genetic change resulting from an interspecific interaction.

We encountered major difficulties, however, in deciding whether the term should be confined to adaptive effects of selection at the one or two population level, or whether it should also include effects arising on a broader scale, i.e., adaptations emanating from interactions amongst many species. Evolutionary processes occupy a continuum ranging from responses to direct selective pressure on one species, such as high or low sternopleural chaetae in selected lines of laboratory *Drosophila*, to cases of highly specific molecular compatibility between a parasite and its host, to diffuse coevolution involving many forms at the ecosystem level. Determining an acceptable degree of latitude for a useful definition of coevolution was a thorny problem. Since species occurring in the same environment always influence each other to some extent, several discussants preferred a broad, minimally stringent definition. Futuyma, however, mused that although it may be true ". . . that you cannot touch a flower without troubling a star," it becomes increasingly difficult to document coevolution as we proceed from two species to N species. One insect species may be attacked by an array of parasitoids, and coevolution may indeed be occurring between all or some of them, but it is very difficult to positively identify reciprocity and to document the coevolution.

Futuyma also introduced the question of time scale. What sort of time lags might there be between the genetic responses of species to each other? What is the maximal time period within which we would call an interaction coevolutionary? Would a geological period be too long?

When considering species interactions, we must take note of potential rates of evolutionary change as well as of time scale. Therefore we discussed the question of relative evolutionary rates among interacting species. Coevolutionary hypotheses, particularly those concerning insect parasitoid-host systems, often imply concurrent selection on organisms with generation times of similar orders of magnitude. Chabora suggested a somewhat paradoxical situation within the parasitoids. In terms of specific life history features, precise behavioral repertoires required for host finding and courtship, and exact developmental cueing with the host, parasitoids are probably selected for evolutionary conservatism. Some parasitoid characteristics are evidence for this idea: 1) Many species

Table 1. Participants

Dr. Peter C. Chabora—Co-organizer
Ph.D. Program in Biology
Graduate School and University Center,
  CUNY
33 W. 42nd Street
New York, NY 10036

Dr. H. Roberta Koepfer—Co-organizer
Department of Biology
Queens College, CUNY
Flushing, NY 11367

Dr. J.J.M. van Alphen
Department of Population Biology
Zoological Laboratory of the University of
  Leiden
P.O. Box 9516, 2300 RA Leiden
The Netherlands

Dr. M. Bouletreau
Génétique des Populations
CNRS-Université Lyon 1
69622 Villeurbanne
France

Lic. Paula Brunner
G.I.B.E., Depto. de Ciencias Biológicas
Facultad de Ciencias Exactas y Naturales
Universidad de Buenos Aires
Ciudad Universitaria, PAB.2
(1428) Buenos Aires
Argentina

Dr. Yves Carton
Laboratoire de Biologie et Génétic Évolutives
C.N.R.S.
91190 Gif-sur-Yvette
France

Dr. Douglas Futuyma
Department of Ecology and Evolution
S.U.N.Y.
Stony Brook, NY 11794

Dr. Paul Gross
Department of Entomology
University of Maryland
College Park, MD 20742

Dr. Marilyn A. Houck
Department of Ecology and Evolutionary
  Biology
University of Arizona
Tucson, AZ 85721

Dr. Arne Janssen
Department of Population Biology
University of Amsterdam
Kruislaan 302
1098 SM Amsterdam
The Netherlands

Dr. Arthur M. Kopelman
Department of Science and Mathematics
F.I.T.
7th Avenue at 27th Street
New York, NY 10001

Dr. Lex Kraaijeveld
Department of Population Biology
Zoological Laboratory, University of Leiden
P.O. Box 9516, 2300 RA Leiden
The Netherlands

Dr. Robert May
Department of Zoology
Oxford University
Oxford OX1 3PS, UK

Dr. Jan G. Sevenster
Department of Population Biology
Zoological Laboratory, University of Leiden
P.O. Box 9516, 2300 RA Leiden
The Netherlands

Dr. Samuel W. Skinner
Biology Department, Jordan Hall
Indiana University
Bloomington, IN 47405

Dr. Richard Stouthammer
Department of Biology
University of Rochester Rochester, NY 14627

Dr. John N. Thompson
Departments of Botany and Zoology
Washington State University
Pullman, WA 99164

Dr. Sara Via
Department of Entomology
Cornell University
Ithaca, NY 14853

Dr. John Werren
Department of Biology
University of Rochester
Rochester, NY 14627

Dr. David Wool
Israel

are parthenogenetic. 2) Sexual species of hymenopterous parasitoids are arrehenotokous; females are diploid and haploid males arise from unfertilized eggs. Malemale competition among parasitoids is common, particularly in species exhibiting sib mating and local mate competition. During such competition, the total genotype of a haploid male is exposed, thereby permitting strict sexual selection by females of genetically determined male courtship and mating behaviors. 3) Available data on allelic frequencies of parasitic hymenoptera suggest heterozygosity levels several times below that found in other insects. However, parasitoid species are very numerous, and herein lies the paradox. Thompson pointed out that if speciation is so common amongst parasitoids, then perhaps they are not so conservative after all. Clearly, this can be experimentally clarified.

There was considerable doubt as to whether coevolutionarv. process are, in fact, a valid area for scientific inquiry. Wool submitted that there is no strong reason for viewing the parasitoid-host relationship as a coevolving system; he sees it as a situation in which the parasitoid simply evolves the capacity to destroy the host. At present, Wool is not convinced that there was ever a strong case for considering coevolution to be a general phenomenon.

Chabora queried as to how much, and what types of, evidence are needed to unequivocally demonstrate the current or previous existence of coevolutionary processes. Do general descriptions of parasitoid-host biology and anecdotal evidence provide adequate confirmation for coevolution? For example, would an examination of the immune responses of various host populations to various strains of parasitoids be enough? Would blood cell data alone be sufficient? Would we need additional support from parasitoid reproductive data? Or ovipositional data? Or host reproductive biology data? To which Kopelman replied, "all the above."

Two major courses of investigation were suggested by Futuyma. In the first, and here he was referring to plant-insect interactions, agricultural evolution might be demonstrated by examining genetic variation within and among populations of each of two interacting species. Local populations that are exposed to a certain herbivore may show a genetically developed defense system that is different from that found in populations not exposed to the predator, and so forth. Of course, one would also have to show concomitant changes in the development of genetically based counter-defenses by the herbivore.

The second approach involves a broad taxonomic survey of the properties of organisms, and would provide the kind of informational base from which speculation about the possibility of coevolution has arisen. Here one might examine a variety of hymenopteran taxa, for example, and delineate a subset of characteristics that vary among the taxa in such a way that they might be interpreted as coevolutionary adaptations. Having established the basic phylogenetic framework, we must then infer which members of that subset are derived characters; if a character is going to be interpreted as an adaptation, it must first be distinguished from a primitive state that might have evolved in the distant past for an entirely different reason. Once a character is shown to be derived, and also appears to show some functional adaptation to a particular feature of life, speculation about coevolutionary patterns may be appropriate. At some point, however, we must also consider the other species in the association, i.e., the host(s) if the initial phylogenetic analysis was determined for the parasitoids. Describing parallel responses within clades of hosts responding to the selective pressures initiated by specific clades of parasitoids adds another level of complexity and effectively multiplies the amount of speculation.

Sex ratio literature provides abundant examples from parasitoid-host relationships, and, therefore, we must ask what, if any, are the selective pressures imposed by sex ratio phenomena. We first noted a fundamental difference between solitary and gregarious parasitoids: local mate competition is remarkably prevalent among gregarious parasitoids, while solitary forms often demonstrate sex ratios approaching unity.

Effects of host size on solitary parasitoids were discussed in terms of oviposition behavior, in which daughter eggs are usually deposited in larger hosts, and sons in smaller ones. Werren and Skinner agreed that the frequency distribution of host sizes may affect the evolution of parasitoid sex ratio, but predictions regarding the reciprocal result—selection on host size distribution—were far more difficult and speculative. Thus, although host selection behavior by the parasitoid may indeed select for changes in some characteristic of the host, it was not at all clear which characteristic that might be; coevolution does not necessarily require a response in a complementary set of characteristics in both parties. Werren did suggest some ways by which parasitoids might influence the evolution of host size. Since it is of no concern to the host whether its mortality is caused by a developing male or female parasitoid, in a direct sense, parasitoid oviposition behavior has negligible influence. Secondary effects could result, however, from increasing parasitic ability to use smaller and smaller hosts for the production of female offspring, which would augment the potential for greater parasitoid population growth rates. This would entail a high percentage of parasitism, a topic to which we refer later. Futuyma used this interchange to adjust our focus back to the question of how we wish to view coevolution: do we wish to think of coevolution as simply involving the existence of features, in each of two species, that evolved in response to the interaction, or do we wish a more narrow definition that involves changes in a given character in one species in response to changes in a given character of another species? Using the latter criterion, it does not appear at all obvious that a character of the host that influences parasitoid sex ratio is going to be responsive to the evolution of sex ratio per se.

Remember that the above exchange considered only cases of parasitism in which the host is attacked after reaching its final size, a pupa, for example. Chabora cautioned that we must also consider cases in which the host is attacked at an early developmental stage and the parasitoid female has no means of predicting final host size, e.g., cynipid parasitoids attacking first and second instar drosophilid larvae. The deposition of a male or female egg is independent of the final host (pupa) size, but the situation does provide a wonderful array of adaptations on which selection may be acting and potential coevolutionary mechanisms may abound. To further develop this example consider that, following oviposition into an early immature larva, the parasitoid egg and early developmental stages must be essentially invisible to the host. The developing parasitoid must not initiate a host immune response nor retard the host's development to an extent that the host becomes uncompetitive. In fact, it has been shown by Bouletreau and his colleagues that parasitized *Drosophila* larvae outcompete unparasitized larvae in experimental situations. Upon completion of host larval development, the parasitoid cues into hormonal changes accompanying host pupation, a period of rapid parasitoid development ensues, and the host tissues are consumed. Although many adaptations are necessary both for the parasitoid and host, none require explanation by processes other than simple natural selection, and apparently no literature exists supporting a coevolutionary explanation for processes involved in the development of endoparasitic larval parasitoids.

van Alphen described a third category of parasitoids, e.g., soft scale insects, which attack hosts that grow for a period and then begin to reproduce while continually growing, eventually producing a population with enormous size variation among reproductives. In cases where parasitoids select larger hosts for female eggs, he hypothesized selection for reduced host size that could go so far that the parasitoid could no longer reproduce on that host because the sex ratio became too male-biased. Research on such parasitoids may provide evidence for selection for host size reduction followed by a parallel reduction in parasitoid size. Yet, here again, simple selection rather than coevolution provides adequate explanations.

Thompson proposed that a more fruitful approach than focusing on the details of morphological, behavioral, or physiological traits of parasitoids and their hosts would be to consider research on natural populations that demonstrate a gene-for-gene coevolution between plants and plant pathogens. When examining plant based systems, investigators do not concentrate on the specific details of coevolutionary mechanisms, but consider instead the observed gene-for-gene pattern of resistance and counter-resistance. The possibility of gene-for-gene systems operating in parasitoid-host systems may be minute, but we do not need unique details to inquire as to whether coevolution has occurred at all.

In spite of such research on parasite-host interactions, little has concentrated on evolutionary responses, and even less on specific gene interactions. Chabora reviewed the suggestion made by Ilse Walker in the early 1960s that a single gene was responsible for observed variations in the immune response of geographically separated *Drosophila* to separated demes of cynipid parasitoids. Subsequent studies have not described such a clear pattern. Numerous investigations of host immune responses indicate considerable within-population variation, suggesting a more complicated and environmentally sensitive system, and these qualities will be discussed below.

One active area of parasitoid-host experimentation concerns associations between parasitoids and microorganisms—bacteria and viruses—that are found within the parasitoid's reproductive tract. These can modify the host's immune response or affect parasitoid offspring sex ratio, sometimes eliminating males. Perhaps those microorganisms that initiate parthenogenesis may ultimately be responsible for the generation of high levels of species diversity. Chabora reported that parthenogenesis is prevalent among the highly diverse complex of cynipid parasitoids attacking drosophilids in the Caribbean region, and suggested that the study of associated microorganisms might provide fruitful areas for investigation. In parthenogenetic species, individual clones are the units under selection, and adaptation to the precise parasitic way of life demands little latitude in variation. Yet, even among sympatric parthenogenetic species there is considerable breadth in host specificity; some species are content to be host specific, while others attack a wide range of hosts. Thus, even a parthenogenetic clone may harbor a considerable capacity for environmental breadth and resilience. We might ask, then, what data would be necessary to implicate microorganisms in coevolutionary processes?

The general consensus was that the examination of sex ratio phenomena alone is not a particularly fertile area for the examination of coevolutionary concepts, but that microorganisms may affect parasitoid population dynamics in ways that would ultimately have profound effects on the host populations. Werren suggested that a switch in emphasis from parasitoid-host to parasitoid-microorganism might provide a more profitable domain in which to look for evidence of coevolution. Information offered by discussants regarding the frequency and distribution of microorganisms within insects, including non-parasitic species, indicated that many insects carry microorganisms spanning a broad taxonomic range, and at least 20 cases are known where sex ratios were affected. Skinner noted that at least five orders of insects, four of which are diploid, were shown to have microorganisms that kill sons and not daughters. Haplodiploidy may facilitate sex ratio effects, but it certainly is not a prerequisite for these interactions to occur. The cellular immune response of hosts following parasitism and the parasitoid's ability to evade this response may provide direct evidence for coevolutionary patterns. When an endoparasitoid deposits an egg into the host's haemocoele, the host may recognize the egg as a foreign object and respond by producing enlarged, flattened blood cells that surround the egg, then melanize to form a capsule. The egg is often killed by suffocation or because its hatching is inhibited. The pattern of observed responses suggests that parasitoids associated with a given host population rarely elicit defense responses, but newly associated species do. Thus, because of the potential for experimental manipulation, particularly by using island populations, immune response studies may be a viable method for examination. The discussants reviewed several examples emanating from biological control programs where immune responses suggested shifts in host gene frequency resulting from the introduction of hosts with encapsulation abilities. There was, however, no feeling that such cases exemplified coevolutionary concepts. Bouletreau then argued that adaptations following species introductions are a global process, and distinguishing between various adaptations is extremely difficult. Whether an adaptation is behavioral, or immunological, or physiological, the degrees of adaptation resulting from environmental responses in general and from coevolutionary processes are difficult to elucidate. Perhaps, he speculated, an ideal experimental methodology would be to artificially destroy adaptations, then attempt to examine the process and mechanism of restoration. In supportive response, several discussants entered a second plea for joint systematic and ecological research in order to separate those adaptations due to environmental constraints, and those directly due to species interactions.

Several points were then considered. First, there may very well be variation among populations that influences the success of a parasitoid or the success of a host in warding off a parasitoid, but such variation may not be adaptive in the coevolutionary sense because it resulted from other causes. Therefore, we must focus on patterns in the distribution of that variation. A research program could be designed that tests predictions about how a particular character will vary if it is indeed an adaptation to interaction with another species. For example, one may examine the distribution of an immune response among populations of hosts along a gradient ranging from areas of severe parasitism to areas where parasitism is negligible. van Alphen and Kraaijeveld then described research from their laboratory which fits that model precisely. A parasitoid that normally attacks *Drosophila subobscura* throughout Europe has also been attacking *D. melanogaster* as it invades northern reaches of Europe during warm months. In the Mediterranean basin the parasitoid and *D. melanogaster* populations interact continuously, but in more northern regions only seasonal interactions occur. Population characteristics involving immune responses determined thus far, appear to fit the predictions, but further selection experiments are planned. Additional examples of *Drosophila*-parasitoid experiments were reviewed by Chabora, but Futuyma stressed that *D. melanogaster* was a poor species for experiments of this nature because of its recent invasion and seasonal distributional changes.

A word of caution and an echo of support was voiced by Wool who observed that difficulties may arise from interpretation of experimental results. He warned that if we believe in the process of coevolution, we tend to interpret our results as cases of coevolution. Perhaps the best approach, Wool suggested, would be to work out the details of one case completely, showing genetic reciprocity of change in both interacting species. At this point, he is convinced that only one case shows coevolution, and that is the fig wasps, because neither species can live without the other. That, however, is not a case of parasitoid-host interaction and is more likely an example of mutualism.

The discussion then turned to a consideration of natural systems where parasitoid-host interactions comprise complex systems incorporating numerous species existing on a set of resources with varying degrees of stability and longevity. Coevolution under these conditions has been referred to as diffuse coevolution. Chabora and Carton portrayed the community of *Drosophila* spp. and its complex of tropical parasitoids, where a single infested fruit may have several host species and as many as 17 parasitoid species. Additional examples were described by Janssen for studies in the Netherlands, where 15 parasitoids attacking 20 drosophilids were described, although only five species of *Drosophila* and three parasitoids were common. The examples were similar in that in both cases parasitoids attacking their usual host failed to elicit immune responses, but an immune response and encapsulation often occurred when the parasitoid oviposited into a "foreign" host. Such observations do not suggest coevolution, but rather efficient adaptation to the host on the part of the parasite. Perhaps we could paraphrase and invoke a "ghost of coevolution past."

The results of extensive field work by Sevenster on drosophilids and their parasitoids in Panama demonstrated an extremely high species diversity for parasitoids, but illustrated one of the complications of such investigations. It is difficult to elucidate natural species associations because species that can be successfully reared are usually generalists, while parasitoids that are host specific are far more difficult to maintain. This is particularly true when the hosts are also substrate specific. Field observations made on naturally occurring substrates show considerably different results than those utilizing baited traps, where highly unusual combinations of hosts and parasitoids abound in unusually high densities.

A second observation was that the frequency of parasitization is generally quite low under natural conditions. This important point is dealt with in detail later, but suggests that the interpretation of experimental laboratory results may imply coevolutionary processes that are considerably different than those actually occurring under natural conditions.

At this point in considering parasitoid-host complexes, further discussions of behavioral aspects of host finding, host selection, and oviposition seemed appropriate. van Alphen expanded earlier comments on the studies of parasitoid host searching behavior by Vet, and related them to Sokolowsky's work on a single gene that influences host larval behavior in that the larvae are either highly mobile or quiescent. Vet demonstrated that *Drosophila* parasitoids utilize three basic strategies in locating hosts within their substrate. In one strategy, parasitoids employ vibrotaxis to locate the host larvae borrowing through the substrate. Some use various combinations of random probing with their ovipositor, while others tap the substrate with their antennae, searching for host spiracles. Clearly, the behavior of host larvae may differentially affect the parasitism rates of parasitoid species utilizing diverse searching strategies. One can imagine that, in a complex community, the frequency of a single gene affecting host larval behavior may alter the abundance of parasites utilizing certain hunting strategies. Carton reported that preliminary data from Canada, Tunisia, and the Lesser Antilles suggest differences in the gene frequencies of the alleles causing the "rover" and "sitter" behavioral morphs in these three *melanogaster* populations.

On a related issue, research on the chemical ecology of insect-plant interactions has provided enticing information on how parasitoid behavior may be influenced either directly or indirectly by plant volatiles. Numerous studies have demonstrated that parasitoids are attracted to certain plants only after the plants are attacked by herbivores, but coevolutionary explanations invoking evolution of the volatiles to attract herbivore parasitoids, and thus protect the plant, would be extraordinarily difficult to prove. Several discussants found this circumstantial evidence quite convincing, although to ascribe coevolution as a causal mechanism is not parsimonious because simple adaptations can describe these phenomena.

The manner in which a parasitoid disperses its eggs among available hosts was suggested as a behavior that might demonstrate integration between parasitoids and their host. Bouletreau pointed out that the egg parasitoid, *Trichogramma*, demonstrates genetic variability in oviposition behavior, where some individuals scatter their offspring among distant host eggs and others tend to clump ovipositions. van Alphen countered with the observation that models of adaptive superparasitism have predicted exactly what had been described, and that ovipositional patterns have nothing to do with coevolution.

In the context of the preceding discussions, Chabora observed that we had been considering scenarios at two very different levels of complexity. When considering any parasitoid-host interaction, one must take into account the resource upon which the host is feeding; those systems where the host is feeding upon an evolutionarily inert substrate, such as a decaying fruit, must be compared to those where the host is feeding upon a resource capable of evolving, a living plant, for example. We seem to know little enough

690

about simple parasitoid-host interactions; attempting to interpret interactions involving three living variables can only make matters still more difficult.

Early models concerned with the coevolution of interacting systems focused on the dynamics of the populations and the evolution of stability. The concept of a "balance in nature," which was so pervasive in ecological and evolutionary thought, influenced the development of theoretical coevolution models comprising population interactions that would eventually produce a balanced system with decreasing fluctuations of abundance. Pimentel's model of a genetic-feedback mechanism was a classical example of this thinking; through decreases in a parasite's reproductive capacity or virulence, and concomitant increases in a host's resistance, a balance was struck that reduced the potential for extinction through a series of dampened population cycles. Computer simulations of simple genetic models illustrating the predictions of the genetic feedback mechanism were supported to a varying degree by laboratory experiments utilizing fly-parasitoid species.

The oft-cited example of interspecific evolution does not concern parasitoid-host species, but rather the *Myxomatoma* virus, mosquito, and rabbit system in Australia. However, Futuyma noted that the biological properties that made the *Myxomatoma* system evolve the way it did were entirely different from those of parasitoid-host systems. Furthermore, it did not seem obvious to him that there should be the evolution of a more stable and less virulent relationship, and he saw no reason why there shouldn't be an evolution toward extinction.

An additional observation by Futuyma, developed from earlier comments, was that host immune response generally tended to be more sensitive to foreign parasitoids than to the usual parasitoid. If anything, this suggests that the parasitoid is leading the host in an arms race, and does not support a scenario in which the host becomes more resistant or reduces the parasitoid's reproductive potential. It appears that the evolution of increasing ability of the parasitoid bespeaks a greater potential for extinction. Further comments on the exceptional efficiency and "over" specialization among parasitoids supported these observations.

A prevalent assumption in the foregoing discussions of parasitoid-host interactions was that the parasitoid always finds the host. May argued that a number of lines of inquiry suggest that the most influential component in parasitoid-host population dynamics rests simply on the parasitoid's ability to find hosts in a patchy and complex environment. The literature on parasite-host dynamics has become increasingly bizarre and elaborate, and replete with excessively complex mathematical models whose demands for data are far beyond those which can be provided. Recently, several investigators attempted to develop simpler, but approximate, questions relating to actual data available on searching behavior of parasitoids. Laboratory research suggests that there are all sorts of elaborate behavioral adaptations that make for more efficient searching. For example, for a short period after finding a host, a parasitoid may take shorter steps and turn more often, as in a drunken walk, searching the immediate area more thoroughly. Yet, in an evaluation of roughly thirty-odd field studies of searching behavior, which is a very complicated story, May and colleagues asked: in what fraction of cases do density dependent behaviors significantly influence the population dynamics? In what fraction of cases is it just density independent randomness in the way hosts are distributed and density independent randomness in the way that parasitoids do, and do not, find them? Analysis disclosed that in the majority of cases, density independent effects were most important.

Two paradoxes emerged from this investigation. First, if it is indeed true, as laboratory studies suggest, that parasitoids possess adaptive searching behaviors, then why do field studies indicate that such behaviors have very little to do with the overall dynamics of the system? And again, if parasitoids are such able searchers, why are they so inefficient in harvesting their hosts? May pointed out that parasitoids in the field typically take a yield that is an order of magnitude below a maximum sustainable yield. This suggests that hosts in the field are not as easy to find as they are in the laboratory. Thoughts about coevolution, then, must be more complicated than simply considering the interaction between individual parasitoids and hosts. The implications of the extent of selective pressures on the interacting populations are also ambiguous and require further elucidation.

With regard to the first paradox, van Alphen commented that one of the major features in the life of a parasitoid is interspecific competition with other parasitoids. Once an egg is deposited into a host, that host becomes available for superparasitization by other parasitoids. Models demonstrate how long individuals should stay in a patch and continue ovipositing after the arrival of other ovipositing females. Such behavior, adaptive for the individual, would generate aggregated distributions needed to establish stable population models.

Chabora suggested that a comparative field study utilizing similar parasitoid-host communities on natural resources of varying longevities might provide a productive area for exploration of some of these questions. The drosophilid-parasitoid communities, for example, might be analyzed on the basis of the type of host resource being utilized. In north temperate regions, for instance, one might compare the dynamics of highly ephemeral communities associated with mushrooms, to those of communities on more substantial natural fruits where more than one generation may be reared, to those of communities on long lasting sap fluxes. Predictions concerning the host searching behavior of parasitoids (and host behavior as well), parasitoid life history and reproductive schedules, ovipositional patterns, parasitoid competition and superparasitism, and major influences on the dynamics of local populations may be examined.

Following a summary of the discussion thus far by Chabora, which emphasized the difficulties in conclusively demonstrating coevolutionary processes in parasitoid-host systems, van Alphen returned the discussion to some of the very general comments presented by May. He considered that populations were, in the long run, rather stable, and that this may be a very important feature of parasitoid-host systems. There just may not be an "arms race," with parasitoids becoming more and more efficient. Referring to the suggestion by May that parasitoids harvest about 10% of their potential yield, van Alphen commented that extant parasitoid species are really not very efficient after all, thereby assuring their future survival by permitting a portion of the host population to survive. This mechanism would resemble a built-in insurance against driving the host to extinction. Futuyma then inquired whether this wasn't a kind of group selectionist provision for the future.

van Alphen responded, "Almost," but said that the general attributes described in observations of the dynamics of parasitoid-host interactions are such that a portion of the host population escapes parasitism. This confers a degree of stability, and most described interactions are stable. Our task is to explain why this happens. May responded by reiterating Futuyma's comment that these situations are different from the case of *Myxomatoma*, and that he saw no reason why populations cannot get into trajectories that carry them to extinction.

Janssen observed that hosts in systems exhibiting long term persistence may be experiencing only low levels of selective pressure, because the parasitoids are not attacking a major proportion of available hosts. This would explain our failure to observe coevolutionary processes there. Coevolution, therefore, may become apparent only in cases of high parasitoid impact, but because of such intense interaction, extinction may be a prevalent outcome. In that case, we would also search in vain for evidence of coevolution.

Over a ten year period of sampling drosophilid-parasitoid communities in tropical regions, Chabora and Carton noted a considerable stability and predictability in the sequence and abundance of parasitoids attracted to host-baited traps. However, following the considerable impact of a major hurricane, species abundance patterns were highly disturbed. Such natural perturbations are capricious and initiate major effects on community interactions. Thompson interjected with two relevant comments regarding community stability and perturbations. First, even if stable systems could be found, it would be an error to concentrate on them in our search for coevolution, because the coevolutionary process itself may be a mechanism for extinction. That would be a very interesting phenomenon indeed. Second, natural perturbations are the circumstance that may produce bouts of coevolution, and may generate precisely the kinds of quick evolutionary changes observed. Essentially this is classical evolutionary biology, except that we are using the context of coevolution.

The fascinating issue of whether or not we can reasonably consider coevolution to be a kind of arms race—implying continual change in each of two interacting species—leading to an infinite evolution without end, was raised by Futuyma. Concepts of coevolution need not be implicated. There are no grounds for believing that such indefinite evolution occurs. As in most other evolutionary systems there is, however, every reason to believe that adaptations arrive at some sort of quasi-equilibrium, which then persists until something alters the context of that interaction. Much of the motivation for talking about the concept of coevolution stems from observation of patterns of what might be adaptations to an interaction. The existence of these interactions doesn't mean that the system is continually evolving, because it could have arrived at a quasi-equilibrium.

The case is not closed, but the context for defining coevolution, establishing experimental requirements, and pursuing the theoretical development. for elucidating parasitoid-host coevolutionary mechanisms has now been far more clearly stated. For this, the co-organizers thank the participants and accept all responsibility for errors of interpretation, omission and commission in the preparation of these remarks.

GENETIC PROCESSES IN EVOLUTION

PLENARY ADDRESS

Systematics and the Study of Evolutionary Processes

Douglas J. Futuyma

Abstract. Despite the accomplishments of the Modern Synthesis, there has been little interchange between the two major realms of evolutionary biology, the study of evolutionary processes and of evolutionary history. This paper explores ways in which phylogenetic studies of evolutionary history may pose questions and provide information of interest or importance to analyses of evolutionary processes. I focus particularly on questions of a genetic nature that are posed by the data of phylogenetic studies. Research programs drawing on and synthesizing these approaches are illustrated by studies of the evolution of host affiliation in herbivorous insects. A phylogenetic study of the chrysomelid beetle genus *Ophraella* suggests conclusions about the likelihood of coevolution among these insects and their host plants, and about the likelihood of constraints on the evolution of new host affiliations. The possibility of constraints is then explored by studies of genetic variation in the responses of *Ophraella* species to their congeners' hosts, framed in an explicitly phylogenetic context. Preliminary results of the genetic studies are described.

The unity of a field of study does not depend on singularity of explanation. As the unity of genetics does not depend on the universality of Mendelian inheritance or of the genetic code, so the unity of evolutionary biology does not require, say, natural selection or allopatric speciation as exclusive and universal mechanisms. The unity of evolutionary biology lies, rather, in the connections, the interrelationships, the interpenetration of the concepts, discoveries, and questions of the several subdisciplines that constitute the field. Thus, paleontologists and population geneticists can each be enriched by the others' discipline; students of phenotypic evolution and of molecular evolution can offer insights to each other, and draw on a common body of principles.

Nevertheless, there exists something of a schism between the two great halves of evolutionary biology: historical studies, chiefly in paleontology and phylogenetic systematics, and what I have termed "synchronic studies," chiefly of microevolutionary processes (Futuyma, 1987). Although the Modern Synthesis drew these fields together and achieved a conceptual *rapprochement* between them (Mayr & Provine, 1980), a truly unified evolutionary biology, drawing on and synthesizing historical and synchronic studies, has been less evident in the 1980s than in the 1950s. I have suggested (Futuyma, 1987) that real union might be sought by identifying questions and research programs to which both sides might make indispensable, reciprocally illuminating contributions. In this paper (also Futuyma, 1990a), I review and suggest a few such avenues, and briefly describe one in which my laboratory is currently engaged.

I will focus on the interplay between evolutionary studies and systematics, because

Dr. Futuyma is with the Department of Ecology and Evolution, State University of New York, Stony Brook, New York 11794, USA.

696

the need here is so obvious. On one hand, some systematists seem to have only an elementary understanding of evolutionary processes and many do not try to derive insights into evolution from their work; they frequently bury in taxonomic monographs observations that would fascinate evolutionary biologists if they were brought to broader attention. A few systematists have even voiced hostility to integrating systematics and evolutionary studies, although happily this attitude is not common.

On the other hand, every American systematist (at least) knows that the prestige of systematics leaves something to be desired. It is common to meet people who proclaim that they are interested in the Question, not the organism, and deny being ornithologists, entomologists, or botanists, as if those were the scarlet letters of biology—the brands of shame. Every systematist knows of faculty positions replaced by non-systematists, and of institutions from which systematics has been aggressively purged.

It is important, then, to reflect on the contributions of systematics to evolutionary and ecological biology. Non-systematists may need to be reminded that every use of the comparative method, every statement about comparative morphology, and every analysis in community ecology rests on an indispensable taxonomic foundation. We take this for granted until, for example, we try to do insect ecology in the Neotropics.

Almost every concept, almost the entire vocabulary, of macroevolution—adaptive radiation, convergence, heterochrony, mosaic evolution, differential evolutionary rates, evolutionary trends—comes to us from earlier generations of systematists (including here paleontologists). More recently, contributions from systematics to evolutionary theory include character displacement (Brown & Wilson, 1956), taxon cycles (Wilson, 1961), coevolution (especially Ehrlich & Raven [1964], and earlier, parasitologists such as Eichler [1948] and Clay [1949]), and an enormous amount of systematic effort that clarified ideas on geographic variation, species, and speciation. During and after the Synthesis, the distinction between systematics and synchronic evolutionary studies was not sharp; it was not uncommon for individuals who identified themselves as ichthyologists, botanists, and so forth, to work on both systematics and evolutionary processes. For example, Edgar Anderson (1937) published on the continuity between intraspecific and interspecific variation, Robert Rush Miller (1950) on the genetics of species differences in pupfish, Charles Michener (1958) on the evolution of sociality in Hymenoptera, and Carl Hubbs (1955) on the genetics and ecology of hybridization in fishes. Theodosius Dobzhansky (1931, 1941) published major studies of beetle taxonomy. Alfred Sturtevant, best known for his fundamental contributions to genetics, published prolifically on the systematics of *Drosophila* (Sturtevant, 1921, 1942), and was knowledgeable about the systematics of Diptera in general.

During and immediately after the Synthesis, a major contribution of systematics to evolution was in the areas of intraspecific variation and speciation. Today, the major new contributions appear to be mostly in the area of phylogenetic inference. A phylogenetic perspective can contribute substantially to our understanding of evolutionary processes, as the following examples and "thought experiments" illustrate (see also Futuyma, 1990a).

In the arena of molecular evolution, the distinction between evolutionary genetics and phylogenetic analysis is becoming very fuzzy, as almost any issue of *Molecular Biology and Evolution* attests. A recent issue of *Genetics*—of all journals perhaps the least likely— included a paper (Easteal, 1990) on the phylogeny of orders of eutherian mammals. Easteal deduces a phylogeny from globin sequences, shows that substantial sequence evolution has occurred along the internodes, and uses this to challenge the assumption of a nearly simultaneous origin of the orders, which has been used (e.g., by Gillespie [1986]) to argue for heterogeneity of rates of sequence evolution. This is only one of many uses of phylogenetic information (e.g., Langley & Fitch, 1974, Goodman et al., 1982) to evaluate the constancy or inconstancy of rates of molecular evolution, an issue of equal interest to population geneticists and students of evolutionary history. Phylogenetic analysis of molecular data has contributed to the analysis of evolutionary processes in many other contexts; examples include concerted evolution of gene families (Dover et al., 1982), the rate of silencing of duplicate genes (Li et al., 1981), the history of population structure (Avise, 1989), and the analysis of gene flow (Slatkin & Maddison, 1989).

Another important area is the use of the comparative method for deducing the adaptive significance of a trait from its correlation with other features such as the organism's ecology or social system. Ridley (1983), Felsenstein (1985), Donoghue (1989), and others have pointed out that a simple count of species introduces correlations between traits that may reflect not adaptation but common ancestry. What is required is a count of the number of independent origins of a trait, in taxa with one or another ecological or social feature, and this requires a great deal of phylogenetic information (Maddison, 1990). A similar principle holds for what will surely prove to be one of the most important applications of phylogenetic analysis, the study of the steps by which complexes of functionally interacting characters are assembled during evolution. For example, distasteful, aposematically patterned insects are often gregarious, which has led to the common belief that a gregarious habit has been a precondition favoring the evolution of aposematism, because it provides the opportunity for kin selection. Sillén-Tullberg (1988) examined a number of lepidopteran groups that include species with both of these features, and for which the phylogeny is well enough known to map the relative age of origin of the two traits. In a majority of clades, aposematism appears to have arisen before the gregarious habit—which calls into question the hypothesis of kin selection (but see Maddison, 1990).

In another example, Huey and Bennett (1987) measured the preferred temperature and the physiologically optimal temperature of a number of Australian skinks, and inferred evolutionary changes in these traits by mapping them onto a phylogeny. Evolution toward both higher and lower values has occurred in both traits, generally in parallel, as expected. But the coadaptation has been only partial, so that the changes in physiology do not exactly match those in preferenda, and in at least one case the preference and the physiological optimum have apparently evolved in opposite directions—an unexpected result that invites more detailed study, perhaps of selection gradients.

At a more macroevolutionary level, Mitter et al. (1988) have made the simple but important observation that because sister groups are by definition equal in age, the difference between them in species richness must be due to a difference not in age but in rate of diversification. One may then test the effect of a feature on diversification rate by seeing whether it is consistently associated with higher diversity in a number of lineages in which the trait has evolved independently, compared to their sister lineages. The habit of herbivory in insects has evolved many times, and Mitter et al. (1988) found that herbivorous lineages are consistently more diverse than their non-herbivorous sister groups. This may be the most explicit evidence to date in favor of Simpson's proposition that invasion of a new adaptive zone enhances the rate of diversification. This approach obviously could be extended to explore the effects on diversity of other so-called key adaptations, or of features such as ecological specialization. It also highlights the very difficult question of why adaptation at the level of individual organisms should promote diversification. For example, Mitter and Farrell (in press) have taken this approach to test Ehrlich and Raven's (1964) suggestion that diversification of plant lineages has been enhanced by the evolution of new defenses against herbivory. They report that plant clades with resin- or latex-bearing canals (which are known to provide defense against insects) are statistically more diverse than sister clades that retain the ancestral condition. It is not obvious why the evolution of a defense should increase the diversity of species rather than merely the abundance of plants.

An area yet to be explored is the genetic basis and population genetics of characters that phylogenetic analysis shows to be especially interesting. For instance, every systematist is familiar with characters that show extreme conservatism, or stasis, and with characters that display apparent irreversibility. For example, most of the thousands of species of chrysomelid beetles, as in beetles generally, have 11 antennal segments, the ancestral number. Among the 80 North American genera in the subfamilies Galerucinae and Alticinae, departures occur in only two genera, in which the number has independently been reduced to ten. In almost all beetle groups with other than 11 segments, the number has been reduced rather than increased. Beetles with long antennae have long segments, not more segments. Why is there such conservatism, and why the directionality? Is there a paucity of genetic variation? Are these characters highly canalized? Do genetic variants generally have severe deleterious pleiotropic effects? Is the distribution of mutational variation asymmetrical? I am aware of virtually no information about the level of genetic variation, or of genetic correlations with other features, in conservative versus rapidly evolving characters. In fact, it is not even clear if they differ in levels of phenotypic variation (Kluge & Kerfoot, 1973; Rohlf et al., 1983).

It has become common to suppose that genetic variation is seldom limiting, and that stasis may be ascribed to stabilizing selection (e.g., Charlesworth et al., 1982). The evidence for this view includes the almost invariable response in *Drosophila* populations to artificial selection on almost any character (Lewontin, 1974; Weber and Diggins, 1990)—the chief exception being consistent asymmetry (handedness) of ocelli (Maynard Smith & Sondhi, 1960). However, some observations on apparent lack of variation suggest that genetic constraints may be important (see also Loeschcke, 1987). Among species of grasses that grow near copper mines, populations from normal soils revealed genetic variation for copper tolerance in those species that have become adapted to mine wastes, but no genetic variation was found in the majority of the species that have not established tolerant populations (MacNair, 1987). Intraspecific variations in the arrangement of carpals in bolitoglossine salamanders represent only a few of the possible patterns, and fixation of new arrangements as the species-typical condition has been limited to these few (Alberch, 1983). Because most features are a "black box" from a developmental point of view, we often cannot say that an imaginable evolutionary change is in fact possible: the requisite developmental pathway may not exist. There cannot be genetic variation in a "character" for which there is no developmental foundation. A search for constraints on genetic variation may therefore not be as quixotic as it may seem.

Interesting questions are provided by characters that have diversified greatly in some groups, but not in others. For example, the morphology of the hind leg of chrysomelid beetles in the subfamily Galerucinae (which merely walk) is monotonously uniform. The flea beetles, subfamily Alticinae, are derived from the galerucines, and have a key adaptation, a metafemoral spring, that enables them to jump (Furth, 1980). Various features of the tibia and tarsomeres are extremely diverse among the flea beetles (Fig. 1). These may be adaptations in these jumping insects affecting purchase on and lift-off from different plant surfaces. The question is, is the genetic variation required for such modifications already immanent in the galerucines, or in those flea beetles that do not have these modifications? How quickly can such coadapted features evolve, once a key adaptation arises? Have different lineages evolved different modifications because of initial differences in genetic variation? This particular example might not be the most suitable for such an exploration, because the functions of these modifications are not well understood, but every systematist knows of comparable examples.

One can think of many questions of a genetic nature, posed by phylogenetic patterns in particular characters, including ecological features. I close with a brief description of our own current work, conceived explicitly as an attempt to join phylogenetic, genetic, and ecological approaches. The context is the host associations of phytophagous insects, and the possible coevolution of insects with their food plants. The stage was set in 1964 by Ehrlich and Raven, who noted that related species of insects, such as the 60 or so species of *Heliconius* butterflies, often feed as larvae on related plants—various species of

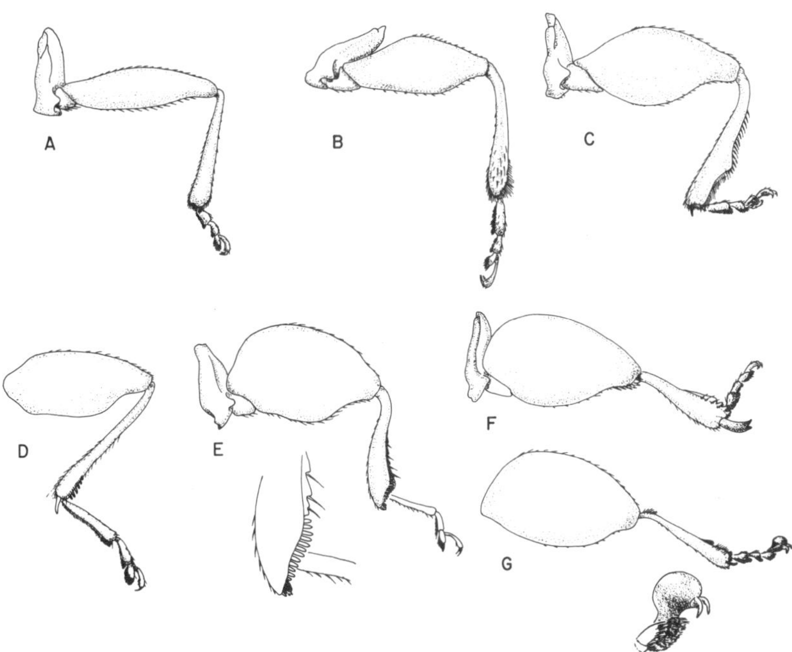

Figure 1. Left hind leg, in ventral view, of one species of Galerucinae (A) and six species of Alticinae (B–G), illustrating morphological diversity in the latter. A. *Ophraella conferta* (LeConte). B. *Altica woodsi* Isely, illustrating the enlarged femur typical of Alticinae, but no other pronounced modifications. C. *Blepharida rhois* (Forester), with a tibial crest and enlarged setae. D. *Longitarsus melanurus* Melsheimer, with elongated basal tarsomere and large tibial spur. E. *Chaetocnema confinis* Crotch, with insert showing tibial crest and modified setae. F. *Dibolia borealis* Chevrolat, with large, bifid tibial spur. G. *Capraita circumdata* (Randall), with insert showing swollen apical tarsomere. Illustrations by the author, from specimens collected in New York State. Not to scale.

Passifloraceae in the case of *Heliconius*. They suggested that conservatism of diet was due to adaptation to the similar secondary compounds of related plants, which, they suggested, had evolved as defenses against herbivores. Some authors have gone further (e.g., Benson et al., 1975, Spencer, 1986), and suggested that plants and their associated insects have diversified simultaneously and in parallel as they exerted selection on each other. Such a pattern of congruent cladogenesis would be reflected in congruent phylogenies of the insects and their hosts. Alternatively (as Ehrlich and Raven apparently supposed), an insect group might diversify later than its host plants, and shift from one plant species to another by a process of colonization, which might be facilitated by chemical similarity of the new host to the insect's ancestral host (Mitter & Brooks, 1983; Mitter & Farrell, in press).

Both these scenarios imply constraints on genetic variation in those features that govern the insect's host association. These include behavioral traits—the insect's willingness, so to speak, to eat and lay eggs on a new plant—and physiological traits that enable it to grow and survive in the face of toxic secondary compounds and other inimical plant features. Although the phylogenetic patterns imply constraints, a panadaptationist might suppose that genetic variation is usually sufficient to adapt to virtually any plant—after all, *Drosophila* populations respond to selection on almost any character, including behavioral characters, and hundreds of species of insects have evolved physiological tolerance to novel toxins, namely synthetic insecticides. It is not clear, then, whether the history of host affiliation displayed by a group of insects has been constrained by the availability of genetic variation, so that for *genetic* reasons the realized history was more probable than

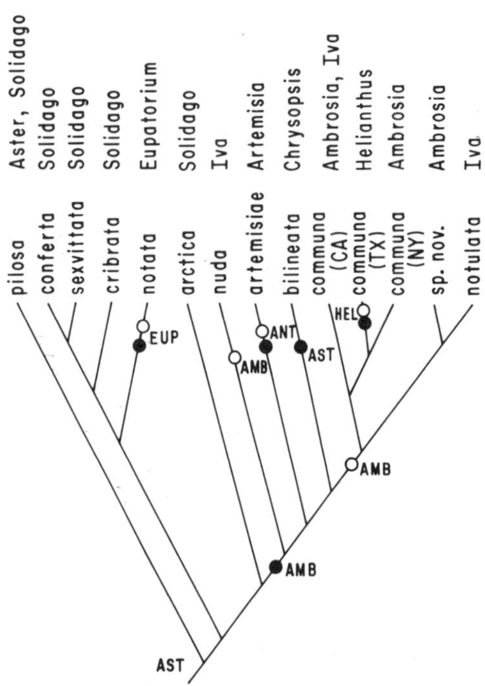

Figure 2. Best current estimate of the phylogeny of *Ophraella* (from Futuyma & McCafferty, 1990). Host genera of each species arrayed at top. Open and filled circles on branches of the phylogenetic tree are two equally parsimonious interpretations of evolutionary shifts in host affiliation, between tribes or subtribes of Asteraceae (AST, Astereae; AMB, Heliantheae, subtribe Ambrosiinae; ANT, Anthemideae; EUP, Eupatorieae; HEL, Heliantheae, subtribe Helianthinae). The ancestral affiliation is inferred to have been with Astereae. *O. communa* and *O. notulata* are associated with different species of *Iva*, and in the widespread *O. communa*, association with *Iva* has been documented only for a small, isolated population. Transition from *Ambrosia* to *Iva* is considered likely in this group.

other histories that we can imagine, but which did not occur—or whether insects have the genetic potentiality to adapt to virtually any plant, so that the realized history, guided perhaps by a unique series of selection events, is only one out of many possibilities that could equally well have transpired.

Our program of research is first, to infer the history of host shifts within an insect group, by parsimoniously mapping host affiliation onto a phylogeny inferred from other characters; and second, to examine genetic variation in a species for its behavioral and physiological responses to plants on which it does not normally feed, but which are the hosts of related species. In particular, we focus on a species that retains a locally ancestral host affiliation (host A), but which has a close relative with a derived affiliation (host B). We search for genetic variation in behavioral and physiological responses (for example, ability to survive) both to host B and to one or more other plants, say X and Y, which are the hosts of more remotely related species, and which have not figured in the history of the insect species under study. If constraints on genetic variation have guided the history of host affiliation, we might expect to find genetic variation in responses to the host of the close relative (plant B), but not to the hosts of the distant relatives (plants X and Y). It is also interesting to ask if, as might be expected (Courtney et al., 1989), species with derived host associations display more genetic variation in responses to locally ancestral hosts than vice versa.

Such tests are required for several species, in order to search for general patterns in the relationship between genetic variation and evolutionary history.

Our current work concerns a rather small genus of exclusively North American chrysomelid beetles, *Ophraella* (Wilcox, 1965, LeSage, 1986, Futuyma, 1990b). Both adults and larvae feed on foliage of Asteraceae (Compositae). The hosts are in four tribes of Asteraceae, and each species of beetle feeds exclusively on certain species within at most three genera in a single tribe. Several feed only on a single plant species.

Our cladistic analysis of *Ophraella* (Futuyma & McCafferty, 1990) was based on both morphological and electrophoretic characters. Parsimony analysis of the two data sets gave substantially similar results, and analysis of the combined data sets yielded the phylogeny displayed in Figure 2, which also presents two parsimonious interpretations of the history of shifts between tribes of asteraceous hosts.

These results bear on the historical aspects of Ehrlich and Raven's scenario. First, does the phylogeny of the beetles conform to that of their hosts? Cladistic analyses of tribal relationships in the Asteraceae have been made by Bremer (1987), based on non-molecular characters, and by Jansen et al. (1991), based on restriction site variation in chloroplast DNA. According to Jansen (pers. comm.), these authors agree that the DNA data provide stronger evidence on the relationships among the tribes that include hosts of *Ophraella*. These relationships differ importantly from the topology of inferred transitions between tribes in *Ophraella* (Fig. 3). Moreover, if we use even the most liberal published estimates of the relation between genetic distance and divergence time, most of the speciation events in *Ophraella* are post-Miocene (Futuyma & McCafferty, 1990)—but there are Miocene pollen records of modern genera in two of the tribes.

I conclude that the phylogenies of the beetles and their host plants are not congruent, that shifts in host affiliation occurred between tribes that had become differentiated much earlier, and that these beetles therefore have not been the agents of selection responsible for the secondary chemical differences that distinguish the tribes. Thus there has been little or no coevolution, at least at this level. This does not exclude the possibility of more recent coevolutionary changes in chemistry, within or among closely related species of host plants.

Our conclusion that there has been little simultaneous cladogenesis of these insects and their host plants agrees with most of the few other such studies (Futuyma & McCafferty, 1990; Mitter and Farrell, in press). Congruent cladogenesis does appear to have occurred in some instances, however (e.g., Farrell & Mitter, 1990); considerably more phylogenetic study will be required to determine which is the more common pattern.

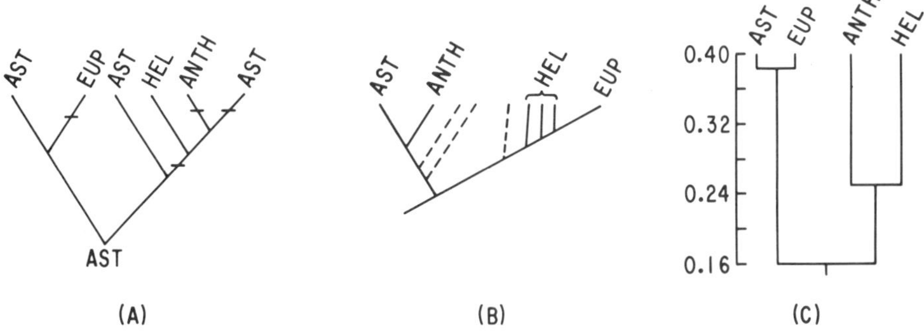

Figure 3. A. A postulated history of shifts in host affiliation in *Ophraella*, among tribes of Asteraceae, from Fig. 2. B. Phylogenetic relationships among these tribes of Asteraceae, based on Jansen et al. (1990). Broken lines indicate position of tribes that do not include hosts of *Ophraella*. C. Phenetic similarity of tribes of Asteraceae that include hosts of *Ophraella*, based on UPGMA analysis of shared classes of secondary compounds. The contrast between the history of host shifts and the phylogeny of plant tribes is equally great for the other, equally parsimonious, interpretation of host shifts indicated in Fig. 2.

Ehrlich and Raven (1964) postulated that insects "colonize" new hosts in evolutionary time on the basis of plants' similarity in secondary compounds. The secondary chemistry of Asteraceae is the subject of a large literature. When we compiled this information and calculated overall phenetic similarity among tribes, we found modest congruence between the topology of host shifts and the topology of the phenogram (Fig. 3). In addition, there have been five shifts between genera in the same tribe and at least two shifts between congeneric plants, which are chemically quite similar. On the whole, then, host shifts seem to be correlated with overall chemical similarity. Comparable analyses appear not to have been reported for other groups of insects and plants.

This phylogenetic analysis has implications for our understanding of evolutionary processes in insects and their host plants. The host associations of these insects appear to be attributable more to "colonization" of and adaptation to pre-existing host plants than to "association by descent" (Mitter & Brooks, 1983). Speciation and cladogenesis in the plants is therefore not immediately responsible for speciation of the insects (as it might be in some more intimate associations of parasites and hosts; cf. Hafner & Nadler, 1988). Because certain groups of plants (e.g., Ambrosiinae) antedate the origin of those beetles that feed on them, these insects cannot have been the selective agents responsible for the evolution of the (presumably defensive) secondary compounds that characterize the plant clades (e.g., complex sesquiterpene lactones in the Ambrosiinae). This conclusion is consonant with the hypothesis of "diffuse coevolution," in which the evolution of plant defenses is stimulated more by a diverse array of herbivores than by any particular herbivore (Janzen, 1980; Fox, 1981; Futuyma, 1983; Strong et al., 1984). The conclusion that host shifts in *Ophraella* have occurred largely among chemically similar plants suggests that physiological, sensory, or other constraints (i.e., "genetic constraints") limit the variety of possible host shifts available to a species at any time. This, of course, is a particular instance of the Darwinian (and modern) doctrine that *natura non facit saltum*.

The phylogenetic analysis was undertaken in order to provide a historical framework for designing and interpreting studies of genetic variation in the responses of beetles to the host plants used by their congeners. We have done a variety of preliminary experiments that show that the potential barriers to colonizing a new host include both larval and adult feeding behavior (simple refusal to feed on a congener's host, even unto death), oviposition response (both maturation of eggs and oviposition behavior are blocked), and probably toxicity (since larvae have greatly reduced growth and survival even if they feed).

Most of our work has been on feeding responses, which would be the first, necessary step in colonization of a new host. Preliminary trials, intended only to establish the probable spectrum of responses, consisted of examining the feeding responses of wild-caught adults to hosts of their congeners, whenever a particular combination of beetle and plant species was available. The numerous trials displayed a wide range of responses (Fig. 4; Futuyma, 1991). In some cases, the phenotypic variation among individuals was great, ranging from considerable consumption to none at all (e.g., feeding on *Solidago bicolor*, a host of *Ophraella pilosa*, by *O. arctica*, which naturally feeds on *S. multiradiata*). In other instances, no individual consumed a test plant at all. For example, we have now tested more than 250 adults and 280 larvae of *O. communa* (the natural hosts of which are in the Ambrosiinae) on *Solidago altissima* (host of *O. conferta*), and have observed no feeding whatever. We can detect no phenotypic variance, hence no expressed genetic variance, in this case.

We are presently engaged in increasing our sample sizes, and in determining whether phenotypic variation in feeding, growth, and survival on various test plants has a genetic component. The details of the data and experimental methods will be published elsewhere, and may be only briefly summarized here. The most extensive data to date are on feeding responses, measured as the leaf area consumed in 24 hours by individual beetles

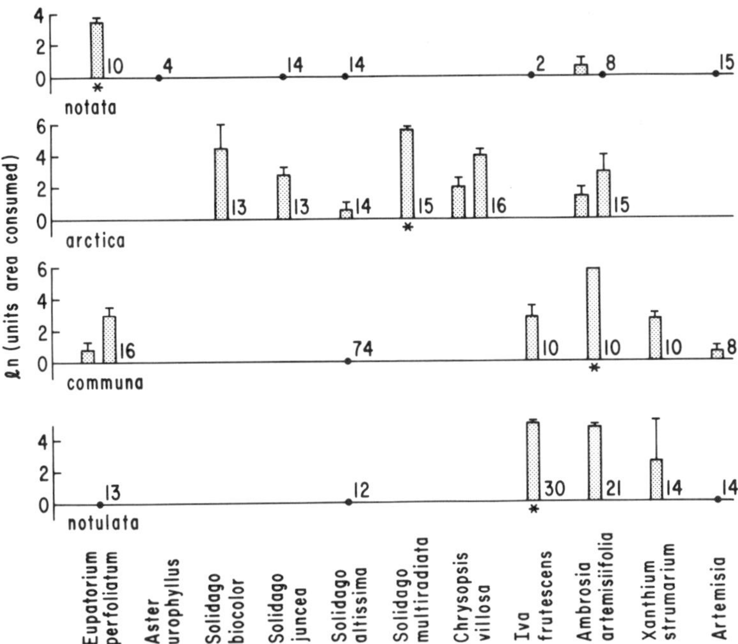

Figure 4. Representative results of feeding trials of wild-caught adult *Ophraella* beetles, tested on their own and congeners' host plants. Extracted and condensed from Futuyma (1991). For each of four species of *Ophraella*, on each of several test plants, the height of the column indicates mean consumption in 24 hours, and the bar the standard error of the mean; associated numbers are sample sizes. The natural host of each beetle species is marked by an asterisk. *Xanthium strumarium* is closely related to *Iva*, and has been recorded as an occasional host of *O. communa*. Two species of *Artemisia* were used: *O. communa* and *O. notata* were tested on *A. vulgaris*, a European species that is closely related to the host of *O. artemisiae* (and is readily consumed by that species); *O. notulata* was tested on *A. Douglasiana*, the host reported for *O. californiana* (not included in the phylogenetic analysis).

in a no-choice test (a beetle is confined with leaf discs of the test plant in a dish with moist filter paper). In *O. communa*, we have data for half-sib families (each sire mated to two dams; design unbalanced because of variable survival among broods) on the feeding responses to *Iva frutescens* and *Eupatorium perfoliatum*. Data on adults for two, and on larvae for three successive generations indicated a significant sire effect in almost every case (Table 1). Data on *O. notulata*'s feeding responses to *Ambrosia artemisiifolia* and *Eupatorium perfoliatum* have not yet been analyzed, but our strong impression is that they too will provide evidence for a genetic component to the variation in responses to both plants. *O. communa* and *O. notulata* are closely related; they feed on *Ambrosia artemisiifolia* and *Iva frutescens* respectively, the association with *Iva* probably being a derived condition with respect to the association with *Ambrosia* (Fig. 2). The phylogenetic analysis implies that association with *Eupatorium* has not figured in the history of the clade to which these species belong.

From our work to date, it appears likely that if a species displays any feeding response to a congener's host, the variation in response is likely to have a genetic component, and so far this appears to be true whether the plant is a host of a close or of a distant relative. However, some plants appear to elicit no response at all; this appears to be more frequently the case when the plant is the host of a distant relative of the test species rather than a close relative. We are still far from being able to say whether patterns of genetic variation bear any relationship to the history of host shifts. Whatever pattern emerges, however, it will bear on the question of whether studies of genetic variation can tell us anything about evolutionary history—and vice versa.

Table 1. Analyses of variance of feeding responses of adult and larval *Ophraella communa* to two test plants (*Iva* and *Eupatorium*). Each sire was mated to two dams; because of mortality, the number scored per brood varied, including zero in some instances. The variate is leaf area consumed in 24 hours, measured under a microscope with an ocular reticle. Each specimen was tested on both plants separately. Analysis was by the PROCGLM routine in SAS. Significant sire effects imply a genetic component of variance. The proportion of variance explained by the model is given by r^2.

		Iva				*Eupatorium*				
	Source	df	F	p	r^2	Source	df	F	p	r^2
1. F_2 adults	model	74	1.41	0.041	0.42	model	74	2.01	0.0002	0.51
	sire	49	1.42	0.057		sire	49	2.28	0.0001	
	dam (sire)	25	1.45	0.091		dam (sire)	25	1.55	0.058	
	error	142				error	142			
2. F_3 adults	model	14	2.27	0.014	0.34	model	11	2.57	0.020	0.49
	sire	11	2.27	0.006		sire	8	2.74	0.021	
	dam (sire)	3	0.99	0.042		dam (sire)	3	2.23	0.105	
	error	61				error	30			
3. F_2 larvae	model	87	2.83	0.0001	0.58	model	87	3.06	0.0001	0.60
	sire	43	4.38	0.0001		sire	43	4.03	0.0001	
	dam (sire)	44	1.33	0.103		dam (sire)	44	2.12	0.0003	
	error	176				error	176			
4. F_3 larvae	model	15	3.40	0.002	0.64	model	14	17.97	0.0001	0.89
	sire	12	3.71	0.002		sire	11	17.75	0.0001	
	dam (sire)	3	2.30	0.099		dam (sire)	3	18.75	0.0001	
	error	29				error	30			
5. F_4 larvae	model	6	4.45	0.006	0.58	model	5	3.18	0.056	0.61
	sire	5	2.76	0.049		sire	4	3.91	0.036	
	dam (sire)	1	12.89	0.002		dam (sire)	1	0.78	0.398	
	error	19				error	10			

In 1944, George Gaylord Simpson wrote that

"Not long ago, paleontologists felt that a geneticist was a person who shut himself in a room, pulled down the shades, watched small flies disporting themselves in milk bottles, and thought he was studying nature," whereas "the geneticists said that paleontology had no further contributions to make to biology, that its only point had been the completed demonstration of the truth of evolution, and that it was a subject too purely descriptive to merit the name 'science.' The paleontologist, they believed, is like a man who undertakes to study the principles of the internal combustion engine by standing on a street corner and watching the motor cars whiz by."

Simpson's words could equally well describe the attitudes of population biologists and systematists, perhaps even more aptly around 1985 than in the 1940s. The Modern Synthesis changed such attitudes to some extent, but not enough. At least sociologically, the Synthesis wasn't complete—or perhaps it became unravelled. It appears likely that we are now entering a new era of synthesis between studies of history and process in evolution. The Fourth International Congress of Systematic and Evolutionary Biology, with "The Unity of Evolutionary Biology" as its theme, has given testimony that the synthesis is well underway.

ACKNOWLEDGMENTS

The research on *Ophraella* described here has been supported by the National Science Foundation (BSR8516316 and BSR8817912). I am grateful to Mark Keese for collaboration on the experiments, and to Sonja Scheffer for assistance. This is Contribution No. 781 in Ecology and Evolution from the State University of New York at Stony Brook.

LITERATURE CITED

Alberch, P. 1983. Morphological variation in the neotropical salamander genus *Bolitoglossa*. *Evolution* 37:906–919.

Anderson, E. 1937. Supra-specific variation in nature and in classification from the view-point of botany. *Amer. Natur.* 71:223–235.

Avise, J. C. 1989. Gene trees and organismal histories: a phylogenetic approach to population biology. *Evolution* 43:1192–1208.

Benson, W. W., Brown, K. S. & L. E. Gilbert. 1975. Coevolution of plants and herbivores: passion flower butterflies. *Evolution* 29:659–680.

Bremer, K. 1987. Tribal interrelationships of the Asteraceae. *Cladistics* 3:210–253.

Brown, W. L., Jr. & E. O. Wilson. 1956. Character displacement. *Syst. Zool.* 5:49–64.

Charlesworth, B., Lande, R. & M. Slatkin. 1982. A neo-Darwinian commentary on macroevolution. *Evolution* 36:474–498.

Clay, T. 1949. Some problems in the evolution of a group of ectoparasites. *Evolution* 3:279–299.

Courtney, S. P., Chen, G. K. & A. Gardner. 1989. A general model for individual host selection. *Oikos* 55:55–65.

Dobzhansky, Th. 1931. The North American beetles of the genus *Coccinella*. *Proc. U.S. Nat. Mus.* 80:1–32.

Dobzhansky, Th. 1941. Beetles of the genus *Hyperaspis* inhabiting the United States. *Smithson. Inst. Publ.* 3642:1–94.

Donoghue, M. J. 1989. Phylogenies and the analysis of evolutionary sequences, with examples from seed plants. *Evolution* 43:1137–1156.

Dover, G., Brown, S., Coen, E., Dallas, J., Strachan, D. & M. Trick. 1982. The dynamics of genome evolution and species differentiation. Pp. 343–372. *In:* G. A. Dover & R. B. Flavell (eds.), *Genome Evolution.* Academic Press, New York.

Easteal, S. 1990. The pattern of mammalian evolution and the relative rate of molecular evolution. *Genetics* 124:165–173.

Ehrlich, P. R. & P. H. Raven. 1964. Butterflies and plants: a study in coevolution. *Evolution* 18:586–608.

Eichler, W. 1948. Some rules in ectoparasitism. *Ann. Mag. Nat. Hist.* 12:588–598.

Farrell, B., and C. Mitter. 1990. Phylogeny of host-affiliation: have *Phyllobrotica* (Coleoptera: Chrysomelidae) and the Lamiales diversified in parallel? *Evolution.* 44:1389–1403.

Felsenstein, J. 1985. Phylogenies and the comparative method. *Amer. Natur.* 126:1–25.

Fox, L. R. 1981. Defense and dynamics in plant-herbivore systems. *Amer. Zool.* 21:853–864.

Furth, D. G. 1980. Inter-generic differences in the metafemoral apodeme of flea beetles (Chrysomelidae: Alticinae). *Syst. Entomol.* 5:263–271.

Futuyma, D. J. 1983. Evolutionary interactions among herbivorous insects and plants. Pp. 209–231. *In:* D. J. Futuyma & M. Slatkin (eds.), *Coevolution.* Sinauer: Sunderland, MA.

Futuyma, D. J. 1987. *Sturm und Drang* and the evolutionary synthesis. *Evolution* 42:217–226.

Futuyma, D. J. 1990a. History and evolutionary processes. *In:* M. Nitecki (ed.), *History and Evolution.* University of Chicago Press: Chicago, IL.

Futuyma, D. J. 1990b. Observations on the taxonomy and natural history of *Ophraella* Wilcox (Coleoptera: Chrysomelidae), with a description of a new species. *J. New York Entomol. Soc.* 98:163–186.

Futuyma, D. J. 1991. The evolution of host specificity in herbivorous insects: Genetic, ecological, and phylogenetic aspects. Pp. 431–454. *In:* P. W. Price, T. M. Lewinsohn, G. W. Fernandes, & W. W. Benson (eds.), *Plant-Animal Interactions: Evolutionary Ecology in Tropical and Temperate Regions.* Wiley: New York.

Futuyma, D. J. & S. S. McCafferty. 1990. Phylogeny and the evolution of host plant associations in the leaf beetle genus *Ophraella* (Coleoptera: Chrysomelidae). *Evolution.* In press.

Gillespie, J. H. 1986. Variability of evolutionary rates of DNA. *Genetics* 113:1077–1091.

Goodman, M., Weiss, M. L., J. Czelusniak. 1982. Molecular evolution above the species level: branching patterns, rates, and mechanisms. *Syst. Zool.* 31:376–399.

Hafner, M.S. & A. Nadler. 1988. Phylogenetic trees support the coevolution of parasites and their hosts. *Nature* 332:258–259.

Hubbs, C. L. 1955. Hybridization between fish species in nature. *Syst. Zool.* 4:1–20.

Huey, R. B. & A. F. Bennett. 1987. Phylogenetic studies of coadaptation: preferred temperatures versus optimal performance temperatures of lizards. *Evolution* 41:1098–1115.

Jansen, R. K., Michaels, H. J. & J. D. Palmer. Phylogeny and character evolution in the Asteraceae based on chloroplast DNA restriction site mapping. *Syst. Bot.* In press.

Janzen, D. H. 1980. When is it coevolution? *Evolution* 34:611–612.

Kluge, A. G. & W. C. Kerfoot. 1973. The predictability and regularity of character divergence. *Amer. Natur.* 107:426–442.

Langley, C. H. & W. M. Fitch. 1974. An examination of the constancy of the rate of molecular evolution. *J. Mol. Evol.* 3:161–177.

LeSage, L. 1986. A taxonomic monograph of the Nearctic galerucine genus *Ophraella* Wilcox (Coleoptera: Chrysomelidae). *Mem. Entomol. Soc. Canada* 133:1–75.

Lewontin, R. C. 1974. *The Genetic Basis of Evolutionary Change.* Columbia University Press: New York.

Li, W.-H., Gojobori, T. & M. Nei. 1981. Pseudogenes as a paradigm of neutral evolution. *Nature* 292:237–239.

Loeschcke, V. (ed.) 1987. *Genetic Constraints on Adaptive Evolution.* Springer-Verlag: Berlin.

MacNair, M. R. 1987. Heavy metal tolerance in plants: a model evolutionary system. *Trends in Ecology and Evolution* 2:354–359.

Maddison, W. P. 1990. A method for testing the correlated evolution of two binary characters: are gains or losses concentrated on certain branches of a phylogenetic tree? *Evolution* 44:539–557.

Maynard Smith, J. & K. Sondhi. 1960. The genetics of a pattern. *Genetics* 45:1039–1050.

Mayr, E. & W. B. Provine (eds). 1980. *The Evolutionary Synthesis: Perspectives on the Unification of Biology.* Harvard University Press, Cambridge, MA.

Michener, C. D. 1958. The evolution of social behavior in bees. *Proc. Tenth Internat. Congr. Entomol.* 2:441–447.

Miller, R. R. 1950. Speciation in fishes of the genera *Cyprinodon* and *Empetrichthys*, inhabiting the Death Valley region. *Evolution* 4:155–163.

Mitter, C. & D. R. Brooks. 1983. Phylogenetic aspects of coevolution. Pp. 65–98. *In*: D. J. Futuyma & M. Slatkin (eds.), *Coevolution.* Sinauer: Sunderland, MA.

Mitter, C. & B. Farrell. Macroevolutionary aspects of insect/plant relationships. *In*: E. A. Bernays (ed.), *Insect-Plant Interactions.* Vol. 2. CRC press, Boca Raton, FL. In press.

Mitter, C., Farrell, B. & B. Wiegmann. 1988. The phylogenetic study of adaptive zones: Has phytophagy promoted insect diversification? *Amer. Natur.* 132:107–128.

Ridley, M. 1983. *The Explanation of Organic Diversity: The Comparative Method and Adaptations for Mating.* Clarendon, Oxford.

Rohlf, F. J., Gilmartin, A. J. & G. Hart. 1983. The Kluge-Kerfoot phenomenon: a statistical artifact. *Evolution* 37:180–202.

Sillén-Tullberg, B. 1988. Evolution of gregariousness in aposematic butterfly larvae: a phylogenetic analysis. *Evolution* 42:293–305.

Simpson, G. G. 1944. *Tempo and Mode in Evolution.* Columbia University Press: New York.

Slatkin, M. & W. P. Maddison. 1989. A cladistic measure of gene flow inferred from the phylogenies of alleles. *Genetics* 123:603–613.

Spencer, K. C. 1986. Chemical mediation of coevolution in the *Passiflora-Heliconius* interaction. Pp. 167–240. *In*: K. C. Spencer (ed.), *Chemical Mediation of Coevolution.* Academic Press: San Diego, CA.

Strong, D. R., Lawton, J. H. & R. Southwood. 1964. *Insects on Plants: Community Patterns and Mechanisms.* Blackwell: Oxford.

Sturtevant, A. H. 1921. The North American species of *Drosophila. Carnegie Inst. Wash. Publ.* 301:1–150.

Sturtevant, A. H. 1942. The classification of the genus *Drosophila* with descriptions of nine new species. *Univ. Texas Publ.* 4213:5–51.

Weber, K. E. & L. T. Diggins. 1990. Increased selection response in larger populations. II. Selection for ethanol vapor resistance in *Drosophila melanogaster* at two population sizes. *Genetics* 125:585–597.

Wilcox, J. A. 1965. A synopsis of North American Galerucinae (Coleoptera: Chrysomelidae). *Bull. New York State Mus. Sci. Serv.* 400:1–226. Albany, NY.

Wilson, E. O. 1961. The nature of the taxon cycle in the Melanesian ant fauna. *Amer. Natur.* 95:169–193.

Phenotypic Plasticity
and the Expression of Genetic Variation

G. de Jong and S. C. Stearns

INTRODUCTION

The Darwinian revolution replaced typological thinking with thinking about variation within populations[1,2], in traits and in reproductive success. Natural selection presupposes phenotypic variation among individuals in reproductive success[3]; a response to selection is only possible in traits where the genetic variation in the trait is correlated with reproductive success. The relation between phenotypic and genetic variation is therefore a key element of evolutionary theory. The origin of the non-genetic part of the phenotypic variation in any trait should not be relegated to oblivion.

Phenotypic plasticity refers to phenotypic variation induced systematically by environmental variation, most clearly documented as variation among individuals of a single clone raised in different environments. Plastic traits are often genetically variable, either in the plasticity itself, in the amount of change with the environment, or in the level at which the variation takes place. If genetic variation for phenotypic plasticity is present in a population, it would seem that no genotype is optimally plastic, indicating that while plasticity may be adaptive (a view supported by Bradshaw[4], Schlichting[5], and Sultan[6]), it is also constrained (a view supported by the results of Berven, Gill, and Smith-Gill[7] and by the extensive literature on genotype x environment interactions in domestic animals and crop plants[8,9]). Genetic variation in phenotypic plasticity affects the expression both of genetic variation for a single trait and of genetic covariation between any pair of traits, within and across environments. The implications of phenotypic plasticity for the expression of genetic variation have been neither explicitly analyzed nor widely appreciated. Several approaches have been suggested recently; they were proposed independently and need to be coordinated. This article presents these approaches and tries to indicate where they correspond. While surveying concepts, we do emphasize one approach and use it to try to unite several others.

TWO APPROACHES TO PLASTICITY: WHAT IS A TRAIT?

Phenotypic plasticity, loosely used, means any phenotypic change across environments, but its exact definition is the change in the mean phenotypic value of a genotype across environments. Because interesting plastic traits are frequently polygenic, quantita-

Dr. de Jong is with the Vakgroep Populatie- en Evolutiebiologie, Padualaan 8, NL-3584 CH Utrecht, the Netherlands. Dr. Stearns is with the Zoologisches Institut, Universität Basel, Rheinsprung 9, CH-4051 Basel, Switzerland.

tive genetics provides an appropriate approach. Quantitative genetics can be developed either from population genetics or as a statistical theory[10], and phenotypic plasticity can be incorporated as another element of the statistical treatment[11,12]. For present purposes, however, we prefer to use the population genetic approach to quantitative genetics.

Within quantitative genetics, two approaches to plasticity are conditioned in part by how the environment is visualized. One approach, Falconer's[13], visualizes discrete environments, exemplified by different host plant species for an herbivorous insect[14] or by pools at different altitudes for a frog[15]. Another approach, Woltereck's[16], visualizes continuous environments, exemplified by temperature and humidity[17]. (Experimental designs frequently impose discreteness on the underlying, natural continuity.)

For two environments, Falconer's approach[13] is straightforward: he defined the realisation of a trait in two environments as two traits and represented the plasticity of the trait as the correlation between the realisation of the trait in the two environments: *the cross-environmental correlation.* For continuous environments, Woltereck's approach[16] through reaction norms is appropriate. A reaction norm, empirically defined, is a plot of the phenotypic mean against the environment for each genotype separately. For a set of discrete experimental environments cut out of a continuum, a two-factor (genotype $\times$ environment) analysis of variance is also appropriate. Each choice of approach carries with it an implicit view of plasticity with consequences for how it is modeled and what its adaptive significance might be.

Falconer's Approach

This is most convenient for several discrete environments that cannot be ordered on a single dimension. The realisation of the trait in each environment is taken as a separate trait (figs. 1a, 2a), and the plasticity of the trait is measured by the cross-environmental correlation. By using known genotypic backgrounds, one can calculate the genetic component of the phenotypic covariance across environments (Box 1). This is impractical for many environments.

Via and Lande[18] extended Lande's[19] general model of simultaneous selection on several quantitative traits by adding Falconer's cross-environment covariance to model the evolution of adaptive plasticity. If selection in each environment is optimizing, if there is an unlimited supply of genetic variants of all sizes (the distribution of allelic effects is normal), and if the genetic variances in both environments and the genetic covariance between environments remain constant during selection, then the optimum mean

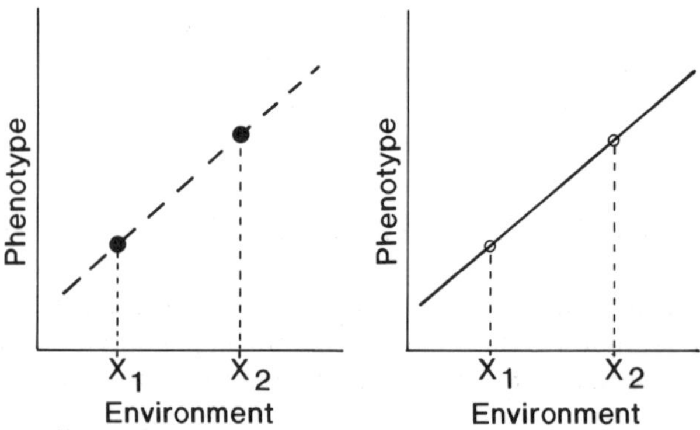

Figure 1. Two phenotypes realized by one genotype in two environments can be regarded as two traits, emphasizing the points (left), or as one trait, i.e., the line representing the reaction norm (right).

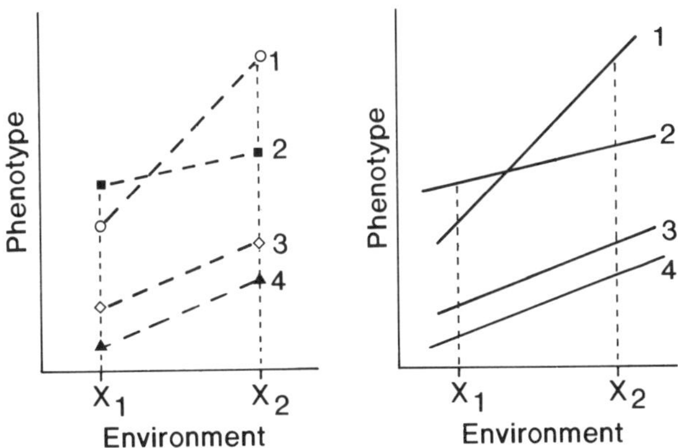

Figure 2. Four genotypes in two environments. Genotype by environment interaction is diagrammed in two ways that correspond to the two ways of defining a trait: as genetically related points in each environment (left) and as reaction norms (right). Only the genotypes 3 and 4 do not show genotype by environment interaction.

phenotype will be achieved in each environment. (With such assumptions, one would expect the optimum to be reached regardless of the model.)

The two conditions under which the optimum is not reached—a cross-environment genetic correlation of +1 or −1—are of special interest. Via and Lande[18] interpret a cross-environmental correlation of +1 as identical plasticity of all genotypes; lack of genetic variation for plasticity, lack of variation in slopes of reaction norms, prevents optimization. A cross-environmental correlation of −1 means that the genotypes reverse rank order between environments in a particular way; all lines connecting the two expressions of each genotype cross in one point. That all lines cross in one point imposes a constraint on selection, for the line connecting the optima might not be present in the bundle of crossing lines.

Woltereck's Approach

Woltereck[16] realized that genetically similar organisms share, not some value for a trait, but rather a common way of responding to environmental heterogeneity. In *Daphnia* reared under different food conditions, he observed that samples from the same clone reacted similarly to changes in food supply, while those from different clones reacted differently. He concluded that the genotype, what is actually inherited, what evolves, is the norm of reaction—a conclusion that has often been repeated[20,21].

A reaction norm can be treated as a function that maps the environment onto the phenotype (fig. 1b). The reaction norm is the function, the environmental value is the argument, and the phenotypic value is the function value (Box 1). Here the reaction norm itself is seen as the trait under selection, a notion most straightforwardly applied to clonal organisms but extendable without difficulty to sexually reproducing genotypes. Different genotypes have different reaction norms; this suggests that genotypic differences are differences in function. Because the simplest functions are straight lines, we first need a quantitative genetics of linear functions. The rest of this paper reviews recent progress in this approach to the genetics of plasticity. How are genetic variances and covariances and the heritability of plasticity defined for quantitative traits that are linear functions of the environment?

THE GENETICS OF LINEAR REACTION NORMS

Genetic and Environmental Effects May Not Be Separable

When a reaction norm is a linear function of an environmental variable, the slope of the line connecting the phenotypic values in the different environments measures plasticity. Lines of identical slope but different intercept represent different genotypes that have different mean values but identical plasticities. Within a genotype, the plasticity variance is the variance in the environmental variable—as experienced by different individuals of the same clone—multiplied by the square of the slope (Box 2).

Genotypic variation in phenotypic plasticity means reaction norms with different slopes; these show up in a two-factor analysis of variance as genotype × environment interaction (fig. 2b). A similar design can be carried out in heterogeneous environments. In that case, for linear reaction norms, the genotype × environment interaction variance is the product of the variance of the environmental variable and the genetic variance in slope. The genetic main effect is the genetic variance at the average environment, and the environmental main effect is the average slope over all genotypes times the variance in the environmental variable (Box 2). Because the genotypic variance deduced from the genetic main effect depends on the mean of the environmental distribution, and because the environmental variance deduced from the environmental main effect depends on a mean value over all genotypes, there is no clean separation of genetic and environmental effects.

The Heritability of Plasticity

The first two concrete suggestions for measuring the heritability of plasticity were based on a two-factor analysis of variance that partitions variation into genetic, environmental, and genotype × environment effects. Scheiner and Goodnight[22] proposed estimating the variance due to plasticity as the sum of the environmental main-effect variance and the variance due to genotype × environment interaction (Box 2). However, the main-effect environmental component contains effects that result not so much from heritable differences in plasticity as from the average slope over all genotypes. Therefore, Scheiner and Lyman[23] proposed using the genotype × environment interaction variance divided by total phenotypic variance as the heritability of phenotypic plasticity. Only if the reaction norms are straight lines does the genotype × environment variance involve a real additive genetic variance[24], and for linear reaction norms, a more precise measure of the heritability of plasticity would be the genetic variance in slope divided by the total variance in slope. If the reaction norms are not linear, there is no good measure for the heritability of phenotypic plasticity. (Using only two environments imposes linearity on the reaction norms.)

Heritabilities and Genetic Covariances Change Across Environments

It has long been recognized that heritabilities and genetic covariances are functions of the environments in which they are measured[10]. In this section we show how both change systematically across the range of an environmental variable under the assumption of linear reaction norms. Such changes have important implications for the potential response to selection.

Heritabilities. Take the same genotypes, at the same frequencies, as before, but picture identical replicate populations sitting at different points on the range of the environmental variable. Within each population, the environment is constant. Differences in slopes of reaction norms will be reflected in different heritabilities of the same trait of measured in each separate population. Can these changes in genetic variance be expressed as a function of the environment?

The answer is straightforward if we use Falconer's[10] treatment of the additive genetic variance per locus. It consists of the variance in gene frequencies times the square of the "average effect of a gene substitution" (Box 3). If the genotypic value is the sum of the allelic effects, the "average effect of a gene substitution" becomes simply the *difference* of the allelic effects. Linear allelic effects lead through linear genotypic values to linear reaction norms, and if the allelic effects are not parallel, their difference is again linear. Crossing linear reaction norms imply a linear "average effect of a gene substitution". Such a linear "average effect of gene substitution" means that the additive genetic variance per locus is a quadratic function of the environment (Box 3). Summing over loci and introducing linkage disequilibrium does not change that. If dominance is present, the dominance variance per locus may also become quadratic.

Genetic covariances. Similarly, the additive genetic covariance of two phenotypically plastic traits determined by crossing reaction norms also becomes a quadratic function of the environment (Box 3). One locus that pleiotropically influences both traits suffices to demonstrate the principle. Consider again the case where genotypic values can be found as the sum of allelic effects, this time for both traits. It is the differences among the allelic effects that produce the "average effects of a gene substitution" for the two traits, and the differences must be taken for both traits with the allelic effects in the same order. Because the additive genetic covariance involves the product of these two linear average effects, it is also a quadratic function of the environment.

Covariances are not constrained, like variances, to remain positive. When should additive genetic covariance change sign across environments? To be of interest, such a sign change should arise from some feature of the biology of the traits, not just from the mathematical formalism. What might such features be? Again, looking at one locus brings some insight. Such a locus has two average effects, one for each trait. They have to be multiplied to get the contribution of that locus to the additive genetic covariance, and if they are linear, that multiplication yields a product that changes sign when one of the average effects changes sign but the other does not (Box 3). For each locus, a sign change of the genetic covariance occurs when the allelic effects cross for one trait but not for the other. But when both average effects change sign simultaneously, in the same environment, their product does not change sign.

Polygenic traits can be dealt with too. When many loci contribute to the two traits through linear reaction norms, a null hypothesis predicts sign change (Box 4): if all slopes and intercepts of the reaction norms are mutually independent, the expected value of the additive genetic covariance changes sign across environments[25]. A sufficiently strong correlation among the average effects for the two traits will prevent the sign change.

A biological interpretation. Sign change in additive genetic covariance should occur when the average effects of gene substitutions are independent because the biochemical paths from the genes to the two traits are not physiologically or functionally connected. We should not expect such a sign change when tight functional links force genetic effects to be correlated, as is the case for allocation of one substrate to two traits[26,27].

Genetic variation in the allocation of one substrate to two traits leads to average effects of a gene substitution that have opposite signs for the two traits. Such tight physiological connections prevent sign changes in the additive genetic covariance of the two traits, which is forced to be negative. Allocation of one substrate to two traits is the clearest example of a biological trade-off. Similar functional trade-offs arising in the epigenetic system would lead to negative additive genetic covariances in all environments. However, ecological trade-offs that do not arise from tight functional constraints are compatible with a sign change in additive genetic covariance across environments.

The additive genetic covariance of two traits determined by linear reaction norms

712

changes sign for a wide range of conditions; lack of sign change is the (mathematical) exception rather than the rule. The exception has a clear interpretation: the two traits are tightly related through constraints on gene action in the epigenetic system. However, loosely constrained gene action is compatible with a sign change, and the observation of a change of sign in genetic covariance across environments indicates that the epigenetic links between the two traits cannot be tight.

The genetic covariance of some traits varies across environments[5,28,29], and Gebhardt and Stearns[17] have documented a sign change in the genetic covariance of developmental time and weight at eclosion in *Drosophila mercatorum*. We could use a catalogue of traits that were so tightly coupled in the epigenetic system that they showed no sign change in additive genetic covariance across environments. Research on this point is just beginning.

DISCUSSION

Modelling the plastic response as a bundle of linear reaction norms brings some unity to the diverse models used to describe phenotypic plasticity and its evolution. One can see what the components of phenotypic variance represent[22,23] and what Falconer's[13] approach would look like (Box 1).

One can also see how Via and Lande's model[18,30] can be connected with that of Gillespie and Turelli[31]. They come to opposite conclusions concerning the maintenance of genetic variation, for their assumptions differ dramatically. In Via and Lande's model, the optimum phenotype is reached in both environments unless the genetic correlation between environments is +1 or −1, and the only genetic variation that remains after selection is maintained by selection-mutation balance. The assumption about linear reaction norms that corresponds to Via and Lande's assumption about normal distributions of allelic effects is that one has an infinite supply of straight lines with all possible slopes and intercepts. Finding the optimum in both environments consists simply of picking out the straight line that goes through the two points required. Given the assumptions, a homozygote with this straight line will exist. If the cross-environment genetic correlation were ±1, the supply of straight lines would be too small to guarantee that one of them would go through the two points required.

What about the maintenance of genetic variation? Lande's[19] model cannot handle selected polymorphism, and the conclusion that the only genetic variation remaining after selection is maintained by selection-mutation balance is more a feature of the model than of biology in general. The linear reaction norm connecting the two optimal points need not represent a homozygous genotype. If it represents a heterozygote, it is quite possible that the reaction norms representing the accompanying homozygotes cross; genetic variation would be maintained. This links the maintenance of genetic variation in a heterogeneous environment to genotype × environment interaction[31], as one would expect from the tradition of models that use G × E interaction in fitness to explain the maintenance of genetic variation[32,33].

The behaviour of the genetic covariance between two traits across many environments informs us about their developmental integration. Covariances functioning as constraints on the response to selection are better viewed as originating in epigenetic mechanisms that respond to environmental heterogeneity than as genetic covariances caused by linkage and pleiotropy. Of course they involve pleiotropy, but pleiotropy itself is not a constraint on selection or a developmental constraint and is compatible with a flexible epigenetic system in which two traits are not tightly coupled by function.

Box 1: What is a Trait?

Consider a phenotypically plastic character expressed in two environments. Falconer[13] proposed regarding the realization of the character in the two environments as two traits (Fig. 1, left). Woltereck[16] suggested that the trait is the way the character reacts to the environmental change (Fig. 1, right). He coined the phrase, "The genotype is the reaction norm."

Corresponding to the two ways to define plastic traits, there are two ways to describe whether the reaction to the environment is the same for both genotypes. Note that genotype by environment interaction means that genotypes react differently to environmental change. Falconer[13] proposed using the cross-environment genetic correlation between the phenotypic values of several genotypes to measure their reaction. For example, genotypes 1 and 2 (Fig. 2) produce a genetic correlation $r_A = -1$; genotypes 3 and 4 produce a genetic correlation $r_A = +1$. In a two-factor analysis of variance for genotypes and environments, the component attributable to genotype by environment interaction corresponds to a measure of the dissimilarity with which the genotypes react to the environment. If we used only genotypes 3 and 4 in such an analysis, there would be no component attributable to genotype $\times$ environment interaction, for both react the same way. Using 1 and 2, or all four, would imply a $G \times E$ component.

Clearly, the cross-environment genetic correlation and the genotype by environment interaction must be related. The linear reaction norm defined by the pair of points for the ith genotype (Fig. 2) has intercept a_i and slope c_i. Both the genetic covariance between the two environments (COV_G) and the genotype by environment interaction variance (V_{GE}) are derived from the same straight lines. The genotypic main effect in the ANOVA is the genetic variance measured in the mean environment, $V_{G \text{ at } \bar{x}}$. Then (de Jong[24] differing from Yamada[11]):

$$COV_G = V_{G \text{ at } \bar{x}} - V_{GE}$$

The genetic variance in the mean environment can be defined in terms of the variances and covariance of slopes and intercepts as

$$V_{G \text{ at } \bar{x}} = V_G(a) + 2\bar{x}\, COV_G(a,c) + \bar{x}^2 V_G(c)$$

The variance in the environmental values over the two environments is

$$\sigma^2_x = 1/4\, (x_1 - x_2)^2.$$

The genotype by environment interaction variance then becomes the variance in the environmental values over the two environments, σ^2_x, times the genetic variance in slopes:

$$V_{GE} = \sigma^2_x\, V_G(c)$$

Under these assumptions, neither the genetic correlation between slopes and intercepts nor the genetic correlation between the slopes and the values of the genotypes in the mean environment is constrained; both can take any value from -1 to $+1$. Thus, using linear reaction norms yields the flexibility necessary for a model of selection.

714

Box 2: Phenotypic and Genetic Variation in Reaction Norms.

In a heterogeneous environment, the population experiences a distribution in the values of an environmental variable. Through the interaction of the environment with each individual over the course of development, the distribution of the environmental variable is mirrored into the distribution of phenotypes for that genotype (Fig. 3, left). The genotype maps the environmental distribution into a phenotype distribution. The reaction norm is the mapping function, the environment is the argument of the function, and the genotypic value is the value taken by the function, to which is added some error variation to get the phenotypic value. A difference between genotypes is a difference in mapping function (Fig. 3, right).

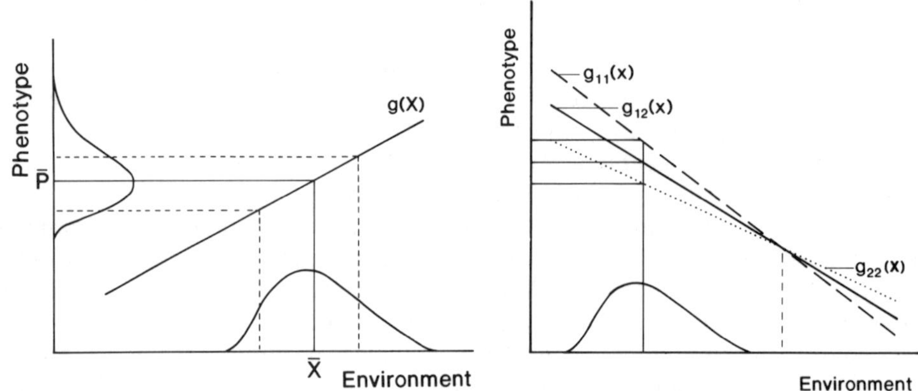

Figure 3. Reaction norms map the distribution of an environmental variable into a phenotypic distribution: one genotype (left), three genotypes (right).

For any type of mapping function, the total phenotypic variance V_P can be partitioned into a genetic variance V_G, an environmental variance V_E, and a genotype by environment interaction variance V_{GE}[22,23]. If the environmental variable x has mean $\bar{x}$ and variance σ^2_x, then for linear reaction norms the partitioning can be specified in terms of slopes, variances, and means. The genetic variance is measured at the mean value of the environmental variable ($V_{G\,at\,\bar{x}}$; Box 1). The environmental variance is the square of the mean slope times the variance of the environmental variable:

$$V_E = (\bar{c})^2 \sigma^2_x$$

The genotype by environment interaction variance is the variance in the environmental variable times the variance in the slopes of the reaction norms[24] (which is strictly genetic):

$$V_{GE} = \sigma^2_x\, V_G(c)$$

Apart from these variance components, there will always be variance contributed by experimental error and environmental noise, V_R. If the error in each environment is independent of the environment and the genotype, the phenotypic variance in slopes will be larger than the genetic variance found by using replicate samples within a clone. The best measure of the heritability in plasticity in such a model is the genetic variance in slope divided by the phenotypic variance in slope rather than divided by total phenotypic variance. Using V_{GE}/V_P places σ^2_x in the numerator and a confusing diversity of things in the denominator[17].

Box 3: The Principle Underlying Changes in Genetic Variances and Covariances Based on Linear Reaction Norms.

Imagine a large number of replicate populations that are identical except that each occupies a different position along the range of an environmental variable. Over the whole range of the variable, the phenotypic value changes for a given genotype as in Fig. 3, and similarly for all genotypes and many characters. How do we expect genetic variance and covariance to change over the range of the environmental variable? The approach is through the single locus treatment of quantitative traits discussed in Falconer[10].

Genotype by environment interaction is a synonym for nonparallel reaction norms. When a genotype by environment interaction exists, genetic variances and covariances change with the environment. Crossing reaction norms for additive genotypes have the heterozygote's reaction norm exactly intermediate to those of the two homozygotes (Fig. 4a, c). They can be thought of as arising from linear allelic effects that themselves cross (Fig. 4b, d). Adding the allelic effects gives the genotypic values; subtracting them gives "the average effect of a gene substitution" α, which also varies linearly with the environment and is zero where the allelic effects cross (Fig. 4b, d). The contribution to the additive genetic variance from this locus is given by the product of the variance in allele frequencies and the square of "the average effect of a gene substitution" α: $2pq\,\alpha_1^2$ for trait 1 and $2pq\,\alpha_2^2$ for trait 2. For a given locus influencing a trait, the additive genetic variance is a quadratic function of the environmental variable (Fig. 4i, j), but a different such function for each trait.

The contribution from this locus to the additive genetic covariance between two traits is given by the product of the variance in gene frequencies with the product of the "the average effect of a gene substitution" for each of the two traits: $2pq\,\alpha_1\alpha_2$. For each locus, the additive genetic covariance is also a quadratic function of the environmental variable (Fig. 4f).

There is no reason why the average effects of a gene substitution at a given locus for two traits should be zero at the same value of the environmental variable. Going from left to right along the axis of the environmental variable, first one average effect changes sign, then the other. The product of the average effects, $\alpha_1\alpha_2$, thus changes sign twice (Fig. 4f). When many such loci contribute to the two traits, the total additive genetic covariance is the sum of all their contributions (Fig. 4g).

Only if the "the average effects of a gene substitution" for the two traits both change sign at the same environmental value for each locus (Fig. 4h) will the additive genetic covariance not change sign across the range of the environmental variable.

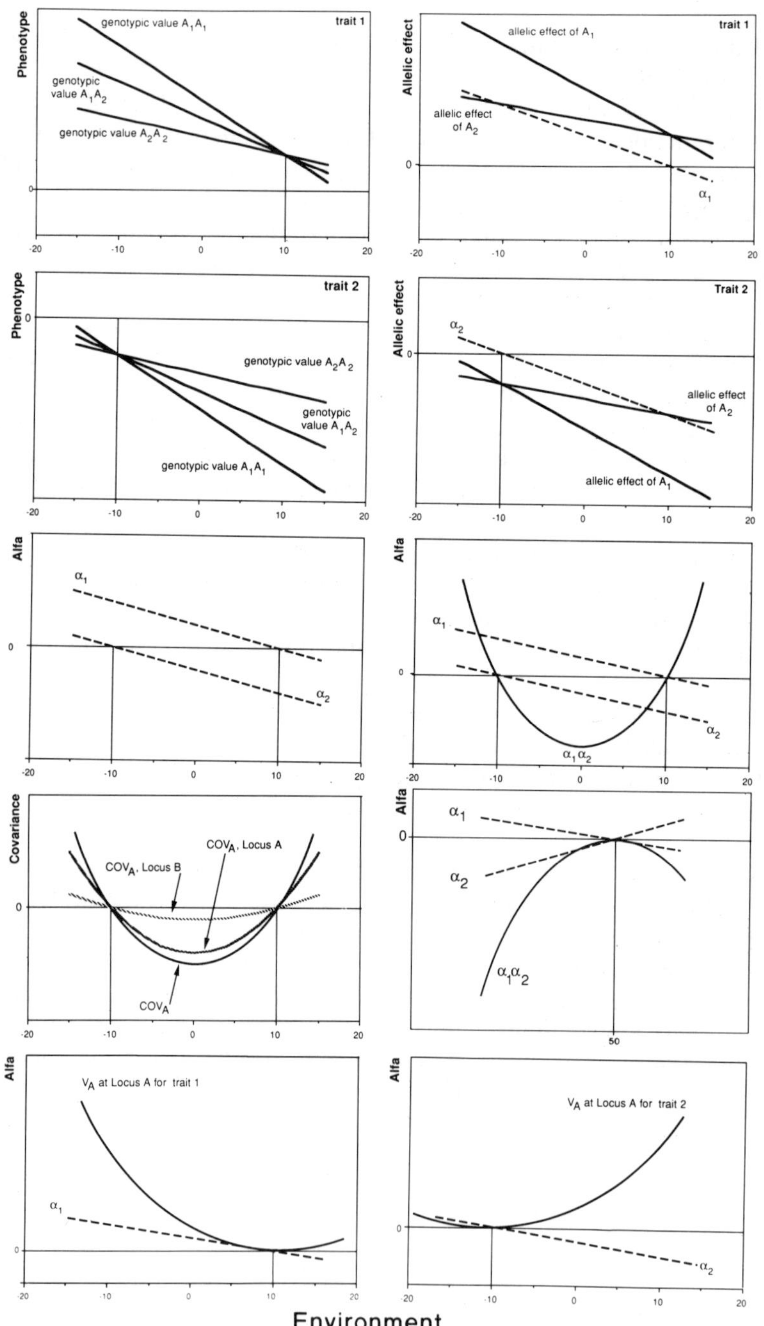
Phenotype
genotypic value A₁A₁
trait 1
genotypic value A₁A₂
genotypic value A₂A₂
0
-20
-10
0
10
20
Allelic effect
allelic effect of A₁
trait 1
allelic effect of A₂
0
α₁
-20
-10
0
10
20
Phenotype
trait 2
0
genotypic value A₂A₂
genotypic value A₁A₂
genotypic value A₁A₁
-20
-10
0
10
20
Allelic effect
Trait 2
α₂
0
allelic effect of A₂
allelic effect of A₁
-20
-10
0
10
20
Alfa
α₁
0
α₂
-20
-10
0
10
20
Alfa
α₁
0
α₂
α₁α₂
-20
-10
0
10
20
Covariance
COV_A, Locus B
COV_A, Locus A
0
COV_A
-20
-10
0
10
20
Alfa
α₁
0
α₂
α₁α₂
50
Alfa
V_A at Locus A for trait 1
α₁
0
-20
-10
0
10
20
Alfa
V_A at Locus A for trait 2
0
α₂
-20
-10
0
10
20
Environment

Figure 4. One locus: linear genotypic values for two traits and their relation to the genetic variances and covariance.

(a, upper left) —Genotypic values for trait 1 as a function of an environmental variable.

(b, upper right) —Allelic effects of alleles A_1 and A_2 and "the average effect of a gene substitution" α_1, for trait 1. The genotypic values in Fig. 4a are found by adding the allelic effects for trait 1; "the average effect of a gene substitution" α_1, is found by subtracting them.

(c, second left) —Genotypic values for trait 2 as a function of an environmental variable.

(d, second right) —Allelic effects of alleles A_1 and A_2 and "the average effect of a gene substitution" α_2, for trait 2. The genotypic values in Fig. 4c are found by adding the allelic effects for trait 2; "the average effect of a gene substitution" α_2, is found by subtracting them.

(e, third left) —The two "average effects of a gene substitution", α_1 for trait 1 and α_2 for trait 2.

(f, third right) —If the "average effects of a gene substitution" α_1 and α_2 do not change sign simultaneously, then their product $\alpha_1\alpha_2$ changes sign each time one of them changes sign.

(g, fourth left) —Excursion to two loci, A and B. The additive genetic covariance as a sum over the per locus contributions. (Here the sign changes for the α's coincide for each trait; the sign change of the total additive genetic covariance occurs at the same environmental value for all gene frequencies and linkage disquilibria).

(h, fourth right) —Another locus. Here the product $\alpha_1\alpha_2$ does not change sign as the "average effects of a gene substition" α_1 and α_2 for the two traits both change sign simultaneously, at the same value of the environmental variable.

(i, lower left) —Locus A again. The "average effect of a gene substitution" α_1 (the same as in Fig. 4b, e on different scaling) and the additive genetic variance $2pq\,\alpha^2_1$ for trait 1 due to genetic variation at locus A. Note that the minimum of the additive genetic variance for trait 1 at locus A and one of the roots of the additive genetic covariance between traits 1 and 2 at locus A coincide, at the environmental value of $x = 10$.

(j, lower right) —Locus A again. The "average effect of a gene substitution" α_2 (the same as in Fig. 4d, e on different scaling) and the additive genetic variance $2pq\,\alpha^2_2$ for trait 2 due to genetic variation at locus A. Note that the minimum of the additive genetic variance for trait 2 at locus A and one of the roots of the additive genetic covariance between traits 1 and 2 at locus A coincide, at the environmental value of $x = -10$.

Box 4: From Single Loci to Polygenes: Changes in Additive Genetic Variance and Covariance

If there is genotype by environment interaction involving linear reaction norms at one locus, the additive genetic variance and covariance are quadratic functions of the environment (Box 3). The same holds for polygenic traits determined by large numbers of loci if at least one locus behaves like the one described in Box 3.

Consider the distribution of average effects over 20 loci (A, B, C . . .) that each influence both trait 1 and trait 2[25]. Then there are 20 average effects for trait 1 ($\alpha_{1,A}$, $\alpha_{1,B}$, $\alpha_{1,C}$ . . .) and 20 for trait 2 ($\alpha_{2,A}$, $\alpha_{2,B}$, $\alpha_{2,C}$ . . .). For each trait the 20 average effects form a distribution with a mean and a variance, and there might be a covariance between them. If the distributions of the average effects for the two traits are independent of each other, the sign of the additive genetic covariance will change over environments (Fig. 5a). If the covariance between α_1 and α_2 is strong enough, it can prevent the sign change (Fig. 5b).

If genetic constraints exist, they have their origin not in the genetic covariance, but in the covariance of the average effects. Covariance in average effects implies strong similarity in the effects of each locus on the two traits. A biological mechanism that would produce such similar effects is a physiological tradeoff. This suggests that any causes of constraint on the response to selection should be sought in the biological realities of the epigenetic system.

718

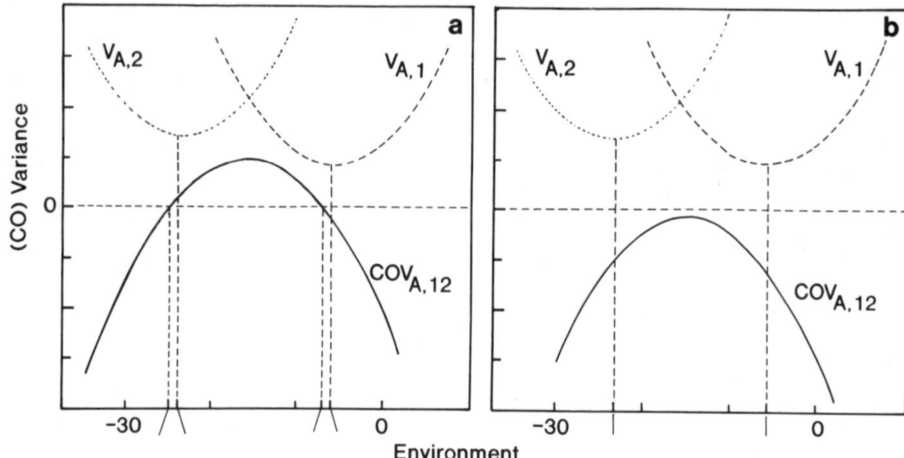

Figure 5. Many loci, each influencing both trait 1 and trit 2; for each locus, there exists an "average effect of a gene substitution" for both traits. For detail, see ref. 25.
a. The "average effects of a gene substitution" over loci are independently distributed for the two traits.
b. The "average effects of a gene substitution" over loci show a tight negative covariance between the two traits.

LITERATURE CITED

1. Mayr, E. (1982). *The Growth of Biological Thought.* Belknap Press.
2. Mayr, E. (1988). *Towards a New Philosophy of Biology.* Belknap Press.
3. Futuyma, D. (1986). *Evolutionary Biology,* 2nd Ed. Sinauer.
4. Bradshaw, A. D. (1965). *Adv. Genet.* 13: 115–155.
5. Schlichting, C. D. (1986). *Ann. Rev. Ecol. Syst.* 17: 667–693.
6. Sultan, S. E. (1987). *Evolutionary Biology* 21: 127–178.
7. Berven, K. A., Gill, D. E. & S. J. Smith-Gill. (1979). *Evolution* 33: 609–623.
8. Pani, S. N. & J. F. Lasley. (1972). *Miss. Agr. Res. Stn. Res. Bull.* 992.
9. Comstock, R. E. & R. H. Moll. (1963). Pp. 164–196. *In:* Hanson, W. D. & H. F. Robinson (eds.), *Statistical Genetics and Plant Breeding.* Washington: National Academy of Sciences.
10. Falconer, D. S. (1989). *Introduction to Quantitative Genetics.* 3rd Edition. Longman.
11. Yamada, Y. (1962). *Jap. Jour. Genet.* 37: 498–509.
12. Freeman, G. H. (1973). *Heredity* 31: 339–354.
13. Falconer, D. S. (1952). *Amer. Natur.* 86: 293–298.
14. Via, S. (1984). *Evolution* 38: 896–905.
15. Berven, K. A. & D. E. Gill. (1983). *Amer. Zool.* 23: 85–97.
16. Woltereck, R. (1909). *Verhlgn. Deutschen Zool. Gesellschaft* 19: 110–172.
17. Gebhardt, M. D. & S. C. Stearns. (1988). *J. Evol. Biol.* 1: 335–354.
18. Via, S. & R. Lande. (1985). *Evolution* 39: 505–522.
19. Lande, R. (1979). *Evolution* 33: 402–416.
20. Dobzhansky, Th. (1937). *The Genetical Theory of Natural Selection.* Columbia University Press.
21. Schmalhausen, I. I. (1949). *Factors of Evolution: The Theory of Stabilizing Selection.* Blakiston Press.
22. Scheiner, S. M. & C. J. Goodnight. (1984). *Evolution* 38: 845–855.
23. Scheiner, S. M. & R. F. Lyman. (1989) *J. Evol. Biol.* 2: 95–107.
24. Jong, G. de. (1990a). *Genetica* 81:171–177.
25. Jong, G. de. (1990b). *J. Evol. Biol.* 3:447–468.
26. Riska, B. (1986). *Evolution* 40: 1303–1311.
27. Noordwijk, A. J. van & G. de Jong. (1986). *Amer. Natur.* 128: 137–142.
28. Murphy, P. A., Giesel, J. T., & M. N. Manlove. (1983). *Evolution* 37: 1181–1192.
29. Service, P. M. & M. R. Rose. (1985). *Evolution* 39: 943–945.
30. Via, S. & R. Lande. (1987). *Genet. Res.* 49: 147–156.
31. Gillespie, J. H. & M. Turelli. (1989). *Genetics* 121: 129–138.
32. Ludwig, W. (1950). *Neue Erg. Probl. Zool. (Klatt-Festschrift)* 516.
33. Levene, H. (1953). *Amer. Natur.* 116: 463–479.

Introduction to the Symposium

Bruce Riska

Due to a tragic accident, Bruce Riska was unable to finish his manuscript for this symposium volume. We (Barry Sinervo and Tim Mousseau) wrote the following contribution for Bruce based on his unfinished manuscript and the talk he presented at ICSEB IV. Any errors, omissions, and inconsistencies are the sole responsibility of B. S. and T. M. We would also like to dedicate this symposium to Bruce, from whose inspiration and insights we have benefited tremendously.

We are accustomed to thinking of the effects of genes on the individual in which those genes occur: the phenotype is the expression of an individual's genotype, and adaptive evolution is possible because an individual's genetic and phenotypic values are positively correlated. This is sometimes an oversimplification. Many different kinds of interactions can occur among related individuals, in which genes present in one individual can affect the phenotype of another, and the correlation between an individual's genetic and phenotypic values can be negative (Cheverud, 1984a; Lynch, 1987). The most common example of inter-individual genetic effects are maternal effects found in many animals and plants, such as body size in mice (Dickerson, 1947; Falconer, 1965; Cheverud, 1984a; Riska et al., 1985; Cowley et al., 1989), offspring size in fish (Reznick, 1981, 1982) and lizards (Sinervo, 1990), seed size and dormancy in plants (Mazer, 1987a,b; Roach & Wulff, 1987; Schaal, 1984), development time, dormancy and body size in insects (Mousseau & Dingle, 1991; Janssen et al., 1988; Rossiter, 1991), and oviposition site choice in reptiles with its associated effects on environmental sex determination (Bull et al., 1982). These maternal effects may be heritable in the mother but are experienced as environmental effects by the offspring.

Maternal effects are characteristic complications of inheritance in mammals (Falconer, 1965, 1981; Eisen, 1970; Riska et al., 1985; Cheverud, 1984a). Although typically strongest during prenatal and pre-weaning postnatal growth, when they may be the largest source of phenotypic variance in body size, maternal effects can also persist to maturity (Cowley et al., 1989). Part of the maternal effect on body size in laboratory mice is negatively correlated with litter size: large mice have large litters that are composed of small mice, which later produce small litters composed of large mice (Falconer, 1965; Eisen, 1970). Since body size and litter size are correlated with many life history traits, maternal effects could be very important sources of variance in fitness. Although biologists sometimes view maternal effects as environmental "noise" to be eliminated by experimental design, maternal effects can be heritable and have been studied in both laboratory and agricultural contexts precisely because of their confusing, constraining effects on selection response

Please address all correspondence to: B. Sinervo, Department of Integrative Biology, University of California, Berkeley, CA 94720, USA.

720

(Dickerson, 1947; Falconer, 1965; Hanrahan, 1976; Cheverud, 1984b). At least three different models, with different assumptions and predictions, have been used to interpret maternal effects on body size and other traits in mammals. Despite their potential importance, these models and their consequences are generally unknown to evolutionary biologists working on maternally influenced traits in mammals and other organisms.

The importance of heritable maternal effects in evolution arises from several properties. First, applied and theoretical work both show that traits influenced by maternal performance can show continued negative or cyclical responses to selection, delayed response to selection after selection stops, and other unusual behaviors (e.g., Kirkpatrick & Lande, 1989). This is partly because of a lag of one or more generations in the expression of genes' effects through maternal performance, so that individuals showing the effects of a particular gene need not possess that gene. Also, if the individual does possess that gene, the gene may affect the individual's phenotype directly, through its effects on growth and development, as well as indirectly through the maternal effect, and these two sources of genetic variance may be pleiotropically correlated, and the correlation can be negative (Dickerson, 1947; Willham, 1972). Second, maternal effects are often strongest in the early stages of development and growth (Reznick, 1981; Schaal, 1984; Sinervo & McEdward, 1988), such as prenatal and early postnatal growth in mammals (Monteiro & Falconer, 1966). They are thus most important during the periods of growth and development that are often thought to contribute most to major evolutionary change (Gould, 1977; Raff & Kaufman, 1983), and might thus make disproportionately large contributions to macroevolution (Martin, 1981; Martin & MacLarnon, 1985; Sinervo & McEdward, 1988). Divergence of body size among mammals appears to occur chiefly by differences in prenatal growth, possibly as a result of differences in maternal investment (Riska, 1989). The earliest stages of life are, in some organisms, the most subject to selection, and resources supplied by the mother can be crucial for survival (Schaal, 1984).

A symposium on maternal effects in evolutionary biology is appropriate at this time, as more researchers discover the importance of maternal influences in many different systems. In addition to recent theoretical developments (Kirkpatrick & Lande, 1989), this is indicated by studies of mammals (Cheverud, 1984a; Cowley et al., 1989, this volume), fish (Reznick, 1981, this volume), birds (Lande & Price, 1989), lizards (Sinervo, 1990, this volume; Sinervo & Huey, 1990), insects (Mousseau & Dingle, 1991, this volume; Janssen et al., 1988), sea urchins (Sinervo & McEdward, 1988), angiosperms (Schaal, 1984; Lacey, this volume), and amphibians (Kaplan, 1989, this volume). The broad taxonomic range of these and other studies is matched by the diversity of research questions, ranging from evolutionary trends in morphology or life history, to "epigenetic" effects of maternal influences on embryogenesis, to evolution of whole-organism performance traits such as sprint speed, to kin selection models and parent-offspring conflict. The theoretical framework for some of this work, however, is varied and confusing. Recent studies encompass those adopting models directly from animal breeding, those with new theoretical treatment, and those reporting experimental results with no reference to existing genetic models.

Animal breeding, the source of most maternal effects models currently used in evolutionary biology, has provided two main models (Falconer, 1965; Dickerson, 1947; Willham, 1972) whose differences, assumptions, and consequences are not widely appreciated, even within the animal breeding community. Willham (1963, 1972) presented a biometrical development of the Dickerson (1947) and related models incorporating a breeding value for maternal performance (Fig. 1a). This is the most widely used model in animal breeding and evolutionary biology. Persistent effects in this model are treated by using a breeding value for each generational comparison (e.g., "grandmaternal performance" trait). Falconer (1965) simplified these models to allow analysis via regression of daughter's body size on mother's, through litter-size effects. This was the first model to incorporate an infinite chain of diminishing effects through maternal ancestors (Fig. 1b). Extrapolations by Cheverud (1984a) and Lynch (1987) were used to investigate kin selection and related models. Aspects of both the Willham's and Falconer's approaches were incorporated by Riska et al. (1985) to provide a more complete model. This model included both genetic inheritance and "persistent environmental" effects (Fig. 1c). Also, a very general class of models including all of the above has been presented with a general analytical methodology aimed at understanding evolutionary consequences of the situations represented by the different models (Kirkpatrick & Lande, 1989). The field is thus developing rapidly and needs a sorting-out of the somewhat confusing and complicated models.

One aim of some recent theoretical work (Kirkpatrick & Lande, 1989) appears to be the incorporation of maternal effects into multiple-trait analyses of selection (Lande & Arnold, 1983), allowing for cross-generational analysis of selection gradients on correlated traits. In so doing, the maternal effect is considered in terms of specific measurable traits in the mother, rather than the tradition all-inclusive "maternal performance" phenotype, which can only be estimated indirectly, from covariances among relatives. This subtle distinction has the advantage of allowing simpler analyses, but the disadvantage of perhaps missing parts of the maternal effect (Falconer, 1965).

The theoretical work also provides a framework for understanding the evolutionary significance of maternal effects in natural populations. Specifically, fitness attributes of an organism such as development time, survival, and reproduction depend on maternally-influenced aspects of the phenotype such as size and time to hatching, sprouting, weaning, or fledging. These and other maternally-influenced traits affecting offspring performance and ecological interactions are the mechanistic links of special importance for the evolution of the life history. Maternal effects can often be thought of as cross generational norms of reaction. Such interactions can limit response to selection and perhaps constrain evolutionary change; they can also provide the raw material for adaptation. Because selection response with maternal effects can depend on traits in other generations, stabilizing selection and antagonistic pleiotropy could arise from interactions between traits in different generations, and attempts to measure selection on multiple traits could be confounded by selection on correlated traits in generations not measured.

How widespread and important are maternal effects and are they genetically heritable? Unfortunately, all the problems familiar to quantitative genetics are associated with the traditional measurement and treatments of maternal affects in that many families and offspring are required to estimate reliably the genetic variance and covariances. Alternatively, can we identify the specific measurable maternal traits influencing offspring phenotypes so that we can apply the theoretical models (Kirkpatrick & Lande, 1989; Lande & Arnold, 1983) being developed to an understanding of the role of maternal effects in the evolution of natural populations? Specifically, how often do the unusual dynamics suggested by the models (Riska et al., 1985; Kirkpatrick & Lande, 1989) arise in natural populations. To answer these questions, we must include maternal effects in theoretical and empirical studies. Fruitful approaches might include careful descriptive studies in combination with manipulative experiments to examine the causal relationships among them. This symposium represents a first attempt to assemble and coordinate contemporary views concerning the importance of maternal effects in evolution.

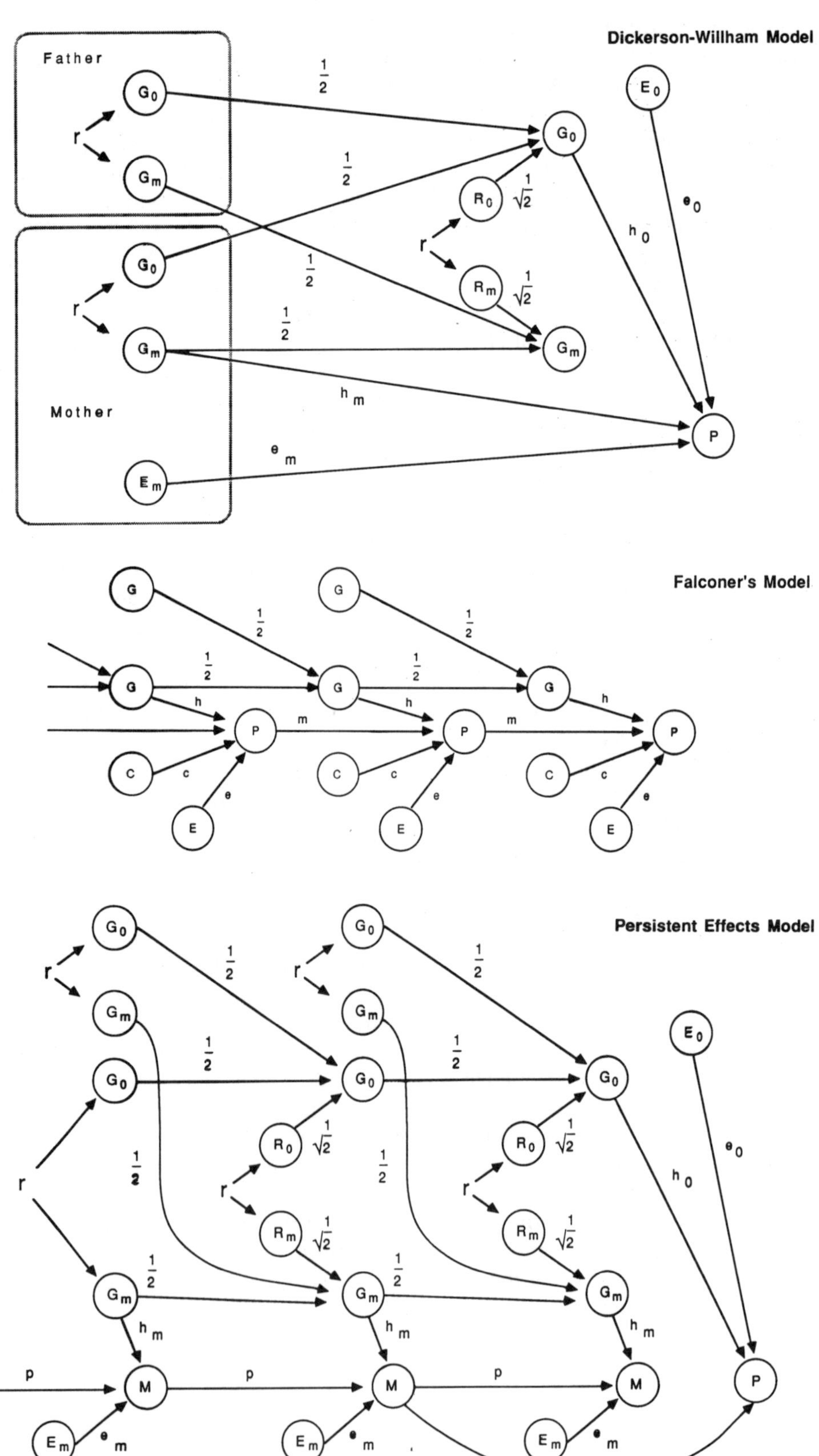
Dickerson-Willham Model
Father
Falconer's Model
Mother
Persistent Effects Model

Figure 1. Path diagrams for three quantitative genetic models of maternal effects (a) The Dickerson-Willham model (Dickerson, 1947; Willham, 1963, 1972). In this model, the phenotype of a character in the offspring, P, is a sum of the direct contribution of the individual's genotype, G_o, environmental effects, E_o, and genetic and environmental maternal effects, G_m and E_m, respectively. Rather than identifying individual characters in the mother that mediate the maternal effect, the overall maternal effect is visualized as stemming from a composite trait called "maternal performance." The genes for maternal performance and the trait under consideration can also be pleiotropically correlated through the two-headed arrows labelled "r." Mendelian inheritance of genetic factors is represented by the paths of 1/2. The paths of $2/\sqrt{2}$ arise because of correlated residuals corresponding to random assortment, accounting for the deviation of offspring genetic value from average of parents. The regression of additive genetic value on the phenotype is $B = h_0^2 + 3/2h_0rh_m + 1/2h_m^2$. (b) The maternal effect model of Falconer (1965). Here the phenotype, P, is a function of genetic (G), environmental (E), common environment (C), and maternal effects (m). Notice that the model differs from the Dickerson-Willham model because the loops that can be traced out for the maternal effect form an infinite series extending back into the maternal lineage. In this model, the regression of additive genetic value on the phenotype is $B = h^2(2+m)/(2-m)$. (c) The Riska persistent effects model (Riska et al., 1985), combining the Dickerson-Willham model and the Falconer model. Thus, the maternal phenotype, M, affects the offspring phenotype through the path labeled "m," and it also affects itself through the path "p." This model results in an infinite series as does the Falconer model in (b). Here, the regression of additive genetic value onto phenotype is $B = h_0^2 + 3/2h_0rh_m + 1/2h_m^2$, (as in the Dickerson-Willham model), $+ ph_m^2 + h_0rh_m[p/(2-p)]$.

ACKNOWLEDGMENTS

We (Barry Sinervo and Tim Mousseau) would like to thank D. Cowley for comments on earlier versions of this manuscript, and for writing the figure legends. We would also like to thank M. Turelli for obtaining Bruce's notes on the manuscript.

LITERATURE CITED

Bull, J. J., Vogt, R. C., & M. G. Bulmer. 1982. Heritability of sex ratio in turtles with environmental sex determination. *Evolution* 36: 333–341.

Cheverud, J. M. 1984a. Evolution by kin selection: a quantitative genetic model illustraded by maternal performance in mice. *Evolution* 38: 766–777.

Cheverud, J. M. 1984b. Quantitative genetics and developmental constraints on evolution by selection. *Journal of Theoretical Biology* 110: 155–171.

Cowley, D. E., Pomp, D., Atchley, W. R., Eisen, E. J., & D. Hawkins-Brown. 1989. The impact of maternal uterine genotype on postnatal growth and adult body size in mice. *Genetics* 122: 193–203.

Dickerson, G. E. 1947. Composition of hog carcasses as influenced by heritable differences in rate and economy of gain. *Iowa Agricultural Experimental Station Research Bulletin* 354: 492–524.

Eisen, E. J. 1970. Maternal effects on litter size in mice. *Canadian Journal of Genetics and Cytology* 12: 209–216.

Falconer, D. S. 1965. Maternal effects and selection response. Pp. 763–774. *In:* S. J. Geerts (ed.), *Genetics Today, Proceedings of the XI International Congress of Genetics,* Vol. 3. Pergamon Press: Oxford.

Falconer, D. S. 1981. *Introduction to Quantitative Genetics,* 2nd ed. Longman: New York. 340 pp.

Gould, S. J. 1977. *Ontogeny and Phylogeny.* Harvard University Press: Cambridge, MA. 501 pp.

Hanarahan, J. P. 1976. Maternal effects and selection response with an application to sheep data. *Anamal Production* 22: 359–369.

Janssen, G. M., de Jong, G., Joosse, E. N. G., & W. Scharloo. 1988. A negative maternal effect in springtails. *Evolution* 42:828–834.

Kaplan, R. H. 1989. Ovum size plasticity and maternal effects on the early development of the frog, *Bombina orientalis* Boulenger, in a field population in Korea. *Functional Ecology* 3: 597–604.

Kirkpatrick, M., & R. Lande. 1989. The evolution of maternal characters. *Evolution* 43: 485–503.

Lande, R. & S. J. Arnold. 1983. The measurement of selection on correlated characters. *Evolution* 37: 1210–1226.

Lande, R. & T. Price. 1989. Genetic correlations and maternal effects coefficients obtained from parent-offspring regression. *Genetics* 122: 915–922.

Lynch, M. 1987. Evolution of intrafamilial interactions. *Proceedings of the National Academy of Science U.S.A.* 84: 8507–8511.

Martin, R. D. 1981. Relative brain size and basal metabolic rate in terrestrial vertebrates. *Nature* 293: 57–60.

Martin, R. D. & A. M. MacLarnon. 1985. Gestation period, neonatal size, and maternal investment in placental mammals. *Nature* 313: 220–223.

Mazer, S. J. 1987a. Parental effects on seed development and seed yield in *Raphanus raphanistrum*: implications for natural and sexual selection. *Evolution* 41: 355–371.

Mazer, S. J. 1987b. The quantitative genetics of life-history characters in *Raphanus raphanistrum*: ecological and evolutionary consequences of seed weight variation. *American Naturalist* 130: 891–914.

Monteiro, L. S. & D. S. Falconer. 1966. Compensatory growth and sexual maturity in mice. *Animal Production* 8:179–192.

Mousseau, T. A. & H. Dingle. 1991. Maternal effects in insect life histories. *Annual Review of Entomology* 36: 511–534.

Raff, R. A. & T. C. Kaufman. 1983. *Embryos, Genes, and Evolution.* Macmillan:N.Y. 395 pp.

Reznick, D. 1981. "Grandfather Effects": the genetics of interpopulational differences in offspring size in the mosquito fish. *Evolution* 35: 941–953.

Reznick, D. 1982. Genetic determination of offspring size in the guppy (*Poecilia reticulata*). *American Naturalist* 120: 181–188.

Riska, B. 1989. Composite traits, selection response, and evolution. *Evolution* 43: 1172–1185.

Riska, B., Rutledge, J. J., & W. R. Atchley. 1985. Covariance between direct and maternal effects in mice, with a model of persistent environmental influences. *Genetical Research* 45: 287–297.

Roach, D. A. & R. D. Wulff. 1987. Maternal effects in plants. *Annual Review of Ecology and Systematics* 18: 209–236.

Rossiter, M. 1991. Maternal effects generate variation in life history: consequences of egg size plasticity in the gypsy moth. *Functional Ecology.* In press.

Schaal, B. A. 1984. Life history variation, natural selection, and maternal effects in plant populations. Pp. 188–206. *In:* R. Dirzo & J. Sarukhan (eds.), *Perspectives in Plant Population Ecology.* Sinauer:Sunderland, MA.

Sinervo, B. & L. R. McEdward. 1988. Developmental consequences of an evolutionary change in egg size: an experimental test. *Evolution* 42: 885–899.

Sinervo, B. 1990. The evolution of maternal investment in lizards: an experimental and comparative analysis of egg size and its effects on offspring performance. *Evolution* 44: 279–294.

Sinervo, B. & R. B. Huey. 1990. Allometric engineering: an experimental test of the causes of interpopulational differences in performance. *Science* 248: 1106–1109.

Willham, R. L. 1963. The covariance between relatives for characters composed of components contributed by related individuals. *Biometrics* 19: 18–27.

Willham, R. L. 1972. The role of maternal effects in animal breeding. III. Biometrical aspects of maternal effects in animals. *Journal of Animal Science* 35: 1288–1293.

Experimental and Comparative Analyses of Egg Size in Lizards: Constraints on the Adaptive Evolution of Maternal Investment per Offspring

Barry Sinervo

Abstract. Elucidating the mechanistic links between maternal effects and fitness is fundamental to an understanding of the evolution of maternal effects. For animals that lay eggs, the amount of yolk allocated to each egg is an ideal maternal effect to focus on in an analysis of the selective consequences of these traits in natural populations. Egg size is related to fitness through its effects on fecundity—the tradeoff between egg size and egg number is a common theme underlying life history variation in many groups of egg-laying organisms. Yolk mass is also likely to have large and measurable effects on offspring survival in nature through its effects on post-hatching yolk reserves and hatchling size.

By experimentally manipulating yolk volume independent of maternal phenotype, it is possible to disentangle the potentially complex causal network of interactions (i.e., correlated maternal effects on other maternal and offspring traits) that govern the evolution of this maternal effect in natural populations. For example, in a complementary set of experimental manipulations clutch size was increased by injections of follicle stimulating hormone, and clutch size was decreased by surgically ablating follicles (removing yolk but otherwise leaving the follicle intact) from the ovary of vitellogenic females, i.e., yolkectomy). Egg size either decreased with increased clutch size or egg size increased with decreased clutch size. These experiments indicate that follicle stimulating hormone is involved in the proximate regulation of clutch size and, in a cascading fashion, egg size. This mechanistic coupling between the regulation of clutch size and egg size that is reflected in the pervasive tradeoff between these traits is a rigid constraint on adaptive evolution. The functional limits on the maximum egg size that is possible in lizards is also discussed.

In addition, I have used two experimental manipulations of yolk mass that permit an analysis of the fitness effects arising from maternal investment. Female side-blotched lizards (*Uta stansburiana*) from the inner Coast Range of California produce large clutches of small eggs early in the season and small clutches of large eggs later in the reproductive season. The fitness consequences arising from the seasonal change in maternal investment were investigated by releasing miniaturized and gigantized hatchlings as well as normal-sized control hatchlings into two study populations located in the Coast Range of California. Experimentally miniaturized eggs and hatchlings can be produced by removing some of the yolk from freshly laid lizard eggs (aspirating yolk with a syringe). Gigantized eggs and hatchlings can be produced by surgically ablating follicles (yolkectomy) from the ovary of vitellogenic females. Eggs are gigantized because the intact follicles remaining in the ovary incorporate the yolk that would normally be allocated to the entire clutch of eggs.

Results from an analysis of one month survivorship of size-manipulated and control neonates released into nature indicated that the seasonal pattern of reproductive allocation in *U. stansburiana* is adaptive—natural selection early in the season favored an optimal egg size that was smaller than the optimal egg size favored later in the season. Moreover, these results also indicate that natural selection acting later in the season favors an optimum egg size that was very close to the functional limits on the maximum size of an egg that can be laid by female side-blotched lizards. Thus, the experimental manipulations provide insights into the mechanistic causes of the underlying the tradeoff between egg size and clutch size, of constraints on maximum egg size, and of the dynamics of natural selection on these traits.

Dr. Sinervo is with the Department of Integrative Biology, University of California, Berkeley, CA 94720, USA.

INTRODUCTION

An understanding of maternal effects is essential to life history theory because of their influences on early development and morphogenesis and cascading effects on the juvenile and adult phases of the life history (Riska, this volume). Moreover, they are significant for more general aspects of evolutionary biology for two reasons. First, they confound our interpretation of variation in physiological, morphological, and life historical traits among geographically distinct populations as well as our analyses of the quantitative genetic bases underlying variation in the traits among individuals within a population. Second, and more importantly, the traits underlying maternal effects may themselves be under direct or indirect selection and thus are an important aspect of the evolving phenotype (Kirkpatrick & Lande, 1989). Offspring size is an example of a pervasive maternal effect that is likely to have considerable ramifications for many aspects of the organismal phenotype (Lack, 1947, 1954).

Egg size, like all maternal effects, is a unique life history trait in that it is simultaneously a maternal and an offspring trait. However, this coupling between maternal and offspring traits confounds genetic and environmentally-induced sources of variation of the maternal phenotype with the offspring phenotype. This presents practical problems in the analysis of the quantitative genetic aspects associated with the evolution of the maternal effects (Kirkpatrick & Lande, 1989; Riska, 1990; Riska et al., 1985). One possibility for addressing these effects would be the route of classical quantitative genetics (Riska, 1990). However, given the difficulty of generating the half-sib designs necessary for a quantitative genetic analysis of maternal effects (Falconer, 1981), it would be useful if an alternative, less time consuming methodology could be brought to bear on the problem. An experimental manipulation of egg size would permit an analysis of the effects of offspring size independent of maternal phenotype (Sinervo, 1990a; Sinervo & Huey, 1990; Sinervo & McEdward, 1988). Moreover, such an experimental manipulation would provide an analysis of the causal bases underlying these maternal effects that would complement future quantitative genetic analysis.

In this paper, I describe and synthesize experimental methodologies for unraveling issues involving the evolution of egg size in reptiles (Sinervo, 1990a; Sinervo & Huey, 1990; Sinervo & Licht, in press). A discussion of the evolution of this maternal effect would not be complete without simultaneously considering clutch size. Not only are egg size and clutch size key life history traits because of their close associations with fitness, but there may exist a strong negative correlation between these traits that is reflected in a fundamental tenet of life history theory—the tradeoff between egg size and egg number (Lack, 1947, 1954; Wilbur, 1977; Williams, 1966). I will first discuss experiments that elucidate the mechanistic bases underlying the physiological tradeoff between offspring size and fecundity. I will then summarize the consequences of this tradeoff for several physiological and morphological traits in hatchling lizards. In so doing, I will illustrate how it is possible to make inferences concerning the role of egg size in shaping trait evolution among geographically distinct populations of lizards. I will then integrate these mechanistic analyses of the physiological and hormonal regulation of egg size and the functional consequences for offspring performance with preliminary results from a study of the fitness consequences of egg size in natural populations. These experiments have been conducted on two species of iguanid lizards, the western fence lizard (*Sceloporus occidentalis*) and the side-blotched lizard (*Uta stansburiana*), that are found across a broad geographic range in western North America (Stebbins, 1985). Across each species' range, egg size and clutch size covary in a fashion that reflects the tradeoff between the two traits (Sinervo, 1990a; Sinervo & Licht, in press).

THE PHYSIOLOGICAL AND HORMONAL CONTROL OF EGG SIZE IN REPTILES AND ITS RELATIONSHIP TO THE EGG SIZE AND EGG NUMBER TRADEOFF

Implicit in many considerations of sets of traits which covary consistently among phylogenetically-related groups of organisms is that these traits are in some fashion constrained. The constraints underlying the egg size and egg number tradeoff seem to result from the relatively straightforward, easily explained constraints involving energetic or functional limitations imposed by the burden of a clutch on females (Bauwens & Thoen, 1981; Shine, 1980; Sinervo et al., in press; Vitt & Congdon, 1978) as well as other potential costs of reproduction (Reznick, 1985). Given these limitations on total clutch mass, it seems intuitively obvious that a female producing large offspring must produce fewer offspring than a female producing small offspring. Until recently, the mechanisms underlying the tradeoff between egg size and egg number have remained unclear. How are egg size and egg number regulated in reptiles? Knowledge of the mechanistic bases of such a fundamental tradeoff of life history theory would be useful in elucidating a more complete set of contraints on reproduction and specifically the maternal effects subsumed by the trait of egg size. For example, ultimately there must be constraints on the maximum size of an egg that can be laid by a lizard. The important question as it relates to life history theory is how often do these functional constraints limit the action of natural selection?

The pattern of covariation underlying the egg size and egg number tradeoff is typified by populations of the iguanid lizard *Uta stansburiana* that are found in the inner coast ranges of California (Sinervo & Licht, in press). In a northern population found just south of the San Francisco Bay area, average egg size is 0.42 g and average clutch size is 4 eggs (Corral Hollow Road). Egg size averages 0.35 g and average clutch size is 6 eggs in a population found 110 km further south located on Billy Wright Road (40 km w of Los Banos). Thus, populations producing large eggs produce small clutches and vice versa. Similarly, the egg size and egg number tradeoff is apparent among females within a population (Sinervo & Licht, in press).

In a complementary set of experiments (Sinervo & Licht, in press), we have shown that the egg size and egg number tradeoff is regulated by a relatively simple hormonal and physiological interaction. By experimentally supplementing vitellogenic females from the large-egged, small-clutched northern populations with exogenous follicle stimulating hormone (FSH) we increased the number of eggs from 4 to 6 eggs per clutch. Moreover, we observed a concomitant drop in average egg size of these northern females from 0.42 g to 0.36 g. In a second series of experiments we surgically reduced the number of eggs laid by females by palping the yolk out of vitellogenic follicles (yolkectomy). The number of eggs (2–3) laid by these females was decreased by the number of follicles removed during the yolkectomy procedure (2 follicles) and eggs (0.53 g) from these clutches were dramatically larger than eggs laid by either unmanipulated or sham-manipulated females (0.44 g, 4–5 eggs per clutch).

These complementary experiments indicate that egg number is regulated by critical levels of FSH during early vitellogenesis and that egg size is largely a function of the number of eggs recruited into the clutch. Thus, egg size is largely constrained by the regulation of clutch size, which is regulated by gonadotropin levels (FSH) —indirect early events which precede the major period of vitellogenesis and yolk incorporation that one would naively assume to be more directly related to average egg size in a clutch (Sinervo & Licht, in press). Results from these experimental manipulations provide the mechanistic basis underlying the pattern of covariation observed among populations and among females within populations of the lizard *Uta stansburiana* (Nussbaum, 1981; Sinervo & Licht, in press) and *Sceloporus occidentalis* (Sinervo, 1990a). Given the generality of this tradeoff among other amniote vertebrates (Ballinger, 1978) and the relatively conserved aspects of hormonal regulation among vertebrates (Licht et al., 1977) these results are likely to be generalizable to other vertebrate life histories.

What about constraints on maximum egg size? Females that naturally lay clutches consisting of two eggs produce eggs that average 0.53 g. This two-egged phenotype occurs with low frequency in these populations of *U. stansburiana*. In a second series of experiments, we produced one-, two- and three-egged clutches using the yolkectomy protocol (Sinervo & Licht, in prep.). If egg size was reaching some sort of functional limit we would expect egg size to be comparable across this range of clutch sizes. If egg size was not constrained we would expect that egg size would still increase with decreasing clutch size (as it does over the natural range of variation—typically 3–9 eggs on the first clutch laid during the reproductive season and 2–8 eggs on subsequent clutches). Over the range of experimentally reduced clutch sizes, one to three eggs, we found no significant increase in egg size. Thus, egg size has apparently reached a plateau at $\sim$0.53 g.

EGG SIZE AND ITS CONFOUNDING EFFECTS ON THE ANALYSIS OF POPULATION DIFFERENTIATION

Given the pervasiveness of clutch size variation among geographically distinct populations of reptiles (Fitch, 1985) and the negative correlation between egg size and clutch size it is reasonable to assume that many species of lizards vary in egg size and thus hatchling size. How does the variation in clutch size and egg size affect offspring performance traits that are likely to be the mechanistic bases of differences in fitness (Arnold, 1983, 1987)? Clutch size and egg size covary among populations of the western fence lizard, *Sceloporus occidentalis*. Northern races from Washington and Oregon have relatively small eggs and large clutches compared to southern races from California. Many other physiological and morphological traits also differ between the northern and southern races of *S. occidentalis*. Notably, populations representative of the southernmost races have higher growth rates (Sinervo, 1990b), longer limbs (Sinervo et al., in press; Sinervo & Huey, 1990), higher thermal preference as juveniles (Sinervo, 1990b), greater burst speed and endurance (Sinervo, 1990b; Sinervo & Huey, 1990; Tsuji et al., 1989; van Berkum et al., 1989) than their northern counterparts.

Direct comparison of juveniles from northern and southern parts of the species distribution with regards to any of these performance traits would be confounded by the dramatic difference in egg and hatchling size found among populations. By removing yolk from freshly laid eggs it is possible to miniaturize the larger southern hatchlings to the same size as the northern hatchlings and compare various physiological and morphological traits without the confounding effects of the difference in hatchling size (Sinervo, 1990a; Sinervo & Huey, 1990).

For example, hatchling lizards from northern populations sprint more slowly and have lower endurance than hatchling lizards from southern populations. Are these differences in locomotor performance simply an allometric consequence of the difference in egg size? By comparing miniaturized southern hatchlings with naturally smaller northern hatchlings we found that all of the differences in sprint speed were an allometric consequence of the size difference. However, only some of the difference in endurance was attributable to size; the majority was attributable to other, as yet, undetermined physiological differences. In a similar fashion, I have shown that the differences in growth rate among populations (Sinervo, 1990a) seem to be related to differences in the thermal physiology of growth (Sinervo, 1990b) and not directly to the differences in hatchling size. Presumably the concordant physiological changes in endurance, growth rate and thermal physiology found among populations of *S. occidentalis* are indicative of evolved changes in the suite of correlated physiological traits related to energy metabolism and utilization.

Based on our experimental and comparative population analyses, it appears that traditional statistical analyses (ANCOVA) of purely comparative data may present a misleading portrait of the role of offspring size in the evolutionary divergence of allometric traits among populations (Sinervo & Huey, 1990). The discrepancies between the experimental and purely comparative approaches are likely due to the complex nature of size-related traits. In some cases most of the variation in the traits is attributable to size (e.g., sprint speed) in other cases, other as yet unidentified traits correlated with size (metabolism, etc.) are likely to influence the allometric trait of interest (e.g., stamina). Nevertheless, the experimental manipulation of offspring size by removal of yolk permits a causal basis to be ascribed to the maternal effects associated with yolk mass that is not possible with the purely comparative approaches. Moreover, the complementary experimental manipulation of offspring size, gigantization via yolkectomy (Sinervo & Licht, in press, summarized above), is also possible in *S. occidentalis* (Sinervo & Licht, unpublished data). The experimental manipulations, when compared to results from comparative analyses, indicate that evolutionary divergence in some offspring traits functionally related to size (e.g., sprint performance) may be more constrained relative to traits that are determined by other aspects of development and physiology (e.g., stamina and growth rate). Thus, these results provide a basis for interpreting variation in the suite of traits involved in the evolution of the tradeoff between egg size and number—adaptive changes governed in a large part by natural selection on offspring size.

NATURAL SELECTION ON EGG SIZE: INFERENCES FROM EXPERIMENTAL MEASUREMENTS

Natural selection on egg size presumably acts in part through the correlations between this trait (Sinervo, 1990a; Sinervo & Huey, 1990) and performance traits more directly related to fitness (Arnold, 1983). Egg size can also affect fitness through its effects on "hatchling quality" that are largely determined by yolk composition and volume (Congdon, 1989; Sinervo, 1990a). Regardless of the precise nature of the cascading effects and causal interactions of egg size with other fitness traits such as clutch size, hatchling performance and survival, an experimental manipulation of yolk mass and measurements of its effects on fitness would subsume all of the direct and indirect effects (Arnold & Wade, 1984; Lande & Arnold, 1983) into the resultant selective force acting on egg size. This greatly simplifies the prediction of the direction of trait evolution under various selection regimes (Arnold, 1987).

The experimental manipulation can be thought of as a controlled environmental manipulation of the maternal effect yolk volume. The experimental manipulation leaves no genetically correlated effects passed by inheritance to offspring but does carry over as potential non-genetic (i.e., controlled environmental effect) maternal effects that may persist from generation to generation (Riska et al., 1985). Presumably, the experimental manipulation of yolk volume would have the same effects on hatchling size and quality that genetic variation in the control of egg size might have. However experimental manipulation does not have any of the pleiotropic affects on other maternal or offspring traits that a genetic effect would have. For example, a genetic change in genes controlling the hormonal regulation of clutch size (e.g., FSH, see above) and egg size might be correlated with other physiological traits associated with reproduction. Direct experimental manipulation of egg size has no such effects. By comparing unmanipulated and manipulated offspring, it is theoretically possible to disentangle these separate genetic and non-genetic maternal effects. However, the quantitative genetic analysis of the maternal effects necessarily requires a half-sib breeding design (Falconer, 1981). The appropriate experimental design would include the experimental manipulation of size nested within each clutch (i.e., split clutch design in which half of the offspring were unmanipulated and half were manipulated). Limitations on the quantitative genetic analysis of such traits dictate that an organism with a large clutch size (>10, such as *S. occidentalis*) and multiple clutches be used in such analyses.

At this point, I would like to raise an important caveat concerning the use of experimental manipulations in analyzing natural selection. If the population is evolving under directional or stabilizing selection it is likely that the experimentally-induced effects will introduce an important artifact with regards to the response to selection—the underlying genotypic effects of egg size would become unscrambled from the phenotypic effects that govern selection on these traits (see Riska et al., [1985]; and Kirkpatrick & Lande [1989] for a theoretical treatment of these effects). On the other hand, such an experimental manipulation may mimic naturally occurring environmentally-induced effects that potentially confound interpretations of levels of quantitative genetic variation found in natural populations (Riska et al., 1985). In either case, experimentally induced effects could serve as a useful probe if an unmanipulated reference population were used in conjunction with the experimentally manipulated population.

Such persistent environmentally-induced effects might result in a complicated approach to evolutionary equilibrium of the maternal effects traits under selection, depending on the precise nature of the maternal effect on offspring and maternal fitness traits (Kirkpatrick & Lande, 1989; Riska et al., 1985). Are such pathways affecting offspring size pervasive in natural populations of reptiles? An example of a naturally occurring, environmentally-induced pathway affecting the maternal traits associated with offspring size of reptiles is related to the interaction of two maternal effects. Oviposition site choice affects offspring size through the effects of egg hydration and the concomitant effects on yolk utilization (Ackerman et al., 1985; Gutzke et al., 1987; Packard & Packard, 1988; Tracy & Snell, 1985). Most reptile eggs take up some water from the substrate throughout the incubation period. If sufficient water is unavailable, all of the yolk that was allocated by the female to individual eggs is not used and juvenile lizards hatch at reduced size (Packard & Packard, 1988; Tracy & Snell, 1985). This interaction would uncouple yolk provisioning and egg size from hatchling size. Given the pervasiveness of this physiological route in reptiles, and the stochastic nature of proper egg hydration in natural populations, it is likely that this environmentally induced effect is widespread in reptiles. An experimental manipulation of yolk mass would greatly facilitate an investigation of the potential influence of this confounding effect on selection on these traits in natural populations.

NATURAL SELECTION ON EGG SIZE: A CASE STUDY

Uta stansburiana has a broad geographic range in western North America and is found from Baja California to Washington State. In the southern parts of its range it produces 3 to 5 clutches throughout the reproductive season, approximately one month apart (Parker & Pianka, 1975). Seasonal changes in egg size have been observed in the three subspecies of *U. stansburiana* that are found in different parts of the geographic range (Ferguson & Snell, 1986; Nussbaum, 1981; Sinervo & Licht, in press). Moreover, Ferguson and Fox (1984) found that larger hatchlings had higher seasonal survival rates, supporting the notion that selection on the seasonal change in egg size was adaptive. However, lack of adequate data on other components of fitness (most notably the tradeoff between egg size and number) prohibit an assessment of whether the pattern of egg size change reflects an optimum allocation of maternal investment per offspring as a function of season.

Recently, we (B. Sinervo, P. Doughty & R. B. Huey, in prep.) have measured natural selection on egg size in populations of the lizard *U. stansburiana* found in the inner coast range of California. Space limitations do not permit a complete description of the study and only a superficial treatment of the results will be given here. We obtained (1989) control, gigantized, and miniaturized eggs and hatchlings from lab-reared and wild-caught females (see Sinervo, 1990a; and Sinervo & Licht, in press for details of husbandry) and released neonates back on the same study plot from which we obtained the females.

An analysis of the one month survivorship of these control and experimentally miniaturized and gigantized hatchlings in nature indicated that directional selection favoring large hatchlings during the second and third clutch was more intense than on the first clutch in both populations studied (along Del Puerto Canyon Rd; and Billy Wright Road, Los Banos). In all clutches, fecundity selection favored females that laid small eggs because they laid more eggs (i.e., a negative correlation between egg size and number, see results described above and Sinervo and Licht, in press).

Combining the fitness effects across these two episodes of selection (i.e., fecundity selection favoring small eggs and survival selection favoring large hatchlings) results in optimizing selection on egg size that typically (3 out of 4 clutch/population analyses) favored a female with an intermediate egg size. That is, females with an intermediate egg size produced more surviving offspring at the one month census than females producing the smallest or the largest eggs. However, the optimum egg size on the first clutch ($\sim$0.31 g in both populations) was smaller than the optimum egg size on the second clutch ($\sim$0.49 g) in both populations studied. Moreover, the optimal egg size on the second clutch ($\sim$0.49 g) was very close to the constraint describing the maximum egg size that is possible in *U. stansburiana* (0.53 g; Sinervo & Licht, in press and Servo & Licht, in prep.). And indeed, some females laying the smallest clutches (2 eggs) produced as large an egg as is functionally possible (0.53 g) in this species.

Thus, the seasonal change in egg size is adaptive in that females are producing larger young later in the reproductive season in response to a harsher selective environment. However, natural selection appears to be bringing the populations close to physiological and morphological constraints that limit the evolution of this maternal effect. It will be interesting to compare results from these populations which have small eggs relative to populations found at the northernmost limit of *U. stansburiana* in the Coast Range of California (e.g., Corral Hollow) that have a larger egg size. Presumably these northern populations are closer to the functional constraints on egg size. Moreover, an analysis of selection on egg size over the long term as revealed from our on going studies will indicate whether the pattern of selection varies among years (and among populations). This detailed, long-term analysis of natural selection should indicate the degree to which the adaptive evolution of these traits is constrained (ultimate constraints) by the physiological and functional limitations governing clutch size and egg size (proximate mechanisms).

DISCUSSION

Experimental manipulation of clutch size and egg size provides insights into the proximate mechanisms governing the tradeoff between these two components of fitness. For example, the complementary experimental manipulations that increase clutch size via injections of follicle stimulating hormone or that decrease clutch size by removal of follicles (i.e., yolkectomy) both have concomitant effects on egg size. Eggs either decreased in size with increased clutch size or increase in size with decreased clutch size. These experiments indicate that follicle stimulating hormone is involved in the proximate regulation of clutch size and, in a cascading fashion, egg size. Moreover, the experimental data indicate that intraspecific and interspecific patterns of egg size and clutch size variation (Ballinger, 1978; Nussbaum, 1981; Sinervo, 1990a; Sinervo & Licht, in press) are indeed a rigid constraint on the adaptive evolution of these results. In addition, experimental reduction of clutch size to the one egg limit (Sinervo & Licht, in prep.) indicates that egg size may have reached a functional constraint ($\sim$0.53 g) in this species.

The fitness consequences of egg size and clutch size have been dealt with at great length in various theoretical treatments over the past two decades (Brockelman, 1975; Kaplan & Cooper, 1984; McGinley et al., 1987; Smith & Fretwell, 1974; Wilbur, 1977). Two basic premises reflected in the counterbalance of two selective forces are common to all models. Because juvenile size can influence survival in certain environments, the evolution of the tradeoff between clutch size and egg size is probably mediated by fecundity selection acting on females that favors large clutches of small eggs, and by survival selection as a function of size that favors larger offspring. Very few empirical data have been obtained to estimate the magnitude of the effects of offspring size on the fitness of individuals in natural populations of reptiles (however, see Ferguson & Fox, 1984). Indeed, adequate emperical data demonstrating that egg size of any reptile population is under stabilizing selection (i.e., optimizing selection) are currently unavailable. This requires fairly precise measurements of the relationships among fecundity, survival, and egg size. Based on our preliminary results (Sinervo, Doughty, & R. B. Huey, in prep.), experimental alterations of offspring size, either gigantization or miniaturization, greatly facilitate the measurement of selection. Such manipulations not only help verify links between these traits and fitness, but also increase the number of individuals in the tails of the frequency distribution for egg size (Endler, 1986; Mitchell-Olds & Shaw, 1987; Schluter, 1988; Sinervo, 1990a). This enhances one's ability to describe the true shape of the fitness function and facilitates the detection of stabilizing selection (Schluter, 1988). In our investigations, experimental manipulation of egg size (and clutch size) verified the shape of the relationship between egg size and fecundity (Sinervo & Licht, in prep.) as well as the shape of the relationship between egg size and offspring survival (Sinervo, Doughty, & R. B. Huey, in prep.).

The actual advantages of large offspring size will necessarily be complex because offspring size is genetically or functionally correlated with numerous offspring traits and capacities (e.g., development rate, locomotor performance, morphology, and feeding capability (Sinervo, 1990b; Sinervo & McEdward, 1988)). Any of these correlated traits—not just size per se—might be the focus of selection. The need for a manipulative approach can be demonstrated with a thought experiment. Imagine that a study demonstrates that large hatchlings are favored in certain environments. Such data cannot distinguish whether large hatchlings have a survival advantage simply because they are large, or because they inherited "high quality" genes from their parents. However, any correlation between size and fitness found in an experimental analysis reflects the causal impact of size and size alone (Sinervo, 1990a; Sinervo & McEdward, 1988).

Moreover, experimental manipulation of offspring size greatly simplifies the statistical analysis of natural selection because the detection of selection on offspring size in an experimental analysis is not confounded with selection on correlated traits as it would be in a comparative analysis of natural selection on egg size (Arnold & Wade, 1984; Lande & Arnold, 1983; Mitchell-Olds & Shaw, 1987; Schluter, 1988). Our preliminary experimental analyses (Sinervo, Doughty, & R. B. Huey, in prep.) indicate that natural selection on egg size in the lizard *Uta stansburiana* favors an optimum egg size, but that the optimum egg size shifts seasonally—females produce larger clutches later in the reproductive season because selection on offspring size in later clutches favors larger offspring relative to selection on offspring size on the first clutch of the reproductive season. Moreover, the optima observed on the later clutches are close to the functional constraints on egg size and some females in the population are producing two-egged clutches that are effectively at these functional limits. Thus, it appears that selection on egg size is constrained not only by the tradeoff between egg size and egg number involved in fecundity selection, but also by the functional constraints on the maximum egg size that is possible in this species of lizard.

These experimental methodologies allow for an investigation into the proximate mechanisms governing the regulation of egg size in reptiles. This mechanistic analysis not only identifies constraints on the adaptive evolution of the suite of traits involved in the tradeoff between egg size and egg number, but the cascading effects of this tradeoff on offspring performance, physiology, and morphology (Sinervo, 1990b; Sinervo & Huey, 1990; Sinervo & Licht, in press; Sinervo & McEdward, 1988). Moreover, these experimental approaches can also be applied to individuals in natural populations to gain an understanding into the dynamics of selection on these traits. An experimental analysis of egg size is possible in lizards as well as many other organisms (see Sinervo & McEdward and Sinervo, 1990a for a partial list of taxa in which offspring size manipulation is possible). Thus, this approach is likely to be generalizable to many other taxa in addition to squamates.

ACKNOWLEDGMENTS

I wish to thank Bruce Riska for engaging discussions on the role of maternal effects in evolution that have profoundly shaped my outlook on these issues. Bruce was the inspiration behind our (B. R. and B. S.) organization of this symposium and his guidance and advice helped me to carry it through to press. Unfortunately, a tragic accident prevented Bruce from seeing it to fruition in this symposium volume. We, the symposium participants, all recognize his outstanding contributions to the symposium. I also wish to thank Steve Adolph and Bob Kaplan for editorial comments on previous versions of this manuscript.

LITERATURE CITED

Ackerman, R. A, Seagrave, R. C., Dmi'el, R., & A. Ar. 1985. Water and heat exchange between parchment-shelled reptile eggs and their surroundings. *Copeia* 1985: 703–711.

Arnold, S. J. 1983. Morphology, performance and fitness. *American Zoologist* 23: 347–361.

Arnold, S. J. 1987. Genetic correlation and the evolution of physiology. Pp. 189–211. *In:* M. E. Feder, A. F. Bennett, W. W. Burggren & R. B. Huey (eds.), *New Directions in Ecological Physiology.* Cambridge University Press: Cambridge, MA.

Arnold, S. J. & M. J. Wade. 1984. On the measurement of natural and sexual selection: theory. *Evolution* 38: 709–719.

Ballinger, R. E. 1978. Variation in and evolution of clutch and litter size. Pp. 789–825. *In:* R. E. Jones (ed.), *The Vertebrate Ovary.* Plenum: New York.

Bauwens, D. & C. Thoen. 1981. Escape tactics and vulnerability to predation associated with reproduction in the lizard *Lacerta vivipera. Journal of Animal Ecology* 50: 733–743.

Brockelman, W. Y. 1975. Competition, the fitness of offspring, and optimal clutch size. *American Naturalist* 109: 677–699.

Congdon, J. D. 1989. Proximate and evolutionary constraints on energy relations of reptiles. *Physiological Zoology* 62: 356–373.

Endler, J. A. 1986. *Natural Selection in the Wild,* Princeton University Press: Princeton, N.J. 336 pp.

Falconer, D. S. 1981. *Introduction to Quantitative Genetics,* 2nd edition. Longman: Great Britain. 340 pp.

Ferguson, G. W. & S. F. Fox. 1984. Annual variation of survival advantage of large juvenile side-blotched lizards, *Uta stansburiana:* its causes and evolutionary significance. *Evolution* 38: 342–349.

Ferguson, G. W. & H. L. Snell. 1986. Endogenous control of seasonal change of egg, hatchling, and clutch size of the lizard *Sceloporus undulatus garmani. Herpetologica.* 42: 185–191.

Fitch, H. S. 1985. Variation in clutch and litter size in new world reptiles. *University of Kansas Museum of Natural History Miscellaneous Publication* No. 76.

734

Gutzke, W. H., Packard, G. C., Packard, M. J., & T. J. Boardman. 1987. Influence of the hydric and thermal environments on eggs and hatchlings of painted turtles (*Chrysemys picta*). *Herpetologica* 43:393–404.

Kaplan, R. H. & W. S. Cooper. 1984. The evolution of developmental plasticity in reproductive characteristics: an application of the adaptive coin flipping principle. *American Naturalist* 123: 393–410.

Kirkpatrick, M. & R. Lande. 1989. Selection, inheritance, and evolution of maternal characters. *Evolution* 43: 485–503.

Lack, D. 1947. The significance of clutch size. *Ibis* 89: 302–352.

Lack, D. 1954. *The Natural Regulation of Animal Numbers*, Clarendon Press: Oxford.

Lande, R. & S. J. Arnold. 1983. The measurement of selection on correlated characters. *Evolution* 37: 1210–1226.

Licht, P., Papkoff, H., Farmer, S. W., Muller, C. H., H. W. Tsui & D. Crews. 1977. Evolution of gonadotropin structure and function. *Recent Progress in Hormone Research* 33: 169–248.

McGinley, M. A., Temme, D. H. & M. A. Geber. 1987. Parental investment in offspring in variable environments: theoretical and empirical considerations. *American Naturalist* 130: 370–398.

Mitchell-Olds, T. & R. G. Shaw. 1987. Regression analysis of natural selection: statistical and biological interpretation. *Evolution* 41: 1149–1161.

Nussbaum, R. A. 1981. Seasonal shifts in clutch size and egg size in the side-blotched lizard, *Uta stansburiana* Baird and Girard. *Oecologia* 49: 8–13.

Packard, G. C. & M. J. Packard. 1988. The physiological ecology of reptilian eggs and embryos. Pp. 523–605. *In:* C. Gans & R. B. Huey (eds.), *Biology of the Reptilia*. A. R. Liss: New York.

Parker, W. S. & E. R. Pianka. 1975. Comparative ecology of populations of the lizard *Uta stansburiana*. *Copeia* 1975: 615–632.

Reznick, D. 1985. Costs of reproduction: an evaluation of the empirical evidence. *Oikos* 44: 257–267.

Riska, B., Rutledge, J. J. & W. R. Atchley. 1985. Covariance between direct and maternal genetic effects in mice, with a model of persistent environmental influences. *Genetical Research Cambridge* 45: 287–297.

Schluter, D. 1988. Estimating the form of natural selection on a quantitative trait. *Evolution* 42: 849–861.

Shine, R. 1980. 'Costs' of reproduction in reptiles. *Oecologia* 46: 92–100.

Sinervo, B. 1990a. The evolution of maternal investment in lizards: an experimental and comparative analysis of egg size and its effects on offspring performance. *Evolution* 44: 279–294.

Sinervo, B. 1990b. The evolution of thermal physiology and growth rate between populations of the western fence lizard (*Sceloporus occidentalis*). *Oecologia* 83: 228–237.

Sinervo, B., Hedges, R., & S. C. Adolph. Decreased sprint speed as a cost of reproduction in the lizard *Sceloporus occidentalis*: variation among populations. *Journal of Experimental Biology.* In press.

Sinervo, B. & R. B. Huey. 1990. Allometric engineering: an experimental test of the causes of interpopulational differences in locomotor performance. *Science* 248: 1106–1109.

Sinervo, B. & P. Licht. The physiological and hormonal control of clutch size, egg size, and egg shape in *Uta stansburiana*: Constraints on the evolution of lizard life histories. *Journal of Experimental Zoology.* In press.

Sinervo, B. & L. R. McEdward. 1988. Developmental consequences of an evolutionary change in egg size: an experimental test. *Evolution* 42: 885–899.

Smith, C. C. & S. D. Fretwell. 1974. The optimal balance between size and number of offspring. *American Naturalist* 108: 499–506.

Stebbins, R. C. 1985. *A Field Guide to Western Reptiles and Amphibians*, 2nd edition. Houghton Mifflin: Boston. 336 pp.

Tracy, C. R. & H. L. Snell. 1985. Interrelations among water and energy relations of reptilian eggs, embryos, and hatchlings. *American Zoologist* 25: 999–1008.

Tsuji, J. S., Huey, R. B., van Berkum, F. H., Garland, Jr., T., & R. Shaw. 1989. Locomotor performance of hatchling fence lizards (*Sceloporus occidentalis*): quantitative genetics and morphological correlates. *Evolutionary Ecology* 3: 240–252.

van Berkum, F. H., Huey, R. B., Tsuji, J. S. & T. Garland, Jr. 1989. Repeatability of individual differences in locomotor performance and body size during early ontogeny of the lizard *Sceloporus occidentalis* (Baird and Girard). *Functional Ecology* 3: 97–105.

Vitt, L. J. & J. D. Congdon. 1978. Body shape, reproductive effort, and relative clutch mass in lizards: resolution of a paradox. *American Naturalist* 112: 595–608.

Wilbur, H. M. 1977. Propagule size, number, and dispersion patterns in *Ambystoma* and *Asclepias*. *American Naturalist* 111: 43–68.

Williams, G. C. 1966. *Adaptation and Natural Selection.* Princeton University Press: Princeton, NJ. 307 pp.

Parental Effects on Life-History Traits in Plants

Elizabeth P. Lacey

INTRODUCTION

In 1987, Roach and Wulff reviewed the literature on maternal effects in plants and concluded that although data show that these effects can strongly influence offspring phenotype for many life history traits, few studies have identified the specific causes of these effects. One specific conclusion that they reached was that little is known about the relative contributions of genetic versus environmental effects. In this paper I summarize what new information has appeared since 1987 that could help determine these contributions. I also address four additional questions: 1) Through which parent are environmental effects transmitted? Although Roach and Wulff reviewed only maternal effects, they did note that paternal effects influence seed size in some crop species. 2) When during the parental (maternal or paternal) life does the environment most strongly contribute to a parental effect? 3) Through what structural tissues of the seed is the parental effect mediated? 4) Why are parental environmental effects evolutionarily important? Although a maternal (or paternal) environmental effect has traditionally been viewed as a factor that can influence the degree to which a trait may respond to selection, it may, itself, undergo selective change.

GENETIC VERSUS ENVIRONMENTAL EFFECTS

Many studies have shown that maternal effects can influence seed, seedling, and, in some cases, adult life history traits (see references in Roach & Wulff, 1987). One can draw only limited conclusions about the cause of these effects, however, because most studies either confound environmental and genetic causes or focus only on one type of cause. Crop scientists have usually focused predominantly on the genetic component alone. The typical plant genetic experiment has been to perform a diallel cross among inbred lines of a crop species and then partition the genetic effect into its nuclear component (also called the general combining ability) and reciprocal component (includes maternal and paternal genetic effects). Seldom, however, have these studies reported the conditions under which the inbred lines were grown or considered the fact that the lines were grown in different environments, sometimes geographically different places (see references in Roach & Wulff, 1987). Geneticists who have paid attention to environmental effects have attempted to test for them by growing progeny of crosses over two generations. If the maternal effect disappears, then the effect is assumed to be environmental; if it persists, it

Dr. Lacey is with the Department of Biology, Eberhart Building, University of North Carolina, Greensboro, NC 27412, USA.

is assumed to be genetic. Went (1959), Durrant (1971), Grun (1976) and Alexander and Wulff (1985) among others have shown that this may be an erroneous assumption. Plant physiological ecologists have focused on environmental effects alone. Their studies have shown that nutrient availability, photoperiod, and temperature can influence offspring seed weight, germination, and even days to flowering (see references in Roach & Wulff, 1987). Temperature can even induce heritable changes (e.g., Durrant, 1971; Cullis, 1977). None of these studies, however, has examined the environmental contribution relative to the genetic contribution.

I am aware of only seven studies that have attempted to assess both genetic and environmental components of maternal, or paternal, effects. Roach and Wulff cited four studies, all showing both parental (maternal or paternal) genetic and environmental effects (Durrant & Timmis, 1973; Sawhney & Naylor, 1979; Alexander & Wulff, 1985; and Antonovics & Schmitt, 1986). Of these, Antonovics and Schmitt's (1986) study most clearly partitioned environmental, genetic nuclear, and genetic non-nuclear causes. They found in the offspring of diallel crosses of *Anthoxanthum odoratum* that only a maternal environmental effect, of the three possible causes, influenced seed size. This contrasts somewhat with my current study of *Plantago lanceolata*, ribwort plantain, in which I have found both a strong environmental effect and a genetic non-nuclear effect for seed size.

In an experiment that I have recently completed, I explored the strength of the temperature effects first reported by Alexander and Wulff (1985) and the transmission of the effects, either maternal or paternal, in *Plantago lanceolata*. Briefly, clonal replicates of 9 genotypes, 4 from one population and 5 from another, were randomly assigned to one of two growth chambers that differed in temperature: 15°C nights/20°C days or 20°C nights/26°C days. Using a Comstock-Robinson type II mating design, I crossed genotypes by bagging together reproductive spikes from different combinations of genotypes. The bags were placed over the maternal spikes just prior to stigma emergence. Seeds were collected from each cross by treatment combination, and a sample was weighed and germinated in petri dishes in the same two growth chambers.

Crosses were made among plants not only within chambers but also between chambers so that I could partition the temperature effects into three components: maternal pre-bagging temperature—the temperature under which the mother was growing before pollination, paternal pre-bagging temperature, and post-bagging temperature—the temperature during gamete maturation, fertilization, and seed development. There were six temperature treatments, described in Figure 1. For two of the treatments (1 and 4), either both parents were grown at high temperature or both parents were grown at low temperature. For two other treatments (2 and 6), either the mother or father was grown at high temperature and the other parent, grown at low temperature, was transferred to the high temperature chamber at the time of bagging. For the last two treatments (3 and 5), either the mother or father was grown at low temperature, and the other parent, grown at high temperature, was transferred to the low temperature at bagging. All six treatments differed in at least one of the three components of parental temperature. Although I will describe the complete experiment and its results elsewhere, I present some preliminary results here.

The data show that when other physical environmental factors that might be confounded with genetic effects are held constant, temperature strongly influenced seed size, with lower temperatures producing heavier seeds (Fig. 2). One can see this most easily by comparing the mean seed size for treatments 1 and 4 in Figure 2. There was also a significant non-nuclear genetic component to size, but an insignificant nuclear component. Whether the non-nuclear component is maternal or paternal in origin remains to be determined.

Two other recent studies have also found environmental effects for several life history traits. Stratton (1989) found that offspring of *Erigeron* parents grown in high fertilizer

TREATMENTS

	1	2	3	4	5	6
	HxH	LxH	HxL	LxL	LxH	HxL
		→	→		←	←
M pre	H	L	H	L	L	H
P pre	H	H	L	L	H	L
post	H	H	L	L	L	H

Figure 1. Experimental design for the growth chamber experiment. Immediately under each of the six treatments is shown the pre-bagging temperature of the mother X the pre-bagging temperature of the father: H = high, L = low. Where relevant, the arrow indicates which parent was moved to a different chamber to produce the cross. Temperatures for each developmental phase of the parents for each treatment are shown in the box: M pre = maternal pre-bagging temperature, P pre = paternal pre-bagging temperature, post = post-bagging temperature.

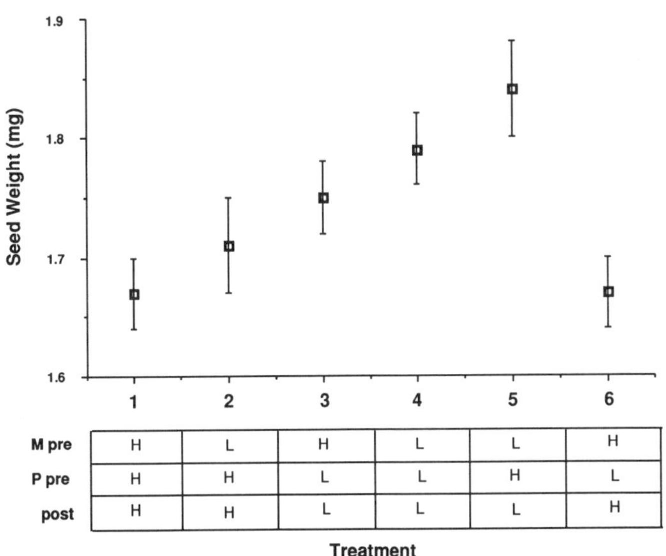

	1	2	3	4	5	6
M pre	H	L	H	L	L	H
P pre	H	H	L	L	H	L
post	H	H	L	L	L	H

Treatment

Figure 2. Mean seed weight ± 2 standard errors for each temperature treatment. Means were determined by individually weighing 12 seeds per cross per treatment, or as many seeds as possible for the few crosses that produced fewer than 12 seeds.

emerge earlier and grow more quickly, especially under competition, than do offspring of parents grown in low fertilizer. Schmitt et al. (unpubl. ms.) have observed in a greenhouse experiment that the light and nutrient environment of the parents in *P. lanceolata* can influence seed weight and germination. Thus, recent studies are confirming that parental effects for life history traits are both genetically and environmentally based; the relative importance of the two components of the effects, however, still remains unclear.

TIMING OF ENVIRONMENTAL EFFECTS

Theoretically a parental environmental effect could be initiated either before gametes are produced, during gamete production, at the time of fertilization, or during seed maturation (i.e., early embryonic development of the offspring). A few physiological ecologists have found significant environmental effects when they have placed parents in different environments for anthesis and seed maturation, suggesting the importance of the environment during that time (e.g., Riddell & Gries, 1958; Koller, 1962; Cook, 1975; Pourrat & Jacques, 1975; Sionit & Kramer, 1977; Gutterman, 1980/81, 1983a, b; Siddique & Goodwin, 1980), but none has considered also the environment of the parent prior to that time.

In the *Plantago* experiment that I have briefly described, I could distinguish among pre-bagging maternal temperature effects, paternal temperature effects, and post-bagging temperature effects. The pre-bagging effects included any temperature effects that were initiated prior to gamete maturation. Post-bagging effects included the period of ovary and pollen maturation, fertilization, and seed development. In general, the data show that the temperature after bagging most strongly influenced seed size and germination. Low post-bagging temperature produced heavier seeds (Fig. 2; compare treatments 3 and 6 or treatments 2 and 5), whereas differences in pre-bagging maternal (compare treatments 1 and 2 or 3 and 4) or paternal (compare treatments 4 and 5 or 1 and 6) temperature produced no significant differences in seed weight. The post-bagging temperature also most strongly affected germination. Seeds that were produced and that matured under low temperature showed lower percent germination after one (Fig. 3) and two weeks than did seeds produced under high temperature. This can be seen by comparing the treatments differing only in post-bagging temperature (Fig. 3; compare treatments 3 and 6 or treatments 2 and 5). At the end of the first week of germination, there was also a pre-bagging paternal temperature effect, the direction of which depended on the background temperatures (Fig. 3; compare treatments 1 and 6 or 4 and 5). Under a high post-bagging temperature, a low pre-bagging paternal temperature increased germination, whereas under low post-bagging temperature, a low pre-bagging paternal temperature reduced germination. This paternal temperature effect disappeared by the second week.

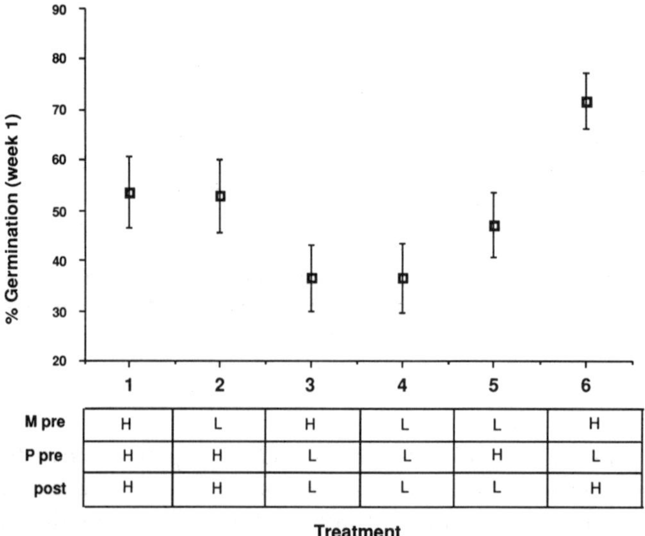

Figure 3. **Mean percent germination by the end of the first week ± 2 standard errors for each temperature treatment.**

Whether the effects of other environmental variables resemble the temperature effects is unknown, for I have found no other study that examines the effect of parental environment as it changes over the life cycle (developmental) phases of the parent. It is possible that the effects may vary. At one end of the spectrum, some variables may produce limited, i.e., immediate, effects during one developmental phase of a life cycle, whereas others, like nutrients, which can accumulate and be transported within a parent, might produce wider-ranging, i.e., longer-term, effects that span several developmental phases.

ROLE OF DIFFERENT SEED TISSUES

The Roach and Wulff (1987) review showed that little is known about the relative importance of the different tissues that make up a seed in mediating a maternal (or paternal) effect. A typical seed is composed of three types of tissues. The embryo, which contains the meristematic cells that give rise to the new plant, is diploid tissue that receives half its complement of chromosomes from the mother and half from the father. The endosperm, food tissue, is triploid, receiving ⅔ of its chromosomes from the mother and ⅓ from the father. The seed is entirely maternally derived tissue and is diploid. Theoretically, a genetic maternal effect could be transmitted through the embryo or endosperm cytoplasm or through the endosperm nucleus, if there is a dosage effect (Roach & Wulff, 1987). An environmental effect could be transmitted through any of the three tissues.

Information about the role of the different tissues in mediating a parental effect is still limited today. The only data of which I am aware come from a small study that an undergraduate and I conducted to complement the larger *Plantago* experiment that I have already briefly described. After *P. lanceolata* seeds germinate, their seed coats often fall off the cotyledons or can be easily removed from them. So, after weighing and germinating individual seeds for 11 crosses from all six treatments, we retrieved as many seed coats as we could, let them dry, and weighed them. By subtracting coat weight from total seed weight, we also calculated embryo/endosperm weight. Although the coat may have con-

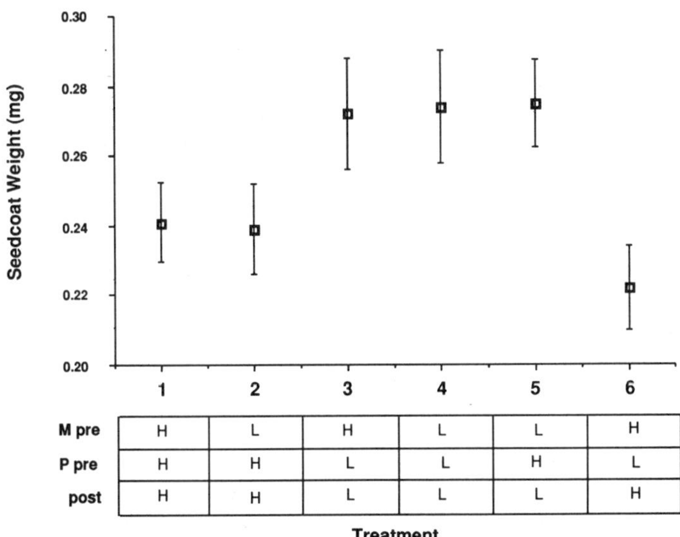

Figure 4. Mean seed coat weight ± 2 standard errors for each temperature treatment. Means were determined by individually weighing 1–3 seed coats per cross per treatment.

740

tained residual endosperm tissue that was not absorbed by the embryo, our examination of the coat under a compound scope suggested that any residual tissue, if present, was negligible. We found no significant difference in total seed weight or endosperm/embryo weight among treatments, but we did observe a significant treatment effect on coat weight alone in these crosses. Thus, coat weight can vary with no effect on total seed weight. Low parental temperatures produced heavier seed coats (Fig. 4), and the treatment effect on coat weight was explained solely by the post-bagging temperature (Fig. 4; compare treatments 1, 2, and 6 with 3, 4 and 5).

The fact that coat weight alone varied among treatments suggests that a parental environmental effect can independently affect different tissues of a seed. This independence may explain why parental environmental effects can alter offspring phenotype for some life history traits and not others. For example, a coat influences, among other traits, dormancy and its counterpart, germination (e.g., see references in Roach & Wulff, 1987). In many species, thicker coats delay germination. The observation that coat weight, and by inference thickness, was greater in seeds maturing at low temperature would suggest that low temperature should, consequently, delay germination. This is, in fact, what I observed when looking at germination patterns over the whole *Plantago* data set. Germination did occur more slowly for the seeds that had matured at low temperature (Fig. 3).

MATERNAL VERSUS PATERNAL EFFECTS

Environmental effects, like genetic effects, have historically been presumed to be transmitted via the mother. Therefore, paternal effects have received little attention. The typical plant genetic experiment has been to perform a diallel cross among imbred lines of crop species and then to analyze the data using a model that contains a variable for reciprocal effects, which are assumed to be maternal in origin. These studies have shown the presence of maternal effects but because no paternal effects were sought, they also were not found. Likewise, many ecologists have collected seeds from open-pollinated mothers grown under different environmental conditions and concluded that any environmental effects are maternally transmitted. The untested assumptions in this interpretation are that all paternal genotypes have contributed equally to the seed output of all mothers and that environmental effects are not transmitted paternally. Both of these assumptions seem quite shaky, if for no reason other than that genotypes usually differ somewhat in flowering time (Rathcke & Lacey, 1985).

Empirical studies are showing that both genetic non-nuclear and environmental effects can be transmitted paternally. A few studies show that paternal genetic effects can be important for some traits in some (e.g., Beddows et al., 1962; Garwood et al., 1970; Fleming, 1975; Aksel, 1977), though not all, species (Hayward & Breese, 1968; Antonovics & Schmitt, 1986; Mazur, 1987). Wolff (1988) recently found within a single population of *P. lanceolata* evidence for a maternal genetic effect for some traits like cotyledon length but a paternal genetic effect for other traits like leaf length. Also, chloroplast DNA inheritance can range from maternal to paternal and even biparental transmission within angiosperms (e.g., Metzlaff et al., 1981; Tilney-Basset & Birky, 1981; Corriveau & Coleman, 1988; Schumann & Hancock, 1989; Boblenz et al., 1990). Other types of studies are demonstrating that the environment can influence pollen success, whether measured by pollen size (e.g., Stanley & Linskens, 1974; Broaddus & Lacey, unpubl. data), germination (Schlichtling, 1986; Schoper et al., 1986), or siring of seeds (Kuo et al., 1981; Young & Stanton, 1990). Thus, evidence for paternal environmental effects is increasing.

The *Plantago* data showed no paternal temperature effect for seed size, but they did show one for early germination (Fig. 3). Pre-bagging paternal temperature influenced percent germination during the first week. Thus, parental environmental effects could hypothetically be transmitted through either the maternal or paternal parent, depending on the particular life history trait. Even though the paternally transmitted temperature effect disappeared after the first week, it may still be ecologically and evolutionarily important. For individuals in natural populations, timing of germination is often critical for seedling establishment, and this paternal temperature effect may determine which seedlings successfully establish and which do not.

EVOLUTIONARY IMPORTANCE OF PARENTAL EFFECTS

Parental environmental effects, regardless of when they are produced and how they are transmitted, have traditionally been viewed as evolutionarily important because they can alter the response to selection in a given generation. Quantitative geneticists have noted that parental contributions to phenotypic variance outside the normal nuclear contribution can alter the rate of response to selection (e.g., Willham, 1963; Van Vleck, 1976; Falconer, 1981; Riska et al., 1985; Kirkpatrick & Lande, 1989). The greater these contributions, the less an individual's genotype contributes to its own phenotype and consequently the weaker the response to selection. Additionally Haldane's (1931) and Wright's (1969) work in population genetics (see also Riska et al., 1985; Kirkpatrick & Lande, 1989) suggests that non-nuclear contributions may influence the direction as well as rate of the response. In all these cases a parental effect is envisioned as a factor that can influence a response to selection for some particular trait like seed size, or growth rate.

A parental environmental effect is, however, also a form of phenotypic plasticity, a trait that can evolve under natural selection. Schmalhausen (1949), Bradshaw (1965), and others have argued that phenotypic plasticity is genetically based and, therefore, can itself evolve, and most biologists accept this idea today. Plasticity has traditionally been defined as an individual's range of phenotypic responses—the norms of reactions—to its immediate environment. Defined this way, plasticity is an "intragenerational" phenomenon because it is the environment to which the individual is exposed during its lifetime that determines the plastic response. Plasticity could include, however, an individual's range of phenotypic responses to environments of previous generations, an "intergenerational" component. In this case, the phenotypic response to environment is manifested in the generation subsequent to the one experiencing the environment. This intergenerational plasticity, or parental environmental effect, may also potentially evolve under selection if it is, at least, partially heritable.

Given the appropriate experimental design, one that partitions parental genetic effects into their nuclear and non-nuclear components, we can look for evidence of a genetic basis for an environmental effect in two ways. First, we can look for a significant reciprocal (non-nuclear) by treatment interaction, which means that reciprocal pairs respond differently to treatments. The responses are manifested in the subsequent generation. Second, if we have determined that the treatment effect occurs prior to fertilization, we can look for a significant nuclear by treatment interaction. If the environmental effect exerts its influence after fertilization, then the interaction may really reflect intragenerational plasticity, i.e., the environment may be acting directly on the offspring in its embryonic stage of development. If, however, offspring differ in their response to the environment of the parent prior to fertilization, then we should be truly observing intergenerational plasticity.

My *Plantago* data provide no evidence for intergenerational plasticity in germination, but they do for seed size. The reciprocal by treatment effect for seed weight was highly significant, whereas the cross by treatment effect was not. Antonovics and Schmitt (1986) also observed a significant reciprocal by environmental interaction for seed size in *Anthoxanthum odoratum*. Durrant and Timmis (1973) observed that lines of flax responded

742

differently in both plant weight and DNA content to changing levels of nitrogen and phosphorus during parental growth and reproduction, and their data strongly suggest a non-nuclear genetic control of this plasticity. Thus, at this point, there is support for both the presence of phenotypic variation in environmental effects and the heritability of these effects. The implication of these results is that a parental environmental effect may not represent just "noise" that interferes with the process of selection for some agronomic or life history trait. Rather it may be a trait in its own right, itself subject to change by natural selection.

CONCLUSION

Given the potential evolutionary implications of maternal and paternal effects, it seems important to characterize them better than they have been to date. Roach and Wulff (1987) raised several questions that are worth investigating in future studies, and I have raised several more. I have limited my questions to the relative importance and causes of environmental effects, specifically when and how they are transmitted and whether they are heritable. As Roach and Wulff have noted, ultimately we need to know, also, what the consequences of these effects are, particularly in natural populations. One should be encouraged by the fact that most of these questions should be answerable through experimental studies.

ACKNOWLEDGMENTS

I thank Robin Hopkins for her help collecting data, Sara Smith for her help collecting and weighing seed coats, J. Antonovics for his comments on an earlier draft of this manuscript, NSF for supporting my work on parental effects, and B. Riska for inviting me to speak at this symposium. This work was also supported partially by an NSF grant #BSR 87-06429 to the Duke University Phytotron.

LITERATURE CITED

Aksel, R. 1977. Quantitative genetically non-equivalent reciprocal crosses in cultivated plants. Pp. 269–280. *In:* A. Muhammed, R. Aksel & R. C. von Borstel (eds.), *Genetic Diversity in Plants.* Plenum Press: New York.

Alexander, H. M. & R. Wulff. 1985. Experimental ecological genetics in *Plantago.* X. The effects of maternal temperature on seed and seedling characters in *P. lanceolata. J. Ecol.* 73:271–282.

Antonovics, J. & J. Schmitt. 1986. Paternal and maternal effects on propagule size in *Anthoxanthum odoratum. Oecologia* 69:277–282.

Beddows, A. R., Breese, E. L. & B. Lewis. 1962. The genetic assessment of heterozygous breeding material by means of a diallel cross. I. Description of parents self- and cross-fertility and early seedling vigour. *Heredity* 17:501–512.

Boblenz, K., Nothnagel, T. & M. Metzlaff. 1990. Paternal inheritance of plastids in the genus *Daucus. Mol. Gen. Genet.* 220:489–491.

Bradshaw, A. D. 1965. Evolutionary significance of phenotypic plasticity in plants. *Adv. Genet.* 33:115–155.

Cook, R. E. 1975. The photoinductive control of seed weight in *Chenopodium rubrum* L. *Amer. J. Bot.* 62:427–431.

Corriveau, J. L. & A. W. Coleman. 1988. Rapid screening method to detect potential biparental inheritance of plastid DNA and results for over 200 angiosperm species. *Amer. J. Bot.* 75:1443–1458.

Cullis, C. A. 1977. Molecular aspects of the environmental induction of heritable changes in flax. *Heredity* 38:129–154.

Durrant, A. 1971. Induction and growth of flax genotrophs. *Heredity* 27:277–298.

Durrant, A. & J. N. Timmis. 1973. Genetic control of environmentally induced changes in *Linum. Heredity* 30:369–379.

Falconer, D. S. 1981. *Introduction to Quantitative Genetics.* Longman: London.

Fleming, A. A. 1975. Effects of male cytoplasm on inheritance in hybrid maize. *Crop Sci.* 15:570–573.

Garwood, D. L., Weber, E. J., Lambert, R. J. & D. E. Alexander. 1970. Effect of different cytoplasms on oil, fatty acids, plant height and ear height in maize. (*Zea mays* L.). *Crop Sci.* 10:39–41.

Grun, P. 1976. *Cytoplasmic Genetics and Evolution.* Columbia University Press: New York.

Gutterman, Y. 1980/81. Influence on seed germinability: phenotypic maternal effects during seed maturation. *Israel J. Bot.* 29:105–117.

Gutterman, Y. 1983a. Phenotypic maternal effect of photoperiod on seed germination. Pp. 67–79. *In:* A. A. Kahn (ed.), *The Physiology and Biochemistry of Seed Development, Dormancy and Germination.* Elsevier: New York.

Gutterman, Y. 1983b. Flowering, seed development, and the influences during seed maturation on seed germination of annual weeds. Pp. 1–25. *In:* S. O. Duke (ed.), *Weed Physiology Vol. 1. Reproduction and Ecophysiology.* CRC Press: Florida.

Haldane, J. B. S. 1931. A mathematical theory of natural and artificial selection. Part VII. Selection intensity of mortality rate. *Proc. Camb. Philos. Soc.* 27:131–136.

Hayward, M. D. & E. L. Breese. 1968. The genetic organization of natural populations of *Lolium perenne* L. III. Productivity. *Heredity* 23:357–368.

Kirkpatrick, M. & R. Lande. 1989. The evolution of maternal characters. *Evolution* 43:485–503.

Koller, D. 1962. Preconditioning of germination in lettuce at time of fruit ripening. *Am. J. Bot.* 49:841–844.

Kuo, C. G., Peng, J. S. & J. S. Tsay. 1981. Effect of high temperature on pollen grain germination, pollen tube growth, and seed yield of Chinese cabbage. *HortScience* 16:67–68.

Mazur, S. J. 1987. Parental effects on seed development and seed yield on *Raphanus raphanistrum:* implications for natural and sexual selection. *Evolution* 41:355–371.

Metzlaff, M., Borner, T. & R. Hagemann. 1981. Variations of chloroplast DNAs in the genus *Pelargonium* and their biparental inheritance. *Theor. Appl. Genet.* 60:37–41.

Pourrat, Y. & R. Jacques. 1975. The influence of photoperiodic conditions received by the mother plant on morphological and physiological characteristics of *Chenopodium polyspermum* L. seeds. *Plant Sci. Letters* 4:273–279.

Rathcke, B. & E. P. Lacey. 1985. Phenological patterns in terrestrial plants. *Ann. Rev. Ecol. Syst.* 16:179–214.

Riddell, J. A. & G. A. Gries. 1958. Development of spring wheat: III. Temperature of maturation and age of seeds as factors influencing their response to vernalization. *Agron. Jour.* 50:743–746.

Riska, B., Rutledge, J. J. & W. R. Atchley. 1985. Covariance between direct and maternal genetic effects in mice, with a model of persistent environmental influences. *Genet. Res. Camb.* 45:287–295.

Roach, D. A. & R. Wulff. 1987. Maternal effects in plants: evidence and ecological and evolutionary significance. *Ann. Rev. Ecol. Syst.* 18:209–235.

Sawhney, R. & J. M. Naylor. 1979. Dormancy studies in seed of *Avena fatua.* 9. Demonstration of genetic variability affecting the response to temperature during seed development. *Can. J. Bot.* 57:59–63.

Schlichtling, C. D. 1986. Environmental stress reduces pollen quality in *Phlox:* compounding the fitness deficit. Pp. 483–488. *In:* D. L. Mulcahy, G. B. Mulcahy & E. Ottaviano (eds.), *Biotechnology and Ecology of Pollen.* Springer-Verlag: New York.

Schmalhausen, I. I. 1949. *Factors of Evolution.* Blakiston Press: New York.

Schoper, J. B., Lambert, R. J. & B. L. Vasilas. 1986. Maize pollen viability and ear receptivity under water and high temperature stress. *Crop Sci.* 36:1029–1033.

Schumann, C. M. & J. F. Hancock. 1989. Paternal inheritance of plastids in *Medicago sativa. Theor. Appl. Genet.* 78:863–866.

Siddique, M. A. & P. B. Goodwin. 1980. Seed vigour in bean (*Phaseolus vulgaris* L. cv. Apollo as influenced by temperature and water regime during development and maturation. *J. Exptl. Bot.* 31:313–323.

Sionit, N. & P. J. Kramer. 1977. Effect of water stress during different stages of growth of soybean. *Agron. J.* 69:274–278.

Stanley, R. G. & H. F. Linskens. 1974. *Pollen: Biology, Biochemistry, Management.* Springer-Verlag: NY.

Stratton, D. A. 1989. Competition prolongs expression of maternal effects in seedlings of *Erigeron annuus* (Asteraceae). *Amer. J. Bot.* 76:1646–1653.

Tilney-Bassett, R. A. E. & C. W. Birky Jr. 1981. The mechanism of the mixed inheritance of chloroplast genes in *Pelargonium. Theor. Appl. Genet.* 60:43–53.

Van Vleck, L. D. 1976. Selection for direct maternal and grandmaternal genetic components of economic traits. *Biometrics* 32:73–181.

Went, F. W. 1959. Effects of environment of parent and grandparent generations on tuber production by potatoes. *Amer. J. Bot.* 46:277–282.

Willham, R. L. 1963. The covariance between relatives for characters composed of components contributed by related individuals. *Biometrics* 19:18–27.

Wright, S. 1969. *Evolution and the Genetics of Populations. II. The Theory of Gene Frequencies.* University of Chicago Press: Chicago.

Wolff, K. 1988. Natural Selection in *Plantago* Species: a Genetical Analysis of Ecologically Relevant Morphological Variability. Ph. D. thesis. University of Groningen, The Netherlands.

Young, H. J. & M. L. Stanton. 1990. Environmental stress influences pollen quality in wild radish. *Science* 298:1631–1633.

Maternal Effects in Insects: Examples, Constraints, and Geographic Variation

Timothy A. Mousseau and Hugh Dingle

Maternal effects are a frequent, and often troublesome, source of environmental resemblance. . . .

Falconer 1989, p. 159

INTRODUCTION

A sentiment shared by many plant and animal breeders has been that maternal effects are often an annoying source of variance that must be analytically dealt with prior to reaching useful conclusions concerning response to selection or evolution. The objective of this paper is to convince the reader that there are sound biological reasons to consider maternal effects. Our central thesis is that many maternal effects in insects have evolved as a special form of phenotypic plasticity that permits adaptive fine tuning or programming of life cycles in response to heterogeneous environments.

In outline, we present a series of examples of maternal effects associated with life cycle regulation. This is followed by a summary of a literature review of 74 species found to display maternal effects. We also present evidence that maternal effects are most often observed between an adult mother and her embryos, suggesting some constraint on the distance between life cycle stages over which maternal effects may act. In the final section of the paper we present data indicating that maternal effects often vary geographically within species, and that this variation is usually correlated to the seasonal characteristics of the habitat from which the population was collected, suggesting genetic divergence and local adaptation of these maternal characters. It should be noted that many of the points presented here are abbreviations or amplifications of those made in Mousseau & Dingle (1991) and Danks (1987).

A BRIEF REVIEW OF INSECT LIFE CYCLES

In general, most insect life cycles can be divided into two parts (Figs. 1 & 2): a period of growth and reproduction corresponding to seasons of favorable meteorological conditions, and a period of dormancy or quiescence which occurs during the unfavorable

Drs. Mousseau and Dingle are with the Department of Entomology and The Center for Population Biology, University of California, Davis CA 95616, USA. Dr. Mousseau's present address is Department of Biological Sciences, The University of South Carolina, Columbia, SC 29208, USA.

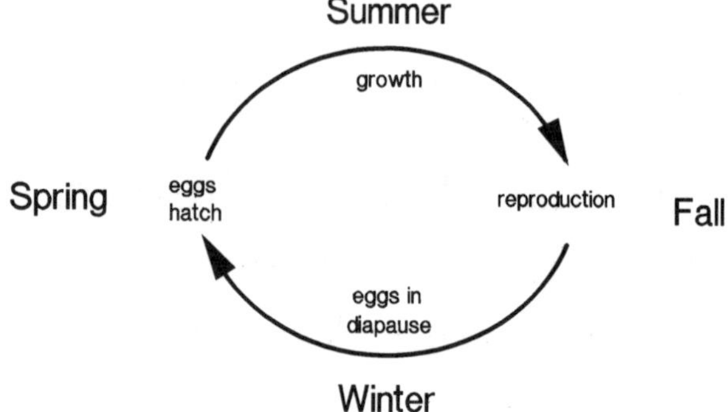

Figure 1. A typical univoltine insect life cycle with an overwintering egg stage. Life cycle shifts occur if other stages overwinter.

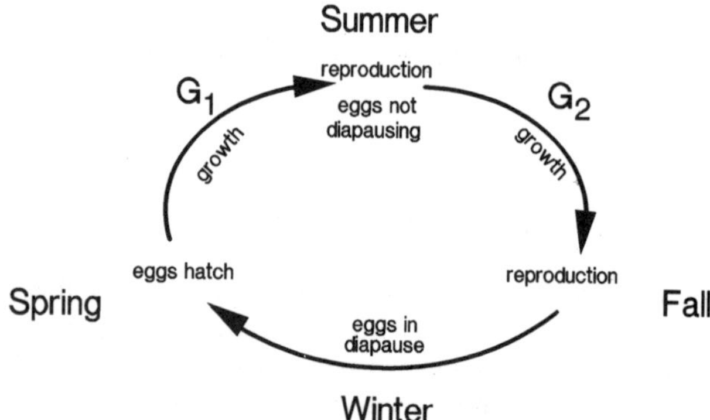

Figure 2. A typical bivoltine insect life cycle with an overwintering egg stage. Life cycle shifts occur if other stages overwinter.

season. In temperate and northern regions dormancy usually occurs during the winter, while in hot dry regions dormancy often occurs during the summer (Danks, 1987; Tauber et al., 1986).

There are three critical aspects of a typical univoltine insect life cycle with egg diapause (Fig. 1). First, the eggs must hatch at an appropriate time. An egg that hatches early risks mortality from a late frost or snowfall, while an egg that hatches late loses valuable time for growth and reproduction. Second, all growth and reproduction must be completed prior to the onset of winter. However, there is also selection for development to proceed for as long a time as is permitted by the local environment, because development time is positively correlated to body size and fecundity in many insects (Roff, 1986). The third critical component to this life cycle is that eggs must be physiologically equipped to survive harsh winter (or summer) conditions, which usually means that they must be in a state of diapause. Many insects exhibit variations on this simple life cycle, although the exact configuration is often rotated such that larvae, pupae, or adults are the diapausing stage. However, the same basic constraints with respect to the coordination of the life cycle to patterns of seasonality always apply.

These constraints also apply to bivoltine (and multivoltine) insects (Fig. 2), except that the first generation must develop and reproduce in the first half of the growing season, leaving sufficient time for the second generation to complete development and reproduction prior to winter. As a result of this constraint, the eggs produced by the first generation must not be in diapause; they must be capable of hatching immediately in order to maximally utilize the time remaining in the season for the production of the second generation.

In summary, the three critical aspects of any insect life cycle are: the timing of egg hatch (or the resumption of development following diapause), the coordination of growth and reproduction to seasons of favorable climate, and the decision to diapause (or not). Most of the maternal effects to be discussed in this paper relate to the mechanisms by which insects regulate these three fundamental components of their life cycles.

A FEW EXAMPLES OF MATERNAL EFFECTS IN INSECTS

Many insect species display maternal effects on the control of offspring diapause. Figure 3 illustrates the results of a series of experiments by D. S. Saunders (1966) where females of the parasitic wasp *Nasonia vitripennis* were subjected to a wide variety of photoperiods, and their offspring were examined for the incidence of diapause. Mothers which experienced short photoperiods (<15hrs L:D), a condition which recreates fall-like conditions in the wild, produced a high proportion of diapausing offspring. However, when females experienced long photoperiods, such as they would encounter in the summer, mothers produced almost exclusively non-diapausing offspring, which in the wild would continue development to produce a second or sebsequent generation. Clearly, this kind of response allows mothers to program an appropriate developmental switch in the offspring, which in this case is to diapause in preparation for over-wintering, or to continue development and produce an additional generation. Similar effects of photoperiod on maternal control of diapause have been reported for more than 60 species (Table 1; Danks, 1987; Saunders, 1982).

There are many other environmental cues that can trigger maternal effects on diapause. For example, in *Nasonia vitripennis* host deprivation will result in an increased incidence of diapausing progeny (Fig. 4). Females that are not deprived of hosts, or are only deprived for a single day, produce exclusively non-diapausing offspring. However, females that are deprived of hosts for four days produce almost entirely diapausing progeny. In this example, host deprivation is a cue which is correlated to deteriorating environmental conditions and is used by the mother to program development in her offspring.

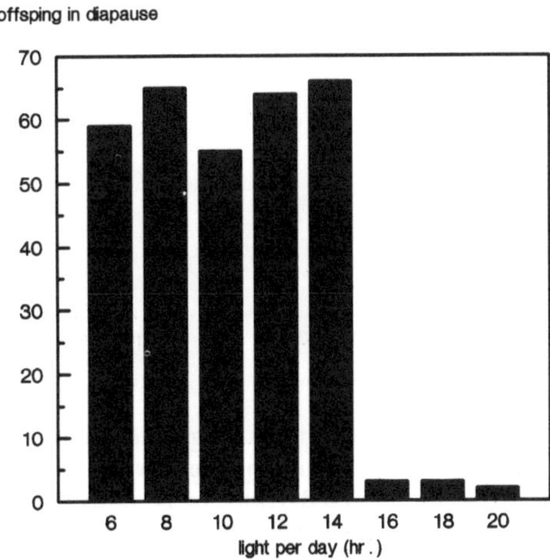

Figure 3. The proportion of offspring in diapause versus the photoperiod experienced by the maternal generation in the parasitic wasp *Nasonia vitripennis*. Photoperiods ≤ 14 hrs day length result in mainly diapausing offspring while photoperiods ≥ 15 hrs day length result in non-diapausing offspring. Data from Saunders (1966).

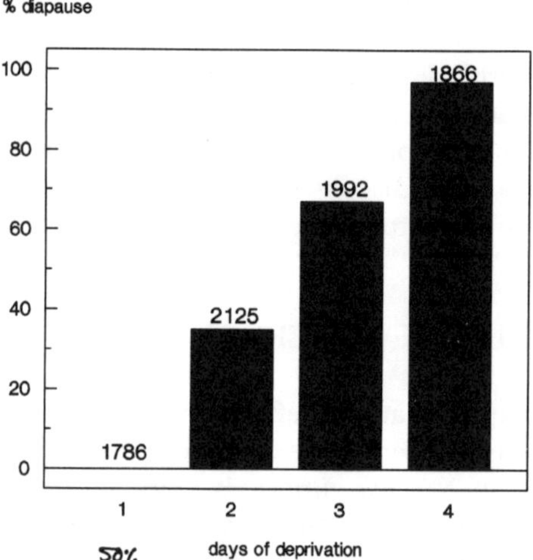

Figure 4. The proportion of diapausing offspring versus the number of days that mothers are deprived of hosts in *Mormoniella* (= *Nasonia vitripennis*). Mothers that are deprived of hosts in which to deposit offspring for one day or less produce exclusively non-diapausing offspring. Mothers deprived of hosts for four days produce almost entirely diapausing offspring. Number above bar indicates total numbers of offspring assayed. Data from Schneiderman and Horwitz (1958).

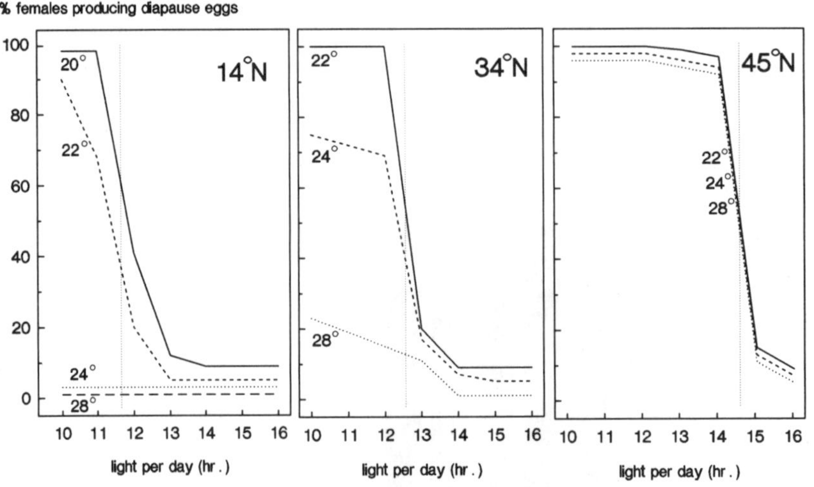

Figure 5. The proportion of females producing diapausing eggs versus photoperiod and temperature experienced by mothers for three populations of the mosquito, *Aedes atropalpus*, collected from three latitudes. In a population collected from 14°N latitude photoperiod effects were observed only when mothers experienced cool temperatures (20° and 22° C). In a northern population (45°N) there was no difference among temperature treatments and photoperiod was the only factor to result in variation in maternal control of offspring diapause. In an intermediate population (34°N) both photoperiod and temperature influenced maternal control of diapause in offspring. The dotted vertical line denotes the critical photoperiod for aversion or induction of diapause. Redrawn from Beach (1978).

Often, it is not a single environmental cue that triggers a maternal effect, but a combination of several. For example, in a study of three populations of the mosquito, *Aedes atropalpus,* Beach (1978) found that in a southern population collected from Puerto Rico (14°N), females that were reared at cool temperatures displayed a significant maternal effect on diapause in response to photoperiod, while females reared at high temperatures displayed no photoperiod induced effect (Fig. 5). On the other hand, in a population collected from 45°N latitude, there was little effect of temperature, and photoperiod was the only cue to result in a significant maternal effect. An intermediate population (34°N) displayed intermediate levels of interaction between photoperiod and temperature as cues associated with maternal control of diapause in offspring. These patterns of maternal susceptibility to environmental cues are consistent with a hypothesis of reliability of the cues as indicators of future meteorological conditions. In the south, photoperiod alone is not a reliable cue since its change with season is very gradual and of small total magnitude. In the north, photoperiod is a very reliable indicator of seasonal position (Danks, 1987).

Maternal age has also frequently been found to influence patterns of development in offspring. For example, in the striped ground cricket, *Allonemobius fasciatus,* individual females from bivoltine populations produce increasing proportions of diapausing eggs as the mother ages (Fig. 6; Mousseau, 1991). This maternal-age effect presumably reflects the decreasing probability of a first generation female producing a second generation as she gets older (and the season progresses). In univoltine populations, however, there is no maternal-age effect on offspring diapause. This is not surprising if we accept the hypothesis that maternal-age effects are a consequence of natural selection. In univoltine populations there is rarely the opportunity to produce a second generation because of the short growing season (Mousseau & Roff, 1989) and thus there would be no selection for maternal-age effects since in the wild all eggs are diapausing (e.g. no phenotypic variation).

There are many characters besides diapause that are subject to maternal effects in insects, although, as an exemplar diapause is particularly attractive because of its obvious

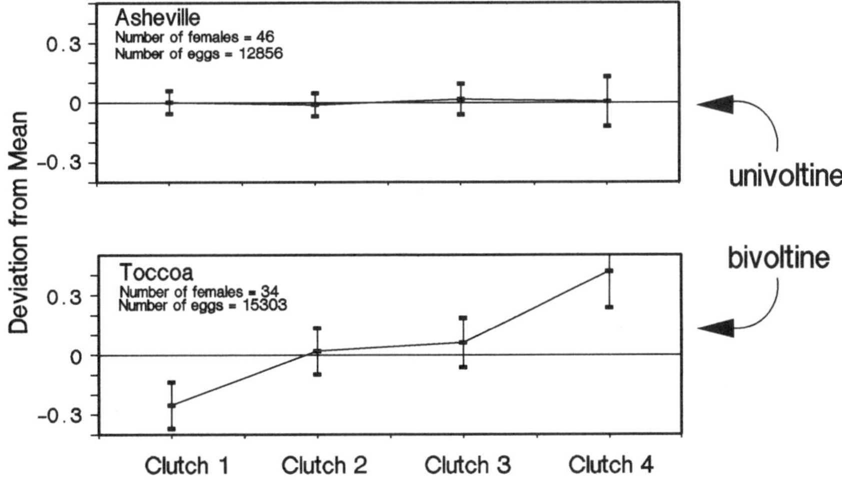

Figure 6. The influence of maternal age on the proportion of diapausing offspring in two populations of the striped ground cricket, *Allonemobius fasciatus.* The data have been scaled so that the mean proportion of diapause for each female is standardized to zero, and deviations from zero reflect changes from the mean with respect to clutch number. In the univoltine population, collected from Asheville, NC, there is no effect of maternal age on the proportion of diapause in offspring. In the bivoltine population (collected from Toccoa, GA) however, the proportion of diapausing offspring increases as females age (indicated by clutch number). Data from Mousseau (1991).

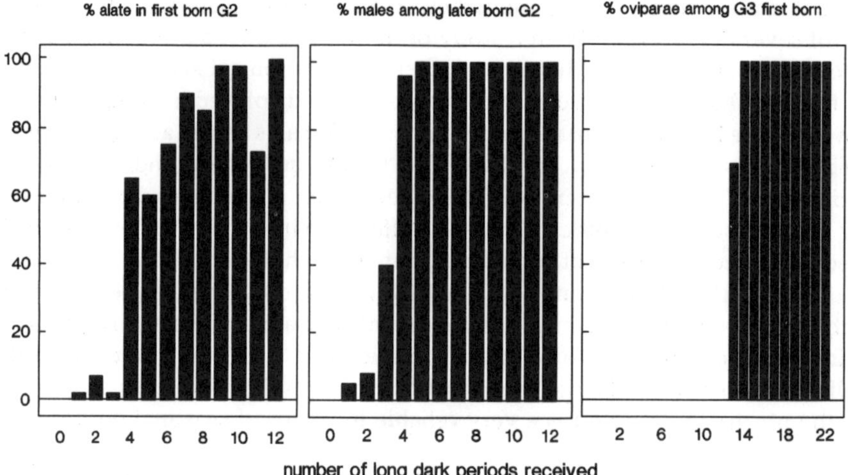

Figure 7. The influence of photoperiod on maternal and grandmaternal control of wing dimorphism, sexual morph production, and ovipary in the aphid, *Myzus persicae.* **Mothers (G1 generation) that experience prolonged exposure to short day lengths switch from the production of wingless, nonsexual morphs to the production of alate and sexual offspring (G2 generation). Similarly, G1 mothers that experience short day lengths produce oviparous G3 grand-offspring. Redrawn from Blackman (1975).**

association with fitness. In aphids, many components of life cycle regulation are subject to maternal control. In *Myzus persicae,* mothers that are subjected to prolonged periods of short photoperiods shift from the production of wingless non-dispersing offspring to the production of alate offspring which can then disperse in search of primary host plants (Fig. 7; Blackman, 1975). These photoperiod effects can also induce a shift from the production of parthenogenic to sexual morphs. Similarly, short photoperiods experienced by grand-mothers can induce a shift from the production of viviparous to oviparous grand-offspring, which are better equipped to handle over-wintering on primary hosts.

SUMMARY OF MATERNAL EFFECTS IN INSECTS

Table 1 lists a total of 74 species that have found to display maternal effects. It must be noted that this list is far from complete; it serves primarily to demonstrate the wide variety of taxa, the large number of characters, and the richness of cues, biotic and abiotic, that are associated with maternal effects in insects. A logical conclusion from this survey is that maternal effects must be considered an important aspect of insect life cycle regulation, and must be included in any study of insect life history.

MATERNAL EFFECTS AS PHENOTYPIC PLASTICITY

In most of the cases reported in Table 1, maternal effects result from exposure to environmental cues like photoperiod or temperature, cues which give a reliable indication of future meteorological or biotic conditions; the developmental and phenotypic effects in offspring that result are often easily interpreted as adaptive switches which increase the fitness of offspring under the predicted conditions. Most of these effects can be thought of as a form of trans-generational phenotypic plasticity whereby a mother that

Table 1. A sample of insect species that display maternal effects. The most commonly reported environmental cues are photoperiod (photo) and/or temperature (p/t).

Species	Character	Cue	Reference
Adelphorcoris lineolatus	diapause	p/t	Ewen 1966
Aedes albopictus	diapause	p/t	Mori et al. 1981
Aedes atropalpus	diapause	p/t	Anderson 1968, Kalpage and Brust 1974, Beach 1978
Aedes caspius	diapause	p/t	Vinogradova 1975a
Aedes caspius dorsalis	diapause	age	Yakubovich 1983
Aedes dorsalis	diapause	p/t	McHaffey and Harwood 1970
Aedes sollicitans	diapause	p/t	Anderson 1970
Aedes togoi	diapause	p/t	Vinogradova 1965, McGinnis and Brust 1982
Aedes triseriatus	diapause	p/t	Love and Whelchel 1955
Aedes vexans	diapause	p/t	Wilson and Horsfall 1970
Allonemobius fasciatus	diapause	age	Sarai 1967, Mousseau 1991
Amphigerontia bifasciata	diapause	p/t	Glinyanaya 1975
Anopheles freeborni	diapause	p/t	Depner and Harwood 1966
Aphidius nigripes	diapause	age	Brodeur and McNeil 1989
Aphis fabae	wing dimorphism, sex ratio	photo, crowding, age	Lees 1983, Hardie 1987, Hardie and Lees 1985
Aphis nerii	wing dimorphism	crowding	Groeters and Dingle 1989
Aphis pisum	wing dimorphism	crowding	Lamb and MacKay 1979
Bacillus rossius	diapause	p/t	Scali 1968
Bathyplectes curculionis	diapause	p/t, age	Parrish and Davis 1978
Bombyx mori	diapause	p/t	Kogure 1933
Calliphora vicina	diapause	p/t	Bogdanova et al. 1978, Khachatryan 1981, Vinogradova 1974, 1975b, Vinogradova and Zinovjeva 1972, Saunders 1987, Saunders et al. 1986
Callosobruchus maculatus	dispersal, size	age, behavior	Messina 1990, Sano-Fujii 1979
Catolaccus aeneoviridis	diapause	age	McNeil and Rabb 1973
Cavelerius saccharivorus	diapause	age	Hokyo et al. 1983
Coecilius flavidus	diapause	p/t	Glinyanaya 1975
Coeloides brunneri	diapause	p/t	Ryan 1965
Cryptus inornatus	diapause	p/t	Simmonds 1946, 1948
Culicoides pulicarius	diapause	p/t, age	Isayev 1975
Delia radicum	diapause	p/t	Read 1969
Drosophila melanogaster	malathion resistance, survival, development	photo, age	Singh and Morton 1981, Barnes 1984, Cadieu 1983, Fleuriet and Vageille 1982, Kerver and Rotman 1987, Giesel 1988, Giesel et al. 1989, Mills and Hartmann-Goldstein 1985
Endria inimica	diapause	p/t	Gustin 1974
Glossina morsitans	size, survival	diet	Langley et al. 1978
Haematobia irritans	diapause	p/t	Depner 1961, 1962
Heliothis punctigera	diapause	p/t	Cullen and Browning 1978
Heliothis virescens	diapause, growth	p/t, diet	Benschoter 1970, Gould 1988
Heliothis zea	diapause	p/t	Adkisson and Roach 1971, Benschoter 1970, Wellso and Adkisson 1966
Hypera postica	diapause	p/t	Rosenthal and Koehler 1968
Hypopteromalus tabacum	diapause	age	McNeil and Rabb 1973
Locusta migratoria	diapause, color, marching	age, crowding	Matthée 1951, Hardie and Lees 1985, Kennedy 1961, Labeyrie 1967

Table 1. Continued.

Species	Character	Cue	Reference
Locusta pardalina	diapause	age	Matthée 1951
Lucilia caesar	diapause	p/t, age	Fraser and Smith 1963, Ring 1967b
Lucilia sericata	diapause	p/t	Cragg and Cole 1952, Saunders et al. 1986
Lymantria dispar	pupal weight, dispersal, development time	diet, age	Rossiter et al. 1990, Rossiter 1991a, 1991b
Megoura viciae	wing dimorphism, sex ratio	photo, age, crowding	Lees 1984
Mesovelia mulsanti	diapause	age	Galbreath 1976
Muellerianella brevipennis	diapause	p/t	Witsack 1971
Musca domestica	survival, development time	age	Rockstein 1957
Myzus persicae	wing dimorphism, sex ratio	photo, age, crowding	Blackman 1975
Nasonia vitripennis	diapause	p/t, age	Saunders 1962, 1965b, 1973, etc., Schneiderman and Horwitz 1958
Neoseiulus fallacis	diapause intensity	photo	Rock et al. 1971
Nomadacris septenfasciata	lipid reserves, development time	crowding	Hardie and Lees 1985, Kennedy 1961, Labeyrie 1967
Nysius groenlandicus	diapause	p/t	Bocher 1975
Oligonychus ununguis	diapause	p/t	Shinkaji 1975
Oncopeltus fasciatus	diapause, development time	p/t, age	Groeters and Dingle 1987, 1988, Phelan and Frumhoff 1991
Orchesella cincta	age at maturity, weight, clutch size	genetic	Janssen et al. 1988
Orgyia antigua	diapause	p/t	Doskoeil 1957, Kind 1965, 1972
Orgyia thyellina	diapause	p/t	Kimura and Masaki 1977, Sato 1977
Pectinophora gossypiella	diapause	age	Raina and Bell 1974
Periphyllus testudinatus	diapause	p/t	Bonnemaison 1956
Peripsocus phaeopterus	diapause	p/t	Glinyanaya 1975
Peripsocus quadrifasciatus	diapause	p/t	Eertmoed 1978
Phanoptera grisea	diapause	p/t	Helfert 1980
Phanoptera nana	diapause	p/t	Helfert 1980
Pseudatomoscelis seriatus	diapause	p/t	Gaylor and Sterling 1977
Psorophora ferox	diapause	p/t	Pinger and Eldridge 1977
Pteronemobius nigrofasciatus	diapause	p/t	Masaki 1973
Pyrrhocoris apterus	wing dimorphism	photo	Honek 1980
Sarcophaga bullata	diapause	p/t, age	Denlinger 1972a, b, Henrich and Denlinger 1982, Rocky et al. 1989, Rockey and Denlinger 1986
Schistocerca gregaria	lipid reserves, development time	crowding	Hardie and Lees 1985, Kennedy 1961, Labeyrie 1967
Spalangia drosophilae	diapause	p/t, age	Simmonds 1946, 1948
Tenebrio molitar	growth rate, size	age	Ludwig and Fiore 1960, 1961
Trichogramma euproctidis	diapause	p/t	Qui and Zaslavsky 1983
Trichogramma evanescens	diapause	p/t	Zaslavsky and Umarova 1982
Zophobas atratus	growth rate, size	age	Tschinkel 1990

experiences environmental cue "a" will produce offspring whose phenotype is appropriate to the conditions predicted by this cue (Fig. 8). Similarly, if this same mother experiences cue "b", she will produce offspring with a phenotype appropriate for the environment predicted by cue "b". In the case of diapause, many species display this adaptation with short photoperiods and/or cool temperatures leading to diapausing offspring, while long photoperiods and/or high temperatures result in non-diapausing offspring.

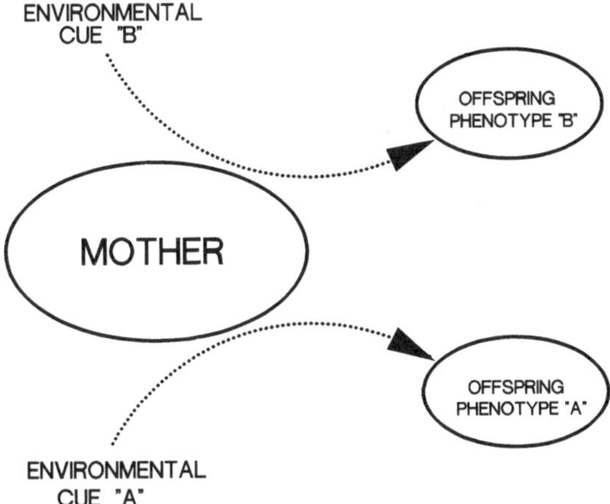

Figure 8. A schematic of maternal effects as a special case of phenotypic plasticity. In many of the examples reported in this paper environmental conditions experienced by mothers result in developmental switches in the offspring. These developmental effects often result in offspring of a phenotype appropriate to the environmental conditions predicted by the cues experienced in the maternal generation.

CONSTRAINTS ON MATERNAL EFFECTS

A pattern to emerge from this survey is that for species that display maternal effects, it is the adult stage of the maternal generation which is most likely to be sensitive to biotic or abiotic cues, and it is the embryonic stage of the offspring which is most often observed to display phenotypic plasticity in response to maternal sensitivity to environmental conditions. Figure 9 illustrates the pattern of interaction between maternal sensitivity and phenotypic variation in offspring for 61 species that exhibit maternal effects on diapause. We have used diapause as our model trait since there is a wealth of data available for this character (mainly extracted from Danks [1987] and Saunders [1982]). In 40 species, it is the adult mother that is sensitive to environmental influences, while in only 17 species is the larval stage of the maternal generation sensitive. Danks (1987) reports only one species (*Bombyx mori*) in which the egg stage of the maternal generation has a significant influence on diapause in the offspring.

Species that diapause as eggs or embryos are much more likely to respond to maternal effects than species that diapause at later stages of development. Forty of the species shown in Figure 9 diapause as eggs or embryos, while only 19 species diapause as larvae (or pupae), and two species diapause as adults. We have lumped species that diapause as pupae with those that diapause as larvae because of the possibility of bias introduced as a consequence of differences between pauro- and hemi- and holometabolous insect life cycles (pauro-, hemimetabolous insects do not pupate).

It is our contention that these patterns of interaction between maternal and offspring life cycle stages are a corollary of a general rule of environmentally modulated phenotypic

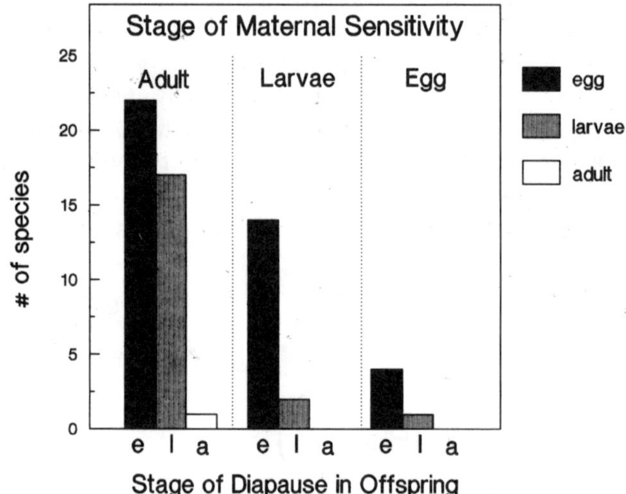

Figure 9. The stage of maternal sensitivity to environmental cues, and the stage of diapause in 61 insect species that exhibit maternal effects on diapause. In 40 species the adult mother is sensitive to environmental cues. Also, in 40 of the 61 species which display maternal effects, the offspring diapause as eggs or embryos. Species which diapause as pupae have been combined with those that diapause as larvae. Data from Danks (1987) and Saunders (1982).

plasticity, whereby the decision to diapause is most influenced by environmental conditions contemporary to the stage at which diapause occurs or the stage(s) immediately prior to it. Figure 10 schematically outlines patterns of interaction of 146 insect species for which these life history details have been reported (Danks, 1987). It can be seen that in 62 species that diapause as adults, it is environmental conditions experienced by the adults themselves that most frequently influence the decision to diapause. In only one species do conditions in the maternal generation influence diapause of the adult offspring. Similarly, for 50 species that diapause as larvae or pupae, it is mainly conditions experienced by the larvae or pupae themselves that influence diapause. However, in ten species (20%), environmental conditions experienced by the previous generation can significantly influence diapause in the offspring. Conversely, in 27 of the 34 species that diapause as eggs, conditions experienced by the maternal generation influence whether or not the egg will diapause.

The above patterns of interaction between life cycle stages strongly support our contention that maternal effects are a subset of a general class of environmentally modulated phenotypic plasticity that occurs both within and among generations. The observed patterns of interaction suggest a constraint over the distance between life cycle stages that phenotypic plasticity can act.

There are two competing hypotheses relating to the evolution of the observed constraint. First, these patterns could simply reflect physiological limits to the distance between life cycle stages that environmental information can be transmitted. This hypothesis is consistent with what little is known concerning maternal regulation of diapause in embryos. In many instances, it is a hormone(s) produced by the mother and transmitted cytoplasmically to the eggs which directly regulates prenatal physiological pathways to control development (Mousseau & Dingle, 1991). Such direct effects would be expected to diminish as the offspring develops and its own biochemical machinery supercedes or dilutes direct maternal effects. An alternative hypothesis is that natural selection has favored short term over long term effects, as a result of the increased uncertainties when prediction is extended chronologically.

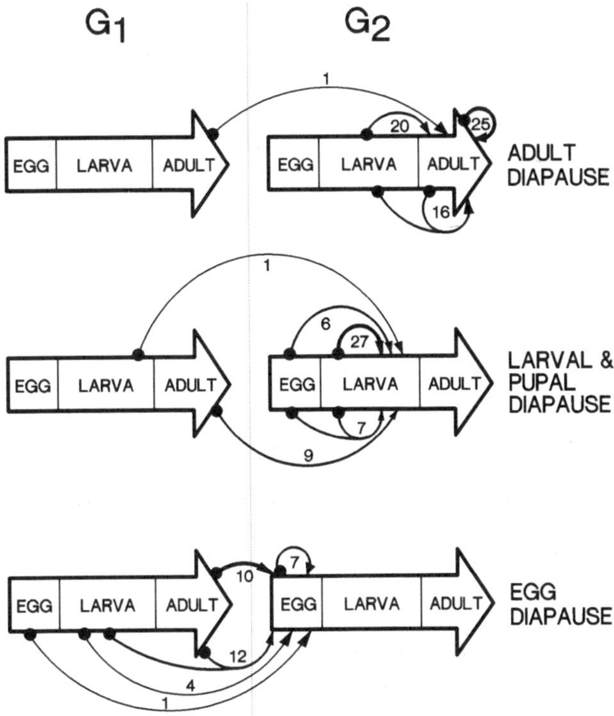

Figure 10. Patterns of phenotypic plasticity in control of diapause. Arrows indicate interactions between stage of sensitivity to environmental cues and stage at which diapause occurs. Numbers associated with arrows indicate number of species displaying the noted pattern of interaction. For species that diapause as adults (top), it is environmental conditions experienced by the adult or the larva that most frequently influence the decision to diapause. In only one species do conditions in the previous generation influence adult diapause. However, for species that diapause as eggs or embryos, it is most often environmental conditions experienced in the maternal generation (G1) that affect diapause. Data from Table 1, and Danks (1987).

At present, we are unable to exclude either hypothesis. However, it is interesting to note that aphids frequently display multi-generational maternal effects, presumably as a result of their vastly telescoped life cycles, with embryos developing within embryos of the grandmaternal generation. Unfortunately, in aphids physiological and chronological interactions are confounded.

GEOGRAPHIC VARIATION FOR MATERNAL EFFECTS

The last point to be addressed in this paper concerns the evidence for geographic variation in maternal effects. This question is of great interest since it points to the causative agents for the evolution of maternal effects, and provides the strongest evidence in support of the adaptive significance of maternal control in nature. Using diapause once again as our model trait, we have found geographic variation in maternal control of offspring development in 17 of the 18 species for which sufficient data could be compiled to address this issue (Table 2 in Mousseau & Dingle [1991]). In most cases, these maternal effects vary geographically in relation to the seasonal characteristics of the location from which the populations were collected.

For example, as was discussed earlier, Beach (1978) found in *Aedes atropalpus* that the interaction between photoperiod and temperature as cues for maternal induction of

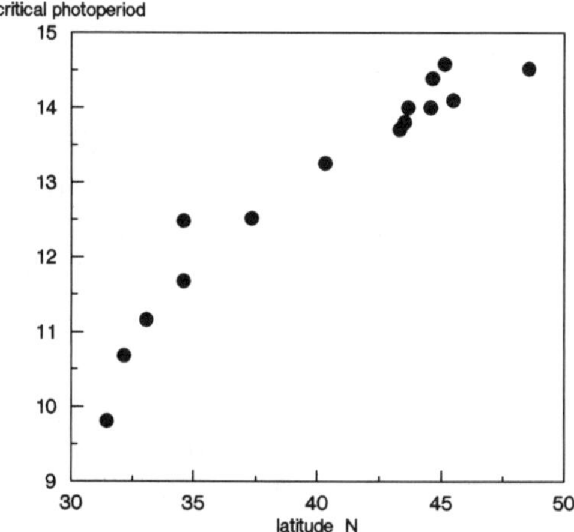

Figure 11. Geographic variation in the critical photoperiod for maternal induction of diapause in the psocid *Peripsocus quadrifasciatus*. **Note that the critical photoperiod increases with latitude. Redrawn from Eertmoed (1978).**

diapause in offspring varied among populations in a manner consistent with that expected, given the reliability of the two cues in the different habitats (Fig. 5). Also, the critical photoperiod for the induction or aversion of diapause varied, with a much longer photoperiod being required in the northern population than in the southern population. Since these experiments were performed under common garden conditions, these differences indicate genetic divergence among populations. Eertmoed (1978) took this question one step further and examined 15 populations of the psocid, *Peripsocus quadrifasciatus*, from a wide variety of latitudes and found that the critical photoperiod for diapause induction varied consistently with latitude, suggesting local adaptation and genetic divergence among populations (Fig. 11). In a study of the ten populations of the striped ground cricket, *Allonemobius fasciatus*, maternal age effects on diapause were most pronounced in bivoltine populations where such age effects would bestow a selective advantage to mothers that displayed them (Mousseau, 1991).

To date there have been no definitive tests for the heritability of maternal effects in insects. However, the observed geographic variation in many species strongly supports the hypothesis that natural selection has acted to produce local adaptations, and genetic divergence among populations. This is good evidence that these traits are indeed heritable, given that most of the geographic variation has probably evolved during the current interglacial period. This finding is of particular interest given the sorts of questions posed by Riska (this volume) and Kirkpatrick & Lande (1989).

CONCLUSIONS

The principle findings of this paper are:

1. Maternal effects are observed in a wide variety of insect taxa, for a large number of adaptive traits, in response to many environmental cues.

2. Many of the maternal effects presented here are a special case of adaptive variation known as phenotypic plasticity in which individual genotypes exhibit variable life histories in response to environmental cues. In almost all species reviewed, maternal effects were correlated with predictable patterns of seasonality and resulted in appropriate switches in life history.

3. We have found that maternal effects on diapause are strongest between an adult mother and her embryos. This observation may be the result of constraints imposed by biochemical machinery and developmental pathways involved in maternal effects; alternatively, natural selection may have favored short term over long term control.

4. We find geographic variation in maternal control of diapause in most species for which sufficient data exist to test for it. This is evidence for genetic divergence and local adaptation of maternal effects.

Thus, rather than simply being a troublesome source of environmental variance, maternal effects, as these data from insects demonstrate, are integral and important aspects of life histories. They allow mothers to influence the future performance of offspring, and they allow circumvention of constraints resulting from genetic correlations among traits (Groeters & Dingle, 1987). Natural selection can therefore act across generations with important consequences (Kirkpatrick & Lande, 1989; Riska, this volume). For these reasons, maternal effects must be considered salient elements of life cycles, and they will need to be incorporated into newly developing theories of life history evolution.

ACKNOWLEDGMENTS

We thank Bruce Riska and Barry Sinervo for organizing the ICSEB IV symposium on Maternal Effects in Evolution. We especially appreciate Bruce's inspiration, intellectual support, guidance, and friendship. Also, we thank the many entomologists who have contributed both published and unpublished data to this review; MaryCarol Rossiter and Peter Frumhoff were especially generous in this regard. We would like to acknowledge Hugh Dank's monograph on dormancy in insects, which kindled the discusssions that ultimately resulted in this paper. This work was supported by an NSERC (Canada) Postdoctoral Fellowship and a UC Davis Center for Population Biology travel award to T.A.M., and NSF grants to H.D.

LITERATURE CITED

Adkisson, P. L. & S. H. Roach. 1971. A mechanism for seasonal discrimination in the photoperiodic induction of pupal diapause in the bollworm *Heliothis zea* (Boddie). Pp. 272–280. *In:* M. Menaker (ed.), *Biochronometry,* National Academy of Sciences.

Anderson, J. F. 1968. Influence of photoperiod and temperature on the induction of diapause in *Aedes atropalpus* (Diptera: Culicidae). *Entomol. Exp. Appl.* 11:321–330.

Anderson, J. F. 1970. Induction and termination of embryonic diapause in the salt marsh mosquito, *Aedes solicitans* (Diptera: Cullicidae). *Bull. Conn. Agric. Exp. Stn.* 711. 22 pp.

Barnes, P. T. 1984. A maternal effect influencing larval viability in *Drosophila melanogaster. J. Heredity* 75:288–292.

Beach, R. 1978. The required day number and timely induction of diapause in geographic strains of the mosquito, *Aedes atropalpus. J. Insect Physiol.* 24:449–455.

Benschoter, C. A. 1970. Culturing *Heliothis* species (Lepidoptera: Noctuidae) for investigation of photoperiod and diapause relationships. *Ann. Ent. Soc. Am.* 63:699–701.

Blackman, R. L. 1975. Photoperiodic determination of the male and female sexual morphs of *Myzus persicae. J. Insect Physiol.* 21:435–453.

Bocher, J. 1975. Notes on the reproductive biology and egg diapause in *Nysius groenlandicus* (Zett.) (Heteroptera: Lygaeidae) *Vidensk. Meddr Dansk. Naturh. Foren.* 138:21–38.

Bogdanova, T. P., Vinogradova, Ye. B. & V. A. Zaslavsky. 1978. The inter-relationship of the maternal influence and reactions governing diapause in *Calliphora vicina* R.-D. (Diptera, Calliphoridae). *Trudy Zool. Inst. Leningr.* 69:62–79.

Bonnemaison, L. 1975. Determinisme de l'apparition des larves estivales de *Periphyllus* (Aphidinae). *C. r. Hebd. Seanc. Acad. Sci., Paris* 243:1166–1168.

Brodeur, J. & J. N. McNeil. 1989. Biotic and abiotic factors involved in diapause induction of the parasitoid, *Aphidius nigripes* (Hymenoptera: Aphidiidae). *J. Insect Physiol.* 35:969–974.

Cadieu, N. 1983. Maternal influence and the effect of heterosis on the viability of *Drosophila melanogaster* during different pre-imaginal stages. *Can. J. Zool.* 61:1152–1155.

Cragg, J. B. & P. Cole. 1952. Diapause in *Lucilia sericata* (Mg.) (Diptera). *J. Exp. Biol.* 29:600–604.

Cullen, J. M. & T. O. Browning. 1978. The influence of photoperiod and temperature on the induction of diapause in pupae of *Heliothis punctigera*. *J. Insect Physiol.* 24:595–601.

Danks, H. V. 1987. *Insect Dormancy: An Ecological Perspective.* Biological Survey of Canada Monograph series No. 1.

Denlinger, D. L. 1972a. Induction and termination of pupal diapause in *Sarcophaga* (Diptera: Sarcophagidae). *Biol. Bull.* 142:11–24.

Denlinger, D. L. 1972b. Seasonal phenology of diapause in the flesh fly *Sarcophaga bullata. Ann. Ent. Soc. Am.* 65:410–414.

Depner, K. R. 1961. The effect of temperature on development and diapause of the horn fly, *Siphona irritans* (L.) (Diptera:Muscidae). *Can. Ent.* 93:855–859.

Depner, K. R. 1962. The effects of photoperiod and ultraviolet radiation on the incidence of diapause in the horn fly, "*Haematobia irritans* (L.) (Diptera: Muscidae)" *Int. J. Biomet.* 5:68–71.

Depner, K. R. & R. F. Harwood. 1966. Photoperiodic responses of two latitudinally diverse groups of *Anopheles freeborni* (Diptera: Culicidae). *Ann. Ent. Soc. Am.* 59:7–11.

Doskoeil, J. 1957. Beitrag zur Kenntnis der Insektendiapause. 2. Einfluss der Beleuchtungslange auf die Entstehung der Diapause der Eier. *Vest. Csl. Spol. Zool.* 21:273–283.

Eertmoed, G. E. 1978. Embryonic diapause in the psocid, *Peripsocus quadrifasciatus:* photoperiod, temperature, ontogeny and geographic variation. *Physiol. Ent.* 3:197–206.

Ewen, A. B. 1966. A possible endocrine mechanism for inducing diapause in eggs of *Adelphocoris lineolatus* (Goeze) (Hemiptera: Miridae). *Experentia* 22:470.

Falconer, D. S. 1989. *Introduction to Quantitative Genetics,* Third Edition. Longman: New York.

Fleuriet, A. & M. C. Vageille. 1982. On the maintenance of the polymorphism at the ref (2)P locus in populations of *Drosophila melanogaster. Genetica* 59:203–210.

Fraser, A. & W. F. Smith. 1963. Diapause in larvae of green blowflies (Diptera: Cyclorrhaphia: *Lucilia* spp.). *Proc. R. Ent. Soc. Lond.* (A) 38:90–97.

Galbreath, J. E. 1976. The effect of the age of the female on diapause in *Mesovelia mulsanti* (Hemiptera: Mesoveliidae). *J. Kans. Ent. Soc.* 49:27–31.

Gaylor, M. T. & W. L. Sterling. 1977. Photoperiodic induction and seasonal incidence of embryonic diapause in the cotton fleahopper, *Pseudatomoscelis seriatus. Ann. Ent. Soc. Am.* 70:893–897.

Giesel, J. T. 1988. Effects of parental photoperiod on development time and density sensitivity of progeny of *Drosophila melanogaster. Evolution* 42:1348–1350.

Giesel, J. T., Lanciani, C. A. & J. F. Anderson. 1989. Effects of parental photoperiod on metabolic rate in *Drosophila melanogaster. Florida Entomol.* 71:499–503.

Glinyanaya, Ye. I. 1975. The importance of daylength in the control of seasonal cycles and diapause in some Psocoptera. *Ent. Rev.* 54(1):10–13.

Gould, F. 1988. Stress specificity of maternal effects in *Heliothis virescens* (Boddie) (Lepidoptera: Noctuidae) larvae. *Mem. Ent. Soc. Can.* 146:191–197.

Groeters, F. R. & H. Dingle. 1987. Genetic and maternal influences on life history plasticity in response to photoperiod by milkweed bugs (*Oncopeltus fasciatus*). *Am. Nat.* 129:332–346.

Groeters, F. R. & H. Dingle. 1988. Genetic and maternal influences on life history plasticity in milkweed bugs (*Oncopeltus fasciatus*): response to temperature. *J. Evol. Biol.* 1:317–333.

Groeters, F. R. & H. Dingle. 1989. The cost of being able to fly in the milkweed-oleander aphid, *Aphis nerii* (Homoptera: Aphididae). *Evol. Ecol.* 3:313–326.

Gustin, R. D. 1974. Termination of diapause in the painted leafhopper, *Endria inimica. Ann. Ent. Soc. Am.* 67:607–609.

Hardie, J. 1987. The photoperiodic control of wing development in the black bean aphid, *Aphis fabae. J. Insect Physiol.* 33:543–549.

Hardie, J. & A. D. Lees. 1985. Endocrine control of polymorphism and polyphenism. Pp. 441–490. *In:* G. A. Kerkut & L. I. Gilbert (eds.), *Comprehensive Insect Physiology, Biochemistry, and Pharmacology.* Vol. 8.

Helfert, B. 1980. Die regulative Wirkung von Photoperiode und Temperatur auf den Lebszklus okologisch unterschiedlicher Tettigoniiden-Arten (Orthoptera, Saltatoria). 2. Teil: Embryogenese und Dormanz der Filialgenera. *Zool. Jb. Abt. Syst. Okol. Geogr. Tiere* 107:449–500.

Henrich, V. C. & D. L., Denlinger. 1982. A maternal effect that eliminates pupal diapause in progeny of the flesh fly, *Sarcophaga bullata. J. Insect Physiol.* 28:881–884.

Hokyo, N., Suzuki, H. & M. Murai. 1983. Egg diapause in the oriental chinch bug, *Cavelerius saccharivorus* Okajima (Heteroptera: Lygaeidae). I. Incidence and intensity. *Appl. Ent. Zool.* 18:382–391.

Honek, A. 1980. Maternal regulation of wing polymorphism in *Pyrrhocoris apterus:* effect of cold activation. *Experientia* 36:418–419.

Isayev, V. A. 1975. Photoperiodic induction of diapause in the egg phase in *Culicoides pulicarius punctatus* Mg. (Diptera, Ceratopogonidae). *Parazitologiya* 9:501–606.

Janssen, G. M., de Jong, G., Joosse, E. N. G. & W. Scharloo. 1988. A negative maternal effect in springtails. *Evolution* 42:828–834.

Kalpage, K. S. P. & R. A. Brust. 1974. Studies on diapause and female fecundity in *Aedes atropalpus. Envir. Ent.* 3:139–145.

Kennedy, J. S. 1961. Continuous polymorphism in locusts. *Symp. Entomol. Soc. London* 1:80–90.

Kerver, W. J. M. & G. Rotman. 1987. Development of ethanol tolerance in relation to the alcohol dehydrogenase locus in *Drosophila melanogaster.* II. The influence of phenotypic adaptation and maternal effect on alcohol supplemented media. *Heredity* 58:239–248.

Khachatryan, A. G. 1981. Structure of the sensitive period in the fly *Calliphora vicina* (Calliphoridae, Diptera) during photoperiodic induction of larval diapause. *Biol. Zh. Arm.* 34:682–687.

Kimura, M. T. & S. Masaki. 1977. Brachypterism and seasonal adaptation in *Orgyia thyellina* Butler (Lepidoptera, Lymantriidae). *Kontyu* 45:97–106.

Kind, T. V. 1965. Neurosecretion and voltinism in *Orgyia antiqua* L. (Lepidoptera, Lymantriidae). *Ent. Rev.* 44(3):326–327.

Kind, T. V. 1972. Endocrine determination of diapause in the moth *Orgyia antiqua* (Lepidoptera, Lymantriidae). Pp. 210–228. *In:* N. I. Goryshin (ed.), *Problems in Photoperiodism and Diapause in Insects.* Leningrad University Press: Leningrad.

Kirkpatrick, M. & R. Lande. 1989. The evolution of maternal characters. *Evolution* 43:485–503.

Kogure, M. 1933. The influence of light and temperature on certain characteristics of the silkworm, *Bombyx mori. J. Dep. Agric. Kyushu Imp. Univ.* 4:1–93.

Labeyrie, V. 1967. Physiologie de la mère et etat de la progéniture chez les insectes. *Bull. Biol.* 101:13–71.

Lamb, R. J. & P. A. MacKay. 1979. Variability in migratory tendency within and among natural populations of the pea aphid, *Acyrthosiphon pisum. Oecologia* 39:289–299.

Langley, P. A., Pimley, R. W., Mews, A. R. & M. E. T. Flood. 1978. Effect of diet composition on feeding, digestion, and reproduction in *Glossina morsitans. J. Insect Physiol.* 24:233–238.

Lees, A. D. 1983. The endocrine control of polymorphism in aphids. Pp. 369–377. *In:* R. G. H. Downer & H. Laufer (eds.), *Endocrinology of Insects.* Liss: New York.

Lees, A. D. 1984. Parturition and alate morph determination in the aphid *Megoura viciae. Entomol. Exp. Appl.* 35:93–100.

Love, G. J. & J. B. Whelchel. 1955. Photoperiodism and the development of *Aedes triseriatus* (Diptera: Culicidae). *Ecology* 36:340–342.

Ludwig, D. & C. Fiore. 1960. Further studies on the relationship between parental age and the life cycle of the mealworm, *Tenebrio molitor. Ann. Entomol. Soc. Amer.* 53:595–600.

Ludwig, D. & C. Fiore. 1961. Effects of parental age on offspring from isolated pairs of the mealworm, *Tenebrio molitor. Ann. Entomol. Soc. Amer.* 54:463–464.

Masaki, S. 1973. Climatic adaptation and photoperiodic response in the band-legged ground cricket. *Evolution* 26:587–600.

Matthée, J. J. 1951. The structure and physiology of the egg of *Locusta pardalina* (Walk.). *Sci. Bull. Dep. Agric. For. Un. S. Afr.* 316. 83 pp.

McGinnis, K. M. & R. A. Brust. 1982. The effect of photoperiod and temperature on the induction of embryonic diapause in *Aedes togoi* (Theobald) (Diptera: Culicidae) and overwintering. *Proc. Ent. Soc. Manitoba* 38: 23.

McHaffey, D. G. & R. F. Harwood. 1970. Photoperiod and temperature influences on diapause in eggs of the floodwater mosquito, *Aedes dorsalis* (Meigen) (Diptera: Culicidae). *J. Med. Ent.* 7:631–644.

McNeil, J. N. & R. L. Rabb. 1973. Physical and physiological factors in diapause initiation of two hyperparasites of the tobacco hornworm, *Manduca sexta. J. Insect Physiol.* 19:2107–2118.

Messina, F. J. 1990. Alternative life-histories in *Callosobruchus maculatus:* environmental and genetic basis. In K. Gujii (ed.), *Proceedings of the Second International Symposium on Bruchids and Legumes.* Junk: Amsterdam.

Mills, A. & I. Hartmann-Goldstein. 1985. Maternal age, development time, position effect variegation in *Drosophila melanogaster. Genet. Sel. Evol.* 17:171–178.

Mori, A., Oda, T. & Y. Wada. 1981. Studies on the egg diapause and overwintering of *Aedes albopictus* in Nagasaki. *Trop. Med. Nagasaki* 23:79–90.

Mousseau, T. A. 1991. Geographic variation in maternal age effects on diapause in a cricket. *Evolution,* 45:1053–1059.

Mousseau, T. A. & H. Dingle. 1991. Maternal effects in insect life histories. *Ann. Rev. Ent.* 36:511–34.

760

Mousseau, T. A. & D. A. Roff. 1989. Adaptation to seasonality in a cricket: patterns of phenotypic and genotypic variation in body size and diapause expression along a cline in season length. *Evolution* 43:1483–1496.

Parrish, D. S. & D. W. Davis. 1978. Inhibition of diapause in *Bathyplectes curculionis,* a parasite of the alfalfa weevil. *Ann. Ent. Soc. Am.* 71:103–107.

Phelan, J. P. & P. C. Frumhoff. 1991. Differences in the effects of parental age on offspring life history between tropical and temperate populations of milkweed bugs. *Evolutionary Ecology,* in press.

Pinger, R. R. & B. F. Eldridge. 1977. The effect of photoperiod on diapause induction in *Aedes canadensis* and *Psorophora ferox* (Diptera: Culicidae). *Ann. Ent. Soc. Am.* 70:437–441.

Qui, M. P. & V. A. Zaslavsky. 1983. Photoperiodic and temperature reactions in *Trichogramma euproctidis* (Hymenoptera, Trichogrammatidae). *Zool. Zh.* 62:1676–1680.

Raina, A. K. & R. A. Bell. 1974. Influence of dryness of larval diet and parental age on diapause in the pink bollworm, *Pectinophora gossypiella* (Saunders). *Envir. Ent.* 3:316–318.

Read, D.C. 1969. Rearing of the cabbage maggot with and without diapause. *Can. Ent.* 101:725–737.

Ring, R. A. 1967a. Photoperiodic control of diapause induction in the larvae of *Lucilia caesar* L. (Diptera: Calliphoridae). *J. Exp. Biol.* 46:117–122.

Ring, R. A. 1967b. Maternal induction of diapause in the larvae of *Lucilia caesar* L. (Diptera: Calliphoridae). *J. Exp. Biol.* 46:123–136.

Rock, G. C., Yeargan, D. R. & R. L. Rabb. 1971. Diapause in the phytoseiid mite *Neoseiulus* (T.) *fallacis. J. Insect Physiol.* 17:1651–1659.

Rockey, S. J., Miller, B. B. & D. L. Denlinger. 1989. A diapause maternal effect in the flesh fly, *Sarcophaga bullata:* transfer of information from mother to progeny. *J. Insect Physiol.* 35:553–558.

Rockey, S. J. & D. L. Denlinger. 1986. Influence of maternal age on incidence of pupal diapause in the flesh fly, *Sarcophaga bullata. Physiol. Entomol.* 11:199–204.

Rockstein, M. 1957. Longevity of male and female houseflies. *J. Gerontology* 12:253–256.

Roff, D. A. 1986. Predicting body size with life history models. *Bioscience* 36:316–323.

Rosenthal, S. S. & C. S. Koehler. 1968. Photoperiod in relation to diapause in *Hypera postica* from California. *Ann. Ent. Soc. Am.* 61:531–534.

Rossiter, M. 1991a. Environmentally based maternal effects: a hidden force in insect population dynamics. *Oecologia,* in press.

Rossiter, M. 1991b. Maternal effects generate variation in life history: consequences of egg size plasticity in the gypsy moth. *Functional Ecology,* 5(3).

Rossiter, M., Yendol, W. G. & N. R. Dubois. 1990. Resistance to *Bacillus thuringiensis* in the Gypsy Moth, (Lepidoptera: Lymantridae): genetic and environmental causes. *J. Econ. Ent.* 83:2211–2218.

Ryan, R. B. 1965. Maternal influence on diapause in a parasitic insect, *Coeloides brunneri* Vier. (Hymenoptera: Braconidae). *J. Insect Physiol.* 11:1331–1336.

Sano-Fujii, I. 1979. The effect of parental age and developmental rate on the production of active form of *Callosobruchus maculatus* (F.) (Coleoptera: Bruchidae). *J. Stored Prod. Res.* 2:187–195.

Sarai, D. S. 1967. Effects of temperature and photoperiod on embryonic diapause in *Nemobius fasciatus* (DeGeer) (Orthoptera, Gryllidae). *Questiones Ent.* 3:107–135.

Sato, T. 1977. Life history and diapause of the white-spotted tussock moth, *Orgyia thyellina* Butler (Lepidoptera: Lymantriidae). *Jap. J. Appl. Ent. Zool.* 21:6–14.

Saunders, D. S. 1962. The effect of the age of female *Nasonia vitripennis* (Walker) (Hymenoptera, Pteromalidae) upon the incidence of larval diapause. *J. Insect Physiol.* 8:309–318.

Saunders, D. S. 1965a. Larval diapause induced by a maternally operating photoperiod. *Nature* 206:739–740.

Saunders, D. S. 1965b. Larval diapause of maternal origin: induction of diapause in *Nasonia vitripennis* (Walk.) (Hymenoptera, Pteromalidae). *J. Exp. Biol.* 42:495–508.

Saunders, D. S. 1966. Larval diapause of maternal origin. II. The effects of photoperiod and temperature on *Nasonia vitripennis. J. Insect. Physiol.* 12:569–581.

Saunders, D. S. 1966. Larval diapause of maternal origin. III. The effect of host shortage on *Nasonia vitripennis. J. Insect Physiol.* 12:899–908.

Saunders, D. S. 1982. *Insect Clocks,* Second Edition. Pergamon Press, Oxford.

Saunders, D. S. 1987. Maternal influence on the incidence and duration of larval diapause in *Calliphora vicina. Physiol. Ent.* 12:331–338.

Saunders, D. S., Macpherson, J. N. & K. D. Cairncross. 1986. Maternal and larval effects of photoperiod on the induction of larval diapause in two species of fly, *Caliphora vicina* and *Lucilia sericata. Expl. Biol.* 46:51–58.

Scali, V. 1968. Biologia riproduttiva del *Bacillus rossius* (Rossi) nei dintorni di Pisa con particolare riferimento all' influenza del fotoperiodo. *Memorie Soc. Tosc. Sci. Nat.* 75:108–139.

Schneiderman, H. A. & J. Horwitz. 1958. The induction and termination of facultative diapause in the chalcid wasps *Mormoniella vitripennis* (Walker) and *Tritneptis klugii* (Ratzeburg). *J. Exp. Biol.* 35:520–551.

Shinkaji, N. 1975. Seasonal occurence of the winter eggs and environmental factors controlling the evocation of diapause in the common conifer spider mite, *Oligonychus ununguis* (Jacobi), on chestnut (Acarina: Tetranychidae). *Appl. Ent. Zool.* 19:105–111.

Simmonds, F. J. 1946. A factor affecting diapause in hymenopterous parasites. *Bull. Ent. Res.* 37:95–97.

Simmonds, F. J. 1948. The influence of maternal physiology on the incidence of diapause. *Phil. Trans. R. Soc. Ser. (B)* 233:385–414.

Singh, R. S. & R. A. Morton. 1981. Selection for malathion resistance in *Drosophila melanogaster*. *Can. J. Genet. Cytol.* 23:355–369.

Tauber, M. J., Tauber, C. A. & S. Masaki. 1986. *Seasonal Adaptations of Insects*. Oxford University Press: New York.

Tschinkel, W. R. 1990. Crowding, maternal age, age at pupation, and the quantitative life history of *Zophobas atratus* (Coleoptera: Tenebrionidae) (Manuscript).

Vinogradova, Ye. B. 1965. An experimental study of the factors regulating induction of imaginal diapause in the mosquito *Aedes togoi* Theob. (Diptera, Culicidae) *Ent. Rev.* 44(3):309–315.

Vinogradova, Ye. B. 1974. The pattern of reactivation of diapausing larvae in the blowfly, *Calliphora vicina*. *J. Insect Physiol.* 20:2487–2496.

Vinogradova, Ye. B. 1975a. The role of photoperiod reaction and temperature in the induction of diapause in the egg stage in *Aedes caspius caspius* Pall. (Dipt., Culicidae). *Parazitologia* 9:385–392.

Vinogradova, Ye. B. 1975b. Intraspecific variation in the reactions controlling larval diapause in *Calliphora vicina*. *Ent. Rev.* 54(4):11–20.

Vinogradova, Ye. B. & K. B. Zinovjeva. 1972. Maternal induction of larval diapause in the blowfly, *Calliphora vicina*. *J. Insect Physiol.* 18:2401–2409.

Wellso, S. G. & P. L. Adkisson. 1964. Photoperiod and moisture as factors involved in the termination of diapause in the pink bollworm, *Pectinophora gossypiella*. *Ann. Ent. Soc. Am.* 57:170–173.

Wilson, G. R. & W. R. Horsfall. 1970. Eggs of floodwater mosquitoes. XII. Installment hatching of *Aedes vexans* (Diptera: Culicidae). *Ann. Ent. Soc. Am.* 63:1644–1647.

Witsack, W. 1971. Experimentelle-ökologische Untersuchungen über Dormanz-formen von Zikaden (Homoptera, Auchenorrhyncha). I. Zur Form und der Embryonaldormanz von *Muellerianella brevipennis* (Boheman) (Delphacidae). *Zool. Jb. Abt. Syst. Ökol. Georgr. Tiere* 98:316–340.

Yakubovich, V.Ye. 1983. Maternal influence on the physiological state of the progeny in the mosquito *Aedes caspius dorsalis*. *Medskaya Prazit.* 52:16–19.

Zaslavsky, V. A. & T.Ya. Umarova. 1982. Photoperiodic and temperature control of diapause in *Trichogramma evanescens* Westw. (Hymenoptera, Trichogrammatidae). *Ent. Rev.* 60(4):1–12.

Prenatal Effects on Mammalian Growth: Embryo Transfer Results

David E. Cowley

Abstract. Maternal effects that mammalian offspring experience prenatally can have a long-lasting impact on postnatal phenotype. Furthermore, they can potentially contribute to evolution because they are in part genetically determined aspects of maternal performance. Reciprocal embryo transfer experiments with mice indicate genetic prenatal maternal effects on embryo survival, pregnancy rate, body weight from birth to ten weeks of age, tail length from birth to six weeks, early and late growth rates, age at ear and eye opening, the amount of fat deposition at 21 days and at ten weeks, testes weight at three weeks of age, and adult skeletal morphology. Maternal physiological factors contributing to these genetic prenatal effects may include interstrain differences in basal metabolic rate, resting oxygen consumption, and systolic blood pressure. In order to understand these maternal effects in an evolutionary context, future studies should focus on estimating the genetic variances for prenatal and postnatal maternal effects, the genetic covariance between them, and their maternal effect coefficients.

It seems clear that the prenatal rate of growth depends largely on the characteristics of the dam as far as it depends on heredity at all.

Wright, 1922, p. 21

Maternal factors constrain the inherent capacity for mammalian growth and development that is conferred by the individual's genome. Various attributes of the mother, including her physiology, anatomy and behavior, are important for early growth and survival of her progeny. Furthermore, these maternal effects can assume an important role in mammalian evolution by selection (Kirkpatrick & Lande, 1989; Atchley & Newman, 1989). The objectives of this paper are to summarize the results of recent embryo transfer experiments aimed at characterizing prenatal maternal effects, to review the diverse literature describing prenatal effects, to examine how these effects might contribute to evolution, and to describe some new experimental designs for estimating the genetic variance for prenatal and postnatal maternal effects in random-bred populations. Because models incorporating maternal effects have generally been developed by animal breeders, the potential exists for confusion to arise over some aspects of terminology. Some of this terminology regarding maternal effects has been outlined in the paper by Riska in this volume and the reader is encouraged to consult it.

The complexity of phenotypic expression of a trait in mammals is sometimes not fully appreciated (Fig. 1). The progeny's nuclear genes; which are inherited equally from both parents, are not always equally expressed. In addition to parental imprinting of chromosomes (Surani et al., 1990), the expression of the progeny's genes can be further modu-

---

Dr. Cowley is with the Department of Genetics, North Carolina State University, Raleigh, NC 27695, USA.

lated through two distinct maternal avenues. Prenatally, factors such as maternal body size, age, parity, uterine litter size (number of fetuses *in utero*) or position in the uterus can have a significant impact on the development and growth of fetal tissues. After birth, a different set of maternal characteristics plays an important role in preweaning development. The quality and quantity of milk, nesting behavior, defense from predators, and other aspects of maternal care are examples of factors that can cause postnatal maternal effects. Whereas these prenatal and postnatal maternal effects are experienced as environmental effects by the offspring, it is important to recognize that the maternal effects themselves can have a genetic basis. Prenatal and postnatal maternal traits that have maternal effects on characteristics in the offspring have been commonly lumped into a composite maternal performance trait. Rarely have attempts been made in the evolutionary biology literature to distinguish between them (Atchley & Newman, 1989).

Development of the mammalian phenotype is further complicated by interactions between the progeny genotype, the prenatal and postnatal maternal factors, and the external environmental conditions. For example, environmental effects are generally visualized as having their effect in the postweaning phase of growth. However, the environment the mother experiences during pregnancy can also have profound indirect consequences on progeny development. Hyperthermia, stress, and nutritional regime are examples of maternal environmental factors which condition the genetic potential of maternal ability and affect fetal development (DeSesso, 1987).

Maternal effects are important in fields as diverse as agriculture, reproductive physiology, toxicology, and evolutionary biology. In toxicology, for example, one is often interested in evaluating developmental responses to substances ingested by the mother during pregnancy or lactation. Much of the interest in human studies revolves around characterizing the potential for fetal damage via trans-placental transfer of toxic effects of foodstuffs, drugs, and chemicals in the environment. In agriculture, crosses between

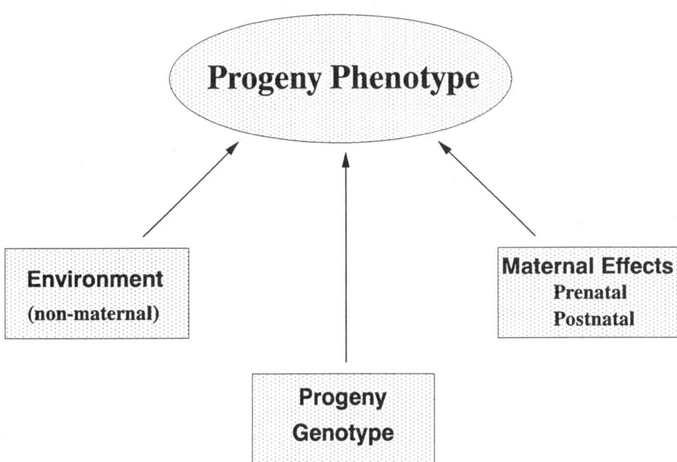

Figure 1. Factors contributing to the phenotypic expression of a quantitative trait in mammals. Environmental effects arise from the external environment, exclusive of maternal effects. Progeny genotype effects stem from Mendelian inheritance of factors influencing a trait. Maternal effects, experienced as an environmental effect by the progeny, can be manifested through expression of the genes in the mother or through environmental conditions the mother experiences during gestation (prenatal) and lactation (postnatal). Myriad interactions between these components are also possible. Recent experiments (Cowley et al., 1989; Pomp et al., 1989; Atchley et al., 1991; Cowley, 1991) have used embryo transfers to estimate prenatal maternal effects on offspring phenotype while holding postnatal maternal effects constant. Other factors that can influence progeny phenotype (not shown) include maternally-transmitted cytoplasmic effects (including mitochondrial DNA) and chromosome imprinting.

strains are often carried out to determine if the performance of individuals from some crosses is superior to the performance of other crosses. By comparing reciprocal crosses, the breeder can ascertain if the contribution of maternal effects (and/or sex-linked effects) is sufficient to warrant using females of a particular strain in the crosses. In evolutionary studies, one may desire to characterize not only direct additive genetic variance arising from the individual's own genotype, but also to estimate additive genetic variance for maternal effects. This information could then be used to predict the response to selection (Kirkpatrick & Lande, 1989).

THE IMPACT OF PRENATAL MATERNAL EFFECTS

The potency of prenatal maternal effects to constrain fetal development can be understood better by examining development of the laboratory mouse. Four stages of prenatal mammalian development (Hafez, 1963) can be recognized conveniently depending on location of the developing offspring in the female reproductive tract and the source of nutrition to the developing embryo or fetus (Table 1). Whereas developmental biologists typically recognize different developmental stages by embryonic or fetal morphology, the stages delineated in Table 1 are more conducive to understanding the impact that prenatal maternal effects can have on progeny development.

Fertilization in the mouse occurs in the oviduct, and the cleaved ovum (Stage 1), nourished by its own cytoplasm and oviductal secretions (Hamner, 1971), attains the blastocyst stage and passes into the uterus at about 3½ days post-fertilization (Stage 2). The embryonic genome becomes active very early (Pedersen & Spindle, 1976) and the rate of cleavage is largely determined by the *Ped* gene, which is closely linked to the MHC *H-2* locus (Warner et al., 1988). Prior to implantation the blastocyst is nourished by uterine secretions (Beier et al., 1971; Daniel, 1971; Aitken, 1979). Implantation of the blastocyst into the uterine wall occurs about 4½ days post fertilization (Theiler, 1983) and for a brief period (Stage 3) the implanted embryo probably obtains nutrients from the trophoblast cells (Snell & Stevens, 1966). Although the embryonic genome is largely responsible for the rate of development, survival of the conceptus through these first three stages is dependent on the internal environment of the oviduct and uterus which is controlled largely by the maternal genome.

Table 1. Stages of prenatal development in the mouse.

			Growth			
Stage	Location	Nutrition	Cell #	Cell Size	Limiting Factors	Duration (%)
Cleaved Ovum	Oviduct	Cytoplasm	+	−	Fetal Genotype	16
		Oviductal Secretions				
Blastocyst	Uterus	Uterine Secretions	+	−	Fetal Genotype	8
Implanted Embryo	Uterus	Trophoblast Cells	+	−	Fetal Genotype	3
Fetus	Uterus	Placenta	+++	+	Fetal Genotype Maternal Genotype Maternal Environment	73

Adapted from Hafez (1963). For cell number and cell size, "+" indicates an increase and "+++" indicates a rapid increase. A "−" denotes a decrease. Limiting factors refers to those factors controlling growth. The maternal genotype and maternal environment are important to embryonic survival during the first three stages, but they are not generally a limiting factor. The approximate duration of each stage is based on figures given by Theiler (1983) assuming a 19 day gestation period.

The mammalian fetus relies entirely on the mother for its essential nutrients during the postimplantation phase of prenatal development (Stage 4). Fetal growth during stage 4 is controlled by fetal genotype (Blakley, 1979), maternal genotype (Cowley et al., 1989), and the external environment the mother experiences (Barr et al., 1970; McLeod et al., 1972). In contrast to the brief preimplantation stages, the fetal stage in the mouse is about 70% of the total gestation period (Theiler, 1983). In a variety of mammalian species, prenatal maternal effects appear to have their greatest impact on fetal growth late in gestation (Gluckman, 1986). In the mouse, this corresponds to the last three or four days of gestation (Snow, 1986).

Prenatal maternal effects have a significant impact on a wide variety of traits in the offspring of mammals, and the most obviously affected trait is birth weight. Prenatal maternal factors are an important influence on birth weight in a wide variety of mammals including humans. Estimates of the percentage of variance in birth weight that is accounted for by prenatal maternal effects range from 25% in lambs (Dickinson et al., 1962), to 40–50% in calves, mice and swine (Donald et al., 1962; Young et al., 1965; El Oksh et al., 1967; Lush et al., 1934), and 75% in guinea pigs (Wright, 1922). Maternal genotype has been estimated to account for 25% of the variation in neonatal birth weight in humans (Robson, 1978). A composite of maternal genotype and maternal environmental effects was reported to account for 40–50% of the variation in human birth weight, and the maternal effect variance increased with parity (Nance et al., 1983).

Through the use of embryo transfers, the influence of prenatal factors on birth weight and subsequent postnatal growth has been demonstrated for a number of mammalian species. Venge (1950) reported that body size of the maternal breed in rabbits was positively correlated with birth weight and postnatal growth. Hunter (1956) and Dickinson et al. (1962) reported similar results for embryos transferred between breeds of sheep differing in body size, and similar results in pigs were reported by Smidt et al. (1967). Hunter (1956) also demonstrated that maternal age and uterine litter size made significant contributions to the birth weight of lambs.

Embryo transfer experiments in mice and deermice have yielded equivocal results regarding the influence of prenatal maternal effects on neonatal body weight. Brumby (1960) reported that prenatal effects were important to the growth of mice from long-term within-litter selection strains. His study showed that prenatal maternal performance of females from the smaller body size strain was generally inferior to females of the larger body size strain. Moore et al. (1970) also reported that selection for increased body size in different mouse strains enhanced prenatal maternal performance. In contrast, Aitken et al. (1977) and Al-Murrani and Roberts (1978) reported that prenatal maternal effects were of very minor importance to fetal growth in strains of mice selected for large or small body size. However, the study by Aitken et al. (1977) examined fetal size at 16 days of gestation, a time when prenatal maternal factors may not have had their full effect (Snow, 1986). Pomp et al. (1989) found significant genetic prenatal maternal effects on birth weight in the inbred mouse strains C3H and SWR and their F1 hybrid C3SWF1, but there was no evidence of nonadditive (heterotic) genetic prenatal maternal effects on birth weight. Roth and Klein (1986) reported that maternal body size had an important effect on progeny birth weight in two Peromyscus subspecies (deermice) that differ substantially in body size. The latter result suggests that the finding of significant prenatal maternal effects on neonatal body size is not unique to laboratory mice.

Generally, experiments using inbred mice have revealed significant prenatal maternal effects on body size of progeny. In contrast, similar experiments conducted on mice from strains selected for increased and decreased body size sometimes fail to show significant prenatal maternal effects. It could be argued that the failure to find significant prenatal maternal effects in lines of mice selected for large and small body size is simply a reflection of selection for rapid or slow postnatal growth, which would be expected to be controlled largely by the individual's genome, rather than from maternal effects experienced prenatally. Furthermore, selection studies in mice are often carried out using within-family selection, a procedure that is expected to minimize the contribution of maternal effects to selection response (Falconer, 1989, p. 233).

Embryo transfer experiments have demonstrated genetic and environmental prenatal maternal effects on a variety of other characters (Table 2). Fekete and Little (1942) demonstrated that female mice of the inbred C57 mouse strain (low mammary tumor incidence) that had been gestated by uterine mothers of the inbred DBA mouse strain (high incidence of spontaneous mammary tumors) had a markedly increased incidence of mammary tumors. The reciprocal transfers of DBA embryos into C57 females resulted in a decreased incidence of mammary tumors in the DBA progeny. The pattern of lobation of the liver was shown to be under at least partial prenatal maternal control in mice, although only female progeny had an altered lobation pattern (Rauch, 1952). The avoidance of alcohol by inbred DBA mice was shown to be affected by the uterine mother's selection/avoidance of alcohol (Randall & Lester, 1975). Cowley et al. (1989) demonstrated significant genetic and environmental prenatal maternal effects on postnatal body weight from birth to nine weeks of age, on tail length from birth to three weeks, on early and late growth rates, and on the ages at ear opening and eye opening in mice. Atchley et al. (1991) found genetic prenatal maternal effects on skeletal morphology of adult mice. A number of authors have reported prenatal maternal effects on embryo survival in mice (Fekete, 1947; Runner, 1951; Moler et al., 1981; Baunack et al., 1986; Pomp et al., 1989; Cowley, 1991).

Other research, not utilizing embryo transfers has implicated prenatal maternal factors in a number of additional progeny characteristics. In humans, there appears to be a prenatal maternal effect on the determination of sole ridge counts (Malhotra et al., 1987). Ottman et al. (1988) reported that offspring of epileptic mothers were more prone to seizures than were offspring of epileptic fathers, suggesting the possibility of a prenatal maternal effect. In rodents, Wolff, et al. (1986) reported that prenatal determination of coat color differentiation, postnatal obesity and tumor susceptibility of A^{vy}/a mice was dependent on the dam's strain and her agouti locus phenotype. In Tabby mice, whisker number of offspring resembles the mother rather than the father (Kindred, 1961). Significant prenatal maternal effects (genetic and environmental) have also been reported for gonad size in male and female mouse fetuses (Argyropoulos & Shire, 1989). An interesting prenatal maternal effect in rodents involves day length. Weaver et al. (1987) reported that day length the mother experiences during pregnancy was communicated to the fetus, via maternal melatonin secretion, during the last few days of gestation. Lee et al. (1989) found that day length the fetuses experience during gestation influences their rate of postnatal development.

One might expect that since postnatal body weight can be strongly influenced by prenatal maternal effects (Cowley et al., 1989), the various tissues of the body might equally respond to the prenatal maternal effects. However, Cowley (1991) reported that variability occurred in the magnitude of prenatal maternal effects. Significant effects were found for testis weight, kidney weight, and the amount of fat accumulated in the subcutaneous and epididymal fat pads in 21 day old mice, as well as the amount of subcutaneous fat in 70 day-old mice. In contrast, postnatal cranial capacity (an index of brain size) and liver weight were unaffected by prenatal maternal genotype. Thus, the sizes of some components of body weight were affected by the genotype of the mother whereas other organs were insensitive to genetic or environmental prenatal maternal effects.

Ontogenetic variability in prenatal maternal effects has also been reported to occur. The prenatal influence of the mother has been shown to extend well into postnatal ontogeny in mice and in deermice, although the relative importance declines with age (El Oksh et al., 1967; Moore et al., 1970; Roth & Klein, 1986; Cowley et al., 1989). El Oksh et al. (1967) reported a gradual decline in the relative importance of prenatal maternal effects,

Table 2. Embryo transfer experiments either demonstrating or failing to find prenatal maternal effects.

Authors	Species	Result
Heape (1890)	rabbit	First report on growth of transferred embryos
Fekete and Little (1942)	mouse	Prenatal maternal effect on the incidence of mammary tumors in female offspring
Fekete (1947)	mouse	Prenatal maternal effects on embryo survival in C57 and DBA inbred strains
Venge (1950)	rabbit	Birth weight positively correlated with body size of maternal breed
Runner (1951)	mouse	Prenatal maternal effect on embryo survival
Rauch (1952)	mouse	Liver lobation pattern of female offspring affected prenatally by mother
Hunter (1956)	sheep	Maternal body size, age and litter size affected birth weight and postnatal growth
McLaren and Michie (1958)	mouse	Number of lumbar vertebrae affected by prenatal maternal factors
Brumby (1960)	mouse	Prenatal maternal effects important to growth of mice from long-term within-litter selection strains, performance of smaller mothers inferior to larger ones
Dickinson et al. (1962)	sheep	Maternal body size had a positive effect on birth weight and postnatal growth of offspring
Smidt et al. (1967)	pig	Body size of maternal strain positively correlated with birth weight and postnatal body size of offspring
McLaren (1970)	mouse	Gestation length negatively correlated with prenatal litter size
Moore et al. (1970)	mouse	Selection for increased body size enhanced prenatal maternal performance, prenatal maternal effects disappeared after about 2 weeks postnatally
Randall and Lester (1975)	mouse	Alcohol avoidance of mice similar to that of mother for inbred DBA mice
Aitken et al. (1977)	mouse	Prenatal maternal effects of minor importance to fetal size at day 16 of gestation in strains selected for large and small body size
Al-Murrani and Roberts (1978)	mouse	Prenatal maternal effects of minor importance in strains selected for large and small body size
Blakley (1979)	mouse	Prenatal maternal environment of large body size line superior to that for small body size line for tibia growth at day 15 of gestation
Moler et al. (1981)	mouse	Maternal genotype more important than fetal genotype in determining embryo survival
Baunack et al. (1986)	mouse	Embryo survival affected by genotype of foster mother
Roth and Klein (1986)	deermouse	Maternal body size affected early progeny development in Peromyscus subspecies, but had no effect on the adult body size difference between subspecies
Iida et al. (1987)	mouse	No prenatal maternal genotype heterosis for embryo survival
Cowley et al. (1989)	mouse	Genetic and environmental prenatal maternal effects on postnatal body size, rate of growth, and age at eye and ear opening. Prenatal maternal heterosis was important for tail length, but not for body weight.
Pomp et al. (1989)	mouse	Genetic prenatal maternal effects on embryo survival and pregnancy rate, but no evidence of prenatal maternal heterosis
Atchley et al. (1991)	mouse	Genetic and environmental prenatal maternal effects on skeletal morphology at 70 days of age, much of the prenatal effect associated with a prenatal maternal effect on overall body size of the offspring
Cowley (1991)	mouse	Genetic prenatal maternal effects on fat deposition and testis weight at 21 days of age

and about 15% of the variance in body weight at 42 days of age was attributable to prenatal maternal effects. Recent studies have demonstrated that even though the relative importance of genetic prenatal maternal effects declined postnatally, they significantly altered genetic expression of body weight and skeletal dimensions in the offspring well into adult ages (Cowley et al., 1989; Atchley et al., 1991). Furthermore, the prenatal maternal effects incurred during gestation can extend into a subsequent F_2 generation (Zamenhof et al., 1971; McLeod et al., 1972).

The experiment conducted by Cowley et al. (1989) can yield information about the *relative* importance of prenatal maternal effects on the variance of traits in laboratory-reared mice. The percentages of total variance for progeny genotype, genotype of the uterine mother, and the interaction between these two genotypes are given in Table 3 for body weight and for tail length. Prior to obtaining the variance components, the data were corrected for sex and for litter size at birth. The reader is cautioned that the percentages in Table 3 give only a rough approximation of the proportion of variance contributed by progeny genotype and by genotype of the uterine mother. This is because the variance components are based on only two degrees of freedom. Furthermore, the experiments utilized only two inbred strains and a hybrid between them. Thus, only three genotypes are represented in the data. Therefore, the reader should focus attention on whether the relative magnitudes are large or small, not on the exact numbers given in Table 3. With this cautionary note, one can observe that genetic prenatal maternal effects are relatively small, perhaps 10% or less of the variance in offspring body weight and tail length. However, for both traits the interaction variance was relatively larger. In general, the percentage of variance attributable to the progeny's genotype was considerably larger than the contribution from prenatal maternal effects.

Table 3. Variance components for progeny (p) and prenatal maternal (uterine, u) genotypes and their interaction expressed as a percent of the total variance.

Trait	Variance Component	Age (days)												
		0	3	6	9	12	21	28	35	42	49	56	63	70
Body Weight	p	48	43	34	15	3	7	3	3	38	42	52	44	49
	u	1	7	2	3	1	4	8	7	5	6	4	3	0
	p × u	7	10	15	17	14	13	8	5	3	4	3	2	7
Tail Length	p	25	48	59	53	55	66	56	52	50	46	43	48	
	u	0	1	3	1	3	<1	1	4	2	2	5	2	
	p × u	7	2	0	4	1	2	6	3	6	6	7	6	

Data were adjusted for sex and the continuous covariate litter size at birth. Data originated from the embryo transfer experiment reported by Cowley et al. (1989).

PRENATAL FACTORS INFLUENCING PROGENY DEVELOPMENT

Many of the prenatal factors that have been recognized to affect fetal development are given in Table 4. A listing such as that in Table 4 is somewhat redundant because some of the factors are interrelated. For example, maternal age and parity are generally positively correlated. Poor maternal nutrition can be considered a form of stress, although stress in Table 4 refers to environmental variables such as hyperthermia and overcrowding. Despite these minor complications, the classification of prenatal factors should be adequate for this discussion.

Jones (1976) stated that fetal growth is regulated by three intertwined factors. First, the ability of the mother, through her own metabolism, to supply nutrients is paramount to fetal growth. Second, the capacity of the placenta to transfer maternally supplied nutrients to the fetus further regulates the capacity of fetal growth. Third, fetal metabolism determines the rate at which the maternally supplied nutrients are assimilated into tissues. A fourth factor, which encompasses various aspects associated with the litter *in utero* (Table 4), could be added to this list.

The mother's ability to supply nutrients to the fetus is a function of her health, nutrition, stress, age, parity, and body size. Although health, nutrition, and stress are largely environmental, their effects on offspring development can be substantial (DeSesso, 1987). Maternal age can be a particularly important source of maternal effect in a population of heterogeneous age composition. Hunter (1956) demonstrated older ewes generally produced larger lambs than did younger ewes. Parity can have an important effect on birth weight in humans (Karn & Penrose, 1951; Karn, 1952; Nance et al., 1983). Nance et al. (1983) reported that prenatal maternal factors accounted for 40% of the variance in birth weight at first parity in humans, and the contribution gradually increased to about 54% at the fifth parity.

Maternal body size has long been recognized as an important contributor of prenatal maternal effects. In a classic experiment, Walton and Hammond (1938) reported that the birth weights of progeny from reciprocal crosses between the large Shire horse and the small Shetland pony were characteristic of birth weights of the maternal breed. Shire mares gave birth to heavier offspring than did Shetland mares, and the birth weights were about the same as for purebred matings. Similar results have been obtained in rabbits (Venge, 1950), sheep (Hunter, 1956; Dickinson et al., 1962), cattle (Joubert & Hammond, 1954; Donald et al., 1962), pigs (Smidt et al., 1967), and deermice (Roth & Klein, 1986). In

Table 4. Intrauterine factors that can affect fetal development.

Factor	Genetic	Environmental	Reference	Species
		Maternal Aspects		
Health	X	X	8, 24	Mouse, human
Nutrition		X	11, 21, 26, 32	Mouse, rat
Age		X	9, 10, 14, 15	Mouse, human
Parity		X	14, 15, 19, 22	Mouse, human
Stress		X	2, 4, 9, 11	Mouse, rat, sheep
Maternal Body Size	X	X	3, 5, 7, 12, 23, 25, 27, 28, 31	Mouse, cow, sheep, human, pig, deermouse, rabbit, horse
Utero-Placental Blood Supply	X	X	4, 10, 13, 18, 20	Mouse, human, rhesus macaque, sheep, rat, guinea pig
		Aspects Associated with the Litter		
Litter Size	X	X	1, 3, 6, 7, 10, 17, 20, 33, 34	Mouse, rat, human, guinea pig, rabbit
Intrauterine Position		X	1, 10, 18, 29, 30	Rat, mouse
Genotype of Other Fetuses	X	X	16	Mouse
Duration of Pregnancy		X	9, 14, 15, 35	Mouse, rabbit, human

1. Barr et al. (1970); 2. Chernoff et al. (1987); 3. Cowley et al. (1989); 4. DeSesso (1987); 5. Donald et al. (1962); 6. Falconer (1960); 7. Falconer (1965); 8. Fekete and Little (1942); 9. Hafez (1963); 10. Healy et al. (1960); 11. Hemm et al. (1977); 12. Hunter (1956); 13. Jones and Robinson (1979); 14. Karn (1952); 15. Karn and Penrose (1951); 16. Kryzanowska (1967); 17. McKeown et al. (1976); 18. McLaren (1965); 19. McLaren (1979); 20. McLaren and Michie (1960); 21. McLeod et al. (1972); 22. Nance et al. (1983); 23. Ounsted and Ounsted (1966); 24. Penrose (1952); 25. Roth and Klein (1986); 26. Shambaugh et al. (1987); 27. Smidt et al. (1967); 28. Venge (1950); 29. Vom Saal and Bronson (1978); 30. Vom Saal and Bronson (1980); 31. Walton and Hammond (1938); 32. Winick and Noble (1966); 33. Wright (1922); 34. Wright (1960); 35. Cowley (1991)

the latter study, although they demonstrated a maternal body size effect on birth weight of offspring, these authors rejected maternal effects as being responsible for maintaining the phenotypic body size difference between the subspecies.

Maternal body size is a complex trait influenced by numerous physiological variables. In the embryo transfer experiments described by Cowley et al. (1989) and Cowley (1991), body size of the maternal strain was an important prenatal maternal factor. Irrespective of the embryo's own genotype, and after correction for differences in prenatal litter size, progeny that had been gestated as embryos by females of the larger body size strain always had greater body weight and growth rates than those gestated by females of the smaller body size strain. The strains used in those experiments, C3HeB/FeJ and SWR/J, differed substantially in several important physiological characteristics. The SWR strain has been shown to have a high basal metabolic rate and high oxygen consumption (Pennycuik, 1967; Storer, 1967), and high systolic blood pressure (Schlager & Weibust, 1967). In contrast, the C3H strain has been shown to have low blood pressure (Schlager & Weibust, 1967), a low oxygen consumption and low metabolic rate (Pennycuik, 1967; Storer, 1967). Cowley (1991) suggested that these physiological differences across strains are probably responsible, in part, for the interstrain difference in body size.

Physiological characteristics that may be involved in determining maternal body size may also be intimately related to the second factor regulating fetal growth, i.e., the ability of the placenta to transfer maternally supplied nutrients to the fetus. Utero-placental factors (Table 4) have been demonstrated to significantly affect prenatal growth in the mouse, rat, guinea pig, sheep, rhesus macaque, and human. Cowley (1991) suggested that in his embryo transfer experiment: 1) interstrain differences in metabolic rate could result in differing nutrient supply to the fetus; 2) differences in oxygen consumption could result in differential oxygenation of the maternal blood and hence limit fetal respiration; and 3) differences in blood pressure could affect the supply of blood to the uterine arteries and placenta.

There is no general consensus regarding the relative importance of fetal genotype, the third factor regulating fetal growth. Fetal genotype is largely the limiting factor in the rate of embryonic growth prior to implantation (Table 1; Warner et al., 1988). Embryo survival has been reported to be more affected by maternal genotype than by progeny genotype (Moler et al., 1981; Pomp et al., 1989). After implantation, the relative importance of fetal genotype varies across species. In humans, prenatal maternal effects have been reported to be a more important determinant of birth weight than fetal genotype (Robson, 1955; Morton, 1955; Mi et al., 1986). Estimates of the percentage of total variance of human birth weight that is attributable to fetal genotype range from 10% to 50% (Robson, 1955, 1978; Magnus, 1984). In livestock, Lush et al. (1934) estimated less than 10% of the birth weight variation in pigs was attributable to progeny genotype. For sheep, the contribution of progeny genotype to variation of birth weight of lambs has been reported to range from 10–20% (Hunter, 1956) to more than 70% (Dickinson et al., 1962). In cattle, about 30% of the variance for birth weight was reported to be accounted for by progeny genotype (Donald et al., 1962). In mice, El Oksh et al. (1967) reported that progeny genotype made an insignificant contribution to birth weight. In contrast, data given in Table 3 indicate that progeny genotype was more important than the mother's genotype in determining variability of neonatal weight and tail length.

The fourth prenatal factor affecting fetal growth includes several components associated with the litter *in utero*. Litter size has been demonstrated to have a marked effect on prenatal growth in a number of species (Table 4). Wright (1960) considered litter size to be the single most important factor affecting perinatal survival and early growth of guinea pigs. In general, litter size is negatively correlated with prenatal growth, i.e., members of large litters are of smaller body size than are members of small litters (Kirkpatrick & Rutledge, 1987; Kirkpatrick et al., 1988; Cowley et al., 1989; Cowley, 1991). Large litters tended to be gestated a slightly shorter time than small litters (McLaren & Michie, 1963; McLaren, 1970; Cowley, 1991).

Intrauterine position of fetuses can have an effect on fetal growth, especially in large litters. In mice, heavier fetuses tend to be located at the ovarian end of the uterine horn, with lighter fetuses near the cervical end (Healy et al., 1960; McLaren, 1965). In contrast, Barr et al. (1970) reported that in rats, fetuses in the middle of the uterine horn were larger than fetuses at either end. Interaction between fetuses of different genotypes can occur (Kryzanowska, 1967), and the sex of adjacent fetuses can also influence postnatal characteristics. Vom Saal and Bronson (1978, 1980) described postnatal reproductive and behavioral masculinization of females that had been located between two male fetuses *in utero*.

EVOLUTIONARY SIGNIFICANCE OF PRENATAL EFFECTS

Undoubtedly one of the most potent forces affecting prenatal and early postnatal survival and growth in mammals is the impact the mother has during the prenatal development of her progeny. The demonstration of genetic determination of prenatal maternal effects has several evolutionary implications. First, because prenatal maternal effects can have a genetic basis (Cowley et al., 1989; Pomp et al., 1989; Atchley et al., 1991; Cowley, 1991), the outcome of multivariate phenotypic selection could be affected (Kirkpatrick & Lande, 1989; Atchley & Newman, 1989; Lande & Kirkpatrick, 1990). Second, environmental modulation of prenatal maternal performance could magnify the effects of genetic drift.

Prenatal effects can potentially contribute to mammalian evolution in several ways. Because these prenatal maternal effects can alter progeny phenotype even at adult ages (Cowley et al., 1989), they have the capacity to significantly affect future reproduction. It has been demonstrated in rats, for instance, that prenatal effects can extend their impact into a second F_2 generation (Zamenhof et al., 1971; McLeod et al., 1972). With this type of effect, the response to selection may be delayed or it may even occur temporarily in a direction opposite to the direction of selection (Cheverud, 1984; Kirkpatrick & Lande, 1989). Since prenatal effects can alter early growth rate (Cowley et al., 1989), selection for an early age at first reproduction, as might occur in a natural population (Conley et al., 1974), could be affected strongly by genetic prenatal effects.

A potentially significant question for mammalian evolution is whether or not genetic prenatal maternal effects can contribute to the covariances among traits in the offspring. This is a significant consideration because of the importance of maternal effects in the equilibrium genetic and phenotypic covariance matrices as well as in the response to directional selection (Kirkpatrick & Lande, 1989). Cowley (1991) reported significant genetic prenatal effects on the among-trait covariances at 21 but not at 70 days of age. Thus, it is possible for prenatal maternal effects to have an impact on the covariances among traits, and hence, they have the potential to affect the multivariate response to selection. As an example, consider body size selection imposed early in postnatal life. The response to selection during this period could be significantly affected by prenatal maternal effects, and the earlier that selection is imposed, the greater is the potential for genetic prenatal maternal effects to contribute to selection response.

A second way that prenatal maternal effects could influence mammalian evolution is through genetic drift. At the population level, there are very important maternal effects on population size and population dynamics that are mediated through female reproductive capacity. Whereas evolutionary biologists frequently think of genetic determination of characters as a requisite for a population to respond to natural selection, one can also envision how environmentally induced prenatal maternal effects might enhance the effects of genetic drift in a population. As an example, consider the sensitivity of female reproductive performance in desert rodents to prevailing environmental conditions (Conley et al., 1974; Reichert, 1979) and how variable female reproductive capacity can increase the probability of population size periodically becoming very small. Reproduction of rodents in a highly unpredictable desert environment is correlated with precipitation and other environmental factors affecting primary productivity. Under adverse conditions female reproductive capacity is depressed and may fail entirely (Conley et al., 1974; Reichert, 1979). Diminished population size brought about by a deterioration in reproductive capacity of the females enhances the effects of genetic drift in local populations and it increases the probability of periodic local extinction. With low migration rates, prenatal maternal effects on population size that are modulated by variable environmental conditions could be an important factor in genetic diversification of local populations.

UNRESOLVED PROBLEMS AND FUTURE INVESTIGATION

Several unexplored questions remain regarding prenatal maternal effects, and answers to these questions are necessary to begin to understand the importance of maternal effects in the evolution of natural populations. First is the covariance between prenatal and postnatal maternal effects. It is presently unknown if postnatal maternal effects reinforce, counteract, or are independent of prenatal maternal effects. There may be cases for each, and without further investigation generalization will be impossible. Second, whereas it has been demonstrated in inbred strains of mice that prenatal maternal effects can have a genetic basis, it is unknown if they are of sufficient magnitude in random-bred populations to actually contribute significantly to selection response.

A common criticism voiced by some ecologists and evolutionary biologists studying natural populations is that studies using domesticated laboratory populations, such as much of the literature reviewed in this paper, have little if any bearing on understanding evolution in natural populations. An alternative view, and one held by this author, is that before one can realistically model and predict evolution in natural populations, one must first study, under carefully controlled laboratory conditions, the relative importance of various sources of genetic variation that can contribute to evolution by selection. The experiments outlined below use random-bred strains of mice to study genetic variation for prenatal and postnatal maternal effects. Given that prenatal effects can have a genetic basis, as demonstrated with studies using completely inbred and isogenic strains of mice, the next logical step is to consider genetic variation in random-bred populations. Estimates of additive genetic variance for prenatal and postnatal maternal effects obtained from studies of random-bred laboratory populations will provide a better understanding of the potential importance of maternal effects in modifying the course of evolution in natural populations.

Using embryo transfers, there are several ways one could estimate the variance for maternal effects in random-bred populations. Two possible embryo transfer experiments utilizing the transfer of inbred embryos into random-bred uterine mothers are diagramed in Figure 2. In the first design (Fig. 2a; Design 1 of Table 5), the pups produced by transfer of isogenic embryos into unrelated recipients are fostered at birth to isogenic nurse mothers. This strategy will standardize the postnatal maternal effects and allow estimation of the variance for prenatal maternal effects. As can be seen in Table 5, the expectation of the variance component for uterine mothers is a composite of additive genetic (Au), dominance (Du) and uterine common environmental (Eu) variance. Unless one is willing to assume a strictly additive genetic model of prenatal maternal effects, this design is of limited utility for estimating genetic prenatal maternal variance. The design would be more useful in preliminary investigations where the experimenter desires to know

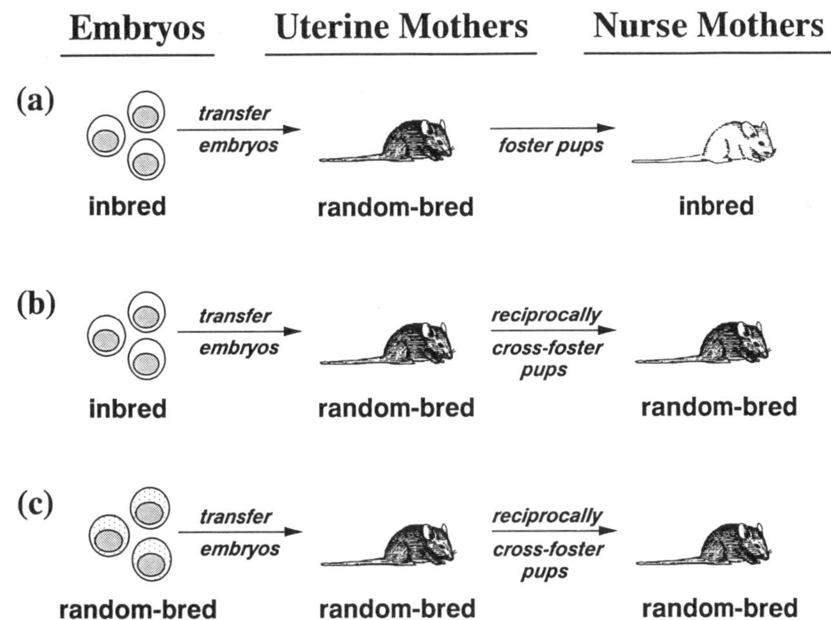

Figure 2. Experimental designs for transfer of embryos from inbred (isogenic) strains into random-bred uterine mothers. Expectations of variance components for these designs are given in Table 5. (a) Utilizing postnatal fostering of the pups to unrelated isogenic females, the experimenter can hold postnatal nursing effects constant. If the recipient females are unrelated, this design will provide an estimate of the total prenatal maternal variance, including additive genetic, dominance, and common environmental components. If the recipient females are paternal half sibs, then one can obtain an estimate of the additive genetic variance for prenatal maternal effects. (b) By reciprocally cross-fostering pups at birth, one can estimate the additive genetic covariance between prenatal and postnatal maternal effects. (c) Transfer of random-bred embryos from a single donor female into pairs of random-bred uterine mothers. It is assumed in this design that the embryos are unrelated to the uterine mothers. At birth, pups are to be reciprocally cross-fostered between the two recipient females. Recipient females could either be related or unrelated. Expectations of the variance components arising from this experimental design are given in Table 5.

whether or not prenatal maternal effects make a significant contribution to variability of offspring.

A variation of the first design (Design 2, Table 5) would be to transfer the inbred embryos into pairs of full sib females. By using related females, the experimenter can isolate genetic sources of variation from the uterine common environmental source of variation. Thus, the variance between replicate pairs (R) of full-sib females receiving inbred embryos includes one-half the prenatal additive genetic variance and one-fourth the prenatal dominance variance (Table 5). Depending on the character under study, this design could yield an informative estimate of prenatal additive genetic variance. For example, prenatal maternal genotype heterosis (a measure of nonadditive or dominance effects) was not an important effect on birth weight and postnatal body weight (Pomp et al., 1989; Cowley et al., 1989). However, tail length was significantly influenced by nonadditive prenatal maternal effects from nine days to nine weeks of age (Cowley et al., 1989). For traits significantly influenced by prenatal dominance effects, the additive genetic prenatal maternal variance could be estimated by transferring the embryos into pairs of paternal half sisters (Design 3, Table 5). The variance between pairs of paternal half sisters estimates one fourth the prenatal maternal additive genetic variance.

Another experiment utilizing the transfer of inbred embryos is shown in Figure 2b. The design here is the same as in Figure 2a except that, within pairs of females receiving

Table 5. Expectations of variance components for some experiments designed for estimating prenatal and postnatal maternal variances and covariances.

Design*	Recipients	Variance Component	σ^2_{Ao}	σ^2_{Do}	σ^2_{Au}	σ^2_{Du}	σ^2_{Am}	σ^2_{Dm}	σ_{AoAu}	σ_{AoAm}	σ_{AuAm}	σ^2_{Eu}	σ^2_{Ec}	σ^2_{e}
1 (2a)	unrelated	σ^2_U	0	0	1	1	0	0	0	0	0	1	0	0
2 (2a)	full sibs	σ^2_R	0	0	1/2	1/4	0	0	0	0	0	0	0	0
		σ^2_U	0	0	1/2	3/4	0	0	0	0	0	1	0	0
3 (2a)	paternal half sibs	σ^2_R	0	0	1/4	0	0	0	0	0	0	0	0	0
		σ^2_U	0	0	3/4	1	0	0	0	0	0	1	0	0
4 (2b)	unrelated	σ^2_R	0	0	0	0	0	0	0	0	1	0	0	0
		σ^2_U	0	0	1	1	0	0	0	0	0	1	0	0
		σ^2_N	0	0	0	0	1	1	0	0	0	0	1	0
5 (2b)	full sibs	σ^2_R	0	0	1/2	1/4	1/2	1/4	0	0	3/2	0	0	0
		σ^2_U	0	0	1/2	3/4	0	0	0	0	0	1	0	0
		σ^2_N	0	0	0	0	1/2	3/4	0	0	0	0	1	0
1–5	all	σ^2_W	0	0	0	0	0	0	0	0	0	0	0	1
6 (2c)	unrelated	σ^2_R	1/2	1/4	0	0	0	0	0	0	1	0	0	0
		σ^2_U	0	0	1	1	0	0	0	0	0	1	0	0
		σ^2_N	0	0	0	0	1	1	0	0	0	0	1	0
7 (2c)	full sibs	σ^2_R	1/2	1/4	1/2	1/4	1/2	1/4	0	0	3/2	0	0	0
		σ^2_U	0	0	1/2	3/4	0	0	0	0	0	1	0	0
		σ^2_N	0	0	0	0	1/2	3/4	0	0	0	0	1	0
8 (3)	unrelated	σ^2_R	1/4	0	0	0	0	0	1/2	1/2	1	0	0	0
		σ^2_U	0	0	1	1	0	0	0	0	0	1	0	0
		σ^2_O	1/4	1/4	0	0	0	0	0	0	0	0	0	0
		σ^2_N	0	0	0	0	1	1	0	0	0	0	1	0
6,7,8	all	σ^2_W	1/2	3/4	0	0	0	0	0	0	0	0	0	1

*The number in parentheses following the Design number is a reference to a figure, e.g., Designs 1 and 2 refer to Figure 2a. Variance components refer to those obtained with analysis of variance. Subscript "R" denotes replicated pairs, "U" represents uterine mother (prenatal), "N" designates nurse mother (postnatal), "O" is ovarian mother, and "W" represents the within-female residual error variance. Causal variance components represent genetic and environmental sources of phenotypic variation. Subscript "Ao" is additive genetic for the offspring genotype, "Do" stands for offspring dominance genetic variance, "Au" is uterine (prenatal maternal) additive genetic, "Du" is uterine dominance, "Am" is additive genetic for postnatal maternal (nursing), "Dm" is postnatal maternal dominance variance, "AoAu" is additive genetic covariance between the progeny's genotype and the prenatal mother's genotype, "AoAm" is additive genetic covariance between offspring genotype and postnatal maternal genotype, "AuAm" is additive genetic covariance between prenatal maternal and postnatal maternal genotypes, "Eu" is environmental effects common to a prenatal sibship, "Ec" is environmental effects common to a postnatal sibship, "e" designates random environmental variation unique to individuals. Refer to figure legends and text for additional details. Note that in design 8, all interaction terms have zero expectation, i.e., $\sigma^2_{UO} = \sigma^2_{UN} = \sigma^2_{ON} = \sigma^2_{UON} = 0$. Thus, the mean squares for each interaction term in Design 8 gives an independent estimate of the residual variance, σ^2_W.

inbred embryos, the resultant pups are reciprocally cross-fostered at birth. When the recipient females are unrelated (Design 4, Table 5), one can estimate the additive genetic covariance between prenatal and postnatal maternal effects (AuAm). As with the design in Figure 2a, this design could also be varied to use uterine mothers that are related. The expectations of the variance components using full-sib recipients is given in Table 5 (Design 5). By comparing analyses from Designs 4 and 5, the difference in variance components for nurses for these two designs estimates the genetic variance for postnatal maternal effects. Coupled with Designs 1, 2, and 3, approximate estimates of the additive genetic variances for prenatal and postnatal effects and the additive genetic covariance between prenatal and postnatal effects can be obtained. The only notable advantage to using Designs 1–4 over designs to be described later is that these designs hold progeny genotype effects constant, allowing one to focus on the maternal effects. By holding progeny genotype effects constant, one may be able to use fewer animals than if progeny genotype varies between individuals.

One could also transfer embryos from a random-bred strain into random-bred uterine mothers (Fig. 2c). It is assumed in this design that embryos are unrelated to the recipient females into which they are transferred. The expectations for variance components are shown in Table 5 for recipients that are unrelated (Design 6) and recipients that are full sisters (Design 7). Inspection of the expected values of variance components suggests that there is no advantage to using these embryo transfer designs, except possibly in conjunction with other designs. None of the variance components provides an estimate of the prenatal or postnatal maternal variance independent of the respective prenatal and postnatal common environmental variances.

An experimental design with the potential to provide a tremendous amount of information is reciprocal transplantation of ovaries between pairs of females (Design 8 of Table 5, Fig. 3). If the females are unrelated and are mated to the same male, the design generates four different variances of individuals, ten different covariances of full sibs, and ten different covariances of paternal half sibs. Five variance components with genetic expectations can be obtained from the analysis of variance (Table 5). Whereas one might think the interaction terms from the analysis of variance would yield estimates of additive genetic covariances among progeny genotype and prenatal and postnatal maternal effects, all of the interaction variance components have an expectation of zero.

Using only the variance components arising from the reciprocal ovarian transplantation, an investigator cannot simultaneously estimate all the causal components shown across the top of Table 5. However, the analysis of this design coupled with two standard cross-fostering experiments (pairs of paternal half sib females and pairs of full sib females, Designs 2 and 3 of Newman et al., 1989) will allow one to estimate all the causal components of variance and covariance. An advantage of this design over the embryo transfer designs shown in Figure 2 is that prenatal and postnatal maternal effects could be studied for lifetime reproduction. The embryo transfer experiments can only be used to study single parities.

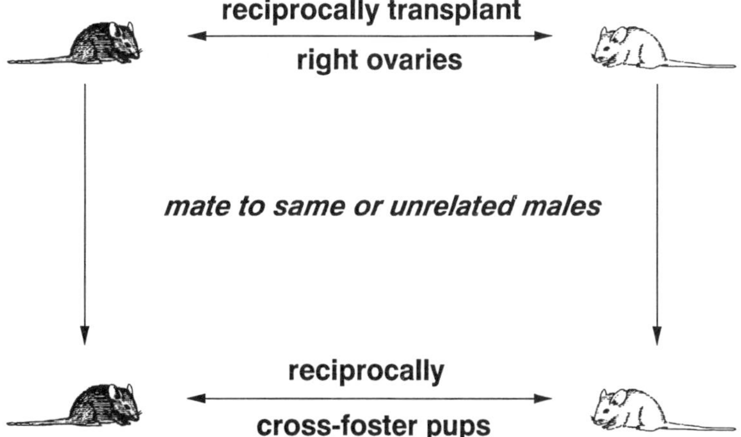

Figure 3. Experimental design for reciprocal transfer of one ovary between females. Females could then be mated to related or unrelated males. Postnatal fostering of pups would be greatly enhanced by using random-bred strains differing in coat color. An advantage of this design over the embryo transfer designs shown in Figure 2 is that the maternal effects could be studied across multiple parities in the same females. Using the analysis of variance of this design, coupled with the analyses of standard cross-fostering between pairs of paternal half sib females and pairs of full sib females (Designs 2 and 3 of Newman et al., 1989), one can simultaneously estimate additive genetic and dominance variances for direct (progeny genotype), prenatal and postnatal effects, the genetic covariances between direct, prenatal, and postnatal maternal effects, as well as prenatal and postnatal common environmental variances (Table 5).

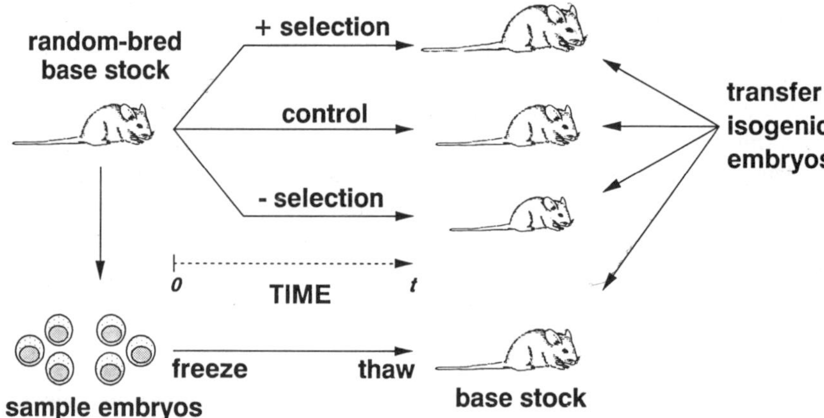

Figure 4. An experimental design for studying how prenatal and postnatal maternal effects change in response to selection. Samples of embryos frozen at the initiation of a selection experiment (Time = 0) would provide a baseline for measuring future maternal performance in selected and control lines. At some future time, t, the embryos could be thawed and reproductively viable females representative of the base stock could be obtained. The performance of females from the base stock could be compared with that of selected females, maintaining both under contemporaneous environmental conditions. Comparison of the maternal performance of base population females with control line females would give an estimate the influence of genetic drift on maternal effects.

Other possible experiments include bisection of embryos followed by transfer and postnatal fostering. The expectations for several designs utilizing embryo bisection have been worked out (Cowley unpublished) but they are not presented here. In general, they would be very difficult to carry out in the laboratory and they would be considerably more expensive than the other designs already mentioned.

One could also use embryo manipulation to address the second issue important to mammalian evolution, i.e., are prenatal maternal effects of sufficient importance to significantly affect the course of selection? Figure 4 gives an experimental design by which one could address this question. At the initiation of a selection experiment (Time = 0), one could freeze a sample of embryos. At the end of selection (Time = t), one could thaw the frozen embryos and rear reproductively viable individuals. One could then ascertain if prenatal effects changed in response to selection by comparing prenatal maternal performance of the base stock and the selected lines under contemporaneous environmental conditions. One might also examine the effects of genetic drift of maternal performance by comparing the performance of base stock females with females from a control line. By transferring inbred embryos into the various recipient females, one could minimize the extraneous variation due to progeny genotype and focus on the prenatal and/or postnatal maternal variance.

Additional experimentation is critical to understanding more completely the relative importance of maternal effects in mammalian evolution. The experiments described above will provide important data that will enhance significantly our understanding of how maternal effects modify the expression and the evolution of Mendelian-inherited genetic factors.

ACKNOWLEDGMENTS

The author is indebted to W. R. Atchley, E. J. Eisen, T. F. C. Mackay, S. Newman, and V. L. Roth for critical comments on the manuscript. The author's research was supported by National Science Foundation grants BSR-8507855 and BSR-8605518 to William R. Atchley.

LITERATURE CITED

Aitken, R. J. 1979. Uterine proteins. *Oxford Reviews of Reproductive Biology* 1:351–382.

Aitken, R. J., Bowman, P. & I. Gauld. 1977. The effect of synchronous and asynchronous egg transfer on foetal weight in mice selected for large and small body size. *J. Embryol. Exp. Morph.* 37:59–64.

Al-Murrani, W. K. & R. C. Roberts. 1978. Maternal effects on body weight in mice selected for large and small size. *Genet. Res.* 32:295–302.

Argyropoulos, G. & J. G. M. Shire. 1989. Genotypic effects on gonadal size in fetal mice. *J. Reprod. Fertil.* 86:473–478.

Atchley, W. R. & S. Newman. 1989. A quantitative genetic model of mammalian development. *Amer. Natur.* 134:486–512.

Atchley, W. R., Logsdon, T. R., Cowley, D. E. & E. J. Eisen. 1991. Uterine effects, epigenetics and postnatal skeletal development in the mouse. *Evolution* 45:891–909.

Barr, M. J., Jr., Jensh, R. P. & R. L. Brent. 1970. Prenatal growth in the albino rat: Effects of number, intrauterine position and resorptions. *Am. J. Anat.* 128:413–428.

Baunack, E., Wieding, B. & K. Gartner. 1986. Prenatal survival of reciprocal F1 hybrids in inbred mice caused by both embryonic factors and genotype of foster mother. *Zuchthygene* 21:115–120.

Beier, H. M., Kühnel, W. & G. Petry. 1971. Uterine secretions as extrinsic factors in preimplantation development. *Adv. Biosci.* 6:165–189.

Blakley, A. 1979. Embryonic bone growth in lines of mice selected for large and small body size. *Genet. Res.* 34:77–85.

Brumby, P. J. 1960. The influence of the maternal environment on growth in mice. *Heredity* 14:1–18.

Chernoff, N., Kavlock, R. J., Beyer, P. E. & D. Miller. 1987. The potential relationship of maternal toxicity, general stress, and fetal outcome. *Teratogenesis, Carcinogenesis, and Mutagenesis* 7:241–253.

Cheverud, J. M. 1984. Evolution by kin selection: a quantitative genetic model illustrated by maternal performance in mice. *Evolution* 38:766–777.

Conley, W., Nichols, J. D. & A. R. Tipton. 1974. Reproductive strategies in desert rodents. Pp. 193–215. *In:* R. H. Wauer & D. H. Riskind (eds.), *Symposium on the Biological Resources of the Chihuahuan Desert Region, United States and Mexico. Proceedings and Transactions Series, National Park Service, No. 3.*

Cowley, D. E. 1991. Genetic prenatal maternal effects on organ size in mice and their potential contribution to evolution. *J. Evol. Biol.* (in press).

Cowley, D. E., Pomp, D., Atchley, W. R., Eisen, E. J. & D. Hawkins-Brown. 1989. The impact of maternal uterine genotype on postnatal growth and adult body size in mice. *Genetics* 122:193–203.

Daniel, J. C. 1971. Uterine proteins and embryonic development. *Adv. Biosci.* 6:191–203.

DeSesso, J. M. 1987. Maternal factors in developmental toxicity. *Teratogenesis, Carcinogenesis, and Mutagenesis* 7:225–240.

Dickinson, A. G., Hancock, J. L., Hovell, G. J. R., Taylor, St. C. S. & G. Wiener. 1962. The size of lambs at birth—a study involving egg transfer. *Anim. Prod.* 4:64–79.

Donald, H. P., Russell, W. S. & St. C. S. Taylor. 1962. Birth weights of reciprocally cross-bred calves. *J. Agric. Sci.* 58:405–412.

El Oksh, H. A., Sutherland, T. M. & J. S. Williams. 1967. Prenatal and postnatal maternal influence on growth in mice. *Genetics* 57:79–94.

Falconer, D. S. 1960. The genetics of litter size in mice. *J. Cell. Comp. Physiol.* 56, Suppl. 1:153–167.

Falconer, D. S. 1965. Maternal effects and selection response. *Proc. XIth Intl. Congr. Genetics* 3:763–774.

Falconer, D. S. 1989. *Introduction to Quantitative Genetics.* Longman Scientific and Technical, Essex, England.

Fekete, E. 1947. Differences in the effect of uterine environment upon development in the dba and C57 black strains of mice. *Anat. Rec.* 98:409–415.

Fekete, E. & C. C. Little. 1942. Observations on the mammary tumor incidence of mice born from transferred ova. *Cancer Res.* 2:525–530.

Gluckman, P. D. 1986. The regulation of fetal growth. P. 85–104. *In:* Buttery, P. J., Haynes, N. B. & D. B. Lindsay (eds.), *Control and Manipulation of Animal Growth.* Butterworths: London.

Hafez, E. S. E. 1963. Maternal influence on fetal size. *Int. J. Fertil.* 8:547–553.

Hamner, C. E. 1971. Composition of oviductal and uterine fluids. *Adv. Biosci.* 6:143–163.

Healy, M. J. R., McLaren, A. & D. Michie. 1960. Foetal growth in the mouse. *Proc. Roy. Soc. London, Ser. B.* 153:367–379.

Heape, W. 1890. Preliminary note on the transplantation and growth of mammalian ova within a uterine foster mother. *Proc. Roy. Soc.* 48:457–458.

Hemm, R. D., Arslanoglon, L. & J. J. Pollock. 1977. Cleft palate following food restriction in mice; association with elevated maternal corticosteroids. *Teratology* 15:243–248.

Hunter, G. L. 1956. The maternal influence on size in sheep. *J. Agric. Sci.* 48:36–60.

Iida, K., Mizuma, Y. & J. Nagai. 1987. Heterosis in survival of transferred mouse embryos. *Z. Versuchstierkd.* 30:99–103.

Jones, C. T. 1976. Fetal metabolism and fetal growth. *J. Reprod. Fertil.* 47:189–201.

Jones, C. T. & J. S. Robinson. 1979. The influence of changes in the utero-placental circulation on the growth and development of the fetus. Pp. 395–409. *In:* D. R. Newth & M. Balls (eds.), *Maternal Effects in Development.* Cambridge University Press: Cambridge.

Joubert, D. M. & J. Hammond. 1954. Maternal effect on birth weight in South Devon × Dexter cattle crosses. *Nature* 174:647–651.

Karn, M. N. 1952. Birth weight and length of gestation of twins, together with maternal age, parity and survival rate. *Ann. Eugen.* 16:365–377.

Karn, M. N. & L. S. Penrose. 1951. Birth weight and gestation time in relation to maternal age, parity and infant survival. *Ann. Eugen.* 16:147–164.

Kindred, B. M. 1961. A maternal effect on vibrissae score due to the Tabby gene. Aust. *J. Biol. Sci.* 14:627–636.

Kirkpatrick, B. W. & J. J. Rutledge. 1987. The influence of prenatal and postnatal fraternity size on reproduction in mice. *Biol. Reprod.* 36:907–914.

Kirkpatrick, B. W., Arias, J. & J. J. Rutledge. 1988. Effects of prenatal and postnatal fraternity size on long-term reproduction in mice. *J. Anim. Sci.* 66:62–69.

Kirkpatrick, M. & R. Lande. 1989. The evolution of maternal characters. *Evolution* 43:485–503.

Kryzanowska, H. 1967. Cell number in relation to heterosis during embryonic growth of mice. *Genet. Res.* 10:235–240.

Lande, R. & M. Kirkpatrick. 1990. Selection response in traits with maternal inheritance. *Genet. Res.* 55:189–197.

Lee, T. M., Spears, N., Tuthill, C. R. & I. Zucker. 1989. Maternal melatonin treatment influences rates of neonatal development in meadow vole pups. *Biol. Reprod.* 40:495–502.

Lush, J. L., Hetzer, H. O. & C. C. Culberson. 1934. Factors affecting birth weights of swine. *Genetics* 19:329–343.

Magnus, P. 1984. Causes of variation in birth weight; a study of offspring of twins. *Clin. Genet.* 25:15–24.

Malhotra, K., Vijayakumar, M., Borecki, I. B., Mathew, S., Poosha, D. V. R. & D.C. Rao. 1987. Resolution of genetic and uterine environmental effects in a family study of new dermatoglyphic measure: sole pattern ridge counts. *Am. J. Phys. Anthro.* 74:103–108.

McKeown, T., Marshall, T. & R. G. Record. 1976. Influences on fetal growth. *J. Reprod. Fertil.* 47:167–181.

McLaren, A. 1965. Genetic and environmental effects on foetal and placental growth in mice. *J. Reprod. Fertil.* 9:79–98.

McLaren, A. 1970. The fate of very small litters produced by egg transfer in mice. *J. Endocr.* 47:87–94.

McLaren, A. 1979. The impact of pre-fertilization events on post-fertilization development in mammals. Pp. 287–320. *In:* D. R. Newth & M. Balls (eds.), *Maternal Effects in Development.* Cambridge University Press: Cambridge.

McLaren, A. & D. Michie. 1958. Factors affecting vertebral variation in mice. 4. Experimental proof of the uterine basis of a maternal effect. *J. Embryol. Exp. Morphol.* 6:645–659.

McLaren, A. & D. Michie. 1960. Control of prenatal growth in mammals. *Nature* 187:363–365.

McLaren, A. & D. Michie. 1963. Nature of the systemic effect of litter size on gestation period in mice. *J. Reprod. Fertil.* 6:139–141.

McLeod, K. I., Goldrick, R. B. & H. M. Whyte. 1972. The effect of maternal malnutrition on the progeny in the rat. *Aust. J. Exp. Biol. Med. Sci.* 50:435–446.

Mi, M. P., Earle, M. & J. Kagawa. 1986. Phenotypic resemblance in birth weight between first cousins. *Ann. Hum. Genet.* 50:49–62.

Moler, T. L., Donahue, S. E., Anderson, G. B. & G. E. Bradford. 1981. Effects of maternal and embryonic genotype on prenatal survival in two selected mouse lines. *J. Anim. Sci.* 51:300–303.

Moore, R. W., Eisen, E. J. & L. C. Ulberg. 1970. Prenatal and postnatal maternal influences on growth in mice selected for body weight. *Genetics* 64:59–68.

Morrison, D. F. 1976. *Multivariate Statistical Methods.* McGraw-Hill: New York.

Morriss, G. M. 1979. Teratogenesis. Pp. 351–373. *In:* D. R. Newth & M. Balls (eds.), *Maternal Effects in Development.* Cambridge University Press: Cambridge.

Morton, N. E. 1955. The inheritance of human birth weight. *Ann. Hum. Genet.* 20:125–134.

Nance, W. E., Kramer, A. A., Corey, L. A., Winter, P. M. & L. J. Eaves. 1983. A causal analysis of birth weight in the offspring of monozygotic twins. *Am. J. Hum. Genet.* 35:1211–1223.

Newman, S., Rutledge, J. J. & B. R. Riska. 1989. Estimation of prenatal maternal genetic effects. *J. Anim. Breed. Genet.* 106:30–38.

Ottman, R., Annegers, J. F., Hauser, W. A. & L. T. Kurland. 1988. Higher risk of seizures in offspring of mothers than of fathers with epilepsy. *Am. J. Hum. Genet.* 43:257–264.

Ounsted, M. & C. Ounsted. 1966. Maternal regulation of intra uterine growth. *Nature* 212:995–997.

Pedersen, R. A. & A. I. Spindle. 1976. Genetic effects on mammalian development during and after implantation. *CIBA Found. Symp.* 40:133–140.

Pennycuik, P. R. 1967. A comparison of the effects of a variety of factors on the metabolic rate of the mouse. *Aust. J. Exp. Biol. Med. Sci.* 45:331–346.

Penrose, L. S. 1952. Data on the genetics of birth weight. *Ann. Eugen.* 16:378–381.

Pomp, D., Cowley, D. E., Eisen, E. J., Atchley, W. R. & D. Hawkins-Brown. 1989. Donor and recipient genotype and heterosis effects on survival and prenatal growth of transferred mouse embryos. *J. Reprod. Fertil.* 86:493–500.

Randall, C. L. & D. Lester. 1975. Alcohol selection by DBA and C57BL mice arising from ova transfers. *Nature* 255:147–148.

Rauch, H. 1952. Strain differences in liver patterns in mice. *Genetics* 37:617 (Abstr.).

Reichert, S. E. 1979. Development and reproduction in desert mammals. Pp. 797–822, *In:* D. W. Goodall & R. A. Perry (eds.), *Arid-land Ecosystems: Structure, Functioning and Management.* Cambridge University Press: London.

Robson, E. B. 1955. Birth weight in cousins. *Ann. Hum. Genet.* Pp. 19:262–268.

Robson, E. B. 1978. The genetics of birth weight. Pp. 285–297. *In:* F. Faulkner & J. M. Tanner (eds.), *Human Growth. 1. Principles and Prenatal Growth.* Plenum Press: New York.

Roth, V. L. & M.S. Klein. 1986. Maternal effects on body size of large insular *Peromyscus maniculatus:* evidence from embryo transfer experiments. *J. Mamm.* 67:37–45.

Runner, M. N. 1951. Differentiation of intrinsic and maternal factors governing intrauterine survival of mammalian young. *J. Exp. Zool.* 116:1–20.

Schlager, G. & R. S. Weibust. 1967. Genetic control of blood pressure in mice. *Genetics* 55:497–506.

Shambaugh, G. E., III, Mankad, B., Derecho, M. L. & R. E. Koehler. 1987. Enzyme markers of maternal malnutrition in fetal rat brain. *J. Nutr.* 117:144–152.

Smidt, D., Steinbach, J. & B. Scheven. 1967. Use of miniature pigs in reproductive biology research. Pp. 45–50. *In: Proceedings of a Symposium on Swine in Biomedical Research, Batelle Memorial Institute,* Richland, WA.

Snell, G. D. & L. C. Stevens. 1966. Early embryology. Pp. 205–245. *In:* E. L. Green (ed.), *Biology of the Laboratory Mouse.* Dover Publications, Inc.: New York.

Snow, M. H. L. 1986. Control of embryonic growth rate and fetal size in mammals. Pp. 67–82. *In:* F. T. Faulkner & J. M. Tanner (eds.), *Human Growth, Vol. 1.* Plenum Press: New York.

Storer, J. B. 1967. Relation of lifespan to brain weight, body weight, and metabolic rate among inbred mouse strains. *Exp. Gerontol.* 2:173–182.

Surani, M. A., Allen, N. D., Barton, S.C., Fundele, R., Howlett, S. K., Norris, M. L. & W. Reik. 1990. Developmental consequences of imprinting of parental chromosomes by DNA methylation. *Phil. Trans. R. Soc. Lond. Ser. B* 326:313–327.

Theiler, K. 1983. Embryology. Pp. 121–136. *In:* H. L., Foster, J. D. Small & J. G. Fox (eds.), *The Mouse in Biomedical Research. Vol. 3. Normative Biology, Immunology and Husbandry.* Academic Press: Orlando, FL.

Venge, O. 1950. Studies on the maternal influence on the birth weight in rabbits. *Acta Zool.* 31:1–148.

Vom Saal, F. S. & F. H. Bronson. 1978. *In utero* proximity of female mouse fetuses to males: effect on reproductive performance during later life. *Biol. Reprod.* 19:842–853.

Vom Saal, F. S. & F. H. Bronson. 1980. Sexual characteristics of adult female mice are correlated with their blood testosterone levels during prenatal development. *Science* 208:597–599.

Walton, A. & J. Hammond. 1938. The maternal effects on growth and conformation in Shire horse-Shetland pony crosses. *Proc. Roy. Soc. London, Ser. B.* 125:311–335.

Warner, C. M., Brownell, M.S. & M. A. Ewoldsen. 1988. Why aren't embryos immunologically rejected by their mothers? *Biol. Reprod.* 38:17–29.

Weaver, D. R., Keohan, J. T. & S. M. Reppert. 1987. Definition of a prenatal sensitive period for maternal-fetal communication of day length. *Am. J. Physiol.* 253:E701–704.

Winick, M. & A. Noble. 1966. Cellular response during malnutrition at various ages. *J. Nutr.* 89:300–306.

Wolff, G. L., Roberts, D. W. & D. B. Galbraith. 1986. Prenatal determination of obesity, tumor susceptibility, and coat color pattern in viable yellow (A^{vy}/a) mice. *J. Hered.* 77:151–158.

Wright, S. 1922. The effects of inbreeding and cross-breeding of guinea pigs. III. Crosses between highly inbred families. *USDA Bull.* 1121.

Wright, S. 1960. The genetics of vital characters of the guinea pig. *J. Cell. Comp. Physiol.* 56, *Suppl.* 1:123–151.

Young, C. W., Legates, J. E. & B. R. Farthing. 1965. Prenatal influences on growth, prolificacy and maternal performance in mice. *Genetics* 52:553–561.

Zamenhof, S., Van Martens, E. & L. Gravel. 1971. DNA (cell number) in neonatal brain: second generation (F_2) alteration by maternal (F_0) dietary protein restriction. *Science* 172:850–851.

Maternal Effects in Fish Life Histories

David N. Reznick

Abstract.   Where sought, maternal effects have often been found in fish. The most common maternal effects involve offspring size, but this is possibly more of a consequence of where investigators have looked than a measure of where maternal effects are most likely to be found. The Forster-Blouin study illustrates that a wider array of effects may be found, but special experiments are required to detect them. A review of the consequences of egg size for subsequent development also suggests that the maternal effect on egg size may be correlated with other aspects of the life history. Specifically, larger eggs tended to produce larger hatchlings, these eggs were more likely to survive early development, the hatchlings survived longer in the absence of food, and they tended to grow faster. Other studies suggest that faster growth and larger body size will, in turn, influence the age and size at maturity. The presence of such effects argues that a more detailed study of maternal effects would be well rewarded and suggests that these effects could strongly influence the short-term response to natural selection.

INTRODUCTION

The goal of theoretical treatments of maternal effects is understanding how such effects can influence the evolution of phenotypes under natural or artificial selection (Kirkpatrick & Lande, 1989). Having established that maternal effects can strongly influence the response to selection (Riska, this volume), the next question is: "How common are such effects and what traits are the most likely to be influenced?". My goal is to summarize some of what is known about maternal effects in fish.

I will first give an example of a maternal genotype effect, then multiple examples of maternal environment effects. These examples are derived from research done by myself and my colleagues and all involve work done on species in the family Poeciliidae. Because most known maternal effects are on egg or offspring size, I will then review what is known of the effects of egg size on future development. This added review will therefore consider the degree to which maternal effects on offspring size can influence other aspects of the offspring's phenotype. Finally, I will consider some possible directions for future research on maternal effects. This is intended as a prospective review, in that there has been little direct interest in maternal effects in the past, so the results summarized here are often incidental to investigations with other goals. This review has been done with the intent of establishing the possible importance of maternal effects and hence to motivate interest in their empirical study.

Dr. Reznick is with the Department of Biology, University of California, Riverside, CA 92521, USA.

EFFECTS OF MATERNAL GENOTYPE—"GRANDFATHER EFFECTS" IN *POECILIA RETICULATA* AND *GAMBUSIA AFFINIS*

Maternal genotypic effects represent genes expressed in the mother with phenotypic consequences in the offspring. I investigated such effects as part of my studies of the genetic basis for interpopulation differences in offspring size (Reznick, 1981, 1982a). In each study, I hybridized laboratory reared fish from populations which produced large versus small offspring (Fig. 1). The results summarized here represent the experiments done on *Poecilia reticulata* (Reznick, 1982a), although the same results and conclusions apply to *Gambusia affinis*. The populations chosen differed in that one produced offspring that were approximately 50% larger in dry weight than the other. The experiments were designed so that I could evaluate the independent contributions of fathers and mothers to offspring size (see Fig. 1 caption for details).

I found, after the first generation of hybridization, that offspring size was determined by the site of origin of the mother and was not affected by the father (Fig. 2). So, for example, mothers from localities where mothers generally produced large offspring also produced large offspring, regardless of who they were mated to.

I performed a second generation of crosses between the four types of females from the first experiment (LL, LS, SL, and SS—see Fig. 1 caption) and a new group of laboratory born males from the same two sites (Fig. 3). The experiment therefore involved eight types of crosses, two types of fathers crossed with four types of mothers. When analysed as a two-way experiment (mothers × fathers), I again found that only mothers influenced offspring size, with no significant contribution coming from the fathers. The effects of fathers are represented by the averages of each row in Figure 3; note that the means for each row are nearly identical. The averages of the columns represent female effects.

So far, the results are the same as in the first experiment; however, the female effect can be further broken down to consider the influence of grandmothers and grandfathers on offspring size (Fig. 4). It is possible to consider the contribution of the grandfathers by averaging the SS and LS results and comparing them to the SL and LL results (Fig. 4). Similarly, it is possible to assess the effects of the grandmothers (Fig. 4). These comparisons reveal a significant contribution of grandmothers and grandfathers to offspring size, as illustrated in the right hand portion of Figure 4. The magnitude of the grandfather

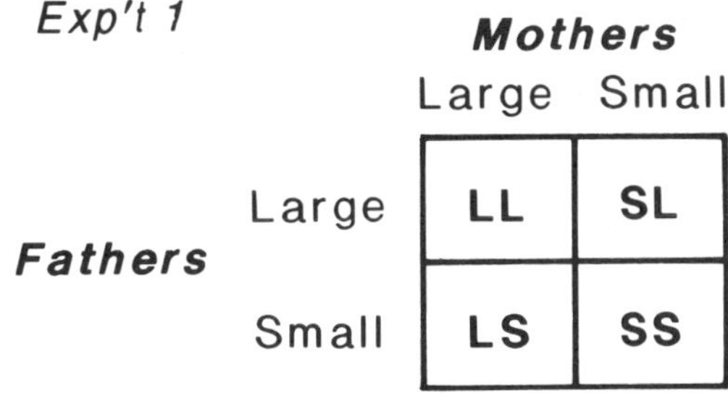

Figure 1. Schematic of the first experiment which describes the genetic basis of interpopulation differences in offspring size. Virgin females from the two types of localities (ones that produce large versus small offspring) were mated to laboratory-born males from the same two types of localities. The experiment is thus designed for a two-way analysis of variance, with mothers and fathers as the main effects and the two localities as the levels of each effect. The offspring from these crosses are denoted by two letters, the first representing the mother's locality, the second the father's locality. "LS" therefore represents the offspring from a mother from a locality that typically produces large offspring crossed to a father from a locality that typically produces small offspring.

782

Experiment 1 Results

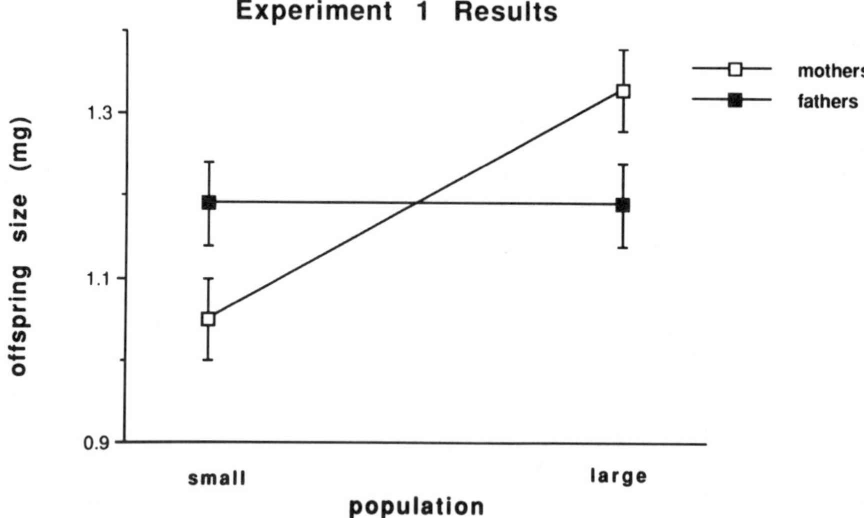

Figure 2. Results for the first generation of crosses in guppies (*Poecilia reticulata*—from Reznick, 1982). The y-axis represents offspring size (dry weight) while the x-axis represents the locality of origin. The plotted values are means plus or minus one standard error. The locality of origin for the females had a strong and significant effect on offspring size. The father did not have a significant effect on offspring size and there was no significant interaction between fathers or mothers.

Exp't. 2 Results

Mothers

Fathers	SS	SL	LS	LL	ave. off. size (by father)
S	0.82	0.94	1.21	1.22	→ 1.07
L	0.94	1.10	1.10	1.42	→ 1.15
ave. off. size (by mother)	0.88	1.02	1.15	1.32	

Figure 3. Results for the second generation of crosses in guppies (*Poecilia reticulata*—from Reznick, 1982). The four types of females are represented by the column headings; these headings correspond to the four crosses described in the caption to Figure 1. The two types of males are the row headings. The entry in each cell is the mean dry weight of the offspring produced by four replicate crosses. The numbers at the end of each row and column represent the average for that entire row or column.

and grandmother contributions did not differ significantly and none of the interactions in this three-way analysis (grandmothers × grandfathers × fathers) were significant.

My interpretation of the combined results of these two experiments is that offspring size is determined by the mothers' genotype. These fish are livebearing and ovoviviparous, or lecithotrophic (Wourms, 1981), which means that eggs are fully provisioned prior to fertilization, and are then simply retained by the female after fertilization with little additional input of resources by the mother. This means that egg size, and hence offspring size, is determined prior to fertilization. However, the father contributes to the genotype of the daughter, and hence contributes to the size of his grandchildren. This effect is therefore a consequence of the maternal genome controlling the characteristics of the offspring prior to when the paternal genome begins to function. These experiments consider genetic differences among populations. To be relevant to the sort of maternal

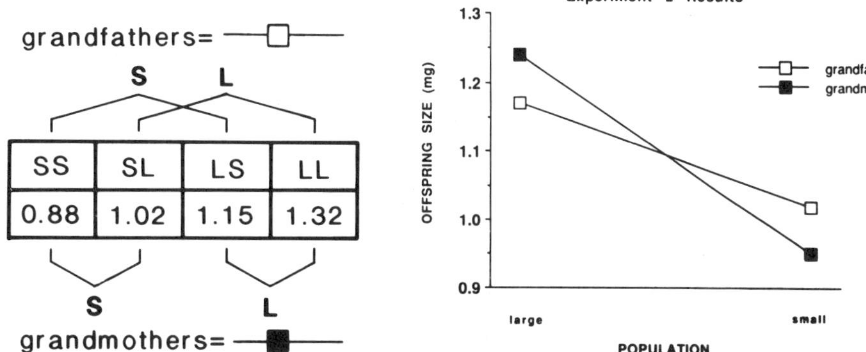

Figure 4. The analysis of grandparent effects in the second generation of crosses in guppies (after Reznick, 1982). The left hand portion of the figure represents the means for the four columns in Fig. 3 and illustrates the values used to estimate the influence of grandmothers and grandfathers on offspring size. The right hand portion of the figure illustrates the significant grandmother and grandfather effects. Plotted values are means plus or minus one standard error.

effects discussed in this symposium, it is necessary to show that the same effect applies to genetic variation within a population. This result should also apply to all egg-laying species, which includes most fish species.

MATERNAL ENVIRONMENT

Factors Influencing Offspring Size

Maternal environment effects represent environmental factors experienced by the mother that influence the phenotype of the offspring. Such effects are usually the consequence of the environment experienced by the mother during egg formation or gestation. Examples presented here include the influence of food availability, salinity, and temperature experienced by the mother during the development of the young. All of these investigations consider factors which are known to vary in the natural environment of these fish. The values chosen for each environmental factor in each study also represent ones normally experienced by these fish.

I also consider the influence of body size, and the age of the mother. These last two factors are generally confounded in fish because they grow throughout their lifespan, hence older fish are also generally larger. Falconer (1981) refers to such effects as "maternal environment, general," meaning that they are attributable to non-genetic variation among mothers, but would apply to all offspring that a female produces. Such an effect is in contrast to those enumerated above, which Falconer refers to as "maternal effects, immediate," meaning that they apply presumably only to a given litter and are attributable to the environment experienced by the female during the formation of that litter.

Food Availability.   I and Tony Yang investigated the influence of food availability for the mother on the nature of her offspring, following a design first used by Hester (1964) (Fig. 5). We began with sexually mature, reproducing females. Females were assigned to either a "high" or "low" level of food availability after producing a litter, then kept on this level of food availability until they produced their second litter (Interval 1). After producing the second litter, they were again randomly assigned to a high or low level of food availability, with the constraint that there be equal numbers in each of four treatments: high-high, high-low, low-high, low-low. They were maintained on this level of food availability until they produced their third litter (Interval 2). Dependent variables asso-

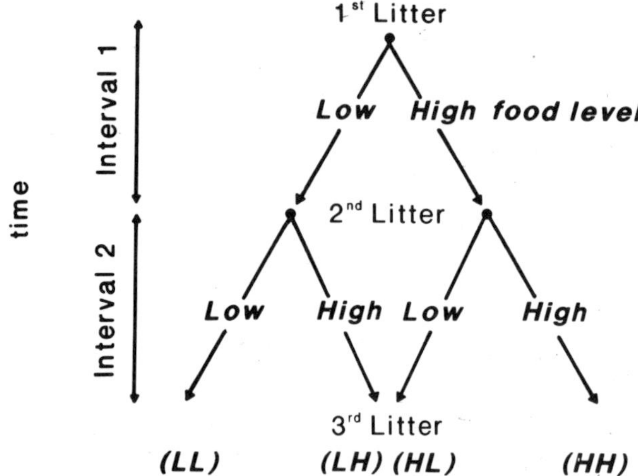

Figure 5. Our modification of an experimental design first used by Hester (1964) to investigate the influence of food availability on reproduction in guppies.

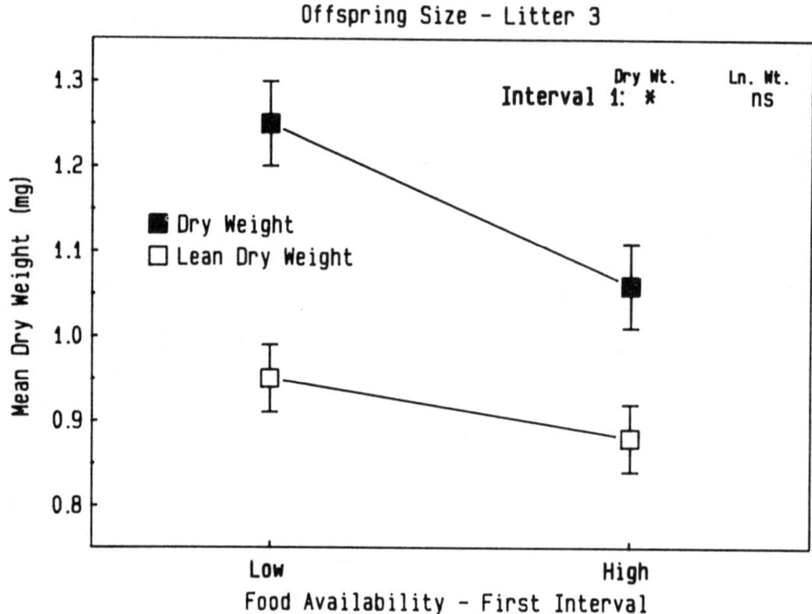

Figure 6. Effects of food availability on offspring weight in *Poecilia reticulata*. The upper line represents the mean effect of food availability during the first time interval on the dry weight of offspring in the third litter. The lower line represents the same comparison, but with the lean dry weight of the offspring as the dependent variable. Error bars represent plus or minus one standard error. Food availability has a significant impact on total dry weight, but not on the lean dry weight.

ciated with this third litter were analysed with two way analyses of variance, with the independent variables being the first and second intervals; however, for the purposes of this discussion, I will consider just the effects of the first time interval.

We found that fish which received "low" food during the first interval produced heavier babies than those that received high food (Fig. 6). The upper line in Figure 6 is the value for all individuals that received either "high" or "low" food during the first interval alone. This means that we have averaged across the second time interval (i.e., the "low"

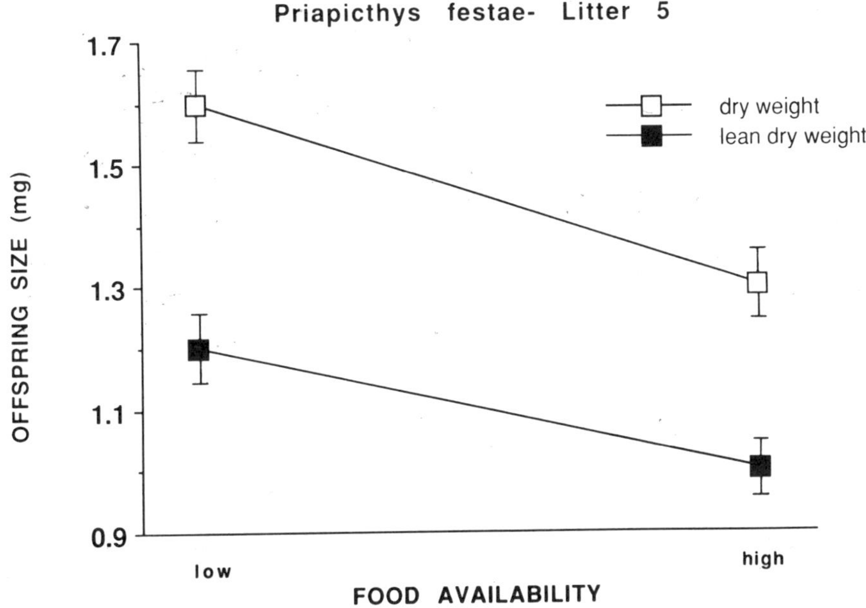

Figure 7. Effects of food availability on offspring weight in *Priaphichthys festae*. The results are for an experiment similar in design to Fig. 5. The data represented here are the same as in Fig. 6. For this species, the effect of food availability is significant on both total and lean dry weight.

value is the average of the low-low and low-high treatments). We then extracted the ether soluble fats from the offspring and compared the lean weights (the lower line in Fig. 6). Most of these size differences disappeared. The low food fish were slightly heavier, but the difference was no longer significant. The size differences between the two treatments are therefore mostly attributable to the differences in fat reserves. Females which receive less food pack more fat into each embryo.

In a similar experiment on the poeciliid *Priapichthys festae*, Ray Llauredo and I again found that low food resulted in larger offspring, independently of the size of the mother (Fig. 7). However, in this case the size difference was equally distributed between the fat and non-fat components of the embryo, so the size differences remained significant after ether soluble fats were extracted (lower line in figure 7). This difference in results between *Poecilia reticulata* and *Priapichthys festae* suggests that offspring size alone is not an adequate description of a maternal effect. The composition of the eggs (embryos) may also be important in the consequences of these size differences.

Temperature and Salinity. Joseph Travis (pers. comm.) has considered the influence of temperature and salinity on the size of offspring produced by sailfin mollies (*Poecilia latipinna*). These are environmental factors that vary considerably within and among populations of this species during the reproductive season. He performed a two-way experiment in which he independently manipulated water temperature and salinity. For both variables, he chose values that were represented in the natural environment (Fig. 8). The combination of higher water temperatures and zero parts per thousand salinity resulted in almost no reproduction, so it is not represented in Figure 8. He found that, at 24°C, offspring size increased as salinity increased from 0 to 12 parts per thousand. He also found that, at the two higher salinities, females kept at 29°C produced larger offspring than those kept at 24°C.

Female Age and Size. Many investigators have considered the combined effects of the age and size of a female on the size of the offspring that she produced. These two

786

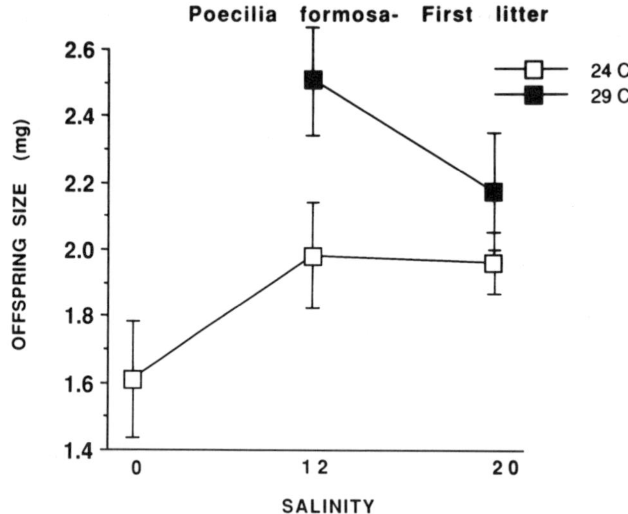

Figure 8. Effects of salinity by temperature on offspring size (dry weight) in the first litter of young produced by *Poecilia formosa*. Plotted values are means plus or minus one standard error. The results were qualitatively the same for the second and third litters (not reported here).

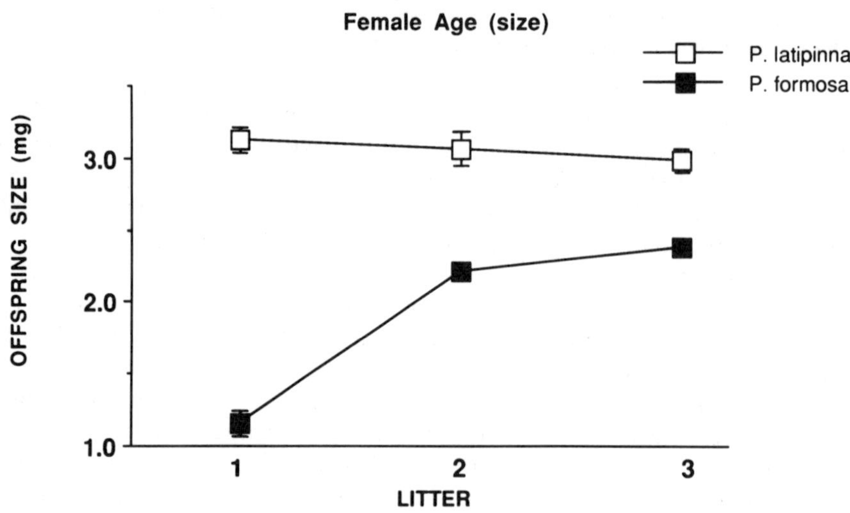

Figure 9. Effect of female age on offspring size in *Poecilia latipinna* and *P. formosa*. The x-axis represents the first, second, and third litters produced by individual females. The y-axis represents the dry weight (means plus or minus one standard error) of offspring produced in each of these three litters.

variables are generally confounded because fish have indeterminate growth, so older females also tend to be larger. I will again illustrate this type of maternal effect with unpublished work by Joseph Travis. He considered the average weight of offspring produced in the first through third litters by *Poecilia latipinna* and *P. formosa*. Offspring size remained constant with age in *P. latipinna,* but increased with age in *P. formosa* (Fig. 9). A similar increase in offspring size with the age(size) of the mother has been reported for a wide variety of fish species (e.g., *Engraulis anchoita*–de Ciechomski, 1966; *Cyprinus carpio*–Tomita et al., 1980; *Poecilia reticulata*–Reznick, 1982b; *Salmo salar*–Thorpe et al., 1984; see review of Gall, 1974 below for an additional example). It is much more common for such an effect to be present, rather than absent.

In conclusion, many environmental factors have been found to influence the size and possibly the composition of offspring. Those considered here—food availability, temperature, salinity, and age (size)—represent common sources of environmental variation for many species. Other environmental factors that influence offspring size have been described. For example, Marsh (1986) has shown that egg size changes on a seasonal basis in the fractional spawning orange-throated darter (*Etheostoma spectabile*). This list is therefore not exhaustive.

Influence on Factors Other Than Offspring Size

Sharon Forster-Blouin (1990), working with *Heterandria formosa* (Poeciliidae), considered the influence of maternal environment on factors other than offspring size. She began her study with juveniles collected from two field localities, then reared to maturity at a single temperature in the laboratory. These females were then mated and were kept at either high or low temperatures during the gestation of the young. The young were then reared at either high or low temperatures. She considered whether the incubation temperature of the mother influenced the critical thermal maxima and minima or the age at maturity of the offspring. She found that, when the mother was kept at a low temperature during incubation, the young were more heat and cold tolerant. She also found that the young attained maturity at a slightly earlier age than their counterparts whose mothers had been kept at high temperatures. The effects of maternal environment on temperature tolerance were substantial, while those on age at maturity were small, accounting for only 1% of the total variance. These results are complicated by interactions, as summarized in full by Forster-Blouin.

This work demonstrates that there might be a diversity of maternal effects on factors other than offspring size. Detecting them will require experiments that evaluate the life history of offspring reared in a controlled environment.

THE INFLUENCE OF OFFSPRING SIZE ON SUBSEQUENT DEVELOPMENT

Because the most commonly reported maternal effect is on offspring size, I have also reviewed what is known about the consequences of differences in offspring size in future development (Table 1). I consider this literature separately because many of these investigators did not consider maternal effects as the source of the initial difference in egg size. This work is of interest because it indicates how a maternal effect on offspring or egg size might influence other aspects of the phenotype. The general trends that I found are similar to those reported for seed size in plants, which are that there are consistent effects of egg size early in development but these become less consistent across studies and less prominent within studies as development proceeds.

These studies were based primarily on species of commercial importance. Most of the studies involve salmonids, including trout, salmon, and charr. The actual methods and dependent variables estimated are quite variable; I will comment on some of these variations where appropriate below.

Larger Eggs Produce Larger Hatchlings

Eighteen out of the 19 investigations that I reviewed reported that larger eggs also produced larger hatchlings. This means that the differences in egg size can most often be attributed to real differences in the quantity of resources per egg and not to some other factor, such as water content.

Table 1. Summary of literature which evaluates the influence of egg size on 1) hatchling size, 2) the duration of survival in the absence of food, 3) probability of survival either to hatching or after hatching, if fed, and 4) growth rate. A '+' entry indicates that the offspring from larger eggs were 1) larger, 2) survived longer, 3) were more likely to survive, or 4) either retained their size advantage or grew faster. A '−' entry means that there was no difference between large and small eggs. No entry means that the comparison was not made.

Reference	Species	1	2	3	4
Bagenal, 1969	*Salmo truta*		+	+	+
Beacham et al., 1985	*Oncorhynchus keta,*				
	O. kisutch	+			
Blaxter and Hempel, 1963	*Clupea harengus*	+	+		+
de Ciechomski, 1966	*Engraulis anchoita*	+	−		−
Elliot, 1984	"chinook salmon"	+			+
Fowler, 1972	"chinook salmon"	+		−*	+,−*
Gall, 1974	"rainbow trout"	+		+	+
Glebe et al., 1979—1,2	*Salmo salar*	+		−	+
Hayes and Armstrong, 1942	"salmon"	+			+
Iwamoto et al., 1984	*O. kisutch*	+			+
Kazakov, 1981—1	*Salmo salar*	+			
Kirpichnikov, 1966—1	*Cyprinus carpio*	+		+	−
Lagomarsino et al., 1988	*Cichlasona citrinellum*	−			−
Lindroth, 1972—2	*Salmo salar*				−
Marsh, 1986	*Etheostoma spectabile*	+	+		−
Millenbach, 1950	"rainbow trout"	+			
Pitman, 1979	"rainbow trout"	+		+	+
Privnolev et al., 1964—3	"salmon"	+			−
Reagan and Conley, 1977	"channel catfish"				−
Springate and Bromage, 1985	*Salmo gairdneri*	+		−	+#
Thorpe et al., 1984	*Salmo salar*	+		−	−
Tomita et al., 1980	*Cyprinus carpio*			−	
Wallace and Aasjord, 1984	*Salvelinus alpinus*	+		+	+
Zonova, 1973—1	*Cyprinus carpio*			−	−
Total:		18:1	3:1	5:6	11:9

*Mortality rates were higher in large eggs. Large eggs retained their size advantage in 1 out of 3 replicates.
#Young from two- vs. three-year old females were compared. Older females produced larger eggs. There was no correlation between egg size and survival within an age class. Comparisons were not made between age classes. The size differences persisted until the end of the experiment, but there were no differences in growth rate.
1—cited by Springate and Bromage, 1985
2—cited by Thorpe et al., 1984
3—cited by Wallace and Aasjord, 1984

Larger Hatchlings Survive Longer

Three out of four investigators reported that, when kept without food, the offspring from larger eggs survived longer. For example, Marsh (1986), working with the orange-throated darter (*Etheostoma spectabile*), evaluated eggs which ranged in mass from approximately 0.2 to 0.5 mg dry weight. When evaluated at low temperatures, the time to starvation for these extremes was approximately 45 and 69 days, respectively.

The one investigation which reports no relationship between egg size and time to starvation (de Ciechomski, 1966) on Argentine anchovies (*Engraulis anchoita*), simply noted that the fry from small eggs survived for 7 to 8 days while those from large eggs survived for 7 to 9 days. No means or statistical analyses were reported.

An increase in survival time in the absence of food suggests higher fitness in the offspring from larger eggs if initial food availability is low.

Larger Eggs Have Greater Viability

Five out of 11 studies reported that the young from larger eggs have higher survival. Five of the remaining studies reported that there were not significant differences in survival between large and small eggs; one reported that small eggs had higher viabilities. Viability was most often evaluated as the probability of survival to the eyed stage, or a stage of development shortly before hatching. In one study on rainbow trout (*Salmo gairdneri;* Pitman, 1979), survival was also evaluated for 200 days after the completion of yolk absorption. The offspring from larger eggs continued to have a survival advantage throughout the experiment.

The Offspring From Larger Eggs Grow More Rapidly

Growth rate was evaluated for periods of time varying from a few weeks in laboratory studies to up to a year in the field. All laboratory studies compared the performance of isolated groups of large versus small eggs. In most cases, the young were fed to satiation. The one field study (*Salmo trutta;* Elliot, 1984) compared average egg sizes and size of hatchlings among years and performed a correlation using yearly averages. He then correlated hatchling size with size at the end of the first year, again using yearly averages as the dependent variables. Some investigators simply considered whether or not initial size differences persisted into later stages of development, while others formally analysed growth rate constants. In all cases, the study ended before the subjects had attained sexual maturity. I considered a result positive if initial size differences remained at the end of the experiment.

Eleven out of 20 studies report that the size differences at hatching persist for periods of up to one year after hatching. The remaining nine studies report non-significant differences between the offspring from large versus small eggs. Many of these investigations reported that the size differences persisted for some interval of time after hatching, then disappeared. It therefore appears that the size and growth advantage associated with larger eggs may be transient.

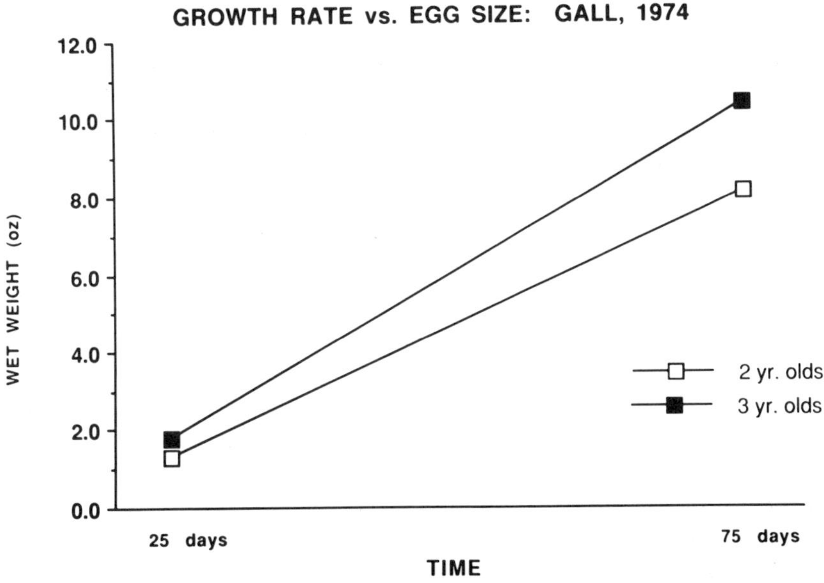

Figure 10. Results from Gall (1974). The x-axis represents the number of days after hatching. The y-axis represents mean offspring size, measured as wet weight. The two lines represent the results for young derived from two versus three year old mothers. Sizes differed significantly at 25 and 75 days.

Gall's (1974) study on rainbow trout (*Salmo gairdneri*) illustrates three of these results for the effects of egg size plus the effect of maternal age(size). He compared the performance of young produced by two versus three year-old females. Three year old females were larger and produced significantly larger eggs than two year-olds. These larger eggs produced larger offspring, were more likely to survive to hatching, and produced more rapidly growing young that were larger at 25 and 75 days after hatching (Fig. 10). The larger hatchlings actually had higher mortality in the fingerling stage, but the overall survival was still greater for the full term of the experiment.

Additional Correlates of Egg Size

Comparisons of the morphology and performance of hatchlings from large versus small eggs have also been made. For example, Bams (1967), working with sockeye salmon (*Oncorhynchus nerka*) found that larger fry, from larger eggs, had better swimming stamina. He, and Mead and Woodall (1968), who also worked with sockeye, found that larger fry were less susceptible to predation. Magnuson (1962), working with medaka (*Oryzias latipes*), found that the fry from larger eggs were more aggressive and competitive. Finally, Blaxter and Hempel (1963), who worked with herring larvae (*Clupea harengus*) found that larger fry had longer jaws and a larger gape; such a size difference should result in larger fry being able to utilize a wider range of food particle sizes.

Such performance variables represent potentially important consequences of maternal effects because they can influence the growth and survival of the young in more natural settings than those used in the above series of experiments. Most of the cited experiments kept each size class of prey in a separate container and fed it to satiation. Such circumstances should minimize the importance of differences in performance because they are a less critical aspect of resource acquisition. When large versus small offspring interact in a more realistic setting, the influence of size on survival and growth rate could be greatly enhanced.

Possible Longer Term Consequences of Egg Size

All of the above experiments ended before the subjects attained sexual maturity; it remains possible that maternal effects extend further into development. Studies of the influence of growth rate and body size on age and size at maturity, and on fecundity, may provide clues to these possible maternal effects (reviewed by Stearns & Crandall, 1984). As one example, I found that more rapid growth (caused by higher levels of food availability) results in earlier maturity at a larger body size, higher size specific fecundity, and smaller offspring size (Reznick, 1982b, 1990). Such results suggest that maternal effects may also have more long-term effects on the phenotype of the offspring and could even extend to subsequent generations, similar to results reported for mice (Falconer, 1965).

RECOMMENDATIONS FOR FUTURE WORK

The predominance of reports of maternal effects on offspring size is more a function of where investigators have looked rather than where maternal effects are most likely to be found. The results of the survey on the effects of egg size on future development yield a partial set of expectations for other possible maternal effects. Forster-Blouin's results illustrate that other effects can be attained, but special experiments are required to detect them.

When considering egg size, it is important to note that size alone may be an incomplete description of a maternal effect. There are different ways for egg size to vary, as illustrated in the differences in the response of *Poecilia reticulata* and *Priapichthys festae* to food availability (Figs. 6 and 7). Such differences in composition may in turn influence the subsequent performance of the offspring. The composition of the eggs may therefore represent an important element of any maternal effect.

When considering maternal effects correlated with egg size, it would pay to have defined growth conditions (e.g., temperature, density, food availability) and replicated treatments. Growth conditions were sometimes poorly defined in the literature and the nature of the conditions may influence the outcome of experiments. Most of the studies cited above had only one replicate with large or small eggs, so egg size was confounded with any uncontrolled differences among the treatment containers. It would further pay to have multiple levels of key environmental factors. To illustrate the virtue of such an experiment, Figure 11 illustrates the hypothetical outcome of an study with two different levels of food availability. This result implies that the differences between egg size are important at low but not high levels of food availability.

Such an outcome could reconcile the differences in survival when the fry are not fed (the extreme case of low food availability) with the variable results for survival and growth when they are fed to satiation. It could also reconcile the inconsistent results obtained by different investigators who worked with the same species (Table 1). For example, Glebe et al. (1979, in Springate and Bromage, 1985) reported that the offspring from large eggs in *Salmo salar* grew more rapidly than offspring from small eggs; Thorpe et al. (1984), who worked with the same species, reported no difference in growth rate. Similarly, Pitman (1979) found that the offspring from large eggs have higher survival, while Springate and Bromage (1985), working with the same species (*Salmo gairdneri*) report no differences in survival. The studies that reported no differences between egg size classes may have been conducted under more favorable circumstances. More generally, the differences in the results may be attributable to differences in culture conditions.

The kind of experiment illustrated in Figure 11 would also help us to evaluate the possible adaptive significance of the response to food availability reported for *Poecilia reticulata* and *Priapichthys festae;* the larger young produced with low resource availability would have an advantage in those conditions. However, if there were an associated reduction in fecundity, then producing larger young could be disadvantageous when food availability is high because there is no longer a difference in performance between the young from large versus small eggs.

The sort of result in Figure 11 also gives a familiar message in a new context; the influence of maternal effects will depend on the environment in which they are investigated.

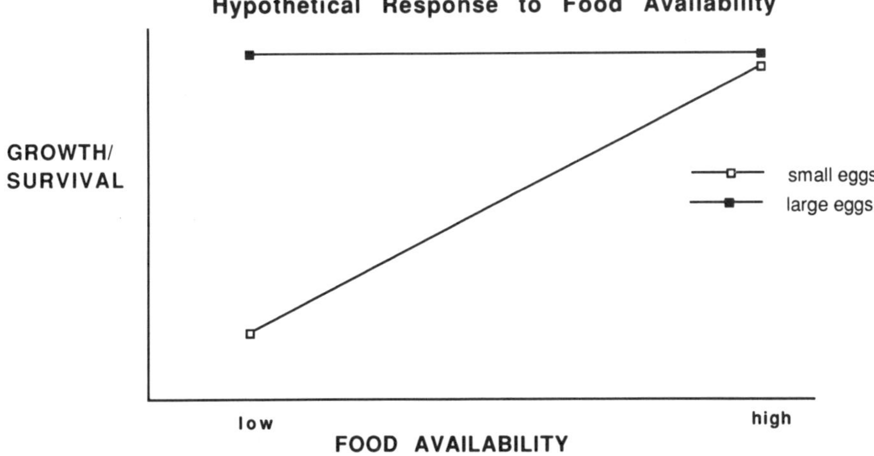

Figure 11. The hypothetical response of offspring from large versus small eggs to different levels of food availability. The x-axis represents food availability and the y-axis the response in some aspect of the offspring, such as growth rate or survival. The two lines represent large versus small eggs.

792

Finally, results could be collected with the goal of applying some of the available models. In the case of the model proposed by Kirkpatrick and Lande (1989), it is assumed that there is some feature of the mother's phenotype which is correlated with the maternal influence on the offspring's phenotype. Most of the above investigations do not consider such correlations between maternal and offspring phenotypes. Furthermore, there do not appear to be any simple correlations between traits in the mother and the maternal effect. For example, one easy candidate would be to seek an association between the size of the mother and the size of her offspring. In the case of age effects, the two are positively correlated. In the case of the maternal genotype effect, the size differences among localities are independent of the size of the female. In the case of the effect of food availability, the low food females which are producing the larger offspring are smaller than the high food females, suggesting an inverse correlation. More generally, I know of no consistent aspect of the female which predicts the maternal effect. Establishing such associations is a challenge for the empirical scientists interested in this phenomenon and is required in order to apply some forms of the theory.

LITERATURE CITED

Bagenal, T. B. 1969. Relationship between egg size and fry survival in brown trout, *Salmo trutta* L. *J. Fish. Biol.* 1:349–353.

Bams, B. A. 1967. Differences in performance of naturally and artificially propagated sockeye salmon migrant fry, as measured with swimming and predation tests. *J. Fish. Res. Bd. Can.* 25:1117–1153.

Beacham, T. D., Withler, F. C. & R. B. Morley. 1985. Effect of egg size on incubation time and alevin and fry size in chum salmon (*Oncorhynchus keta*) and coho salmon (*Oncorhynchus kisutch*). *Can. J. Zool.* 63:847–850.

Blaxter, J. H. S. & G. Hempel. 1963. The influence of egg size on herring larvae (*Clupea harengus*). *J. Cons. Int. Explor. Mers.* 28:211–240.

de Ciechomski, J. D. 1966. Development of the larvae and variations in the size of eggs of the Argentine anchovy, *Engraulis anchiota* Hubbs and Marini. *J. Cons. Int. Explor. Mer.* 30:281–290.

Elliot, J. R. 1984. Growth, size, biomass and production of young migratory trout, *Salmo trutta*, in a Lake District stream. *J. Anim. Ecol.* 53:979–994.

Falconer, D. S. 1981. *Introduction to Quantitative Genetics.* Longman Group, Ltd. London.

Falconer, D. S. 1965. Maternal effects and selection response. Pp. 763–774. *In:* S. J. Geerts (ed.) *Genetics Today,* Vol. 3, Proceedings of the Eleventh International Congress of Genetics. Pergamon Press: Oxford.

Forster-Blouin, S. 1990. Genetic and Environmental Components of Thermal Tolerance in the Least Killifish, *Heterandria formosa*. PhD. dissertation, Florida State University.

Fowler, G. 1972. Growth and mortality of fingerling chinook salmon as affected by egg size. *Prog. Fish. Cult.* 34:66–69.

Gall, G. A. E. 1974. Influence of size of eggs and age of female on hatchability and growth of rainbow trout. *Calif. Fish Game* 60:26–35.

Glebe, B. D., Appy, T. D. & R. L. Saunders. 1979. Variation in Atlantic salmon (*Salmo salar*). *I.C.E.S. Cn.* 1979/M 23. 11 pp.

Hayes, F. R. & F. H. Armstrong. 1942. Physical change in the constituent parts of developing salmon eggs. *Can. J. Res.* 20:99–114.

Hester, F. J. 1964. Effects of food supply on fecundity in the female guppy, *Lebistes reticulatus* (Peters). *J. Fish. Res. Bd. Canada* 21:757–764.

Iwamoto, R. N., Alexander, B. A. & W. K. Hershberger. 1984. Genotypic and environmental effects on the incidence of sexual precocity in coho salmon (*Oncorhynchus kisutch*). *Aquaculture* 43:105–121.

Kirkpatrick, M. and R. Lande. 1989. The evolution of maternal characters. *Evolution* 43:485–503.

Lagomarsino, I. W., Francis, R. C. & G. W. Barlow. 1988. The lack of correlation between size of egg and size of hatchling in the Midas cichlid, *Cichlasoma citrinellum. Copeia* 1988:1086–1089.

Magnuson, J. J. 1962. An analysis of aggressive behavior, growth and competition for food in medaka (*Oryzias latipes*) (Pisces, Cyprinodontidae). *Can. J. Zool.* 40:313–363.

Marsh, E. 1986. Effects of egg size on offspring fitness and maternal fecundity in the Orange-throat Darter, *Etheostoma spectabile* (Pisces: Percidae). *Copeia* 1986:18–30.

Mead, B. W. & W. L. Woodall. 1968. Comparison of sockeye salmon fry produced by hatcheries, artificial channels and natural spawning areas. *Int. Pac. Salmon Fish. Comm. Prog. Rep.* No. 20.

Millenbach, C. 1954. Rainbow brood-stock selection and observations on its application to fishery management. *Progressive Fish Culturist* 12:151–152.

Pitman, R. W. 1979. Effects of female age and size on growth and mortality in rainbow trout. *Prog. Fish. Cult.* 41:202–204.

Reagan, R. E. & C. M. Conley. 1977. Effect of egg diameter on growth of channel catfish. *Prog. Fish. Cult.* 39:133–134.

Reznick, D. N. 1981. "Grandfather effects": the genetics of interpopulation differences in offspring size in the mosquito fish *Gambusia affinis*. *Evolution* 35:941–953.

Reznick, D. N. 1982a. The genetics of offspring size in the guppy (*Poecilia reticulata*). *Am. Nat.* 120:181–188.

Reznick, D. N. 1982b. The impact of predation on life history evolution in Trinidadian guppies: the genetic components of observed life history differences. *Evolution* 36:1236–1250.

Reznick, D. N. 1990. Plasticity in age and size at maturity in male guppies (*Poecilia reticulata*): an experimental evaluation of alternative modes at development. *J. Evol. Biol.* 3:185–203.

Springate, J. R. C. & N. B. Bromage. 1985. Effects of egg size on early growth and survival in rainbow trout (*Salmo gairdneri* Richardson). *Aquaculture* 47:163–172.

Stearns, S.C. & R. E. Crandall. 1984. Plasticity for age and size of sexual maturity: a life-history response to unavoidable stress. Pp. 13–33. *In:* G. Potts & R. J. Wooton (eds.) *Fish Reproduction.* Academic Press: London.

Thorpe, J. E., Miles, M.S. & D. S. Keary. 1984. Developmental rate, fecundity, and egg size in Atlantic salmon, *Salmo salar* L. *Aquaculture* 43:289–305.

Tomita, M., Iwahashi, M. & R. Suzuki. 1980. Number of spawned eggs and ovarian eggs and egg diameter and percent eyed eggs with reference to the size of female carp. *Bull. Jpn. Soc. Fish* 96:1077–1081.

Wallace, J. C. & D. Aasjord. 1984. An investigation of the consequences of egg size for the culture of Arctic char (*Salvelinus alpinus*). *J. Fish. Biol.* 24:427–435.

Wourms, J. P. 1981. Viviparity: the maternal-fetal relationship in fishes. *Amer. Zool.* 21;473–515.

Zonova, A. S. 1973. The connection between egg size and some of the characters of female carp (*Cyprinus carpio* L.). *J. Ichthyol.* 13:679–689.

Developmental Plasticity and Maternal Effects in Amphibian Life Histories

Robert H. Kaplan

Abstract. Sources of variation in birth weight in viviparous organisms can be partitioned into factors that are associated with the genotype and the environment of the offspring (Falconer, 1989). Since the environment is the female parent this component of the variation is all maternal in origin. In amphibians with external fertilization, the offspring's genotype has not expressed itself at the time of oviposition, therefore, all variation in the offspring's phenotype at the time of birth (i.e., egg size) is the result of a maternal effect. These maternal effects can be further partitioned into those that result from the genotype of the female, from environmental variation associated with the female during her development, from the environmental variation that immediately affects the female during the egg making process, and from other factors such as parity and age. In the work discussed here only one type of maternal effect is emphasized, i.e., the effect of the immediate environment during vitellogenesis on ovum size. Clutch size does not necessarily negatively covary with ovum size in amphibians and is thus not concomitantly treated. This work relates theories and studies on the evolution of parental investment in offspring with theories and studies on the evolution of developmental plasticity and maternal effects.

INTRODUCTION

Over the past six years I have addressed issues concerning the evolution of life history strategies and developmental plasticity in the oriental fire bellied toad, *Bombina orientalis*. This frog lives in eastern temperate Asia and has the advantage over many temperate zone amphibians of breeding in the laboratory at regular intervals throughout the year. It has a typical amphibian life cycle in which heterogeneous environments play a large role. The uncoupling of environmental heterogeneity between larval and adult phases allows for the exploration of some interesting issues. For example, environmental variation that induces predictable variation in egg size during the fall and winter, may exert its influences on offspring development in an unpredictable environment during the spring and summer. Thus, there is potential for interaction between environmental unpredictability at one stage of the life cycle with environmental unpredictability at a subsequent stage.

What follows is a discussion of the results of some experiments that address these issues. I will first show that environmental variation does in fact influence ovum size. Then I'll show that this variation in ovum size has further ramifications for the development of offspring in ways which should affect offspring fitness. Finally, I'll show that the fitness of offspring that develop from eggs of different sizes is dependent on an interaction with the thermal environment in which they develop. This will be followed by a discussion of the theoretical implications of this line of investigation.

Dr. Kaplan is with the Department of Biology, Reed College, Portland, Oregon 97202, USA.

THE ENVIRONMENTAL SENSITIVITY OF THE VITELLOGENIC PROCESS

Variation in propagule size in general and ovum size in particular has often been treated as generally unimportant in nature as a result of theory which focuses on the evolution of optimal parental investment in offspring. It is becoming increasingly obvious that such variation is common and even quite large (i.e., Capinera, 1979; Kaplan & Cooper, 1984; Tessier & Consolatti, 1989). As more attention is drawn to the quantification of this variation, questions about the sources of the variation arise. It has become clear in our laboratory that much of the variation in ovum size of amphibians is environmental in nature. Laboratory experiments have shown that varying food levels and temperatures associated with females for the 10 weeks prior to oviposition readily influences the sizes of eggs that they produce. Cold temperatures and high food supply yield the largest eggs and warm temperatures and diminished food supply yield the smallest (Kaplan, 1987). It should be emphasized that this plastic response is highly predictable and repeatable.

Over five years of field work on *Bombina orientalis* in Korea have shown that there is much variation among individuals each year in each of three populations. Three types of evidence indicate that environmental sensitivity of the vitellogenic process is responsible for a significant portion of this variation. First, over five years in two populations, mean ovum size changes significantly from year to year (in prep). Individuals in these populations are long lived and it is unlikely that genetic changes in the population are responsible for these shifts. The shifts are not in the same direction in the two populations, one being from a low elevation rice paddy population and the other from a high elevation mountainous region. In addition, mean body weights also change significantly from year to year in each of the populations in ways which parallel the results of the laboratory experiment discussed above. When females lose weight they also produce eggs of smaller size, suggesting fluctuation in food availability as a source of the variation in the field.

A second line of evidence implicating environmental plasticity as common in the field involves the repeated breeding in the laboratory under environmentally uniform conditions, of individuals whose ovum size was first measured in the field. If we assume that all the variation in ovum size in the field is due to the sum of female genetic factors, general environmental factors associated with the female's development, and special environmental factors associated with the female during vitellogenesis (Falconer, 1989) then breeding in the laboratory under uniform environmental conditions can only diminish the special environmental factors leaving the inherent female specific factors untouched. In the three populations we examined there was substantial reduction in variance in ovum size among individuals in the laboratory. The proportion of the field variance eliminated in the lab ranged from 28% to 71%. In addition to providing strong evidence that vitellogenesis in the field is subject to environmental perturbation, these values also set upper limits on the heritabilities of ovum size.

A third line of evidence showing that a single female produces eggs of different sizes during different years comes from our ongoing mark/recapture work. To date, five females in one population have been caught breeding in the field during successive years. In each case the mean ovum size that they produced had changed significantly. In three of the five ovum size decreased and in two it size increased. In addition to providing the most direct evidence that maternal investment in offspring is subject to environmental effects, these data also indicate that micro-environmental heterogeneity among females could result in a situation where the mean ovum size in the population as a whole does not change from year to year, but each female's ovum size does change.

796

THE RAMIFICATIONS OF OVUM SIZE VARIATION DURING OFFSPRING DEVELOPMENT

Evidence from both the laboratory and the field indicates that ovum size variation has important ramifications for offspring development within a great many species. In amphibians, these effects are now well recognized. In *Bombina orientalis*, for example, laboratory studies (in prep.) have shown that offspring developing from larger eggs are larger at the hatching stage, stage of first feeding, and metamorphosis. In addition, these studies have also indicated that individuals developing from larger eggs metamorphose in less time.

Field experiments which incorporate thermal heterogeneity of the environment have also shown that ovum size variation impacts growth and development. Differences in ovum size did not affect developmental time to the hatching stage, but there were large effects associated with micro-climatic temperature variation both within and between years. In 1987, for example, the embryos in some breeding pools hatched in 2.5 days while others, in different pools, hatched in 8.0 days. In addition, temperature variation affected the size of larvae at hatching, with those that developed in cold environments hatching at a larger size. Despite these environmentally induced effects, the maternal effect (i.e., ovum size) continued to influence size of larvae at the stages of hatching and first feeding, and for several weeks beyond this when the experiments were terminated (Kaplan, 1989).

THE INTERACTION OF OVUM SIZE PLASTICITY WITH THE SUBSEQUENT OFFSPRING ENVIRONMENT

In many amphibians with complex life cycles, dividing developmental plasticity into two components (one relating to vitellogenesis, the other to larval development) offers a unique opportunity to study the potentially adaptive nature of plasticity induced at both stages and, most importantly, their interaction. Recent laboratory studies on *Bombina orientalis* have addressed these issues by considering the effect of different egg sizes on development in environments that differ both in the density of larvae reared in individual containers and the amount of food provided to them. These studies (in prep.) have shown that under uncrowded, food-rich conditions larvae that develop from large eggs metamorphose at a larger size and in less time, as discussed above. But, as conditions become more severe, as a result of both crowding and food limitation, larvae developing from large eggs metamorphose at a smaller size and take more time. An earlier result in the salamander, *Taricha torosa*, also showed an interaction between initial egg size and larval density (Kaplan, 1985). In that case, however, the interaction was somewhat different. In the high density situation, larvae that developed from large eggs still required more time to metamorphose but they metamorphosed at a larger size. Berven & Chadra's (1988) recent result on *Rana sylvatica* also show an interaction between initial egg size, crowding, and food limitation. But in this case it was larvae from small eggs that metamorphosed at a larger size when conditions were good, and those from large eggs that metamorphosed at a larger size when conditions were severe.

The main point of these three laboratory studies, however, is that they make it clear that increasing parental investment in offspring does not necessarily lead to an increase in those traits which we usually assume to be correlated with offspring fitness. Thus, the fitness outcome of plasticity induced during vitellogenesis that affects egg size can only be evaluated by taking into consideration the pattern of environmental variation during subsequent embryonic development. The fact that the nature of the interactions differs among the three species discussed above is an interesting observation which will require future work.

A recent field experiment (in review) on *Bombina orientalis* has evaluated initial ovum size variation by looking at the probability of larvae surviving predation after developing from eggs of different sizes in environments of different temperatures. Naturally occurring interspecific tadpole predators (*Rana amurensis coreana*) were allowed to predate on newly hatched *Bombina* larvae. The first result was not unexpected. Logistic regression analysis showed that larger larvae had a higher probability of surviving. All prior work had indicated that larger eggs give rise to larger larvae so the connection seemed clear. But, when ovum size was brought into the logistic model the result indicated that at a given size larvae that developed from large eggs had a lower probability of surviving predation than those that developed from small eggs. This was unexpected and indicated that egg size was contributing to variation in something other than simple total length.

A path analytic approach was used to evaluate the simultaneous effects of egg size and developmental temperature on total length. It was found that developmental temperature greatly influenced total length and therefore the probability of surviving predation. Embryos that developed in colder environments attained substantially greater total lengths than those that developed in warmer environments. It was found, however, that total length was not the most appropriate predictor of survival because some important dynamics were overlooked. Specifically, temperature affected total length only by influencing the length of the tail and not the snout-vent length. Thus, the source of increased total length with decreased developmental temperature is accompanied by a change in the tail length to body length ratio. In addition, egg size had little effect on tail length, but exerted most of its influence on total length through its effect on snout-vent length.

This phenomenon resulted in the unexpected result discussed above. A logistic regression analysis indicated that the probability of surviving predation at the hatching stage was negatively associated with developmental temperature and positively associated with egg size. Most interesting, however, was the significant interaction between egg size and temperature. At the coldest developmental temperatures larvae that developed from large eggs had a higher probability of surviving predation than those that developed from small eggs. At the warmest temperatures the reverse was true, with larvae that developed from large eggs actually suffering a higher probability of predation.

The path analytic approach helps explain the result by showing that at warm temperatures larvae hatched at smaller sizes and larvae that developed from large eggs had tails comparable in size to those that developed from small eggs. They are nevertheless larger due to the influence of egg size on snout-vent length. Thus, at warm temperatures the ratio of tail length to snout-vent length is smaller in larvae that develop from large eggs. This could result in problems with locomotion. At colder temperatures, when larvae hatch at larger size the tail to snout vent length ratios are equivalent in larvae that develop from large and small eggs; larvae from larger eggs are therefore able to capitalize on their larger size. Thus, plasticity in ovum size induced during vitellogenesis, coupled with thermally induced plasticity during early development, and thermal heterogeneity during the first 24 hours after hatching, results in an unpredictability about the adaptive utility of a particular egg size at the time when initial plasticity in egg size is induced in the female. Larvae that develop from large eggs have greater fitness than those that developed from small eggs (when assessed as the probability of surviving predation during this early stage of life) when they develop in colder environments, and larvae that develop from small eggs have greater fitness than those that develop from large eggs when they develop in warmer environments.

EMPIRICAL AND THEORETICAL IMPLICATIONS

The results discussed above join two other studies in showing that, by considering development in heterogeneous environments, the usual assumption of optimal parental investment (i.e., that increased investment in individual offspring is positively related to offspring fitness) is not necessarily true. In 1985 Kaplan showed that in larvae of the salamander, *Taricha torosa*, the consequences of increased parental investment per offspring were contingent upon food availability during larval development (see above). Similarly, Berven & Chadra (1988) working with larvae of the frog, *Rana sylvatica*, found a similar interaction between egg size and food availability.

Given that maternal investment in offspring in temperate zone amphibians is greatly influenced by environmental factors (Kaplan, 1987) the question arises as to whether such plasticity can be adaptive. Phenotypic plasticity in offspring size has only recently been recognized in many organisms (e.g., Tessier & Consolatti 1989). While a number of explanations are emerging that attempt to explain such variation (Capinera 1979, Crump 1981, Kaplan & Cooper 1984, McGinley et al. 1987), it is clear that more information is needed about the implications of variation in offspring size for offspring fitness. It is argued here that one of the first assumptions of many models of optimal parental investment (i.e., that offspring fitness increases with investment per offspring) is not universally applicable. The studies on *Bombina orientalis* demonstrate that environmental variation can induce predictable plasticity in ovum size. This plasticity can be uncoupled from subsequent environmentally induced, predictable plasticity in early development. Thus, a female cannot anticipate the environment in which her offspring will develop at the time when allocation of energy to offspring is made. This uncoupling of the environmental sources of variation that influence two distinct phases of the life cycle provide a mechanism for the model of adaptive coin-flipping that we developed several years ago (Cooper & Kaplan, 1982; Kaplan & Cooper, 1984). In that model an optimal solution to dealing with environmental uncertainty associated with characteristics that are related to fitness is to allow for the development of phenotypic variability (that does not anticipate the future and therefore appears random) at an earlier stage in the life cycle. The environment during vitellogenesis which results in a predictable response in egg size is actually a coin-flip with respect to the environment that a larva encounters later in the life cycle. In general, the plasticity in maternal investment per offspring that exists among amphibians offers a valuable tool for testing the growing theory (Kaplan & Cooper 1984, Bull 1987, Schultz 1989) which indicates that environmental uncertainty can result in the selection of developmental plasticity in one stage of the life cycle even though the fitness ramifications are realized in an unpredictable environment at a subsequent developmental stage.

ACKNOWLEDGMENTS

This paper has benefited from comments by Barry Sinervo. It was written while continually wishing that Bruce Riska gets the miracle that he deserves.

LITERATURE CITED

Berven, K. A. & B. G. Chadra. 1988. The relationship among egg size, density, and food level on larval development in the wood frog (*Rana sylvatica*). *Oecologia* 75:67–72.
Bull, J. J. 1987. Evolution of phenotypic variance. *Evolution* 41:303–315.
Capinera, J. L. 1979. Qualitative variation in plants and insects: effect of propagule size on ecological plasticity. *American Naturalist* 114:350–361.

Cooper, W. S. & R. H. Kaplan. 1982. Adaptive "coin-flipping": a decision-theoretic examination of natural selection for random individual variation. *Journal of Theoretical Biology* 94:135–151.

Crump, M. L. 1981. Variation in propagule size as a function of environmental uncertainty for tree frogs. *Am. Nat.* 117:724–737.

Falconer, D. S. 1989. *Introduction to Quantitative Genetics.* Longman, New York.

Kaplan, R. H. 1985. Maternal influences on offspring development in the California newt, *Taricha torosa. Copeia* 1985:1028–1035.

Kaplan, R. H. 1987. Developmental plasticity and maternal effects of reproductive characteristics in the frog, *Bombina orientalis. Oecologia* 71:273–279.

Kaplan, R. H. 1989. Ovum size plasticity and maternal effects on the early development of the frog, *Bombina orientalis,* in a field population in Korea. *Functional Ecology* 3:597–604.

Kaplan, R. H. & W. S. Cooper. 1984. The evolution of developmental plasticity in reproductive characteristics: an application of the "adaptive coin-flipping" principle. *American Naturalist* 123:393–410.

McGinley, M. A., Temme, D. H. & M. A. Geber. 1987. Parental investment in offspring in variable environments: theoretical and empirical considerations. *American Naturalist* 130:370–398.

Schultz, D. L. 1989. The evolution of phenotypic variance with iteroparity. *Evolution* 43:473–475.

Smith, C. C. & S. D. Fretwell. 1974. The optimal balance between size and number of offspring. *American Naturalist* 108:499–506.

Tessier, A. J. & N. L. Consolatti. 1989. Variation in offspring size in *Daphnia* and consequences for individual fitness. *Oikos* 56:269–276.

CELLULAR AND MOLECULAR LEVELS OF EVOLUTION

Leeuwenhoek in Lilliput

Peter H. A. Sneath

Abstract. An overview is presented of the contribution of microbiology to systematics, both directly to rationale and data-handling, and indirectly through molecular biology. Microbiology had no historical dogma of taxonomy. This allowed systematics to be re-evaluated from fundamentals that apply to all organisms, and led to the concepts of information-rich groups, overall resemblance based on numerous equally-weighted characters, and the reconstruction of phylogeny from phenetic analyses, for which additional assumptions about evolution are needed. These ideas (associated with numerical taxonomy in the broad sense) have led to the greatest advance in systematics for a century or more. Phenetics is not theory-free, but founded on sound theory in logic, philosophy, and information science. The importance of reliability is stressed, with illustration of the considerable uncertainty in most phylogenetic trees. Examples are given of the use of these ideas in identification, biotechnology, and interpretation of evolutionary events. The importance of nucleic acids in museum specimens is stressed; this is a vast heritage that we may one day be able to exploit.

Three hundred years ago, a Dutch draper succeeded in making the first microscope, a simple lens in a brass plate—and opened up a new world to his admiring contemporaries (Dobell, 1932). The man was Antonie van Leeuwenhoek. The newly-formed Royal Society of London eagerly welcomed the accounts of his observations. His simple microscope was soon followed by the familiar compound microscope, due largely to Robert Hooke. These instruments gave entry to the world of small insects and protozoa. Jonathan Swift was thinking of this world when he wrote his well-known epigram (*Oxford Dictionary of Quotations*, 1972):

> So, naturalists observe, the flea
> Hath smaller fleas than on him prey;
> And these have smaller fleas to bite 'em
> And so proceed *ad infinitum.*

But Swift was writing of plagiarism among poets, and the sting is in the tail, in the less often quoted couplet:

> Thus every poet in his kind,
> Is bit by him that comes behind.

Every science has many examples of unacknowledged plagiarism,—even systematics.

The size of humans is a meter or two, small insects are measured in millimeters, and bacteria in micrometers, so Leewenhoek's invention allowed study of a world about a

Dr. Sneath is with the Department of Microbiology, Leicester University, Leicester, LE1 7RH, UK; and Garvin Visiting Professor 1990–1991, Virginia Polytechnic Institute and State University, Blacksburg, Virginia 24061, USA.

million times smaller than our familiar surroundings. Advances in electron microscopy have carried this still further.

This address aims to give an overview of how new techniques affect systematic and evolutionary biology, illustrating this from the part that microbiology has played in recent advances in systematics. This is appropriate to our main Congress theme, the Unity of Evolutionary Biology, because bacteria are living organisms, not just birds and butterflies—not to mention begonias. They all have genomes composed of DNA and are subject to natural selection. Systematics must embrace them all if it is to produce a philosophy of taxonomy and evolution that satisfies the needs of all scientists. Most illustrations will be from bacteria, but there has been parallel work on viruses, microfungi and protista. All kinds of systematists face very similar, and varied, problems, and systematics must be able to solve these in every group of organisms.

Microbiology has made many contributions to systematic and evolutionary biology, both directly to the rationale of systematics and the handling of complex data, and indirectly through pioneering the development of molecular biological techniques. It should also be said that the contribution toward a new system of nomenclature for bacteria, by Victor Skerman of the University of Brisbane (Lapage et al., 1975) is now attracting much interest in botany and zoology, where nomenclature is becoming unwieldy.

Perhaps the biggest contribution of microbiology to systematics has been a healthy skepticism of previous taxonomic dogma (Vernon, 1988). As in the Renaissance, the discarding of many opinions that had been repeated parrot-fashion over the years was a prerequisite to new thinking. There were a number from other disciplines who also felt the time had come for re-thinking the logic of systematics (e.g., Beckner, 1959; Cain & Harrison, 1958; Gilmour, 1951; Gregg, 1954; Michener & Sokal, 1957; Williams & Lambert, 1959), and here I particularly associate my colleague Robert Sokal, who brought his own sense of skepticism from his background in statistics. For my own part, because there was no established method for classifying bacteria, no fossil record of bacteria, and no agreed-upon phylogenies, it was possible to start thinking from fundamentals: homology, character-weighting, overall similarity, information content, clustering. What is homology when two bacteria ferment glucose? How many times more important is motility than filamentous branching? What is a bacterial species? Much was swept away. The Method of Division from Above, the Hierarchy of Characters, the Primacy of Differential Characters, are now little heard of (Sokal & Sneath, 1963; Sneath & Sokal, 1973).

The other major contribution was an emphasis on quantitation and reliability. Taxa are not formed by arbitrary division on a few characters given extreme weights, but on grouping together organisms on the extent to which they share large numbers of common properties; that is, on overall similarity without excessive differential character weights. These stages lead to phenetic groupings, but this is not the essence of the logic. The critical factor is the view that classifications are highly multivariate, subsuming large numbers of variables, leading to information-rich polythetic groups.

The common theme in all this is one of information science. Such methodology is essential to reconstructing phylogenies, although it is not sufficient for this because further assumptions and techniques are required. It is also essential for analysis of molecular data, for numerical identification, and for systematic data bases. This is evident because none of these disciplines can operate without phenetic steps (which are always present, though not always acknowledged: the phylogenetic method of Hennig, 1966, implies one can correctly choose *a priori* characters that define clades, but this is feasible neither in logic nor in practice; it is a close parallel to the dogma of division into taxa from above).

Many of these advances came with what is known as numerical taxonomy in the broad sense (Sokal & Sneath, 1963). It was fortunate in its timing, just as computers were becoming available, but it would have appeared nonetheless: the early numerical classifications were made with pencil and paper or optical matching devices. There is, I believe, no doubt that numerical taxonomy will be seen as the greatest advance in systematics since Darwin, and perhaps—because Darwin has had far less influence on systematic method than is often thought—since Linnaeus. Without the painstaking exploration of fundamental concepts in logic, philosophy, and information theory carried out by numerical taxonomy—simply because they had to apply to bacteria as well as birds and begonias—the later developments would have had no foundation. Of course, we all stand on the shoulders of giants, and if the work had not been done then it would have been done later or by others. But I speak of what is, not what might have been. An open mind was a *sine qua non.*

There are also different sorts of open mind; many critics of numerical taxonomy have had the inestimable advantage of approaching it with a completely open mind: they have not read what is already known before writing about it.

If we turn to some examples of these techniques applied to bacteria, it is surprising that we see very similar phenetic patterns in bacteria as in eukaryotes. Bacteria and other prokaryotes are predominantly haploid, asexual, and have a small genome. They are capable of very rapid multiplication in favorable conditions, but of very slow multiplication, or long dormancy, when conditions are unfavorable. Yet phenetic analysis shows a similar taxonomic structure to that in other organisms: there are fairly tight clusters of individuals (in this case bacterial strains) which are separated by distinct gaps from nearby clusters, and these primary clusters are traditionally equated with species. The species clusters form groups which are equated with genera, and in some genera one sees more complex arrangements similar to the swarms of highly similar species in some plants and animals.

Above genus level the taxonomic structure is less clear in bacteria and it is particularly due to the work of Carl Woese and his colleagues (Woese, 1987; Woese et al., 1975, Woese & Fox, 1977) on ribosomal RNA sequences that a way forward has been found. The various ribosomal sequences, but particularly the 16S and 5S sequences, have allowed study at higher taxonomic levels, and they have shown that there is a major division in the prokaryotes between the extreme thermophils, halophils, and methanogens—the Archaebacteria—and other bacteria plus the blue-green algae (now commonly referred to as the cyanobacteria)—the Eubacteria. Although some aspects are still not entirely clear, these two form two of the major groups of organisms, Superkingdoms one may call them, with the Eukaryotes—protists, plants, fungi, and animals—forming a third Superkingdom (Woese & Fox, 1977).

The sequence analysis is also a remarkable confirmation of the view, expressed most lucidly by Lynn Margulis (Margulis, 1970) for the symbiotic origin of the eukaryotes, because the ribosomes of mitochondria and chloroplasts show unmistakable affinities to bacterial ribosomes. The chloroplasts are clearly affiliated to the blue green algae. We now have a broad map of all important groups of living organisms, most clearly shown by the work of Hori and Osawa on 5S rRNA (Hori & Osawa, 1986).

It may be noted in passing that these analyses, though most impressive, still pose some questions. They reflect only a small portion of the genome, and a portion that is concerned only with the synthesis of proteins. They are subject to uncertainty because the sequences are relatively small. And there are some incongruences that will be mentioned later. Though often interpreted as phylogenies—as is quite legitimate provided that it is realized that this involves further assumptions, that is, on how evolution occurs—they are the result of what is essentially phenetic analysis, overall similarities between large numbers of observed character states with little differential weighting. This can be seen from the fact that no phenetic analyses of these levels had been done before; the assumption of traditional botanists was that blue-green algae belonged to the plants, and their placement turned on arguments over whether photosynthetic pigments or nuclear structures were the earlier to evolve—typical pre-phenetic issues (Humphries & Richardson, 1980); the contrast between the old and new could not be more striking.

I have referred to the question of reliability of classifications. It was the hope of numerically minded systematists that their classifications would in some sense be more reliable than those made by earlier methods. Of course, in a formal sense, the question does not arise for arbitrary classifications: if flowering plants must have chloropyll, and fungi must not, then anything that lacks it, like the bird's nest orchid, is simply placed in the fungi. Much debate is occurring on the relative reliability of different taxonomic methods. Yet little is said on the inherent uncertainly in small sets of characters.

If in comparing two organisms we observe four differences between them in a set of ten characters, we cannot expect that there will be exactly four differences between them in a second comparable set of characters; it might be three, or six. This sampling error affects both phenetic and cladistic or phylogenetic analyses. This has been recognized in phenetic work from the beginning (Sokal & Sneath, 1963) and sensible judgements are made without much fuss. But attempts to find the one true phylogeny have led to arguments over methods without considering this critical point: that analyses are based on finite, and usually small, sets of characters.

A tree can be presented in a form that shows the uncertainty using plausible statistical arguments (Sneath, 1986; Sneath & Radbourne, 1991). Figure 1 shows a cladogram of species in two genera with the numbers of evolutionary changes shown on the internodes of the tree in the upper part, and the uncertainty due to sampling error in the lower part. Almost the only secure conclusions are (1) the genera do differ significantly, and (2) perhaps each genus contains two groups of species that differ, but most of the remaining detail is almost valueless. The 25 characters employed are quite insufficient to give a sound cladogram of the 17 species. Figure 2 shows another example from a thorough study of

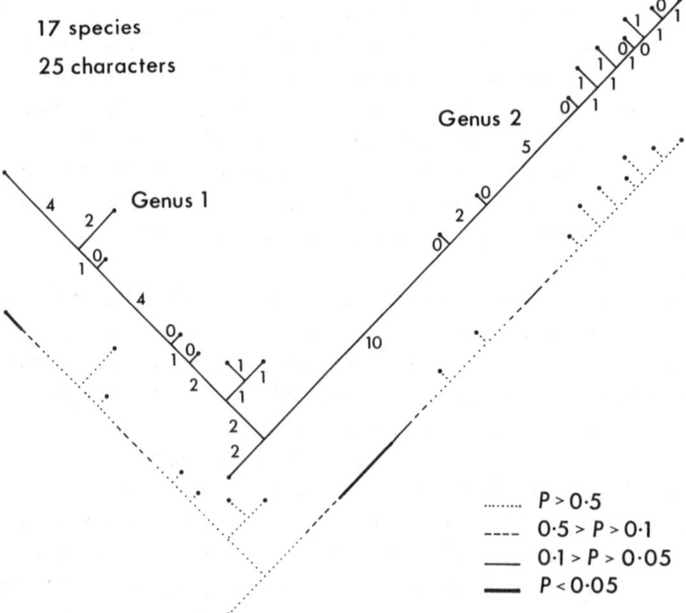

Figure 1. A cladistic tree from the literature (by no means atypical) showing the considerable uncertainty in trees based on relatively few characters. The stem lengths are proportional to the numbers of changes (zero change is symbolized by half a change to facilitate representation). The version above shows the numbers of changes. The version below shows the confidence of the tree using the method of Sneath (1986). Dotted regions indicate where a node might move (in a tree from a second comparable set of characters) with a probability of 50%. Other probabilities are shown by the other symbols. Thus the thick line indicates uncertainty of 5% or confidence of 95%.

higher groups of Plecoptera by Nelson (1984). This shows six equally parsimonious trees each with 154 steps, and it is thus evident that the 113 characters used cannot reliably indicate the details of the tree topology for the 22 taxa. Figure 3 shows another almost identical topology, but with stem lengths proportionally to numbers of differences as indicated by Nelson's data table, and then drawn to indicate uncertainty. It can be seen that the uncertainty extends far beyond that implied by six parsimonious trees. One must question whether it is worth going to great lengths to construct numerous alternative trees, or to devise alternative algorithms, for data that can support only such weak conclusions. It has

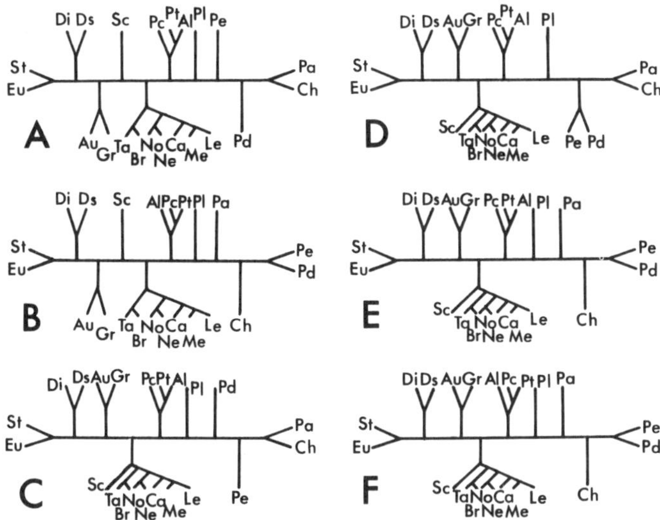

Figure 2. Six minimum-length cladograms of higher groups of stoneflies (Plecoptera) from Nelson (1984) based on 113 binary characters. The six trees all have 154 steps, but somewhat different topologies (stem lengths are not proportional to numbers of changes). The higher groups are mostly families and tribes (e.g., St = Stenoperlinae; details in Nelson, 1984).

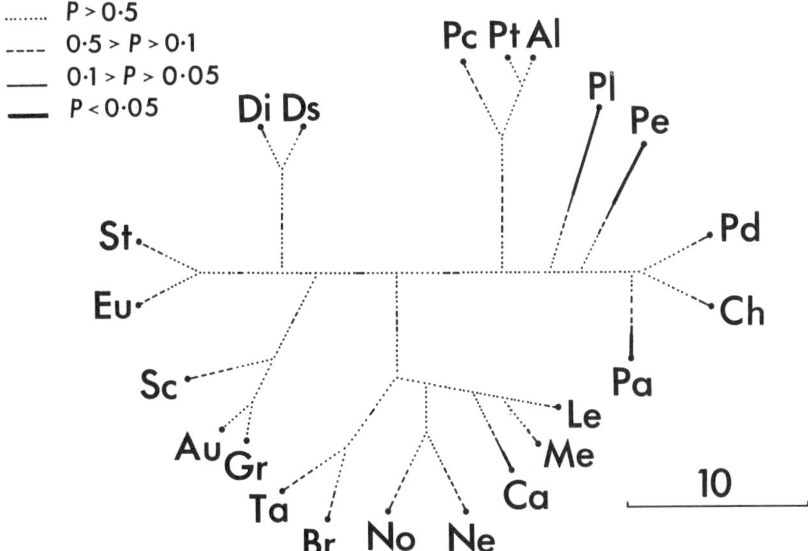

Figure 3. Taxa in Fig. 2 displayed to show tree confidence (conventions as in Fig. 1, Sneath 1986, method 1). The tree is based on average linkage clustering of data in Nelson (1984) with stem lengths proportional to numbers of changes as shown by scale bar. The topology is almost the same as A in Fig. 2, and though not directly comparable, the total number of steps implied is about 156.

808

to be said, quite firmly, that arguments over phylogenetic reconstructions that ignore sampling error are largely a waste of time.

Further, phylogenetic reconstruction must require critical assumptions on how evolution occurs, which are not raised by phenetics. Consider the minimum-length three in Figure 4 based on distances from over 50 characters. Does this imply evolution of these OTUs in the "clades" seen here? In a Congress where any tree-like diagram is likely to be uncritically called a cladogram the temptation is to say that it does. But Figure 4 is a non-cladistic cladogram, and the OTUs are chemical elements (Sneath, 1988). The concept of homology that was used was certainly not evolutionary, but was based on a more fundamental concept of correspondence of properties. Yet the phenetic groups are real enough in the sense that they represent groups of high information content and predictivity; they are natural groups in the sense of the late John Gilmour whose contribution to taxonomic logic (Gilmour, 1937, 1940) will be long remembered. Chemists use these groups every day; halogens, noble gases, and so on. Further, such groups do, as was postulated by the philosopher J. S. Mill (1886), owe their existence to underlying common causes which it is the duty of the scientist to uncover; the cause is the pattern of orbital electrons of the atoms. But their origin, in nuclear reactions in stars, is entirely unrelated to their phenetics. The elements do not have a genome in the same sense as living organisms.

We see here four different theoretical aspects intertwined: theory of homology, i.e., what should be compared with what; information theoretic concepts of groups; theory of underlying electronic causes; and theory of historical nuclear transformations. It is thus perverse to speak of phenetics as theory-free (it concerns the first three), or phylogeny as theoretical, without explaining that the theory is evolutionary theory. It also shows that one must have a model of how evolution actually operates. It is, regrettably, philosophers of science who are the main offenders on both counts (Hull, 1988; Sober, 1988; Scott-Ram, 1990).

It cannot be said too forcefully that the stubborn refusal of many systematists to acknowledge no other goal of systematics than reconstructing phylogeny—to deny that any other activity is worthwhile—cuts them off from other scientists, and can easily lead to the self-deception that everyone else in the world is wrong. Their bitter quarrels (recounted by Hull, 1988; Scott-Ram, 1990) do not further the cause of systematics.

The newer ideas are leading in several directions. One is toward the establishment of taxonomic data banks. A good deal is occurring in microbiology, because of two urgent needs, accurate diagnosis and biotechnology. There are, of course, large banks of molecular data, but most of these are not arranged in comparative or taxonomic format.

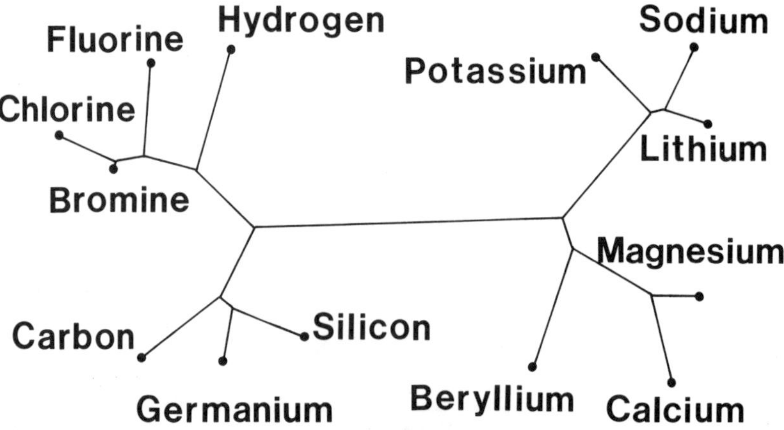

Figure 4. Minimum length tree for selected chemical elements, based on 54 properties (modified from Sneath, 1988).

Numerical identification or diagnosis is very well established in bacteriology (Wilcox et al., 1980; Goodfellow et al., 1985). It follows the simple logic of numerical taxonomy: (1) choice of entities and attributes; (2) calculation of resemblance between entities; (3) assembling of entities into groups or taxa; and (4) identification of new entities with these taxa. Traditional diagnostic keys have many limitations in bacteriology, because of intraspecies variation and technical problems. Instead, probabilistic or distance models are much used. The basic idea is simple. Once the species have been defined, by numerical taxonomy of numerous bacteria isolates or strains, these species can be represented by roughly spherical (or hyperspherical) clusters of strains. The clusters or spheres can then be described as the percent of positive test reactions shown by strains of a given species in a defined set of tests. These percentages define the position and radius of each species.

When an unknown strain is run through the tests the results imply its position in the phenetic space (Fig. 5). If the unknown lies within a sphere it is identified as a member of that species. Other outcomes occur: (1) strains may be intermediate between two close species; (2) strains may be just outside the periphery of a sphere, and are then assumed to be aberrant members of that species; (3) strains may lie alone in space away from all spheres, and these are unidentifiable strains which may be the nuclei of species which have not been described, or are not in the system; or (4) possibly such strains are extreme genetic variants. The distances in the phenetic space can be expressed as probabilities, so that the likelihood of correct identity can be assessed, and a warning can be given if further testing is needed. Yet more sophisticated statistics can also be used (Sielaff et al., 1982). These systems are now found in automated equipment in microbiology and give excellent results. Indeed they can be simplified into profile registers that record the test pattern and identity of all but the most aberrant strains. An example of a few entries, annotated from a register produced at our institution, is shown in Table 1.

But these data-bases are of great potential in biotechnology, in the search, for example, for new antibiotics. They may suggest where to look; genera that produce many antibiotics may produce many more if studied further. But other uses exist. If one wishes to isolate selectively strains of a particular species one can use so-called computer-designed isolation media, that is, media containing the best nutrients for the desired species, and containing those chemicals that inhibit unwanted species (Vickers et al., 1984). These ideas are now much used in microbiology.

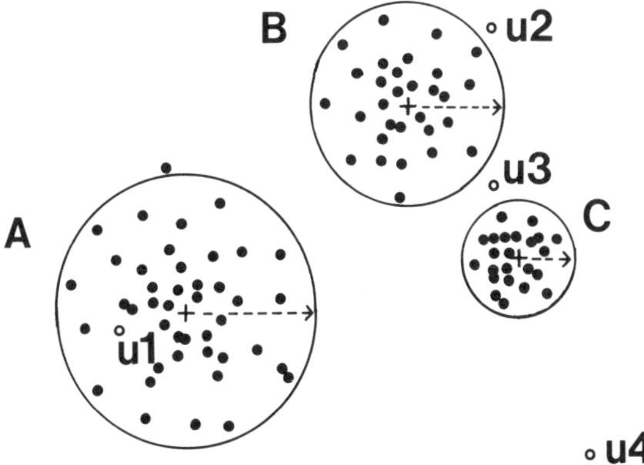

Figure 5. Schematic representation of identification space. Three species of bacteria, A, B and C are shown, based on known strains (solid circles). The test results define the positions of the strains, and hence also the centre and radius of the species. Open circles indicate unknown strains whose positions indicate their probable identity: u1 is a member of species A; u2 is an atypical member of species B; u3 is an intermediate between species B and C; u4 may be a strain of an unrecorded species or perhaps an extreme genetic variant.

810

Table 1. Example of entries in a profile register of bacteria of the genus *Staphylococcus*. The test results are coded octally to give the profile number, and all common profiles are listed in order.

Profile	Taxon	Probability	Further Tests
(test pattern)	(species of *Staphylococcus*)	(of identity)	(or comments)
23320	S. epidermidis	.9735	
23322	S. epidermidis	.8860	WARNING!
	S. haemolyticus	.1017	RECHECK TESTS
23326	S. epidermidis	.6782	Coagulase
	S. aureus	.2383	(extra test advised)

There are also many interfaces between microbial systematics and other disciplines. Some peculiarities of bacteria have been mentioned, which must be relevant in their population genetics. Their ecological niches are peculiar: they may be fractions of a millimeter across in a soil grain, yet extend for thousands of kilometers. They may be alternately exposed to intense r-selection as opportunists in favorable conditions, and intense k-selection when eking a living on traces of nutrients. They have, it seems, a core of essential genes on their main chromosome, and numerous smaller "accessory chromosomes" or plasmids, carrying less essential genes which exchange readily with the main chromosome and move from one bacterium to another (Richmond, 1970). These, and bacterial viruses, are important vectors of genes for antibiotic resistance, and probably have a major part to play in prokaryote evolution.

Viruses are now also attracting interest as possible vectors of genes in eukaryotes. Arthur Mourant suggested many years ago that the sudden appearance of calcified skeletons in many invertebrate phyla at the start of the Cambrian Period might have been due to virus transfer of an essential enzyme such as an alkaline phosphatase (Mourant, 1971; we now have methods for testing this hypothesis). Gene transfer occurs from *Agrobacterium* to flowering plants, and other examples are turning up (Sikorski et al., 1990). In addition, viruses afford almost the only examples of evolution occurring before our eyes. The influenza virus has been studied for some decades (Skehel et al., 1982), and it may be noted that at least superficially its evolution exhibits considerable zig-zag behavior (Fig. 6).

Leewenhoek's microscope and its successors has been the workhorse of systematics. Now we are close to other developments. We are close to instruments that can read molecular detail at almost 10^{-10} metres. They are like Sam Weller's "patent double million magnifyin' gas microscope of hextra power" in Dicken's *Pickwick Papers* (*Oxford Dictionary of Quotations*, 1972). A new world, which Leewenhoek would have delighted in, is opening up—the world of molecular biology. Of course, we cannot yet visualize the molecular details directly; our studies must be made with physicochemical techniques that yield sequences of proteins or nucleic acids that must be analyzed by computer; but it is a new world nevertheless, and at this Congress its impact is very evident. It is one of the great themes for the future, and one of the workhorses will be the computer.

The contributions of molecular biology to systematics are now becoming substantial. The 5S rRNA map of life has been mentioned. What has not been sufficiently emphasized is the extraordinary evolutionary record present in the genome, and the impressive concordance with non-molecular and fossil evidence. All creatures carry a fossil record within themselves. An example is the tree from alpha-globins of vertebrates. The groupings are much as one would expect (subject to the inevitable sampling error). They are not haphazard. Too much attention can be paid to a few aberrancies, and we may lose the excellent overall picture. There are, it is true, problems in knowing how to compare the various sequences in the genome, because of the extraordinary complexity of the genome, which systematists of the future will have to master.

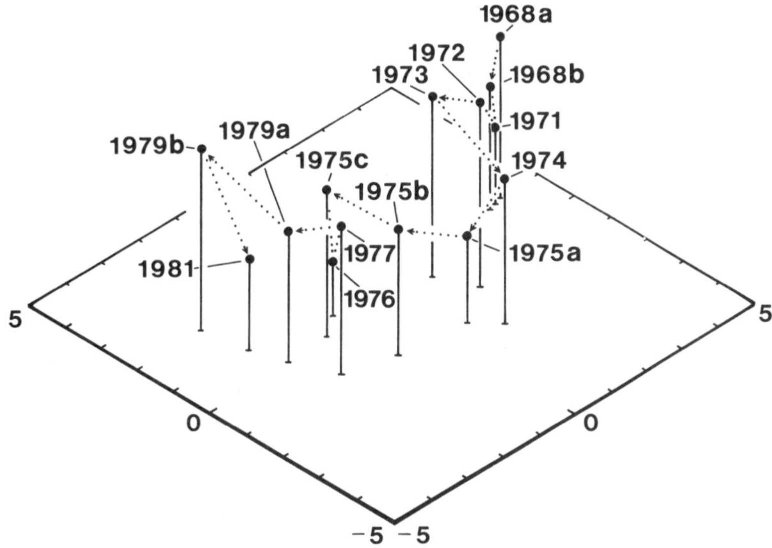

Figure 6. Evolution in influenza A virus. Principal coordinate analysis of immunological distances of the virus haemagglutinin from Skehel et al. (1982). Dotted lines connect temporally adjacent virus strains isolated from 1968 to 1981.

But there is a way round, though at the price of some assumptions about the physical chemistry of the genome itself. This is to compare the physicochemical binding of single strand DNA from two organisms when it reassociates to form the double-stranded Watson-Crick helix. One assumes that the disturbing influences largely cancel out after removing repetitious DNA, and the method relies, in a sense, on safety in numbers. Indeed, it is a point meet for philosophers to debate: whether the similarity in entire genomes is a sample, and if so, of what. Like all new techniques, it should not be pressed too far, but it has given some convincing answers in microbiology.

The outstanding application, however, has been made in ornithology by Charles Sibley and his team, in one of the major advances in systematics of the past decade. The pairing of single-copy DNA has yielded a wealth of information that has now emerged in a comprehensive new classification of the birds (Sibley et al., 1988). It is of course now being critically evaluated, but it looks very sound to me. Sampling error is relatively very small because of the large numbers of differences (many millions) implied by DNA-DNA physicochemical dissimilarity. There is a little experimental biological error (not yet evaluated in detail), but this is unlikely to be serious. As an example, one of the first studies may be cited, on the ratite birds (Sibley & Ahlquist, 1981). Sampling error is (theoretically) negligible, and its place is taken by experimental or biological error (Fig. 7), which is small in the work on birds, though uncomfortably large in work on bacteria (Hartford & Sneath, 1990).

It may be noted in passing that the molecular variation within a typical bacterial species seems as much as that within all birds; and we do not know the significance of this.

The concordance of molecular and other data has been stressed. Where there are two comparable molecular sequences does one find discordances that are greater than could be explained by sampling error? There are a few instances where 16S and 5S rRNA give results that seem too different to be dismissed (Sneath, 1989). The positions of origin of the flowering plants and yeasts are above the Protista on the 16S tree and below the Protista on the 5S tree (Fig. 8). The cellular slime moulds also have a rather different origin

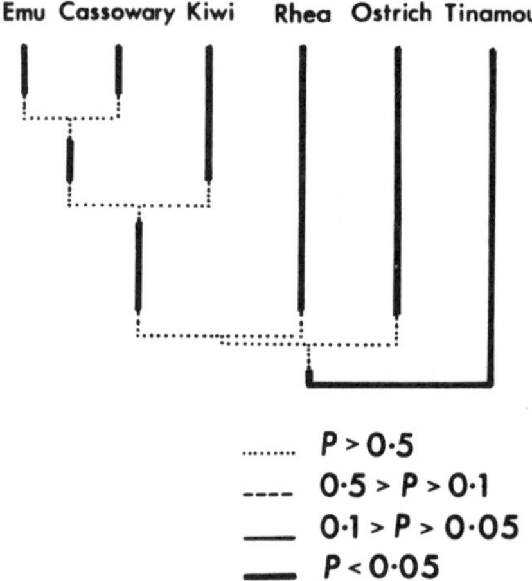

Figure 7. Cladogram of ratite birds from DNA pairing data of Sibley & Ahlquist (1981), modified from Sneath (1986). The uncertainty indicates that due to experimental error of the DNA technique (not sampling error). The only region of appreciable uncertainty is the order of branching of the rhea and the ostrich. Conventions as in Fig. 1.

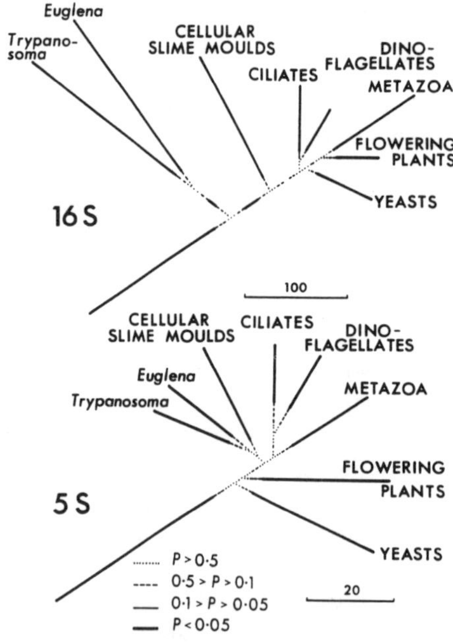

Figure 8. Relations between major eukaryote groups as shown by 16S and 5S rRNA, drawn to show uncertainty due to sampling error. Conventions as in Fig. 3. 16S data from Woese (1987), 5S data from Hori & Osawa (1986). The lower stems lead to the prokaryotes. Scales show numbers of base pair differences. Reproduced with permission from Sneath (1989).

in the two. A few more might be cited, and recent work hints at further discrepancies (Van den Eynde et al., 1990).

A good illustration is seen in plant viruses. The geminiviruses fall into two major groups, A and B. Group A viruses attack monocotyledons, are transmitted by leafhoppers, and have an unsegmented genome; group B viruses attack dicotyledons, are transmitted by whiteflies, and have a segmented genome. Howarth & Vandemark (1989) constructed two minimum-length phylogenetic trees based on the sequences of, respectively, the

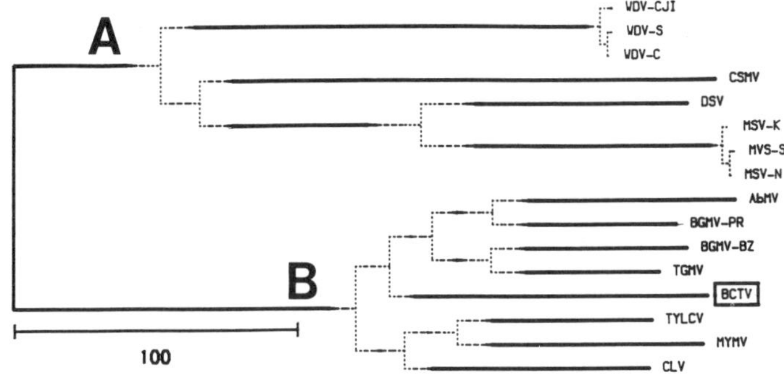

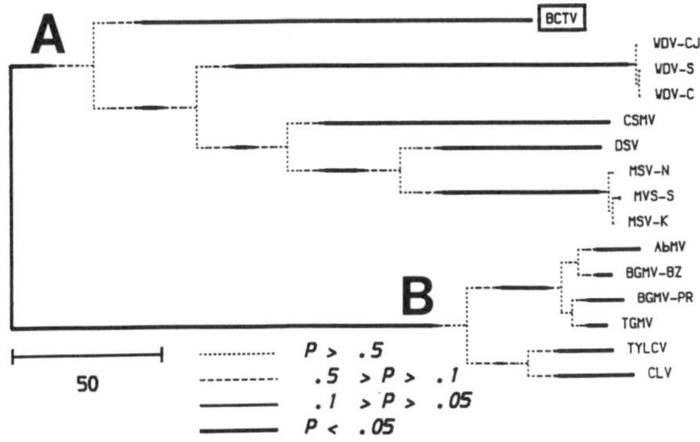

Figure 9. Relations of geminiviruses based on virus proteins; minimum-length trees from Howarth & Vandemark (1989) drawn to indicate tree confidence by the computer program of Sneath & Radbourne (1991). Above, tree from the replication-associated protein; below that from the coat protein. The virus groups A and B are shown. Note the discrepancy in the position of beef curly top virus (BCTV) in the two trees (see text). Conventions as in Fig. 3. Virus names are abbreviated (e.g. WDV = wheat dwarf virus; TGMV = tomato golden mosaic virus).

replication-associated protein and the coat protein of these viruses. The trees are broadly concordant, showing the division into groups A and B. There is, however, one anomalous virus, beet curly top virus (BCTV). This attacks dicotyledons, is transmitted by leafhoppers, and has an unsegmented genome; it belongs to group A on the tree from the coat protein and group B on the tree from the replication-associated protein. When the tree is drawn to express uncertainty it is seen that the difference is highly significant (Fig. 9); it is exceedingly unlikely that sampling error would explain the degree of movement of BCTV that this would imply. These authors considered that evolution of one of the proteins was constrained by the need to permit transmission by one or other type of insect vector, and I believe that a likely explanation is that BCTV is a hybrid between two geminiviruses, one from group A and one from group B. This example illustrates the power of these numerical methods in molecular data, and indicates how unusual phenomena may lead to reevaluation of our concepts of evolution.

Molecular probes are now becoming important. These are short lengths of nucleic acid, labelled in some way, which, when single-stranded, can recognize the complementary strand in a mixture of many nucleic acid molecules. They can be used directly in the environment. Rita Colwell and her colleagues here at the University of Maryland were able to extract bacterial nucleic acid directly from inland rock samples in the Antarctic, and to demonstrate that this "fossil" nucleic acid was highly similar to the nucleic acid of marine vibrios (Colwell et al., 1989). Recent reports (Giovanni et al., 1990; Ward et al., 1990) show that there are many unstudied microorganisms in waters, only recognized at present by unique nucleic acids. Probes can also be tailored to recognize one bacterial species or all members of a bacterial family (Regensberger et al., 1988).

This leads to a final topic. Microbiology may have little at present to contribute to conservation of the environment, though it has more to say on global effects. But through its pioneering methodology it has much to say on gene banks. These are usually thought of as seed stores and the like. But DNA libraries can be prepared consisting of all the genome (though in fragments) inside bacteria plasmids. The ability to extract DNA from dried or spirit-preserved museum material is another important pointer: DNA of the extinct quagga has been extracted and parts have been sequenced (Higuchi et al., 1984). The systematics of the future may rest heavily on the DNA in museum specimens.

Our museums thus contain a priceless heritage; they are not old just bones and skins and insect fragments—they are, in a sense, still alive. Soon we will learn how to exploit this heritage. Through developmental biology we may learn to introduce genes into tissue cultured cells. Perhaps we can backcross such cells to "fossil" gene banks, at first in plants, later (after many practical and ethical problems have been solved) in animals. Perhaps we can induce development of extinct organisms; perhaps the mammoth will shake the earth again.

LITERATURE CITED

Beckner, M. 1959. *The Biological Way of Thought.* Columbia University Press: New York. 200 pp.

Cain, A. J. & G. A. Harrison. 1958. An analysis of the taxonomist's judgement of affinity. *Proceedings of the Zoological Society of London* 131:85–98.

Colwell, R. R., MacDonell, M. T. & D. Swartz. 1989. Identification of an Antarctic endolithic microorganism by 5S rRNA sequences analysis. *Systematic and Applied Microbiology* 11:182–186.

Dobell, C. 1932. *Antonie van Leeuwenhoek and his "Little Animals".* Bale & Danielsson: London. 430 pp.

Gilmour, J. S. L. 1937. A taxonomic problem. *Nature* 139:1040–1042.

Gilmour, J. S. L. 1940. Taxonomy and philosophy. Pp. 461–474. *In:* J. Huxley (ed.), *The New Systematics.* Clarendon Press: Oxford. 583 pp.

Gilmour, J. S. L. 1951. The development of taxonomic theory since 1851. *Nature* 168:400–402.

Giovanni, S. J., Britschgi, T. B., Moyer, C. L. & K. G. Field. 1990. Genetic diversity in Sargasso Sea bacterioplankton. *Nature* 345:60–63.

Goodfellow, M., Jones, D. & F. G. Priest (eds.). 1985. *Computer-assisted Bacterial Systematics.* Academic Press: London. 443 pp.

Gregg, J. R. 1954. *The Language of Taxonomy.* Columbia University Press: New York. 71 pp.

Hartford, T. & P. H. A. Sneath. 1990. A review. Experimental error in DNA-DNA pairing: a survey of the literature. *Journal of Applied Bacteriology* 68:527–542.

Hennig, W. 1966. *Phylogenetic Systematics.* University of Illinois Press: Urbana. 263 pp.

Higuchi, R., Bowman, B., Freiberger, M., Ryder, O. A. & A. C. Wilson. 1984. DNA sequences from the quagga, an extinct member of the horse family. *Nature* 312:282–284.

Hori, H. & S. Osawa. 1986. Evolutionary change in 5S rRNA secondary structure and a phylogenetic tree of 352 5S rRNA species. *BioSystems* 19:163–172.

Howarth, A. J. & G. J. Vandemark. 1989. Phylogeny or geminiviruses. *Journal of General Virology* 70:2717–2727.

Hull, D. L. 1988. *Science as a Process: An Evolutionary Account of the Social and Conceptual Development of Science.* University of Chicago Press: Chicago. 586 pp.

Humphries, C. J. & P. M. Richardson. 1980. Hennig's method and phytochemistry. Pp. 353–378. *In:* F. A. Bisby, J. G. Vaughan & C. A. Wright (eds.), *Chemosystematics, Principles and Practice.* Academic Press: London.

Lapage, S. P., Sneath, P. H. A., Lessel, E. F., Skerman, V. B. D., Seeliger, H. P. R. & W. A. Clark (eds.). 1975. *International Code of Nomenclature of Bacteria.* American Society for Microbiology: Washington, D.C. 180 pp.

Margulis, L. 1970. *Origin of Eukaryotic Cells.* Yale University Press: New Haven. 349 pp.

Michener, C. D. & R. R. Sokal. 1957. A quantitative approach to a problem in classification. *Evolution* 11:130–162.

Mill, J. S. 1886. *A System of Logic.* Longmans, Green: London, 622 pp.

Mourant, A. E. 1971. Transduction and skeletal evolution. *Nature* 231:466–467.

Nelson, C. H. 1984. Numerical cladistic analysis of phylogenetic relationships in Plecoptera. *Annals of the Entomological Society of America* 77:466–473.

Oxford Dictionary of Quotations, 2nd ed. 1972. Oxford University Press: London. 1003 pp.

Regensburger, A., Ludwig, W. & K. H. Schleifer. 1988. DNA probes with different specificities from a cloned 23S rRNA gene of *Micrococcus luteus. Journal of General Microbiology* 134:1197–1204.

Richmond, M. H. 1970. Plasmids and chromosomes in prokaryotic cells. *Symposia of the Society for General Microbiology* 20:249–277.

Scott-Ram, N. R. 1980. *Transformed Cladistics, Taxonomy and Evolution.* Cambridge University Press: Cambridge. 238 pp.

Sibley, C. G. & J. E. Ahlquist. 1981. The phylogeny and relationships of the ratite birds as indicated by DNA-DNA hybridization. Pp. 301–335. *In:* G. G. E. Scudder & J. L. Reveal (eds.), *Evolution Today.* Carnegie Mellon University: Pittsburgh.

Sibley, C. G., Ahlquist, J. E. & B. L. Munroe, Jr. 1988. A classification of living birds of the world based on DNA-DNA hybridization. *Auk* 105:409–423.

Sielaff, B. H., Matsen, J. M. & J. E. McKie. 1982. Novel approach to bacterial identification that uses the Autobac system. *Journal of Clinical Microbiology* 15:1103–1110.

Sikorski, R. S., Michaud, W., Levin, H. L., Boeke, J. D. & P. Heiter. 1990. Trans-kingdom promiscuity. *Nature* 345:581–582.

Skehel, J. J., Douglas, A. R., Wilson, I. A. & D. C. Wiley. 1982. Antigenic variation in the influenza A (Hong Kong) viruses. *Symposia of the Society for General Microbiology* 33:215–226.

Sneath, P. H. A. 1986. Estimating uncertainty in evolutionary trees from Manhattan-distance triads. *Systematic Zoology* 35:470–488.

Sneath, P. H. A. 1986. The phenetic and cladistic approaches. Pp. 252–273. *In:* D. L. Hawksworth (ed.), *Prospects in Systematics.* Clarendon Press: Oxford.

Sneath, P. H. A. 1989. Analysis and interpretation of sequence data for bacterial systematics: the view of a numerical taxonomist. *Systematic and Applied Microbiology* 12:15–31.

Sneath, P. H. A. & J. C. Radbourne. 1991. DOTDND: a FORTRAN-77 program for showing graphically the confidence or uncertainty in phylogenetic trees. *Computers & Geosciences* 17:689–718.

Sneath, P. H. A. & R. R. Sokal. 1973. *Numerical Taxonomy. The Principles and Practice of Numerical Classification.* W. H. Freeman: San Francisco. 473 pp.

Sober, E. 1988. *Reconstructing the Past: Parsimony, Evolution, and Inference.* MIT Press: Cambridge, MA. 265 pp.

Sokal, R. R. & P. H. A. Sneath. 1963. *Principles of Numerical Taxonomy.* W. H. Freeman: San Francisco. 359 pp.

Van den Eynde, H., Van de Peer, Y., Perry, J. & R. De Wachter. 1990. 5S rRNA sequences of representatives of the genera *Chlorobium, Prosthecochloris, Thermomicrobium, Cytophaga, Flavobacterium, Flexibacter* and *Saprospira* and a discussion of the evolution of eubacteria in general. *Journal of General Microbiology* 136:11–18.

Vernon, K. 1988. The founding of numerical taxonomy. *British Journal for the History of Science* 21:143–159.

Vickers, J. C., Williams, S. T. & G. W. Ross. 1984. A taxonomic approach to selective isolation of streptomyces from soil. Pp. 553–561. *In:* L. Ortiz-Ortiz, L. F. Bojahil & V. Yakoleff (eds.), *Biological, Biochemical and Biomedical Aspects of Actinomycetes.* Academic Press: Orlando, FL.

Ward, D. M., Weller, R. & M. M. Bateson. 1990. 16S rRNA sequences reveal numerous uncultured microorganisms in a natural community. *Nature* 345:63–65.

Willcox, W. R., Lapage, S. P. & B. Holmes. 1980. A review of numerical methods in bacterial identification. *Antonie van Leeuwenhoek* 46:233–299.

Williams, W. T. & J. M. Lambert. 1959. Multivariate methods in plant ecology. 1. Association-analysis in plant communities. *Journal of Ecology* 47:83–101.

Woese, C. R. 1987. Bacterial evolution. *Microbiological Reviews* 51:221–271.

Woese, C. R. & G. E. Fox. 1977. Phylogenetic structure of the prokaryotic domain: the primary kingdoms. *Proceedings of the National Academy of Sciences of the USA* 74:5088–5090.

Woese, C. R., Fox, G. E., Zablen, L., Uchida, T., Bonen, L., Pechman, K., Lewis, B. J. & C. R. Woese. 1975. Conservation of primary structure in 16S ribosomal RNA *Nature.* 254:83–86.

Chloroplast DNA Evolution and Phylogenetic Relationships in *Chlamydomonas*

Monique Turmel, Eric Boudreau, Jean Boulanger, Jean-Patrick Mercier,
Christian Otis, and Claude Lemieux

Abstract. Our current view of the evolutionary patterns of the chloroplast genome is very biased in favor of land plants. We wish to broaden this view by providing insights into the evolutionary history of chloroplast DNA (cpDNA) in *Chlamydomonas,* a large and morphologically diverse group of unicellular green algae whose relationships are not well understood. Studies of *Chlamydomonas* cpDNA have focused so far on two pairs of interfertile species, *C. eugametos/C. moewusii* and *C. reinhardtii/C. smithii.* These algal cpDNAs are structurally similar to their land plant counterparts, but their size and arrangement show as much variation as that observed among all of the land plant cpDNAs examined to date. Except for deletions/additions, the cpDNAs are colinear within each pair, whereas they are extensively rearranged between the two pairs. To assess the significance of these rearrangements, we recently reconstructed a cpDNA phylogeny for 17 *Chlamydomonas* species representing 14 of the 15 sporangial autolysin groups *sensu* Schlösser and examined the chloroplast gene organization of nine of these species. The cpDNA phylogeny was inferred from large subunit (LSU) rRNA gene sequences. It features multiple evolutionary lineages, most species being allied in a lineage with *C. eugametos* or in a lineage with *C. reinhardtii.* Mapping of 32 genes on the cpDNAs of nine species representing the various lineages disclosed a great diversity of chloroplast genome arrangements. Only two sets of closely linked genes were found to be shared by *C. eugametos/C. moewusii, C. reinhardtii,* and the six other algae examined. Comparison of the two most related cpDNAs indicated that, as in land plants, changes in gene order occur by inversions of sequences and that the large inverted repeat region varies in size by spreading or shrinkage. A special feature of *Chlamydomonas* chloroplast LSU rRNA genes is their great variability in intron composition. A total of 39 group I introns are inserted at 12 distinct positions within the LSU rRNA genes of the 17 algae examined. No correlation was noted between the chloroplast LSU rRNA-based phylogeny and the distribution of the introns, suggesting that these are or were once mobile elements.

INTRODUCTION

Although comparisons of rRNA sequences support the view that the chloroplast DNAs (cpDNAs) of green algae and land plants have a common origin (see Gray, 1989), there is growing evidence that the evolutionary trends of these DNAs are not necessarily the same. Extensive characterization of cpDNAs from a large number of angiosperms and several representatives of the other land plant groups has revealed that most of these

Drs. Turmel, Boudreau, Boulanger, Mercier, Otis, and Lemieux are with the Département de Biochimie, Faculté des Sciences et de Génie, Université Laval, Québec (Québec) G1K 7P4, Canada. Correspondence should be sent to Dr. Turmel.

DNAs are very conserved at both the levels of their primary sequence and their gene order (see Palmer, 1985). The majority of land plant chloroplast genes are organized into multicistronic operons, several of which are remarkably similar to their homologues in cyanobacteria, the modern representatives of the presumed ancestor of chloroplasts in the green algal-land plant lineage. In contrast, the limited studies on green algal cpDNAs published to date have disclosed a highly variable gene organization with only a few conserved gene clusters, suggesting that the evolutionary pattern of green algal cpDNAs is less conservative than that of their land plant counterparts (see Manhart et al., 1989). Extensive sequence rearrangements have been reported even within *Chlamydomonas* (Lemieux & Lemieux, 1985; Lemieux et al., 1985a), the only green algal genus in which the cpDNAs of more than one species have been studied in detail by physical and gene mapping.

The genus *Chlamydomonas* is an assemblage of unicellular flagellates sharing a pair of apical flagella, a cell wall closely appressed to the protoplast, a single large chloroplast with a variable shape, and one or more pyrenoids (see Harris, 1989; and Trainor & Cain, 1986). Over 450 *Chlamydomonas* species have been described (Ettl, 1976), most of which have not been found to exhibit a sexual cycle and thus have been distinguished from one another on the basis of a few morphological characteristics that are visible in the microscope. It is largely for this reason that the validity of many *Chlamydomonas* species has been questioned (Lewin, 1975). The most recent and extensive work on the taxonomy of *Chlamydomonas* has been reported by Ettl (1976). This author described nine distinct groups of species which differ primarily by the chloroplast morphology and the number and location of pyrenoids. Groups of related *Chlamydomonas* species have also been identified on the basis of sensitivity to sporangial wall autolysins. Using this character, Schlösser (1984) classified 65 *Chlamydomonas* strains into 15 autolysin groups.

Studies of *Chlamydomonas* cpDNAs have focused on two pairs of interfertile species from distinct autolysin and morphological groups: *C. eugametos/C. moewusii* and *C. reinhardtii/C. smithii* (see Table 1 for more details on their classification). Whether these four species should be considered as separate species remains controversial (see Trainor & Cain, 1986). Although *C. eugametos* and *C. moewusii* mate readily to form zygotes, there is considerable lethality among the hybrid progeny; in contrast, *C. reinhardtii* and *C. smithii* are clearly interfertile. The four *Chlamydomonas* cpDNAs have been physically mapped (Harris, 1989; Lemieux et al., 1985b; Rochaix, 1978; Turmel et al., 1987) and compared by cross-hybridizations using large cpDNA fragments as probes (Lemieux & Lemieux, 1985; Lemieux et al., 1985a). Their structure is very similar to that of their land plant homologues, but their size (195–292 kbp) and gene arrangement show as much variation as that found among all of the land plant cpDNAs examined to date. Except for deletions/additions, the cpDNAs are colinear within each of the two pairs of interfertile algae, whereas they are extensively rearranged between the two algal pairs. As *C. eugametos* and *C. reinhardtii* display important differences at the level of the primary sequence of their cpDNAs as well as at the morphological, physiological, and reproductive levels, the numerous rearrangements could be explained simply by a great evolutionary distance separating these two algae. Alternatively, they could suggest that *Chlamydomonas* cpDNAs rearrange at a fast rate. In an attempt to distinguish between these two possibilities and to gain insights into the mode of structural evolution of *Chlamydomonas* cpDNAs, we have recently undertaken a detailed sequence analysis of the chloroplast large subunit (LSU) rRNA gene from several *Chlamydomonas* species and investigated the overall chloroplast gene arrangement of representative species. Our sequence analysis allowed us to determine not only the relationships among the species examined, but also the distribution of introns in their chloroplast LSU rRNA gene.

RELATIONSHIPS AMONG 17 *CHLAMYDOMONAS* SPECIES DEDUCED FROM COMPARISONS OF CHLOROPLAST LSU rRNA GENE SEQUENCES

We wished to examine the cpDNAs of representative species of *Chlamydomonas* whose evolutionary distances are well defined. Because the current taxonomy of the genus did not provide precise information on the evolutionary distances separating species, we set out to assess the phylogenetic relatedness of the 17 species listed in Table 1 by comparing their chloroplast LSU rRNA gene sequences. With the exception of *C. indica, C. eugametos, C. moewusii,* and *C. starii,* the species selected were not expected to be very closely related because they belong to different autolysin groups (Table 1). They represent 14 of the 15 autolysin groups described by Schlösser (1984) and four of the nine morphological groups described by Ettl (1976).

The entire coding region of each of the 17 *Chlamydomonas* chloroplast LSU rRNA genes was sequenced by the dideoxy chain termination method; the universal M13 primer was used for *C. eugametos* and *C. moewusii,* while a collection of oligodeoxynucleotides specific to highly conserved rRNA regions was employed for the remaining algae. Before this work was initiated, the only reported *Chlamydomonas* chloroplast LSU rRNA gene sequences were partial sequences from *C. reinhardtii* (see Lemieux et al., 1989). Our sequence data were found to represent the entire length of the LSU rRNA gene from five species, *C. eugametos, C. moewusii, C. reinhardtii, C. gelatinosa,* and *C. pischmanni;* the other algal gene sequences were partial due to the presence of introns of more than 200 bp (see Fig. 3) which could not be completely sequenced with the primers used.

Figure 1 presents the organization of the *C. eugametos* chloroplast LSU rRNA gene deduced from both DNA and RNA analyses (Turmel et al., 1991). This gene reveals two special features: 1) six group I introns and 2) three internal transcribed spacers, designated ITS1, ITS2, and ITS3. The finding of six introns is in sharp contrast to the situation in *C. reinhardtii* where a single group I intron (Rochaix et al., 1985) is inserted three nucleotides upstream of *C. eugametos* intron 6, and also to the situation in land plants where no interrupted LSU rRNA genes have been identified. Prior to our study, the maximal number of introns known in the LSU rRNA gene was three (in the nucleus of the Carolina strain of *Physarum polycephalum;* Muscarella & Vogt, 1989). All six *C. eugametos* LSU rRNA introns are inserted within highly conserved regions of primary sequence and/or rRNA

Table 1. *Chlamydomonas* **species examined in this study.**

Species	Autolysin group[a]	Morphological group[b]	Source[c]
reinhardtii Dangeard	1	Euchlamydomonas	SAG 11-32b
komma Skuja	2	Euchlamydomonas	SAG 26.72
peterfii Gerloff	3	Chlamydella	SAG 70.72
mexicana Lewin	4	Agloë	SAG 11-60a
iyengarii Mitra	5	Euchlamydomonas	SAG 25.72
humicola Lucksch	7	Chlorogoniella	SAG 11-9
species 66.72	8	ND	SAG 66.72
frankii Pascher	9	Euchlamydomonas	SAG 19.72
pallidostigmatica King	10	ND	SAG 9.83
gelatinosa Korshikov	11	Euchlamydomonas	SAG 69.72
indica Mitra	12	Chlamydella	SAG 11-11
eugametos Moewus	12	Chlamydella	UTEX 9
moewusii Gerloff	12	Chlamydella	UTEX 97
starii Ettl	12	Chlorogoniella	SAG 3.73
pitschmanni Ettl	13	Chlorogoniella	SAG 14.73
geitleri Ettl	14	Chlorogoniella	SAG 6.73
zebra Korshikov	15	Euchlamydomonas	SAG 10.83

[a]Schlösser (1984) [b]Ettl (1976); ND: not determined [c]SAG: from the Sammlung von Algenkulturen Göttingen; UTEX: from the Culture Collection at the University of Texas at Austin.

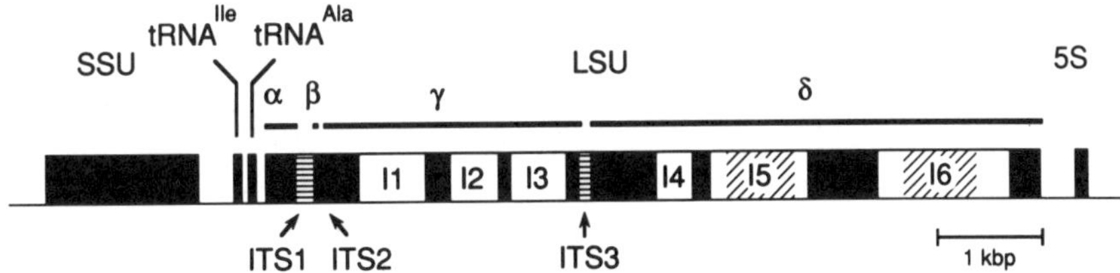

Figure 1. Organization of the *C. eugametos* chloroplast rDNA operon as deduced from DNA and RNA sequence analyses. This operon contains the genes for the small subunit (SSU; Durocher et al., 1989) and LSU (Turmel et al., 1991) rRNAs, 5S rRNA (A. Gauthier & C. Lemieux, unpublished results), tRNA^Ile and tRNA^Ala (Turmel et al., 1991). Filled areas denote coding regions for mature RNA species, while open areas denote introns (I1-I6). Four distinct RNA species, termed α, β, γ and δ, make up the LSU rRNA. Horizontally hatched areas represent the three ITS sequences (ITS1-ITS3) that separate the LSU rRNA coding regions, whereas cross-hatched areas denote intron ORFs.

secondary structure, and represent novel intron insertion sites compared to those that have been described so far (Turmel et al., 1991). Each of the two largest introns (introns 5 and 6) contains a long open reading frame (ORF) potentially encoding a basic protein of more than 200 amino acids. The intron 5 ORF belongs to a family of intron ORFs that have been identified in the mitochondrial DNAs of *Podospora anserina* and *Neurospora crassa*, whereas the intron 6 ORF is unrelated to any known ORFs (Turmel et al., 1991). From a genetic standpoint, the intron 5 is the most interesting of the *C. eugametos* LSU rRNA introns; it is absent from *C. moewusii* (A. Gauthier & C. Lemieux, unpublished results) and is mobile during interspecific crosses between the two algae (Lemieux & Lee, 1987). The molecular basis of this mobility is discussed in the following section. We have recently tested the self-splicing capacity of the six *C. eugametos* introns and found that only intron 5 cannot self-splice *in vitro* (M.-J. Côté, M. Turmel & C. Lemieux, unpublished results).

Unlike the introns, the three ITSs in the *C. eugametos* chloroplast LSU rRNA gene are located within highly variable regions of primary sequence and/or rRNA secondary structure. All 17 *Chlamydomonas* chloroplast LSU rRNA genes analyzed were found to display ITSs at common positions. These sequence elements are usually less than 300 nucleotides in size and are excised post-transcriptionally from a precursor RNA to yield four mature rRNA species, designated α, β, γ and δ (Fig. 1). The 7S and 3S rRNAs reported by Rochaix and Darlix (1982) in the *C. reinhardtii* chloroplast correspond to species α and β. We strongly suspect that green algae from other genera share with *Chlamydomonas* ITSs at the same positions. For instance, in *Chlorella ellipsoidea*, an insert that does not have the characteristics of an intron has been reported at the same position as ITS3 (Yamada & Shimaji, 1987). Three ITSs have been identified in the maize chloroplast LSU rRNA gene (Kössel et al., 1985), but at positions different from those of *Chlamydomonas*; one of them accounts for the 4.5S rRNA species.

The alignment of the various *Chlamydomonas* cpDNA sequences encoding the four LSU rRNA species revealed a total of 418 phylogenetically informative sites. Phylogenetic trees with essentially the same topology were constructed using these characters and various methods. Figure 2 presents a global chloroplast phylogeny including land plant, *Chlamydomonas* and other algal taxa, which was reconstructed with the aid of the distance matrix method. It can be seen that there exists more than one evolutionary lineage in *Chlamydomonas*. Two main lineages can be distinguished: one features seven species, including *C. eugametos,* and the other the remaining ten species, including *C. reinhardtii.* Within each of these main lineages, there are additional lineages that form sister groups. The positions of all *Chlamydomonas* taxa, with the exception of four taxa, are very well supported by the sequence data. The ambiguous positions observed are those of *C. reinhardtii, C. zebra,* and *C. gelatinosa* relative to one another, and that of *C. humicola* relative to the other

820

groups; the branchings associated with these taxa (Fig. 2) were found in less than 75% of replications during bootstrap analyses.

Our chloroplast rRNA gene sequence comparison is, in many respects, consistent with the classifications established by Ettl (1976) and Sclösser (1984) (see Fig. 2 and Table 1). It differs mainly from these classifications in the case of two taxa, *C. peterfii* and *C. starii*. Ettl places *C. peterfii* with *C. eugametos, C. moewusii,* and *C. indica* in the Chlamydella group, whereas our rRNA sequence analysis allies *C. peterfii* with *C. mexicana* (Ettl's Agloë group). Although Schlösser's classification places the latter species in different autolysin groups, this classification is consistent with the rRNA sequence data in that *C. peterfii* produces an autolysin that can break down the sporangial wall of *C. mexicana* cells (Schlösser 1984). Ettl allies *C. starii* with *C. pitschmanni* and *C. geitleri* in the Chlorogoniella group; however, the sequence data suggest that *C. starii* is allied with the "*C. reinhardtii*" clade (Euchlamydomonas group). In Schlösser's classification, *C. starii* is allied with *C. eugametos, C. moewusii,* and *C. indica* in autolysin group 12.

From Fig. 2, it is also evident that *Chlamydomonas* chloroplast LSU rRNA genes display extensive sequence variation. Together, the various *Chlamydomonas* lineages reveal at least

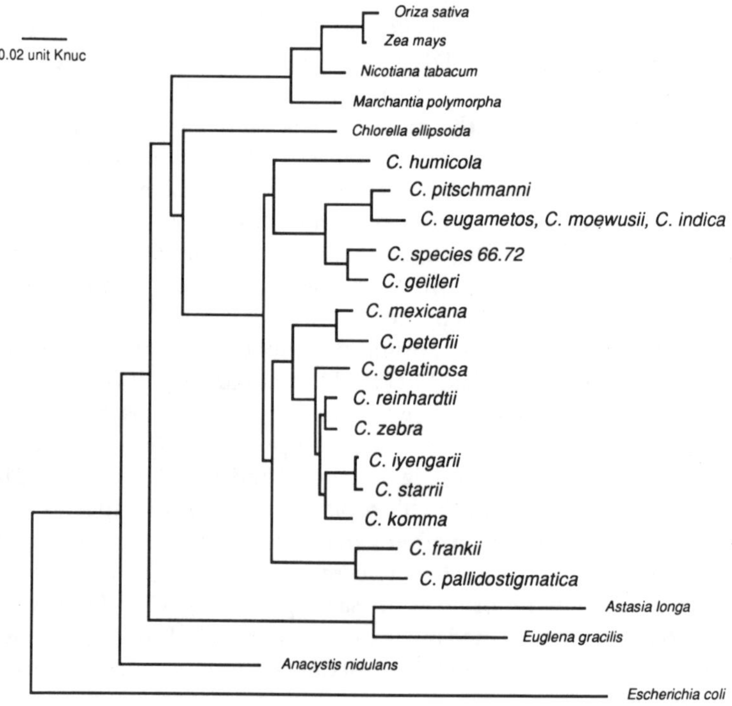

Figure 2. Relationships among 17 *Chlamydomonas* species as deduced from comparisons of chloroplast LSU rRNA gene sequences. The positions of the *Chlamydomonas* taxa relative to a green alga from another genus (*Chlorella ellipsoidea*), four land plants (*Oriza sativa, Zea mays, Nicotiana tabacum* and *Marchantia polymorpha*), two euglenophytes (*Euglena gracilis* and *Astasia longa*) and a cyanobacterium (*Anacystis nidulans*) are also presented. The *E. coli* 23S rRNA gene sequence was used as an outgroup. The phylogenetic tree is drawn to scale and is based on a distance matrix derived from an alignment of 2,565 partially conserved nucleotides. Computation of the corrected number of nucleotide differences (Kimura's K_{nuc} values) and construction of the tree were carried out using the DNADIST and FITCH programs of the Felsenstein-PHYLIP package (version 3.2), respectively. The LSU rRNA sequences were taken from: *C. eugametos,* Turmel et al., 1991; *C. moewusii,* A. Gauthier & C. Lemieux, unpublished results; *C. reinhardtii,* Lemieux et al. (1989); the remaining 14 *Chlamydomonas* species, M. Turmel, C. Otis, J.-P. Mercier & C. Lemieux, unpublished results; *Astasia longa,* Siemeister & Hachtel (1990); and other organisms, see the references cited by Gutell et al. (1990).

twice the range of sequence variation seen in all land plants. The level of sequence divergence is also considerable within distinct lineages: in at least three lineages (including *C. eugametos*, *C. reinhardtii*, and *C. frankii*), this level is greater than that found between a monocot and a dicot, two highly divergent land plant groups. This wide range of sequence variation suggests that the various *Chlamydomonas* lineages include distinct genera and represent distinct families and even orders. An independent cladistic analysis carried out in collaboration with one of us clearly supports this interpretation in the case of the "*C. humicola*" and "*C. reinhardtii*" lineages and reveals that *Chlamydomonas* is not a monophyletic or natural genus (Buchheim et al., 1990).

In this study, Buchheim et al. reconstructed a phylogeny from partial nuclear small subunit rRNA sequences of 13 of the 17 *Chlamydomonas* species listed in Table 1 as well as from the rRNA sequences of *C. callosa* and three flagellates from genera regarded as close relatives of *Chlamydomonas*, i.e. *Chlorogonium elongatum*, *Haematococcus lacustris*, and *Volvox aureus*. The various algal nuclear rRNA sequences were obtained from RNA templates and their alignment provided 92 phylogenetically informative characters. The phylogeny derived from cladistic analysis of these characters is congruent with that based on the chloroplast LSU rRNA gene sequences in displaying three main lineages of *Chlamydomonas* species represented by *C. eugametos*, *C. humicola*, and *C. reinhardtii*. However, the relationships among the *Chlamydomonas* taxa within a given lineage, with the exception of the "*C. eugametos*" lineage, are less resolved in the nuclear phylogeny than in the chloroplast phylogeny. Remarkably, the three non-*Chlamydomonas* taxa were found to fall within two distinct *Chlamydomonas* clades, indicating that the current taxonomy of this genus does not reflect natural or monophyletic groups. *Chlorogonium elongatum* and *H. lacustris* are allied with *C. humicola*, whereas *V. aureus* is part of the "*C. reinhardtii*" lineage. This non-monophyletic concept of the *Chlamydomonas* genus is well supported by the sequence data.

DISTRIBUTION OF GROUP I INTRONS IN THE CHLOROPLAST LSU rRNA GENE OF *CHLAMYDOMONAS*

Our sequence analysis of various *Chlamydomonas* chloroplast LSU rRNA genes disclosed a great diversity of introns. As shown in Fig. 3, a total of 39 introns representing 12 distinct insertion sites within domains II, IV and V of the LSU rRNA secondary structure were identified in the LSU rRNA genes of the 17 species listed in Table 1. Although all of these introns have not been completely sequenced and their RNA secondary structure analyzed, it is clear that they belong to the group I intron family because their 5′ terminal sequence is preceded by a U residue and can pair with the upstream exon, and also because their 3′ terminal sequence ends with a G residue. Moreover, some of the partial intron sequences determined were found to display one or more of the conserved sequence elements P, Q, R, and S that allow the formation of the characteristic group I core structure.

The intron composition of *Chlamydomonas* chloroplast LSU rRNA genes is extremely variable and does not agree with the phylogeny derived from the coding sequences of these genes (Fig. 3). Similar variations in intron composition have been noted for two additional *Chlamydomonas* chloroplast genes (small subunit rRNA gene: Durocher et al., 1989; *psbA*: see Turmel et al., 1989) and several genes from other systems (see Dujon, 1989). This situation has been explained by the ability of group I introns to excise from, and integrate into, genes (see Dujon, 1989). Evidence for two types of intron integration events has been reported. One type of events, called intron homing, allows the maintenance of a group I intron at a specific site and is evidenced in crosses between intron-plus and intron-minus strains. The other type of intron mobility involves the trans-

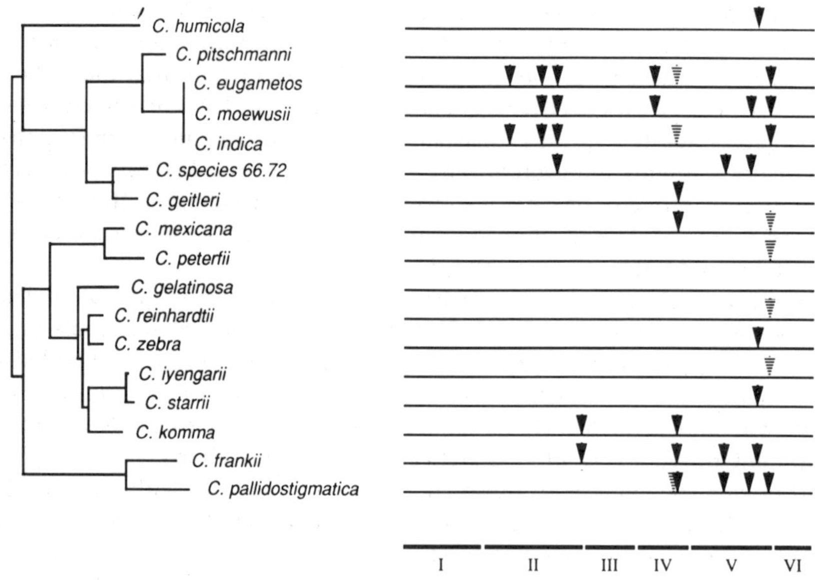

Figure 3. Distribution of group I introns in the 17 *Chlamydomonas* chloroplast LSU rRNA genes analyzed in this study. Arrowheads denote the intron insertion sites. The introns inserted within very closely linked sites were distinguished by solid and dashed arrowheads. Thick lines labelled with roman numerals refer to the six major LSU rRNA domains defined by long range base-pairing interactions (Noller, 1984). The various algal LSU rRNA sequences were not drawn to scale; the *E. coli* 23S rRNA was used as a model and variations in length of the variable regions were not taken into account. To the left of the sequence alignment is shown the phylogenetic tree presented in Fig. 3.

fer of a group I intron from one genomic site to another.

The molecular mechanism underlying intron homing has been recently elucidated for two yeast mitochondrial introns, two bacteriophage T4 introns, and a nuclear intron of *Physarum polycephalum* (see Lambowitz, 1989). This same mechanism is responsible for the mobility of the *C. eugametos* chloroplast LSU rRNA intron 5 (CeLSU·5) during reciprocal crosses with *C. moewusii* strains that lack the intron (Lemieux & Lee, 1987). In such interspecific crosses, the CeLSU·5 intron is acquired by all intronless sites of the LSU rRNA gene and is transmitted to the entire progeny, whereas most of the other poly-morphic cpDNA markers are transmitted mainly from the mating-type plus parent (Lemieux et al., 1988). Like the aforementioned mobile introns, CeLSU·5 codes for a site-specific DNA endonuclease that makes a double-strand break at or near the homing site of the intron in the recipient DNA (Gauthier et al., 1991). The *C. eugametos* endonuclease activity was identified by overexpressing the CeLSU·5 ORF of 218 codons in *Escherichia coli*, using plasmid constructs containing the homing site; as a result of this expression, the plasmid constructs were found to be linearized in close proximity to the intron homing site. In hybrid zygotes, therefore, the *C. eugametos* endonuclease can cleave specifically the *C. moewusii* cpDNA at this site. Subsequent repair of the resulting double-strand break, most probably by the gap repair mechanism, results in the insertion of the intron into virtually all intronless sites of the LSU rRNA gene present in the chloroplast.

Although it has not been directly demonstrated that group I introns can move from one genomic site to another, several circumstantial lines of evidence support this phenom-enon (see Dujon, 1989). Two mechanisms have been proposed to explain this transposi-tion. One of them occurs at the DNA level and involves intron-encoded endonucleases (Dujon, 1989), while the other occurs at the RNA level and involves the insertion of a group I intron into a foreign RNA by reversal of the self-splicing reaction, followed by reverse transcription of the recombined RNA and integration into the genomic DNA by homologous recombination (Woodson & Cech, 1989). In the case of the rRNA genes whose mature products exist predominantly as RNA-protein complexes in the ribosomes, the first mechanism predicts that group I introns would insert more or less randomly following a T residue within any region of these gene sequences, whereas the second mechanism predicts that they would insert preferentially or exclusively within RNA domains located at the surface of the ribosome. Consistent with such locations, the *E. coli* rRNA regions corresponding to most, if not all, of the insertion positions of *Chlamydomonas* chloroplast LSU rRNA introns were found to be protected from degradation by chemical probes after binding of tRNAs (Moazed & Noller, 1989), antibiotics (Moazed & Noller, 1987), and elongation factor EF-G (Moazed et al., 1988) to the large ribosomal subunit. It is thus likely that some of the multiple *Chlamydomonas* intron insertion sites arose by the gain of introns at the RNA level. When the sequences of a sufficient number of introns at each of the insertion sites become available, it will be interesting to see if they have a common origin.

CHLOROPLAST GENOME ORGANIZATION IN *CHLAMYDOMONAS*

We have investigated the chloroplast gene organization of *C. eugametos/C. moewusii, C. frankii, C. gelatinosa, C. humicola, C. iyengarii, C. peterfii, C. pitschmanni*, and *C. reinhardtii*. These nine species represent the various *Chlamydomonas* lineages found in the chloroplast LSU rRNA based-phylogeny shown in Fig. 2. Figure 4 indicates the positions of the 32 genes studied on the physical map of the *C. moewusii* cpDNA. Several of these genes were mapped in the course of recent sequencing studies, while the others were identified by heterologous hybridizations using *C. reinhardtii* or land plant gene-specific probes (Turmel et al., 1988; E. Boudreau & M. Turmel, unpublished results). *Chlamydomonas* gene-specific probes are now available for most of the 32 genes; the advantage of these probes is discussed below. The chloroplast gene organization of *C. moewusii* differs dramatically from the consensus gene order found in land plants, the rDNA operon being the only common gene cluster. An interesting feature of the *C. moewusii* cpDNA, which has also been observed in *C. reinhardtii* (Kück et al., 1987), is the presence of the three exons making up the coding region of the *psaA* gene in widely scattered regions. Like its *C. reinhardtii* homologues (Choquet et al., 1988), the *C. moewusii psaA* exons are probably transcribed independently and the resulting precursors assembled in *trans*.

To study the gene organization of the six *Chlamydomonas* cpDNAs whose physical maps were not available, we decided to construct simultaneously the physical and gene maps of these cpDNAs by hybridizing various gene-specific probes to Southern blots containing several digests of each algal cpDNA. Basically, this approach consists of a linkage analysis of chloroplast genes. Provided that the genes examined are numerous enough and well dispersed throughout the chloroplast genome, the hybridization data should allow us to deduce the relative order of all genes and construct a physical map of each algal cpDNA. Restriction endonucleases that generate few cpDNA fragments were selected to maximize the chances of finding two genes on a given fragment. It also became obvious that *Chlamydomonas* gene probes were required as several land plant chloroplast genes hybridize only weakly or not at all to *Chlamydomonas* cpDNAs.

With the gene probes employed, we have been able to construct a gene map covering the total length of the *C. pitschmanni* cpDNA and, for each of the five remaining species, we have associated most of the genes to separate DNA segments representing together 50 to 90% of the chloroplast genome. Although the relative orientation of these segments has not been determined, our data clearly indicate that the various *Chlamydomonas* lineages

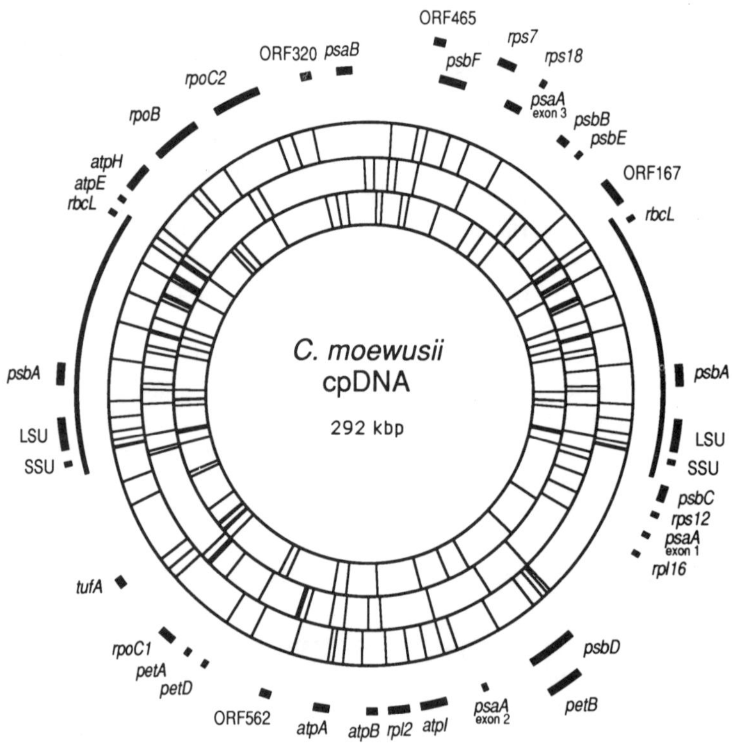

Figure 4. Physical and gene maps of the *C. moewusii* cpDNA. The three circles from the inside to the outside represent the *Eco*RI, *Bst*EII, and *Ava*I restriction maps, respectively (Turmel et al., 1987). The positions of 32 genes are denoted by dark boxes. The genes whose functions have not been determined were identified by sequence analysis and designated as those of *Marchantia poly-morpha* (Ohyama et al., 1986). The thick lines connecting the loci for the rRNA genes, *psbA* and *rbcL*, delimit the inverted repeat sequence. Only one of the two possible orientations of the single-copy regions relative to one another is shown.

feature distinct chloroplast gene arrangements. The nine *Chlamydomonas* species examined have in common only two gene clusters consisting of the rDNA operon-*psbA* and *petA-petD*. The significance of these gene clusters remains unclear as the physical maps of all algal cpDNAs are not available. They could represent diagnostic features of *Chlamydomonas* chloroplast genomes and the *petA-petD* pair could constitute an operon. In support of the latter hypothesis, our sequence analysis of the *C. moewusii petA* and *petD* genes (A. Bergeron & M. Turmel, unpublished results) revealed that these genes are very closely linked and encoded on the same DNA strand. Interestingly, similar observations have been reported for the *petA* and *petD* genes of *Scenedesmus obliquus* (Kück, 1989; see also Kück et al., 1990).

Figure 5 presents our comparative analysis of the chloroplast gene maps of *C. eugametos/C. moewusii, C. reinhardtii,* and *C. pitschmanni.* The individual events that led to the extensively rearranged *C. eugametos/C. moewusii* and *C. reinhardtii* cpDNAs could not be resolved in this analysis, even though the fragments we used as probes were much smaller than those employed by Lemieux and Lemieux (1985). Given that *C. eugametos* and *C. reinhardtii* represent the two main *Chlamydomonas* lineages observed in the phylogeny inferred from chloroplast rRNA sequence comparisons (see Fig. 2), it is not surprising to find that their cpDNAs differ extensively in gene organization. In contrast, the chloroplast gene maps of *C. eugametos/C. moewusii* and *C. pitschmanni,* three species from the same lineage, show only three noticeable differences, which are due to inversions and to the

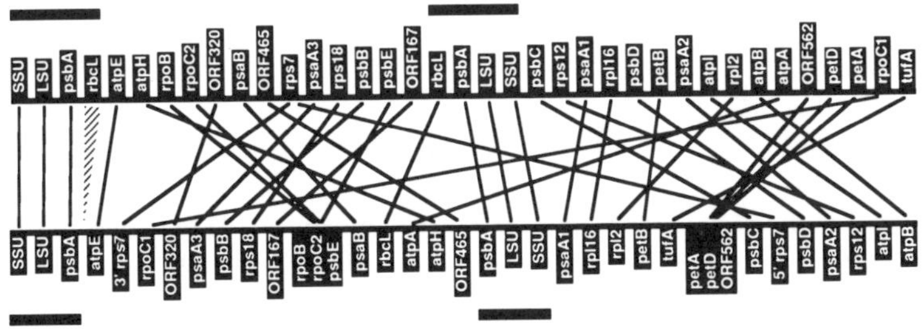

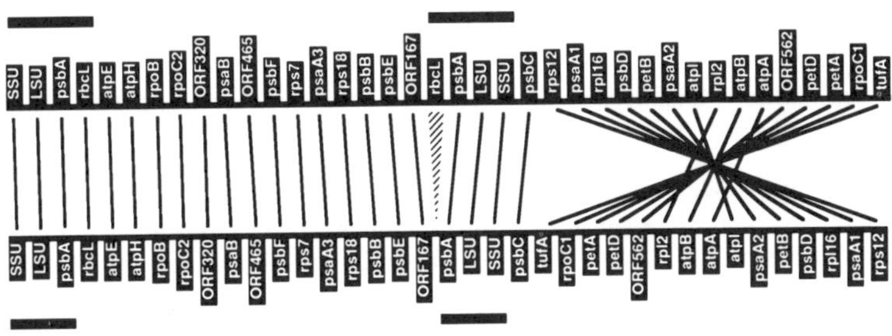

Figure 5. Compared gene maps of the *C. eugametos/C. moewusii*, *C. reinhardtii*, and *C. pitschmanni* cpDNAs. These three algal gene maps were linearized at a similar position to allow direct comparison of the *C. eugametos/C. moewusii* cpDNA with the *C. reinhardtii* and *C. pitschmanni* cpDNAs. In each map, dark boxes denote the positions of the 32 genes presented in Fig. 4. Note that the relative order of the genes assigned to a common box has not been determined and that the physical distances separating contiguous genes are not indicated. The thick lines outside the gene maps delimit the inverted repeat sequence. In each of the two cpDNA comparisons, the shaded triangle denotes a difference in the gene content of the inverted repeat involving *rbcL*, and, except for this gene, all genes having identical functions are connected by solid lines.

contraction or shrinkage of the inverted repeat sequence. A single-copy region displays two inversions, one of which involves the segment extending from *rps12* to *tufA* and the other the internal region containing *rpl2, atpB*, and *atpA*. In *C. eugametos/C. moewusii, rbcL* lies within the inverted repeat sequence, while, in *C. pitschmanni*, it is immediately adjacent to this sequence. Such differences were not unexpected as, in land plant chloroplast genomes, occasional changes in gene order are usually caused by inversions, and variations in genome size are due to the expansion or shrinkage of the inverted repeat (see Palmer, 1985). In future studies, it will be interesting to see if the frequency of cpDNA rearrangement is similar within distinct *Chlamydomonas* lineages and if some of the ancestral operons conserved in land plants are found in a few *Chlamydomonas* species.

ACKNOWLEDGMENTS

We thank J. E. Boynton, N. W. Gillham, E. H. Harris and J. D. Palmer for their gifts of chloroplast gene probes and U. G. Schlösser for providing *Chlamydomonas* strains. This research was supported by the Natural Sciences and Engineering Research Council of

Canada and by the Fonds pour la Formation de Chercheurs et l'Aide à la Recherche. C. L. is a Scholar and M. T. an Associate in the Evolutionary Biology Program of the Canadian Institute for Advanced Research.

LITERATURE CITED

Buchheim, M. A., Turmel, M., Zimmer, E. A. & R. L. Chapman. 1990. Phylogeny of *Chlamydomonas* (Chlorophyta) based on cladistic analysis of nuclear 18S rRNA sequence data. *J. Phycol.* 26:689–699.

Choquet, Y., Goldschmidt-Clermont, M., Girard-Bascou, J., Kück, U., Bennoun, P. & J. D. Rochaix. 1988. Mutant phenotypes support a *trans*-splicing mechanism for the expression of the tripartite *psaA* gene in the *C. reinhardtii* chloroplast. *Cell* 52:903–913.

Dujon, B. 1989. Group I introns as mobile genetic elements: facts and mechanistic speculations—a review. *Gene* 82:91–113.

Durocher, V., Gauthier, A., Bellemare, G. & C. Lemieux. 1989. An optional group I intron between the chloroplast small subunit rRNA genes of *Chlamydomonas moewusii* and *C. eugametos*. *Curr. Genet.* 15:277–282.

Ettl, H. 1976. Die Gattung *Chlamydomonas* Ehrenberg *Beih. Nova Hedwigia* 49:1–1122.

Gauthier, A., Turmel, M. & C. Lemieux. 1991. A group I intron in the chloroplast large subunit rRNA gene of *Chlamydomonas eugametos* encodes a double-strand endonuclease that cleaves the homing site of this intron. *Curr. Genet.* 19:43–47.

Gray, M. W. 1989. The evolutionary origins of organelles. *Trends in Genet.* 5:294–299.

Gutell, R. R., Schnare, M. N. & M. W. Gray. 1990. A compilation of large subunit (23S-like) ribosomal RNA sequences presented in a secondary structure format. *Nucl. Acids Res.* 18:2319–2330.

Harris, E. H. 1989. *The* Chlamydomonas *Sourcebook*. Academic Press: San Diego. 780 pp.

Kössel, H., Natt, E., Strittmatter, G., Fritzsche, E., Gozdzicka-Jozefiak, A. & D. Przybyl. 1985. Structure and expression of rRNA operons from plastids of higher plants. Pp. 183–198. *In:* L. Vloten-Doting, G. S. P. Groot & T. C. Hall (eds.) *Molecular Form and Function of the Plant Genome*. Plenum Press: New York.

Kück, U. 1989. The intron of a plastid gene from a green alga contains an open reading frame for a reverse transcriptase-like enzyme. *Mol. Gen. Genet.* 218:257–265.

Kück, U., Choquet, Y., Schneider, M., Dron, M. & P. Bennoun. 1987. Structural and transcription analysis of two homologous genes for the P700 chlorophyll α-apoproteins in *Chlamydomonas reinhardtii*: evidence for *in vivo trans*-splicing. EMBO J. 6:2185–2195.

Kück U., Godehardt, I. & U. Schmidt. 1990. A self-splicing group II intron in the mitochondrial large subunit rRNA (LSU rRNA) gene of the eukaryotic alga *Scenedesmus obliquus*. *Nucl. Acids Res.* 18:2691–2697.

Lambowitz, A. M. 1989. Infectious introns. *Cell* 56:323–326.

Lemieux, B. & C. Lemieux. 1985. Extensive sequence rearrangements in the chloroplast genomes of the green algae *Chlamydomonas eugametos* and *Chlamydomonas reinhardtii*. *Curr. Genet.* 10:213–219.

Lemieux, B., Turmel, M. & C. Lemieux. 1985a. *Chloroplast* DNA variation in *Chlamydomonas* and its potential application to the systematics of this genus. *BioSystems* 18:293–298.

Lemieux, C., Turmel, M., Seligy, V. L. & R. W. Lee. 1985b. The large subunit of ribulose–1, 5-bisphosphate carboxylase-oxygenase is encoded in the inverted repeat sequence of the *Chlamydomonas eugametos* chloroplast genome. *Curr. Genet.* 9:139–145.

Lemieux, C. & R. W. Lee. 1987. Nonreciprocal recombination between alleles of the chloroplast 23S ribosomal RNA gene in interspecific *Chlamydomonas* crosses. *Proc. Natl. Acad. Sci. USA* 84:4166–4170.

Lemieux, B., Turmel, M. & C. Lemieux. 1988. Unidirectional gene conversions in the chloroplast of *Chlamydomonas* interspecific hybrids. *Mol. Gen. Genet.* 212:48–55.

Lemieux, C., Boulanger, J., Otis, C. & M. Turmel. 1989. Nucleotide sequence of the chloroplast large subunit rRNA gene from *Chlamydomonas reinhardtii*. *Nucl. Acids Res.* 17:7997.

Lewin, R. A. 1975. A *Chlamydomonas* with black zygospores. *Phycologia* 14:71–74.

Manhart, J. R., Kelly, K., Dudock, B. S. & J. D. Palmer. 1989. Unusual characteristics of *Codium fragile* chloroplast DNA revealed by physical and gene mapping. *Mol. Gen. Genet.* 216:417–421.

Moazed, D. & H. F. Noller. 1987. Chloramphenicol, erythromycin, carbomycin and vernamycin B protect overlapping sites in the peptidyl transferase region of 23S ribosomal RNA. *Biochimie* 69:879–884.

Moazed, D., Robertson, J. M. & H. F. Noller. 1988. Interaction of elongation factors EF-G and EF-Tu with a conserved loop in 23S RNA. *Nature* 334:362–364.

Moazed, D. & H. F. Noller. 1989. Interaction of tRNA with 23S rRNA in the ribosomal A, P, and E sites. *Cell* 57:585–597.

Muscarella, D. E. & V. M. Vogt. 1989. A mobile group I intron in the nuclear rDNA of *Physarum polycephalum.* Cell 56:443–454.

Noller, H. F. 1984. Structure of ribosomal RNA. *Ann. Rev. Biochem* 53:119–162.

Ohyama, K., Fukuzawa, H., Kohchi, T., Shirai, H., Sano, T., Sano, S., Umesono, K., Shiki, Y., Takeuchi, M., Chang, Z., Aota, S. I., Inokuchi, H. & H. Ozeki. 1986. Chloroplast gene organization deduced from complete sequence of liverwort *Marchantia polymorpha* chloroplast DNA. *Nature* 322:572–574.

Palmer, J. D. 1985. Comparative organization of chloroplast genomes. *Ann. Rev. Genet.* 19:325–354.

Rochaix, J. D. 1978. Restriction endonuclease map of the chloroplast DNA of *Chlamydomonas reinhardii. J. Mol. Biol.* 126:597–617.

Rochaix J. D. & J. L. Darlix. 1982. Composite structure of the chloroplast 23 S ribosomal RNA genes of *Chlamydomonas reinhardii.* Evolutionary and functional implications. *J. Mol. Biol.* 159:383–395.

Rochaix, J. D., Rahire, M. & F. Michel. 1985. The chloroplast ribosomal intron of *Chlamydomonas reinhardii* codes for a polypeptide related to mitochondrial maturases. *Nucl. Acids. Res.* 13:975–984.

Schlösser, U. G. 1984. Species-specific sporangium autolysins (cell-wall-dissolving enzymes) in the genus *Chlamydomonas.* Pp. 409–418. *In:* D. E. G. Irvine & D. M. John (eds.), *Systematics of the Green Algae.* Academic Press: New York.

Siemeister, G & W. Hachtel. 1990. Organization and nucleotide sequence of ribosomal RNA genes on a circular 73 kbp DNA from the colourless flagellate *Astasia longa. Curr. Genet.* 17:433–438.

Trainor, F. R. & J. R. Cain. 1986. Famous algal genera. I. *Chlamydomonas.* Pp. 81–127. In: F. E. Round & D. J. Chapman (eds.), *Progress in Phycological Research, Volume 4.* Biopress Ltd.: Bristol.

Turmel, M., Bellemare, G. & C. Lemieux. 1987. Physical mapping of differences between the chloroplast DNAs of the interfertile algae *Chlamydomonas eugametos* and *Chlamydomonas moewusii. Curr. Genet.* 11:543–552.

Turmel, M., Lemieux, B. & C. Lemieux. 1988. The chloroplast genome of the green alga *Chlamydomonas moewusii:* Localization of protein-coding genes and transcriptionally active regions. *Mol. Gen. Genet.* 214:412–419.

Turmel, M., Boulanger, J. & C. Lemieux. 1989. Two group I introns with long internal open reading frames in the chloroplast *psbA* gene of *Chlamydomonas moewusii. Nucl. Acids Res.* 17:3875–3887.

Turmel, M., Boulanger, J., Schnare, M. N., Gray, M. W. & C. Lemieux. 1991. Six group I introns and three internal transcribed spacers in the chloroplast large subunit ribosomal RNA gene of the green alga *Chlamydomonas eugametos. J. Mol. Biol.* 218:293–311.

Woodson, S. A. & T. R. Cech. 1989. Reverse self-splicing of the *Tetrahymena* group I intron: implication for the directionality of splicing and for intron transposition. *Cell* 57:335–345.

Yamada, T. & M. Shimaji. 1987. An intron in the 23S rRNA gene of the *Chlorella* chloroplasts: complete nucleotide sequence of the 23S rRNA gene. *Curr. Genet.* 11:347–352.

Pattern and Process in
Plant Mitochondrial Genome Evolution

Michael B. Coulthart

Abstract. Recently published data strongly suggest that the mitochondrial genomes of angiosperms evolve in a highly distinctive manner, with a very high rate of rearrangement but a very low rate of nucleotide substitution in comparison with chloroplast genomes of the same plant lineages. The factual basis for this claim is reviewed. Then, after a brief review of certain elementary theoretical concepts of evolutionary genetics, an analysis of possible mechanistic explanations is presented, with reference to specific features of plant mitochondria and mitochondrial DNA (i.e., cellular behavior, mutation rates, and cryptic genetic variation) that have been suggested to contribute to the evolutionary pattern.

INTRODUCTION

Genome evolution can be defined simply as molecular evolution at levels of genetic organization above that of the single gene. Its study begins, as do analogous evolutionary studies carried out at other levels of biological organization, with the use of the comparative method to describe interspecies *differences* and, by appropriate inference, to chronicle evolutionary *change*. When enough data of this sort have accumulated, it is appropriate to begin a search for *patterns*, defined here as regularities that emerge from observation of multiple instances of evolutionary change. With this prodigious task of elucidating history and defining pattern well under way, it is quite natural to begin to ask questions about *process*. In this phase, established evolutionary theory, much of it framed in the language of molecular, transmission, and population genetics, is used in attempts to generate and test explanations of observed patterns. In some cases, it is actually possible to use the results of such confrontations between fact and theory to generate new theoretical insight into evolutionary processes. One of the main theoretical goals in the study of genome evolution is to understand something of how the organization and function of the genome itself—the "genetic system"—impose constraints on and furnish opportunities for evolution. This goal is consonant with current emphasis on hierarchy in evolution (e.g., Gould & Eldredge, 1977; Brandon & Burian, 1984; Doolittle, 1988).

The diversity with respect to structural organization and specifics of genetic function that exists among mitochondrial genomes of eukaryotes is enormous; it is now beginning to appear that evolutionary patterns and processes in these genomes may be equally diverse, with animals, fungi, protists and plants each displaying unique or characteristic features (Gray, 1989). Especially when combined with their comparatively small size and

---

Dr. Coulthart is with the Department of Biochemistry, Dalhousie University, Halifax, Nova Scotia, B3H 4H7, Canada.

genetic simplicity, this diversity suggests that mitochondrial genomes provide an excellent opportunity to approach the above-mentioned theoretical goal of understanding the evolutionary importance of genetic systems. The purpose of the present chapter is to discuss evolutionary features unique to or characteristic of plant mitochondrial genomes, while attempting to place the current state of knowledge in this area within the context of the history/pattern/process/theory paradigm outlined above. By way of a disclaimer, however, I should point out that because data on non-angiospermous plant mitochondrial DNAs (mtDNAs) are almost vanishingly scarce, the present discussion of "plant" mtDNA evolution is based on data from angiosperms alone.

In the following discussion I will first define, and then present some illustrative supporting data for, a tentative evolutionary pattern that has emerged from recent comparative studies of plant mtDNA arrangement and nucleotide sequence. Then, after a brief digression to review some relevant theoretical concepts, I will sketch and critically examine three recently published examples of speculations about possible mechanistic underpinnings of the pattern. After each of these two major sections (one on pattern, one on process), I will venture some ideas as to what kinds of experimental data might help to confirm and extend current knowledge in this area.

THE PATTERN

To define the evolutionary pattern in question, some preliminary groundwork should be laid. First, it should be emphasized that the physical organization of plant mtDNA is complex and somewhat mysterious, and is still a subject of active investigation (see, for example, the analysis of native plant mtDNA by pulsed-field gel electrophoresis reported by Bendich & Smith, 1990). Despite this uncertainty, it has proven possible in every case examined so far to reconstruct circular restriction maps from cloned fragments of a given plant mtDNA. Thus, the currently best-accepted physical model for plant mtDNA places the entire sequence complexity of the genome in a single, large "master circle"; the sizes of the dozen or so circular maps available to date vary widely between 200 and ca. 600 kilobase pairs (kbp) (reviewed by Lonsdale et al., 1988). However, complexity is introduced because in almost all plant species these master circles contain families of relatively large (1–15 kbp), dispersed repeated sequences that appear capable of homologous recombination to generate excised subgenomic circles as well as rearranged master circles (Falconet et al., 1984; Lonsdale et al., 1984; Palmer & Shields, 1984). The main structural consequence for the plant mitochondrial genome would seem to be fragmentation ("multicircularity"). This complex organizational situation appears, nevertheless, not to be incompatible with a considerable degree of compositional stability of the mtDNA within species (Oro et al., 1985; Palmer, 1988). Note that, with the evolutionary rearrangements in plant mtDNAs mentioned below, it is the master-circle maps that are compared between taxa.

Second, I make a distinction between *genome rearrangements* and *nucleotide substitutions*. Although this distinction can be somewhat arbitrary (e.g., how are small insertions/deletions of one or a few bases classified?), it serves well enough for the data under consideration here. What is more important is the usefulness of the distinction for thinking about the possible existence of different classes of molecular mechanisms that underlie evolutionary change in genomes. Finally, I would like to point out that the pattern under discussion is a *quantitative* one, having to do with rates of evolutionary change in plant mitochondrial genomes, rather than a *qualitative* pattern based solely on the *kinds* of change that are seen. Plant mtDNA does exhibit interesting qualitative evolutionary patterns (for instance, its tendency to incorporate segments of exogenous DNA from the plastid and nuclear genomes—see, e.g., Nugent & Palmer, 1988; Schuster & Brennicke,

830

1987). However, these are not the main subject of discussion here.

The nature of this quantitative pattern is succinctly summarized in the title of a recent paper by Palmer and Herbon (1988)—"Plant mitochondrial DNA evolves rapidly in structure, but slowly in sequence." In this study it was shown that, in a sample of six species of *Brassica* and the closely related genus *Raphanus*, anywhere between 3 and at least 14 major rearrangements (inversions) could be detected between the genomes of a given pair of species. Although this degree of genome rearrangement is low enough so that individual rearrangements can be discerned, it stands in striking contrast to the extreme structural conservatism of the genomes of, for example, chloroplasts, which have co-existed with mitochondria in the same plant cell and organism lineages for hundreds of millions of years (Martin et al., 1989). To illustrate graphically the magnitude of the difference, it is sufficient to note that the chloroplast genomes of two species belonging to different divisions of the plant kingdom—a bryophyte, *Marchantia*, and an angiosperm, *Nicotiana*—differ by only a few readily discernible rearrangements (Ohyama et al., 1986, Shinozaki et al. 1986). What is more, the choroplast genomes of the vast majority of angiosperm species examined, including the six studied by Palmer and Herbon, have exactly the same gene order (Palmer et al., 1983; Palmer, 1985).

To return to the above-mentioned paper, Palmer and Herbon (1988) also determined the degree of restriction site divergence between mitochondrial genomes of all possible pairs of the six *Brassica* and *Raphanus* species examined, and compared these values with those for restriction site divergence between the chloroplast genomes of the same species. The remarkable result was that for this group, despite a rate of blockwise sequence rearrangement in the mitochondrial genome that is apparently orders of magnitude higher than it is in the chloroplast genome, the rate of restriction site divergence (and by inference the rate of nucleotide sequence divergence) over the entire genome appears to be about threefold *lower* in mtDNA than it is in chloroplast DNA (cpDNA).

The evidence for this type of pattern is not restricted to *Brassica*, or to dicotyledonous plant species. The most spectacular published example of rapid rearrangement between two completely mapped plant mtDNAs is that of the N- and T-cytoplasm lineages of maize (Fauron & Havlik, 1989). Despite being part of the gene pool of a single plant species (*Zea mays* L.), these mtDNAs differ by at least 30 separate rearrangements. Although restriction site divergence between these maize mtDNAs was not compared with that between cpDNAs of the same varieties, only 3 site differences were detected with a complement of 3 restriction enzymes between the N and T mtDNAs in the approximately 500 kbp of sequence common to the two genomes, confirming the *Brassica* results. A final set of observations to illustrate this evolutionary pattern was published by Wolfe et al. (1987, 1989). In these studies, estimates of relative rates of divergence by synonymous nucleotide substitution in protein-coding genes of mtDNA, cpDNA, and nuclear DNA (nDNA) were obtained between representative monocots and dicots. The results are in interesting agreement with those obtained from *Brassica*—the estimated rate of synonymous substitution in plant mitochondrial protein-coding genes ($0.8–1.1 \times 10^{-9}$ per site per year) is about three times lower than it is in protein-coding genes of chloroplasts (i.e., $2.1–2.9 \times 10^{-9}$ per site per year), which in turn accumulate synonymous substitutions about four times more slowly than do plant nuclear genes. Although sequence data for plant nuclear genes are still relatively scarce, there does not appear to be a large difference in rate estimates between this group of plant genes ($5.1–7.1 \times 10^{-9}$ per site per year) and genes in animal nuclei ($2–8 \times 10^{-9}$ per site per year) (Wolfe et al., 1989). Thus, plant mitochondrial genomes, although they appear to suffer very few constraints on gross structure, have perhaps the lowest rate of nucleotide substitution known among eukaryotic cellular genomes.

Before proceeding to a discussion of possible mechanisms, a few comments on this putative pattern are in order. First and perhaps most important, its validity should be confirmed and extended with new data from other plant species, since extensive data are at present only available for *Brassica* and maize. The maize and *Brassica* data already suggest that rate variation may exist between plant taxa, at least for rearrangements. If the overall pattern is in fact widespread, it may be especially helpful to identify more examples of such variations and exceptions within groups of close relatives (e.g., between genera within families), as these cases may embody differences in key factors that generate the pattern. For instance, it would be interesting if relatively small and simple plant mtDNAs like those of *Brassica* species (ca. 220 kbp, one pair of recombining repeats; Palmer & Herbon, 1988) proved to be less prone to evolutionary rearrangement than are larger, more complex mtDNAs like those of maize (ca. 550 kbp with 5 sets of recombining repeats; Fauron & Havlik, 1989).

Second, as with any quantitative analysis, care must be given to the nature of the comparisons made and to the significance of the specific metrics employed in making those comparisons. As emphasized in discussions by Lonsdale et al. (1988) and Palmer (1990), the most appropriate genomes to compare with plant mitochondrial genomes in evolutionary-rate studies at low taxonomic levels would seem to be chloroplast genomes of the same plant lineages. This type of comparison is more likely to generate focused and rigorous explanatory hypotheses than, say, a comparison between plant and animal mtDNAs, between which differences in many fundamental molecular, cellular and population parameters likely exist. Also, there is at present no reason to expect that the aspect of rapid rearrangement is in any way causally connected with the low rate of nucleotide substitution; it may be advisable to treat the two features independently for the time being.

Finally, a sampling problem exists with nucleotide sequence data; it stems from the fact that different sequence "sectors" that occur in all genomes (coding regions, intergenic regions, introns, different codon positions, etc.) typically evolve at different rates. This can make extrapolation from divergence rates in a small sample of sequence to divergence rates for the entire genome difficult. Thus, if differences in whole-genome sequence divergence rates are of interest (as they seem to be for plant mtDNA and cpDNA), and if localized sequence data rather than genome-wide restriction maps are used, the choice of a sequence data set for rate comparison must be made with care. I will comment further on this point at the end of the chapter.

POSSIBLE PROCESSES

Assuming the validity of the evolutionary pattern described above, what can be said in genetic terms about the mechanistic processes that underlie it? Before presenting and evaluating some possible answers to this question, it is first worth sketching a few well-known elementary principles of evolutionary genetics. More detailed discussion of these principles, as well as excellent general treatments of basic evolutionary genetics theory, can be found, for example, in Hartl and Clark (1989) and Maynard Smith (1989). A particularly enlightening discussion of theory as it pertains to evolution of plant organelle genomes is given by Birky (1987).

The rate of genetic evolution, E, in a population or lineage can be conveniently decomposed into (i) the rate of mutation to new variants per unit time (M), and (ii) the probability of ultimate fixation of a new variant in the lineage (F). This leads to the equation $E = M \times F$, which condenses a vast complexity of processes into a very simple form. The M term represents the combined contributions of processes of DNA damage, replication, repair, and recombination in their respective cellular contexts, to the total mutation rate. M can also be written as Nu, where N is the number of genetic units (genes, genomes, etc.) in the population and u is the rate of appearance of mutant genetic units per unit, per unit time.

832

With reference to plant mtDNA evolution, it is worth noting that chromosome rearrangements can be accommodated within this scheme as readily as can "point mutations." The term F includes the probabilities of fixation by stochastic processes (random genetic drift) and by natural selection. It depends, therefore, on the distribution of selection coefficients, s, for new variants (often assumed to have a slight bias toward small, negative values). However, F also varies inversely with effective population size (N_e—a type of average that reflects deviations of the population from ideal evolutionary behavior), which influences the degree to which either stochastic forces or selective ones dominate the fate of a variant.

Two more points are relevant to the current discussion. First, for the class of variants that are selectively neutral, within which all variants are (by the definition of neutrality) equally likely to be fixed, $F = 1/N$, so that $E = Nu \times 1/N = u$, the neutral mutation rate. Thus, for truly neutral variants the rate of evolution at the genetic level can be quite high— on the order of the mutation rate to that type of variant. Second, since the population behavior of many variants (notably those with small, negative values of s) tends more to randomness the smaller the effective population size, if a population suffers even occasional restrictions of population size the proportion of *effectively* neutral variants may actually be quite large. To state the evolutionary consequences of these two points as a principle, one might say that evolution at the genetic level consists, to a variable extent, of a "core" process of neutral fixation occurring at a stochastically constant rate; this rate can in some cases be so large as to approach the order of the total mutation rate. This, of course, does not mean that natural selection does not play a critical role in adaptive evolution; it simply means that at the genetic level the bulk of fixation of new variants in a lineage may take place independently of any selective process. Deviations from predominance of the core process result from strong directional (positive) or constraining (negative) effects of selection on individual variants, mediated by large selection coefficients or large population size.

This framework is useful in thinking critically about possible mechanisms underlying the unusual evolutionary rates seen in plant mtDNA. If we assume for the moment that most evolutionary rearrangements and nucleotide substitutions in both mtDNA and cpDNA are fixed as selectively neutral variants, this suggests that the differences in evolutionary rates are underlain by differences in mutation rates, probabilities of random fixation, or perhaps both. Palmer (1990), for instance, suggests that since mtDNA in plants has more families of dispersed-repeat sequences than does cpDNA, and seems to have a very active homologous-recombination system, recombination between such repeats in plant mtDNA helps to explain its high evolutionary rate of rearrangement. In other words, there may be an effect of a higher total mutation rate (to rearrangements) for mtDNA in comparison with cpDNA. This suggestion is made more intriguing by the association of repeated sequences with the ends of certain rearrangements that do occur in some cpDNAs, and by the seeming increase in rate of fixation of rearrangements in some cpDNA lineages that have higher numbers of dispersed repeats (Palmer et al., 1987; Tsai & Strauss, 1989). Certainly, the suggestion is also parsimonious in its appeal solely to basic aspects of genome organization and function, rather than to complex and relatively inaccessible factors such as population structure.

However, as Palmer (1990) notes in his discussion, another clear basic difference between plant mtDNA and cpDNA is a large difference in the proportion of intergenic spacer DNA (about 90% for mtDNA *versus* about 10% for cpDNA). The implication is that what we observe at the level of a difference in evolutionary rates of rearrangement between the two genomes (as well as between slowly and rapidly evolving cpDNAs) may more accurately be seen as an effect of a difference in *neutral*, rather than *total*, mutation rates. In fact, to a first approximation the chance that both ends of a mutationally rearranged DNA segment fall within nonessential DNA can be considered to vary directly with the square of the fraction of nonessential DNA in the genome. This is not merely a fine point of theory: the proposal of a difference in total mutation rate carries with it specific assumptions about molecular mechanisms (i.e., more frequent homologous recombination in mtDNA than in cpDNA), and is therefore concrete and testable. In this connection it may also be salutary to point out that direct evidence for frequent, ongoing homologous recombination in plant mitochondria has yet to be obtained, despite the now-plentiful evidence alluded to above that recombinant junctions do in fact exist within plant mtDNA populations.

Another, considerably more complex, hypothesis is that put forward by Lonsdale et al. (1988). These authors draw attention mainly to a single key difference between plant mitochondria and chloroplasts: the former organelles are presumed to undergo frequent fusion (to a degree where they should be considered a single "dynamic syncytium" within a plant cell lineage), whereas the latter do not engage in such behavior and thus constitute a population of discrete, clonally propagating organelle lineages within a cell lineage. Combined with other important features of the mtDNA and the cpDNA per se, this key difference is then postulated to have two ultimate effects on evolutionary rates—one on the rate of rearrangement, and one on the rate of base substitution.

First, it is noted that plant mtDNA is multicircular and lacks an obvious mechanism for equable partitioning of subgenomic molecules between daughter cells during cell division. These factors, combined with mitochondrial nondiscreteness, should increase the probability of a daughter cell randomly receiving a genetically unbalanced or even incomplete complement of mtDNA. What is more, Lonsdale et al. (1988) reason that, if an aberrant mtDNA complement were received by a cell, selection to restore viability in the lineal descendants of such a cell might favor the amplification of rearranged variant molecules pre-existing at low frequency within the mtDNA population, if such amplification helped to restore balanced genetic function. And second, if homologous recombination takes place frequently within the mitochondrial syncytium, it is suggested that an efficient process of "copy correction" could be operating at the level of the mtDNA sequence, presumably as a side effect of the strand interactions and/or repair processes associated with homologous recombination. In other words, the authors seem to be proposing that frequent homologous recombination within a mitochondrial syncytium would effectively lower the total point mutation rate for mtDNA. In contrast, the absence of fusion between chloroplasts would mean that, even if recombination is frequent within these organelles, because of their mutual isolation individual cpDNA lineages will be able to accumulate more nucleotide substitutions with a consequently higher evolutionary rate for cpDNA in the plant lineage as a whole.

This is an interesting proposal, since it explicitly attempts to take into account factors operating at the population level. Organelle DNAs of multicellular eukaryotes actually exist in multiple, nested population/lineage contexts (molecules within organelles, organelles within cells, cells within organisms, organisms within populations). This adds to the evolutionary dynamics of variant fixation in organelles a complexity that must be considered when attempting to understand evolutionary patterns. However, the proposal offered by Lonsdale et al. is seriously flawed in at least two important aspects. With respect to the fixation of rearranged molecules, it is rather difficult to imagine how rare variants of this kind could be systematically favored by random genetic drift via aberrant mtDNA segregation, since it is the rare members of any population that are most likely to be stochastically lost when the population suffers a decrease in numbers. The idea of viability selection for increase of a rare molecule in a cell lineage with aberrant mtDNA composition also seems to rely heavily on unlikely events—specifically, the accumulation of favorable mutations in an initially rare component of the mtDNA population, before lowered cell viability leads to extinction of the cell lineage.

Logical difficulties also exist with the postulated effects on nucleotide substitution

834

rates. The invocation of copy correction does not in itself appear to help very much, because it leaves open the question how such a process could lead to a net correction bias in favor of ancestral sequences. Indeed, a new variant is, as a result of copy correction, just as likely to increase in frequency as to decrease. Thus, it is fairly readily shown how sequence *uniformity* can be maintained within a mtDNA lineage through concerted evolution driven by a copy-correction process (Birky & Skavaril, 1976, Ohta, 1977), but not so readily how a significant lowering of effective mutation rate could be achieved by the same process. Neither is it easy to see how mitochondrial fusion *per se* should remove this difficulty. In fact, fusion would appear essentially to remove one nested level of population structure for mtDNA (i.e., organelles within cells). This should *increase* the rate of nucleotide substitution by facilitating fixation of a mtDNA variant within a cell lineage, which is one necessary step in the multilevel process of genetic frequency change leading to fixation in the organismal lineage as a whole.

A final example of current thinking on evolutionary mechanisms in plant mtDNA is that offered by Small et al. (1989). These authors address the question of rapid plant mtDNA rearrangement only, and do not explicitly deal with comparative evolutionary rates. However, the main elements of their scheme, derived from elegant genetic and molecular analysis of certain mtDNA rearrangements in maize, are highly relevant to the current discussion. Specifically, they suggest that plant mitochondria contain a persistent subpopulation of individually rare (but collectively perhaps quite abundant), extensively rearranged mtDNA species, which they term "sublimons." In the maize example that they analyze in detail, the rearranged molecules arise by infrequent recombination between short, dispersed repeats, and seem able to persist at low ("subliminal") levels in the mtDNA for long periods of time. The critical evolutionary role envisaged for sublimons is their collective capacity to serve as a reservoir of standing variation in a mtDNA population, capable of quickly responding to changes in "fixation pressure" (the combined effect of random drift and selection on the tendency to fix new variants in a lineage). A key point seems to be that the instantaneous rate of evolution of plant mtDNA arrangement may fluctuate greatly over time. Thus, what we observe as a high overall evolutionary rate of rearrangement for this genome may have less to do with the rate of mutation *per se* (neutral or otherwise) or ongoing differences in the opportunity to fix mutations, than with an ability of plant mtDNA to accumulate cryptic variation and then occasionally to rapidly fix some of that variation when conditions change.

One reason why this model is interesting is that it provides an explicit, substantiated molecular model for processes leading to the origin of evolutionarily significant genetic variants in a particular case. It is also exciting because it deviates somewhat from the class of evolutionary rate model on which much of the foregoing discussion was founded. Instead of viewing evolution on the genetic level strictly in terms of a quasi-steady state process with stochastically constant inputs and outputs, additional, biologically interesting components—stored variation and fluctuating environments—are introduced, and enliven the discussion considerably, even if they tend to complicate theory. If sublimon-like molecules prove to occur widely in plant mtDNA and to play a major part in its evolution, investigation of the underlying molecular, cellular and population mechanisms could contribute significantly to a better general understanding of the processes of organelle genome evolution.

SUMMARY AND CONCLUSIONS

The foregoing is intended as a brief illustrative outline of the stage to which our understanding of plant mitochondrial genome evolution has progressed. Enough comparative data are now available to make statements about evolutionary pattern, in quantitative terms relating to rates of blockwise genome rearrangement and of nucleotide substitution. It seems that the former rate is "unusually" high, at least in comparison with that of cpDNA in the same plant lineages, and that the latter rate is "unusually" low, in comparison not only with cpDNA and nDNA in the same plant lineages but also with other eukaryotic cellular genomes. It is appropriate at this point, then, to make some comments and suggestions as to the general form future investigations might most profitably take.

I previously mentioned that care should be exercised in the choice of plant mtDNA sequence data used to infer genome-wide evolutionary rates, and therefore to further investigate the validity of the evolutionary pattern. This is especially true because the sequence subset most commonly used for comparative rate estimates, synonymous codon positions within genes, cannot be safely assumed to be evolving entirely free of selective constraints (Sharp & Li, 1987). Furthermore, the distinct possibility remains that variation at translationally synonymous positions is under more stringent constraint in plant mtDNA than in cpDNA, even though Wolfe et al. (1987) point out that the extent and nature of codon usage bias within genes in plant mtDNA and cpDNA are not detectably different. For example, there may be special sequence requirements for mRNA editing in plant mitochondrial genes (Covello & Gray, 1989). It would be most helpful to have estimates of rates of sequence divergence in a subset of "minimally constrained" sequences such as that represented by pseudogenes in nuclear genomes (Gojobori et al., 1982). For these sequences, because base changes can be assumed to a close approximation to be selectively neutral, their evolutionary rate of nucleotide substitution can be assumed to be a measure of the total mutation rate, at least for the local regions of the genome in which they occur (see above discussion). Such a sequence subset is almost certainly available in plant mtDNAs, among "promiscuous" sequences transferred from other cellular compartments, especially the chloroplast (Stern & Lonsdale, 1982; Nugent & Palmer, 1988). Some effort should be directed, therefore, towards identifying homologous transferred segments that have been retained by related plant mtDNAs and are presently nonfunctional in the mitochondrion, and towards estimating their rates of sequence divergence relative to other mtDNA segments.

With respect to the investigation of evolutionary mechanisms, processes at several organizational levels should be taken into account. At the molecular level, it would be very useful to understand plant mtDNA recombination systems in more detail. Is the dominant process one of homologous exchange, dependent for its operation on extensive sequence homology and proceeding through essentially the same molecular steps as, say, the RecBCD pathway of *Escherichia coli* (Smith, 1988)? Or are site-specific mechanisms an important possibility, as suggested by the results of Joyce et al. (1988)? Furthermore, is the answer to this question the same for all angiosperms? One of the most crucial sets of questions from the current perspective concerns the frequency and timing of recombination in plant mtDNA. Whether we can understand the rapid rate of rearrangement in plant mtDNA as partly the result of a higher rate of mutational input to the fixation process hinges largely on how this question is answered. In the first instance, detailed and comprehensive analysis of DNA sequences surrounding naturally occurring rearrangement breakpoints in plant mtDNA seems one obvious path to follow. The results of this kind of work might also provide some clue as to the nature of the molecular mechanism(s) underlying the low sequence divergence rate, which at present remains quite a puzzle.

At the level of cells and cell lineages, our understanding of the transmission genetics of plant mtDNA is exceedingly primitive. How is genetic integrity of a multicircular DNA population maintained through somatic and sexual cell generations, if there is no mechanism for ensuring proper partitioning? If plant mtDNA is in fact compositionally stabilized by specific cellular mechanisms, an understanding of the way in which these normally operate in a stabilizing manner might suggest a role for them in promoting the origin of mutant cells under destabilizing circumstances. To complicate this issue, data on

836

the amount and the structure of intraspecies mtDNA variation at the population level in plants are almost nonexistent; we cannot at present even properly distinguish between the possibilities of actual intraspecies stability, and rapid evolutionary turnover leading to widespread identity by recent common descent. Such variation data should be widely collected, with special attention paid to the multilevel population structure of organelle DNAs, and to sublimon-type variants. The results may help to distinguish between a relatively simple model of molecular evolution such as that based mainly on neutral variation and random drift, and more complex ones such as those involving sequestered variation and episodes of rapid evolution.

In closing, I think it is fairly safe to say that however the blanks are eventually filled in for the specifics of the mechanism in plant mtDNA evolution, the uniqueness of mitochondrial genomes in general and plant mitochondrial genomes in particular ensure that they will continue to assist at the birth of significant new insights into genetic aspects of the evolutionary process.

ACKNOWLEDGMENTS

I thank Pat Covello for helpful discussions. This work was supported by a post-doctoral fellowship (Biotechnology Training Program) from the Medical Research Council of Canada to M. B.C., as well as by an MRC operating grant (MT-4124) to Dr. Michael W. Gray.

LITERATURE CITED

Bendich, A. J. & S. B. Smith. 1990. Moving pictures and pulsed-field gel electrophoresis show linear DNA molecules from chloroplasts and mitochondria. *Curr. Genet.* 17:421–425.

Birky, C. W., Jr. 1987. Evolution and variation in plant chloroplast and mitochondrial genomes. Pp. 23–53. *In:* L. D. Gottlieb & S. K. Jain (eds.), *Plant Evolutionary Biology.* Chapman and Hall: New York. 300 pp.

Birky, C. W. Jr. & R. V. Skavaril. 1976. Maintenance of genetic homogeneity in systems with multiple genomes. *Genet. Res.* 27:249–265.

Brandon, R. N. & R. M. Burian (eds.). 1984. *Genes, Organisms, Populations: Controversies over the Units of Selection.* MIT Press: Cambridge, MA. 300 pp.

Covello, P. S. & M. W. Gray. 1989. RNA editing in plant mitochondria. *Nature* 341:662–666.

Doolittle, W. F. 1988. Hierarchical approaches to genome evolution. *Can. J. Philosophy* 14 (suppl.):101–133.

Falconet, D., Lejeune, B., Quetier, F. & M. W. Gray. 1984. Evidence for homologous recombination between repeated sequences containing 18S and 5S ribosomal RNA genes in wheat mitochondrial DNA. *EMBO J.* 3:297–302.

Fauron, C. & M. Havlik. 1989. The maize mitochondrial genome of the normal type and the cytoplasmic male sterile type T have very different organization. *Curr. Genet.* 15:149–154.

Gojobori, T., Li, W.-H. & D. Graur. 1982. Patterns of nucleotide substitution in pseudogenes and functional genes. *J. Mol. Evol.* 18:360–369.

Gould, S. J. & N. Eldredge. 1977. Punctuated equilibria: the tempo and mode of evolution reconsidered. *Paleobiology* 3:115–151.

Gray, M. W. 1989. Origin and evolution of mitochondrial DNA. *Ann. Rev. Cell Biol.* 5:25–30.

Hartl, D. L. & A. G. Clark. 1989. *Principles of Population Genetics* (2nd ed.). Sinauer: Sunderland, MA. 682 pp.

Joyce, P. B. M., Spencer, D. F. & M. W. Gray. 1988. Multiple sequence rearrangements accompanying the duplication of a tRNA^{Pro} gene in wheat mitochondrial DNA. *Plant Mol. Biol.* 11:833–843.

Lonsdale, D. M., Brears, T., Hodge, T. P., Melville, S. E. & W. H. Rottmann. 1988. The plant mitochondrial genome: homologous recombination as a mechanism for generating heterogeneity. *Phil. Trans. Roy. Soc. London Series B* 319:149–163.

Lonsdale, D. M., Hodge, T. P. & C. M.-R. Fauron. 1984. The physical map and organisation of the mitochondrial genome from the fertile cytoplasm of maize. *Nucl. Acids Res.* 12:9249–9261.

Martin, W., Gierl, A. & H. Saedler. 1989. Molecular evidence for pre-Cretaceous angiosperm origins. *Nature* 339:46–48.

Maynard Smith, J. 1989. *Evolutionary Genetics*. Oxford University Press: New York. 325 pp.

Nugent, J. M. & J. D. Palmer. 1988. Location, identity, amount and serial entry of chloroplast DNA sequences in crucifer mitochondrial DNAs. *Curr. Genet.* 14:501–509.

Ohta, T. 1977. On the gene conversion model as a mechanism for maintenance of homogeneity in systems with multiple genomes. *Genet. Res.* 30:89–91.

Ohyama, K., Fukuzawa, H., Kohchi, T., Shirai, H., Sano, T., Umesono, K., Shiki, Y., Takeuchi, M., Chang, Z., Aota, S.-I., Inokuchi, H. & H. Ozeki. 1986. Chloroplast gene organization deduced from complete sequence of liverwort *Marchantia polymorpha* chloroplast DNA. *Nature* 322:572–574.

Oro, A. E., Newton, K. J. & V. Walbot. 1985. Molecular analysis of the inheritance and stability of the mitochondrial genome of an inbred line of maize. *Theor. Appl. Genet.* 70:287–293.

Palmer, J. D. 1985. Evolution of chloroplast and mitochondrial DNA in plants and algae. Pp. 131–239. *In:* R. J. MacIntyre (ed.), *Molecular Evolutionary Genetics.* Plenum: New York.

Palmer, J. D. 1988. Intraspecific variation and multicircularity in *Brassica* mitochondrial DNA. *Genetics* 118:341–351.

Palmer, J. D. 1990. Contrasting modes and tempos of genome evolution in land plant organelles. *Trends Genet.* 6:115–120.

Palmer, J. D. & L. A. Herbon. 1988. Plant mitochondrial DNA evolves rapidly in structure, but slowly in sequence. *J. Mol. Evol.* 28:87–97.

Palmer, J. D., Osorio, B., Aldrich, J. & W. F. Thompson. 1987. Chloroplast DNA evolution among legumes: loss of a large inverted repeat occurred prior to other sequence rearrangements. *Curr. Genetics* 11:275–286.

Palmer, J. D. & C. R. Shields. 1984. Tripartite structure of the *Brassica campestris* mitochondrial genome. *Nature* 307:437–440.

Palmer, J. D., Shields, C. R., Cohen, D. B. & T. J. Orton. 1983. Chloroplast DNA evolution and the origin of amphidiploid *Brassica* species. *Theor. Appl. Genet.* 65:181–189.

Schuster, W. & A. Brennicke. 1987. Plastid, nuclear and reverse transcriptase sequences in the mitochondrial genome of *Oenothera*: is genetic information transferred between organelles via RNA? *EMBO J.* 6:2857–2863.

Sharp, P. M. & W.-H. Li. 1987. The rate of synonymous substitution in enterobacterial genes is inversely related to codon usage bias. *Mol. Biol. Evol.* 4:222–230.

Shinozaki, K., Ohme, M., Tanaka, M., Wakasugi, T., Hayashida, N., Matsubayashi, T., Zaita, N., Chunwongse, J., Obokata, J., Yamaguchi-Shinozaki, K., Ohto, C., Torazawa, K., Meng, B. Y., Sugita, M., Deno, H., Kamogashira, T., Yamada, K., Kusuda, J., Takaiwa, F., Kato, A., Tohdoh, N., Shimada, H. & M. Sugiura. 1986. The complete nucleotide sequence of tobacco chloroplast genome: its gene organization and expression. *EMBO J.* 5:2043–2050.

Small, I., Suffolk, R. & C. J. Leaver. 1989. Evolution of plant mitochondrial genomes via substoichiometric intermediates. *Cell* 58:69–76.

Smith, G. R. 1988. Homologous recombination in procaryotes. *Microbiol. Rev.* 52:1–28.

Stern, D. B. & D. M. Lonsdale. 1982. Mitochondrial and chloroplast genomes of maize have a 12-kilobase sequence in common. *Nature* 299:698–702.

Tsai, C. H. & S. H. Strauss. 1989. Dispersed repetitive sequences in the chloroplast genome of Douglas-fir. *Curr. Genet.* 16:211–218.

Wolfe, K. H., Li, W.-H. & P. M. Sharp. 1987. Rates of nucleotide substitution vary greatly among plant mitochondrial, chloroplast, and nuclear DNAs. *Proc. Nat. Acad. Sci. USA* 84:9054–9058.

Wolfe, K. H., Sharp, P. M. & W.-H. Li. 1989. Rates of synonymous substitution in plant nuclear genes. *J. Mol. Evol.* 29:208–211.

Phylogenetic Implications of the Plasmid-like Features of Mitochondrial DNA

Howard T. Jacobs

Abstract. Mitochondrial DNA exhibits significantly more features in common with the extrachromosomal as opposed to the chromosomal genomes of prokaryotes. In this paper, I review aspects of mitochondrial genome organization, transcription, and replication that support this view, as well as evidence for relics of plasmid maintenance systems in mtDNA. These features suggest that mtDNA may have originated as an extrachromosomal genome of the prokaryote ancestor of mitochondria. The horizontal mobility of such replicons raises the possibility that the present-day mitochondrial genome is a genetic mosaic, or that it is phyletically distinct from the chromosomal genome of the original prokaryotic endosymbiont that harboured it, and whose genes were eventually incorporated into nuclear DNA. Conversely, mitochondria may themselves be polyphyletic, even if their modern-day genomes are not. The possibility of horizontal transfer also renders unsafe any phylogenetic conclusions drawn from studies of mtDNA in isolation. In an extreme view, it could also invalidate the idea of a physical union of cells as the original endosymbiotic event. The physical maintenance of mtDNA as a semi-independent genome over evolutionary time may be due to an aggressive mechanism inherent in the properties of the plasmid-like replicon from which it originated.

INTRODUCTION: THE ENDOSYMBIOTIC HYPOTHESIS

The endosymbiotic hypothesis, formulated in its modern form by Margulis[1,2], proposes that mitochondria arose as a prokaryotic endosymbiont, which was engulfed by the primitive eukaryote, to mutual advantage. This relationship is supposed to have provided the two partners with reciprocal nutritional opportunities, the proto-mitochondrion metabolising, via respiratory pathways, the secondary metabolites provided by the 'host' cell. The intimacy of the relationship is assumed to have favoured the eventual transfer of genetic information from the endosymbiont to the host nuclear chromosomes, leaving behind only a fragmentary genome encoding a small number of respiratory chain polypeptides, plus the RNA components of a translation apparatus. A similar history is proposed for the other DNA-containing organelle of eukaryotes, the chloroplast: the process of gene transfer to the nucleus may have gone a stage further in the case of some other organelles, such as peroxisomes, which no longer have detectable genomes[3].

Recent analysis of the nucleotide sequences of the large and small subunit rRNAs, which are held in common among all prokaryotes, eukaryotes, mitochondria, and chloroplasts, has provided strong evidence in support of an endosymbiotic origin for

Dr. Jacobs is with the Institute of Genetics, University of Glasgow, Church Street, Glasgow GI1 5JS, Scotland.

mitochondria[4]. Mitochondrial rRNAs are markedly more similar to eubacterial than to nuclear-encoded eukaryotic rRNAs[4], and are most closely related to those of a particular group of prokaryotes (the alpha-subdivision of the purple nonsulfur bacteria), suggesting a specific phyletic origin of the mitochondrial genome[5,6].

Whilst these studies have established, with some degree of certainty, the phyletic origin of (at least the RNA components of) the mitochondrial translation system, they do not permit the phyletic origin of the eukaryotic respiratory chain, nor of mitochondria *per se* to be extrapolated with certainty, unless it is assumed that mitochondria and, by implication, their free-living ancestors, are/were themselves monophyletic. This assumption rests on the premise that the extant mitochondrial genome is the end result of a 'subtractive' process, starting from an originally complex eubacterial endosymbiont genome, from which the majority of the chromosomal genes were either transferred to the nucleus or discarded entirely, because their functions were redundant.

In this paper I review various kinds of circumstantial evidence for an *extrachromosomal* origin for the mitochondrial genome. If this alternate view were proven to be correct, the simplest version of the endosymbiont hypothesis and many of its ramifications would be open to serious question because of the evident horizontal mobility of extrachromosomal DNAs in prokaryotes.

PLASMID-LIKE FEATURES OF mtDNA

Although mtDNA encodes components of a translation apparatus, a property not shared with any known prokaryotic extrachromosomal replicon, the mitochondrial genomes of animals, fungi and eukaryotic protists exhibit many features which bear a superficial resemblance to those of bacterial episomes. (Plant mtDNA remains somewhat less clearly understood and will not be discussed specifically in this context.) Although all these resemblances could individually be ascribed to convergent evolution, based on adaptation to a reduced coding capacity and a requirement for propagation and maintenance systems distinct from those of chromosomal (nuclear) DNA, the number of such features held in common warrants closer examination of a possible direct evolutionary relationship between mtDNA and prokaryotic episomes. In particular, the likelihood of convergent evolution can be assessed by considering the number of shared features against the diversity of evolutionary solutions to a given problem. The fact that almost no other eukaryotic extrachromosomal DNAs share these features strongly suggests that these homologies are not co-incidental. Features of chloroplast genome organization and gene expression are, by contrast, much more like those of eubacterial chromosomes[7].

The phage- or plasmid-like features of mtDNA, several of which will be discussed in greater detail, may be summarized as follows:

1. compact genome organization, including elimination of spacer DNA, the occurrence of overlapping genes and the presence of only minimal control information[8-10].
2. predominantly multicistronic, but 'non-operon' transcription, with regulation operating *within* rather than *between* transcription units[11-14].
3. formation of a long RNA primer for replication, by transcriptional initiation at an 'ordinary' promoter, and 3' RNA processing by a specific RNase, rather than through the use of an origin-specific primase[15,16].
4. formation of a D-loop triplex as an early replication intermediate[15,17].
5. 'extensively' continuous, unidirectional DNA synthesis on both strands during replication [15,18,19].

840

6. lagging-strand initiation by a distinct mechanism, requiring exposure of one or more sites in ssDNA[15,20].
7. transcription by a 'phage-type' RNA polymerase[16,21].
8. possible relics of at least two plasmid maintenance systems similar to those used in bacteria (post-segregational killing and site-specific recombination)[22].

Not all of the above points are supported by convincing or universal evidence, as will be discussed below, though most of them are well documented in at least some systems (commonly either in fungi or in metazoa, but not always in both). Therefore some attention needs to be paid to establishing their validity and generality, before their significance can be ascertained. Work in our own system, sea urchin mtDNA, has provided support for points (1–5) and (8), which are now discussed in greater detail. In a number of respects echinoderm mtDNA appears to 'bridge the gap' between fungal and vertebrate mtDNAs, and we have therefore suggested that it represents a particularly conservative mitochondrial genome[23].

PLASMID-LIKE REPLICATION MECHANISMS IN mtDNA

The mechanism of replication of mtDNA in metazoans bears a striking resemblance to that of plasmids found in both Gram-positive and Gram-negative bacteria, notably colE1 and its relatives. It differs in a number of important respects from that of chromosomal DNAs in bacteria or in the eukaryotic nucleus, or from that of other extrachromosomal (e.g. viral) DNAs in eukaryotes. A striking feature is the absence of a specific primase. The origin region appears to adopt a non-standard conformation that permits persistent primer-template base-pairing, under as yet ill-defined conditions. Both in mtDNA and in prokaryotic plasmids, this structure forms adjacent to a GC homopolymer run of variable length[24,25]. In sea urchin mtDNA, there is evidence that the formation of a structure with single-stranded character, adjacent to the GC homopolymer, is dependent upon DNA topology[24], and I have hypothesized that the topological constraint may be provided by the association of an anchorage-providing protein with the template, coupled with transcription from the upstream promoter at which the synthesis of the primer initiates[24,26]. As in plasmids such as colEl[27,28], the primer RNA is of substantial length. The properties of this primer RNA have not been investigated in detail in echinoderms, but in mammals its function appears to be influenced by competing intra-and inter-molecular RNA base-pairing[29,30], which again resembles aspects of the regulation of the colE1 primer[27,28].

The 3' end of the primer RNA is created in both colEl and mammalian mtDNA by a specific RNA processing event, catalysed, respectively, by RNase H (Ref. 31) and RNase MRP (Ref. 32), though these enzymes cut by rather different mechanisms. In both cases, the 3' end of the primer is extended unidirectionally, and for a limited distance only, by a DNA polymerase which thus generates the D-loop triplex intermediate[33,34]. Further extension of the D-strand, leading to productive replication, is a potential regulatory step for both mtDNA and for colEl. In colEl, D-strand synthesis and extension are carried out by different DNA polymerases (PolI and PolIII, respectively)[27,34]; in mitochondria the mechanism of D-strand extension is not properly understood, except that the D-loop region appears to be bounded by short, imperfect sequence reiterations, which resemble protein-binding sites[24,33]. Probably the most 'diagnostic' feature of mtDNA replication shared with that of eubacterial phages[35] and plasmids[27,36,37] is continuous DNA synthesis on both strands, which contrasts with the semi-discontinuous synthesis employed by prokaryotic and eukaryotic chromosomal DNAs. Initiation on the displaced strand is delayed, and occurs by an entirely distinct mechanism, involving a primase active only on ssDNA[15,20,35–38], and with a greater or lesser degree of sequence specificity[15,16,19,35–38]. In sea urchin mtDNA, lagging-strand synthesis initiates at multiple positions[19,39], as may also be the case in yeast[16] and some prokaryotic plasmids[40], but the units of synthesis are large compared with the overall size of the genome.

TRANSCRIPTION OF mtDNA BY A PHAGE-TYPE POLYMERASE

The RNA polymerase employed to transcribe mtDNA is structurally and functionally distinct from the family of RNA polymerases that transcribe eukaryotic nuclear and bacterial chromosomal genes, as well as chloroplast DNA. The latter group all exhibit a prototypic $\alpha\beta\beta'$ subunit structure[41], to which are added additional subunits allowing discrimination of particular sets of promoters and terminators. The major subunits of $\alpha\beta\beta'$ polymerases from phylogenetically disparate sources are clearly related to one another. By contrast, the mitochondrial RNA polymerase is structurally unrelated to the $\alpha\beta\beta'$ family, but resembles the single-subunit RNA polymerases of the T-odd bacteriophages (T3, T7, SP6 etc.), as shown by sequence analysis of the gene which encodes this enzyme in yeast[16,21]. It also functions in a rather similar manner. The T-odd phage RNA polymerases discriminate their promoters without the aid of other proteins, though the mitochondrial enzyme from various sources appears to require guidance from proteins that interact in a sequence-specific manner with the template[16,42]. Unlike the promoters for the $\alpha\beta\beta'$ polymerases, however, which generally comprise one or more recognition elements positioned in a precise manner over tens of base-pairs upstream (or downstream) of the initiation site, the T-odd/mitochondrial polymerases use short promoters, typically of 9–12 bp, which just overlap the start site[16,41]. Interestingly, the only other known RNA polymerases of this family are encoded by mitochondrial plasmids found in plants[43] and fungi[44], many of which are associated with pathological states.

RELICS OF PLASMID MAINTENANCE SYSTEMS IN mtDNA

All prokaryotic plasmids, especially those at low copy number, deploy a multiplicity of elaborate mechanisms for their own maintenance, in the absence of strong selection for the functions of the genes that they encode[45,46]. This is of obvious advantage not only to the plasmid, but also to its host, in allowing it to remain able to respond to an environmental condition which is not itself continuously maintained (e.g., the presence of an antibiotic or of an unusual substrate). These systems, by maintaining the plasmid in an extrachromosomal form, may also facilitate its efficient transfer, and enable amplification of the copy number when required.

Typical maintenance mechanisms employed by plasmids include active partitioning systems associated with the cell division machinery[45,47], site-specific recombination systems to resolve plasmid multimers to monomers[48] (which maximizes the chance of both daughter cells acquiring at least one copy of a low copy-number plasmid), and so-called post-segregational killing (PSK) systems[49], which bring about the programmed cell death of cells that fail to inherit the plasmid, in favour of their sister cells which do. I shall now discuss evidence that mitochondrial DNA contains relics of at least one PSK system, as well as the possible remnants of a site-specific recombination system.

PSK IN PROKARYOTIC PLASMIDS

The prototypic PSK system is encoded by the so-called *hok* (host-killing) gene of *E. coli* plasmid RI (Refs. 49–51). This gene specifies a hydrophobic, 52 amino-acid toxin, whose synthesis is translationally repressed by an antisense transcript (designated *sok*). This transcript is maintained in excess by a high rate of transcription in cells that retain the plasmid. In daughter cells which lose the plasmid at cell division the supply of *sok* RNA cannot be replenished, and, because of the inherent instability of this RNA, *sok* repression of *hok* mRNA translation is overcome, leading to cell death[51].

The *hok* gene product acts rapidly at the cell membrane by uncoupling oxidative phosphorylation and inhibiting respiration[50]. Over half of all plasmids in Gram-negative bacteria which have been sequenced encode a *hok*-related polypeptide (K. Gerdes, pers. comm.). In addition, chromosomal genes with structural and functional resemblances to *hok* have been detected in all eubacteria examined. Thus, in *E. coli*, the chromosomal *relF* gene codes for a 51 amino-acid respiratory toxin, which kills cells in a very similar manner to *hok*[50]. The precise function of this gene is unknown, though its presence in an operon, other genes of which function in response to starvation, suggests that it may be needed for programmed cell death under extreme starvation conditions, to permit the survival of neighbouring cells that remain viable. Despite their functional homology, *hok* and *relF* exhibit only limited primary sequence similarity, though they share a very similar hydropathy profile.

IDENTIFICATION OF A *hok*-RELATED GENE IN mtDNA

Since mtDNA has been sequenced from a wide variety of taxa, one may reasonably ask whether any genes with a recognizable degree of similarity with *hok* have been identified. Superficially this does not seem a promising approach, as all of the common mtDNA-encoded gene products have been previously identified as *bona fide* respiratory chain components, either genetically, biochemically, or both[52]. Nevertheless, a PSK system based on crippling respiration would seem to be an attractive mechanism for maintaining an extrachromosomal genome in mitochondria and/or the mitochondrial ancestor. One might also postulate that, if PSK function has been discarded subsequently, the ancient *hok*-like gene of mitochondria might have evolved into a respiratory chain polypeptide.

Consideration of the respiratory chain polypeptides encoded in mtDNA leads to the conclusion that virtually all of them have indisputable counterparts in the corresponding enzyme complexes of prokaryotes and chloroplasts. The single, clear exception is subunit 8 of ATP synthase (A8). This polypeptide is encoded in the mtDNA of all fungi and metazoa (except nematodes, where the gene remains unidentified), and has also been identified in some protozoan mitochondrial genomes. A8 is a hydrophobic, integral membrane protein of about 52 amino acids in length (48 in *Aspergillus*, 51 in *Drosophila*, 54 in sea urchins, 55 in *Xenopus*), which exhibits very little primary sequence similarity between distant taxa, but which exhibits a moderately conserved hydropathy profile. Comparison of the hydropathy profiles of *hok* and A8 from a variety of taxa[22] have revealed striking similarities (Fig. 1). The *Drosophila* A8 profile matches very closely to that of *hok*, and sea urchin A8 also matches well; in fact, if a seven amino acid gap is introduced into the sea urchin sequence between the hydrophobic N-terminal and hydrophilic C-terminal domains, the profile can be almost perfectly superimposed onto that of *hok*. An A8 consensus profile built from the sequences from five diverse taxa also matches that of *hok*, except at its extreme C-terminus which is, in any case, the most variable portion of A8, being truncated and much less hydrophobic in fungi, but considerably extended and more hydrophilic in mammals.

Comparisons of the hydropathy profiles of A8 and members of the *hok* family by a numerical method[22] have revealed a statistically significant similarity, when judged against the background scores obtained in comparing polypeptides of the *hok* family with unrelated, integral membrane polypeptides of similar length and overall hydrophobicity, selected from the databases. This analysis will be presented in greater detail elsewhere[22].

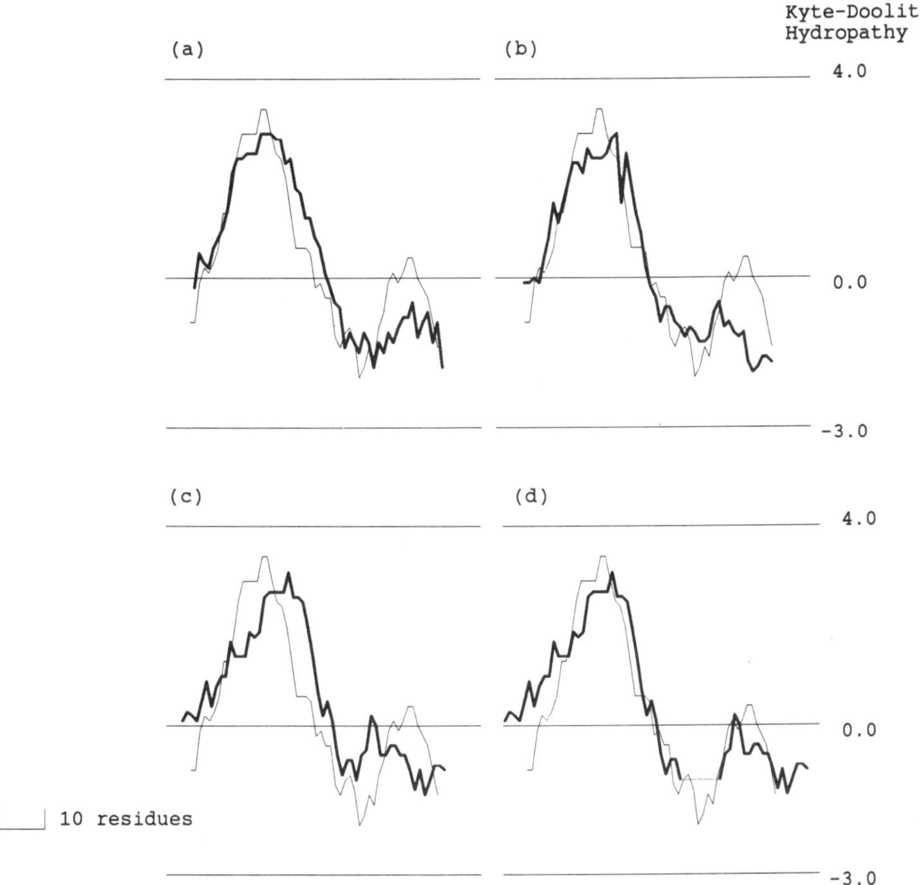

Figure 1. Kyte-Doolittle hydropathy profiles[72] for *hok* (Refs. 50, 51; faint line, 11-residue window), compared with (bold lines) those for (a) *Drosophila* A8 (Ref. 73; 11-residue window, shifted by one residue to maximize alignment); (b) a consensus hydropathy profile for A8 sequences of *Drosophila*[73], *Aspergillus*[74], sea urchin[10], cow[57] and *Xenopus*[75]. Mean Kyte-Doolittle hydropathies were calculated over a 7-residue window, at each position, for the 5 sequences aligned individually with *hok*; (c) sea urchin[10] (11-residue window, shifted by one residue to maximize alignment); (d) sea urchin[10], with the N- and C-terminal halves arbitrarily interrupted (dotted line), to maximize alignment with the *hok* profile. Positive values indicate hydrophobic character, negative values hydrophilic character.

UNUSUAL PROPERTIES OF A8 CONSISTENT WITH ITS DERIVATION FROM A *hok*-LIKE PROTEIN

Evidence that present-day A8 functions in a manner analogous with *hok* is lacking. However, A8 has a number of properties that suggest that its synthesis must be precisely controlled, and that incorrect expression can have damaging effects on ATP synthase assembly or activity, which would be fully consistent with the proposition that it was recruited to its present role from a polypeptide able to interact destructively with the proton channel of ATP synthase. How this may have come about is discussed below.

As indicated, A8 has no counterpart in prokaryotic or choloroplast proton-pumping ATP synthases, but is, at least in yeast, an essential subunit of the enzyme, as determined by genetic criteria. Mutants in the *aap1* gene, which codes for A8 in yeast, fail to incorporate not only A8 but also subunits 6 and 9 (A6, A9) into the inner mitochondrial

844

membrane[53], suggesting that A8 is a key polypeptide required for the assembly of the F_O portion of ATP synthase. This is supported by the observation that antisera directed against mammalian A8 immunoprecipitate at least two other newly synthesized mitochondrial translation products (A6 and ND1), in addition to A8 (Ref. 54). Although the ND1 result may be due to immunological cross-reaction[55], the reactivity with A6 suggests that nascent or newly synthesized A6 and A8 interact directly. Experiments in which the yeast A8 gene was engineered for expression in nuclear DNA, and used to complement a mitochondrial mutation in *aap1*, did not restore full wild-type levels of ATP synthase activity[56]; a possible explanation for this is that over-expression of A8 in nuclear DNA may lead to a partial uncoupling of oxidative phosophorylation (P. Nagley, pers. comm.).

An unusual feature of the A8 gene in all metazoan and fungal mtDNAs is its translational coupling with A6. The two polypeptides are translated from different reading frames of a single mRNA and, in metazoa, the two reading frames overlap by a greater or lesser distance[10,53,57]. This suggests that their synthesis must be coupled in some way, consistent with the notion that over-expression of either subunit is highly deleterious to the cell.

POSSIBLE RELICS OF A SITE-SPECIFIC RECOMBINATION SYSTEM IN mtDNA

Metazoan mtDNA lacks a generalised recombination mechanism, although a variety of recombinases appear to act on fungal mtDNA. One or more of these may be derived from an ancient site-specific recombination system to maintain a plasmid in the monomeric state. Specifically, the recombinases that catalyze the site-directed movement of introns which contain them, by introducing specific double-strand nicks in target DNA[58–60] (which then serve to promote biased gene conversion), are good candidates. Intriguingly, yeast mtDNA contains other open reading frames that exhibit sequence similarity with these intron-encoded recombinases, but which are found as 'free-standing' (i.e. non-intronic) genes[61]. Intronic ORFs of this family are also found in some eubacterial phages[62].

PHYLOGENETIC IMPLICATIONS OF AN EPISOMAL ORIGIN FOR mtDNA

The plasmid-like features of mtDNA, especially its mode of replication and transcription, and the presence of a relic of a PSK system, suggest that it may have originated as a prokaryotic episomal genome in the eubacterial endosymbiont, rather than as its chromosomal genome, as conventionally believed. This has a number of potential implications for phylogeny and for the evolutionary relationship between nuclear and mitochondrial DNA. First, a PSK system offers at least a partial explanation for the evolutionary maintenance of mtDNA. Second, if derived from a horizontally mobile genetic element, the mitochondrial genome may itself be a genetic mosaic: the mitochondrial translational apparatus, the various portions of the respiratory chain encoded in mtDNA and the 'nuts and bolts' of the replication and transcription machinery of mtDNA may have quite different phyletic origins and histories from one another, and from the original mitochondrial endosymbiont. Third, if the mitochondrial genome were maintained for a long period as an extrachromosomal element in the endosymbiont, subsequent to its incorporation into the primitive eukaryotic cell, whilst the chromosomal genes of the endosymbiont were being progressivly lost to the nucleus, it may have retained 'horizontal mobility' long after the original endosymbiotic event. Hence, phylogenies derived from analyzing nuclear and mitochondrial genes in diverse eukaryotes need not agree with one another—or, put another way, discrepancies in such phylogenies do not necessarily indicate an erroneous methodology. Finally, in an extreme view, the endosymbiotic origin of mitochondria need not be seen as a *physical* union of two cells, but could reflect a horizontal transfer merely of *genetic* information; for example, in the form of a phage genome, coding for the respiratory chain plus a translation apparatus, which gained access to an internalized membrane compartment of a primitive eukaryotic cell already possessing an endomembrane system. These implications will now be considered individually in greater detail.

PSK AND THE EVOLUTIONARY MAINTENANCE OF mtDNA

The retention of mtDNA with an essentially invariant gene content over vast periods of evolutionary time remains an unsolved puzzle. The mitochondrial genome of metazoa, for example, has remained unchanged, except for a few genomic rearrangements, since the divergence of the metazoan phyla[52]. Traditional explanations for this persistence, based on the hydrophobicity of the gene products encoded in mtDNA, or the supposed unlikelihood of gene transfer to the nucleus, have had to be abandoned in the face of clearly contradictory data. Thus, the argument that the hydrophobic mitochondrial translation products need to be synthesized at the site of their incorporation into the membrane has been disproved by the creation of synthetic versions of such genes, expressible in nuclear DNA, which, with the appropriate transit information, have been successfully expressed and used to complement the corresponding mitochondrial mutations[53,56,63]. Furthermore, a number of hydrophobic polypeptides integral to, or closely associated with, the respiratory chain are now encoded in nuclear DNA in eukaryotes, even though the corresponding genes are found in prokaryotes in the same operons as genes for respiratory chain subunits that are encoded in mtDNA in eukaryotes. Obvious examples are cytochrome c_1 (co-transcribed with cytochrome *b* in bacteria[64]); at least two hydrophobic 'assembly' factors for cytochrome oxidase, present in the *cox* operon alongside COI, COII and COIII in *Paracoccus*[65] (M. Saraste, pers. comm.), but in nuclear DNA in yeast (A. Tzagoloff, pers. comm.), and ATP synthase subunit 9 (nuclear-encoded in most fungi and all metazoa[52] but part of the *unc* operon in *E. coli*).

Gene transfer from organelles to the nucleus is clearly possible over evolutionary time and has occurred frequently, as documented by the widespread occurrence of pseudogenes of mitochondrial origin in nuclear DNA[66-68]. More startlingly, it has recently been demonstrated to occur at a measurable frequency (estimated at about 2×10^{-5} per cell per generation) in the laboratory, by experiments in which nuclear markers, not expressible in the mitochondrial genetic system, were introduced into mitochondrial DNA by experimental DNA transformation, and then observed to give rise spontaneously to derivatives in which the test gene had been transferred to the nucleus[69]. Gene transfer *per se* is therefore not unlikely. It may also be noted that a large proportion (typically 5%) of random DNA sequences from *E. coli* or mammalian genomes are capable of conferring mitochondrial targeting information on a polypeptide which lacks it[70], suggesting that mitochondrial genes that relocated to the nucleus early in eukaryote evolution would not have experienced severe difficulties in having their products re-directed to mitochondria.

Proponents of the view that 'real' gene transfers were unlikely events are obliged to fall back on the argument that even if 'DNA transfers' did occur, they were not likely to result in productive relocation of a functional gene, due to the incompatibilities in the systems of gene expression in the two compartments. Whilst this argument is sustainable today, it does little to explain why 95% of the presumed genes of the endosymbiont were able to make the jump during early eukaryote evolution, leaving behind a few key components of the respiratory chain. Furthermore, the idiosyncracies of mitochondrial gene expression (RNA editing, genetic code differences, unusual modes of splicing and RNA processing, translation-regulatory sequences within coding regions etc.) are largely taxon-specific, suggesting that they are a *consequence*, rather than the *cause* of the long period of isolated evolution of genomes in the two compartments.

The observation that mtDNA appears to possess a relic of a PSK system offers a radically different kind of explanation for the persistence of mtDNA and the genes it encodes. If one presumes that some or all of the genes for the respiratory chain were acquired by a plasmid genome of the original endosymbiont, possessing an aggressive PSK mechanism, whose function was to prevent loss of the plasmid or its acquisition by transcription units of a foreign replicon, then one may easily see how such a plasmid could be retained as an independent genome *within* the endosymbiont, right up to the point at which the endosymbiont surrendered the last vestiges of its genetic autonomy, by loss of its own chromosomal genome. One may argue, similarly, that subsequent transfer of the DNA containing the *hok*-like gene to nuclear DNA would be strongly disfavoured by the negative effects on the whole cell of its unregulated expression. Transfer of other genes from mtDNA to the nucleus would also be disfavoured over time, because of the likelihood of such sequences acting to facilitate the incorporation of the remainder of the mitochondrial genome, including the *hok*-like gene, into an active nuclear transcription unit, by homologous recombination. The high rate of promiscuous transfer of sequence information observed in yeast suggests this would be very likely to eliminate the descendents of any cell that stably incorporated mtDNA sequences into the nuclear genome, prior to the effective genetic isolation of the two compartments.

Additionally, one may view the negative consequences of transfer of *hok* sequences into nuclear DNA as providing a powerful selection pressure in favour of the divergence of the nuclear and mitochondrial genetic systems to the point of incompatibility. The idiosyncracies of mitochondrial gene expression being taxon-specific, as already mentioned, implies that the disappearance of the remains of the chromosomal genome of the endosymbiont may have occurred relatively late in eukaryotic evolution, and at a different time and in a slightly different manner in different taxa.

EVOLUTION OF A *hok*-LIKE PROTEIN INTO A SUBUNIT OF ATP SYNTHASE

Obviously a major problem with the hypothesis that A8 is a relic of an ancient PSK system for mtDNA is to explain how and why a *hok*-like protein involved in PSK has evolved into a functional subunit of ATP synthase. The genetic isolation of nuclear and mitochondrial DNA, which may have been accelerated by the presence of a *hok*-like gene in mtDNA, would seem to render the *hok* function superfluous. A possible clue comes from the evidence, discussed above, of interactions between A8 and A6. This would suggest the possibility that protein level detoxification may, at some stage, have replaced translational repression by *sok* as the means of controlling the activity of the *hok*-like gene product. This assumes that the *hok*-like protein may have a natural affinity for that part of the proton channel of ATP synthase provided by A6—which is testable, and which is fully consistent with its physiological action—and that it may have been possible for the cell to detoxify it by a large excess synthesis of A6. Loss of the plasmid encoding both *hok*/A8 and A6, coupled with an inherent instability of the non-membrane-incorporated A6 polypeptide, and/or greater relative stability of the *hok* mRNA over the A6 mRNA, would have led to programmed cell (organelle) death. This natural interaction may, in a subsequent stage, have allowed the *hok*-like protein to evolve for function, for example as a regulator of the activity of the ATP synthase proton channel, or of ATP synthase assembly, without loss of its primary 'destructive' role as a maintenance system for the genome. With loss of the need for the latter function, the gene would nevertheless have been retained, due to its established regulatory role, perhaps evolving new mechanisms to prevent its inappropriate expression, including the translational coupling to A6. The rapid sequence diversification of A8, by far the most evolutionarily divergent mtDNA-encoded polypeptide, may represent its acquisition of, and adaptation, to new roles subsequent to the loss of its original function.

CONSEQUENCES OF A POLYPHYLETIC ORIGIN
FOR THE MITOCHONDRIAL GENOME

Plasmid-borne genes are commonly those required for rapid adaptation to a harshly varying environment, such as those conferring resistance to antibiotics, or the ability to metabolize exotic substrates or cope with huge fluctuations in the external conditions (nutrient levels, pH, etc.). The acquisition by a plasmid of most or all of the genes for the respiratory chain would have enabled any cell into which it was transferred to colonize, in a single step, an aerobic, nutrient-poor environment from which it would previously have been barred. One might ask what would be the advantages for such a cell in retaining respiratory chain genes in an extrachromosomal form. Although a central premise of the foregoing discussion is that such an outcome is a result of inherent properties of the plasmid itself, it is more plausible to regard the situation as a mutually advantageous symbiosis.

There is no obvious answer to this question. However, it is noteworthy that many eukaryotic organisms quite capable of utilising respiratory pathways to generate ATP will, under defined conditions, 'revert' to glycolytic/anaerobic ATP generation. This is particularly well illustrated by yeast, which represses mitochondrial respiration in the presence of glucose. If it is assumed that the genes which are retained on plasmids are those for which there is a potential requirement for high copy-number (for example, to synthesize sufficient amounts of a product required for effective detoxification), then such an argument may also have applied in the primitive eukaryotic cell, where rapid synthesis of a large amount of the respiratory chain polypeptides might have been necessary to divert reduced cofactors into the chain. In other words, the 'host' cell's pre-existing metabolism may have been seriously maladapted to the use of the extraneously grafted respiratory pathways. The high copy-number of at least some of the respiratory chain genes is, in fact, maintained today, by the multi-copy nature of mtDNA, compared with the nuclear-encoded subunits of the chain and other mitochondrial polypeptides, which are almost invariably specified by single-copy genes.

This (admittedly speculative) argument need not detract from the main idea, however, that an extrachromosomal locale would have allowed the genes for respiration to spread rapidly throughout the prokaryotic world and permit the colonization of previously hostile environments. One may even speculate that the mitochondrial respiratory chain itself may have been compiled in stages, by successive transfers of genes carried on extrachromosomal elements, between cells that had each acquired partial and relatively inefficient means of reoxidising the reduced cofactors generated catabolically.

The genes for the mitochondrial respiratory chain, seen in this light, may have originated in one or more different lines of prokaryote, distinct from the actual endosymbiont that entered the primitive eukaryotic cell. Possibly the only genetic remnant of the endosymbiont's chromosomal genome may be the genes for those parts of its translation apparatus acquired aggressively, during the final stages of its genetic degeneration, by mtDNA. This means that phyletic deductions based on sequence analysis of mitochondrial rRNA or tRNA genes, or of mitochondrial translation factors now encoded in nuclear DNA, may not be safely applicable to the phyletic origin of the mitochondrial respiratory chain, either in whole or in part. Alternatively, the mitochondrial translation machinery and respiratory chain may have had a *common* phyletic origin distinct from that of the *other* genes of the endosymbiont, some or all of which have found their way into nuclear DNA. A phage carrying genes for a translation apparatus may be regarded as a rather extreme mechanism for transferring genes across vast phyletic distances. Much attention has been devoted to the possibility that the mitochondrial genomes of different groups of eukaryotes may have been independently acquired[71], in separate or successive endosymbiotic events. This would go some way to explaining discrepancies between 'nuclear' and 'mitochondrial' phylogenies. Whilst this cannot be ruled out, an extrachromosomal origin for mtDNA raises two other interesting possibilities: 1) that multiple endosymbiotic events, in which primitive eukaryotes acquired prokaryotic partners of quite diverse origins and metabolic virtuosities, nevertheless deposited an essentially identical mitochondrial genome; or 2) that endosymbiosis may have occurred only once, but that horizontal passage of the mitochondrial genome occurred countless times until quite late in eukaryote evolution.

It should be stressed that an extrachromosomal origin for mtDNA does not *necessarily* invalidate conventional phylogenies—it merely requires that they be considered cautiously in the light of a wider data set. Of considerable value in examining this question will be phylogenetic analysis of genes more clearly derived from the endosymbiont chromosomal genome, such as those coding for the biochemically distinct mitochondrial isoforms of enzymes found also in the cytosol.

ENDOSYMBIOSIS: A CELLULAR OR GENETIC EVENT?

The membrane topologies of the major eukaryotic organelles, as well as their biochemistry, make endosymbiosis a highly attractive idea. In the case of chloroplasts, sequence analysis of cpDNA, which has shown it to be more or less equivalent to a 'pared down' cyanobacterial genome, renders this idea almost inescapable. However, the 'extrachromosomal' features of mtDNA, discussed above, leave some doubt as to whether mitochondria also arose by a physical union of two distinct pre-existing cells.

Two alternate hypotheses can be considered. First, horizontal transfer of the genes now carried in the mitochondrial genome, for example, in the form of a phage, could have supplied respiratory function to a long established (non-respiratory) organelle of the early eukaryotic cell, which itself arose endosymbiotically. An obvious candidate would be the chloroplast; such a possibility raises numerous phylogenetic problems, and would require the resurrection of the currently discredited view that the acquisition of the photosynthetic organelle was an early event in eukaryote evolution. It differs from the older view by postulating that a plasmid-like genome, carrying the genes for the respiratory chain (plus a translation apparatus), which had been acquired from a purple bacterium, hijacked an ancient plastid and transformed it into a respiratory organelle, with loss of its photosynthetic capability and the expulsion of its original genome. The presence of genes of cyanobacterial origin in nuclear DNA, coding for some mitochondrial isozymes in animals, fungi or protozoa, would be a clear and testable prediction of such a hypothesis.

The alternate view, already mentioned earlier, would be that the progenitor of the mitochondrial genome gained access to a membrane compartment of the early eukaryotic cell which did not itself arise endosymbiotically. In this scheme, the endosymbiotic origin of mitochondria would resemble a cultural myth: spiritually (genetically) true, but historically false.

TESTING THE PREDICTIONS OF AN EVOLUTIONARY SPECULATION

Like all novel evolutionary speculations, the above can only be evaluated in the light of the validity of its predictions. Fortunately, a number of these are clearly testable, and it will become possible to address others as data accumulate. The central premise, that mtDNA derives from an extrachromosomal genome, is strongly suggested but not proven by existing data. Detailed elucidation of the mechanism and enzymology of transcription and replication of mtDNA from diverse taxa should clarify to what extent the apparent episomal features of these processes are genuine. The derivation of A8 from a *hok*-like protein, currently supported largely by statistical arguments, can be usefully investigated by studying the precise physiological consequences of over-expression and ectopic expression of A8, in bacteria as well as in mitochondria, both alone and in combination with other subunits of ATP synthase. A8 and *hok* are sufficiently short polypeptides to enable a detailed investigation of structure-function relationships to be undertaken, using reverse genetics. The construction of chimeric genes, for example, containing portions of A8 replaced with the corresponding sequences from *hok*, should prove useful in determining whether all or part of A8 is functionally, as well as structurally, related to *hok*. The over-expression of A8 (and eventually A6) in bacteria should also make it possible to investigate whether these polypeptides interact directly with their common mRNA to regulate their own or each other's synthesis. It should also prove feasible to set up a variety of *in vitro* systems for studying other aspects of the translational control of A8 and A6 synthesis, with a view to uncovering relics of a translational regulator analogous with *sok*, if such exists.

The broader phylogenetic questions, relating especially to the monophyly of mitochondria, will obviously be answered by the gathering of much more comprehensive sequence data on nuclear, nuclear-coded mitochondrial, mitochondrial, and bacterial genes, and by refinement of the methods in use for constructing meaningful phylogenies.

ACKNOWLEDGMENTS

I am grateful to numerous colleagues for discussions and suggestions, especially to Dave Sherratt, Phil Nagley, Tom Fox, Kenn Gerdes, Tom Cavalier-Smith and Matti Saraste. My research is supported by funding from the Royal Society, MRC, the Wellcome Trust, the European Community, and NATO.

LITERATURE CITED

1. Margulis, L. M. 1970. *The Origin of Eukaryotic Cells.* Yale University Press: New Haven, CT.
2. Margulis, L. M. 1981. *Symbiosis in Cell Evolution.* W. H. Freeman: San Francisco, CA.
3. Cavalier-Smith, T. 1987. *Ann. N.Y. Acad. Sci.* 503:55–71.
4. Gray, M. W., Sankoff, D. & R. J. Cedergren. 1984. *Nucl. Acids Res.* 12:5837–5852.
5. Cedergren, R. J., Gray, M. W., Abel, Y. & D. Sankoff. 1989. *J. Mol. Evol.* 28:98–112.
6. Yang, D., Oyaizu, Y., Oyaizu, H., Olsen, G. J. & C. R. Woese. 1985. *Proc. Natl. Acad. Sci. USA* 82:4443–4447.
7. Gray, M. W. 1988. *Biochem. Cell Biol.* 66:325–348.
8. Anderson, S., Bankier, A. T., Barrell, B. G., de Bruijn, M. H. L., Coulson, A. R., Drouin, J., Eperon, I. C., Nierlich, D. P., Roe, B. A., Sanger, F., Schreier, P. H., Smith, A. J. H., Staden, R. & I. G. Young. 1981. *Nature* 290:457–465.
9. Zimmer, M., Lückerman, L., Lang, B. F. & K. Wolf. 1984. *Mol. Gen. Genet.* 196:473–481.
10. Jacobs, H. T., Elliott, D. J., Math, V. B. & A. Farquharson. 1988. *J. Mol. Biol.* 202:185–217.
11. Levens, D., Ticho, B., Ackerman, E. & M. Rabinowitz. 1981. *J. Biol. Chem.* 256:5226–5232.

12. Christianson, T. W. & D. A. Clayton. 1986. *Proc. Natl. Acad. Sci. USA* 83:6277–6281.
13. Elliott, D. J. & H. T. Jacobs. 1989. *Mol. Cell. Biol.* 9:1069–1082.
14. Jacobs, H. T. 1989. *BioEssays* 11:27–34.
15. Clayton, D. A. 1982. *Cell* 28:693–705.
16. Schinkel, A. H. & H. F. Tabak. 1989. *Trends in Genet.* 5:149–154.
17. Kasamatsu, H., Robberson, D. L. & J. Vinograd. 1971. *Proc. Natl. Acad. Sci. USA* 68:2252–2257.
18. Goddard, J. M. & D. R. Wolstenholme. 1980. *Nucl. Acids Res.* 8:741–759.
19. Mayhook, A. G., Rinaldi, A.-M. & H. T. Jacobs. 1991. *Nucl. Acids Res.* In press.
20. Martens, P. A. & D. A. Clayton. 1979. *J. Mol. Biol.* 135:327–351.
21. Masters, B. L., Stohl, L. L. & D. A. Clayton. 1987. *Cell* 51:89–99.
22. Jacobs, H. T. 1990. *J. Mol. Evol.* (In press).
23. Jacobs, H. T., Elliott, D. J., Mayhook, A. G., Segreto, D. T. & A.-M. Rinaldi. 1990. *In:* Pp. 145–151. E. Quagliariello et al. (eds.), Structure, Function and Biogenesis of Energy Transfer Complexes. Elsevier: Amsterdam.
24. Jacobs, H. T., Herbert, E. R. & J. Rankine. 1989. *Nucl. Acids Res.* 17:8949–8965.
25. Masukta, H. & J.-I. Tomizawa. 1990. *Cell* 62:331–338.
26. Jacobs, H. T. 1991. *Bull. Mol. Biol. Med.* In press.
27. Scott, J. R. 1984. *Microbiol. Rev.* 48:1–23.
28. Tomizawa, J. & T. Itoh. 1982. *Cell* 31:575–583.
29. Chang, D. D., Fisher, R. P. & D. A. Clayton. 1987. *Biochim. Biophys. Acta* 909:85–91.
30. Chang, D. D. & D. A. Clayton. 1987. *EMBO J.* 6:409–417.
31. Itoh, T. & J. Tomizawa. 1980. *Proc. Natl. Acad. Sci. USA* 77:2450-2454.
32. Chang, D. D. & D. A. Clayton. 1989. *Cell* 56:131–139.
33. Doda, J. N., Wright, C. T. & D. A. Clayton. 1981. *Proc. Natl. Acad. Sci. USA* 78:6116–6120.
34. Kingsbury, D. T. & D. R. Helinski. 1970. *Biochem. Biophys. Res. Commun.* 41:1538–1544.
35. Arai, K. & A. Kornberg. 1981. *Proc. Natl. Acad. Sci. USA* 78:69–73.
36. Van der Ende, A., Teerstra, R. & P. J. Weisbeek. 1983. *J. Mol. Biol.* 167:751–756.
37. Veltkamp, E. & A. R. Stuitje. 1981. *Plasmid* 5:76–99.
38. McMacken, R., Ueda, K. & A. Kornberg. 1977. *Proc. Natl. Acad. Sci. USA* 74:4190–4194.
39. Matsumoto, L,. Kasamatsu, H., Piko, L. & J. Vinograd. 1974. *J. Cell Biol.* 63:146–159.
40. Arai, K. I., Low, R. & A. Kornberg. 1981. *Proc. Natl. Acad. Sci. USA* 78:707–711.
41. Rowland, G. C. & R. E. Glass. 1990. *BioEssays* 12:343–346.
42. Fisher, R. P. & D. A. Clayton. 1988. *Mol. Cell. Biol.* 8:3496–3509.
43. Kuzmin, E. V., Levchenko, I. V. & G. N. Zaitseva. 1988. *Nucl. Acids Res.* 16:4177.
44. Chan, B. S.-S. & H. Bertrand. 1988. *Genome* 30 (Suppl I):318.
45. Sherratt, D. J. 1986. *In:* Pp. 239–250. I. Booth & C. Higgins (eds.), *Symposium of the Society for General Microbiology. Regulation of Gene Expression.*
46. Nordström, K. & S. J. Austin. 1989. *Ann. Rev. Genet.* 23:37–69.
47. Ogura, T. & S. Hiraga. 1981. *Cell* 32:351–360.
48. Austin, S., Ziese, M. & N. Sternberg. 1981. *Cell* 25:729–736.
49. Gerdes, K., Rasmussen, P. B. & S. Molin. 1986. *Proc. Natl. Acad. Sci. USA* 83:3116–3120.
50. Gerdes, K., Bech, F. W., Jørgensen, S. T., Løbner-Olesen, A., Rasmussen, P. B., Atlung, T., Boe, L., Karlstrom, O., Molin, S. & K. Meyerburg. 1986. *EMBO J.* 5:2023–2029.
51. Gerdes, K., Helin, K., Christensen, O. W. & A. Løbner-Olesen. 1988. *J. Mol. Biol.* 203:119–129.
52. Chomyn, A. & G. Attardi. 1987. *Curr. Topics in Bioenergetics* 15:295–329.
53. Nagley, P. 1988. *Trends in Genet.* 4:46–52.
54. Chomyn, A., Mariottini, P., Cleeter, M. N. J., Ragan, C. I., Doolittle, R. F., Matsuno-Yagi, A., Hatefi, Y. & G. Attardi. 1985. *In:* Pp. 259–275. E. Quagliariello et al. (eds.), Achievements and Perspectives of Mitochondrial Research, Vol III. Elsevier: Amsterdam.
55. Chomyn, A., Mariottini, P., Gonzalez-Cadavid, N., Attardi, G., Strong, D., Trovato, D., Riley, M. & R. F. Doolittle. 1983. *Proc. Natl. Acad. Sci. USA* 80:5535–5539.
56. Nagley, P., Farrell, L. B., Gearing, D. P., Nero, D., Meltzer, S. & R. J. Devenish. 1988. *Proc. Natl. Acad. Sci. USA* 85:2091–2095.
57. Fearnley, I. M. & J. E. Walker. 1986. *EMBO J.* 5:2003–2008.
58. Jacquier, A. & B. Dujon. 1985. *Cell* 41:383–394.
59. Macreadie, I. G., Scott, R. M., Zinn, A. R. & R. A. Butow. 1985. *Cell* 41:395–402.
60. Wenzlau, J. M., Saldanha, R. J., Butow, R. A. & P. S. Perlman. 1989. *Cell* 56:425–430.
61. Quirk, S. M., Bell-Pedersen, D. & M. Belfort. 1989. *Cell* 56:455–466.
62. Séraphin, B., Simon, M. & G. Faye. 1987. *J. Biol. Chem.* 262:10146–10153.
63. Banroques, J., Delahodde, A. & C. Jacq. 1986. *Cell* 46:837–844.
64. Kurowski, B. & B. Ludwig. 1987. *J. Biol. Chem.* 262:13805–13811.
65. Raitio, M., Jalli, T. & M. Saraste. 1987. *EMBO J.* 6:2825–2833.

66. Fox, T. D. 1983. *Nature* 301:371–372.
67. Jacobs, H. T., Posakony, J. W., Roberts, J. W., Grula, J. W., Xin, J.-H., Britten, R. J. & E. H. Davidson. 1983. *J. Mol. Biol.* 165:609–632.
68. Fukuda, M., Wakasugi, S., Tsuzuki, T., Nomiyama, H., Shimada, T. & T. Miyata. 1985. *J. Mol. Biol.* 186:257–266.
69. Thorsness, P. E. & T. D. Fox. 1990. *Nature* 346:376–379.
70. Baker, A. & G. Schatz. 1987. *Proc. Natl. Acad. Sci. USA* 84:3117–3121.
71. Gray, M. W. 1989. *Trends in Genet.* 5:294–299.
72. Kyte, J. & R. F. Doolittle. 1982. *J. Mol. Biol.* 157:105–132.
73. Clary, D. O. & D. R. Wolstenholme. 1985. *J. Mol. Evol.* 22:252–272.
74. Netzker, R., Kochel, H. G., Basak, N. & H. Kuentzel. 1982. *Nucl. Acids. Res.* 10:4783–4794.
75. Roe, B. A., Ma, D.-P., Wilson, R. K. & J. F. H. Wong. 1985. *J. Biol. Chem.* 260:9759–9774

Alternative Theories of Molecular Evolution

James F. Crow

Abstract. The neutral theory of molecular evolution has strong and weak forms. The strong form asserts that the great majority of molecular fixations, including amino acid substitutions, are neutral (i.e., so nearly neutral as to be dominated by mutation and random drift). The weak form accepts neutrality for the bulk of DNA, but not for amino acid substitutions. Many who oppose or are dubious about the strong form accept the weak form. Evidence for some specific instances of selection are given, but the overall conclusion is that the strong theory is consistent with more facts than the various selective alternatives. Other topics discussed are: male-driven evolution, the mutation spectrum, age of a polymorphism, Darwinian gradualism as an extension of neutrality, and the origin of new genes.

My assignment in this symposium is to give an overview of molecular evolution from the viewpoint of a population geneticist. In particular, I am asked to consider neutralist and selectionist theories.

Most DNA evolution is now widely believed to be driven by mutation and dominated by random processes—the Neutral Theory of Kimura (1968, 1983) and King & Jukes (1969). Yet, organismic evolution is dominated by natural selection, and this must surely be reflected at the molecular level. How can these be reconciled?

That evolutionary change involves both selection and random changes goes all the way back to Darwin. For example, very early in "The Origin of Species" (Darwin, 1872, p. 40), he says:

> I am inclined to suspect that we see, at least in some of these polymorphic genera, variations which are of no service or disservice to the species, and which consequently have not been seized on and rendered definite by natural selection.

In more recent times, A. H. Sturtevant (1921) noted that:

> Such genes [those whose phenotypic effect appears in species crosses] are perhaps to be thought of as having arisen by mutation and having been perpetuated through chance. Since they do not produce any significant effect in the genetic complex in which they arose, they are not eliminated by natural selection, and some of them will sooner or later happen to be incorporated into the race.

Of course Darwin and Sturtevant were thinking of morphological structures, whereas this symposium is concerned with molecular changes. The rules are quite different. Much of organismic selection is concerned with polygenic traits, stabilizing selection, elimina-

Dr. Crow is with the Genetics Department, University of Wisconsin, Madison, WI 53706, USA.

tion of harmful mutations, sexual selection, maintenance of polymorphisms, short-term environmental changes, changes in competing species, and selection at different levels. I think it likely that most such selection (at least in sexual species) involves frequency changes of alleles that have been segregating in the population for some time, rather than incorporation of new mutations. Most of the time in phenotypic evolution, we are not keeping track of the individual genes involved, and often we don't care. In contrast, with molecular evolution we *are* tracking individual genes, or DNA regions, but we often don't know what they are doing—particularly in areas outside the coding region. For this symposium I shall concentrate on molecular changes, thereby brushing aside a world of interesting problems in morphological, functional, and behavioral evolution.

MOLECULAR EVOLUTION RATES

I shall first discuss the rate of molecular evolution. Much of what I shall say comes directly from the work of Motoo Kimura (1983). Since the theory was first advanced (Kimura, 1968; King & Jukes, 1969), most of its general features have been repeated so often that it seems supererogatory to repeat it. I briefly summarize parts of the theory here for the benefit of newcomers and old-timers with the least retentive memories.

It is convenient to distinguish between strong and weak forms of the theory. The strong form (Kimura, 1983) asserts that the great majority of molecular changes, including amino acid substitutions, are neutral (i.e., so nearly neutral as to be dominated by mutation and random drift). The weak form accepts neutrality for the bulk of DNA, but not for amino acid substitutions.

The rate of evolution is the number of mutant fixations that occur in a very long time period—long relative to the time required for an individual substitution to take place. Of course most mutations will be lost in a few generations, whether they are harmful, neutral, or beneficial. Only the few that are lucky enough to get through the hazard of early random extinction have any chance of persisting, as first shown quantitatively by R. A. Fisher (1930, p. 73). If we look at a sufficiently long time period, the time required for an individual fixation becomes irrelevant; only the number of such events is counted.

The rate of substitution, k, per unit time (generation or year) is given by

$$k = 2N\mu P \tag{1}$$

in which N is the population number at the time the mutant occurs, μ is the mutation rate per gene (or nucleotide, or codon) per time unit, and P is the probability that the mutation will ultimately be fixed. The factor two comes from diploidy. The formula for P is the result of successive refinements by five of the great figures in population genetics, Haldane, Fisher, Wright, Malécot, and Kimura (see Kimura 1983, p. 43).

$$P = \frac{1-e^{-2N_e s/N}}{1-e^{-4N_e s}} \sim \frac{2N_e s}{N(1-e^{-4N_e s})} \tag{2}$$

in which s is the selective advantage of the mutant heterozygote and N_e is the variance effective population number (Crow, 1954; Crow & Denniston, 1988). Thus

$$K \sim \frac{4N_e\mu s}{1-e^{-4N_e s}} \tag{3}$$

This simple formula assumes that the mutation is partially or fully dominant; in particular that it is not overdominant. It also assumes that only one substitution at this site is in transit at a time (or if more than one, they can be individually tracked). That is to say, the fixation time is short relative to the interval between successive successful mutations. The last

854

assumption appears to be realistic, especially if we are observing individual nucleotides or amino acids.

If the mutation is neutral (s → 0) k = μ; the long-time rate of substitution in the population is the same as the mutation rate per generation. This is also obvious from simple symmetry, since the probability that the ultimately successful site will be the new mutant is simply the frequency of that mutant, which, if it is represented only once, is P = 1/2N (King & Jukes, 1969). Fisher (1930) showed that fitness increases at a rate measured by the variance of the least squares estimate of the fitnesses, and called his creation "The Fundamental Theorem of Natural Selection." Perhaps we should name the principle, k = μ, the "Fundamental Theorem of Neutral Evolution."

Equation 3 is illustrated in Figure 1 for various values of N_e and s. The graph shows that when the mutation is favored (s > 0) the rate of substitution increases with N_e. When s is small and $N_e s > 1$ Equation 3 becomes $k \sim 4N_e s\mu$. The formula obviously applied only to mutations that are favorable from the time of initial occurrence.

Even if the conditions are not strictly met, k should increase with N_e. This is reasonable intuitively, since the larger the population the more often will a favorable mutation arise; this enhances the chance of one of them surviving the risky early generations after its occurrence. The fact that the numerous sets of data show no enhanced substitution rate with increasing population size (for example, ungulates evolve molecularly no faster than carnivores, or insects than mammals) argues against any measurable fraction of molecular changes being successive incorporation of favorable mutations.

When the mutation is deleterious (s < 0) the rate is greater in small populations, as is intuitively obvious, since the smaller the population the more a deleterious mutation behaves like a neutral one. But unless s is very near to zero, the substitution rate is extremely low (Figure 1).

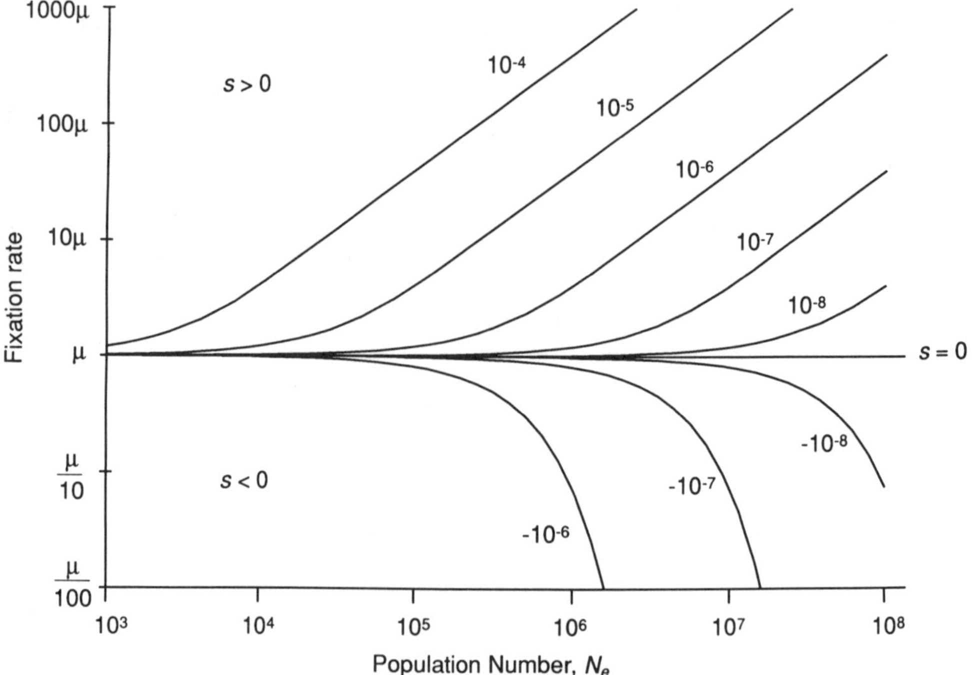

Figure 1. The rate of evolution (number of mutation fixations per generation) as a function of effective population number, N_e, for different values of the selective advantage of the mutant heterozygote, s. The mutation rate per generation (or other time unit) is μ.

CRITICISMS OF THE NEUTRAL THEORY

The neutral theory has been strongly and uncritically opposed by several evolutionists and accepted, equally uncritically, by a number of molecular biologists. A theory with such important ramifications deserves the most thoughtful scrutiny and rigorous testing. Since the theory is mathematical and makes quantitative predictions, it can sometimes be tested with experimental or observational data. The 1970s brought a plethora of papers testing various predictions of the theory, often introducing new statistical methods. Some supported the theory, while others contradicted it. The theory itself underwent several modifications, some because of criticisms.

Ohta (1976, 1977; see also Kimura, 1979) suggested that molecular evolution consists mainly of the random fixation of very slightly deleterious mutations. This had two virtues: it helped resolve the dilemma that substitution rates are more nearly constant per year than per generation, whereas mutation rates are thought to be adjusted to generation length; and, it helped explain why heterozygosity is small and doesn't increase with population size in the way strict neutrality would predict. Nei & Graur (1984) explained the second problem by assuming a widespread population size bottleneck. The year-generation difficulty is mitigated by data showing faster yearly rates in rodents and slower in long-lived higher primates (Wu & Li, 1985, Li et al., 1990). Koop et al. (1989) and Goodman (this symposium) report that globin pseudogenes in the higher primates evolve at about 1/5 the yearly rate of prosimians.

The most persistent and detailed mathematical criticism has come from Gillespie (1986a, b, 1987). He subjects several predictions of the theory to detailed analysis and finds troublesome inconsistencies. He doesn't regard either the Ohta or the Nei and Graur modifications as convincing. Most important is his belief that the rates of amino acid changes in different lineages are not as constant as the neutral theory would predict; rather, the molecular clock is overdispersed. He (Gillespie, 1989) notes that the observed index of dispersion (ratio of variance to mean) for replacement substitutions is about 7.5, which he regards as much too variable for a mutation-driven process. He proposed an alternative selection model in which: (1) the environment changes at intervals that are short relative to the duration of the lineage being examined; (2) the environmental challenge can be met by any of a large number of essentially equivalent mutations; and (3) mutations two steps away from the prevailing type are very rare. Such a model generates variability in rates that appear consistent with observed values. Gillespie also argues that the greater evolutionary rate of unimportant changes, which is expected on the neutral hypothesis, is also expected in a selection hypothesis; for, as Fisher (1930, p. 38) noted, the smaller the effect of a mutation, the more likely it is to be beneficial. More recently Ohta and Tachida (1990) have considered a model in which there is spatial variation in selection coefficients, which are weak enough to be influenced by both selection and random drift.

As I mentioned earlier there is now a rather wide consensus (including Gillespie) in favor of the weak form of the theory, that evolution of pseudogenes, introns, synonymous codons, and the bulk of non-coding DNA follows neutral kinetics. This is supported by the observation that synonymous codons, pseudogenes, and other presumably nonfunctional DNA regions evolve at a rapid rate that is roughly constant over the genome. If these were selectively favored, they would hardly all be favored to the same degree.

Although there is no doubt that some of the predictions of the strong neutral theory are contradicted by empirical observations, I nevertheless believe that the neutral theory of molecular evolution in it general features—i.e., the view that *most* (but not all) amino acid substitutions are mutation-driven and so nearly neutral that random drift is the major determinant of individual mutation fates—is consistent with more facts than any selective alternative so far presented. Here are a few reasons, largely following Kimura:

856

1. As mentioned earlier, there is no correlation of evolutionary rate and population size. A theory that assumes that substituted mutations are favorable from the time of occurrence predicts a more rapid rate in large populations. The alternative, which seems unlikely to me, is a stream of favorble mutations at multiple loci occurring at such a high rate as not to be limiting.

2. There is no correlation of molecular evolution rates with the rates of morphological changes, which are selection-driven. Molecular evolution proceeds at the same rate in slowly evolving lineages (e.g., some fish) as in rapidly evolving ones (mammals). If molecular rates were determined by selection, surely molecular and phenotypic rates would have some common causal factors and show some dependence.

3. Molecular evolution rates show no correlation with rates of change of any known environmental variables. On a selection theory some correlations should surely emerge, for example in comparisons of land and sea animals.

4. In now-classical experiments Cox and Yanofsky (1967) measured the rate of mutation caused by the Treffers mutator gene in *Escherichia coli,* and found an astonishing rate of some 7 changes from AT to GC pairs per bacterium per cell division. There was no obvious change in phenotype. In a stable population the maximum selection intensity is 50 percent; in an expanding population it is somewhat less. It is hard to escape the conclusion that the bulk of these mutations is neutral. And, since there is very little nongenic material in *E. coli* the mutations must have been mainly in coding and regulatory regions.

5. When mutation rates are very high, as in retroviruses and mitochondria, the substitution rates increase roughly in proportion, and the ratio of synonymous to nonsynonymous rates is again about 3.

The arguments for greater than Poisson variability in rates is persuasive, but there is no reason to think that average mutation rates are the same in different lineages. I am more impressed by the wide differences in rates between proteins within lineages and the relative constancy within proteins between lineages, than by the variability of the latter.

I regret that these arguments are largely qualitative, whereas a main strength of the neutral theory is supposed to be its being quantitative. Further quantitative study is needed, and is sure to be done. But we shouldn't expect perfect agreement of data and theory; there are too many complications and exceptions.

IS NEUTRAL EVOLUTION MALE-DRIVEN?

It is likely that a large fraction of mutations of individual nucleotides are errors of replication. If so, we should expect a much higher rate in males than females because of the much larger number of cell divisions between zygote and sperm than between zygote and egg. This is supported by data on mutation rates of some X-linked genes in man (Crow & Denniston, 1985). A prediction of mutation-driven molecular evolution would then be a different rate of evolution between genes on the X and Y chromosomes and those on the autosomes. Specifically, if the male mutation rate is much higher than the female, the X chromosome should evolve at about ⅔ the autosomal rate and the Y chromosome at twice the autosomal rate.

Evolutionary data for sex chromosomes are limited, but published values clearly point in this direction. Miyata et al. (1990), considering synonymous sites, found a ratio of 0.58 for the X-autosome ratio, and for a Y-linked pseudogene, 2.2 times the rate of its autosomal counterparts. In contrast, theories based on selection predict that the X should evolve faster than autosomes and, in particular, X and Y rates should deviate from autosomal rates in the same direction (Charlesworth et al., 1987). If the findings of Miyata et al. are supported by further work, they would strongly support male-driven molecular evolution. Especially revealing would be evidence on the Y (or W) chromosome in birds, which on this hypothesis should have a very low rate of molecular evolution.

EVIDENCE FOR POSITIVE SELECTION IN MOLECULAR EVOLUTION

This symposium is concerned with selection in molecular evolution. I have given reasons for thinking that most molecular evolution is mutation-driven. Yet there must surely be some positive selection, even if this constitutes a small proportion of all molecular change. Although there is abundant evidence for purifying selection, there is very little for positive selection. What kind of evidence would support this?

A strong candidate would be a gene, or DNA region, that evolves faster than a pseudogene, or one in which amino-acid replacement substitutions are faster than synonymous ones. Inhibitors of serine proteinases appear to have this property, offering strong evidence for selection (Laskowski et al., 1987; Hill & Hastie, 1987; Laskowski, this symposium). Particularly important is the identification of specific highly variable regions at active contact points. This suggests that it may be easier to find evidence of selection in parts of a gene than in the whole, where a rapidly changing minority may be swamped out by a slower majority.

Positive selection is usually difficult to prove by rate data. Positively selected mutations are likely to be rarer than neutral ones. Letting the subscript f stand for favorable and n for neutral, we have the following substitution rates: favorable mutations, $k_f = 4N_e s \mu_f$; neutral mutations, $k_n = \mu_n$. Thus $k_f > k_n$ if $\mu_f/\mu_n > 1/4N_e s$. For example, if $4N_e s = 10$, the substitution rate of a positively selected mutation will exceed that of a neutral allele only if its mutation rate is larger than $1/10$ the neutral rate.

A sudden increase in the rate of substitution following a gene duplication or a changed environment is sometimes cited as evidence for positive selection. But it is ordinarily not possible to distinguish between positive selection and decreased selective constraint. For example, the amino acids of the lens protein of blind mole rats has changed more rapidly since they went underground (Hendriks et al., 1987; Lubsen & de Jong, this symposium). But since the rate is less than the pseudogene or third position rate, this does not constitute proof of positive selection.

A second kind of evidence is offered by long-lasting polymorphisms, a topic to which I shall return.

Evidence for a selectively maintained polymorphism can involve specific knowledge of the function, or evidence from the geographical pattern. The ADH polymorphism (Kreitman, this symposium) offers strong evidence because of a latitudinal cline that is found in widely separated longitudes. The selection is clearly directed at, or very near, the amino acid responsible for the charge difference by which the alleles are identified.

AN ARGUMENT FOR MOST POINT MUTATIONS
BEING EFFECTIVELY NEUTRAL

I'll assume that pseudogenes follow neutral kinetics and use their rate of substitution as a measure of the average mutation rate over the time the process has been occurring. The rates are in reasonable agreement with what little information there is on human mutation rates.

The mammalian pseudogene evolution rate is estimated as about 5×10^{-9} per nucleotide per year (Li et al., 1985; Li et al., 1990). Data from Koop et al. (1989) and Goodman (this symposium) show a considerably slower rate for recent human ancestry: 1.1×10^{-9}, or 22×10^{-9} per 20-year human generation. With 6×10^9 base pairs per diploid cell, this means that a zygote has 132 new mutations.

Ohta (1990) has shown that the mutation load is less in multigene families. A stronger claim has been made by Kondrashov (1988). If all selection is by truncation, or something approximating this, it is mathematically possible for an arbitrarily large number of deleterious mutations to be maintained in mutation-selection balance with a reasonable mutation load. Therefore, Kondrashov argues, the genomic mutation rate is probably considerably larger than one, and may be orders of magnitude larger. But for selection to remove 10^2 deleterious mutations each generation requires that virtually all selection be by truncation. This seems very unlikely to me, but only future research will settle the question. At the same time, it is reasonable, without postulating any particular form of selection, for most nonsynonymous mutations in coding regions to be deleterious.

THE MUTATION SPECTRUM

Geneticists have long known that mildly deleterious mutations greatly exceed those with drastic effects. For example, mutations having a mild effect on viability occur at least 20 times as frequently as lethals (Mukai et al., 1972), which in turn are more common than those with a visible effect. Mutations of small effect appear to become increasingly frequent until they approach a value too small to detect.

Whether those of still smaller effect are *still* more frequent is not clear from empirical data. Nevertheless, for reasons given above, I believe that the great majority of nucleotide substitutions over the genome as a whole are essentially neutral. At the same time, most amino acid replacement mutations may well be deleterious. In a highly conserved protein like some of the histones, almost all mutations are harmful, whereas for less conserved ones, such as fibrinopeptide, neutral changes may predominate. For hemoglobins, which evolve at about .1 to .2 of the pseudogene rate, this would argue that some 80–90% of the mutations are deleterious, assuming that favorable mutations are a negligibly small fraction.

What is the molecular nature of the very numerous, mildly deleterious mutations that grade imperceptibly into zero effect? Mukai (1990, and earlier) has noted that these probably have a variety of molecular causes. Transposable elements are responsible for some, for strains known to have such an element have a greatly increased polygenic viability-mutation rates (Mukai, 1990). Mukai and his colleagues have also measured the spontaneous rate of amino acid-altering point mutations (Mukai & Cockerham, 1977; Mukai, 1990). These seem to be insufficient, by a factor of about 30, to account for the variance generated by viability-reducing mutations. Mukai and Cockerham argue that this must mean that most such mutations are outside the coding region. Many years ago K. Mather (1944) suggested that many polygenic mutations reside in the heterochromatin, i.e., are not ordinary protein-coding genes. The Mukai-Cockerham suggestion is cut from the same cloth. I think it is also quite possible that the bulk of the mildly deleterious mutations studied in *Drosophila* are not base substitutions. The situation cries out for molecular analysis.

HETEROZYGOSITY AND AGE OF A POLYMORPHISM

Consider first neutral mutations at an individual nucleotide. The observed heterozygosity per nucleotide in mammals is about 0.002 (Kazazian et al., 1983; Nei, 1987, p. 267). The fraction of heterozygous nucleotides at equilibrium between mutation and random drift in a population of effective size N_e with k allelic states (for nucleotides, k = 4) and mutation rate μ is

$$H = \frac{4N_e\mu(k-1)}{k-1 + 4N_e\mu k} \sim 4N_e\mu \tag{4}$$

Kimura (1983). The heterozygosity is 2p(1-p) where p is the frequency of the mutant allele, so $p \sim 0.001$. The mean age of a neutral allele of frequency p in a population of effective size N_e is

$$A = -4N_e[p \log p + (1-p)\log(1-p)] \tag{5}$$

generations (Kimura & Ohta, 1973). The brackets enclose two terms. The first is the contribution from rare mutations that have drifted upward to 0.001; the second is from common ones that started near fixation and drifted downward. The latter group consists of some that are quite old, with an age of the order of the population number, but these are extremely rare.

Thus, from (5), when $p = 0.001$ the mean age, A, is $0.0316N_e$. From (4) $N_e = H/4\mu$ where $H = 0.002$ and μ, estimated from the pseudogene evolution rate, is 2.2×10^{-8}/generation. So $N_e = 22,727$ and $A = 718$ generations. 720 human generations is roughly 20,000 years, so the average age of a neutral nucleotide polymorphism is quite small in evolutionary terms. The age of a favored or deleterious mutation is still less.

In contrast, if there is balancing selection the polymorphism can be much older. If a polymorphism has existed for a time very much longer than the population number, it must be due to some form of balancing selection, or some other balancing process, such as meiotic drive (Wu & Hammer, 1990).

Recent years have brought important advances in the genealogical approach to population genetics (Watterson, 1975; Kingman, 1982; Griffiths, 1980). In this approach one looks at the alleles in the present population and traces the coalescing paths back to a common ancestral allele. I can illustrate this by simulation results from a recent article by Takahata (1990) based on work by Takahata and Nei (1990).

Consider first a neutral locus. If $N_e\mu = 0.01$, the expected heterozygosity is 0.04 (from Equation 4). An individual locus is monomorphic most of the time and the average age of a heterozygous site is small. Even if the mutation rate is unrealistically high, e.g., $N_e\mu = 1$, leading to a heterozygosity comparable to that of MHC loci (about 0.9) the mean number of generations back to the common ancestor of a random pair of alleles (the separation time) is roughly the population number. If the mutation is beneficial, the time is much less for a comparable level of heterozygosity.

The simulations for deleterious alleles are as expected. Mutations are even more quickly eliminated. Heterozygosity remains low. The substitution rate is extremely low unless the allele is nearly neutral. Those mutations existing in the population are rare, and this can easily account for the excess of very rare alleles found in numerous observations of the distribution of allele frequencies. This must surely be the prevailing form of selection for most functional genes.

Since most mutations with very small effects exhibit partial dominance, in any reasonably large population the average frequency of a mutant allele is given by the deterministic equation μ/s, where s is the selection against the mutant heterozygote. The mean persistence of a mutation (roughly $1/s$), based on *Drosophila* data (Crow 1979), is on the order of 100 generations.

Takahata's simulations with overdominance offer a striking contrast. Here the polymorphism can be very old, hundreds or thousands of times the population number. The pattern generated in the simulations is very much like that seen for the MHC loci. Strikingly, one rodent polymorphism has persisted from a time before the evolutionary separation of mouse and rat.

The very long persistence of the polymorphism argues forcefully for some form of balancing selection in the evolution of these loci. A similarly long persistence of t alleles in mice is evidence for a balance between meiotic drive and selection.

THE IMPOSSIBILITY OF DISTINGUISHING SOME FORMS OF SELECTION

I want to emphasize, however, that when frequency-dependent selection is permitted it is often impossible, from gene frequency kinetics or equilibria, to distinguish various forms of selection (Denniston & Crow, 1990). Strikingly different fitness patterns, such as those with dominance reversed, can lead to the same equations of allele frequency change. In particular, an overdominant model (A) and a frequency-dependent model (B) lead to the same equation for allele frequency change. Therefore they can't be distinguished by any procedure based on expected allele-frequency changes.

Genotype	Fitness A	Fitness B
A_iA_i	$1 - s_i$	$(1 - s_ip_i)^2$
A_iA_j	1	$(1 - s_ip_i)(1 - s_jp_j)$

This means that the genealogical data on the MHC loci does not distinguish between these two selection patterns, patterns that differ greatly in their biological meaning. Other information is required. Nei has given biological reasons for preferring overdominance.

UTILIZING NEUTRAL MARKERS TO STUDY NATURAL SELECTION

Molecular studies—at first isozymes, then RFLPs, and now sequencing and PCR—have provided a wealth of information of potential use. Important issues in evolution hinge on detailed knowledge of population structure. The acceptability of Sewall Wright's "shifting balance" theory (Wright, 1988 and earlier) depends on, among other things, the frequency with which the type of population structure that he postulates actually exists. There are abundant molecular polymorphisms now available that are nearly enough neutral to serve as indicators of the degree of population subdivision, migration, and random drift. Likewise, a structured population with partially isolated groups offers the possibility of development of cooperative and altruistic traits within groups, by an extension of the principle of kin selection. For this, too, molecular markers can provide clues as to whether the relevant structure exists (Crow & Aoki, 1984).

DARWINIAN GRADUALISM

It is sometimes said that neutralism and natural selection are antagonistic, that the acceptance of neutral evolution weakens the role of natural selection. I believe the contrary. If there are neutral alleles, there must also be very weakly selected alleles. And Darwinian gradualism depends on genes with very small effects, especially many genes with small effects on a quantitative character.

So, to me a reasonable interpretation is this: the bulk of the individual nucleotide substitutions throughout the vertebrate genome are essentially neutral, driven by mutation and random drift. There are also a number of mutations, especially in coding and regulatory regions, that are deleterious and held in mutation-selection balance. Many are very mild; they can be amino acid substitutions, changes in regulatory regions, or other alterations with effects that are more quantitative than qualitative. In addition there is a small fraction of mutations that have small beneficial effects. They need not be favorable at the time of occurrence. Some may cause quantitative changes in traits with an intermediate optimum and in balance between mutation and stabilizing selection. These permit the continuous fitness changes that are required to keep up with an ever-changing, and often hostile environment of which the major element is other species evolving themselves.

I should like to sing a paean to polygenes, and restate some old principles. Multiple factors have a number of evolutionarily desirable attributes: (1) The smaller the individual effects, the more nearly additive the genes become. This is to be expected theoretically, but it is also true experimentally. The more nearly additive, the greater the heritability, and the greater the response to selection. (2) The smaller the individual effects, the better is the opportunity for evolutionary fine-tuning. If adjusting the eye lens to the retina requires a precision of 1/10 mm, genes that change the quantity by this amount or more are too coarse-grained. (3) With many factors there are the variance-conserving properties of Mendelism, without the noise that arises from segregation at individual loci. With a large number of loci, opposite random effects largely cancel; the Mendelian variance is reduced to something more like the variance of a mean. (4) Mendelian recombination permits each allele to be tested in many different combinations, so that those that do best on the average ("good mixers") are selected. (5) And, most important, multiple Mendelian factors have a maximum of potential variance with a minimum of standing variance. This is shown by numerous artificial selection experiments in which continuing selection has produced strains with means several standard deviations away from the original population values. In some cases it has been shown that new mutations are not the explanation because of the failure to obtain the same result by selecting in homozygous strains.

I don't, of course, want to assume that all natural selection is based on additive variance, although I think a great deal is. True evolutionary novelty, however, may well come from other sources, such as unusual gene interactions (as Wright, 1988 and earlier, has forcefully argued), genomic changes, and new gene functions.

WHERE DO NEW GENES COME FROM?

The evolution of a major new gene function demands some process that permits acquisition of a new function while retaining the old. Geneticists have long been aware that Mendelian heredity, including mutation, does not permit addition of alleles, only substitution. A mutated gene, in the process of acquiring a new function, ordinarily loses the old one. The obvious solution to the dilemma is to duplicate the gene, whereupon one of the two copies is free to mutate to a new function while the other preserves the old one, letting the organism have its cake and eat it. In this way the system can become more complex. There is abundant evidence that this is indeed what has happened.

Although this view has had much discussion in recent years, it is worth mentioning that it is not a new idea. The early Drosophilists, especially Bridges and Muller, emphasized this from the time of the earliest evidence for gene duplication. One of the clearest statements is from Muller (1936), just after the independent proof by Bridges and himself that the *Bar* eye mutation in *Drosophila* is a duplication.

> We consider the point of chief interest in the Bar case to be its illustration of the manner of origination of extra genes in evolution. Bar had for a long time offered the best case yet known for the idea that genes could arise de novo. . . . [There is now] no reason to doubt the application of the dictum "all life from preexisting life" and "every cell from a pre-existing cell" to the gene; "every gene from a pre-existing gene." We need at present make an exception here only of those very special conditions under which life itself, as a naked gene, originates.

What is new, however, is the plethora of new mechanisms discovered in recent years. Muller knew of the classical cytogenetic processes of inversion, duplication, translocation, and polyploidy. But he could have had no idea of the richness of the genomic repertoire brought about by gene conversion, replicative transposition, retroviruses, and the various other means of gene transposition and amplification, along with possible horizontal transmission. The importance of gene amplification in molecular evolution is apparent. Repeat units are uniquitous and the great excess of seemingly non-functional, often highly repetitious DNA is abundant evidence of DNA's ability to multiply. The genome is far from static. For the most part, I suspect, all this shuffling and multiplication isn't doing the organism any good. There is a constant conflict between wayward DNA, ever tending to increase itself, and stodgy organisms, trying vainly to keep this from happening. DNA has all sorts of methods for increasing its amounts, at rates that are very rapid compared to ordinary mammalian evolution. Until this produces substantial fitness effects the species simply tolerates it; there is little else it can do.

But, clearly, repeated and transposed DNA has created new and important functions. Sorting these out and distinguishing between DNA that is serving only its own ends and that which has been co-opted by the organism for *its* benefit is a major task for future research in molecular evolution. Wu (this Congress; Wu & Hammer, 1990) has shown a specific function for satellite DNA, in this case the *responder* locus in *Drosophila melanogaster*. Not only is it involved in the meiotic drive system, but it also has a measurable fitness effect on its own.

MOLECULAR EVOLUTION IN BACTERIA AND VIRUSES

There is a rich and rapidly growing field of molecular evolution in microorganisms. In fact, the most specific antecedents of the neutral theory are found in the work of Freese (1962) and Sueoka (1962). Each of them independently postulated mutation pressure of selectively equivalent base changes to account for the diversity of DNA amounts in different bacterial species. More recently, bacteria with greatly different base compositions have shown how the genetic code can be modified by mutation pressure (Jukes, 1985). John Maynard Smith, in this Congress, has urged more population genetic and evolutionary research on microorganisms.

Finally, I shall mention but not discuss the exciting research on virus evolution, particularly the very rapid evolution of retroviruses and the novel ways it has for generating variability (Pathak, 1990). Here, as in the many orders of magnitude slower rates of molecular evolution in eukaryotes, there appears to be a predominance of mutation-driven changes. Interesting problems in molecular evolution are clearly not confined to eukaryotes.

FINAL REMARKS

The neutral theory has forced evolutionists to take stochastic processes into account. I see no dichotomy between probabilistic and deterministic treatments of evolutionary theory. All evolution, if examined in sufficient detail, is stochastic. Deterministic equations are approximations, often very good ones. We can neglect stochastic processes for strong selection in large populations, just as a physicist ignores quantum uncertainty in most practical problems.

An understanding of the role of selection in molecular evolution needs, and is getting, a case-by-case analysis. This may not be easy, for as I have already emphasized it may be very difficult in individual instances to demonstrate that selection has occurred. Only in special circumstances, such as amino acid replacements at higher than the neutral rate, long persisting polymorphisms, and correlation with specific environmental factors, can selection be demonstrated. And we need the quantitative theory. As Gould has emphasized in this Congress, the study of evolution is both contingent and lawful. We study the cases, but we expect the theory to provide guidance.

The great merit of the neutral theory, in which Kimura has played the starring role along with several supporting actors and many bit players, is that it is quantitative. It has provided a wealth of testable predictions (and retrodictions). It has the requisites of a good scientific theory: it explains puzzling facts, it provides connections between otherwise disparate observations, it makes predictions, and it is heuristic; in addition it has elegance, and it brought that element of surprise that characterizes some of the nicest theories. It also shows the power of mathematical methods, sure to be of increasing value in the study of molecular evolution. For the first time in evolutionary study, we have—thanks to molecular biology—sufficiently detailed and abundant data to justify such a quantitative theory.

Almost 60 years ago, in January 1931, J. B. S. Haldane gave a lecture at the Prifysogol Cymru, Aberystwyth (in Wales, I needn't say), and said this (Haldane, 1932):

> The permeation of biology by mathematics is only beginning, but unless the history of science is an inadequate guide, it will continue.

ACKNOWLEDGMENTS

I am indebted to Bill Engels, John Gillespie, Motoo Kimura, Alexey Kondrashov, and Michael Turelli for many useful comments. This does not imply that they are in agreement with all that I have said. This is paper number 3168 from the Genetics Laboratory, University of Wisconsin.

LITERATURE CITED

Charlesworth, B., Coyne, J. A. & N. H. Barton. 1987. The relative rates of evolution of sex chromosomes and autosomes. *Amer. Natur.* 130:113–146.

Cox, E. C. & C. Yanofsky. 1967. Altered base ratios in the DNA of an *Escherichia coli* mutator strain. *Proc. Natl. Acad. Sci. USA* 58:1895–1902.

Crow, J. F. 1954. Breeding structure of populations. II. Effective population number. Pp. 543–556. *In:* O. Kempthorne, T. A. Bancroft, J. W. Gowen & J. L. Lush (eds.), *Statistics and Mathematics in Biology.* Iowa State College Press: Ames, Iowa.

Crow, J. F. 1979. Minor viability mutants in Drosophila. *Genetics* 92:s165–172.

Crow, J. F. & K. Aoki. 1984. Group selection for a polygenic behavioral trait: estimating the degree of population subdivision. *Proc. Natl. Acad. Sci. USA* 81:6073–6077.

Crow, J. F. & C. Denniston. 1985. Mutation in human populations. Pp. 59–123. *In:* H. Harris & K. Hirschhorn (eds.), *Advances in Human Genetics.* Plenum Publishing Company: New York.

Crow, J. F. & C. Denniston. 1988. Inbreeding and variance effective population numbers. *Evolution* 42:482–495.

Darwin, C. 1872. *The Origin of Species.* 6th ed. Modern Library: New York.

Denniston, C. & J. F. Crow. 1990. Alternative fitness models with the same allele frequency dynamics. *Genetics* 125:201–205.

Fisher, R. A. 1930. *The Genetical Theory of Natural Selection.* Oxford University Press: Oxford. Second edition, 1958. Dover Publications: New York.

Freese, E. 1962. On the evolution of base composition of DNA. *J. Theor. Biol.* 3:82–101.

Gillespie, J. H. 1986a. Natural selection and the molecular clock. *Mol. Biol. Evol.* 3:138–155.

Gillespie, J. H. 1986b. Variability of evolutionary rates of DNA. *Genetics* 113:1077–1091.

Gillespie, J. H. 1987. Molecular evolution and the neutral allele theory. Pp. 10–37. *In:* P. H. Harvey & L. Partridge (eds.), *Oxford Surveys in Evolutionary Biology.* Oxford University Press: Oxford.

Gillespie, J. H. 1989. Lineage effects and the index of dispersion of molecular evolution. *Mol. Biol. Evol.* 6:311–315.

Griffiths, R. C. 1980. Lines of descent in the diffusion approximation of neutral Wright-Fisher models. *Theor. Pop. Biol.* 17:37–50.

Haldane, J. B. S. 1932. *The Causes of Evolution.* Harper: New York.

Hendriks, W., Leunissen, J., Nevo, E., Bloemendal, H. & W. W. de Jong. 1987. The lens protein αA-crystallin of the blind mole rat, *Spalax ehrenbergi:* evolutionary change and functional constraints. *Proc. Natl. Acad. Sci. USA* 84:5320–5324.

Hill, R. E. & N. D. Hastie. 1987. Accelerated evolution in the reactive centre regions of serine protease inhibitors. *Nature* 326:96–99.

Jukes, T. H. 1985. A change in the genetic code in *Mycoplasma capricolum.* *J. Mol. Evol.* 22:361–362.

Kazazian, H. H., Chakravarti, A., Orkin, S. H. & S. E. Antonarakis. 1983. DNA polymorphism in the human β globin gene cluster. Pp. 137–146. *In:* M. Nei & R. K. Koehn (eds.). *Evolution of Genes and Proteins.* Sinauer, Sunderland, MA.

Kimura, M. 1968. Evolutionary rate at the molecular level. *Nature* 217:624–626.

Kimura, M. 1979. Model of effectively neutral mutations in which selective constraint is incorporated. *Proc. Natl. Acad. Sci. USA* 76:3440–3444.

Kimura, M. 1983. *The Neutral Theory of Molecular Evolution.* Cambridge University Press: Cambridge.

Kimura, M. & T. Ohta. 1973. The age of a neutral mutant persisting in a finite population. *Genetics* 75:199–212.

King, J. L. & T. H. Jukes. 1969. Non-Darwinian evolution. *Science* 164:788–798.

Kingman, J. F. C. 1982. The coalescent. *Stoch. Proc. Appl.* 13:235–248.

Kondrashov, A. S. 1988. Deleterious mutations and the evolution of sexual reproduction. *Nature* 336:435–440.

Koop, B. F., Tagle, D. A., Goodman, M. & J. L. Slightom. 1989. A molecular view of primate phylogeny and important systematic and evolutionary questions. *Mol. Biol. Evol.* 6:580–612.

Laskowski, M., Kato, I., Ardelt, W., Cook, J., Denton, A., Empie, M. W., Kohr, W. J., Park, S. J., Parks, K., Schatzley, B. L., Schoenberger, O. L., Tashiro, M., Vichot, G., Whatley, H. E., Wieczorek, A. & M. Wieczorek. 1987. Ovomucoid third domains from 100 avian species: Isolation, sequences, and hypervariability of enzyme-inhibitor contact residues. *Biochemistry* 26:202–221.

Li, W.-H., Gouy, M., Sharp, P. M., O'Huigin, C. & Y.-W. Yang. 1990. Molecular phylogeny of rodentia, lagomorpha, primates, artiodactyla, and carnivora and molecular clocks. *Proc. Natl. Acad. Sci. USA* 87:6703–6707.

Li, W.-H., Luo, C.-C. & C.-I. Wu. 1985. Evolution of DNA sequences. Pp. 1–94. *In:* R. J. MacIntyre (ed.), *Molecular Evolutionary Genetics.* Plenum Publishing Company: New York.

Mather, K. 1944. The genetical activity of heterochromatin. *Proc. Roy. Soc. Lond. B* 132:308–332.

Miyata, T., Kuma, K., Iwabe, N., Hayashida, H. & T. Yasunaga. 1990. Different rates of evolution of autosome-, X chromosome-, and Y chromosome-linked genes: hypothesis of male-driven molecular evolution. Pp. 341–357. *In:* N. Takahata & J. F. Crow (eds.), *Population Biology of Genes and Molecules.* Baifukan: Tokyo.

Mukai, T., Chigusa, S. I., Mettler, L. E. & J. F. Crow. 1972. Mutation rate and dominance of genes affecting viability in *Drosophila melanogaster. Genetics* 72:335–355.

Mukai, T. 1990. Viability polygenes in populations of *Drosophila melanogaster.* Pp. 199–218. *In:* N. Takahata & J. F. Crow (eds.), *Population Biology of Genes and Molecules.* Baifukan: Tokyo.

Mukai, T. & C. C. Cockerham. 1977. Spontaneous mutation rates at enzyme loci in *Drosophila melanogaster. Proc. Natl. Acad. Sci. USA* 74:2514–2517.

Muller, H. J. 1936. Bar duplication. *Science* 83:528–530.

Nei, M. 1987. *Molecular Evolutionary Genetics.* Columbia University Press: New York.

Nei, M. & D. Graur. 1984. Extent of protein polymorphism and the neutral mutation theory. *Evol. Biol.* 17:73–118.

Ohta, T. 1976. Role of very slightly deleterious mutations in molecular evolution and polymorphism. *Theor. Pop. Biol.* 10:254–275.

Ohta, T. 1977. Extension of the neutral mutation drift hypothesis. Pp. 148–167. *In:* M. Kimura (ed.), *Molecular Evolution and Polymorphism.* National Institute of Genetics, Mishima, Japan.

Ohta, T. 1990. Some new aspects of population genetics arising from gene multiplicity. Pp. 169–180. *In:* N. Takahata & J. F. Crow (eds.), *Population Biology of Genes and Molecules.* Baifukan: Tokyo.

Ohta, T. & H. Tachida. 1990. Theoretical study of near neutrality. I. Heterozygosity and rate of mutant substitution. *Genetics* 126:219–229.

Pathak, V., Hu, W.-H. & H. M. Temin. 1990. Retrovirus variation and hyper-mutation. *In:* E. J. Steele (ed.). *Somatic Hypermutation in V-regions.* CRC Press: Cleveland.

Sturtevant, A. H. 1921. Genetic studies on *Drosophila simulans.* III. Autosomal genes. General discussion. *Genetics* 6:179–207.

Sueoka, N. 1962. On the genetic basis of variation and heterogeneity of DNA base composition. *Proc. Natl. Acad. Sci. USA* 48:582–592.

Takahata, N. 1990. Allelic genealogy and MHC polymorphisms. Pp. 267–286. *In:* N. Takahata & J. F. Crow (eds.), *Population Biology of Genes and Molecules.* Baifukan: Tokyo.

Takahata, N. & M. Nei. 1990. Allelic genealogy under overdominant and frequency-dependent selection and polymorphism of major histocompatibility complex loci. *Genetics* 124:967–978.

Watterson, G. A. 1975. On the number of segregating sites in genetical models without recombination. *Theor. Pop. Biol.* 7:256–276.

Wright, S. 1988. Surfaces of selective value revisited. *Amer. Natur.* 131:115–123.

Wu, C.-I. & W.-H. Li. 1985. Evidence for higher rates of nucleotide substitution in rodents than in man. *Proc. Natl. Acad. Sci. USA* 82:1741–1745.

Wu, C.-I. & M. F. Hammer. 1990. Molecular evolution of ultraselfish genes. Pp. 177–203. *In:* R. K. Selander, T. Whittam & A. G. Clark (eds.). *Evolution at the Molecular Level.* Sinauer: Sunderland, MA.

An Hypothesis on Molecular Evolution that Combines Neutralist and Selectionist Views

EXTENDED ABSTRACT

Morris Goodman

An important premise of the neutral theory (Kimura, 1983) is that the rate of accumulation of base substitutions in a lineage over tens of millions of years of time might approach but cannot be greater than the rate of occurrence of mutations in the organisms of that lineage. This premise receives support from the finding that the maximum rate of base substitutions within genomes occurs where mutations are least likely to be selected against, such as at third positions (silent sites) of four-fold degenerate codons and in pseudogenes. In turn, it follows that the cumulative effects of natural selection have drastically slowed rates of molecular evolution. This happened in a most basic way by selection of adaptive mutations that improved the molecular machinery of life. Those adaptations that led to the evolutionary emergence of the network of interacting enzymes behind the machinery of DNA replication and repair greatly reduced the rates of occurrence of mutations. Those selected adaptive mutations that improved the machinery of life in so doing probably increased the density of lock and key type interaction sites among the encoded functionally active macromolecules. If so, these selected changes increased the proportion of sites in exons where new mutations were likely to be detrimental, thus selected against, and thereby slowing the nonsynonymous substitution rates. The reasoning behind this conclusion was first presented some years ago in an hypothesis that held the fastest rates of molecular evolution occur when the likelihood was greatest that newly occurring mutations were neutral to natural selection (Goodman, 1961, 1963).

That neutralist and selectionist views on molecular evolution can be complementary rather than conflicting is apparent on considering how major evolutionary advances arise. In the neutralist view, as expressed by Ohno (1970): "big leaps in evolution required the creation of new gene loci with previously nonexistent functions. Only the cistron which became redundant was able to escape from the relentless pressure of natural selection, and by escaping, it accumulated formerly forbidden mutations to emerge as a new gene locus." In the selectionist view, there is a synergism between gene duplication and selection. If the "redundant cistron" is to become a new gene and not a pseudogene, positive Darwinian selection at the new locus must favor the mutant alleles that encode a new beneficial function. Evidence for such Darwinian evolution comes from the pattern of rate changes in duplicated genes that acquire new functions; initially rates accelerate but later they decelerate as the natural selection that was positive becomes purifying.

Dr. Goodman is with the Department of Anatomy and Cell Biology, Wayne State University, School of Medicine, Detroit, MI 48201, USA.

Globins and their encoding genes exemplify this selection driven acceleration-deceleration pattern of substitution rates. Phylogenetic trees constructed for homologous globin sequences have revealed that the amino acid substitution rate of vertebrate globins speeded up during earlier stages of the vast adaptive radiation that saw the rise of the jawed and boney vertebrates and the first tetrapod invasions of the continents from the sea (Goodman et al., 1975; Goodman, 1981; Goodman et al., 1987). These studies also revealed that later in the vertebrate radiation, in the lineages to warm blooded amniotes (birds and mammals), rates slowed-down. The speeded-up rates in the early vertebrates occurred when selection on duplicated genes first shaped separate myoglobin and hemoglobin branches of the jawed vertebrate globin family and then, just before divergence of the Chondrichthyes and Osteichthyes, divided the hemoglobin branch into the α- and β-loci that encode heterotetrameric $(2\alpha2\beta)$ hemoglobin molecules. The fastest evolving sequence positions in nascent α and β chains were emerging $\alpha_1\beta_2$ contact and Bohr effect sites (Goodman, 1981), the sites which brought about cooperative subunit interactions in heterotetrameric hemoglobin and more efficient delivery of oxygen to respiring tissues. Selection first perfected these important functional sites and then preserved them, as evident from the fact that, from the bird-mammal ancestor to the present, they along with heme contact sites were the slowest evolving positions. Similarly, in the jawed vertebrate myoglobin branch, certain functional sites (e.g., salt bridges which stabilize monomeric structure) that were fast evolving in the early vertebrates became very slow evolving sites from the bird-mammal ancestor to the present (Goodman, 1981).

Closer to the present, globin evolution again exhibited a speed-up in rates followed by a slow-down. The γ-hemoglobin genes of primates provide a dramatic example. Phylogenetic reconstructions show that originally the γ genes functioned in embryonic life as if they were extra ε-hemoglobin genes, while β genes functioned in both fetal and postnatal life; however, in the stem-lineage to the simian primates, not only were there genic regulatory changes that delayed γ expression from embryonic to fetal life and β from fetal to postnatal life but also, in the globin chains encoded by γ genes, there were amino acid changes that permitted oxygen to bind more firmly to fetal hemoglobin than to adult hemoglobin (Goodman et al., 1987; Tagle et al., 1988). By ensuring that the flow of oxygen would always be from mother to fetus, the emergence of such a unique fetal hemoglobin may have been a precondition for the developmental changes that lengthened gestation and intrauterine fetal life in simian primates. The role of selection (positive and then purifying) is evident from the fact that both the absolute rate of nonsynonymous substitutions and the ratio of nonsynonymous to synonymous substitutions sharply increased in the stem-simian γ gene line and then sharply decreased in later simian γ gene lines such as those of hominoids and Old World monkeys (Goodman et al., 1987; Tagle et al., 1988).

Recently Fitch et al. (1990) have further elucidated the evolutionary history of primate γ-hemoglobin genes. They have been able to establish that early in the stem-simian lineage, when it still had only a single γ-hemoglobin locus, a member of the family of transposons called Lines inserted itself downstream of the γ locus while a second closely related Line inserted itself upstream of the γ locus. An unequal homologous crossover between these two Line sequences then produced the duplicate γ genes found in simian primates. This tandem gene duplication doubled the chances that one or the other duplicate would escape the selective constraints of an embryonically functioning gene and thus be freer to evolve into a fetally functioning gene. If positive selection for emergence of fetal γ-hemoglobin genes occurred (as it apparently did), then gene conversions between the paired γ hemoglobin genes allowed selection to fix in both genes the mutations that initially transformed one or the other embryonic γ gene into a fetal gene.

This selectionist view is compatible with the theory of nearly neutral mutations as presented in the recent study of Ohta and Tachida (1990) but is not necessarily incompatible with the strictly neutral theory of Kimura (1983). In the strictly neutral theory, an accelerated substitution rate after a gene duplication results from relaxation of selective constraints, which allows a larger proportion of mutations to be effectively neutral and open to either fixation or loss by random drift. If mutations fixed by random drift happen to shape a functionally different gene that at some later date turns out to be beneficial for its bearers, these randomly fixed mutations will have in effect shaped a new set of selective constraints and then, with this increased likelihood that new mutations will now be detrimental rather than neutral, purifying selection will reduce the substitution rate. However, this neutralist view which attributes almost nothing to positive selection but quite a lot to purifying selection, can be reconciled with the selectionist view of the preceding paragraph. All one has to do is consider the fate of the population (either an evolving deme or evolving species) in which mutations that were randomly fixed turn out at a later date to be advantageous for the members of the deme or species. The natural selection which is purifying within this population group preserves genes that may give the members of the group an advantage over those of other competing demes or species and, in this larger context, natural selection acts pretty much as Darwin envisioned it, i.e., competition is not just between individuals belonging to the same species but is also between individuals belonging to different species.

As single events, many nonsynonymous substitutions may have been neutral, or may have had an extremely small selection coefficient over relatively limited periods of time. However, when the potential for multiple interactions among many single substitutions over millions of years is considered it becomes apparent that selection for coadaptive interactions among molecular sites determines the course of evolution. A mutation which was neutral when it occurred and became fixed in its deme or species by random drift, if expressed in a protein, could probably not survive over the long time spans of macroevolution unless it became part of a beneficial network of molecular interactions as a result of subsequent mutations at other molecular sites. It is likely that many extinct taxa littering the fossil record were species which failed to evolve advantageous genic complexes as fast as the surviving species did.

Although alternative theories on molecular evolution disagree on the respective roles of random drift and positive selection in governing the rate of nonsynonymous substitutions, there is general agreement that mutation pressure and random drift primarily govern synonymous substitution rates in exons as well as base substitution rates in pseudogene and other noncoding, nonfunctional DNA (Crow, this volume). Yet even these synonymous and noncoding base substitution rates, which largely escape the dampening effects of purifying selection, provide evidence of the powerful role of natural selection in organismal evolution. This is illustrated by the substitution rates in primates over the $\psi\eta$-globin pseudogene locus and other noncoding DNA of the genomic domain which contains the $\psi\eta$ locus as well as ε, γ, δ, and β globin genes. These substitution or drift rates in "neutral" DNA, which should mirror the rates of occurrence of spontaneous mutations, correlate with the trends in primate evolution towards lengthened periods of gestation and fetal life, prolonged care of neonates that favors learned behavior, increased generation lengths, and longer life spans—all trends that were especially pronounced in the primate lineage to humans. Over time in descent from the early lower primates to higher primates (the present day simians), neutral drift rates that ranged from 3.5 to 6 × 10^{-9} substitutions per site per year in the earlier ancestors of present day primates decreased to rates in the range of 1 to 2 × 10^{-69} in simian lineages, with the hominoid lineage to human having the lowest rate (1×10^{-9}) (Koop et al., 1989; Bailey et al., 1990). The selectionist explanation for this pattern of fast "neutral" drift rates in early primates followed later in primate evolution by slowed "neutral" drift rates that became especially slow in hominoids is that selection brought about reduced per year mutation rates in later primates. To some extent this would be an automatic consequence of the longer generation times and slower rates of reproduction of the higher primates (humans having the

868

most K-selected life history strategy). It would also make evolutionary sense if the lineage to such long lived primates as humans had evolved improved DNA repair mechanisms. Such mechanisms would reduce per year mutation rates and the lower rates would be advantageous for these primates.

ACKNOWLEDGMENTS

This research was supported by National Science Foundation Grant No. BSR-8607202 and by National Institutes of Health Grant. No. HL33940.

LITERATURE CITED

Bailey, W. J., Fitch, D. H. A., Tagle, D. A., Czelusniak, J., Slightom, J. L. & M. Goodman. 1990. Molecular evolution of the $\psi\eta$-globin gene locus: gibbon phylogeny and the hominoid slowdown. *Mol. Biol. Evol.* In press.

Fitch, D. H. A., Bailey, W. J., Tagle, D. A., Goodman, M., Siew, L. & J. L. Slightom. 1990. Duplication of the g-globin gene mediated by repetitive L1 LINE sequences in an early ancestor of simian primates (submitted to *Proc. Natl. Acad. Sci. USA*).

Goodman, M. 1961. The role of immunochemical differences in the phyletic development of human behavior. *Hum. Biol.* 33:131–162.

Goodman, M. 1963. Man's place in the phylogeny of the primates as reflected in serum proteins. Pp. 204–234. *In:* S. L. Washburn (ed.), *Classification and Human Evolution.* Aldine Press: Chicago, IL.

Goodman, M. 1981. Decoding the pattern of protein evolution. *Prog. Biophys. Mol. Biol.* 37:105–164.

Goodman, M., Moore, G. W. & G. Matsuda. 1975. Darwinian evolution in the genealogy of haemoglobin. *Nature* 253:603–608.

Goodman, M., Czelusniak, J., Koop, B. F., Tagle, D. A., & Slightom, J. L. 1987. Globins: a case study in molecular phylogeny. *Cold Spring Harbor Symp. Quant. Biol.* 52:875–900.

Kimura, M. 1983. *The Neutral Theory of Molecular Evolution.* Cambridge University Press: Cambridge.

Koop, B. F., Tagle, D. A., Goodman, M. & J. L. Slightom. 1989. A molecular view of primate phylogeny and important systematic and evolutionary questions. *Mol. Biol. Evol.* 6:580–612.

Ohno, S. 1970. *Evolution by Gene Duplication.* Springer-Verlag: New York. 160 pp. (Quote is from Preface of this book.)

Ohta, T. & H. Tachida. 1990. Theoretical study of near neutrality. I. Heterozygosity and rate of mutant substitution. *Genetics* 126:219–229.

Tagle, D. A., Koop, B. F., Goodman, M., Slightom, J. L., Hess, D. L. & R. J. Jones. 1988. Embryonic ε and γ globin genes of a prosimian primate (*Galago crassicaudatus*): nucleotide and amino acid sequences, developmental regulation, and phylogenetic footprints. *J. Mol. Biol.* 203:439–455.

Distinguishing the Forces of Evolution with Molecular Data: An Example in *Drosophila*

Margaret A. Riley

Abstract. A summary of two surveys of nucleotide variation for the xanthine dehydrogenase locus (*Xdh*) in *Drosophila pseudoobscura* is presented. These data, which include a four-cutter restriction map survey and complete DNA sequence survey, allow the examination of the importance of the forces of evolution, including migration, mutation, recombination, natural selection and random genetic drift, in producing and maintaining genetic variation in natural populations of *Drosophila*.

INTRODUCTION

Population geneticists have spent a considerable effort trying to distinguish among the forces controlling genetic variation in natural populations. Much of this effort has concentrated on the use of electrophoretically determined protein variation to examine the relative roles of natural selection, genetic drift, migration, mutation, and recombination in generating and maintaining genetic variation within and between populations and species. Extensive surveys of protein variation have accumulated over the past twenty years. It has been clearly shown by those studies that natural populations of most organisms harbor a very large amount of genetic variability (eg., Nevo et al., 1984). However, in most cases, these data have failed to distinguish between alternative hypotheses regarding the production and maintenance of this variability (reviewed in Nevo et al., 1984; Fuerst et al., 1977; Lewontin, 1974).

It has been argued that surveys of DNA sequence polymorphism will provide the sort of data required to test hypotheses regarding the forces operating on genetic variation (Lewontin, 1985; Kreitman, 1988). The power of this approach lies in the detail afforded with complete DNA sequence information. Nucleotide positions can be examined that are closely linked but functionally distinct, eg., synonymous vs. non-synonymous positions. By comparing the levels of polymorphism and divergence at these functionally distinct positions with levels predicted by the neutral mutation theory (Kimura, 1983), it will be possible, in many cases, to distinguish between neutral vs. selective forces operating to produce the observed variation (see for example Hudson, et al., 1987). In addition, examination of the patterns of nucleotide polymorphism will often allow the importance of recombination and migration, in molding the variation observed, to be assessed (Riley et al., 1989).

Dr. Riley is with the Museum of Comparative Zoology, Harvard University, Cambridge, MA 02138, USA.

A large survey of DNA sequence variation at the alcohol dehydrogenase locus (*Adh*) in *Drosophila melanogaster* clearly reveals the power of this approach. A combinatin of complete DNA sequences for 11 *Adh* genes (Kreitman, 1983), one sequence from a closely related species, *D. simulans* (Kreitman and Aguade, 1986b), and four-cutter restriction mapping for a large number of additional *Adh* genes (Kreitman and Aguade, 1986a, b; Simmons et al., 1989) have provided strong evidence for the role of natural selection in maintaining a single segregating amino acid as a balanced polymorphism (Hudson et al., 1987).

The question remains, however, whether ADH is truly representative of the majority of protein polymorphisms. ADH was initially chosen for study by Kreitman (1980) because it displayed a widespread two-allele amino acid polymorphism in populations of *D. melanogaster*. A repeated clinal distribution for these two major variants of ADH argued a priori for the importance of positive selection in producing the observed allozyme distribution.

I present here a summary of two surveys of nucleotide variation for the xanthine dehydrogenase locus (*Xdh*) in *Drosophila pseudoobscura* (Riley et al., 1989; Riley et al., submitted). These studies are part of a large survey of sequence variation at the locus designed to discriminate among the forces operating at *Xdh*. In contrast to ADH, the encoded enzyme, xanthine dehydrogenase is highly polymorphic in all species of *Drosophila* where it has been examined electrophoretically (Coyne, 1976; Buchanon and Johnson, 1983; Keith et al., 1985). In a sample of 185 genomes from two natural populations of *D. pseudoobscura*, 20 allozyme classes were distinguished (Keith et al., 1985). In addition, the two populations, while separated by over 500 Km, were shown to be statistically indistinguishable in allozyme identity and frequency distribution.

The XDH electrophoretic data raise two questions. 1) Why is XDH such a highly polymorphic enzyme? Are the XDH allozyme classes the result of positive selection or, rather, the random drift of selectively neutral alternatives? 2) Why do the two widely separated populations have identical frequency distributions of XDH allozymes? Do the populations share a common gene pool or do they experience similar selective pressures, or both?

The combined use of four-cutter restriction mapping and complete DNA sequence analysis has allowed some success in addressing these questions. I report here a summary of our investigations to distinguish the forces of evolution acting at the *Xdh* locus in *Drosophila pseudoobscura*. The data I will discuss come from two separate surveys of nucleotide variation at the *Xdh* locus (Riley et al. 1989, Riley et al. ms submitted). All of the analysis and much of the discussion that I present come directly from those investigations.

FOUR-CUTTER RESTRICTION MAPPING AT *Xdh*

Kreitman and Aguade (1986a) introduced the methods for high resolution restriction mapping to survey genetic variation. This technique employs restriction enzymes with a four base pair recognition sequence. These sites occur, on average, every 200 base pairs in the genome. Thus, with a battery of 10 enzymes and with an efficiency of site detection of about 85%, it is possible to screen 1 in every 7 bases in the region of interest.

We have employed this technique to survey restriction site and insertion/deletion variability at the *Xdh* locus (Riley et al., 1989). Fifty-eight genomes were included in this study, representing 11 allozyme classes from the Keith et al. (1985) electrophoretic survey of XDH. The motivation for this work was to reveal population- or allozyme-specific levels or patterns of variability. Essentially, we asked whether high resolution restriction data can 1) distinguish if the two populations sampled by Keith et al. (1985) share a common gene pool and 2) if the allozyme classes are the product of positive selection.

Seven four-cutter enzymes were employed. These enzymes detected 135 restriction sites, of which 49% were polymorphic, plus 12 insertion/deletion length polymorphisms. The locations of the 78 polymorphisms are given in Figure 1. From these data it is possible to estimate the proportion of base pairs polymorphic in this sample of genes using the formula derived from Hudson (1982), modified to take account of multiple hits in each four base sequence (Riley et al., 1989). About 1 in every 12 base pairs is polymorphic in this sample for the *Xdh* region, given the simplifying assumption that polymorphisms are randomly distributed across the region probed.

Figure 2 shows the joint distribution of the + allele frequencies (the presence of a recognition site is scored as a +) in the two populations sampled. This figure reveals an extreme similarity in the allele frequencies in the two populations (r = 0.99, P < 0.001). The populations, and indeed the allozyme classes as well, are also indistinguishable in terms of the average number of sites differing per genome (Table 1).

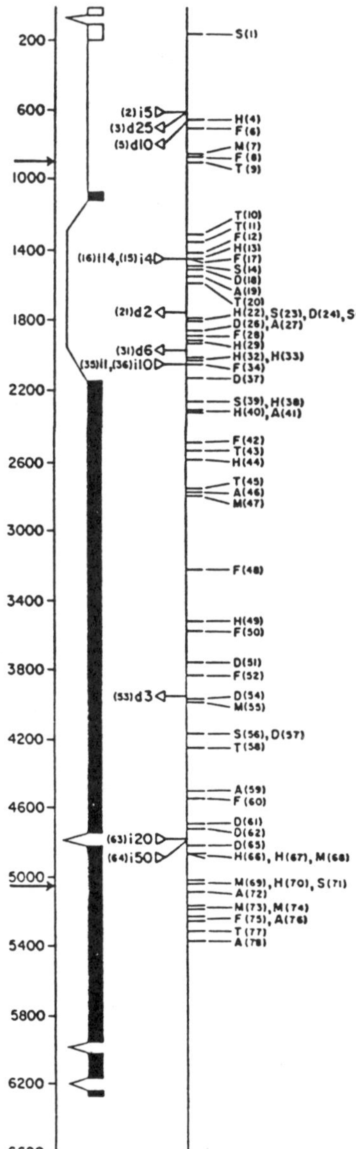

Figure 1. Structure of the *Xdh* gene in *Drosophila pseudoobscura* and the location of the polymorphic restriction sites found in this study. Center: Black boxes indicate *Xdh* exons, clear boxes indicate 1(3)s12 exons. Depressed lines indicate introns. Below: Scale is in base pairs with arrows delimiting the location of the probe employed. Above: Polymorphic restriction sites above the line. S = *Sau*96A, A = *Alu*I, D = *Dde*I, H = *Hae*III, F = *Hin*fIII, M = *Msp*I, T = *Taq*.

872

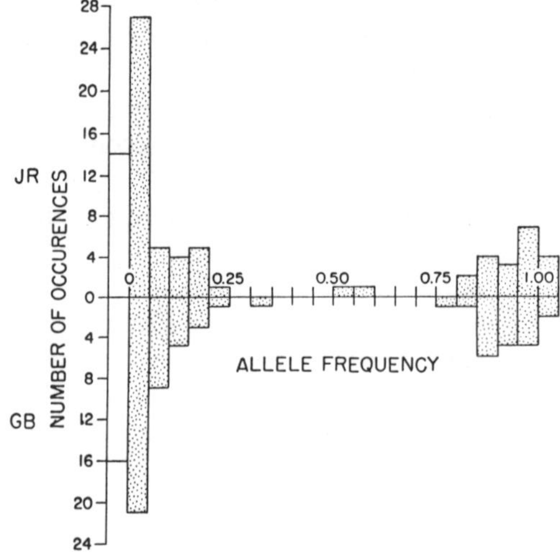

Figure 2. Distribution of + allele frequencies (q) for the 78 polymorphisms in the two populations. JR populations above the line. GB populations below the line. Open boxes are the polymorphisms absent or fixed in either population.

Table 1. Number of sites differing per genome, within and between localities and electromorphs.

	Mean	Variance	N	SE
Within populations	9.0086	4.9748	734	0.0823
Between populations	9.2524	25.0162	919	0.1650
Within morphs	8.5509	86.0773	216	0.6313
Between morphs	9.2742	5.6438	1437	0.0627
In pops. in morph	8.0612	80.4018	98	0.9058
In pops. bet. morph	9.2469	97.2181	636	0.3910
Bet. pops. in morph	8.9576	94.1574	118	0.8933
Bet. pops. bet. morph	9.2959	14.9233	801	0.1365
Total data	9.1796	16.1170	1653	0.0987

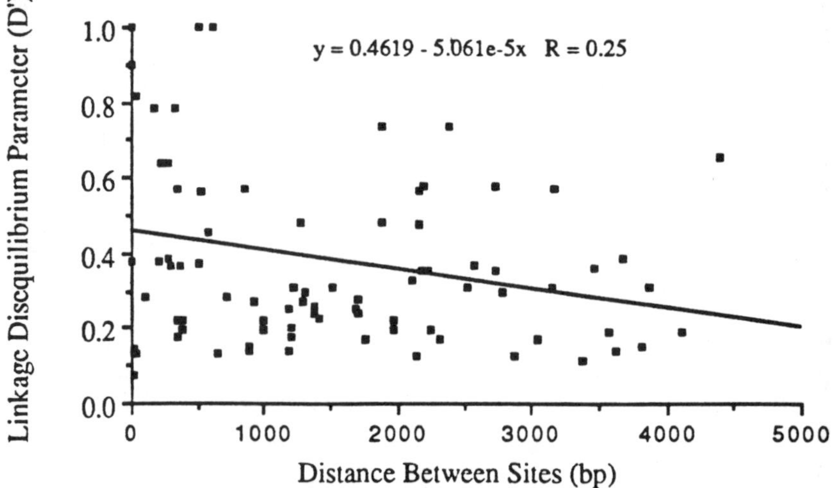

Figure 3. Standard linkage disequilibrium parameter, [D'], as a function of distance between sites, for sites that are significantly associated at the 0.05 level of significance.

High haplotype diversity, 0.993 in this sample of genes, and the absence of distinctive levels of polymorphism within allozyme classes or populations suggest that recombination may be important in shuffling variation when the polymorphisms reach appreciable frequencies in the populations. Recombination was examined by testing for the significance of linkage association between all pairs of polymorphic restriction sites by a Fisher's exact test for a 2×2 table (for details see Riley et al., 1989). Figure 3 shows the average absolute value of D' for the pairs of sites that have a significant association in the Fisher's test, as a function of distance between the sites. Clearly there is very little linkage disequilibrium even for sites very close together.

The four-cutter data provide some indication of the age and size of the populations sampled. The very high level of haplotype diversity, the high frequency of singly represented polymorphic sites and the inferred importance of recombination, argue that the populations sampled are quite large and have remained large for a long time. Estimates of the effective population size from these data range from 3.8×10^5 to 3.2×10^6. In addition, the lack of population-specific patterns of polymorphism, the identity of the frequency distribution of sites between the populations, and the presence of identical haplotypes in both populations, argue quite strongly that migration may account for the identity of the allozyme distribution noted by Keith et al. (1985).

If we compare the four-cutter diversity directly with the frequency distribution of electrophoretic alleles there seems to be an immense discrepancy (see Figure 4). The electrophoretic data can be interpreted as indicating considerable selective constraint on amino acid sequence. Yet, at the nucleotide level there is no haplotype that characterizes a particular allozyme class. The electrophoretic distribution indicates a phenotypic distinction between, but not within, allozyme classes, while the restriction site distribution suggests overall homogeneity.

Two alternative scenarios could explain this apparent discrepancy. The first is that positive selection is maintaining particular amino acid sequences, or classes of sequences, in the populations. The electrophoretic distribution would be revealing this selective constraint, while the restriction site distribution would be revealing neutral polymorphism accumulating unlinked to the site(s) of selection. Alternatively, the two distributions could differ simply because electrophoresis has not resolved all amino acid variability segregating for XDH. If the allozyme classes consist of multiple amino acid sequences that are indistinguishable electrophoretically, then the four-cutter data would be revealing the true distribution of variation and the null hypothesis of neutrality would best fit the *Xdh* data. The four-cutter survey does not resolve the question of selection at XDH. A second stage of screening for genetic variation was required.

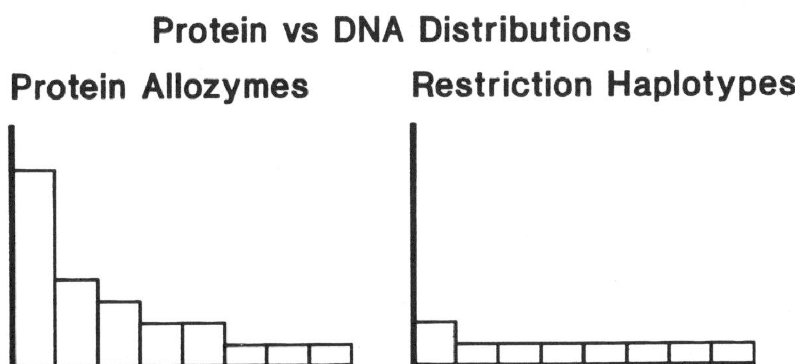

Figure 4. Schematic representation of the XDH protein electrophoretic distribution versus the four-cutter restriction site haplotype distribution. Frequency of each haplotype is given on the Y-axis. Each box indicates a haplotype.

DNA SEQUENCE ANALYSIS AT *Xdh*

In an attempt to distinguish the importance of positive selection for allozyme variability in XDH, Riley et al. (ms. submitted) have obtained complete DNA sequences for *Xdh* in seven genomes from the original population sample of Keith et al. (1985). The sample includes two representatives of the major allozyme class and one representative each of 5 additional allozyme classes. Figure 5 indicates the 185 nucleotide polymorphisms and six insertion/deletion variants observed.

Two genes chosen at random differ, on average, at 68 of the 5464 sequenced base pairs. Synonymous positions are, by far, the most highly polymorphic, with an average of 35 sites out of 964 (3.6%) differing between any pair of sequences. Table 2 provides a summary of the amino acid and nucleotide polymorphism by region. Table 3 details the 28 segregating amino acid residues.

The most striking result from this analysis is the extremely high level of amino acid variation revealed at the *Xdh* locus. Two percent of the residues are polymorphic in this sample of seven sequences, with an average of nine amino acid differences between any pair of allozymes. From the DNA sequence data it is possible to estimate the degree of purifying selection experienced by XDH. This requires an assumption that synonymous positions, experiencing little selective constraint, accurately reflect the level of neutral polymorphism for a coding region. At *Xdh*, 9% of silent sites are polymorphic. Thus, roughly 276 replacement sites (0.09×3062 base pairs) are expected to be segregating if the protein were under no functional constraint. Clearly, XDH, in this light, is highly constrained. Only 10% of the predicted replacement mutations are observed. However, when compared to *Adh*, a locus for which there is comparable polymorphism data (Kreitman, 1983), in which only 2.6% of the expected replacement polymorphisms are observed, XDH appears to experience a reduced level of purifying selection.

Given the estimate for effective population size provided from the four-cutter restriction survey, it is possible to predict the upper limit of the selection coefficient for neutral mutations at *Xdh*. This relationship is given by Kimura (1983) as $S < \frac{1}{2}Ne$, where S is the selection coefficient for neutral mutations and Ne is the effective population size. In *Drosophila pseudoobscura* an effectively neutral mutation at *Xdh* will experience a selection coefficient of 1×10^{-7}. In other words, in such a large population, purifying selection will govern the fate of most mutations, even those with slightly deleterious effects. And yet, at

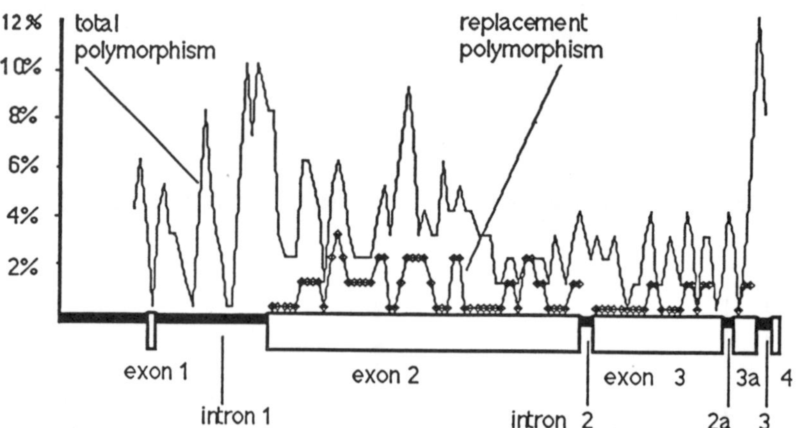

Figure 5. Percent nucleotide polymorphism at the *Xdh* locus in *Drosophila pseudoobscura*. The physical map of the region is indicated along the X-axis, percent polymorphism, per one hundred nucleotides, is given along the Y-axis (a sliding window of 50 nucleotides was employed). Total polymorphism (intron, silent, replacement) and replacement site polymorphism are given separately.

XDH there is a substantial amount of protein polymorphism. This argues that either the high level of protein polymorphism is being actively maintained through positive selection or that all or most of this polymorphism represents neutral alternatives that do not effect protein function, and therefore are not subjected to the action of purifying selection.

There is one striking difference between the four-cutter data and the complete sequence data. The restriction data show no less heterogeneity within than between allozyme classes. However, with complete sequence information the two representatives of the major allozyme class are shown to be, by far, more similar that any other pair of sequences. As is seen in Table 3, the major allozyme sequences are identical at the amino acid level and differ by only nine nucleotide polymorphisms in total. Recall that the average number of nucleotide differences between a pair of alleles is 68 and the average number of amino acid differences is nine.

Table 2. *Xdh* sequence polymorphism.

		Nucleotide Sites			Amino Acid Residues		
Region	#bp	#sites polym	%sites polym	i/d* polym	#amino acids	#polym	%polym
5′	196	7	3.6				
Exon 1	54				18		
Intron 1	1024	43	4.2	5			
Exon 2	2613	88/91 !	3.4		871	21/23	2.4
Intron 2	62	3	4.8				
Exon 3	1143	21/22	1.8		381	3	0.8
Intron 3	69	5	7.7				
Exon 4	168	3	1.8		56	1	1.8
Intron 4	67	8	11.9	1			
Exon 5	48	2	4.2		16	1	6.3
3′	12						
Summary: Nucleotides							
Exons	4026	114/118	2.8				
Silent	964	87/90	9.0				
Replace	3062	27/28	0.9				
Intron	1222	59/60	4.7	6			
5′ + 3′	208	7	3.4				
Total	5456	180/185	3.3	6			
Total Amino Acids					1342	26/28	1.9

* i/d = insertion/deletion variants
! #s preceding a / = # of *sites* polymorphic, #s following a / = total
# of base pairs or amino acids segregating.   % polym = polymorphism

Table 3. Polymorphic amino acids at XDH.

Position ruler (tens markers over the units digits): `5,` over column 5, `1` over column 9, `20,` over column 19.

	1	2	3	4	6	7	8	9	0	1	2	3	4	5	6	7	8	9	1	2	3	4	5	6	7	8
CONSENSUS	L	S	D	V	T	P	E	P	L	A	T	H	I	A	E	S	T	N	S	L	S	L	R	M	C	E
JR48					A		K		F	S	K			T				H								
GB336			G	E	P											R			G		R			T		
GB50		A			A			L					O		G							M				
GB369													V													D
GB87													V													D
JR436		A			A														N						F	
GB361	P					A									C					V			H			
MELAN!	L	A	—	E	K	D	E	K	L	S	L	H	I	A	E	S	S	N	T	L	K	L	R	M	C	P

! denotes *D. melanogster* residues

Are the two unique amino acids shared by the major allozyme representatives under positive selection? The two DNA sequences could be similar simply because they share a more recent common ancestor, i.e., by chance we chose two closely related flies from the population sample. Alternatively, the two DNA sequences could be alike because one or both of the shared amino acid polymorphisms are under some form of positive selection. The four-cutter data, which includes ten representatives from the major allozyme class, argue strongly that the identity is due to descent. Members of the major class are not significantly more similar in haplotypes than any randomly chosen set of alleles. Clearly additional sequence data is required to resolve this issue. A relatively few additional representatives of the major allozyme class should be sufficient. This work is currently in progress (Susanne Kaplan, personal communication).

Putting aside the question of selection within the major allozyme class, several lines of evidence strongly suggest that the remaining polymorphism, both at the nucleotide and amino acid level, is selectively neutral. 1) The high level of amino acid polymorphism observed supports a preliminary conclusion that no particular amino acid sequence is being positively selected. This assertion is perhaps more convincing when one recalls the lack of differentiation within and between allozyme classes observed in the four-cutter survey. Positive selection would both generate linkage disequilibrium between sites and lead to classes of homogeneous sequences. 2) The frequency distribution for replacement polymorphisms is indistinguishable from that observed for silent and intron sites. The majority of all polymorphisms, irrespective of site function, are unique. These results argue that similar selective pressures (purifying or positive selection), or an absence of selection is experienced equally by the three classes of nucleotide sites. 3) Synonymous and replacement polymorphisms map to the pattern revealed in a between species comparison of *Xdh* (Riley et al., 1989). The neutral theory predicts a positive correlation between levels and patterns of divergence and polymorphism at a neutrally evolving locus (Kimura, 1983). The predicted positive correlation is observed at *Xdh*. 4) A statistical test of the neutral theory has been developed by Tajima (1989). This test examines the neutral prediction for the relation between the number of segregating sites and the average number of nucleotide differences in a population sample. *Xdh* does not deviate significantly from the relationship predicted for neutrality.

Each of the four points just raised does not individually rule out the possibility that positive selection has molded levels and patterns of polymorphism at *Xdh*. However, I would argue that the composite picture of nucleotide variation that has emerged most strongly supports the view that *Xdh* polymorphism conforms to predictions from the neutral theory.

CONCLUSIONS

The combined use of four-cutter restriction mapping and complete DNA sequence determination has provided a detailed picture of the levels and patterns of genetic variation segregating at the *Xdh* locus in *Drosophila pseudoobscura*. These data have allowed us to distinguish the importance of several evolutionary forces, such as migration, recombination, and purifying selection, in the production and maintenance of the observed variation. In addition, although the data do not allow a resolution regarding the importance of positive selection for amino acid sequence identity within the major allozyme class, they do provide specific and easily testable predictions. What is required is a few additional sequences of *Xdh* from members of the major allozyme class.

Surveys of nucleotide polymorphism at *Xdh* have served to illustrate the power of molecular data in addressing population genetic questions. Perhaps most important is the fact that *Xdh* tells a very different story about the evolutionary forces acting on variation than was determined from the first such survey, which reveals a balanced polymorphism at the *Adh* locus in *D. melanogaster* (reviewed in Kreitman, 1988). It is clear that many surveys of nucleotide polymorphism, in particular surveys of additional loci within the same species, will be required before general patterns of evolution at the population level will be revealed.

ACKNOWLEDGMENTS

The work summarized in this review was supported by a grant to R. C. Lewontin from the National Institutes of Health (GM-29301). I thank Susanne Kaplan, Mary Ellen Hallas and Richard Lewontin for the extensive contributions they have made in the collection and analysis of these data.

LITERATURE CITED

Buchanon, B. A. & D. L. E. Johnson. 1983. Hidden electrophoretic variation at the xanthine dehydrogenase locus in natural populations of *Drosophila melanogaster*. *Genetics* 104:301–315.

Coyne, J. A. 1976. Lack of genetic similarity between two sibling species of *Drosophila* as revealed by varied techniques. *Genetics* 84:593–607.

Fuerst, P. A., Chakraborty, R. & M. Nei. 1977. Statistical studies on protein polymorphism in natural populations. I. Distribution of single locus heterozygosity. *Genetics* 86:455–483.

Hudson, R. R. 1982. Estimating genetic variability with restriction endonucleases. *Genetics* 100:711–719.

Hudson, R. R., Kreitman, M. & M. Aguade. 1987. A test of neutral molecular evolution based on nucleotide data. *Genetics* 116:153–159.

Keith, T. P., Brooks, L. D., Lewontin, R. C., Martinez-Cruzado, J. C. & D. L. Rigby. 1985. Nearly identical allelic distributions of xanthine dehydrogenase in two populations of *Drosophila pseudoobscura*. *Mol. Biol. Evol.* 2:206–216.

Kimura, M. 1983. *The Neutral Theory of Molecular Evolution*. Cambridge University Press, Cambridge.

Kreitman, M. 1980. Assessment of variability within electromorphs of alcohol dehydrogenase in *Drosophila melanogaster*. *Genetics* 95:467–475.

Kreitman, M. 1983. Nucleotide polymorphism at the alcohol dehydrogenase locus of *Drosophila melanogaster*. *Nature* 304:412–417.

Kreitman, M. 1988. Molecular population genetics. *Oxford Surv. Evol. Biol.* 4:38–60.

Kreitman, M. & M. Aguade. 1986a. Genetic uniformity in two populations of *Drosophila melanogaster* as revealed by filter hybridization of four-nucleotide-recognizing enzyme digests. *Proc. Natl. Acad. Sci. USA* 83:3562–3566.

Kreitman, M. & M. Aguade. 1986b. Excess polymorphism at the *Adh* locus in *Drosophila melanogaster*. *Genetics* 114:93–110.

Lewontin, R. C. 1974. *The Genetic Basis of Evolutionary Change*. Columbia University Press: New York. 346 pp.

Lewontin, R. C. 1985. Population genetics. *Ann. Rev. Gen.* 19:81–102.

Nevo, E. A. Beiles & R. Ben-Shlomo. 1984. The evolutionary significance of genetic diversity: ecological, demographic and life history correlates. *Lect. Notes Biomath.* 53:1–213.

Riley, M. A., Hallas, M. E. & R. C. Lewontin. 1989. Distinguishing the forces controlling genetic variation at the *Xdh* locus in *Drosophila pseudoobscura*. *Genetics* 123:359–369.

Riley, M. A., S. R. Kaplan, M. Vieux. DNA sequence polymorphism at the *Xdh* locus in *Drosophila pseudoobscura*. Submitted to *Molecular Biology and Evolution*.

Simmons, G. M., Kreitman, M., Quattlebaum, W. F. & N. Miyashita. 1989. Molecular analysis of the alleles of alcohol dehydrogenase along a cline in *Drosophila melanogaster*. I. Maine, North Carolina and Florida. *Evolution* 43:393–409.

Tajima, F. 1989. Statistical method for testing the neutral mutation hypothesis by DNA polymorphism. *Genetics* 123:597–601.

Polymorphism Due to Balancing Selection at the Major Histocompatibility Complex Loci

Masatoshi Nei and Austin L. Hughes

Abstract.   Classical MHC loci are extremely polymorphic, and there are several features that are unique to MHC polymorphism. 1) The heterozygosity is as high as 90% in many human and mouse populations. 2) The number of nucleotide differences between alleles is unusually high. 3) In the antigen-binding site of the gene, the rate of nonsynonymous substitution is much higher than that of synonymous substitution. 4) Polymorphic alleles (allelic lineages) persist in the population for tens of millions of years. To explain the above observations, some form of balancing selection is necessary. One of the most reasonable forms of balancing selection to explain them is overdominant selection (heterozygote advantage). The hypothesis of overdominant selection is well supported by the function of MHC molecules and the pattern of nucleotide substitution in MHC genes. However, at the present time, certain models of frequency-dependent selection cannot be ruled out, though experimental evidence is weak. It is not clear how realistic these models are for the case of MHC polymorphism. It seems that more careful experimental studies are necessary to distinguish between the overdominance and frequency-dependent selection hypotheses.

INTRODUCTION

The major histocompatibility complex (MHC) loci in mammals are known to be exceptionally polymorphic, and the maintenance of polymorphism at these loci has been studied for more than two decades. Yet no consensus has been reached.

In recent years, however, new biochemical techniques such as gene cloning and DNA sequencing have been applied to the study of the genomic organization of the MHC, and this has contributed greatly to an understanding of the maintenance of polymorphism as well as of the function of the genes at the molecular level (Klein, 1986; Lawlor et al., 1990). Several new hypotheses, including the gene conversion hypothesis (Weiss et al., 1983), have also been proposed for explaining MHC polymorphism. Bjorkman et al. (1987a, b) recently identified the antigen recognition site (ARS) of the class I MHC molecule, using X-ray crystallography. This identification led Hughes and Nei (1988) to reason that if there is any positive Darwinian selection at the MHC loci, the selection must occur at the ARS. Examining the pattern of nucleotide substitution at the ARS and the remaining regions of the genes, they then showed that positive selection is indeed operating at the ARS.

In this paper we first examine the nature and pattern of MHC polymorphism and then discuss the mechanisms of its maintenance.

Drs. Nei and Hughes are with the Institute of Molecular Evolutionary Genetics and Department of Biology, Pennsylvania State University, University Park, PA 16802, USA.

EXTENT OF MHC POLYMORPHISM

MHC loci show an extraordinarily high degree of polymorphism, and no other protein coding loci in the mammalian genome match them in this regard. For example, the human MHC loci, *HLA-A* and *-B*, show a heterozygosity of about 90% in many populations; the number of alleles detected in a sample of about 200 individuals ranges approximately from 15 to 30 (see Roychoudhury & Nei, 1988). Similarly, the mouse MHC loci, *H2-K* and *-D*, are known to be extremely polymorphic. In other organisms the extent of allelic polymorphism is not well studied, but experiments on skin grafting, serological tests, and restriction fragment analyses suggest that many vertebrate organisms are highly polymorphic for both class I and class II loci (Table 1).

It should be noted, however, that there are organisms in which all the MHC loci are virtually monomorphic. An extreme example is the cheetah, in which even skin grafting can easily be made between different individuals (O'Brien et al., 1985). Another organism in which a low degree of polymorphism is observed is the Syrian hamster (McGuire et al., 1985). Although the number of individuals examined is relatively small, no polymorphic alleles have been detected at the class I loci, and the number of alleles detected at the class II loci is also small (Streilein, 1987). The fin whale also shows a low level of MHC polymorphism, as do mouse populations living on small islands in the North Sea (Figueroa et al., 1986).

In recent years DNA sequences have been determined for many different alleles at the polymorphic MHC loci in humans and mice. One can therefore examine the extent of DNA variation by using nucleotide diversity (π), i.e., the average number of nucleotide differences per site between two randomly chosen genes (Nei, 1987). The π values estimated for the *HLA-A* and *H2-K* loci are 0.04 and 0.08, respectively. These values are much higher than those for other loci in the mammalian genome, where π is usually less than 0.01 (Nei, 1987). This indicates that at the MHC loci not only the number of polymorphic alleles is large but the extent of nucleotide divergence between alleles is also high. Note that the $\pi = 0.04 \sim 0.08$ is much higher than the average number of nucleotide differences per site (0.016) for the η globin pseudogene region between the human and the chimpanzee (Miyamoto et al., 1987).

Table 1. MHC polymorphism in vertebrate species.

Species	Method	MHC Polymorphism Class I	Class II	References
Chicken	SE	High	High	Simonsen et al. (1982)
	RFLP	N.A.	High	Warner et al. (1989)
Cheetah	SG	Absent	N.A.	O'Brien et al. (1985)
Cottontop tamarin	IEF	Low	Moderate	Watkins et al. (1988)
Balkan mole rat	RFLP	Low	Absent	Nizetic et al. (1988)
Israeli mole rat	RFLP	High	N.A.	Nizetic et al. (1985)
Fin whale	RFLP	Low	Low	Trowsdale et al. (1989)
Sei whale	RFLP	Low	Low	Trowsdale et al. (1989)
Syrian hamster	RFLP	Low	Moderate	McGuire et al. (1985)
	SE	Low	Moderate	Streilein (1987)
Rat	RFLP	High	N.A.	Palmer et al. (1983)
Domestic cat	RFLP	Moderate	Moderate	Yuhki & O'Brien (1989)
	SE	Moderate	Moderate	Winkler et al. (1989)
Pig	RFLP	High	High	Chardon et al. (1985)
Mouse	All methods	High	High	Klein (1986)
Human	All methods	High	High	Klein (1986)

Methods of detection: SG = Skin grafting; IEF = Iso-electric focusing electrophoresis; RFLP = Restriction fragment method; SE = Serology. N.A.; data not available.

PATTERN OF NUCLEOTIDE SUBSTITUTION

One way of studying the nature and pattern of natural selection operating at the DNA level is to examine the number of synonymous substitutions per synonymous site (d_S) and the number of nonsynonymous substitutions per nonsynonymous site (d_N). If there is no selection for a DNA region, d_S and d_N are more or less the same. If there is purifying selection, d_S is expected to be higher than d_N because selection would occur at the amino acid level. By contrast, if there is positive Darwinian selection, d_N is expected to be higher than d_S. As mentioned earlier, the antigen specifity of the MHC molecule is determined by the amino acids in the ARS, which is located in the α_1 and α_2 domains in the class I MHC. Therefore, if there is positive selection at the MHC loci, it should occur at the ARS region of the gene. We therefore examined the d_S and d_N values for the ARS region and the remainder region separately (Hughes & Nei, 1988, 1989a).

Table 2 shows the mean d_S and d_N values for all sequence comparisons at each class I polymorphic locus. In the case of the human *HLA-A* locus, d_N is much higher than d_S in the ARS, the difference being significant at the 0.1% level. In the other regions of the α_1 and α_2 domains and the α_3 domain of the MHC, d_S is higher than d_N. The same pattern of nucleotide substitution is also observed for the *HLA-B* and *-C* loci. It is interesting to see that d_S is more or less the same for all three regions examined, whereas d_N in the ARS is significantly higher than d_S in any region. This clearly indicates that amino acid substitutions in the ARS are enhanced by positive Darwinian selection. By contrast, d_N is much smaller than d_S in the non-ARS regions, suggesting that amino acid substitutions are generally subject to purifying selection in these regions.

Table 2 also includes the results of allelic comparisons at the mouse *H2-K* and *-L* loci. The d_N in the ARS is again substantially higher than the d_S in any region. Therefore, the pattern of nucleotide substitution for the mouse MHC loci is essentially the same as that for the human MHC loci. (The unusually low value of d_S for α_3 may have been caused by interlocus genetic exchange; Hughes & Nei, 1988).

We conducted a similar statistical analysis for class II MHC loci. Class II MHC molecules are composed of the α and β chain polypeptides, and the ARS is located in domain 1 (D1) of both α and β chains (α_1 and β_1). Table 3 shows that in both mouse and human poly-

Table 2. Mean numbers of nucleotide substitutions per synonymous site (d_S) and per nonsynonymous site (d_N) expressed as percentages between alleles from the same and different class I MHC polymorphic loci in humans (*HLA-A, B,* and *C*) and mice (*H2-K,* and *D*).

Locus (No. sequences)	Antigen recognition site (ARS) ($N = 57$)		Remaining codons in α_1 & α_2 ($N = 124, 125$)[a]		Domain α_3 ($N = 92$)	
	d_S	d_N	d_S	d_N	d_S	d_N
Human						
A (5)	3.5	13.3***	2.5	1.6	9.5	1.6**
B (4)	7.1	18.1**	6.9	2.4*	1.5	0.5
C (3)	3.8	8.8	10.4	4.8	2.1	1.0
Overall means	4.7	14.1***	5.1	2.4	5.8	1.1**
Mouse						
K (4)	15.0	22.9	8.7	5.8	2.3	4.0
L (4)	11.4	19.5	8.8	6.8	0.0	2.5**
Overall means	13.2	21.2*	8.8	6.3	1.2	3.6**

The d_S and d_N values were computed by using Nei and Gojobori's (1986) method I.

The difference between d_S and d_N is significant at 5% level (*), at 1% level (**), or at 0.1% level (***) (see Hughes & Nei, 1988, for details). N = number of codons compared. [a]N is 124 for the mouse and 125 for the human.

morphic loci, d_N is again much higher than d_S in the ARS, though the difference is not necessarily significant because of the small number of codons involved. In the conserved domain (α_2 or β_2 domain) region, however, d_N is always smaller than d_S. Therefore, class II polymorphic loci show the same pattern of nucleotide substitution as that of class I loci.

TRANS-SPECIFIC POLYMORPHISM

Theoretically, a pair of neutral alleles at a locus may persist for $2N$ generations in a population but not much longer, where N is the effective population size (Takahata & Nei, 1990). However, McConnell et al. (1988) have shown that two polymorphic allelic lineages at the $A\beta 1$ locus are shared by three species of mice *(Mus)*, which apparently diverged about three million years ago. Figueroa et al. (1988) showed that two poly-morphic allelic lineages marked with insertions/deletions at the $A\beta 1$ locus are shared by mice and rats, which diverged at least 10 million years ago. Similar shared poly-morphisms have also been reported for the class I and class II MHC loci in humans and chimpanzees, which diverged probably 5∿7 million years ago (Fan et al., 1989; Lawlor et al., 1988; Mayer et al., 1988).

Figure 1 shows the phylogenetic tree of human (*HLA-A* and *-B*) and chimpanzee (*CHA-A* and *-B*) MHC alleles. The human allele *H-A11* is closer to the chimpanzee allele *C-A108* than to any other human alleles at the *A* locus, and the human allele *H-A24* is closer to the chimpanzee allele *C-A126* than to any other human alleles. Here we can clearly see that a pair of polymorphic allelic lineages are shared by humans and chimpanzees. The *B* locus also shows a similar shared polymorphism. These polymorphic lineages apparently coexisted in both human and chimpanzee populations for a long time. For example, the allelic pair *H-A11* and *H-A24* in humans or *C-A108* and *C-A126* in chimpanzees is esti-mated to have been polymorphic for more than 20 million years if we use the rate of synonymous substitution estimated by Li et al. (1987) for hominoids. This indicates that polymorphic allelic lineages persist in the population through speciation events. Klein (1987) called this the trans-specific mode of evolution.

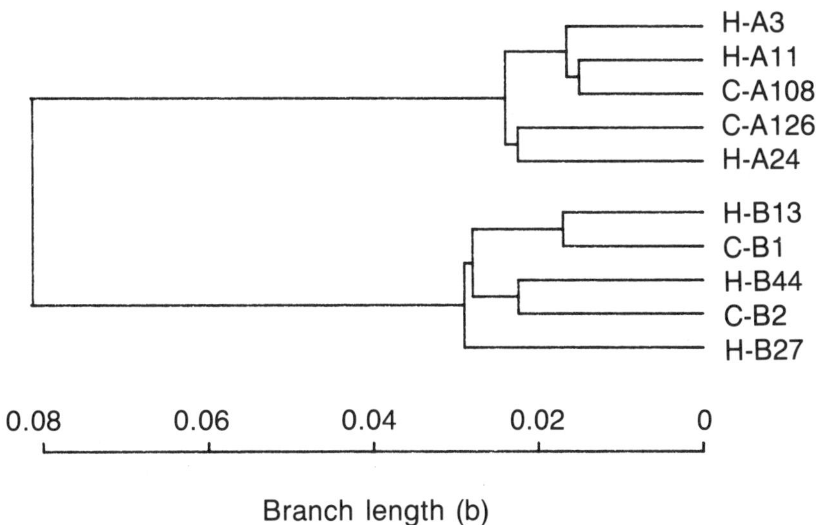

Figure 1. Phylogenetic tree for alleles from the class I MHC loci *A* and *B* in the human (*HLA-A* and *-B*) and the chimpanzee (*CHA-A* and *-B*). This tree was obtained by UPGMA, and b represents the branch length measured in the number of nucleotide substitutions per site. H: human. C: chimpanzee.

MECHANISMS OF MAINTENANCE OF MHC POLYMORPHISM

It is now clear that MHC polymorphism is characterized by several unique features and that any hypothesis for the mechanism of its maintenance must be able to explain all these observations. The unique features are: 1) The extent of polymorphism is extraordinarily high. 2) The number of nucleotide differences between alleles is unusually high. 3) In the antigen-recognition site of the gene, the rate of nonsynonymous substitution is higher than the rate of synonymous substitution. 4) Polymorphic alleles (allelic lineages) persist in the population for tens of millions of years.

As mentioned earlier, many different hypotheses for explaining MHC polymorphism have been proposed in the past. The major ones are: 1) a high mutation rate (Bailey & Kohn, 1965), 2) neutral mutations (Klein, 1987), 3) gene conversion (Lopez de Castro et al., 1982), 4) maternal-fetal incompatibility (Clarke & Kirby, 1966; Hedrick & Thomson, 1988), 5) mating preference (Yamazaki et al., 1976), 6) overdominant selection (Doherty & Zinkernagel, 1975; Hughes & Nei, 1988), and 7) frequency-dependent selection (Snell, 1968; Bodmer, 1972). Let us now examine these hypotheses in the light of present knowledge about the MHC.

Neutral Mutation and Gene Conversion

The hypothesis of a high mutation rate is certainly capable of explaining the higher degree of polymorphism, but this hypothesis can be rejected because it cannot explain the relationship of $d_N > d_S$ in the ARS region of the gene. Furthermore, it is now known that the rate of synonymous nucleotide substitution in the MHC genes is not particularly high compared with that of other nuclear genes (Klein, 1986). For example, the number of synonymous substitutions between the human and mouse class I genes is about 0.32, which is less than half the average d_S for other genes (Hayashida & Miyata, 1983).

Klein (1987) proposed the hypothesis of neutral mutations to explain the transspecific mode of evolution. (He actually proposed that positive selection may occur when a drastic change in environment occurs but that MHC alleles are neutral through most of evolutionary time.) In practice, however, this hypothesis is incapable of explaining the long persistence of polymorphic alleles in rodents and hominoids. As mentioned above, the coexistence of a pair of neutral alleles in a population rarely persists for more than $2N$ generations. The extent of protein polymorphism in humans suggests that the long-term effective population size (N) of this species is of the order of 10^4 (Nei & Graur, 1984). If we assume that the average generation time in the human lineage was 10 years in the past, $2N$ generations correspond to 2×10^5 years. This is clearly too short compared with the observed persistence time of MHC polymorphic alleles. Note also that this hypothesis cannot explain the relationship of $d_N > d_S$ in the ARS.

Another hypothesis that does not invoke any selection is that of gene conversion. There is indirect evidence that unequal crossover, exon shuffling, or gene conversion generates genetic exchange between different alleles at the same locus (Holmes & Parham, 1985) or at different loci (Weiss et al., 1983; Hughes & Nei, 1989b). Therefore, intragenic or intergenic recombinatioin seems to be a source of genetic variation. However, the frequency of occurrence of these events seems to be quite low (Hughes & Nei, 1989b; Klein et al., 1990; Lawlor et al., 1990), and they cannot be a major factor in MHC polymorphism. Furthermore, this hypothesis cannot explain the observation $d_N > d_S$ in the ARS.

Maternal-Fetal Incompatibility

This hypothesis was first proposed by Clarke and Kirby (1966), who attempted to explain James' (1965) observation that fetal growth in mice was enhanced when the mother and father were of different MHC haplotypes. The idea behind the hypothesis is that a histoincompatible fetus stimulates the mother's immune response, which in turn makes the fetus more vigorous. This postulated maternal-fetal interaction is expected to produce a net heterozygous advantage (Hedrick & Thomson, 1988). However, this hypothesis is not based on a solid body of experimental data. Subsequent studies involving the rat (Palm, 1974) and the mouse (Hamilton et al., 1985) have failed to replicate James' original finding.

Certain human data have been taken as support for the maternal-fetal incompatibility hypothesis. For example, some authors (e.g., Gill, 1983) have found a high rate of spontaneous abortion when the mother and father share the same MHC alleles. These data, however, can be explained equally well by the hypothesis of deleterious recessive genes at one or more loci that are closely linked to the MHC (Schacter et al., 1984). Therefore, they do not really constitute evidence for the hypothesis. Perhaps the most powerful argument against this hypothesis is the fact that the chicken has a high degree of MHC polymorphism even though there is no maternal-fetal interaction.

Mating Preference

It is often stated that mice mate disassortatively with respect to MHC genotype and that such a mating pattern might promote MHC polymorphism (e.g., Partridge, 1988). In practice, evidence for such disassortative mating is quite meager. Certainly, there is evidence that mice and rats can discriminate by olfaction between individuals that apparently differ only at MHC loci (Yamazaki et al., 1979; Singh et al., 1987). Yamazaki et al. (1983) have shown that mice can discriminate by urine odor between congenic individuals homozygous for the K^b allele and those homozygous for the K^{bml} allele. The MHC molecules encoded by alleles K^b and K^{bml} are known to differ only in three amino acid positions. However, showing that mice have the ability to make such discriminations is not the same as showing that they actually use this ability in nature, much less that they use it as a basis for mate choice. Furthermore, individual odors of mice in natural conditions are likely to be influenced by many factors—including protein products of many non-MHC loci, diet, and so forth—in addition to MHC products in the urine. Note also that mating pattern in man is hardly affected by MHC genotypes (Rosenberg et al., 1983). For these reasons, this hypothesis is not very attractive (see Nei & Hughes (1991) for further discussion of this problem).

Overdominant Selection

The overdominance hypothesis has a solid biological basis. Showing that different MHC molecules bind different foreign antigens, Doherty and Zinkernagel (1975) argued that a heterozygote having two different MHC molecules should be more resistant to infectious diseases than a homozygote having one. Since then, many authors have shown that different allelic products differ in their ability to bind and present foreign peptides to T cells in both class I and class II loci (Berzofsky et al., 1979; Buus et al., 1986; Enssle et al., 1987; Sette et al., 1989; Celis & Karr, 1989). Probably because resistance to viral disease is a complex phenomenon, involving not only the class I and class II MHC but also the antibody system, immunologists have so far found relatively few cases where resistance to a natural viral pathogen is correlated with the MHC genotype of the host. The best documented case available is that of resistance to Marek's disease virus, a pathogen of the domestic fowl (Longenecker et al., 1976). In this case, different MHC haplotypes of the fowl show several gradations of resistance to this virus; the B^{21} haplotype is most strongly resistant, several other haplotypes show moderate resistance, while still others are highly susceptible (Longenecker & Mosmann, 1981). In the mouse, it has recently been shown that the $H-29$ haplotype confers resistance to a mouse-acquired immunodeficiency syndrome virus, whereas the *H-26* haplotype is highly susceptible (Hamelin-Bourassa et al., 1989). In humans, statistical associations have been reported between the presence of certain MHC alleles and resistance or susceptibility to a wide variety of infectious agents (Tiwari & Terasaki, 1985). The cheetah, which is virtually monomorphic for the MHC loci, is known to be susceptible to certain pathogens (O'Brien et al., 1985), as is the cottontop tamarin, which also has low MHC polymorphism (Watkins et al., 1988).

As we have seen in Tables 2 and 3, the rate of nonsynonymous nucleotide substitution is enhanced in the ARS apparently because of positive Darwinian selection $(d_N > d_S)$. This enhancement of nonsynonymous substitution is exactly what is expected under overdominant selection (Maruyama & Nei, 1981). Overdominant selection enhances the rate of nonsynonymous (amino acid) substitution, because a new mutant allele is almost always in heterozygous condition and thus enjoys a selective advantage over more common alleles that would exist in both heterozygous and homozygous conditions.

The overdominance hypothesis is also capable of explaining the other features of MHC polymorphism mentioned above. That is, the high heterozygosity *(H)* and the large number of alleles (n_a) observed can easily be explained by this hypothesis, since these values increase as Nv and Ns increase, where N, v, and s are the effective population size, mutation rate per locus, and selective advantage of heterozygotes over homozygotes (Table 4; see also Kimura & Crow, 1964; Wright, 1966; Maruyama & Nei, 1981). For example, when $Nv = 0.01$ and $Ns = 1000$, the expected values $(\bar{H}$ and $\bar{n}_a)$ of H and n_a become 0.95 and 20, respectively, for a sample of 1000 individuals. When $Nv = 0.01$ and $Ns = 10,000$, $\bar{H}$ and $\bar{n}_a$ become 0.98 and 56 (Wright, 1966). If a species is subdivided into subpopulations as in man and mice, $\bar{n}_a$ becomes even higher.

The long-term persistence of polymorphic alleles can also be explained by overdominant selection, as can be seen from a computer simulation presented in Figure 2. This figure shows the phylogeny of different alleles sampled at an evolutionary time (present). All alleles eventually go back to the common ancestral allele. The time at which all the alleles converge to this ancestral allele is called the coalescence time for polymorphic alleles. This coalescence time can be used to obtain an idea of how long polymorphic alleles persist in the population. Figure 2 shows that the coalescence time is about $900N$ generations. As mentioned above, the long-term effective population size for humans is

Table 3. **Mean numbers of nucleotide substitutions per synonymous site (d_S) and per nonsynonymous site (d_N) expressed as percentages between alleles at mouse and human class II MHC loci.**

Locus (No. sequences)	Putative recog. site (ARS) (N=15–20)		Domain 1 excluding ARS (N=64–78)		Domain 2 (N=94)	
	d_S	d_N	d_S	d_N	d_S	d_N
		β Chain loci				
Mouse						
$A\beta1$ (7)	0.0	30.0***	4.0	6.7	7.3	1.3*
$E\beta1$ (4)	11.6	41.5*	1.8	5.2	0.9	0.6
Human						
$DP\beta1$ (3)	3.9	19.0	2.4	2.8	5.3	0.6
$DQ\beta1$ (8)	13.7	26.5	8.5	6.7	5.6	1.6
$DR\beta$ (14)[a]	15.0	45.7**	8.0	4.5	8.3	3.3**
		α Chain loci				
Mouse						
$A\alpha$ (6)	3.2	23.7***	2.8	2.6	7.2	0.7**
Human						
$DQ\alpha1$ (4)	21.7	27.0	8.0	4.3	4.0	2.4

The difference between mean d_S and d_N is significant at 5% level (*), at 1% level (**), or at 0.1% level (***) (see Hughes & Nei, 1989a, for details). N = number of codons compared. [a]$DR\beta1$ and $DR\beta3$ are treated as a single locus. For ARS codons, see Brown et al. (1988).

about 10^4. Since $900N$ generations are about 9×10^4 generations or about 90 million years, there is no problem for explaining the long persistence of MHC polymorphic alleles. Table 4 shows that the long persistence of polymorphic alleles occurs whenever Ns is large and the mutation rate is relatively low.

Earlier we mentioned that in some organisms or populations the extent of MHC polymorphism is very low. In these organisms, however, the long-term effective population size seems also to be quite low. For example, the population size of the cheetah has been estimated to be 1,500 to 25,000 and seems to have experienced several bottlenecks in the past (O'Brien et al., 1985). Therefore, the low level of MHC polymorphism in this organism is most probably caused by the small effective population size. Similarly, the low level of MHC polymorphism in the mouse populations in the North Sea area can be explained by the small effective population size. These observations suggest that the selective advantage (s) of heterozygotes over homozygotes is quite small, probably of the order of $0.01 \sim 0.1$. If the selection coefficient is small, overdominant selection is effective in increasing heterozygosity only in large populations (Table 4).

Table 4. Means of heterozygosity $(\bar{H})$, number of alleles $(\bar{n}_a)$, coalescence time $(\bar{T}_c)$, and rate of amino acid substitution $(\bar{\alpha})$ for neutral mutations $(Ns = 0)$ and overdominant selection $(Ns > 0)$. These results were obtained by computer simulation (Takahata & Nei, 1990).

Nv	Ns	$\bar{H}$	$\bar{n}_a$	$\bar{T}_c$	$\bar{\alpha}$
0.001	0	(0.004)	(1.0)	—	(1.0)
	33	0.698	3.6	387.0	10.8
	100	0.812	5.9	468.0	10.2
	10,000	(0.977)	(49)	—	—
0.01	0	0.056	1.3	1.5	0.8
	33	0.755	4.6	50.4	6.2
	100	0.842	7.2	80.8	7.7
	10,000	(0.980)	(56)	—	—
0.1	0	0.216	2.5	2.4	0.9
	33	0.814	7.3	13.1	3.3
	100	0.877	10.9	19.3	4.3
	10,000	(0.984)	(73)	—	—

N = effective population size. v: mutation rate per locus. s: selective advantage of heterozygotes over homozygotes. $\bar{T}_c$ is given in terms of N generations. $\bar{\alpha}$ is given relative to the expected value for neutral mutations, i.e., $\bar{\alpha} = v$. All values in this table are the means for 20 replications. It is clear that overdominant selection enhances all of $\bar{H}$, $\bar{n}_a$, $\bar{T}_c$, and $\bar{\alpha}$ compared with the case of neutral mutations. The values in parentheses were obtained by analytical formulas; in these cases the simulation was not conducted because of the large computer time required. Sample size was assumed to be 100 individuals except for the cases where $\bar{H}$ and $\bar{n}_a$ were determined analytically; in these cases sample size was assumed to be 1000.

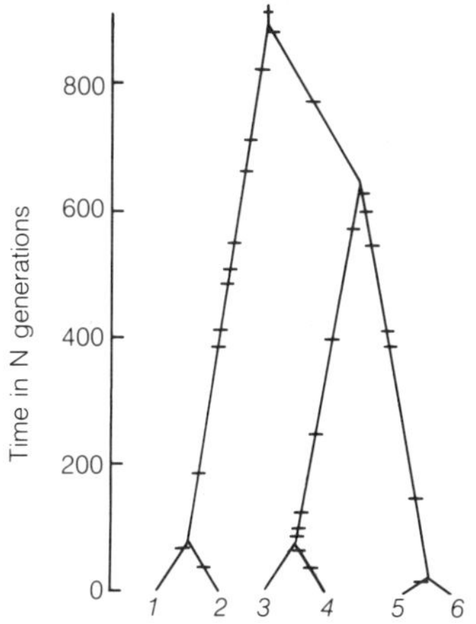

Figure 2. Allelic genealogy for overdominant alleles. This genealogy was obtained by computer simulation (Takahata & Nei, 1990). The $-$ symbol denotes the occurrence of a mutation. $Nv = 0.001$ and $Ns = 100$ were assumed.

Frequency-Dependent Selection

The hypothesis of frequency-dependent selection has been advocated by many authors during the last two decades. Surprisingly, however, this hypothesis is ill-defined, and the arguments for the hypothesis are highly speculative. The most popular model of frequency-dependent selection is to assume that host individuals carrying a recently arisen mutant allele have a selective advantage because pathogens will not have had time to evolve the ability to infect host cells carrying a new mutant antigen (Snell, 1968; Bodmer, 1972). This will result in a constant turnover of alleles in the population because old alleles lose resistance to pathogens. While this model generates a higher value of d_N than d_S, it can explain neither the high degree of polymorphism nor the long-persisting polymorphism at the MHC loci (Hughes & Nei, 1988; Takahata & Nei, 1990).

There are some other forms of frequency-dependent selection that potentially maintain a high degree of polymorphism. Mathematically, one can show that selection due to minority advantage can enhance the extent of polymorphism as much as overdominant selection (Takahata & Nei, 1990). The only problem with this model is that there is no evidence that this type of selection really occurs.

Nevertheless, it seems difficult to reject the hypothesis of frequency-dependent selection, particularly minority advantage, at present, because experimental data on this issue are too few and because overdominant selection and minority advantage often produce similar population dynamics of genes. It is hoped that in the future more experimental or statistical studies will be conducted to distinguish between the hypotheses of overdominant selection and frequency-dependent selection. One such study would be to examine the fitness of a homozygote for a rare allele at a locus. If the frequency-dependent selection hypothesis is correct, this genotype should have a higher fitness than a heterozygote for a pair of common alleles. By contrast, if the overdominance hypothesis is correct, the reverse would be true.

ACKNOWLEDGMENT

This study was supported by research grants from the National Institutes of Health and the National Science Foundation.

LITERATURE CITED

Bailey, D. W. & H. I. Kohn. 1965. Inherited histocompatibility changes in progeny of irradiated and unirradiated inbred mice. *Genet. Res.* 6:330–340.

Berzofsky, J. A., Richman, L. K. & D. J. Killion. 1979. Distinct *H-2*-linked *Ir* genes control both antibody and T cell responses to different determinants on the same antigen, myoglobin. *Proc. Natl. Acad. Sci. USA* 76:4046–4050.

Bjorkman, P. J., Saper, M. A., Samraoui, B., Bennett, W. S., Strominger, J. L. & D. C. Wiley. 1987a. Structure of the human class I histocompatibility antigen, *HLA-A2. Nature* 329:506–512.

Bjorkman, P. J., Saper, M. A., Samraoui, B., Bennett, W. S., Strominger, J. L. & D. C. Wiley. 1987b. The foreign antigen binding site and T cell recognition regions of class I histocompatibility antigens. *Nature* 329:512–518.

Bodmer, W. 1972. Evolutionary significance of the *HL-A* system. *Nature* 237:139–145.

Brown, J. H., Jardetzky, T., Saper, M. A., Samraoui, B., Bjorkman, P. J. & D. C. Wiley. 1988. A hypothetical model of the foreign antigen binding site of class II histocompatibility molecules. *Nature* 322:845–850.

Buus, S., Colon, S., Smith, C., Freed, J. H., Miles, C. & H. M. Grey. 1986. Interaction between a "processed" ovalbumin peptide and Ia molecule. *Proc. Natl. Acad. Sci. USA* 83:3968–3971.

Celis, E. & R. W. Karr. 1989. Presentation of an immunodominant T-cell epitope of hepatitis B surface antigen by the *HLA-DPw4* molecule. *J. Virol.* 63:747–752.

Chardon, P., Vaiman, M., Kirszenbaum, M., Geffrotin, C., Renard, C. & D. Cohen. 1985. Restriction fragment length polymorphism of the major histocompatibility complex of the pig. *Immunogenetics* 21:161–171.

Clarke, B. & D. R. S. Kirby. 1966. Maintenance of histocompatibility polymorphisms. *Nature* 211:999–1000.

Doherty, P. C. & R. M. Zinkernagel. 1975. Enhanced immunological surveillance in mice heterozygous at the *H-2* gene complex. *Nature* 256:50–52.

Enssle, K.-H., Wagner, H. & B. Fleischer. 1987. Human mumps virus-specific cytotoxic T lymphocytes: quantitative analysis of *HLA* restriction. *Human Immunol.* 18:135–149.

Fan, W., Kasahara, M., Gutknecht, J., Klein, D., Mayer, W. E., Jonker, M. & J. Klein. 1989. Shared class II MHC polymorphisms between humans and chimpanzees. *Human Immunol.* 26:107–121.

Figueroa, F., Günther, E. & J. Klein. 1988. Mhc polymorphism pre-dating speciation. *Nature* 335:265–267.

Figueroa, F., Tichy, H., Berry, R. J. & J. Klein. 1986. MHC polymorphism in island populations of mice. *Curr. Top. Microbiol. Immunol.* 127:100–105.

Gill, T. J. 1983. Immunogenetics of spontaneous abortion in humans. *Transplantation* 35:1–6.

Hamelin-Bourassa, D., Skamene, E. & F. Gerrais. 1989. Susceptibility to a mouse acquired immunodeficiency syndrome is influenced by the *H-2*. *Immunogenetics* 30:266–272.

Hamilton, B. L., Hamilton, A. & M. S. Hamilton. 1985. Maternal-fetal disparity at multiple minor histocompatibility loci affects the weight of the feto-placental unit in mice. *J. Reprod. Immunol.* 8:257–261.

Hayashida, H. & T. Miyata. 1983. Unusual evolutionary conservation and frequent DNA segment exchange in class I genes of the major histocompatibility complex. *Proc. Natl. Acad. Sci. USA* 80:2671–2675.

Hedrick, P. W. & G. Thomson. 1988. Maternal-fetal interactions and the maintenance of *HLA* polymorphism. *Genetics* 119:205–212.

Holmes, N. & P. Parham. 1985. Exon shuffling in vivo can generate novel *HLA* class I molecules. *EMBO J.* 4:2849–2854.

Hughes, A. L. & M. Nei. 1988. Pattern of nucleotide substitution at major histocompatibility complex class I loci reveals overdominant selection. *Nature* 335:167–170.

Hughes, A. L. & M. Nei. 1989a. Nucleotide substitution at major histocompatibility complex class II loci: evidence for overdominant selection. *Proc. Natl. Acad. Sci. USA* 86:958–962.

Hughes, A. L. & M. Nei. 1989b. Ancient interlocus exon exchange in the history of the *HLA-A* locus. *Genetics* 122:681–686.

James, D. A. 1965. Effects of antigenic dissimilarity between mother and foetus on placental size in mice. *Nature* 205:613–614.

Kimura, M. & J. F. Crow. 1964. The number of alleles that can be maintained in a finite population. *Genetics* 49:725–738.

Klein, J. 1986. *Natural History of the Major Histocompatibility Complex.* Wiley: New York.

Klein, J. 1987. Origin of major histocompatibility complex polymorphism: the trans-species hypothesis. *Human Immunol.* 19:155–162.

Klein, J., Kasahara, M., Gutkneckt, J. & F. Figueroa. 1990. Origin and function of Mhc polymorphism. *Chem. Immunol.* 49:35–50.

Lawlor, D. A., Ward, F. E., Ennis, P. D., Jackson, A. P. & P. Parham. 1988. *HLA-A* and *B* polymorphisms predate the divergence of humans and chimpanzees. *Nature* 335:268–271.

Lawlor, D. A., Zemmour, J., Ennis, P. D. & P. Parham. 1990. Evolution of class I MHC genes and proteins: from natural selection to thymic selection. *Ann. Rev. Immunol.* 8:22–63.

Li, W.-H., Tanimura, M. & P. M. Sharp. 1987. An evaluation of the molecular clock hypothesis using mammalian DNA sequences. *J. Mol. Evol.* 25:330–342.

Longenecker, B. M. & T. R. Mosmann. 1981. Structure and properties of the major histocompatibility complex of the chicken: speculations on the advantages and evolution of polymorphism. *Immunogenetics* 13:1–23.

Longenecker, B M., Pazderka, F., Gavora, J. S., Spencer, J. L. & R. F. Ruth. 1976. Lymphoma induced by herpes virus: resistance associated with a major histocompatibility gene. *Immunogenetics* 3:401–407.

Lopez de Castro, J. A., Strominger, J. L., Strong, D. M. & H. T. Orr. 1982. Structure of crossreactive human histocompatibility antigens *HLA-A28* and *HLA-A2*: possible implications for the generation of *HLA* polymorphism. *Proc. Natl. Acad. Sci. USA* 79:3813–3817.

Maruyama, T. & M. Nei. 1981. Genetic variability maintained by mutation and overdominant selection in finite populations. *Genetics* 98:441–459.

Mayer, W. E., Jonker, M., Klein, D., Ivanyi, P., van Seventer, G. & J. Klein. 1988. Nucleotide sequences of chimpanzee MHC class I alleles: evidence for trans-species mode of evolution. *EMBO J.* 7:2765–2774.

McConnell, T. L., Talbot, W. S., McIndoe, R. A. & E. K. Wakeland. 1988. The origin of MHC class II gene polymorphism within the genus *Mus. Nature* 332:651–654.

McGuire, K. L., Duncan, W. R. & P. W. Tucker. 1985. Syrian hamster DNA shows limited polymorphism at class I-like loci. *Immunogenetics* 22:257–268.

Miyamoto, M. M., Slightom, J. L. & M. Goodman. 1987. Phylogenetic relations of humans and African apes from DNA sequences in the $\psi\eta$-globin region. *Science* 238:369–373.

Nei, M. 1987. *Molecular Evolutionary Genetics.* Columbia University Press: New York.

Nei, M. & T. Gojobori. 1986. Simple methods for estimating the numbers of synonymous and nonsynonymous nucleotide substitutions. *Mol. Biol. Evol.* 3:418–426.

Nei, M. & D. Graur. 1984. Extent of protein polymorphism and the neutral mutation theory. *Evol. Biol.* 17:73–118.

Nei, M. & A. L. Hughes. 1991. Polymorphism and evolution of the major histocompatibility complex loci in mammals. Pp. 222–247. *In:* R. K. Selander, T. Whittam & A. G. Clark (eds.), *Evolution at the Molecular Level.* Sinauer Associates: Sunderland, MA.

Nizetic, D., Figueroa, F., Nevo, E. & J. Klein. 1985. Major histocompatibility complex of the mole-rat. *Immunogenetics* 25:55–67.

Nizetic, D., Stevanovic, M., Soldatovic, B., Savic, I. & R. Crkvenjakov. 1988. Limited polymorphism of both classes of MHC genes in four different species of the Balkan mole rat. *Immunogenetics* 28:91–98.

O'Brien, S. J., Roelke, M. E., Marker, L., Newman, A., Winkler, C. A., Meltzer, D., Colly, L., Greumann, J. F., Bush, M. & D. E. Wildt. 1985. Genetic basis for species vulnerability in the cheetah. *Science* 227:1428–1434.

Palm, J. 1974. Maternal-fetal histocompatibility in rats: an escape from adversity. *Cancer Research* 34:2061–2065.

Palmer, M., Wettstein, P. J. & J. A. Frelinger. 1983. Evidence for extensive polymorphism of class I genes in the rat major histocompatibility complex *(RT1). Proc. Natl. Acad. Sci. USA* 80:7616–7620.

Partridge, L. 1988. The rare-male effect: what is its evolutionary significance? *Phil. Trans. R. Soc. Lond.* B 319:525–539.

Rosenberg, L. T., Cooperman, D. & R. Payne. 1983. *HLA* and mate selection. *Immunogenetics* 17:89–93.

Roychoudhury, A. & M. Nei. 1988. *Human Polymorphic Genes: World Distribution.* Oxford University Press: New York.

Schacter, B., Weitkamp, L. R. & W. E. Johnson. 1984. Parental *HLA* compatibility, fetal wastage, and neural tube defects: evidence for a *T/t*-like locus in humans. *Am. J. Hum. Genet.* 36:1082–1091.

Sette, A., Buus, S., Appella, E., Smith, J. A., Chestnut, R., Miles, C., Colon, S. M. & H. M. Grey. 1989. Prediction of major histocompatibility complex binding regions of protein antigens by sequence pattern analysis. *Proc. Natl. Acad. Sci. USA* 86:3296–3300.

Simonsen, M., Crone, M., Koch, C. & H. Hála. 1982. The MHC haplotypes of the chicken. *Immunogenetics* 16:513–532.

Singh, P. B., Brown, R. E. & B. Roser. 1987. MHC antigens in urine as olfactory recognition cues. *Nature* 327:161–164.

Snell, G. D. 1968. The *H-2* locus of the mouse: observations and speculations concerning its comparative genetics and polymorphism. *Folia Biologica* 14:335–358.

Streilein, W. 1987. Studies on an MHC exhibiting limited polymorphism. Pp. 379–395. *In:* G. Kelsoe & D. H. Schulze (eds.), *Evolution and Vertebrate Immunity.* University of Texas Press: Austin.

Takahata, N. & M. Nei. 1990. Allelic genealogy under overdominant and frequency-dependent selection and polymorphism of major histocompatibility complex loci. *Genetics* 124:967–978.

Tiwari, J. L. & P. I. Terasaki. 1985. *HLA and Disease Associations.* Springer-Verlag: New York.

Trowsdale, J., Groves, V. & A. Arnason. 1989. Limited MHC polymorphism in whales. *Immunogenetics* 29:19–24.

Warner, C., Gerndt, B., Xu, Y., Bourlet, Y., Auffray, C., Lamont, S. & A. Nordskog. 1989. Restriction fragment length polymorphism analysis of major histocompatibility complex class II genes from inbred chicken lines. *Anim. Genet.* 20:225–231.

Watkins, D. I., Hodi, F. S. & N. L. Letvin. 1988. A primate species with limited major histocompatibility complex class I polymorphism. *Proc. Natl. Acad. Sci. USA* 85:7714–7718.

Weiss, E. H., Mellor, A. L., Golden, L., Fahrner, K., Simpson, E., Hurst, J. & R. A. Flavell. 1983. The structure of a mutant *H-2* gene suggests that the generation of polymorphism in *H-2* genes may occur by gene conversion-like events. *Nature* 301:671–674.

Winkler, C., Schultz, A., Cevario, S. & S. O'Brien. 1989. Genetic characterization of *FLA*, the cat major histocompatibility complex. *Proc. Natl. Acad. Sci. USA* 86:943–947.

Wright, S. 1966. Polyallelic random drift in relation to evolution. *Proc. Natl. Acad. Sci. USA* 55:1074–1081.

Yamazaki, K., Beauchamp, G. K., Egorov, I. K., Bard, J., Thomas, L. & E. A. Boyse. 1983. Sensory distinction between *H-2^b* and *H-2^{bml}* mutant mice. *Proc. Natl. Acad. Sci. USA* 80:5685–5688.

Yamazaki, K., Boyse, E. A., Miké, V., Thaler, H. T., Mathieson, B. J., Abbott, J., Boyse, J., Zayas, Z. A. & L. Thomas. 1976. Control of mating preferences in mice by genes in the major histocompatibility complex. *J. Exp. Med.* 144:1324–1335.

Yamazaki, K., Yamaguchi, M., Baranoski, L., Bard, J., Boyse, E. A. & L. Thomas. 1979. Recognition among mice: evidence from the use of a Y-maze differentially scented by congenic mice of different major histocompatibility types. *J. Exp. Med.* 150:755–760.

Yuhki, N. & S. J. O'Brien. 1988. Molecular characterization and genetic mapping of class I and class II MHC genes of the domestic cat. *Immunogenetics* 27:414–425.

Generation of Allelic Polymorphism at the HLA Class II DRB Loci of Primates: Intra-exon Exchange Within and Between Loci

Ulf B. Gyllensten, Mats Sundvall, and Henry A. Erlich

Abstract. The evolution of polymorphism at loci encoding the β-chains of the major histocompatibility complex class II DR antigens in primates was studied by DNA amplification. Phylogenetic analysis was performed on 63 nonhuman primate DRB sequences and 53 human sequences, and indicates the presence of five DRB loci in primates. Segments encoding the β-sheet and the α-helix of the first domain, separated by a putative recombination signal located within the DRB second exon, appear to have different evolutionary histories. The polymorphisms in the β-pleated sheet appear to reflect the ancestral relationships among DRB alleles, while the polymorphisms in the α-helix may have been recombined into the framework of ancestral DRB alleles. Thus, the polymorphisms in the α-helical domain could be of more recent origin and under selection for variability, while polymorphisms in the α-sheet are older and conserved by selection.

INTRODUCTION

The HLA-D or class II region on chromosome 6 consists of the three subregions HLA-DR, DQ and DP, each of which encodes one α- and at least one β-chain (1,2). These chains form a highly polymorphic integral membrane protein that binds peptide fragments derived from processed antigens (Ag). The peptide fragments are thought to be located within an antigen binding cleft formed by two α-helices and a β-pleated sheet in the N-terminal outer domain of the heterodimer (3,4). The HLA-Ag complex is recognized by the T-cell receptor, leading to T lymphocyte activation. Susceptibility to a number of autoimmune diseases has been found to be associated with specific alleles of the DQB1 and DRB1 loci in this region (5–11). Analyses of disease susceptibility (5–11), as well as peptide binding studies (12) have implicated specific polymorphic residues in the α-helical domain as being functionally important.

The loci encoding these antigens are among the most polymorphic genes in higher organisms, as revealed by immunologic and molecular analyses (1–6). Most of the allelic sequence diversity resides in the second exon, which encodes the amino terminal domain of the antigen, and is characterized by: a) a large number of alleles per locus, b) a large diversity among the alleles at a locus, with individual alleles differing at numerous polymorphic positions including short deletions. c) a distribution of allele frequencies that is significantly different from that of most nuclear genes, with many alleles at intermediate

Drs. Gyllensten and Sundvall are with the Department of Medical Genetics, Biomedical Center, University of Uppsala, Box 589, S-751 23 Uppsala, Sweden. Dr. Erlich is with the Department of Human Genetics, Cetus Corporation, 1400 Fifty-Third Street, Emeryville, CA 94608, USA.

frequencies, rather than the J-shaped allele frequency distribution characteristic of most polymorphic genes, and d) loci encoding β-chains consistently being more polymorphic than loci encoding α-chains.

The origin and maintenance of this extensive allelic diversity has been the subject of considerable controversy. The diversity is generally believed to be maintained by selection (13) although some argue that the polymorphism is neutral (14–17). Two hypotheses have been advanced to explain the origin of the sequence polymorphism; either the polymorphism could have been generated during the lifetime of the human species or, alternatively, it predates speciation and has been maintained in the contemporary human population by selection (18). Recently, allelic types at HLA-DQA1 (19), DQB1 (20) and the DRB (21–23) loci in primates have been shown to predate the divergence of hominoids, and in some cases even that of Old World monkeys, indicating that the polymorphism at some class II loci has been maintained over 5–20 million years. Similarly, the polymorphism at class I loci has been shown to predate speciation, both in primates and rodents (24–26). The maintenance of allelic types over such time spans is inconsistent with the polymorphism being neutral (19,20). In fact, analyses have shown that individual allelic types at the DQA1 and DQB1 loci accumulate change at different rates, presumably as an effect of relative differences in selection pressure (19). Thus, the evolutionary studies provide strong support for selection acting to maintain the polymorphism as well as for an ancient origin of the polymorphism.

A more controversial issue is the mechanism responsible for generating the allelic diversity. It has been suggested that the extensive polymorphism has been generated by gene conversion or segmental transfer between alleles (27). However, if new alleles were generated by combining segments of different preexisting alleles, and this process was rapid relative to the accumulation of point mutations, the similarity of alleles within species would increase and the similarity of allelic types between species would eventually disappear. Thus, one would predict that if allelic types were generated mainly by segmental transfers, they would not be conserved over evolutionary periods of time. The conservation of allelic types at DQA1 and DQB1 over 5–20 million years thus does not suggest that segmental transfers are a main source of allelic diversity at these loci. However, two out of the 28 DQA1 sequences examined could have been generated by intra-exom recombination (19), indicating that these events may occur, but only at a low frequency.

The patchwork pattern of allelic diversity of other HLA class II loci has been taken to indicate that allelic diversity is generated *mainly* by recombination, or segmental transfer. For instance, the similarity between HLA-DRB1 and DRB5 sequences on the DR2 haplotype was postulated to be due to a reciprocal exchange (28), and, similarly, sequences from the second exon of HLA-DRB3 on a DRw6 haplotype were suggested to have been donated by gene conversion to the DRB1 locus (29). However, the presence of a shared polymorphism between alleles could, in principle, be due to common ancestry, to convergent evolution, or to sequence exchanges between alleles; these alternative explanations can be evaluated when the age of the polymorphism is known. To obtain a more detailed understanding of the antiquity of the allelic polymorphism at the DRB1–5 loci as well as the mechanisms for generation of the polymorphism, we analyzed second exon nucleotide sequences in primates by enzymatic amplication (PCR), using oligonucleotide primers known to amplify all HLA-DRB genes (30).

EVOLUTIONARY ANALYSIS USING PCR

The polymerase chain reaction (PCR, 31–33) provides the most powerful technique so far developed for studying the evolutionary process of closely related taxa at the DNA level. Oligonucleotide primers based on the sequence of regions of the DNA that are con-

892

served or evolving at a normal rate has successfully been applied to species that have been isolated for 40–80 million years. Assuming a rate of neutral substitutions of 10^{-9} per basepair per year, two 40 bp segments (the average length of two PCR primers) derived from a common ancestral sequence will accumulate on the average 1 silent difference in $1/(10^{-9} \times 40) = 25$ million years. This corresponds to an average of a total of one change in about 6 million years (including the replacement changes). Two random 20 bp sequences can therefore be expected to accumulate a total of about 5 changes per 30 million years, assuming no selection. Most of these mutations will be compatible with PCR, with the exception of some nucleotide changes in the 3' end of the oligonucleotide primers, which could prevent the DNA polymerase from extending the template-primer complex. Since all templates synthesized in the PCR will contain sequences perfectly matched to the priming oligonucleotides, mismatches between the primer and the template will be eliminated after the first cycle of PCR. Primers based on human Mhc sequences will, in our hands, amplify the homologous gene in species that have been separated from 30–40 million years (DQA1, DPB1) to 60–80 million years (DQB1, DRB1) using the reaction conditions found to be optimal for amplification of human genes. However, the priming sites in some of these genes are in conserved regions of the coding sequences (exon-intron borders of exon 2) and may be accumulating changes slower than the neutral rate due to conserving selection.

With PCR primers designed to amplify the polymorphic second exon of the HLA DRB1–5 loci (30) we have studied the homologous DNA segments from regular chimpanzee (*Pan troglodytes*), pygmy chimpanzee (*Pan paniscus*), gorilla (*Gorilla gorilla*), baboon (*Papio leucophaeus*) and rhesus (*Macaca mulatta*) (22,23). The approximate divergence times from the human lineage of these species rise gradually, from approximately 5 million years for *Pan* and *Gorilla* to nearly 20 million years for *Papio* and *Macaca* (34–36). Subsequent to the PCR, we have employed a denaturating gradient gel system (37,38) (Figure 1) to (a) score individuals as homo- and heterozygous and (b) purify preparatively the individual allelic templates. Amino acid sequences of individual alleles were deduced from the nucleotide sequence of cloned PCR products, or from PCR fragments that were gel-purified and sequenced directly using a procedure for generating single-stranded DNA in the PCR (39).

The similarity of nucleotide sequences was studied by parsimony analysis (40,41), using both the PHYLIP program package (Joe Felsenstein, University of Washington, Seattle, WA) and the PAUP program (David Swofford, Illinois Natural History Survey, IL).

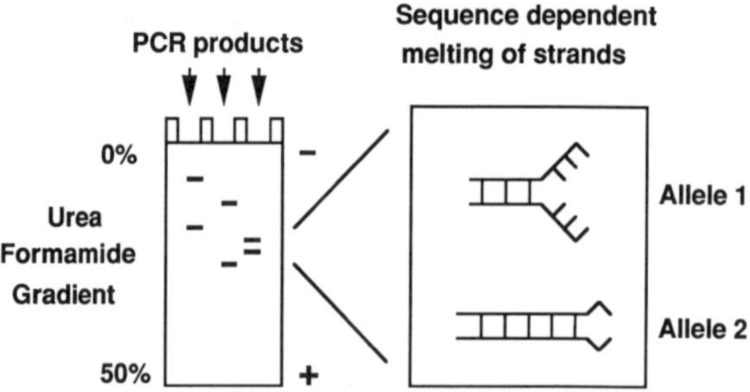

Figure 1. Experimental procedure for analysis of the allelic spectrum at primate Mhc loci. Separation of PCR amplified allelic variants by denaturating gradient electrophoresis. a) The principle for separation of allelic templates in a gradient of increasing denaturant (urea and formamide). The PCR products are electrophoresed into a linear gradient until they melt partially, and due to the formation of trifurcated DNA molecules slow down in migration. Since the melting point is determined by the nucleotide composition of the DNA molecules, alleles will start to melt at different points in the gradient.

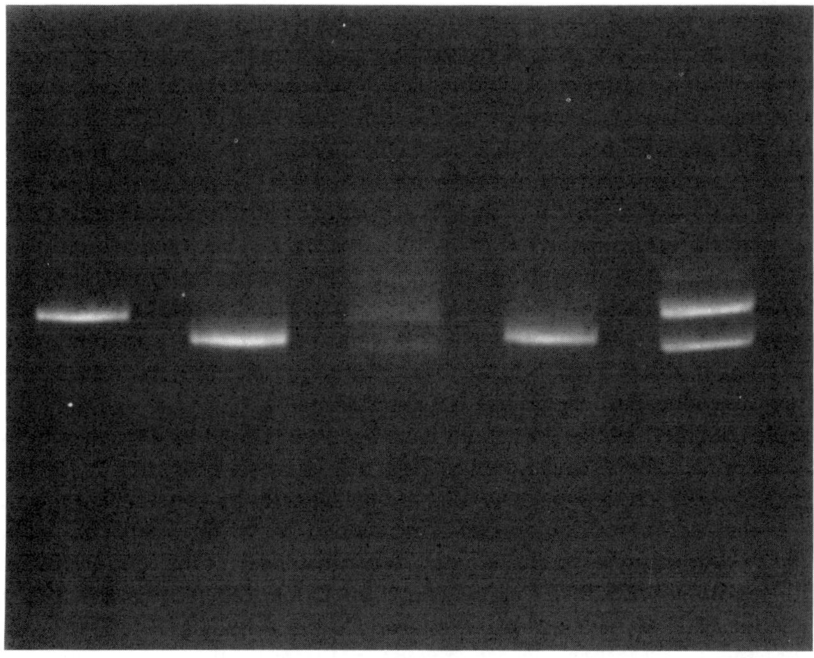

Figure 1b. Separation of PCR amplified allelic variants of the HLA-DQA locus by denaturating gradient gel electrophoresis. PCR products from each of five reactions were loaded onto a 20×20 cm, 12.5% polyacrylamide gel carrying a linear, vertical gradient of between 25–60% denaturant. The gel was submerged in buffer heated to 60°C, and run at 100mA, 150V, for 7 hr. The DNA was visualized by staining with ethidium bromide. The genotypes of the five samples are given above the lanes. The third band in the heterozygote individuals (c,e) is a heteroduplex formed by annealing one strand from each of the two alleles.

GENERATION OF ALLELIC DIVERSITY AT THE DRB1 LOCUS

A total of 39 human DRB1 alleles and 21 alleles obtained by PCR amplification from nonhuman primates was used to study the mechanisms for generating polymorphism at this locus (22). In general, phylogenetic analysis is based on the assumption that sequence similarities reflect ancestral relationships. Similarities due to gene conversion (segmental exchange) can yield branching patterns which are not related to phylogeny. If sequences from one part of the exon can be shown to have a branching order similar to that obtained by independent analysis, such as RFLP analysis or serological typing, this tree is likely to reflect phylogenetic relationships. A different branching order for a second fragment would suggest that this segment was inserted by, for example, gene conversion into the sequences whose phylogeny is revealed by the tree for the first segment. Thus, lack of consistency in the branching order derived from sequences from different parts of the exon is a strong indication of recombinational mechanisms in generating the contemporary allelic sequences. To test whether different parts of the second exon of the DRB1 locus reflect the same evolutionary history, phylogenetic analysis of primate sequences was therefore performed on two discrete parts of the exon.

894

Although multiple alternative breakpoints in the exon can be postulated, it has been noted (28) that codons 51–55 have a conserved nucleotide sequence (AC)G GAG CTG GGG GGG, which differs at only 3 out of 13 nucleotides from the consensus χ-like recombination sequence observed in the minisatellite 33.15 (GGAGGTGGGCAGG) (42), and could represent a general breakpoint for exchange of sequence elements among alleles. The structural model for the class II molecule suggests that this position represents a transition point between sequences encoding the β-pleated sheet and those encoding the α-helix of the first domain (4). This sequence is conserved in all HLA-DRB genes as well as all nonhuman primate DRB sequences, except DRB2 (22,23). In addition, the observation of two such χ-like sequences in the second exon of the DPB1 locus (43,44) at positions flanking the α-helix, as well as in class I loci, is consistent with this hypothesis (23). While a number of other breakpoints in the exon are possible, i.e., between sequences encoding different β-strands, our analysis (below) indicates that gene conversion (segmental exchange) at a single point, without recombination of flanking markers, could account for the diversity of many DRB1 alleles. This does not preclude the possibility of exchange within the β-sheet, as has been indicated for some sequences (45). However, these events may be less frequent than exchanges between domains.

Examples of DRB1 alleles that could have been generated by exchange between two other contemporary alleles are shown in Figure 2. The possibility that these alleles were created by sequence exchange was evaluated statistically by considering the number of positions in each of the two domains favoring each of the three possible branching orders of the putative donor and recipient sequences. For instance, for the *0404, *1302 and *1402 alleles the topology with *0404 as an outgroup is favored for the β-sheet, while the topology with *1302 as an outgroup is favored for the α-helix (Table 1). For each of the examples tested the hypothesis that the three topologies for the α-sheet are equally likely can be statistically rejected and, similarly, in four out of the five cases a particular topology is favored for the α-helix (Table 1). Since the trees favored are different for the two domains, sequences encoding the β-pleated sheet and those encoding the α-helix must reflect different evolutionary histories.

This result is not likely to be due to clustering of several mutations in the same codons. Analysis of the third positions in the codons, that are independent of each other and not subjected to the same selective constraint, similarly favor different topologies for the two different domains (Table 1).

In the phylogenetic trees for the β-pleated sheet of all the DRB1 alleles, sequences cluster according to DR specificities (Figure 3a). This pattern corresponds to the

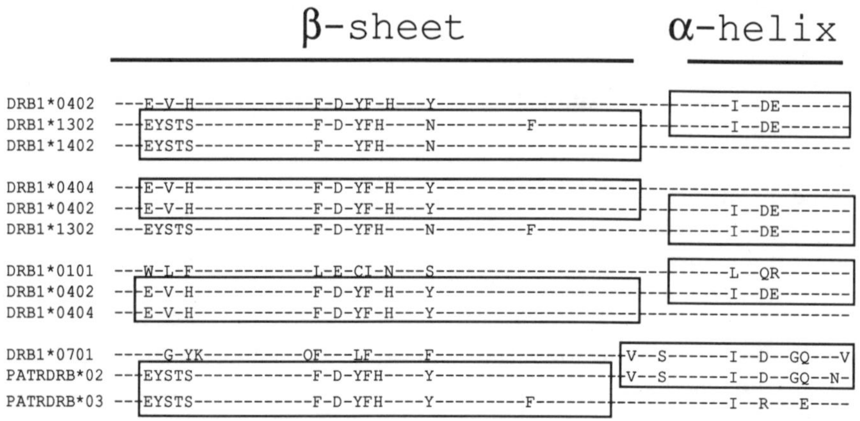

Figure 2. Putative recombination events in the evolution of DRB1 alleles.

serological "public" specificities DRw52 (associated with the haplotypes DR3, 5, 6 and 8) and DRw53 (associated with the haplotypes DR4, 7, and 9). The branching order of the β-sheet sequences is also consistent with the RFLP patterns for the DR haplotypes (46) as well as with the phylogeny based on human third exon sequences (22,47). By contrast, in the tree based on codons 54–78, encoding the α-helix (Figure 3b), the DR specificities are mixed and there is a tendency for sequences to cluster by species.

Table 1. Statistical comparison of the distribution of sites in the β-sheet and α-helix favoring each of three possible branching orders of three DRB1 sequences. The numbers to the left of the dash represent total number of nucleotide differences and the number to the right differences in the third position of the codons.

Sequences compared	Domain	Topology (A+B) vs C	(A+C) vs B	(B+C) vs A	Probability (a)
A. DRB1*0404					
B. DRB1*1302	β-sheet	1/1	1/0	13/4	0.01 / >0.05
C. DRB1*1402	α-helix	0/0	6/2	0/0	0.05 / >0.05
A. DRB1*0402					
B. DRB1*0404	β-sheet	14/4	0/0	0/0	0.001 / >0.05
C. DRB1*1302	α-helix	0/0	6/0	0/0	0.05 / —
A. DRB1*0101					
B. DRB1*0402	β-sheet	0/0	0/0	14/5	0.001 / 0.05
C. DRB1*0404	α-helix	0/0	6/2	0/0	0.05 / >0.05
A. DRB1*0101					
B. DRB1*0103	β-sheet	14/5	0/0	0/0	0.001 / 0.05
C. DRB1*0402	α-helix	0/0	0/0	6/2	0.05 / >0.05
A. DRB1*0701					
B. PATRDRB*02	β-sheet	1/0	0/0	18/0	0.001 / —
C. PATRDRB*03	α-helix	9/3	2/1	3/1	>0.05 / >0.05

(a) Binomial probabilities.

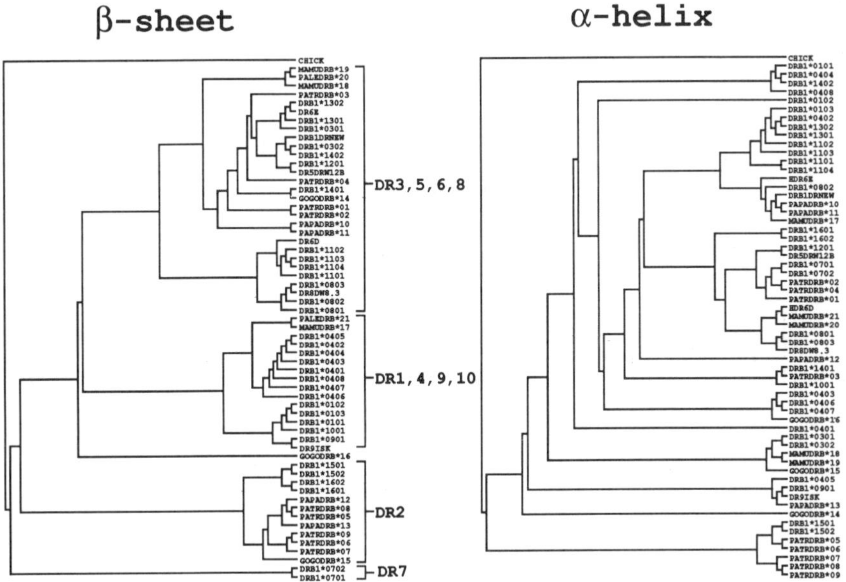

Figure 3. Parsimony trees based on the nucleotide sequences of the second exon of DRB1. Separate trees were generated for codons 6–53 (a), encoding the β-pleated sheet and codons 54–78 (b) encoding the α-helix. Branch lengths are not proportional to the amount of change. Phylogenetic trees were constructed as described in (22).

Thus, allelic diversity at DRB1 appears to be generated in part by combining the left part (encoding the β-pleated sheet) and right part (encoding the α-helix) of different ancestral alleles. However, the similarity among alleles in the β-sheet could, in principle, be due to convergent evolution resulting in similar amino acid sequences of alleles with different evolutionary origins (48). To evaluate this possibility, phylogenetic trees based on the nucleotide sequence variation at the third position of the codons were constructed. Comparison of the topology of the trees for the β-sheet, based on the third position of the codons, was found to be significantly different from the tree based on all codon positions in the α-helix domain, but not from the tree based on all codon positions in the β-pleated sheet (22). This indicates that the β-sheet tree reflects ancestral relationships rather than convergent evolution.

To determine the extent of the DRB1 locus participating in the sequence exchanges, phylogenetic trees for the second and third exon sequences were compared. Human sequences for the third exon are only available for some haplotypes (47,49). For the second exon, separate trees were generated for the β-pleated sheet (Figure 4a), and the α-helix (Figure 4b). Comparisons indicate no significant difference in topology between the tree based on the third exon (Figure 4c) and that based on the region of the second exon encoding the β-pleated sheet. However, the topology of the third exon is significantly different from that of the second exon region encoding the α-helix (22). Thus, we conclude that the 5' part of the second exon and the third exon appear to reflect a similar evolutionary history, while sequences in the 3' part of the second exon have changed independently of the surrounding exons.

The results thus indicate that the two functional domains of the second exon have different evolutionary histories. Sequence motifs in the part of the exon encoding the β-pleated sheet (e.g., EYSTS in Figure 2) appear to reflect a common ancestry of that subset of sequences (DR3, 5, and 6), while shared sequences encoding the α-helix (e.g., I-DE in Figure 2) may have been recombined into the framework of different alleles by segmental transfer after the divergence of ancestral allelic types (or lineages). The β-sheet epitope EYSTS, located on the first β-strand, is present in all primate species examined, suggesting that it is more than 20 myr old (22). By contrast, the polymorphic epitopes of the α-helical domain, particularly those of the third hypervariable region (codons 68–72), are not generally shared among different species.

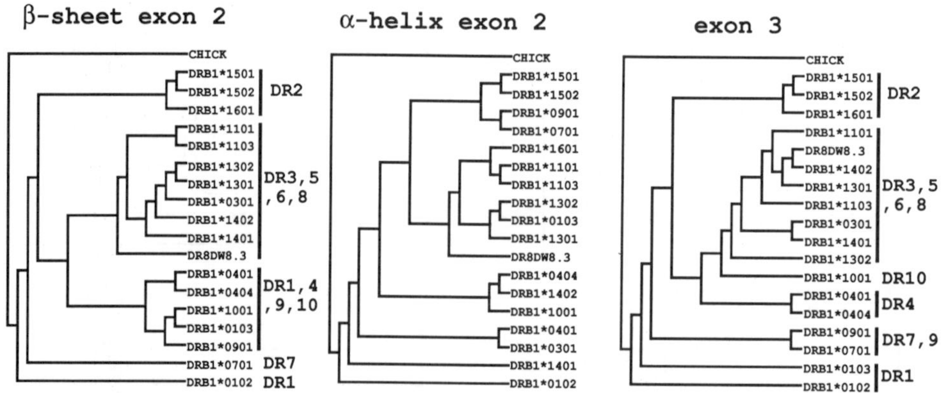

Figure 4. Parsimony trees for human nucleotide sequences from the third exon (14) (a) and the two parts of the second exon of the DRB1 locus, encoding the β-pleated sheet and the α-helix. Separate trees were generated for codons 6–53 (a), encoding the β-pleated sheet and codons 54–78 (b) encoding the α-helix of the second exon. Branch lengths are not proportional to the amount of change.

GENETIC EXCHANGES AMONG DRB LOCI

In contrast to the DQ and DP regions with only one expressed β-chain locus and one pseudogene each, in the DR region human haplotypes have variable numbers of β-chain loci. Most haplotypes contain two expressed β-chain loci, and, in addition, one or several presumed pseudogenes (2,22,47).

To examine the extent of the putative intra-exon exchange between DRB loci, separate phylogenetic trees were generated for the β-sheet and the α-helix of all 63 nonhuman primate sequences and, in addition, of 53 human sequences from the DRB1–5 loci (23). The pattern of polymorphism in some of the DRB alleles suggests that they could have been generated by exchange between two different DRB loci (Figure 5). The sequences of most of these alleles are consistent with the point of exchange being close to the transition between the β-sheet and the α-helix (23). The possibility that these alleles were created by sequence exchange was evaluated, similar as for DRB1 sequences, by considering the number of positioins in each of the two domains favoring each particular branching order of the putative donor and recipient sequences. For example, for the DRB1*0404 allele, representing a putative exchange between a *0402-like β-sheet and a *1302-like α-helix, *1302 is favored as an outgroup for the β-sheet and *0402 as an outgroup for the α-helix (Table 2). Similarly, for the PATRDRB*01, *03 and *55 alleles the topology with *55 as an outgroup is favored for the β-sheet, while the topology with *01 as an outgroup is favored for the α-helix (Table 2). For all of the tested DRB alleles, the hypothesis that the three topologies for the β-sheet are equally likely can be statistically rejected. The topologies favored for the α-helix were, in five out of the six cases discussed, different than that for the β-sheet, although the three topologies were only statistically different in one of these tests.

The extent of this putative intra-exon exchange within and between DRB loci was then studied by computing separate phylogenetic trees for the β-sheet and the α-helix of 116 sequences from all five DRB loci (Figure 6). The large number of sequences included in the analysis and the relatively small number of phylogenetically informative characters, especially for the α-helix, resulted in a number of equally likely trees. A consensus tree was therefore computed from the alternative individual trees for each of the two parts of the exon. Most of the alternative trees differ only in the position of sequences at the terminal twigs. In the tree for the β-sheet, the sequences form clusters representing different loci or, in the case of DRB1, different serological specificities. By contrast, in the tree for the α-helix, most of the major clusters each contain representatives from different loci or, in the

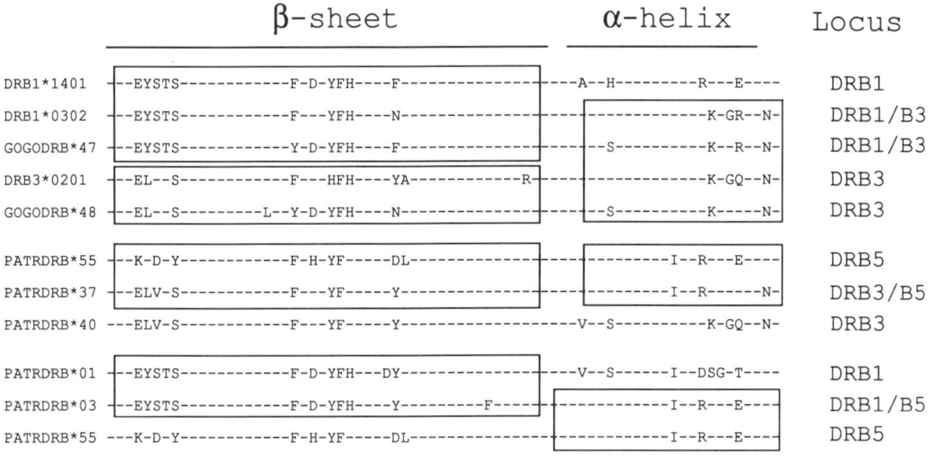

Figure 5. Putative recombination events in the evolution of alleles from DRB1–5.

898

Table 2. Statistical comparison of the distribution of sites in the β-sheet and α-helix favoring each of three possible branching orders of three DRB1–5 sequences.

Sequences compared	Domain	Topology (A+B) vs C	(A+C) vs B	(B+C) vs A	Probability (a)
I. Putative exchanges between DRB1 alleles					
A. DRB1*0402					
B. DRB1*0404	β-sheet	14	0	0	0.001
C. DRB1*1302	α-helix	0	6	0	0.05
II. Putative exchanges between DRB1 and DRB5					
A. DRB1*1302					
B. DRB1*DR6E	β-sheet	15	0	0	0.0001
C. DRB5*0102	α-helix	3	0	3	>0.20
A. PATRDRB*01					
B. PATRDRB*03	β-sheet	13	1	1	0.01
C. PATRDRB*55	α-helix	0	0	11	0.001
III. Putative exchanges between DRB3 and DRB5					
A. PATRDRB*55					
B. PATRDRB*37	β-sheet	0	0	10	0.01
C. PATRDRB*40	α-helix	8	1	2	>0.20
IV. Putative exchanges between DRB1 and DRB3					
A. DRB1*1401					
B. DRB1*0302	β-sheet	11	0	1	0.01
C. DRB3*0201	α-helix	1	1	8	>0.20
A. GOGODRB*47					
B. GOGODRB*48	β-sheet	0	9	0	0.05
C. GOGODRB*14	α-helix	4	0	2	>0.20

(a) Binomial probabilities.

case of the DRB1 locus, different serological specificities encoded by the DRB1 locus. By comparing the position of each of the DRB sequences in the tree for the β-sheet and the α-helix, genetic exchanges can be inferred. For instance, the sequence DRB1*0302, that clusters with DRB1 sequences with respect to the β-sheet epitope, is found in the cluster with DRB3 alleles in the α-helix tree. To the extent that selection is acting on the recombinant alleles, the frequency of inferred exchanges, based on the observed alleles, may differ substantially from the actual rate of generation of recombinant alleles. The number of inferred exchanges was highest between alleles at the DRB1 locus, followed by exchanges among DRB1 and either DRB3 or DRB5, and lowest between DRB1 and DRB2 (23). By contrast, the DRB2 sequences form a monophyletic group in both β-sheet and α-helix trees indicating that this locus does not participate to the same extent in the sequence exchange. It is noteworthy that the DRB2 sequences in the putative recombination region are less conserved than the other DRB loci (23).

EVOLUTION OF DRB LOCI

The presence of sequences from nonhuman hominoids in each of the seven major clusters of sequences in Fig. 6a indicate that the duplication of loci into the contemporary DRB1–5 loci occurred at least 5 million years ago (34–36). In addition, sequences from the rhesus monkey and the baboon were found to cluster with human DRB1 (DR 3, 5, 6 and DR4), DRB2, and DRB5 alleles, indicating that the duplication, and in some cases allelic diversification into some DRB1 allelic types, must have been initiated prior to the divergence of the rhesus monkey and the human, more than 20 million years ago (34–36). The branching order in the phylogenetic tree for the β-sheet (Fig. 5a) can also be used to infer

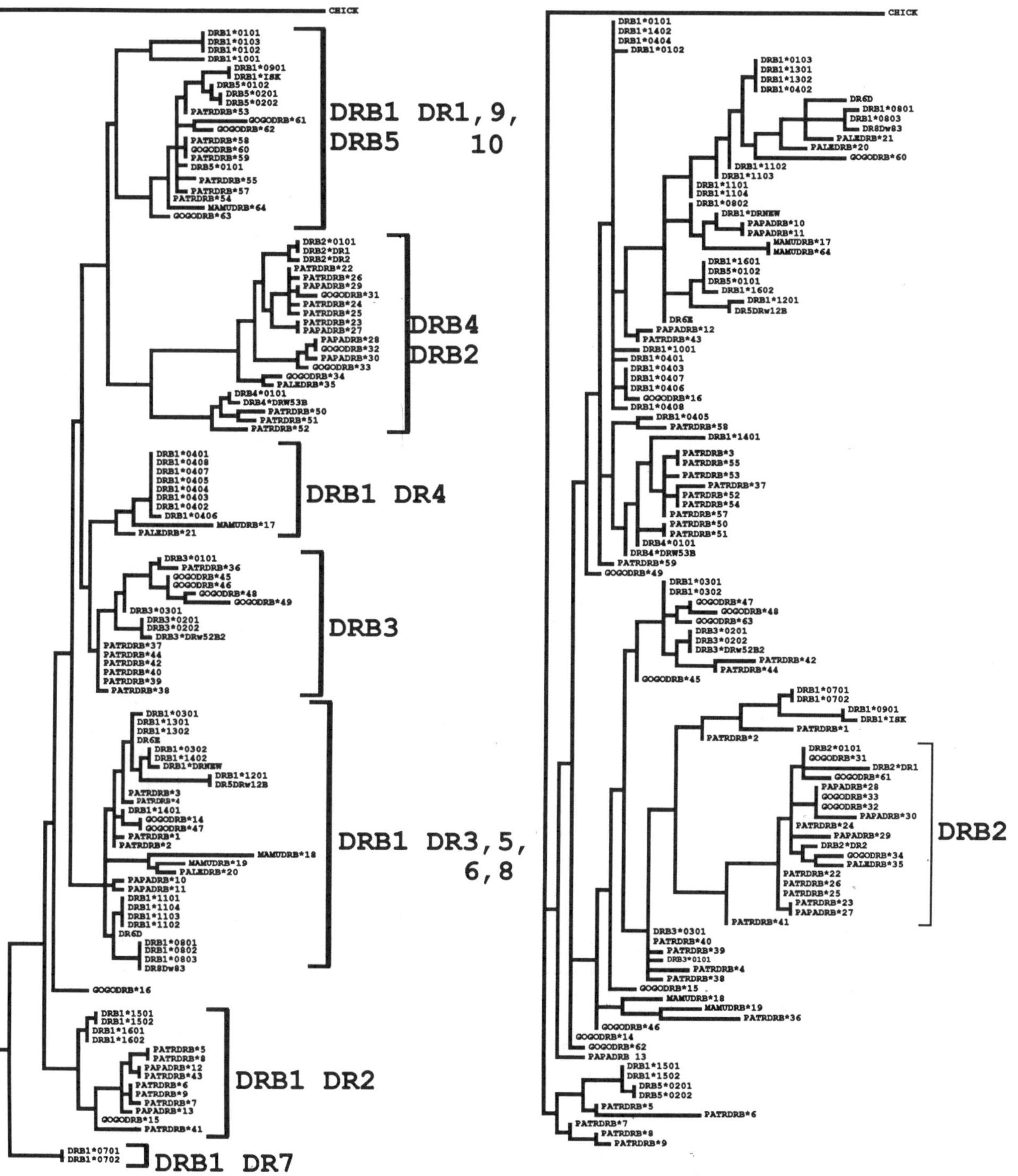

Figure 6. Parsimony trees of 63 nonhuman primate and 53 human DRB nucleotide sequences. Separate trees were constructed for the β-sheet (codons 6–53) (a) and the α-helix (codons 54–74) (b). Phylogenetic trees were constructed as described in (23).

the order by which the contemporary DRB loci, or allelic types, were generated from a hypothetical ancestral DRB gene by gene duplications. The position of the DRB1 DR1, 2, 7, 9 and 10 sequences closest to the root (domestic chicken) and outside the other groups indicates that the primordial DRB gene was DRB1-like. The DRB1 sequences do not form a monophyletic cluster, indicating that the duplication of loci in some cases have occurred after the split of some allelic types. The DRB2, DRB4 and DRB5 loci all appear to have been derived from ancestral DRB1 sequences, the most similar contemporary DRB1 sequences of which are found on the HLA-DR1, 9, and 10 haplotypes. The DRB3 sequences appear to be derived from the same ancestral locus as some DRB1 lineages (on DR4 haplotypes, Fig. 4a). However, alternative topologies for positioning the DRB3 sequences require from two (DRB3 with DRB1 DR3, 5, 6) to six (DRB3 with DRB1 DR7) additional mutations, and can therefore not be statistically rejected. Similarly, alternative positions for the DRB1 DR4 sequences require one (DR4 with DRB2–5) to four (DR4 with DR7) additional changes.

EVOLUTION OF DRB POLYMORPHISM

In general, specific polymorphic epitopes in the first hypervariable region (e.g., EYSTS at positions 9–13 of the first DRB1 β-strand) appear to be ancient since they are found in all primate species within a particular lineage (e.g., the DRB1 DR3, 5, 6-like clade). Thus, the generation of allelic diversity in this region clearly predated speciation. In the tree for the whole β-sheet sequence, however, there is a strong tendency for sequences from each locus, or in the case of the DRB1 locus, from each DR-type (e.g., DR4, DR3, 5, 6, 8; Fig. 4a) to cluster by species. This pattern is consistent with a model of evolution where certain epitopes encoded by expressed genes are conserved, and only minor modifications (by point mutations) within a species are tolerated. The exceptions are sequences from the DRB2–5 loci, where mixing of different species is more pronounced.

In the tree for the sequences encoding the α-helix, allelic types at the DRB1, as well as sequences from other DRB loci, are to a larger extent mixed (with the exception of DRB2), but nevertheless some clustering of sequences by species is maintained. This pattern of sequence clustering may indicate that sequences in the α-helix are exchanged among alleles, and occasionally also between loci, as noted previously, but the polymorphism in this part of the exon is to a larger extent generated *after* the divergence of species. The polymorphism in the first domain may thus be subjected to two different types of selection; the β-sheet for conservation of a limited set of epitopes and the α-helix for increased variability, mediated in part by intra-exon transfer of segments.

Analysis of the DQA1 locus has shown that most of the allelic diversification preceded speciation and could be explained without invoking gene conversion (19). Similarly, the phylogenetic analysis of DQB1 indicated that, while the allelic types DQB1*02, 03, 04, 05 and 06 at the DQB1 locus have been maintained for over 20 myr, considerable allelic diversification within these major types or lineages has occurred *after* the species diverged (20).

In the evolution of DRB loci, however, intra-exon recombination appears to have played a more important role in the generation of polymorphism. The conservation in all primate species examined of a consensus recombination signal sequence (at codon 51 to 55) between the β-sheet and α-helix regions may be functionally related to this putative recombinational mechanism of generating diversity. Recently, a minisatellite repeat sequence, virtually identical to the DRB recombination signal, was found to be the target of a specific DNA binding protein, possibly involved in the recombination mechanism (50).

Genetic exchange between different DRB loci can also be inferred the the β-sheet and α-helix phylogenetic trees in Fig. 6. Wu et al. (28) have reported a reciprocal exchange between sequences of the human DRB1 and DRB5 loci on the DR2 heplotype and noted the consensus χ-like sequence located between the recombined segments. The DRB2 and DRB4 loci, based on the monophyletic clustering in the β-sheet and the α-helix phylogenetic tree (fig. 6a and 6b), do not participate to the same extent in such inter-locus exchange (23). Interestingly, the consensus recombination signal sequence is not conserved in DRB2.

Recombination seems also to have played a role in the evolution of polymorphism at other MHC loci. The generation of new class I alleles (which contain two highly polymorphic exons) by inter-exon recombination has been suggested recently (51). Here, as in the case of DRB allelic diversity, recombination between alleles appears to have been more frequent than between loci. Specific polymorphic segments encoding the β-sheet have been maintained virtually unchanged over more than 20 million years. The finding of alleles that have been generated by similar genetic exchanges (e.g., between DRB1 and DRB3) in different species suggest that such exchanges may be relatively infrequent, and that some of the observed alleles resulting from putative recombination events may predate the divergence of hominoid lineages. Presumably, selection for the new allelic variants (13,17), generated in part by these recombinational events, has maintained them over evolutionary time periods.

ACKNOWLEDGMENTS

We thank Allan C. Wilson, Oliver Ryder and Karen Garner for providing DNA samples and Corey Levenson, Dragan Spasic and Lauri Goda for synthesis of oligonucleotides. U. B. G. was supported by a fellowship from the Knut and Alice Wallenberg Foundation and a grant from the Swedish Natural Science Research Council.

LITERATURE CITED

1. Trowsdale, J., Young, J. A. T., Kelly, A. P., Austin, P. J., Carson, S., Meunier, H., So, A., Erlich, H. A., Spielman, R., Bodmer, J. & W. F. Bodmer. 1985. Structure sequence and polymorphism in the HLA-D region. *Immunol. Rev.* 85:5–43.
2. Kappes, D. & J. L. Strominger. 1988. Human class II major histocompatibility complex genes and proteins. *Ann. Rev. Biochem.* 57:991–1028.
3. Bjorkman, P. J., Saper, M. A., Samraoui, B., Bennett, W. S., Strominger, J. L. & D. C. Wiley. 1987. Structure of the human class I histocompatibility antigen HLA-A2. *Nature* 329:506–512.
4. Brown, J. H., Jardetzky, T., Saper, M. A., Samaraoui, B., Bjorkman, P. J. & D. C. Wiley. 1988. A hypothetical model of the foreign antigen binding site of class II histocompatibility molecules. *Nature* 332:845–850.
5. Horn, G. T., Bugawan, T. L., Long, C. & H. A. Erlich. 1988. Allelic variation of HLA-DQ loci: relation to serology and to insulin-dependent diabetes susceptibility. *Proc. Natl. Acad. Sci. USA* 85:6012–6016.
6. Todd, J. A., Bell, J. I. & H. O. McDevitt. 1987. HLA-DQβ gene contributes to susceptibility and resistance to insulin-dependent diabetes mellitus. *Nature* 329:599–604.
7. Scharf, S., Friedman, A., Brautbar, C., Szafer, F., Steinman, L., Horn, G., Gyllensten, U. & H. A. Erlich. 1988. HLA class II allelic variation and susceptibility to *Pemphigus vulgaris. Proc. Natl. Acad. Sci. USA* 85:3504–3508.
8. Sinha, A. A., Brautbar, C. Szafer, Friedman, A., Tzfoni, E., Todd, J. A., Bell, J. I. & H. O. McDevitt. 1988. A newly characterized HLA-DQβ allele associated with *Pemphigus vulgaris. Science* 239:1026–1029.
9. Bugawan, T. L., Angelini, G., Larrick, J., Auricchio, S., Ferrara, G. B. & H. A. Erlich. 1988. A combination of a particular HLA-DPβ allele and an HLA-DQ heterodimer confers susceptibility to coeliac disease. *Nature* 339:470–473.
10. Begovich, A., Bugawan, T. L., Nepom, G., Klitz, W., Nepom, B. & H. A. Erlich. 1989. A specific HLA-DPβ allele is associated with pauciarticular juvenile rheumatoid arthritis but not adult

rheumatoid arthritis. *Proc. Natl. Acad. Sci. USA* 86:9489–9493.

11. Scharf, S. J., Friedman, A., Steinman, L., Brautbar, C. & H. A. Erlich. 1989. Specific HLA-DQβ abd HLA-DRβ1 alleles confer susceptibility to *Pemphigus vulgaris*. *Proc. Natl. Acad. Sci. USA* 86:6215–6219.

12. Robbins, P. A., Lettice, L. A., Santos-Aguado, J., Rothbard, J., McMichael, A. J. & J. L. Strominger. Definition of a discrete location within HLA-A2 for an influenza B virus nucleoprotein peptide. Submitted for publication.

13. Hughes, A. & M. Nei. 1989. Nucleotide substitution at major histocompatibility complex class II loci: evidence for overdominant selection. *Proc. Natl. Acad. Sci. USA* 86:958–962.

14. Gustafsson, K., Wiman, K., Emmoth, E., Larhammar, D., Bohme, J., Hyldig-Nielsen, J. J., Ronne, H., Peterson, P. A. & L. Rask. 1984. Mutation and selection in the generation of class II histocompatibility antigen polymorphism. *EMBO J.* 3:1655:1661.

15. Gorski, J. 1989. Second domain polymorphism reflects evolutionary relatedness of alleles and may explain public serologic epitopes. *J. Immunol.* 143:329–333.

16. Serjeantson, S. W. 1989. The reasons for Mhc polymorphism in man. *Tran. Proc.* 21:598–601.

17. Klein, J. & F. Figueroa. 1986. Evolution of the major histocompatibility complex. *CRC Crit. Rev. Immunol.* 6:295–386.

18. Klein, J. 1987. Origin of the major histocompatibility complex polymorphisms: The trans-species hypothesis. *Hum. Immun.* 19:155.

19. Gyllensten, U. & H. A. Erlich. 1989. Ancient roots for polymorphism at the DQα locus of primates. *Proc. Natl. Acad. Sci. USA* 86:9986–9990.

20. Gyllensten, U., Lashkari, D. & H. A. Erlich. 1990. Allelic diversification at the class II DQB locus of the mammalian major histocompatibility complex. *Proc. Natl. Acad. Sci. USA* 87:1835–1839.

21. Fan, W., Kasahara, M., Gutknecht, J., Klein, D., Mayer, W. E., Jonker, M. & J. Klein. 1989. Shared class II Mhc polymorphism between human and chimpanzees. *Hum. Immunol.* 26:107.

22. Gyllensten, U., Sundvall, M. & H. A. Erlich. 1991. Allelic diversity is generated by intra-exon exchange at class II Mhc DRB1 locus of primates. *Proc. Natl. Acad. Sci. USA.*

23. Gyllensten, U., Sundvall, M., Ezcurra, I. & H. A. Erlich. 1991. Genetic diversity at class II DRB loci of the primate histocompatibility complex. *J. of Immunology.*

24. Figueroa, F., Günther, E. & J. Klein. 1988. Mhc polymorphism predating speciation. *Nature* 335:265–268.

25. Lawlor, D. A., Ward, F. E., Ennis, P. D., Jackson, A. P. & P. Parham. 1988. HLA-A and B polymorphism predate the divergence of human and chimpanzees. *Nature* 335:268–271.

26. McConnell, T. J., Talbot, W. S., McIndoe, R. A. & E. K. Wakeland. 1988. The origin of Mhc class II gene polymorphism within the genus *Mus*. *Nature* 332:651–654.

27. Nathenson, S. G., Geliebter, J., Pfaffenbach, G. M. & R. A. Zeff. 1986. Murine major histocompatibility complex class I mutants: molecular analysis and structure function implications. *Ann. Rev. Immunology* 4:471.

28. Wu, S., Saunders, T. & F. Bach. 1986. Polymorphism of human Ia antigens generated by reciprocal exchange between two DRβ loci. *Nature* 324:676–679.

29. Gorski, J. & B. Mach. 1986. Polymorphism of human Ia antigens: gene conversion between two DRb loci results in a HLA-D/DR specificity. *Nature* 322:67–70.

30. Erlich, H. A. & T. L. Bugawan. 1989. HLA class II gene polymorphism: DNA typing, evolution and relationship to disease susceptibility. *In*: PCR Technology Principles and Applications for DNA Amplifications. Pp. 193–208. Stockton Press: New York, London.

31. Mullis, K. B. & F. Faloona. 1987. Specific synthesis of DNA in vitro via a polymerase catalyzed chain reaction. *Meth. Enzyml.* 155:335–350.

32. Saiki, R., Scharf, S., Faloona, F., Mullis, K., Horn, G., Erlich, H. A. & N. Arnheim. 1985. Enzymatic amplification of β-globin genomic sequences and restriction site analysis for diagnosis of sickle cell anemia. *Science* 230:1350–1354.

33. Saiki, R. K., Gelfand, D. H. Stoffel, S., Scharf, S., Higuchi, R., Horn, G., Mullis, K. B. & H. A. Erlich. 1988. Primer-directed enzymatic amplification of DNA with a thermostable DNA polymerase. *Science* 329:487–491.

34. Sakoyama, Y., Hong, K.-J., Byun, S. M., Hisajima, H., Ueda, S., Yaoita, Y., Hayashida, H., Miyata, T., & T. Honjo. 1987. Nucleotide sequences of immunoglobin e genes of chimpanzee and orangutan: DNA molecular clock and hominoid evolution. *Proc. Natl. Acad. Sci. USA* 84:1080–1084.

35. Pilbeam, D. 1984. The descent of hominoids and hominids. *Scientific Ameican* 250, 84–96.

36. Sarich, V. M. & A. C. Wilson. 1967. Immunological time scale for hominoid evolution. *Science* 158:1200–1203.

37. Fisher, S. G. & L. S. Lerman. 1983. DNA fragments differing by single base-pair substitutions are separated in denaturing gradient gels: correspondence with melting theory. *Proc. Natl. Acad. Sci. USA* 80:1579–1583.

38. Myers, R. M., Sheffield, V. C. & D. R. Cox. 1988. Detection of single base changes in DNA: ribonuclease cleavage and denaturation gradient gel electrophoresis. Pp 95–141. *In:* Davies K. E. (ed.): *Genome Analysis: A Practical Approach.* IRL Press, Oxford.

39. Gyllensten, U. & H. A. Erlich. 1988. Generation of single stranded DNA by the polymerase chain reaction and its application to direct sequencing of the HLA-DQA locus. *Proc. Natl. Acad. Sci. USA* 85:7652–7656.

40. Farris, S. J. 1970. Methods for computing Wagner trees. *Syst. Zool.* 19:83–92.

41. Fitch, W. M. 1977. Towards defining the course of evolution: minimum change for a specific tree topology. *Syst. Zool.* 20:406–415.

42. Jeffreys, A. J., Wilson, V. & S. L. Thein. 1985. Hypervariable "minisatellite" regions in human DNA. *Nature* 314:67–73.

43. Bugawan, T. L., Horn, G. T., Long, C. M., Mickelson, E., Hansen, J. A., Ferrara, G. B., Angelini, G. & H. A. Erlich. 1988. Analysis of HLA-DP allelic sequence polymorphism using the in vitro enzymatic DNA amplification of DP-α and DP-β loci. *J. Immunol.* 141:4024–4030.

44. Bugawan, T. L., Begovich, A. B. & H. A. Erlich. 1990. Rapid HLA-DPβ typing using enzymatically amplified DNA and non-radioactive sequence specific oligonucleotide probes. *Hum. Im.*

45. McClure, G. R., Ruberti, G., Fathman, C. G., Erlich, H. A. & A. B. Begovich. 1990. DRB1*LY10—A New DRB1 allele and its haplotypic association. *Immunogenetics.* In press.

46. Simons, M. J., Wheeler, R., J-M. Lalouel & B. Dupont. (ed.). Pp. 959–1023. *In: Immunobiology of HLA.* Springer Verlag.

47. Gorski, J. 1989. Second domain polymorphism reflects evolutionary relatedness of alleles and may explain public serologic epitopes. *J. Immunol.* 143:329–333.

48. Stewart, C.-B., Shilling, J. W. & A. C. Wilson. 1987. Adaptive evolution in the stomach lysozymes of foregut fermentors. *Nature* 330:401–404.

49. Andersson, G., Larhammar, D., Widmark, E., Servenius, B., Peterson, P. A. & L. Rask. 1987. Class II genes of the human major histocompatibility complex. Organization and evolutionary relationship of the DRβ genes. *J. Biol. Chem.* 262:8748–8758.

50. Collick, A. & A. J. Jeffreys. 1990. Detection of a novel minisatellite-specific DNA-binding protein. *Nucleic Acids Research* 18:625–629.

51. Holmes, N. & P. Parham. 1985. Exon shuffling in vivo can generate novel class I molecules *EMBO J.* 4:2849–2854.

Fight for Sight in Evolution:
The Eye Lens Proteins and Their Genes

Nicolette H. Lubsen and Wilfried W. De Jong

Abstract. The abundant water soluble structural proteins of the vertebrate eye lens, the crystallins, have proven to be a rich source of evolutionary studies. Here we outline two aspects of the evolution of these proteins. First, the various steps in the acquisition of lens-specific expression as exemplified by the taxon-specific crystallins and secondly, the cause (in the human γ-crystallin gene family) and consequence (for the αA-crystallin of the blind mole rat) of loss of function.

INTRODUCTION

The vertebrate eye lens contains only two cell types, the cuboidal epithelial cells, which form an anterior layer, and the long ribbon-like fibre cells, which make up the bulk of the lens. The primary fibre cells, found in the nucleus of the lens, arise through differentiation of the posterior cells of the lens vesicle, early in development. The lens subsequently grows through addition of the secondary fibre cells to the outside of the lens. These secondary fibre cells arise from differentiation of the epithelial cells at the equator of the lens. The oldest lens cells are thus found in the centre of the lens, while the younger cells form the cortex (McAvoy, 1978a,b; Piatigorsky, 1981).

The requirement for transparency of the lens has resulted in some unusual properties of the lens fibre cells and its constituent proteins. First of all, the fibre cells are very protein rich (up to 50% in the lens nucleus). The bulk of the fibre protein is made up of the crystallins, the water soluble structural lens proteins. To ensure transparency of the lens, the crystallins must remain soluble even at these high protein concentrations. Secondly, the fibre cells do not die nor does the protein turn over in the (normal) fibre cells. Hence the crystallins must be sufficiently stable to persist throughout the life span of the animal. Finally, the tertiary structure of the crystallins must be such that they can be packed in a short range order, again a requirement set by the optical properties of the lens (Delaye & Tardieu, 1983). Three protein families have evolved to suit these requirements, the α-, β-, and γ-crystallins (for review, see Piatigorsky, 1984; Bloemendal, 1985; De Jong & Hendriks, 1986; Wistow & Piatigorsky, 1988). Members of these three protein families, the ubiquitous crystallins, are found in all vertebrate lenses. The best characterized in evolutionary sense is αA-crystallin, a member of the α-crystallin family (see, for example, De Jong et al., 1985; De Jong & Goodman, 1988). This protein is highly conserved (see also below). The available information indicates that the same is true for the β- and γ-crystallins (for review, see Lubsen et al., 1988).

Dr. Lubsen (Department of Molecular Biology) and Dr. De Jong (Department of Biochemistry) are with the University of Nijmegen, Toernooiveld, 6525 ED Nijmegen, the Netherlands. Please address correspondence to Dr. Lubsen.

Table 1. Vertebrate crystallins.

Crystallin	Occurrence	Amount	Related proteins
α	all vertebrates	up to 50%	small heat shock proteins, *Schistosoma* and *Mycobacterium* antigens
β	all vertebrates	up to 70%	*Myxococcus xanthus* protein S, *Physarum polycephalum* 3a
γ	all vertebrates	up to 40%	*Myxococcus xanthus* protein S, *Physarum polycephalum* 3a
δ	birds, reptiles	up to 70%	$\delta 2$ = argininosuccinate lyase
ε	many birds, crocodiles	up to 23%	= lactate dehydrogenase B4
η	elephant shrew	24%	aldehyde dehydrogenases
ς	guinea pig	10%	alcohol dehydrogenases
λ	rabbit, hare	8%	hydroxyacyl-CoA dehydrogenases
μ	wallaby, kangaroo	up to 10%	?
ρ	frogs (genus *Rana*)	12%	aldose and aldehyde reductases, prostaglandin F-synthase
τ	lamprey, some fishes, birds, reptiles	up to 10%	= α-enolase

For references, see De Jong et al., 1989, Wistow, 1990, and Wistow et al., 1990. Related proteins are indicated with = when the protein is encoded by the same gene, ? indicates that no related protein has yet been found.

Some species contain additional crystallins besides the ubiquitous α-, β- and γ-crystallins (for review, see Wistow & Piatigorsky, 1988; De Jong et al., 1989; Piatigorsky and Wistow, 1989). These additional crystallins, the so-called taxon-specific crystallins, are sometimes specific for a limited group of related species, other taxon-specific crystallins are more widespread in certain vertebrate classes (see Table 1). Most of the taxon-specific crystallins have now been identified as related or identical to a housekeeping enzyme which, for as yet unknown reasons, accumulates in the lens of some, but not all, species.

The lens system thus presents a paradox: on the one hand the expectation of stringent constraints is realized when one looks at the ubiquitous crystallins, on the other hand the lens is apparently able to accept the accumulation of a number of household proteins without adverse optical effects. It is the challenge for the protein chemists to find a common denominator between the diverse enzymes/crystallins. For the evolutionary biologist this system presents a fascinating glance at genes on their way to tissue-specificity. The steps that can be discerned at this time are briefly outlined below.

THE ACQUISITION OF CRYSTALLINS

The One Gene–Two Functions System

The identity of a number of taxon specific crystallins is known. Some of these, such as ε- and τ-crystallin, have been shown to be identical to a known housekeeping enzyme. For example, ε-crystallin is identical to lactate dehydrogenase B4 (Wistow et al., 1987; Hendriks et al., 1988) while τ-crystallin is indistinguishable from α-enolase (Wistow et al., 1988). Other taxon specific crystallins have been identified as belonging to a certain class of enzymes but the exact enzymatic function of the presumed housekeeping equivalent is still obscure (see Table 1).

ε- or τ-crystallin are encoded by single copy genes (Hendriks et al., 1988; Wistow et al., 1988). Hence a single gene supplies the crystallin as well as the enzymatic function. Clearly the sequence must be under dual selective pressure, pressure originating from its enzymatic function and pressure from its function as a crystallin. Intuitively one would predict that these dual pressures would tend toward conservatism, i.e., that the shared gene would be less likely to accept changes than its non-shared orthologs in other species.

This prediction is not borne out by the (few) data available for the ε/LDH B sequences: the duck and chicken sequences are more distant from the ancestral sequence than mammalian (human and pig) sequences (Hendriks et al., 1988). These data point towards another possible scenario, namely a period of rapid change in LDH B either directly preceding or directly postdating its acceptance as a lens protein. These changes would then have been directed towards modelling the LDH B protein as a crystallin without unduly influencing its enzymatic activity. One piece of evidence that supports this scenario is that LDH B from most birds and higher reptiles is more thermostable than its mammalian counterpart (Wilson et al., 1964). Thermostability is indicative of protein stability, a requirement for the long-lived lens proteins.

τ-Crystallin is, properly speaking, not a taxon-specific crystallin as it is found in such disparate groups as lampreys, certain fishes, birds, and reptiles. If the recruitment of τ-crystallin as a lens protein was a one-time event, it must have occurred even before the divergence of the cyclostome and gnathostome lineages, more than 400 Myrs ago. Subsequent loss of high lens expression in some taxa must then be invoked to explain the present patchy occurrence of τ-crystallin. Similarly, the recruitment of lactate dehydrogenase B4 as a crystallin must date from the common archosaurian ancestor of birds and crocodiles. Yet today not all birds do contain ε-crystallin and the distribution of ε-crystallin in the avian orders does not correspond to their taxonomic relationships (Wistow et al., 1987). The pattern of occurrence of τ- or ε-crystallins can be explained by assuming that their presence is selectively neutral and that their lens expression has been randomly lost in time. However, their distribution fits equally well that of a character of which the advantages outweigh the disadvantages only in certain habitats. At present there are too few data to distinguish between these alternatives. A possible disadvantage of the enzyme-crystallins could be that the high concentration of the particular enzyme disturbs the intermediate metabolism (note, however, that α-enolase—τ-crystallin—is inactivated in the lens by a post-translational modification; Wistow et al., 1988). To our knowledge, no experiments that address this question have been reported. The possible advantage of the enzyme-crystallins is obscure as well. Some of the enzyme-crystallins identified thus far (i.e., ε, λ, ρ and ς) are oxidoreductases and their presence correlates with an increased level of NADH or NADPH (and their oxidized forms) in the lens (Rao & Zigler, 1990; Zigler & Rao, 1990). The high concentration of these pyridine nucleotides may be beneficial in combating the chronic oxidative stress to which lenses are exposed. There is no obvious correlation, however, between habitat, expected oxidative stress, and the occurrence of enzyme-crystallins. A different explanation has been offered in the case of ε-crystallin. As its presence correlates with water or high altitude habitats, it has been suggested that the NADH bound by ε-crystallin could serve as a UV filter and reduce glare (Wistow et al., 1987).

The Case of the Duplicated Housekeeping Gene

If a single gene encodes the enzyme as well as the crystallin, then the optimal protein sequence must be a compromise between the needs of the lens and that of the housekeeping system. If the housekeeping gene duplicates, then one copy can be used as crystallin while the other copy can serve the housekeeping needs. Such a system has evolved for δ-crystallin, a crystallin found in birds and reptiles. It has long been known that the chicken genome has two closely linked copies of the δ-crystallin gene. It was puzzling, however, that only one of these genes contributes significantly to the δ-crystallin in the lens: 99% of the δ-crystallin transcripts in chicken lens derive from the δ1 gene (for review, see Wistow and Piatigorsky, 1988). A second puzzling property of the δ-crystallin genes was that their expression is not quite lens specific; low but significant levels of δ-crystallin mRNA are found in other tissues (for review, see Clayton et al., 1986). These problems were resolved once δ-crystallin was identified as very closely related to argininosuccinate lyase. Presumably the "lens-specific" δ1-crystallin gene is a duplicate of the argininosuccinate lyase gene (the δ2-crystallin gene) being adapted to its role as crystallin. The few amino acid replacements between the chicken δ1 and δ2 polypeptides have led to loss of most of the enzymatic activity of the δ1 protein. Duck lenses, however, do show a very high endogenous argininosuccinate lyase activity (Piatigorsky et al., 1988). Hence the enzymatic activity itself does not appear to interfere with the function of lens cells.

δ-Crystallin is found in reptiles as well as birds, and must have been recruited before the divergence of reptiles and birds. Unfortunately, the organization of the δ-crystallin gene(s) has not yet been studied in reptiles. Hence it is not known whether the gene duplication pre- or postdates the recruitment of the gene. The very high sequence similarity between the two chicken δ-crystallin genes, which includes all but three of the sixteen introns, suggested that the gene duplication was a relatively recent event (Nickerson et al., 1986). It is perhaps more likely that this sequence similarity is maintained by gene conversion, as a duplicate δ-crystallin gene now has also been found in the duck (Piatigorsky et al., 1987).

The Ubiquitous Crystallin Genes: Diverged Copies of Housekeeping Genes

The first inkling that crystallins were closely related to housekeeping proteins actually came from sequence studies of one of the ubiquitous crystallins, αA-crystallin. This sequence was found to be homologous with the small heat shock proteins (Ingolia & Craig, 1982). In fact, the human small heat shock protein is somewhat more closely related to αA-crystallin (and its sibling αB-crystallin) than to the *Drosophila melanogaster* small heat shock proteins (De Jong et al., 1988). The heat shock origin of the α-crystallins may explain the fact that αB-crystallin (but not αA-crystallin) is expressed in other tissues as well as the lens (Bhat and Nagineni, 1989; Dubin et al., 1990).

The vertebrate ancestor of the other super-family of ubiquitous crystallins, the β- and γ-crystallins, has not yet been found. That there is likely to be such an ancestor is suggested by the fact that the distinctive β- and γ-crystallin protein fold is also found in the *Myxococcus xanthus* protein S (a spore protein; Wistow et al., 1985) and *Physarum polycephalum* spherulin 3a (Wistow, 1989).

Is the Recruitment of Housekeeping Proteins a Lens-specific Phenomenon?

The frequent recruitment of housekeeping proteins as lens crystallins begs the question whether this is a more general phenomenon or whether there is something about the regulation of gene expression in the lens that predisposes it to overexpression of housekeeping proteins. One possible explanation for the close relationship between crystallins and housekeeping proteins is a trivial one. Lens development is a relatively recent acquisition. Hence there may have been insufficient time to obscure the relation between a crystallin and its more generally expressed ancestor. The occurrence of taxon-specific crystallins (e.g., λ-, ρ- and ς-crystallin; see Table 1), however, suggests that the recruitment is an ongoing process in the lens.

Recruitment could be a general phenomenon that is rejected by most tissues because of the disturbance of the cellular metabolism by the resulting enzyme imbalance. The low water content of the terminally differentiated lens allows little metabolic activity and could thereby permit recruitment. A more interesting, though very speculative, possibility is that recruitment is indeed a reflection of the manner in which gene expression is regulated in the lens. Lens is one of the few tissues in vertebrates that can be regenerated. Indeed transdifferentiation to lens can occur from a bewildering variety of tissues, such as iris, cornea, retina, brain and adenohypophysis (for review, see Clayton et al., 1986). The basis for the ease with which tissues can be induced to transdifferentiate to lens is unknown, but the mere fact that it happens suggests that there is something special about lens induction and lens-specific gene expression.

LOSS OF FUNCTION OF THE CRYSTALLINS

The optical properties of the lens are determined by the concentration of the various crystallins. Changes in habitat may alter the visual requirements of a species and thereby alter the selective pressure on the crystallins.

The Case of the Mole Rat

The blind mole rat (*Spalax ehrenbergi*) has adapted to subterranean life during some 25 Myrs. Unlike moles, these animals do not show any light-dark response and only a rudimentary eye is found. The "lens" in this eye consists of an irregular mass of apparently undifferentiated cells. These cells still express α-crystallin as assayed by immunofluorescent staining (Quax-Yeuken et al., 1985). The αA-crystallin gene of this species has been isolated and sequenced (Hendriks et al., 1987). Not unexpectedly, the rate of change of the mole rat αA-crystallin is much higher than that found in other rodents. The mole rat αA-crystallin has nine amino acid changes, while none are found in its nearest relatives, the mouse, hamster, gerbil, and rat. It is reasonable to assume that the amino acid changes in the mole rat αA-crystallin have only been accepted after the loss of the need for vision. The mole rat diverged from the rodent lineage some 40 Myrs ago and fossil evidence shows a complete subterranean life about 25 Myrs ago. Taking the latter time, the average rate of change would be 0.9×10^{-9} subst/site/year or about 20% of the rate observed in pseudo-genes. Hence either the loss of vision occurred considerably later than indicated by the fossil evidence or αA-crystallin is still not free to drift. Although the mole rat does not show a light-dark response, the animals do have a photoperiod perception. The retina, which is well differentiated (Sanyal et al., 1990), is probably involved in this response. The development of the retina necessarily involves the induction of a lens vesicle, which in turn could require the expression of at least partially functional αA-crystallin. A study of the rate of evolution of the β- or γ-crystallins, which are expressed only during terminal lens cell differentiation, could resolve this issue.

The Case of the Human γ-crystallin Genes

The change from aboveground to underground living provides a rather obvious and drastic change in the visual requirement of an animal. More subtle changes in habitat can have consequences for the distribution and complexity of the crystallins as well. One example is provided by the human γ-crystallins. The amount of γ-crystallin is inversely correlated with the water content of the lens. The hard, round rodent lenses contain six different γ-crystallins (γA—γF), which together comprise about 40% of the total crystallins (Siezen et al., 1988), while in the soft human lenses only 11% of the crystallins is γ-crystallin (Thomson & Augusteyn, 1985). The human γ-crystallin fraction contains only two major species (γC and γD; Siezen et al., 1987). However, evolutionary evidence indicates that the common ancestor of man and rat had six active γ-crystallin genes (Aarts et al., 1988). These genes are still all active in the rat but in man four must have been at least partially inactivated.

The molecular basis for the inactivation of two genes, the γE and γF genes, became apparent as soon as the coding regions were sequenced (Meakin et al., 1985). These genes contain an in frame stop codon in the second exon and can thus not yield a full length protein. In addition the γF gene lacks the expected promoter sequences. (Note that the human γF gene is not the ortholog of the rat γF gene but rather a recent duplicate of the human γE gene. The ortholog of the rat γF gene in man is the γG gene. This gene has been partially deleted in the primate lineage and is represented in man by a quarter-gene fragment; Brakenhoff et al., 1990). In contrast to the γE and γF genes, the human γA and γB genes show no aberration at the sequence level (Den Dunnen et al., 1985, Meakin et al., 1985). Model studies showed that the promoter of the human γA gene is poorly active and the relatively low level of the γA protein in the human lens is likely to be due to a low rate of transcription. The promoter of the human γB gene is very active, however, its RNA product appears to be unstable (Brakenhoff et al., 1990). The cause of the instability of this transcript could be traced to the sequence of the third and last exon. By analogy to other systems (Brawerman, 1989), the multiple copies of the AUUUA motif in the 3' non-coding region are predicted to be the destabilizing element.

The human γ-crystallin gene family thus provides examples of gene inactivation at three different levels: the genomic level (inactivation of the coding region or deletion; γE, γF and γG), the transcriptional level (decrease in promoter activity; γA) and the post-transcriptional level (mRNA instability; γB). This inactivation in the primate lineage has presumably been permitted by a lesser need for γ-crystallin in the water-"rich" human lens. Interestingly, the decreasing importance of γ-crystallins during human evolution coincides with an accelerated rate of possibly adaptive changes in the αA-crystallin molecule (De Jong & Goodman, 1988).

POSTSCRIPT

The genes for lens crystallins provide not only a model for the gain and loss of (tissue-specific) expression, they also show a broad range of other molecular evolutionary phenomena, such as concerted evolution, intron loss, alternative splicing, splice site sliding, multiple translation initiation sites and capture of simple sequence DNA (see De Jong & Hendriks, 1986; Lubsen et al., 1988). We expect that the lens system will continue to be a fruitful source of evolutionary studies in the future.

ACKNOWLEDGMENTS

Our work has been carried out under the auspices of the Netherlands Foundation for Chemical Research (SON) and with financial aid from the Netherlands Organization for the Advancement of Pure Research (NWO).

LITERATURE CITED

Aarts, H. J. M., den Dunnen, J. T., Leunissen, J., Lubsen, N. H. & J. G.G. Schoenmakers. 1988. The γ-crystallin gene families: sequence and evolutionary patterns. *J. Mol. Evol.* 27:163–172.

Bhat, S. P. & C. N. Nagineni. 1989. αB Subunit of lens-specific protein α-crystallin is present in other ocular and non-ocular tissues. *Biochem. Biophys. Res. Comm.* 158:319–325.

Bloemendal, H. 1985. Lens research: from protein to gene. *Exp. Eye Res.* 41:429–448.

Brakenhoff, R. H., Aarts, H. J. M., Reek, F. H., Lubsen, N. H. & J. G.G. Schoenmakers. 1990. The human γ-crystallin genes: a gene family on its way to extinction. *J. Mol. Biol.* 216:519–532.

Brawerman, G. 1989. mRNA decay: finding the right targets. *Cell* 57:9–10.

Clayton, R. M., Jeanny, J.-C., Bower, D. J. & L. H. Errington. 1986. The presence of extralenticular crystallins and its relationship with transdifferentiation to lens. *Curr. Topics Dev. Biol.* 20:137–151.

De Jong, W. W., Zweers, A., Versteeg, M., Dessauer, H. C. & M. Goodman. 1985. α-Crystallin A sequences of *Alligator mississippiensis* and the lizard *Tupinambis teguixin*: molecular evolution and reptilian phylogeny. *Mol. Biol. Evol.* 2:484–493.

De Jong, W. W. & W. Hendriks. 1986. The eye lens crystallins: ambiguity as evolutionary strategy. *J. Mol. Evol.* 24:121–129.

De Jong, W. W. & M. Goodman. 1988. Anthropoid affinities of *Tarsius* supported by lens αA-crystallin sequences. *J. Human. Evol.* 17:575–582.

De Jong, W. W., Leunissen, J. A. M., Leenen, P. J. M., Zweers, A. & M. Versteeg. 1988. Dogfish α-crystallin sequences. Comparison with small heat shock proteins and *Schistosoma* egg antigen. *J. Biol. Chem.* 263:5141–5149.

De Jong, W. W., Hendriks, W., Mulders, J. W. M. & H. Bloemendal. 1989. Evolution of lens crystallins: the stress connection. *TIBS* 14:365–368.

Delaye, M. & A. Tardieu. 1983. Short-range order of crystallin proteins accounts for eye lens transparency. *Nature* 302:415–417.

Den Dunnen, J. T., Moormann, R. J. M., Cremers, F. P. M. & J. G.G. Schoenmakers. 1985. Two human

910

γ-crystallin genes are linked and riddled with Alu-repeats. *Gene* 38:197–204.

Dubin, R. A., Wawrousek, E. F. & J. Piatigorsky. 1989. Expression of the murine αB-crystallin gene is not restricted to the lens. *Mol. Cell. Biol.* 9:1083–1091.

Hendriks, W., Leunissen, J., Nevo, E., Bloemendal, H. & W. W. De Jong. 1987. The lens protein αA crystallin in the blind mole rat, *Spalax ehrenbergi*: evolutionary change and functional constraints. *Proc. Nat. Acad. Sci. USA* 84:5320–5324.

Hendriks, W., Mulders, J. W. M., Bibby, M. A., Slingsby, C., Bloemendal, H. & W. W. De Jong. 1988. Duck lens ε-crystallin and lactate dehydrogenase B4 are identical: selective pressure from two distinct functions acting on a single copy gene. *Proc. Nat. Acad. Sci. USA* 85:7114–7118.

Ingolia, T. D. & E. A. Craig. 1982. Four small *Drosophila* heat shock proteins are related to each other and to mammalian α-crystallin. *Proc. Nat. Acad. Sci. USA* 79:2360–2364.

Lubsen, N. H., Aarts, H. J. M. & J. G.G. Schoenmakers. 1988. The evolution of lenticular proteins: the β- and γ-crystallin super gene family. *Prog. Biophys. Molec. Biol.* 51:47–76.

McAvoy, J. W. 1978a. Cell division, cell elongation and distribution of α-, β- and γ-crystallins in the rat lens. *J. Embryol. Exp. Morph.* 44:149–165.

McAvoy, J. W. 1978b. Cell division, cell elongation, and the co-ordination of crystallin gene expression during lens morphogenesis in the rat. *J. Embryol. Exp. Morph.* 45:271–281.

Meakin, S. O., Breitman, M. L. & L.-C. Tsui. 1985. Structural and evolutio-nary relationships among five members of the human γ-crystallin gene family. *Mol. Cell. Biol.* 5:1408–1414.

Nickerson, J. M., Wawrousek, E. F., Borras, T., Hawkins, J. W., Norman, B. L, Filpula, D. R., Nagle, J. W., Ally, A. H. & J. Piatigorsky. 1986. Sequence of the chicken δ2 crystallin gene and its intergenic spacer. Extreme homology with the δ1 crystallin gene. *J. Biol. Chem.* 261:552–557.

Piatigorsky, J. 1981. Lens differentiation in vertebrates. *Differentiation* 19:134–153.

Piatigorsky, J. 1984. Lens crystallins and their gene families. *Cell* 38:620–621.

Piatigorsky, J., Norman, B. & Jones, R. E. 1987. Conservation of δ crystallin gene structure between ducks and chickens. *J. Mol. Evol.* 25:308–317.

Piatigorsky, J., O'Brien, W. E., Norman, B. L., Kalumuck, K., Wistow, G. J., Borras, T., Nickerson, J. M. & E. F. Wawrousek. 1988. Gene sharing by δ-crystallin and argininosuccinate lyase. *Proc. Nat. Acad. Sci. USA* 85:3479–3483.

Piatigorsky, J. & G. J. Wistow. 1989. Enzyme/crystallins: gene sharing as an evolutionary strategy. *Cell* 57:197–199.

Quax-Jeuken, Y., Bruisten, S., Bloemendal, H. & W. W. De Jong. 1985. Evolution of crystallins: expression of lens-specific proteins in the blind mammals mole (*Talpa europaea*) and mole rat (*Spalax ehrenbergi*). *Mol. Evol. Biol.* 2:279–288.

Rao, P. V. & J. S. Zigler, Jr. 1990. Extremely high levels of NADPH in guinea pig lens: correlation with ς-crystallin concentration. *Biochem. Biophys. Res. Comm.* 167:1221–1228.

Sanyal, S., Jansen, H. G., De Grip, W. J., Nevo, E. & W. W. De Jong. 1990. The eye of the blind mole rat, *Spalax ehrenbergi*: rudiment with hidden function? *Invest. Ophthalmol. Vis. Sci.* In press.

Siezen, R. J., Thomson, J. A., Kaplan, E. D. & G. B. Benedek. 1987. Human lens γ-crystallins: isolation, identification and characterization of the expressed gene products. *Proc. Nat. Acad. Sci. USA* 84:6088–6092.

Siezen, R. J., Wu, E., Kaplan, E. D., Thomson, J. A. & G. B. Benedek. 1988. Rat alens γ-crystallins. Characterization of the six gene products and their spatial and temporal distribution resulting from differential synthesis. *J. Mol. Biol.* 199:475–490.

Thomson, J. A. & R. C. Augusteyn. 1985. Ontogeny of human lens crystallins. *Exp. Eye Res.* 40:393–410.

Wilson, A. C., Kaplan, N. O., Levine, L., Pesce, A., Reichlin, M. & W. S. Allison. 1964. Evolution of lactic dehydrogenase. *Fed. Proc. Am. Soc. Exp. Biol.* 23:1258–1266.

Wistow, G., Summers, L. & T. Blundell. 1985. *Myxococcus xanthus* spore coat protein S may have a similar structure to vertebrate βγ-crystallins. *Nature* 315:771–773.

Wistow, G. J., Mulders, J. W. M. & W. W. De Jong. 1987. An enzyme as a structural protein: lactate dehydrogenase in avian and crocodilian lenses. *Nature* 326:622–624.

Wistow, G. J. & J. Piatigorsky. 1988. Lens crystallins: the evolution and expression of proteins for a highly specialized tissue. *Ann. Rev. Biochem.* 57:479–504.

Wistow, G. J., Lietman, T., Williams, L. A., Stapel, S. O., De Jong, W. W., Horwitz, J. & J. Piatigorsky. 1988. τ-Crystallin/α-enolase: one gene encodes both an enzyme and a lens structural protein. *J. Cell Biol.* 107:2729–2736.

Wistow, G. 1990. Evolution of a protein superfamily: relationships between vertebrate lens crystallins and microorganism dormancy proteins. *J. Mol. Evol.* 30:140–145.

Wistow, G., Anderson, A. & H. S. Kim. 1990. Taxon-specific crystallin expression in mammalian lenses. *Invest. Ophthalmol. Vis. Sci.* 31 (suppl.):162.

Zigler, J. S. Jr. & P. V. Rao. 1990. Nucleotide-binding enzyme/crystallins lead to extremely high levels of pyridine nucleotides in the ocular lens. Submitted.

Adaptive Features in Digestive Enzymes in Mammals

Jaap J. Beintema and Adriana Furia

Abstract. The digestive enzymes, pancreatic ribonuclease and stomach lysozyme, show several similarities in mammals:
1. There is a high level of expression of ribonuclease in the pancreas, and of lysozyme in the stomach, of ruminants and species with ruminant-like digestion. However, both enzymes are also expressed in other tissues.
2. Pancreatic ribonucleases and stomach lysozymes in ruminants and species with ruminant-like digestion are less basic than homologous enzymes in other species and tissues. The explanation for this feature is different for ribonucleases and lysozymes.
3. Several gene duplications occurred in ancestral ruminants, with the less basic enzymes expressed in the pancreas (ribonuclease) or stomach (lysozyme) and the more basic ones expressed in other tissues.
4. The evolutionary rates of pancreatic ribonuclease and stomach lysozyme slowed down in ruminants.

Pancreatic ribonuclease is one of the most intensively investigated digestive enzymes in mammals (Beintema et al., 1988b). However, it differs from other digestive enzymes in having highly different levels of expression in different species (Barnard, 1969; Beintema et al., 1973) and in being found in other tissues and body fluids as well, indicating that the enzyme has other physiological functions besides its role in digestion. High levels of pancreatic ribonuclease occur in ruminants, in species with ruminant-like digestion, and in several mammals with cecal digestion. Barnard (1969) proposed that these elevated levels are a response to the necessity of digesting large amounts of ribonucleic acid derived from the symbiotic microflora of the stomach or cecum of these herbivorous mammals.

Elevated levels of lysozyme have been found in the stomach of several ruminants and species that have a ruminant-like digestion (Dobson et al., 1984) and in the intestine of species with cecal digestion (Hammer et al., 1987). These elevated levels are explained by the necessity to digest cell walls from the symbiotic bacterial microflora in these mammals. The recruitment of ribonuclease and lysozyme in the digestive systems of herbivores to hydrolyze components from the symbiotic microflora has not only resulted in adaptations in the levels of expression of these enzymes, but also in convergence to similar properties in different taxa.

During our studies on the primary structures of about 40 mammalian ribonucleases, we have found several examples of selective amino acid replacements that influence the

Dr. Beintema is with the Biochemisch Laboratorium, Nijenborgh 16, 9747 AG Groningen, the Netherlands. Please address correspondence to him. Dr. Furia is with the Dipartimento di Chimica Organica e Biologica, Università di Napoli, Via Mezzocannone 16, 80135 Napoli, Italy.

912

properties of the protein molecule. These concern glycosylation and charge characteristics.

Pancreatic ribonucleases differ in the presence or absence of covalently attached carbohydrates to Asn residues in Asn-X-Ser/Thr sequences that are part of highly variant sequences at the surface of the molecule. The glycosylation state of ribonucleases could be correlated with the type of fermentation used: species with cecal digestion, such as pig, horse and several rodents, have ribonucleases with large carbohydrate moieties attached to several positions at the surface of the molecule. Therefore, it was suggested that the presence of carbohydrate protects ribonuclease from absorption in the gut, enabling it to be transported to the large intestine where it should hydrolyse the ribonucleic acid from the cecal microflora (Beintema et al., 1976). Ruminants and species with ruminant-like digestion (camel, hippopotamus, sloth, kangaroo) have less carbohydrate attached to their ribonucleases. Therefore, it seems that glycosylation of pancreatic ribonuclease may not be advantageous for species with stomach fermentation.

Similar glycosylation requirements for ribonucleases from different taxa may explain convergences in functional and structural characteristics. The pancreatic ribonuclease of giraffe is positioned with pronghorn and not with deer in the most parsimonious tree (Fig. 1) as a consequence of a number of rather rarely observed amino acid replacements. However, since these shared replacements occur at internal positions, the relationship does not come to clear expression in many of the properties of these enzymes. Thus, giraffe ribonuclease is very similar to moose (a deer species) ribonuclease in glycosylation and charge properties. It may be significant that in a classification of ruminants based on feeding ecology and stomach structure (Hofmann et al., 1976), moose and giraffe are grouped together as large "concentrate selectors" with a diet containing relatively little cellulose fibre, and with a simple (primitive) stomach.

Another striking similarity is the large number of identical amino acid residues in 2 large surface regions (about one third of the sequence) of pancreatic ribonucleases of pig and guinea pig B, including two glycosylation sites, as a result of parallel evolution. This has been attributed to an adaptation to cecal digestion (Beintema et al., 1977).

A similar example of parallelism has been described by Stewart et al. (1987), who demonstrated that the stomach lysozyme of langur, an Old-World monkey with ruminant-like digestion, shares many amino acid replacements with ox lysozyme. Ox and langur lysozyme are less basic than human lysozyme and have a lower pH optimum of enzymatic activity as an adaptation to the low pH of the stomach.

Pancreatic ribonucleases of ruminants and species with ruminant-like digestion are also less basic than ribonucleases which have no or only a minor function in the digestion of microfloral RNA from the stomach or cecum, such as the pancreatic ribonucleases from man and whale, and bovine seminal and brain ribonuclease (see below). The pancreatic ribonuclease of langur differs from human pancreatic ribonuclease at 14 amino acid positions. Eight of these involve changes in charge, with the result that the langur possesses 5 positive charges less than the human enzyme (Beintema, 1990).

Ribonucleases act in the intestine and do not need as low a pH optimum as lysozyme. Here the excess of positive charges may serve to degrade double-stranded viral RNAs (Libonati et al., 1976) in other tissues than the pancreas.

If ribonucleases expressed in the pancreas and in other tissues have different functions, and have different structural adaptations to these separate functions, the use of a single gene product in a species may be considered as a compromise between different requirements. In man, the same ribonuclease gene product has been found in several tissues and body fluids—although large differences in organ-specific glycosylation have been observed (Beintema et al., 1988a). The situation is, however, different in ruminants. Three homologous bovine secretory ribonucleases have been isolated and sequenced (Carsana et al., 1988, Beintema et al., 1988b). These are the enzymes from the pancreas and

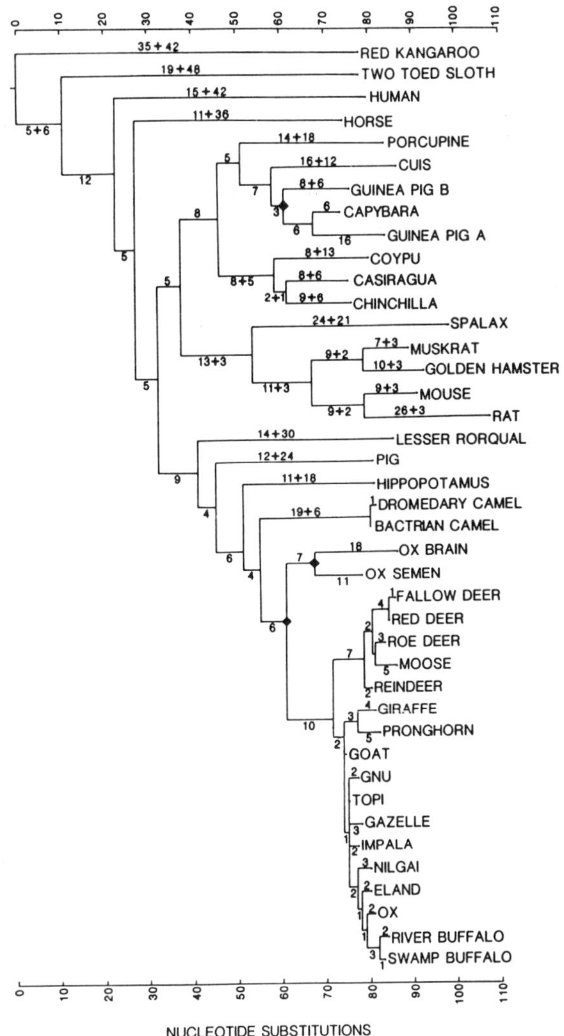

Figure 1. Evolutionary tree of ribonucleases. The most parsimonious tree observed is shown. Horizontal lengths are proportional to the corrected number of nucleotide substitutions, which are shown, rounded to the nearest integer, in the form $a + b$, where a is the number of uncorrected substitutions demanded by parsimony and b is the number of additional substitutions required by a correction procedure for unseen substitutions. No b is shown if no correction is required (Fitch & Beintema, 1990). Vertical distances are solely to separate lineages. The diamonds represent three demonstrated gene duplications. The taxa used in the study were: red kangaroo, *Macropus rufus;* two-toed sloth, *Bradypus infuscatus;* human, *Homo sapiens;* horse, *Equus caballus;* porcupine, *Hystrix cristata;* cuis, *Galea musteloides;* guinea pig, *Cavia porcellus;* capybara, *Hydrochoerus hydrochoeris;* coypu, *Myocastor coypus;* casiragua, *Proechimys quairae;* chinchilla, *Chinchilla brevicaudata;* spalax, *Spalax ehrenbergi;* muskrat, *Ondatra zibethica;* golden hamster, *Mesocricetus auratus;* mouse, *Mus musculus;* rat, *Rattus norvegicus;* lesser rorqual, *Balaenoptera acutorostrata;* pig, *Sus scrofa;* hippopotamus, *Hippopotamus amphibius;* dromedary camel, *Camelus dromedarius;* bactrian camel, *Camelus bactrianus;* ox, *Bos taurus;* fallow deer, *Dama dama;* red deer, *Cervus elaphus;* roe deer, *Capreolus capreolus;* moose, *Alces alces;* reindeer, *Rangifer tarandus;* giraffe, *Giraffa camelopardalis;* pronghorn, *Antilocapra americanus;* goat, *Capra hircus;* gnu, *Connochaetes taurinus;* topi, *Damaliscus korrigum;* gazelle, *Gazella thomsoni;* impala, *Aepyceros melampus;* nilgai, *Boselaphus tragocamelus;* eland, *Taurotragus oryx;* and buffalo (swamp and river), *Bubalus bubalis.*

the already mentioned basic enzymes produced in the seminal vesicles and the brain, respectively. Analysis of the sequences to derive the most parsimonious tree (Fig. 1) indicates that these ribonucleases are products of gene duplications that occurred rather recently, in the ancestors of the ruminants after divergence from the other artiodactyls (Beintema et al., 1988b).

More direct evidence for these gene duplications was obtained by southern blot analysis (Fig. 2). Genomic DNA samples (10 μ) from several artiodactyls were digested with restriction endonucleases (70–80 units; Boehringer) at 37°C for 2–4 hours. For each species four different digests were made with EcoRI, PvII, HindIII and PstI, respectively. Samples were electrophoresed in 0.8% agarose gel, transferred to Hybond-N membranes and hybridized to a ^{32}P nick-translated probe which contained the coding sequence and the 3' untranslated region of the bovine pancreatic ribonuclease gene (Carsana et al., 1988). Hybridization was performed in a medium consisting of 6xSSC, 5x Denhardt's solution and 0.5% SDS. It consisted of prehybridization during 1 h with 200 μg/ml

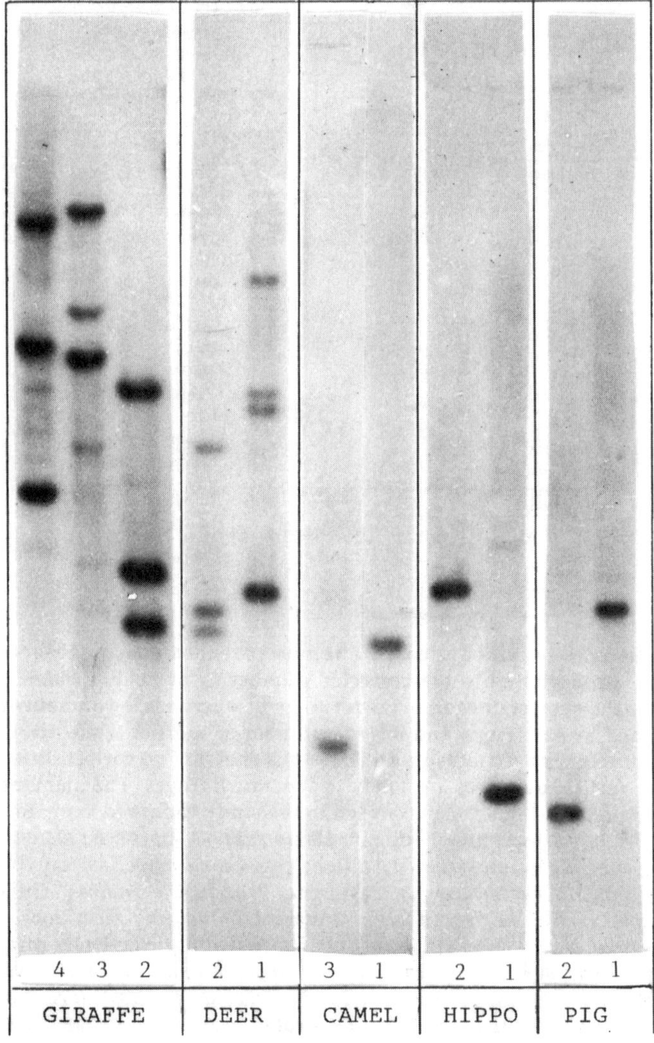

Figure 2. Southern blot analysis of genomic DNAs of several artiodactyls. See text for more information about species and experimental conditions. Restriction enzymes: 1, EcoRI; 2, PstI; 3, PvII; 4, Hind III.

denaturated salmon sperm DNA, followed by overnight hybridization with the probe, both at 60°C. Filters were washed in 2xSSC, 0.1% SDS, 5 mM EDTA at 60°C, and exposed to Kodak films with intensifying screens at −70°. The restriction patterns obtained in non-ruminant genomes (pig, *Sus scrofa;* peccary, *Tayassu tajacu;* hippopotamus, *Hippopotamus amphibius;* camel, *Camelus dromedarius;* llama, *Lama glama*) exhibited only one cross-hybridizing band, suggesting the presence of a single ribonuclease gene. (Genes of distantly related members of the ribonuclease superfamily like the nonsecretory ribonuclease and angiogenin (Beintema et al.;1988b) probably deviate too much in sequence to be detectable.) In the restriction patterns of ruminants (ox, *Bos taurus;* water buffalo, *Bubalus bubalis;* sheep, *Ovies aries;* goat, *Capra hircus;* pronghorn, *Antilocapra americanus;* giraffe, *Giraffa camelopardalis;* black-tailed deer, *Odocoileus hemionus*), on the other hand, a number of bands were observed, indicating that gene duplications occurred before divergence of the ruminants. This is in agreement with the indirect evidence from the most parsimonious tree (Fig. 1). The most likely number of cross-hybridizing ribonuclease genes in the investigated ruminant species as derived from the restriction patterns is three. These may be the orthologues of the bovine pancreatic, seminal and brain ribonuclease genes, respectively.

Similar observations have been made on the number of lysozyme genes in different mammalian species. Generally, only one or a few lysozyme genes are found, but in ruminants as many as 8 to 10 genes have been demonstrated (Irwin et al., 1989). Several of them, with a digestive function, are expressed in the stomach while others are found elsewhere and have an antibacterial function.

Most parsimonious trees of ribonuclease (Fig. 1) and lysozyme (Jollès et al., 1989) show increased evolutionary rates of the two enzymes at the origin of the ruminants, when the enzymes acquired more specific functions after the gene duplications had taken place. Later on, when the enzymes were adapted to their functions, the presence of stable symbiotic systems in the rumen caused very low evolutionary rates during the more recent stages of ruminant evolution (Fig. 1; Jollès et al., 1989).

Both ribonuclease and lysozyme have been recruited as digestive enzymes in ruminants and species with ruminant-like digestion, and adaptive structural and functional changes and gene duplications have led to a versatile system for both enzymes in the ruminants. Since separate mammalian taxa underwent similar adaptive changes of their digestive system, enzymes involved in this process have turned out to exhibit a rich repertoire of examples of convergent and parallel evolution.

ACKNOWLEDGMENTS

We thank Dr. D. M. Irwin, Dept. of Biochemistry, University of California, Berkeley, for his very generous gift of DNA samples of several artiodactyls, The Progetto Bilaterale, Consiglio Nazionale delle Ricerche for support of these studies and Dr. R. N. Campagne for critically reading the manuscript.

LITERATURE CITED

Barnard, E. A., 1969. Biological function of pancreatic ribonuclease. *Nature* 221:340–344.

Beintema, J. J., 1990. The primary structure of langur (*Presbytis entellus*) pancreatic ribonuclease: adaptive features in digestive enzymes in mammals. *Mol. Biol. Evol.* 7:470–477.

Beintema, J. J., Blank, A., Schieven, G. L., Dekker, C. A., Sorrentino, S & M. Libonati. 1988a. Differences in glycosylation patterns of human secretory ribonucleases. *Biochem. J.* 225:501–505.

Beintema, J. J., Gaastra, W., Lenstra, J. A., Welling, G. W. & W. M. Fitch. 1977. The molecular evolution of pancreatic ribonuclease. *J. Mol. Evol.* 10:49–71.

Beintema, J. J., Gaastra, W., Scheffer, A. J. & G. W. Welling, 1976. Carbohydrate in pancreatic ribonucleases. *Eur. J. Biochem.* 63:441–448.

Beintema, J. J., Scheffer, A. J., van Dijk, H., Welling, G. W. & H. Zwiers. 1973. Pancreatic ribonuclease: distribution and comparisons in mammals. *Nature New Biol.* 241, 76–78,

Beintema, J. J., Schüller, C., Irie, M. & A. Carsana. 1988b. Molecular evolution of the ribonuclease superfamily. *Prog. Biophys. Mol. Biol.* 51:165–192.

Carsana, A., Confalone, E., Palmieri, M., Libonati, M. & A. Furia. 1988. Stucture of the bovine pancreatic ribonuclease gene: the unique intervening sequence in the 5' untranslated region contains a promotor-like element. *Nucleic Acids Res.* 16:5491–5502.

Dobson, D. E., Prager, E. M. & A. C. Wilson. 1984. Stomach lysozymes of ruminants. I. Distribution and catalytic properties. *J. Biol. Chem.* 259:11607–11616.

Fitch, W. M. & J. J. Beintema. 1990. Correcting parsimonious trees for unseen nucleotide substitutions: the effect of dense branching as exemplified by ribonuclease. *Mol. Biol. Evol.* 7:438–443.

Hammer, M. F., Schilling, J. W., Prager, E. M. & A. C. Wilson, 1987. The recruitment of lysozyme as a major enzyme in the mouse gut: duplication, divergence and regulatory evolution. *J. Mol. Evol.* 4:272–279.

Hofmann, R. R., Geiger, G. & R. König. 1976. Vergleichend-anatomische Untersuchungen an der Vormagen-Schleimhaut von Rehwild und Rotwild. *Z. Säugetierk.* 41:167–193.

Irwin, D. M., Sidow, A., White, R. T. & A. C. Wilson. 1989. Multiple genes for ruminant lysozymes, pp. 73–85. *In:* S. Smith-Gill & E. Sercarz (eds.), *The Immune Response to Structurally Defined Proteins: The Lysozyme Model.* Adenine: Schenectady, NY.

Jollès, J., Jollès, P., Bowman, B. H., Prager, E. M., Stewart, C.-B. & A. C. Wilson. 1989. Episodic evolution in the stomach lysozymes of ruminants. *J. Mol. Evol.* 28:528–535.

Libonati, M., Furia, A. & J. J. Beintema. 1976. Basic charges on mammalian ribonuclease molecules and the ability to attack double-stranded RNA. *Eur. J. Biochem.* 69:445–451.

Stewart, C.-B., Schilling, J. W. & A. C. Wilson. 1987. Adaptive evolution in the stomach lysozymes of foregut fermenters. *Nature* 330:401–404.

Functional Characteristics of the Calcium Modulated Proteins Seen from an Evolutionary Perspective

Robert H. Kretsinger, Susumu Nakayama, and Nancy D. Moncrief

Abstract. We have constructed dendrograms relating 178 proteins that contain from two to eight tandemly repeated homolog domains called EF-hands or calmodulin folds. There are 14 distinct subfamilies and 13 individual proteins that may be the first representatives of new subfamilies. In Figure 1 we indicate the 27 subfamilies, the number of EF-hand domains, observed or inferred calcium binding, and the relationships among the subfamilies expressed in minimal mutation distances. Nodal sequences for each of the 14 subfamilies emphasize the essential features of those subfamilies. The existence of numerous isotypes with several subfamilies cautions against using the EF-hand proteins to establish phylogenies of organisms. We have also constructed dendrograms based on pairs of EF-hand domains (Figure 3). The distribution of intron positions and phases are consistent within subfamily but show little interpretable relationship among subfamilies. Three results—sensitivities of the dendrogram relating subfamilies, lack of congruence between dendrograms using entire sequences and pairs of domains, and lack of congruence between dendrograms based on amino acid sequences and that based on introns—all support the conclusion that the rates of amino acid sequence change were much slower after the functions of the subfamilies were defined and their domain structures established. Further, we suggest that the complex patterns of gene duplications and splicings that occured prior to the establishments of these subfamilies may not be interpretable with present data and analytical procedures.

INTRODUCTION

Kretsinger (1975) proposed that calcium functioning as a cytosolic messenger is bound by proteins and that these calcium modulated proteins are members of one homolog family. The members of this family contain several tandem copies of the calmodulin fold or EF-hand. It consists of an α-helix, calcium binding loop, α-helix. The canonical calcium binding EF-hand is easily recognized by the 1975 mnemonic:

```
----------- α-helix E -----------                   ----------- α-helix F -----------
                    ----- calcium binding loop -----
                        1 1 1 1 1 1 1 1 1 1 2 2 2 2 2 2 2 2 2 2
1 2 3 4 5 6 7 8 9 0 1 2 3 4 5 6 7 8 9 0 1 2 3 4 5 6 7 8 9
E n * * n n * * n X * Y * Z G # I-X * *-Z n * * n n * * n  (1975)
```

The first residue is Glu; "n" refers to hydrophobic residues—Val, Ile, Leu, Met, Phe, Tyr, Trp—the inner aspects of the α-helices; X, Y, Z, -X, and -Z refer to calcium coordinating

Drs. Kretsinger, Nakayama, and Moncrief are with the Department of Biology, University of Virginia, Charlottesville, VA 22901, USA. Dr. Moncrief is also with the Virginia Museum of Natural History, Martinsville, VA, 24112, USA.

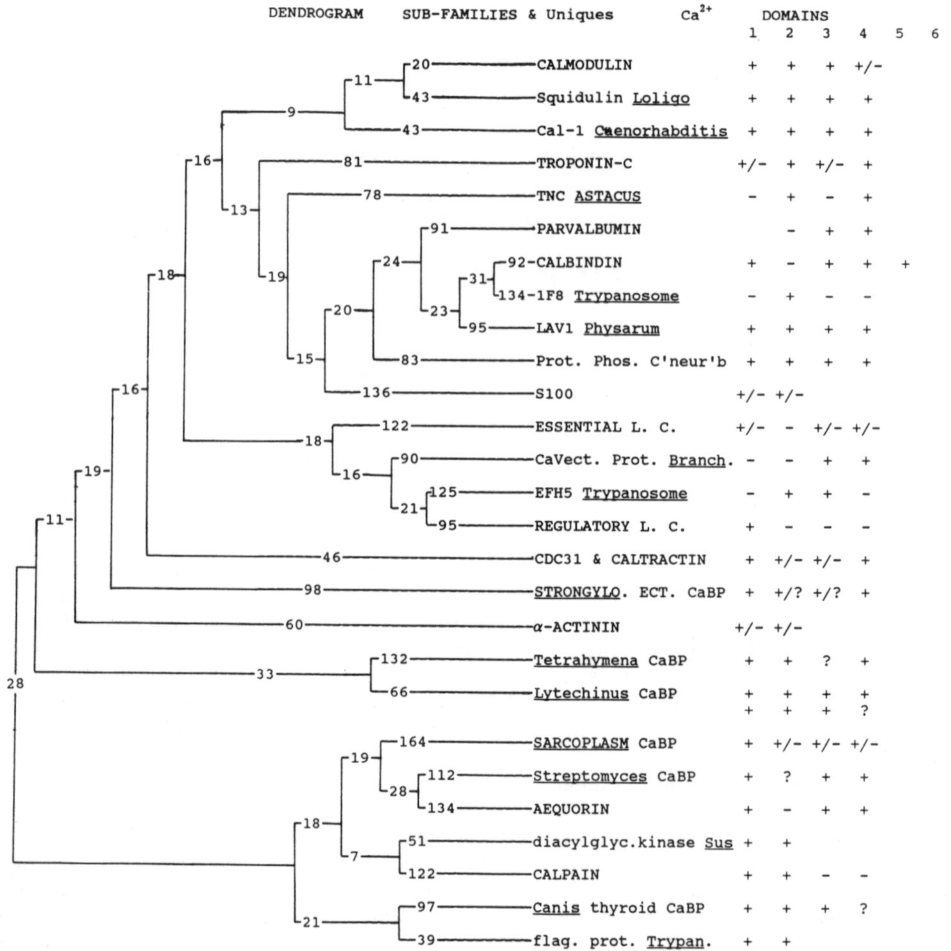

Figure 1. Dendrogram of Twenty-seven EF-hand Subfamilies. We have identified 14 subfamilies and 13 unique proteins, probably the first representatives of separate subfamilies; the uniques are indicated in lower case script. The numbers of EF-hand domains are indicated as well as their (inferred) ability to bind calcium. "+/−" indicates that some domains do and some do not bind calcium; "?" indicates uncertainty. The protein from *Lytechinus* has eight domains. Minimum mutation distances are indicated; for proteins having other than four domains distances are normalized to four. Distances to the 14 terminal nodes of subfamilies are weighted averages of all members. The root of the dendrogram has not been determined; the lower five subfamilies appear most distant from the upper 17 uniques and subfamilies.

amino acids—Asp, Asn, Ser, Thr, Glu, Gln— at the vertices of an octahedron; Gly occurs at a sharp bend and uses dihedral angles restricted to the other 19 amino acids; "I'"—Ile, Leu, Val—provides a hydrophobic side chain to the core of the molecule. The "#" at position 16, the -Y vertex, can be occupied by any amino acid; here calcium is coordinated with the carbonyl oxygen atom. The "*" indicates that any amino acid is acceptable at those twelve positions. Our immediate goal is to identify all sequenced EF-hands and to distinguish them for numerous similar analogs. Second, we have classified the proteins that contain EF-hands. Finally, we draw conclusions about the relationships of evolutionary processes and functional characteristics.

CHARACTERISTICS OF DOMAINS

There are 647 EF-hand domains in the 178 homologous proteins. Twenty of these are the unique "S100" domain found as the first of the two domains of members of the S100 subfamily. The first domains of the S100 proteins have two amino acids inserted into the calcium binding loops. Instead of coordinating calcium with five side chains and one carbonyl oxygen at vertex -Y (#), the S100 domain employs four carbonyl oxygen atoms "=O" and one side chain, Glu, at the -Z vertex. Loop positions are numbered from the -Y residue with carbonyl oxygens at 0, -3, -5, and -8. Although the helix, loop, helix motif is recognized, the consensus for the S100 domain is significantly different from that of the canonical EF-hand.

E n * * n n * * n X * Y * Z G # I-X * *-Z n * * n n * * n (1975)
                                =O
* n * * n F * * n -8 * * E G D * * # L * K * E L K * L n * * E (S100)
              =O =O =O =O

Of the remaining 627 domains 371 are judged to bind calcium; 16 are questionable; and 240 probably do not bind calcium. Those not binding calcium have a variety of substitutions and insertions or deletions. Kretsinger et al., (1990) developed and evaluated several algorithms for recognizing and aligning these more divergent domains. There are also many examples of cysteines at calcium coordinating postions, of amino acids other than glycine at position 15, and of prolines at positions in which their dihedral angle of $-55°$ could not be accommodated in the four EF-hand proteins of known crystal structure.

PROCEDURES

As described by Kretsinger et al., (1990) several alignment procedures and query patterns were used to identify 178 proteins that contain from two to eight EF-hands in tandem repeat and to distinguish these homologs from several very similar analogs. The maximum parsimony methods used to generate dendrograms were described by Moncrief et al., (1990). The internodal distances of figure 1 were normalized to four domain long proteins. The distances to terminal nodes of subfamilies represent weighted averages of all members of those subfamilies. The dendrogram is unrooted. The seven subfamilies at the bottom seem most distant from the twenty at the top of the dendrogram.

DISTRIBUTION

The calmodulins are found in all four kingdoms—Fungi, Protists, Plants, and Animals. Most other EF-hand proteins are restricted to animals, excepting CDC31 and caltractin found in fungi and protists, α-actinin and LAV1 of slime mold, the essential light chain of myosin found in fungi, the three proteins—1F8, flagellar protein, and EFH5—of trypanosomes, and the *Tetrahymena* calcium binding protein. Several have been found only in invertebrates: the sarcoplasm calcium binding proteins, squidulin, cal–1, calcium vector protein, *Strongylocentrotus* ectodermal protein, *Lytechinus* calcium binding protein, and aequorin. The *Streptomyces* calcium binding protein is the only representative from a prokaryote; whether it functions as a second messenger remains to be determined. S100, calcineurin B, and the thyroid protein have been found only in mammals. A typical mammal would contain EF-hand proteins from at least 11 subfamilies and from within each subfamily multiple isotypes.

Few simple patterns emerge. Excepting the first domain of S100, there seems to be no

920

special subset of domains associated with any subfamily. As noted in the preceding paragraph some subfamilies have, to date, been found only in certain kingdoms or even in a single species; even so, we see no patterns of amino acid sequences associated with specific groupings of organisms. Although one usually associates calcium binding with the EF-hand, we cannot infer with certainty whether the primordial domain or pair of domains bound calcium.

RELATIONSHIPS WITHIN DOMAINS

Dendrograms were constructed for each subfamily, for example, that of parvalbumin as shown in Figure 2. These dendrograms are relatively robust, in contrast to the dendrogams relating subfamilies; the topologies change little upon addition or subtraction of individual members. The rates of evolution within subfamilies are relatively constant and, not surprisingly, the inferred characteristics, such as calcium binding and deletions, are quite constant within subfamilies. As illustrated for parvalbumin, one organism often contains several isotypes. It can be difficult to distinguish orthologs from paralogs. This multiplicity of isotypes, as well as of subfamilies, makes it difficult to use the EF-hand proteins as markers to establish organismal phylogenies.

PAIRS OF DOMAINS

We constructed dendrograms based on pairs of domains, for instance, domains 1,2 and domains 3,4 of calmodulin. With only a few minor exceptions the relationships within subfamilies are the same whether constructed from the entire proteins or pairs of domains.

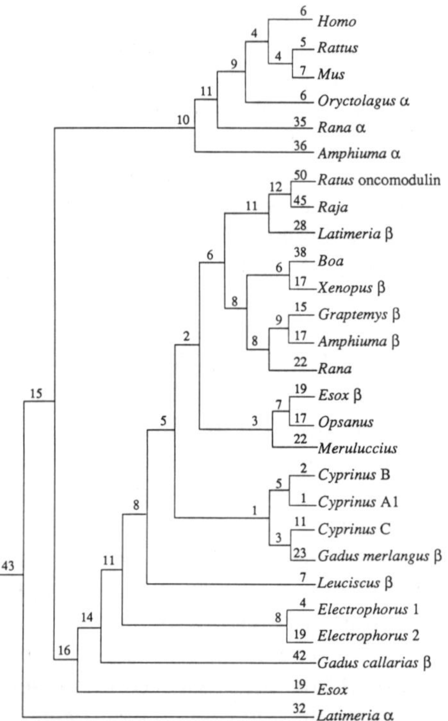

Figure 2. Dendrogram of the Parvalbumin Subfamily. The genus of the source of parvalbumin is indicated as well as the author's designation of a or b subfamily, if given. Minimum mutation distances between nodes are indicated for this three domain protein.

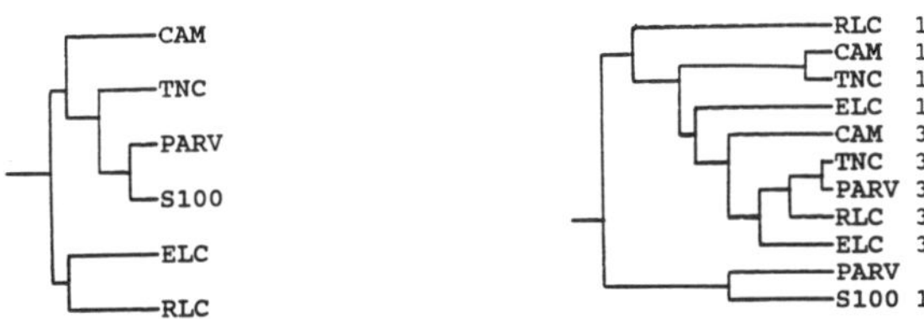

Figure 3. Relationships among Pairs of Domains vs Entire Molecules. The six subfamilies—calmodulin, troponin C, parvalbumin, S100, enzymatic light chain, and regulatory light chain of myosin—contain 109 of the 178 EF-hand containing proteins of known amino acid sequence. The relationships among the six subfamilies are shown on the left. On the right is the dendrogram calculated for pairs of domain plus the residues connecting domains 1 and 2 and those connecting domains 3 and 4. The relationships within subfamilies (not shown) are similar for either pairs of domains and for the entire molecules. The S100 subfamily contains only domains 1 and 2; parvalbumin has only domain 2 of the 1,2 pair. The lack of congruence between the two dendrograms is tentatively interpreted as indicting several gene duplication and splicing events during the evolution of these subfamilies.

In marked contrast significant changes in the dendrograms relating subfamilies often alter the overall score by only a few additional mutation units. This sensitivity, or lack of robustness, is also seen when addition of new sequences alters previous relationships among, but not within, domains. Tentatively we interpret these findings to indicate that the rate of change of these proteins was much greater prior to the establishments of the subfamilies than after each subfamily had assumed an important function in the precursor organism. The relationships among subfamilies computed from pairs of domain show significant deviations from congruence. That is, all pairs 1,2 do not form a dendrogram identical to the dendrogram of all pairs 3,4 nor identical to the dendrogram of complete molecules. Although it is difficult to assign a statistical significance to these results, we suggest that this lack of congruence reflects complex patterns of gene duplications and recombinations.

The six subfamilies—calmodulin, troponin C, parvalbumin, S100, essential light chain, and regulatory light chain of myosin—contain 109 (24 + 13 + 27 + 20 + 25 + 17) of the 178 proteins. They are also the best characterized biochemically. Figure 3 shows the relationships among the 11 pairs of domains. The relationships for pairs of domains within subfamilies were permitted to vary from those found optimal for entire sequences. If the relationships within subfamilies are constrained to be the same as those for entire sequences, the relationships among the 11 pairs of domains are similar to those of the unconstrained dendrogram. The important point is that using the data for 109 proteins one finds significant lack of congruence between dendrograms based on pairs of domains and that based on entire sequences.

INTRON DISTRIBUTION

Genomic DNA sequences are available for 23 EF-hand proteins representing seven different subfamilies and two uniques. The 106 different introns have been classified, positions 1 through 25, by their positions within or between (or preceding) EF-hand domains and by their phase relative to encoding triplets. For calmodulin, enzymatic light chains, and regulatory light chains there are gDNA sequences available from both *Drosophila* and *Rattus*. For all three subfamilies the introns in *Drosophila* are also present in *Rattus*; for all three there are additional introns in *Rattus*. These seven subfamilies and two uniques are shown in Figure 4 in the same cladistic relationship as in Figure 1 and their intron positions are also indicated. The dendrogram based on amino acid sequences does not optimally describe the relationships based on intron distribution.

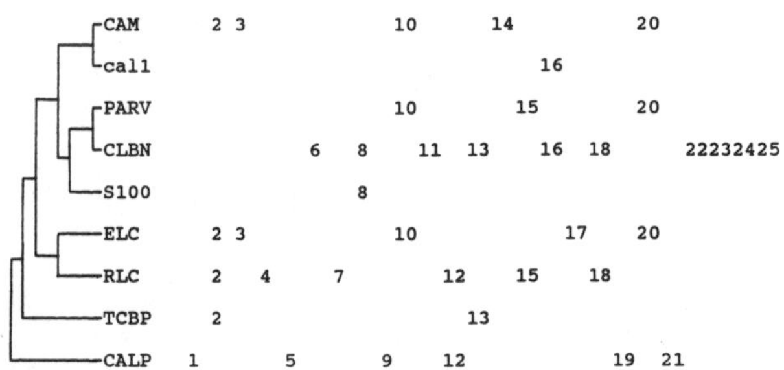

Figure 4. Summary of intron positions and types within subfamilies. Genomic DNA sequences are available for nine of the known subfamilies and uniques. The cladistic relationship of these nine is shown as extracted from Figure 1. There are 25 different classes of introns within the 26 gDNA sequences in these nine subfamilies. The occurences of the classes are indicated for each subfamily.

CONCLUSIONS

We have established a data base of the EF-hand proteins, confirmed its accuracy, aligned the EF-hand domains, and constructed dendrograms using the criterion of minimum mutation distance. The proteins are classified into 14 subfamilies and 13 unique proteins. This classification of subfamilies accords well with known physiological and biochemical characteristics of the proteins. When account is taken of the multiple isotypes, the dendrograms within subfamilies accord with relationships between species.

Excepting the first domains of the S100 subfamily, the numerous variations in the EF-hand homolog domain, such as insertions and deletions, calcium binding ability, and special amino acid substitutions, do not fall into discernible patterns. Dendrograms based on complete amino acid sequences are congruent within domains with those based on pairs of domains. In contrast there are significant deviations from congruence among domains when comparing dendrograms based on entire sequences with those based on pairs of domains, and with those based on distribution of introns. We interpret this lack of congruence as indicating that there were numerous gene duplication and splicing events prior to the establishment of functions for the subfamilies. After these establishments there was strong selective pressure to restrict alterations to point mutations.

We anticipate the discovery of many more subfamilies of EF-hand homolog proteins. As more functional characteristics are found we should be able to relate the evolution of these subfamilies to their functional constraints.

ACKNOWLEDGMENT

NASA grant NAGW-1233 to R.H.K.

LITERATURE CITED

Kretsinger, R. H. 1975. Hypothesis: Calcium Modulated Proteins Contain EF Hands. Pp. 469–478. *In: E. Carafoli, Clementi, F., Drabikowski, W. & A. Margreth. (eds.), Calcium Transport in Contraction and Secretion.* North-Holland Publishing Co., Amsterdam.

Kretsinger, R. H., Tolbert, D., Nakayama, S. & W. Pearson. 1991. The EF-Hand, Homologs and Analogs *In:* C. Hiezmann (ed.), *Novel Calcium Binding Proteins.* 17–38.

Moncrief, N. D., Goodman, M. & R. H. Kretsinger. 1990. Evolution of EF-Hand Calcium-Modulated Proteins I. Relationships Based on Amino Acid Sequences. *J. Mol. Evol.* 30:522–562.

Tests to Distinguish Between Phylogenetic Information and Random Noise in Nucleotide Sequence Data

James W. Archie

Abstract. Because of the nature of DNA sequence evolution, levels of randomness (homoplasy) in DNA sequence data can be expected to be relatively high, making the analysis of these data problematical for use in phylogenetic inference studies. As the level of homoplasy increases in these data, trees of equal and nearly equal length become more disparate. Computer simulations have shown that the consensus of estimated trees with the true tree decreases as the level of homoplasy increases. A randomization test has been developed to determine if there is significant nonrandom evolutionary information in a set of sequence data. The test distribution is derived by permuting the nucleotides among taxa within sites and deriving the minimum length tree from the randomized data. The length of the original tree can then be compared to this distribution. An associated index, the homoplasy excess ratio (Archie, 1989b), can be used to measure the overall level of randomness and identify problematical data sets. Two other homoplasy indices, the consistency index (Kluge and Farris, 1969) and retention index (Farris, 1989) are shown to be inappropriate for this purpose. The randomization test can be used to evaluate particular subsets of taxa or sequence domains. Ribosomal DNA sequences (Field et al., 1988) of a series of representative animal phyla along with plant, fungi, slime mold, and ciliate taxa are evaluated for their homoplasy levels. Overall, a very high level of homoplasy is present in these data and data for particular subsets of taxa are shown to contain no phylogenetic information.

INTRODUCTION

Recent technological advances are rapidly increasing the biological systematist's ability to collect comparative nucleotide sequence data for use in phylogenetic analysis. Although there is undoubtedly a tremendous amount of information on phylogenetic history in these data, three major problems confront systematists trying to use this information for phylogenetic inference—1) computational complexity, 2) sequence alignment, and 3) repetitive or redundant sequence change (homoplasy). The availability of faster computers and more efficient computer programs are aiding the complexity problem, although solutions for many inference problems will always be only approximate. The sequence alignment problem is related to the first but is more biological in nature. The third problem is intricately related to and dependent on the others—depending on the particular alignment chosen, homoplasy levels inferred from the data will either increase or decrease. In this paper, I address the problem of the estimation of homoplasy levels in nucleotide sequence data and the importance of having an adequate

Dr. Archie is with the Department of Biology, California State University, Long Beach, CA 90840, USA.

924

estimate of homoplasy level in evaluating the quality of and expected confidence in both the data and the resulting estimates of phylogenetic relationship derived from the data.

ESTIMATING LEVELS OF RANDOMNESS IN NUCLEOTIDE SEQUENCES

There are two approaches to estimating the levels of randomness or repetitive sequence change in nucleotide sequences. The first is derived from the theoretical expectation of the accumulation of repetitive changes correlated with sequence divergence. Nei (1987) provides a recent review. As the percentage of sequence divergence increases, as estimated by aligning and comparing any two sequences, the expected proportion of repetitive changes increases. The difficulty with this approach is that the estimated amount of repetitive change associated with a particular level of divergence depends on the initial nucleotide frequencies before divergence, and on the substitution probabilities exhibited by the sequences during evolution and that the investigator is willing to assume is reasonable. For example, with equal frequencies and substitution probabilities, the saturation level of divergence is 0.75, while with substantially unequal starting frequencies and a high variance in substitution probabilities for codon groups, the saturation level of divergence may be as low as 0.33 (Lewontin, 1989). Estimated levels of repetitive change can only be derived relative to these saturation levels. As a result, the interpretation or estimation of amount of randomness in nucleotide sequences from the observation of percent sequence divergence depends more on intrinsic, and in many cases unknowable, properties of the biological system (initial nucleotide frequencies and substitution probabilities) than on empirical observation.

The second approach to the estimation of levels of randomness in nucleotide sequences is to scale an estimate of the observed amount of repetitive change to the worst possible case. A lower bound on the amount of repetitive sequence change (randomness or homoplasy) can be estimated directly from the observed sequence data using the principle of parsimony derived from the field of cladistics (Farris, et. al., 1970). In this approach, a tree of relationships is estimated from the data by weighting each site equally, allowing each nucleotide to change to any other nucleotide with equal probability, and by treating each site independently. Under these assumptions, the best estimate of relationships (and the tree that provides a lower bound on the level of repetitive sequence change) is the minimum length tree, i.e., the one requiring the fewest hypotheses of sequence change. The worst possible case, i.e., the maximum amount of repetitive change for the observed sequences, is then obtained by the estimation of the distribution of lengths of minimum length (ML) trees for *comparable* but *phylogenetically random* sequence data. The degree of randomness in observed sequences can then be estimated by comparing the length of the ML tree from the original data to the distribution of tree lengths for the phylogenetically random data. The difficulty is in the estimation of the distribution of tree lengths for phylogenetically random sequence data.

A RANDOMIZATION TEST FOR PHYLOGENETIC INFORMATION

To examine the distribution of steps in random phylogenetic data, the observed distribution of states within sites must be maintained, but all hierarchical correlations between sites must be eliminated. Perfect hierarchical correlations between sites implies that there have been no repetitive changes. Although it is possible to create random sequence data easily using a random number generator from a computer to assign nucleotide states to sequences, it would be necessary to choose an arbitrary probability model to generate such data; for example, equal frequencies of all four nucleotides.

Table 1. Formulae for three available homoplasy indices.

1. Consistency Index (Kluge and Farris, 1969)

$$CI = \frac{\Sigma(\text{Number of States in Character} - 1)}{\text{Character Lengths on ML Tree}}$$

$$= \frac{\text{Minimum Possible Changes}}{\text{Length of Minimum Length Tree}}$$

$$= \frac{\text{Minimum Possible Changes}}{\text{Min. Changes} + \text{Homoplastic Changes}}$$

2. Retention Index (Farris, 1989)

$$RI = \frac{\text{Max. Possible Changes} - \text{Observed Changes}}{\text{Max. Possible Changes} - \text{Minimum Changes}}$$

Minimum Changes $= \Sigma$ (Number of States in Character $- 1$)

Max. Possible Changes — Determined from the distribution of character states among the taxa

3. Homoplasy-Excess Ratio (Archie, 1989)

$$HR = \frac{1.0 - \text{Observed Homoplasy on ML Tree}}{\text{Maximum Possible Homoplasy}}$$

$$= \frac{\text{Max. Possible Changes} - \text{Observed Changes}}{\text{Max. Possible Changes} - \text{Minimum Changes}}$$

Minimum Changes $= \Sigma$ (Number of States in Character $- 1$)

Maximum Possible Changes $=$ mean number of steps on minimum length tree for randomized data.

However, it is well known that codon bias exists for many, if not most, sequences and Archie (1989a) has shown that the distribution of lengths of minimum length trees differs for different character state distributions. As a result, it is necessary to maintain the state distributions *within each site.* Such random sequence data can be obtained by randomly permuting the nucleotide state assignments within each site. A distribution of tree lengths for minimum length trees for random sequence data can be obtained by successively permuting the site data and estimating the minimum length tree for the randomized data some arbitrary number of times (e.g., $r > 100$). The length of the tree calculated from the original (non-randomized) data can then be compared to this distribution. An hypothesis of no phylogenetic information in the data can be tested by determining if the length of the original data tree is less than 95% or 99% of the lengths of the randomized data trees. An example of these distributions will be given below.

INDICES OF LEVELS OF HOMOPLASY

Three indices have been introduced as measures of the amount of homoplasy (repetitive change). These are the consistency index (Kluge and Farris, 1969), the homoplasy-excess ratio (Archie, 1989b), and the retention index (Farris, 1989)[1]. Specific formulae for these indices are given in Table 1. The main difference between the consistency index (CI) and the other two is that the CI compares the observed amount of homoplasy for the data (actually, the minimum number of changes plus the number of homoplastic changes) to the length of a tree that would be obtained if there was no homoplasy, while the

---

[1]Farris (1989, 1990) showed that the retention index is the same as the homoplasy-excess ratio maximum introduced by Archie (1989b).

926

homoplasy-excess ratio (HER) and retention index (RI) compare the observed amount of homoplasy to the maximum possible homoplasy for the data. The difference between the latter two indices is in the way that the maximum possible homoplasy for the data is determined. The two formulae are identical otherwise.

In deriving the formula for the HER, Archie (1989b) estimated the maximum possible homoplasy from the mean number of steps on minimum length trees for randomized data, as described above. In contrast, Farris (1989; see also Archie [1989b]) determined the maximum possible homoplasy from the distribution of character states among the taxa. This can be seen by referring to Fig. 1. For a single binary character, such as that in Fig. 1, the number of steps on a "bush" is the maximum possible number that could be obtained on any tree. The actual number is determined by choosing the state of the ancestor. Since we are interested in the number of steps on minimum length trees, as the state of the ancestor we choose the state with the highest frequency among the taxa, in this case 0. The maximum number of steps is simply the number of 1 states among the taxa. The maximum number of steps on a tree can be determined for nucleotide sequence data (4 possible states) by determining the most frequent state at each site and subtracting this value from the number of taxa, t. The overall retention index is calculated by summing the maximum number of steps at each site (determined in this way), across all sites. Clearly, the RI is an easier statistic to calculate than the homoplasy ratio. The question is which of these, or the consistency index, is a "better" estimate of the amount of homoplasy in the data. Each index achieves a value of 1.0 if there is no homoplasy in the data and decreases towards 0.0 as the amount of homoplasy in the data increases. One would expect that if there is no phylogenetic information in the data, i.e., they exhibit phylogenetic randomness, that each index should equal 0.0, otherwise, a particular value would be difficult to interpret. In addition, an appropriate index for comparative purposes should be scaled to account for different numbers of taxa and characters as well as character state distributions. Two examples of the use of these statistics and the randomization procedure, described above, will now be discussed.

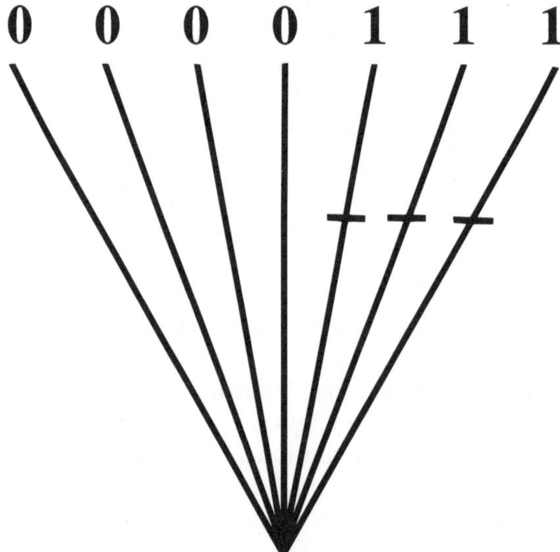

Figure 1. Derivation of the maximum number of steps possible on any minimum length tree. The bars represent character state changes. The state 0 is assumed to have been assigned to the ancestral node.

AN EXAMPLE USING PROTEIN SEQUENCES ON PLANT FAMILIES

In an analysis of the relationships of plant families, Bremer (1988) analyzed the unambiguous, informative nucleotide sequence sites derived from amino acid sequences for four proteins: ribulose biphosphate carboxylase, plastocyanin, cytochrome c, and ferredoxin. As the ferredoxin sequences were available for only six of the families, he performed two separate analyses, one with six taxa and one with nine. Bremer's basic observations from the results of his analyses were that trees that differed by only two or three steps from the minimum length tree for the data, differed substantially in inferred topological relationships. In fact, for the analysis with nine taxa, if a consensus tree (Rohlf, 1982) was made of all trees that differed from the minimum length tree by at most 3 steps (151-154 steps), there was no structure to the consensus tree. Similarly, for the six taxa study, a consensus tree of trees that differed from the minimum length tree by at most two steps (161–163 steps) had no structure. The consistency indices for the minimum length trees were 0.583 and 0.689, neither of which appeared alarmingly low. Archie (1989c) presented an analysis of these data using the randomization technique described here to elaborate on Bremer's results. As shown in Table 2, the range of tree lengths (r = 200 and 100, respectively) for nine and six taxa included the length of the minimum length tree, and Archie (1989c) showed that the hypothesis of no phylogenetic information in these data cannot be rejected. In contrast to the CI, the homoplasy-excess ratios for these data are 0.046 and -0.008, agreeing completely with the conclusion of no phylogenetic information.

Table 2. Results of the homoplasy analysis of plant families based on nucleotide sequences derived from proteins. The data are from Bremer (1988).

	9	6
Number of Taxa	9	6
Number of Sites	64	84
Number of states in data	88	111
Length of minimum length tree	151	161
Consistency Index (CI)	0.583	0.689
Number of Minimum Length Trees	2	1
Number of trees 1 step longer	3	1
Number of trees 2 steps longer	9	3
Number of trees 3 steps longer	24	—
Mean Tree Length for Randomized Data	155.0	160.61
Range of tree lengths	(146–162)	(148–166)
Homoplasy Excess Ratio	0.046	−0.008

AN EXAMPLE USING RIBOSOMAL RNA SEQUENCES

Field et al. (1988) presented an analysis of sequences of 18S ribosomal RNA for 24 animal taxa from diverse animal phyla plus four non-animal taxa (Table 3). Using the Fitch and Margoliash (1967) distance based method they derived a phylogenetic tree for these 28 taxa. The relationships inferred from this study (Fig. 2) have been particularly controversial for two reasons: 1) their analysis implies that the kingdom Animalia and, also, the Metazoa, are polyphyletic, with the Cnidaria more closely related to the Fungi, Plantae, and a ciliate protozoan than to the rest of the animal kingdom, and 2) their analysis implies that the Arthropoda is a relatively early derivative from the primary animal lineage, having branched off the main lineage prior to other Protostomia, including the Annelida. Traditional interpretations have the Mollusca and Annelida being derived before the Arthropoda within the Protostomia.

Table 3. Taxa used by Field et al., (1988) in their phylogenetic analysis of animal phyla.

Protista		
	Ciliate	— *Oxytricha nova*
	Cellular slime mold	— *Dictyostelium discoideum*
Fungi		
	Yeast	— *Saccharomyces cerevisiae*
Plantae		
	Corn	— *Zea mays*
Animalia		
	Acoelomates	
	Cnidaria	
	Hydra	— *Hydra* sp.
	Anemone	— *Metridium senile*
	Platyhelminthes	
	Planaria (Flatworm)	— *Dugesia tigrina*
	Coelomates	
	Protostomia	
	Annelida	
	Polychaete	— *Chaetopteris* sp.
	Oligochaete	— *Lumbricus* sp.
	Mollusca	
	Clam	— *Mya arenaria*
	Clam	— *Spisula solidissima*
	Chiton	— *Cryptochiton stelleri*
	Nudibranch	— *Anisodoris nobilis*
	Arthropoda	
	Fruit fly	— *Drosophila melanogaster*
	Brine shrimp	— *Artemia salina*
	Millipede	— *Spirobolus marginatus*
	Horseshoe crab	— *Limulus polyphemus*
	Sipuncula	
	Peanut worm	— *Golfingia gouldii*
	Pogonophora	
	Vent worm	— *Riftia pachyptila*
	Lophophorata	
	Brachiopoda	
	Lamp shell	— *Lingula reevi*
	Deuterostomia	
	Echinodermata	
	Starfish	— *Asterias forbesi*
	Brittlestar	— *Ophiocoma wendtii*
	Urchin	— *Arbacia punctulata*
	Crinoid	— *Lamprometra palmata*
	Chordata	
	Tunicate	— *Styela clava*
	Amphioxus	— *Branchiostoma californiense*
	Human	— *Homo sapiens*
	Frog	— *Xenopus laevis*

Although these data have been reanalyzed extensively (Ghiselin, 1988; Patterson, 1989; Lake, 1990) with certain of the controversial results reinterpreted, it is enlightening to examine these data using Wagner parsimony and the randomization technique (described here) along with the three homoplasy indices (Archie, in preparation). The results from these analyses are presented in Tables 4–8. In each analysis, I first used either PAUP (Swofford, 1985) or HENNIG86 (Farris, 1988) to determine the length of the minimum length tree for the particular subsample of taxa chosen. This provided a value for the consistency index and retention index. I derived a value for the homoplasy-excess ratio using 25 or more randomizations of the original data followed by estimation of the

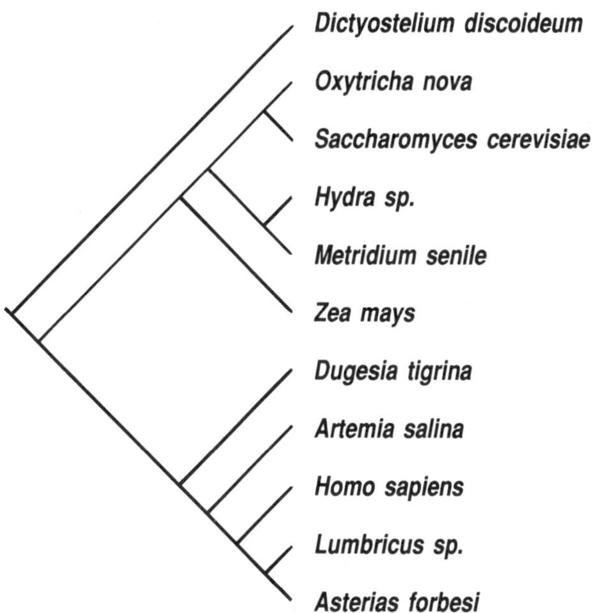

Figure 2. Diagram of relationships among major taxonomic groups for 28 taxa derived using Fitch-Margoliash optimization of a genetic distance matrix as depicted in Field et al. (1988).

length of the minimum length tree using PAUP. In most cases I used global branch swapping with PAUP to try to get a better estimate of the ML tree. In marginal cases, I used the branch-and-bound option of PAUP to guarantee a most parsimonious solution. In addition to determining the length of the ML tree for the randomized data, in each case I also calculated the value of the consistency index and the retention index. The average values from the replicates provide estimates of the minimum possible value achievable for each of these statistics assuming that there is no phylogenetic information in the data. Using this reasoning, the average minimum-value for the homoplasy-excess ratio will be equal to 0.0.

For all 28 taxa and 346 informative sites, the analysis yielded two minimum length trees with 1420 steps (CI=0.44; RI = 0.50). Figure 3 shows one of the two minimum length trees. In this tree the Metazoa and Animalia are monophyletic, but the Arthropoda are paraphyletic and the Mollusca are polyphyletic. Using 180 informative sites (due to program limitations), I obtained a tree with 735 steps (CI= 0.437, RI=0.504, HER = 0.349; Table 4). The range of steps on randomized data trees was 943–964. Clearly, the ML tree length is well below that for the randomized data trees, allowing rejection of the hypothesis of no phylogenetic information. However, the low values of all three indices indicate that there is a substantial amount of homoplasy in these sequence data. An important observation (see below), is that the mean values for the CI and RI derived from the randomized data were not 0.0, but were found to be 0.336 and 0.358, respectively. HER for these data is 0.0, by definition.

Because of the large amount of homoplasy and the controversial results from the analysis of the original data by Field et al., (1988), I performed several subsequent analyses to determine if I could identify particular parts of the data that might be considered problematical or exhibit exceptionally high average levels of homoplasy. Four additional analyses were performed (Tables 5–8). The first analysis (11 taxa total) included all Deuterostome taxa analyzed by Field et al. (1988) plus *Artemia*, an oligochaete, and Planaria, the last three taxa added for rooting purposes. For this analysis, the CI increased

930

Table 4. Results from randomization analysis of entire Field et al., (1988) data set.

Number of Taxa	28
Number of Informative Sites	180
Number of States in Data	501
Maximum Possible Number of Steps	1156
Mean Tree Length for Randomized Data	954.2
Range of Tree Lengths for 50 Replicates	(943, 964)
Minimum Length Tree	735
Consistency Index	0.437
[CI = (501—180)/735]	
Mean CI for Randomized Data	0.336
Retention Index	0.504
[RI = (1156—735)/(1156—735)]	
Mean RI for Randomized Data	0.358
Homoplasy-Excess Ratio	0.349
[HR = (954.2—735)/(954.2—321)]	
Mean HER for Randomized Data	0.00

Table 5. Results from analysis relationships of Deuterostome taxa including four Echinodermata (*Arbacia, Asterias, Ophiocoma,* and *Lamprometra*), four Chordata (*Styela, Branchiostoma, Homo, Xenopus*), Arthropoda (*Artemia*), Annelida (*Lumbricus*), and Platyhelminthes (*Dugesia*).

Number of Taxa	11
Number of Informative Sites	138
Number of States in Data	346
Maximum Possible Number of Steps	509
Mean Tree Length for Randomized Data	429.3
Range of Tree Lengths for 25 Replicates	(419, 439)
Minimum Length Tree	347
Consistency Index	0.599
[CI = (346—138)/347]	
Mean CI for Randomized Data	0.485
Retention Index	0.538
[RI = (509—347)/(509—138)]	
Mean RI for Randomized Data	0.218
Homoplasy-Excess Ratio	0.372
[HR = (429.3—347)/(429.3—138)]	
Mean HER for Randomized Data	0.00

Table 6. Results from analysis of relationships of eucoelomate protostome taxa (Nudibranch, *Mya, Spisula,* Chiton, Brachipoda, Pogonophora, Oligochaete, Polychaete, Sipunculid, *Artemia,* Human, Starfish, Planaria)

Number of Taxa	11
Number of Informative Sites	180
Number of States in Data	488
Maximum Possible Number of Steps	737
Mean Tree Length for Randomized Data	625.5
Range of Tree Lengths for 25 Replicates	(615, 636)
Minimum Length Tree	531
Consistency Index	0.580
[CI = (332—132)/363]	
Mean CI for Randomized Data	0.493
Retention Index	0.480
[RI = (737—531)/(737—308)]	
Mean RI for Randomized Data	0.096
Homoplasy-Excess Ratio	0.298
[HR = (625.5—531)/(625.5—308)]	
Mean HER for Randomized Data	0.00

Table 7. Analysis of Arthropods (*Limulus*, Millipede, *Artemia*, *Drosophila*, Oligochaete, Human, Starfish, Planaria)

Number of Taxa	8
Number of Informative Sites	141
Number of States in Data	348
Maximum Possible Number of Steps	419
Mean Tree Length for Randomized Data	361.6
Range of Tree Lengths for 150 Replicates	(352, 376)
Minimum Length Tree	353
Consistency Index	0.586
[CI = (346—138)/347]	
Mean CI for Randomized Data	0.573
Retention Index	0.311
[RI = (509—347)/(509—138)]	
Mean RI for Randomized Data	0.271
Homoplasy-Excess Ratio	0.055
[HR = (429.3—347)/(429.3—138)]	
Mean HER for Randomized Data	0.00

Table 8. Results from analysis of deepest relationships (Ciliate, Slime Mold, Yeast, *Zea*, Hydra, Planaria)

Number of Taxa	6
Number of Informative Sites	136
Number of States in Data	335
Maximum Possible Number of Steps	361
Mean Tree Length for Randomized Data	314.8
Range of Tree Lengths for 150 Replicates	(304, 329)
Minimum Length Tree	304
Consistency Index	0.655
[CI = (332—132)/363]	
Mean CI for Randomized Data	0.633
Retention Index	0.333
[RI = (361—304)/(361—199)]	
Mean RI for Randomized Data	0.285
Homoplasy-Excess Ratio	0.067
[HR = (314.8—304)/(314.8—199)]	
Mean HER for Randomized Data	0.00

substantially over the entire data set while the RI and HER increased only slightly. In comparison, the mean CI for the randomized data also is higher (0.485 vs. 0.336) while the mean RI has actually decreased (0.218 vs 0.358). In any event, the amount of homoplasy for these taxa appears to be somewhat less than for the complete data set.

The species involved in the third analysis primarily represent the species of eucoelomate protostomes (t=11) but the results are quite different (Table 6). The consistency index for these data is only very slightly lower than for the deuterostomes and still higher than for the complete data set. However, the retention index is somewhat lower than for the prior two analyses (RI = 0.480) and the homoplasy-excess ratio has dropped to 0.298. In this case both the RI and HER indicate that there is somewhat greater average homoplasy among these taxa than for either the entire data set or the deuterostomes alone. Note that the difference between the observed CI and the mean CI for the randomized data sets is less than for either of the other two analyses.

The final two analyses address the problem from the complete data set of 1) the placement of the arthropods and 2) the relationship of the cnidarians to the other animal taxa. Recall that these relationships conflict substantively with prior placement of these groups. The results for these analyses are again quite different than for any of the prior analyses. In the analysis of the relationships of the Arthropoda (4 taxa) to several distantly related taxa

932

(oligochaete, human, starfish, and Planaria), the CI is similar in value (0.586) to that from the previous two subsamples (Table 7). The RI is lower than seen in the other subsample analyses and is comparable in value to that seen in the analysis of the complete data set (RI = 0.311). In contrast to both of these, however, the HER value for this analysis is close to 0.0, HER = 0.055. In addition, the length of the minimum length tree (L = 353) falls in the range of values observed for lengths of randomized data trees (352–376, r = 100). Both of these results indicate that there is essentially no information in these data on the relationships of the Arthropoda to these other taxa. Although this result is not apparent from the observed values of the CI or RI, note that the observed values are very similar in magnitude to the mean values obtained for these two statistics for the randomized data trees.

The final analysis uses six taxa (t=6) to investigate the deepest relationships addressed by these data—the relationships of representatives of the non-animal kingdoms plus the cnidarians and a single representative of the other animal phyla. The results are presented in Table 8. The CI for these data is the highest value observed for any of the analyses presented, CI = 0.655. The retention index is only slightly higher than in the analysis of the Arthropoda. Again, however, HER is close to 0.0, HER = 0.067, and the minimum length tree (L = 304) was equal to that observed for at least one of the randomized data trees (range = 304–329, r = 150). As with the arthropod analysis, the mean CI and RI for the randomized data differ very little from the observed values, even though the observed CI was higher than in any other analysis. The interpretation of these results is again that because of the large amount of homoplasy these sequence data contain no significant residual information on the deepest relationships addressed in the analysis, including the relationship of the cnidarians to the non-animal.

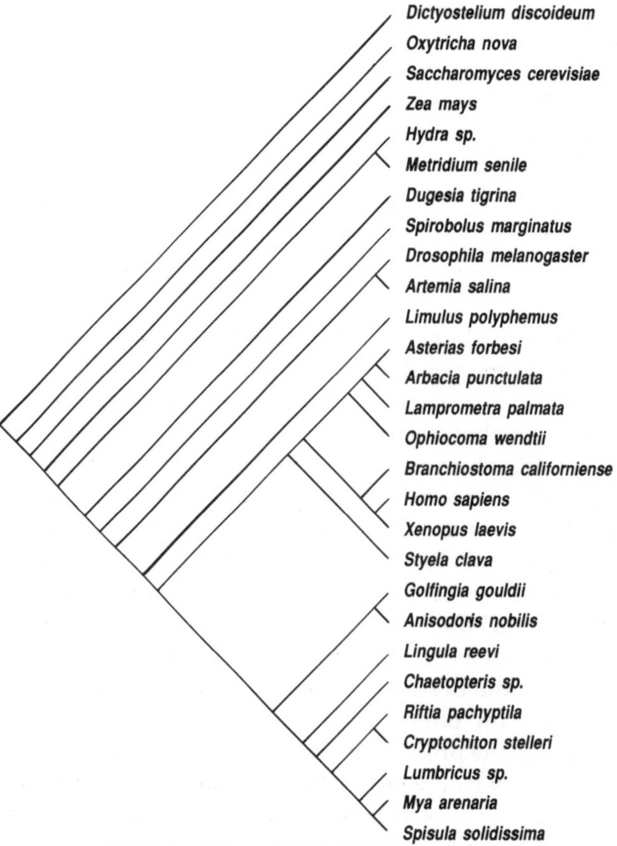

Figure 3. Minimum length tree for 28 taxa in Field et al., (1988) using the same alignments as those presented by Field et al.

DISCUSSION

Clearly, levels of homoplasy are important in evaluating the data and results from a phylogenetic analysis. As has been shown here, high homoplasy levels affect both the stability and reliability of estimated relationships. For phylogenetically informative data, the optimal case would be where the minimum length tree has substantially fewer steps than the next shortest tree. The usual case, however, is that there will be two to several trees that are either of minimum length or near minimum length. In these cases, the common observation is that these trees resemble one another in tree shape (topology). For sequence data which exhibit little residual phylogenetic information, however, trees that differ by only one or a few steps can be expected to differ drastically from one another in topology. In fact, as longer sequences are gathered, trees of nearly equal length would be expected to differ from each other more and more.

It should be clear that in order to evaluate the level of homoplasy in a set of data, an appropriate index for its measurement must be used. Although the consistency index has been used for the past 20 years as a comparative measure of homoplasy, until recently (Archie, 1989a, b; Sanderson and Donoghue, 1989) the properties of this statistic had not been investigated. It has now been shown (Archie, 1989b) that CI is directly negatively correlated with both the number of taxa, t, and the number of characters, N, used in a phylogenetic analysis. In addition, it is also affected by character state distributions, an aspect of the data uncontrollable by the investigator. Finally, the minimum achievable value for the CI is directly related to these same factors. As I have shown here and elsewhere (Archie, 1990) for the retention index, although uncorrelated with t, it as well as its minimum achievable value are affected by both N and character state distributions. In contrast, because of the manner in which it is calculated, Archie (1989b) showed that the HER is unaffected by either t, N, or character state distributions, and its minimum expected value is always 0.0. It is directly and appropriately affected, however, by the absolute level of apparent or estimated homoplasy determined by the number of repetitive changes on the minimum length tree.

One contradiction seems to be present in the use of HER as a measure of homoplasy. Although the expected value of HER for phylogenetically random data is 0.0, as was seen in the analysis of the plant family data, HER can take on negative values. Rather than being a contradiction, because of the formulation of HER, this is actually to be expected for finite length sequence data. Instead of using the maximum observed tree length from the randomized data, the formula uses the mean from the r repetitions. Whereas the maximum from a series of repetitions is very sample size dependent, the mean is a stable statistic and has been observed to have a very low variance among replicates. If the data are truly random, if tree lengths from the randomized data are symmetrically distributed about the mean, then approximately 50% of all trees derived from such random data should be negative, even though with infinitely long sequences the expected value of HER is 0.0.

The analyses that I have presented demonstrate several possible uses of the character randomization procedure and the homoplasy-excess ratio as tools for investigating the amount and variability in homoplasy levels among a series of taxa. The most important point from these results is to identify the necessity for biologists to evaluate the level of homoplasy in their data set as a whole, and to investigate the variability of inferred evolutionary rates and estimated homoplasy in different parts of their data. This test is expected to be quite powerful and easy to reject. For example, following the final analysis presented, if the anemone is added to the data set and reanalyzed, the null hypothesis is easily rejected and the HER increases to 0.1841 (the CI actually decreases; CI = 0.649). The residual information in the data identified by the test and HER is thus restricted primarily to the two relatively closely related cnidarian taxa.

There is clearly room for additional development of the randomization technique,

934

and, in particular, there is a need to more precisely define the nature of the statistical tests being tested. For example, Faith (1990) analyzed a set of sequence data originally analyzed by Thomas et al., (1989). He showed that the length of the minimum length tree that they obtained did not differ significantly from lengths of minimum length trees for randomized data. However, Thomas et al., (1989) had found that certain of their groups were well supported by the data using the bootstrapping technique of Felsenstein (1985a, b). Faith and Cranston (submitted) have found several other such data sets for which the null hypothesis tested by the randomization procedure could not be rejected. Clearly these results seem in conflict, but they can be explained by the small number of informative sites used by Thomas et al., (1989).

ACKNOWLEDGMENTS

I wish to thank the organizer of this symposium, David Hillis, for inviting me to participate. This research was partially supported by a grant from the National Science Foundation (USA), BSR-8918042.

LITERATURE CITED

Archie, J. W. 1989a. A randomization test for the presence of phylogenetic information in systematic data. *Systematic Zoology* 38:239–252.

Archie, J. W. 1989b. Homoplasy excess ratios: new statistics for measuring levels of homoplasy in systematic data. *Systematic Zoology* 38:253–269.

Archie, J. W. 1989c. Phylogenies of plant families: a demonstration of randomness in DNA sequence data derived from proteins. *Evolution* 43:1796–1800.

Archie, J. W. 1990. Homoplasy excess statistics and retention indices: a reply to Farris. *Systematic Zoology* 39:169–174.

Bremer, K. 1988. The limits of amino acid sequence data in Angiosperm phylogenetic reconstruction. *Evolution* 42:795–803.

Faith, D. 1990. Chance marsupial relationships. *Nature* 345:393–394.

Farris, J. S. 1988. *Hennig86*. Port Jefferson, NY.

Farris, J. S. 1989. The retention index and the rescaled consistency index. *Cladistics* 5:417–419.

Farris, J. S. 1990. The retention index and homoplasy excess. *Systematic Zoology* 38:406–407.

Farris, J. S., Kluge, A. G. & M. J. Eckhardt. 1970. A numerical approach to phylogenetic systematics. *Systematic Zoology* 19:172–189.

Felsenstein, J. 1985a. Confidence limits on phylogenies with a molecular clock. *Systematic Zoology* 34:152–161.

Felsenstein, J. 1985b. Confidence limits on phylogenies: an approach using the bootstrap. *Evolution* 39:783–791.

Field, K. G., Olsen, G. J., Lane, D. J., Giovannoni, S. J., Ghiselin, M. T., Raff, E. C., Pace, N. R. & R. A. Raff. 1988. Molecular phylogeny of the animal kingdom. *Science* 239:748–753.

Fitch, W. M. & E. Margoliash. 1967. Construction of phylogenetic trees. *Science* 155:279–284.

Ghiselin, M. T. 1988. The origin of molluscs in the light of molecular evidence. Oxford *Surveys in Evolutionary Biology* 5:66–95.

Kluge, A. G. & J. S. Farris. 1969. Quantitative phyletics and the evolution of anurans. *Systematic Zoology* 18:1–32.

Lake, J. A. 1990. Origin of the Metazoa. *Proc Natl. Acad. Sci., USA* 87:763–766.

Lewontin, R. 1989. Inferring the number of evolutionary events from DNA coding sequence differences. *Molecular Biology and Evolution* 6:15–32.

Nei, M. 1987. *Molecular Evolutionary Genetics* Columbia: New York. 512 pp. Patterson, C. 1989. *In:* B. Fernholm, K. Bremer, and H. Jornvall (eds.), *The Hierarchy of Life*. Excerpta Medica: Amsterdam, the Netherlands.

Rohlf, F. J. 1982. Consensus indices for comparing classifications. *Mathematical Bioscience* 59:131–144.

Sanderson, M. J. & M. J. Donoghue. 1989. Patterns of variation in levels of homoplasy. *Evolution* 43:1781–1795.

Swofford, D. L. 1985. *PAUP—Phylogenetic Analysis Using Parsimony*, Version 2.4. Illinois Natural History Survey, Champaign.

Thomas, R. H., Schaffner, W., Wilson, A. C. & S. Paabo. 1989. DNA phylogeny of the extinct marsupial wolf. *Nature* 340:465–467.

Experimental Phylogenies
from T7 Bacteriophage

Mary E. White, J. J. Bull, Ian J. Molineux, and David M. Hillis

Abstract.   One of the major impediments to assessing methods of phylogenetic inference is the almost complete absence of known phylogenies. Using bacteriophage T7, we have devised an experimental system for the rapid creation of phylogenies. The research design involves growing bacteriophage in the presence of the mutagen nitrosoguanidine, dividing the mutated phage into lineages, and studying the relationships among these lineages. When the desired number of phage lineages has been obtained, the mutated phage can be analyzed using molecular techniques such as restriction enzyme mapping and DNA sequencing. Different methods can be used to infer relationships from the molecular data, and inferred phylogenies can then be compared to the actual phylogeny. In a preliminary study involving eight taxa, cross-contamination of the lines prevented such comparisons of methods of phylogenetic inference. However, we present data showing the feasibility of the approach, and discuss methods for the prevention and detection of contamination. We also discuss the consequences and limitations of the experimental design. Although alternative approaches have some advantages, the T7-nitrosoguanidine system appears to be an effective approach for the creation and study of experimental phylogenies.

INTRODUCTION

A serious obstacle to the evaluation of methodologies of phylogenetic inference is the almost complete absence of phylogenies whose topology is known *a priori*. A few partially known phylogenies exist for domesticated and laboratory animals and plants (e.g., Baum, 1984; Fitch & Atchley, 1985a, b, 1987), but these phylogenies are of limited use for comparing phylogenetic methods because they exhibit limited divergence, reticulate evolution, or simplistic tree topologies. In addition, such phylogenies were produced over tens or hundreds of years, so experimental replication and control are impractical. Our purpose is to describe a simple experimental system that can be used to produce a known phylogeny of virtually any topology over a matter of months.

The system we describe uses the bacteriophage T7 grown in the presence of the mutagen N-methyl-N'-nitro-N'-nitrosoguanidine (hereafter called nitrosoguanidine). The use of T7 for studies in experimental evolution was suggested by Studier (1980):

> . . . it might be possible to follow divergence of phage DNAs in the laboratory. Because of its rapid growth rate, many generations of T7 can be grown in a relatively short time. If the natural mutation frequency is high enough, or if it could be increased by growing the phage in the presence of a mutagen, changes might accumulate rapidly enough to be detected by restriction analysis, and ultimately it might be possible to produce phages as divergent as [other wild-type T phages].

Drs. White, Bull, and Hillis (Department of Zoology) and Dr. Molineux (Department of Microbiology) are with the University of Texas, Austin, TX 78712, USA. Please address correspondence to Dr. Hillis.

936

Studier (1980) showed that T7 phage accumulated mutations rapidly when grown in the presence of nitrosoguanidine. Starting with wild-type T7 phage, he grew 45 consecutive lysates in cultures of *Escherichia coli* that contained this mutagen, and he then analyzed plaque-purified isolates from the final lysate by restriction analysis. He estimated that the mutated T7 differed from wild-type T7 by approximately one nucleotide substitution in every 400 base-pairs, as well as by some deletions.

The T7 system described by Studier (1980) can easily be modified for experimental studies of phylogeny. In this paper we will describe our modifications to Studier's methods and present some early results. In addition, we will discuss potential problems and limitations of this approach to experimental phylogeny construction.

MATERIALS AND METHODS

T7 phage were grown in cultures of *Escherichia coli* W3110. For each cycle of growth, phage from a single isolated plaque were eluted in 250 μl of SM buffer (Sambrook et al., 1989). An aliquot (100 μl) of this eluate was added to 900 μl of LB-broth (Sambrook et al., 1989) containing bacteria and 20 μg of nitrosoguanidine. The cultures were incubated at 38–40°C until lysis was complete. A portion of this lysate was plated with bacteria in the absence of nitrosoguanidine to form plaques representing clonal isolates. Prior to plating, each agar plate was premarked with a spot; the isolated plaque that appeared closest to this mark was selected for the next cycle of growth. The remainder of the SM eluate from the original plaque was stored at 4°C as the "ancestral" stock.

The experimental design of the phylogeny is shown in Figure 1. Each internode distance consisted of 35 cycles of plaque-lysate-plating as described above. After every 35 cycles, each lineage was split by using phage obtained from the same isolated plaque to infect two cultures containing nitrosoguanidine. The process was continued until eight lineages had been produced, each of which had been grown through a total of 105 cycles in the presence of the mutagen.

DNA was isolated from the eight terminal lineages and from wild-type T7 phage using the glycerol method described by Sambrook et al. (1989). These lineages were each assigned a letter (R through Z exclusive of U) in a manner unrelated to their relationships, so that the investigators who analyzed the samples could not be biased by *a priori* knowledge of relationships. Each DNA sample was cleaved with 24 restriction enzymes, each of which cleaved at a different 5- or 6-bp recognition sequence (*Bam* HI, *Bcl* I, *Bgl* II, *Bst* BI, *Bst* EII, *Bst* NI, *Cla* I, *Dra* I, *Eco* RI, *Eco* RV, *Hind* III, *Kpn* I, *Nde* I, *Nhe* I, *Pvu* II, *Pst* I, *Sal* I, *Sca* I, *Spe* I, *Ssp* I, *Stu* I, *Xba* I, *Xho* I, and *Xmn* I). Restriction fragments were separated by electrophoresis on 0.8% agarose gels along with appropriate size standards and were visualized by staining with ethidium bromide.

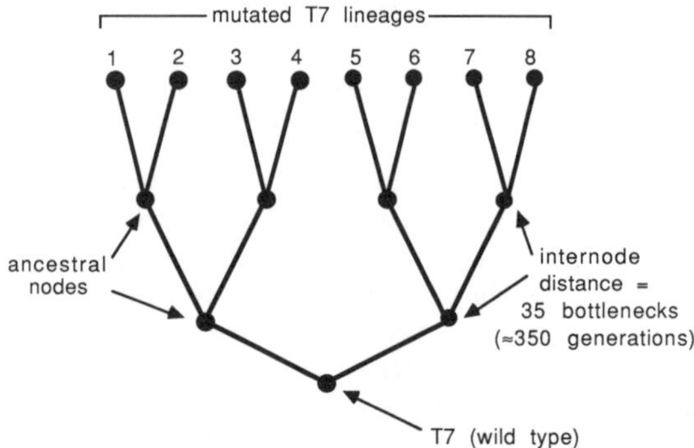

Figure 1. Experimental design of phylogeny, starting with wild-type T7 phage. See text for details.

Restriction site gains and losses (relative to wild-type T7 phage) were mapped by double digestion with one or more enzymes that had a number of conserved restriction sites (*Cla* I, *Dra* I, *Kpn* I, and *Xba* I). Deletions were detected by parallel changes in the lengths of restriction fragments produced by each enzyme, as well as by the loss of restriction sites in the deleted region. Site gains and losses, as well as the gains of deletions, were used to infer the phylogenetic relationships among the lineages using Swofford's (1990) Phylogenetic Analysis Using Parsimony (PAUP) program. We examined unrooted trees of the eight terminal lineages, as well as trees rooted using wild-type T7 as the ancestor. Approximate confidence limits for internal branches were obtained by bootstrapping: characters were sampled with replacement to produce 1,000 pseudoreplicate data matrices (Felsenstein, 1985), which were also analyzed with PAUP.

RESULTS

Terminal Taxa and an Unrooted Tree

The distribution of site gains, losses, and deletions (all relative to wild-type T7 phage) is shown in Figure 2. The unrooted tree is shown in Figure 3, where it is compared with the experimental design. Resolved branches on the inferred tree contained highly variable numbers of changes, from 18 changes along the lineage connecting taxa T and V to a single change connecting taxa R and W. Although the unrooted tree matched the experimental design for all branches that were resolved, the actual phylogeny may have differed from the experimental design for reasons discussed in the next section.

The distribution of observed restriction site gains and losses reflects our choice of enzymes used for screening the DNA. Indeed, it is possible to choose enzymes that are especially likely to reveal readily-detectable changes of a particular type (i.e., site gains or losses). For example, it is obvious that for any restriction enzyme failing to cut wild type, the only observable changes are site gains. Furthermore, nitrosoguanidine tends to cause the mutations G → A and C → T (see *Limitations of the system,* below). Some enzymes have many "one-off" sites in the T7 genome—regions that require only a single G → A or C → T transition in order to become a recognition site. These enzymes are prone to reveal site gains more readily than enzymes with few "one-off" sites. For example, the 4-base sequence GATC is cut by *Mbo* I and *Sau* 3AI. There are only six GATC sites in wild-type T7, but there are 416 "one-off" sites (GGTC or its complement, GACC), so we expect to observe several *Mbo* I site gains in these phylogenies. This expectation is, in fact, realized. The ideal combination for the detection of site gains is therefore few recognition sites but many one-off sites in the wild-type sequence. Conversely, recognition sites that are GC-rich are expected to be relatively prone to loss of sites. However, all recognition sequences are so short that the probability of loss is low per site, hence site losses are expected only for GC-rich enzymes that cut wild-type T7 many times. The restriction pattern of any

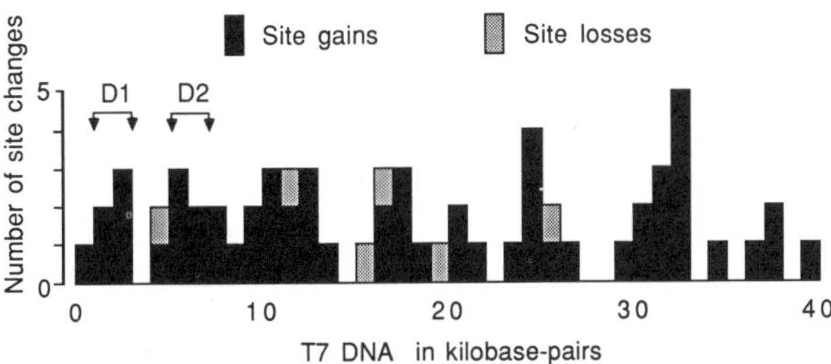

Figure 2. Distribution within the T7 genome of restriction site gains and losses, as well as two deletions (D1 and D2), summed across the experimental lines.

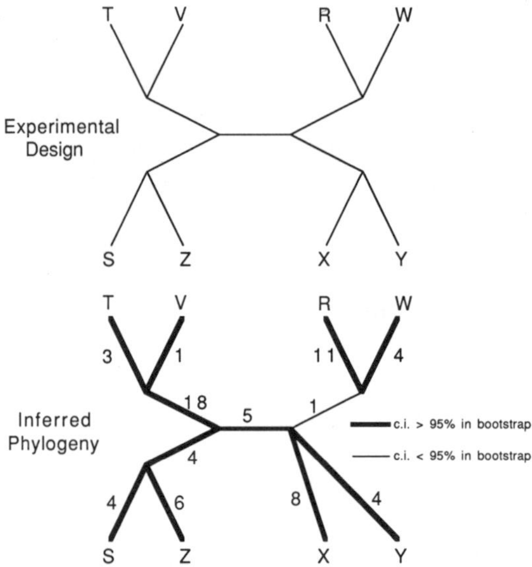

Figure 3. Comparison of the topologies of the experimental design phylogeny (above) and the inferred phylogeny (below). Numbers along branches indicate inferred site changes and deletions.

enzyme with many cuts is difficult to analyze, so that there is no parallel basis for choosing enzymes prone to reveal easily-scored site losses as there is for choosing enzymes with easily-scored site gains.

Ancestral Taxa and a Rooted Tree

When ancestral T7 (wild-type) was added to the tree for purposes of rooting, the inferred root of the tree was along the lineage uniting taxa (T, V) to the remaining taxa (Fig. 4). However, the root in the experimental design was between taxa ([T, V], [S, Z]) and ([R, W], [X, Y]) (see Fig. 4). To examine the reasons for this difference, we extracted DNA and analyzed restriction site variation from several ancestral taxa. Examination of ancestors of (S, Z) revealed an apparent regain of a deleted region of DNA, and a simultaneous convergence of the restriction patterns to the ([R, W], [X, Y]) lineage. The extreme unlikelihood of these apparent simultaneous convergences (especially the "regain" of previously deleted DNA) indicates that at least one contamination event occurred between these experimental lineages. Upon discovering this contamination in our experimental lines, we abandoned further phylogenetic analysis of these lineages and began refining procedures to avoid and detect contamination during future experiments.

DISCUSSION

The likelihood of cross-contamination of the lineages in this study precludes use of the present data to test alternative methods of phylogenetic inference. Nonetheless, this study demonstrates the feasibility of using *in vivo* mutagenesis of phage for work in experimental phylogeny construction. Since successful production of experimental phylogenies depends on the elimination of cross-contamination, we will emphasize the methods we have devised for detection and avoidance of contamination in future studies of phage evolution. We will also discuss the limitations of the T7 system and alternative designs for experimental studies of molecular evolution.

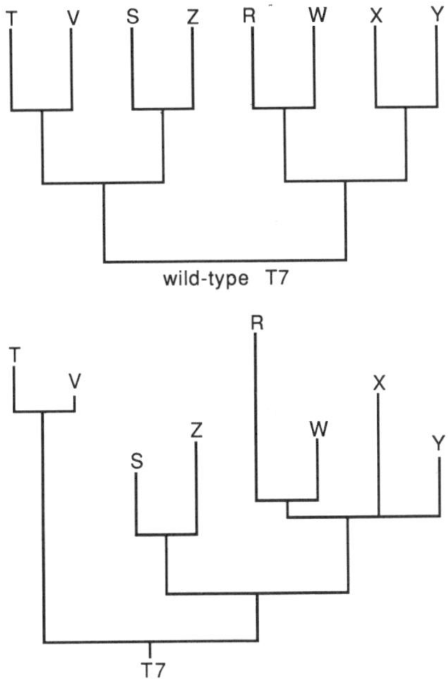

Figure 4. Comparison of the experimental design phylogeny (above) and the rooted inferred phylogeny (below). The branch lengths correspond to the inferred amount of change.

Contamination: Detection and Avoidance

We suspect that contamination was most likely due to aerosols resulting from the use of negative displacement pipetters with disposable sterile tips for transferring phage lysates. Now all our phage growth, handling, and storage procedures are carried out with glass apparatus, which can be sterilized in its entirety, or platinum wires and loops, which can be flamed immediately prior to use. Because most of the phage handling occurs at the plating step, we have modified the protocol (see below) to reduce the number of platings needed to create an extensive phage lineage. Regardless of the extent of precautions, however, it is necessary to assume that, at some level, contamination will occur, and to devise criteria for detection and elimination of contaminated lineages. These criteria must be stringent enough that real reversals or convergences between lines are not mistaken for contaminations; otherwise, the system would be biased against homoplasy.

Limitations of the System and Alternative Designs

Any biological system used to construct phylogenies must include two processes: mutation and incorporation of mutations. It is well established that the rate and the spectrum of mutations vary according to environmental conditions and even vary according to the organism; the extent to which mutations are incorporated into the population depends on the fitness effects of those mutations (which in turn depends on the types of mutations and on the organism) and on the structure of the population in which the mutations arise. All of these factors are influenced by the experimental design, so the investigator is confronted with a wide array of possible designs for the experimental construction of phylogenies. Each of these points will be discussed below.

MUTATION

The intrinsic mutation rate of most self-replicating DNA molecules is quite low per nucleotide, although the rate for RNA molecules is substantially higher. In addressing both the spectrum and the rate of mutation, there is a large selection of different molecules (bacterial chromosomes, plasmids, viruses, naked molecules propagated *in vitro*) and of different mutagens that may be considered for this work.

T7

Our choice of bacteriophages for creation of phylogenies rested primarily on their short generation times, as well as the ease of their purification and of nucleic acid isolation. Phages with either RNA or DNA genomes are common, and both types are represented as single- or double-stranded nucleic acids. The RNA phages are potentially useful organisms for a phylogenetic study because their spontaneous mutation rates are high enough (e.g., $\geq 10^{-4}$ per base per generation) that it may not be necessary to employ chemical mutagens. However, the single-stranded RNA phages have a severe disadvantage in this type of study in that they are not rapidly inactivated upon drying, a property that increases the risk of contamination in any long-term evolutionary study. Furthermore, assays of sequence divergence in RNA phages most likely would necessitate sequencing, since fingerprinting partial RNase digests provides only limited information that may be difficult to interpret. In contrast, DNA phages can be analyzed by restriction enzyme mapping as well as by sequencing—analyses that are complementary so long as neither method is used exhaustively.

Among the DNA phages, T7 has a relatively high spontaneous mutation frequency since T7 DNA replication is largely independent of host replication and repair functions that have high levels of fidelity. Its latent period is among the shortest of all phages (~ 13 min at 42°C) and visible plaques form in 2–3 hours. The T7 genome is double-stranded DNA, allowing restriction enzyme mapping of DNA isolated from phage particles as a screen for the acquisition of mutations. Furthermore, the distribution of restriction enzyme sites is highly skewed from statistical expectations; many 6-base recognition sites are rare, or even absent, in the 39,937 base-pair molecule, although there are many sequences that match five of the six positions. The T7 genome has been completely sequenced (Dunn & Studier, 1983; Moffat et al., 1984) and there is a large body of information on the biology of the phage (Studier & Dunn, 1982). Nearly half of the 51 genes of the phage, corresponding to almost 25% of the total genetic information, are non-essential for growth on most laboratory strains of *E. coli*. This allows an easy comparison of the rate of mutation in genes having different levels of constraint to divergence.

Two disadvantages of T7 are that its spontaneous mutation rate is too low to conveniently construct phylogenies without recourse to chemical mutagenesis, and that the large size of the genome likely restricts the maximum tolerable mutation rate. To elaborate on this last point, if a fraction P of the molecule is constrained so that any mutation (of a particular type) is lethal, hence $Q=1-P$ of the molecule can tolerate mutation, the chance that n mutations to a single molecule will not be lethal is Q^n. Thus, in comparing a long and a short molecule, each with the same value of Q, each molecule will be able to tolerate the same average *number* of mutations. However, the per-base mutation *rate* will be lower for the longer molecule; if data are to be obtained by DNA sequencing methods, more bases will need to be surveyed from the longer molecule than from the shorter molecule to obtain the same number of changes. The advantage of shorter DNA molecules in this respect applies only to the extent that the fraction of bases tolerating mutation (Q) is similar between long and short molecules, and does not apply to data obtained from restriction enzyme patterns.

Mutagen

The mutagenic effects of various chemical and physical agents have been explored over the last few decades. Mutagens have been found to vary in their spectrum of changes, both with respect to the ratio of lethal:non-lethal changes and in the types of changes they induce in the nucleotide sequence. Studier's choice of nitrosoguanidine was based on the relatively high frequency of viable mutations obtained with this mutagen (Studier, pers. comm.). The spectrum of mutations from nitrosoguanidine is chiefly GC → AT transitions, and although a broader spectrum may be desirable, the spectrum of certain natural mutation processes is known to be similarly narrow (Cambareri et al., 1989). Furthermore, the transition bias produced by nitrosoguanidine mirrors the transition bias observed in many studies of molecular phylogenetics (e.g., Brown et al., 1982).

Alternative Systems

In addition to the T7 nitrosoguanidine system, we have contemplated a few alternative systems for possible use in phylogenetic reconstruction. One obvious and desirable alternative is an RNA virus such as $Q\beta$. RNA viruses have intrinsic mutation rates high enough that use of artificial mutagens may not be necessary (e.g., 10^{-4} per base per generation). However, $Q\beta$ in particular is not killed easily, thus enhancing the possibility of contamination. Furthermore, any assay of sequence divergence would require sequencing.

A second possible alternative system involves the use of a single-stranded DNA phage (e.g., M13) with a heterologous DNA fragment inserted into its genome, the latter providing a selectively neutral target for any sequence changes. Such an insert can be specifically targeted for *in vitro* mutagenesis by forming gapped duplex molecules which are then either attacked with single-strand-specific mutagens (e.g., sodium bisulfite) or are repaired using DNA polymerases under conditions in which accuracy of DNA synthesis is low (Kunkel et al., 1987; Pine & Huang, 1987). Enormous mutation rates of the neutral target DNA can be achieved under such conditions (the functional part of the phage molecule is protected during mutagenesis), whereby a single round of mutation provides sufficient change for an entire branch of the phylogeny. The main drawback of this system would seem to be the confinement of mutations to essentially inert regions, in which all changes are selectively neutral; in addition, the spectrum of mutations obtained with sodium bisulfite is also atypically narrow, being entirely GC → AT. Nonetheless, these systems are unique in providing extremely high rates of change in a short period of time.

INCORPORATION OF MUTATIONS

In general, the rate at which mutations are incorporated in a DNA sequence depends on the mutation rate, the fitness effects of the mutations, and the population structure. The relevance of experimental design to mutation rate was discussed above. The fitness effects of any mutation depend both on the nature of the mutation/mutagen and on the function of the sequence in which it occurs. For example, it was noted above that nitrosoguanidine yields a greater ratio of non-lethal:lethal mutations than do many other mutagens. Similarly, the preceding paragraph described the targeting of mutations to sequences inserted into filamentous phages, these sequences being able to tolerate enormous mutation rates because they have no useful function to the phage.

A less obvious impact of experimental design on the incorporation of mutations concerns population structure. At one extreme, the phage (or other molecule) may be propagated in mass culture, whereby the minimum population size is a million or greater; Studier (1980) used such a design, transferring 0.1 ml of a lysate to initiate each new lysate.

942

At the other extreme—and as used here—the phage may be "bottlenecked" each cycle to a population size of one, choosing a single plaque to initiate each new lysate. The impact of these alternative designs, and of intermediate designs, lies in the extent to which deleterious mutations are incorporated. (It needs to be emphasized that bottlenecking to a single clone is necessary at least at every node in the phylogeny, so that one is assured that each pair of sister lineages share a common ancestor at the specified point in the phylogeny. However, there is considerable latitude as to how many additional bottlenecks may be employed along each branch.)

Except for targeted *in vitro* mutagenesis methods, it may be assumed that a nontrivial fraction of mutations will have deleterious consequences for the phage; in addition, some mutations will likely be selectively neutral, or nearly so. From principles established by population genetics models (Crow & Kimura, 1970), the population structure of the experimental design should have a substantial impact on the rate of incorporation of deleterious and advantageous mutations, although no effect on the average rate of incorporation of neutral ones. For strictly neutral mutations, the number expected to arise in a generation is $\mu{\cdot}N$, where μ is the mutation rate and N is the population size. The chance that the population in the future will come to be descended from an arbitrarily chosen mutation in the current population is $1/N$, so the rate of incorporation of neutral mutations is $(m{\cdot}N)(1/N) = \mu$. Thus the population structure (as reflected in N) should have no effect on the average rate of incorporation of neutral changes.

However, this conclusion fails when the mutations affect fitness. To obtain a qualitative understanding of these principles, consider first the case of large population size. In a nearly infinite population, deleterious alleles are maintained at low frequency by a balance between mutation (which inputs them) and selection (which eliminates them); conversely, advantageous mutations spread to fixation. Small populations lessen the impact of selection on allele frequencies by increasing the role of chance in the fixation of alleles. Thus, deleterious alleles can be fixed and advantageous alleles lost in small populations.

Under our experimental design, every lineage was bottlenecked to a population size of 1 each cycle. This design should therefore have facilitated the fixation of deleterious mutations (without affecting the incorporation of neutral ones), yielding a net incorporation of mutations greater than would be expected from a design employing large population sizes continually (such as that used by Studier, 1980). Indeed, a substantial incorporation of deleterious mutations seems to have occurred under our design, because the time required for lysis of a culture increased dramatically over the course of the experiment (from two to over ten hours). The main drawback of this outcome was an increase in the amount of time required to complete the phylogeny construction. Work currently underway suggests that a design in which cultures are bottlenecked to a single plaque after 5 cycles of lysis incorporates a similar level of mutation as observed in this first study yet does not suffer nearly so high a load from deleterious mutations.

ACKNOWLEDGMENTS

We thank F. William Studier for advice and information on the T7 nitrosoguanidine system. This work was supported by NSF grant BSR-8657640 (to DMH), NIH grant GM-32095 (to IJM), and a Texas Advanced Technology Program grant (to JJB).

LITERATURE CITED

Baum, B. R. 1984. Application of compatibility and parsimony methods at the infraspecific, specific, and generic levels in Poaceae. Pp. 192–220. *In:* T. Duncan & T. F. Stuessy (eds.), *Cladistics:*

Perspectives on the Reconstruction of Evolutionary History. Columbia University Press: New York.

Brown, W. M., Prager, E. M., Wang, A. & A. C. Wilson. 1982. Mitochondrial DNA sequences of primates: tempo and mode of evolution. *J. Mol. Evol.* 18:225–239.

Cambareri, E. B., Jensen, B.C., Schabtach, E. & E. U. Selker. 1989. Repeat-induced G-C to A-T mutations in *Neurospora. Science* 244:1571–1575.

Crow, J. F. & M. Kimura. 1970. *An Introduction to Population Genetics* Theory. Harper and Row: New York.

Dunn, J. J. & F. W. Studier. 1983. Complete nucleotide sequence of bacteriophage T7 DNA and the locations of genetic elements. *J. Mol. Biol.* 166:477–535.

Felsenstein, J. 1985. Confidence limits on phylogenies: An approach using the bootstrap. *Evolution* 39:783–791.

Fitch, W. M. & W. R. Atchley. 1985a. Evolution in inbred strains of mice appears rapid. *Science* 228:1169–1175.

Fitch, W. M. & W. R. Atchley. 1985b. Rapid mutations in mice? *Science* 230:1408–1409.

Fitch, W. M. & W. R. Atchley. 1987. Divergence in inbred strains of mice: a comparison of three different types of data. Pp. 203–216. *In:* C. Patterson (ed.), *Molecules and Morphology in Evolution: Conflict or Compromise?* Cambridge University Press: Cambridge, UK.

Kunkel, T. A., Roberts, J. D. & R. A. Zakour. 1987. Rapid and efficient site-specific mutagenesis without phenotypic selection. *Meth. Enzymol.* 154:367–382.

Moffat, B. A., Dunn, J. J. & F. W. Studier. 1984. Nucleotide sequence of the gene for bacteriophage T7 RNA polymerase. *J. Mol. Biol.* 173:265–269.

Pine, R. & P. C. Huang. 1987. An improved method to obtain a large number of mutants in a defined region of DNA. *Meth. Enzymol.* 154:415–430.

Sambrook, J., Fritsch, E. F. & T. Maniatis. 1989. *Molecular Cloning: A Laboratory Manual,* 2nd Ed. 3 Volumes. Cold Spring Harbor Laboratory: Cold Spring Harbor, NY.

Studier, F. W. 1980. The last of the T phages. Pp. 72–78. *In:* N. H. Horowitz & E. Hutchings, Jr. (eds.), *Genes, Cells, and Behavior: A View of Biology Fifty Years Later.* W. H. Freeman: San Francisco, CA.

Studier, F. W. & J. J. Dunn. 1982. Organization and Expression of Bacteriophage T7 DNA. *Cold Spring Harbor Symp. Quant. Biol.* 42:999–1007.

Swofford, D. L. 1990. Phylogenetic Analysis Using Parsimony, version 3.0. University of Illinois: Urbana.

METHODS AND RESOURCES

Assessing the Methods of Quantitative Vicariance Biogeography: A Simulation Study of Phenetic Similarity Analysis

Edward F. Connor

Abstract. Four quantitative methods have been proposed for developing and testing hypotheses concerning the historical relationships between geographical regions; 1) phenetic similarity analysis, 2) generalized track analysis, 3) cladistic biogeographic inference of area cladograms, and 4) cladistic character inference of area cladograms. I outline an approach to assessing the relative abilities of each of these four methods to actually recover the "true" historical relationships between areas, using hypothetical biotas subjected to known sequences of vicariance.

Preliminary results for phenetic similarity methods indicate that indices based on both the shared presences and joint absence of species are better than those based simply on the number of shared presences, or on the number of lineages shared. For the hypothetical biotas simulated, the ability of the Simple Matching Coefficient to correctly identify the "true" area cladogram was greater when a small number of regions was examined, and increased when more taxa were included in the simulated biota.

The results presented here suggest that each of these techniques has some ability to correctly identify actual historical relationships between geographical regions, but I anticipate that methods that use some information on the evolutionary relationships between taxa will require data on fewer taxa to be as accurate as methods that use no data on evolutionary relationships between taxa.

INTRODUCTION

Four quantitative methods have been proposed for developing and testing hypotheses concerning the historical relationships between geographical regions; 1) phenetic similarity analysis, 2) generalized track analysis, 3) cladistic biogeographic inference of area cladograms, and 4) cladistic character inference of area cladograms (Connor, 1988).

Phenetic similarity analysis involves calculating an index of the percent similarity between the biotas of regions on a pair-wise basis, and using a clustering algorithm to produce a dendrogram of relationships between regions. The inference that is drawn via this method is that regions that share many taxa have strong historical relationships. This method has been criticized because it is ad hoc, because similarity indices are sensitive to differences between regions in the number of resident taxa, and because it equates strong historical relationships between regions with sharing many widespread taxa (Nelson &

Dr. Connor is with the Department of Environmental Sciences, Clark Hall, University of Virginia, Charlottesville, Virginia 22903, USA, and The Blandy Experimental Farm, P.O. Box 175, Boyce, Virginia 22620, USA.

Platnick, 1978, Connor, 1988). This method has been used since J. D. Hooker applied it in analyzing the historical relationships of the Galapagos flora (Hooker, 1846), and is still commonly used (McCoy and Heck, 1987, Nichols, 1988).

Generalized track analysis is based on Croizat's panbiogeography (Croizat et al., 1974), and contends that regions that share many closely related taxa have a strong historical connection. This method focuses on more narrowly distributed taxa, even endemic taxa, and measures the strength of an historical relationship between regions as the number of higher taxa with sub-taxa present in a set of geographical regions. The major difference between this approach and those based on phenetic similarity analysis is that it employs information on the evolutionary relationships between taxa to infer the historical relationships between regions. This method has also been criticized as being ad hoc, lacking statistical rigor, and lacking statistical power relative to methods based on cladistic character inference of area cladograms (Ball, 1975, McDowall, 1978, Simberloff et al., 1981, Connor, 1988). Although both Page (1987) and Connor (1988) suggest approaches that may overcome some of these criticisms, this method has never been used in analyzing the historical relationships between geographical regions.

Cladistic biogeographic inference of area cladograms was developed by Legendre (1986) and uses information on the presence or absence of taxa as "characters" in a cladistic analysis of the relationship between regions. It is similar to phenetic similarity analysis in that it employs no information about the evolutionary relationships between taxa. As with cladistic methods applied to phenotypic character states, this approach considers regions that share taxa with restricted or narrow distributions (derived characters) to be closely related. This technique has only been applied in one instance (Legendre, 1986).

Cladistic character inference of area cladograms was first proposed by Rosen (1978), but later elaborated by Platnick and Nelson (1978), Nelson and Platnick (1981), Simberloff et al. (1981), Brooks (1981, 1985, 1990), Rosen (1985), Zandee and Roos (1987), Page (1988), and Wiley (1987, 1988). This method involves performing a cladistic analysis of phenotypic character states on taxa in several monophyletic lineages, and combining this information on the evolutionary relationships among taxa with data on geographical distribution to construct parsimonious or consensus trees of the relationships between areas (area cladograms). As defined here, cladistic character inference of area cladograms encompasses the "component analysis" of Nelson and Platnick (1981), the "parsimony" or "Brooks parsimony analysis" of Brooks (1981, 1985) and Zandee and Roos (1987), and variations on these methods (Page, 1988, Brooks, 1990). Much of the discussion of cladistic character inferences of area cladograms has revolved around the treatment of widespread taxa and the treatment of lineages with taxa missing from some of the geographic regions whose relationships are of interest (Page, 1988). However, these methods have also been criticized for being ad hoc or lacking a statistical basis (Simberloff et al., 1981, Page, 1988). Because cladistic character inference methods require the greatest amount of information of all of the methods described above, they have been applied in a limited number of instances. However, computer packages that perform these analyses are now or will soon be available.

The debate concerning the appropriateness, suitability, and accuracy of each of these methods for inferring historical relationships between geographical regions has paralleled the debate in systematics about the relative merits of phenetics and cladistics. All arguments for or against a particular approach rest entirely on theory and conjecture rather than on an empirical assessment of the abilities of each method to recover known relationships between geographical regions. I report here on a preliminary approach to assessing the ability of these four methods to recover known sequences of vicariance between hypothetical regions using simulated biotas, and illustrate this approach through an assessment of phenetic similarity methods.

ASSESSING THE METHODS OF VICARIANCE BIOGEOGRAPHY

To assess the relative abilities of phenetic similarity, generalized track, cladistic biogeographic, and cladistic character inference methods to depict the actual historical relationships between geographical regions, one could apply these methods to the biota of a set of geographical regions with a *known* history, and then determine which method matches most closely the true relationships between regions. Rather than assume that we know the true historical relationships between any actual set of geographic regions, I propose to use computer simulations of known sequences of vicariance as a test of the relative abilities of these methods to recover known historical relationships.

Simulating Vicariant Biotas

To simulate known sequences of vicariance, I developed a computer algorithm that begins with a large ancestral geographical region throughout which all taxa are ubiquitously distributed and then subjects this region and its biota to a vicariant event (Fig. 1). Each taxa in the ancestral biota has a probability equal to 1/3 of dispersing to both daughter regions without cladogenesis, dispersing with cladogenesis, or not dispersing (extinction). Each daughter region may subsequently be subjected to additional vicariant events with their respective biotas experiencing dispersal, cladogenesis, or extinction again with each probability equal to 1/3. The specific sequence of vicariance to be simulated is then set by which daughter regions are subjected to additional vicariant events.

The geographical distribution and evolutionary relationships within each lineage is preserved in an r × c matrix where each of the r - rows represents a lineage, and each of the c - columns represents a geographical region. Successively higher integers indicate more recently evolved taxa within a lineage, and zeros indicate that no members of a lineage are distributed in that region.

The Correspondence Between "True" and Estimated Area Cladograms

To examine the correspondence between the "true" area cladogram and that estimated by applying one of the four methods of quantitative vicariance biogeography, one of the methods is applied to the simulated biota, and whether or not it matches exactly the "true" are cladogram is determined. Any discrepancy between the true and estimated area

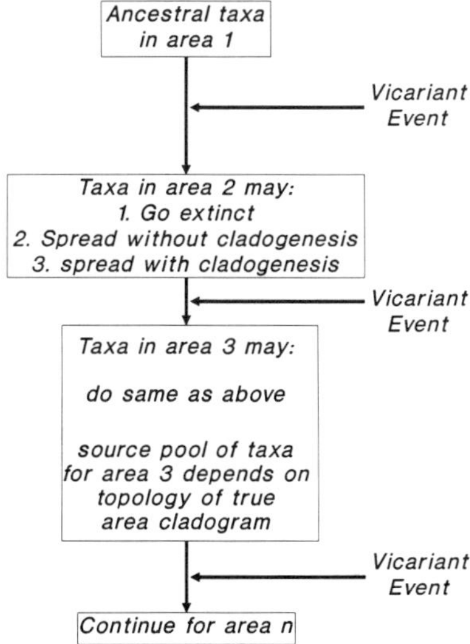

Figure 1. Algorithm to generate hypothetical biota subject to known sequence of vicariance.

950

cladogram in defining the historical relationships between geographical regions is grounds for declaring the estimated cladogram a mismatch. When examining the historical relationship between several regions (>5) it may be better to measure the match between "true" and estimated area cladograms using an index of cladogram congruence such as outlined by Page (1988).

PHENETIC SIMILARITY ANALYSIS: AN ILLUSTRATION AND PRELIMINARY ASSESSMENT

To illustrate this approach to assessing the relative accuracy of the methods of quantitative vicariance biogeography, I simulated 1000 sequences of vicariance for each of the possible topological relationships of 3, 4, and 5 geographic regions (Figs. 2–4), applied phenetic similarity methods to estimate relationships between geographical regions, and compared the estimated relationships to the actual to determine the percentage of estimated area cladograms matching the true area cladogram.

I used both Jaccard's index and the Simple Matching Coefficient to represent the behavior of phenetic indices since Jaccard's index is based on the number of joint occurrences of taxa in a pair of regions, and the Simple Matching Coefficient is based on the sum of the number of joint occurrences and the number of joint absences (Sneath & Sokal, 1973). Estimated area cladograms were then generated using the UPGMA technique applied to the matrix of pair-wise similarity values (Sneath & Sokal, 1973).

When phenetic methods are applied to a hypothetical biota containing 100 lineages, the Simple Matching Coefficient, calculated using the occurrences of species, correctly identifies the "true" relationships between geographical regions more than 90% of the time (Fig. 5). Jaccard's Index was no better than 50% accurate for 3-member area cladograms, and recovered less than 5% of the true relationships between regions for 5-member area cladograms. If, rather than counting the number of species shared or jointly absent between sites, we count the number of lineages shared or jointly absent between sites, the Simple Matching Coefficient performs more poorly, but is still considerably better than Jaccard's Index (Fig. 6). The Simple Matching Coefficient also has the desirable property that as more taxa are included in the analysis the percentage of "true" area cladograms that are correctly identified increases whereas the opposite is true for Jaccard's Index (Fig. 7).

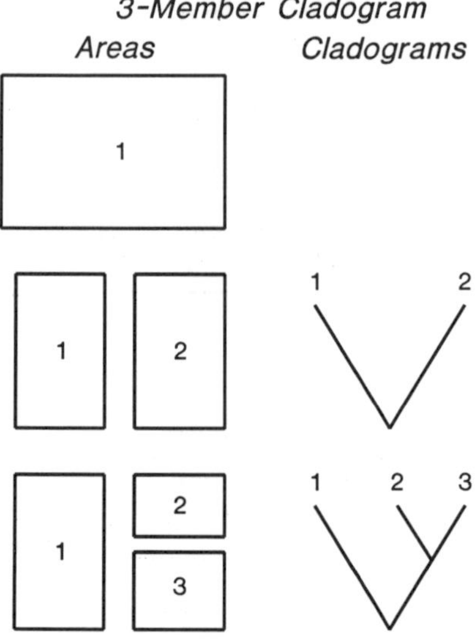

Figure 2. Vicariance sequence and area cladogram for 3-member cladograms.

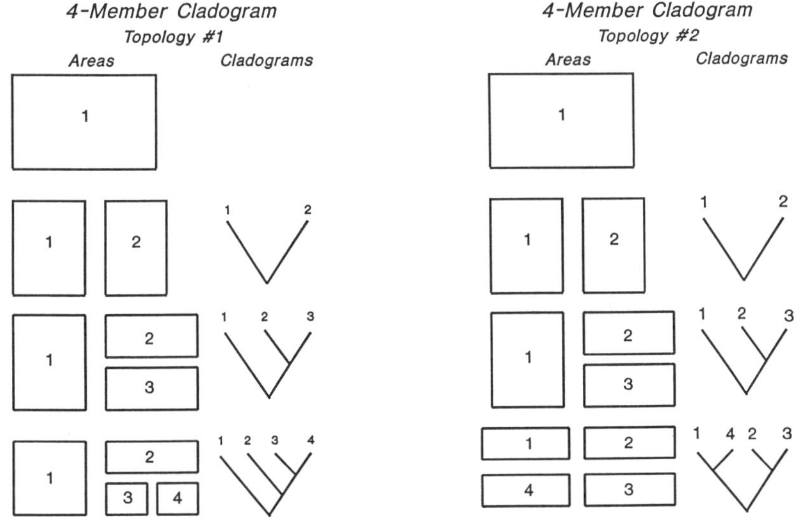

Figure 3. Vicariance sequence and area cladogram for 4-member cladograms.

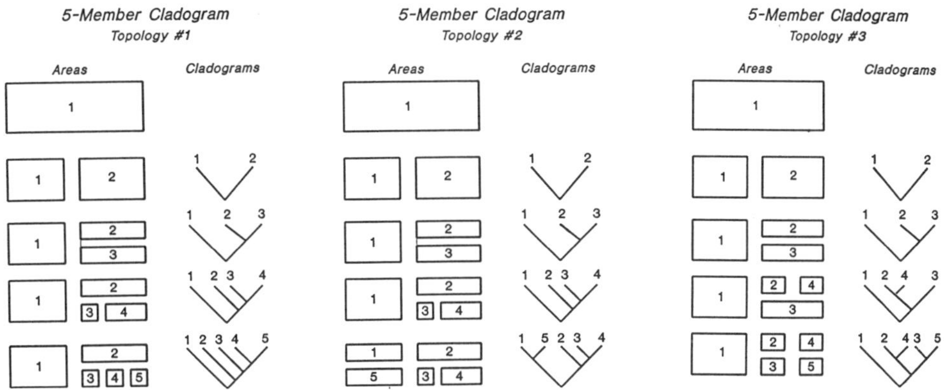

Figure 4. Vicariance sequence and area cladogram for 5-member cladograms.

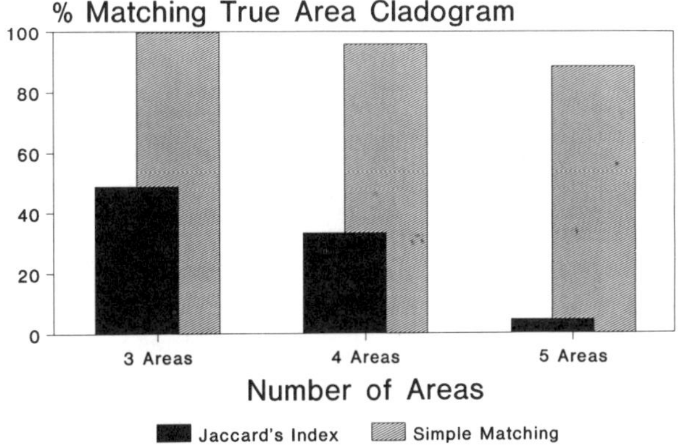

Figure 5. Percent of area cladograms estimated using phenetic similarity methods that match "true" area cladogram. This analysis is based on the number of species shared or jointly absent between each pair of regions. Each value was estimated using 1000 simulated biotas consisting of 100 lineages. For 4 and 5 member cladograms 1000 simulated biotas were examined for each possible cladogram topology.

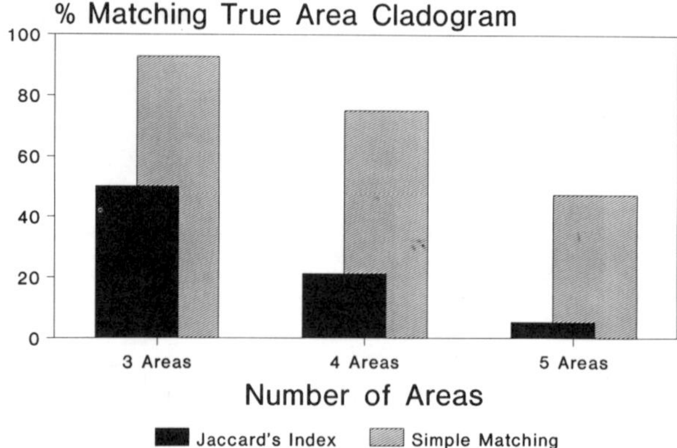

Figure 6. Percent of area cladograms estimated using phenetic similarity methods that match "true" area cladogram. This analysis is based on the number of lineages shared or jointly absent between each pair of regions. Each value was estimated using 1000 simulated biotas consisting of 100 lineages. For 4 and 5 member cladograms, 1000 simulated biotas were examined for each possible cladogram topology.

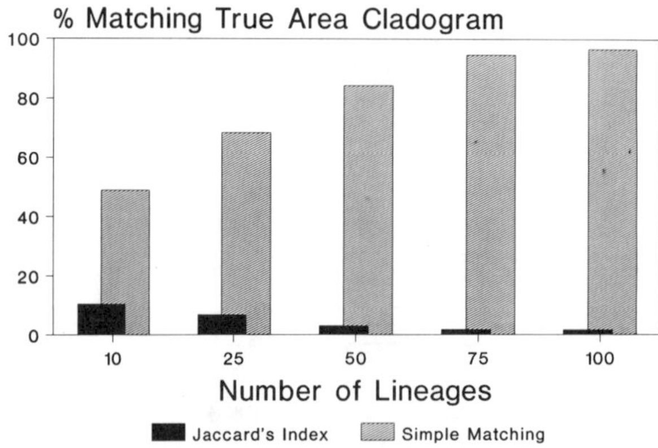

Figure 7. Effect of the number of taxa examined on the ability of phenetic similarity methods to correctly identify the "true" area cladogram. This analysis is based on the number of species shared or jointly absent from each pair of regions. These values were estimated using 1000 simulated biotas for each of the possible 4-member cladogram topologies.

DISCUSSION

The protocol outlined here holds the promise of providing an empirical assessment of the relative abilities of the four methods of quantitative vicariance biogeography. A similar approach could be used to assess generalized track, cladistic biogeographic, and cladistic character methods, and to assess the sensitivity of each of these methods to long-distance dispersal, higher rates of extinction, fossil evidence, ghosts (sensu Simberloff et al., 1981), adaptive radiation within regions, and other forces that have shaped biogeography.

Of the four methods: phenetic similarity, generalized track, cladistic biogeographic, or cladistic character inference, I would define the "best" method as the one that produces the most accurate picture of the true historical relationships between geographical regions, using the least data. This definition is analagous to the statistical concept of efficiency, where a more efficient statistical test requires a smaller sample size to have equal accuracy to that of the less efficient test. However, since some of the methods require different kinds of information (i.e., only two of the methods require information on the evolutionary relationships of the subject taxa), it is not possible to calculate the efficiency of these methods in the strictest sense.

While the assumptions made in the vicariance algorithm presented above were not intended to be a completely realistic simulation, they do embody the essential elements of a vicariating biota. Hence, the results I obtained regarding the ability of phenetic methods to depict "true" historical relationships between regions show that phenetic methods may be quite useful, particularly in regions or with taxa where data on evolutionary relationships may not be available or may be difficult to obtain.

I would anticipate that those methods of quantitative vicariance biogeography that use some information of the evolutionary relationships of taxa would be better methods, inasmuch as they may be more efficient. However, the results presented here for phenetic similarity methods suggest that none of these approaches should be viewed as inappropriate, uninformative, or incorrect, since equal accuracy may be obtained by increasing the number of taxa involved in an analysis of historical relationships.

I propose to expand the analysis presented here to all four methods of inferring historical relationships between regions that are used in quantitative vicariance biogeography. I also propose to examine the sensitivity of any conclusions reached in the assumptions made in simulating the effects of hypothetical sequences of vicariance, and in calculating congruence between "true" and estimated historical relationships between geographical regions.

LITERATURE CITED

Ball, I. R. 1975. Nature and formulation of biogeographical hypotheses. *Systematic Zoology* 24: 407–430.

Brooks, D. R. 1981. Hennig's parasitological method: a proposed solution. *Systematic Zoology* 30: 229–249.

Brooks, D. R. 1985. Historical ecology: a new approach to studying the evolution of ecological associations. *Annals of the Missouri Botanical Garden* 72:660–680.

Brooks, D. R. 1990. Parsimony analysis in historical biogeography and coevolution: methodological and theoretical update. *Systematic Zoology* 39: 14–30.

Connor, E. F. 1988. Fossils, phenetics, and phylogenetics: inferring the historical dynamics of biogeographic distributions. Pp. 254–269. *In:* J. K. Liebherr (ed.) *Zoogeography of Caribbean Insects.* Cornell University Press: Ithaca, NY.

Croizat, L., Nelson, G., & D. Rosen. 1974. Centers of origin and related concepts. *Systematic Zoology* 23: 265–287.

Hooker, J. D. 1846. On the vegetation of the Galapagos archipelago, as compared with that of some other tropical islands and the continent of America. *Transactions of the Linnean Society of London* 20: 235–262.

Legendre, P. 1986. Reconstructing biogeographic history using phylogenetic tree analysis of community structure. *Systematic Zoology* 35: 68–80.

McCoy, E. D. & K. L. Heck. 1987. Some observations on the use of taxonomic similarity in large-scale biogeography. *Journal of Biogeography* 14:79–87.

McDowall, R. M. 1978. Generalized tracks and dispersal in biogeography. *Systematic Zoology* 27: 8–104.

Nelson, G. & N. I. Platnick. 1978. The perils of plesiomorphy: widespread taxa, dispersal, and phenetic biogeography. *Systematic Zoology* 27: 474–477.

Nelson, G. & N. I. Platnick. 1981. *Systematics and Biogeography: Cladistics and Vicariance.* Columbia University Press: New York.

Nichols, S. W. 1988. Kaleidoscopic biogeography of West Indian Scaritinae (Coleoptera:Carabidae). Pp. 71–120. *In:* J. K. Liebherr (ed.) *Zoogeography of Caribbean Insects.* Cornell University Press: Ithaca, NY.

Page, R. D. M. 1987. Graphs and generalized tracks: quantifying Croizat's panbiogeography. *Systematic Zoology* 36: 1–17.

Page, R. D. M. 1988. Quantitative cladistic biogeography: constructing and comparing area clado-
grams. *Systematic Zoology* 37:254–270.
Platnick, N. I. & G. Nelson. 1978. A method of analysis for historical biogeography. *Systematic Zoology* 27: 1–16.
Rosen, D. E. 1978. Vicariant patterns and historical explanation in biogeography. *Systematic Zoology* 27: 159–188.
Rosen, D. E. 1985. Geologic hierarchies and biogeographic congruence in the Caribbean. *Annals of the Missouri Botanical Garden* 72: 636–659.
Simberloff, D., Heck, K. L., McCoy, E. D., & E. F. Connor. 1981. There have been no statistical tests of cladistic biogeographic hypotheses! Pp. 40–63. *In:* G. Nelson & D. E. Rosen (eds), *Vicariance Biogeography: A Critique.* Columbia University Press: New York.
Sneath, P. H. A. & R. R. Sokal. 1973. *Numerical Taxonomy.* W. H. Freeman and Company: San Francisco, CA.
Wiley, E. O. 1987. Methods in vicariance biogeography. Pp. 283–306. *In:* P. Hovenkamp (ed.) *Systematics and Evolution: A Matter of Diversity.* Institute of Systematic Botany: Utrecht.
Wiley, E. O. 1988. Parsimony analysis and vicariance biogeography. *Systematic Zoology* 37: 271–290.
Zandee, M. & M. C. Roos. 1987. Component-compatibility in historical biogeography. *Cladistics* 3:305–332.

Literature Databases

Maureen C. Kelly

Abstract. Literature databases serve as valuable references to the published research literature. They constitute massive archives, spanning over 20 years and containing millions of entries. Modern computer retrieval techniques make it possible to use these databases in new ways. Literature databases are useful for identifying what research has been conducted and for tracking trends in research. These databases contain a wealth of information on the organisms reported on in the literature. A project is described which uses a literature database (BIOSIS Previews®) as a resource for building a register of taxonomic names.

INTRODUCTION

Good afternoon. Thank you for this opportunity to speak today on the topic of 21st century literature databases. Growing up as I have in the 20th century, I've become accustomed to thinking of the 21st century as a strange and distant future, an exotic combination of Buck Rogers and the movie 2001. Well, the 21st century is no longer very distant; and space ships and high-tech, talking computers no longer seem very exotic.

The 21st century is now less than a decade away. And it takes on a kind of familiarity when you think of it as part of a continuous process of change, built on the steady progress of technology and the adaptation of our lifestyles to that technology.

Information systems are also susceptible to changes in technology and lifestyles. Having watched the progress of information processing and transfer over the past 20+ years, I have come to believe that the future of information systems can be anticipated from a careful reading of the past and present.

For this reason, I will begin my talk with a review of literature databases and their role in the dissemination of scientific information. Naturally, my observations are influenced by my perspective as a representative of a large abstracting and indexing service, BIOSIS.

BIOSIS

For those of you who may not know us, BIOSIS is an independent, not-for-profit publisher of printed and computer-readable reference products in the life sciences. Our products include *Biological Abstracts®* and the *BIOSIS Previews®* and *BioBusiness®* databases. We also publish *Zoological Record®* in conjunction with the Zoological Society of London.

We have been serving the information needs of biologists since 1926. Our Board of Trustees is drawn from both the academic and industrial sectors, and they provide us with an important link to the biological community. We have an ongoing commitment to fostering the growth and dissemination of biological knowledge.

Ms. Kelly is Director, BIOSIS, Document Analysis Division, 2100 Arch Street, Philadelphia, PA 19103, USA.

CONTENT OF LITERATURE DATABASES

Size and Coverage

In fulfillment of this commitment, we have amassed a 64-year archive of over 8.8 million printed literature references. Our computer-readable archive spans 21 years and contains over 6.7 million references. In 1989, we monitored approximately 9,000 journals, books, and conference proceedings from over 100 countries worldwide.

Our coverage reflects the tremendous growth which has taken place in the published record of biology (Fig. 1). In 1926 when we began, we published 14,500 references. This year, we will publish 593,000 references, a 40-fold increase. While it took us 30 years to publish our first million references, it took only 10 years to publish our second million. We are now publishing over a million references every two years.

Zoological Record has a much longer history; it has been published in printed form for 125 years. Its annual coverage has grown from 3,000 references in 1864 to over 70,000 (70,446) references in 1989, a 23-fold increase. *Zoological Record Online*® has been building for 11 years now and contains 790,000 references.

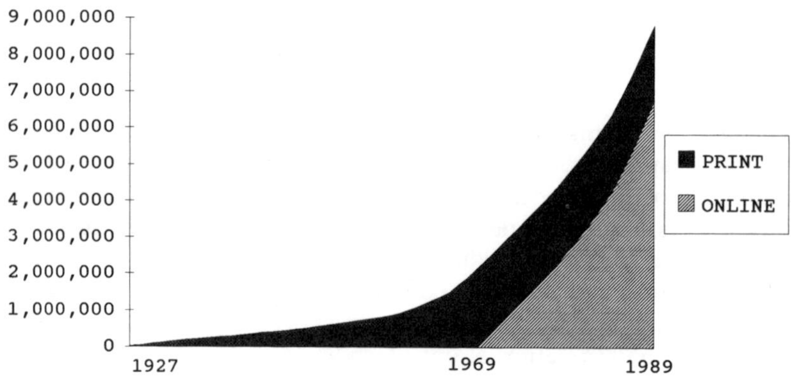

Figure 1. Growth in BIOSIS coverage.

Record Structure

Literature databases such as *BIOSIS Previews* and *Zoological Record Online* serve as inventories of scientific research. They offer abbreviated records of vast arrays of research reports put together in such a way as to facilitate access.

They contain bibliographic citations; often they include abstracts or summaries of the original work; and, most important of all, they include indexing information. These databases and the print products they duplicate were designed to help users identify the information they need.

The databases are often called "secondary" databases because they provide a secondary form of access to the primary literature, i.e., the journal articles and books in which research is originally published. Now it is becoming more common to see "primary" literature databases as well. These contain the full text of journal articles and citations; they may or may not contain indexing. Because these databases quickly become very large, however, they sometimes suffer from slow retrieval times.

USE OF LITERATURE DATABASES

Database Usage Patterns

Literature databases may be used alone or as complements to scientific data files such as *GenBank* or *GRIN* (Germplasm Resources Information Network). Traditionally, literature databases and their printed counterparts have been used for preparing bibliographies and for identifying the extent and nature of research that has been conducted in a particular field.

A recent study at Cornell University shows that close to 70% of their academics and researchers turn to literature databases prior to embarking on new research (69.8%) and when writing papers or reports (67.7%). They also use literature databases during research (40.6%) and when preparing grant proposals (34.4%).

Only 29% of the group studied, however, used literature databases to keep current in their field. The vast majority (70%) still rely on the primary literature and on colleagues to remain current. Apparently, the convenience of the printed medium and the chemistry of the personal interaction still have advantages to offer in this high tech world of ours.

Tracking Research Trends

Literature databases have utility beyond the preparation of bibliographies and grant proposals. In the broadest sense, they serve as records of the scientific discovery process. From them, much can be learned about the directions that scientific research has taken over the past 20 years.

Approximately 20% of *Zoological Record* and 5% of references covered in *BIOSIS Previews* pertain to systematics and evolutionary biology. Based on *BIOSIS Previews* coverage, research in evolutionary biology has increased in recent years after a decline between 1980 and 1983 (Fig. 2). Our coverage of systematics research shows a similar pattern over the past decade (Fig. 3).

By studying literature databases, we can detect shifts in the type of research being conducted. For example, in recent years we see an increase in research that uses the techniques of molecular genetics and biotechnology in conjunction with evolution (Fig.4). The overlap of systematics with molecular properties research shows a somewhat different pattern (Fig. 5).

Most recently, we have begun to see a small number of research reports which address evolution in conjunction with global warming and conservation biology. As is often the case for new areas of research, approximately 70% of these reports are from meetings (Fig. 6).

Tracking Nomenclature Trends

Literature databases also provide a useful record of recent nomenclatural usage. As part of our quality control process, BIOSIS tracks all terms coming into the literature for the first time. We also track the first occurrences in the database of the scientific names of organisms. This information has been maintained in a Master Vocabulary File since 1975.

In its first year, the file contained nearly 300,000 terms. Over 50% of that file was made up of organism names. By 1989, the file had grown to over five times its 1975 size. It now contains over 1.5 million terms, including nearly 730,000 unique organism names (49% of the file). In recent years, this translates into an average of one unique organism name for every 16 papers published (Fig. 7).

BIOSIS Previews can also be used to track the volume of literature that reports new taxa. The number of references recording new taxa has remained relatively stable since 1975, averaging 5,400 references per year. Since 1975, over 80,000 documents have been published that contained records of new taxa (Fig. 8). Approximately 80% of these documents recorded new species.

The indexing in *Zoological Record Online* makes it possible to quantify the actual number of new taxa (as opposed to documents containing new taxa, as is the case for *BIOSIS Previews*). In 1989, ZR recorded nearly 25,000 new species and subspecies.

Building & Maintaining Nomenclature Databases

BIOSIS has been able to use its literature database to build and maintain a register of bacterial nomenclature. In the case of bacteria, the nomenclature is well established. A comprehensive and exclusive *List of Approved Names* was published in 1980, setting a new starting point for bacterial nomenclature. The rules for valid publication of new bacterial names are also tightly controlled.

Even with this orderly nomenclature, literature databases can serve a useful role. BIOSIS has taken each unique name that has occurred in the literature since 1980 and added it to this register along with the approved names. There are currently 12,600 names on file, including genera, species, and subspecies. Synonymy and classification are included, and all names are coded to indicate whether they have been validly published. The broad international coverage of *BIOSIS Previews* is useful for identifying new names which have been published outside of the authorized mechanisms. This sometimes occurs for literature originating in the Soviet Union and China.

Members of the staff producing *Zoological Record* have used their resources to build a partially machine-readable authority file consisting of 200,000 generic and subgeneric names, including synonyms. They are planning to cooperate with the Zoological Society of London on the production of the next volume of *Nomenclator Zoologicus* (1970–1980). They are also evaluating the feasibility of using optical character recognition technology to create a database of the earlier volumes of *Nomenclator Zoologicus*. This comprehensive database might then be usable as a starting point for production of a list of names in common use in zoology, in cooperation with the International Commission on Zoological Nomenclature. (This is similar to a project now underway to produce lists of botanical names.)

CURRENT LIMITATIONS

As I have just described, literature databases can be used in a variety of ways to facilitate scientific research. However, there are limitations to their utility. A major limitation is the scope of the literature covered. Although 20 years of data and 6.7 million references make for a sizeable database, it still lacks the historical substance needed to support archival systematics research, where crucial records may date back over 100 years.

Our coverage of the current literature is also limited. In spite of the fact that BIOSIS published 550,000 references in 1989 and monitored over 9,000 journals as well as books and conference proceedings from over 100 countries, there is still a wealth of literature we are unable to cover. Production budgets are, after all, finite accounts. In the case of botany, for example, we estimate that we are able to cover just over 50% of the existing literature sources. What tends to be omitted is the popular literature, newsletters, trade journals, locally published documents, and the "grey" literature. Although accessibility of the literature is sometimes a factor, the cost of processing the literature is the primary limitation on coverage.

Cost affects the users of as well as the producers of literature databases. In the past, when print products were the main reference tools, the library purchased a subscription and made it available at no charge to its patrons. Now, however, it is not uncommon to see

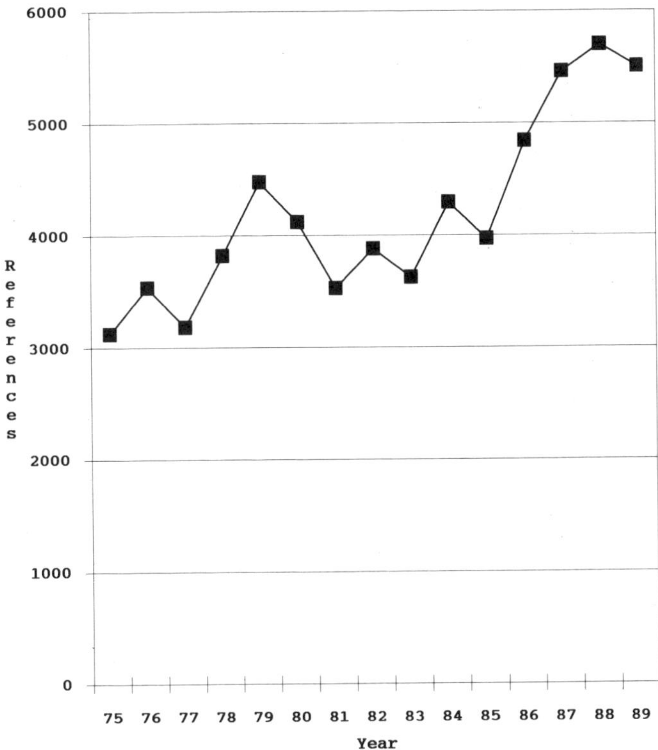

Figure 2. Evolution—General. 1975–1989

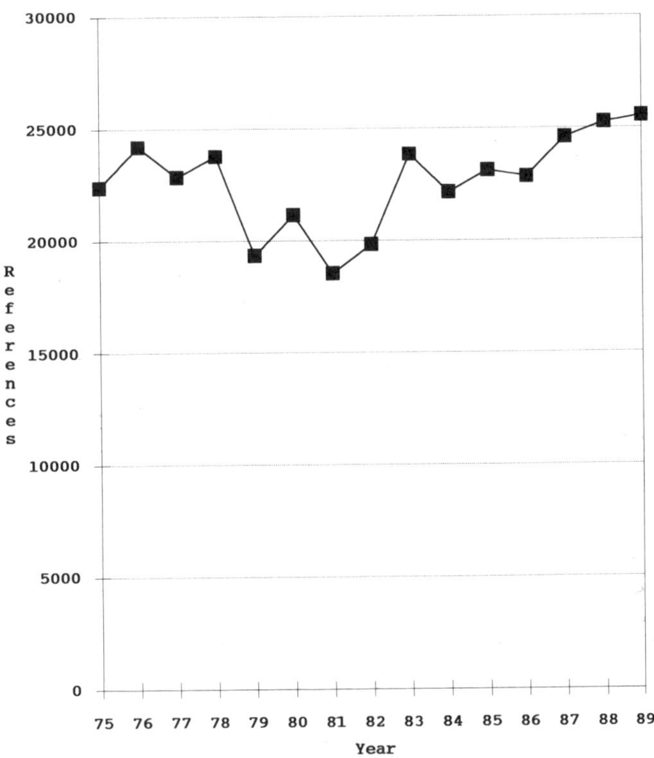

Figure 3. Systematics—General. 1975–1989

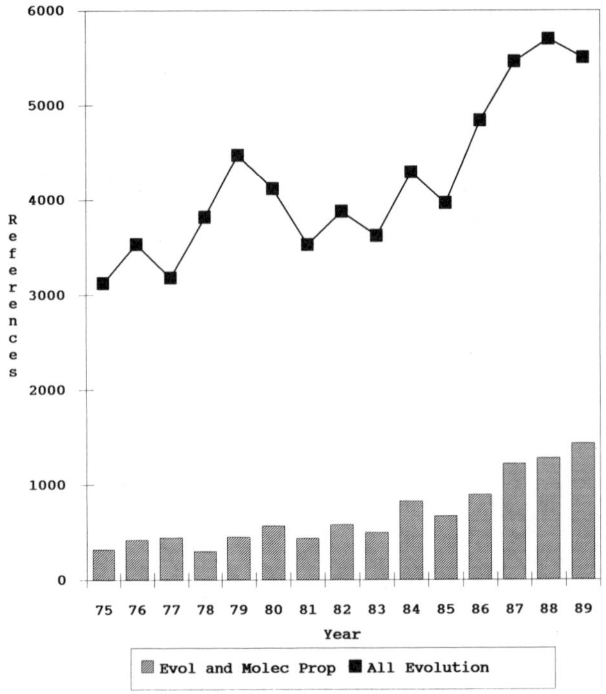

Figure 4. Evolution and molecular properties. 1975–1989

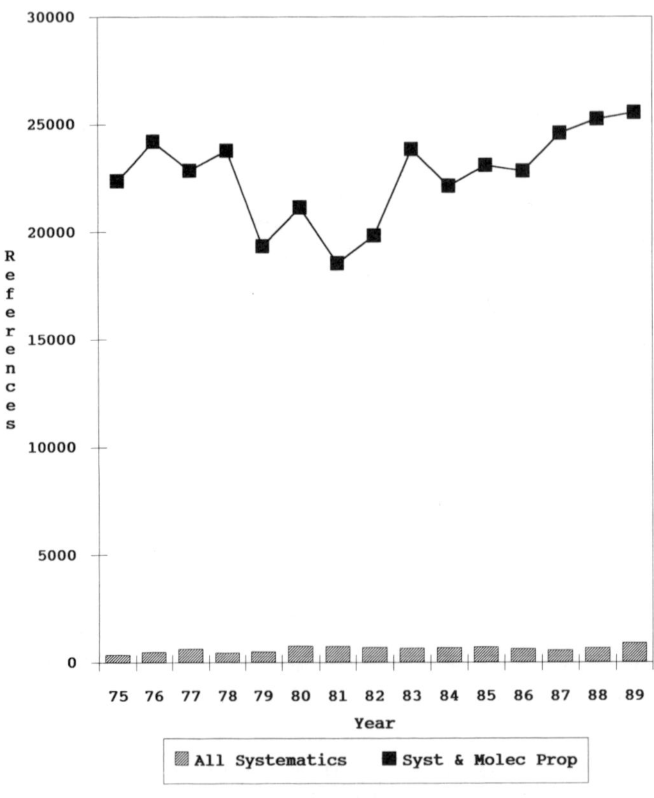

Figure 5. Systematics and molecular properties. 1975–1989

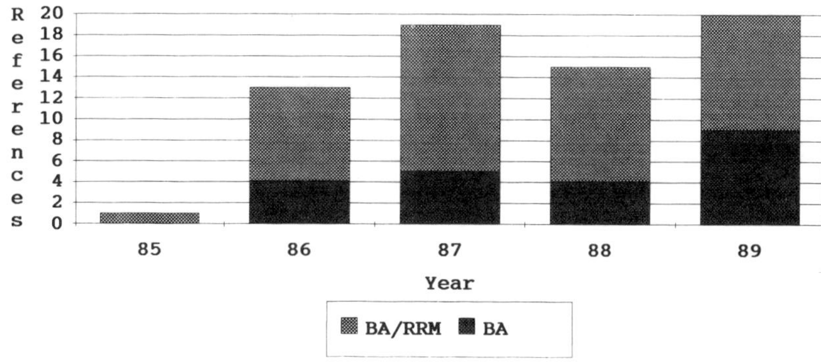

Figure 6. Evolution and conservation BIOSIS Previews, 1985–1989

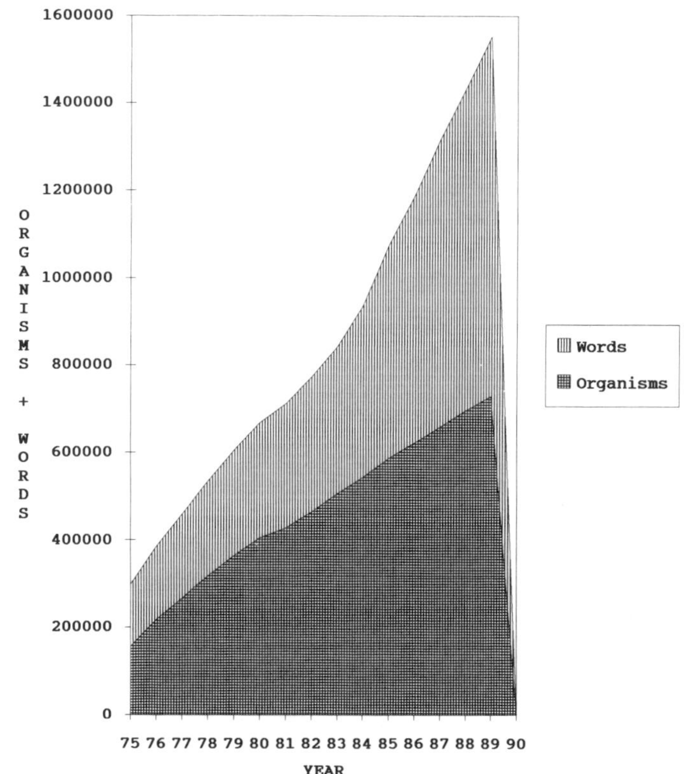

Figure 7. Master vocabulary file totals.

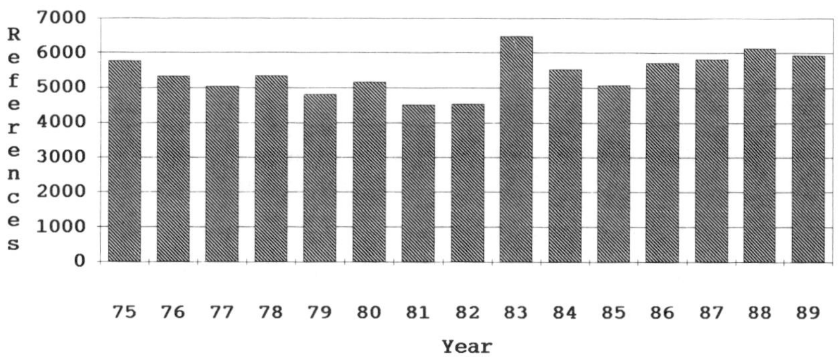

Figure 8. New taxa references. BIOSIS Previews, 1975–1989

962

libraries pass on to users the cost of online database searching. Furthermore, charges for online database usage are typically based on how long you use them and how successful you are in finding information. This "metering" of usage tends to discourage browsing and reduces the chances for serendipitous discovery of useful information. Fortunately, this situation is beginning to improve with new delivery media such as compact disks and new pricing options such as fixed price leases for databases.

FUTURE OPPORTUNITIES

Sadly, I expect that the economics of information processing and transfer will continue to present obstacles for us into the next century. Although this is considered by many to be the "era of biology," basic biological research still suffers from a shortage of funds. This is especially true for disciplines like systematics and evolutionary biology.

On the positive side, I envision a future replete with opportunities to overcome some of these limitations and to achieve new levels of service. Technological advances will offer opportunities for improvements both in the ways in which literature databases are produced and also in the ways in which they are delivered and used. Another, very important opportunity for improvement will come from cooperative partnerships; from working together to achieve significant advances in scientific information systems and services.

Technologies for Database Production

Technology has already had a substantive impact on the way in which literature databases are built and maintained. Computers are now an accepted part of database production, especially in supporting data entry. Although optical scanning technology failed to live up to expectations in the 1980s, it holds renewed promise for the next decade. Scanners are now available that work in two stages. First, a document is scanned quickly to produce a digitized image, analogous to the image transmitted by a FAX machine. Next, sophisticated programs are used to transform this image into a standard (ASCII) computer file. This software can interpret multiple fonts, format the data into prescribed fields, and make use of customized dictionaries to ensure accurate translation. This separation of the physical scanning from the character recognition analysis offers significant efficiencies for paper handling and staff utilization.

Another key area for technology will be in support of indexing. Computer-aided-instruction is just beginning to be used to train indexers and catalogers. The National Agriculture Library is working with Apple Computer Inc. to develop one such system. Artificial intelligence systems also hold promise for reducing the labor needed to index databases. The National Library of Medicine has long been a leader in informatics research of this type. Both computer-aided-instruction and artificial intelligence could help to control the costs of database production and to improve the options for data retrieval.

Technologies for Database Delivery and Use

Technology will have its most conspicuous impact in the area of database delivery and use. There are three key elements here: storage, processing, and transmission.

Data Storage. Optical storage techniques have already begun to have an impact. A CD-ROM holds about 550 megabytes of data and offers a convenient alternative to both print and online data storage for relatively large archival files. One year of *Biological Abstracts*®, containing some 275,000 citations with abstracts and indexing, just about fills a single compact disk. These disks can be stamped out or "printed" in a very cost effective manner. (By contrast, the data on a magnetic disk or tape must be laid on sequentially; a much slower process.) The CD-ROM readers can be used in conjunction with microcomputers and are available for $700–$800.

A limitation of optical disks has been that their contents, once printed, is fixed. The newer, erasable-optical disks use a combination of magnetic and optical technologies to crowd over 650 megabytes of data onto one side of a 5¼ inch disk. (It works by using a laser to change the electrical composition of the disk.) Sony Corporation sells a magneto-optical drive priced from $3,000–$6,000, depending on the specifications. Panasonic has recently announced a new erasable optical drive based on a faster, phase-change technology. (It works by changing the reflective surface of the disk and requires only a single pass over the disc to erase and rewrite.) It can store up to 1 gigabyte (1,000 megabytes) of information, equivalent to about 500,000 typed pages. The drive is priced at $5,000.

All three optical technologies take about five times longer than the hard disk on a microcomputer to search out information. (85–90 milliseconds vs. 20 milliseconds or less) However, their phenomenal storage capacities and transportability make it possible for users to store databases locally rather than relying on online access to remote databases. Plus, erasable-optical disks make it possible for users to customize literature databases with the addition of their own data.

Data Processing.   Another factor which makes it possible for users to maintain large local data collections is the availability of affordable but powerful local workstations. Computers with radically new architectures and efficient processing methods are giving rise to desktop supercomputers. These technologies are just becoming more accessible. SUN Microsystems Inc. has begun advertising a high-power RISC workstation (i.e., reduced instruction set computer) for under $5,000. These machines are designed to function as nodes in a local area network.

Powerful workstations and erasable-optical media give users the opportunity to change the way they interact with literature databases. Databases with the full text of articles (rather than only citations and abstracts) can be maintained and searched. Much faster and more sophisticated retrieval software can be supported, along with expert systems and artificial intelligence. Literature records can be consolidated with research records and data files. High resolution graphics, with their demanding storage requirements, can also be accommodated.

Data Transmission.   When you combine these enhancements in storage and processing with new opportunities for high speed, high capacity data transmission, you really begin to change the way people work with information. It has been said that "the real gains in productivity will come from giving knowledge workers direct and immediate access to knowledge." When the Association for Computing Machinery asked a group of experts to predict the impact of information technology trends, they agreed that "what is going to count . . . is the ability to move high volumes of information in multi-media forms . . . back and forth between people and places where data is stored."

Just such a high performance data highway has been proposed by Senator Albert Gore and the administration's Office of Scientific and Technical Policy. The proposed network goes by the name of NREN: the National Research and Education Network. Gore's Senate bill 1067 would authorize $400 million over five years to establish a network capable of transmitting three gigabits of data per second (i.e., 3,000 megabits per second). The National Science Foundation's NSFNET currently transmits data at the rate of 1.5 million bits (or megabits) per second. The NREN system is being designed to support researchers in their increasingly data-intensive pursuits. Formidable demands for data storage, computation, and communications arise with projects such as NASA's Earth Observing System and the National Institutes of Health's human genome mapping efforts.

NREN is seen as one component of a national information environment that includes digital data libraries and a national knowledge bank. Robert Kahn of the Corporation for

964

National Research Initiatives is one of the pioneers of NREN. Kahn is busy planning an information future for us filled with "tireless knowbots." These invisible, software-equivalents of knowledge robots will "make endless journeys through information space," collecting and interpreting knowledge from a myriad of resources, translating the user's question into meaningful results.

Technology Caveats.   This idea so engages my 10-year old son that he's looking forward to the time when he can have his own knowbot. However, not everyone agrees that such a future would be in the best interests of all. An article on this topic appeared earlier this year in the Society for Scholarly Publishers newsletter. It cautioned that the high costs of supporting the NREN network may result in "a societally intolerable gap between the information rich and the information poor," especially if printed products are abandoned, and public information is made available only through this network.

Indeed, a gap already exists between those who have access to technology and those who do not. An item in the *New Scientist* puts information technology in perspective by describing conditions in the Sudan, where frequent major power failures combined with heat and dust make the use of computers very difficult. Throughout the less developed countries technology is limited, power supplies are unreliable, and funds are very scarce. With four-fifths of the world's five billion people living in less developed countries, we would do well to keep this technology gap in mind when working toward our information future.

Literature Databases as Part of the 21st Century Biological Knowledge Base

As I reviewed the literature in preparation for this talk, I came across many different forecasts of what the 21st century will hold for knowledge users. The projections range from dystopian through incrementalist to utopian. I suppose I would characterize myself as a sort of optimistic incrementalist. Crystal balls are fun to look into, but, in my experience, the future is best built on hard work.

There is no avoiding the impact of technology on our future. The computer and telecommunications technology have been and will continue to be engines for change in our society. New technologies have brought about changes in our information lifestyles; these new lifestyles have whetted our appetite for ever new technological solutions to our information problems. But we have a responsibility to look beyond the technology. Biological knowledge, widely distributed and wisely used, can improve our lives and our environment. The literature databases are a tool for sharing the knowledge of our discoveries, enabling us to build our future with an improved awareness of the experiences of the past. To be fully effective in that role, literature databases must be skillfully integrated with the other knowledge resources at our disposal. This is going to be one of the hard parts.

We have a choice. We can be caretakers, carefully tending our individual databases. Or we can be innovators. We can reach out in partnership; we can cooperate and compromise and build the bridges that will bring our knowledge resources successfully into the 21st century.

LITERATURE CITED

Anon. Designs for a Global Plant Species Information System. *CODATA Newsletter* May 1990: 4.
Anon. Hands-off Experience. *New Scientist* 31 March 1990: 67.
Anon. 1990. Infobits: Sony Corp. *IDP (Information & Database Publication) Report* 11(7): 11.
Anon. Panasonic Introduces Rewritable Optical Drive. *Information Industry Bulletin* 7 June 1990: 3–5.
Anon. 1990. Release 63. *News From GenBank* 3(1): 1.

Gilna, P. & J. Ryals. 1990. Moving Data From the Laboratory to GenBank. *News From GenBank* 3 (1): 1–2.

Johnson, S. W. 1990. Perspective on public libraries: considering the future. *Information Retrieval & Library Automation* 26(1): 1–6.

Martin, S. K. 1990. Proposed national research and education network: first impressions. *SSP Letter* 12(1): 1–3.

National Research Council. 1988. *Toward a National Research Network.* National Academy Press. Washington, DC. 55pp.

Palca, J. 1990. Getting together bit by bit. *Science* 248: 160–162.

Penn, E. Fossil record aids in predictions of global warming's consequences. *The Scientist* 14 May 1990: 5, 10.

Schulz, W. 1990. Molecular biology adds new depth to systematics research. *Smithsonian Institution Research Reports* (60): 2, 5.

Seiter, C. Erasable opticals: new light on data. *MacWorld* March 1990: 152–159.

Straub, D. W. & J. C. Wetherbe. 1989. Information technologies for the 1990s: an organizational impact perspective. *Communications of the ACM* 32(11): 1328–1339.

U.S. Congress, Office of Technology Assessment. 1989. *High Performance Computing & Networking For Science—Background Paper.* U.S. Government Printing Office: Washington, DC. 36pp.

Verity, J. W. 1990. The next frontier is the text frontier. *Business Week* (3165) June 18, 1990: 178–179.

Waldrop, M. M. 1990. Learning to drink from a fire hose. *Science* 248: 674–675.

Weber, R. 1990. Libraries without walls? *Publishers Weekly* 237 (23): 20–22.

Electronic Research Collections: Perspectives and an Example from the Evolution of Terrestrial Ecosystems Consortium

John Damuth

Abstract. Computerized database technology makes increasingly feasible the development of large, general compilations of data and references built primarily to serve as novel research tools in a particular subject area. Such databases treat as data both purely observational data and scientific conclusions and interpretations relevant to the chosen subject. They thus summarize the state of current knowledge in a specific area, entail basic research in their compilation, and are to a great extent "problem-oriented." Data generation and compilation are the major time-consuming activities. (This contrasts sharply with other important computerized database projects whose aim is to increase the availability of, and efficiency of access to, traditional data sources, or to manage traditional research collections more effectively. In the latter case, rate of data-entry is the primary limitation on growth.) The construction of an electronic research database in biology and paleontology is more akin to the establishment, maintenance, and development of the material collections of a natural history museum, although in this case the data items that constitute the collections are non-material ones. The analogy with a museum collection is an apt one. The database represents a collection resource amassed over a period of time by many workers. Work has been done to seek out and compile the data, interpret them, organize them, and store them in a form in which they can be retrieved and used in research. Many different pieces of information must be kept together, or their significance is lost. While the ultimate purpose of the database is its use as a research tool, data can be added without any specific research questions in mind, knowing that eventually these data may form the basis of comparative research projects. Some compiling of data may be done intentionally to answer specific research questions, but once the data are entered they may prove useful for other projects. In all these ways a non-material research collection resembles and functions similarly to a traditional museum collection. The construction of a large non-material research collection in paleobiology has been initiated by researchers in the Evolution of Terrestrial Ecosystems ("ETE") Consortium (Smithsonian and UCSB). The ETE database is designed to allow broad-scale comparisons of the paleobiology and paleoecology of terrestrial ecosystems and their plant and animal communities. In particular, interest focuses on making detailed comparisons on a locality-by-locality basis of paleocommunity structure, paleovegetation, and paleoenvironments, and on tracing patterns in biological structural change in ecosystems throughout geologic time. Information will come from heterogeneous sources—the literature, museum collections, and original research. It is recommended that such electronic databases be recognized as a type of collection-resource, and their construction the legitimate long-term objective of independent research groups.

INTRODUCTION—RESEARCH DATABASES

The development of computer technology has made it progressively easier to compile and manage large amounts of biological data. At present many, if not most, biologists

Dr. Damuth is with the Department of Biological Sciences, University of California, Santa Barbara, CA 93106, USA.

use some kind of computer on a daily basis. Further, the computing power necessary to handle large database projects is now no longer exclusively in the hands of institutional mainframe computers and their operators, but can be put on a desktop. Powerful workstations, formerly used primarily in engineering, now appear in increasing numbers in biological science departments. These offer huge increases in speed, memory, and mass storage over the PC-based systems introduced a few years ago. At the same time the "high-end" versions in traditional personal computer product lines have begun to achieve the performance levels of advanced workstations. These developments allow smaller institutions, research groups, and even individuals to contemplate long-term database projects that would have been impractical before.

The most obvious applications of database technology in biology involve the computerization of vast databases that we already have in another medium— for example, bibliographic information, where the information resides on catalogue cards or in specialized bibliographies, ultimately referring to library collections. Or, to take another important example, the information associated with specimens in museum natural history collections, where the information has traditionally resided on a combination of labels, index cards, and ledger books.

Such computerization projects are of the utmost importance. They will vastly increase the availability of the information contained in such databases. In addition, computerization of library and museum collections-information is of considerable benefit in the efficient management of such physical resources.

However, I want to discuss a distinctly different type of electronic database—a kind of collection of facts and inferences that computer technology may have made possible for the first time. This is a collection of facts and inferences specifically intended to be the direct object of research activities and analysis. By this I mean that the information in the database has been compiled from diverse sources and organized there such that it can be analyzed in a variety of ways without recourse to other data or collections. It is such an organized collection of information that I will refer to as a *research database.*

An example that might be familiar to some is the extensive compilation of marine fossil taxa that Sepkoski and his colleagues have built (e.g., Raup & Boyajian, 1988; Sepkoski, 1986). This data set consists of the geologic age-ranges recorded for many thousands of marine fossil genera and represents a synthesis of a large amount of literature. The database has been used in a variety of studies of changes in diversity of marine organisms over geologic time and in the study of patterns of mass extinction.

The important point for the present discussion is that when these studies are performed, it is the contents of the database itself that are analyzed as the "raw" data— researchers do not return to the original sources. In a real sense, the relevant research involving those sources has already been done in the process of compiling the database.

In contrast to the way that a research database is used, non-research databases (such as computerized bibliographic or museum-collection databases) are usually intended as indexes to *another* collection whose contents need to be consulted—for example, books, articles, or specimens. While it may sometimes be possible to answer a research question directly from such a database, its primary function is to locate research resources.

Non-research databases allow one to *find* the data of interest—research databases already *contain* the information of interest.

The data in a research database are intended to be analyzed as they are represented in the database, and the form and content of the data embody scientific research decisions. Some data are partly analyzed, standardized, or combined with other data; some potential data are even deemed of no interest and omitted. This means that the data in the database cannot be used for any research purposes for which the original research or data collecting effort was not appropriate (the same applies, for example, to museum collections). Research databases are thus *problem-oriented.* Of course, from a research database we can always go back to the original sources to verify our data, just as we can go back to nature and collect or make observations to verify what is in a museum. However, we are not obliged to do so if the data in the database are adequate to the investigation at hand—and this is the great value of a research database.

Research databases are already familiar to most scientists. Most researchers compile small research databases in the course of their normal research activities. Any data set intended for analysis is, in some sense, a research database. But typically these data sets are compiled for only a single study—they are of limited lifespan and are created ad hoc. The significant and perhaps novel situation arises with the idea that sometimes such a database can be maintained as a general research resource. It can be extended, updated and built upon indefinitely. It can be used for a variety of research purposes by many workers, who will return to it again and again for the data they need.

One of the most important characteristics of such an electronic research database is that it can integrate information from a wide variety of sources (Fig. 1). Consider a hypothetical research database involving vertebrate species of some taxon. We might have locality data and field measurements from the collections-data of several museums; measurements made on specimens from several museums; behavioral observations from the field for each species; results of physiological experiments performed on individuals of the various species in the lab; information from reports about the behavior or ecology of the species from the literature; and systematic information from an as-yet unpublished cladogram provided by a colleague. All of this information would be standardized and in one place, ready to form the basis of comparative analyses.

In contrast, non-research databases usually refer to only one type of resource, and often are restricted as well to a single collection or institution.

In addition to the data directly documented in the sources (for example, geographic locality data from a museum collection), a research database can contain anything else we wish to put in it. This can include syntheses of diverse data and scientific interpretations that are far removed from raw data. It will be useful to distinguish data of two types—though the distinction is not always clear-cut: *non-inferential data,* which are subject to little or no scientific interpretation (e. g., "hard" empirical facts, direct observations, catalogue numbers), and *inferential data,* which are conclusions or interpretations based upon the results of scientific research (e.g., systematic affinities, diets of dinosaurs, or productivity of habitats). Non-research databases contain primarily non-inferential data. By contrast, research databases may contain large numbers of inferential data.

RESEARCH DATABASES AS ELECTRONIC MUSEUMS

The frequent references to museum collections throughout the first section are intentional. In this section I will draw an extended analogy between museum collections and the computerized research databases that I have just described. I will argue that computerized research databases are so similar to museum collections that we can think of their design, building, and maintenance in very similar terms. This perhaps somewhat surprising correspondence may help clarify the value of both.

Note that I am comparing museum collections with research databases, *not* with computerized databases describing a museum's own collections. The latter are non-research databases—and are just a re-representation of the label and catalogue data of the museum collection. A *research database* might not include any data derived from museum collections.

Often the construction, maintenance, and study of natural history collections are regarded as merely descriptive activities devoid of theoretical content. In fact, the construction of museum collections, and the construction of research databases, share the

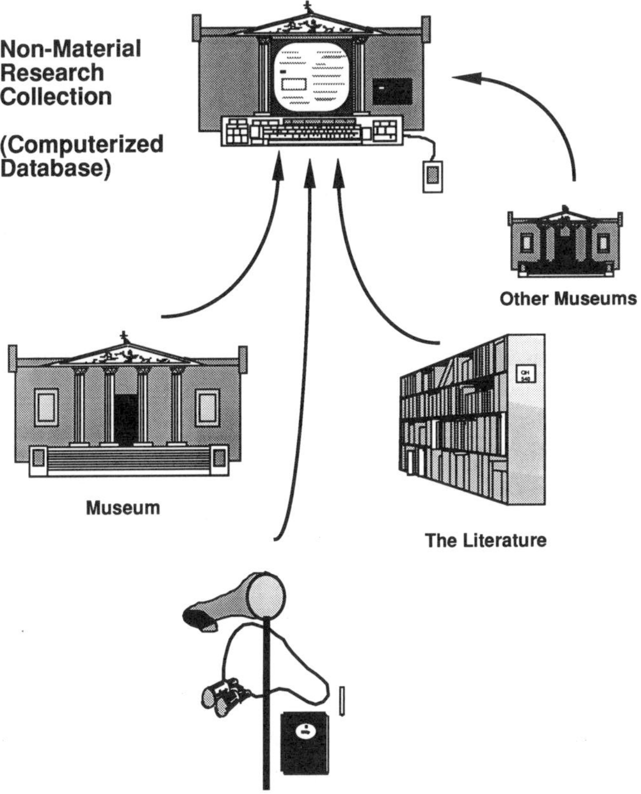

Figure 1. Computerized databases intended to be non-material research collections can integrate data from many different sources (see text).

same type of bold conjectures about what may be important to preserve and compare, and what kinds of hypotheses will be fruitful to test. These are hardly theoretically neutral issues.

Theory and theoretical motivation may be more evident today in the design of computerized research databases than in that of natural history museums, because much of the theoretical foundation of natural history collections was worked out in the past and has occasioned little discussion in recent years. Nevertheless, the organization of museums and their collections involves theoretical decisions. Griesemer (1990) has described in some detail how Joseph Grinnell designed the Museum of Vertebrate Zoology at Berkeley specifically so that its collections could test his theories about the relationship between ecological factors and variation and speciation. Indeed, Griesemer makes a strong case for our regarding the museum collection and its associated data as a kind of theoretical model of nature—a simplified representation that we can study directly in lieu of nature. In fact, perhaps the broadest definition of a "model" of nature would be any representation of nature that can be the object of scientific study in its own right. A research database is also a representation of nature, arranged for the purpose of study. One could also regard it, then, as a simplified model of nature, built to allow investigation of particular natural processes—exactly the function of a museum collection.

Museum collections can be seen to be types of large, expensively maintained scientific devices whose function is to record and represent in a manageable and comparable form particular aspects of nature—aspects that we know or suspect that we will have repeated interest in studying. Their existence allows us to make observations that would otherwise be either prohibitively expensive or impossible. In this way collections are facilities shared by the biological community that are similar to the telescopes, particle accelerators, and remote-sensing satellites of other scientific communities.

Museum collections contain less information than nature, and can be used to pursue only certain kinds of research questions. A museum collection is thus as problem-oriented as is a research database.

We can summarize the important similarities between museum collections and research databases. Characteristics of museum collections and their use that are shared with computerized research databases include:

1. The collection allows the organized preservation of the accumulated work of previous collectors.
2. This database can be built upon indefinitely.
3. The collection affords researchers the opportunity to juxtapose material and data from widely different sources (geographical regions, time periods, etc.). For example, a researcher who has an inspiration about how rodents living in African versus South American forests should differ in morphology need not gather the resources for extensive transcontinental field work in order to test this hypothesis. It is likely that most of the data are already available in one or more major museum collections.
4. This affords the researcher the opportunity to test efficiently ad hoc hypotheses about nature.
5. Comparative study of the collection can also suggest patterns in nature that can form the basis for new theory or hypotheses.

Fundamental aspects of the process of building a museum collection that are shared with that of *building* a research database include:

1. Obtaining the data (specimens) in proper form—performing collecting expeditions, preparing specimens and field notes—is more arduous and time consuming than is cataloguing per se. For a research database, the primary effort is in obtaining or processing data and getting them into the proper form for entry, not data entry itself.
2. Scientific work has to be done to seek out and compile the data, interpret them, organize them, and store them in a form in which they can be retrieved and used in research.
3. Many different pieces of information must be kept together, or their significance is lost. Information must be managed, and effort must be expended to maintain the collection.
4. Collecting may be undertaken to answer specific research questions, but once the specimens are collected they may prove useful for other investigations. Specimens (records) can be added without any specific research questions in mind; later, these data may form the basis of comparative research projects.
5. Choices of data to collect, as well as what to store and how storage is arranged, and what retrieval mechanisms are available, derive from (or embody) biological theory and a particular purpose for the collection (i.e., the problem(s) it is designed to address).

Thus, electronic research collections can be understood to be *non-material* research collections analogous to *material* (natural history) collections of museums.

THE EVOLUTION OF TERRESTRIAL ECOSYSTEMS CONSORTIUM

The construction of a large non-material research collection in paleobiology has been initiated by researchers in the Evolution of Terrestrial Ecosystems ("ETE") Program of the National Museum of Natural History and in the Department of Biological Sciences, University of California at Santa Barbara. Paleontologists at these two institutions have formed a consortium, whose purpose is to share the work of compiling the data, and to share the resulting research database.

The ETE database is designed to allow broad-scale comparisons of the paleobiology and paleoecology of terrestrial ecosystems and their plant and animal communities. In particular, interest focuses on making detailed comparisons on a locality-by-locality basis of paleocommunity structure, paleovegetation, and paleoenvironments, and on tracing patterns in biological structural change in terrestrial ecosystems throughout geologic time.

To support this activity, we use a commercially available, full-featured, relational database management system (*Ingres*, Ingres Corporation, Alameda, California). The entry and modification of data are accomplished through a customized forms-based utility program. This program performs sophisticated error-detection and transparently executes a number of required data transformations. It also automatically documents changes made to existing data fields. The system runs on a local network of Hewlett-Packard/Apollo UNIX workstations. Identical databases are maintained in Washington, D.C. and in California.

The basic entity in the database is a fossil locality—a particular geographic location from which a significant collection of fossil specimens has been obtained. For each locality, in addition to its name, age and location, there are a number of fields of sedimentological, taphonomic, and paleoecological data (including inferences about the vegetation and physical environment). In addition, for each locality there is a species list, and for each species in the database a set of 25–30 morphological and ecological descriptors (depending upon the type of species). For vertebrates, the data include attributes such as esti-mated body mass, inferred dietary preference, locomotor category, and sexual dimorphism. For plant macrofossils, the entries include such things as leaf size, leaf margin, life form, and trunk diameter. There are additional tables for recording literature references and the update history of entries. In all there are 17 base tables and 130 data fields.

The database may be queried directly using SQL (the industry-standard database query language), or may be accessed through a graphics interface which displays fossil localities on maps. The system allows researchers to see the localities that satisfy arbitrarily complicated (boolean) conditions specified by the user. Data sets for statistical analysis can be generated from the graphics interface using a menu-based utility, and common statistical analyses can be performed directly from the interface. The large number of ecological inferences contained in the database will allow one to compare aspects of com-munity structure (such as body-mass distribution, or percentage of woody vs. herbaceous species) across paleocommunities whose contained species may be only very distantly related (Damuth et al., in press). It is of course possible to pursue more traditional "taxon-based" studies with the contents of the ETE database as well.

It is expected that eventually all significant terrestrial plant and animal localities will be recorded in the ETE database. This ambitious, long-term goal means in practice the involvement of numerous researchers as members of smaller, manageable projects to deal with the parts of the fossil record that they know best. The compilation projects will be managed by members of the small research group that initiated the database, who will be responsible for the comparability and quality of the resulting data.

Much of the information will be taken from the literature and some will come directly from museum collections. The large number of inferential data make it likely that much of

972

the data compilation will involve original, unpublished research on the part of compilers. Most of this research will involve work with museum specimens, but in some cases it may also include new observations from the field. The ETE database will represent an unprecedented compilation of all aspects of the terrestrial fossil record, in a manageable form.

As an example of a research database, the ETE database exhibits all of the major features such compilations share with museum collections. Such electronic databases, intended to be maintained beyond the scope of a single research project, should be clearly recognized by the scientific community as a type of collection-resource, and their construction the legitimate long-term objective of independent research groups.

LITERATURE CITED

Damuth, J., Jablonski, D., Harris, J. A., Potts, R., Stucky, R., Sues, H.-D. & D. B. Weishampel. Taxonfree characterization of faunas. *In:* The Evolution of Terrestrial Ecosystems Consortium (eds.), *Evolutionary Paleoecology of Terrestrial Plants and Animals.* University of Chicago Press: Chicago. In press.

Griesemer, J. R. 1990. Modeling in the museum: on the role of remnant models in the work of Joseph Grinnell. *Biology and Philosophy* 5: 3–36.

Raup, D. M. & G. E. Boyajian. 1988. Patterns of generic extinction in the fossil record. *Paleobiology* 14: 109–125.

Sepkoski, J. J., Jr. 1986. Global bioevents and the question of periodicity. Pp. 47–61. *In:* O. Walliser (ed.), *Global Bio-events. Lecture Notes in Earth Sciences* 8.

Beyond the Hardcopy:
Databasing Flora of North America Information

Nancy R. Morin

Abstract. Documentation and study of organisms—their characteristics, distributions, and genetic and ecological relationships—is urgently needed for purposes of conservation, resource management, and applied and basic research. Flora of North America is designed to make accessible to a wide range of users standardized information on plants. Computers are used in Flora of North America as mechanical tools for gathering, editing, and formatting the information; as a conceptual tool for defining appropriate use of terms and for categorizing characteristics of any kind; and as devices for making the data accessible. The programs being used for these applications (TROPICOS, developed at the Missouri Botanical Garden) result in a relational database of names, characteristics, distributions, and other features. Collaboration with other projects on national and international levels is ongoing.

INTRODUCTION

Flora of North America is a collaborative effort among some 20 universities, colleges, and museums to prepare, edit, and publish a synoptic flora of the vascular plants of North America north of Mexico: the continental United States, Canada, Greenland, and St. Pierre and Miquelon Islands; and to develop a computerized database therefrom. No such work has ever been completed for this area. The flora will synthesize available knowledge about the plants found in the region and will delineate taxa and geographical areas in need of further, more critical, study. It will serve as a systematic conspectus of the North American flora; not only for theoretical work in taxonomy, ecology, and phytogeography but also for practical use and general reference in biology, environmental sciences (including wildlife management, forestry, range management, horticulture, and agriculture).

Flora of North America will be published in 12 volumes issued over a period of 12 years. The treatments of the plant groups will comprise illustrations, pertinent synonymies, identification keys, comparative descriptions, summaries of habitat and geographic ranges, chromosome numbers, phenological information, and other significant biological observations. Volume 1 will contain introductory material, including brief chapters on history of floristic botany, physiography, evolution of the flora, and phytogeography of the flora area; and treatments of seedless vascular plants and gymnosperms. This is scheduled for publication in 1991. A comprehensive, consolidated bibliography and complete index to the flora will be published in the last volume. About 17,000 species will be covered (Soil Conservation Service, 1982), which is approximately 7% of the world's total species of vascular plants. The treatments of species are being

---

Dr. Morin is with the Missouri Botanical Garden, P. O. Box 299, St. Louis, MO 63166, USA.

974

written and reviewed by experts from throughout the systematic botanical community based on original observations of herbarium specimens supplemented by a critical review of the literature.

Flora of North America does not stop with the hard copy volumes: a computerized database composed of data from the treatments together with additional information not appropriate for publication in the flora is being developed. Further information will be added to the database after the flora itself has been completed, and the database will be maintained as a permanent resource. It will be made available through a variety of media, making the information accessible in new ways to a wide range of users.

The database of the Flora of North America project will provide, for the first time, names, synonymies, descriptions, and geographic distributions—prepared by specialists, rigorously reviewed, and presented in a consistent and comparable manner. The database is entered into and maintained on computers so that for the first time plant characteristics can be analyzed and compared in the context of their hypothesized phylogenetic relationships and geographical and ecological distributions. Correlations can be made, for example, among morphological characters, between these characters, and with aspects of habitat, ecology, chemistry, genetics, etc.

The data format is structured to meet two primary goals: comparison of characteristics across all taxa, and production of detailed, precise, computer-generated descriptions. The software and data structure provide an advanced multiple-entry key function for detailed comparisons and analyses that will assist in the writing of such keys. The production of printed, sequential keys will also be possible.

FLORISTICS FOR THE 21st CENTURY

Floras have been written by botanists for at least the past three centuries, always as "hard copy" (although describing the often exquisitely illustrated books this way seems hardly right). In any event, the works were limited by the constraints typical of books, and this in turn limited who used the information and how they used it. Traditionally, a 'flora' may contain anything from a simple list of the plants occurring in an area to a very detailed account of those plants. A flora almost always contains scientific names, and it may also include vernacular (or "common") names, literature references, descriptions, habitats, geographical distribution, illustrations, flowering times, or notes. Less often, floras include chemical information, chromosome numbers, population occurrences, etc. Sometimes the plants are listed alphabetically, and sometimes they are presented within a classification system. Floras often also include keys for identification.

Flora of North America (FNA) sought to determine the needs of those who use floras, and specifically what non plant-taxonomists need from a flora. To do this, FNA, with the American Society of Plant Taxonomists and the American Institute of Biological Sciences, held a workshop attended by about 70 professionals from a wide range of disciplines (Morin et al., 1989). The participants themselves identified a very extensive list of potential flora users (Table 1). Furthermore, workshop participants made it clear that floristic information is not sufficiently available or useful for their purposes, that they need more information, that it must be directly comparable, presented with a minimum of jargon, and presented within in a *classification system* to facilitate investigation of characteristics and distributions of related species. Finally, they concluded that users want to have access to the information in all possible forms: hard copy reports, on diskettes, on CDs, on-line, etc.

DATA SOURCES

The first challenge for meeting the needs of such users is to locate and capture the information needed. Floristic information resides in herbarium collections, published and unpublished literature, and in the minds of people, mostly botanists, who have observed the plants in the field or laboratory. Flora of North America is designed to locate, involve, and guide the many botanists who are knowledgeable about North American plants in order to synthesize this information into treatments that meet the needs of our users. Ninety percent of the treatments in the first three volumes will be written by relevant specialists. To locate more specialists, and to help those already signed up, we have developed preliminary files both on specialists and on taxa that will be treated. For instance, we have computerized files on names accepted in 25 major floras, names that authors will be required to account for.

Information on plants, their characteristics and their distributions, can be found in a vast number of published scientific research papers, which include revisionary or monographic studies on specific groups, and other floristic studies. The Flora of North America project in an incomplete survey has identified over 800 relevant monographic or revisionary articles published between 1971 and 1984, and hundreds of floristic projects of varying sizes and scopes are currently underway. Floristic information may also be found in ecological studies, which may encompass analyses of life-forms, phenology, relations to specific abiotic factors, form and function of plant structures, and descriptive analyses of plant communities.

Available sources of unpublished floristic information include Master's theses, Ph. D. dissertations, electronic databases, collectors' field notebooks, data files in local and regional herbaria, and notes compiled by private and public conservation agencies. Additional unpublished information, accessible only on an individual basis, remains associated with archival herbarium collections throughout North America.

FNA authors are expected to examine as many herbarium specimens as possible. The original plan of the project did not include gathering precise locality information. Several factors argue that we will have to find mechanisms to do this nonetheless. First, much of the power of the data depends on being able to tie it to physical parameters (Illinois Natural History Survey Reports 281–282, 1988). Ideally, it should be possible to determine whether specific features or suites of features are correlated with any physical elements, such as soil or climate. Second, anticipating such applications, the botanical community, encouraged by the National Science Foundation, is beginning to plan strategies for databasing all or most of the 60 million plant specimens held in North American herbaria. It makes sense for FNA to be involved in specimen computerization as much as possible because many of the FNA authors will be recording specimen information in their own databases. This is an ideal time to make a concerted effort to capture that information because many of the specimens will have been recently and authoritatively annotated as well.

DATA MANAGEMENT

The organizational center for Flora of North America is located at the Missouri Botanical Garden, and the FNA database is an integral part of the Garden's larger database, TROPICOS. TROPICOS is a series of interrelated files about vascular plants and mosses worldwide. The programs were written by R. E. Magill in collaboration with M. R. Crosby. This system has been used for a variety of projects at the Garden, as well as for curatorial and bibliographic procedures. The hardware and software work extremely well for the main purposes of Flora of North America: managing data on names, descriptions, and distributions of plants.

The files are maintained on an Applied Digital Data Systems, Inc. (a subsidiary of NCR) MENTOR 5300 computer with two 300-megabyte hard disks and one megabyte of working memory with a Pick-based operating environment. It uses 9-track, 1600/3200-bits-per-inch, phase encoded, ½ inch tape drives, with an American Standard Code for Information Interchange (ASCII) and EBCDIC character sets and four 1200-baud modems. Additionally, the system can accommodate 2400- to 19200-baud modems. The Mentor uses a virtual memory system that expands RAM by assigning a portion of available hard disk space as virtual memory. This memory is directly addressable, as is RAM, resulting in an extremely powerful and fast operating system. Flora of North America alone may require on the order of 10 GB of storage capacity.

Unlike most other microcomputer database management systems such as dBASE and R:Base, which run on top of the computer's operating systems such as MS-DOS, Pick is designed to use available disk storage to its maximum by not requiring a fixed field-length format. Data files occupy only the space of the data entered plus system delimiters. The operating system also allows on-line expansion or contraction of a field's data or even addition or removal of fields from records or files. The ADDS Mentor as a database management machine is designed for relational file manipulation. The Pick-based operating system provides report management software and data processing functions that use English statements for sorting or selecting data for output. Files are accessed through user-defined dictionaries and system verbs that together can produce formatted reports on a variety of media. Report production is quick and simple using dictionary pointers to requested data fields. These dictionaries provide the data format and position to the operating system and allow for great flexibility in report generation by allowing the user to set the parameters. Inter-and intra-file sorts are possible and selected lists of requested records can be stored for repeated use. Complicated reports can be performed on-screen or stored by a remote processor for future use. Thus programming required for most operations is limited to data input and correction tasks or special output needs. Pick in various versions runs on many computers, including McDonnell-Douglas's Microdata, DEC, Prime, and some large IBMs. In addition to mini and mainframe releases, Pick has been available for the IBM XT/AT and compatible microcomputers with at least 10 megabytes of hard disk since 1984.

The TROPICOS software can also be used with Revelation (Revelation Technologies, Inc.), which runs on top of the disk operating system (DOS) and has the same data structure as PICK. Revelation is compatible with all IBM PCs, XT/AT, PS/2, and 100% compatibles. It requires 512 K of RAM, MS-DOS or PC-DOS, and a hard disk.

TROPICOS DATA FIELDS

The NAMES file is the core of the TROPICOS system. It is a purely nomenclatural file, containing Latin names and the citation of their original place of publication. NAMES makes use of the authority files AUTHORS, TITLES, and BIBLIOGRAPHY. There is a separate NAMES data file for each plant family, as well as a file for mosses and one for ferns. Currently, more than 400,000 names of species and infraspecies are in NAMES.

NAMES are placed in the correct genus using the GENUS INDEX, based on the Index Nomenum Genericorum (ING), kindly supplied to the Garden through the Smithsonian Institution as an updated subset of the ING project. It is a pointer file that places each of the 38,000 or so generic names for mosses and vascular plants in its proper subfile under NAMES.

SYNONYMS is a system for recording synonymy from the literature. Each recorded synonym is attributed to a literature reference, which is cataloged in BIBLIOGRAPHY. Thus it is easy to trace the source of a taxonomic or nomenclatural judgment.

DISTRIBUTIONS records the geographical distributions of plants at three levels of resolution, country, state, and county, or the appropriately named geographical subdivisions in that hierarchical structure. The information in DISTRIBUTIONS can be literature-based, in which case citations are recorded in BIBLIOGRAPHY, or specimen based, in which case vouchers are in the specimen files.

The EXSICCATAE file contains information found on herbarium specimen labels. It draws upon NAMES and uses the distribution menus.

A file called BIBLIOGRAPHY is actually a series of bibliographies for particular subsets of the TROPICOS system. It allows one to flag any word in the title as a key word, enter additional key words, and enter notes that can be designated for printing (e.g., for an annotated bibliography).

Several authority files save space, standardize data, and aid entry of further data. AUTHORS is a file of authors of names of plants based on the Kew Draft List and supplementary information, particularly names of authors that have appeared in the literature since the Kew list was published. The entries in AUTHORS are coded in such a way that information that does not appear in the Kew list can be retrieved easily and forwarded to Kew for updating their list.

TITLES contains information concerning the titles of periodicals and books pertinent to TROPICOS, now numbering about 13,000 entries. Abbreviations are included in this file, following the recommendations of Botanico-Periodicum-Huntianum and Taxonomic Literature. AUTHORS and TITLES are particularly important to NAMES, where unabbreviated versions of authors' names and titles of publications are being replaced by numerical codes that correspond to the alphabetic information in these files. This assures that these names will be consistently abbreviated (or spelled out in full) as they appear in various products of the TROPICOS system.

The Hunt Institute for Botanical Documentation maintains a database of literature pertinent to the Flora of North America. These files are also relational and allow cross-checking. This database documents the place in which every reference appears, whether it is for a nomenclatural entry or not, how it should be listed in the index, etc.

DESCRIPTIONS: A SPECIFIC CHALLENGE

Developing a comprehensive, parallel, and searchable descriptions database for plants is a very tall order. Several excellent programs already exist for capturing and manipulating such data (Dallwitz, 1984), one of which (TROPICOS) was developed at the Missouri Botanical Garden and is being used for Flora of North America. Developing the character and character state list is a major challenge (Allkin, 1984). Each of the programs available for writing descriptions requires the user to design his or her own character list. This is an extremely time-consuming and tricky task, and one that not many people persist in, usually because they lack the training or because it is an untraditional approach. Furthermore, individual character lists are not easily cross-comparable, because the characteristics may have been treated differently. Flora of North America has, for the past several years, been assembling a master character list that would be applicable to any family. The only practical way to do this, however, has been to start with a general list and then focus on individual large groups or families. The master character list now covers ferns, gymnosperms, and many of the monocotyledonous plants.

Ultimately, the descriptions database will be an enormous, practical editorial tool. It will help speed up and improve the work of the project enormously. It is not yet making that contribution. The very exercise of putting together the character lists, however, has been an excellent mechanism for encouraging authors to think about what needs to be included in their treatments and to agree among themselves about how to treat characters and what terminology to use.

Within TROPICOS, descriptions are entered via menu-driven screens, either by the author or based on the author's treatment. Measurements are entered as numbers and calibrations, stored as millimeters. Instructions are given as to what is to be printed for each character. Branch questions—those that determine the next set of questions, are built in, as are jumps, e.g., if leaves are absent, further questions relating to leaves are skipped. Major categories of descriptors, such as shape, are held in glossaries. Ultimately definitions and illustrations will be available on-line. For data entry, an author can specify which characters and character states are required for the genus or family in question, and these are presented in menus. Authors who do not want to use the TROPICOS programs are welcome to use worksheets instead or simply to submit manuscript. It is important to emphasize that the descriptions program helps greatly in insuring that the treatments are complete and parallel.

Because Flora of North America authors will be entering data at their own institutions, ideally using the FNA/TROPICOS programs, the problems of checking and merging incoming data will not be trivial. Few projects that have off-site data capture have dealt specifically with this problem. The International Legume Database and Information Service (ILDIS) provides an excellent model, however (Bisby, 1989).

Volume 1 information will be available on-line by the time it is published. Special programming effort will be required to make the information as easily accessible as possible. Because of the precision of the descriptions, searches will need to be phrased carefully so that all permutations within a more general state are located, and programs will be developed to assist with this. For instance, the answer to a query about which plants have red flowers needs to take into account that the user may not be differentiating between colored perianth parts and colored bracts, and that many different perianth parts may be scored as being entirely or partially red. The benefit gained from such precision is clearly the level of information content held in the database.

To help in production of the volumes, the computer programs are designed to draw together coded information from nomenclature, synonymy, references, descriptions, and distributions. Printer's codes can be embedded automatically so that tapes can be sent to the printer after a minimal amount of word processing.

The Missouri Botanical Garden's computing staff are working with M. J. Dallwitz at the CSIRO Division of Entomology toward being able to import and export data from DELTA, the system based on comparative descriptions that he developed in collaboration with L. Watson (the Australian National University). DELTA provides translation of encoded descriptions into various formats including those for classificatory analysis, interactive identifications, data retrieval, and key generation. Links between database programs, as in the case of TROPICOS and DELTA, broaden the facility and flexibility of each system and foster cooperation.

Flora of North America anticipates that an increasing number of authors will want to prepare their treatments using descriptions programs on personal computers. Coordinating this will require additional technical support, because authors will need to be able to add characters and character states. The project benefits greatly in a variety of ways by having authors use the programs directly, and to encourage this it will be necessary to make the programs more "user friendly." Unless the programs suit the user's needs almost perfectly, they are not likely to become a practical research tool.

The programs must allow extraction of floristic treatments of specific regions from the database, at different political and ecological levels, e.g., states, provinces, territories, parks, vegetation zones. Other kinds of reports, such as regional lists of poisonous plants, or lists of plants containing certain chemical compounds may also be needed. The project hopes to widely publicize among the "laboratory" subdisciplines of plant biology the predictive value of the database, to demonstrate the potential for experimental use.

Connections among databases need to be established using either existing networks such as BITNET or dedicated links between host institutions. Many databases already in existence are clearly related to the elements in the primary database, e.g., census, climate, soil, and wildlife data.

Flora of North America interacts with the many regional floras underway, some of which include databasing activities. For example, the Flora of Pennsylvania project is managed by the Morris Arboretum in collaboration with the University of Pennsylvania, the Carnegie Museum of Natural History, and the Philadelphia Academy of Natural Sciences. The project has received funding jointly with Flora of North America to explore the requirements for using Geographical Information Systems applications for specimen databases. This is particularly important in view of the development of several regional consortia to computerize specimen information in herbaria. The California herbaria consortium and the southeastern herbaria consortium are just two examples of a growing trend.

A great deal of botanical research is undertaken in federal agencies, and these agencies are also developing botanical databases with which the FNA database should interact. Examples are the Soil Conservation Service Plant List of Accepted Nomenclature Taxonomy and Symbols (PLANTS); National Park Flora, Germplasm Resource Information Network (GRIN), Canadian Heritage Information Network (CHIN)—Natural History Museum of Canada; the Plant Information Network (PIN) Data Base for Colorado, Montana, North Dakota, Utah, and Wyoming; and the Multi-State Fish and Wildlife Information Systems (CFWIS). Other continent level databases, such as that held by The Nature Conservancy, are also relevant.

Networking on a global level is also very important. Some projects with which Flora of North America collaborates are also headquartered or have editorial centers at Missouri Botanical Garden and similarly use the TROPICOS programs. These include floristic projects in Mesoamerica, Colombia, Ecuador, Peru, Bolivia, and other areas of Latin America; Madagascar; Flora of China; and a botanical inventory of Taiwan. In addition, Flora of North America has direct collaboration with the revision of Flora Europaea and Flora USSR and anticipates collaboration with Flora of the Philippines. The Species Plantarum Project, which has its secretariat at the Royal Botanic Gardens, Kew, hopes to produce family treatments on a global level in both printed and computerized form, and Flora of North America would play an important role in this for taxa that are primarily North American.

Such collaboration, which is absolutely essential given the enormity of the overall task of managing data about the world's plants, requires facile exchange of data, which in turn requires a much greater community commitment to designing and accepting minimum data standards. Although hardware and software compatibility are not absolute requirements for data exchange, we must not underestimate the amount of staff time and expertise that may be required to convert files before they can be shared.

It is clear that there is an immediate and urgent need for information about plants of North America for applied scientists and conservationists as well as by for a wide array of biologists. Computerized databases provide a means of very efficiently capturing and presenting such information, making the best use of the time and energies of botanists who are contributing to the project. Automated databases such as Flora of North America offer remarkable potential for managing existing data about plants and for stimulating diverse research interests, especially enhancing interdisciplinary studies, in which the data are manipulated, analyzed, and made available in a variety of often novel forms.

LITERATURE CITED

Allkin, R. 1984. Handling taxonomic descriptions by computer. Pp. 263–278. *In:* R. Allkin & R. A. Bisby (eds.), *Databases in Systematics.* Academic Press: London. Bisby, F. A. 1989. Databases, information systems, and legume research. Pp. 811–825. *In:* C. H. Stirton & J. L. Zarucchi (eds.), *Advances in Legume Biology. Monographs in Systematic Botany from the Missouri Botanical Garden* 29.

Dallwitz, M. J. 1984. Automatic typesetting of computer-generated keys and descriptions. Pp. 279–290. *In:* R. Allkin & R. A. Bisby (eds.), *Databases in Systematics.* Academic Press: London.

Illinois Natural History Survey Reports 281–282. 1988.

Morin, N. R., Whetstone, R. D. Wilken, D. & K. Tomlinson. 1989. Floristics for the 21st Century. *Monographs in Systematic Botany from the Missouri Botanic Garden.* Vol. 28. 163pp.

Soil Conservation Service, 1982. *National List of Scientific Plant Names.* Vols. 1 and 2. U.S. Department of Agriculture: Washington, D.C.

From Data Bases to Knowledge Bases

Theodore J. Crovello

Abstract. Continued data collection is necessary but insufficient for effective progress in systematics and evolutionary biology. Creation and expansion of explicit, computer-compatible knowledge bases also is needed to summarize the large amounts of data and knowledge already acquired in these fields and to determine what knowledge is most needed to move the fields ahead. Knowledge bases are a major component of artificial intelligence, a field with roots in cognitive science and computer science. Data and knowledge are distinct. *Data* include basic facts, etc. that people or sensing equipment can observe or otherwise note (e.g., presence of a plant species in a given sample). *Knowledge* requires higher levels of cognition and often involves analysis and synthesis of data to determine trends, rules, etc. (e.g., "generalized tracks" in biogeography are "knowledge chunks"). Several types of knowledge are valuable in biology. *Declarative, domain-dependent* knowledge summarizes insights relevant to a given domain (e.g., IF the shrub has one bud scale AND is dioecious THEN it is likely *Salix*). *Procedural, domain-dependent* knowledge describes part of a process that is specific to a field (e.g., IF species diversity is low but the area of a region is very large, THEN it still can be a major floristic classification unit). *Procedural domain-independent* knowledge involves knowledge useful in many fields. Immediate needs include development of standards and formats to "represent" systematic and evolutionary knowledge in computer knowledge bases and procedures to extract knowledge from current sources (e.g., data bases and the literature). Current expert system "shells" can represent only some types of knowledge effectively. More generally, researchers must expand their philosophical approach to include computer-compatible knowledge bases. Perhaps a first step will be that journal articles will include a statement of important knowledge chunks emerging from the research reported, just as such articles now contain an abstract and key words.

INTRODUCTION

Systematists, evolutionary biologists, and biogeographers for centuries have wished for additional data on which to base their conclusions. This continues to the present time and likely will continue forever since the "expanding frontier of knowledge" has an associated increase in the border between what we know and what we do not know. While more data always will be a desideratum, the huge magnitudes of data already accumulated by researchers in the above fields have made understanding and retrieval of such data difficult. An early use of computers by biologists was the creation of data bases to help store, retrieve and process such expanding data bases (e.g., Crovello, 1972).

But even with the rapidly expanding creation and use of data bases in these areas, we still are not making the most effective use of such data, and thus make decisions with less foundation than we should, and also waste scarce resources by again obtaining data that has already been accumulated but is not readily retrievable. A possible solution is to turn to computer based knowledge, to move from just the creation of data bases to the creation of knowledge bases.

Dr. Crovello is with California State University, 714 Administration Building, 5151 State University Drive, Los Angeles, CA 90032, USA.

Knowledge bases, a major component of Artificial Intelligence (AI), may be as important an advance to systematics and evolution as the use of computers themselves. Neither computers nor AI is a panacea, but the wise use of computers has helped many researchers achieve their goals. The same can be expected from AI. The essential modifier in the above sentence is "wise."

This paper has several purposes: to describe basic AI concepts; to suggest why AI is particularly timely today; to provide some examples of how AI can be used in our disciplines; and to discuss the different ways in which knowledge bases can be created using today's program packages.

Effective use of AI requires a clear understanding of what the goals, procedures and knowledge of our fields are. This last statement may seem unnecessary, since we know our own field and what we are doing. While we presumably know these things, it is essential that such knowledge be made *explicit*, i.e., able to be understood clearly by others and now also by a computer. In addition, the ultimate goal is to have such explicit knowledge at a level of detail sufficient to permit completion of aspects of analyses without further human consultation.

Every practicing biologist is an expert, someone very knowledgeable about their field. Such knowledge has been acquired through study and practical experience. An expert's "knowledge base" can be viewed as a model of how that particular researcher organizes and understands the field. Just as it is important to test other types of hypotheses and models, so too is it important to test an expert's knowledge base. The process of making the knowledge base explicit is also the way to evaluate, correct and expand such expertise. Once such a knowledge base has been evaluated and corrected to an acceptable level, other workers can use it with confidence. In contrast, use of a knowledge base with serious errors in it will confuse more than contribute.

To summarize, the goal of AI in systematics, evolution and biogeography is to make the knowledge of its experts explicit enough so that it can be stored and used by a computer in ways that might be described as intelligent.

CONCEPTS OF ARTIFICIAL INTELLIGENCE

As a separate discipline AI is only several decades old. Since its growth has been exponential, many of its concepts and results are quite recent and still undergoing revision. AI is an overlap of the fields of computer science and cognitive science (Gardner, 1986). AI workers have adopted as their major paradigm the information processing model. This model dominates cognitive psychology, one of the supporting disciplines of cognitive science. As long as other aspects and factors of intelligence are excluded from the AI model, it may be difficult for AI to duplicate higher aspects of intelligence (e.g., creativity), but useful knowledge bases are still possible to construct in fields like phytogeography today.

AI can be viewed as a set of problem-solving procedures. A more widely used definition is that it allows computers to exhibit behaviors that if shown by a human being would be considered intelligent. This begs the question of what intelligent behavior is. Some feel that there is a clear, unambiguous separation between intelligent and nonintelligent behavior. But, at least operationally, it seems best to view intelligence quantitatively and multidimensionally. Perhaps the best general definition of AI is one that I have used for the computer itself; AI is simply an extension of the user's mind. Such a definition is positive, open-ended and not threatening to our sense of uniqueness.

DATA VERSUS KNOWLEDGE

AI distinguishes between data and knowledge, so the uses of each term must be understood and not used interchangeably. "Data" are defined as basic facts which most people or sensing equipment can observe or otherwise note. Examples include the presence or absence of certain species of fish in a sample, the altitude at a given geographic location, and replacement lists of species in a study of community succession.

In contrast, knowledge requires higher levels of cognition. It often involves analysis or synthesis of data to determine trends, generalizations, rules, laws, etc. For example, the "generalized tracks" of vicariance biogeography as well as estimates of floristic or faunistic similarities between OGUs each are "knowledge chunks" of varying size. An OGU, or Operational Geographic Unit, is a unit of geography chosen for study (Crovello, 1981). It is analogous to Operational Taxonomic Unit. They and other knowledge chunks usually result from the data processing in combination with an expert's previous knowledge and methods of "knowledge processing." The latter includes evaluation, comparison and integration with what one's brain already knows.

Several types of knowledge are valuable in biology:

1. Declarative, domain-dependent knowledge summarizes rules or laws relevant to a given domain. For example, IF foodweb diversity increases THEN species survival should increase. (Throughout this paper pieces of knowledge presented as IF . . . THEN . . . rules will be capitalized as indicated, for emphasis.)

2. Procedural, domain-dependent knowledge describes part of a process that is specific to a given field. For example, IF species diversity is low but the area of a region is very large, THEN it still can be a major floristic classification unit.

3. Procedural, domain-independent knowledge involves knowledge useful during activities in many fields. For example, IF data sources differ in reliability THEN use the most reliable one available.

KNOWLEDGE BASED SYSTEMS

The area of AI with the most immediate value to phytogeography focuses on so-called expert or knowledge based systems (these terms will be used synonymously in this paper). Waterman (1986) provides a thorough and readable introduction to this topic, including many examples.

Buchanan and Duda (1983) define a knowledge based or expert system and also describe three of its important properties. "An expert system is a computer program that provides expert-level solutions to important problems and is:

1. Heuristic, i.e., it reasons with judgmental knowledge as well as with formal knowledge of established theories
2. Transparent, i.e., it provides explanations of its line of reasoning, and answers to queries about its knowledge
3. Flexible, i.e., it integrates new knowledge incrementally into its existing store of knowledge."

An expert system has two essential components: the domain knowledge base and the inference engine. The latter will be described later. Using the definitions of the previous section, a domain knowledge base contains knowledge of a specific domain, e.g., biology. In actuality no current knowledge base completely includes such a wide domain or even a major subdiscipline like phytogeography. Knowledge bases exist for very narrow segments of a domain. Examples possible today include procedures to choose taxa for use in a floristic study; choice of the best procedure for vegetation or floristic classification; and identification of individual organisms.

The domain knowledge base has several types of entities. It contains data, much like that found in any phytogeographic data base. For example, it may have data on the distribution of taxa in a series of OGUs, or about climatic or geological data for each OGU, etc. But in addition to data it also contains knowledge.

To understand the contents of a knowledge base and to underscore the diffence between data and knowledge, consider a species by OGU phytogeographic data matrix. The item in each cell is one piece of data, i.e., the presence or absence of a particular species in a particular OGU. But phytogeographic data matrices are not sets of random data. Rather, analysis of such matrices usually reveals patterns or structures among the data that are important to phytogeography. For example, certain phylogenetically related taxa may share similar geographic distribution patterns. Or, geographically distant OGUs may be found via multivariate analysis to be very close in terms of the taxa found in each (e.g., Crovello 1982). The resulting phytogeographic knowledge base would contain statements ("knowledge chunks") about which specific OGUs are closest to each other, etc. Such derived knowledge is based on data and data analysis, including interpretation of the results in the context of previous knowledge.

As another example, a phytogeographer may discover that taxa she knows to have a certain leaf shape also have a similar distribution pattern. This last situation is an example of an expert combining prior knowledge (what taxa possess a certain leaf shape) with new data to create new knowledge. This example also emphasizes that not everyone presented with the same sets of data and knowledge can use it effectively or can create new knowledge or insights. So we would not expect a biochemist, or anatomist, or even another phytogeographer viewing the same series of phytogeographic data bases to always derive the same kinds and amounts of phytogeographic knowledge from them.

Another major and more challenging goal of AI and more specifically of expert systems research is to develop procedures to allow computers to "process knowledge" just as a domain expert would. The concept of knowledge processing includes all stages, from the planning of knowledge processing to its actual accumulation, representation, verification, storage, retrieval, use, modification and replacement. To continue an earlier example, Takhtajan (1986) describes his floristic classification of the world. How was the classification produced? Most likely he repeatedly, consistently, and cleverly applied a number of pieces of domain-procedural knowledge to the huge amount of floristic data accumulated over his productive career. One such rule might have been the following: IF family endemism is high THEN the OGU should be considered as a possible floristic kingdom.

SOME USES OF KNOWLEDGE BASE SYSTEMS

In other fields knowledge bases already have been created to carry out one or more tasks. The same uses are relevant here, although some of the names may seem peculiar. The following list is based primarily on Waterman (1986):

–Design (configure objects under constraints, e.g., resource allocation studies)

–Implementation (put a design into action, e.g., carry out steps in a phytogeographic study)

–Interpretation (knowledge from sensory data, e.g., from remote sensing of laboratory instruments)

–Classification (creation of classes of objects, e.g., hierarchies in systematics or floristics)

–Identification, or Diagnosis (determination of membership in a class or group, determination of system malfunctions, etc., e.g., to which plant community type a site belongs, or why a biological preserve is not continuing to serve as a critical habitat)

–Debugging (choice of remedies for malfunctions, e.g., creating a new management plan)

–Repair (execution of plans to administer the remedy, e.g., carrying out the new management plan)

–Monitoring (comparison of observations to expectations, e.g., monitoring changes in environmental characters at a series of environmental sites)

–Control (government of overall system behavior, e.g., making decisions during a phytogeographic study)

–Prediction (infer likely consequences, e.g., the likely results of a proposed phytogeographic experiment)

–Machine learning (the computer acquisition of more knowledge, including knowledge of relationships, tasks, etc. within biogeography, e.g., building a computer based semantic network among phytogeographic concepts)

–Education (teaching people; e.g., learning via computer about new phytogeographic knowledge produced from others' research)

Not all expert systems act autonomously. Human interaction during the decision making process can be frequent, e.g., to allow additional data input, to decide which hypothesis to explore next, or to provide additional knowledge that the expert system does not yet possess.

KNOWLEDGE REPRESENTATION IN AN EXPERT SYSTEM

Decisions about how to represent knowledge is an important step in creation of a knowledge base, and thus of an expert system. As the term suggests, it concerns ways in which knowledge can be represented in a computer, and begs the question of how it might best be represented to achieve a given function. Analogies exist with data representation and organization in the design of maximally effective data bases.

Knowledge can be represented in many ways: prose statements; an equation such as a regression equation relating number of species and island area; a graph or picture; etc. People can process knowledge that is represented in many diverse ways, but computers are not yet so versatile. For the most part, current knowledge bases can not accommodate ambiguity, lack common sense, and have difficulty integrating knowledge from sources that represent it in different ways. Consequently, knowledge representation in AI most frequently takes the form of rules presented in an "IF . . . THEN" format, as in the following simple example:

IF grass level is high, AND

IF water levels at water holes are adequate, AND

IF lion number is low,

THEN Thompson's gazelle has a high survival value (CF = .85)

Each rule represents one knowledge chunk. Rules consist of two parts: one or more premises (or conditions), and one or more conclusions (or actions). When an expert system is being used in a given situation, IF all of the premises of the rule being evaluated are true THEN the conclusion is true. CF= .85 describes the value of a Certainty Factor, an indication of the expert's estimate of the reliability of the conclusion based on those premises.

Even the above elementary example can reveal the power of knowledge chunks and their representation as rules. Students and researchers may form such a rule based on many observations of that specific ecosystem, but of even more importance is the development of a more general rule that integrates analogous sets of specific rules. If similar rules

986

were derived from multiple studies in a diversity of habitats, then the following more general rule would be valid and have even more power for understanding the field or making resource management decisions:

IF food resources are high, AND

IF water resources are adequate, AND

IF predator number is low,

THEN the prey species have a high survival value (CF = .85)

Several rules also can be linked together to form a rule chain, defined as a set of rules wherein the conclusion of one rule is a premise of the next rule. When the conclusions of several rules form the premises of another rule, a rule network is created.

This concept of an inference network emphasizes how the inference engine in an expert system can use a knowledge base that represents its knowledge as rules. For example, in the rule just given above, a user may ask the expert system if species i is in OGU q. This is referred to as the "goal rule." The inference engine would "chain backward," through a "chain" or network of rules to determine if all the premises of the goal rule are true. When all the premises have been determined as being true then the goal is considered true (or false if one or more of the premises are false). The rule format is natural and common in our normal thinking, as emphasized by the many IF . . . THEN . . . statements in this paper.

To paraphrase the process of backward chaining, the inference engine first would search for rules that have the user's question as a conclusion. Why? Because IF its premises can be shown to be true THEN its conclusion also is true. IF it finds a rule with the user's question as a conclusion, THEN the next task is to determine the correctness of each premise of that rule. This requires a search of the rule base to find a rule that has one of the goal rule's premises as a conclusion. That rule's premises then must be verified, with the program chaining backward up through as many levels of the rule network as needed.

IF a premise of some rule in a chain relevant to the user's current question can not be verified from the knowledge base THEN the program may pause and ask the user to supply the information. IF the user can not supply the information THEN the program will continue to reason, but may indicate that the resulting final answer has less certainty than if such information were known.

IF in its search of the knowledge base the inference engine discovers more than one rule with the same conclusion THEN it must determine which rule to evaluate first. Procedures for such conflict resolution also often take the form of sets of rules themselves. So rules can be found both in the domain knowledge base and in the inference engine.

This introduction to basic concepts of expert systems of necessity omits many aspects, such as rule generalization, rule hierarchies, introspection, and other types of knowledge representation and reasoning. Interested readers should consult some of the many recent books on expert systems.

WHAT SYSTEMATISTS, ECOLOGISTS, AND BIOGEOGRAPHERS EXPLICITLY DO

Before AI can contribute to a field, the goals, procedures and knowledge of that field must be stated explicitly; that is, unambiguously and in sufficient detail such that someone (or some intelligent computer) could carry out the required tasks successfully without further human assistance.

The general goals of the above fields include the following: to describe the patterns of variation in nature; explain how these patterns arose and maintain themselves; predict future patterns; and suggest management strategies for biota and geographic areas. These goals can apply both to natural patterns and to ecosystems that have been strongly influenced by society (e.g., via farming or forestry).

To accomplish these general goals we engage in many diverse activities, each of which contains subgoals. Such activities include all or most of the following: theory formation, classification, identification, historical reconstruction, and character analysis. Crovello (1970, 1981) suggested that every study involves a multistage decision process. That is, it involves many stages and substages, and each step requires choices or decisions among options. These are based both on knowledge about the subject area and on general procedural knowledge. This is an ideal context for expert systems.

Young researchers often have a relatively small set of options to consider. Interestingly, experienced workers may have thought about many options over their professional lives but have forgotten some of them. The computer need never forget (this also has disadvantages but they can be overcome).

ARTIFICIAL INTELLIGENCE AND IDENTIFICATION

To focus on just one use of AI, consider the goal of creating an expert system to identify to which floristic region a specific geographic site might belong through use of the detailed floristic classification of the world in Takhtajan (1986). Alternatively, one may prefer to think about creation of an expert system to identify to which biome or formation an OGU belongs. Although identification is one of the easier tasks we encounter, it still is challenging and can illustrate basic concepts of expert systems.

Several papers already have appeared in the biological literature that focus on taxonomic identification using expert systems. Atkinson & Gammerman (1987) describe several uses of EXPERT KEY, an expert system program that accomodates reasoning with uncertainty. The authors demonstrate the increased efficiency of the expert system over conventional identification programs. Mascherpa and Pellegrini (1987) introduce concepts of AI relevant to biological identification and provide examples of possible uses, such as with the flora of Switzerland. Woolley & Stone (1987) discuss a prototype knowledge base consisting of 114 rules. It was created for the identification of individuals belonging to one species group of the insect genus, *Signophora*.

A standard, published, dichotomous key contains knowledge derived from analysis of large amounts of data and thus can be considered a knowledge base. The procedure to use that knowledge for the goal of identification is straightforward: begin at the start of the key and work through it until the goal is reached (e.g., determination of the floristic province to which an OGU belongs). A knowledge base represented in the form of a published key can be directly transformed into a rule based representation since each part of a couplet can be stated as an AI rule. For example,

"Trees present....................2,"

can be restated as, "IF trees are present THEN go to rule 2."

Computer based identification in taxonomy introduced several alternatives to the dichotomous key (e.g., Pankhurst 1975, 1978). Representation of identification knowledge inside the computer can vary. For example, the program might simply have the equivalent of a printed key available as data. The sequence of the questions asked by the computer would be fixed, just as occurs in published keys.

Alternatively, the computer could "dynamically" choose the next character about which to request information. This usually requires the presence of a full data matrix within the computer. It also requires a procedure to determine which question the computer is to ask next. For example, choose that character among those not yet asked whose states will divide the remaining taxa into two groups, each with more or less the same number of taxa.

If a complete, reliable data matrix is available within the computer, then AI is not needed. But if data for certain characters are missing for some formation, or if the variation in its character states is not fully understood, then the AI rule-based approach is useful to determine the best identification. In other words, if complete, absolute knowledge is not available, then the judgmental knowledge that an expert system can accomodate may contribute to the correct identification.

Given representation of identification knowledge in a rule format, it is clear how a set of rules derivable from a key can become a rule chain or network. In fact an alternative, succinct definition of a phytogeographic or other key is the following: knowledge for the purpose of identification that is represented as rules and forms an inference network.

It is important not to dismiss AI in phytogeography simply as a different way to represent knowledge of dichotomous keys. This would be like equating all use of AI in medical diagnosis only with the ability to distinguish among nine easily identifiable childhood diseases. Such uses are "low-end" AI. As the biogeographer's identification goal becomes more challenging due to incomplete information, the need for expertise, for judgemental knowledge, increases. Unlike computer based keys that rely only on data and some decision rules to choose what data to request next, AI can accommodate a biogeographer's heuristics, i.e., the rules of thumb that the phytogeographer has found useful in previous identifications, even if totally similar situations have not been encountered before. In such cases an expert's domain of independent, procedural knowledge also can be incorporated into the knowledge base.

Simple examples useful in many domains when the goal is identification include the following heuristics, represented here as rules:

IF several samples from the same object (e.g., an OGU) are available THEN evaluate all samples.

IF some characters are more reliable than others THEN first use those with higher reliability.

Such "metalevel" rules are obvious once they are read, and seem to be simply common sense. They are obvious because they are so ingrained in our problem solving procedures that we take them for granted. Yet such presumed knowledge must be explicitly stated before it can be entered into a computer knowledge base.

In summary, like other computer approaches to identification, an expert system has several advantages over a book of expertise. An expert system is interactive, allowing the user to more quickly obtain an answer, and allowing the computer to dynamically optimize each interaction during a session. Such programs can be easily duplicated to allow distribution to other research and service laboratories. In addition, unlike most other computer identification programs, the rule format of knowledge representation in an expert system is easily understood by the user. In addition, it can be readily understood by other computer programs and permits direct access to and use of data bases. The knowledge base can be electronically updated quickly, without editing the program itself (because the knowledge is not integrated into the program code). An expert system is testable for inconsistencies in its procedural knowledge, and allows easy revision of it. An expert system can explain its reasoning procedures, and only to the level of detail desired by each user.

HOW AN EXPERT SYSTEM IS CREATED AND MAINTAINED

An expert system can be created in various ways, depending on the level of sophistication of the program product used. At one extreme no expert system program product would be involved. Based on the purposes of the expert system, the designer would devote considerable time to the organization and contents of the inference engine as well as of the domain knowledge base. Decisions would include what data and knowledge to include in the knowledge base and how to represent it. Design of the inference engine would involve the ability to recognize rule formats, to determine when all the premises of a rule are correct, to determine what rules are relevant at different moments in a given session and how to resolve rule conflicts. In addition, the program should be able to explain its reasoning to the user and allow easy modification of the rule base. Finally, additional programming is required to permit efficient, high level communication between the computer and the user. No biologist would do all of the above.

While an expert system can be created using any well known computer language (e.g., BASIC or Pascal) or special AI languages such as LISP or PROLOG, few biologists will choose to build an expert system from scratch using such languages. Most will opt to use an expert system "shell." Shells are "utility" programs allowing users to enter only knowledge and data relevant to their particular use. That is, they do not have to create an inference engine, a user interface, etc. Familiar examples of utility programs are data base, word processing, and statistical programs, since the user only enters words or numbers in an appropriate format and the program carries out the required analysis. Expert system shells available for microcomputers range in price from $100 to over $10,000.

The trend in expert system shells is to allow compatibility with major commercially available data base programs as well as with such special programs as HYPERCARD. Used on Apple Computer Company's Macintosh computer, this linkage with HYPERCARD allows easy access to graphics and pictorial screens, including maps, photographic images of OGUs, etc.

TRANSFORMING DATA BASES INTO KNOWLEDGE BASES

Many useful computerized data bases already exist. One is, "GEOECOLOGY," a county level environmental data base for the conterminous United States (Olson et al., 1980). It contains information not only on distribution of many taxa, but also on climatic, topographic, geological and other characteristics. A major thrust in many disciplines in the next decade will be to convert data bases to knowledge bases. This will occur first via direct interaction of domain experts as well as AI experts (called knowledge engineers). But work is under way to create intelligent programs that will automatically create knowledge bases from data bases. Naturally, any such programs that are successful will have been provided with a considerable amount of a human expert's knowledge about how to accomplish such a challenging task.

To understand how data bases can be transformed into knowledge bases, consider again the difference between data and knowledge. Most data bases about the real world are not random. That is, their contents reveal repeated patterns of variation. This should be expected since patterns of variation are the rule in nature. For centuries biologists have been identifying such patterns. The question is: how do they do it? If the data base consists of concrete observations then some procedures are obvious. For example, one procedure might be as follows:

1. Choose certain pairs of characters that are of biological interest (e.g., number of species and latitude; number of species and island area).
2. Subject such series of paired observations to regression analyses.
3. IF some regressions are significant THEN formulate rules summarizing the relationship. For example, IF island area increases THEN number of species increases as twice the square of island area.
4. Have such rules reviewed by a biogeographic expert.
5. Compare rules accepted in step 4 with those already in the knowledge base to determine rule conflicts.

6. IF rule conflicts occur THEN resolve them EITHER with a conflict resolution program AND/OR with human expert intervention.

The above procedure obviously is only one of many possible, and is incomplete. For example, it does not allow consideration of the relative reliability of each data base or of its parts. But as an example it illustrates that today's data bases can become even more valuable as major sources of knowledge bases. Interestingly, this was the same rationale used to convert data stored in museums, data about ecological relevés, etc. into computerized data bases.

CONCLUSIONS

The wise use of AI in systematics, evolution and biogeography promises to be as valuable as the use of computers themselves. Expert systems is the area of AI that will be most useful. Effective development and use of expert systems will be possible only after researchers reflect on how we do what we do, and are able to make both our procedural and declarative knowledge sufficiently explicit to be processed by computer. AI thus offers an opportunity for knowledge and expertise both to increase and to live longer.

Neither computers nor expert systems are panaceas. But as a logical extension of the use of computers in biology, AI can be a valuable addition. Just as important, the design, development, and use of expert systems can produce deeper insights into both our concepts and procedures. For some this may be the most valuable result of AI.

There is nothing mysterious about AI. It will be useful because relevant knowledge (not just data) of one or more experts will have been entered into the computer. The advantage is that every subsequent use of such expertise will not require the presence of the expert.

More specifically, since AI can electronically clone more expertise than previously possible, the resulting knowledge bases can have the following results:

–An expanded lifetime for such knowledge. The expertise need not be lost when people retire.

–Increased simultaneous use. The expertise easily can be duplicated electronically and distributed to those who need it. This is similar to the effects of the printing press, but exceeds it since it has the advantages of being in an interactive mode, of dealing with knowledge instead of just data, and of being in a form that can quickly focus on that small amount of knowledge that is relevant to the task at hand.

–Increased performance levels by both experts and paraprofessionals.

–Expanded and better use of knowledge by more people, be they academic or applied biologists or others.

–Emphasis on the centrality and ubiquity of problem-solving. The AI paradigm clearly shows that even deciding to what floristic category a site belongs is problem-solving. It involves repeated use of the "generate a hypothesis and test it" cycle until the correct answer is obtained.

–AI's requirement that knowledge be made explicit will increase understanding of what we do, and possibly suggest ways of how it might be done better.

Viewing expert systems as intelligent assistants, associates and colleagues provides a positive, open-ended perspective on AI. AI's success in fields similar to ours suggests that the time researchers devote to experimenting with knowledge based systems will contribute significantly to advancing research and understanding.

LITERATURE CITED

Atkinson, W. D. & A. Gammerman. 1987. An application of expert systems technology to biological identification. *Taxon* 36:705–714.

Buchanan, B. G. & R. O. Duda. 1983. Principles of rule-based expert systems. *Advances in Computers* 22:163–216.

Crovello, T. J. 1970. Analysis of character variation in ecology and systematics. *Ann. Rev. Ecol. & Syst.* 1:55–98.

Crovello. T. J. 1972. Computerization of the Edward Lee Greene herbarium at Notre Dame. *Brittonia* 24:131–141.

Crovello, T. J. 1981. Quantitative biogeography: an overview. *Taxon* 30:563–575.

Crovello, T. J. 1982. Floristic similarities among 51 regions of the Soviet Union based on the Brassicaceae. *Taxon* 31:451–461.

Gardner, H. 1986. *The Mind's New Science: A History of the Cognitive Revolution.* Basic Books Inc.: New York. 423 pp.

Mascherpa, J-M. & C. Pellegrini. 1987. Approche par les systèmes experts de la determination des familles de la flora de suisse. Cahiers de la Faculté des Sciences (Univ. de Geneve) 14:7–21.

Olson, R. J., Emerson, C. J. & M. K. Nungesser. 1980. *Geoecology: a county-level environmental data base for the conterminous United States.* Oak Ridge National Laboratory. Environmental Sciences Division Publication 1537. Oak Ridge, TN.

Pankhurst, R. (ed.). 1975. *Biological Identification with Computers.* Academic Press: New York. 333 pp.

Pankhurst, R. 1978. *Biological Identification.* Edward Arnold: London. 104 pp.

Takhtajan, A. 1986. *Floristic Regions of the World.* University of California Press: Berkeley. 522 pp.

Waterman, D. A. 1986. *A Guide to Expert Systems.* Addison-Wesley Publishing Co: Reading, MA. 419 pp.

Woolley, J. B. & N. D. Stone. 1987. Application of artificial intelligence to systematics: SYSTEX—a prototype expert system for species identification. *Syst. Zool.* 36:248–267.

Microbiological Databases
in the 21st Century

Micah I. Krichevsky

NATURE OF MICROBIOLOGICAL DATA

Microbiologists concern themselves with a wide variety of phenomena, with the common thread being the size of the organisms under study. Consequently, the form of data used to describe and analyze the phenomena are equally diverse. Much of the data is simple incidence data. Does something occur or not? Does the organism possess a particular attribute or not? Is the organism found in a particular place or not?

Numerical data describe growth and death, levels of enzyme activity, sizes, chemical composition, biological potential and other continuously varying phenomena. Statistical descriptions and models are constructed from both types of data. In many cases, these descriptions are turned into word descriptions as in taxonomic treatises.

INTERDISCIPLINARY TOWER OF BABEL

A partial list of the disciplines that are confluent with various aspects of the broad discipline of microbiology includes engineering, ecology, epidemiology, biochemistry, genetics, molecular biology, geology, oceanography, geography, medicine, pharmacy, pharmacology, pathology, anatomy, zoology, botany, nutrition, food technology, toxicology, immunology and both traditional and modern biotechnology. Each of these disciplines has its own specialized vocabulary.

Wide variations within and among the various disciplines exist in the degree of vocabulary standardization and ready accessibility of those standards. Databases which must link these disciplines must also conform to the vocabulary standards developed in each field. The future of publicly-accessible databases will be predicated on the development of a system for communicating and updating the standards to the database producers.

We now see the beginnings of standard formats for transmitting data among database hosts. This, too, will continue and facilitate database construction at all levels of complexity. There is no reason to assume that complete standardization of transfer protocols will ever be, or even should be, accomplished. As with the current proliferation of utility programs for interconversion of data in proprietary formats for spread sheets, word

Dr. Krichevsky is with the Microbial Systematics Section, Epidemiology and Oral Disease Prevention Program, National Institute of Dental Research, National Institutes of Health, Bethesda, MD, USA.

processors, and the like, there will likely be development of powerful "truth table" conversion programs that will ask for the logical correspondence of elements of each format and generate a procedure to accomplish the translation.

TRENDS IN STANDARDIZATION OF VOCABULARY IN MICROBIOLOGY

As a paradigm for the standardization of vocabulary in the various fields impinging on microbiology, we may consider the state of standardization in microbiology as such. The trends we shall consider fall into two broad classes, the conventions for naming organisms and the conventions for describing attributes of organisms. Nomenclature standardization of microorganisms is highly structured and officially sanctioned by internationally recognized scientific organizations. The vast majority of microbiologists adhere to the standards as set by these bodies.

The converse is true in the case of descriptors of attributes of microorganisms. The scientific establishment in microbiology, under the aegis of the International Council of Scientific Unions (ICSU) does not have generally recognized standards for terminology of descriptors of microorganisms.

NAMES OF ORGANISMS

Perhaps the most pervasive element in microbiological databases is, and will continue to be, the "name" given to the taxonomic place of the organism. Usually this will revolve around the species name. Within each of the major groups of microbiological entities, good standardization of rules of nomenclature exist. However, the rules for each group are maintained by, and are effectively isolated from, the several nomenclatural bodies set up within two of the Unions adhering to ICSU: International Union of Biological Societies (fungi, algae, and protozoa) and International Union of Microbiological Societies (fungi, bacteria, and viruses). The relationships between groups and the codes are shown in Table 1.

The various specialists in each of these microbiological disciplines are reasonably cognizant of the rules of nomenclature applying to their subjects of study. Clearly, those concerned with algae must make up their minds as to which code they follow. Even within a discipline, the rules allow scientific conflict. Unfortunately for the database producer, a clear set of authoritative names is not available from the nomenclaturists in most cases. With the exception of the International Committee on Taxonomy of Viruses, the committees responsible for establishing the naming rules do not judge scientific validity of the names, only the correct etymology in some cases.

This situation is changing and will affect future databases. Mycologists are seriously discussing lists of the "names in common use" which seems a euphemism for an approved list of names. In the bacteria, the International Committee on Systematic Bacteriology published, first in 1980, the Approved Lists of Bacterial Names (Skerman et al., 1989). All names not on that list were declared invalid. Those wishing to conserve an older name or

Table 1. Coverage of Codes of Nomenclature and the Responsible Unions.

Organisms	Codes of Nomenclature (Informal name)	ICSU Union
Viruses	Viral code	IUMS
Bacteria	Bacteriological code	IUMS
Fungi, lichens, algae	Botanical code	IUBS
Protozoa, algae	Zoological code	IUBS

propose a new name must publish the recommendation in the International Journal of Systematic Bacteriology. Where there is reason for adjudication on the name, the International Judicial Commission will make the required judgement (but only on the correct application of the Code, not on scientific merit).

Thus, for the bacteria at least, the database producer can go to a single source, the International Journal of Systematic Bacteriology, for valid names. Further, all names published prior to 1980 can be ignored if not on the list. There is one obvious exception to this. The names used in the older literature still cannot be ignored in abstract and similar databases.

A continuing problem faced by database producers is the lack of availability of such "official" lists in publicly-accessible, machine readable form (Cameron, 1990). The publishers of these lists (and many other lists of standards) often maintain these lists in computer files. However, the prevailing view of these publishers is that they wish to sell hard copy (books or journals) rather than have people make and use electronic copies. In some cases, these lists have been retyped into databases and made available to the public as value-added databases. The Taxonomic Reference File of BIOSIS is an example of the bacterial Approved Lists being available. I have been told that there is another version of the Approved Lists available online in Germany. Presently, each database producer is faced with the task of either retyping each list needed into the computer or recording the list online. The legal copyright issues involved in all of this could be the subject of an entire essay. Questions of accessibility of ancillary standards for use in building databases obscures prediction of some aspects of future trends in database building.

ATTRIBUTES OF ORGANISMS

Accessibility to terminology standards for descriptors of microorganisms is essentially nonexistent. The reason is simple: nothing to access. There are almost no ICSU sanctioned standard setting organizations that have the same "official" status as the nomenclature committees. There have been some attempts at setting standards, such as that of the ICTV for viruses (Atherton, et al., 1983). For the most part, the compendia of descriptors have been private efforts, often contained as glossaries within larger works. There are major efforts to compile descriptions of taxa in the viruses, bacteria, yeasts, fungi, protozoa, etc.

The terminology contained within each of these compendia largely is internally consistent. The terms often are obscure to those outside of the discipline. Further, different compendia in the same discipline often will not be consistent with each other. The choices for the database producer are to choose one "authority" or to construct systems of apparent synonymy.

Database builders of the past and present microbiological databases model the data structures on the example of the printed book. Descriptions of organisms are largely at the species rather than strain level. Usually only positive attributes are listed. No distinction is made between unknown and negative.

A considerable portion of data recorded by microbiologists is in the form of presence/absence (yes/no). Any coding system must allow for missing observations to capture all the information. Thus, the data type is not truly binary but ternary in nature. Two states are the complement of a positive response in microbiology. That is, the complement of positive is the combination of negative and missing responses, i.e., NOT POSITIVE = NEGATIVE AND MISSING. All three responses (positive, negative, and missing) should be allowed and be searchable in a database at the strain level. Databases are beginning to be constructed in this way and will be quite common in the future.

Species descriptions should be in the form of numerical frequencies of occurrence of the positive states. This is seldom done. Word qualifiers are used to indicate approximate frequency. Attributes are stated as "usually positive," "most often absent," "occasionally present," etc.

WORDS

The database producer first must compile a list of terms that will be used in the database. Natural language (free form) databases exist and are even common in open literature abstract databases. Even in these, key word lists are used to facilitate searching.

As with any language, words acquire new meanings by being adapted to fit new situations. Thus, the meaning of words may vary across disciplines. A "spore" may be a reproductive structure, a survival structure, or both. Consider the meanings of the word "field" in microscopy, information retrieval, agriculture, and as a profession. Future databases will have glossaries included so that the user can obtain a definition of a word in the midst of interacting with the database. These glossaries will be "multimedia" in implementation. Initially, and already coming on stream, will be combinations of words and pictures. Later, sound will be added. The conceptual framework and even the programs to allow implementation exist so the prediction is not speculative.

Such high technology will not help too much if the present situation of lack of standardization of terms continues. Another trend is the increasing awareness of biologists that there is an obligation to communicate the results of their labors to others not facile in the jargon of their field. To do so, an understandable, standardized vocabulary must be available. Of course, such standardization will aid communication within a field as well. We often do not recognize that the Babel phenomenon applies even within a field.

PHRASES AND SENTENCES

Single words rarely suffice to define a data element. Often, single words are meaningless in isolation. Consider the word "mannose" and limit the consideration to simple concepts about mannose metabolism.

Examples, abstracted from Rogosa, et al., (1986), are shown in Table 2. A statement, in the form of one or more numerically indexed sentences, represents the descriptive concept. For understandable communication, we need such full statements. Very few statement-oriented vocabularies exist. In most cases, we leave it to the reader to surmise what the word denotes as a feature.

In most other forms of communication, we convey meaning through linked combinations of words. The proximity, order, and context of the words in combination all control the effectiveness of communication of meaning.

The following three sentences illustrate the importance of these factors in conveying meaning.

We treated the animals with **disease.**
We treated the animals with **penicillin.**
We treated the **diseased** animals with **penicillin.**

There is no difference in sentence structure between the first two sentences. The quite different meanings must be gleaned from interpretation of the context. The last sentence combines the two concepts. The importance of order and proximity, and the difference between them is illustrated by moving the position of one word:

We treated the animals **diseased** with **penicillin.**

The consideration of context reduces the sentence to nonsense.

Increasingly, databases are built and accessed across disciplinary lines. The construction of controlled, statement-oriented vocabularies will command increasing attention to facilitate understanding. We see such trends in a number of areas of microbiology even now.

996

Table 2. RKC Codes for Mannose Features.

024506: D-Mannose is catabolized.
024507: D-Mannose is catabolized aerobically.
024508: D-Mannose is catabolized anaerobically.
025022: D-Mannose is utilized.
025080: D-Mannose is oxidized.
025138: D-Mannose is reduced.
025196: Acid is produced from D-mannose.
025254: Gas is produced from D-mannose.
025312: D-Mannose can be used as the sole source of carbon.

DATABASES

Publicly-accessible databases fall into three main categories: single builder, single host (e.g., Taxonomic Reference File of BIOSIS); multiple builders, combined into a single database (e.g., the CODATA/IUIS Hybridoma Data Bank, International Working Group on Mycobacteria Taxonomy cooperative study on slowly growing mycobacteria); and multiple builder, multiple host distributed networks (e.g., the Microbial Strain Data Network (MSDN) central directory and online culture collection catalogues linked to the World Data Center on Microorganisms). Whatever the structure of the database, the trend is to increase the number and complexity of databases. In addition to the aforementioned construction of more precise and accessible vocabularies, other changes are predictable.

As new database-building techniques become available to the microbiologist and the cost of computational facilities decreases, larger personal databases will be built. Combining of such databases will be another driving force to induce use of data transfer standards. We see the growth of institutional database structures to supplement the personal databases. The installation of Local Area Networks (LAN) facilitates this trend. (Although LANs are mixed blessings, they do inherently force communication.)

Finally, the telecommunications facilities have achieved a geographic coverage and a cost that makes electronic digital communication accessible to most of the world's microbiologists. Countries that are not presently connected to the communications networks either will be soon or individuals and institutions will be able to access satellite communications at reasonable cost.

The MSDN provides a paradigm of the future developments in microbiological databases (Hill & Krichevsky, 1985). The MSDN has users in about forty countries at present. This distributed network of users is not unusual. What is unusual is the distributed nature of the databases accessible on the system. While some of the databases are installed on the system computers (in both the USA and UK) directly, databases hosted and maintained in about six other countries are connected by electronic gateways. Thus, the basic structure provides a mechanism for easily (and cheaply) making databases available to the public.

Lest the reader get the impression that informational Nirvana has arrived, we must note that the ease of connection of databases through gateways does not include uniform data structures or use conventions. The ease of use of these disparate databases varies widely. However, at a cost of $600–1100 per connection, new databases can be made available to microbiologists. The only requirement is that the host computer already be accessible to the packet switching services that abound in the world today.

There are systems that attempt to impose standards on the database structures and conventions for use. There are other attempts at placing a uniform use or translation shell on top of disparate databases. These techniques work well only when the databases have underlying similarities and functions. Combining a numerical database with a literature database through either strategy requires that the single standard must be rich enough to cope with both data structures to avoid loss of power.

The "one-stop shopping" that is currently available on the MSDN ranges from quite specifically microbiological databases (culture collection catalogues, directories of collections, microbiologists with their special interests, properties and availability of monoclonal antibodies, etc.) to general purpose databases for science and business (through a gateway to DataStar, a general purpose public database utility). The trend in the future will be to take advantage of this type of easy connectivity to develop a wide variety of distributed databases in various branches of microbiology.

LITERATURE CITED

Atherton, J. G., Holmes, I. R. & E. H. Jobbins. 1983. ICTV code for the description of virus characters. *Monographs in Virology* 14:1–154.

Cameron, G. 1990. Terminology in biological databases—delivery systems. *Binary—Computing in Microbiology* 2:111–117.

Hill, L. R. & Krichevsky M. I. 1985. *Needs and Specifications for an International Microbial Strain Data Network. Proceedings of a Workshop held in Brussels, Belgium 15–17 November 1983 and Executive Summary of the Working Group Meeting, Bangkok, Thailand, 23–25 November 1984.* UNEP: Nairobi. 58 pp.

Rogosa, M., Krichevsky, M. I. & R. R. Colwell. 1986. *Coding Microbiological Data for Computers.* Springer-Verlag: New York. 299 pp.

Skerman, V. B. D., McGowen, V. & P. H. A. Sneath. 1989. *Approved Lists of Bacterial Names (Amended Edition)*, and Moore, W. E. C. & L. V. H. Moore. 1989. *Index of the Bacterial and Yeast Nomenclatural Changes.* (2 vol. set). American Society for Microbiology: Washington, D.C. 188 pp. & 72 pp.

Standardization of Terminology
for Access to Biological Data

Lois Blaine

Abstract. The advent of new and highly specific factual biological databases demands a unified and standardized set of descriptors for biological entities. Cross-disciplinary retrieval of data becomes increasingly difficult as the "vocabulary of biology" proliferates. There are numerous organizations involved in classifying objects and developing nomenclature for the classes. All major branches of biology (zoology, botany, bacteriology, virology) have nomenclatural committees dealing with the naming of taxa at the species level and above. A far more complex aspect of biological nomenclature is the development of a well defined and standardized terminology for describing the morphological, physical, and biochemical components and characteristics of these organisms. The work of these groups is often not known or recognized outside of their own disciplines. This situation results in the duplication of effort and the creation of thesauri which have little meaning to the scientific community. Steps are now being taken by governmental and private organizations which show promise of addressing problems of biological nomenclature on a global and interdisciplinary scale.

INTRODUCTION

The Committee on Data for Science and Technology (CODATA) of the International Council of Scientific Unions (ICSU) is an interdisciplinary scientific organization that facilitates access to scientific and technical data and seeks to improve its quality and reliability. CODATA holds international biannual meetings to review projects and to provide an opportunity for interdisciplinary communication. The meetings also allow the CODATA General Assembly to evaluate proposals for new initiatives to fulfill the organization's mission.

During the 11th CODATA Conference, held in September 1988 in Karlsruhe, FRG, a "Commission on the Terminology and Nomenclature of Biology" was created as the result of a Workshop on the Vocabulary and Nomenclature of Biology. It was the consensus of participants that, while there are several outstanding examples of widely publicized and accepted sets of standard terms for describing biological entities, these well known "authorities" are the exception rather than the rule. It is clear that much work is being done by special groups and committees within the Bio-Unions. However, knowledge of the results is often confined to the originating sub-discipline. There are also areas of biology where standards of nomenclature do not exist.

The major purpose of the CODATA Commission is to focus the attention of the biological sciences and bioinformatics communities on the work of the Bio-Unions' standards setting bodies. It is not the intent of the Commission to set standards for biological

Ms. Blaine is with the American Type Culture Collection, Rockville, MD, USA.

nomenclature, rather but to facilitate communication among the subcommittees of the Bio-Unions and to ensure that the scientific community is aware and able to take advantage of their efforts. To emphasize this point, the name of the Commission has been changed to "Commission on Standardized Access to Biological Data."

The stated goal of the Commission is "to improve international access to and use of information resources in biology and related disciplines by promoting international and interdisciplinary cooperation in the development, enhancement, and use (in various languages) of terminologies, controlled vocabularies, nomenclatures, and classifications."

The need for standardized terminology within the biological sciences has been recognized for many years. This need is growing more critical because of the recent proliferation of automated systems for storage and retrieval of biological information.

Database developers are facing problems in choosing standard descriptors for biological entities. The problem is more pronounced due to the recent increase in interdisciplinary research. Cross disciplinary retrieval of data becomes increasingly difficult as the "vocabulary of biology" proliferates. Traditionally, nomenclature committees have existed within various biological fields. Their function has been to provide advice and set standards for terminology and nomenclature within subspecialties of the field. Several organizations are now addressing the problem from an interdisciplinary point of view.

The International Council of Scientific Unions' (ICSU) Biological Unions and the Committee on Data for Science and Technology (CODATA) are major contributors to the establishment and dissemination of standards for biological terminology and nomenclature. Others, such as the Matrix of Biological Knowledge Group (Morowitz & Smith 1987), are working on the development of a "biological matrix" which will aid in drawing generalizations from the vast biological knowledge base. This will require a common data dictionary of biological terms that are standardized and defined.

The U.S. National Library of Medicine (NLM) has an online directory of biotechnology information resources (DBIR) which includes biological nomenclature committees and their publications. NLM's National Center for Biotechnology Information (NCBI), in cooperation with NLM's Bibliographic Services Division, is involved in the development of an online hierarchical dictionary of standardized biological terms. This list represents one activity in NCBI's mission to coordinate efforts to collect and disseminate biotechnology information on a worldwide basis. The International Information Centre for Terminology (INFOTERM) works in liaison with the International Organization for Standardization (ISO) to coordinate terminology standardization. These and other activities, described in more detail below, should increase accessibility of standardized terminology across disciplines.

THE DEVELOPMENT OF STANDARDIZED TERMINOLOGY IN BIOLOGY: EVOLUTION OF REQUIREMENTS

Historically, the various components of ICSU, specifically the Biological Unions such as the International Union of Biological Sciences (IUBS), International Union of Biochemistry (IUB), International Union of Immunological Societies (IUIS), International Union of Microbiological Societies (IUMS) and others have appointed committees to recommend standardized nomenclature and terminology in defined subdisciplines. The IUB, for example, has a committee which sets standards for enzyme nomenclature. Their publication, *Enzyme Nomenclature* (Webb, 1984), is widely recognized and used by authors publishing in the biological literature, by database developers building thesauri of biological terms, and by searchers of biological databases. Unfortunately, the situation regarding enzyme nomenclature is unique. Although other groups have managed to

1000

develop internationally recognized nomenclature systems within specific areas of biology, the existence of standards is the exception rather than the rule.

There are numerous organizations involved in classifying objects and developing nomenclature for the classes. All major branches of biology (zoology, botany, bacteriology, virology) have nomenclatural committees that deal with the naming of taxa at the species level and above. Two major problems exist in utilization of the products of these committees: a) disagreement among various factions that make up the committees, resulting in confusion for potential users of the nomenclature and b) lack of reliable pathways to the products of these committees.

Developing standards for the naming of organisms, the traditional role of taxonomists, is only the beginning of the problem. A far more complex aspect of biological nomenclature is the development of well defined and standardized terminology for describing the morphological, physical, and biochemical components and characteristics of these organisms.

In addition to sanctioned committees, there are literally hundreds of ad hoc groups working on standardized terminology for extremely narrow and well defined areas—ranging from "nomenclature of leukocyte surface antigens" (Nomenclature Committee of the 4th International Workshop on Human Leukocyte Differentiation Antigens. 1989) to "standardized terminology for the description and analysis of equine locomotion." The work of these groups is often not known or recognized outside of their own disciplines. This situation results in duplication of effort and the creation of thesauri by database developers which have little meaning to the scientific community.

Database developers such as BIOSIS, Excerpta Medica, Chemical Abstracts, UNESCO, and NLM have devoted major resources to develop and update thesauri and lists of controlled vocabulary terms for building their databases. While these tools are invaluable for use with their corresponding databases, there is little uniformity in the content and form of these thesauri. Furthermore, most of these lists have been developed for use with bibliographic databases, i.e., they are composed of "index terms" to aid information scientists in the storage and retrieval of data on particular biological entities. None were designed to specifically describe each and every component of biological entities. Most bibliographic databases are supplemented by complete abstracts of the material contained in the papers recorded in the databases. Specific descriptors, therefore, appear as "free text" in the abstracts and are not subject to any attempt at standardization in contrast to the broader "index terms."

The advent of factual databases for biologists, such as the sequence databanks (GenBank, EMBL, PIR), the Hybridoma Databank, molecular structure databanks, and databases describing the characteristics of microbial strains demands a more precise and comprehensive tool that encompasses all of the terminological work produced by the Biological Unions and other sanctioned groups.

The realization of a system that would give cross disciplinary access to the standardized, well defined and documented terminology requires much preliminary work. This work will include at least the following steps: a) the identification of existing committees and ad hoc groups sanctioned by the appropriate components and affiliates of ICSU working on vocabulary and nomenclature problems within specific disciplines, b) publication and wide international dissemination of information on the work of these committees, c) development of computer systems for storage and retrieval of comprehensive data on the source, definition, history, and hierarchical relationships of biological terms, and d) in-depth study of the principles of terminology and standards development. All of these aspects are being addressed by existing organizations and should provide impetus for achieving the long-term goal. The work of these organizations requires support and input from the entire biological community.

THE CODATA COMMISSION ON STANDARDIZED TERMINOLOGY FOR ACCESS TO BIOLOGICAL DATA

Initially, the Commission is seeking to identify private and public producers of databases and publications that depend upon terminology, nomenclature, and classification of biological entities. Organizations, groups, and individuals active in the building and maintenance of terminology, and of controlled vocabularies, are also being sought. Sub-disciplines where such constructs are needed, but not available, will be identified and Commission resources will be used for assisting in these areas.

The Commission activities are contributing to a directory of major "producers" of sanctioned standardized lists of terminology and nomenclature (see description of DBIR above) and are providing forums for communication among relevant "producers" to focus on problem areas and thus facilitate linkages among related vocabularies.

Figure 1 illustrates a few aspects of a specific biological question for which the investigator would be required to consult numerous data sources, i.e., "What is known about development of a retroviral vaccine?" Bibliographic databases such as BIOSIS, MEDLINE, Chemical Abstracts, and EMBASE; factual databases such as GENBANK, Protein Information Resource, EMBL, Hybridoma Data Bank; and specialized databases such as World Patent Index would all provide essential information to resolve this query. The investigator not only has to be familiar with the search commands required by each of these data systems, but also must be cognizant of the terminology selected by each database producer to codify the information. If standardized sets of descriptors were available for biological concepts, the terminology of the databases could be mapped to these descriptors. The CODATA Commission is encouraging the development of such descriptors by the appropriate scientific societies and their sanctioned standards committees.

Members of the Commission are Biological Union representatives appointed by the Unions. Representatives from other organizations such as database producers, scientific journals, biological networks, CODATA Task Groups, etc. are joining as liaison members. Funding for the initial work of the Commission is being provided by U.S. and European government organizations. Because of the international and interdisciplinary interest in the work of the Commission, expansion of funding will be sought from international sources. Endorsement of the work of the Commission and expressions of willingness to participate in its work have been received from most of the Biological Unions of ICSU.

DEVELOPMENT OF RETROVIRUS VACCINE

1) PHYSICAL PROPERTIES
   - infectivity, pathogenicity
   - species differences

2) PATENTS

3) NUCLEIC ACID
   - properties
     - synthesized?
     - single stranded?

4) PROTEINS
   - properties
     - binding sites
   - sequences

5) PROTEIN STRUCTURE

6) IMMUNOLOGICAL PROPERTIES
   - immunoglobulins
     - class
     - sequence
     - binding sites

7) MONOCLONAL ANTIBODIES
   - defined epitopes
   - cross reactivities

8) CULTURE COLLECTIONS
   - cells for growing
   - viral receptors

Figure 1. Biological data types covered by separate and distinct databases.

TERMINOLOGY: THE SCIENCE OF CONCEPTS AND TERMS

All of the efforts described above and the work of numerous other groups attempting to derive standards for the vocabulary of biology must adhere to the basic principles of lexicography. These are organizations which exist solely for that purpose.

The American Society for Testing and Materials (ASTM), for example, has a subcommittee on Terminology which has just revised its guidelines on form and style of Terminology. These guidelines will be published in 1990 and are intended for use by ASTM in publishing its "standards" documents. However, these guidelines contain basic principles that will be invaluable to those working on standardization of biological terminology. Issues such as order, definition, hierarchy, preferred terms, descriptors, abbreviations, symbols and acronyms are addressed by the guidelines. ASTM documents will, of course, be available to all. It would be of great benefit, however, if some of those involved in designing these standards could work with the Bio-unions and CODATA to assist in applying the guidelines specifically to biological nomenclature and terminology problems.

One such organization affiliated with ICSU which has agreed to collaborate with the CODATA Commission is the International Information Centre for Terminology (INFOTERM). This organization, established in 1971, is sponsored by UNESCO within the framework of the UNISIST programme. INFOTERM works in liaison with the technical committee on "Terminology: Principles and Coordination" of the International Organization for Standardization. INFOTERM's International Institute for Terminology Research provides a forum for theoretical and applied terminology research. The Institute is involved in the promotion and coordination of basic research in terminology, including methods of international unification; knowledge theory; study of scientific terminologies; and design of knowledge bases and training courses in theory of science, knowledge theory, knowledge engineering, information science, information management and special language research.

It is clear that the biological community will have to agree upon and accept a set of standards that apply to the science of terminology if there is to be consistency in a system which will be assembled almost like pieces of a puzzle. The pieces, in this case, will come from a great diversity of contributors—all of whom speak different "languages." In order to bring the "language of biology" into a cohesive and unified system, we biologists and computer scientists have much to learn from those who have devoted themselves to the "science of Terminology."

This brief overview of interdisciplinary efforts to standardize the terminology and nomenclature of biology should, if nothing else, provide clues to the enormity of the task before us.

LITERATURE CITED

Morowitz, H. & T. Smith. 1987. *Report of the Matrix of Biological Knowledge Workshop.* Santa Fe Institute. Santa Fe.

Webb, E. C. 1984. *Enzyme Nomenclature.* Academic Press. New York.

Nomenclature Committee of the 4th International Workshop on Human Leukocyte Differentiation Antigens. 1989. Towards a Better Definition of Human Leukocyte Surface Molecules. *Immunology Today* 10(8):253–258.

Systematics Resources, Training, and Jobs

K. Elaine Hoagland

Nearly one hundred people attended the workshop on "Resources, Training, and Job Placement of Systematists and Evolutionists on a Worldwide Scale." In addition to the scheduled speakers, audience participation added information and suggestions from many countries in Europe and Latin America. The group agreed to maintain contact through representatives of several countries, and to plan training courses and develop other mechanisms to further systematics, particularly in developing countries.

LATIN AMERICAN CONCERNS

The session began with a review of resources and opportunities in Brazil, presented by Dr. P. E. Vanzolini, Director of the Museu de Zoologia, Sao Paulo. He cited the need for systematists, given Brazil's efforts at industrialization and power generation. "We have no entomologists, so have given up on insect identifications and are concentrating on vertebrates to understand biological diversity," he said. There are resources to train vertebrate biologists in-country, although the universities are not oriented towards systematics but rather towards molecular biology. Vanzolini noted a recent increase in interaction between systematists and biochemists, and increased economic incentives to do systematics. He thought that it is possible to create new systematics labs at universities, but it is doubtful that new museums can be created, because it is too difficult to assemble the necessary collections and library.

The University of Sao Paulo has a Ph.D. program in systematics, supported by some computer capability and a good library. The National Museum of Rio is renovating its collections. There is also a small museum in Bene, in the Amazon, which is well curated.

Only in the last 30 years has Peru begun to train systematists. The subject is very popular with students, according to Gerardo Lamas of the Museo de Historia Natural, National University of San Marcos, Lima. Students are using the modern tools of biochemistry, statistics and cladistic analysis, along with collections and field work. Advanced training is possible only in Lima, but courses are given in the interior at times. There is government support through a Research Council, despite social and economic difficulties. The main problem is job opportunities. Dr. Lamas suggested looking to private companies using natural products to endow teaching and research positions, in return for tax deductions. Private foundations and non-governmental organizations should also take part in the plan. Museums and educational organizations must work on public education to improve the status of systematics.

Dr. Hoagland is the Executive Director, Association of Systematics Collections, 730 11th Street N.W., Washington, D.C. 20001, USA.

Dr. Tila Perez spoke of the diverse habitats and high endemism in her native Mexico. She works at the National University in Mexico City, where there are systematics collections. Vera Cruz also has a center for systematics. Although interest in systematics is increasing in Mexico, funds are declining due to the ongoing financial crisis, which itself has caused a decline in standard of living and an increase in pollution. The faculty in Mexico City are implementing a graduate program in systematics, but there is a need to send some students abroad for training. There is insufficient opportunity for direction of thesis projects on the local biota, and many habitats, such as the dry forests, are in need of study. But the government is creating natural reserves, and these may provide opportunities in the future. Dr. Perez called for a national plan to focus on key habitats for study, national collections development, and new storage facilities.

Jack Schuster of the University of the Valley, Guatemala, also spoke of the remarkable diversity of habitats, from desert to rain forest, in the very small area of Guatemala. In 1977, Guatemala graduated its first biologist in the country at the BS level. There are no graduate programs, and no indigenous taxonomists. Dr. Schuster has a small museum, and there is a national museum. There are plans for a diagnostic network in entomology, with links to Costa Rica and the rest of Central America. Many students are interested in systematics but there is little support, except for some international funds from US AID and other agencies. "Time is our major problem," he said, "for studying the biota of Guatemala."

From the audience, Jeremy Holloway of CAB International, an agricultural organization centered in Britain, commented on the need to build resources and traditions in developing countries. He pointed to Malaysia as a country that has a tradition of field work and collections, but no national museum. Peter Raven commented that there is actually a decline in the functioning of governmental institutions in Latin America, but a developing tradition of non-governmental organizations. "There were 70 NGO's in 1980, and there are over 3,000 now; competent people are fleeing from government." It is difficult to find people to manage parks and run ministries. He placed some blame on failure of the United States to support Latin America and worried that other countries are withdrawing support as international attention moves to Europe. Bush's recent statements in support of economic ties with Latin America are encouraging and might make debt swaps easier, he suggested.

Raven stressed the need to present a context for systematics that governments can understand, such as has been done by INBio in Costa Rica. Dr. Portez of the University of Puerto Rico agreed, and cautioned for scientists not to impose models of research on developing countries. "New types of aid are needed to train local people and allow the researcher to mix with the people," she said.

Dr. Gamez, Director of the National Biodiversity Institute of Costa Rica (INBio), was asked to present his model plan for systematics in that country, one which involves the training of parataxonomists to do identifications and monitor biodiversity. Dr. Vanzolini disagreed with the approach, for it does not produce fully-trained systematists. However, Gamez and others pointed out that the method not only produces an immediate practical result and gives the local people a stake in biodiversity, but helps to identify persons who are qualified for further training. The INBio model uses the slogan, "To save biodiversity, we have to use it. To use it, we have to know it."

Another member of the audience commented on the situation in Argentina. Salaries and support for science are down. Once there were excellent museums, but there is no money for maintenance, and there is an exodus of qualified personnel and students. "There is no prestige for biologists in Argentina, only for industry and pollution."

THE DEVELOPED COUNTRIES

The second part of the program featured speakers from Britain, the USSR, and the United States. In Britain, said Brian Clarke of the University of Nottingham, the problem is the number of jobs available. "There is enough training for the few qualified posts, but by the criteria of national and international needs, not enough infrastructure to support research in biodiversity." Professor Clarke decried the division of biochemistry and evolutionary/population biology, just at a time when interesting things are happening at the boundaries. Recently, the British Museum had to hire an American in the field of molecular/evolutionary biology, because no British scientists could be found to qualify. "Senior scientists who advise governments don't believe [the importance of systematics]," he said. We must first convince these senior scientists.

Vladimir Sokolov of the USSR Academy of Sciences spoke about the situation in the Soviet Union. Although the USSR has a good tradition in taxonomy, the number of practitioners has declined to the point where "taxonomists might need to go into the red databook [of endangered species]," he quipped. Training is available at a few universities, including Kiev, Moscow, and Leningrad. There are several good museums and botanical gardens in major cities and provincial areas, but there is a restriction on hiring new staff. Although science has privilege, this privilege does not extend to general biology. "A bus driver has a better salary than a biologist," he lamented. Decision-makers do not understand the need to protect the biosphere. "We are preparing a document for the government to explain these needs, and our needs," he said. There is danger that the public could lose confidence in science, because the newspapers sometimes blame scientists for not improving the situation in the country, as they are expected to do.

Robert Hoffmann, Assistant Secretary for Research at the Smithsonian Institution, spoke positively about the resources for systematics in the U.S.: good collections, mentors, and funding for students (at least compared with other countries). He saw jobs as the greatest problem. Collections are underutilized for training and research. As in other countries, the shift in focus of universities away from systematics is of concern. Curriculum requirements in systematics and related subjects can help prevent polarization at universities, he suggested. "We've failed to make alliances with molecular biological disciplines. We must persuade these biologists how we can help each other." Dr. Hoffmann praised cooperative programs between universities and museums. He also proposed that we encourage the fundamental interest of people in systematics who are working in other fields. One possible source for support of systematics is industry, but Hoffmann was not under the illusion that it would be easy to tap this resource.

Walter Reid of the World Resources Institute spoke about the Strategy for Biodiversity being developed by his organization, IUCN, UNEP, and others. The project is to develop a strategy for local, national, and international action to save, study, and sustainably use the world's biological diversity. It is clear that systematics must play a role in such a strategy. By working internationally but on a local level, the WRI hopes to identify types of action needed and present the strategy to policy-makers. Reid cited Terry Erwin's work on insects in the tropics as evidence that policy-makers can understand that conservation decisions require "basic data on where and what."

Dr. Reid noted that the conservation and development communities are coalescing, as evidenced by INBio in Costa Rica, where conservation priorities are related to sustained use of species and the development of cottage industries based on natural resources. According to Reid, an advantage of systematics research is that it develops a scientific infrastructure that carries immediate benefits but is low in capital outlay.

Michael Mares concluded the formal part of the program by describing the Fulbright Program for scholarly exchanges between the United States and 120 countries. The purpose of the Fulbright program is to encourage internationalism. In addition to year-long exchanges, split grants (e.g., two summers) or short grants of ten days to two weeks are possible. Proposals in biology, including wildlife biology, conservation biology, and systematics, are eagerly sought.

Dr. Mares gave advice to those considering submission of a grant. The four-page applications are first reviewed by a panel in the area of study, and then go to a general panel with only one scientist on it. Therefore, the proposals must be understandable to the intelligent layman. People with no foreign experience have an edge in the application process, but it is helpful to contact an in-country host before submission.

Mares answered some earlier speakers who worried about the loss of trained persons who failed to find jobs in systematics. It is a positive trend if these persons are at the Masters Degree level and take their understanding of systematics to another related field such as conservation, he claimed.

From the audience, several individuals had comments about other countries. The French Society of Systematics has developed a book to send to ministers and other policy-makers on the importance of systematics. Systematics is not being taught at many universities, and recently an American was hired to fill a post in France. In Germany, jobs are hard to find and several systematists have emigrated to the United States to find jobs. One idea broached at the workshop was to enlist a network of European systematists to lobby the European parliament on the importance of systematics.

COMMON PROBLEMS

It is clear from the individual accounts of the various countries that there is not so much difference between so-called developed and developing countries in the status of systematics. Although the infrastructure is much stronger in developed countries, all face the problem of loss or lack of jobs in systematics, and loss of programs from universities.

William Anderson of the University of Michigan claimed that the biggest problem is with faculty in universities and even in biology departments; they make the decisions on what will be taught, and "This is where the battle is being fought, at least in the United States." Anderson's comment reinforced Brian Clarke's call to educate other scientists who are themselves policy-makers, and Robert Hoffmann's urging to seek out biochemists for collaboration and intellectual exchange.

Peter Raven, however, said that it was a mistake to lament the loss of departments in universities, because ". . . universities don't have to study certain things, and it is anti-intellectual to say so." He added that too strident an emphasis on conservation instead of the intellectual framework in systematics and evolutionary biology is also a mistake. "Systematics is not just of use for something else. The why of systematics is that the biota is fundamental to human progress. Universities pursue intellectually challenging things, and this changes. The structure of universities must be flexible enough to bring together groups of people at the fringes of fields." Raven called for a global system of sharing of systematists specializing in different taxa.

Raven, Clarke, and Hoffmann agreed that we must emphasize the excitement of the fundamental unknown to policy-makers and the public. Hoffmann used the example of the National Plan for Astronomy and Physics as an appeal to curiosity and intellect. Mares added that policy-makers also respond to the importance of biotic resources. Donald Duckworth summed up the discussion by commenting that we must be "for something, not just about something."

SOLUTIONS

The most specific proposals revolved around training in developing countries. Many participants were impressed with the INBio model and proposed expanded programs for training paraprofessionals and re-training of scientists in other fields to give them a background in systematics. Networking of organizations and institutions in developing countries was seen as a key step in expanding the INBio model to other countries. The Latin American Botanical Network headquartered in Santiago, Chile, was cited as another organization to be emulated.

Training students was not seen as just bringing them to the U.S. or to other developed countries, although Smithsonian Institution programs of that type were seen as beneficial. Some persons felt that industry and private foundations should be approached for help to endow faculty or museum positions, while others felt that such attempts would fail unless they were well within the mission of the organization being approached. Dr. Gamez said that we must sell a package to government—the components of collections, databases, and personnel all go together to provide the information package that they need.

Some attendees felt that US AID was a good source of support, while others were skeptical. Raven said that if one can get a national plan adopted that will call for particular work being done, such as study and protection of biodiversity, then AID or the World Bank or others would go for it. Gamez's package presented through the vehicle of INBio is such a national plan. Vicki Funk of the Smithsonian Institution commented that the US too needs a national plan that involves both ecologists and systematists as information providers.

The group felt that it might be worthwhile to explore the possibility of holding a workshop for persons in Latin America interested in setting up institutions similar to INBio. They also wished to institute short-term courses bringing instructors to developing countries to do training while also conducting field work. Dr. Funk commented that much of the energy in small projects is burned up in scrambling for funds, hence an umbrella funding mechanism for a series of short courses would be superior to individual efforts.

At the conclusion of the workshop, a group of people agreed to communicate further about these ideas and form a network for information on training, jobs, and infrastructure. Members of the group are Douglass Miller, Vicki Funk, Jeremy Holloway, Bruce Collette, Jack Schuster, Chris Darling, and the speakers who gave presentations at the workshop. Others who wish to be kept informed on these issues and who have suggestions for programs may write to ASC.

ACKNOWLEDGMENTS

We thank The Nature Conservancy, in cooperation with the U.S. Fish & Wildlife Service, for making funds available to assist in travel support for our Latin American speakers.

APPENDIXES

Program Committee

Chair:
Dr. Marjorie L. Reaka-Kudla
Department of Zoology
University of Maryland, College Park, MD

Dr. Craig Black
Director
Los Angeles County Museum of Natural
  History
Los Angeles, CA

Dr. Michael Braun
Laboratory of Molecular Systematics
Smithsonian Institution
Washington, D.C.

Dr. Jonathan Coddington
Department of Entomology
Smithsonian Institution
Washington, D.C.

Dr. Rita R. Colwell
Director
Maryland Biotechnology Institute
University of Maryland, College Park, MD

Dr. Douglas Futuyma
Department of Ecology and Evolution
State University of New York
Stony Brook, NY

Dr. Stephen J. Gould
Museum of Comparative Zoology
Harvard University
Cambridge, MA

Dr. Anson Hines
Smithsonian Environmental Research Center
Edgewater, MD

Dr. Alan Kohn
Department of Zoology
The University of Washington
Seattle, WA

Dr. K. June Lindstedt-Siva
Manager, Environmental Sciences
Atlantic Richfield Company
Los Angeles, CA

Dr. Mary Mickevich
Department of Entomology
University of Maryland
College Park, MD

Dr. Douglass Miller
Systematic Entomology Laboratory
U.S. Department of Agriculture
Beltsville, MD

Dr. Charles Mitter
Department of Entomology
University of Maryland
College Park, MD

Dr. Peter Raven
Director
Missouri Botanical Garden
St. Louis, MO

Dr. James E. Rodman
Systematic Biology Program
National Science Foundation
Washington, D.C.

Dr. David Wake
Director
Museum of Vertebrate Zoology
University of California
Berkeley, CA

Dr. Edward O. Wilson
Museum of Comparative Zoology
Harvard University
Cambridge, MA

Dr. Carl Woese
Department of Microbiology
University of Illinois
Champagne-Urbana, IL

Plenary and Special Lectures

Dr. Peter Raven, Director, Missouri Botanical Garden, P.O. Box 299, St. Louis, MO 63166, USA. *Biodiversity in an age of extinction: what is our responsibility?*

Professor Robert M. May, FRS, Department of Pure and Applied Biology, Imperial College, Prince Consort Road, London SW7 2BB, United Kingdom. *Community patterns and their importance for understanding and conserving diversity.*

Dr. Erick Greene, Department of Avian Sciences, University of California, Davis, CA 95616, USA. Dobzhansky Lecture: *Evolutionary aspects of a developmental polymorphism in a caterpillar.*

Dr. Richard Leakey, Director, Kenya Wildlife Service, P.O. Box 40241, Nairobi, Kenya. *An overview of the evidence for African origins.*

Dr. Margaret B. Davis, Regent's Professor of Ecology, Department of Ecology, Evolution and Behavior, University of Minnesota, Minneapolis, MN 55455, USA. *New understanding of the Ice Age: implications for the study of evolution.*

Dr. William B. Provine, Section of Ecology and Systematics, Cornell University, Ithaca, NY, USA. The Wilhelmina E. Key 1990 Invitational Lecture of The American Genetics Association: (Introduced by Professor Bruce Wallace, Department of Genetics, Virginia Polytechnic and State University, Blacksburg, VA, USA.). *Motoo Kimura, the Neutral Theory of Molecular Evolution and the Disunity of Modern Evolutionary Biology.*

Dr. Douglas J. Futuyma, Department of Ecology and Evolution, Division of Biological Sciences, State University of New York, Stony Brook, NY, USA. *Systematics and the study of evolutionary processes.*

Professor John Maynard Smith, FRS, School of Biological Sciences, University of Sussex, Falmer, Brighton BN1 9QG, United Kingdom. *The evolution of prokaryotes: does sex matter?*

Dr. Peter H. A. Sneath, Department of Microbiology, University of Leicester, P.O. Box 138, Medical Sciences Building, University Road, Leicester LE1 9HN, United Kingdom. *Leeuwenhoek in Lilliput.*

Dr. Eugenie Clark, Department of Zoology, University of Maryland, College Park, MD 20742, USA. *Sea monsters and deep sea sharks.*

Dr. Stephen J. Gould, Museum of Comparative Zoology, Harvard University, Cambridge, MA 02138, USA. *Darwin's unrecognized appeal to species selection.*

APPENDIX 3

Congress Symposia and Organizers

A. Evolution in Perspective: Biodiversity, Conservation, Biotechnology and Global Change

1. *Critical Issues in Biological Diversity*
   Organizers: Dr. V. A. Funk and Dr. Stanwyn Shetler, Department of Botany and Office of the Director, National Museum of Natural History, Smithsonian Institution, Washington, D.C. 20560, USA

2. *Evolution in a Rapidly Changing Environment: Global Warming*
   Organizer: Dr. Robert Buddemeier, Chief, Geohydrology Section, Kansas Geological Survey, University of Kansas, Lawrence, KS 66046, USA

3. *UV-B Radiation as an Evolutionary Stress Factor*
   Organizer: Dr. George Bean, Department of Botany, University of Maryland, College Park, MD 20742, USA

4. *Systematics and the Release of Genetically Engineered Organisms*
   Organizer: Dr. Ramon J. Seidler, U.S. Environmental Protection Agency, 200 Southwest 35th Street, Corvallis, OR 97333, USA

5. *Systematics, Biogeography and Evolutionary Significance of Hydrothermal Vents and Vent-Related Seeps: The Emerging Global Pattern*
   Organizer: Dr. Meredith Jones, 2283 Mitchell Bay Road, Friday Harbor, WA 98250, USA

6. *Evolution in Island Archipelagos: The Emerging Picture*
   Organizer: Dr. Scott Miller, Department of Entomology, Bishop Museum, Honolulu, HI 96817, USA

7. *Biotic Exchange: How Has It Influenced Subsequent Evolution?*
   Organizer: Dr. Geerat Vermeij, Department of Geology, University of California, Davis, CA 95616, USA

8. *The Evolution and Ecology of Small Populations*
   Organizer: Dr. Fred W. Allendorf, Division of Biological Sciences, University of Montana, Missoula, MT 59812, and Population Biology and Physiological Ecology, National Science Foundation, 1800 G Street, Washington, D.C. 20550, USA

9. *Conservation in Evolutionary Perspective: Madagascar, A New Arena*
   Organizer: Dr. Porter P. Lowry II, Missouri Botanical Garden, St. Louis, MO 63166, USA and Laboratoire de Phanerogamie, Museum National d'Histoire Naturelle, 16, rue Buffon, Paris, France

10. *Extinction and Evolution*
    Organizer: Dr. David Jablonski, Department of Geophysical Sciences, University of Chicago, Chicago, IL 60637, USA

11. *Long and Short-Term Views of Ecosystem Stability: Implications for Evolution*
    Organizers: Dr. William A. DiMichele and Dr. Scott L. Wing, Department of Paleobiology, National Museum of Natural History, Smithsonian Institution, Washington, D.C. 20560, USA

12. *The Role of Systematics and Evolution in Biotechnology*
    Organizer: Dr. Rita R. Colwell, Director, Maryland Biotechnology Institute, University of Maryland, College Park, MD 20742, USA

1014

B. Evolutionary Mechanisms and Processes

13. *Diversification: Patterns, Rates, Causes, and Consequences*
    Organizers: Dr. Alan Kohn, Department of Zoology, University of Washington, Seattle, WA
    98195; Dr. Charles Mitter and Brian Farrell, Department of Entomology, University of Maryland, College Park, MD 20742, USA

14. *Hybrid Zones and Evolutionary Process*
    Organizer: Dr. Richard Harrison, Section of Ecology and Systematics, Cornell University,
    Ithaca, NY 14853, USA

15. *Genetic Constraints in Evolution: Do They Occur and How Do They Work?*
    Organizer: Dr. Stephen C. Stearns, Director, Zoological Institute, University of Basel, Basel,
    Switzerland

16. *Maternal Effects in Evolutionary Biology*
    Organizers: Dr. Bruce Riska, Department of Genetics, University of California, Davis, CA
    95616, and Dr. Barry Sinervo, Department of Integrative Biology, University of California, Berkeley, CA 94720, USA

17. *Functional Morphology, Biomechanics and Evolutionary Process*
    Organizer: Dr. Marvalee Wake, Department of Integrative Biology, University of California, Berkeley, CA 94720, USA

18. *Developmental Processes and Evolutionary Change*
    Organizer: Dr. Jerome C. Regier, Center for Agricultural Biotechnology, Maryland Biotechnology Institute and Department of Entomology, University of Maryland, College Park, MD 20742, USA

19. *Multiple Levels of Selection in Relation to Evolutionary Theory*
    Organizer: Dr. Elisabeth S. Vrba, Department of Geology and Geophysics, Yale University,
    New Haven, CT 06511, USA

20. *The Role of Multigene Families in Molecular Evolution*
    Organizers: Dr. Gabriel A. Dover, Department of Genetics, University of Cambridge, Cambridge CB2 3EH, United Kingdom, and Dr. Russell Doolittle, Center for Molecular Genetics, University of California, La Jolla, CA 91238, USA

21. *Natural Selection in Molecular Evolution*
    Organizers: Dr. Morris Goodman, Department of Anatomy and Cell Biology, Wayne State University School of Medicine, Detroit, MI 48201, and Dr. Martin Kreitman, Biology Department, Princeton University, Princeton, NJ, USA

C. Systematics and Phylogenetic Reconstruction

22. *Phylogenetic Analysis of Nucleotide Sequence Data: Methods, Comparisons, and Applications*
    Organizer: Dr. David Hillis, Department of Zoology, University of Texas, Austin, TX 78712,
    USA

23. *Molecular and Paleontological Perspective on the Evolution of Modern Humans*
    Organizer: Dr. G. Phillip Rightmire, Department of Anthropology, State University of New
    York, Binghamton, NY, USA

24. *A Critical Reappraisal of Theories of Character Evolution in Phylogenetic Inference*
    Organizers: Dr. Mary Mickevich, Department of Entomology, University of Maryland, College Park, MD 20742, and Dr. Richard Holmquist, Space Sciences Laboratory, University of California, Berkeley, CA 94720, USA

25. *Origin and Evolution of Mitochondrial and Plastid Genomes*
    Organizer: Dr. Michael W. Gray, Department of Biochemistry, Sir Charles Tupper Medical Building, Dalhousie University, Halifax, Nova Scotia, Canada B3H 4H7

26. *Toward a Phylogeny of the Protistans*
    Organizer: Dr. Mitchell L. Sogin, Center for Molecular Evolution, Marine Biology Laboratory, Woods Hole, MA 02543, USA

27. *Early Life*
    Organizer: Dr. J. William Schopf, Center for the Study of Evolution and the Origin of Life, Institute of Geophysics and Planetary Physics, Geology Building, University of California, Los Angeles, CA 90024, USA

28. *Mode and Tempo of Viral Evolution*
    Organizer: Dr. Wen-Hsiung Li, Center for Demographic and Population Genetics, University of Texas Health Science Center, Houston, TX 77225, USA

Congress Special Interest Symposia/Workshops and Organizers

29. *Resources, Training and Job Placement of Systematists and Evolutionists on a World-Wide Scale: Their Significance in the Global Biodiversity Crisis*
    Organizer: Dr. K. Elaine Hoagland, Executive Director, Association of Systematics Collections, 730 11th Street, N.W., Washington, D.C. 20001, USA

30. *21st Century Data and Knowledge Bases in Systematics and Evolutionary Biology*
    Organizers: Dr. Theodore J. Crovello, Graduate Studies and Research Office, California State University, Los Angeles, CA 90032; and Dr. Candace McManus, Microbial Systematics Section, National Institute of Dental Research, National Institutes of Health, Bethesda, MD 20892, USA

31. *The Latest Tools for Cladistic Analysis*
    Organizers: Dr. Estelle Russek-Cohen, Department of Animal Sciences, University of Maryland, College Park, MD; Dr. James M. Carpenter, Department of Entomology, Museum of Comparative Zoology, Harvard University, Cambridge, MA, USA; Dr. Wilentina H. deWeerdt, Institute of Taxonomic Zoology, Zoologische Museum, University of Amsterdam, Amsterdam, The Netherlands; and Dr. Roderic Page, Department of Zoology, University of Auckland, Private Bag, Auckland, New Zealand

Round Table Discussion Groups and Organizers

1. *Evolution and Phylogeny of Protistan Groups*
    Organizers: Dr. Robert A. Andersen, Center for Culture of Marine Phytoplankton, Bigelow Laboratory for Ocean Sciences, McKown Point, West Booth Bay Harbor, ME 04575, USA, and Dr. John O. Corliss, Albuquerque, NM 87153, USA

2. *The Origin of the Metazoa*
    Organizer: Dr. Claus Nielsen, Zoologisk Museum, Universietetsparken, DK 2100 Kobehavn, Denmark

3. *Hybrid Zones*
    Organizer: Dr. Richard Harrison, Section of Ecology and Systematics, Cornell University, Ithaca, NY 14853, USA

4. *The Impact of Biomechanics and Functional Morphology on Studies of Evolution*
    Organizer: Dr. Marvelee Wake, Department of Integrative Biology, University of California, Berkeley, CA 94720, USA

5. *Co-Evolution: Insects/Parasitoids*
    Organizers: Dr. Peter C. Chabora, The Graduate School and University Center, City University of New York, NY 10036, and Dr. H. Roberta Koepfer, Department of Biology, Queens College, CUNY, Flushing, NY 11367, USA

6. *Symbiosis in Evolution*
    Organizer: Dr. Mary Beth Saffo, Institute of Marine Sciences, University of California, Santa Cruz, CA 95064, USA

7. *Nomenclature*
    Organizer: Dr. Christopher Thompson, Department of Entomology, National Museum of History, Smithsonian Institution, Washington, D.C. 20560, USA

8. *High Diversity Marine Ecosystems: Advanced Research Aspects* (Sponsored by UNESCO and IUBS/IABO)
    Organizer: Dr. Pierre Lasserre, Station Biologique de Roscoff, University of Paris, and C.N.R.S., Roscoff, France; Dr. Frederick Grassle, Institute of Marine and Coastal Sciences, Cook College, Rutgers University, New Brunswick, NJ, USA; and Dr. G. Carleton Ray, Department of Environmental Sciences, University of Virginia, Charlotteville, VA, USA

9. *The Relationship of Development to Morphological Evolution*
    Organizer: Dr. David B. Wake, Museum of Vertebrate Zoology, University of California, CA 94720, USA

10. *Evolution on Islands and Conservation: West Indies*
    Organizer: Dr. Gregory Mayer, Division of Amphibians and Reptiles, National Museum of Natural History, Smithsonian Institution, Washington, D.C., USA

11. *Evolution on Islands and Conservation: Phillippines*
    Organizer: Dr. S. H. Sohmer, Assistant Director, Research and Scholarly Studies, Bishop Museum, P.O. Box 1900A, Honolulu, HI 96817, USA

12. *Energy and Community Evolution*
    Organizers: Dr. Leigh Van Valen and Dr. Virginia Maiorana, Department of Ecology and Evolution, Whitman Laboratory, University of Chicago, 915 East 57th Street, Chicago, IL, USA

13. *Homeosis and the Evolution of Plants*
    Organizers: Dr. U. Posluszny, Department of Botany, College of Biological Sciences, University of Guelph, Guelph, Canada, and Dr. Rolf Sattler, Biology Department, McGill University, Montreal, Quebec, Canada

14. *The Influence of Molecular Biology on Evolutionary Theory*
    Organizer: Dr. Gabriel A. Dover, Department of Genetics, University of Cambridge, Cambridge CB2 3EH, United Kingdom

15. *Evolutionary Consequences of Genetic Engineering and Conservation Practices* (Graduate Student Discussion Group)
    Organizer: Allison E. L. Colwell, Department of Biology, Washington University, St. Louis, MO 63130, USA

16. *Biodiversity, Conservation, and Global Change* (Graduate Student Discussion Group)
    Organizer: John Ware, Department of Zoology, University of Maryland, College Park, MD 20742, USA

17. *Conceptual Issues in Systematics and Phylogenetic Reconstruction* (Graduate Student Discussion Group)
    Organizer: Bryan Dutton, Department of Botany, University of Maryland, College Park, MD 20742, USA

Society Sponsored and Special Symposia and Workshops

A. Affiliated Society Symposia

1. *Rates and Weights: Rates of Evolution and Character Weighting* (Willi Hennig Society)
   Organizer: Dr. Christopher J. Humphries, Botany Department, British Museum, Cromwell Road, London SW7 5BD, United Kingdom

2. *The Phylogeny of Behavior* (Willi Hennig Society)
   Organizer: Dr. John W. Wenzel, Department of Entomology, University of Georgia, Athens, GA 30602, USA

3. *Comparing Trees: Measures of Congruence and Coevolution* (Willi Hennig Society)
   Organizer: Dr. Susan J. Weller, Department of Entomology, National Museum of Natural History, Smithsonian Institution, Washington, D.C. 20560, USA

4. *Young Investigator Symposium* (American Society of Naturalists)
   Organizer: Dr. Barbara Bentley, Department of Ecology and Evolution, State University of New York, Stony Brook, NY 11794, USA

5. *Sensory Drive: Does Sensory Biology Bias or Constrain the Direction of Evolution?* (Vice-Presidential Symposium, American Society of Naturalists)
   Organizer: Dr. John A. Endler, Department of Biological Sciences, University of California, Santa Barbara, CA, USA

6. *Quantitative Approaches to the Study of Evolution* (Numerical Taxonomy Association; Co-Sponsored by the Society for the Study of Evolution)
   Organizer: Dr. George Estabrook, Herbarium and Department of Biology, University of Michigan, Ann Arbor, MI 48109, USA

7. *Molecular Evolution of Ultraselfish Genes* (The Society for the Study of Evolution)
   Organizer: Dr. Chung-I Wu, Department of Biology, University of Rochester, Rochester, NY 14627, USA

8. *Host-Parasite Interactions and the Evolution of Reproductive Characters* (The Society for the Study of Evolution)
   Organizers: Dr. Samuel W. Skinner and Dr. Keith Clay, Biology Department, Indiana University, Bloomington, IN 47405, USA

9. *Pattern Versus Process: Causal Explanation in Evolutionary Biology* (The Society of Systematic Zoology)
   Organizer: Dr. Joel Cracraft, Department of Anatomy and Cell Biology, University of Illinois, Chicago, IL 60612, USA

10. *Evolutionary Genetics of Aging* (The Society for the Study of Evolution)
    Organizer: Dr. Michael R. Rose, Department of Ecology and Evolutionary Biology, School of Biological Sciences, University of California, Irvine, CA 92717, USA

B. Affiliated Society Workshops and Special Sessions

11. *Linnaean Plant Name Typification Project* (Linnaean Society)
    Organizers: Dr. Charles Jarvis, Department of Botany, The Natural History Museum, London, England, and Dr. James L. Reveal, Department of Botany, University of Maryland, College Park, MD 20742, USA

12. *Molecular Evolution of Archaebacteria* (University of Maryland Center for Marine Biotechnology; Co-Sponsored by the Association of Systematic Collections)
> Organizer: Dr. Frank Robb, Center of Marine Biotechnology, University of Maryland, Baltimore, MD, USA

13. *Toward More Stable Biological Nomenclature: Proposed Lists of Names in Current Use*
> Organizers: Dr. D. L. Hawksworth, CAB International Mycological Institute, Kew, United Kingdom; Dr. W. Greuter, Botanischer Garten and Botanischer Museum, Berlin, Germany, and Dr. J. McNeill, Royal Ontario Museum, Toronto, Canada.

14. *T-24* (Numerical Taxonomy Association)

Contributed Paper Sessions and Their Co-Chairs

1. *Sex and Sex Ratios.* Chair: Dr. Lin Chao, Department of Zoology, University of Maryland, College Park, MD 20742
2. *Molecular and Morphological Relationships Among Populations I.* Co-chairs: Mr. F. X. Villablanca, Museum of Vertebrate Zoology, University of California, Berkeley, CA 94720; Dr. Richard Highton, Department of Zoology, University of Maryland, College Park, MD 20742
3. *Plant Mating Systems.* Co-chairs: Dr. Paul R. Neal, Department of Ecology and Evolution, State University of New York, Stony Brook, NY 11794; Dr. Linda F. Delph, Department of Biology, Indiana University, Bloomington, IN 47405
4. *Functional Morphology, and Patterns of Natural Selection.* Co-chairs: Dr. Sharon Emerson, Department of Biology, University of Utah, Salt Lake City, UT 84112; Dr. D. W. McShea, Committee on Evolutionary Biology, University of Chicago, Hinds Geophysical Sciences Building, 5734 South Ellis Avenue, Chicago, IL 60637
5. *Developmental and Morphological Patterns in Evolution.* Co-chairs: Dr. Erick Greene, Department of Avian Sciences, University of California, Davis, CA 95616; Mr. Andres Collazo, Museum of Vertebrate Zoology and Department of Integrative Biology, University of California, Berkeley, CA 94720
6. *Molecular and Morphological Relationships Among Populations II.* Co-chairs: Dr. Nancy Knowlton, Smithsonian Tropical Research Institute, APO Miami, FL 34002; Dr. Darryll Felder, Department of Biology, University of Southwestern Louisiana, Lafayette, LA 70504
7. *Genetic Constraints, and Levels of Selection.* Co-chairs: Dr. Mark Dybdahl, Friday Harbor Laboratories, University of Washington, Friday Harbor, WA 98250; Mr. Sean H. Rice, Department of Ecology and Evolutionary Biology, University of Arizona, Tucson, AZ 85721
8. *Evolution on Islands.* Co-chairs: Dr. Jonathan B. Losos, Museum of Vertebrate Zoology, University of California, Berkeley, CA 94720; Dr. Thomas E. Reimchen, Department of Zoology, University of Alberta, Edmonton, Canada T6G 2E9
9. *Genetic Structure of Populations I.* Co-chairs: Ms. Sabine S. Loew, Department of Ecology and Evolution, State University of New York, Stony Brook, NY 11794; Dr. Joseph E. Neigel, Biology Department, University of Southwestern Louisiana, Lafayette, LA 70504
10. *Rates of Evolution, and Analysis of Phylogenetic Patterns I.* Co-chairs: Dr. Harilaos A. Lessios, Smithsonian Tropical Research Institute, Apartado 2072, Balboa, Republic of Panama; Dr. Margaret F. Smith, Museum of Vertebrate Zoology, University of California, Berkeley, CA 94720
11. *Analysis of Phylogenetic Patterns II.* Co-chairs: Dr. C. Riley Nelson, Department of Zoology, University of Texas, Austin, TX 78712; Dr. Daphne G. Fautin, Department of Systematics and Ecology, University of Kansas, Lawrence, KS 66045, and Kansas Geological Survey, University of Kansas, Lawrence, KS 66046
12. *Genetic Structure of Populations II.* Co-chairs: Dr. James L. Hamrick, Department of Botany, University of Georgia, Athens, GA 30602; Dr. David C. Culver, Department of Biology, American University, Washington, D.C. 20016
13. *Analysis of Phylogenetic Patterns III, and Life History Evolution I: Reproduction and Allocation.* Co-chairs: Mr. Keith Karoly, Committee on Evolutionary Biology, University of Chicago, Chicago, IL 60637; Ms. Louisa A. Stark, Department of Environmental, Population and Organismic Biology, University of Colorado, Boulder, CO 80309
14. *Hybrid Zones and Speciation.* Co-chairs: Ms. Patricia L. Sawaya, Department of Biological Sciences, University of Cincinnati, Cincinnati, OH 45221; Dr. Thomas Parsons, Laboratory of Molecular Systematics, Museum Support Center, Smithsonian Institution, Washington, D.C. 20560
15. *Sexual Selection.* Co-chairs: Dr. Gerald Borgia, Department of Zoology, University of Maryland, College Park, MD 20742; Dr. Anne E. Houde, Department of Biology, Program in Ecology, Evolution and Behavior, Princeton University, Princeton, NJ 08544

16. *Mechanisms of Isolation, Speciation and Hybridization.* Co-chairs: Ms. Nancy L. Reagan, Department of Ecology and Evolution, University of Chicago, 940 East 57th Street, Chicago, IL 60637; Dr. Judy Rhymer, Laboratory of Molecular Systematics, Museum Support Center, Smithsonian Institution, Washington, D.C. 20560

17. *Behavior and Evolution.* Co-chairs: Dr. Lynne Parenti, Department of Vertebrate Zoology, Smithsonian Institution, Washington, D.C. 20560; Ms. Rosemary J. Smith, Department of Ecology and Evolutionary Biology, University of Arizona, Tucson, AZ 85721

18. *Speciation and Diversification.* Co-chairs: Dr. Warren Douglas Allmon, Department of Geology, University of South Florida, Tampa, FL 33620; Ms. Jane Masterson, Committee on Evolutionary Biology, University of Chicago, 915 East 57th Street, Chicago, IL 60637

19. *Evolution of Genes and Genomes I.* Co-chairs: Dr. Wolfgang Stephan, Department of Zoology, University of Maryland, College Park, MD 20742; Dr. Marc Epstein, College of Agriculture and Maryland Institute of Biotechnology, University of Maryland, College Park, MD 20742, and Department of Entomology, Smithsonian Institution, Washington, D.C. 20560

20. *Phenotypic Plasticity and Quantitative Genetics.* Co-chairs: Dr. Martin D. Gebhardt, Department of Genetics, University of Georgia, Athens, GA 30602; Dr. Susan J. Mazer, Department of Biological Sciences, University of California, Santa Barbara, CA 93106

21. *Historical Processes, Biogeography, Community Stability and Diversity.* Co-chairs: Dr. Robyn J. Burnham, New Mexico Museum of Natural History, Albuquerque, NM 87194; Dr. Charles W. Thayer, Department of Geology, University of Pennsylvania, Philadelphia, PA

22. *Evolutionary Interactions Between Species I.* Co-chairs: Mr. Evan D. Brodie III, Department of Ecology and Evolution, University of Chicago, Chicago, IL 60637; Dr. Douglas Gill, Department of Zoology, University of Maryland, College Park, MD 20742

23. *Evolutionary Interactions Between Species II, and Evolutionary Ecology.* Co-chairs: Dr. G. A. Allen, Department of Biology, University of Victoria, Victoria, British Columbia, Canada V8W 2Y2; Ms. Deborah J. Morrin, Department of Zoology and Program in Marine and Estuarine Environmental Sciences, University of Maryland, College Park, MD 20742

24. *Evolution of Genes and Genomes II.* Co-chairs: Dr. Patricia Gentili, Department of Entomology, Smithsonian Institution, Washington, D.C. 20560; Dr. Brook G. Milligan, Department of Botany, University of Texas, Austin TX 78713

25. *Character Analysis, Phylogenetic Inference and Methodology I.* Co-chairs: Dr. Carey Krajewski, Laboratory of Molecular Systematics, Smithsonian Institution, Washington, D.C. 20560; Dr. Anthony H. Bledsoe, Department of Biological Sciences, University of Pittsburgh, Pittsburgh, PA 15260

26. *Evolution of the Prokaryotes and Multicellular Eukaryotes: Molecular Approaches.* Co-chairs: Mr. James Sniezek, Department of Zoology, University of Maryland, College Park, Maryland 20742, USA; Dr. Katherine G. Field, Department of Microbiology, Oregon State University, Corvallis, Oregon 97331, USA.

27. *Evolutionary Genetics of Populations.* Co-chairs: Dr. Stephen R. Palumbi, Department of Zoology, University of Hawaii at Manoa, Honolulu, Hawaii 96822, USA; Mr. Paul Jivoff, Department of Zoology, University of Maryland, College Park, Maryland 20742, USA.

28 *Character Analysis and Methodology II: Education and Policy.* Co-chairs: Dr. Peter Houde, Department of Biology, Princeton University, Princeton, New Jersy 08544, USA; Dr. Susan Weller, Department of Entomology, Smithsonian Institution, Washington, D.C. 20560, USA.

29. *Life History Evolution II: Growth, Generation Time, and Natural Selection.* Co-chairs: Dr. William J. Etges, Department of Zoology, University of Arkansas, Fayetteville, Arkansas 72701, USA; Dr. George Roderick, Department of Entomology, University of Maryland, College Park, Maryland 20742, USA.

Poster Sessions

1. Evolutionary Ecology
2. Sex, Breeding Systems, Sexual and Natural Selection
3. Life History Evolution
4. Evolutionary Genetics
5. Molecular and Chromosomal Evolution
6. Phylogenetic Relationships and Diversity
7. The Role of Environmental, Genetic, Developmental and Morphological Factors in Evolution
8. Historical Processes and Biogeography

Local Organizing Committee

Chair:
Dr. Rita R. Colwell
Director
Maryland Biotechnology Institute
University of Maryland, College Park

Dr. George Bean
Department of Botany
University of Maryland, College Park

Dr. Eugenie Clark
Department of Zoology
University of Maryland, College Park

Dr. John O. Corliss
University of Maryland, Emeritus
Albuquerque, New Mexico

Dr. Pierre-Marc Daggett
Congress Co-ordinator
Department of Microbiology
University of Maryland, College Park

Dr. Elizabeth C. Dudley
Special Editor, Congress *Proceedings*
Dioscorides Press

Dr. Theodore R. Dudley
U.S. National Arboretum
Washington, D.C.

Dr. Vicki A. Funk
Department of Botany
National Museum of Natural History
The Smithsonian Institution
Washington, D.C.

Dr. Richard Highton
Department of Zoology
University of Maryland, College Park

Dr. K. Elaine Hoagland
Executive Director
Association of Systematics Collections
Washington, D.C.

Dr. Mary Mickevich
Department of Entomology
University of Maryland, College Park

Dr. Douglass Miller
Systematic Entomology Laboratory
Agriculture Research Service
U.S. Department of Agriculture
Beltsville, Maryland

Dr. Sidney K. Pierce
Department of Zoology
University of Maryland, College Park

Dr. Arthur Popper
Chairman
Department of Zoology
University of Maryland, College Park

Dr. Marjorie L. Reaka-Kudla
Department of Zoology
University of Maryland, College Park

Dr. James Reveal
Department of Botany
University of Maryland, College Park

Dr. James E. Rodman
Program Officer
Systematic Biology Program
National Science Foundation
Washington, D.C.

Dr. Estelle Russek-Cohen
Department of Animal Science
University of Maryland, College Park

Ms. Louise Salmon
Meetings Manager
American Institute of Biological Sciences
Washington, D.C.

Dr. Stanwyn Shetler
Office of the Director
National Museum of Natural History
Smithsonian Institution
Washington, D.C.

Officers and Members of the Past Council and International Committee of the International Congress of Systematic and Evolutionary Biology

Officers

C. Barry Cox (UK), Co-President
James L. Reveal (USA), Co-President
Nils Stenseth (Norway), Secretary
    General/Treasurer

Council Members (including the above)

Mary K. Arroyo (Chile)
Thomas Cavalier-Smith (Canada)
William G. Chaloner (UK)
Rita R. Colwell (USA)
Joel Cracraft (USA)
Werner Greuter (Germany)
Shoichi Kawano (Japan)
Geoffrey G. E. Scudder (Canada)
Beryl B. Simpson (USA)

International Committee

T. N. Anathakrishnan (India) (1996)
M. E. Barkworth (USA) (1990)
S. A. Bengtson (Norway) (1990)
R. Berry (UK) (1996)
B. Briggs (Australia) (1996)
D. A. Brothers (South Africa) (1990)
C. Coetzee (Namibia) (1996)
P. Crane (USA) (1996)
J. Crisci (Argentina) (1996)
B. Dommee (France) (1996)
W. W. de Jong (Netherlands) (1996)
A. Dubois (France) (1990)
J. Felsenstein (USA) (1996)
O. Frota-Pessoa (Brazil) (1990)
K. Bachmann (Netherlands) (1996)
A. Bell (UK) (1996)
F. Bernini (Italy) (1996)
L. Borgen (Norway) (1996)
D. Brncic (Chile) (1990)
P. Calow (UK) (1996)
D. Cohen (Israel) (1996)
R. Crewe (South Africa) (1996)
I. S. Dareusky (USSR) (1990)

J. Doyle (USA) (1996)
F. di Castri (France) (1996)
F. Ehrendorfer (Austria) (1996)
W. Fitch (USA) (1996)
I. Fukuda (Japan) (1996)

Council Members

F. R. Ganders (Canada) (1990)
W. F. Grant (Canada) (1996)
W. Greuter (Germany) (1990)
D. Heppell (UK) (1996)
H. Hoenigsberg (Chile) (1990)
D. Jablonski (USA) (1996)
L. A. S. Johnson (Australia) (1996)
K. Joysey (UK) (1996)
M. Kimura (Japan) (1996)
P. Lavalle (France) (1996)
C. Lindsey (Canada) (1996)
S. Lovtrup (Sweden) (1996)
L. Margulis (USA) (1996)
J. McNeill (Canada) (1990)
M. Monasterio (Venezuela) (1996)
G. Mulligan (Canada) (1996)
K. Pedersen (Denmark) (1996)
K. Pirozynski (Canada) (1996)
P. H. Raven (USA) (1990)
D. Ride (Australia) (1996)
J. Sarukhan (Mexico) (1990)
A. K. Sharna (India) (1990)
L. Slobodkin (USA) (1996)
D. Stone (USA) (1996)
A. Takhtajan (USSR) (1996)
D. Tomback (USA) (1996)
K. Urbanska (Switzerland) (1996)
J. Vassal (France) (1996)
G. Vida (Hungary) (1996)
G. Wagner (Austria) (1996)
Wu Cheng-yih (PRC) (1990)
M. Goodman (USA) (1996)
S. W. Greene (UK) (1996)

D. Hawksworth (UK) (1996)
L. J. Hickey (USA) (1990)
C. Humphries (UK) (1996)
C. Jeffrey (UK) (1996)
K. Jones (UK) (1996)
R. W. Kiger (USA) (1990)
B. Lanza (Italy) (1990)
G. M. Lechon (France) (1996)
M. Littler (USA) (1996)
Ma Shi-jun (PRC) (1990)
A. Marshall (UK) (1996)
H. Y. Mohanram (India) (1990)
D. Moore (UK) (1996)
V. Novak (Czechoslovakia) (1996)

D. Penny (New Zealand) (1996)
G. Prance (UK) (1990)
O. Reig (Venezuela) (1990)
G. Sarmiento (Venezuela) (1990)
W. Schwemmler (Germany) (1996)
J. Slipka (Czechoslovakia) (1996)
O. T. Solbrig (USA) (1996)
E. Szathmary (Hungary) (1996)
L. Thaler (France) (1996)
S. Ulfstrand (Sweden) (1990)
P. E. Vanzolini (Brazil) (1990)
C. Vazquez-Vanez (Mexico) (1996)
E. Vrba (South Africa) (1996)
M. Wake (USA) (1996)

Resolutions Passed at ICSEB IV

1. Resolution of gratitude for organization and support of ICSEB IV

At the conclusion of the Fourth International Congress of Systematic and Evolutionary Biology, the Congress:

- **Recognizes with gratitude** the great success and intellectual stimulus of the meetings and the opportunities which they have provided to scientists from many countries for international collaboration in numerous fields.

The Congress:

- **Records its thanks** to the Regents, the Chancellor, and President of the University of Maryland at College Park, and to the Regents and Secretary of the Smithsonian Institution, joint hosts of the Congress, and in particular, to the staff of the University, and of the U.S. National Museum of Natural History, for their part in making the Congress so welcoming and successful.

The Congress:

- **Records special thanks** to Dr. Rita R. Colwell and the members of the Local Organizing Committee of the Congress, and to Dr. Marjorie L. Reaka-Kudla, Chair of the Program Committee, without whose efforts the Congress would not have happened at all.

The Congress also:

- **Gratefully acknowledges** the support which it has received from corporate sponsors, the co-sponsoring societies and organizations that provided support for participants, especially: Alfred P. Sloan Foundation, Commission of European Communities, The Smithsonian Institution, The Nature Conservancy, United Nations Educational, Scientific and Cultural Organization, United States Fish and Wildlife Service Office of International Affairs, The United States National Science Foundation, and The University of Maryland.

2. Resolution welcoming participation in International Union of Biological Sciences (IUBS)

The Congress:

- **Records its Appreciation** of the work of the Officers and Council of ICSEB in successfully completing the formalities for the admission of ICSEB as a Scientific Member of the International Union of Biological Sciences.

The Congress:

- **Looks Forward**, as the IUBS Commission on Systematic and Evolutionary Biology, to participating in the international programs of the Union in collaboration with other members.

3. Resolution on free movement of scientists and international collaboration

Welcoming the recent dramatic changes in relations between the countries of Western and Eastern Europe and, in particular, the opportunities that these changes afford for scientists to face together many global problems, the Congress:

- **Urges** national bodies throughout the world to facilitate the free movement of scientists between countries and to encourage their participation in international programs.

4. Resolution on Biodiversity

Emphasizing the global threat currently facing ecosystems, species, and populations;
Confirming that the need to understand fundamental aspects of biodiversity and its maintenance,

and factors of global change currently threatening it has become the most urgent issue in environmental science today;

Urging that the preservation of biodiversity be adopted as a major focus of international and national bodies, and

Recognizing that issues of maintaining global biodiversity are inseparable from those concerning the use of natural resources by humanity, the distribution of resources, and development, the Fourth International Congress of Systematic and Evolutionary Biology:

- **Warmly endorses** the initiative of the United Nations Environmental Program (UNEP) in promoting an International Convention on Biological Diversity.
- **Applauds** the resolution of the Preparatory Committee, appointed by the General Assembly of the United Nations to prepare for the forthcoming United Nations Conference on the Environment and Development, to specify in its mandate protection of the oceans and other waters as well as protection of the atmosphere and land resources, including biodiversity.
- **Urges** national and scientific members of the International Union of Biological Sciences to actively participate in the international Program on Biodiversity adopted by IUBS6, at its most recent General Assembly.

5. *Resolution emphasizing urgent need for support for inventory studies*

Recognizing that 20 to 25 per cent of the total species of plants, fungi, and microorganisms are likely to become extinct in the next three decades if current trends continue, and

Asserting the urgent need for meaningful biological inventories in most parts of the world, the Congress:

- **Commends** the efforts of countries, institutions and people engaged in, and supporting, such studies; and
- **Emphasizes** to funding agencies and employers of biologists the urgency to expand such work especially of field surveys and eco-taxonomic studies at all levels.

6. *Resolution emphasizing support for systematics*

Recognizing that the result of inventory studies only become meaningful within systematic frameworks and that the current taxonomic infrastructure probably provides for no more than 20 per cent of species of the world biota; and

Recognizing also that support for systematic studies (including basic taxonomy) has often been lacking both from scientists and organizations, the Congress:

- **Urges** *scientists* and those supporting research to expand efforts in basic taxonomy and, in particular, to increase taxonomic effort and effectiveness by developing new approaches to training, collecting and analyzing, and to making available internationally the results of research in the field.

7. *Resolution urging increased stability in the names of organisms*

Considering that the results of inventory studies, taxonomic and systematic research and biological studies of all kinds cannot be communicated effectively and easily unless there is stability and uniformity in the names of organisms, the Congress:

- **Commends** the Standing Committee on Biological Nomenclature of IUBS and the Sections of Zoological Nomenclature and Botanical Nomenclature of IUBS for their efforts since ICSEB III to promote the compilation and publication of lists of names in current use (and the bases of such names) in support of the resolution passed by that Congress.
- **Encourages** IUBS, and its national and scientific members, to continue and increase support for these efforts and, in particular, the programs creating and publishing lists and data bases of Botanical and Zoological names referred to in this resolution.
- **Urges** the forthcoming International Botanical Congress, and in particular its Nomenclature Section, to devise and implement such changes to the International Code of Botanical Nomenclature as may be necessary to bring the lists of names compiled by its working groups into effective use.
- **Commends** BIOSIS for successfully compiling and bringing into service the BIOSIS Register of Bacterial Nomenclature within the TRF system.
- **Commends** the International Commission on Zoological Nomenclature, meeting at the Fourth ICSEB Congress, for deciding to proceed (in collaboration where appropriate, with BIOSIS) to prepare and adopt lists of names in Zoological Nomenclature at all levels.

8. *Resolution requesting museums with international collections to maintain and expand systematic activities*

Viewing with concern the tendency of governments in developed countries to reduce governmental support for taxonomic research (and systematic research generally), with the consequence that institutions are liable to reduce international commitments;
Holding that never before has the need for these institutions to maintain and increase their diverse and world wide effort in systematic biology been so urgent, or the opportunities so great;
The International Congress of Systematic and Evolutionary Biology:

- **Urges** the directors and policy-making boards and councils of museums with international collections to resist the pressures to reduce the amount and coverage of systematic research and to do all that they can to promote their international commitments as represented by the worldwide collections they have made and now hold in trust for science.

9. *Resolution on exploring the use of new techniques for preserving genetic material*

Recognizing that the threat to species and populations requires special efforts to be made to preserve genetic material;
Drawing attention to new techniques permitting laboratory conservation in gene banks of genetic material important to biotechnology, nature conservancy, preservation of threatened species, agriculture and fundamental biology;
The Congress:

- **Recommends** to bodies providing funds in support of biological research that they promote studies of the potential of such techniques and that they develop initiatives in this field.

10. *Resolution welcoming decision to accept invitation to Budapest, Hungary in 1996*

The Fourth International Congress of Systematic and Evolutionary Biology:

- **Warmly welcomes** the invitation of the Hungarian Academy of Sciences and the Eötvös University in Budapest to hold the Fifth Congress of ICSEB in Budapest.

The Congress:

- **Thanks** those Hungarian scientists responsible for promoting the invitation and looks forward to cooperating with them, through its Council and Officers, to make the Fifth Congress as successful and memorable as the Fourth has been.

Officers and Members of the Next Council and International Committee of the International Congress of Systematic and Evolutionary Biology

C. Barry Cox (UK), Co-president
Rita R. Colwell (USA), Co-president
R. W. Kiger (USA), Secretary General
D. Hawksworth (UK), Treasurer

Council Members (Including Above)

T. Cavalier Smith (Canada)
W. Greuter (Germany)
J. L. Reveal (USA) (*Ex. Off.*)
E. Szathmary (Hungary) (*Ex. Off.*)
N. Stenseth (Norway) (*Ex. Off.*)
H. E. Cogger (Australia)

International Committee

V. A. Albert (USA)
B. R. Baum (Canada)
B. C. Clarke (UK)
J. Cracraft (USA)
G. A. Dover (UK)
D. Fautin (USA)
K. Fredga (Sweden)
A. J. Gibbs (Australia)
E. Gittenberger (Netherlands)
J. D. Holloway (UK)
O. Kraus (FRG)
P. Lasserre (France)
P. T. Lehtinen (Finland)
A. Minelli (Italy)
C. Nielson (Denmark)
J. W. Schopf (USA)
R. Schuster (Australia)
R. J. Seidler (USA)
U. Simidu (Japan)
P. H. Sneath (UK)
S. C. Stearns (Switzerland)
F. J. R. Taylor (Canada)
W. Vader (Norway)
G. Zweers (Netherlands)
B. Tan (Philippines)
A. V. Hall (South Africa)
U. Posluszny (Canada)
W. Ford Doolittle (Canada)
C. E. Jarvis (UK)
S. R. P. Holley (New Zealand)
M. Ragan (Canada)
S. J. Mazer (USA)

INDEXES

Index of Author Names

Index of Organism Names

Page numbers in italics denote entries in a figure or table.

THE UNITY
OF
EVOLUTIONARY BIOLOGY

THE UNITY OF EVOLUTIONARY BIOLOGY

Proceedings of the Fourth International Congress of Systematic and Evolutionary Biology

VOLUME I

University of Maryland
College Park, USA
July 1990
Co-hosted by the Smithsonian Institution

Edited by
Elizabeth C. Dudley

DIOSCORIDES PRESS
Theodore R. Dudley, Ph.D., General Editor
Portland, Oregon

ISBN 0-931146-19-4 (two-volume set)
Printed in Hong Kong

DIOSCORIDES PRESS
9999 S.W. Wilshire, Suite 124
Portland, Oregon 97225

Library of Congress Cataloging-in-Publication Data

International Congress of Systematic and Evolutionary Biology (4th :
  1990 : University of Maryland)
    The unity of evolutionary biology : proceedings of the Fourth
International Congress of Systematic and Evolutionary Biology,
University of Maryland, College Park, USA, July 1990 / co-hosted by
the Smithsonian Institution ; edited by Elizabeth C. Dudley.
        p.   cm. -- (Ecology, phytogeography & physiology series : v.
3)
    Includes bibliographical references and index.
    ISBN 0-931146-19-4
    1. Evolution--Congresses.   2. Biological diversity--Congresses.
I. Dudley, Elizabeth Corning.   II. Smithsonian Institution.
III. Title.   IV. Series.
QH359.I58   1990
575--dc20   90-27811
                                                          CIP

Contents

6

EVOLUTION ON ISLANDS

PHYLOGENETIC PROCESSES

VOLUME II

POPULATION AND COMMUNITY EVOLUTION

EVOLUTION AND ECOLOGY OF SMALL POPULATIONS—SYMPOSIUM

10

CELLULAR AND MOLECULAR LEVELS OF EVOLUTION

PLENARY ADDRESS

ORIGIN AND EVOLUTION OF MITOCHONDRIAL AND PLASTID GENOMES—SYMPOSIUM

NATURAL SELECTION IN MOLECULAR EVOLUTION—SYMPOSIUM

INDEXES

Toward The Next Century and Beyond: The Unity of Evolutionary Biology and Its Importance

Marjorie L. Reaka-Kudla and Rita R. Colwell

BACKGROUND: THE HISTORY OF ICSEB

The International Congress of Systematic and Evolutionary Biology was established in 1973 to facilitate international exchange and communication among systematists and evolutionary biologists interested in diverse groups of organisms, particularly plants and animals. About 1,800 individuals attended ICSEB-I, which was held August 4–12, 1973, at Boulder, Colorado (Short, 1973). Representatives from more than 30 countries attended. Three books, including *Co-evolution of Animals and Plants* by Gilbert and Raven (1975) and *Origin and Early Evolution of Angiosperms* (Beck, 1976), resulted from that Congress. In addition, 23 papers were published in *Systematic Botany* and 8 in *Taxon*. ICSEB-II was held July 17–24, 1979, at Vancouver, British Columbia, Canada (Simpson & Brinck, 1981). Attendance included 953 participants from 31 countries. At least one book, *Evolution Today* (Scudder & Reveal, 1981), resulted from this Congress. Held at Brighton, England, July 4–10, 1985, ICSEB-III was attended by 660 participants from 34 countries (Simpson & Scudder, 1986); 31% were from the United Kingdom, 35% from the United States, and 36% from 32 other countries.

We believe that the decline in attendance at the Congress reflected reduced funding available for systematic-evolutionary research as well as movement away from teaching and research in classical systematics. There has been a recent resurgence of interest in systematics because of the threat (and reality) of global environmental change and the loss of biodiversity, and because of exciting new developments in molecular systematics and evolution. ICSEB-IV, hosted by the University of Maryland and the Smithsonian Institution, was held from June 30–July 7, 1990, was organized to reverse the downward trend in participation in systematics-evolution, to capture and magnify recently renewed interest in these fields, to focus on advances in systematic-evolutionary biology since the first Congress was held in 1973, and to promote a resynthesis of the theory of evolutionary biology.

Dr. Reaka-Kudla is with the Department of Zoology, University of Maryland, College Park, MD 20742, USA. Dr. Colwell is President of the Maryland Biotechnology Institute and Professor in the Department of Microbiology, University of Maryland, College Park, MD 20742, USA.

ICSEB-IV

ICSEB-IV is the only major meeting on systematics and evolution that combines a strong inter-taxonomic, interdisciplinary, and international emphasis. Whereas ICSEB was founded to increase communication among researchers working on different groups of organisms (plants and animals) and thereby provide a forum not available in meetings of societies devoted to particular taxonomic groups, we explicitly broadened the inter-taxonomic emphasis to include microorganisms and, thus, all living biota. ICSEB always has differed from most societal meetings in its international orientation, but we, in view of the critical need for scientific communication and cooperation on systematic-evolutionary problems in a rapidly changing global environment, purposefully accentuated the international focus of ICSEB-IV and took specific steps to incorporate under-represented groups of scientists as a way of enhancing the field (see below).

Underscoring the degree of international and interdisciplinary interaction sought, the Congress was co-sponsored by 11 scientific organizations (the Willi Hennig Society, the American Society of Naturalists, the Numerical Taxonomy association, the Society for the Study of Evolution, the Society of Systematic Zoology, the Entomological Society of America, the Linnaean Society of London, the Association of Systematics Collections, the International Union of Biological Sciences, the Phycological Society of America, and the International Union of Microbiological Societies). Five of these (the Willi Hennig Society, the American Society of Naturalists, the Numerical Taxonomy association, the Society for the Study of Evolution, and the Society of Systematic Zoology) held their annual meetings jointly with ICSEB-IV. The program also included symposia, workshops and other meetings of the International Commission on Nomenclature, UNESCO and IUBS/IABO, the Linnaean Society, and the International Union of Biological Sciences. In addition, the International Society for Evolutionary Protistology, the Phycological Society of America, the Society of Protozoologists, and the Executive Board of the International Union of Biological Sciences met at the University of Maryland immediately prior to ICSEB-IV.

The Congress opened on June 30, 1990, with welcoming remarks by Dr. William Kirwin, President of the University of Maryland, College Park, and Dr. Robert Hoffman, Assistant Secretary for Research at the Smithsonian Institute. Opening remarks by Dr. Rita Colwell, Chair of the Local Organizing Committee, Dr. Marjorie Reaka-Kudla, Chair of the Program Committee, and Drs. Barry Cox and James Reveal, Co-Presidents of ICSEB, were followed by the Congress Opening Plenary address, which was delivered by Dr. Peter Raven.

About 1600 persons attended (1413 registrants), 371 individuals agreed to participate as organizers and speakers in Congress Symposia and Discussion Groups, 58 individuals co-chaired Contributed Paper Sessions, 382 speakers spoke in 29 Contributed Paper Sessions, and 90 individuals presented posters in 8 Poster Sessions. Scientists from 38 countries, including 20 developing nations, were represented.

Theme and Major Topics

Except during the interlude of the New Synthesis, there has been limited communication historically among the disciplines of evolutionary biology, particularly between students of evolutionary history (paleontologists and systematists) and those of molecular, population, and organismal biology. There has been increasing realization that barriers between these subfields must be overcome if a complete theory of evolution and systematics is to be forged (see Futuyma, 1988, and this volume). While embracing the diversity of evolutionary biology and systematics, ICSEB-IV sought to further their integration. The program, designed to foster the overall theme *"The Unity of Evolutionary Biology"*, sought to synthesize historical and mechanistic approaches to evolutionary and systematic problems.

Congressional events were focused around three major topics: 1) Evolution in perspective: biodiversity, conservation, biotechnology and global change; 2) Evolutionary mechanisms and processes; and 3) Systematics and phylogenetic reconstruction. These process-oriented topics were selected to foster exchange among researchers using different approaches on different groups of organisms, emphasize currently important issues, re-evaluate previous concepts, and target, to the best of our ability, emerging ideas that are likely to play an important role in the future of the field. The success of the scientific program was due to the ideas and generous participation of the Program Committee (Appendix 1). The important contributions of these individuals to the Congress are gratefully acknowledged.

The Program

Plenary and Special Lectures. Special presentations addressing a diversity of topics occurred during the Congress as Plenary Lectures (Appendix 2). Speakers included Drs. Eugenie Clark, Margaret B. Davis, Douglas Futuyma, Stephen J. Gould, Richard Leakey, Robert May, Peter Raven, John Maynard Smith, and Peter Sneath. Two lectures, those of Drs. Leakey and Clark, were open to the public as well as to Congress participants. The Congress also was honored to be Dr. William Provine's chosen forum for the delivery of the Wilhelmina E. Key 1990 Invitational Lecture (American Genetics Association). In addition, Dr. Erick Greene, the recipient of the Dobzhansky Prize (Society for the Study of Evolution), delivered his address at ICSEB-IV. Other special lectures included the American Society of Naturalists Presidential Address (Dr. Lee Erhmann) and the Society for the Study of Evolution Presidential Address (Dr. Stephen Gould).

Congress Symposia. As indicated in Appendix 3, 28 Congress Symposia (21 half day, 7 full day) were organized. Note that many of the symposia relate to more than one Congress topic, but were listed within one area for convenience. Organizers were requested explicitly to include women and minorities, students, and scientists from developing countries in their symposia, and to embrace diverse philosophies, approaches, and taxonomic groups to foster cross-disciplinary exchange. Our success in fostering these efforts is discussed in greater detail below.

Congress Special Interest Symposia/Workshops. Three Congress Special Interest Symposia/Workshops were established in topics considered to be critical for the field of systematic-evolutionary biology as well as for policy-making bodies in the 21st century (Appendix 4). These fora were conducted in a discussion format with the express purpose of facilitating maximal exchange among participants, and one included a hands-on workshop. Organizers of these Workshop/Symposia specifically sought international participation and representation from developing countries.

Congress Discussion Groups. In an attempt to maximize the potential for exchange and synthesis, a number of individuals were asked to organize Discussion Groups in an informal round table format (Appendix 5). Also, some organizers of Congress Symposia requested the opportunity to pursue ideas generated in the more formal symposium context in ensuing Discussion Groups. The first announcement solicited names and topics from prospective participants, providing another mechanism by which novel topics and under-represented groups could be included. Additionally, three Discussion Groups addressing the three major topics of the Congress were organized by graduate students. The vigorous attendance and debate, comments from participants during the Congress, and the end-of-Congress review by officers and organizers all demonstrated that these Graduate Student Discussion Groups were among the most successful events of the Congress. It was decided that these fora should become a permanent component of the International Congress.

18

Organizers of Discussion Groups were requested to circulate prospective discussion questions at least twice to potential participants well in advance of the meeting, providing the opportunity for interaction and sharpening of focus on controversial and synthetic issues. Key questions for each Discussion Group were listed in the Program so that they could be considered by all participants of the Congress. Discussion Groups were scheduled to follow related symposium topics whenever possible. The Program Committee encouraged minimal formal presentations, a strong emphasis on discussion and synthesis, and publication of the results in the *Proceedings* of the conference.

Sponsored Symposia and Workshops.   The Congress Program was enriched by 14 additional Society Sponsored or Special Sponsored Symposia/Workshops (Appendix 6). The Willi Hennig Society sponsored three Symposia, the American Society of Naturalists two, the Society for the Study of Evolution three, the Numerical Taxonomy association two, and the Society of Systematic Zoology one. In addition, UNESCO and IUBS/IABO sponsored a Special Symposium/Discussion Group Session, and the Linnaean Society, the Center for Marine Biotechnology of the Maryland Biotechnology Institute, and the Association of Systematic Collections sponsored workshops.

Contributed Paper/Poster Sessions.   Congress participants also made presentations in Contributed Paper and Poster Sessions that were arranged by topic (Appendixes 7,8). Some co-sponsoring societies chose to have their own Contributed Paper Sessions, while others integrated their papers within ICSEB Contributed Paper Sessions according to topic. The participation of society officers in organizing their society's contributions, particularly Dr. Michael Bell of the Society for the Study of Evolution, was greatly appreciated. Many other individuals (including Drs. Michael Braun, Carey Krajewski, Theodore Dudley, Elizabeth Dudley, James Reveal, Pierre-Marc Daggett, Ms. Josephine Klapper, and others too numerous to name) who assisted in the final stages of program preparation also deserve special recognition. In addition, we sincerely thank those individuals who agreed to co-chair Contributed Paper Sessions (Appendix 7).

Administration and Organization

Program Committee.   Under development for more than two years, the Program was co-ordinated by a Program Committee (Appendix 1) that included leaders in the field who provided a broad range of scientific expertise. The group targeted currently important topics, ideas needing re-evaluation in light of new information, and areas most fruitful for the future direction of the field. With this meeting, we hoped to inaugurate a decade of work and thought that must prepare us for the 21st century.

Local Arrangements.   A Local Organizing Committee (Appendix 9) co-ordinated space, technical facilities, funding, travel arrangements, food, and other aspects of the Congress. Dr. Pierre-Marc Daggett was Congress Co-ordinator, and oversaw the functions of the Secretariat and Congress logistics. A Society Committee (Drs. Stanwyn Shetler [chair], Vicki Funk, Mary Mickevich, Marjorie Reaka-Kudla) co-ordinated the participation of other societies that met with ICSEB. An Exhibit Committee (Ms. Louise Salmon, AIBS Meetings Manager [chair], Drs. Pierre-Marc Daggett, Estelle Russek-Cohen) organized book, computer and scientific software, and other appropriate displays. We thank all of these individuals and the many others who helped before, during, and after the Congress for their dedication and assistance.

Funding

The University of Maryland provided substantial support for administrative costs, and the Smithsonian Institution provided and funds for mailing, and covered expenses incurred at the Smithsonian during the Congress, as well as participant support for some symposia. Additional funds were received from the U.S. National Science Foundation, the Alfred P. Sloan Foundation, the United Nations Education and Scientific Organization, the Commission of European Communities, the International Union of Microbiological Societies, the Nature Conservancy, the American Type Culture Collection, and Biospherics. To all of these organizations we extend our deepest appreciation.

DISCUSSION: WHAT MADE ICSEB-IV UNIQUE?

International Focus, with Special Emphasis on Participation by Under-represented Groups (including women, minorities, young scientists, and scientists from developing countries)

In the face of a rapidly changing global environment and deteriorating biodiversity that will affect us and future generations, we need to understand past patterns of evolution (through paleontology, phylogenetic reconstruction, analysis of geophysical and ecosystem dynamics over geological time) and we need to predict, through an understanding of the mechanisms and different levels of selection, the responses of the earth's living biota to changing environmental conditions. Rapid solutions to basic questions of how environmental change relates to mechanisms of evolutionary change and to large scale patterns in evolution and systematics will require all of the human resources and talents that we can mobilize. One of the important goals of this Congress was to include individuals who might otherwise not have the opportunity to participate in the solution of these major problems.

A retrospective analysis shows that we were successful, although it is always possible to do better. In 1985 and 1987, 18.1 and 18.7% of all recipients of Ph.D. degrees in Science and Engineering in the United States were women, and 24.3 and 31.1% of all Ph.D. recipients in Science and Engineering were from countries other than the United States (AAAS Observer, 1989). Women were involved as organizers in 29.0% and scientists from outside the U.S. in 16.1% of our 31 Congress Symposia/Workshops. Although we know that some organizers made strong efforts to recruit such individuals, only 13.7% of the speakers invited by Symposium organizers were women. On the other hand, 26.7% of the speakers invited by Symposium organizers were from outside the United States. Speakers in Congress Symposia and Workshops came from 22 countries (including Costa Rica, Argentina, Chile, Peru, Brazil, Pakistan, and Madagascar). Of our 17 Discussion Groups, women were involved as organizers in 29.4% and scientists from outside the U.S. in 23.5% of these events. Of the 48 Congress Symposia/Workshops and Discussion Groups combined, women had leadership roles in 29.2% and non-U.S. scientists in 18.8% of these fora. Two of our 9 (22%) Plenary Speakers were women (three were invited), and three (33%) were from other countries. Women, graduate students and scientists from other countries comprised 24.9%, 23.1%, and 21.9% of overall attendance.

The Program Committee also targeted the development of several Symposia and Workshops with a distinctly international focus. In particular, the Symposium on Conservation in Madagascar highlighted a developing country that faces an environmental/biological/human crisis whose resolution will be of intense interest. Similarly, the Congressional Workshop on Resources, Training, and Job Placement of Systematists and Evolutionists on a World-wide Scale was developed explicitly as an international forum with a strong emphasis on how human resources in developing countries can be enhanced (and galvanized) to assist with the global environmental/biodiversity crisis. In the Biodiversity Symposium, Dr. Gomez's (Costa Rica) address in the Biodiversity Symposium was extremely well attended and received. Numerous other Symposia and

Workshops, including those on the Role of Systematics and Evolution in Biotechnology, Island Biogeography, Evolution and Biogeography of Deep Sea Vents and Seeps, High Diversity Marine Ecosystems, and Genetic Constraints in Evolution, provided good examples of extensive international participation.

Young scientists were encouraged to participate through the development of three Graduate Discussion Groups, the availability of low registration fees, the opportunity for waiver of registration fees in exchange for one day's audio-visual assistance at the Congress, and other financial aid. We were encouraged that 23.1% of the registrants were graduate students, and felt that the extremely successful participation of students in their own Discussion Groups and in other aspects of the Congress was a most promising omen for the future.

Interdisciplinary Emphasis, Including Molecular, Population, Organismic, Paleontological, and Phylogenetic Reconstruction Approaches

In accordance with the theme of the Congress, "The Unity of Evolutionary Biology," organizers were asked specifically to include an interdisciplinary approach. Good examples of this were found in the Symposia on Diversification (which included paleontological, ecological, biogeographical, phylogenetic reconstruction, and population genetical perspectives), Multiple Levels of Selection (with presentations from philosophical, genetic, organismal, population, phyletic, and paleontological points of view), the Evolution of Modern Humans (with talks using molecular, ecological, phylogenetic, and paleontological approaches), Early Life (including presentations based on chemical, molecular, ecosystem, and paleontological evidence), and others.

Intertaxonomic Emphasis, Including Prokaryotes and Protistans

Organizers were encouraged to include a comparative taxonomic approach. In contrast to previous ICSEB meetings, studies within single taxonomic groups were not emphasized, so as to enhance the potential for cross-fertilization of ideas. Examples of events emphasizing cross-taxonomic approaches included, among others, the Symposia on Biodiversity (with presentations on coral reefs, nematodes, plants, and spiders), Conservation in Small Populations (with presentations centered on flies, ungulates, mice, fish, reptiles, and plants), and Developmental Processes (with talks using archosaurs, sea urchins, insects, maize, and phytoflagellates as subjects).

Format Structured to Maximize Accessibility, Discussion, Cross-fertilization of Ideas, and Synthesis (among taxa, disciplines, and scientists from countries with different intellectual traditions)

The earliest manifestation of our dedication to synthesis was the development of our theme, "The Unity of Evolutionary Biology," inspired by the work of Douglas Futuyma (1988) and previous authors (see Mayr & Provine, 1980). The drama of this debate was heightened when, in his Wilhelmina E. Key 1990 Invitational Lecture of the American Genetics Association, Dr. William Provine challenged the view that evolutionary biology could be explained by a single unifying theory. Futuyma responded, in his elegant style, to this idea in his Plenary Lecture and paper (this volume; see also papers by Goodman and Crow). In addition, the Program Committee hoped to incorporate several areas of emerging national and international needs into the multifaceted field of evolutionary biology. The Plenary Lectures by Drs. Peter Raven, Margaret Davis and Robert May; The Symposia on Global Change, Biodiversity, Extinction, Evolution and Ecology of Small Populations, the Role of Systematics and Evolution in Biotechnology, Systematics and the Release of Genetically Engineered Organisms, UV-B Radiation, and Conservation in

Madagascar; the Workshops on 21st Century Data Bases, Training and Job Placement of Systematists and Evolutionists, and High Diversity Marine Ecosystems; the Graduate Student Discussion Groups on Biodiversity, Global Change, Genetic Engineering, and Conservation; and the Discussion Groups on Conservation on Islands, all fostered these connections. Other sessions, including the Symposia on Evolution in Modern Humans (which integrated paleontological, morphological and molecular approaches), Evolution of Protistan Groups (which, along with its associated Discussion Group, drew on multidisciplinary information), Phylogenetic Analysis, and Nomenclature targeted topics of intense current interest and debate. Still other areas (including Island Biogeography, Deep Sea Vents and Seeps, Hybrid Zones, The Role of Development in Evolution) have been the source of much interest and have made major contributions to the field in the past; ICSEB provided a forum in which new developments could be incorporated and the importance of these topics in modern evolutionary biology could be reevaluated.

In addition to developing a theme that would promote synthesis and requesting that organizers incorporate multi-taxonomic and multi-disciplinary approaches as well as a diversity of participants, the Program Committee constructed a format that would maximize these exchange processes. Seventeen major Discussion Groups were established. Questions were circulated among potential participants, honed through interaction, published in the Program, then opened to all for discussion and synthesis (with a galvanizing goal of producing a publishable conclusion) during the meeting.

In addition to the organized Discussion Groups, rooms were available continuously for "spontaneous" Discussion Groups that arose out of discussions and controversies during the Congress. Providing still another outreach mechanism to potential participants, the Program Committee offered to organize contributed paper/poster or other sessions around topics that respondents wished discussed. For example, one group of scientists provided a novel viewpoint on evolution with a Contributed Symposium on "A Thermodynamic Perspective of Evolution." Topics suggested by 116 respondents to the first Congress flier were incorporated into the Program wherever possible. Individuals desiring to participate in discussion of particular topics were put in communication with appropriate Discussion Group and Symposium organizers. Three Congress Workshops/Special Interest Symposia were organized in an informal panel and discussion group format (one with hands-on workshop components). Organizers of Symposia and Discussion Groups were specifically requested to include representatives of opposing views and contrasting approaches, and in several cases the Program Committee asked individuals with different approaches and backgrounds to be co-organizers. Discussion Groups and Contributed Paper/Poster Sessions were scheduled to follow Symposia on related topics, allowing for greater access and exchange among participants and speakers. Social events were scheduled throughout the Congress so that participants could meet each other informally before the final Congress banquet; the mid-Congress 4th of July crab feast and fireworks display, as well as the Smithsonian's open house for researchers, were much enjoyed. Society-sponsored events were integrated among all Congress events, if the society so desired, to maximize exchange and communication among disciplines. Communication between scientists, policy makers, and the public was facilitated in several Symposia (where policy makers were included as participants), by international press coverage, and by two Plenary Lectures that were open to the public.

OFFICERS, COUNCIL MEMBERS, INTERNATIONAL COMMITTEE MEMBERS, AND RESOLUTIONS

The Officers and Members of the Council and the International Committee for ICSEB-V are given in Appendix 10. A number of resolutions were proposed and passed by the membership at the Congress Closing Session on Saturday, July 7, 1990 (Appendix 11). In addition, Dr. Barry Cox presented the Engler Medal to Dr. Peter Taylor.

The Officers, Council Members, and International Committee Members for the 5th International Congress of Systematic and Evolutionary Biology—which will be held in Budapest, Hungary, in 1995—are given in Appendix 12.

LITERATURE CITED

Beck, C. B. (ed.) 1976. *Origin and Early Evolution of Angiosperms.* Columbia Unviersity Press: New York.

Futuyma, D. J. 1988. *Sturm und Drang* and the evolutionary synthesis. *Evolution* 42:217–226.

Gilbert, L. E. & P. H. Raven (eds.). 1975. *Co-evolution of Animals and Plants.* University of Texas Press: Austin.

Mayr, E. & W. B. Provine (eds.). 1980. *The Evolutionary Synthesis: Perspectives on the Unification of Biology.* Harvard University Press: Cambridge, MA.

Scudder, G. G. E. & J. L. Reveal (eds.). 1981. *Evolution Today: Proceedings of the Second International Congress of Systematic and Evolutionary Biology.* Hunt Institute for Botanical Documentation: Carnegie-Mellon University: Pittsburgh, PA.

Short, L. L. 1973. The First International Congress of Systematic and Evolutionary Biology. *Taxon* 22:641–646.

Simpson, B. B. & P. Brinck. 1981. The Second International Congress of Systematic and Evolutionary Biology. Pp. 455–461. In: G. G. E. Scudder & J. L. Reveal (eds.), *Evolution Today Proceedings of the Second International Congress of Systematic and Evolutionary Biology.* Hunt Institute for Botanical Documentation; Carnegie-Mellon University: Pittsburgh, PA.

Simpson, B. B. & G. G. E. Scudder. 1986. Third International Congress of Systematic and Evolutionary Biology, Brighton, England, 4–10 July 1985. *Taxon* 35:464–467.

BIODIVERSITY

Biology in an Age of Extinction:
What is Our Responsibility?

Peter H. Raven

Before we can realistically appraise our potential role as students of systematic and evolutionary biology, we must first consider the context in which we operate. Citizens of rich, industrialized nations such as the United States and Canada have every reason to feel extraordinarily privileged. In such countries, we enjoy the highest standard of living the world has ever known, and possess individual and collective opportunities that would have seemed unthinkable just a decade or two ago. Even though we who have long enjoyed its benefits sometimes seem not to care much about our responsibilities as citizens, evidently preferring a sort of comfortable apathy when it comes to politics, democracy has broken out throughout the world, and the prospect of world peace seems brighter than ever. What should there be to worry us, then, and how should we view our common responsibilities?

Regardless of how fortunate we may be individually—and we should all be humble enough, and realistic enough, to recognize our good fortune—the world is in desperate condition. Our population has more than doubled over the past 40 years to the record level of more than 5.3 billion people, and is continuing to grow at the rate of some 95 million people annually. This is more than the entire population of Mexico, with almost all of the growth in numbers taking place in the less developed countries of the tropics and subtropics. Nearly one in four of us, all in these regions, lives in a condition that the World Bank has defined as absolute poverty—a condition of existence that the Bank has defined as being below the lowest acceptable standard of human existence. Individuals who live in absolute poverty are unable to be sure of providing food, shelter, and clothing to themselves and their families on a day-to-day basis. At the same time, perhaps 500 million people (one in ten of us, or twice the population of the United States) receive less than 80% of the U.N.-recommended amount of calories—which simply means not only that their minds and bodies cannot develop normally, but also that they are simply wasting away every day. Against a background of these figures, and with tens of thousands of babies under the age of four starving to death every day, fools still tell us that we have happily avoided the famine that was predicted in the 1960s, and that we shall be able to continue into the future much as we have in the past. The fact that anyone would make, much less believe, such statements is scarcely to be believed.

Right now, we are consuming, re-directing, or wasting—for example, by eating it, building pastures, or simply burning it up—an estimated 40% of the total net photosynthetic productivity that takes place on land (Vitousek et al., 1986). Although global grain production has increased 2.6 times since 1950, we have lost an estimated 20%

Dr. Raven is with the Missouri Botanical Garden, P.O. Box 299, St. Louis, MO 63166, USA.

of the topsoil on the world's agricultural lands during the same period of time, an estimated 22 to 24 billion tons per year. No one foresees bringing any substantial amount of additional land into cultivation, and yet extension agents in the United States, and presumably elsewhere, still talk about "allowable" losses of topsoil as a base line from which to calculate what they consider to be sound farming practices! This is just one example of the overwhelming effects that human beings are having on the planet Earth today.

What about the future? At the present rate, the global human population would double in 39 years. This clearly will not occur, however, because too many of us will die along the way. Still, something approaching a billion people will be added to the total during the 1990s—as many people as populated the entire world less than two hundred years ago, at the start of the Industrial Revolution. The addition of these people makes the prospects of alleviating the widespread poverty and starvation that afflict so many of us remote, even if we redouble our efforts to accomplish these aims.

It is a national disgrace for us who live in the United States that, having provided leadership in the worldwide family planning efforts that are underway throughout the less developed world now, we have turned our back on these efforts and refuse to provide funds to those very multilateral agencies that could best administer them.

It is a national disgrace that, although we indulge ourselves by consuming at a rate 20 to 30 times that characteristic of most of the less developed countries of the world, we have never even considered adopting a population policy for ourselves. The impossibility of putting in place a appropriate immigration policy without having a national population policy ought to be obvious to any biologist.

It is a national disgrace that, on a per capita basis, we are the meanest donors of foreign development assistance of any industrialized country. Although we have no trouble indulging ourselves by spending $150 or $200 billion dollars beyond our available resources to maintain the conditions of our lives at what we have come to consider a suitable level; although we have currently begun the process of spending perhaps $500 billion or more to bail out savings and loan institutions that have been gutted by predatory and self-serving individuals, we nonetheless have a great deal of trouble in providing a few billion dollars per year, in a $1.3 *trillion* budget, for such assistance. The difficulty of increasing our expenditures is made obvious by the fact that nearly three quarters of us believe that foreign assistance is the least desirable part of our national budget, and the part that ought to be cut first.

But perhaps these are mainly problems of the United States, ones that the rest of the industrialized world does not share. Sadly, that is not the case. We are just starting to recognize the fact that many ecological problems are global, not national, in nature. Although we have long considered the oceans to be a kind of common property that should be protected, no one seems to be responsible for the estimated 2000 to 3000 oil spills of all sizes that occur every year. We read in the papers that the major oil companies are now seeking to solve the problem by a simple strategy: they are looking for third parties to carry their oil, so that they will not be financially liable for the spills. Presumably, they can't afford them, but someone else can! Wouldn't it be simpler to practice conservation, and to invest in alternative forms of energy? The Law of the Sea has never been adopted by the United States, despite more than a decade of inspired effort, and most of the signatories to the Antarctic Treaty, including the United States, deplore the efforts of Australia, France, and now New Zealand to set up an international park there before it is too late to save that pristine wilderness, one which surely constitutes a vital part of our global patrimony.

As students of ecology, we understand that the greenhouse effect, based on the presence of gases such as carbon dioxide, methane, and now nitrogen oxides and CPCs in the atmosphere, makes the earth a relatively warm planet—much warmer than those that lack an atmosphere. Instead of a frigid $-15°C$ average temperature, we have one of approximately $+15°C$. Isn't it incredible that there is a controversy in the media, reflected in public opinion, not only about whether pumping more of those same gases into the atmosphere will warm global climates, but even whether there *is* a greenhouse effect or not? Perhaps the only more surprising matter is the ingenuity of the hypotheses that individuals have invented for miraculous interventions of hitherto unsuspected natural phenomena that will keep the world from getting warmer no matter what we do. In contrast, the depletion of the ozone layer, a globally-generated problem with serious implications for human health and the health of the biosphere, is now widely accepted at the political level as needing attention urgently; strong steps are being taken to protect the ozone layer from further destruction.

At any rate, these matters have created a kind of global consciousness of the environment, and made industrialized nations recognize their interdependence when it comes to managing the atmosphere and the oceans. The recent creation of a world fund to alleviate the ozone depletion problem, involving as it does both rich and poor countries, signals the kind of compromise that must be repeated many times if the problems of the world are to be solved. But what about biological diversity, the subject of this conference? How should we understand it; how should it be viewed by the public; and what should we do about it as scientists?

Sadly, the loss of biological diversity is not widely recognized as a global problem. That may sound strange to this audience, which is so aware of the problem, but the world's leaders, as you may have noticed, are not clamoring to put in place initiatives to deal with this issue. To some extent, I think that we systematic and evolutionary biologists have been at fault. As "global change" has become a widely acceptable theme—the U.S. has proposed spending $1 billion in this area in the coming fiscal year—we have, I would suggest, tried too hard to link biological diversity with this movement, and thus to secure additional funds to support our efforts. The net effect has been to make the problem of the loss of biological diversity appear as a kind of minor component—one of the unfortunate side consequences—of the atmospheric changes that "everybody" understands. The problem of the loss of biodiversity has also been linked conceptually with the loss of tropical forests, which does contribute substantially to the accumulation of carbon dioxide in the atmosphere, and "the biological diversity problem" is often equated in the political mind with "the tropical forest problem," both then being relegated to a side category of a major effort to model global change.

The problem is also, I think, one of style, in terms of approaches within the respective fields. Despite the evident difficulties, the atmosphere can be modeled, and one can, in principle, predict both wide-scale and local changes from these models. It is clearly worth supporting their development, and that is what we in the United States and other nations are doing on a major scale. Comparable models dealing with the loss of species, in contrast, are neither particularly refined nor certain in application. This clearly seems to have been one of the factors in regarding the loss of species as a less substantial problem than, say, global warming, despite the fact that it is of far more fundamental importance, with permanent consequences that are likely to be disastrous for human beings in the future.

It is a sad commentary on the selfishness of human nature that some economists will still argue that there isn't any real problem, so let's get on with business as usual. In this way, many business journals, for example, simply ignore these issues, even though the CEO of many major companies would be dismissed for incompetence if he or she were to take as simple-minded an approach to environmental problems as is presented as gospel on a daily basis in the major trade journals. As long as we regard 2.5% growth as barely acceptable, applaud rising populations in our cities, states, and nations and deplore declines, and fall back on the oxymoron "sustainable development" as the Holy Grail that we seek, we clearly shall continue to destroy the sustainability of the world as surely as if we had adopted a general plan to accomplish this aim.

What I suggest that we need to do is to find a new way of thinking about biodiversity: one that celebrates the plants, animals, fungi, and microorganisms of the world because of their beauty; because we depend on them individually as our primary source of sustainable productivity (and on the communities in which they function for the global stability that makes possible our continued existence), and simply because we have no right to destroy the organisms that share this world with us. What can we say about the dimensions of this problem?

Tropical lowland rain forests remained largely undisturbed while most other kinds of tropical vegetation were being decimated over the past few centuries. The climates where rain forests grow are less hospitable than some other tropical climates, the forests are able to grow on relatively infertile soils, and their often gigantic trees were more difficult to cut before the advent of chain saws and other modern technologies. Especially during the past century, however, people began to move in greater numbers into these forests, which now support several hundred million of us. They have been reduced to about half of their original extent, and now constitute some 6% of the world's land area.

Over the past few years, the estimated remaining total of about 3 million square miles of tropical rain forest—about the size of the 48 contiguous states of the United States—has been reduced by between 62,500 and 80,000 square miles per year—roughly the size of Michigan or Minnesota (Myers, 1989; World Resources Institute, 1990). Large tracts of tropical rain forest in the northern and western Brazilian Amazon, in the interior of the Guyanas of South America, in the Zaire Basin, and perhaps in parts of New Guinea will survive longer than most areas of such forest, so that much of the rest—in Latin America, Africa, and all of Asia—will be reduced to small patches, surely less than 10% of its original size, over the next 20 to 30 years.

Greatly oversimplifying, the forests are lost because of population pressure; because the greater proportion—some 85%—of the world's wealth is concentrated in industrialized countries, which are home to a rapidly shrinking 25% of the world's population; because since 1984 there has been a positive cash flow, amounting to tens of billions of dollars per year, from poor to rich countries, most of which, broadly speaking, have been unwilling to provide substantial amounts of aid except for military and political purposes; and because economically depressed countries often can see no other options to using up the forests as a short-term resource, a stop-gap measure that is often reinforced by inappropriate, but trivially expeditious, policies and laws.

I cannot resist adding that the management of the forests in most developing countries is roughly equivalent to the way we manage them in the Pacific Northwest and in Alaska—cut off the mature trees and never mind that tourism, presumably not devoted primarily to seeing logged over hillsides, is a more important source of revenue for the areas concerned than the timber industry; sell them off for a quick profit; ship them to Japan as unsawn logs for the quickest possible profit, with no value added for the local economies; and be forced within a few years to make the transition to plantation forestry anyway, but with the mature forests permanently lost. When will we come to our senses, and realize that it is simply time to stop?

To the extent that there is a will to reverse the process of forest destruction; to the extent that the industrialized countries join the less-developed ones to promote global sustainability, as proposed so eloquently by the World Commission on the Environment and Development, the Brundtland Commission; or, more broadly, to the extent that we realize that we are all citizens of one small, finite planet, which we must manage for our common benefit, the process of forest destruction might be slowed or even reversed locally. Since the period we are discussing, 20 or 30 years, is one in which the global population is likely to grow by 50%—the equivalent of the population of the entire world in 1930, with almost all growth taking place in the tropics and subtropics—and since the population that exists now is destroying tropical rain forests at the rate of approximately

1.5% per cent per year, it is difficult to visualize changes in human nature so profound that there would be much tropical rain forest, or any other kind of tropical vegetation, left by the middle of the next century. The United Nations has estimated that, even if we maintain a high level of effort in family planning, so that regional and global population stability may be achieved as rapidly as possible, our numbers are likely to grow to at least 11 billion people a century from now before stability can be attained. What do you think the world will be like then? Nevertheless, human societies have exhibited an amazing ability to adapt in the past, and we should certainly resolve to do what we can to preserve as much as possible of the marvelous biological diversity that we enjoy in the early 1990s.

Without such extraordinary and unprecedented changes, it seems likely that 20 to 25% of all species of plants, animals, fungi, and microorganisms may vanish during the next 30 years or so, and that fully half of the total number of species may disappear before the close of the 21st century. To indicate the basis for these predictions, consider plants. In this relatively well-known group, approximately half of the world's species live in or near forested areas that will be reduced to less than a tenth of their current extent over the next 30 years. The species/area relationships predicted by the theory of island biogeography suggest that half of them will be at risk when the forests are decimated, and extrapolations from these numbers into the future are frighteningly simple to make. Although the loss of perhaps a quarter of the world's plant and vertebrate species during our lifetimes is frightening enough, an even higher proportion of the total number of species may be lost, since much higher proportions of groups such as beetles and ants than of plants, occur in the tropics. Now that we have the ability to move genes from one kind of unrelated organism to another, the loss of a single species implies more than the loss of that individual evolutionary masterpiece, with whatever potential it possessed for human benefit, and its role in an aggregate of organisms providing ecosystem services (including the protection of the atmosphere, soils, water, and the like). It also implies the loss of tens of thousands of individual genes that might themselves be of human benefit.

Given the challenge posed by this series of problems, I would like to direct the remainder of these remarks to the responses that the world community of systematic and evolutionary biologists and ecologists might make. We who know and love the world's biological diversity should be in the most favorable position to make suggestions about how to use it intelligently, and how to take action to conserve it. What, then, should we be doing?

First, we must work for the political and general recognition of the loss of biodiversity as a global problem of the greatest importance, absolutely worthy in its own right of the attention of the world's leaders. Proceeding much more rapidly than more widely recognized problems, such as global warming and the depletion of the ozone layer, and completely irreversible, the unparalleled loss of biodiversity will have the most tragic consequences for human beings in the future. We live in a time of extinction that has had no equal in the past 65 million years of earth history, and we are supposed to have the brains to do something about it. Biologists have an obligation to inform their fellow citizens about this problem, so that it can be understood clearly and addressed adequately.

The United Nations Environment Program, taking the lead in suggesting plans for the U.N. family of agencies in implementing the recommendations of the Brundtland report, is attempting to put in place a global convention of biodiversity. Meetings are being held to attempt to generate a set of principles that will lead to the protection of biodiversity throughout the world. The United States and other nations should cooperate in making this convention a reality, because it will provide a legal instrument for the protection of the organisms on which we depend. The implications of elevating biodiversity to the status of a recognized world problem cannot be overemphasized.

Second, we must urgently join forces with our scientific colleagues in less developed countries, and indeed in all countries, to address the major problems of ecology, systematics, and evolutionary biology. Abdus Salaam, that great scientist who has performed such a valuable service in emphasizing the necessity for additional trained people in less developed countries, has estimated that, with three-quarters of the world population, these countries are home to only about 6% of the world's scientists and engineers. They are also home to a minimum of 80%, and possibly much more, of the world's biodiversity. This indicates clearly that a very substantial augmentation of local efforts will be necessary to learn much about the organisms that occur in these regions while there is still time to do so.

How can this problem be addressed in the context of economies, many of which have deteriorated substantially over the past ten years, and government services that are continuing to disintegrate? One very promising approach is that of the Instituto Nacional de Biodiversidad (INBio) in Costa Rica, a relatively prosperous, democratic, literate Central American country about the size of West Virginia. Costa Rica has recognized that pursuing a national inventory of the organisms that inhabit the country represents a strategy that is both scientifically sound and potentially economically productive. Ably headed by Rodrigo Gamez, INBio has gotten off to a very strong start over the past year, consolidating the national collections, surveying parts of the country for various groups of organisms, starting the consolidation of a national data bank based on the existing Nature Conservancy-initiated effort, and initiating the provision of assessment and information services in connection with development projects. To provide an estimate of the magnitude of the problem, Costa Rica has an estimated 500,000 to 1 million species of organisms, which INBio hopes to catalog over the next decade; no more than 40,000 species of all groups have been recorded from the country already, and *no more than 400,000 from the tropics of the entire world—Asia, Africa, and Latin America combined!*

The training of parataxonomists and other technicians employed for the purposes of INBio represents one of the most innovative aspects of the program. Costa Rica, however, has one of the more outstanding and better supported groups of scientists in Latin America, so that the program was able to get off to a faster start than would have been possible elsewhere. Nonetheless, the realization that all of the operations of taxonomy do not have to be carried out by Ph.D.-level scientists is important for us. If we orient our thinking according to objectives, and not the traditions of the past, we may well find better ways to achieve what we want.

Comparable national biological inventories should be undertaken, when possible, in all of the countries of the world. Collectively they would comprise an effective strategy for dealing with the world's biological diversity. Modern data-banking methods make their interconvertibility easily attainable, and should be exploited fully for the construction of a world network of information about biodiversity. The overall effort could readily be adopted as a global strategy for dealing with biodiversity, particularly since it puts each country completely in charge of its own biodiversity, which it would be exploring, learning to use, and conserving for its own purposes. Improving the management capabilities for a country's biodiversity is easily seen as a logical part of the national planning effort, since a major part of the prosperity of a given country will ultimately depend on how good a job they do in this area. In countries like the United States, where hundreds of millions of dollars are spent annually on biodiversity in one way or another, the need for a national biological inventory is obvious; our failure to implement one can logically only be viewed as amounting to a collective lack of understanding of the problem.

Another outstanding approach to developing science and engineering in less developed countries is the network approach, exemplified by the Third World Academy of Sciences and the institutes that are being developed in Trieste and elsewhere for advanced training, and by the Latin American Botanical Network. That network was generated by Latin American plant scientists for their own purposes and is effectively linking institutions throughout the area in imaginative training programs.

A third area that I would like to explore has to do with the nature of biological inventory and the ways in which efforts to catalogue the world's biodiversity are generally viewed by the world at large, by many of our biological colleagues, and possibly even by some of you who are in this room. For better or for worse, science is organized about the construction of hypotheses, the more recently devised the better. The job of cataloguing and understanding biological diversity is often not regarded as "science," whatever "science" may be, and has traditionally fared poorly in competition with the testing of currently fashionable hypotheses or the investigations of new kinds of data. There is no difficulty in understanding why these approaches, hypotheses, and methods are important, but there is a real problem if they cause us to neglect the perhaps mundane, but fundamental, task of learning about the organisms that share this world with us while they are still around.

To illustrate the problem by an analogy, imagine that we had just 30 years to study the biota present 75 million years ago, in the age of dinosaurs. After 30 years at least a quarter of all the organisms present, including some major groups, would be extinct, and our chance of learning about them would be gone forever. If a limited number of people, and a limited amount of resources, were available to apply to the problem—and that is inevitably the case—would you rather have those people spend their time testing a few hypotheses that had occurred to them, out of a potentially unlimited number that might be conceived, or would you rather that they sampled the overall dimensions of the biota, preserving samples and recording carefully what the organisms were like, and where they occurred? The question has no simple answer, but it is clear that a major effort of the latter kind would be highly desirable, and that it would produce results of permanent interest that could not be provided by what could be regarded as highly sophisticated approaches. More simply, we'd certainly like to know what was there, in as exact terms as possible.

The real difficulty of securing funding for inventory studies becomes even more apparent when we recall that they are usually not funded in the course of development work, because they are "too scientific!" On the one hand, we can't be supported in learning about the broad outlines of the world's biota because scientists regard the work as too pedestrian, even though their ultimate conclusions must rest on the interpretation of the patterns, and even though we have surveyed a maximum of only 15% of the world's biota, conservatively estimated at 10 million species, so far. On the other hand, we can't afford to learn about organisms in the course of development efforts, because they and the ecosystem services they provide are scientific—a matter of basic research—and presumably not relevant to the sustainability of our efforts. What a remarkable world we inhabit! Perhaps we need to try harder to change it.

Will our current hypotheses really be seen as so refined and productive that our scientific successors will thank us for testing them in preference to describing the world's organisms more broadly? The question is really a false one, as you will readily recognize, but it does illustrate the dilemma of getting on with the inventory work that many of us believe to be attainable only in the near future unless it is recognized as intrinsically valuable in its own right. The implications of understanding, using wisely, and conserving biodiversity are so obvious in connection with sound, sustainable development that we simply should not allow the neglect of efforts such as national biological inventories by the development community to persist.

Fourth, I would like to offer a few suggestions about how we might address the problem of inventory. There are more species in the world today than there will be tomorrow; far more than there will be a year from now; and many more than will be in existence by the end of the decade, by which time perhaps 5 to 10% of the total species that exist now will have disappeared. In my view, we shall never attain a complete survey of the world's insects, mites, or nematodes. We would be better served to plan the activities that we can carry out as carefully as we can in order to pass on as much information, as many rationally collected specimens as possible to future generations of scientists.

It does appear feasible to complete the world inventory of plants, vertebrates, butterflies, mosquitoes, and a few other relatively well-known groups. I believe that we need to depend on them to reveal the broad outlines of biogeography, and that we should emphasize making our surveys of these groups as complete as possible as soon as possible (National Research Council, 1980). Completing the overall picture of their distribution patterns is intrinsically worthwhile, and ought to be funded and pursued vigorously. The information gained should be applied immediately to conservation, as in the design of protected areas and the selection of organisms for *ex situ* preservation. Such applications will also help greatly in the preservation of species of other groups of organisms that will inevitably remain less well known.

What about groups of organisms that are less well known? How shall we deal with them? There are currently 12,000 described species of nematodes and 30,000 described species of mites, yet specialists have estimated that each group may contain more than a million species. Only a handful of scientists throughout the world is capable of dealing professionally with non-parasitic nematodes—the vast majority of the group—and only a few dozen with most groups of mites. If there are a million species or more in each group, and no one has the foggiest idea of the relationship between the comparative numbers of species that exist in temperate regions versus the tropics, then there is literally no hope of cataloguing a high proportion of them before they disappear. What of insects at large, with about 750,000 described species, and estimates of the total number ranging from 2 million to 80 million? Would it really do us any good to prepare encyclopedias of all these species, recording brief descriptions and a locality or two, while they continue to disappear rapidly? My view is that this approach has some utility, but not much.

What I suggest we should be doing instead is setting up intelligent plans to estimate how many species of nematodes, mites, and insects exist, and what the major evolutionary and biogeographical patterns in these groups are. We should be encouraging specialization in these organisms, and supporting specialists throughout the world; they, in turn, should be generating plans to sample the diversity in their groups while it still exists. Why is no single student of nematodes, or a group of students, proposing to sample the diversity in the Amazon Basin in some logical way, perhaps through a grid of localities, so that, when the organisms are all gone, we could have some rational basis for knowing what the patterns had been like? It stands to reason that we would like to know, like to appreciate just what kind of diversity this group represents now, but we are not taking imaginative steps to try to find out. Similar strategies should be adopted by INBio and other comparable national efforts, since attempting to describe all of the free-living nematodes of Costa Rica is clearly an impossible task, given the difficulties of collecting, recognizing, and classifying the members of this group.

It would be particularly important, as Robert May (1988) and others have suggested, and as our National Research Council Committee on research priorities in tropical biology pointed out a decade ago (National Research Council, 1980), to obtain comprehensive samples of all organisms from particular small sites. These might well include comprehensive ecosystem sites, since we would then understand the patterns of biogeography, evolution, and adaptation in the individual groups better, and also be able, in principle, to dissect the ecological relationships among them at particular sites.

Fifth, and in a similar vein, we systematic and evolutionary biologists should keep in mind that whatever features of groups of organisms, particularly tropical ones, that we want to study will be more accessible now than ever in the future. For example, if we want to understand the biosystematics of tropical trees, we should select a particular group of trees and go to work on it now; so many of the individual populations, races, and species will be gone in the near future that the picture will always be tragically incomplete. If we want to compare the pattern of evolution in *Swartzia* with that in *Quercus*, or *Psychotria* with

Ceanothus, we need to remember that a quarter of the species in the two tropical groups is likely to be gone within 30 years or so, and get on with the job now. Of the 1094 native taxa of Hawaiian plants, 423 (38%) are presumed to be extinct or threatened to some degree, with 107—a tenth of the native taxa—apparently already gone (Wagner, Herbst, & Sohmer, 1990). Unfortunately, that is the kind of pattern that we and our students are going to find increasingly during the decades to come, and we must think carefully about what to do so as to make the best possible contribution while the opportunities are still there.

By the same token, "unusual" specimens—ones preserved in unusual ways, frozen, specially dried, or prepared for particular purposes—will be of unusual value in the future. Systematists and evolutionists should try to decide, in the context of their groups, what species are most likely to yield specimens that will be interesting for future investigations, and to attempt to obtain such specimens, particularly in areas such as Madagascar, the Atlantic forests of Brazil, Hawaii, or the Philippines, where.the biota is disappearing rapidly right before our eyes. Once collected, such specimens will be of special importance, and biologists will need to pay much more attention to the preservation of their materials, since the time-honored assumption that one can simply go back to the field for more is clearly no longer valid.

Sixth, and finally, systematic and evolutionary biologists must see that their activities result in gains for the utilization and, especially, the conservation of the organisms of interest to them. Consider plants, for example: there are about a quarter of a million species, a fifth of which will be at risk during the next 30 years. In principle, it is perfectly possible to monitor all of these plants, to find out whether they are in protected areas or not, to bring them into seed banks and cultivation, and to save them for the future. We who know the earth's biological diversity best ought to love it the most, and to act in such a way as to save it. There is really no excuse for letting even one additional species of plant or vertebrate go extinct, if we have the collective will to save it.

In a country like the United States, with all of its problems, we live in a democracy, and our individual views are influential if we care enough to express them. The environment is a matter of wide concern to our officials, and to the officials of countries all over the world. If they fail to understand our complete and utter dependency on biodiversity, that failure must be shared with us. If we who know better take a narrow view, and simply get on with business as usual, are we really any better than the politicians and businessmen whom we are always ready to blame for their lack of environmental sensitivity? Events are simply moving too rapidly, and the world needs us too much, for us to neglect the opportunities that are available now, and will decrease so rapidly in the future.

In a very real sense, the human race unwittingly has become the proprietor of a sort of gigantic, dispersed Noah's ark, with all of the responsibilities that this entails. When the global human population reaches stability, and our great-grandchildren can think again about re-populating and re-creating fields and forests, the particular kinds of plants, animals, and microorganisms that are available to them will depend to a very great extent on what we all decide to do during the remainder of our lives. You have probably all seen the "Far Side" cartoon that shows a group of dinosaurs lamenting the fact that all they have in the face of impending extinction are brains the size of walnuts; can we really claim to be doing all that much better? All of the knowledge that is ever going to be available about the marvelously intricate patterns that have resulted from billions of years of evolution over large stretches of the world's ecosystems will be gained during the next few years and decades. The times are incredibly challenging, and our actions are fabulously important. Let us resolve to work together to realize these accomplishments, and to take actions that are worthy of us and the great opportunities that we share.

LITERATURE CITED

May, R. M. 1988. How many species are there on earth? *Science* 241:1441–1449.

Myers, N. 1989. *Deforestation Rates in Tropical Forests and Their Climatic Implications.* Friends of the Earth: London.

National Research Council. 1980. *Research Priorities in Tropical Biology.* National Academy of Sciences: Washington, D.C.

Vitousek, P. M., Ehrlich, P. R., Ehrlich, A. H. & P. M. Matson. 1986. Human appropriation of the products of photosynthesis. *BioScience* 36(6):368–373.

Wagner, W. L., Herbst, D. R. & S. H. Sohmer. *Manual of the Flowering Plants of Hawai'i.* 2 vols. Bishop Museum Special Publication 83. University of Hawaii Press/Bishop Museum Press: Honolulu, HI.

World Resources Institute, *World Resources 1990–91.* Oxford University Press: New York. 383 pp.

PLENARY ADDRESS

Nature Reserves and Biodiversity

Richard E. Leakey

I find it particularly relevant in my new job as Director of Wildlife in Kenya to look at some problems with the perspective that comes from 22 years of involvement with prehistory and looking at change.

There is absolutely no question at all that in a continent such as Africa change has always been the rule rather than the exception where climate is concerned. And while climatic changes in a continent so large have expressed themselves in many different ways, they have certainly been a powerful overall dynamic behind the great diversity of species that have come and gone through time.

One of my concerns is that today there is a great effort to set aside parts of the planet as nature reserves and sanctuaries—places where, hopefully, human interference will not impose change, where we will not put undue pressure on the habitat, and where the existing biodiversity can sustain itself—I think most of us assume—indefinitely. In the tropics, particularly in my part of the world, it is very clear that major changes in climate take place quite frequently, and consequently I am often worried by the concept of these national parks. Because they are islands, they are fixed in terms of their peripheries. They are isolated communities of plants and animals surrounded by agricultural settlement. The people living around the national parks in our country have access to piped water, irrigation, chemical fertilizers, seedlings, plants, and so on. There is probably a greater likelihood of habitat survival and continuity of life external to the parks than actually exists within the parks themselves. When we have a dry year, water will continue to be piped in and people will continue to grow their corn. In the national parks, however, because they are supposed to be set aside as pristine areas, the vegetation will dry out in a dry year and a number of animal species will probably disappear because they cannot move from the parks to areas to which they might previously have migrated to get access to food and resources that would have kept them going.

We do face a dilemma therefore, in terms of the protection and maintenance of systems, and it is a dilemma that calls for more research and for ethical and moral judgements—judgements that we are not necessarily ready to make but which, nonetheless, have to be considered. We are already discussing the culling of antelope and the culling of elephants in southern Africa, culling because there are too many—too many because the land area they exist on is too small to cope with the population growth that has come about as a result of some form of protection. That sort of limited intrusive management is accepted to a certain extent, but we also have to consider intrusive management in terms of the greater biodiversity and consider, for example, whether we should be starting to replant stands of vegetation, where we should reintroduce trees and plants no longer

---

Dr. Leakey is with the Kenya Wildlife Service, P.O. Box 40241, Nairobi, Africa. These remarks are excerpted, with his permission, from Dr. Leakey's full speech.

there, and whether we should protect pieces of forest that are otherwise going to disappear because the elephant—or the buffalo—or the giraffe—is destroying them.

I think we've got to know a great deal more about biodiversity and a great deal more about how things work, but I don't think there is any doubt at all that change will come about—that the country will go through drought years and will go through very wet years. In the process these isolated communities of natural plants and natural wildlife are going to suffer. And I have no doubt that the greatest challenge to the future is to work out ways of management that will sustain some of the communities that we've all grown to take so much for granted and that are, to a certain extent, what conservation is all about.

Consider, for example, the regeneration of forests along rivers such as the Tana, which drains off Mt. Kenya and goes down to the Indian Ocean. Down towards the delta are magnificent gallery forests, and it would appear that these forests establish themselves by regenerating on their periphery. The forest basically shifts and islands develop; these islands shift and further and further islands develop. The study of these patches of forest suggests that there is virtually no regeneration of the major trees within the forest itself; seed dispersal on the periphery leads to this movement. Today, these natural forests are abutting against rice fields and going nowhere, and the forests are dying. In such situations should we be planting trees? Should we be creating biosystems that we know once existed but that are disappearing before our eyes? I think these are issues in judgement that are of more relevance to the future than whether or not the single species, such as the elephant, is going to survive. I think we can resolve the elephant's problem; whether we can solve the bigger problem in our time is more problematic.

Let me say that I believe one of the challenges for the paleobotanists and paleontologists of tomorrow is to develop a far greater knowledge of the way in which evolution has occurred in the context of habitat and climate. And I would suggest that more time should be spent on investigating what has caused species to go extinct; the magnificent fossil sites of East Africa are yielding enormous amounts of data that could be relevant to such investigations. Turning our attentions in this direction might well enable us to learn more about one of the major problems that conservationists today are deeply concerned with— the extinction of species in modern ecosystems.

Biological Diversity:
Are We Asking the Right Questions?

Stanwyn G. Shetler

INTRODUCTION

In conservation circles "biodiversity" has become the byword of our time. Thanks to the media, even the general public seems to be conscious, if only vaguely, of the concept of diversity in nature and of the worth of having it. Saving it is good, losing it is bad, they have been conditioned to believe. But few of the public can define biodiversity, identify or measure it when they see it, or understand how and why to save it. Indeed, the level of biodiversity literacy—i.e., ability to distinguish the kinds of plants and animals or any other units of biological diversity in nature—is quite low in the populace. If the preservation of biodiversity is vital to the future of our planet and an essential aim of conservation efforts, then the first line of defense must be the education of the public. The central question is: how do we go about raising the general level of biodiversity literacy?

"Biodiversity," short for either "biological diversity" or "biotic diversity," has gained almost universal currency as the convenient term for the diversity of life-forms on earth since it was first used in 1986 in the National Forum on BioDiversity and then in the resulting book edited by E. O. Wilson and F. M. Peter (1988), titled simply *Biodiversity.* The Forum, held in Washington, D.C., and jointly sponsored by the National Academy of Sciences and the Smithsonian Institution, marked a turning point in the study of biological diversity. In many ways it represented the beginning of a new discipline of biodiversity.

Wilson, in his Foreword, credits Walter Rosen, then a Senior Program Officer of the National Research Council/National Academy of Sciences, with introducing the term "biodiversity." As a member of the Program Committee for the Forum with Rosen and others, I would like to add an anecdotal footnote. Some of us from the Smithsonian insisted on using the conventional term *"biological* diversity," but editorial types at the Academy kept insisting that *"biotic* diversity" was the more correct term. Back and forth it went. Just when Walt Rosen finally despaired of the impasse and coined the eminently sensible contraction "biodiversity," I never knew, but one thing is certain: the new term caught on with a vengeance. Probably few words have entered the vocabulary of science and attained widespread acceptance with such ease and speed. I doubt that Madison Avenue could have done better. (Perhaps Walt Rosen will not remember it exactly this way, but my story is not apocryphal!)

Despite the success of biologists and conservationists in defining and highlighting the global biodiversity crisis in the public mind and in making the case for preserving

Dr. Shetler is with the National Museum of Natural History, Smithsonian Institution Washington, DC 20560, USA biological diversity even to politicians and entertainment personalities, the biodiversity battle has only begun. It is a battle that can never truly be won. Beyond the platitudes and cliches lies an open-ended series of unasked and unanswered scientific questions and a conservation task that defies imagination. Of the various active and passive conservation measures that have been used or proposed, none guarantees success. The loss of diversity is inevitable and relentless.

The biotic bleeding is worldwide (Shetler, in press). We cannot stop the mortal ebb of life-forms, only slow it. Given the scale of the crisis, all conservation measures ultimately are but trivial palliatives. This means that we must be clearheaded about our aims and priorities in research and conservation, so as to make the most of our efforts. We must be careful to pursue the right questions and take actions that make rational use of the answers.

Are we asking the right questions to understand biological diversity and take the right actions? Obviously, it all depends on what we are trying to accomplish. The purpose of this symposium, "Critical Issues in Biological Diversity," is to examine some of the premises and perspectives of biodiversity research and conservation, and some of the questions posed by investigations at the molecular, organismic, and ecological levels in different taxonomic groups and with different methods. The intent is to stimulate thinking and discussion about the right questions, not to assert dogmatically what the right questions are. I begin the symposium by raising a few philosophical questions from my own point of view.

As a higher plant systematist, my own experience has been almost entirely in temperate regions. Usually, for reasons I'll discuss, the biodiversity crisis is viewed mainly as a tropical crisis. As a non-tropical biologist, I will not venture to review the now well-known tropical biodiversity concerns. Rather, I will bring to bear some temperate-zone perspectives that I think are instructive considerations for any part of the globe.

IS THE BIODIVERSITY CRISIS PRIMARILY A TROPICAL CRISIS?

Because the tropical regions harbor the lion's share of the earth's life-forms, the tropics have become the center of attention in the race to study and preserve biological diversity. It has, in fact, become conventional wisdom that the battle to save biodiversity is largely a battle to save the tropics, especially the tropical rain forests, where, according to E. O. Wilson and others, half the world's species are concentrated. The alarming rate at which these forests are being destroyed promotes the view that the biodiversity battle is inconsequential by comparison on all other fronts.

In terms of the scale of the problem, i.e., the sheer numbers of species and the complexity of the ecological interactions, no one can dispute the crucial significance of the tropics. But conservation is not simply an academic matter of comprehending and ranking the scientific worth of various parcels of nature, but also a matter of human motive and a function of how people behave. It ultimately depends, therefore, on social more than scientific factors. Human behavior, such as the greed behind unmitigated economic exploitation, is not diversity-dependent nor geographically or ecologically correlated. This is to say that the people who exploit the species-rich tropical regions are motivated by the same forces as the people who exploit the relatively species-poor temperate regions. In fact, the high standards of living and the consumptive lifestyles of the developed countries are what the developing countries try to emulate when they seek to exploit the wealth of their land, in the process destroying their forests.

The biodiversity crisis is not so much a crisis of species as a crisis of lifestyles. The proximate cause may be the cutting of rain forests, but the primary cause is to be found in the consumption of the highly developed societies found largely in the temperate regions. It is to these regions, then, that we must look for answers and solutions. In a real sense, therefore, the biodiversity crisis is as much, if not more, a temperate as a tropical crisis. We in the developed temperate regions can expect to have ever-diminishing influence and moral force on the tropical scene unless and until we can achieve a breakthrough in reducing consumptive behavior. It must be a breakthrough resulting in a stewardship ethic that stems the marching tide of land conversion and urbanization by placing the intrinsic values of our own natural habitat above the market values.

Though it may be trite to repeat, it is nonetheless true that we in the United States have a rather sorry past record and continue to set a poor example—though, nowadays, our conservation rhetoric can hardly be faulted. We, for example, have long since plowed up nearly all of our natural prairies and cut down 96%–98% of our original forests. Yet every year more forested and other natural land is converted to agriculture or is developed, and urbanization generally continues to shrink natural habitat and threaten or destroy species.

A prime example of our continuing failure in the United States to stop the tree-cutting that we so deplore in the tropics is provided by the now familiar case of the northern spotted owl and the old-growth forests of the Pacific Northwest. This unprotected portion of the remnant forest will be completely gone in another 10–15 years at the present rate of cutting.

Loggers and environmentalists literally have been at loggerheads with each other over the fate of the owl and the forests. At the heart, it is a social issue of economics and jobs. The only way to save the spotted owl is to save the forest, as much of it as possible, and this means shutting down logging in much if not most of the remaining unprotected old-growth stands. This not only pits the timber industry against the environmental community but also pits the two against the government and government agencies against each other.

The timber industry and its political allies in government would reduce the argument to the simple choice between the survival of a bird and the survival of thousands of people, their jobs, livelihoods, and way of life. The hour is already too late for the bird, some argue, because with as few as 4,000 spotted owls remaining it is doubtful that the population is still above the minimum viable size for indefinite survival. In any event, to put the spotted owl above the people is elitist, serving only the interests of the radical few, industry would say.

The issue is far broader, however. The spotted owl is but a symbol, a flagship species speaking for the ancient forest as a whole, including the bird's cohort of a hundred other species that cannot survive without the forest. Can we save the forest for the trees, for its collective biodiversity? Should the survival of some thousands of jobs that can only last another two decades at best take precedence over the survival of the last remnants of a primeval forest and all its organisms? Continued logging to save jobs really places the survival of short-term jobs over the long-term welfare of an entire nation. Selling a birthright so cheaply for the few seems elitist in the extreme. Surely it would be far cheaper in the long-run to find alternative jobs for those who will be dispossessed than to let them continue on their suicide mission. The real fault lies, of course, with all of us who use and consume wood products and derivatives as though there were no limits.

We, in highly developed countries, tend to worship the womb of the material environment and the technological life-support systems that sustain us (Shetler, in press). We try to consume all, yet save all, and the marketplace would have us believe that we can have it both ways. Because too many of us believe this, our efforts to contribute to the solution often are merely adding to the problem, as when we do things that increase or at least do not decrease the consumption of our natural heritage. The deceptive ambiguity of the market hype is epitomized by the latest "green revolution," the new business-and-industry fad of selling so-called "green" products. The superficial notion of the "green" product as environmentally sound begs more questions than it answers, and it completely distracts from the core issue: how do we cut consumption and, thus, the demand for the disappearing forests and other natural resources that take with them the natural habitats and their species? Until less consumptive lifestyles are adopted in the wealthy nations of the temperate regions, the carrying capacities of the countries in the tropical regions will continue to be overtaxed, and the ultimate fate of the rain forests and tropical biodiversity will be sealed. Given this predicament, it is hard to be optimistic.

Is the biodiversity crisis primarily a tropical crisis? Yes, if we focus on the wealth of the species per se, but, no, if we look beyond to the human factors that are the causes of the crisis. If we do this, we look right into our own backyards here in North America where we seem impotent to stop the raging destruction of the bulldozers and the runaway, throwaway, consumptive lifestyles that at some point will threaten every life-form on earth, including our own.

There are, of course, hopeful national policies and practices that are beginning to influence personal stewardship ethics and behavior. Unless those who are the most advantaged set workable examples, nothing will stop the ever more rapid loss of habitat and species. As scientists and conservationists, we have a long way to go in establishing an adequate base of knowledge on which to act. So far, we lack sufficient unity and organization to marshal the essential manpower and resources to answer even the rudimentary biodiversity questions of what and where. Surely, the place to start is here in North America where the access is easy, the necessary resources are obtainable, and the hour for saving the natural environment is very late. We need to get behind the present frail movements toward concerted national efforts to survey and monitor the continent's biotic resources and to establish a national center and network for biodiversity research, as well as appropriate institutes to deal with policy, research, and practice in conservation biology and environmental protection generally.

WHY WASTE TIME ON THE ORDINARY LIVES AND PATCHES OF NATURE?

In the atmosphere of urgency that has surrounded the issue of disappearing diversity, attention has been focussed, not surprisingly, on the most visibly endangered biotic elements in the environment. This has usually meant the most beautiful, majestic, amazing, cuddly, or otherwise charismatic creatures of the realm. I would not for a minute want to belittle the vital endangered species approach nor underestimate the power of a charismatic species, as symbol and indicator, to carry the weight of a whole habitat or ecosystem. But it is an approach with severe limitations, and its weaknesses become more and more apparent as the pace of loss quickens and the scale of destruction expands.

The endangered species focus is vulnerable on several counts. In the first instance, too much emphasis is placed on the intrinsic worth of the species itself, especially in the public mind. Assigning value is very subjective and can be treacherous when there is inadequate knowledge of, and appreciation for, natural systems. In the calculus of the argument, the species tends to get transformed from an intangible value into a tangible commodity, and suddenly economic worth is weighed in cold cost/benefit terms. Few organisms have the broad general appeal of a bald eagle, grizzly bear, or California redwood to rise above the cost/benefit approach. Few others are able to survive the politics and the economics of commodity analysis; witness the difficulties, for instance, of the northern spotted owl, snail darter, and Furbish lousewort. The endangered species approach can be a biological version of the bait and switch game. Environmentalists bait the public into concern about disappearing species with flagship examples, and developers switch the attention to economics and win the argument.

The second problem, which follows from the first, is that all attention is lavished on a few high-profile organisms quite unevenly distributed in the organic kingdom. Given human predilections, some birds, mammals, and, far behind, higher plants get the spotlight. The great majority of the plants and animals, especially the insects and other invertebrates, that make up the bulk of the biomass and the fabric of the ecosystem remains silent and invisible on the biodiversity stage.

Thirdly, focus on species as commodities, rather than as elements of a natural system and process whose immeasurable value as a natural heritage is largely lost out of context, tends to foster a biotechnological, fix-it approach of *ex situ* conservation and propagation. The unstated message is, transplant the endangered species or bring it under perpetual, managed care, then cut the forest, fill the marsh, or build the dam, and everybody wins. Invaluable as *ex situ* methods of conservation may be as tools for research and for preserving priceless gene pools, such artificial means of maintenance, propagation, rehabilitation, and restoration do not save natural systems and tend to confuse public understanding and divert attention from preserving whole habitats and their species *in situ*. Certainly, heroic propagation and/or rehabilitation measures are sometimes justifiable, and there have been dramatic successes, as, for example, with the peregrine falcon. But they are enormously expensive per species and completely out of the question as a general solution. Gardens, zoos, and their hybrids are not the answer to preserving *natural* systems in any event.

As a veteran of countless local skirmishes to save habitats in their natural state, I know how difficult it is to find "handles" for argument before the various public forums. Every conservationist trying to protect a habitat grasps for an "Endangered" or "Threatened" species, with a capital "E" or "T." When one succeeds, it is a powerful tool, like magic—but also like a lightning rod to attract opposition. Rarely is one so lucky, however. Almost always one is dealing with the unheralded, ordinary lives and patches of nature. Thousands of ordinary wildlands are disappearing from our neighborhoods every year, especially to suburban subdivisions, while we stand defenselessly by.

Extinction begins with local extirpation in the ordinary corners of our lives, and for local endemics extirpation *is* extinction. General extinction usually is not perceived until it is too late, but the tragedy accrues in increments. A species may die a thousand deaths across its range before disappearing (Shetler, in press). A declining species dies out and its range shrinks to the point of no return inexorably but imperceptibly over unmeasured time. One person's weed is another person's rare wildflower, and to the one who is about to lose a flower from his or her environment that flower is no less precious just because it is common somewhere else. We must begin to look at extinction as a never-ending gradient and find ways to put drama and meaning into the local scene, where it all begins. Like a great gathering river, the biotic bleeding will become a hemorrhage that drains away the biotic lifeblood of continents. Hardly a species of North America is safe, and we are inextricably interlocked with other continents, especially South America. The fate of one is ultimately the fate of the other.

It's here at home, in the suburbs and in the greater process of urbanization, where the battle for biodiversity really is being lost, daily, because it is here that society's real values and attitudes are being played out. Public ignorance and prejudice about the natural realm are colossal, and the economic forces of our lives are overwhelming and governing. In the subdivision is where a whole new conservation ethic is desperately needed if there is to be any hope. We need an ethic that gives ecological value a competitive edge or at least parity with economic value, as determined by the usual density-of-development value of land. This means better criteria and mechanisms for weighing private against public interests. Overriding all prescriptions, however, is the requirement for biologists and ecologists to do a far better job than before of educating the general public about biodiversity and ecosystems.

SHOULD BIODIVERSITY BE MANAGED?

Bill McKibben, in his recent book (1989), speaks of "the end of nature." I have called it the "Postnatural" era in America in my recent article on the "unnaturalizing" of America (Shetler, in press). What we are speaking of is a planet and a continent so dominated by the human species that unadulterated nature has no place to hide beyond the reach of man. Even the weather and the seasons are no longer tamper-proof, as we are learning from the global warming alarms. Postnatural America is an America of ever more disturbed natural systems and a growing range of artificial systems; it is an America with a nature of synthetic landscapes, fabricated biomes, and remade ecosystems; and it is an America of managed commons, from "crown-vetched" or artificially "wildflowered" roadsides to highly engineered and controlled parks and wilderness areas. There are no sharp boundaries, only a progressive attenuation of nature primeval and a growing human tyranny.

The mind of Postnatural America, conditioned by the triumphs of science and the self-conscious attempts of modern environmentalism to restore original nature, too readily accepts as real the many forms of counterfeit nature. The great natural commons of our past—the forests, prairies, wetlands—are being destroyed or desecrated. More and more we turn of necessity to the artificial, managed commons of our future, from zoos and gardens to reconstructed ecosystems and theme parks. From the point of view of basic science, especially evolutionary biology, they all thwart and falsify historical process and cut us off from the natural, evolutionary history of the land.

With this backdrop, it is not surprising, therefore, if there are mixed feelings and some skepticism about the most prescriptive and managed approaches to conservation biology.

Should biodiversity be managed? Once again it is a matter of goals. If for example, the aim is to save genes or simulate natural systems, then biodiversity management makes sense and may be essential. If the aim is to preserve substitutes of natural systems, then surely management is a desperate last resort, perhaps having amenity value but very dubious scientific value. The natural landscape is a patchwork of interrelated histories— geological and environmental, evolutionary and ecological. There are histories of the land, the species, and the communities; histories of migrations, dispersals, invasions, human disturbances, extinctions. Thus every patch of nature tells a story of then and now; every patch is an expression and record of a long evolutionary stream of history. The extirpation or extinction of a species forever destroys a strand of history. Though the species is reintroduced to the same habitat, the continuity is broken. It now is a species devoid of history. Likewise, when a foreign species is deliberately added to the habitat, it adulterates the natural process and order and thereby dilutes, distorts, or even falsifies the natural history of the habitat. Of course, there is no absolute purity. Virtually all seemingly primeval landscapes have been compromised by human disturbance and artificial manipulation; virtually all natural histories have been garbled by human histories.

"A natural ecosystem is more than the sum of its parts, especially in its historical continuity, and it cannot be concocted like a cake from a mix of set ingredients," I wrote recently in connection with butterfly conservation (Shetler 1990). "Sowing, transplanting, or releasing plant and animal species indiscriminately across the land is neither sufficient nor necessarily wise as a conservation strategy." Surely, the urge must be resisted to turn the whole world into a zoo or botanical garden without traceable history. Even a native *wild*flower is an exotic if it is purposely planted, as, for instance, in beautifying roadsides.

CONCLUSION

There are two keys to the future of biological diversity on the planet: 1) saving natural habitat, and 2) educating the younger generations. If there is a universal strategy for saving species for posterity, it is the preservation of habitat on every hand. To achieve this, people must understand diversity and believe in its values and the wisdom of preserving it. The public is largely illiterate about biodiversity because the schools teach little or nothing about species and natural history. A child can go through 12 grades and literally not learn how to tell one plant or animal from another. How can anyone who is ignorant of the diversity in nature be expected to understand, appreciate, and defend it? What is needed is old-fashioned natural history training, not merely science education. The pitfall of science education is that it usually is devoid of real contact with nature and tends to engender blind faith in scientific progress. To society at large, science too easily becomes a paradigm or philosophy for manipulating nature to achieve directed ends.

LITERATURE CITED

McKibben, Bill. 1989. *The End of Nature.* Random House: New York.

Shetler, Stanwyn G. 1990. Butterfly Gardening and Conservation. Pp. 107–109. *In: Butterfly Gardening: Creating Summer Magic in Your Garden* (created by the Xerces Society in association with the Smithsonian Institution). Sierra Club Books: San Francisco. 192 pp.

———. In press. Three Faces of Eden. *In:* H. J. Viola & C. J. Margolis (eds.), *Seeds of Change: Quincentennial Commemoration.* Smithsonian Institution Press: Washington, D.C.

Wilson, E. O. (ed.) & Frances M. Peter (assoc. ed.). 1988. *Biodiversity.* National Academy Press: Washington, D.C. 521 pp.

Designing and Testing Sampling Protocols to Estimate Biodiversity in Tropical Ecosystems

Jonathan A. Coddington, Charles E. Griswold, Diana Silva Dávila, Efrain Peñaranda, and Scott F. Larcher

Abstract. Sampling methods to estimate total species richness of a defined area (conservation unit, national park, field station, "community") will play an important role in research on the global loss of biodiversity. Such methods should be *fast*, because time is of the essence. They should be *reliable* because diverse workers will need to apply them in diverse areas to generate comparable data. They should also be *simple* and *cheap*, because the problem of extinction is most severe in developing tropical countries where the scientific and museum infrastructure is often still rudimentary. In the past, two scientific fields have been mainly responsible for providing such data: systematics and community ecology. Simplistically summarized, samples collected by systematists better represent the species richness in an area but are intractable statistically, whereas the analysis of samples collected to answer ecological questions is usually straightforward but often poorly represents the total fauna. The two approaches are complementary and we propose a set of methods that seeks the optimal compromise. We applied these methods to sample and estimate species richness of Araneae in Bolivia, but we present criteria so that the methods can be modified for other broadly similar groups. We describe the methods used, present preliminary analyses of the effect of four variables (site, collecting method, collector, and time of day) on total number of adult individuals taken, and discuss analytical approaches that employ such data to estimate total species richness. We also present data from Peruvian canopy fogging samples to show that estimates of species richness of diverse tropical arthropod taxocenes are obtainable in principle.

INTRODUCTION

The alarmingly rapid extinction of species currently underway is mainly a tropical phenomenon. While many authors have decried this loss and suggested sweeping policy initiatives to reverse the trend (Wilson & Peter, 1988), rather less attention has been given to the development of pragmatic inventory methodologies that will aid in addressing the problem. Obviously the cause of accelerated rates of extinction is complex and its amelioration requires diverse approaches. However, we presume that fast, reliable, and comparable estimates of total species richness in areas of interest will be important data on which to base conservation decisions, allocation of resources, and land use planning. Relative richness of endemics might be an even more useful datum, but it is far more difficult to obtain. Scientists traditionally involved in the description and inventory of communities

Drs. Coddington and Griswold are with the Department of Entomology, National Museum of Natural History, Smithsonian Institution, Washington, DC 20560, USA. Dr. Dávila is with the Museo de Historia Natural de la Universidad de San Marcos, Av. Arenales 1256, Apto. 14 0434, Lima 14, Perú. Dr. Peñaranda is with the Instituto de Ecologia, Casilla 10077, La Paz, Bolivia. Please address correspondence to Dr. Coddington.

and biotas can make significant contributions by designing and testing such sampling protocols.

In the past, estimating and describing total species richness of an area has been the purview of two scientific disciplines: systematics and community ecology. Systematists, particularly those associated with museums, are often expert in efficient inventory techniques, where "efficient" basically means maximizing the number of species discovered in a given sampling area per unit effort. Usually museum personnel sample as many habitats as possible, switching emphasis as soon as a good series of each species is obtained. Some museum personnel use quantitative techniques in which quantity of habitat sampled and sampling effort are monitored, but many do not. Results of such latter efforts, hereafter termed "museum collecting," is often no more elaborate than a list of species encountered. Not only relative abundance but the proportion of the fauna as yet undiscovered remains unknown and unknowable through museum collecting as it is often implemented (at least within any reasonable time frame for any reasonably diverse arthropod fauna). Moreover, museum collecting also often fails to take account of two variables fundamental to any quantitative sampling program: sampling effort and the possibility of biased representation of the true relative abundances of species in the area or community sampled. Museum collections doubtless underestimate the true relative abundance of the most common species.

For example, the seminal paper on estimating species richness was partly inspired by a museum collection of Malaysian butterflies (Corbet, in Fisher et al., 1943). R. A. Fisher, who invented the log series in order to cope with these data, evaded the question of biased relative abundance by *assuming* the collector was unbiased in his sampling behavior, but had stopped after collecting 20 individuals of each species. However, he could only ignore the problem of differential sampling effort (e.g. by date, taxon, or microhabitat) because no data were available. As such the collection of butterflies had to be regarded as a single large sample; the potential analytical insight and power that would have resulted from having a series of replicate samples representing equivalent sampling effort was lost.

Accurate estimates of total species richness in complex habitats have not received the attention they deserve from community ecologists. It must be admitted that the total number of species present is an abstract and perhaps not very interesting quantity, ecologically speaking. That situation will probably change, given that species richness is fast disappearing, that the trend can be slowed but not stopped, and that in the complex nexus of "what to save, when, how, at what cost," an important datum, species richness or biodiversity, is typically lacking.

Although the work of Preston (1948 and later papers) and, especially, C. B. Williams (1964) showed that total species richness could be estimated, most workers of the time were more interested in what the results implied about ecological process than in the estimates of that bald number. Interest in this approach waned after May (1975) suggested that the ubiquitously observed lognormal distribution of species abundances was due not to ecological process or law, but mostly to the "law of large numbers" or the central limit theorem. Be that as it may, and regardless of later arguments that the lognormal distribution is not entirely artifactual (Sugihara, 1980), the generally good fit of the lognormal to empirical data remains a striking observation (Taylor, 1978; Magurran, 1988), although largely untested in tropical ecosystems. If anything, the explanation of the fit as an effect of large numbers is a prescription, not a proscription, of its use for estimating total species richness in tropical ecosystems.

From the point of view of practical work on biodiversity, however, this body of ecological theory and method has several limitations. First, one ought to be as interested in the confidence interval associated with any estimate of total species richness as with the estimate itself. However, an analytical derivation of the confidence interval associated with the lognormal estimate of species richness is still lacking (Pielou, 1975). Bootstrapping techniques may help, although we know of no implementations of the boot-strap easily applicable to such species abundance distributions. Other theoretical distributions, such as the negative binomial, provide such confidence intervals, but generally do not fit the empirical data as well. Completely different methods, such as the non-parametric jackknife (Heltsche & Forrester, 1983) are also available, provide confidence intervals, and show promise. The jackknife thus far has been applied only to quadrate based sampling methods. Its extension to plotless methods remains untested. Finally, one may simply plot the accumulation of species as a function of sample number or individuals taken. While straightforward and revealing, this method is ad hoc and it is unclear how to estimate the asymptote and confidence interval. However, fitting such curves to models that have an asymptote (e.g., hyperbolic functions) may be justified. If the assumption of applicability is made, the confidence interval could be estimated by a least squares procedure. Because the set of analytical approaches is still fairly small and relatively untried on tropical data, we advise the use of all techniques applicable to a given data set. After all, convergence of estimates based on completely different theoretical approaches might be evidence that each (or all) is measuring the same quantity. Tractability to different analytical approaches should be a chief design criterion for the sampling protocol. By this we mean, for example, that merely because the lognormal can be fit to data viewed as a single large sample, large samples should still be composed of many smaller replicate samples. Applicability of the lognormal is not thereby sacrificed, and other methods that do assume replicate samples, such as the jackknife or species accumulation curves, become applicable.

Despite the feasibility of analysis, ecologically-oriented sampling has some drawbacks. The emphasis on statistical tractability predisposes workers to prefer methods of rigorously controlled and known bias, such as automated traps, in rigorously defined sampling units, such as plots or quadrats (Southwood, 1978). Automated methods may minimize variability but they can also sample taxa differentially or be unsuitable for some taxa. In addition, setting up, using, and maintaining plots in remote tropical field sites that may be visited only once (typical for systematists) is time-consuming, difficult, and expensive. Therefore the need exists to develop plotless sampling techniques that are better suited to the salvage or "hit-and-run" style of inventory work that the biodiversity crisis is likely to necessitate.

In sum, therefore, several considerations suggest a blend of these two traditional approaches to estimating biodiversity. We perceive the following as important criteria for sampling protocols:

1. The usual complement of collecting methods used by museum personnel involved in inventories (whatever that may be for particular groups), should be modified as little as possible in order to yield analytically tractable data. We start with museum techniques because museum personnel have the most expertise in actually accessing or sampling the total fauna in a region. In the decade to come most of the inventory work will probably be initiated by museums or analogous institutions.
2. The number of collecting methods used should be as few as possible, and the classification of microhabitats should be as simple as possible, in order to minimize the complexity of the sampling protocol. Methods should be chosen based on their efficiency and independence (= low overlap with) other methods. For the purpose of estimating total species richness, microhabitats that are especially difficult to sample yet productive of relatively few species can probably be ignored except in exhaustive inventory projects.
3. The sampling protocol should work well in both plot-based and plotless sampling situations. Time spent sampling may be a good choice as the measure of the sampling unit because of its simplicity and universality. While time is not the most

exact measure for all collecting methods, it is appropriate for many favored by systematists and is easy to measure in the field. In those methods for which another unit is better suited, such as discrete number of events or area, the sample size of these can be at least be adjusted so that total sampling effort (e.g., hours spent doing it) is in some sense comparable over all methods.

4. The sample unit should be large enough to yield adequate numbers (individuals and/or species taken) within samples, but small enough so that the total number of samples taken is large enough for between-factor statistical comparison.

5. Data should be collected so that variation can be estimated and analyzed. Contributing factors at a minimum include: site, individual collector, time of day (and season), and sampling method. Although inventory work is not undertaken to test hypotheses about the effects of these factors on variation in number of individuals or species taken, these exploratory analyses do reveal the quality and structure of data used to achieve estimates of species richness, especially if the analytical method assumes replicate samples.

6. Data on numbers of individuals and species taken should be able to be combined to produce species abundance distributions. These can be used to estimate species richness in three broad ways. They can be fitted to appropriate parametric distributions (negative binomial, lognormal); they can be subjected to the non-parametric jackknife estimate; and they can be used to produce species accumulation, or collector's, curves.

7. Finally, if possible, the analytical approach should yield confidence intervals on the estimates. Although the confidence interval on any empirical estimate is an important scientific consideration in its own right, it is especially important in the present context in which a rough but fast and reliable method to estimate a complex parameter is sought.

In this report we present results from a preliminary trial of the above principles to estimate species richness in a variety of tropical evergreen forests. We focused on arthropods because they comprise nearly all the species richness of any tropical evergreen site and specifically on spiders (Araneae). Besides the obvious fact that we are all araneologists, spiders are a good choice for several reasons. First, they are poorly known taxonomically, and thus typical of many extremely diverse groups. Second, they are among the six or seven most diverse ordinal taxa in terms of species described or estimated to exist (Coddington, 1990, Coddington et al., 1990). As a rough guess, a hectare of typical tropical forest probably supports 300–800 spider species. More than 1000 genera occur in the Neotropics, about a third of the known world genera. Third, spiders must generally be caught by hand, in contrast with other groups such as mites, Lepidoptera or parasitic Hymenoptera for which efficient mass sampling techniques have been devised. Fourth, spiders are generally a good mix of apparent and cryptic species—not so apparent as butterflies, nor so cryptic as centipedes. Fifth, as obligate and abundant arthropod predators near the top of the invertebrate food chain, spiders are an important component of any ecosystem. Insofar as much of the work on tropical biodiversity has concerned herbivorous arthropods or groups of mixed feeding strategy (Farrell & Erwin, 1988; Erwin & Scott, 1980, Erwin, 1979; Broadhead, 1983; Mound & Waloff, 1983), data on obligate carnivores may help to round out the emerging picture on the distribution of species richness across body size and trophic level (May, 1988).

48

METHODS

Each Bolivian sample collected was classified according to four main factors:

Site

We collected in three sites: the Estación Biologica Beni (Biosphere Reserve) (elevation 100m: ca. 14°47'S:66°15'W); a site near the intersection of the Rio Tigre and the La Paz-Coronavi road (500m: ca. 15°23'S:66°59'W); and a forested site on the slope of Cerro Uchumachi, near Coroico (1900m: 16°15'S:67°21'W). These are all evergreen tropical forest sites. A priori we expected maximal species diversity at the intermediate elevation.

Collector

Four of us are experienced field araneologists, although our particular interests range from fossorial mygalomorphs through vagabond hunting spiders to minute web spinners. The fifth, although an experienced field worker, knew much less about the particular techniques and relevant aspects of spider natural history. A priori we expected a significant effect on numbers of individuals taken due to collector.

Time of Day

Naturalist tradition states that many spider species are nocturnal. On the other hand, spiders resting by day are not uncommon. Consequently, we sampled during the day (roughly 0700–1200) and during the evening (1900–2400). A priori we expected more individuals and more species during the evening.

Method

Our initial classification of methods and microhabitats was elaborate (Fig. 1, above). We designed it to access all relatively diverse components of the fauna in all relatively accessible microhabitats. Method and microhabitat were matched so that the sampling protocol would yield a relatively complete picture. The basic sampling unit was one hour of constant collecting, timed with a stopwatch. Collectors agreed to take all putatively adult spiders encountered, without exception, during that hour. "Adult" meant any animal that was or might possibly be adult; only if the collector was certain that the animal was juvenile was it skipped. Spider taxonomists widely agree that con- or heterospecificity can only be judged reliably from sexually mature animals. Skipping juveniles makes collecting more efficient; also they are practically impossible to identify reliably. The use of the stopwatch meant that a given sample need not be strictly continuous—one could turn off the watch to attend to personal details, converse, eat, move from one area to another, etc. Therefore, each sample represented one full hour of attentive application of a particular method.

Using stopwatches provided another benefit. A collector with two stopwatches can use at least three methods simultaneously—two hour-based samples, and a technique based on counts or area. Collectors were free to switch between methods, as long as the tally for each was independent. Being able to choose is important because the collector can thereby maximize the number of species sampled; it helps to maintain the efficiency inherent in museum collecting. However, it may introduce a bias if collectors quit when the hunting is poor.

After two to three days in the field, however, it became clear that the initial habitat classification was inefficient and overly complex. It was inefficient because one might be condemned to spend several hours searching in and for small holes, which might happen to contain nothing of interest. It was too complex to maintain a fully factorial design, as a

Traditional methods	Microhabitats:
Hand-searching	1. Herb layer
Beating trays	2. Shrub layer
Sweep nets	3. Tree bark/surface and beneath
Pitfall traps	4. Leaf litter
Litter sifting/extraction	5. Big holes (burrows, hollows, streambanks)
Bark/log fragmentation	6. Little holes (tubes in soil)
	7. In/under logs, rocks
	8. Forest canopy

Looking up.—Hand searching	accesses 1–3.	Unit = 1 hour.
Looking down.—Hand searching	accesses 4–7.	Unit = 1 hour.
Beating foliage.—Beating tray	accesses 1,2.	Unit = 25 beats.
Litter sifting.—Funnel/sheet	accesses 4.	Unit = 2 sq. meters

Figure 1. (Above) Initial match of collecting methods against classification of microhabitats for sampling of Araneae. (Below) Methods that turned out to be practical in the field, and their relation to the initial design.

collector had to employ too many different methods in different places at different times, and get replicates of each. Since collectors rarely spend more than, say, three weeks at a given site, the entire protocol should be simple enough to be carried out within that time. Thereafter we restricted our methods to four (Fig. 1, below):

Looking up. This calls for hand-picking while the collector moves slowly along on his or her feet, searching any vegetation or structures above knee height. It roughly translates to collecting while walking about.

Looking down. This calls for hand-picking while the collector is on his or her knees or stomach, intensively searching the soil, leaf litter, forest floor debris, and shortest vegetation. It roughly translates to collecting while crawling about.

Beating. A beating "event" was defined as placing a standard sized tray (ca. 0.5 m²) beneath a suitable unit of vegetation (branch, bush, sapling, vine, etc.) and tapping it until no more spiders fell down. Twenty-five such events constituted one sample. A beating sample required about an hour to complete, and so required about the same amount of collector effort as methods 1 and 2.

Litter sifting. Although we utilized this method infrequently, it is doubtless an essential way to access the spider fauna. We gathered up 2 m² of litter, and sifted and sorted it on a white sheet to find the spiders. The area sampled required on average about an hour's effort for one person. Results from this method are not compared here.

Because none of these methods is plot-based, we cannot give accurate figures for the area or volume of habitat sampled per sample unit (except, of course, in the case of sifting). Sampling Araneae is slow work, and in the course of an hour with methods 1–3, no one covered much more or much less than 50 linear meters of ground in the areas we sampled. Assuming that a person samples one meter on either side of that line, each sample represents perhaps 100 m². These figures are offered as a rough comparison only, and the precision of the method could be improved if one took the time to get a rough measure of the area covered. Certainly we did not collect every individual in the area sampled, so our results are lower bounds on the abundance or diversity of the area sampled by each sampling unit. For each site the total area sampled was probably about a hectare.

The Peruvian canopy samples were collected according to methods described by Erwin and Scott (1983) from the Tambopata Reserved Zone and Manu Reserved Zone, Madre de Dios, Perú. Two samples are analyzed, one from an isolated Manu canopy

50

including two individual trees of different species and several liana species, and a second series of five Tambopata canopies including a variety of tree and liana species. In the former, all animals that fell from the canopy were collected but in the latter a series of 45 funnels (1 m^2) were arranged to subsample the falling arthropods. Although these five canopies were from distinctly different forest types, in order to investigate analytical techniques we are treating them as a series of 225 replicate samples.

Analysis

For the Bolivian samples, two variables, numbers of adult individuals per sample and numbers of species per sample, can be analyzed in much the same way. Here we present preliminary analyses of the first variable only. We used SYSTAT version 4.2 (Wilkerson, 1988) for statistical analysis and graphs. We emphasize that we are not trying to test hypotheses about the effects of various factors on numbers of individuals per sample, but rather simply to explore the structure of the data that will be used to estimate richness. In particular, one can anticipate substantial variability between samples, given our imprecise definition of sampling methods and units. The Peruvian data was fitted to a lognormal distribution according to the procedures outlined by Ludwig and Reynolds (1988). Jackknife estimates of species richness were computed using equations presented in Heltsche & Forrester (1983).

RESULTS

In the 35 day Bolivian expedition, we spent roughly ten full days collecting (Table 1). The remaining time was taken up with obtaining permits and visas, obtaining field equipment and transportation, getting to or moving between sites, maintaining vehicles and equipment, and dealing with various emergencies. We used eight days at the end of the trip to label and organize samples adequately, and to sort the samples to family. Although one could have spent those days in further sampling, we felt it was more efficient to get the bulk of the sorting and labelling done, rather than leaving it until some indefinite future date. Experience bears this out; while we knew in eight days how many adults of each family had been collected and had collected the abundance data for the factorial design, seven months later we still have not completed allocating the specimens to species. There is a considerable delay in sorting once collections join a museum "backlog." Sorting at the site or before the end of the trip may be worth the extra cost. The Manu canopy sample, which comprised roughly 95,000 arthropods (T. L. Erwin, pers. comm.) required roughly 23 person-days to sort but included only about 2% spiders (calculated from Erwin, 1989). The five Tambopata samples came into our hands already separated from the rest of the canopy fauna so we have no data on time required to sort it to that level. However, for the five canopy collections, sorting the samples to morphospecies required about 14 working days, collating and relating those morphospecies (= developing final morphospecies concepts) required another 23 days, and data entry and proofing required an additional seven days, or about 44 days total on top of field time and the initial sort to order.

Numbers of samples and adults taken at each Bolivian site are summarized in Table 1. We averaged about three samples per person per day; the average number of adults per sample was 16.4 ± 10 including all sites visited. Some of the samples were taken at sites or by methods not discussed here, and so subsequent analyses treat at most 126 of the total 184 samples.

Figure 2 presents box and whisker plots of the number of adults taken by site, collector, method, and time of day (n = 126). Significant effects on abundance due to collector, method, and time of day are apparent.

Table 1. Summary of numbers of samples and number of individuals taken by site.

Site	Days	Samples	Samp./Pers./Day	Ind./Sample	Total Ind.
El Trapiche	4	67	3.4	17.42 ± 10.4	1167
Rio Tigre	4	72	3.8	15.47 ± 9.7	1114
Cerro Uchumachi	2	45	4.50	16.38 ± 10.1	737
Summary	13	191	3.9	16.4 ± 10.0	3018

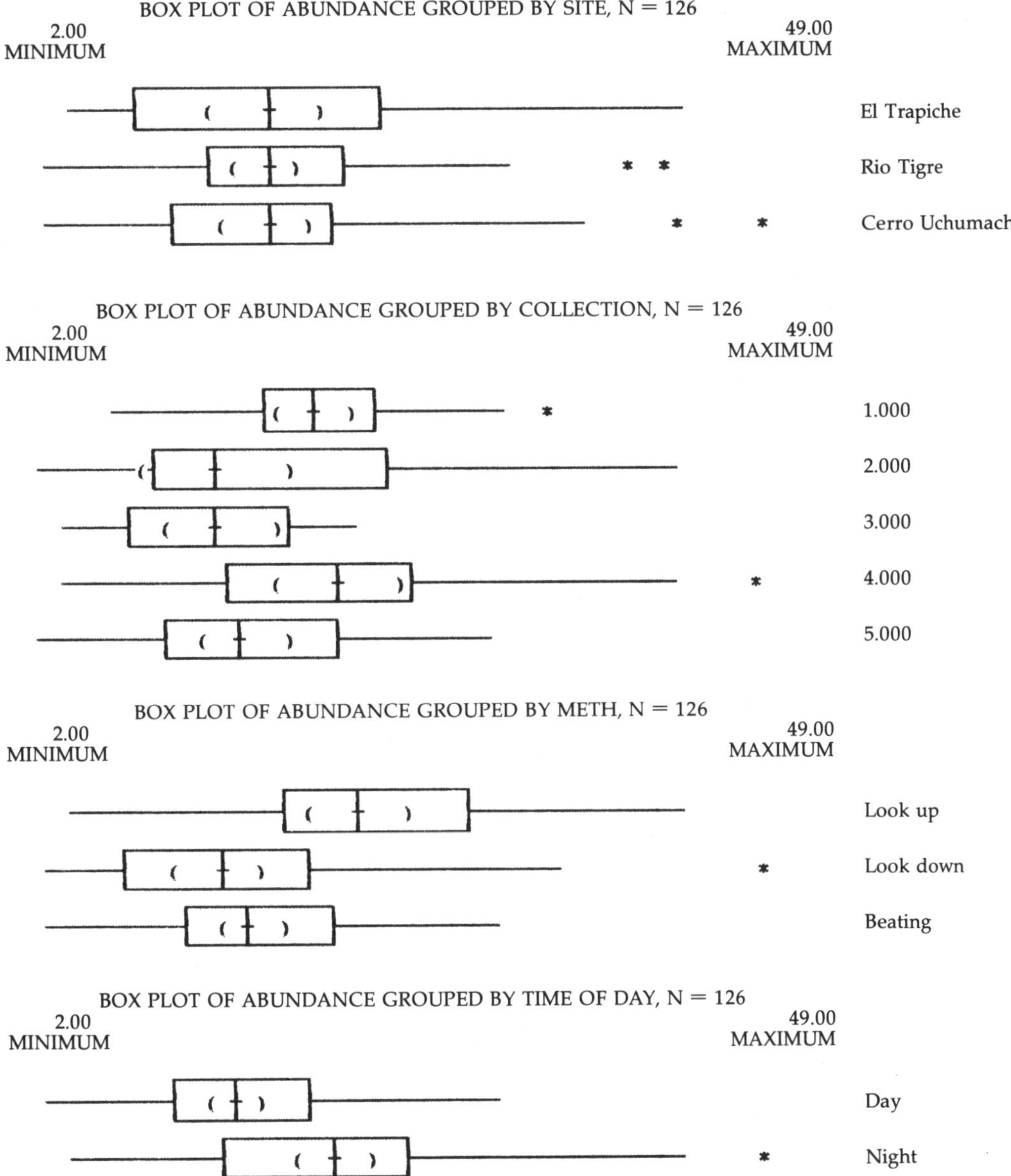

Figure 2. Box and whisker plots of number of adults collected, grouped by site, collector, method, and time of day. Vertical lines marked by "+" are medians, left and right sides of boxes are first and third quartile boundaries, "whiskers" on each side encompass the extreme quartile plus 1.5 times the middle interquartile spread. Parentheses define the 95% confidence interval about the median. Outlier points lying beyond the whiskers are indicated by a "*" and far outliers by "0."

52

Figure 3 presents box and whisker plots of the number of adults taken by each collector, grouped by site. Collectors 1, 2, and 4 did not appear to differ significantly. Collector 3 (the less-experienced person) appeared significantly different at site 1 and 2 but not at site 3. Collector 5 appeared significantly different at site 3, but not 1 or 2.

Figure 4 presents box and whisker plots of the number of adults taken by each method, grouped by site. *Looking up* was most productive at all sites.

Figure 5 presents box and whisker plots of the number of adults taken by time of day, grouped by site. Collecting at night appeared more productive at all sites, although when grouped in this way the difference is not significant at any site.

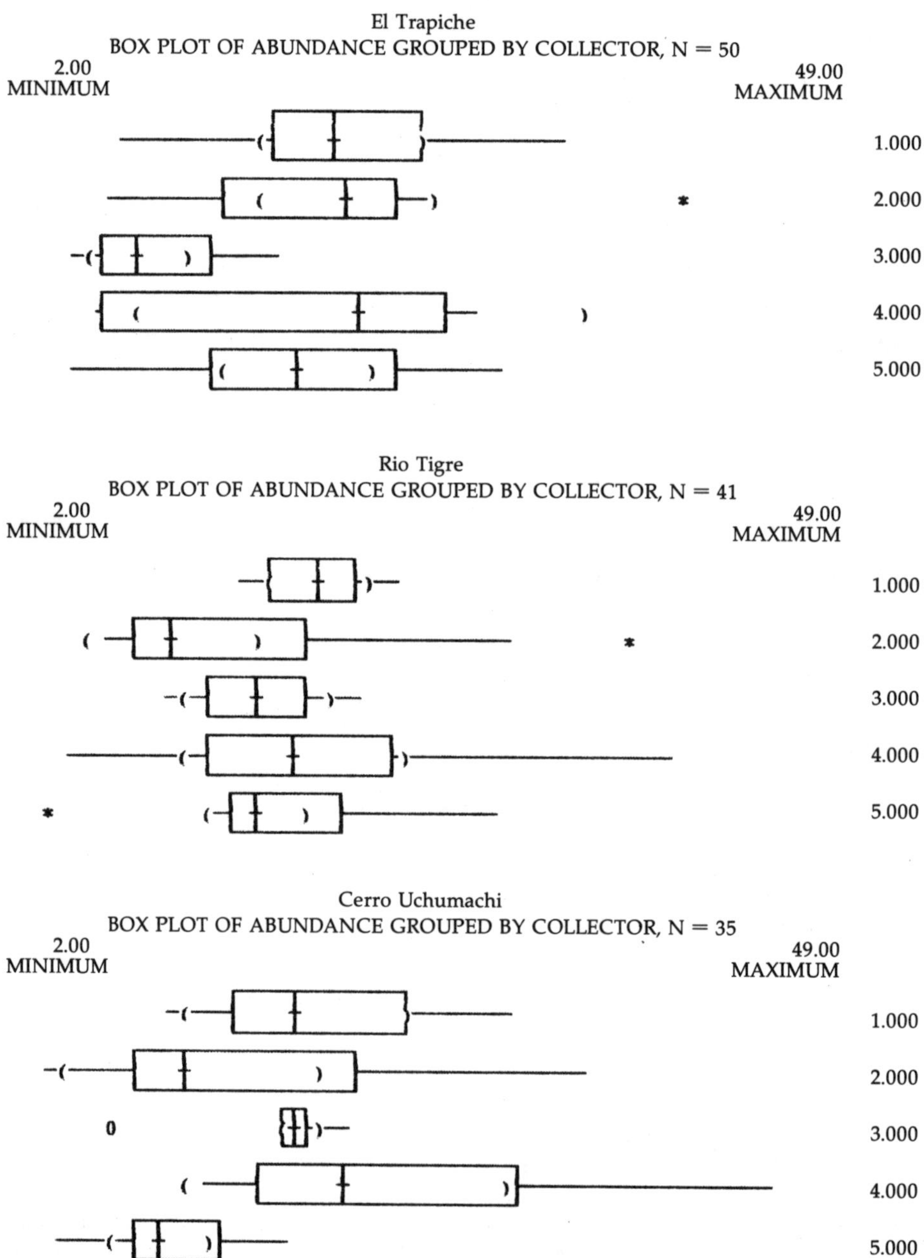

Figure 3. Box and whisker plots of number of adults collected by each collector, grouped by site. See Fig. 2 legend for explanation of plot.

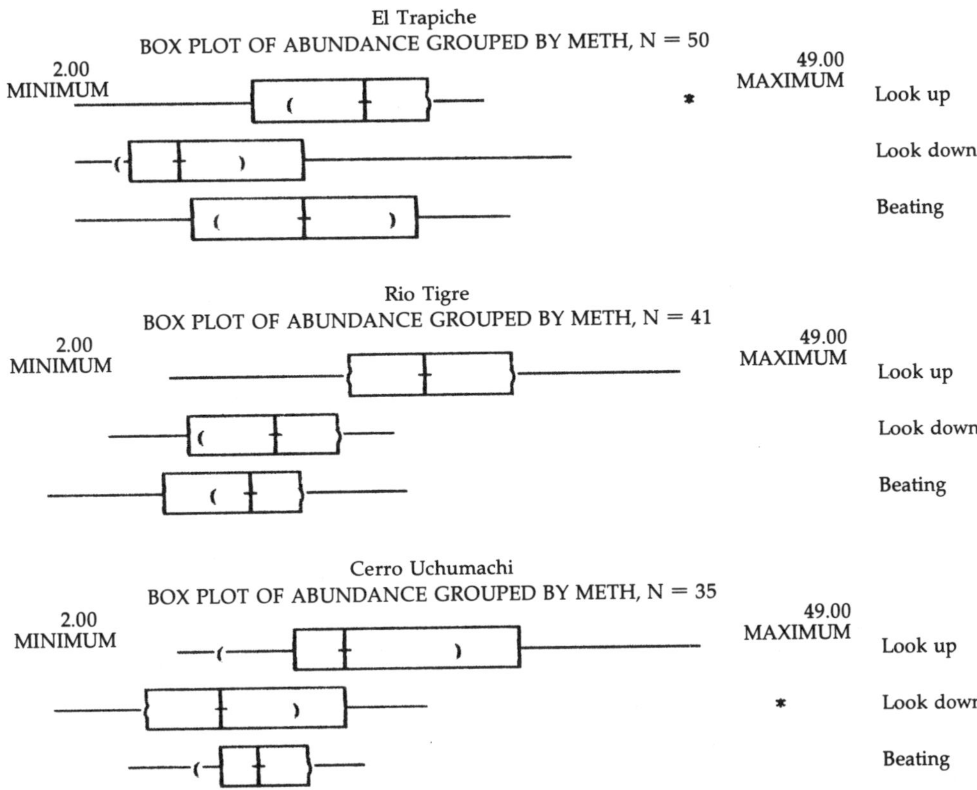

Figure 4. Box and whisker plots of number of adults collected by each method, grouped by site. See Fig. 2 legend for explanation of plot.

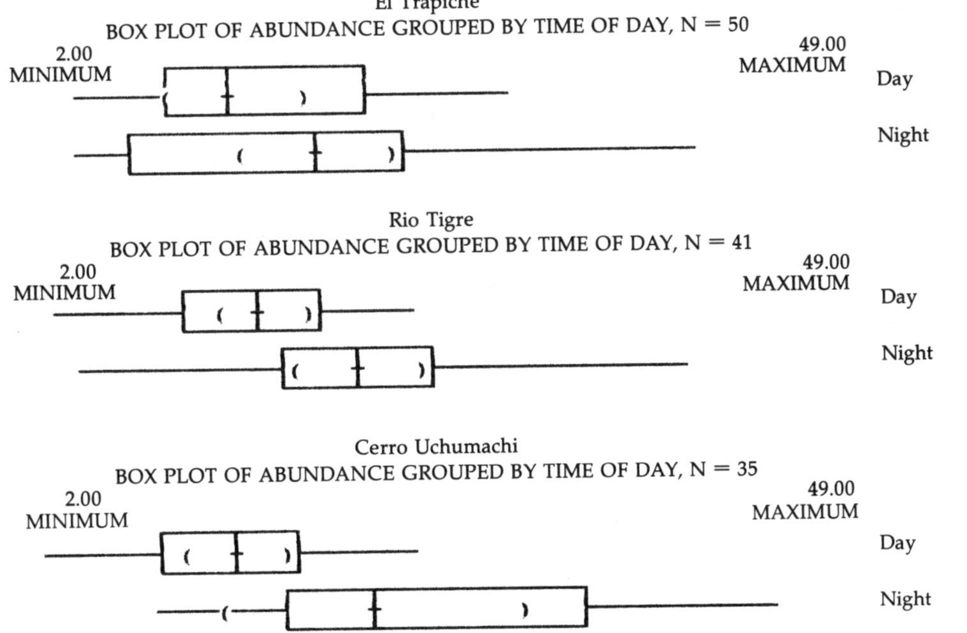

Figure 5. Box and whisker plots of number of adults collected by day and by night, grouped by site. See Fig. 2 legend for explanation of plot.

54

Inspection of the data presented in Figures 3–5 suggested that some of the data was heteroscedastic, especially that grouped by collector (Fig. 3). Bartlett's test for homogeneity of variances on the data grouped by each main factor confirmed significant heterogeneity in data grouped by collector (p< 0.001), and time of day (p< 0.000). In addition, most plots of grouped data disclosed some outlier points (more than 1.5 times the interquartile distance from the median value). Seven samples comprising abundances of 2, 3, 40, 42, 43, and 49 adults per sample were responsible for these outliers. A histogram of the abundance data confirmed a longish left tail and a group of unusually productive samples at the right-hand tail. The highest abundance sample was taken by collector 4 from the lower surface of a wind-thrown tree trunk during a bout of *"looking up"* at night. This habitat differs sufficiently from that normally encountered during *"looking up"* collecting to be regarded as unusual or different. Perhaps the other unusual samples were also the result of collecting in "aberrant" microhabitats. Deletion of these data points eliminated most of the outliers, and the heteroscedasticity in the data grouped by collector (p< 0.156), but not that due to time of day (p< 0.048). Subsequent box and whisker plots of this reduced data (n=119) set did not change most of the results. Significant differences still appeared at El Trapiche and Rio Tigre between collectors 1 and 3, and between *looking up* and *looking down* at El Trapiche and between *looking up* and *beating* at Rio Tigre. Tukey HSD tests on the abundances grouped by collector, method, and time of day (as in Fig. 2) confirmed that the only significant difference was between collectors 1 and 3 (p< 0.013), that *looking up* was significantly more productive than either of the other two methods tested (p< 0.000 from *looking down* and p< 0.011 from *beating*), and that, overall, collecting at night was more productive than during the day (p< 0.017).

Figure 6 presents the observed species abundance distributions for the two Peruvian canopy samples. Raw abundance distributions were transformed logarithmically (log base 2; Preston, 1948; Ludwig & Reynolds, 1988), and plotted as bar charts. Overlaid on each analysis is a curve connecting the expected values for each octave of abundance under the lognormal model.

The Manu single canopy sample (Fig. 6, above) included roughly 95,000 arthropods (Erwin, pers. comm.). From it 613 spiders (222 adults, 391 juveniles) were classified among 187 species. Because the sample arrived as a single collection rather than a series of equivalent samples, the jackknife could not be applied. The range of abundances spanned seven octaves, with an observed mode at 63.5 species.

Expected numbers of species per octave were obtained in two ways: a Chi-square goodness of fit and a least square fit to the lognormal equation ($S[R] = S^0\exp[-a^2R^2]$). Because the Chi-square tests the fit to the lognormal model, we used a non-linear estimation program to minimize the Chi-square ($[O-E]^2/E$) as a criterion for the best-fit estimates of the lognormal parameters. This yielded an expected mode at 54.9 species and a variance of 0.414 ($X^2 =8.19$, df=5, p< 0.1). On this basis, we accept the null hypothesis that the observed distribution is lognormal. The area under the curve, interpreted as the total species richness (S^*), including those species too rare to be included in the sample, is 235 species (calculated as $S^* = 1.77(S^0/a)$; Ludwig & Reynolds, 1988). As pointed out above, no rigorous way exists to compute the confidence interval on this estimate. The number of species observed is obviously a lower "bound." The upper "bound" is unknown, but probably 273 = $1.77(S_{0max}/a_{min})$, where S_{0max} is the upper confidence limit on S_0 and a^{min} is the lower confidence limit on a, is an overestimate (Fig. 6, legend).

The second set of five fogging samples (Fig. 6, below), each from a different group of canopy tree species, is amenable to both the lognormal and the jackknife. In this case, 1834 individuals (354 adults; 1480 juveniles) were classified among 426 species, as above. Nine octaves of abundance were observed.

Judged by Chi-square goodness of fit, the fit to a lognormal distribution is poor (X^2=22.36, df= 7, p< 0.001). Nevertheless, using these estimated parameters, S^* is 560

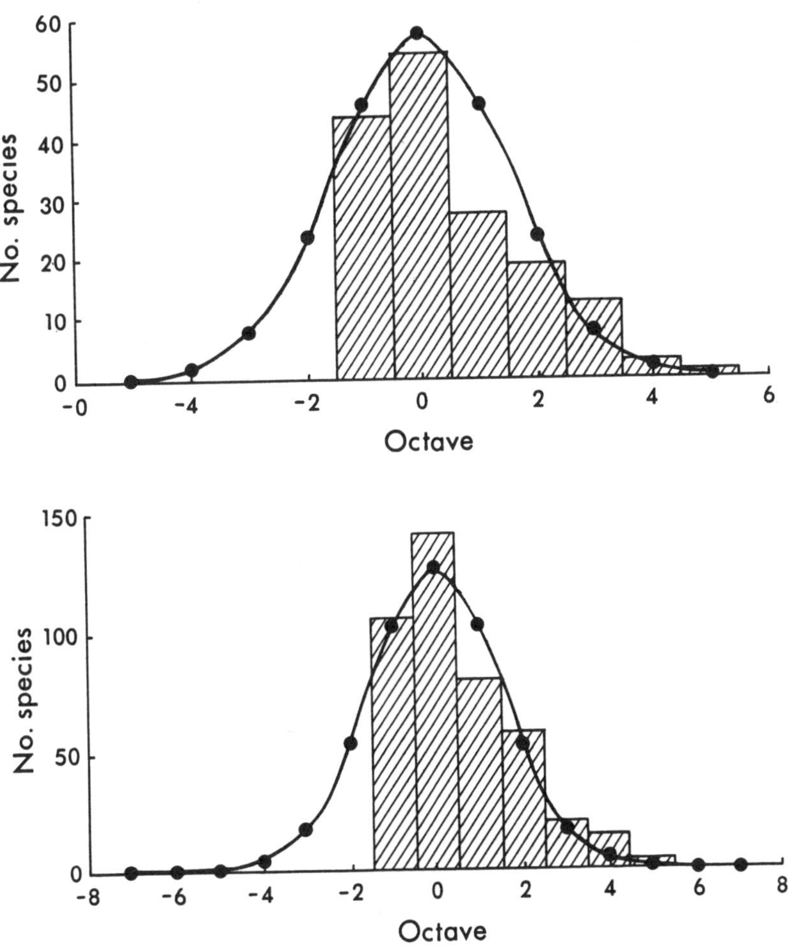

Figure 6. (Above) Estimation of total species richness for fauna of a single canopy. Chi-square estimate of the mean, S.MDSD/0, is 54.9 ± 15.1, and of the variance, a, 0.414 ± 0.082 (X^2 = 8.19, p< 0.1). S* is 235 species (bounds ∿ 142–373 species). Least-square estimates of the same parameters are S^0 = 57.2 ± 16.8; a = 0.473 ± 0.234 (r^2 = 0.91); S* = 214 (bounds ∿ 101–548 species, see text for explanation). (Below) Estimation of total species richness for fauna of five canopies. Chi-square estimate of S^0 = 115.1 ± 25.5; a = 0.366 ± 0.047; (X^2 = 22.36, p< 0.001); S* = 560 species (bounds ∿ 384–780). Least-squares estimates of the same parameters are S^0 = 128.3 ± 21.3, a = 0.469 ± 0.127 (r^2 = 0.96); S* = 484 (bounds = 318–774 species).

species. As above, combining the extreme confidence limits to get a rough idea of the upper bound on the total species richness estimate based on these data, one obtains 426—780 species.

The jackknife estimate ($S^*=S_0$ + k(n–1)/n) works by augmenting the number of species actually observed (S_0) by the number of species (k) unique to any sample, weighted by the number of samples. Its variance (var(S^*)= (n–1)/n$[\Sigma j^2 f(j)-k^2/n]$) is a function of the number of samples f(j) having j unique species. The jackknife estimate for these data therefore yields 647 ± 438 species, which broadly overlaps with the lognormal estimate. It is evident that the estimate of S* from these data is not as precise as that for the single canopy data, which is to be expected given the poor Chi-square fit. If the data are treated instead as five samples (one from each forest type) rather than as 225 samples, the jackknife estimate is 400 ± 1035 species.

56

DISCUSSION

Despite long hours in the field, we averaged rather few samples per person per day. We attribute these surprisingly low numbers to our unit of measurement—the time literally spent searching for animals. Our results may prove typical for similar inventory projects; a surprising amount of time in the field is not actually spent searching for animals.

Fewer significant effects due to collector appeared than we expected. Those effects that proved durable apparently were due to an initial disparity in experience between one collector and the rest. As expected, collectors vary widely in numbers of adults per sample (Fig. 3), and some data points are outliers. These outliers present some analytical problems but, at least in this case, interpretation was not too difficult. Thus, despite our wide variety of taxonomic interests, the experienced collectors performed about equally if judged by total number of individuals taken. Collector 3, the less-experienced individual, got significantly better over the course of the trip (Fig. 3), as one would expect. If typical, this is good news for inventory projects. It suggests that inexperienced personnel in the company of experienced collectors can learn collecting techniques (i.e., become statistically indistinguishable) in a very short time. It also suggests that fairly diverse collectors (at least from the fine-grained point of view of taxonomic speciality), if given explicit instructions, perform about the same. As a cautionary note, the low numbers of individuals taken by Collector 5 at site 3 (Fig. 3) remain unexplained.

Method was significant at all sites and overall (Fig. 4), but not as we had expected. Simple searching in the herb/shrub/sapling zone seems the most efficient technique. Ground-searching (*looking down*) is less effective, but that habitat is less complex, less accessible to humans, and possibly less rich. Beating the former zone was better than searching the ground, but not as productive as searching. A priori, we expected that beating tray samples would be most productive, perhaps because each beating event typically produces multiple individuals. However, judged per hour of collector time, it is not the most productive. The most productive method seems to be *looking up*.

Our classification of methods obviously was coarse. We were satisfied with *beating, looking up*, and *sifting* as a "natural" classification of collecting techniques. They seem to represent consistent collecting behaviors that are repeatable across sites and other variables, given a certain amount of common sense on the part of the collector. "*Looking down*," on the other hand, is heterogeneous. Activities such as tearing apart rotted logs take a lot of time but are not as productive of individuals or species as is hand-searching. However, they yield portions of the total species richness not accessible in other ways. In the future, *looking down* should be restricted to activities exactly comparable to that in *looking up*, and some other category elaborated for different kinds of activities.

On the whole, spiders are more abundant at night (Fig. 2), although the difference is not always significant in reduced data sets (Fig. 5). This increased abundance is probably due both to the nocturnality of the fauna, as well as the simpler and less confusing illumination of a headlamp as compared to dappled sunlight and color during the day.

Surprisingly, numbers of individuals taken per sample at each site did not differ significantly (Fig. 2). Within evergreen tropical forest in Bolivia, spider abundance apparently does not vary much even at sites that vary widely in elevation and vegetation structure. The sampled sites covered an altitudinal range (100 m–2700 m) within which other workers have found significant differences in arthropod abundances. Whether this result will be corroborated by similar studies remains to be seen.

With regard to the expense and efficiency of estimating species richness, our data suggest several interesting points. If one makes the reasonable assumption that the average tropical "site" (however one wants to define site) has about 300–800 species of spiders at any one time, it is possible to calculate best/worst scenarios for how long it would take to get a sample of such a fauna adequate to estimate the total species richness at the site. For the lognormal parameter a > 0.3 in communities of the above size, roughly 10 times as many individuals as species will suffice to estimate reliably species richness, whether by parametric distributions, non-parametric methods, or graphically. According to our results, the best (worst) scenario is that 119(468)—303(1250) samples would have to be completed in order to obtain 3000—8000 identifiable individuals (calculated from Table 1). If the sustainable range of samples per worker per day is taken to be 3–5, one person would need 23(156) field days in a low diversity site and 60(416) days in a high diversity site. Five people might require only a fifth of those ranges. If the sorting time is triple the field time, the total study time for one person would be 92(624) days for a low diversity site or 240(1664) days for a high diversity site, not counting travel, city time, or other such factors. These estimates ignore the obvious advisability of repeating the study in different seasons, although the generality of the importance of seasonality is still unknown. However, Erwin and Scott (1980) found that species turnover between wet and dry seasons was 97%. Given that the field time estimated above for a single person might span significant seasonal change, multi-person teams per taxon are advisable. Nevertheless, in our experience duplication of taxon expertise beyond a pair of people on inventory teams is rare. Insofar as the adequacy of a single sampling effort is a goal either met or not, and if not met, cannot be fixed by a repeat trip without adding yet another factor (season and/or year) to an already complex analysis, it seems well worth the money to ensure that the initial effort is sufficient unto itself.

Similar analyses should be performed on the distribution of numbers of species per sample, categorized by the same main factors as in the above analysis. Because the time-consuming job of sorting animals to species within site is not complete, we cannot yet perform that analysis. Because numbers of species per sample will be much lower than numbers of individuals, differences among factors and treatments are less likely to be significant. This also means that we cannot yet construct species abundance distributions from these data, and therefore cannot yet estimate total species richness for any of the Bolivian sites, or assess the available analytical methods on these data.

The Peruvian canopy samples, kindly provided by Dr. T. L. Erwin, suggests that both the lognormal and the jackknife can be used on tropical faunas. These disparate methods roughly agree in the one case in which we have been able to compare them thus far. However, for both collections we included juveniles in our counts. Without juveniles, the single canopy sample would have provided only 222 classifiable individuals and the five canopy sample 354 individuals—both inadequate samples on which to estimate total species richness. Using juveniles increases sample size but the identification are ambiguous, and studying them tremendously increases sorting time.

For the single canopy sample, the S* estimate was 235 species. This is not at all out of line with our hunch that tropical spider faunas from a single site range between 300 and 800 species, given that this sample was from a single canopy. Estimates such as this, 187 species observed ranging to 273 species maximum as a likely upper "bound," rough as they are, could be useful for conservation planning.

The set of five canopy samples is more problematic because in fact each was from a different forest type. Certainly they represent an increase in area sampled relative to the above, and therefore this is not more intense sampling of the same community, but rather sampling over a re-defined and larger area. Pooling the different forest types may have been responsible for the poor fit to the lognormal distribution.

We followed Ludwig and Reynolds' (1988) recommendations for fitting a lognormal because it meshes well with statistical packages. However, both of our examples may point out a weakness in their approach. No matter how large the number of singletons (species represented by one specimen), if there are doubletons and the octave boundaries are integers, the observed data will always have an apparent mode. Octave 0–1 will always contain fewer species than octave 1–2. Pielou (1975) and Magurran (1988) use logarithmic

58

groupings that avoid this effect on the first two octaves. Hughes (1986) also commented on apparent modes in logged data as effects of the way data are grouped. Insofar as the lognormal requires estimation of the mode, it appears that tropical faunas exhibiting species abundance relations in which singletons are most numerous will always appear lognormal if grouped by Ludwig and Reynolds' method, and this may be a problem with our data. The asymmetry between the first and third octaves in Figure 6 may indicate that singletons are "piling up" in the first and second octaves. Observing two octaves to the left of the mode would give greater confidence that the mode had been obtained. At any rate, a benefit of the lognormal is that it does not require replicate samples. Data collection is therefore easier. On the other hand, the lack of a confidence interval on the estimate is a serious drawback.

In the legend to Figure 6 we also report estimates of the lognormal parameters using a least squares procedure. Although it tells us that a significant amount of the variation is explained, that is hardly surprising. The use of a Chi-square expression as the "loss function" minimized by the nonlinear estimation procedure seems preferable because it better tests the null hypothesis of a lognormal distribution.

The jackknife is not without problems either. Robert Edwards (pers. comm.) points out that the jackknife estimate has an upper limit of about twice the number of species observed, which would be obtained if all observed species were uniques. If less than half the total species in the community are represented in the sample, the jackknife will underestimate the real species richness. Treating the five canopy sample either as 225 replicate samples or 5 replicate samples shows that the accuracy and precision of the jackknife depends greatly on the number of (replicate) samples available.

Based on evidence from other groups studied at Tambopata, it is one of the richest sites on earth (Erwin, 1985). The total number of spider species known to date from non-canopy habitats at Tambopata Forest Reserve is about 500–600 (Silva, in prep.). Without knowing to what extent the faunas of the canopy and non-canopy overlap, we cannot easily compare these numbers. However, based on the possible extremes of no or total overlap, the number of spider species at Tambopata ranges from the known fauna of slightly more than 500 to a high of about 1200. As no other site known to us is comparably diverse in spider species, Tambopata's "megadiversity" reputation is sustained by these data.

Finally, we would like to make three points about future work in this area. All methods that estimate the total species richness of a community make assumptions about the data used in the estimates. We have shown that care in organization of sampling procedures can assess these assumptions at relatively little extra cost. We can say something about the magnitude of the effect of collector experience, collecting method, site, and time of day on relative abundance of individuals in samples. Number of species per sample can be analyzed in the same way as number of adult individuals. Both analyses seem worthwhile because they reveal the structure and quality of the data that underlie estimates of species richness. For example, egregious heterogeneity in the data may indicate that different communities have been inadvertently combined or considered as one. In any study such as ours the number of factors involved will generally be at least four, each factor having several levels. Strictly speaking, this implies a four-way analysis of variance with 90 cells in this case. Adequate replication in each cell implies 270 samples, a number we did not come close to attaining at each site in this preliminary study. Other taxocene-oriented efforts will face the same serious problem. Addition of other factors (e.g., by microhabitat, trophic level, or season) will complicate the analysis. However, if each sample yields about 17 individuals, 270 samples implies some 4600 specimens. This may be about the right sampling intensity for a fauna of 500 species, coincidentally the median value in our guess as to the range of typical species richness of spiders in evergreen tropical sites. In other words, if one is going to need 270 samples anyway, one may as well collect them in a structured fashion. If a collector can average 5 samples per day the sampling could be completed by three collectors in 18 days, which is about the time systematists like to spend at a single, productive site. With respect to the analysis of variance, reducing collectors by two would allow two additional methods to be used, probably a better way to apportion the available cells in the analysis. On the other hand, because inventories are not primarily designed to test hypotheses about collector or method effects, there is no real need to fulfill all the requirements that a strict experimental design imposes.

Secondly, a serious drawback of methods dependent on relative abundance of species is the necessity in most groups to continue collecting individuals of common species during the search for rare ones. Assuming that the lognormal curve for each Bolivian site will look about like that for the Peruvian canopy samples, the cost of disclosing an additional octave of the species abundance distribution at each site would have been an additional 1000 specimens (Table 1), most of which would have been known already from previous samples. Yet another octave implies an additional 4000 specimens. This is a high price, and yet the accuracy and precision of the species richness estimate clearly depends on the number of octaves (data points) observed. Benefits rise linearly but costs rise exponentially. An obvious solution is to truncate intentionally the collection of abundant species and therefore to accept "veil lines" at both tails rather than just the left. As opposed to automated traps, field collectors could collect in this manner if the most common species can be recognized accurately. If conservation research is to have minimum impact on the fauna, this is a strong argument for the kind of sampling protocol done here, as opposed to more automated techniques. On the other hand, the amount of work to get nine octaves is terrific in the tropics, and the rightmost ones are the easiest.

Third, the largest task in this work is the separation of morphospecies. While recognizing the number of species within any sample is easy, collating species across hundreds of samples is difficult. Collating them reliably across different sites becomes increasingly unrealistic without preparing detailed notes, drawings, and, especially diagnoses of hundreds of morphospecies. There comes a time when the "synoptic collection" approach breaks down. That point will differ for different groups and different workers. While conservation biology and ecology tend to see this as a problem in "data management," an older and simpler term exists—alpha taxonomy. Alpha taxonomy is a discipline in decline, and one which the response to the biodiversity crisis evidently hopes to sidestep by ignoring the issue of Latin names. However, our work shows that the need for alpha taxonomy does not grow out of concern with names, but rather out of prosaic data management. We predict that simply to do a competent job of making a comparative inventory of tropical sites, researchers will have to generate somewhere, whether in their notes, computers, or minds, a functional alpha taxonomy of the group under consideration. Therefore we ask, if one has to do all the hard parts anyway, why not go the extra half-step and publish the revision?

ACKNOWLEDGMENTS

We have many people and institutions to thank. The Biological Diversity Program of the Smithsonian Institution funded the work in Bolivia, and the Lowland Tropical Ecosystems Program funded the sorting of the samples from Perú. Harmut Döbel and Edward Barrows gave invaluable advice during the initial stages of protocol design; however, the responsibility for those flaws that remain is ours. Carmen Miranda, Abelardo Sandoval, and Marsha Sitnik helped with logistics. Dr. Mario Baudoin and his staff of the Instituto de Ecologia in La Paz provided equipment and a comfortable place to work. Terry Erwin very kindly allowed us to work on his samples of Peruvian canopy arthropods and we thank Jan Scott for sorting them out of the general collections. The comments of James Carpenter, Robert Colwell, Robert Edwards, Terry Erwin, and Robert Robbins improved

60

the manuscript considerably. This paper is contribution no. 10 from the Biological Diversity in Latin America (BIOOLAT) Project.

LITERATURE CITED

Broadhead, E. 1983. The assessment of faunal diversity and guild size in tropical forests with particular reference to the Pscoptera. Pp. 25–41. *In*: S. L. Sutton, T. C. Whitmore, & A. C. Chadwick (eds.), *Tropical Rain Forest: Ecology and Management*. Blackwell Scientific Publishing, Oxford.

Coddington, J. A., Larcher, S. F. & J. C. Cokendolpher. 1990. The Systematic status of Arachnida, exclusive of Acarina, in North America north of Mexico (Arachnida: Amblypygi, Araneae, Opiliones, Palpigradi, Pseudoscorpiones, Ricinulei, Schizomida, Scorpiones, Solifugae, Uropygi). Pp. 5–20. *In*: M. Koztarab & C. W. Schaeffer (eds.), *Systematics of the North American Insects and Arachnids: Status and Needs*. Virginia Agricultural Experiment Station Information Series 90–1, Virgina Polytechnic Institute and State University: Blacksburg.

Coddington, J. A. 1990. Review of *Advances in Spider Taxonomy 1981–1987: a supplement to Brignoli's A Catalog of the Araneae described between 1940 and 1981*, by Norman I. Platnick, 1989. (Edited by P. Merrett). Manchester University Press. *J. Arachnology*. 18(2): 000-000.

Erwin, T. L. 1983. Beetles and other insects of tropical forest canopies at Manaus, Brazil, sampled by insecticidal fogging. Pp. 59–75. *In*: S. L. Sutton, T. C. Whitmore, & A. C. Chadwick (eds.), *Tropical Rain Forest: Ecology and Management*. Blackwell Scientific Publishing, Oxford.

Erwin, T. L. 1985. Tampobata Reserved Zone, Madre de Dios, Perú. History and description of the Reserve. *Rev. Per. Ent.* 27:1–8.

Erwin, T. L. 1989. Sorting tropical forest canopy samples (an experimental project for networking information. P. 8. *In*: R. J. McGinley (ed.), *Insect Collection News*, vol. 2(1). Department of Entomology, National Museum Natural History, Smithsonian Institution: Washington, D.C.

Erwin, T. L. & J. C. Scott. 1980. Seasonal and size patterns, trophic structure, and richness of Coleoptera in the tropical arboreal ecosystem: the fauna of the tree *Luhea seemanii* Triana and Planch in the Canal Zone of Panama. *The Coleopterists Bull.* 34(3):305–322.

Farrell, B. D. & T. L. Erwin. 1988. Leaf-beetle community structure in an Amazonian rainforest community. Pp. 73–90. *In*: P. Jolivet, E. Petitpierre & T. H. Hsaio (eds.), *Biology of Chrysomelidae*, Kluwer Academic Publishers.

Fisher, R. A., A. S. Corbett & C. B. Williams. 1943. The relation between the number of species and the number of individuals in a random sample of an animal population. *J. Anim. Ecol.* 12(1):42–58.

Heltsche, J. F. & N. E. Forrester. 1983. Estimating species richness using the jackknife procedure. *Biometrics* 39:1–11.

Hubbell, S. P. & R. B. Foster. 1983. Diversity of canopy trees in a neotropical forest and implications for conservation. Pp. 5–41. *In*: S. L. Sutton, T. C. Whitmore & A. C. Chadwick (eds.), *Tropical Rain Forest: Ecology and Management*, Blackwell Scientific Publishing, Oxford.

Hughes, R. G. 1986. Theories and models of species abundance. *Amer. Nat.* 128:879–899.

Ludwig, J. A. & J. F. Reynolds. 1988. *Statistical Ecology: A Primer on Methods and Computing*. Wiley Interscience, New York.

Magurran, A. E. 1988. *Ecological Diversity and its Measurement*. Princeton University Press: Princeton, N.J. 179 pp.

May, R. M. 1975. Patterns of species abundance and diversity. Pp. 81–120. *In*: M. L. Cody & J. M. Diamond (eds.), *Ecology and Evolution of Communities*, Belknap Press, Harvard: Cambridge, MA.

May, R. M. 1988. How many species are there on Earth? *Science* 241:1441–1443.

Mound, L. A. & N. Waloff (eds.) 1978. *The Diversity of Insect Faunas* (Symposium of the Royal Entomological Society No. 9). Blackwell Scientific Publishing: Oxford.

Pielou, E. C. 1975. *Ecological Diversity*. Wiley Interscience: New York, 165 pp.

Pielou, E. C. 1977. *Mathematical Ecology*. Wiley Interscience: New York, 385 pp.

Preston, F. W. 1948. The commonness and rarity of species. *Ecology* 29:254–283.

Southwood, T. R. E. 1978. *Ecological Methods: With Particular Reference to the Study of Insect Populations*, 2nd ed. Chapman and Hall: New York.

Sugihara, G. 1980. Minimal community structure: an explanation of species abundance patterns. *Amer. Nat.* 116:770–787.

Taylor, L. R. 1978. Bates, Williams, Hutchinson — a variety of diversities. Pp. 1–18, *In*: L. A. Mound & N. Waloff (eds.), *The Diversity of Insect Faunas* (Symposium of the Royal Entomological Society No. 9), Blackwell Scientific Publishing: Oxford.

Wilkerson, L. 1988. *SYSTAT: the system for statistics*. SYSTAT Inc.: Evanston, IL. 822 pp.

Williams, C. B. 1964. *Patterns in the Balance of Nature*. Academic Press: London.

Wilson, E. O. & F. M. Peter. 1988. *Biodiversity*. National Academy Press: Washington, D.C.

Processes Regulating Biodiversity in Coral Reef Communities on Ecological vs. Evolutionary Time Scales

Marjorie L. Reaka-Kudla

Abstract This work tests the hypothesis that processes which structure communities in ecological time are relevant to patterns of biodiversity generated over evolutionary time. The work focuses on the solitary motile cryptofauna of coral reefs, which represents some of the highest undocumented biodiversity and biomass within the coral reef community. Manipulative field experiments showed that strong biotic interactions (scramble competition for space, benthic adult-juvenile predation, fish predation)—occasionally interrupted by severe storm disturbance—are the dominant ecological processes structuring this community. Fish predation intensifies rather than negates competition in this refuge-limited system. A separate study tested the hypotheses that competition, behavioral complexity, predation, habitat type, temperature, latitude, depth, species body size, and dispersal potential altered rates of evolutionary change (morphological divergence, speciation, extinction) in populations of stomatopod crustaceans that had been geographically isolated at least 3 my. Of these, only body size and dispersal potential were found to be consistent and statistically significant agents of evolution. Thus, competition, complexity of behavior, and type of habitat occupied are ecologically but not evolutionarily important (except as these factors are correlated with body size and dispersal potential). Fish predation, however, drives this community in ecological time and was responsible for the evolution of high diversity on coral reefs, since bioeroders bored into the reef (providing 3-dimensional crypts into which radiated a vast assemblage of other benthic reef biota) to escape predation. The critical point for the evolution of diversity on coral reefs is that the bioeroded holes constrained the subsequent cryptofauna to small body sizes, limited dispersal, and thus high rates of evolution. The small and cryptic species of the world represent the largest reservoir of diversity still unknown to science; it is these species (both aquatic and terrestrial) to which we should turn our attention for studies of biogeographical patterns, for indications of environmental degradation, and for conservation of untapped genetic, biochemical, and organismal diversity.

INTRODUCTION

Understanding patterns of biodiversity within and among communities of organisms has been a major focus of ecological research for more than 30 years. Ecologists have sought to explain how a given community contains a certain number of species, whether and how much this number varies over time (i.e., whether or not the community is in "equilibrium"), and why other communities (often along environmental gradients) contain fewer or more species. Much of the study of community structure and function has been undertaken with either the direct or indirect objective of understanding questions of

---

Dr. Reaka-Kudla is with the Department of Zoology, University of Maryland, College Park, Maryland 20742, USA.

62

species diversity. One major difficulty, however, has clouded the interpretation of studies of ecological diversity: do community processes that occur on an ecological time scale (one to a few years, or in some cases a few tens of years) have anything to do with patterns of diversity over evolutionary time? One way to approach this problem is to examine (preferably both observationally and experimentally) the ecological processes that control the structure and diversity of a community and then test (using phylogenetic or paleontological methods) the hypothesis that the factors found to be ecologically relevant also influence the historical patterns of diversification and extinction of component taxa within the community. If the two methods do not produce congruent results, then studies of community structure and function, while valid in their own right, tell us little about biogeographical patterns of diversity or trends in diversity over geological time, and provide low predictive power for the long term trajectory of biodiversity in the future.

I will use the above method to evaluate the ecological and evolutionary processes that are responsible for diversity in one of the two most diverse ecosystems on earth—coral reefs. Despite the fact that coral reefs have been the object of considerable public attention because of their beauty and the dramatic adaptations of some of their component organisms, this community has received surprisingly little attention, compared to rainforests, in the international effort to analyze and document biodiversity in the face of a rapidly changing global environment.

ECOLOGICAL PROCESSES OPERATING IN CORAL REEF COMMUNITIES

Coral reefs are comprised of suprabenthic and benthic/cryptic fishes, sessile epibenthic organisms (primarily soft and hard corals, sponges, coralline and fleshy algae), and the cryptofauna (including sessile and motile organisms that bore holes into the calcium carbonate substrate or live in the vacated burrows of previous bioeroders). Information gathered in my laboratory (Reaka-Kudla, Morrin, Severeyn, unpublished data) shows that, of these three major components, the crytofauna represents a large amount of the biomass and certainly the greatest reservoir of biodiversity (largely undocumented) within this community.

Many of the solitary reef cryptofauna (e.g., stomatopods, crabs, snapping shrimps) exhibit elaborate agonistic and territorial behavior, bright color patterns, and armored bodies and appendages. My first question was whether competition was driving the evolution of these characteristics. A series of field manipulations tested the hypothesis that population densities of stomatopod crustaceans were limited by either spatial or food resources. When pieces of rubble containing unoccupied burrows were added experimentally, and when hurricanes obliterated all of the burrows in some study sites, numbers of stomatopods were directly related to the availability of spatial refuges (Reaka, 1985, 1987a; Moran & Reaka-Kudla, 1990). The numbers and sizes of holes in reef substrate across habitats also was positively correlated with the number and sizes of stomatopods present in the field (Moran & Reaka, 1988). Stomatopods reach their largest sizes and reproduce only in the relatively few large holes of the intertidal fringing reef habitat (Reaka, 1987a, Reaka et al., 1989). Following brooding, the larvae emerge and migrate reefward during nocturnal sojourns into the water column (Robichaux et al., 1981, Reaka, 1987a). Adult stomatopods, crabs, and snapping shrimps each prey on their own and each others' juveniles, inhibiting recruitment of stomatopods in nearshore environments and forcing the larvae to settle in deeper reefward habitats and migrate inshore at larger sizes (only larger individuals have a chance of acquiring the few but large and fiercely contested inshore holes; Reaka, 1987a). These reciprocal adult-juvenile competitive-predatory interactions may be of considerable general importance in systems that include several predators whose feeding habits are size dependent. This study also indicated that strong biotic processes operating in only a segment of the population (the nearshore habitat) only part of the time (infrequently interrupted by severe storms; Moran & Reaka-Kudla, 1990) can provide sufficiently intense selection to explain the dramatic adaptations in color, morphology, and behavior in these organisms.

However, densities of stomatopods were consistently unaffected by food in several different field experiments (including superabundant, normal, half normal, low, and very low densities of prey: Reaka, 1980a; Wolf et al., 1983; Reaka, 1985, 1987a; Reaka-Kudla, 1988; see also Dominguez & Reaka, 1988, and below). Field observations on individual stomatopods (Dominguez & Reaka, 1988) and laboratory studies on behavioral preferences for cavities with systematically varied architectures (Reaka-Kudla & Smith, unpublished data) showed that stomatopods accept almost any cavity that they can acquire and defend. Several lines of experimental evidence suggested that, when many different taxa are severely limited by a single resource that cannot be easily subdivided (e.g., space), these species often adopt generalized patterns of resource utilization that promote apparently stochastic recruitment patterns by different species into given refuge sites. Intra- and interspecific competition for spatial resources therefore appears to be a dominant factor in the biology of both motile (above) and sessile (e.g., Jackson & Buss, 1975, Buss, 1979, Jackson, 1977, 1979) reef cryptofauna. Resource limitation usually is important only in conditions that do not suffer strong predation or physical disturbance. My results, however, suggest that, in refuge-limited organisms such as motile cryptobionts, predation intensifies rather than reduces competition for spatial resources (below).

Coral reef ecosystems are productive oases in nutrient-poor waters due to efficient recycling of nutrients (Muscatine & Porter, 1977), suggesting that coral reefs ultimately may be limited by nutrients. However, Grigg and his co-workers concluded that reef communities are regulated from the top down by pervasive predation (Grigg et al., 1984). They note that, despite relatively high productivity, yield at the top of the food web is low because of high internal predation and trophic complexity. However, few studies have addressed the intensity of predation on different components of the reef fauna quantitatively. Despite their interesting results, the above authors were forced by the paucity of published quantitative data to rely on limited or estimated values for some of the interaction terms in their model. Moreover, Jackson and his co-workers (above) have argued that predation is relatively unimportant for sessile reef cryptofauna. Bakus (1981) also notes that sessile epibenthic reef taxa are more likely to contain toxic defensive compounds than cryptic taxa, implying that the latter may encounter relatively little predation. Supporting the hypothesis that fish predation on motile cryptobionts is rare, my initial caging experiments detected no significant difference between densities of stomatopods in caged, partially caged, or open plots (Reaka, 1985). We further tested the hypothesis, however, that new individuals in space-limited populations might immediately colonize sizes opened by predation in the open and cage control plots. When we imposed barriers to immigration in all three experimental conditions, significantly more stomatopods were found in the caged than the control plots (Reaka, 1985). This experiment suggested that the results of many conventional caging studies should be re-evaluated if they treat motile benthic prey. Such methods can seriously underestimate the intensity of predation due to differential turnover rates of prey in caged vs. control plots, and these errors will be especially exaggerated if the prey are space-limited. We instituted similar studies on artificial reefs comprised of fish habitat (cinderblocks), invertebrate habitat (holey rubble), or fish and invertebrate habitat; these reefs were relatively isolated from sources of immigration by a sand flat. These studies showed that some invertebrate taxa (e.g., stomatopods) were negatively influenced by the presence of fishes, and that fish colonization was enhanced by the presence of invertebrate prey (Wolf et al., 1983, Reaka, 1985). Both types of experiments (cages, artificial reefs) yielded an estimate of one successful predation event about every 7–8 months on a given piece of rubble. In contrast, tethered stomatopods in the same sites survived only a few minutes. These results suggest that occupation of a cavity in the protective reef substrate confers considerable protection against fish predators in an environment that is subject to potentially extremely intense predation. This supports the conclusions of Jackson and others (above) that predation by fishes on sessile/relatively sessile cryptofauna is minimal, but indicates that predation intensity increases with greater motility of the cryptofauna (Reaka, 1985).

We also examined temporal activity patterns of reef stomatopods (Dominguez & Reaka, 1988). These crustaceans seal themselves inside burrows overnight and show significant peaks of activity during the early morning and late afternoon hours. We hypothesized that these activity patterns resulted from the availability of prey, activities of fish predators, or visual capabilities of reef stomatopods. Observations on feeding behavior, the volume of food in the stomach, and the rate at which it was processed, indicated that these predators did not feed selectively at a particular time of day. Quantification of benthic prey at different times of the day in the field confirmed that the availability of prey did not determine the crepuscular activity patterns. Field observations, laboratory experiments, and gut contents indicated that stomatopods eat only 7–10 prey/day, whereas each individual's territory includes as many as 1000 available prey (Reaka, 1985, 1987a; Reaka-Kudla, 1988 and in prep.; Dominguez & Reaka, 1988). In a study on O_2 consumption in stomatopods, we calculated that this number of prey provides at least the number of calories required for field activity levels (Reaka-Kudla, 1988 and in prep., White & Reaka-Kudla ms.). Thus, several independent lines of evidence reaffirmed the experimental data discussed above, which suggested that food does not limit these stomatopods. Our data showed that reef stomatopods were not active during crepuscular periods because they could see to feed and avoid predators better at twilight than at midday, and that they were not differentially responding to the observer during midday (their eyes are physiologically and morphologically specialized for bright rather than dim light conditions; Schiff et al., 1986). However, the activity peaks were inversely correlated with the number and sizes of invertivorous fishes present within 1 m of their burrow, suggesting again that potential predation is a major ecological factor molding the biology of reef cryptobionts such as stomatopods.

Following the impact of hurricanes David and Frederick in our study sites in St. Croix, we analyzed the effects of severe episodic physical disturbance upon the cryptofauna (Moran & Reaka-Kudla, 1990). Habitat destruction and disruption to cryptofaunal populations was inversely correlated with depth over five habitats. Several cryptic taxa, including stomatopods, had not recovered by the end of the two year study, especially in intertidal habitats where reboring of holes apparently was retarded by the physically rigorous environment. In areas of intermediate depth and disturbance, however, most cryptofaunal populations were enhanced, with burrowers and then nestlers increasing far above pre-hurricane densities. Many populations were still increasing at the end of the study. These enhanced population densities probably occurred because of accelerated boring in newly available substrate (see Rutzler, 1975), which provided subsequent living space for colonizing nestlers. It is also likely that the impact of the two closely sequenced hurricanes caused a burst of terrigenous run-off around the island, thus increasing planktonic productivity and the survival, recruitment and growth of swimming invertebrate larvae and benthic filter feeders such as many bioeroders (see Birkeland, 1982). This would result in the delayed and relatively long term effects observed in the community. These long term perturbations (> two yr) in such an important component of the reef community are significant for our understanding of the structure, trophic dynamics, and stability of reef communities. Like our research on the extent of bioerosion in dead coral substrate from five habitats in St. Croix (Moran & Reaka, 1988), this study showed that bioerosion, which provides the cryptic spatial refuges that limit the rest of the benthic reef biota, is one of the key elements in understanding the ecology of high diversity coral reef environments.

EVOLUTIONARY PROCESSES OPERATING IN CORAL REEF COMMUNITIES

An independent but related line of work examined the relation of life history patterns and other ecological factors to rates and patterns of evolution in stomatopods on a larger geographic scale (Reaka, 1978, 1979a,b, 1980b, 1986a,b, 1987b; Reaka & Manning, 1981, 1987a,b), since the behavior, ecology, morphology, and systematics of this manageable group (about 450 species world-wide) are relatively well known compared to most groups of organisms. The central issue was whether or not factors operating in ecological time (above experimental studies) were important in molding the evolution of cryptofaunal lineages. These studies examined reef-dwelling gonodactyloid stomatopods (115 subgeneric taxa world-wide; e.g., Reaka, 1980b) and Atlanto-East Pacific stomatopods from all superfamilies and habitats (157 subgeneric taxa; e.g., Reaka & Manning 1981, 1987a). Rates and patterns of evolution (phyletic divergence, species multiplication, extinction) within lineages were examined with respect to species body size and dispersal potential as well as to a number of variables that have been tied to enhanced evolutionary rates in the literature—competition for resources, habitat (reef/rocky vs. soft bottom) and associated biotic pressures (especially predation), complexity of behavior, environmental temperature ($<$ or $>20°C$), latitude (tropical, temperate), and depth (0 to >300 m). Even with large sample sizes that allowed rigorous statistical testing, only body size and dispersal potential proved to be consistent and highly significant agents of evolution. Large species within lineages grow faster, produce more and larger eggs with longer dispersal stages, reproduce more frequently, experience more intense competition for refuges, exhibit more complex fighting and reproductive behavior, suffer lower size-dependent predation, are more brightly colored, have broader geographic ranges, and have lower rates of phyletic divergence, speciation, and extinction than smaller relatives within the same lineage (Reaka, 1979a,b, 1980b, 1986a; Reaka & Manning, 1981, 1987a). Small species within lineages are particularly notable for their small geographic ranges and endemism, major ecological innovations (e.g., invasion of new types of habitat), susceptibility to extinction, and unusually high rates of species multiplication compared to the rest of their lineage. Species on reefs are significantly smaller in adult body size than those on level bottoms (Reaka, 1986a), although most measures of rates of morphological evolution are not significantly different between taxa that occupy these two habitat types (Reaka & Manning, 1981, 1987a).

We hypothesized that taxa of small adult body size evolve rapidly primarily because of restricted dispersal and predation. Small adult body size is strongly associated with abbreviated development and restricted dispersal in marine invertebrates in general (Strathmann & Strathmann, 1982) as well as in stomatopods (Reaka, 1980b, 1986a; Reaka & Manning, 1987a). In addition, rapid morphological divergence in members of small lineages may be driven by predation. Our field observations showed that small individuals are subject to much more intense predation from benthic invertebrates and epibenthic fishes than larger stomatopods (Reaka, 1987a, Dominguez & Reaka, 1988, and unpublished data). We suspect that strong directional selection, facilitated by restricted dispersal, may cause major ecological shifts and subsequent reproductive isolation (e.g., some of the smallest species of the sand-dwelling lysiosquilloids have invaded calcareous worm tubes and coral crevices; some of the smallest taxa in the mud-dwelling squillids inhabit holes in rock and reef substrates; and some of the smallest gonodactylids, which normally occupy shallow reef habitats, are found on relatively level bottom habitats in deep water). The largest stomatopods (to almost 400 mm in body length) are immune to all but the largest fish predators, and these lineages tend to be morphologically and ecologically conservative. In contrast to taxa of small adult body size, many lineages of large body size are conspecific or show little morphological variation between the East Atlantic, West Atlantic, or East Pacific oceans. Even though these regions have been isolated at least 3 million years (Phillips & Forsyth, 1972; Kiegwin, 1978, 1982), morphological differentiation apparently has been retarded by extensive genetic exchange among habitat patches within geographic regions via the long-lived larvae (up to 9 months) in these taxonomic lineages. The Order Stomatopoda is tropical—warm temperate in distribution, and individuals become catatonic at cooler temperatures; thus, larval exchange around the tips of continents at high latitudes is not possible now or (according to our analyses of post-Cretaceous continental geography) in the past.

An alternative (and not mutually exclusive) hypothesis is that small taxa may proliferate because there are more different types of small than large habitats available for invasion. However, in contrast to the above hypotheses on dispersal and predation, this hypothesis is difficult to test. Furthermore, it does not address the correlation between small body sizes and high rates of apparent extinction in stomatopods.

Body size within lineages declines from populations of large adult body size (which produce large numbers of large eggs and long lived larvae) on the western margin of the Pacific toward populations that are progressively smaller in adult body size (which produce fewer and smaller eggs that have more restricted dispersal potential and hence are trapped as endemics) toward the East Central Pacific (Reaka, 1986b; in prep.). Endemism, speciation, and extinction potential are greatest on the eastern periphery of the Indo-West Pacific realm. Populations on atolls are smaller in body size and more likely to be endemic than those on high islands, suggesting that productivity (e.g., Birkeland, 1982) may influence evolutionary patterns of body size. Analysis of fossil and relict groups and current size distributions of species (Fig. 1) suggests that evolution of stomatopods has proceeded from large toward small body sizes over time, the reverse of Cope's Law (Reaka, 1987b; in prep.). I believe that this is due to their proliferation in the bioeroded holes of reefs since the Mesozoic (below).

DISCUSSION

Relation of the above, ecological studies to those on evolutionary processes suggests that competition for limiting resources and the associated complexity of agonistic/reproductive behavior (Reaka & Manning, 1981) are important in ecological time but, interestingly, not over evolutionary time. Competition for food does not appear to be important in the ecology of these predatory reef species; however, studies on bioerosion, hurricanes, and the biogeography of Pacific atolls and high islands show that planktonic productivity and food levels may influence evolutionary trajectories by mediating bioerosion and the availability of large cavities, and thus body size of cryptobionts. Whereas the nature of the habitat (reefs vs. soft bottoms) is very important in ecological time, habitat is evolutionarily important only to the extent that it molds body size and possibly modifies the intensity of predation.

The strong potential predation documented above undoubtedly has had a strong influence on the evolution as well as the ecology of cryptic reef organisms. The following evolutionary scenario for the role of predation in the evolution of high diversity on coral reefs is envisioned. As fast efficient deposit feeders began ejecting mounds of sediment (Thayer, 1979), the stage was set for the more extensive development of framework organisms such as scleractinian corals, which, through their skeletal depositions, could rise above the increased siltation. During the Mesozoic, scleractinian corals became much more extensive (Newell, 1971) and modern teleost reef fish evolved (Tyler, 1980, Vermeij,

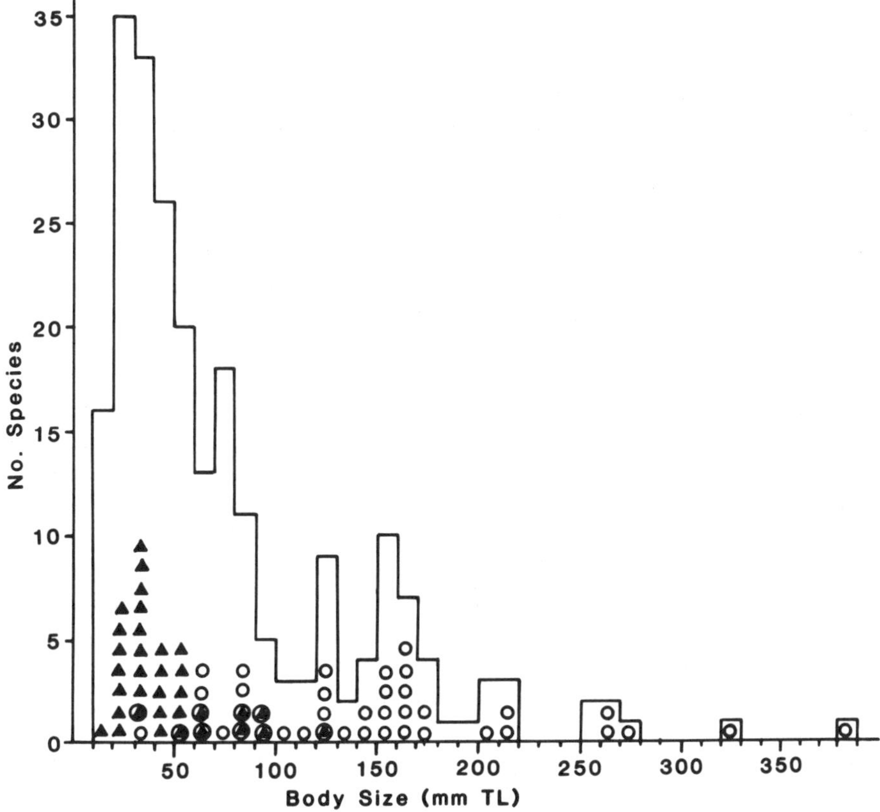

Figure 1. Size frequency distribution of all Atlanto-East Pacific stomatopods. Open circles indicate species with long lived larvae. Closed triangles represent species that have undergone major habitat shifts relative to their ancestral lineage. Body sizes of the bathysquillids (relict species with morphological similarities to the extinct sculdid stomatopods) are > 200 mm in length, and most fossil stomatopods appear to have been moderate to large. These data show that most current stomatopods are small and that body sizes have shifted from large toward small body sizes, at which sizes major ecological and morphological innovations have occurred. The large species, however, are resistant to extinction.

1987). Increasing predation pressures favored organisms that lived in crevices or bored into the reef, and burrowers of hard substrates diversified during the Mesozoic (Palmer, 1982, Vermeij, 1987). The evolution of bioeroders produced a three-dimensional environment, into which radiated a vast assemblage of benthic reef biota (Moran & Reaka, 1988), whose biology in turn was molded by predation and thence by competition for space. A critical feature of this cryptic environment was that it constrained the benthic assemblage to small body sizes, limited dispersal and high evolutionary rates. Thus, predation and, ultimately, bioerosion provide key insights for the evolution of high diversity on coral reefs.

These results, like those of Erwin (1982, 1983, 1988) and May (1988), suggest that the Lilliputian species of the world represent one of science's great, still largely uninvestigated unknowns. Most of the world's known taxa are small (Fig. 1; see also May, 1988), and a still larger number of the world's unknown taxa probably are small. In some cases we know why so many taxa are small: they are constrained to minute sizes in order to fit into crenulated surfaces that protect them from predation (most benthic marine organisms, epiphytic terrestrial organisms, and ectoparasites), or they have invaded a similarly protected but especially nutrient rich environment (endoparasites). In other taxa, we still do not fully understand why there are so many small species (May, 1988, and ICSEB Plenary Lecture). In almost all cases, however, small body size is associated with exceedingly high diversity. Most of the world's species richness resides in these tiny, difficult to study, specialized, and often narrowly distributed species. It is in this group that the greatest loss of genetic, biochemical, and organismal diversity will occur due to environmental changes imposed by increased human populations over the next century. Because of the number of species and individuals, changes in these cryptic communities can be used as sensitive indicators of environmental change. Biogeographical studies of cryptobionts can identify especially vulnerable areas in need of attention from conservationists and policy makers. For example, in marine organisms, mid-Pacific atoll environments represent one of the most fragile ecosystems. These repositories of narrowly distributed species are extremely susceptible to human disturbance, anthropogenic changes such as global warming, and even natural disasters such as storms or El Niño events. We have witnessed the apparent extinction due to a severe storm of a morphologically distinct population that, as far as we can determine, inhabited only one islet of Enewetak Atoll (Reaka, 1980b, Manning & Reaka, 1981, Reaka & Manning, 1987b). Glynn (1989) reported the apparent extinction of several populations and possibly two species of corals in the East Pacific due to the severe 1982–83 El Niño Southern Oscillation event. Whereas tropical marine populations have experienced often lethal increases in temperature due to periodic El Niño climatic oscillations for millenia (Glynn, 1988), these populations live near the upper limit of their temperature tolerance (Jokiel & Coles, 1977). Mean monthly temperature increases of 1–5°C above the long term average during the 1982–83 El Niño resulted in 70–99% mortality of coral populations in the East Pacific (Glynn, 1988, 1989; pers. comm.). The increases in temperature encountered during El Niño episodes are remarkably similar to those predicted for tropical waters over the next decades by global warning models (Mitchell, 1988). When superimposed upon anthropogenic warming trends, the effects of future El Niños upon the diverse and often endemic cryptobiota may be truly catastrophic. This biota deserves our scientific attention.

ACKNOWLEDGMENTS

Parts of this research were funded by the U.S. National Science Foundation (Biological Oceanography), the U.S. National Oceanic and Atmospheric Administration (Office of Undersea Research), the Roger Tory Peterson Research Foundation, the Center for Field Research, and the University of Maryland (Dean of the College of Life Sciences and the Graduate School). I thank my colleagues (especially Dr. Raymond B. Manning, Mr. David Moran, Ms. Jane Dominguez, Ms. Valerie Wesner, Ms. Stephanie Smith, Dr. Nancy Wolf, Dr. Eldredge Bermingham, Ms. Gwen White) for sharing ideas and work.

LITERATURE CITED

Bakus, G. J. 1981. Chemical defense mechanisms on the Great Barrier Reef, Australia. *Science* 211:497–499.

Birkeland, C. 1982. Terrestrial runoff as a cause of outbreaks of *Acanthaster planci* (Echinodermata: Asteroidea). *Mar. Biol.* 69:175–185.

Buss, L. W. 1979. Bryozoan overgrowth interactions—the interdependence of competition for space and food. *Nature* 231:475–477.

Dominguez J. H. & M. L. Reaka. 1988. Temporal activity patterns in reef-dwelling stomatopods: a test of alternative hypotheses. *J. Exper. Mar. Biol. Ecol.* 117:47–69.

Erwin, T. 1982. Tropical forests: their richness in Coleoptera and other arthropod species. *Coleopt. Bull.* 36:74–75.

Erwin, T. 1983. Beetles and other arthropods of the tropical forest canopies at Manaus, Brasil, sampled with insecticidal fogging techniques. Pp. 59–75. *In:* S. Sutton, T. Whitmore & A. Chadwick (eds.), *Tropical Rain Forests: Ecology and Management.* Blackwell: Oxford.

Erwin, T. 1988. The tropical forest canopy: the heart of biotic diversity. Pp. 123–129. In: E. O. Wilson & F. M. Peter (eds.), *Biodiversity.* National Academy Press: Washington, D.C.

Glynn, P. W. 1988. El Niño-Southern Oscillation 1982–83: nearshore population, community and ecosystem responses. *Ann. Rev. Ecol. Syst.* 19:309–345.

Glynn, P. W. 1989. Coral mortality and disturbances to coral reefs in the tropical Eastern Pacific. *In:* P. W. Glynn, (ed.), *Global Ecological Consequences of the 1982-83 El Niño-Southern Oscillation.* Elsevier Press: New York.

Grigg, R. W., Polovina, J. J. & M. J. Atkinson. 1984. Model of a coral reef ecosystem. III. Resource limitation, community regulation, fisheries yield and resource management. *Coral Reefs* 3:23–28.

Jackson, J. B. C. 1977. Competition on marine hard substrata: the adaptive significance of solitary and colonial strategies. *Amer. Nat.* 111:743–767.

Jackson, J. B. C. 1979. Overgrowth competition between encrusting cheilostome ectoprocts in a Jamaican cryptic reef environment. *J. Anim. Ecol.* 48:805–823.

Jackson, J. B. C. & L. W. Buss. 1975. Allelopathy and spatial competition among coral reef invertebrates. *Proc. Nat. Acad. Sci.* 72:5160–5163.

Jokiel, P. L. & S. L. Coles. 1977. Effects of temperature on mortality and growth of Hawaiian reef corals. *Mar. Biol.* 43:201–208.

Kiegwin, L. D. 1978. Pliocene closing of the Isthmus of Panama, based on stratigraphic evidence from nearby Pacific Ocean and Caribbean Sea cores. *Geology* 6:630–634.

Kiegwin, L. D. 1982. Isotopic paleoceanography of the Caribbean and East Pacific: role of Panama uplift in Late Neogene time. *Science* 217:350–353.

Manning, R. B. & M. L. Reaka. 1981. *Gonodactylus aloha,* a new stomatopod crustacean from the Hawaiian Islands. *J. Crust. Biol.* 1:190–200.

May, R. M. 1988. How many species are there on Earth? *Science* 241:1441–1449.

Mitchell, J. F. B. 1988. Local effects of greenhouse gases. *Nature* 332:399–400.

Moran, D. P. & M. L. Reaka. 1988. Bioerosion and availability of shelter for benthic reef organisms. *Mar. Ecol. Prog. Ser.* 44:249–263.

Moran, D. P. & M. L. Reaka-Kudla. 1990. The effects of disturbance: disruption and enhancement of coral reef cryptofaunal populations by hurricanes. *Coral Reefs 9.* In press.

Muscatine, L. & J. Porter. 1977. Reef corals: mutualistic symbioses adapted to nutrient-poor environments. *Bioscience* 27:454–460.

Newell, N. D. 1971. An outline history of tropical organic reefs. *Amer. Mus. Novit.* 2465:1–37.

Phillips, J. D. & D. Forsyth. 1972. Plate tectonics, paleomagnetism and the opening of the Atlantic. *Bull. Geol. Soc. Amer.* 83:1579–1600.

Palmer, T. I. 1982. Cambrian to Cretaceous changes in background communities. *Lethaia* 15:309–323.

Reaka, M. L. 1978. The effects of an ectoparasitic gastropod, *Caledoniella montrouzieri,* upon molting and reproduction of a stomatopod crustacean, *Gonodactylus viridus. The Veliger* 21:251–254.

Reaka, M. L. 1979a. The evolutionary ecology of life history patterns in stomatopod Crustacea. Pp. 235–260. In: S. Stancyk, (ed.), *Reproductive Ecology of Marine Invertebrates,* Belle W. Baruch Lib. Mar. Sci., University of South Carolina Press: Columbia, SC.

Reaka, M. L. 1979b. Patterns of molting frequencies in coral-dwelling stomatopod Crustacea. *Biol. Bull.* 156:328–342.

Reaka, M. L. 1980a. Resource limitation in mobile cryptic species: food or space? *Amer. Zool.* 20:885 (abstr.).

Reaka, M. L. 1980b. Geographic range, life history patterns, and body size in a guild of coral-dwelling mantis shrimps. *Evolution* 34:1019–1030.

Reaka, M. L. 1985. Interactions between fishes and motile benthic invertebrates on reefs: the significance of motility vs. defensive adaptations. *Proc. 5th Internat. Coral Reef Congr.* 5:439–444.

Reaka, M. L. 1986a. Biogeographic patterns of body size in stomatopod Crustacea: ecological and evolutionary consequences. Pp. 209–235. *In:* R. H. Gore & K. L. Heck (eds.), *Biogeography of the Crustacea.* Balkema Press: Rotterdam.

Reaka, M. L. 1986b. Island endemism, life histories and evolution of Pacific crustaceans. *2nd Internat. Symp. Indo-Pacific Mar. Biol. Prog./Abstr.* 30.

Reaka, M. L. 1987a. Adult-juvenile interactions in benthic reef crustaceans. *Bull. Mar. Sci.* 41:108–134.

Reaka, M. L. 1987b. Cope's Rule and patterns of evolution in Crustacea. *Amer. Zool.* 27:142 (abstr.).

Reaka-Kudla, M. L. 1988. The feeding habits of macro-benthic invertebrate predators, and their relationship to the trophic structure of reefs. *Proc. 6th Internat. Coral Reef Symp.* (abstr.).

Reaka, M. L. & R. B. Manning. 1981. The behavior of stomatopod Crustacea, and its relationship to rates of evolution. *J. Crust. Biol.* 1:309–327.

Reaka, M. L. & R. B. Manning. 1987a. The significance of body size, dispersal potential, and habitat for rates of morphological evolution in stomatopod Crustacea. *Smith. Contr. Zool.* 448:1–46.

Reaka, M. L. & R. B. Manning. 1987b. The stomatopod Crustacea of Enewetak Atoll. Pp. 181–190. *In:* D. M. Deveney, E. S. Reese, B. L. Burch & P. Helfrich (eds.), *The Natural History of Enewetak Atoll,* 2. Office Sci. Tech. Info., U.S. Dept. Energy: Washington, D.C.

Reaka, M. L., R. B. Manning & D. L. Felder. 1989. The significance of macro- and microhabitat for reproduction in reef-dwelling stomatopods from Belize. Pp. 183–190. *In:* E. A. Ferrero (ed.), *Biology of Stomatopods.* Collana UZI, Muchi Editore: Modena, Italy

Robichaux, D. M., A. C. Cohen, M. L. Reaka & D. Allen. 1981. Experiments with zooplankton on coral reefs, or will the real demersal plankton please come up? *Mar. Ecol.* 2:77–94.

Rogers, C. S. 1979. The effect of shading on coral reef structure and function. *J. Exp. Mar. Biol. Ecol.* 41:269–288.

Rutzler, K. 1975. The role of burrowing sponges in bioerosion. *Oecologia* 19:203–216.

Schiff, H., R. B. Manning & B. C. Abbott. 1986. Structure and optics of ommatidia from eyes of stomatopod Crustacea from different luminous habitats. *Biol. Bull.* 170:461–480.

Strathmann, R. R. & M. F. Strathmann. 1982. The relationship between adult size and brooding in marine invertebrates. *Amer. Nat.* 119:91–101.

Thayer, C. W. 1979. Biological bulldozers and the evolution of marine benthic communities. *Science* 203:458–461.

Tyler, J. C. 1980. Osteology, phylogeny and higher classification of the fishes of the order Plectognathi (Tetradontiformes). *NOAA Tech. Rept. Nat. Mar. Fish. Serv. Circ.* 434:1–422.

Vermeij, G. J. 1987. *Evolution and Escalation. An Ecological History of Life.* Princeton University Press: Princeton, N.J. 527 pp.

Wolf, N. G., E. B. Bermingham & M. L. Reaka. 1983. Relationship between fishes and mobile benthic invertebrates on coral reefs. Pp. 69–78. *In:* M. L. Reaka (ed.), *The Ecology of Deep and Shallow Coral Reefs,* Symp. Ser. Undersea Res., Nat. Oceanic Atmos. Admin. Undersea Res. Prog., Rockville, MD.

Biological Diversity
from the Perspective of Molecular Biology

V. R. Ferris, J. M. Ferris, and J. Faghihi

Abstract. For small invertebrates with simple morphology like the Nematoda, access to the tools of molecular biology has the potential for changing our thinking about numbers of species, their identification and classification, and the way we ascertain relationships. The prevailing opinion at present among nematode taxonomists is that only morphological data should be required for descriptions of new species, because these are the data that are accessible to most practitioners around the world, and morphological data have been the basis of taxonomic descriptions in the past. Protein and nucleic acid data from several laboratories suggest that the conserved morphology of nematodes can be misleading regarding species identities and proximity of relationships, and molecular data may prove to be less equivocal than traditional approaches for identifying cryptic species. If the technologies can be further simplified and the costs reduced, molecular methods will increasingly augment traditional approaches to nematode taxonomy and systematics and, in some cases, may supplant them.

INTRODUCTION

The term biological diversity (or biodiversity) encompasses the variety of life forms and their genetic differences (Murphy, 1988). Access to the tools of molecular biology is changing our thinking about numbers of species and their identification, as well as classification and the way we ascertain relationships. Because we study invertebrates of the order Nematoda, our examples will come from that group of animals, but we hope our observations will have application to a broader spectrum of organisms.

MOLECULAR DATA IN NEMATODE SYSTEMATICS

As compared with insects, for example, nematodes have a simple morphology. Although we now believe that nematodes probably invaded land during the Paleozoic, the simple but effective body plan has probably changed little over the millenia (Ferris et al., 1976; Poinar, 1988). A widespread conviction exists among nematologists that taxonomic descriptions should require only morphological data obtained with a light microscope, mainly because this is the traditional way. In addition, these are the data believed to be accessible to most people, and to require more exotic kinds of data would be to "restrict taxonomic growth and freedom" (Jairajpuri, 1988). Although the value of biochemical methods for evaluating similarities and differences between species is generally recog-

Drs. Ferris, Ferris, and Faghihi are with the Department of Entomology, Purdue University, West Lafayette, IN 47907, USA.

nized, the use of such techniques for "practical taxonomy" is widely perceived to be limited (Jairajpuri, 1988; Triantaphyllou, 1988). For these reasons, nematode descriptions depend heavily on classical morphometric data, preferably from specimens taken from the entire range of the species habitat; arguments persist as to how many measurements are enough and what kinds of morphological difference have significance for taxonomy.

Data from 2-D Protein Patterns

In our laboratory we have been evaluating our concept of species in the Nematoda for the past decade using protein patterns obtained by two-dimensional polyacrylamide gel electrophoresis (2-D PAGE) (Ferris et al., 1989). Similar work has been carried out in the Netherlands (Bakker & Bouwman-Smits, 1988). When we examine multiple geographic isolates, we generally find 90% similarity or more in the protein patterns if those isolates belong to the same species. For example, in a study of geographic isolates from Hawaii, Fiji, and Florida of a large, free-living, predatory soil nematode found worldwide and assigned on the basis of morphology to a single species (*Labronema pacificum* Cobb), we found the protein patterns from the two South Pacific isolates to be nearly 90% similar, but the Florida isolate to share only about one-third of the proteins with the other two. Clearly, we had been misled by morphology and the Florida isolate has to be considered a separate species (Ferris et al., 1987; Ferris & Ferris, 1988).

In a study of two "strains" of a nominal single species (*Heterodera avenae* Woll.) of a cereal cyst nematode of economic importance in many countries, we found the strains to differ by about 75% of the proteins in our patterns. Although a few small morphological differences between these two strains had been observed for years, the balance of informed taxonomic opinion was that these differences were not large enough to warrant separate species status. Therefore, the two strains have been retained in a single species, despite the fact that they differ markedly in their pathogenicity to important host cereal plants. Again, the evolution of morphological differences has lagged behind the macromolecular changes.

Some years ago Russian workers (Krall' & Krall', 1978) proposed a new generic name (*Bidera*) for the cereal cyst nematodes on the basis of host relationships plus morphological differences that they considered sufficient to differentiate them from the rest of the cyst nematodes of the genus *Heterodera*. This action was never accepted by the leaders of the international taxonomic community who found it "difficult to establish clear differentiation between this group and other *Heterodera* species at the generic level" (Luc et al., 1988). Although we have not seen protein patterns for all of the species in the genus, based on those we have seen the basic protein pattern of the cereal cyst nematode isolates differs markedly from that of other members of the genus *Heterodera* to which they are now assigned. Since major differences in macromolecules do appear to exist between these groups of cyst nematodes, the genus *Bidera* will probably be accepted by nematode taxonomists at some future time.

Two potato cyst nematode species have been the focus of protein studies in several laboratories. Until recently, the potato cyst nematodes were considered to comprise a single species because they are nearly indistinguishable morphologically. They do, however, have clear differences in the genetics of their resistance to potatoes. Workers in the Netherlands, where the potato cyst nematodes are important economic plant pests, found the two species to differ by 70% of the proteins in their 2-D patterns, and we reached a similar conclusion using entirely different techniques to obtain the 2-D patterns (Bakker & Bouwman-Smits, 1988; Ferris et al., unpublished). Again, it is evident that despite their conserved morphology, these nematodes are really very different at the level of their molecules.

Data from DNA

The techniques of molecular biology and recombinant DNA are being applied to problems of systematics in many laboratories at the present time. Often, the new molecular data reveal an unexpected amount of diversity among entities thought previously to be very much alike. The first revelations of this sort for nematodes came from persons working with *Caenorhabditis elegans,* the free-living nematode species vaulted to prominence as an important experimental animal for molecular biology by Sydney Brenner and his followers (Brenner, 1974). As early as 1979, Emmons et al. reported that *C. elegans* and a morphologically almost identical "twin" species differed as much in their DNAs as might be expected for species of other kinds of animals that had been separated for tens of millions of years. In addition, two strains of *C. elegans* showed a degree of nucleotide difference comparable to that observed between distinct species for other kinds of animals, despite the fact that the strains are morphologically indistinguishable and are cross fertile. Molecular techniques using nuclear and/or mitochondrial DNA are now being employed in many nematology laboratories, including our own, to evaluate molecular similarities among morphologically similar taxa. Results from a number of interesting studies have been published, and many more are on the way (Curran & Webster, 1987; Pableo & Triantaphyllou, 1986; Powers et al., 1986). Doubtless, the new data will have a marked effect on taxonomy of the future.

Effect of Molecular Data on Our Concepts of Systematic Relationships

An important issue that must be considered briefly is what, if anything, molecular data can contribute to classification and understanding of the biodiversity we discover. Controversy still exists as to whether classifications based on phylogenetic relationships are even possible for the Nematoda. Some of the most prominent nematode systematists today continue to insist that the absence of fossils dooms our attempts to understand the evolution of nematodes to little more than "intellectual games" (Luc et al., 1987; Jairajpuri, 1988). Among biologists generally, however, a realization has occurred that every organism carries a record of its history encoded in its DNA and that this record may prove to be less cryptic than the history revealed by either morphology or the fossil record (Clark, 1988; Patterson, 1988; Sibley & Ahlquist, 1988). If credible phylogenies can be constructed from ribosomal RNA for prokaryotes, which have even less morphological structure than do nematodes and no fossil record (Lake, 1988; Woese, 1988), the view recently expressed by a prominent nematode systematist that it will not be possible to work out the evolution of nematodes "in the foreseeable future," seems unduly pessimistic (Jairajpuri, 1988). Many examples now exist showing that molecular data can be substituted for morphological data in most of the modern procedures for reconstructing the evolutionary history of organisms, despite continuing debate concerning the best methodologies to employ in the basic algorithms (Patterson, 1988; Ferris & Ferris, 1987, 1988).

THE PRACTICALITY OF MOLECULAR VERSUS MORPHOLOGICAL DATA

Practical Applications of Molecular Data Today

Among nematologists a strong conviction exists that molecular data will always be secondary to morphological data for practical taxonomy and systematics. The techniques for obtaining molecular data are widely perceived to be more complicated and esoteric and not suitable for routine identification and description of new species (Jairajpuri, 1988; Triantaphyllou, 1988).

Examples are beginning to accumulate where biochemical data have proved to be more useful than the traditional approach for routine identifications. In many diagnostic laboratories esterase phenotypes are proving to be the method of choice for distinguishing among the common species of root knot nematodes (*Meloidogyne* species). Although these species, which are of great economic importance worldwide, are very difficult to sort out morphologically, their crop host differences make it imperative that they be diagnosed to species. Rapid electrophoresis on small pre-cast gels has proven to be a simple, efficient, economical, and accurate method for diagnosis of this ubiquitous group of plant pests and the technology is likely to spread quickly to all parts of the world (Esbenshade & Triantaphyllou, 1985; Triantaphyllou, 1988).

Several years ago we were sent two isolates each of the two potato cyst nematode species mentioned above, from a laboratory that supplied breeders over the world with nematodes of these species for genetic studies. To our surprise the two isolates of the first species (*Globodera rostochiensis* Woll.) gave us two different 2-D protein patterns; the two isolates of the second species (*G. pallida* Stone) gave us the same two (different) patterns. Clearly, we had two distinct patterns, but one isolate of each species had been misidentified. We requested more samples of each isolate and obtained the same result. One of the "morphological" differences between the two species is that one (*G. rostochiensis*) passes through a gold stage as the young white female becomes a brown cyst, whereas the second species (*G. pallida*) does not pass through a gold stage. Because we assumed that the laboratory that sent them knew which turned gold and which did not, we had to choose between two alternatives: either the gold color criterion to distinguish the two species of potato cyst nematode was not valid, or the laboratory that sent them to us had experienced a mixup in their species collections at some point in history. We checked the first possibility by growing the isolates again and holding them past the young white female stage at which we usually collect our samples for protein analysis, to see which isolates went through a gold phase. Despite the misleading tags on the sample vials, the two that turned gold were isolates with the same protein pattern, and the two that did not turn gold had the second protein pattern (Fig. 1). We therefore notified the laboratory of the mixup in their isolates.

Although the 2-D PAGE procedure is somewhat cumbersome for routine laboratory use, it did alert us to the identity problems with the potato cyst nematodes, and was critical in a quick resolution of the matter. More recently, two DNA probes have been found by nematologists in England that differentiate the two species from each other by means of a simple dot blot test (Burrows & Perry, 1988). A non-radioactive biotin label for the probes is said to be sensitive to as little as one small juvenile nematode. Thus, the potential for a simple diagnostic test for routine identification of these widely studied species now exists (Burrows, 1988).

Practical Applications of Molecular Data in the Future

Although some progress has been made, we are only at the threshold of the exploitation of molecular techniques for use in the systematics of small invertebrates like the Nematoda. We consider it likely that the relatively new technique of polymerase chain reaction (PCR) will give impetus to more widespread use of molecular methods for systematics and rapid diagnosis (White et al., 1989). As a body of literature accumulates regarding the specific DNA sequences that are best for revealing relationships at a given taxonomic level or for rapid diagnosis for groups of widespread interest, PCR will provide the means for producing large amounts of the desired DNA from very small numbers of the target organisms. Many of the cloning steps, now necessary to produce useful probes, can be circumvented, as well as the need to grow large numbers of nematodes to provide ample quantities of genomic DNA for analysis. The small amounts of DNA needed for

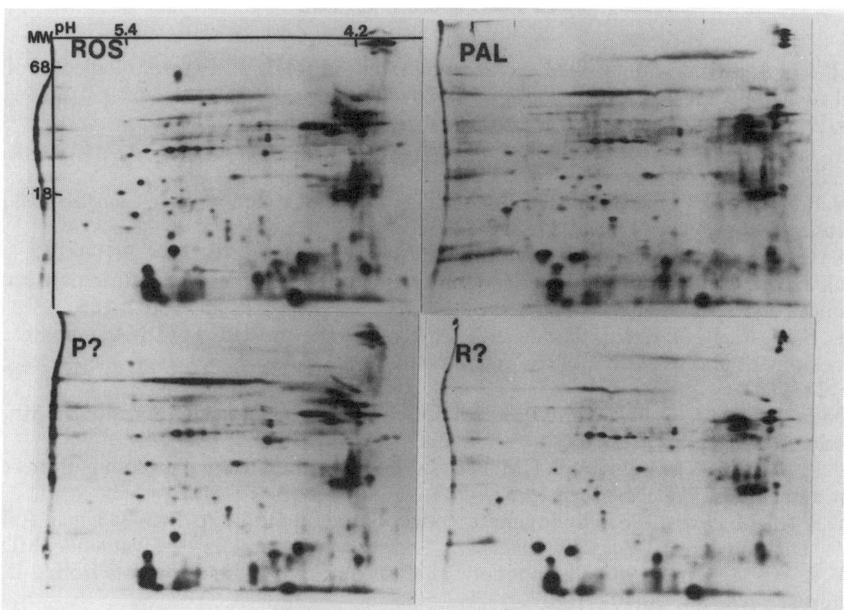

Figure 1. Protein patterns from two-dimensional polyacrylamide gel electrophoresis for the potato cyst nematodes, *Globodera rostochiensis* (left) and *G. pallida* (right). The isolates from which the bottom patterns were prepared were mislabeled, but the protein patterns confirmed their true identity (see text). Molecular weights are given in thousands.

amplification can be produced from a few specimens preserved in alcohol, making import permits for living specimens unnecessary. The amplified DNA sequences can be preserved indefinitely in a frozen state and provide the substrate for infinite amplifications through time.

In due time, as many laboratories become versed in using the tools of molecular biology, it may become commonplace to consider DNA characteristics of primary importance, even for species descriptions, and to place secondary importance on morphological data, including morphometrics. Microscopic analysis will probably always be the most efficient way to provide the first cut in placing a specimen in a broad taxonomic category, but after that biochemical methods may prove to be more efficient and informative. At the present time, the costs of doing molecular biology are prohibitive for many laboratories. Any widespread shift to the use of molecular data rather than morphological data in the systematics of nematodes will require cheaper reagents and laboratory equipment. The production of easy to use economical kits may alleviate some of the cost, or it may be that nematologists will wish to send specimens to special laboratories for molecular data analysis. Whatever happens, it is unlikely that morphology will be retained as the basis of nematode systematics simply because it has historically had that role (Jairajpuri, 1988). Nematode systematics will evolve and change along with the tools to do it in the most efficient and informative manner. Molecular and morphological data will be used together in a process of reciprocal illumination. All of our cherished preconceived ideas regarding numbers of species, numbers of categories, classification, and relationships will have to stand the test of new data as they develop.

ACKNOWLEDGMENTS

Scientific paper, Purdue Agricultural Experiment Station. Research supported in part by NSF grant BSR 8706759 and USDA grants 87-CRCR-1-2289 and 89-37231-4492.

LITERATURE CITED

Bakker, J. & L. Bouwman-Smits. 1988. Contrasting rates of protein and morphological evolution in cyst nematode species. *Phytopathology* 78:900–904.

Brenner, S. 1974. The genetics of *Caenorhabditis elegans. Genetics* 77:71–94.

Burrows, P. R. 1988. The differentiation of *Globodera pallida* from *G. rostochiensis* using species specific DNA probes. *Nematologica* 34:260.

Burrows, P. R. & R. N. Perry. 1988. Two cloned fragments which differentiate *Globodera pallida* from *G. rostochiensis. Revue Nématol.* 11:441–445.

Clark, A. 1988. Review of molecular evolutionary genetics. *Molec. Biol. Evol.* 5:109–112.

Curran, J. & J. M. Webster. 1987. Identification of nematodes using restriction fragment length differences and species specific DNA probes. *Canadian J. Plant Pathol.* 9:162–166.

Emmons, S. W., Klass, M. R. & D. Hirsh. 1979. Analysis of the constancy of DNA sequences during development and evolution of the nematode *Caenorhabditis elegans. Proc. Natl. Acad. Sci. USA* 76:1333–1337.

Esbenshade, P. R. & A. C. Triantaphyllou. 1985. Use of enzymes for identification of *Meloidogyne* species. *J. Nematol.* 17:6–20.

Ferris, V. R., Faghihi, J., Ireholm, A. & J. M. Ferris. 1989. Two-dimensional protein patterns of cereal cyst nematodes. *Phytopathology* 79:927–933.

Ferris, V. R. & J. M. Ferris. 1987. Phylogenetic concepts and methods, Pp. 346–343. *In:* J. A. Veech & D. W. Dickson (eds.), *Vistas on Nematology.* Society of Nematologists: Hyattsville, MD.

Ferris, V. R. & J. M. Ferris. 1988. Phylogenetic analysis in dorylaimids using data from 2-D protein patterns. *J. Nematol.* 20:102–108.

Ferris, V. R., Ferris, J. M., Murdock, L. L. & J. Faghihi. 1987. Two-dimensional protein patterns in *Labronema, Aporcelaimellus,* and *Eudorylaimus* (Nematoda:Dorylaimida). *J. Nematol.* 19:431–440.

Ferris, V. R., Goseco, C. G. & J. M. Ferris. 1976. Biogeography of free-living soil nematodes from the perspective of plate tectonics. *Science* 193:508–510.

Jairajpuri, M. S. 1988. Taxonomy of nematodes, Pp. 101–111. *In:* M. A. Maqbool, A. M. Golden, A. Ghaffar, & L. R. Krusberg (eds.), *Advances in Plant Nematology.* Shamim Printing Press: Karachi, Pakistan.

Krall', E. L. & Kh.A. Krall'. 1978. [Revision of the plant nematodes of the family Heteroderidae on the basis of the trophic specialization of these parasites and their co-evolution with their host plants]. *Fitogel'mintol.* Issledovaniya, Moscow, USSR, "Nauka":36–56.

Lake, J. A. 1988. Origin of the eukaryotic nucleus determined by rate-invariant analysis of rRNA sequences. *Nature* 331:184–186.

Luc, M., Maggenti, A. R. & R. Fortuner. 1988. A reappraisal of Tylenchina (Nemata). 9. The family Heteroderidae Filip'ev & Schuurmans Stekhoven, 1941. *Revue Nématol.* 11:159–176.

Luc, M., Maggenti, A. R., Fortuner, R., Raski, D. J. & E. Geraert. 1987. A reappraisal of Tylenchina (Nemata) 1. For a new approach to the taxonomy of Tylenchina. *Revue Nématol.* 10:127–134.

Murphy, D. D. 1988. Challenges to biological diversity, Pp. 71–76. *In:* E. O. Wilson (ed.), *Biodiversity.* National Academy Press: Washington, D.C.

Pableo, E. C. & A. C. Triantaphyllou. 1986. DNA complexity and relationship of the genome of *Meloidogyne* species *J. Nematol.* 18:626.

Patterson, C. 1988. Introduction, Pp. 1–22. *In:* C. Patterson (ed.), *Molecules and Morphology in Evolution: Conflict or Compromise.* Cambridge University Press: Cambridge.

Poinar, G. O. 1988. The impact of nematodes on mankind in historical perspective. *Nematologica* 34:249–301.

Powers, T. O., Platzer, E. C. & B. C. Hyman. 1986. Species-specific restriction site polymorphism in root-knot nematode mitochondrial DNA. *J. Nematol.* 18:288–293.

Sibley, C. G. & J. E. Ahlquist. 1988. Avian phylogeny reconstructed from comparisons of the genetic material, DNA, Pp. 95–121. *In:* C. Patterson (ed.), *Molecules and Morphology in Evolution: Conflict or Compromise.* Cambridge University Press: Cambridge.

Triantaphyllou, A. C. 1988. Biochemical approaches in nematode taxonomy and identification, Pp. 113–121. *In:* M. A. Maqbool, A. M. Golden, A. Ghaffar, & L. R. Krusberg (eds.), *Advances in Plant Nematology.* Shamim Printing Press: Karachi, Pakistan.

White, T. J., Arnheim, N. & H. A. Erlich. 1989. The polymerase chain reaction. *Trends in Genetics* 5:185–189.

Woese, C. R. 1988. Macroevolution in the microscopic world, Pp. 177–203. *In:* C. Patterson (ed.), *Molecules and Morphology in Evolution: Conflict or Compromise.* Cambridge University Press: Cambridge.

Extinction and Evolution:
The Fossil Record

EXTENDED ABSTRACT

David Jablonski

Extinction has long been recognized as an integral part of the evolutionary equation, but only recently have paleontologists begun to focus on the evolutionary role of extinction, particularly mass extinctions. Recent results suggest that mass extinctions are more important to large-scale evolutionary pattern and process than previously thought, in terms of both eliminating adaptations or clades not previously at risk, and in providing opportunities for diversification among survivors. However, the details of this complex process, played out across several levels in the ecological and evolutionary hierarchies and across broad temporal and geographic scales, are still poorly understood.

THE DATA

The geological timescale was originally built on the basis of local changes in sedimentary rock composition and, more commonly, changes in the fossil content of the rocks (Berry, 1987). Certain biological changes proved to be very widespread or even global in extent, and these biostratigraphic boundaries are the underpinning of today's finely divided and hierarchically arranged timescale of geologic eras, periods, stages, substages, and zones.

Local outcrops and cores are the ultimate source of all the data used to analyse mass extinctions. There are pitfalls, however, to taking the fine structure of local data at face value, and outcrop-scale studies have suffered from massive overinterpretation during the last few years. The problem seems straightforward, but is subtle and far-reaching in its implications: environmental change and breaks in deposition are the rule in virtually all local sedimentary sequences, so that patterns may be generated that are unrelated to global extinctions. Thus breaks in sedimentation will artificially truncate ranges so that extinctions will appear grouped at a particular time horizon; if extinction is roughly constant, the longer the break, the larger the apparently abrupt, synchronous extinction (e.g., Birkelund & Hakansson, 1982). However, if local environments change more gradually, or if sampling density or quality changes through time, an artificially gradual extinction will be seen in the local section (Signor & Lipps, 1982; Raup, 1986). Such artificial extinction is most easily detected when some of the lineages return to the fossil record once favorable

Dr. Jablonski is with the Department of Geophysical Sciences, University of Chicago, 5734 South Ellis Avenue, Chicago, IL 60637, USA.

78

conditions return. This return from apparent extinction has been termed the Lazarus Effect, and can be used to calibrate observed extinctions against demonstrably artificial ones (Jablonski, 1986a; also Raup, 1986). Raup (1989) has shown that these artificially drawn-out extinctions are in fact more likely to seem step-like than smoothly declining, because most fossil lineages are somewhat sporadically distributed through sediments, or differ in their relative abundances. The normal, discontinuous nature of fossil distributions will yield artificial stepwise extinction prior to even the most instantaneous extinction event (see also Koch, 1987).

One response to such problems has been to work at supraspecific levels and sum extinction data over many outcrops and many regions. This approach inevitably sacrifices stratigraphic resolution and taxonomic precision, but provides a generalized global picture that has proven extremely valuable in recognizing or confirming major events and in testing hypotheses on magnitudes and timing of extinctions (e.g., Sepkoski, 1986, 1989; Raup & Sepkoski 1986). The most productive approach has been one that draws in reciprocal fashion on the strengths of studies across a spectrum of scales, from the local outcrop to the province to the global scale, to test the robustness of the observed patterns.

Species-level extinctions can be estimated from observed genus- or family-level losses, on the basis of present-day frequency distributions of species within higher taxa. Thus, the huge Permo-Triassic extinction would have had to remove about 95% of the species to produce the observed losses of about half the marine families and 84% of the genera (Raup, 1979; Sepkoski, 1989, who tabulates extinction magnitudes for the major episodes of the past 250 million years). These back-calculations require simplifying assumptions, not least of which is the generality of taxonomic frequency distributions within the Recent echinoids, which provide the species-family distributions. Further, as Simberloff (1986) points out, the model assumes that congeneric species are no more likely to suffer similar fates than unrelated species. A more realistic approach, that species in the same genus have similarly high or low extinction probabilities, might yield slightly lower species-level extinction estimates as back-calculated from genera or families. However, such refinements are unlikely to change the broad outlines of the story, in terms of relative rankings or approximate absolute magnitudes of events.

SURVIVORSHIP

Mass extinctions might have one of three possible evolutionary effects (Jablonski 1986a,b,c; Raup, 1986). 1) Intensification of survivorship patterns seen during times of background extinction, so that taxa that are usually extinction-prone suffer most severely and the normally extinction-resistant taxa show relatively mild losses. This process will do little more than accelerate background processes. 2) Random removal of taxa, with survival essentially a matter of chance, not biology. Such extinctions would profoundly disrupt evolutionary processes of background times. 3) Changes in the rules of survivorship, so that survivorship is neither random nor determined entirely by the same factors important during background times. This would also disrupt background processes, but might allow for more continuity between the two regimes, e.g., whenever taxa happen to carry traits that are favorable under both background and mass extinctions regimes.

Background and mass extinction patterns in Late Cretaceous mollusks of North American and Europe most closely fit the last of these alternatives (Jablonski, 1986c, 1989a). Species-rich clades are generally more extinction-resistant than species-poor ones, but during the end-Cretaceous extinction, there was no signficant difference in survivorship between the two categories in European and North American bivalves and gastropods (e.g., among bivalves, 45% of species-poor genera and 47% of species-rich genera survived in North America, and 55.4% and 54.5% respectively in Europe). During background times, genera composed of widespread species tend to be more extinction-resistant than genera composed mainly of geographically restricted species, but geographic range at the species level had no effect on genus-level survivorship in North American mollusks (Jablonski, 1986c; data not yet available for European genera).

The positive interaction between species-level geographic range and species richness that prevails during background times is also lost during the end-Cretaceous mass extinction: species-rich genera comprising widespread species, normally the most extinction-resistant category, show no greater survivorship than the species-poor genera comprising localized species, normally the most extinction-prone category (Jablonski, 1986c). Comparable data for other taxa or extinction events are sparse, but species richness is evidently no advantage for Late Cambrian trilobites (Westrop, 1989), end-Ordovician bryozoans (Anstey, 1986), late Devonian corals (Sorauf & Pedder 1986), late Devonian ammonoids (House, 1985; see Ward & Signor, 1983, on the general relation between genus-level and species-level patterns in ammonoids), or end-Cretaceous echinoids (McKinney, 1988). In the one exception I have encountered, Erwin (1989), found species richness to enhance gastropod survivorship during the end-Permian mass extinction.

Despite the apparent disappearance of several buffers against extinction, generic survivorship is not random during the end-Cretaceous and other mass extinction events. While geographic range at the species level has no significant effect on clade survivorship, widespread genera (regardless of the ranges of constituent species) preferentially survive the end-Cretaceous extinction, in both European and North American mollusks (Jablonski, 1986c, 1989a). Preferential survival of widespread genera is recorded for other mass extinctions as well, from Late Cambrian trilobites (Fortey, 1983, Westrop, 1989) to end-Ordovician brachiopods (Sheehan & Coorough, 1990) and bryozoans (Anstey, 1986), to end-Permian gastropods (Erwin, 1989) and bivalves from all of the major extinction events from the end-Ordovician to the end-Triassic (Bretsky, 1973). This effect is difficult to detect during background times, presumably in part because of the timescales involved, but also because it is overwhelmed by other determinants of survivorship (see Jablonski, 1986b). During mass extinctions, however, geographic range at the clade level emerges as one of the clearest predictors of survival.

Jablonski (1986a,b,c, 1989a) has discussed the implications of the change of survivorship patterns between background and mass extinction regimes. Traits that favor survival during mass extinctions need have little correlation with those that enhance survival and diversification during background times, so mass extinctions can have unpredictable and lasting evolutionary effects. Taxa and morphologies might be lost not because they are maladaptive as measured under background conditions (which of course constitute the bulk of geologic time), but because they happened to lack the appropriate biogeographic deployment or other features necessary to weather the mass extinction. Not only do mass extinctions remove adaptations or clades honed under background regimes, they create important ecological and evolutionary opportunities by removing incumbent, dominant taxa and enabling other taxa to radiate in the aftermath of the extinction event (Sheehan, 1982 and this volume; Benton, 1987 and this volume; Hallam, 1987).

Evolutionary rebounds from mass extinctions are clearly important evolutionary phenomena, but they are still poorly known. Accelerated evolutionary rates following mass extinctions are well known (Sheehan, 1982; Hallam, 1987; Miller & Sepkoski, 1988), but rebounds are protracted events when viewed on ecological time scales—as is particularly evident in the history of reefs, where mass extinctions are often followed by five to ten million years before refurbished reef communities appear (Sheehan, 1985; Copper, 1988; Talent, 1988). Jablonski (1989b) found significant differences in post-Cretaceous rebound patterns in the molluscan faunas of Europe and North America. For example, Hansen's (1988) "bloom taxa," which diversify immediately after the extinction in North America, show little such tendency in Europe. The differences could represent geographic variation in faunal vulnerability, geographic variation in the severity of the end-Cretaceous perturbation, or both. These disparities add another level of complexity to the study of post-extinction rebounds, and suggests that no one region can provide an adequate picture of post-extinction recovery.

This is not to say that mass extinctions dominate the entire evolutionary process. Mass extinctions may remove adaptations, and may allow previously obscure groups to move to center stage, but they do not build adaptations. Adaptations like wings or eyes evolve under background regimes—although mass extinctions may have played a significant part in determining how many winged species existed at any point in geologic time. Further, some large-scale evolutionary patterns transcend mass extinctions (Jablonski, 1986b; McKinney & Jackson 1989), although why some trends cross extinction boundaries and others do not is very much an open question. Detailed comparisons of global and clade-specific extinction intensities, within and among extinction episodes, may help to address these problems.

CONCLUSIONS

1. Mass extinctions should be studied at a variety of temporal and geographic scales, from the local outcrop to the global compilation; no one scale gives an adequate picture.

2. Mass extinctions are non-constructive relative to background extinction regimes; selectivity as well as extinction intensity changes during mass extinction intervals. This shift in the rules of survivorship regime can channel evolution in unexpected directions.

3. Evolutionary rebounds may be geologically rapid, but they are not instantaneous and are extremely slow when viewed over ecological timescales. These rebounds may be geographically heterogeneous, adding further complexity to the potential evolutionary consequences of mass extinctions.

ACKNOWLEDGMENTS

I thank S. M. Kidwell, D. M. Raup, and J. J. Sepkoski, Jr., for many illuminating discussions. The research reported here was supported by NSF grants EAR84-17011 and INT86-2045.

LITERATURE CITED

Anstey, R. L. 1986. Bryozoan provinces and patterns of generic evolution and extinction in the Late Ordovician of North America. *Lethaia* 19:33–51.

Benton, M. J. 1987. Progress and competition in macroevolution.*Biol. Rev.* 62:305–338.

Berry, W. B. N. 1987. *Growth of a Prehistoric Time Scale.* 2nd Ed. Blackwell: Palo Alto, CA.

Birkelund, T. & E. Hakansson. 1982. The terminal Cretaceous extinction in Boreal shelf seas—A multicausal event. *Geol.Soc. Am. Spec. Paper* 190:373–384.

Bretsky, P. W. 1973. Evolutionary patterns in the Paleozoic Bivalvia: documentation and some theoretical considerations. *Geol. Soc. Am. Bull.* 84:2079–2096.

Copper, P. 1988. Ecological succession in Phanerozoic reef ecosystems: Is it real? *Palaios* 3:136–151.

Erwin, D. H. 1989. Regional paleoecology of Permian gastropod genera, southwestern United States and the end-Permian mass extinction. *Palaios* 4:424–438.

Fortey, R. A. 1983. Cambrian-Ordovician trilobites from the boundary beds in western Newfoundland and their phylogenetic significance. *Spec. Pap. Palaeontol.* 30:179–211.

Hallam, A. 1987. Radiations and extinctions relative to environmental change in the marine Lower Jurassic of northwest Europe. *Paleobiology* 13: 152–168.

Hansen, T. A. 1988. Early Tertiary radiation of marine molluscs and the long-term effects of the Cretaceous-Tertiary extinction. *Paleobiology* 14:37–51.

House, M. R. 1985. Correlation of mid-Palaeozoic ammonoid evolutionary events with global sedimentary perturbations. *Nature* 313:17–22.

Jablonski, D. 1986a. Causes and consequences of extinction: a comparative approach. Pp. 183–229. *In* D. K. Elliott (ed.), *Dynamics of Extinction*. Wiley: New York.

Jablonski, D. 1986b. Evolutionary consequences of mass extinctions. Pp. 313–329. *In*: D. M. Raup & D. Jablonski (eds.), *Patterns and Processes in the History of Life*. Springer-Verlag: Berlin.

Jablonski, D. 1986c. Background and mass extinctions: the alternation of macroevolutionary regimes. *Science* 231:129–133.

Jablonski, D. 1989a. The biology of mass extinction: a palaeontological view. *Phil. Trans. Roy. Soc. London* B 325:357–368.

Jablonski, D. 1989b. Cretaceous-Tertiary bivalves of Europe and North America: comparison of patterns of extinction and rebound (Abstr.). *Geol. Soc. Am. Abstr.* 21:A208.

Koch, C. F. 1987. Prediction of sample size effects on the measured temporal and geographic distribution patterns of species. *Paleobiology* 13:100–107.

McKinney, F. K. & J. Jackson. 1989. *Bryozoan Evolution*. Unwin Hyman: Boston. 238 pp.

McKinney, M. L. 1988. Extinction selectivity: a key to macroevolutionary processes (Abstr.) *Geol. Soc. Am. Abstr.* 20:A205.

Miller, A. I. & J. J. Sepkoski, Jr. 1988. Modeling bivalve diversification: the effect of interaction on a macroevolutionary system. *Hist. Biol.* 1:251–273.

Raup, D. M. 1979. Size of the Permo-Triassic bottleneck and its evolutionary implications. *Science* 206:217–218.

Raup, D. M. 1986. Biological extinction in Earth history. *Science* 231:1528–1533.

Raup, D. M. 1989. The case for extraterrestrial causes of extinction. *Phil. Trans. Roy. Soc. London* B 325:421–435.

Raup, D. M. & J. J. Sepkoski, Jr. 1986. Periodic extinction of families and genera. *Science* 231:833–836.

Sepkoski, J. J., Jr. 1986. Phanerozoic overview of mass extinction. Pp. 277–295. *In*: D. M. Raup & D. Jablonski (eds.), *Patterns and Processes in the History of Life*. Springer-Verlag: Berlin.

Sepkoski, J. J., Jr. 1989. Periodicity in extinction and the problem of catastrophism in the history of life. *J. Geol. Soc. London* 146:7–19.

Sheehan, P. M. 1982. Brachiopod macroevolution at the Ordovician-Silurian boundary. *Proc. 3rd N. Amer. Paleont. Conv.* 2:477–481.

Sheehan, P. M. 1985. Reefs are not so different—they follow the evolutionary patterns of level-bottom communities. *Geology* 13: 46–49.

Sheehan, P. M. & P. J. Coorough. 1990. Brachiopod zoogeography across the Ordovician-Silurian extinction event. *Geol. Soc. London Mem.* 12:181–187.

Signor, P. W. III & J. H. Lipps. 1982. Sampling bias, gradual extinction patterns, and catastrophes in the fossil record. *Geol. Soc. Am. Spec. Paper* 190:291–296.

Simberloff, D. S. 1986. Are we on the verge of a mass extinction in tropical rain forests? Pp. 165–180. *In* D. K. Elliott (ed.), *Dynamics of Extinction*. Wiley: New York.

Sorauf, J. E. & A. E. H. Pedder. 1986. Late Devonian rugose corals and the Frasnian-Famennian crisis. *Can. J. Earth Sci.* 23:1265–1287.

Talent, J. A. 1988. Organic reef-building: episodes of extinction and symbiosis? *Senckenbergiana Lethaea* 69:315–368.

Ward, P. D. & P. W. Signor III. 1983. Evolutionary tempo in Jurassic and Cretaceous ammonites. *Paleobiology* 9:183–198.

Westrop, S. R. 1989. Macroevolutionary implications of mass extinction—evidence from an Upper Cambrian stage boundary. *Paleobiology* 15:46–52.

Taxonomy and Extinction Patterns

R. A. Fortey

Abstract.   The importance of the taxonomic base in the analysis of extinction patterns has not been sufficiently acknowledged. The terminations of taxa that are good clades correctly record extinction events. The disappearance of paraphyletic taxa is ambiguous. Paraphyletic taxa, as erected by evolutionary systematists in the past, are characteristic of times of accelerated cladogenesis. The "disappearance" of such taxa merely reflects the acquisition of advanced characters, and is no extinction. Because the Cambrian to Ordovician time period was one of intense evolutionary activity it is necessary to beware of such pseudoextinctions. The aftermath of the Cambrian evolutionary "explosion" has been considered the mass winnowing out of groups of high taxonomic status—but this entirely depends on the interpretation of the phylogenetics of these animals. The Cambrian-Ordovician boundary exhibits similar problems. However, by all phylogenetic criteria the end Ordovician marks a true mass extinction.

INTRODUCTION

There is now an overwhelming literature concerning periods of mass extinction, and most of that literature depends to some considerable extent on taxonomy. This was as true of the pioneering studies (Newell, 1967) as it is of those treatments using sophisticated statistical techniques for analysis of pattern in the fossil record (Raup & Sepkoski, 1986). There has, perhaps, been a progression from higher to lower taxonomic level in analysis (from superfamilies to families, and now to genera, each taken as a surrogate for species) as more and more data have become available. There are now innumerable "case histories"—narrative accounts of this event or that as recorded by the disappearance of species through particular rock sections (Donovan, 1989). There is no occasion here to add to these accounts, which have been ably summarised elsewhere (Jablonski, 1986). However, discussions of the influence of taxonomic method on the recognition of patterns of extinction have been far fewer (Smith & Patterson, 1988; Patterson and Smith, 1987; Fortey, 1989). This is surprising, given the fundamental role of taxonomy in all the historical narratives of extinction. It may be that taxonomic method seems, somehow, less a matter for debate than statistical methodology to those considering extinction problems. To redress this imbalance it may be of some service to briefly consider these taxonomic influences on extinction studies, especially with regard to some of the major events claimed in the Lower Palaeozoic.

NATURAL GROUPS AND EXTINCTION PATTERNS

Extinction patterns are based on the termination of taxa. If all supraspecific taxa were natural groups (monophyletic groups of species descended from a common ancestor) there would be no problem about the reality of an event recorded by the demise of such a

---

Dr. Fortey is with The Natural History Museum, Cromwell Road, London SW7 5BD, England.

group. Unfortunately, not all taxa conform to this ideal, because taxa were named for a variety of historical reasons. The simplest kind of error happens because stratigraphic boundaries have been *assumed* to be taxonomic boundaries also—different workers have their fields of expertise on either side of such a boundary, and apply their own taxonomies, without regard to the possibilities of phylogenetic connections between earlier and later taxa (an example would be scleractinian corals before and after the K-T boundary). One does not have to be a cladist to see how misleading such taxonomy can be, because "termination" in such cases may be no more than an arbitrary change of name. At higher (e.g., family) levels some fossil taxa are known from a single species of doubtful relationships; the status of such a family represents no more than taxonomic uncertainty and its "extinction" means no more than the termination of a species. Smith and Patterson (1988) showed that some of the data used by Raup & Sepkoski (1984, 1986) in their analysis of post-Permian mass extinctions was flawed in such taxonomic detail. But perhaps the most important source of confusion, because the most subtle, is the meaning of "extinction" of paraphyletic groups. A surprising number of Raup & Sepkoski's extinctions proved to be paraphyletic taxa according to Patterson and Smith (1987)—is it a coincidence that the disappearances of such taxa are concentrated at horizons of alleged mass extinction?

PARAPHYLETIC GROUPS AND EXTINCTION PATTERNS

Paraphyletic groups include those derived from a single ancestor but *not* including all its descendant species (Eldredge & Cracraft, 1980). Such groups were erected on the basis of shared similarities, which phylogenetic analysis later revealed to be primitive characters. The recognition of paraphyly is one of the achievements of cladistics, but it is not necessary to be a cladist to appreciate the relevance of paraphyletic groups in the extinction debate. It is a characteristic of the early phases of evolutionary radiation (major cladogenesis) that early members of separate clades tend to be "lumped" together in paraphyletic taxa (Fortey, 1989). This is because the majority of their characters are still those of the common ancestor: they share a general resemblance which led earlier workers to classify them together. The result may be paraphyletic families united by symplesiomorphies alone—and within those families paraphyletic genera ("grade groups" in more traditional parlance). Among the planktic graptolites, for example, the early family Anisograptidae is paraphyletic and includes within it all the genera described for the first 10 million years of the history of the group. The extinction of such families is, of course, no extinction at all, but merely records the evolutionary *acquisition* of a distinctive advanced character which is the basis of a "good" clade (Fig. 1). This is a most important difference from a true extinction (it is a taxonomic pseudoextinction in the terminology of Briggs et al., 1988). Only in those cases where the paraphyletic group includes species that persist to a younger horizon than that of the derivation of the true clades could the termination of such a group act as a measure of true extinction in the sense of Raup and Sepkoski—but such cases are not explicit from the taxonomy alone.

The implication is that at times of accelerated cladogenesis apparent extinction rates may rise. If so, paraphyletic groups as conceived by traditional evolutionary taxonomists should register a rise at the same time. To many cladists paraphyletic groups have no meaning, but it *is* of more than passing interest to record the historical reason for their concentration at certain horizons. Because accelerated cladogenesis is often supposed to follow on from major extinction, the two conflated together may account for peaks in generic or familial last appearances; real extinction and taxonomic pseudoextinction cooperate to elevate the peaks of last appearances. But only modern taxonomic analysis on a cladistic basis will reveal satisfactorily what is the case (Patterson & Smith, 1989). Lest this seem excessive shroud-waving on behalf of taxonomy (and taxonomists in critical mood often seem as welcome as the spectre at the feast) the three examples discussed will show

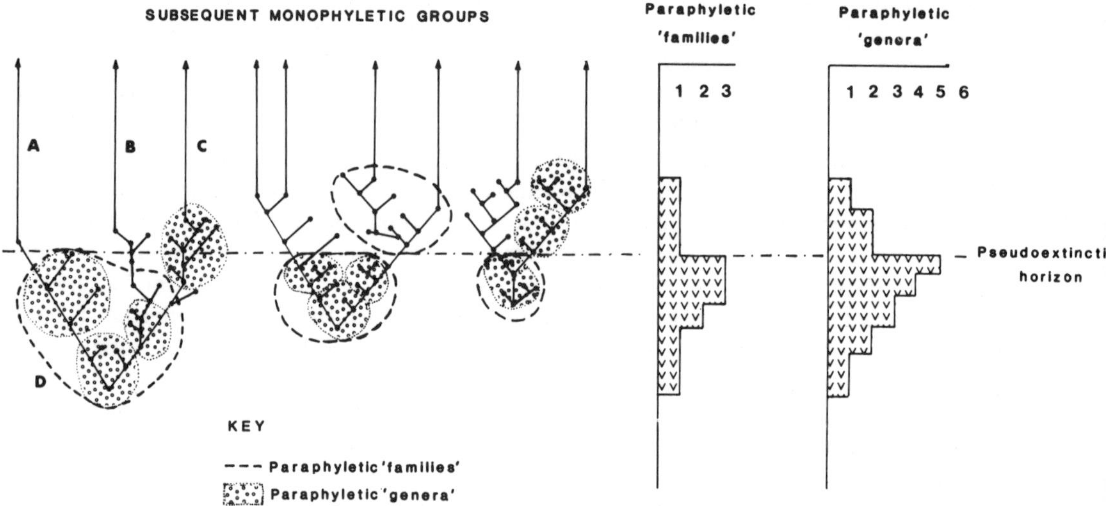

Figure 1. How paraphyletic families and genera can produce pseudoextinctions. Early phases of cladogenesis (species represented by dots, time vertical) are prone to having taxa lumped within paraphyletic genera and families, prior to their separation into 'good' clades (arrows). If such cladogenesis is concentrated within short periods horizons of exceptional apparent 'extinction' result. After Fortey (1989).

how the taxonomic debate cannot be avoided when considering important episodes of extinction.

The Cambrian Evolutionary "Explosion"

The Cambrian fossil record introduces nearly all our living phyla. Recent accounts of the Lower Cambrian have emphasised the supposedly "explosive" nature of the radiation at the base of the Cambrian, and nobody would question the sudden appearance at that level of many kinds of animals with preservable hard parts. Thanks to the success of Gould's (1989) book on the Burgess Shale (Middle Cambrian) the curious animals from that locality, some of which are claimed to have designs as distinctive as any of those of the living phyla, are now widely known. The implication is that more phyla existed in the Cambrian than there are now: "as many as 100 phyla may have existed in the Cambrian, and only 5 per cent or less show any evidence of a Precambrian history" (McMenamin & McMenamin, 1989, p. 168). Burgess Shale type fossils are now known in the Lower Cambrian of China and Greenland. Many of the curious shelly fossils are known *only* in the Lower Cambrian. Gould makes much of this alleged early disparity, claiming that it overturns our view of the history of life—quite literally, because the early richness of designs indicates subsequent impoverishment, and a reversal of the notion of "ever-upward diversification" that was seen as a pervasive paradigm. This may be disputable, but what is not disputable is that the understanding of the taxonomic status of these early forms is crucial to understanding whether or not Gould's view is correct. Because very few of these Cambrian-style "phyla" are known to persist into the Ordovician, we must assume an intra-Cambrian extinction (or several extinctions). Presumably, if it involves the elimination of a great range of phyla, as is claimed, then this must be accounted the greatest extinction of them all (Fig. 2)—the "winnowing out" that set the course of subsequent evolutionary history. After all, the subsequent mass extinctions terminated no groups of phylum status.

The question, then, of how we are to understand the taxonomic status of the Cambrian 'oddballs' is central to our ideas of radiation of living phyla, and to whether there was, or

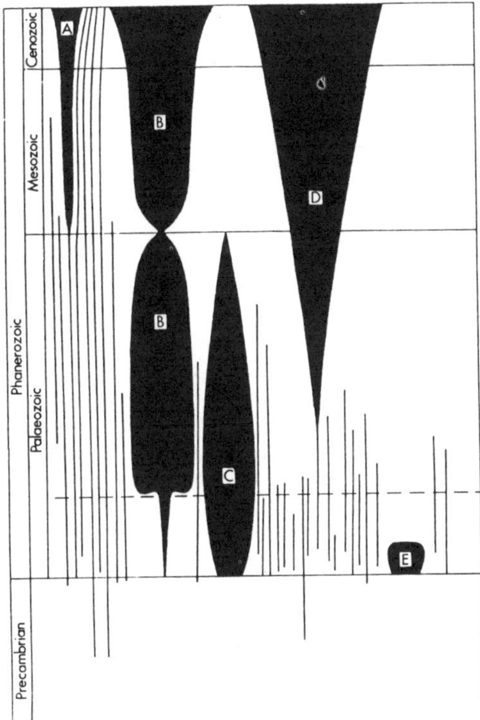

Figure 2. The Cambrian 'radiation' of metazoans according to Whittington (1981); a similar pattern has been claimed for arthropod groups. The product of extremely rapid cladogenesis at the Precambrian-Cambrian boundary, this view eschews phylogenetic connections based on shared morphological characters, accords high taxonomic status to all the separate groups (polyphyly being important), and would regard the early 'culling' of impersistent groups as an important phase of extinction.

was not, a most significant major extinction event(s) in the Cambrian. A popular view, and certainly the most dramatic, is the "100 phyla" view and this must imply at least one major Cambrian extinction event, one at the end of the Lower Cambrian (reduction of small shelly faunas), and probably another one later in the Cambrian, because the majority of "Burgess-type" elements are not known from younger soft-bodied biotas.

The alternative view would lay greater emphasis on the influence of the way taxonomy has been done upon these Cambrian curiosities. It would claim that the over-enthusiastic application of high level taxonomic status to these early animals has been misleading, laying too great an emphasis upon their alleged disparity, to create a taxonomic 'boom'—and a corresponding 'bust' when taxa disappear from the record, not really a mass extinction. This view would not deny the distinctiveness of some of these animals, only that the step from *incertae sedis* to phylum is a serious one to take, and that the greater effort might be directed at placing the enigmas in a phylogenetic context.

These contrasts in approach may, perhaps, be best illustrated by fossils claimed for the Arthropoda. The Cambrian faunas contain a wealth of arthropods (including the familiar trilobites), and their position in relation to living taxa has been problematic. Some of the researchers on the Burgess Shale have stressed the differences (what Gould terms "disparity") from living arthropods, to the extent that a selection of Burgess animals are supposed to have had "separate origins" from other and more familiar arthropods (Whittington, 1981). Any one of these groups would merit high (Class, Subphylum) recognition taxonomically, and it could be argued that their subsequent extinction was an event of some importance. However, a converse view is possible (Briggs & Fortey, 1989). In this

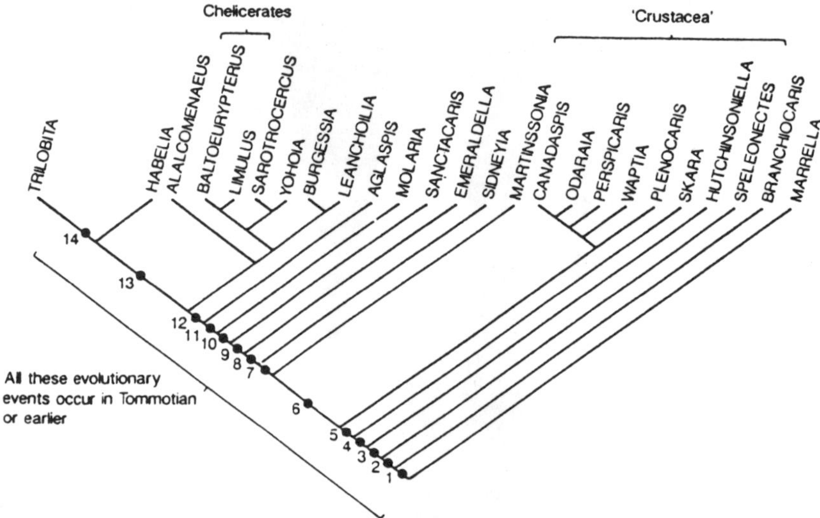

Figure 3. Alternative view of arthropod radiation, after Briggs & Fortey (1989). This view emphasises shared morphological characters (1–14) in classification (making these arthropods monophyletic), tending to reduce the taxonomic status of the 'Cambrian style' arthropods, and hence diminishing the importance of early extinction events.

view the common characters shared by the arthropods indicate common ancestry, less emphasis being placed upon the individual peculiarities of this taxon or that. Under this view it is possible to regard the majority of the Burgess Shale-type arthropods (Fig. 3) as sister groups of forms still living (indeed some living taxa appear to be as primitive as most of the Burgess ones) —some of them represent early states of assembly of arthropods we would readily recognise. For example, the number of appendages incorporated in the head region may have accreted in stages, and the specialisation of certain limbs is assuredly secondary (Manton, 1977). Whether or not one then chooses to place a high taxonomic rank on the early forms becomes debatable—but it is perfectly clear that you cannot claim them as separate "classes" because they serve to connect groups of living arthropods often accorded class status. A strictly cladistic approach would probably designate them as plesions without formal status. In any case, this view of the taxonomy would weaken the case for major Cambrian extinction.

It is worth stressing that this brief account neglects the real difficulties encountered in evaluating the significance of arthropod characters, and the arrangement of arthropods in these phylogenetic schemata will change as better methods of character analysis become available. For other groups with fewer characters (especially the strangest of the Cambrian animals) the problems are still more acute because they have so few characters. So far nobody has given an unequivocal, quantifiable way of assessing morphological divergence, and Raup (1983) has described how the early origin of major biologic groups may be an artefact of the geometry of the evolutionary tree. But the account just given *does* serve to demonstrate the importance of taxonomic stance in evaluating a fundamental problem in what constitutes a 'major' extinction event(s). The ambiguities recognized may account for the differing evaluations of mass extinction within the Cambrian (e.g., end Botomian; Gilinsky & Hubbard, this symposium) and at the Cambrian-Ordovician boundary.

Cambrian-Ordovician Boundary

The boundary between the Cambrian and Ordovician has long been recognised as a watershed in the history of life (Newell, 1967) especially because higher taxa which were to dominate the rest of the Paleozoic appear at that level (Sepkoski, 1981), and some groups still living (corals, nautiloids, bryozoans) first appear thereabouts. It has also been recognised as an extinction horizon, and as a biomere boundary (Stitt, 1975). Biomeres are stratigraphic 'packages' widely recognised in the later Cambrian of the USA, each bounded at its upper limit by an extinction event, which especially affects "platform" taxa (Palmer, 1984). The taxonomic question is this. Because we know that the Cambrian-Ordovician boundary interval was a time of rapid cladogenesis it seems possible that its elevated extinctions might be partly accounted for by 'taxonomic pseudoextinction' associated with an unusual abundance of paraphyletic groups (above). Most Ordovician trilobite families appear early in the Ordovician (in the Tremadoc) but their Cambrian sister taxa are not easy to identify; they must have had them. When they are identified this may classify some of the late Cambrian extinctions as pseudoextinctions. Similarly, the nautiloids underwent a massive evolutionary radiation in the latest Cambrian to earliest Ordovician (Flower, 1964; Chen et al., 1979). It is certain that some of their higher taxa (the order Ellesmeroceratida, for example) which both originate and become apparently extinct within this interval are paraphyletic taxa that should be analysed in terms of later clades. So far as I know the necessary phylogenetic analysis has not been done on the nautiloids, and is just beginning to be done on the trilobites, which implies that we have some way to go before we have a true picture of the Cambrian-Ordovician 'event'. There may be similar phylogenetic problems in other groups (conodonts, brachiopods, clams) where the Ordovician marks a changeover in taxonomy. None of the foregoing should be taken as implying that there were *no* true extinctions among platform taxa—there assuredly were. But it does imply that effort would be as well directed towards taxonomic studies as toward narrative accounts bed-by-bed of rock successions spanning the C-O boundary interval. The interplay of cladogenesis and extinction would be particularly fruitful to investigate at this level; sound taxonomy would provide the key to a proper understanding of the Cambrian-Ordovician 'event'.

Ordovician-Silurian Boundary

Sepkoski (1986) and many others have recognised the Ordovician-Silurian boundary as one of the major extinction horizons, whether measured by last appearances of genera or higher taxa. The extinction interval is within the Ashgill Series, and is generally conceded as coinciding closely with a major glacial episode, although there is much debate about the relative roles of glaciation, anoxia, habitat destruction, and the like, in the extinction of one group or another (Brenchley, 1984). It has been noted (Chatterton & Speyer, 1987; Fortey, 1989) that trilobite and other taxa with oceanic habitats or planktic life cycles were particularly affected. The major clades extinguished at this time are good, natural taxa, so it does not seem likely that the problem of pseudoextinction is prevalent. Furthermore, the succeeding Llandovery Series at the base of the Silurian was not a period of major cladogenesis (other than in the graptolites) and hence the problems associated with paraphyly are not likely to be unusually important. Some of the groups that became extinct at the Ordovician-Silurian boundary (e.g., Agnostida in the trilobites) had a very long previous history from the Cambrian, having survived several previous 'biomere-type' events affecting platform faunas. The kind of doubts which could be entertained about Cambrian or basal Ordovician extinctions as influenced by taxonomic procedure, do not, therefore, apply to the end Ordovician mass extinction to anything like the same extent. This shows that taxonomic critiques can be used constructively as well as critically.

One point that can be made about the end Ordovician mass extinction relates to the poverty of Llandovery faunas. Paul (1982) showed that many cystoid families were "Lazarus taxa" which disappeared from the record late in the Ordovician only to reappear after the Llandovery (or in some cases even later). Clearly there had to be an early Silurian refugium to allow for such phylogenetic continuity. Because cystoids are quite a small field

88

of study, and morphologically complex, Paul had the necessary breadth of expertise to make the phyletic connection between the later and earlier species. There are also trilobites that exhibit a similar Lazarus effect. It does seem possible that critical taxonomic review of other groups well to either side of the end Ordovician extinction will reveal further connections—obscured by the comparative poverty and uniformity of known Llandovery shelly faunas, and by the separation of specialists to either side of the boundary. Revisionary taxonomy can complement fieldwork in the recognition of lacunae in the fossil record.

LITERATURE CITED

Brenchley, P. J. 1984. Late Ordovician extinctions and their relationship to the Gondwana glaciation. Pp. 291–315. *In:* P. J. Brenchley (ed.), *Fossils and Climate.* John Wiley & Sons: New York.

Briggs, D. E. G., Fortey, R. A. & E. N. K. Clarkson. 1988. Extinction and the fossil record of the arthropods. Pp. 171–209. *In:* G. P. Larwood (ed.) *Extinction and Survival in the Fossil Record.* Clarendon Press: Oxford.

Briggs, D. E. G. & R. A. Fortey. 1989. The early radiation and relationships of the major arthropod groups. *Science* 246:241–243.

Chatterton, B. D. E & S. E. Speyer. 1987. Trilobite larval ecology and the Ordovician-Silurian (Ashgill) extinction. *Abstracts with Programs, Geological Society of America* 19:618.

Chen, J.-Y., Zou, X.-P., Chen, T.-E. & D.-L. Qi. 1979. Late Cambrian cephalopods of North China— Plectronocerids, Proactinocerida (ord. nov.) and Yahecerida (ord. nov.). *Acta Paleontologica Sinica* 18:1–23.

Donovan, S. K. (ed.) 1989. *Mass Extinctions: Processes and Evidence.* Belhaven: London. 266 pp.

Eldredge, N. & J. Cracraft. 1980. *Phylogenetic Patterns and the Evolutionary Process.* Columbia University Press: New York. 349 pp.

Flower, R. H. 1964. The nautiloid order Ellesmeroceratida (Cephalopoda). *Memoirs of the New Mexico Institute of Mining and Technology* 12:1–234.

Fortey, R. A. 1989. There are extinctions and extinctions: examples from the Lower Palaeozoic. *Philosophical Transactions of the Royal Society of London, series B:* 325, 327–355.

Gould, S. J. 1989. *Wonderful Life: The Burgess Shale and the Nature of History.* Norton & Company: New York.

Jablonski, D. 1986. Causes and consequences of mass extinctions: a comparative approach. Pp. 183– 227, *In:* D. K. Elliott. (ed.), *Dynamics of Extinction.* John Wiley: New York.

Manton, S. M. 1977. *The Arthropoda: Habits, Functional Morphology and Evolution.* Oxford University Press: Oxford. 527 pp.

McMenamin, M. A. S. & D. L. S. McMenamin. 1989. *The Emergence of Animals: The Cambrian Breakthrough.* Columbia University Press: New York. 217 pp.

Newell, N. D. 1967. Revolutions in the history of life. *Special Papers of the Geological Society of America* 89:63–91.

Palmer, A. R. 1984. The biomere problem: evolution of an idea. *Journal of Paleontology* 58:599–611.

Patterson, C. & A. B. Smith. 1987. Periodicity of extinction: a taxonomic artefact? *Nature* 330:248– 252.

Patterson, C. & A. B. Smith. 1989. Periodicity in extinction: the role of systematics. *Ecology* 70:802– 11.

Paul, C. R. C. 1982. The adequacy of the fossil record. Pp. 75–118. *In:* K. A. Joysey & A. E. Friday (eds.) *Problems of Phylogenetic Reconstruction.* Special Volume of the Systematics Association, Number 21.

Raup, D. M. 1983. On the early origins of major biologic groups. *Paleobiology* 3:107–115.

Raup, D. M. & J. J. Sepkoski. 1984. Periodicity of extinctions in the geologic past. *Proceedings of the National Academy of Sciences* 81:801–805.

Raup, D. M. & J. J. Sepkoski. 1986. Periodic extinction of families and genera. *Science* 231:833–836.

Sepkoski, J. J. 1981. A factor analytic description of the fossil record. *Paleobiology* 7:36–53.

Sepkoski, J. J. 1986. Phanerozoic overview of mass extinction. Pp. 277–295, *In:* D. M. Raup & D. Jablonski (eds.) *Patterns and Processes in the History of Life.* Springer Verlag: Berlin.

Smith, A. B. & C. Patterson. 1988. The influence of taxonomic method on the perception of patterns of evolution. *Evolutionary Biology* 25:127–216.

Stitt, J. H. 1975. Adaptive radiation, trilobite palaeoecology and extinction, Ptychaspid biomere, late Cambrian of Oklahoma. *Fossils & Strata* 4:381–390.

Whittington, H. B. 1981. Cambrian animals: their ancestors and descendants. *Proceedings of the Linnean Society of New South Wales* 105:79–87.

Extinction, Biotic Replacements, and Clade Interactions

Michael J. Benton

INTRODUCTION

During the history of life, major groups of plants and animals have waxed and waned; once-dominant groups have disappeared, and others have risen to take their place. Faunal and floral replacements, or biotic replacements, are a pervasive feature in the long-term evolution of all groups, and they are often seen as punctuation marks in the narrative history of a particular group, ecological realm, or geographic area. Many individual examples have been studied in a general way, and two or three cases have been examined in great detail by paleobiologists. However, little effort has been made to generalise about the role of biotic replacements in the diversification of life, nor to examine the processes that cause them. Indeed, a general attitude has been to assume that they were all broadly competitive, part of the supposed progress of life from primitive to advanced forms. Further, such replacement phenomena have been linked with mass extinction events only in an ad hoc way, and there is no clear view about the broader relationships of the two in a general theory of macroevolution.

There are some fundamental questions to be asked about biotic replacements:

1. Do they exist as discrete events, or are they artefacts of our rather clouded perception of the ecology of major changes in the past?
2. If they exist, how do we define them and categorise them?
3. What is their role in the history of life? Are they unusual rarities, disturbing a more normal, gradual kind of piecemeal species-by-species replacement over time; or, are they a key feature?
4. What causes biotic replacements of different kinds; how do they relate to macro-evolutionary models of competition and opportunism; and how do they relate to extinctions and mass extinctions?
5. How should they be studied in order to develop testable models of patterns and processes?

EXAMPLES

Many biotic replacements have been identified and studied in one way or another (summarised in Benton, 1987). Some of the better-known examples are:

Dr. Benton is with the Department of Geology, University of Bristol, Bristol BS8 1RJ, UK.

90

	Age	Myr
1. progymnosperms vs. pteridosperms	U. Dev.–L. Carb.	20–50
2. gymnosperms vs. angiosperms	L. Cretaceous	10–50
3. Cambrian vs. Paleozoic invertebrates	U. Camb.–U. Ord.	40–80
4. Paleozoic vs. 'modern' invertebrates	Perm.–Trias.	10–90
5. brachiopods vs. bivalves	Paleozoic	50–330
6. hybodonts vs. modern sharks	U. Jur.–L. Cret.	40–70
7. holosteans vs. teleosts	Jur.–Cret.	40–140
8. mammal-like reptiles vs. archosaurs	Triassic	5–40
9. dinosaurs vs. mammals	U. Cret.	1–30
10. multituberculates vs. placentals	L. Tertiary	10–60
11. perissodactyls vs. artiodactyls	U. Tertiary	5–35
12. Great American Interchange	U. Tertiary	1–6

This list illustrates some of the taxic and stratigraphic ranges of the well-known examples of biotic replacement, as well as the category level of the groups involved (mainly orders to phyla), and the total durations (1–50 Myr for minimum duration estimates, and 6–320 Myr for maximum estimates).

TERMINOLOGY AND SCALE

Biotic replacements have been termed ecological or evolutionary relays (Simpson, 1953; Newell, 1963), but these terms contain an implication about the processes involved. The terms faunal and floral replacement refer specifically to the zoological and botanical realms respectively, so that a general equivalent term is proposed here, namely *biotic replacement*, which has no connotations of the processes involved. Terms that explicitly indicate the hypothetical processes are introduced later.

Biotic replacements may be divided into two broad categories, depending upon whether there was a major shift in the geographic distribution of the replacing group(s) or not. In turn, each of these two categories may be further subdivided according to whether extinctions took place or not (Fig. 1):

1. *in situ radiations* (global or smaller-scale), in which there was no substantial change in geographic distribution; the groups involved both/all existed worldwide, or in the same regions(s):
   a. *replacement radiation*, where one group, or set of groups, dies out, and the replacing group(s) radiate into vacated ecospace;
   b. *expansion radiation*, where there are no extinctions involved, and the new group(s) radiate into new ecospace;
2. *invasion radiations*, where the new groups have entered the area because of the destruction of a barrier or because of transport:
   a. replacement radiation, where extinctions occur and the invaders radiate into vacated ecospace;
   b. expansion radiation, where the invaders occupy previously untenanted ecospace; includes *insinuation* (Marshall, 1981), where the invading species find niches that do not seriously affect the species already present.

The scale of biotic replacements is an important topic. As noted in the examples above, the taxa involved typically range from orders to phyla, and the duration of the replacements from 1–330 Myr. Biotic replacements may be discerned at all distances of observation of the history of life, from the global scale involving virtually all marine animals—the

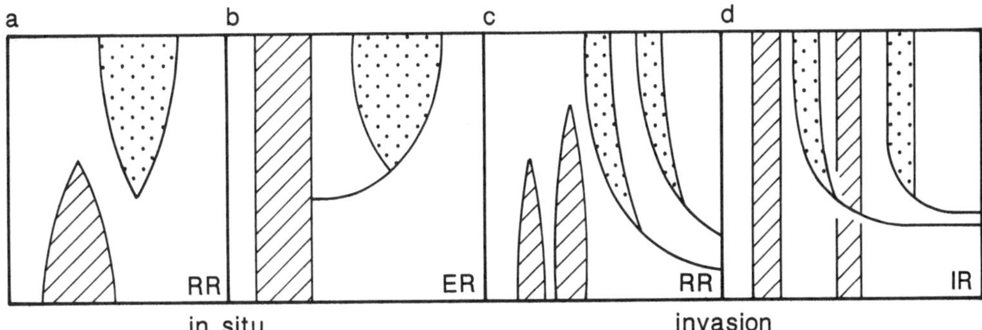

Figure 1. The main kinds of biotic radiations, those involving no geographic shift (a, b), and those involving invasion (c, d). In both cases, the radiation may be a replacement radiation (RR), or an expansion radiation (ER). In the case of an invasion, an expansion radiation has been called an insinuation radiation (IR), after Marshall (1981), since immigrating taxa *insinuate* into the faunas without necessarily causing extinction.

"evolutionary faunas" of Sepkoski (1984)—to geographically restricted events involving selected species, such as the genera of mammals involved in the Great American Interchange. It is likely that these different scales of biotic replacement do not form a restricted set of discrete levels, each displaying different patterns. Rather, the scaling may be more fractal in nature. That is, however close or however distant the observer is from the great multidimensional phylogenetic tree of all living things over the past four billion years, the patterns of branching, extinction, and replacement would show similar degrees of complexity. There are upper and lower limits to the scaling; namely, all organisms that have ever lived on earth at the upper limit, and individual organisms at the lower limit, but similar scale limits apply to the more familiar physical examples of fractal phenomena (for example, the complexity of the coastline of Britain, or turbulent water). The nature of the scaling of *patterns* of biotic replacement does not, of course, necessarily imply a continuity in the *processes* involved (see later).

In the present study, the examples of biotic replacements are at suprageneric level (generally orders to phyla), and they span millions of years (1–330 Myr; mainly in the range 20–50 Myr). An important issue that can fundamentally distort our view of biotic replacements is the phylogenetic status of the taxa involved. This will be discussed first, before the patterns and processes of such replacements are addressed.

THE PROBLEMS OF PARAPHYLY

The taxa involved in postulated biotic replacements must be monophyletic. Indeed, this is doubtless true for the majority of cases noted above, and for many other well-known examples. However, monophyly of the taxa under study must be established before any attempts at macroevolutionary study can be made. This has probably been accepted as an implicit rule over the years, but it has only been expressed explicitly with the full development of cladistic techniques (e.g., Cracraft, 1981; Patterson & Smith, 1987; Smith & Patterson, 1988; Benton, 1988, 1989; Fortey, 1989). This issue was brought to the notice of many people by Patterson and Smith's (1987) survey of parts of the Sepkoski (1982) data set that has been widely used: they found that up to 75% of the families of echinoderms and fishes in this listing were "invalid" for one reason or another, most being non-monophyletic.

Monophyletic taxa, those that contain all the descendants from a single common ancestor, have natural boundaries. In other words, a monophyletic group, or clade, is phylogenetically defined by the possession of one or more unique derived characters, and its termination is precisely defined by the extinction of the included species, a real event or sequence of events.

Non-monophyletic taxa include polyphyletic groups, those that arose from more than one ancestor, and paraphyletic groups, those that contain only some of the descendants of the common ancestor, the other descendants having been removed to another group by systematists. The termination of the group then is not natural. Fortunately, polyphyletic groups have nearly always been avoided in evolutionary studies, but the same is not true of paraphyletic groups. A well-known paraphyletic group is the Class Reptilia as it is generally understood. It includes all of the animals that evolved from a single common ancestor in the Carboniferous, except for the hairy reptiles (Class Mammalia), and the feathered reptiles (Class Aves, the birds). These latter two groups are in themselves monophyletic, but their extraction from the Class Reptilia renders it incomplete. This can be further emphasised in two ways.

Firstly, although the Classes Mammalia and Aves are both monophyletic, having evolved from a single common ancestor, there is debate at present about where the line between 'reptile' and mammal and 'reptile' and bird should be drawn. Wherever the line is drawn, the Classes Mammalia and Aves will always be monophyletic, but the 'Class Reptilia' remains paraphyletic, and it shrinks and expands according to the systematist's whim.

The second, related, aspect that illustrates the uselessness of paraphyletic groups for macroevolutionary study can be illustrated with the 'Class Reptilia'. Suppose one were to plot rates of evolution for the group over the entire span of its existence. These show initially relatively rapid evolutionary rates, with new groups appearing, but after the Triassic and Jurassic periods, the rates appear to stagnate (Benton, 1988). Is this a valid observation or not? The answer is no, since the stagnation in rates is simply an artefact of the extraction of mammals and birds from the evolving lineages. Systematists are manipulating the macroevolutionary findings when they create paraphyletic groups. The solution is to study rates of evolution in the appropriate monophyletic group, in this case the 'Superclass' Amniota (i.e., reptiles + birds + mammals).

Some paleobiologists and evolutionists have argued in favor of using some paraphyletic groups in macroevolutionary studies (e.g. Van Valen, 1978, 1985; Charig, 1982; Sepkoski, 1984, 1987). Reasons include:

1) the groups are well-known and traditional; 2) the groups may be well defined in ecological and adaptive terms, and it is sensible to exclude members of the clade that have adopted entirely different lifestyles (for example, birds are strictly members of the 'Infraclass' Dinosauria, since they arose from theropod dinosaurs, so we cannot be so dogmatic as to say that the dinosaurs have never died out); 3) the groups are more accurate surrogates of species than are monophyletic higher taxa; and 4) in large-scale studies, the effects of summing paraphyla will cancel out, so the detail is unimportant.

The first argument is not relevant to the present discussion: there is no harm at all in talking about reptiles, since everyone knows what is meant, but the group can be avoided in macroevolutionary studies. The second argument might seem to carry more weight. However, it is not clear how such groups are to be recognised (Smith & Patterson, 1988), and they probably introduce circularities into evolutionary studies. If the termination of a paraphyletic group corresponds with a mass extinction event or to a stratigraphic boundary, as is usually the case, then statistical analysis of the stratigraphic distributions of such families is bound to emphasise the events and boundaries. We must use phylogenetic (cladistic) criteria only in defining taxa for evolutionary studies, since only then can they be used to test for extinctions and other events. As for the dinosaurs, all 20 or so major families of dinosaurs are monophyletic, and they have all died out. Birds are included in the Dinosauria, and they have survived. There was indeed a major extinction of several clades of dinosaurs towards the end of the Cretaceous period. The third argument, that paraphyla are the best higher-level representatives of species in macroevolution, is hard to sustain in view of evidence that extinctions of paraphyla often correspond to gaps in the fossil record, rather than to any real species-level events (Smith & Patterson, 1988). The fourth argument, that summing paraphyla will tend to cancel out any errors introduced, is probably also incorrect since the terminations of such unnatural groups often correspond with mass extinction events or major environmental shifts, and errors introduced will tend to be systematic and to reinforce the circularities just noted.

All cases of global, or other in situ, replacements must be demonstrated to involve pairs of clades. Invasion replacements, such as the Great American Interchange, involve at first sight two polyphyletic groups, the edentates, notoungulates, litopterns, and rodents of South America, and the horses, elephants, and carnivores of North America. If this is the appropriate scale of analysis, then that would be the case. However, this famous invasion was not a single event: the swapping of species took several million years, and the interactions are more appropriately considered at the species level. Hence, the taxa are monophyletic.

In conclusion of this section, detailed phylogenetic analyses are a necessary prelude of macroevolutionary studies in order to establish monophyletic groups so that: 1) the earliest representatives of groups can be determined truly to belong to them (avoiding the problems of "ancestor hunting"; Benton, 1988); and 2) true terminations and patterns of decline are determined, not the artificial cut-offs of paraphyletic groups.

HOW TO RECOGNISE A BIOTIC REPLACEMENT

Several general observations of patterns from the fossil record are necessary in order to identify a biotic replacement: 1) two or more clades must be involved; 2) the 'before' and 'after' clades must occupy demonstrably similar adaptive zones (i.e., sets of niches); and 3) there must be an approximate coincidence in timing of the decline and/or disappearance of the 'before' clade(s) and the radiation of the 'after' clade(s), according to a variety of patterns:

i. both over a long time;
ii. both rapidly;
iii. rapid decline/ extinction of one and long-term radiation of the other;
iv. long-term decline of one, and rapid radiation of the other.

In order to determine these basic facts, the fossil record in question has to be adequate, and in particular it ought to possess these qualities: 1) accurate stratigraphic dating throughout; 2) abundant finds of fossils throughout the time-span in question; 3) fossils that may be identified readily to their clades, so that there is no confusion between the 'before' and 'after' taxa; 4) fossils and sediments that give detailed information on ecology and adaptations; 5) clear geological evidence for geographic shifts over time, if an invasion radiation is suspected; and 6) a strongly founded cladistic phylogeny of the taxa of interest.

A well-studied example, the Great American Interchange of the past 6 million years (Marshall, 1981; Marshall et al., 1982; Stehli & Webb, 1985), appears to fulfil these criteria (Fig. 2).

1. **Dating of the sediments** containing mammal remains from the last 6 Myr in South America is reasonably good: six geological formations have been determined, each containing a distinctive fauna of mammals, and there are a variety of radiometric dates and magnetostratigraphic indicators for most of these formations (Marshall et al. 1983, 1984).

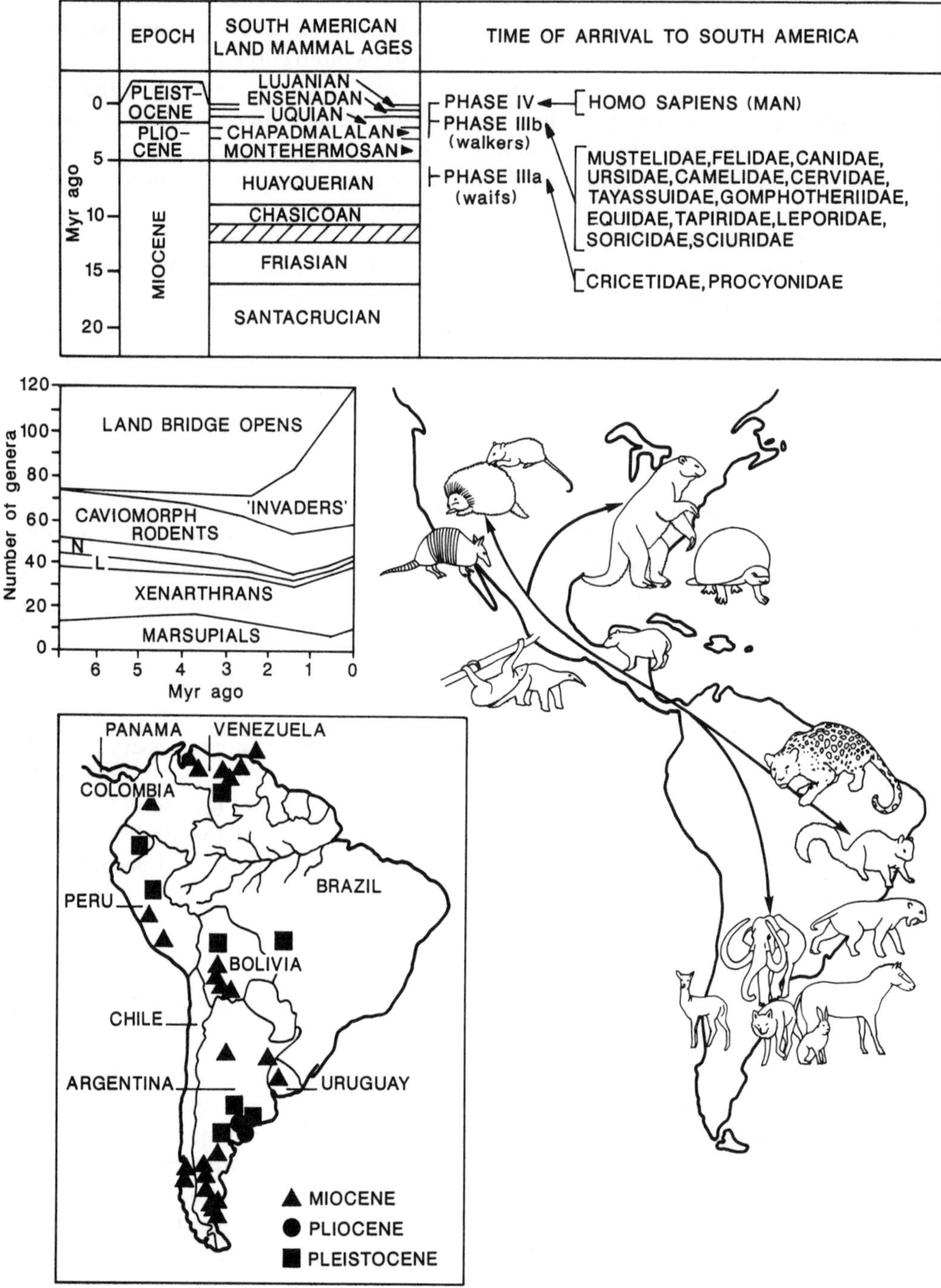

Figure 2. The Great American Interchange, showing some of its major characteristics. The stratigraphic timescale (a) is detailed, and well dated in parts: the interchange itself involves three phases of immigration (termed IIIa, IIIb, and IV). A typical pattern of generic diversity of mammals in South American over the past 6 Myr (b) shows that the North American 'invaders' mainly insinuated, since there is relatively little extinction among the native South American forms (shown shaded). Terrestrial faunas with mammals of Miocene, Pliocene, and Pleistocene age are known from various parts of South America, particularly Argentina (c). The passage of South American mammals into Central and North America and of North American mammals into South America was roughly evenly balanced (d), although the picture looks very different today because of subsequent extinctions. (Based on Marshall 1981, and other sources.)

2. **Fossils are abundant** in all of the formations. Each of the six formations has yielded 72–120 genera (mean 87.8), representing 30–39 families (mean 34.3). Living totals are 75 genera and 35 families, so that the Pliocene and Pleistocene fossil record of mammals would appear to be rather complete compared to the present day, unless past diversities were strikingly higher than they are today (they were in fact higher in the Pleistocene, particularly at the generic level).

3. **The fossils are readily identifiable** to species, genus, and family. Even isolated jaws and teeth, the commonest fragmentary fossils, may be so identified, because of the great variation seen in mammalian teeth, their rich, contained, phylogenetic information, and the enormous experience of mammalian paleontology. In many cases, of course, the mammalian remains in question are virtually complete, to the extent of preservation of hair in some later Pleistocene examples!

4. **The ecology and adaptations may be determined** to a considerable extent, since the teeth and bones give a great deal of specific information on diets and modes of locomotion, since most taxa have closely related living relatives, and since the associated sediments and fossils give extensive information on the habitats.

5. **The geography is relatively well established.** The present location of finds is close to their ancient distribution since the time of interest is within only the past 6 Myr. Independent geological evidence documents the major stages in the opening of the Central American land bridge. Most of the relevant fossil faunas are, unfortunately, from Argentina alone (Marshall 1981), at the southern end of the distribution of most groups, so the geographic story for the whole of South America is unknown.

6. **The phylogenies are essentially cladistic.** Most of the groups have been revised recently, and cladograms developed. The status of the families and orders as clades is now mainly established, although there may be dispute in the cases of some genera, and in placing the families and orders into the framework of a broader cladogram of all mammals.

THE ROLE OF BIOTIC REPLACEMENTS

If the taxa of interest in a macroevolutionary study are all monophyletic, then it is axiomatic that all radiations of new taxa, at whatever scale, must involve either an expansion radiation into new ecospace or a replacement radiation or biotic replacement. The only other way that evolution could proceed in order to avoid this inevitability would be if there were extensive lineage swapping between supraspecific clades, and no clearcut replacement of one by the other. However, this does not appear to be the case, or at least not after a branching event has taken place, so that all clade radiations that are not into previously unoccupied ecospace are to be treated as biotic replacements.

The role of biotic replacements in the history of life may be assessed by the analysis of substantial parts of the fossil record. While it would be best to attempt to compare all parts of the history of plants and animals in order to determine how common and how significant biotic replacements have been, this would represent an enormous undertaking. In part, it might prove impossible, especially where the fossil record is poor, or where the systematic basis for classification is not demonstrably cladistic. Hence, the fossil record of tetrapods has been selected as a test case.

This fossil record is better than most in terms of its cladistic resolution, since vertebrate paleontologists have applied such techniques for a number of years now, and since the taxa lend themselves particularly well to detailed character-based phylogenetic analysis. These arguments are developed more fully elsewhere (Benton, 1988, 1989).

A preliminary analysis of data on families of tetrapods from the Devonian to the Triassic shows that expansion radiations dominated in 'normal' times (ratio, 3:1), while replacement radiations dominated in mass extinction times (ratio, 1:2). This question will be further investigated (Benton, in prep.).

THE CAUSES OF BIOTIC REPLACEMENTS

Is Competition the Cause?

It has already been noted that the tacit assumption by many that major biotic replacements in the past have been mediated broadly by competition, in a way like scaled-up individual/individual or species/species interactions, may be incorrect (Gould & Calloway, 1980; Benton, 1983a, b, 1986, 1987; Gould, 1985, 1988). There are a number of levels of assumptions in many of the story-like accounts of biotic replacements that have gained currency, and these ought to be teased apart.

Firstly, the pattern of replacement must be appropriate for competition to be invoked as the guiding principle. True long-term competition would produce a double-wedge pattern of replacement, in which taxon A (the loser) declines progressively over a measurable geological time span (say, 1–100 Myr in typical postulated examples), while taxon B (the winner) increases in diversity at an equivalent rate. The double-wedge pattern is distinguished from the mass-extinction pattern (Fig. 3) in which taxon A declines and disappears *before* the radiation of taxon B. Typically, the extinction of taxon A will appear to be a rapid event, but it need not be.

After establishing that the fossil record shows a double-wedge pattern for any particular case, the paleobiologist must determine whether the taxa are interacting. This may not always be the case on closer inspection. A rather ludicrous example is the matching double-wedge pattern of the decline of the ichthyosaurs and the rise of the angiosperms which lasted for much of the Cretaceous period, a time span of 70–80 Myr. In this case, it is obvious that the decline of a group of dolphin-like marine reptiles could not have been caused by the rise of flowering plants. But, in other quoted cases, the evidence is more subtle. For example, a rapid look at the Great American Interchange appears to show a gradual increase in dominance of 'North American'-type mammals in South America after 3–4 Myr ago, and a matching decline in numbers of endemic forms. However, in specific cases, the competition argument is hard to make since many of the animals had very different lifestyles and would not have competed. For example, the endemic giant ground sloths and the invading elephants had very different diets and modes of feeding, and it is unlikely that they would have competed for any particular resource. Other evidence indeed confirms (Marshall, 1981; Marshall et al., 1982) that the two groups lived side by side for 2–3 Myr before both died out at the same time in the late Pleistocene.

Thirdly, and most importantly, even if a double-wedge pattern is established, and the two taxa appear to have had near-identical ecologies and to have interacted, competition

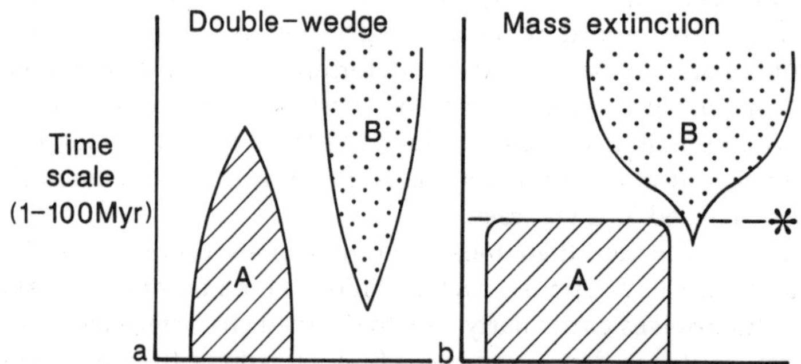

Figure 3. Two distinctive patterns of biotic replacement: (a) the double-wedge pattern (Gould & Calloway, 1980), in which group B waxes, while group A wanes; and (b) the mass extinction pattern, in which group B radiates only after the extinction of group A (the extinction event is indicated by a dashed horizontal line and an asterisk). (After Benton, 1987.)

may not be the only, or even the most likely, explanation. Connell (1980) presented a set of three criteria that must be satisfied in order to demonstrate that competition has caused a particular replacement pattern: 1) divergence in adaptation or mode of life must have occurred; 2) competition must be demonstrated between living members of the taxa involved, and its evolutionary effects must be shown; and, 3) the divergence must have a genetic basis. These criteria apply especially to ecological studies, where competition is cited as the likely cause of changes in adaptations or distribution for living taxa. These criteria are too stringent for application to most macroevolutionary examples, where there may be no closely related living representatives, so that competition in action and genetic coding cannot be assessed (criteria (2) and (3)).

Many ecologists would argue that it is futile even to contemplate testing theories of competition in the distant past. This is strictly correct. It is impossible to apply the critical tests cited by Connell (1980) to any fossil example, and indeed it has proved very hard to do so for modern examples. However, competition has been so all-pervasive in paleobiological writings, and in many well-known macroevolutionary theories (reviewed, Benton, 1987), that some effort must be made to put the subject on a firmer footing. The criteria of Connell (1980) for the demonstration of the role of competition in evolution must be weakened substantially in order to have any applicability to typical paleontological examples: 1) evidence of a double-wedge pattern of biotic replacement; 2) evidence of the likelihood of competition, based on similarities of postulated adaptations, and coincidence of timing and geography; and 3) probability of a genetic basis for the key competitive characters of the taxa involved. Satisfactory tests of all of these points can only be said to point to the *possibility* of competition applying in particular paleontological examples; it can never be seen as an adequate proof. This is an opposite viewpoint to the common assumption of pervasive competition in macroevolution, and it emphasises that the burden of proof rests with those proposing competition, rather than with those criticising it.

Alternatives to Competition

Double-wedge patterns of biotic replacement could be produced by a variety of causes, as reviewed by Benton (1987):

1. **Differential response to predation.** Many marine examples in particular have been investigated in which the relative fates of two or more similar groups appear to have been determined by predator pressure. For example, the evolution of bivalves appears to have been profoundly influenced by the appearance of new kinds of predators at different times. Bivalves that live buried in soft sediment, the endobyssate bivalves, dwindled in importance from the late Paleozoic onwards. This has been ascribed (Stanley, 1977) to increasing predation. A second, better known, burst of new predators arose in the late Mesozoic—voracious crabs, teleost fishes, and carnivorous snails—and these seem to have further reduced the importance of endobyssate bivalves (Stanley, 1977; Vermeij, 1977). The most abundant modern bivalves are those that have reduced the chances of predation by living in rock crevices, by growing thick or spiny shells, by living in the inter-tidal zone, or by being able to swim fast (the scallops).

2. **Differential environmental response.** Many double-wedge replacement patterns may be caused by the differential response of two clades to a major change in the physical or biotic environment. Changes in climate, sea level, topography, geography, and other physical phenomena can profoundly alter the relative survivorship of different groups. Changes in the distribution of plants, or in the major groups present, could produce a double-wedge pattern among dominant herbivores. A well-known example is the so-called 'competitive' model for the replacement of dinosaurs by mammals (e.g., Sloan et al., 1986) in which the decline of the dinosaurs and the rise of the mammals, at the end of the Cretaceous period, is explained by cooling atmospheres and the spread of temperate

98

floras. Differential response to predation (1) is a special case of differential environmental response (2).

3. **Chance.** As already noted, a double-wedge pattern may arise wholly by chance, in that there is no causal connection between the two clades, or between them and a specific causative factor. In paleontological cases where competition or differential response to predation or to environmental change is posited, the pattern might always have arisen by chance.

The non-competitive models for double-wedge patterns of biotic replacement which have just been outlined would loosely be called competition by many researchers—the 'diffuse competition' of Van Valen (1980). However, it is not competition in the strong sense, which must involve a limiting resource, and damage to one group produced by the success of another.

The Role of Competition in Biotic Replacement

Many of the problems with the study of the role of competition in the study of biotic replacement have been engendered by the use of value-laden terminology, or imprecise terminology. The basic patterns of biotic replacements have been noted above, but some further comments on the assumptions implied by commonly-used terms should be noted here.

The portion of a biotic replacement in which competition is demonstrated, or more commonly assumed, is the radiation phase of the 'successful' group. This radiation is usually called an *adaptive radiation,* which immediately implies the possession of a feature, or set of features (*key adaptations*), that ensured success. Hence, for 'adaptive radiation', read, 'radiation in which the group out-competed another and succeeded because of its special feature(s)'. Cracraft (1985) has already rightly suggested that adaptive radiations should be known simply as *radiations.* The term key adaptation involves so many assumptions about process that it should be avoided altogether if possible. It will be used in the following list of models of competition-induced biotic replacements because it is a traditional part of such models. However, as Kemp (1985) has pointed out, it is most unlikely that any radiation was based on the possession of a single character; organisms are much too complex for that.

There is not a single model for a competitive replacement, but at least five, as Benton (1987) noted (Fig. 4):

1. Clade B possesses a key adaptation which allows it to compete successfully with, and replace, clade A. Clade B has demonstrated the competitive superiority conferred upon it by the key adaptation, by causing the extinction of clade A.

2. There is a mass extinction event, and survivors of clade B (with its key adaptation) compete successfully in the disturbed post-extinction ecosystems with the survivors of clade A (which lack the key adaptation) and clade B prevails. Again, clade B has demonstrated its competitive superiority.

3. There is a mass extinction event and clade B resists extinction. Clade A lacks the ability to resist extinction, and it dies out. Clade B has demonstrated its competitive superiority in terms of its ability to resist extinction. That ability may or may not be related to the possession of a key adaptation that mediates the radiation of clade B after the crisis time has passed. Adaptations that favour groups in times of mass extinction stress may be different from those that favour survival in normal selection regimes (Jablonski, 1986).

4. There is a mass extinction event during which many A and B organisms die out, and, by chance, only a few B organisms survive. They evolve into clade B, and they have a key adaptation that ensures a successful radiation. Clade B has not demonstrated its competitive superiority over clade A.

5. There is a mass extinction event during which all or most of clade A, and most of clade B die out by chance. The survivors of clade B radiate, but there is no particular key

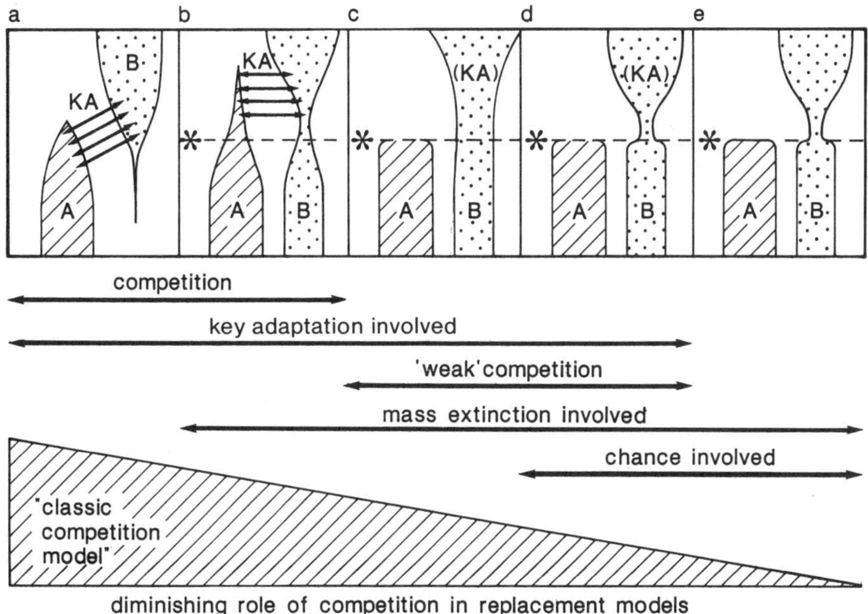

Figure 4. The role of competition, key adaptations, and mass extinction (opportunism) in five models of biotic replacement. The role of competition diminishes from left to right, from a strong, directly-competitive model of replacement (a) to an opportunistic model, in which group B has no particular competitive advantage over group A (e). The intermediate cases involve direct competition after a mass extinction event (b), and competitive advantages in group B which prevent it from becoming extinct (c), or which aid it in its recovery after being depleted (d). In the four models that do not involve strong competition throughout (b-e), an extinction event is an essential trigger (shown by dashed line and asterisk).

adaptation that ensures the success of that radiation. This is a stochastic model for a radiation, and any postulated key adaptations are imaginary.

Most cases of biotic replacement have been assumed to be type 1, that is, fully competitive. However, study of a number of well-known examples (Benton, 1987) has shown that many do not even show a double-wedge pattern, and those that do could just as plausibly be the result of differential environmental response and the conquest of new adaptive space. Examples where simple type 1 competition might apply include the major replacements of plant groups, cheilostome bryozoans with intrazooidal budding by those with zooidal budding (Lidgard, 1986), and multituberculates by rodents (Krause, 1986). However, certain paleobotanists (e.g., DiMichele et al., 1987) argue that large-scale competition in plant evolution is unimportant, and that major floral replacements have more to do with differential responses to stressful changes in the physical environment. Similar arguments might apply to the other two 'competitive' examples.

The Problems of Invoking Large-scale Competition

There are many fundamental problems in postulating competition in a simplistic way as the cause of major biotic replacements. These are noted only briefly here since they have been discussed more extensively by Gould & Calloway (1980), Gould (1985, 1988), Benton (1986, 1987), and others.

1. **Confusion of pattern and mechanism.** In the non-analytical, or 'story-telling', mode of narrative in paleontology, the word 'competition' has often been used as a kind of shorthand for 'patterns of change' or 'trends'. Instead of saying simply that, for example, the mammal-like reptiles declined over time, and the dinosaurs radiated and replaced

them in the late Triassic, certain paleontologists have stated that the two groups competed, and the dinosaurs won. As already discussed, these two sentences are not equivalent, and the latter involves a large number of hypotheses that must be tested at all levels. In fact, this case, like many others, appears to fail even the most basic of tests, since there is no evidence even for a double-wedge pattern of replacement, let alone direct aggressive competition between the two clades.

2. **Over-simplification.** The prevalent confusion of pattern and mechanism just noted is related to the problem of over-simplification of ecological and evolutionary concepts. To imagine that major clades can wax and wane because of the possession or absence of a particular key adaptation is surely simplistic in the extreme. The search for the specific feature of dinosaurs that enabled them to vanquish the mammal-like reptiles, small and large, carnivore and herbivore, globally distributed and endemic, is likely to end in failure. Indeed, in this, and many other particular examples, there is strong evidence for the involvement of a complex of biotic, abiotic, and stochastic factors in the replacement.

3. **Lack of proof.** Competition cannot be treated as the null model against which all other proposals are to be tested. Part of the narrative attitude in paleontology has been that the fossil record tells us the story of a progressive march from the primeval slime to the haughty intelligence of humans today: progress and competition are implicit. This view is seriously flawed (Gould, 1985; Benton, 1987).

4. **Incorrect scaling of concepts.** The postulate that biotic replacements were caused by competition involves a major scaling-up of the concepts of ecology. There are several serious problems with this notion. First, of course, is the question of whether an ecological interaction that takes place between individuals, or species, over periods of years, and in restricted areas, can truly be expanded to apply to supraspecific clade interactions, over periods of millions of years, and often in a global context. Ecological competition can cause the death of individual organisms and the local extinction of species. It could also, arguably, cause global species extinction, shifts in geographic distribution, changes of habits, and genetic changes in the characters of a species. However, even these modest attempts to scale up the postulated effects of observable competition at the present day, have met with considerable criticism (reviewed, Strong et al., 1984; Price et al., 1984; Benton, 1987). It is not clear how local small-scale ecological phenomena can be scaled up even further to the paleontological cases of major biotic replacements. The second problem of scaling concerns the interpretation of 'adaptations of higher taxa'. How does a key adaptation work at supraspecific levels? Do families and orders possess monolithic adaptations that can be compared at clade level, or do all individuals in a clade have an equivalent advantage over all (or most) individuals in another? Surely, as Gould and Calloway (1980) pointed out, the advantageous adaptation might be significant in the first members of a clade, but it is inconceivable that the same feature would continue to provide the same measure of success in interactions between later members of the 'competing' lineages. The third problem of scaling concerns the time scales involved. Major biotic replacements typically spanned times of 1–50 Myr (see above), and hence many millions of generations. Dawkins (1982) and others have noted that it is impossible to imagine how a competitive advantage could be maintained for so long, and still express itself as one group slowly replacing another. The level of genetic advantage per generation would be statistically nil.

CONCLUSIONS

1. Biotic radiations have occurred frequently in the past. They should be divided into expansion radiations, those in which the radiating taxon moves into previously unoccupied adaptive space, and replacement radiations, in which the new group replaces a pre-existing one. Radiations also involve geographic factors, and a distinction should be

made between those that involve no geographic shift, and those that involve invasion.

2. A test example, based on terrestrial tetrapods, suggests that most biotic radiations are expansion radiations, but replacement radiations dominate after episodes of mass extinction.

3. The simple competitive key adaptation model is not the only cause of biotic replacement, and indeed it may only rarely be the cause. A variety of other models, some of them weakly competitive, some of them non-competitive, apply in studied examples.

4. The notion of competition in macroevolution is problematic at various levels: it is hard to demonstrate, and it has been inappropriately scaled in many formulations.

5. The role of mass extinctions in triggering episodes of replacement radiations needs to be investigated further. Certainly, the study of individual radiations should continue, but more general analyses may throw light on broader patterns underlying the long-term diversification of life, and the continual turnover of taxa that has occurred through geological time.

ACKNOWLEDGMENTS

I thank the Leverhulme Trust for funding of part of this work, Miss Anne Blacker for compiling data on tetrapod distributions and ecologies, and Mrs. Pam Baldaro for drafting the diagrams.

LITERATURE CITED

Benton, M. J. 1983a. Large-scale replacements in the history of life. *Nature.* 302:16–17.

Benton, M. J. 1983b. Dinosaur success in the Triassic: a noncompetitive ecological model. *Quarterly Review of Biology,* 58:29–55.

Benton, M. J. 1986. The Late Triassic tetrapod extinction events. Pp. 303–320. *In:* K. Padian (ed.). *The Beginning of the Age of Dinosaurs: Faunal Change Across the Triassic-Jurassic Boundary,* Cambridge University Press: Cambridge.

Benton, M. J. 1987. Progress and competition in macroevolution. *Biological Reviews,* 62:305–338.

Benton, M. J. 1988. Mass extinctions and the fossil record of reptiles: paraphyly, patchiness, and periodicity. Pp. 269–294. *In:* G. Larwood (ed.), *Extinction and Survival in the Fossil Record,* Clarendon Press: Oxford.

Benton, M. J. 1989. Mass extinctions among tetrapods and the quality of the fossil record. *Philosophical Transactions of the Royal Society of London. Series B,* 325:369–386.

Charig, A. J. 1982. Systematics in biology: a fundamental comparison of some major schools of thought. Pp. 363–440. *In:* K. A. Joysey & A. E. Friday (eds.), *Problems of Phylogenetic Reconstruction.* Academic Press: London.

Connell, J. H. 1980. Diversity and the coevolution of competitors, or the ghost of competition past. *Oikos,* 35:131–138.

Cracraft, J. 1981. Pattern and process in paleobiology: the role of cladistic analysis in systematic paleontology. *Paleobiology,* 7:456–468.

Cracraft, J. 1985. Biological diversification and its causes. *Annals of the Missouri Botanical Garden.* 72:794–822.

Dawkins, R. 1982. *The Extended Phenotype.* Oxford University Press: Oxford. 307pp.

DiMichele, W. A., Phillips, T. L. & R. G. Olmstead. 1987. Opportunistic evolution: abiotic environmental stress and the fossil record of plants. *Review of Palaeobotany and Palynology,* 50:151–178.

Fortey, R. A. 1989. There are extinctions and extinctions: examples from the lower Palaeozoic. *Philosophical Transactions of the Royal Society of London. Series B,* 325:327–355.

Gould, S. J. 1985. The paradox of the first tier: an agenda for paleontology. *Paleobiology,* 11:2–12.

Gould, S. J. 1988. Trends as changes in variance: a new slant on progress and directionality in evolution. *Journal of Paleontology,* 62:319–329.

Gould, S. J. & C. B. Calloway. 1980. Clams and brachiopods—ships that pass in the night. *Paleobiology,* 6:383–396.

Jablonski, D. 1986. Causes and consequences of mass extinctions: a comparative approach. Pp. 183–

229. *In:* D. K. Elliott (ed.), *Dynamics of Extinction.* John Wiley & Sons: New York.

Kemp, T. S. 1985. Synapsid reptiles and the origin of higher taxa. *Special Papers in Palaeontology,* 33:175–184.

Krause, D. W. 1986. Competitive exclusion and taxonomic displacement in the fossil record: the case of rodents and multituberculates in North America. *Contributions to Geology, University of Wyoming, Special Paper,* 3:95–117.

Lidgard, S. 1986. Ontogeny in animal colonies: a persistent trend in the bryozoan fossil record. *Science,* 232:230–232.

Marshall, L. G. 1981. The Great American Interchange—an invasion-induced crisis for South American mammals. Pp. 133–229. *In:* M. H. Nitecki (ed.), *Biotic Crises in Ecological and Evolutionary Time.* Academic Press: New York.

Marxhall, L. G., Berta, A., Hoffstetter, R., Pascual, R., Reig, O. A., Bombu, M. & A. Mones. 1984. Geochronology of the continental mammal-bearing Quaternary of South America. *Palaeovertebrata. Mémoire extraordinaire.* 1984:1–76.

Marshall, L. G., Hoffstetter, R. & R. Pascual. 1983. Geochronology of the mammal-bearing Tertiary of South America. *Palaeovertebrata. Mémoire extraordinaire.* 1983:1–93.

Marshall, L. G., Webb, S. D., Sepkoski, J. J., Jr., & D. M. Raup. 1982. Mammalian evolution and the Great American Interchange. *Science,* 215:1351–1357.

Newell, N. D. 1963. Crises in the history of life. *Scientific American.* 208:76–92.

Patterson, C. & A. B. Smith. 1987. Is periodicity of mass extinctions a taxonomic artefact? *Nature.* 330:248–251.

Price, P. W., Slobodchickoff, C. N. & W. S. Gaud. 1984. *A New Ecology.* Wiley Interscience: New York. 515pp.

Sepkoski, J. J., Jr. 1982. A compendium of fossil marine families. *Milwaukee Public Museum. Contributions in Biology and Geology.* 51:1–125.

Sepkoski, J. J., Jr. 1984. A kinetic model of Phanerozoic taxonomic diversity. III. Post-Paleozoic families and mass extinction. *Paleobiology.* 10:246–267.

Sepkoski, J. J., Jr. 1987. [Reply to Patterson and Smith]. *Nature.* 330:252.

Simpson, G. G. 1953. *The Major Features of Evolution.* Columbia University Press: New York. 434pp.

Sloan, R. E., Rigby, J. K., Jr., Van Valen, L. M. & D. Gabriel. 1986. Gradual dinosaur extinction and simultaneous ungulate radiation in the Hell Creek Formation. *Science.* 252:629–633.

Smith, A. B. & C. Patterson. 1988. The influence of taxonomic method on the perception of patterns of evolution. *Evolutionary Biology.* 23:127–216.

Stanley, S. M. 1977. Trends, rates, and patterns of evolution in the Bivalvia. Pp. 209–250. *In:* Hallam, A. (ed.), *Patterns of Evolution as Illustrated by the Fossil Record.* Elsevier: Amsterdam.

Stehli, L. G. & S. D. Webb. 1985. *The Great American Biotic Interchange.* Plenum: New York. 532pp.

Strong, D. R., Jr., Simberloff, D., Abele, L. G. & A. B. Thistle. 1984. *Ecological Communities: Conceptual Issues and the Evidence.* Princeton University Press: Princeton, N.J. 614pp.

Van Valen, L. 1978. Why not to be a cladist. *Evolutionary Theory.* 3:285–299.

Van Valen, L. M. 1980. Evolution as a zero-sum game of energy. *Evolutionary Theory.* 4:289–300.

Van Valen, L. 1985. A theory of origination and extinction. *Evolutionary Theory.* 7:133–142.

Vermeij, G. J. 1977. The Mesozoic marine revolution: evidence from snails, predators and grazers. *Paleobiology.* 3:245–258.

Patterns of Synecology During the Phanerozoic

Peter M. Sheehan

Abstract. Phanerozoic synecology was characterized by nine long intervals when organisms remained in established ecologic associations. The long intervals of stasis in community organization lasted tens of millions of years and most intervals were terminated by extinction events. Following each extinction event, geologically brief intervals of community reorganization lasted a few million years and gradually gave way to new intervals of stasis. These intervals of stasis and community reorganization have been termed Ecologic Evolutionary Units (EEUs) by Boucot (1983). During the Stasis-EEUs evolution and speciation were constrained and lineages of organisms tended to retain existing ecologic roles. Reorganizational-EEUs followed extinction events which removed synecologic constraints, allowing lineages to evolve new adaptations and move into ecologic settings they had not occupied previously. The Reorganizational-EEUs were characterized by repeated, nonevolutionary replacements of communities, progressively increasing diversity within communities, and increasingly dense packing of communities into the ecospace. Thus, extinction events were responsible not only for dramatic changes in the history of lineages of organisms, but extinctions also terminated long standing ecologic associations and allowed massive restructuring of the entire synecologic organization of life.

INTRODUCTION

Five major extinction events appear to have opened ecospace in both marine and terrestrial environments, thereby allowing diversification through bursts of macroevolution (see summaries by Jablonski, 1986a, 1989, herein, and Benton, 1987). The history of community associations through time was punctuated by the same extinction events which were so important in phylogenetic evolution. Not only did many new groups of organisms appear during intervals of macroevolution following extinction events, but the entire synecologic structure of life was reorganized.

The phylogenetic history of organisms has been a focus of research since the early days of paleontology. As a result, the study of patterns of macroevolution in the fossil record has an enormous data base upon which to draw (Sepkoski, 1981a, 1982; Sepkoski et al., 1981; Valentine, 1985). On the other hand, the study of community patterns in the history of life is a relatively recent field of research. Major advances in our understanding of community patterns through geologic time were not possible until substantial numbers of fossil communities had been described.

In a seminal article on fossil communities, Ziegler (1965) described the early Silurian marine community associations in the Welsh Borderland. Ziegler's community studies

Dr. Sheehan is with the Department of Geology, Milwaukee Public Museum, 800 W. Wells St., Milwaukee, WI 53233, USA.

demonstrated to paleontologists that the synecology of fossil assemblages could be examined over broad regions and through significant intervals of geologic time.

Shortly after Ziegler's pioneering work, Bretsky (1968, 1969) synthesized a data base of about 30 published studies of Paleozoic communities. Bretsky (1968, p. 1231) set lofty goals for community studies when he stated, ". . . it seems probable that adaptive shifts, radiations and extinctions may have been directly related to changing community structures." For the next fifteen years numerous studies applied Ziegler's general methodology of community analysis, and a wealth of information on Phanerozoic communities became available (see summaries in Gray et al., 1981 and Boucot, in press).

In his presidential address to the Paleontological Society, Boucot (1983) divided the Phanerozoic into 12 consecutive intervals (Fig. 1), and each of the intervals was characterized by a distinct assemblage of gradually evolving communities. He termed these intervals Ecologic-Evolutionary Units (EEUs). Boucot noted that, during EEUs, species tended to evolve in concert with other species in their community and tended not to invade ecospace occupied by other communities. Groups of organisms that survived from one EEU to the next often invaded environments from which they had previously been excluded.

After an extensive evaluation of environmental changes associated with the boundaries between EEUs, Boucot (1983) could find no consistent types of changes which might have caused the transition from one EEU to the next. Some boundaries were characterized by environmental perturbations, but other boundaries had no perturbations. And some boundaries coincided with major extinction events, but others did not.

In a study relating reef to level bottom community evolution, Sheehan (1985) recognized a pattern of alternating short and long duration EEUs (Fig. 1) in the data set of Boucot (1983). Moreover, most of the long EEUs were terminated by the major extinction events in the fossil record, including those recognized by Raup and Sepkoski (1982). The short duration EEUs followed extinction events, and the transition from the short to the subsequent long EEUs were gradual and not marked by extinction events.

Sheehan (1982, 1985) suggested that the short EEUs were times of community reorganization following the ecologic disruption caused by extinction events. During these short intervals of reorganization many groups of organisms were evolving into new niches. The transitions from short intervals to long intervals occurred gradually as new ecologic associations became established, and stasis in community structure set in once more. The long intervals of stasis persisted until the next extinction event.

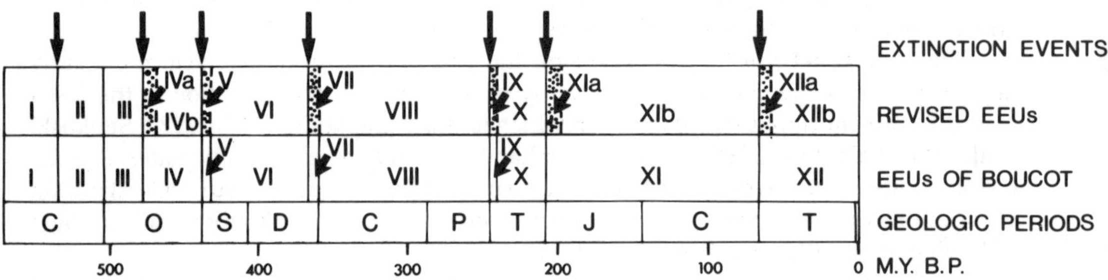

Figure 1. Revised EEUs compared to the EEUs of Boucot (1983). In the post-Cambrian the long, Stasis-EEUs were terminated by extinction events (arrows). See discussion in text concerning extinction events at the end of EEUs I–III. Short, Reorganizational-EEUs (stippled) followed the extinction events. The boundaries between reorganizational and stasis EEUs were gradational.

AN EXAMPLE OF EEUs IN THE TERRESTRIAL ENVIRONMENT

Most of the changes between EEUs are obvious only to workers who specialize in the particular time intervals being considered. However, the change from dinosaurian dominated communities in the Late Mesozoic to mammalian dominated communities in the Cenozoic is one example of change between EEUs that will be familiar to all.

The Miocene community in Nebraska (Fig. 2) provides an example of the kind of stasis in community structure that characterizes EEUs. Although the actual species in Figure 2 are not familiar to any but specialists, many basic groups of animals such as horses, deer, wolves, elephants, hawks, etc. are immediately recognizable. Evolution during the long intervals of stasis is at a level where new species evolve, but the basic groups of animals retain their ecological roles in communities. The community illustrated in Figure 2 is familiar to us because it is from the EEU in which we are living.

Stepping back to an earlier EEU, Figure 3 depicts a community from 66 million years ago. Although we recognize the dinosaurs, they are clearly not part of our modern scene. This is a community which we comprehend by drawing analogies with modern mammalian community structures. Mesozoic terrestrial communities throughout Boucot's EEU 11 were dominated by dinosaurs. Mammals were present only as very small-bodied

Figure 2. Reconstruction of a scene in Nebraska during the Miocene which is part of the present EEU. Most animals can be recognized as belonging to extant families with ecologic roles similar to those of living representatives. Evolutionary changes since the Miocene are of the kind that are common during the long, Stasis-EEUs. Milwaukee Public Museum photo by Mel Scherbarth.

Figure 3. A diorama depicting a scene in Montana during the latest Cretaceous, based on collections from the Hell Creek Formation. Animals from Stasis-EEU XIb are very different from the living fauna. However, ecologic roles such as herbivores, carnivores and scavengers can be determined by analogy with the morphology of living organisms. Comparison of the relatively familiar Miocene scene (Fig. 2) with this dinosaur scene emphasizes the distinctiveness of faunal assemblages in different EEUs. The extinction event which terminated Stasis-EEU XIb allowed development of the modern terrestrial fauna. Milwaukee Public Museum photo by Mel Scherbarth.

animals; most were insectivores and some were herbivores (Lillegraven, 1979). Mesozoic mammals are a typical example of animals locked into their ecologic roles during intervals of stasis. The extinction event at the end of the Mesozoic caused a drastic change in community organization. The extinction cleared dinosaurs from the large body size niches, allowing mammals to radiate into this unoccupied ecospace. During the reign of the dinosaurs mammals had been locked in stasis for 130 million years (MY). Mesozoic mammals had limited roles relative to the modern fauna. Within a few MY after the extinction event mammals radiated to the plethora of large- and small-bodied herbivores, carnivores, and omnivores that characterize the current EEU. The Miocene and Cretaceous scenes provide a striking example of the dramatic differences between EEUs. The sudden changes in community organization that occur between one EEU and the next are a fundamental feature of the history of life.

FOSSIL COMMUNITIES

Bambach (1986), in an outstanding summary of fossil community work, has shown that, consistent with modern community investigations, fossil communities did not exist independently of their constituent organisms. Fossil communities had no genetic continuity, and they changed as evolution operated. But communities were the context in which evolution occurred.

Most fossil community studies have accepted a definition of the community concept that emphasizes descriptive aspects, such as recurrent assemblages of organisms which were thought to have been associated in life. As in modern community studies (May, 1984; Pimm, 1984), most fossil community work has focused on a limited number of taxonomic groups, commonly at only one or two trophic levels. Because of this focus many paleontologists use the term "association" instead of "community," but the use of the term "community" by biologists (see above) is quite similar to the paleontologic usage. As in modern studies, fossil community analysis that considers a cross section of trophic levels improves understanding of community structures.

But fossil communities are distinct from modern communities. The most important difference is that fossil communities are time-averaged (Bambach, 1986; Fursich & Aberhan, 1990) assemblages of species that accumulated at particular horizons in the rock over long intervals, commonly including many seasons and numerous generations. Often substantial changes of the local environment have occurred, and the abundances of populations of the various species may have fluctuated greatly. In essence, the geologic horizons that represent very brief intervals of geologic time were formed over very long intervals of ecologic time.

The introduction of sophisticated analytical techniques has refined paleocommunity analysis (Springer & Bambach, 1985). Most fossil communities had gradational boundaries, and gradient analysis (Springer & Bambach, 1985) permits direct comparison of faunal distributions with different aspects of the physical and biotic environment. Hopefully, future standardization of methods of describing communities will make it easier to compare communities recognized by different workers.

THE HISTORY OF EEUs

Although Boucot (1983) recognized the basic history of EEUs, additional patterns are present. In the post-Cambrian there is a pattern of alternating long and short duration EEUs. In addition, the long EEUs are terminated by major extinction events, while the short EEUs pass gradually to the following EEU without an extinction event (Fig. 2). Short intervals lasted from 3 to 8 MY, while long intervals ranged from 35 to 142 MY.

Boucot noted that several of these short-duration EEUs (V, VII, and IX) had low diversity and were essentially recovery intervals following extinction events. There were similar recovery intervals after each of the extinction events, although Boucot did not recognize the recovery intervals that follow the Late Triassic and the terminal Cretaceous extinction events (See for example, Hansen & Sheehan, 1989; Johansen, 1987, 1989).

Boucot recognized that there were periods of reconstruction of community structure at the transitions between EEUs. He believed reconstruction took two to three MY, possibly much less than one MY. However, the interval of community reorganization actually took the entire duration of the short EEUs which followed extinction events. Viewed in this way EEUs V, VII and IX are not anomalous in having unusually low diversity or in not being terminated by extinction events. The low diversity is expected while the fauna recovered following an extinction event. Rather than being terminated by an extinction event, the short duration EEUs ended gradually as the fauna recovered from the extinction and was reorganized into the next long interval of stasis in community structure. Because there are two basically different EEUs, the short intervals of communities are referred to as Reorganizational-EEUs, and the long intervals of stable community structure are referred to as Stasis-EEUs.

Recognition of the reorganizational intervals in the Middle Ordovician, Early Jurassic, and Paleocene (Sheehan, 1985) brings the total number of EEUs to 15. As discussed below EEUs I, II, and III are not well understood, but the other units are couplets of reorganization and stasis.

The Reorganizational-EEUs which were not recognized by Boucot (1983) need to be designated, and this is done by adding "a" to the Roman Numeral for the reorganizational stage and "b" to the Roman Numeral for the stasis interval (Fig. 1). Ultimately, it may be best to designate all the couplets of reorganization and stasis as subdivisions of nine distinct EEUs. But at present the Cambrian and Early Ordovician EEUs are too poorly known to be certain that couplets existed.

THE PATTERN OF REORGANIZATION—EEU V AS AN EXAMPLE

Boucot (1983) has provided an extensive discussion of the processes of community change during the long intervals of stasis. In addition, Lesperance and Sheehan (1988) examined a sequence of Early Devonian communities to further elucidate changes in communities during an interval of stasis. However, the tumultuous pattern of community change during the Reorganizational-EEUs needs more careful examination. As an example, events during Reorganizational-EEU V, which followed the terminal Ordovician extinction event, will be reviewed. Intervals following other extinction events have similar patterns (see for example, Erwin, 1989; Newell, 1962; Hansen, 1986; Hansen & Sheehan, 1989; Johansen, 1987, 1989).

Sheehan (1975) showed that following the terminal Ordovician extinction, community instability and reorganization persisted into the Upper Llandovery (C_2), the entire extent of EEU V. This was an interval approximately 7 MY in length.

During the Early Silurian, speciation brought diversity back to levels that were comparable to those of the Late Ordovician (Sheehan, 1979, 1982; Walker, 1985; Baarli & Harper, 1986). The increasing numbers of new species dictated that significant changes in community organization must have occurred. As diversity increased, within-community diversity increased also (Sheehan, 1975, 1980; Watkins & Boucot, 1975; Pickerill & Hurst, 1983; Hurst & Pickerill, 1986).

In addition to simple increase in species diversity, the number of communities in given ecologic settings increased. As a result, individual communities occupied progressively less extensive habitats during the Early Silurian (Sheehan, 1975, 1980). Not only were more species being packed into communities, but more communities were being packed into the environment.

The macroevolutionary appearance of new morphologic forms (e.g. new families) following an extinction event (Sepkoski, 1981a) also dictated substantial changes in community organization. This was much like the transformation of terrestrial communities cited previously.

The reorganization and recovery of communities in the early Silurian included macroevolutionary events, increased species packing of communities, and increasingly more environmentally restricted communities. But the recovery was more than what has been described above. In fact, if the Reorganizational-EEUs were merely times when continuously evolving groups remained in the same environment, the short EEUs would be little different from the Stasis-EEUs. However, during the reorganization many groups invaded new ecologic settings. Rather than becoming temporally stable lineages, the newly established stocks commonly were replaced by invaders that had evolved into similar habitats elsewhere.

In addition to the appearance of new families, which unequivocally records stocks moving into new ecospace, many of the surviving groups of organisms also invaded new ecospace following the extinction (Watkins & Boucot, 1975; Boucot 1983; Sheehan, 1982). Sheehan (1977) drew an analogy with ecologic release at the species level when modern species invade new environments that lack competitors and the species immediately expand their niches. Late Ordovician brachiopod species of *Cryptothyrella* and *Streptis* in Baltica were ecologically restricted reef dwellers that evolved into new habitats following the extinction. In the case of *Cryptothyrella* (Sheehan, 1977) post-extinction species dominated many level-bottom communities in open-marine, within-wave-base environments. Post-extinction species of *Streptis* (Hints, 1986) became common in below-wave-base communities.

Another example is provided by the virgianid brachiopods which were small shelled and lived in relatively deep water prior to the Ordovician extinction. When the extinction vacated ecospace in shallow water habitats during the Hirnantian Stage, the virgianids invaded shallow water, expanded dramatically and became large-shelled dominants of these habitats (Hurst & Sheehan, 1982). Boucot (1983) emphasized that such ecologic changes were uncommon during the intervals of stasis.

A characteristic feature of the Reorganizational-EEUs is that community sequences were short lived. It was common for communities or species invading a new region to replace previously existing communities only to be replaced themselves by other invading groups. For example, in the shallow, agitated waters of North American epicontinental seas a series of unrelated, brachiopod-dominated communities sequentially occupied the same physical habitat during Reorganizational-EEU V (Fig. 4). Following eustatic sea level rise at the beginning of the Silurian, one of the initial brachiopod dominated communities in eastern and central North American epicontinental seas was the *Cryptothyrella* Community (Ziegler & Boucot, 1970) derived from Baltica (Sheehan, 1977). At the same time that the *Cryptothyrella* Community radiated into several distinct but related communities in the east, a *Virgiana* Community dominated the western and northern margins of North America (Boucot, 1975; Sheehan, 1980). *Virgiana* originated in Siberia, and the *Virgiana* Community invaded the midwestern states in the latter part of the Lower Llandovery (Fig. 4). The *Virgiana* Community radiated into several closely related communities that dominated the habitats throughout North American epicontinental seas during the Middle Llandovery (Boucot, 1975; Ziegler & Boucot, 1970).

The *Pentamerus* Community replaced the *Virgiana* communities at the beginning of the Upper Llandovery (Fig. 4). However, *Pentamerus* was not derived from *Virgiana*. *Pentamerus* evolved from either *Nondia* (according to Boucot & Chiang, 1974) or from *Borealis* (according to Mork, 1981). In either case, the *Pentamerus* Community displaced, the *Virginana* communities or at least moved into the region they once occupied.

Thus, during Reorganizational-EEU V three distinct and unrelated communities

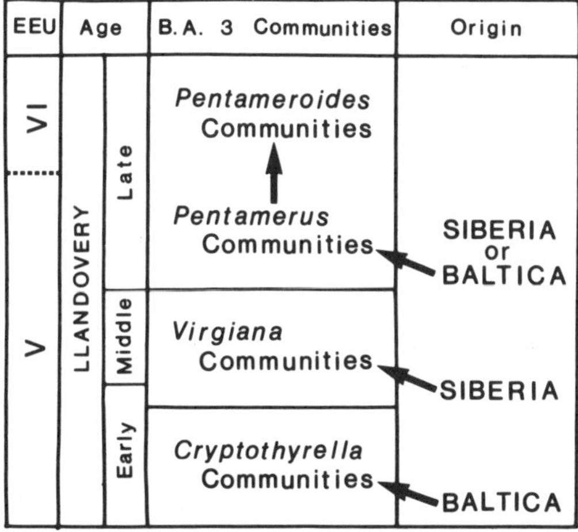

Figure 4. Community replacements characteristic of Reorganizational-EEUs. In shallow, within-wave-base environments or Benthic Assemblage (B.A.) 3 of Boucot (1983) a series of low-diversity communities sequentially replaced one another during Reorganizational-EEU V in North American epicontinenal seas (Sheehan, 1975). Following the terminal Ordovician extinction, the athyridacid dominated *Cryptothyrella* communities, with ancestry in Baltica, migrated to North America in the earliest Silurian (Early Llandovery). The *Cryptothyrella* communities were replaced by the unrelated pentameracid *Virgiana* communities which initially evolved in Siberia. The *Virgiana* communities were replaced by unrelated *Pentamerus* communities at the beginning of the Upper Llandovery. Subsequent community changes were through in-place evolution rather than by outside invasion. The *Pentameroides* communities of the Late Llandovery evolved directly from a *Pentamerus* Community, and Stasis-EEU VI had begun.

110

sequentially dominated rough-water, within wave-base environments in North America. Subsequently, during Stasis-EEU VI species of *Pentamerus* succeeded each other in this environment. New species of *Pentamerus* evolved, but they simply replaced the ancestral species in the communities. The interval of stasis had begun. In the Late Llandovery (C_{4-5}) a new genus, *Pentameroides,* evolved from the plexus of species of *Pentamerus. Pentameroides* became a dominant in this environment during the uppermost Llandoverian and early Wenlockian.

Morphologically the transition from *Pentamerus* to *Pentameroides* was marked by fusing of outer plates of the dorsal valves. These morphologic changes probably provided increased efficiency of muscle action of the brachiopods, but this was only evolutionary fine tuning of the morphology for life in the same ecologic setting. Thus, community evolution during EEU VI stood in stark contrast with EEU V. The earlier pattern of repeated invasions was replaced by a pattern of gradual evolution of new species that remained in essentially the same ecologic setting as their ancestors.

The transitions from recovery to stasis EEUs were gradual, and there is no precise boundary between these two types of EEUs. Different investigators will probably choose slightly different positions for the boundaries. For example, in the transition from EEU V to EEU VI Boucot (1983) placed the boundary at the end of the C_2 interval after *Pentamerus* had become established. The boundary also could have been placed at the end of the Middle Llandovery when the eventually dominant *Pentamerus* Community first was established.

EEUs AND EVOLUTIONARY FAUNAS

The EEUs of Boucot are imposed upon a more basic pattern of three Phanerozoic Evolutionary Faunas (Fig. 3) described by Sepkoski (1981a, 1982, 1984). Each EEU was composed of animals that reflected the current Evolutionary Fauna.

The Evolutionary Faunas were recognized by determining the ranges of skeletonized marine families during the Phanerozoic. The Evolutionary Faunas were found to have progressively higher plateaus of familial diversity (Fig. 5). Sepkoski (1981a) and Bambach (1983, 1985, 1986) emphasized that each succeeding Evolutionary Fauna was more complex ecologically than its predecessor.

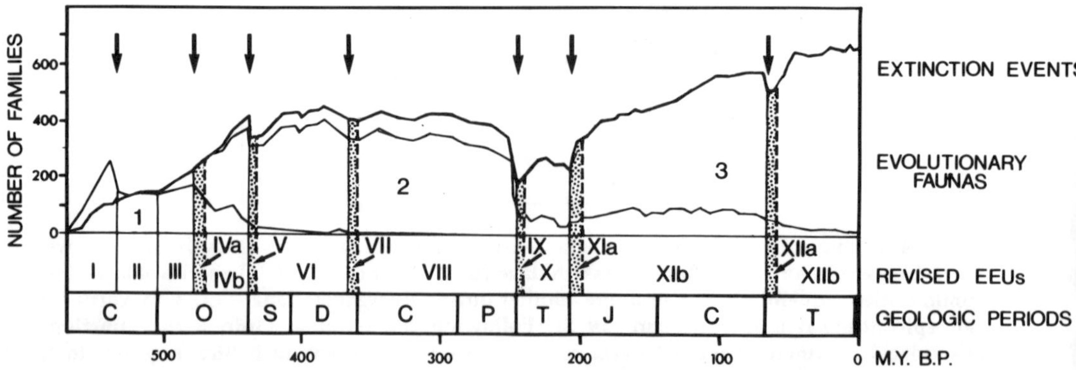

Figure 5. Sequence of EEUs compared to the history of well-skeletonized marine families during the Phanerozoic (after Sepkoski, 1981, 1984). In the chart of Evolutionary Faunas, standing diversity of families is shown by the uppermost dark line. Finer lines separate standing familial diversity of the three Evolutionary Faunas (numbered 1, 2, 3) of Sepkoski. Reorganizational-EEUs are stippled. Arrows designate extinction events discussed in text.

The revised history of EEUs presented here permits integration of the EEUs with the Evolutionary Faunas (Fig. 5). The Reorganizational-EEUs correspond almost precisely with times when diversity of families was rapidly increasing following extinction events. Most Stasis-EEUs correspond to times when global familial diversity remained relatively constant (Sepkoski, 1981a; Raup, 1986), although the intervals when new Evolutionary Faunas were radiating are notable exceptions, as will be discussed.

The Evolutionary Faunas and the diversity patterns are based on global diversity of families, while EEUs are based on the history of community organization of local groups of species. That data from such different levels of organization produce such similar patterns, supports the robustness of these results.

Some level of ecologic integration that constrains macroevolution is implied by both the diversity patterns and the history of communities. The most likely structuring agents are a combination of competition and predator-prey relationships (Sepkoski, 1981a; Van Valen, 1984; but see Flessa & Jablonski, 1985). Boucot (1983) alluded to these constraints simply as "ecologic factors." Establishing the actual structuring agents may be more easily addressed in modern ecologic studies than in the fossil record, where interactions between species are difficult to examine. The importance of the interaction of species within communities remains an unresolved issue in ecology (compare Paine, 1980, 1988; Schoener, 1983, 1985, 1989; Strong, 1983; Connell, 1983; Menge & Sutherland, 1987).

The pattern found in this study, that massive reorganizations of synecologic structure occur only after extinction events terminate intervals of stable ecologic structure or during the transition between Evolutionary Faunas, needs to be addressed. Walker and Valentine (1984) find adaptive space to be relatively open at the species level, but still they (p. 897) fall back on the necessity for freedom from "biotic interference" for the evolution of new species. Walker and Valentine (1984) recognize that adaptive zones available to different organisms are only as densely occupied as speciation and extinction rates of those particular organisms permit. Movement from one adaptive zone to another may be facilitated by extinction events that remove entire groups of organisms from adaptive zones, providing unoccupied space into which surviving groups may evolve. For example, as in the occupation of large body-size niches by dinosaurs but not mammals in the Mesozoic, the critical factor was which group invaded the ecospace first, not which group was capable of most effectively occupying that ecospace.

Evolutionary Faunas have distinct taxonomic compositions and at the family level attained progressively higher diversity plateaus (Sepkoski, 1979, 1981a; Fig 5). Sepkoski (1979, p. 212) suggested that the progressively higher levels of diversity reflect "the early success of ecologically generalized taxa and their later replacement by more specialized taxa." Increasing diversity was accompanied both by within habitat diversity increase (Bambach, 1977, 1983; Ausich & Bottjer, 1985; Sepkoski, 1988) and by occupation of previously unoccupied marine habitats (Bambach, 1983; Sepkoski, 1988). Thus, comparison of communities in particular environments is more difficult between Evolutionary Faunas than within them. The remarkable similarities in composition that Walker and Laporte (1970) found between Ordovician and Devonian communities would not be found between communities of the Cambrian and Paleozoic Evolutionary Faunas. The differing diversities of the Evolutionary Faunas imposed little-understood limits on the complexity of possible faunal interaction. For example, in the low diversity of the Cambrian Fauna, an organism would have faced a more limited variety (number of families) of predators or competitors than it would have in the same environment later in the Phanerozoic.

The changeovers between the Cambrian and Paleozoic Evolutionary Faunas (numbered 1 & 2 in Fig. 5) was gradual and occurred during EEUs III and IV. The changeover involved an onshore toward offshore retreat of the Cambrian Evolutionary Fauna and replacement by the Paleozoic Evolutionary Fauna (Sepkoski & Sheehan, 1983; Lockley, 1983). Global family diversity was increasing during EEUs III and IV and the pat-

112

terns of community changes need further investigation. The Stasis-EEUs during the radiation of the Paleozoic Evolutionary Fauna may have been different from those of other EEUs.

The changeover from the Paleozoic to the Mesozoic-Recent Evolutionary Fauna (numbered 2 & 3 on Fig. 5) occurred more rapidly than did the transition from the Cambrian to the Paleozoic Evolutionary Faunas. The rapidity of change can be attributed to the largest extinction of the Phanerozoic, the Permo-Triassic Extinction Event. Diversity increase of the Modern Evolutionary Fauna continued throughout the Mesozoic and into the Cenozoic (Sepkoski, 1981a; Fig. 5).

EEUs OF THE CAMBRIAN EVOLUTIONARY FAUNA

During EEUs I–III in the Cambrian and Early Ordovician, the pattern of a brief recovery interval followed by a long interval of stasis is not obvious. This may be due to a lack of study, or it may be a fundamental pattern. The Cambrian Evolutionary Fauna is unique in having much lower family diversity than was present during the remainder of the Phanerozoic (Sepkoski, 1981a, b). A clear implication of the pattern outlined for community evolution is that faunal interaction constrains evolution, and once a certain level of community organization is attained, stasis develops. Extinctions depress diversity and disrupt the organization of communities, thus opening environments to recolonization. The low diversity of the Cambrian Evolutionary Fauna may have dictated different rules during the early Phanerozoic. Diversity may not have reached high enough levels to constrain evolution.

An extinction at the end of EEU I, though not recorded as a major extinction event by most workers, did eliminate a very diverse phylum, the archaeocyathids. Sepkoski (1979, 1981a) excluded the archaeocyathids, and thereby eliminated a major decline in family diversity at the end of EEU I. If the archaeocyathids are included (top line of Figure 5 during the Early Cambrian) an extinction event is recorded on the diversity curve. Thus, it is possible that EEU I was terminated by an extinction.

No clear extinction event terminated EEU II, though a series of small extinction events effected trilobites near the close of the Cambrian (Palmer, 1979; Westrop, 1989). Family diversity was actually increasing at the end of EEU III. But the increasing diversity was produced by the radiation of the Paleozoic Fauna (Sepkoski, 1979; Fig. 5). The radiation of the Paleozoic Fauna marks the end of the Cambrian Evolutionary Fauna. This radiation masks the drastic decline or extinction of the Cambrian Evolutionary Fauna (see Fig. 5). Thus, there may be extinctions associated with the termination of EEUs I–III, but they are poorly understood. More community studies are needed before it will be possible to determine whether or not a pattern of stasis-extinction-reorganization is present in EEUs I–III.

COMMUNITY EVOLUTION DURING TRANSITIONS
BETWEEN EVOLUTIONARY FAUNAS

The changes between Evolutionary Faunas are not as abrupt as between EEUs. However, as shown by Sepkoski and Sheehan (1983) for the Cambrian-Paleozoic transition and Sepkoski and Miller (1985) for the Paleozoic-Mesozoic transition, the new Evolutionary Faunas were isolated ecologically and spatially from the older Evolutionary Faunas. Thus, communities of the separate Evolutionary Faunas also must have been separated during these transitions. During the radiations of the Modern Evolutionary Fauna in the Mesozoic, familial diversity increased gradually, nearly doubling the Paleozoic level (Sepkoski, 1981a). Communities during this time were not stable in the sense used previously in this manuscript or by Boucot (1983). The long duration EEUs Xb and XIb had increasingly diverse communities rather than the stasis that characterized other EEUs of long duration. The increase of species level diversity was so large (Signor, 1985, Sepkoski, 1988) that it can be explained only partly by the occupation of new habitats (Bambach, 1985) or by compartmentalization within guilds (Bambach, 1983; Sepkoski, 1988).

The radiation of the Modern Evolutionary Fauna took the entire Mesozoic, and radiation of the Paleozoic Evolutionary Fauna took most of the Ordovician (Sepkoski, 1981a, 1984). During these radiations there may have been sufficient ecologic space within communities to allow continuous increase in diversity of the communities. Until the Evolutionary Fauna attained a diversity plateau, macroevolution may not have been constrained entirely at the community level. New groups appeared in communities throughout the Mesozoic, unlike during more normal times. Mesozoic communities tended to increase in diversity by adding new groups to the communities, rather than by being constantly restructured (McKerrow, 1978), as occurs during Reorganizational-EEUs. Thus, even though diversity was increasing and new goups were being added to communities, there does appear to be a significant level of continuity and stasis in groups already living in particular communities during EEUs XIb and XIIb.

The pattern of reef evolution mimicked the level bottom faunas in the Mesozoic and Ordovician, much as during intervals of stasis previously in the Phanerozoic (Sheehan, 1985). This is further evidence that some level of stasis was present in communities during the replacement of Evolutionary Faunas.

SYNECOLOGY AND EXTINCTION EVENTS

Extinction events have been identified by examining the extinction of genera, families, or even higher taxonomic categories (for examples see Raup & Sepkoski, 1982; Sepkoski, 1986; Benton, 1985). Interest in extinction events has focused on the role they have played in resetting the evolutionary stage (Newell, 1967; Jablonski, 1986a). Because the extinction of a high level taxonomic unit (such as a family) generally requires that many lower level taxonomic units (such as genera or species) became extinct (Raup, 1979), species level mass extinctions can easily be underestimated by analyzing only higher level taxonomic patterns. It is possible that, during early stages of the Phanerozoic, low levels of family diversity may not allow recognition of extinctions. The changeover between EEU II and III may mark an extinction event that does not appear at the family level.

On the other hand, community patterns change markedly at extinction events, which also reset the *synecologic* stage. Synecologic patterns more emphatically reflect extinction events than do high-level taxonomic patterns. As an example, while less than one quarter of the brachiopod families living at the end of the Ordovician became extinct in the terminal Ordovician extinction (Sepkoski, 1982), the worldwide endemism of Late Ordovician brachiopod faunas switched to extreme cosmopolitanism after the extinction (Sheehan & Coorough, 1990). This extinction also ended the Stasis-EEU IV and allowed entirely new community associations to evolve in the Silurian. The synecologic changes were more marked than the extinction was at the family level of analysis.

Similarly, Benton (1985), when examining ranges of families of terrestrial tetrapods, found that the extinction rates of families at the Cretaceous-Tertiary extinction event were not above the background extinction rate. The abrupt disruption and subsequent reorganization of communities at the boundary reflects dramatic events that were not detectable using global diversity of families.

TERRESTRIAL COMMUNITIES

Boucot (1983) emphasized that marine and terrestrial community patterns were decoupled. Radiations of land plants, for example, did not coincide with the boundaries of marine EEUs. However, the radiation of land plants is more similar to the establishment of Sepkoski's Evolutionary Faunas than to the EEUs of Boucot. As indicated previously, the pattern of community evolution is imposed upon the more basic pattern of Evolutionary Faunas. The radiation of the land biota is akin to the radiation of metazoans. Although some boundaries of marine and terrestrial faunas did not coincide, others did. The origin and initial radiation of dinosaurs coincided with EEU XIa, and followed an extinction that terminated a major vertebrate biota dominated by therapsid reptiles and corresponding to EEU Xb (see Benton, 1987). The extinction at the end of the Cretaceous that terminated EEU XIb was synchronous within geologic time scales in marine and terrestrial realms.

In addition, the Paleocene was an interval of evolutionary experimentation and reorganization of communities in both marine and terrestrial environments. In the Paleocene (Reorganizational-EEU XIIa) birds and lizards competed with mammals for dominance. Some archaic groups such as multituberculates, condylarths, and plesiadapiform primates radiated in the Paleocene only to be replaced by other mammals (Krause, 1984). These mammals became dominant during the interval of stasis corresponding to EEU XIIb. Although the synecologic history of terrestrial organisms is still poorly understood, long intervals of stasis appear to have been disrupted by major extinction events which were followed by short intervals when communities were reorganizing.

DURATION OF THE LONG INTERVALS OF STASIS

In order, beginning in the Middle Ordovician, Stasis-EEUs were about 35, 65, 115, 35, 142 and 58 MY in length. The short Reorganizational-EEUs were no more than 8 MY in length. The durations of the Stasis-EEUs depended upon when the major extinction events occurred, as it was the extinction events that terminated long Stasis-EEUs. The relatively short Stasis-EEU Xb in the Triassic (35 MY long) was short only because an extinction event happened to terminate the EEU soon after it was established.

CONCLUSIONS

The Phanerozoic history of communities follows a repetitive pattern of long duration Stasis-EEUs which were terminated by extinction events. The extinction events were followed by short duration Reorganizational-EEUs during which communities gradually established the associations which dominated the subsequent Stasis-EEU.

The Reorganizational-EEUs following extinction events were characterized by community changes which included evolutionary replacement of species and many higher taxonomic-level groups. This produced inevitable changes in community structure. As the recovery progressed, species became more closely packed in communities, and subsequent communities became more closely packed in the environment. Beyond these changes, which necessarily were associated with increased diversity following an extinction event, many new families evolved and some surviving families evolved new ecologic capacities. Communities during Reorganizationzal-EEUs were also characteristically unstable and were replaced repeatedly by unrelated communities, rather than gradually being modified through time as was characteristic during the Stasis-EEUs. In many respects the effects of extinction events are more apparent when synecologic structures such as communities and zoogeographic patterns are examined than when high-level taxonomic patterns are considered.

The ecologic structure of communities appears to have constrained macroevolution during the Stasis-EEUs. Major adaptive radiations in environments that are already occupied by organisms appear to be dependent on extinctions events clearing the ecologic landscape. The brief pulses of macroevolution after the extinction events are followed by the gradual establishment of new synecologic associations which once more begin to constrain macroevolution. The transition from Reorganizational-EEU to Stasis-EEU reflects the gradual development of persistent synecologic patterns. Evolution during Stasis-EEUs is basically fine tuning of organisms to each other and to their physical environment.

The pattern of community evolution has broad implications for our understanding of the history of life. Extinction events were critical intervals during which clearing of ecospace allowed macroevolution and the reorganization of synecologic associations including communities and zoogeographic patterns.

ACKNOWLEDGMENTS

This work was supported generously by grants from the National Science Foundation (EAR 79-01013 and subsequent renewal). Ms. Jayne Miller kindly provided editorial assistance.

LITERATURE CITED

Ausich, W. I. & D. J. Bottjer. 1985. Phanerozoic tiering in suspension-feeding communities on soft substrata: Implications for diversity. Pp. 255–274. *In:* J. W. Valentine (ed.), *Phanerozoic Diversity Patterns—Profiles in Macroevolution.* Princeton Univesity Press: Princeton, NJ.

Baarli, B. G. & D. A. T. Harper. 1986. Relict Ordovician brachiopod faunas in the Lower Silurian of Asker, Oslo Region, Norway. *Norsk Geologisk Tidsskrift* 66:87–98.

Bambach, R. K. 1977. Species richness in marine benthic habitats through the Phanerozoic. *Paleobiology* 3:152–167.

Bambach, R. K. 1983. Ecospace utilization and guilds in marine communities through the Phanerozoic. Pp. 719–746. *In:* M. J. S. Tevesz & P. L. McCall (eds.), *Bioic Interactions in Recent and Fossil Benthic Communities.* Plenum Press: New York.

Bambach, R. K. 1985. Classes and adaptive variety: The ecology of divesification in marine faunas through the Phanerozoic. Pp. 191–253. *In:* J. W. Valentine (ed.), *Phanerozoic Diversity Patterns—Profiles in Macroevolution.* Princeton University Press: Princeton, NJ.

Bambach, R. K. 1986. Phanerozoic marine communities. Pp. 407–428. In: D. M. Raup & D. Jablonski (eds.), *Patterns and Processes in the History of Life.* Dahlem Konferenzen. Springer-Verlag: Berlin.

Benton, M. J. 1985. Mass extinction among non-marine tetrapods. *Nature* 316:811–814.

Benton, M. J. 1987. Progress and competition in macroevolution. *Biological Reviews* 62:305–338.

Berry, W. B. N. 1979. Graptolite biogeography: a biogeography of some lower Paleozoic plankton. Pp. 105–116. *In:* J. Gray & A. J. Boucot (eds.), *Historical Biogeography, Plate Tectonics, and the Changing Environment.* Oregon State University Press: Corvallis.

Bottjer, D. J. & D. Jablonski. 1988. Paleoenvironmental patterns in the evolution of Post-Paleozoic benthic marine invertebrates. *Palaios* 3:540–560.

Boucot, A. J. 1975. *Evolution and Extinction Rate Controls.* Elsevier, Amsterdam, 427p.

Boucot, A. J. 1983. Does evolution take place in an ecological vacuum? II. *Journal Paleontology* 57:1–30.

Boucot, A. J. Final report of Project Ecostratigraphy. *International Union of Geological Societies.* In press.

Boucot, A. J. & K. K. Chiang. 1974. Two new Lower Silurian virgianinid (Family Pentameridae) brachiopods from the Nonda Formation, northern British Columbia. *Journal Paleontology* 48:63–73.

Bretsky, P. W. 1968. Evolution of Paleozoic marine invertebrate communities. *Science* 159:1231–1233.

Bretsky, P. W. 1969. Evolution of Paleozoic benthic marine invertebrate communities. *Palaeogeography, Palaeoclimatology, Palaeoecology* 6:45–59.

116

Connell, J. H. 1983. On the prevalence and relative importance of inter-specific competition: Evidence from field experiments. *American Naturalist* 122:661–696.

Erwin, D. H. 1989. The End-Permian mass extinction: what really happened and did it matter? *Trends in Ecology and Evolution* 4:225–229.

Flessa, K. W. & D. Jablonski. 1985. Declining Phanerozoic background extinction rates: effect of taxonomic structure? *Nature* 313:216–218.

Fursich, F. T. & M. Aberhan. 1990. Significance of time-averaging for palaeocommunity analysis. *Lethaia* 23:143–152.

Gray, J., A. J. Boucot & W. B. N. Berry (eds.), 1981. *Communities of the Past.* Hutchinson Ross Publishing: Stroudsburg, PA.

Hansen, T. A. 1986. Recovery and radiation of molluscan assemblages after the Cretaceous-Tertiary mass extinction. P. A19. *Abstracts with Program. Fourth North American Paleontological Convention.* University of Colorado: Boulder.

Hansen, T. A. & P. M. Sheehan. 1989. Rebounds from the Late Ordovician and Cretaceous extinctions: a comparison. *Geological Society America Abstracts with Programs* 21:A32.

Hints, L. 1986. Rod *Streptis* (Triplesidae, Brachiopoda) iz Ordovika i Silura Estonii. *EESTI NSV Teaduste Akadeemia Toimetised. Geoloogia* 35:20–26.

Hurst, J. M. & R. K. Pickerill. 1986. The relationship between sedimentary facies and faunal associations in the Llandovery siliciclastic Ross Brook Formation, Arisaig, Nova Scotia. *Canadian Journal Earth Sciences* 23:705–726.

Hurst, J. M. & P. M. Sheehan. 1982. Pentamerid brachiopod relationships between Siberia and East North Greenland in the Late Ordovician and Early Silurian. P. 482. *In:* B. Mamet & M. J. Copeland (eds.), *Third North American Paleontological Proceedings, v.2.*

Jablonski, D. 1986a. Background and mass extinctions: the alternation of macroevolutionary regimes. *Science* 231:129–133.

Jablonski, D. 1986b. Causes and consequences of mass extinctions: a comparative approach. Pp. 183–229. *In:* D. K. Elliott (ed.), *Dynamics of Extinction.* John Wiley and Sons: New York.

Jablonski, D. 1989. The biology of mass extinction: a palaeontological view. *Philosophical Transactions Royal Society London* B325:357–368.

Johansen, M. B. 1987. Brachiopods from the Maastrichtian-Danian boundary sequence at Nye Klov, Jylland, Denmark. *Fossils and Strata* 20:1–100.

Johansen, M. B. 1989. Background extinction and mass extinction of brachiopods from the Chalk of Northwest Europe. *Palaios* 4:243–250.

Johnson J. G. 1974. Extinction of perched faunas. *Geology* 2:479–482.

Krause, D. W. 1984. Mammalian evolution in the Paleocene: beginning of an Era. Pp. 87–109. *In:* P. D. Gingerich & C. E. Badgley (eds.), *Mammals. Notes for a Short Course.* University of Tennessee Department of Geological Sciences Studies in Geology, 8.

Kreisa, R. D. 1981. Storm-generated sedimentary structures in subtidal marine facies with examples from the Middle and Upper Ordovician of Southwestern Virginia. *Journal Sedimentary Petrology* 51:823–848.

Lesperance, P. J. & P. M. Sheehan. 1988. Faunal assemblages of the Upper Gaspé Limestones, Early Devonian of eastern Gaspe, Quebec. *Canadian Journal Earth Sciences* 25:1432–1449.

Lillegraven, J. A. 1979. Introduction. Pp. 1–6. *In:* J. A. Lillegraven, Z. Kielan-Jaworowska & W. A. Clemens (eds.) *Mesozoic Mammals—The First Two-Thirds of Mammalian History.* University of California Press, Berkeley.

Lockley, M. G. 1983. A review of brachiopod dominated palaeocommunities from the type Ordovician. *Palaeontology* 26:111–145.

May, R. M. 1984. An overview: real and apparent patterns in community structure. Pp.3–16. *In:* D. R. Strong, D. Simberloff, L. G. Abele, & A. B. Thistle (eds.), *Ecological Communities. Conceptual Issues and the Evidence.* Princeton University Press: Princeton, NJ.

McKerrow, W. S. 1978. *The Ecology of Fossils.* The MIT Press: Cambridge, MA. 384 pp.

Menge, B. A. & J. P. Sutherland. 1987. Community regulation: variation in disturbance, competition, and predation in relation to environmental stress and recruitment. *American Naturalist* 130:730–757.

Mork, A. 1981. A reappraisal of the lower Silurian brachiopods, *Borealis and Pentamerus. Palaeontology* 24:537–554.

Newell, N. D. 1962. Paleontologic gaps and geochronology. *Journal Paleontology* 36:592–610.

Newell, N. D. 1967. Revolutions in the history of life. *Geological Society America Special Paper* 89:63–91.

Paine, R. T. 1980. Food webs: linkage, interaction strength and community infrastructure. *Journal Animal Ecology* 490:667–685

Paine, R. T. 1988. Food webs: road maps of interactions or grist for theoretical development? *Ecology* 69:1648–1654.

Palmer, A. R. 1979. Biomere boundaries re-examined. *Alcheringa* 3:33–41.

Petersen, C. G. 1913. Valuation of the sea. II. The animal communities of the sea bottom and their importance for marine zoogeography. *Reports Danish Biological Station* 21:1–44.

Pickerill, R. K. & J. M. Hurst. 1983. Sedimentary facies, depositional environments, and faunal associations of the Lower Llandovery (Silurian) Beechill Cove Formation, Arisaig Nova Scotia. Canadian Journal Earth Sciences 20:1761–1779.

Pimm, S. L. 1984. Food chains and return times. Pp. 397–412. *In:* D. R. Strong, D. Simberloff, L. G. Abele & A. B. Thistle (eds.), *Ecological Communities. Conceptual Issues and the Evidence.* Princeton University Press: Princeton, NJ.

Raup, D. M. 1979. Size of the Permo-Triassic bottleneck and its evolutionary implications. *Science* 206:217–218.

Raup, D. M. 1986. Biological extinction in Earth history. *Science* 231:1528–1533.

Raup, D. M. and J. J. Sepkoski. 1982. Mass extinctions in the marine fossil record. *Science* 215:1501–1503.

Schoener, T. W. 1983. Field experiments on interspecific competition. *American Naturalist* 122:240–285.

Schoener, T. W. 1985. Some comments on Connell's and my reviews of field experiments on interspecific competition. *American Naturalist* 125:730–740.

Schoener, T. W. 1989. Food webs from the small to the large. *Ecology* 70:1559–1589.

Sepkoski, J. J. 1979. A kinetic model of Phanerozoic taxonomic diversity: II. Early Phanerozoic families and multiple equilibria. *Paleobiology* 5:222–251.

Sepkoski, J. J. 1981a. A factor analytic description of the Phanerozoic marine fossil record. *Paleobiology* 7:36–53.

Sepkoski, J. J. 1981b. The uniqueness of the Cambrian fauna. Pp. 203–207. *In:* M. E. Taylor (ed.), Short Papers for the Second International Symposium on the Cambrian System. *U.S. Geological Survey Open-File Report* 81–743.

Sepkoski, J. J. 1982. A compendium of fossil marine families. *Milwaukee Public Museum Contributions Biology Geology* 51.

Sepkoski, J. J. 1984. A kinetic model of Phanerozoic taxonomic diversity. III. Post-Paleozoic families and mass extinctions. *Paleobiology* 10:246–267.

Sepkoski, J. J. 1986. Phanerozoic overview of mass extinction. Pp. 277–295. *In:* D. M. Raup & D. Jablonski (eds.), *Patterns and Processes in the History of Life.* Dahlem Konferenzen. Springer-Verlag: Berlin.

Sepkoski, J. J. 1988. Alpha, beta, or gamma: where does all the diversity go? *Paleobiology* 14:221–234.

Sepkoski, J. J., Bambach, R. K., Raup, D. M., & J. W. Valentine. 1981. Phanerozoic marine diversity and the fossil record. *Nature* 293:435–537.

Sepkoski, J. J. & A. I. Miller. 1985. Evolutionary faunas and the distribution of Paleozoic marine communities in space and time. Pp. 153–190. *In:* J. W. Valentine (ed.), *Phanerozoic Diversity Patterns—Profiles in Macroevolution.* Princeton University Press: Princeton, NJ.

Sepkoski, J. J. & P. M. Sheehan. 1983. Diversification, faunal change, and community replacement during the Ordovician radiations. Pp. 673–717. *In:* M. J. S. Tevesz & P. L. McCall (eds.) *Biotic Interactions in Recent and Fossi Benthic Communities.* Plenum Press: New York.

Sheehan, P. M. 1975. Brachiopod synecology in a time of crisis (Late Ordovician-Early Silurian). *Paleobiology* 1:205–212.

Sheehan, P. M. 1977. Late Ordovician and earliest Silurian meristellid brachiopods in Scandinavia. Journal Paleontology 51:23–43.

Sheehan, P. M. 1979. Swedish Late Ordovician marine benthic assemblages and their bearing on brachiopod zoogeography. Pp. 61–74. *In:* J. Gray & A. J. Boucot (eds.) *Historical Biogeography, Plate Tectonics, and the Changing Environment.* Oregon State University Press: Corvallis.

Sheehan, P. M. 1980. Paleogeography and marine communities of the Silurian carbonate shelf in Utah and Nevada. Pp. 19–37. *In:* T. D. Fouch & E. R. Magathan (eds.), *Paleozoic Paleogeography of the West-Central United States.* Rocky Mountain Section, Society Economic Paleontologist Mineralogists, West-Central United States Paleogeography Symposium 1.

Sheehan, P. M. 1982. Brachiopod macroevolution at the Ordovician-Silurian boundary. Pp. 477–481. *In:* B. Mamet. & M. J. Copeland (eds.), *Third North American Paleontological Convention. Proceedings,* V.2.

Sheehan, P. M. 1985. Reefs are not so different—they follow the evolutionary pattern of level-bottom communities. *Geology* 13:46–49.

Sheehan, P. M. & T. A. Hansen. 1986. Detritus feeding as a buffer to extinction at the end of the Cretaceous. *Geology* 14:868–870.

Sheehan, P. M. & P. J. Coorough. 1990. Brachiopod zoogeography across the Ordovician-Silurian extinction event. Pp. 181–187. *In:* W. S. McKerrow & C. R. Scotese (eds.) *Palaeozoic*

Palaeogeograph and Biogeography. Geological Society of London Memoir 12.

Signor, P. W. 1985. Real and apparent trends in species richness through time. Pp. 129–150. *In:* J. W. Valentine (ed.) *Phanerozoic Diversity Patterns—Profiles in Macroevolution*. Princeton University Press: Princeton, NJ.

Springer, D. A. & R. K. Bambach. 1985. Gradient versus cluster analysis of fossil assemblages: a comparison from the Ordovician of southwestern Virginia. *Lethaia* 18:181–198.

Strong, D. R. 1983. Natural variability and the manifold mechanisms of ecological communities. *American Naturalist* 122:636:660. Valentine, J. W. (ed.) 1985. *Phanerozoic Diversity Patterns—Profiles in Macroevolution*. Princeton University Press: Princeton, NJ. 441 pp.

Van Valen, L. M. 1984. A resetting of Phanerozoic community evolution. *Nature* 307:50–52.

Walker, K. R. & L. F. Laporte. 1970. Congruent fossil communities from the Ordovician and Devonian carbonates of New York. Journal *Paleontology* 44:928–944.

Walker, T. D. 1985. Diversification functions and the rate of taxonomic evolution. Pp. 311–334. *In:* J. W. Valentine (ed.), *Phanerozoic Diversity Patterns in Profiles in Macroevolution*. Princeton University Press: Princeton, NJ.

Walker, T. D. & J. W. Valentine. 1984. Equilibrium models of evolutionary species diversity and the number of empty niches. *American Naturalist* 124:887–899.

Watkins, R. & A. J. Boucot. 1975. Evolution of Silurian brachiopod communities along the southeastern coast of Acadia. *Geological Society America Bulletin* 86:243–254.

Westrop, S. R. 1989. Macroevolutionary implications of mass extinction—evidence from an Upper Cambrian stage boundary. *Palaeobiology* 15:46–52.

Ziegler, A. M. 1965. Silurian marine communities and their environmental significance. *Nature* 207:270–272.

Ziegler, A. M. & A. J. Boucot. 1970. North American Silurian animal communities. Pp. 95–106. *In:* W. B. N. Berry & A. J. Boucot (eds.). Correlation of the North American Silurian rocks. *Geological Society America Special Paper* 102.

A Phylogenetic Perspective on Articulate Brachiopod Diversity and the Permo-Triassic Extinctions

Sandra J. Carlson

Abstract. Why did the Permo-Triassic extinction event have such a profound effect in shaping articulate brachiopod evolutionary history? Why did the articulates suffer such tremendous losses in this event, while some other groups (gastropods and bivalves, for example) suffered relatively little? And why, unlike the molluscs, has articulate brachiopod recovery from the extinction perturbation lacked substantial, sustained diversification? In an attempt to answer these questions, I investigated patterns of taxonomic, phylogenetic, and ecological selectivity at the Permo-Triassic boundary among articulate brachiopods. Patterns of taxonomic selectivity identify strophomenides as particularly vulnerable to extinction, spiriferides resistant to extinction (at the end-Permian, but not the end-Triassic), and terebratulides capable of recovering from extinctions. Despite this apparent taxonomic selectivity, there is no evidence to suggest that the Permian and Triassic extinction events exhibited an interpretable pattern of phylogenetic selectivity, based on the preliminary phylogenetic analysis presented here. The pattern of apparent ecological selectivity at the Permian extinction event among brachiopods with interlocking and noninterlocking hinge structures, documented by this study, is entirely consistent with Thayer's hypothesis of biological bulldozing (1979). A phylogenetic approach to the analysis of extinctions and originations is valuable because it can provide an explicit framework for the identification and comparison of clades and grades, and character homology and analogy. Evaluating changes in the relative frequency of sets of characters (both homologous and nonhomologous) can yield considerable insight into the evolutionary dynamics of mass extinctions and their long term effects on the subsequent recovery of clades such as the articulate brachiopods.

INTRODUCTION

The dramatic drop in taxonomic diversity at the end of the Permian period represents something of an evolutionary hallmark for the articulate brachiopods. Because they were the most diverse and abundant marine invertebrates throughout most of the Paleozoic era (Fig. 1), paleontologists have long been puzzled by the profound effect this extinction perturbation had in shaping articulate brachiopod evolutionary history. Many causal hypotheses have been proposed, among them the loss of suitable habitat resulting from the formation of Pangaea (Valentine & Moores, 1972), loss of stable soft sediment substrate because of increased bioturbation (Thayer, 1979), lack of a "competitive edge" with respect to molluscs, both anatomically and physiologically (Stanley, 1979; Steele-Petrovic, 1979), and lack of developmental flexibility in nonplanktotrophic larval ecology (Valentine & Jablonski, 1983). Despite the abundance of hypotheses proposed, com-

Dr. Carlson is with the Department of Geology, University of California, Davis, California 95616, USA.

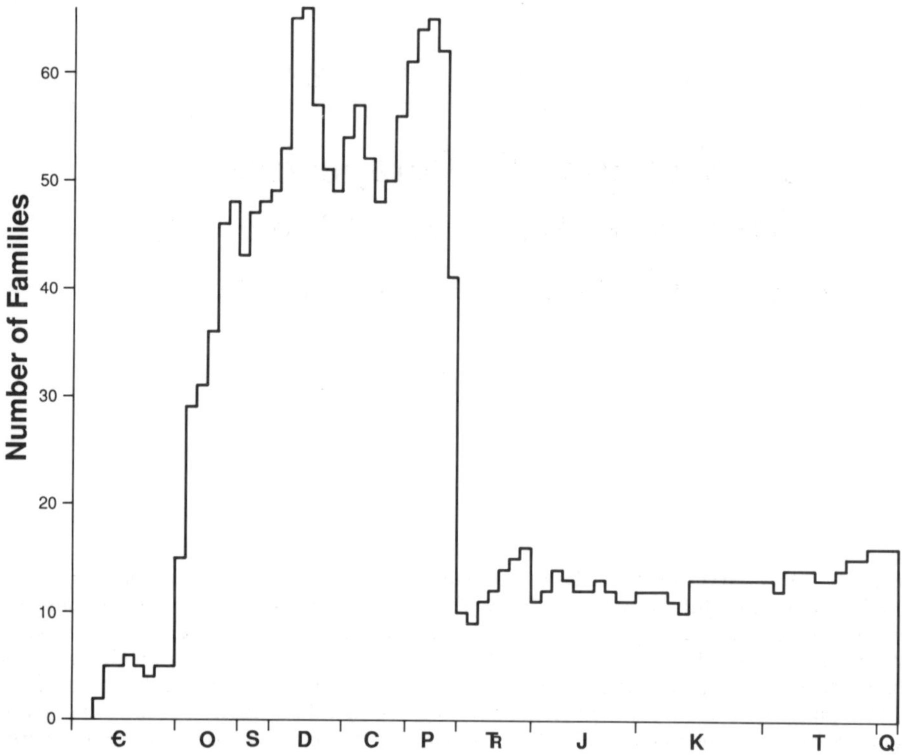

Figure 1. Pattern of family-level diversity of articulate brachiopods through the Phanerozoic (compiled from Sepkoski, 1982, 1983, 1986, 1988).

pelling and well-documented explanations to account for both the severity of the extinctions among the brachiopods, as well as the lingering effects of the extinction over the subsequent 200 million years, are still lacking. Why did the articulates suffer such tremendous losses in this event, while some other groups (gastropods and bivalves, for example) suffered relatively little? And why, unlike the molluscs, has articulate brachiopod recovery from the extinction perturbation lacked substantial, sustained diversification?

To begin to answer these questions, I will focus on three aspects of the pattern of articulate brachiopod diversity.

1. The pattern of taxonomic (superfamilial) diversity through the Phanerozoic. Taking a closer look at the Permian and Triassic periods, what is the pattern of familial and generic diversity within superfamilies? It is clear that some sort of phylum-level taxonomic selectivity occurred at the end-Permian; brachiopods and echinoderms were affected far more than molluscs. Does an interpretable pattern of taxonomic selectivity emerge at lower taxonomic levels within the articulate brachiopods? I will examine the taxonomic composition of three groups of brachiopods: those extinct by the end of the Permian period, those crossing the boundary (occurring in both the Permian and Triassic periods), and those originating after the Permian.

2. The phylogenetic integrity of these three groups, designated relative to the extinction boundary. Since it is frequently a mistake to assume that taxa (names in an organizational hierarchy) are clades (monophyletic groups) (Patterson & Smith, 1987), it is reasonable to ask whether patterns of taxonomic selectivity have phylogenetic significance. Did the Permian extinction event have an effect at the clade level, eliminating some monophyletic groups but not others? Are the survivors closely or only distantly related to one another? I will discuss the taxonomic patterns identified above in the context of two

preliminary phylogenetic reconstructions of articulate brachiopod superfamilies (Carlson, 1991).

3. The pattern of extinction and origination of morphological characters. What insights, if any, can be gained from a character-based approach, in addition to a taxon-based approach to the study of extinction events? I will argue that it is possible to make inferences about ecological selectivity by studying and comparing the characteristics of Paleozoic and post-Paleozoic brachiopods. This is by no means a novel idea; paleontologists have long been comparing, in a general way, the morphologies of "winners and losers" at extinction boundaries. However, by comparing morphology in an explicitly phylogenetic framework, attempting to distinguish homologous from nonhomologous and derived from primitive characters, the evolutionary significance of those patterns comes into sharper focus. Did the Permian extinction event have a grade-level effect, eliminating certain ecological or functional groups but not others? I will examine the distribution of three sets of characters relating to brachiopod shell structure, lophophore support, and habitat or mode of life and discuss their taxonomic and phylogenetic significance.

It is particularly important to keep in mind the distinction between evolutionary pattern and evolutionary process when analyzing extinction boundaries. Although it presents something of a chicken-and-egg dilemma, the patterns whose cause we seek to explain must be clearly delineated before testable hypotheses of cause and effect can be generated. The entities forming the patterns must be identified clearly—are they taxa (if so, at what level?), clades (are they nested or not?), characters (what kinds?), or are they some other type of entity? With respect to which model(s)—statistical, phylogenetic, ecological, or other—are the patterns being compared and analyzed? This paper concerns the analysis of evolutionary patterns primarily, but discusses some possible implications for inferring evolutionary process.

MATERIALS AND METHODS

The ordinal, superfamilial, familial, and generic classification of articulate brachiopods referred to here follows (with a few exceptions noted in Table 1) Williams & Rowell (1965). Geological ranges of families and genera were extracted largely from Sepkoski (1982, plus updates), with modifications based on additional data in Williams & Rowell (1965), Grant (1970), Cooper & Grant (1974–1975), and Ross & Ross (1979). A simple count of the number of families present in each stage from the Upper Permian through the Triassic was plotted to produce Figure 3, noting those families first appearing and last appearing in each stage.

The phylogenetic analyses were performed as described in Carlson (1991). Briefly, I compiled morphological information on the distribution of over 150 characters of soft and hard anatomy, embryology, and development among the 39 superfamilies recognized in the Treatise on Invertebrate Paleontology (Williams & Rowell, 1965). These data were analyzed using the heuristic search option in the computer algorithm PAUP (Swofford, 1989). Five inarticulate superfamilies were used as outgroups in the first analysis; the two geologically oldest articulate superfamilies (billingselloideans and orthoideans) were used as "outgroups" in the second analysis. Fifty percent majority-rule consensus cladograms were calculated when more than a single most-parsimonious tree was found. I acknowledge that the use of consensus trees is controversial (Carpenter, 1988), but given the preliminary nature of these results, they appeared to provide the most economical (although, in some sense, less informative) way to represent conflicts in tree topology that could not yet be resolved unambiguously. Subsequent analyses, in which the characters were weighted according to the Rescaled Consistency Index of Farris (1989), yielded

122

single most-parsimonious cladograms with topologies virtually identical to the consensus diagrams in Figures 4 and 5. The consistency indices of these cladograms with unevenly-weighted characters were typically twice as high as the multiple equally-parsimonious cladograms with evenly-weighted characters. Topological differences and differences in consistency index are noted in the figure captions.

The three characters or character complexes chosen for analysis were selected from the data matrix compiled for phylogenetic analysis, described above. Because this study focuses on patterns emerging at the superfamily level, and seeks to test the phylogenetic validity of the nine orders of articulates currently recognized (Williams & Rowell, 1965), the characters are tallied at the superfamily level. Samples of brachiopods in the Extinct, Cross, and Originate groups are necessarily small (10, 13, and 7, respectively), because the phenomenon of interest is the largest extinction event in brachiopod evolutionary history. Examining lower taxonomic levels (e.g., families or genera) as a way of increasing sample sizes is unfortunately not successful; for example, four genera in 13 families in 13 superfamilies cross the boundary (Table 1). To compare differences in the frequency of nominal character states distributed among the three taxonomic/temporal samples (brachiopods going extinct, crossing the boundary, originating after the boundary), Chi-square tests of independence (contingency tests) were performed. Because of the unavoidably small sample sizes involved, the biological significance of statistical significance values should be interpreted generously.

Table 1. Generic list of superfamilies: A—extinct by the end of the Permian period; B—crossing the boundary; C—originating in the Triassic. Superfamily endings have been adjusted from -acea to -oidea, according to the current revision of the *Treatise on Invertebrate Paleontology* (Williams, pers. comm. 1990). Several additional families and genera, in superfamilies that cross the boundary, are extinct by the end of the Permian period; they are not listed here by name, because the superfamily level is the focus of attention of this study. These data on lower taxonomic levels can be retrieved relatively easily, however, from the references listed below. An asterisk following a genus name in (A) indicates that specimens have been collected above the base of the Triassic, but are considered to be reworked from Upper Permian strata (see also discussion in Kummel, 1979); thus, they are listed in (A) rather than (B). Stage (in A and B) refers to the final stage in which the genus occurs, according to the references cited; Stage (in C) refers to the first stage in which the genus occurs. Numbered references refer to: 1—Sepkoski, 1982 (1983, 1986, 1988); 2—Williams & Rowell, 1965; 3—Ross & Ross, 1979; 4—Grant, 1970; 5—Boucot, 1975; 6—Waterhouse & Bonham-Carter, 1975; 7—Cooper & Grant, 1974–1975; 8—Baker, 1990; 9—Hoover, 1979; 10—Brunton & MacKinnon, 1972. Full citations are provided in the Literature Cited section. Reference numbers in parentheses indicate uncertainty in the accuracy of the reference to the data, or to the data itself.

A. Superfamilies extinct by end of Permian period	Stage	Reference
Order Orthida		
ORTHOIDEA		
Orthidae		1
One unidentified genus	Dzul	6
ENTELETOIDEA		
Enteletidae		1, (3)
*Enteletes**	Dzul	2, 4
Enteletella	Up. Perm	2
Enteletina	Up. Perm	2
*Orthotichia**	Dzul	2, 4
Rhipidomellidae		1, 3, 6
Rhipidomella	Dzul	2, 4
Order Strophomenida		
DAVIDSONIOIDEA		
Meekellidae		1, 3
*Orthothetina**	Dzul	(2), 4

Table 1. Continued

A. Superfamilies extinct by end of Permian period	Stage	Reference
(?DERBYOIDEA) Orthotetellidae		1, 7
Genus uncertain	Dzul	4, 7(?)
Orthotetidae		1, 3
Derbyia	Guad	4, 5
Schuchertellidae		1, 3
?Kiangsiella	Dzul	(2), 4, 5
CHONETOIDEA		
Chonetidae		1, 3
Dyoros	Dzul	2
Neochonetes	Dzul	2, 4
Waagenites	Up. Perm	2
RICHTHOFENIOIDEA		
Cyclanthariidae		1, 7
Cyclantharia	Guad	7
Sestropoma	Guad	7
Richthofeniidae		1, 3
Richthofenia	Dzul (Guad)	2, 4
STROPHALOSIOIDEA		
(?AULOSTEGOIDEA) Aulostegidae		1, (3)
Aulosteges	Dzul	2
?Institella	Up. Perm	2
?Strophalosina	Up. Perm	2
(?CHONETOIDEA) Chonetellidae		1, (?4)
Chonetella	Dzul	2
(?AULOSTEGOIDEA) Cooperinidae		1, 7
Cooperina	Guad	7
Strophalosiidae		1, 3
Dasyalosia	Guad	2
Orthothrix	Guad	2
Craspedalosia	Guad	2
Teguliferinidae		1, 3
?Teguliferina	Guad	2
(?AULOSTEGOIDEA) Tschernyschewiidae		1, 3, 6
Tschernyschewia	Dzul	2
Order Uncertain		
Suborder Dictyonellidina		
EICHWALDIOIDEA		
Isogrammidae		1, 3
Megapleuronia	Guad	2
Order Rhynchonellida		
RHYNCHOPOROIDEA		
Rhynchoporidae		1, 3
Rhynchopora	Guad	2
STENOSCISMATOIDEA		
Atriboniidae		1, 3
Camarophorina	Dzul	2
Camarophorinella	Dzul	2
Cyrolexis	Guad	2, 4
Stenoscismatidae		1, (3)
Stenoscisma	Dzul	2, 4
Order Terebratulida		
STRINGOCEPHALOIDEA		
Centronellidae		1, 5
Genus uncertain	Dzul	5(?)
Mutationellidae		1, 3
Glossothyris	Guad	2

124

Table 1. Continued

B. Superfamilies crossing the Permo-Triassic boundary	Stage	References
Order Strophomenida		
LYTTONIOIDEA		
Lyttoniidae		2
No genera cross; 6 out Perm, *Bactrynium in Rhaet		2
PRODUCTOIDEA		
Marginiferidae		(1), 3
Spinomarginifera	Indu	(2), 4
Order Rhynchonellida		
RHYNCHONELLOIDEA		
Wellerellidae		1, 2, 3
No genera cross; *Pseudowellerella, Denticuliphora, Gerassimovia* out Up. Perm, *Euxinella, Robinsonella* in Trias		2
Order Spiriferida		
CYRTIOIDEA		
Ambocoeliidae		1
Crurithyris	Indu	4
ATHYRIDOIDEA		
Athyrididae		1, 3
Athyris	Rhaet	2
RETICULARIOIDEA		
?Martiniidae		2, 3
?*Mentzelia*	Rhaet	2
?Reticulariidae	Rhaet	2
No genera cross; ?*Spirelytha* out Perm, ?*Triadispira* in Trias		2
ATHYRISINOIDEA		
Athyrisinidae	Nori	1
No genera cross; *Uncinella* out Perm, *Misolia* in Trias		2
RETZIOIDEA		
Retziidae	Rhaet	1
No genera cross; *Hustedia* out Perm, *Neoretzia* in Trias		2
SPIRIFERINOIDEA		
Spiriferinidae		1, 3
No genera cross; *Crenispirifer, Odontospirifer* out Up. Perm, *Spiriferinoides* in M. Tr.		2
SUESSIOIDEA		
Cyrtinidae	Carn	1, 3
No genera cross; *Cyrtina* out Perm, ?*Hirsutella* in M. Tr.		2
SPIRIFEROIDEA		
Spiriferidae	Indu	1, 6
One unidentified genus present in Dienerian, ?1 in Tr. (unclear if crossed boundary or not)		6
Order Terebratulida		
DIELASMATOIDEA		
Dielasmatidae		1, 3
No genera cross; *Yochelsonia* out Up. Perm, *Adygella, Coenothyris, Cruratula* in M. Tr.		2
CRYPTONELLOIDEA		
Cryptonellidae		1, 2
No genera cross; *Cryptonella, Heterelasma* out Perm, *Obnixia* in Up. Tr.		9

Table 1. Continued

C. Superfamilies originating in the post-Permian	Stage	References
Order Strophomenida		
DAVIDSONIOIDEA		
Thecospiridae		1
Thecospira	Carn	2
(Spiriferida) (THECOSPIROIDEA) (Thecospiridae) (*Thecospira*)		8
CADOMELLOIDEA		
Cadomellidae		1
Cadomella	Pliens	2
(Spiriferida) (KONINCKINOIDEA) (Koninckinidae) (*Cadomella*)		10
Order Spiriferida		
KONINCKINOIDEA		
Koninckinidae	Rhaet	1
Five genera in Trias		2
Order Uncertain		
(Spiriferida)THECIDEOIDEA		8
Thecideidae	Toar	1
8 genera orig.		2
(20 genera orig.)		8
Thecidellinidae	Rhaet	1
4 genera orig.		2
(7 genera orig.)		8
(Spiriferida) (THECIDEOIDEA) (Bactryniidae) *Bactrynium*		8
Order Terebratulida		
TEREBRATELLOIDEA		
Dallinidae	Nori	1
Aulacothyropsis	?mid. Tr.	2
TEREBRATULOIDEA		
Terebratulidae	Ladi	1
Plectoconcha	Trias	2
ZEILLERIOIDEA		
Zeilleriidae	Olen	1
Five genera in Trias		2

TAXONOMIC SELECTIVITY

The tremendously rich morphological and taxonomic diversity exhibited by the articulate brachiopods can be best organized and understood in the context of major elements of the articulate brachiopod Bauplan (sensu McGhee, 1980). Articulate brachiopods are bivalved lophophorates whose valves grow by hemiperipheral accretion of shell material. The valves articulate posteriorly in a tooth and socket hinge mechanism. Unlike bivalved molluscs, the mantles and lophophore filaments are unfused. These brachiopods live epifaunally, either attached to a hard substrate with a nonmuscular pedicle, cemented by their ventral valves, or free-living on soft sediment substrates. Valve size, shape, ornament, and internal characters are all quite variable, as even a cursory examination of a sample of Recent and fossil brachiopods will reveal. The Permian fauna, in particular, is characterized by a number of bizarre, aberrant forms including the conical, reef-dwelling richthofeniaceans, the extreme concavo-convex productaceans, and the odd, plate-like lyttoniaceans.

The morphological diversity in the Class Articulata has traditionally been organized into nine orders and 39 superfamilies (Williams & Rowell, 1965). The stratigraphic ranges of numbers of families per superfamily in each order are illustrated in Figure 2. Several

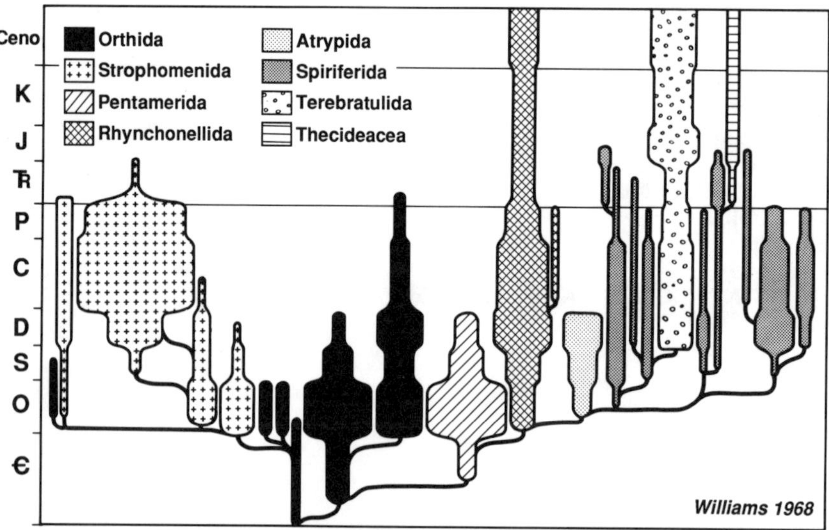

Figure 2. Geologic ranges of families within superfamilies in each order of articulate brachiopods, through the Phanerozoic, adapted from Williams, 1968. Superfamilies currently classified in the same order are shaded in the same pattern, as indicated in the key. Hypothesized phylogenetic relationships among the superfamilies (Williams, 1968) are indicated also.

generalizations about taxonomic diversity patterns may be noted. Paleozoic diversity is far greater than post-Paleozoic diversity; representatives of three orders are extant. The end-Permian and end-Triassic extinction events together significantly reduced superfamilial diversity. While some superfamilies go extinct at the end of the Permian (orthides), others cross the boundary represented by only one or two families (spiriferides), and still others appear to cross with relatively minor effect (rhynchonellides). Phylogenetic relationships among the superfamilies were proposed by Williams (1968; see Fig. 2), and are strongly influenced by the relative first appearance of these taxa in the fossil record. Several orders were considered by Williams to be paraphyletic, notably the orthides and spiriferides.

General observations on patterns of ordinal level diversity in the Permian and Triassic often center on two groups, the strophomenides and the spiriferides (Fig. 3A). Strophomenides, which had the highest familial diversity of any order in the Permian, lost four of six superfamilies by the end-Permian. If *Bactrynium*, which makes a brief appearance in the Rhaetian, is more appropriately classified with the thecideoideans in the Spiriferida, as Baker (1990) proposes, the figure increases to five out of six superfamilies. Easily the most morphologically bizarre groups occurring in the Permian are classified as (aberrant) strophomenides. Their morphological diversity is extraordinary, and is commonly attributed to the ontogenetic loss of a functional pedicle, which is related to their free-living mode of life. The adoption of this particular mode of life is thought to have eliminated many of the "constraints" shaping more typical brachiopod morphology. The scenario of strophomenide vulnerability to extinction comfortably accommodates the wide range of morphological and thus ecological specialization exhibited by the Permian strophomenides.

Spiriferides, as opposed to strophomenides, are considered to be relatively extinction resistant. Representatives of all eight spiriferide superfamilies present in the Upper Permian crossed the boundary and occur in the Triassic (Fig. 3B). The range of morphological diversity among the Permian spiriferides is also quite broad, as evidenced by the large number of superfamilies recognized. Why were the spiriferides essentially unaffected by events presumably resulting in the near elimination of the strophomenides?

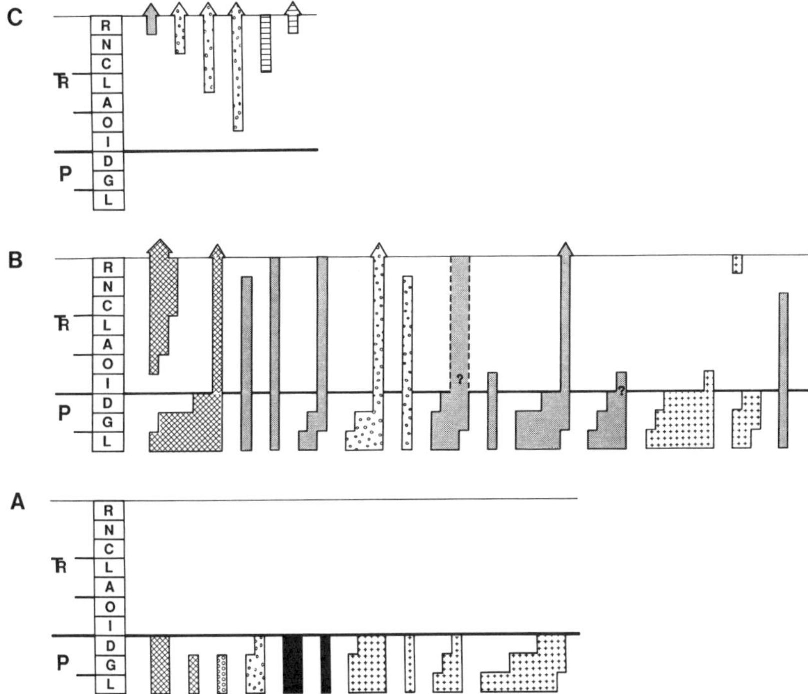

Figure 3. Geologic ranges of families within superfamilies, focusing on the Permian and Triassic periods. Geologic stages are identified by the first letter of the stage name (e.g., D = Dzulfian). Each geometric shape represents a different superfamily. The shapes are shaded according to their superfamilial assignment (see Key in Fig. 2). A—Superfamilies extinct by the end of the Permian period, showing decline in constituent families through the final three stages of the Permian. B—Superfamilies with at least one constituent family crossing the Permo-Triassic boundary, also indicating familial decline in the Permian. In most cases, the genera assigned to those families crossing the boundary are different on each side of the boundary (see Table 1). C—Superfamilies originating in the Triassic period. *Cadomella*, originating in the Jurassic, is omitted from the figure.

Perhaps the single characteristic that they all possessed, a spiral brachidium, or calcareous lophophore support, provided them with some kind of adaptive advantage. If so, then why were seven of these eight superfamilies extinct by the end of the Triassic? The simplest explanation would seem to be that the two extinction events, by affecting these two different groups of brachiopods very differently, were substantially different in kind. It is not entirely clear that the data currently in hand are sufficient to support such a hypothesis. Also, why were the rhynchonellides, which lack an elaborate calcareous lophophore support, able to rediversify successfully after the end-Permian (see Ager, 1987)? Perhaps calcareous lophophore supports are an attribute of incidental importance in the apparent extinction resistance of the spiriferides. This hypothesis remains as yet untested.

Several types of issues can obscure patterns of taxonomic extinction. First, changes in diversity at lower taxonomic levels (species, genera, even families) may be masked by changes at higher levels (superfamilies and orders). Second, the geological interpretation of the position of an extinction boundary, problems with lithostratigraphic and biostratigraphic correlation, the reworking of older sediments and fossils into younger strata, the difficulty of reconciling differences among local sections to propose a single global interpretation, and, the fact that many of the important details from the rock and fossil record may be inadvertently overlooked in analyses of taxonomic extinction—all are potentially obscuring factors. Third, the problem of phyletic pseudoextinction (Fortey,

128

1989), where experts working on two adjacent time periods give two different names to what is in fact the same taxon, will artificially generate an extinction and an origination event where evolutionarily neither existed. Fourth, the problem of nonmonophyletic taxa (Patterson & Smith, 1987; Fortey, 1989)—what evolutionary significance does the "extinction" of a paraphyletic or polyphyletic group have? It might have considerable evolutionary significance (see Ecological Selectivity), but not necessarily phylogenetic significance.

To begin to address the first two issues just raised, I will briefly discuss the evidence supporting the geological ranges of each of the Permian and Triassic superfamilies (Fig. 3, Table 1). Each of the three "extinction-related" groups will be discussed in turn: brachiopods extinct by the end of the Permian, brachiopods crossing the boundary into the Triassic, and brachiopods originating after the Permian period.

Extinct

At least some representatives of three articulate brachiopod orders became extinct through the Late Permian; all the remaining taxa in two orders (Orthida and Dictyonellidina, order Uncertain) became extinct. The Spiriferida was the only order present in the Permian that completely escaped superfamilial extinction. Eight of the 13 superfamilies in which at least some representatives crossed the boundary also experienced considerable familial extinctions in the Guadalupian and Dzulfian stages (Fig. 3B). 43% (10 of 23) of the superfamilies and 82% (59 of 72) of the families present in the Leonardian were extinct by the end of the Dzulfian; the percentage for genera has not been calculated, but is almost certainly over 95%. However, only the stenoscismatoideans, enteletoideans, strophalosioideans, davidsonioideans, and chonetoideans lost more than one genus in the Dzulfian extinction event. Three of those five superfamilies are strophomenides. Thus, as noted earlier, it appears that of all the articulates, strophomenides were particularly affected by the Permian extinctions at the generic, familial, and superfamilial levels.

Cross

In most of the superfamilies (11 of 13), only a single family crossed the boundary, and in those families, very few genera (3 of 13) were actually recorded as occurring on both sides of, or "crossing," the boundary. In other words, most of the occurrences of families crossing the boundary are based on the occurrence of one genus in that family that goes extinct at the end of the Permian and another, different genus in the same family that originates in the Triassic (and typically goes extinct by the end of the Triassic). In the rhynchonelloideans, not a single genus actually crosses the boundary. One new genus, placed in a family that also occurred in the Permian, appears in the Induan. This boundary thus marks a complete turnover in the rhynchonelloidean fauna. Other rhynchonelloidean taxa appeared and diversified throughout the Triassic (Fig. 3B). The terebratulides also turn over completely; no genera cross the boundary. The extension of the strophomenides across the boundary is remarkably tenuous, although it may be difficult to appreciate that fact in compilations such as these. One specimen of *Spinomarginifera* is sufficient to extend the range of the Productoidea into the Triassic (see Table 1). The current problems associated with assigning *Bactrynium* to a higher taxon were mentioned previously. The majority of superfamilies crossing the boundary are spiriferides (8 of 13; 9 families cross), although only one of those eight continues into the Jurassic. Most of the superfamilies (69% or 9 of 13) crossing the Permian boundary are extinct by the end of the Triassic period.

Originate

Seven superfamilies originate in the post-Paleozoic (Fig. 3C), six in the Triassic, and one in the Jurassic. Terebratulides experience a fairly substantial radiation in the Triassic; three superfamilies make their first appearance at this time, two of which persist to the present day. One unusual spiriferide superfamily, the Koninckinoidea, first appears in the Late Triassic and is extinct by the end of the Lower Jurassic. The assignment of the remaining three superfamilies to higher taxa is controversial (Table 1), and thus has considerable significance in the interpretation of patterns of higher taxonomic diversity over this time interval. The classification in the Treatise (Williams & Rowell, 1965) places the monogeneric families Thecospiridae and Cadomellidae in two different strophomenide superfamilies, and leaves the ordinal assignment of the Thecideoidea uncertain. Brunton and MacKinnon (1972) suggested removing *Cadomella* to the Koninckinoidea (Spiriferida). Baker (1990), in a revision of the classification of thecideidine brachiopods, places *Bactrynium* in the Thecideoidea, groups Thecospiridae with the Thecideoidea under the suborder Thecideidina, and places both in the Spiriferida.

Summary of Results of Taxonomic Selectivity

1. Strophomenides appear to suffer the greatest taxonomic losses; four (possibly five) of six superfamilies ended with the Permian. Not a single spiriferide superfamily went extinct in the Permian; all eight crossed the boundary, but seven of these were extinct by the end of the Triassic. Rhynchonellides, at the family level, and terebratulides, at the superfamily level, experienced complete replacement of Permian taxa by Triassic taxa. These broad scale, ordinal level patterns have led to generalizations about extinction vulnerability and resistance within the articulate brachiopods.

2. The Permian extinctions effectively reduced lower level (genus and family) taxonomic diversity. The Triassic extinctions reduced the remaining higher level (superfamily) diversity, which had been "weakened" by the loss of significant numbers of lower level constituent taxa.

3. Family-rich superfamilies (e.g., strophalosioideans) and family-poor superfamilies (e.g., eichwaldioideans, rhynchoporoideans) seemed to fare equally poorly at the end-Permian and equally well in crossing the boundary. The apparent irrelevance of within-taxon richness as a predictor of extinction resistance or vulnerability has been noted at lower taxonomic levels as well (Jablonski, 1986).

4. When dealing with higher taxa that are represented by a single genus, or even a single species, whose relationship to other higher taxa (i.e., other genera in other higher taxa) is uncertain, major differences in taxonomic tallies will occur depending on where the taxa are classified. The high degree of taxonomic uncertainty surrounding some groups must be kept in mind in when interpreting patterns of taxonomic selectivity at extinction boundaries.

PHYLOGENETIC SELECTIVITY

To investigate the phylogenetic validity of the taxonomic patterns outlined above, I performed two phylogenetic analyses of articulate brachiopod superfamilies. In one analysis, I used the outgroup method of polarity determination; in the other, I used a version of the stratigraphic method of polarity determination. In order to choose appropriate outgroups, some kind of preliminary phylogenetic analysis must be performed to identify likely sister groups to the ingroup taxa, in this case, the articulate superfamilies. Because brachiopods are currently organized into two classes, Inarticulata and Articulata, it seemed logical to search among the inarticulates for appropriate outgroups. I performed two preliminary analyses to provide a basis for choosing outgroups from among the inarticulate taxa. First, I analyzed the seven extant brachiopod superfamilies, both inarticulate and articulate, using the other two lophophorate phyla (phoronids and bryozoans) as outgroups. Total ingroup monophyly was preserved (the seven brachiopod superfamilies formed a clade), and the three inarticulate and four articulate superfamilies each clustered together, forming monophyletic groups. Second, I analyzed relationships among the eight inarticulate superfamilies and the articulates, as a single monophyletic taxon, also using phoronids and bryozoans as outgroups. Once again, the monophyly of the inarticulates was indicated. From these preliminary analyses, five inarticulate super-families were chosen as outgroups (Fig. 4).

The results of this first phylogenetic analysis can be summarized as follows. Two major clades emerge, with a single primitive "outlier," the triplesioideans, a group of aber-rant orthides. The orthides, strophomenides, thecideoideans, and one group of spiriferides (the suessioideans) form one clade; all other superfamilies cluster together in

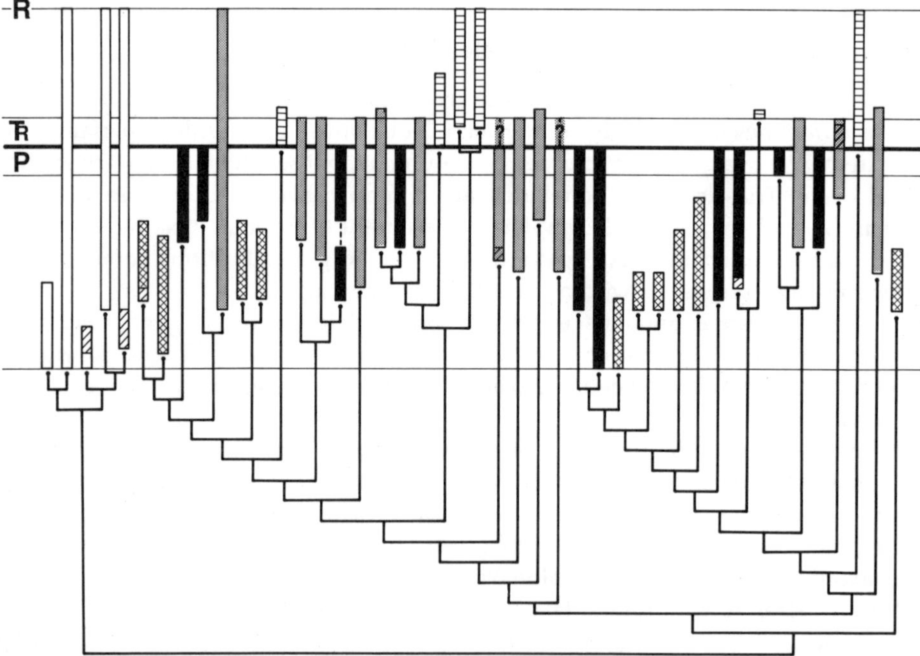

Figure 4. Hypothesized phylogenetic relationships among the 39 articulate brachiopod super-families, using five inarticulate taxa as outgroups. This is a 50% majority rule consensus diagram of 6 cladograms of length 478, with an overall Consistency Index = .370. Rectangles denote the known stratigraphic ranges of the superfamilies; those in black are taxa going extinct at the end of the Permian; those shaded cross the boundary; those originating in the post-Paleozoic are marked with horizontal lines; those with a diagonal grid pattern are extinct before the Permian. The identity of the taxa, listed in order from left to right on the diagram, is as follows: Paterinoidea, Linguloidea, Kutorginoidea, Discinoidea, Cranioidea; Pentameroidea, Porambonitoidea, Stenoscismatoidea, Rhynchoporoidea, Rhynchonelloidea, Atrypoidea, Dayioidea, Koninckinoidea, Athyrisinoidea, Retzioidea, Eichwaldioidea, Athyridoidea, Dielasmatoidea, Stringocephaloidea, Cryptonelloidea, Zeillerioidea, Terebratelloidea, Terebratuloidea, Reticularioidea, Cyrtioidea, Spiriferinoidea, Spiriferoidea, Enteletoidea, Orthoidea, Billingselloidea, Gonambonitoidea, Clitambonitoidea, Plectambonitoidea, Strophomenoidea, Davidsonioidea, Chonetoidea, Cadomelloidea, Richtho-fenioidea, Productoidea, Strophalosioidea, Lyttonioidea, Thecideoidea, Suessioidea, Triple-sioidea. Weighting characters according to methods described in Farris (1989), yielded a single, most parsimonious cladogram, with a C.I.=.616. The only topological difference between this cladogram and Fig. 4 is that the eichwaldioideans move to a position as sister group to the terebratulides.

the second clade. Pentamerides, atrypides, and terebratulides are each monophyletic; rhynchonellides and strophomenides are paraphyletic. Orthides, with the sole exception of the triplesioideans, also form a clade. Spiriferides, of all the currently recognized orders, are the most highly fragmented, phylogenetically. Two paraphyletic clusters of spiriferides emerge, a strophic (straight-hinged) group and an astrophic (curved-hinged) group. This relative pattern of relationships is, in many ways, similar to the pattern proposed by Williams (1968; Fig. 2). The major difference, however, and it is a significant one, is that the stratigraphically most primitive taxa, the orthides and pentamerides, emerge as the most derived taxa in this analysis of morphology independent of stratigraphic position.

At least three interpretations of this perplexing phylogenetic pattern are plausible. First, the pattern could be misleading because the inarticulate outgroups share so few characters with the ingroup taxa that they have relatively little power in structuring the articulate relationships. Second, homoplasy among characters and character complexes could be so pervasive among the articulates (as has been noted by brachiopodologists for the past century) that the power of the homologous characters to structure the diagram is swamped by the overabundance of homoplasies. Third, the pattern might indicate that

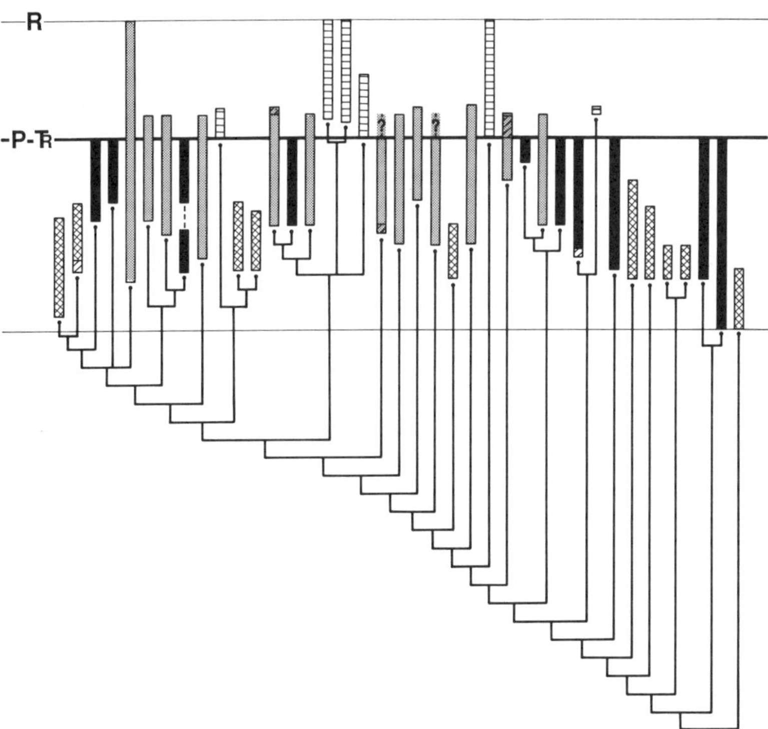

Figure 5. Hypothesized phylogenetic relationships among the 39 articulate brachiopod superfamilies, using the two groups appearing first in the fossil record (Billingselloidea and Orthoidea) as outgroups. This is a 50% majority rule consensus diagram of 4 cladograms of length 371, with an overall Consistency Index = .320. The known geologic ranges of each superfamily, shaded as in Fig. 2, are plotted above the branching pattern of relationship. The identity of the taxa, listed in order from left to right on the diagram, is as follows: Porambonitoidea, Pentameroidea, Stenoscismatoidea, Rhynchoporoidea, Rhynchonelloidea, Athyrisinoidea, Retzioidea, Eichwaldioidea, Athyridoidea, Koninckinoidea, Atrypoidea, Dayioidea, Dielasmatoidea, Stringocephaloidea, Cryptonelloidea, Terebratelloidea, Terebratuloidea, Zeillerioidea, Reticularioidea, Cyrtioidea, Spiriferinoidea, Spiriferoidea, Triplesioidea, Suessioidea, Thecideoidea, Lyttonioidea, Richthofenioidea, Productoidea, Strophalosioidea, Chonetoidea, Cadomelloidea, Davidsonioidea, Strophomenoidea, Plectambonitoidea, Gonambonitoidea, Clitambonitoidea, Enteletoidea, Orthoidea, Billingselloidea.

132

heterochrony, or developmental alterations, particularly paedomorphic changes, have had a significant affect in brachiopod evolution. In other words, the geologically oldest taxa may possess fewer characters shared with geologically younger taxa (and are thus relatively derived) because the geologically younger taxa may have truncated developmental pathways and thus possess a relatively greater number of more general or more primitive characters. Of course, this final suggestion remains entirely speculative until evidence on the ontogenies of these taxa is revealed. Nevertheless, the pattern itself is permissive of such a testable interpretation. Obtaining evidence leading to the rejection of one or all of these interpretations is the focus of my current investigations.

In the second phylogenetic analysis, I eliminated the inarticulate taxa as outgroups and instead used the two articulate superfamilies that first appear in the fossil record in the Lower Cambrian—the orthoideans and the billingselloideans (Fig. 5). The traditional view of brachiopod phylogeny has relied quite heavily on the relative order of appearance in the fossil record to polarize the direction of character transformation (see Fig. 2). Taxa appearing earliest are considered, on that basis, to possess the most primitive morphologies. In a crude attempt to incorporate at least some stratigraphic information into an otherwise strictly morphological phylogenetic analysis, I choose to follow the procedure described above. This is not a very satisfactory method to use as a way of incorporating relative stratigraphic position into a phylogenetic analysis. Ideally, the relative stratigraphic position of each taxon would be evaluated along with, but in a manner different from, the morphological character information. An algorithm to accomplish this kind of evaluation is currently being developed by Fisher, Maddison, and Maddison (MacClade ver. 3.0; Fisher, in prep.), but is not yet available.

Rather than forming two clades, as in the previous analysis, the articulates form one large clade when the orthoideans and billingselloideans are the designated outgroups. The taxa included within the first clade in the previous analysis form the primitive foundation of this single large clade; the taxa appear in the same relative order as before but in reverse polarity. This is not too surprising, since the two taxa used as outgroups here appeared as two of the most derived taxa in the first analysis. The second clade in the first analysis forms the most derived portion of this large clade, with the taxa in nearly the same relative order and polarity as before. Pentamerides, atrypides, and terebratulides are again monophyletic; rhynchonellides, strophomenides and orthides (once again with the exception of the triplesioideans) are paraphyletic. Spiriferides are fragmented again and appear to form two roughly paraphyletic clusters, as before.

This pattern is somewhat more consistent with the order of first appearance of these taxa in the fossil record, not surprising given the rationale for the chosen outgroups. The pentamerides are still an anomaly, however. They are among the earliest brachiopods to appear, and yet they are consistently the most derived morphologically. Brachiopodologists have long recognized that many pentamerides are remarkably "modern" in their overall morphology and share a number of similarities with Recent terebratulides and rhynchonellides. Future research will be directed toward understanding the disparities between these two patterns (Figs. 4 and 5) of relationship, particularly with respect to the position of the pentamerides and the "spiriferides."

Summary of Results of Phylogenetic Selectivity

1. No clear pattern of phylogenetic selectivity exists, with respect to the two preliminary hypotheses of superfamilial phylogenetic relationships among articulate brachiopods. The group of taxa extinct by the end of the Permian is scattered haphazardly across both cladograms, as is the group of taxa crossing the boundary, and the group originating after the Permian. The stratigraphic cladogram suggests that the remaining

most primitive groups were removed by the end-Permian, but at least as many derived taxa were also removed.

2. The spiriferides clearly do not form a monophyletic group. The Spiriferida is frequently referred to as a "clade," or phylogenetically coherent unit, because it exists as a distinct order in the current classification of brachiopods (Williams & Rowell, 1965), defined by a single character (the spiralium) thought to be a shared, derived character. The results of these phylogenetic analyses indicate that the possession of a spiral brachidium is either a shared primitive feature that has transformed in several different ways over the course of brachiopod evolution, or that it is a feature that has arisen convergently many times. Because it is not a clade, the order Spiriferida cannot be considered a particularly extinction resistant taxon.

3. Perhaps the characters that structure the phylogenetic hypotheses illustrated in Figures 4 and 5 are related to mode of life and attachment of the brachiopod to a substrate. In other words, patterns of massive character convergence, rather than character homology, are structuring the phylogenetic diagram. Just as Gardiner (1982) has argued that birds and mammals are sister groups on the basis of an abundance of character similarities, all of which are related to endothermy (Gauthier et al., 1988), it is possible that these phylogenetic hypotheses reflect large scale, ecological convergence among more distantly related brachiopod taxa. Possibly the weakness of the outgroups chosen, because of the few characters shared among outgroups and the ingroup taxa, fails to unambiguously resolve the ingroup relationships. This phylogenetic exercise has highlighted very clearly the most important direction of future research on the project—a resolution of these conflicting results.

ECOLOGICAL SELECTIVITY

A character-based approach to the study of extinction events, in addition to a taxon-based and clade-based approach, can yield valuable insights into the functional and ecological dynamics of extinctions and originations. By looking in some detail at specific morphological features and their distribution among Permian and post-Permian brachiopods, it is possible to begin to bridge the gap between the analysis of evolutionary pattern and the inference of evolutionary mechanism. The morphological characters chosen for analysis here represent several significant aspects of the articulate brachiopod Bauplan, discussed earlier. Variation in shell structure, in calcareous lophophore supports, and in characters relating to mode of life and substrate preference among superfamilies have been tallied for each of the three temporal groups.

Brachiopods, as bivalved lophophorates, mineralize their valves in a manner very similar to bivalved molluscs. The mantle outer epithelial cells secrete an organic periostracum, upon which a low magnesium calcite primary layer, and then a fibrous secondary layer are deposited (Williams & Rowell, 1965; Williams, 1956, 1990). The process is frequently referred to as terminal accretion, because the addition of shell material is most pronounced along the anterior commissure of the valves. In fact, a very thin layer of shell material is deposited over the entire internal surface of the valves as well. The valves of many brachiopods possess punctae, or cylindrical cavities extending through the thickness of the shell. These cavities contain extensions of the mantle tissue, called caecae, which are considered to be important to the brachiopod in respiration (Peck et al., 1986) and possibly in the storage of excess nutrients and as a deterrent to boring predators (Curry, 1983). Up to 50% of the total soft tissue of some punctate brachiopods is contained within the shell itself, in the caecae (Curry & Ansell, 1986). Caecae clearly play an important role in the physiology of punctate brachiopods. However, not all brachiopods have punctae; in fact, punctae appear to have arisen several times indepen-

134

dently from impunctate ancestors, which lack these structures entirely (Williams & Rowell, 1965; Carlson, 1991). Pseudopunctate shell structures are also known; they consist of conical deflections of secondary shell material that point inwardly and anteriorly, forming tubercles on the internal valve surface. The function of pseudopunctae is not yet clear; they may represent some sort of constructional "artifact" of growth, and/or they may increase the biomechanical strength of the shell fabric.

The distribution of these three different shell fabrics among Permian and post-Permian brachiopods is illustrated in Figure 6 and Table 2. Punctate brachiopods fare identically in each of the three groups; pseudopunctate forms also vary relatively little. More impunctate brachiopods cross the Permian boundary, but all except one is extinct by the end-Triassic. The distribution of characters of shell structure in the Cross group is not statistically significantly different from the Extinct and Originate groups (p = 0.662). No clear pattern of character selectivity over this time interval emerges.

Possession of a lophophore, a filamentous organ used for feeding and respiration, is a particularly characteristic feature of brachiopods that is shared with bryozoans and phoronids. Three developmental pathways, with diverging lophophore geometries, can be observed in Recent brachiopods (Williams, 1956; Rudwick, 1970). Plectolophe, spirolophe, and ptycholophe lophophores are the most common adult geometries, although adults with trocholophe, schizolophe, and zygolophe geometries may also be found. A wide variety of wonderfully complex mineralized structures that provide greater or lesser support to the arms of the lophophore has evolved. These calcified structures are mineralized in a manner comparable to the valves themselves, by an infolding or rolling of the outer epithelium, forming long, ribbon-like structures internal to the lophophore base. Four major types of mineralized lophophore supports are recognized. Crura are usually fairly short, prong-like processes that extend from the cardinalia; they are characteristic of rhynchonellides, which possess spirolophe lophophores. Loops are loop-shaped processes that extend through most of the base of the lophophore; they are characteristic of terebratulides, many of which possess plectolophe lophophores, as well as a few other unusual brachiopods. Spiralia are spiral-shaped processes that extend throughout the

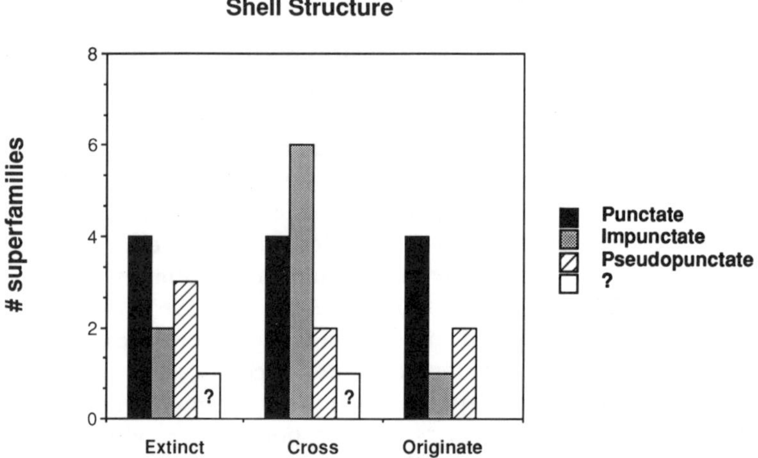

Figure 6. Histogram of number of articulate brachiopod superfamilies possessing one of three different shell structures and their distribution relative to the Permo-Triassic extinction events. The Extinct group includes 10 superfamilies; the Cross group, 13; the Originate group, 7. Open rectangles with question marks indicate the number of taxa for which no character state could be unambiguously assigned; the character may be presently unknown, inapplicable and thus unknowable, or polymorphic at lower taxonomic levels. This last situation is designated by "?poly" in Table 2.

Table 2. List of selected morphological characters and their distribution among superfamilies.

Superfamily	Shell structure	Calcareous lophophore support	Hinge line	Hinge structure	Attachment to substrate
Extinct:					
Orthoidea	Impunctate	??	Strophic	Noninterlock	??
Enteletoidea	Punctate	??	Strophic	Noninterlock	??
Davidsonioidea	?poly	Ridges	Strophic	Noninterlock	Cemented
Chonetoidea	Pseudo	??	Strophic	Noninterlock	Free-living (adults)
Richthofenioidea	Pseudo	Ridges	Strophic	??	Cemented
Strophalosioidea	Pseudo	Ridges	Strophic	Noninterlock	Cemented
Eichwaldioidea	Punctate	??	?poly	Noninterlock	??
Rhynchoporoidea	Punctate	Crura	Astrophic	Interlock	Pediculate
Stenoscismatoidea	Impunctate	Crura	Astrophic	Interlock	Free-living
Stringocephaloidea	Punctate	Loops	Astrophic	Interlock	Pediculate
Cross:					
Lyttonioidea	Pseudo	?poly	??	Noninterlock	Free-living
Productoidea	Pseudo	Ridges	Strophic	Noninterlock	Free-living
Rhynchonelloidea	Impunctate	Crura	Astrophic	?poly	Pediculate
Cyrtioidea	Impunctate	Spiralia	Strophic	Noninterlock	Free-living
Athyridoidea	Impunctate	Spiralia	?poly	?poly	Pediculate
Reticularioidea	Impunctate	Spiralia	Strophic	Noninterlock	Free-living
Athyrisinoidea	Impunctate	Spiralia	Astrophic	?poly	Pediculate
Retzioidea	Punctate	Spiralia	Astrophic	Noninterlock	Pediculate
Spiriferinoidea	?poly	??	Strophic	Noninterlock	??
Suessioidea	Punctate	??	Strophic	Noninterlock	Free-living
Spiriferoidea	Impunctate	Spiralia	Strophic	Noninterlock	Pediculate
Dielasmatoidea	Punctate	Loops	Astrophic	Interlock	Pediculate
Cryptonelloidea	Punctate	Loops	Astrophic	Interlock	Pediculate
Originate:					
Thecospiroidea	Pseudo	Spiralia	Strophic	Interlock	Cemented
Cadomelloidea	Pseudo	Spiralia	Strophic	Interlock	Free-living
Koninckinoidea	Impunctate	Spiralia	Strophic	Noninterlock	Pediculate
Thecideoidea	Punctate	Ridges	Strophic	Interlock	Cemented
Terebratelloidea	Punctate	Loops	Astrophic	Interlock	Pediculate
Terebratuloidea	Punctate	Loops	Astrophic	Interlock	Pediculate
Zeillerioidea	Punctate	Loops	Astrophic	Interlock	Pediculate

entire base of certain spirolophe lophophores; no extant brachiopods possess spiralia, but they were quite common in many extinct forms, and serve to define the order Spiriferida. Ridges are ridge-shaped extensions of shell material on the internal surface of the dorsal (brachial) valve; they occur in thecideides, which possess ptycholophe lophophores.

The distribution of these various lophophore supports among Permian and post-Permian brachiopods is illustrated in Figure 7 and Table 2. No crural supports originated after the Paleozoic, but the rhynchonellides (with crura) are important elements in the Paleozoic and post-Paleozoic faunas. The Extinct group includes more ridge-bearing taxa than any other group; no spire-bearing superfamilies went extinct at the end of the Permian. Most of the Cross group consists of spire-bearers; however, all are extinct by the end-Triassic. The pattern of character distribution among the three groups together does not reveal statistically significant differences in any one group relative to the other two (p = 0.102), although the Extinct and Originate groups alone are significantly different from one another (p = 0.029). Thus, the brachiopods extinct by the end of the Permian have a

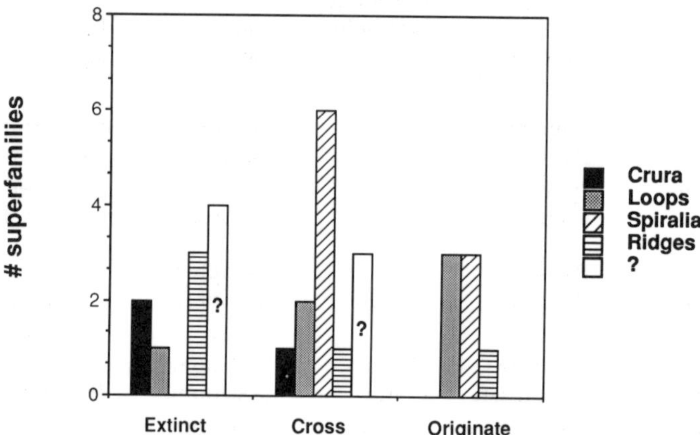

Figure 7. Histogram of the number of superfamilies possessing one of four different calcareous (mineralized) lophophore supports. Other elements of the histogram as described for Figure 6.

significantly different distribution of calcareous lophophore supports than the brachiopods that originate after the Permian; this result bears further examination.

The last character to be investigated is actually a complex of three characters that relate to the attachment of articulate brachiopods to the substrate, which varies considerably today, as in the past. Most brachiopods are pediculate, and attached to a hard substrate (a rock, or another organism with a hard shell, frequently another brachiopod) by a tough, nonmuscular pedicle. Some extant inarticulates and extinct articulates lived with their ventral (pedicle) valves cemented directly to a hard substrate. Although there are no well-documented cases of extant brachiopods living unattached on a soft sediment substrate, many extinct brachiopods (including pentameroideans and productoideans) adopted this mode of life. Some possessed pedicles as juveniles, which were lost later in ontogeny, resulting in free-living adults. Many of these brachiopods possessed broad, flat interareas or an extensive array of spines on the ventral valve, which provided support to the animal living on a soft, unconsolidated substrate.

At least two other morphological characters are associated with the mode of substrate attachment—the nature of the hinge line and valve articulation. Two classes of hinge line geometry are recognized: straight (strophic) hinge lines, and curved (astrophic) hinge lines. Straight hinge lines, in which the axis of rotation of the two valves coincides with the hinge line itself, are characteristic of orthides and strophomenides. Curved hinge lines, in which the hinge axis intersects the hinge line at only two points, are characteristic of rhynchonellides and terebratulides. The configuration of the hinge structures can be characterized as either noninterlocking, where the teeth simply rest in the sockets, or interlocking, where the teeth and sockets grow simultaneously through mineralization and resorption to form a tightly interconnected structure (Jaanusson, 1971; Carlson, 1989). Valves with interlocking articulations can be separated only by breaking the hinge structures.

The distribution of these characteristics relating to mode of life among Permian and post-Permian brachiopods are illustrated in Figure 8 and Table 2. The Extinct group appears to be a pretty equal mixture of pediculate, free-living, and cemented forms. The Cross group is dominated by pediculate and free-living forms; no cemented articulate brachiopods crossed the Permian boundary. Those taxa originating in the Triassic also

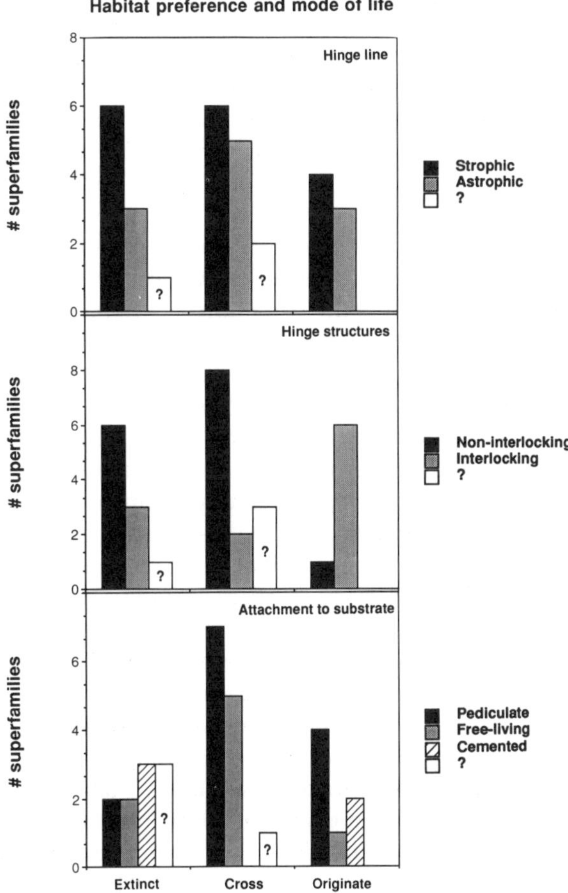

Figure 8. Histogram of the number of superfamilies possessing either strophic or astrophic hinge lines and interlocking or noninterlocking hinge structures, and those superfamilies that are either pediculate, free-living, or cemented to hard substrates. Other elements of the histogram as described for Figure 6.

represent a mixture of each of the three modes of attachment. No significant difference in the distributions is apparent (p = 0.134), but the Extinct and Cross groups alone suggest differences (p = 0.067) that probably merit further investigation. Interestingly enough, the only three Paleozoic groups cemented to the substrate were all present in the Permian and all three went extinct at the end of the Permian. Most of these brachiopods possessed strophic hinge lines and noninterlocking hinge structures. Brachiopods with straight and curved hinge lines fared similarly in the three temporal groups (p = 0.825). Consistently, a few more brachiopod taxa with strophic than astrophic hinge lines were extinct by the end of the Permian, crossed the boundary, and originated after the Permian.

The most striking difference in the characterization of these three temporal groups resides in the nature of valve articulation. The Cross and Originate groups are highly significantly different (p = 0.009), while the Extinct and Cross groups are not (p = 0.573); all three groups together also exhibit significant differences (p = 0.032). Thus, the Permian and Triassic extinction events effectively removed virtually all brachiopods with noninterlocking hinge structures; only those taxa with interlocking hinge structures rediversify (although modestly) following these extinction perturbations.

Summary of Results of Ecological Selectivity

1. In part because of the homoplasious distribution of a punctate shell structure, I was interested in investigating how punctate brachiopods fared, relative to brachiopods with other, different shell structures, throughout these extinction perturbations. If punctae have evolved several times independently, it is reasonable to hypothesize that they might have afforded some adaptive advantage to those brachiopods possessing them. Nothing in these data supports this hypothesis, however. Perhaps such a pattern could only be revealed in the distribution of this character state at lower taxonomic levels.

2. Similarly, if the spiriferides (defined by the possession of a spiralium) comprise a polyphyletic group, indicating that the spiralium is either a shared, primitive feature or a homoplasious feature, of what evolutionary significance is the conclusion that the spiriferides represent an "extinction resistant" taxon? If spiralia are a shared primitive or homoplasious feature of the spiriferides, then the retention of this character by the members of the group has apparently conferred some evolutionary advantage to them, at least in weathering the end-Permian perturbation, if not the end-Triassic perturbation. Thus, the focus of attention should more appropriately be diverted from the taxon Spiriferida, which is apparently nonmonophyletic, to the distribution of the character "defining" the taxon, which, as a primitive or homoplasious character, may well possess functional or adaptive significance that deserves further scrutiny in subsequent evolutionary investigations.

4. The most compelling pattern that emerges in the distribution of these selected characteristics over the Permian and Triassic focuses on the apparent extinction vulnerability of brachiopods with noninterlocking hinge structures, and the extinction "recoverability" of brachiopods with interlocking hinge structures. The status of these characters as homologues is still unclear, although noninterlocking hinge structures appear to characterize a paraphyletic group, while interlocking structures may characterize a monophyletic group. Only thecideoideans, which are problematic for a number of additional reasons, do not fit this pattern; they may well have evolved this feature convergently.

DISCUSSION

Taxonomic Selectivity

Patterns of taxonomic selectivity identify strophomenides as particularly vulnerable to extinction, spiriferides resistant to extinction (at the end-Permian, but not the end-Triassic), and terebratulides capable of recovering from extinctions. The Permian extinctions appear to have reduced lower level (familial) taxonomic diversity, while the Triassic extinctions eliminated the remaining low diversity higher taxa (superfamilies). High lower-level (familial) diversity does not appear to provide a buffer against significant reductions in superfamilial diversity during mass extinctions.

The ordinal assignment of a few "key" taxa has great influence on the pattern of ordinal level taxonomic diversity following the Permian extinction event, and on the interpretation of taxonomic selectivity at this extinction event. According to the Treatise (Williams & Rowell, 1965), strophomenides diversify significantly (three new families; two new superfamilies), if only briefly, in the early Mesozoic. New strophomenide genera (*Bactrynium, Cadomella,* and *Thecospira*) representative of morphological divergence at the superfamilial level, appear in the fossil record after a hiatus of approximately 15 million years (comprising the Olenekian, Anisian, and Ladinian stages) after the last strophomenides went extinct. Mesozoic spiriferides consist solely of the short-lived koninckinoideans. In a very different interpretation, according to Baker (1990) and

Brunton & MacKinnon (1972), the strophomenides are extinct by the end of the earliest Triassic. The spiriferides, on the other hand, experience a substantial radiation (four new superfamilies) in the Mesozoic, including one group that persists to the present day. Such conflicts in taxonomic assignment must be resolved in order to interpret this pattern with respect to taxonomic selectivity in the Permian and Triassic periods.

Phylogenetic Selectivity

Based on the preliminary phylogenetic analysis presented here, there is no evidence to suggest that the Permian and Triassic extinction events exhibited phylogenetic selectivity, despite the fact that some degree of taxonomic selectivity is indicated. The degree of disparity between an established classification scheme and a robust phylogenetic hypothesis is a good indicator of the potential problems involved in using taxonomy as a proxy for phylogeny. If the disparity is high, named taxa may represent either monophyletic or nonmonophyletic groups. Using both as evolutionarily "equivalent" entities in investigations of macroevolutionary phenomena will result in evolutionary patterns that are altered or obscured by nonevolutionary patterns. In order to distinguish phylogenetic (genealogical) macroevolutionary patterns from nonphylogenetic (adaptational or other) macroevolutionary patterns, taxa defined on the basis of shared derived characters should be identified, distinguished from, and analyzed separately from, taxa defined on the basis of shared primitive or homoplasious characters.

The implication that the strophomenide "clade" is all but eliminated at the end of the Permian, while the spiriferide "clade" flourishes into the Triassic is not accurate. Strophomenides appear to be a "clean" paraphyletic group, so taxonomic patterns in this instance can serve at least as an interpretable, if somewhat obscured, proxy for phylogenetic patterns. The effects of spiriferide polyphyly are much more difficult to interpret, however. Did spiralia (as either a primitive or convergent character) play a role in the success of the spiriferides in surviving the Permian extinction event? Or were they just an incidental characteristic of the survivors of this perturbation? Why did spiriferides fail to rediversify in the Triassic and then suffer substantially in the Norian extinction event? Examining in much greater detail the similarities and differences between these two extinction events and their effect in fundamentally restructuring brachiopod faunas between the Paleozoic and Mesozoic promises to be a fruitful area of future research.

Ecological Selectivity

The most striking pattern that emerges in an investigation of characteristics particularly relevant to the articulate brachiopod Bauplan is in the apparent selectivity of hinge structures. Noninterlocking hinge structures indicate extinction vulnerability; interlocking hinge structures indicate, if not extinction resistance, then diversification "permission." Brachiopods with interlocking hinge structures are ecologically quite versatile, in that they can live on almost any type of substrate, in almost any orientation with respect to gravity. This capability apparently conferred considerable ecological and evolutionary significance to brachiopods with interlocking hinge structures. All living brachiopods have interlocking hinge structures. Brachiopods with noninterlocking hinge structures must expend muscular energy to maintain valve articulation in any orientation except one—when the commissural plane is parallel to a flat substrate and the dorsal valve is uppermost. The lack of orientational flexibility, and consequent restriction to a soft sediment substrate, apparently prevented brachiopods with noninterlocking hinge structures from rediversifying after the Permian and Triassic extinction events.

Why did articulate brachiopods suffer such a dramatic decrease in diversity at the end of the Paleozoic era? Why did articulate brachiopods fail to rediversify in the post-Paleozoic? Habitat specificity is implicated because of the pattern of decrease in taxonomic

140

diversity and the temporal distribution of morphological characters relating to mode of life and attachment to the substrate. The extinctions are not instantaneous, but occur throughout the entire Late Permian and Triassic periods. This is more consistent with an ecological, rather than a catastrophic, causal explanation involving the interaction of sessile, epifaunal brachiopods with other benthic marine organisms.

Thayer (1979) introduced the concept of biological bulldozing to account for patterns of declining diversity through the Phanerozoic of taxa restricted to soft sediment substrates. He argued that the increasing diversification of infaunal, burrowing organisms (molluscs, in particular) from the late Paleozoic onward resulted in increased disruption of the soft sediment habitat. Many free-living Paleozoic brachiopods were restricted to precisely this habitat. Because of bulldozing, stable soft sediment substrates were reduced dramatically, preventing free-living brachiopods from rediversifying. Only brachiopods capable of living on hard substrates could find suitable substrates for attachment in the post-Paleozoic benthos. For this reason, rhynchonellides and terebratulides experienced a modest diversification in the Triassic; this level of diversity has been maintained through the Mesozoic and Cenozoic, and up to the present day. The lack of sustained diversification after the Triassic might be explained by different types of interactions between brachiopods and the rest of the marine fauna that do not involve habitat specificity directly (e.g., predation, competition, etc); such explanations remain purely speculative at this time, however. Nevertheless, it is fair to say that the pattern of ecological selectivity at the Permian extinction event among brachiopods with interlocking and noninterlocking hinge structures, documented by this study, is entirely consistent with Thayer's hypothesis of biological bulldozing (1979).

CONCLUSIONS

While the current classification of articulate brachiopods has influenced our interpretations of strophomenide vulnerability and spiriferide resistance to extinct perturbations, neither of these groups is monophyletic, and evolutionary conclusions based on them must, therefore, be reevaluated. A phylogenetic approach to the analysis of extinctions and originations is valuable because it can provide an explicit framework for the identification and comparison of clades and grades, and character homology and analogy. Without such an analysis, it is more difficult to distinguish between morphological similarities that are shared by virtue of functional "constraints" and those shared as a result of common ancestry. Both types of similarity are evolutionarily important, but are important for very different reasons, and under different circumstances. Therefore, failing to distinguish between clade-level taxa and grade-level taxa in analyses of patterns of taxonomic diversity has the potential to conflate and thus obscure different types of evolutionarily interesting patterns of variation. Evaluating changes in the relative frequency of sets of characters (both homologous and nonhomologous) can yield considerable insight into the evolutionary dynamics of mass extinctions and their long term effects on the subsequent recovery of clades such as the articulate brachiopods.

With respect to the admittedly preliminary hypotheses of phylogenetic relationship discussed here, brachiopod superfamilies extinct by the end of the Permian period and those crossing the Permo-Triassic boundary are not clustered in any easily interpretable phylogenetic pattern. The majority of superfamilies originating in the Triassic are classified in the Terebratulida, an order that appears to be undoubtedly monophyletic.

No matter whether this time period is analyzed from a taxonomic or phylogenetic perspective, the end-Permian to end-Triassic marked a time of major transition within the articulate brachiopods. Permian extinctions substantially reduced lower level (familial) diversity within the class; taxon-rich superfamilies appear to have fared no better nor worse than taxon-poor groups. The end-Triassic extinctions eliminated most of the remaining Paleozoic low-diversity higher taxa; Triassic originations provided the foundations of the Mesozoic-Cenozoic (and Recent) brachiopod fauna.

Having identified compelling patterns of phylum-level selectivity at extinction horizons (Jablonski, 1986), it is appropriate to begin to examine these patterns in far more detail taxonomically, phylogenetically, and ecologically. This study represents a preliminary attempt to identify patterns of taxonomic diversity within the articulate brachiopods, to evaluate the phylogenetic significance of those patterns, and to investigate patterns of character extinction and origination and their ecological implications. Patterns of greater biological and paleobiological significance will emerge at these deeper levels of analysis, bringing us closer to understanding the effects of extinction perturbations not only on patterns of diversity, but on the history and evolution of life.

ACKNOWLEDGMENTS

This research was supported by a grant from the National Science Foundation: BSR 8717424. I thank P. Koch, T. Early, and S. Rudman for providing valuable comments on early versions of this manuscript. I particularly thank R. E. Grant and A. J. Rowell for their comments. I alone am responsible for any errors that might remain.

LITERATURE CITED

Ager, D. V. 1987. Why the rhynchonellid brachiopods survived and the spiriferids did not: a suggestion. *Palaeontology* 30(4):853–857.

Baker, P. G. 1990. The classification, origin and phylogeny of thecideidine brachiopods. *Palaeontology* 33(1):175–191.

Boucot, A. 1975. *Evolution and Extinction Rate Controls*. Elsevier: Amsterdam. 427 pp.

Brunton, C. H. C. & D. I. MacKinnon. 1972. The systematic position of the Jurassic brachiopod *Cadomella. Palaeontology* 15(3):405–411.

Carlson, S. J. 1989. The articulate brachiopod hinge mechanism: morphological and functional variation. *Paleobiology* 15(4):364–386.

Carlson, S. J. 1991. Phylogenetic relationships among brachiopod higher taxa. Pp. 3–10. *In:* D. I. MacKinnon (eds.), *Brachiopods Through Time*. Proceedings of the Second International Brachiopod Congress. Dunedin, 1990. A. A. Balkema: Netherlands.

Carpenter, J. M. 1988. Choosing among multiple equally parsimonious cladograms. *Cladistics* 4(3):291–296.

Cooper, G. A. & R. E. Grant. 1974–1975. Permian Brachiopods of West Texas, II & III. *Smithsonian Contributions to Paleobiology* nos. 15 & 19.

Curry, G. B. 1983. Brachiopod caeca—a respiratory role? *Lethaia* 16:311–312.

Curry, G. B. & A. D. Ansell. 1986. Tissue mass in living brachiopods. Pp. 231–241. *In:* P. R. Racheboeuf & C. Emig (eds.), *Les Brachiopodes Fossiles et Actuels*, Actes du le Congres International sur les Brachiopodes, Brest, 1986. Biostratigraphie du Paleozoique 4, Universite de Bretagne: Brest, France.

Farris, J. S. 1989. The retention index and the rescaled consistency index. *Cladistics* 5:417–419.

Fisher, D. C. In prep. Appendix: Stratigraphic data in phylogenetic inference. *MacClade version 3.0, Documentation.* Distributed by the authors, D. and W. Maddison, available from D. Maddison, Museum of Comparative Zoology, Harvard University, Cambridge, MA 02138.

Fortey, R. A. 1989. There are extinctions and extinctions: examples from the Lower Palaeozoic. *Philosophical Transactions of the Royal Society of London* B 325:327–355.

Gardiner, B. 1982. Tetrapod classification. *Zoological Journal of the Linnean Society* 74:207–232.

Gauthier, J., Kluge, A. G. & T. Rowe. 1988. Amniote phylogeny and the importance of fossils. *Cladistics* 4:105–209.

Grant, R. E. 1970. Brachiopods from Permian-Triassic boundary beds and age of Chhidru Formation, West Pakistan. Pp. 117–151. *In:* B. Kummel & C. Teichert (eds.), *Stratigraphic Boundary Problems: Permian and Triassic of West Pakistan*. Department of Geology Special Publication 4, The

142

University Press of Kansas: Lawrence, KS.

Hoover, P. R. 1979. Early Triassic terebratulid brachiopods from the Western Interior of the United States. *U.S. Geological Survey Professional Paper* 1057:1–20.

Jaanusson, V. 1971. Evolution of the brachiopod hinge. *Smithsonian Contributions to Paleobiology* 3:33–46.

Jablonski, D. 1986. Background and mass extinctions: the alternation of macroevolutionary regimes. *Science* 231:129–133.

Kummel, B. 1979. Triassic. Pp. A351–A389. *In:* R. C. Moore (ed.), *Treatise on Invertebrate Paleontology,* Part A. Introduction—Biogeography and Biostratigraphy. Geological Society of America and University of Kansas: Boulder, Colorado & Lawrence, Kansas.

McGhee, G. R. 1980. Shell form in the biconvex articulate Brachiopoda: a geometric analysis. *Paleobiology* 6(1):57–76.

Patterson, C. & A. B. Smith. 1987. Is the periodicity of extinctions a taxonomic artifact? *Nature* 330:248–251.

Peck, L. S., Morris, D. J. & A. Clarke. 1986. Oxygen consumption and the role of caeca in the Recent Antarctic brachiopod *Liothyrella uva notorcadensis* (Jackson, 1912). Pp. 349–355. *In:* P. R. Racheboeuf & C. Emig (eds.), *Les Brachiopodes Fossiles et Actuels,* Actes du le Congres International sur les Brachiopodes, Brest, 1986. Biostratigraphie du Paleozoique 4, Universite de Bretagne: Brest, France.

Ross, C. A. & J. R. P. Ross. 1979. Permian. Pp. A291–A350. *In:* R. C. Moore (ed.), *Treatise on Invertebrate Paleontology,* Part A. Introduction—Biogeography and Biostratigraphy. Geological Society of America and University of Kansas: Boulder, Colorado & Lawrence, Kansas.

Rudwick, M. J. S. 1970. *Living and Fossil Brachiopods.* Hutchinson University Library: London. 199 pp.

Sepkoski, J. J., Jr. 1982. A compendium of fossil marine families. *Milwaukee Public Museum Contributions to Biology and Geology* 51:1–125. Unpublished updates 1983, 1986, 1988.

Stanley, S. M. 1979. *Macroevolution.* W. H. Freeman & Co.: San Francisco, CA. 332 pp.

Steele-Petrovic, H. M. 1979. The physiological differences between articulate brachiopods and filter-feeding bivalves as a factor in the evolution of marine level-bottom communities. *Palaeontology* 22:101–134.

Swofford, D. 1989. *PAUP (Phylogenetic Analysis Using Parsimony),* ver. 3.0. Distributed by the author, Illinois Natural History Survey, 607 East Peabody Drive, Champaign, IL 61820.

Thayer, C. W. 1979. Biological bulldozers and the evolution of marine benthic communities. *Science* 203:458–461.

Valentine, J. W. & D. Jablonski. 1983. Larval adaptations and patterns of brachiopod diversity in space and time. *Evolution* 37:1052–1064.

Valentine, J. W. & E. M. Moores. 1972. Global tectonics and the fossil record. *Journal of Geology* 80:167–184.

Waterhouse, J. B. & G. F. Bonham-Carter. 1975. Global distribution and character of Permian biomes based on brachiopod assemblages. *Canadian Journal of Earth Sciences* 12(7):1085–1146.

Williams, A. 1956. The calcareous shell of the Brachiopoda and its importance to their classification. *Biological Reviews* 31:243–287.

Williams, A. 1968. Evolution of the shell structure of the articulate brachiopods. *Special Papers in Paleontology* 2:1–55.

Williams, A. 1990. Biomineralization in the lophophorates. Pp. 67–82. *In:* J. G. Carter (ed.), *Skeletal Biomineralization: Patterns, Processes, and Evolutionary Trends.* Van Nostrand Reinhold: New York.

Williams, A. & A. J. Rowell. 1965. Evolution and phylogeny. Pp. H164–H199. *In:* R. C. Moore (ed.), *Treatise on Invertebrate Paleontology,* Part H—Brachiopoda (1). Geological Society of America and University of Kansas: Boulder, Colorado & Lawrence, Kansas.

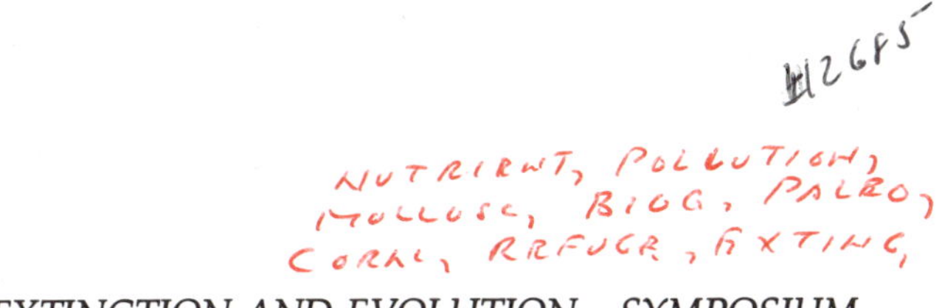

Marine Extinctions
and Their Implications for Conservation
and Biogeography

Geerat J. Vermeij

Abstract. Many marine regions serve as geographical refuges for taxa that had broader geographical distributions during the Paleogene and Neogene periods. For Recent molluscs, these regions include the tropical eastern Pacific (especially the west coast of Mexico), the north and east coasts of South America, the west coast of Africa, the Indo-Malayan area of the tropical Indo-West Pacific, and the cool-temperate northwestern Pacific and northwestern Atlantic Oceans. High planktonic productivity, which characterizes large parts of these regions, is believed to be responsible for the persistence of many molluscan taxa in these refuges. Human-caused eutrophication of the world's coastal waters, though detrimental to corals and other organisms that require clear waters and low planktonic productivity, may be benefiting other marine species as long as the increase in nutrients is not accompanied by overexploitation or by poisoning (including anoxia). Inferences of biogeographical history can be misleading if only Recent distributions of species are considered, because many historical changes in distribution leave no trace in the living biota.

INTRODUCTION

Paleobiology offers conservation biologists a unique perspective on the long-term consequences of environmental change. Through the detailed study of organisms with a good fossil record, we can gain insight into which kinds of species have been especially prone to extinction, which factors ameliorate the causes of extinction, and which circumstances favor speciation as a means of replenishing lost diversity. The essential dimension that is unique to the fossil record is time. Just as the study of human history can enlighten us about the present political state of the world, so the study of fossils has much to offer the biologist seeking to understand patterns of adaptation, ecology, and biogeography.

This contribution is especially clear in the study of extinction. Many extinctions today and in the geological past are local or regional rather than global; that is, the taxa in question die out in one region but survive elsewhere in geographical or habitat refuges. The characterization of refuges can reveal much about the causes, consequences, and dynamics of extinction; and it may point to factors that can prevent local extinctions from becoming global. In this paper, I summarize what I believe are the salient features of geographical refuges in the modern marine biota. Some implications for marine conservation of species and for biogeographical inference will also be discussed briefly.

Dr. Vermeij is with the Department of Geology, University of California, Davis, CA 95616, USA.

TROPICAL REFUGES

Extinctions during the Paleogene and Neogene periods have profoundly affected marine biotas, but some regions were more affected than others. In tropical America, the magnitude of post-Miocene molluscan extinction was about twice as high in the western Atlantic as it was in the eastern Pacific. Woodring (1966) was the first to draw attention to the fact that many molluscs living on both sides of tropical America during the Miocene and early Pliocene became restricted to either the eastern Pacific or the western Atlantic after the middle Pliocene, when the Central American isthmus that now separates these two oceans came to be fully formed. He showed that Paciphiles (taxa that became restricted to the Pacific side) greatly outnumbered Caribphiles (taxa that became restricted to the Atlantic side). Taxonomic work since Woodring's compilation has led to a substantial expansion and modification of the list of geographically restricted molluscs, but Woodring's conclusion remains sound. By my current estimate, there are 101 Paciphiles at the level of subgenus or species, and only 23 Caribphiles. The fact that Paciphiles outnumber Caribphiles by more than 4 to 1 implies strongly that the tropical eastern Pacific serves as a major refuge for molluscs that during the Late Neogene were distributed in both the eastern Pacific and western Atlantic (see also Vermeij & Petuch, 1986). Heck and McCoy (1978), on the other hand, have demonstrated that the situation is the opposite for corals, with the western Atlantic serving as the more important refuge in the Recent biota of tropical America.

Within the western Atlantic, the north and east coasts of South America act as important refuges for molluscan taxa that during the Neogene were more widespread in the tropical western Atlantic (Petuch, 1981, 1982, 1988; Vermeij & Petuch, 1986). My preliminary tally shows that there are at least 23 such taxa, including a number of Caribphiles (e.g., *Muracypraea*, *Dalium*, and *Panamurex*).

Deposits from late Miocene or Pliocene age in Ecuador have yielded a number of molluscan genera known living today only in cool to warm-temperate portions of the northern Pacific. These include *Kelletia*, *Ceratostoma*, and *Aforia* (see Olsson, 1964; Vokes, 1988). Together with the existence of several genera with disjunct distributions in the temperate northeastern and southeastern Pacific (see Lindberg, 1988; Santelices, 1980), these finds suggest that the tropical eastern Pacific was cool enough from time to time to permit the penetration of temperate species. Although this region served as a refuge for tropical species, it clearly played the opposite role for temperate molluscs.

No one has previously commented on the possibility that there might be areas of the tropical eastern Pacific that serve as refuges within a refuge. A continuing analysis of the distributions of eastern Pacific species whose closest relatives occur in the Recent or Late Neogene western Atlantic is showing, however, that the Gulf of California and the open Pacific coast of Mexico south to Guatemala harbor many molluscs that once had much broader distributions. There are also species from west Mexico that as fossils are recorded from Ecuador or Panama. Thus far, I have recognized 18 such Mexican-restricted molluscs, but continuing work will doubtless greatly increase this number. Among these taxa are several Paciphiles, including *Eupleura muriciformis*, *Muricopsis pauxillus*, *Trajana*, and *Semele guaymasensis*. Further work may reveal that other areas of the eastern Pacific, such as the Galápagos Islands and the mainland coast of Ecuador and northern Peru, are also refuges within the eastern Pacific; but, as with west Mexico, careful documentation is needed.

The West African tropics appear to be refugial for two kinds of taxa, those known during the Miocene or Pliocene in the Mediterranean (see Freneix et al., 1988) and those known formerly also in the western Atlantic. In the latter category are several molluscan taxa that within tropical America are categorized as Paciphiles (e.g., *Purpurellus* and *Harpa*) (see Vermeij, 1986).

Although tropical West Africa is a refuge for some warm-loving species, it was just as clearly a locus of extinction for other species. Many warm-temperate molluscs today have a discontinuous distribution in Europe and South Africa, with a conspicuous gap in tropical West Africa. It is reasonable to infer that this disjunction indicates a formerly continuous distribution, perhaps during the glacial episodes of the Pleistocene, and that subsequent warming eliminated the temperate species from low latitudes. The biogeographical history of West Africa thus parallels that of the tropical eastern Pacific.

The vast tropical Indo-West Pacific region may contain several important refuges. The only one that has been at all well documented is the so-called Indo-Malayan area, the complex of continental and high-island shores from mainland southeast Asia to New Guinea. Many molluscan taxa previously found further east in the islands of Micronesia and Polynesia during the Miocene, Pliocene, and early Pleistocene have since become restricted to the Indo-Malayan area.

The Indo-West Pacific as a whole contains at least 21 molluscan taxa at the level of subgenus that were living in tropical America during the Eocene, Miocene, or Pliocene, but that are now extinct there (see also Vermeij and Petuch, 1986). This number is certain to increase as fossil American molluscs are analyzed more critically (e.g., Vokes, 1989, for a careful study of Muricidae). The Indo-West Pacific is also home to a large number of taxa that occurred in Europe during the Paleogene and Miocene (e.g., Robba, 1987). These contractions in range apply to corals as well (Heck and McCoy, 1978).

TEMPERATE REFUGES

In a previous analysis of geographical restriction among Late Neogene cool-water molluscs, I suggested that the northwestern Pacific and northwestern Atlantic Oceans are important refuges for cool-water species that occurred on the eastern margins of the Pacific and Atlantic during the Late Neogene or Pleistocene (Vermeij, 1989). At the time, I recognized 15 taxa at the subgenus or species level that became restricted to a northwestern Pacific distribution from a formerly amphi-Pacific range, and one possible case (*Mytilus californianus*) of restriction to the northeastern Pacific. I have since learned that Kafanov (1987) showed that *M. (Pacifimytilus) californianus* is, and always has been, a strictly eastern Pacific lineage, and that the supposed northwestern Pacific fossil representative is referrable to the independent amphi-Pacific subgenus *Tumidimytilus*.

I previously recorded six molluscan taxa that apparently became restricted from a Pleistocene amphi-Atlantic distribution to a Recent northwestern Atlantic refuge. Further analysis indicates that there are at least 25 such molluscs. About 15 of these lived in the North Sea basin during the late Pliocene and early Pleistocene and persist today in the Arctic waters of Europe and North America as well as in cool-temperate eastern North America; the other ten, also known from the Pliocene and Pleistocene of the North Sea basin, today occur only in the American sector of the Atlantic. There appears to be no case of restriction to the northeastern Atlantic.

A few molluscs with distributions in both the North Pacific and North Atlantic Oceans during the Pliocene or early Pleistocene are found today only in the North Pacific. In my 1989 paper I estimated there to be five such molluscs, but this number has now dropped to four. The buccinid gastropod genus *Searlesia*, which has generally been regarded as containing one or more Pliocene European species and several Oligocene to Recent North Pacific members, now appears to consist of two independent lineages, one of the Pacific and an extinct one in the northeastern Atlantic (Vermeij, in preparation).

Molluscs that have undergone range contractions in the Atlantic are unrepresentative with respect to their biogeographical histories when compared to the North Atlantic molluscan fauna as a whole. At least 20 of the 25 species that were eliminated from the

146

temperate European fauna but that still persist in the Arctic and temperate western Atlantic are the products of trans-Arctic interchange. They either originated in the North Pacific or were derived from invaders from the Pacific. The same is true for three of the four species that persist in the North Pacific and became extinct in the Atlantic. Pacific invaders or their descendants comprise about 13% of the cool-temperate molluscan fauna of Europe, 33% of the cool-water molluscs of the western Atlantic, and 17% of the cool-temperate North Atlantic molluscan fauna as a whole (see Vermeij, 1991). These percentages lie far below the 75 to 80% observed among Atlantic taxa that underwent geographical restriction. I do not have any explanation for why taxa of Pacific origin are so well represented among geographically restricted Atlantic molluscs.

South-temperate molluscs undoubtedly have also undergone contractions in range, but these historical patterns have not been investigated systematically. Southern Australia and New Zealand seem to be important refuges for taxa that during the Paleogene and early Neogene inhabited warm-water areas in Europe and southeast Asia (see Vermeij, 1986; Beu and Maxwell, 1990). Further analyses of these and other patterns of restriction should prove most interesting.

DISCUSSION

What do these refuges have in common, and what do they tell us about the causes of extinction and the conditions necessary for biological renewal? One answer seems to be that all refuges contain large areas of high planktonic primary productivity (Birkeland, 1977; Vermeij, 1978, 1986, 1989; Vermeij and Petuch, 1986). This is especially clear in the case of the eastern Pacific as compared to most of the tropical western Atlantic, but the pattern is also obvious in a comparison of mainland northern and eastern South America with the less productive island shores of the West Indies. These areas, as well as the Indo-Malayan region, West Africa, and the northwestern coastal areas of the Pacific and Atlantic Oceans, contain extensive areas of upwelling, and receive land-derived nutrients from large rivers (Walsh, 1988). High planktonic productivity enables suspension-feeders to grow rapidly and to large size, permits high densities of suspension-feeders to be maintained, and promotes high outputs and high recruitment success of planktotrophic larval stages (Birkeland, 1977, 1982, 1990; Highsmith, 1980; Vermeij, 1989, 1990).

It remains unknown if the refuges increased in planktonic productivity, if the areas from which species were lost decreased in productivity (temporarily or permanently), or if a combination of these phenomena occurred. Analyses of annual growth increments in the shells of fossil and living pelecypods from within and outside refuge areas should provide the answers to these questions and make possible the test of the hypothesis that changes in productivity were responsible for the observed historical changes in distribution. At this point it is noteworthy that, as would be expected from this hypothesis, scleractinian corals show a pattern of geographical restriction in tropical America that is opposite to that of molluscs. These corals require clear waters and low planktonic productivity in order to make maximum use of algal symbionts and to be competitively successful against suspension-feeders such as barnacles, pelecypods, and sponges that do not contain symbionts and that grow rapidly in more productive waters (Glynn and Stewart, 1973; Birkeland, 1990).

Several areas of major upwelling have not yet been recognized as refuges. These include especially the west coasts of temperate South America and southern Africa, as well as the Antarctic region. It will be interesting to examine closely the past distributions of molluscs now living in these regions.

Humans have enriched coastal waters around the world with nutrients thanks to urbanization, agricultural run-off, and industrial pollution (Walsh, 1988). This enrichment is undoubtedly detrimental to reef-building organisms, which thrive on low nutrient levels, but it may be benefiting other marine species, as long as the increased nutrition is not accompanied by overexploitation or by poisoning (including oxygen depletion) of waters and sediments. It is important to understand how nutrient enrichment affects species at different trophic levels, how it influences larval recruitment and adult trophic interactions, and whether the human-caused eutrophication of coastal waters is similar in its effects to the natural nutrient enrichment near the mouths of large rivers and in areas of upwelling.

Changes in temperature and perhaps seasonality may also have had a hand in creating the observed pattern of Recent geographical refuges. The elimination of many taxa from the Miocene and Pliocene Mediterranean and their persistence in West Africa and the Indo-West Pacific may be ascribed partly to climatic cooling and to the salinity crisis in the Mediterranean at the end of the Miocene (see also Por, 1989). The disappearance of temperate molluscs from the west coasts of tropical America and Africa, and the persistence of these taxa in the northern and southern temperate zones, is likely to be the consequence of Pliocene and Pleistocene cool episodes. Stanley (1986) provides additional biogeographical evidence of range truncations of warm-water species, which during the Pliocene extended further north in the northern hemisphere than they do today. I agree with Stanley that these restrictions are related to falling temperatures and perhaps to increasing seasonality at higher latitudes during the late Pliocene and Pleistocene. It must be emphasized that the pattern of geographical restriction interpretable as resulting from changes in productivity is quite distinct from restriction resulting from changes in temperature.

From the data summarized here, it is obvious that many molluscs have undergone major changes in their geographical distributions. We know this because we have available to us an excellent fossil record. Some of the restrictions can be surmised from Recent disjunct distributions; if a taxon is known from two regions that lie far apart and not from the intervening area of potentially suitable habitat, the reasonable inference may be drawn that it also lived in the intermediate region at one time. Many of the restrictions, however, have left no trace in the form of a disjunction in the Recent fauna. This circumstance should serve as a warning to those biogeographers who seek to infer distributional histories on the basis of present-day species ranges alone. Unless a group has a well documented fossil record and a well understood phylogeny, such inferences are apt to be wrong or misleading.

In summary, the study of geographical distributions through time may yield insights into the causes of extinction and into the conditions necessary for conserving species threatened by particular kinds of environmental change. Paleobiology offers the essential ingredient of time with which to infer the history of geographical distributions of fossilizable species.

ACKNOWLEDGMENTS

I thank Janice Cooper, Edith Zipser, and Elizabeth Dudley for technical assistance; and David Jablonski for critically reading the manuscript.

LITERATURE CITED

Beu, A. G. & P. A. Maxwell. 1990. Cenozoic Mollusca of New Zealand. *New Zealand Geological Survey Paleontological Bulletin* 58:1–518.
Birkeland,C. 1977. The importance of rate of biomass accumulation in early successional stages of

148

benthic communities to the survival of coral recruits. *Proceedings of the Third International Coral Reef Symposium* 1:15–21.

Birkeland, C. 1982. Terrestrial run-off as a cause of outbreaks of *Acanthaster planci* (Echinodermata: Asteroidea). *Marine Biology* 69:175–185.

Birkeland, C. 1990. Geographic comparisons of coral-reef community processes. *Proceedings of the Sixth International Coral Reef Congress.* In press.

Freneix, S., Saint Martin, J.-P. & P. Moissette. 1988. Huîtres du Messinien d'Oranie (Algérie occidentale) et paléobiologie de l'ensemble de la faune de bivalves. *Bulletin du Museum National d'Historie Naturelle, Section C* (4) 10:1–21.

Glynn, P. W. & R. H. Stewart. 1973. Distribution of coral reefs in the Pearl Islands (Gulf of Panama) in relation to thermal conditions. *Limnology and Oceanography* 18:367–379.

Heck, K. L. Jr. & E. D. McCoy. 1978. Long-distance dispersal and the reef-building corals of the eastern Pacific. *Marine Biology* 48:349–356.

Highsmith, R. C. 1980. Geographic patterns of coral bioerosion: a productivity hypothesis. *Journal of Experimental Marine Biology and Ecology* 46:177–196.

Kafanov, A. I. 1987. Podsemeistvo Mytilinae Rafinesque 1815 (Bivalvia, Mytilidae) v Kainozoe severnoi Patsifiki. Pp. 65–103. *In: Fauna i Raspredelenie Mollyuskov: Severnaya Patsifika i Polyarnui Bassein.* Akademiya Nauk SSSR: Vladivostok.

Lindberg, D. R. 1988. Systematics of the Scurriini (new tribe) of the northeastern Pacific Ocean (Patellogastropoda: Lottiidae). *Veliger* 30:387–394.

Olsson, A. A. 1964. *Neogene Mollusks from Northwestern Ecuador.* Paleontological Research Institute: Ithaca, NY. 256 pp.

Petuch, E. J. 1981. A relict Neogene caenogastropod fauna from northern South America. *Malacologia* 20:307–347.

Petuch, E. J. 1982. Geographical heterochrony: contemporaneous coexistence of Neogene and Recent molluscan faunas in the Americas. *Palaeogeography, Palaeoclimatology, Palaeoecology* 37:277–312.

Petuch, E. J. 1988. *Neogene History of Tropical American Mollusks: Biogeography and Evolutionary Patterns of Tropical Western Atlantic Mollusca.* Coastal Education and Research Foundation: Charlottesville, VA. 217 pp.

Por, F. D. 1989. *The Legacy of Tethys: An Aquatic Biogeography of the Levant.* Kluwer: Dodrecht, The Netherlands. 214 pp.

Robba, E. 1987. The final occlusion of Tethys: its bearing on Mediterranean benthic molluscs. Pp. 405–426. *In:* K. G. McKenzie (ed.), *Shallow Tethys 2.* Balkema: Rotterdam.

Santelices, B. 1980. Phytogeographic characterization of the temperate coast of Pacific South America. *Phycologia* 19:1–12.

Stanley, S. M. 1986. Anatomy of a regional mass extinction: Plio-Pleistocene decimation of the western Atlantic bivalve fauna. *Palaios* 1:17–36.

Vermeij, G. J. 1978. *Biogeography and Adaptation: Patterns of Marine Life.* Harvard University Press: Cambridge, MA. 332 pp.

Vermeij, G. J. 1986. Survival during biotic crises: the properties and evolutionary significance of refuges. Pp. 231–246. *In:* D. K. Elliott (ed.), *Dynamics of Extinction.* Wiley: New York.

Vermeij, G. J. 1989. Geographical restriction as a guide to the causes of extinction: the case of the cold northern oceans during the Neogene. *Paleobiology* 15:335–356.

Vermeij, G. J. 1990. Tropical Pacific pelecypods and productivity. *Bulletin of Marine Science.* In press.

Vermeij, G. J. 1991. Anatomy of an invasion: the Trans-Arctic Interchange. *Paleobiology.* In press.

Vermeij, G. J. & E. J. Petuch. 1986. Differential extinction in tropical American molluscs: endemism, architecture, and the Panama land bridge. *Malacologia* 27:29–41.

Vokes, E. H. 1988. Muricidae (Mollusca: Gastropoda) of the Esmeraldas Group, northwestern Ecuador. *Tulane Studies in Geology and Paleontology* 21:1–50.

Vokes, E. H. 1989. Neogene paleontology in the northern Dominican Republic 8. The family Muricidae (Mollusca: Gastropoda). *Bulletins of American Paleontology* 97:5–94.

Walsh, J. J. 1988. *On the Nature of Continental Shelves.* Academic Press: San Diego, CA. 520 pp.

Woodring, W. P. 1966. The Panama land bridge as a sea barrier. *American Philosophical Society Proceedings* 110:425–433.

Long-term Biotic Effects of Major Climatic Perturbations: The Plant Fossil Record

EXTENDED ABSTRACT

Jack A. Wolfe

A, if not the, major driving force in plant evolution and on the development of plant ecosystems and communities during the Late Cretaceous and Tertiary was climate. During the Late Cretaceous, lower to middle latitude climates were generally warm and aseasonally subhumid (Upchurch & Wolfe, 1987; Wolfe & Upchurch, 1987). Such climate selects for open-canopy woodlands, trees of generally low stature, and trees and shrubs that have deep root systems and the capability of rapid water conduction. At lower middle latitudes, with high warmth and high water stress, angiosperms displaced most non-coniferous gymnosperms and ferns by the early part of the Late Cretaceous (Crane, 1987). Like the gymnosperms they replaced, many of the angiosperms were wind-pollinated (Kedves, 1989) and small-seeded (Tiffney, 1985), characteristics that are also selected in open-canopy woodlands. This was an environment in which large herbivores could flourish. At higher middle and lower high latitudes, gymnosperms and ferns slowly declined but maintained a strong minority position until near (but not at) the end of the Cretaceous (Upchurch & Wolfe, *in press*); deciduous angiosperms, probably because of the light regime (Wolfe, 1987), were an important part of the vegetation, which was also open-canopy woodland. At high latitudes, the light regime selected deciduous plants that also had plasticity in leaf sizes (Wolfe, 1987), and again angiosperms dominated. Cretaceous angiosperm evolution appears to be characterized by rapid origination of families in the Cenomanian and then to have proceeded at a steady but slower pace, with many extant orders (Muller, 1981), most extant superorders, and all but one extant subclass recognizable by the end of the period.

Events at the Cretaceous-Tertiary boundary had profound effects on, especially, lower latitude ecosystems (Wolfe & Upchurch, 1987). The initial impact winter was particularly devastating on broad-leaved evergreens; plants that had facultative dormancy mechanisms survived this event relatively unscathed, as did both plants and animals in the aquatic ecosystem (Wolfe & Upchurch, 1986). However, the greenhouse effect following the impact winter was probably more significant: temperature increased by about 10°C at mid-latitudes (and probably at high latitudes), concomitant with a four-fold increase in precipitation, and these levels persisted for an estimated 0.5-1.0 Ma (Wolfe, 1990). At low

---

Dr. Wolfe is with the U.S. Geological Survey, MS-919, Denver, CO 80225, USA.

middle latitudes, the new climate favored development of high-statured, closed-canopy rain forest, i.e., selection involved trees that could either grow high (and thus form a closed canopy) or those that could tolerate shade; these were new factors and few Late Cretaceous angiosperms would have been pre-adapted. Insect pollination also appears to have increased (Crepet & Friis, 1987), as might be expected in closed-canopy rain forest. High middle latitudes had megathermal temperatures, but few surviving megathermal broad-leaved evergreens at lower latitudes would have been pre-adapted to the light regime. This led to selection of deciduous angiosperms and gymnosperms that could tolerate megathermal temperatures, to a high-productivity, closed-canopy forest, and to gradual adaptation of broad-leaved evergreens to the light regime. As the closed-canopy forests became high-statured, tetrapods of arboreal habit would be favored. Whether large herbivores such as dinosaurs (and their predators) became extinct precisely at the Cretaceous-Tertiary boundary is a moot point; closely spaced, high-statured trees and high precipitation in the early Paleocene were basically unsuited to large, herbivorous tetrapods. Local plant diversity remained low at middle latitudes during the earliest Paleocene (Wolfe & Upchurch, 1987). At high latitudes, however, extinction appears to have been minimal at the boundary (Fredericksen, 1989), although precipitation appears to have increased and thus presumably resulted in a denser, but still largely deciduous, forest. Many of the same deciduous angiosperms found at lower latitudes were also dominant at high latitudes. The various ecosystems appear to be floristically blurred during the early Paleocene, a blurring that persisted on a diminished scale into the late Paleocene and Eocene (Wolfe, 1987). Adaptation to new climatic conditions at middle (and presumably even lower) latitudes led to more than doubling the origination rate of families of angiosperms during the Paleocene. Rain-forest ecosystems probably provided many more ecologic opportunities than did Late Cretaceous woodland ecosystems. The apparently gradual development of seasonally dry, megathermal, and mesothermal environments during the early Tertiary (Wolfe, 1985) also provided new opportunities, although few familial originations occurred. These drier climates produced forests that were at least partially open-canopy, and large herbivorous tetrapods could find suitable habitats.

Microthermal ecosystems were, during the Late Cretaceous, restricted to polar and subpolar regions and were of generally low diversity, even at the generic level (Wolfe, 1987). The thermal maximum of the early Eocene areally restricted microthermal climates and created a bottleneck for ancient microthermal lineages. Tectonism and volcanism in western North America, concomitant with global cooling following the early Eocene, provided topographically diverse microthermal environments at middle latitudes by the middle Eocene. Numerous clades, especially those of deciduous habit, entered and diversified in these new environments; with few exceptions, this diversification was at the subgeneric level (Wolfe, 1987, 1988). Whereas Late Cretaceous microthermal vegetation was largely deciduous, in which wind-pollination would be predominant, the canopy of Eocene microthermal vegetation was dominated by conifers; the angiospermous understory of this Eocene vegetation was dominated by groups that are today largely insect-pollinated, and these groups underwent the greatest amount of diversification during the Eocene (Wolfe, 1988).

A major, rapid cooling in the earliest Oligocene (33 Ma) restricted megathermal and mesothermal climates to their smallest extent since the Late Cretaceous. The result, especially in North America, was major extinction in megathermal, and moderate extinction in mesothermal, clades. The Oligocene deterioration, however, represented not just lowering of global temperature but also a shift to markedly seasonal temperature at middle and high latitudes (Wolfe, 1985). Higher seasonality in comparison to the Eocene favored development of a deciduous rather than a coniferous canopy in microthermal climates. In these climates, the woody flora suffered few extinctions; deciduous clades were presumably pre-adapted to temperature extremes. At a subgeneric level, some groups that had been of low diversity in less extreme microthermal climates during the Eocene diversified during the Oligocene and earlier half of the Miocene (Wolfe, 1987). Part of this diversification is probably related to these wind-pollinated groups assuming dominance in the canopy. Although many deciduous clades of woody plants were well adapted to seasonality, whatever plants occupied groundcover habitats during the Eocene apparently were not. The Oligocene and Neogene saw a major increase in origination of, and/or diversification in, angiosperm families that are now dominantly or exclusively terrestrial herbs (Muller, 1981). This increase emphasizes the fundamental change that occurred in climates during and following the early Oligocene.

LITERATURE CITED

Crane, P. R. 1987. Vegetational consequences of angiosperm diversification. Pp. 107–144. *In:* E. M. Friis, W. G. Chaloner & P. R. Crane (eds.), *The Origins of Angiosperms and their Biological Consequences.* Cambridge University Press, Cambridge.

Crepet, W. L. & E. M. Friis. 1987. The evolution of insect pollination in angiosperms. Pp. 181–203. *In:* E. M. Friis, W. G. Chaloner & P. R. Crane (eds.), *The Origins of Angiosperms and their Biological Consequences.* Cambridge University Press, Cambridge.

Fredericksen, N. O. 1989. Changes in floral diversities, floral turnover rates, and climates in Campanian and Maastrichtian time, North Slope of Alaska. *Cretaceous Research* 10:249–266.

Kedves, M. 1989. Evolution of the Normapolles complex. Pp. 1–8. *In:* P. R. Crane & S. Blackmore (eds.), *Evolution, Systematics, and Fossil History of the Hamamelidae,* Vol. 2. Clarendon Press, Oxford.

Muller, J. 1981. Fossil pollen records of extant angiosperms. *Botanical Review* 47:1–142.

Tiffney, B. H. 1985. Seed size, dispersal syndromes, and the rise of the angiosperms: evidence and hypothesis. *Annals Missouri Bot. Garden* 71:551–576.

Upchurch, G. R. & J. A. Wolfe. 1987. Mid-Cretaceous to Early Tertiary vegetation and climate: evidence from fossil leaves and woods. Pp. 75–105. *In:* E. M. Friis, W. G. Chaloner & P. R. Crane (eds.), *The Origins of Angiosperms and their Biological Consequences.* Cambridge University Press, Cambridge.

Upchurch, G. R. & J. A. Wolfe. Cretaceous vegetation of the Western Interior. *Geological Assoc. Canada Special Publ.* In press.

Wolfe, J. A. 1985. Distribution of major vegetational types during the Tertiary. Pp. 357–375. *In:* E. T. Sundquist & W. S. Broecker (eds.), *The Carbon Cycle and Atmospheric CO_2: Natural Variations, Archean to Present.* American Geophysical Union (Mon. 32): Washington, D.C.

Wolfe, J. A. 1987. Late Cretaceous-Cenozoic history of deciduousness and the terminal Cretaceous event. *Paleobiology* 13:215–226.

Wolfe, J. A. 1988. An overview of the origins of the modern vegetation and flora of the northern Rocky Mountains. *Annals Missouri Bot. Garden* 74:785–803.

Wolfe, J. A. 1990. Paleobotanical evidence for a marked temperature increase at the Cretaceous-Tertiary boundary. *Nature* 343:153–156.

Wolfe, J. A. & G. R. Upchurch. 1986. Vegetation, climatic and floral changes at the Cretaceous-Tertiary boundary. *Nature* 324:148–152.

Wolfe, J. A. & G. R. Upchurch. 1987. North American nonmarine climates and vegetation during the Late Cretaceous. *Palaeogeography, Palaeoclimatology, Palaeoecology.* 61:33–77.

Extinction in an Extinction-Resistant Clade: The Evolutionary History of the Gastropoda

Douglas H. Erwin and Philip W. Signor

Abstract. Marine prosobranch gastropods diversified continuously, although at varying rates, throughout the Phanerozoic and are now one of the dominant components of modern marine faunas. This dramatic increase in taxic diversity is mirrored by an increase in trophic roles, particularly following the Cretaceous radiation of predaceous caenogastropods. The diversification of the clade appears to result in equal parts from increased resistance to extinctions during mass extinctions and enhanced survival during the intervals between mass extinctions. However, the diversification masks a high degree of reorganization at subclade levels: the clades that dominated the lower Paleozoic were replaced by new groups in the upper Paleozoic, a pattern which repeated between the Permian and Triassic and between the Triassic and Jurassic.

INTRODUCTION

Gastropods are one of the most important components of modern marine biotic diversity; they constitute the most species-rich clade of marine invertebrates and are particularly diverse in warm shallow seas. The gastropods diversified unsteadily but nearly continuously from the Upper Cambrian to the present (Fig. 1). This protracted diversification has had its ups and downs as mass extinctions intervened, older groups dwindled and in some cases became extinct, and newly evolved clades achieved sudden dominance. The overall diversity history is the summation of origination and extinction of species, and the manifestation of these events at the level of genera and families. In this contribution we are interested in causes of the long-term diversification of gastropods. We advance the hypothesis that the increasing dominance of the Gastropoda is due, in part, to the clade's greater extinction-resistance. Our emphasis here is on taxic diversity patterns through time, using genera at the resolution of geologic stage as the best available proxy for species diversity patterns.

Few of the taxa discussed here have been subject to rigorous phylogenetic analysis, although such work is in progress by a number of workers, including one of us (DHE). Thus our analysis is subject to the criticisms leveled by Smith (1988) and Smith & Patterson (1988). They maintain that markedly different patterns emerge when holophyletic taxa are employed in analyses of taxonomic diversity and turnover. However, while some of the groups involved are either paraphyletic or polyphyletic, the overall patterns of

Dr. Erwin was with the Department of Geological Sciences, Michigan State University, E. Lansing, MI 48824, USA. However, Dr. Erwin's present address is Dept. of Paleobiology, Room E-206, USNMNH, Smithsonian Institution, Washington, D.C. 20560. USA.
Dr. Signor is with the Department of Geology, University of California, Davis, CA 95616, USA.

diversity change appear to reflect genuine changes at the species level. Taxic diversity may also be effected by monographic bias introduced by particularly well-studied intervals and the vagaries of taxonomic practice. In our collection of the data we have attempted to account for such problems. Finally, we emphasize that this is a preliminary note from our collaborative research, and we will publish a longer treatment of Phanerozoic gastropod history in the near future.

Long-term survival and the opportunity to diversify are the rewards of clades that avoid extinction. Clades that possessed the greatest capacity for rapid diversification (e.g., archaeocyathans, trilobites, ammonoids) have tended to become extinct. In fact, gastropod genera have among the greatest longevity of any marine genera (Stanley, 1979). Average generic longevity during the Paleozoic exceeds 25 my and reaches 80 my by the Permian. We have identified four possible explanations for the increased extinction-resistance of gastropods, which are not mutually exclusive. These are 1) greater longevity during background times, perhaps produced by broader trophic/ecologic tolerance and/or greater biogeographic distribution; 2) greater resistance to mass extinctions; 3) taxonomic bias, i.e., a lack of sufficient morphologic characters so that gastropod genera are not equivalent to genera in other clades (e.g., Schopf et al., 1975; but see Eldredge, 1976); and 4) the apparent extinction resistance is not real, with extinction-resistance at the family level masking high extinction and recovery rates at lower (generic) levels. We will first consider the extinction history of gastropods during the Phanerozoic before returning to these problems.

MATERIALS AND METHODS

The data used here come from several sources. The original basis of the family data is Sepkoski's compendium of marine families (1982). We have been individually assembling generic data bases, initially using such secondary sources as Wenz (1938) and the Treatise on Invertebrate Zoology (Knight et al., 1960). However, both data sets are now based primarily on the primary literature and (for the Paleozoic) on museum collections. DHE is responsible for the Cambrian-Triassic interval while PWS is responsible for the post-Triassic. The Paleozoic-Triassic data include 702 genera, of which six are resolved only to period and were not considered further. Genera which were resolved only to series level were assigned a stage level resolution proportionate to the underlying pattern of fully resolved data. This assignment had no effect on the diversity patterns. The other data set includes approximately 4,200 genera, including over a thousand genera that apparently have no fossil record. Few of the genera have a resolution only to the level of system or series and the vast majority are resolved to the stage level.

PALEOZOIC GASTROPOD HISTORY

Archaeogastropods comprise the bulk of Paleozoic diversity, with mesogastropods, bellerophonts (at least some of which were likely monoplacophorans) and other groups (including opisthobranchs) constituting minor components (Fig. 1). During the Paleozoic, family diversity rose steadily and archaeogastropods reached a Paleozoic diversity plateau by the Early Devonian. Mesogastropods continued to diversity until the Early Carboniferous. This general pattern was punctuated by sharp drops during the mid-late Devonian and the late Permian.

Generic diversity follows a trajectory similar to familial diversity, but shows greater variability, as expected (Fig. 2). There are pulses of origination throughout the Ordovician, during the Late Silurian, during the early Devonian, during the Visean stage of the

154

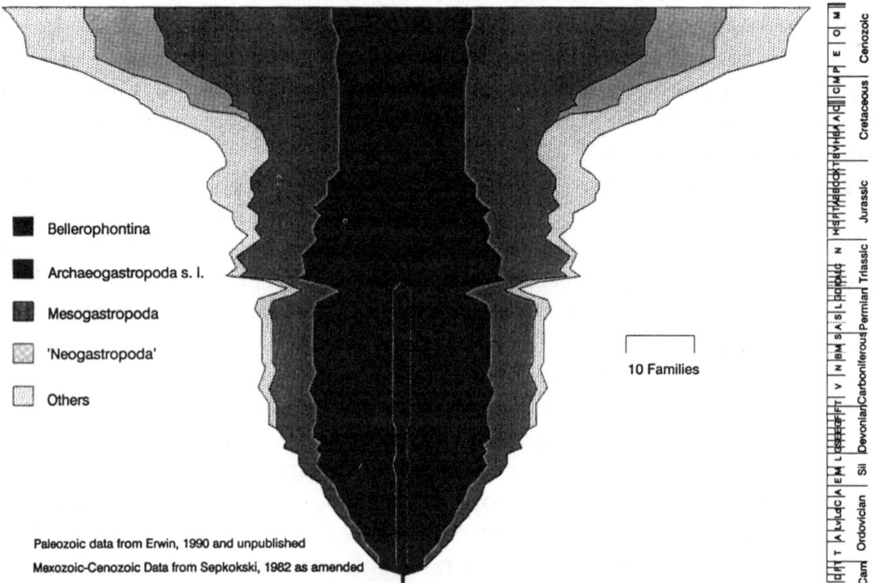

Figure 1. Phanerozoic family diversity, by stage, showing the diversity of the Bellerophontina (which some place in the Monoplacophora), the Archaeogastropoda, *sensu lato,* and the Caenogastropoda (as the Mesogastropoda and Neogastropoda). Paleozoic data from Erwin 1990a; Mesozoic and Cenozoic data from Sepkoski, 1982. The time scale in all figures is based on Palmer (1983).

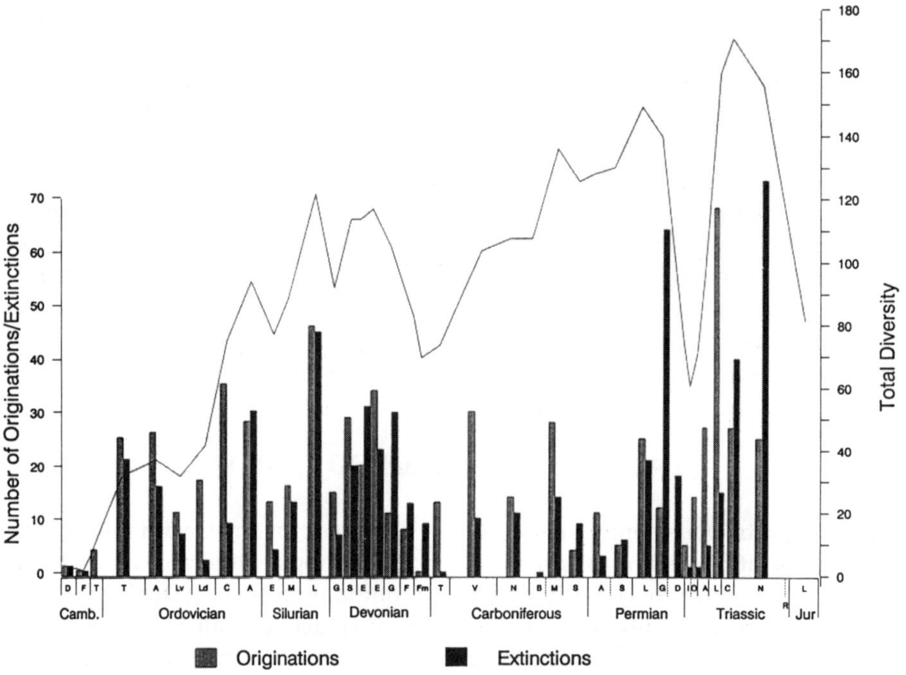

Figure 2. Cambrian-Triassic generic originations, extinctions and total diversity, by stage. Data is from the primary literature and museum collections, as described in Erwin (1990a). The Late Silurian origination and extinction peaks appear to be a result of monographic bias, and the origination and extinction lows during the Bashkirian stage of the Carboniferous is due to preservational bias.

Carboniferous, and during the Triassic as faunas recovered from the end-Permian mass extinction. Increased absolute numbers of extinctions occur during the end-Ordovician, end-Silurian, Middle Devonian, Late Permian, and end-Triassic. These same origination and extinction events are apparent when the rate and proportion of origination and extinction are calculated (Fig. 3). Each of the extinction events is seen in the marine record as a whole (Sepkoski, 1984, 1990) except for the Middle Devonian episode when the total marine fauna shows an extinction peak near the end of the period. The increased extinction of middle Devonian gastropods is probably linked to a replacement of lower Paleozoic groups by several new groups. Extinction rates were high during the Ordovician, as were origination rates, but extinctions and originations per my were fairly low. Thus the high extinction rates were simple a consequence of low overall diversity.

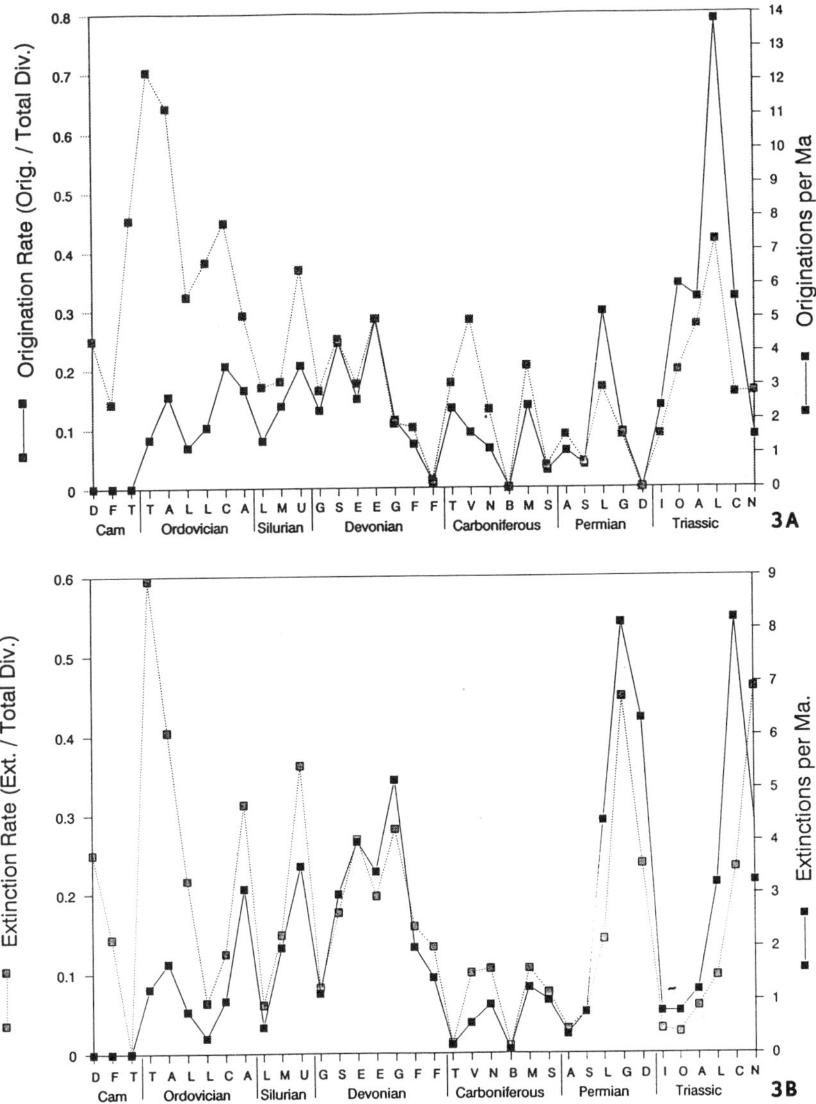

Figure 3. Cambrian-Triassic generic origination (3A) and extinction (3B) rates and proportions. The higher origination and extinction rates in the Lower Paleozoic are a result of low overall diversity. Rates are calculated as the number of originations or extinctions as a proportion of standing diversity. Originations and extinctions per Ma are calculated as the absolute number of events recorded in each stage divided by the duration of each stage.

156

One striking aspect of Figure 2 is that several of the origination and extinction episodes occur during the same interval, rather than in the more customary set of mass extinction followed by a recovery. Post-extinction recoveries are extended affairs, often requiring up to 10 my (Sepkoski, 1984; Erwin et al., 1987). The Early Ordovician, Late Silurian and Leonardian clearly constitute such radiation-extinction events (and the Middle Devonian arguably does as well) and are different from other intervals. One could argue that the originations and extinctions are separate but the data are insufficient to resolve them, or that they represent monographic bias. For the Leonardian and lower Ordovician this is not the case, since the genera are well-established and generally widely distributed, and more detailed data shows coincident originations and extinctions (Erwin, unpublished data). The Late Silurian is more complex and may be artificially inflated by Lindstrom's description of the magnificent Bornholm fauna. Thus at least two of the events represent a complicated turnover of the gastropod assemblage.

One possibility that could account for juxtaposed origination and extinction events is that the two are causally linked. We can suggest two types of linkage that could account for the pattern. The first is a biological interaction such as competitive exclusion in geologic time (the evolution of a new group of predators or parasites could also generate the observed pattern). Previous workers have argued that competitive interactions are responsible for the replacement of one clade by another in geologic time (e.g., Krause, 1986). A second possibility is that taxonomists have arbitrarily defined paraphyletic taxa that are stratigraphically restricted across boundaries where high rates of origination and extinction are observed.

A crucial concern in identifying the factors leading to diversification of the gastropods is distinguishing between the different contributions of mass extinctions, background extinctions, and the radiation-extinction events described above. Quantitatively, background extinctions account for 57% of all Paleozoic extinctions (Table 1) which would seem to suggest that these events, and the associated long-term adaptive changes, are responsible for much of gastropod evolution, as Vermeij has suggested (1987). Mass extinctions account for only 26% of all generic extinctions. At issue, however, is not the relative magnitude of the different classes, but their importance in determining the composition of the biota.

Table 1. Magnitude of extinction episodes, both absolute number of extinct genera, and extinction as a percentage of standing diversity during the interval. Ordovician-Triassic data from Erwin, unpublished data.

	Extinctions	
	Total Genera	% Diversity
Mass Extinctions		
End-Ordovician	31	31
End-Devonian	45	38
End-Permian	84	58
Total	160	26
Radiation-Extinction		
Lower Ordovician	39	61
Upper Silurian	46	36
Leonardian	22	14
Total	107	17
Background Extinctions	349	57

Another way of analyzing the relative significance of background and mass extinctions is to study diversity patterns within subclades. Figure 4 illustrates Late Cambrian through Triassic generic diversity in six different gastropod clades. These clades were identified as independent, monophyletic units in a preliminary phylogenetic analysis of Paleozoic gastropods (Erwin, unpublished). These spindle diagrams show the significance of both mass and background extinctions. The Bellerophontina, Macluritina and Euomphalina, and the Neritina are important parts of Ordovician-Lower Devonian diversity but all decline in both diversity and importance during the Middle Devonian reorganization of gastropod assemblages. Archaeogastropods undergo a steady increase throughout the Paleozoic, albeit punctuated by marked declines during extinction episodes. The history of the caenogastropod diversity is more equivocal. They are a minor part of Paleozoic diversity, but do increase in importance after the Visean.

A detailed analysis of the end-Permian mass extinction suggests just the opposite, however (Erwin 1990a). In that case the change in the dominance relationships between superfamily-level groups was clearly a consequence of the extinction and of differential success during the subsequent recovery. There is no evidence that trends established before the extinction contributed to success.

In a study of Permian gastropod assemblages from the southwestern US (Erwin, 1990b) there was no evidence for the alternation of macroevolutionary regimes observed in Cretaceous gastropods and bivalves (Jablonski, 1986, 1989). The Zygopleuridae and Bellerophontina illustrate the importance of clade history to recovery following a mass extinction. The Zygopleuridae sensu lato includes the present families Paleozygopleuridae, Psudozygopleuridae, and Zygopleuridae and is a monophyletic assemblage. Members of the clade are high-spired caenogastropods that first appear in the Early Devonian and radiate during the Late Carboniferous. Generic diversity drops from 12 to 1 during the Late Permian. The clade recovered during the Triassic and became one of the dominant components of Triassic gastropod assemblages. Two other important Triassic gastropod families, the Coelostylinidae and Spirostylinidae, may be descended from the zygopleurids.

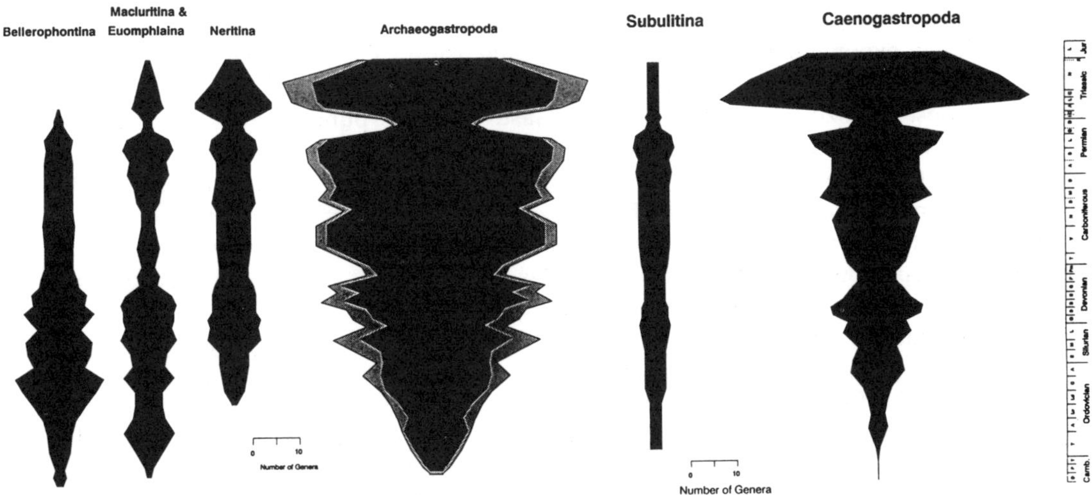

Figure 4. Spindle diagrams showing Cambrian-Triassic generic diversity for the six major gastropod clades. Note that the upper Paleozoic and Triassic bulge in the Macluritina & Euomphalina clade may include genera unrelated to earlier taxa. The lightly stippled area in the Archaeogastropod clade includes groups doubtfully assigned to the Archaeogastropoda. The Subulitina appear to be independent of all other caenogastropods and are listed independently here. Data from Erwin, 1990a and Erwin, unpublished.

158

This pattern contrasts with that of the Bellerophontina, an important group during the early Paleozoic, but one which declines in importance in the late Paleozoic. During the late Paleozoic bellerophont genera have long ranges but comparatively few new genera appear, so the net growth of the clade (originations minus extinctions) is negative during much of the interval. However, during the end-Permian mass extinction bellerophonts experienced the least loss of genera (c. 30%) of any major gastropod clade. But, despite this relative success, bellerophonts failed to radiate during the Triassic. They were swamped by the continuing diversification of other groups and died out in the mid-Triassic. Present evidence suggests that while the Zygopleuridae suffered greater losses during the mass extinction the clade was still spawning new genera at a sufficient rate to re-radiate during the Triassic. Bellerophonts, in contrast, had not been producing new genera and failed to do so after the extinction, ultimately leading to the disappearance of the clade.

Diversity is a function of the historical variation of cladogenesis and extinction within a clade. The extinction of the Bellerophontina (if, indeed, the clade is extinct: see McLean, 1984) followed from a failure to evolve new genera. The task is to determine how general this pattern is, and what characteristics are important in the success or failure of a clade. Analyses of families within clades (Gilinsky and Bambach, 1987) indicate that a reduction in the rate of production of new taxa is a common pattern in the fossil record.

In summary, Paleozoic gastropod clades had a more complex history of subclade replacement than is apparent in a superficial analysis. Groups dominant during the early Paleozoic declined and were replaced by other groups. This suggests that while gastropods may appear relatively extinction-resistant at subordinal and higher taxonomic levels, this masks a more complex dynamic at lower levels.

POST-PALEOZOIC GASTROPOD EXTINCTION HISTORY

The Mesozoic and Cenozoic history of the Gastropoda is a story of diversification, when the clade left its status as a minor component of marine communities and became the most diverse clade of durably-skeletonized marine invertebrates. The diversification, beginning in the mid-Cretaceous, primarily involved predaceous caenogastropods (Taylor et al., 1980). It is difficult to identify the precise start of the diversification because the toe of an exponential diversification is a subtle feature. Certainly, the diversification was well underway by the Barremian Stage of the Early Cretaceous.

The post-Paleozoic diversification of the Gastropoda was punctuated by two intervals of notably increased extinction. Forty-one percent of extant genera became extinct at the end of the Cretaceous (Maastrichtian Stage), a percentage exceeded by the gastropods only once in the Phanerozoic. In the final stage of the Eocene 26% of extant genera were lost. Both intervals are recognized as times of mass extinction and involve general reductions in the taxonomic diversity of marine clades. At other times the percentages of genera becoming extinct remain less than 15% (Fig. 5).

Both, intervals of increased extinction are followed immediately by intervals of increased origination. The first stage of the Cenozoic has a 39% increase in the number of genera and the first stage of the Oligocene has a 19% increase. As in the Paleozoic, intervals of extinction are tightly coupled with intervals of origination (but not necessarily the converse), suggesting a causal relationship between origination and extinction. The nature of this coupling will be the subject of future research.

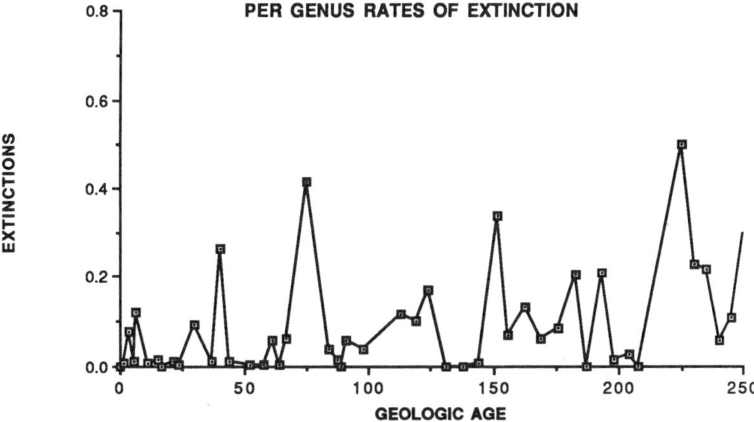

Figure 5. Changes in the frequency of extinction through time. There are two significant peaks in extinction after the Triassic, one at the end of the Cretaceous and one following the Eocene.

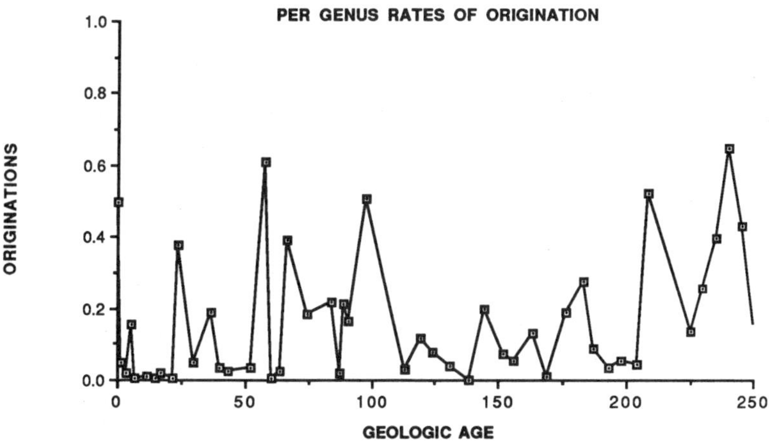

Figure 6. Changes in the frequency of origination through time. Both major peaks in extinction (Fig. 5) are closely followed by stages with increased originations.

CONCLUSIONS

In part the persistence and diversification of gastropods reflect their much lower extinction rates during major mass extinctions compared with other class-level groups. For example, during the end-Permian mass extinction, the most severe of the Phanerozoic, gastropod family diversity declined by only 10% while the average for all marine invertebrates was 54% and articulate brachiopods declined by 84%. Similarly, 41% of the genera become extinct during the last stage of the Mesozoic, a figure that is high, but low in comparison to the 48% of all invertebrate genera that became extinct at that time. Yet the success of gastropods may also be related to greater survival during the intervals between mass extinction, so-called background extinction.

160

ACKNOWLEDGMENTS

We thank Dave Jablonski for inviting us to participate in the symposium on extinctions. DHE acknowledges support from the National Science Foundation, Systematic Biology Program. PWS acknowledges support from NSF EAR 88-04798.

LITERATURE CITED

Eldredge, N. 1976. Differential evolutionary rates. *Paleobiology* 2:174–177.

Erwin, D. H. 1990a. Diversity patterns in Mississippian-Triassic gastropod genera and the end-Permian mass extinction. *Paleobiology* 16:187–203.

Erwin, D. H. 1990b. Regional Paleoecology of Permian gastropod genera, Southwestern United States and the end-Permian mass extinction. *Palaios* 4(5):424–438.

Erwin, D. H., Valentine, J. W. & J. J. Sepkoski, Jr. 1987. A comparative study of diversification events: the early Paleozoic versus the Mesozoic. *Evolution* 41:1177–1186.

Gilinsky, N. L. & R. K. Bambach. 1987. Asymmetrical patterns of origination and extinction in higher taxa. *Paleobiology* 13:427–445.

Jablonski, D. 1986. Background and mass extinctions: the alternation of macroevolutionary regimes. *Science* 231:129–133.

Jablonski, D. 1989. The biology of mass extinction: a palaeontological view. *Phil. Trans. R. Soc. Lond.* B 325:357–368.

Knight, J. B., Cox, L. R., Keen, A. M., Batten, R. L., Yochelson, E. L. & R. Robertson. 1960. *Treatise on Invertebrate Paleontology, I. Mollusca.* Geological Society of America; Lawrence, KS.

Krause, D. W. 1986. Competitive exclusion and taxonomic displacement in the fossil record: the case of rodents and multituberculates in North America. *University of Wyoming Contributions to Geology Special Paper* 3:95–117.

McLean, J. H. 1984. A case for the derivation of the Fissurellidae from the Bellerophontacea. *Malacologia* 25:3–20.

Palmer, A. R. 1983. The Decade of North American Geology 1983 Geologic Time Scale. *Geology* 11:503–504.

Schopf, T. J. M., Raup, D. M., Gould, S. J. & D. S. Simberloff. 1975. Genomic versus morphologic rates of evolution: influence of morphologic complexity. *Paleobiology* 1:63–70.

Sepkoski, J. J., Jr. 1982. *A Compendium of Fossil Marine Families.* Milwaukee Public Museum Contributions in Biology and Geology 51:125.

Sepkoski, J. J., Jr. 1984. A kinetic model of Phanerozoic taxonomic diversity. III. Post-Paleozoic families and mass extinctions. *Paleobiology* 10:246–267.

Sepkoski, J. J., Jr. 1989. Periodicity in extinction and the problem of catastrophism in the history of life. *Journal of the Geological Society of London* 146:7–19.

Smith, A. B. 1988. Patterns of diversification and extinction in Early Paleozoic echinoderms. *Palaeontology* 31:799–828.

Smith, A. B. & C. Patterson. 1988. The influence of taxonomic method on the perception of patterns of evolution. *Evolutionary Biology* :127–216.

Stanley, S. M. 1979. *Macroevolution.* W. H. Freeman; San Francisco, CA.

Taylor, J. D., Morris, N.J., & C. N. Taylor. 1980. Food specialization and the evolution of predatory prosobranch gastropods. *Palaeontology* 23:375–409.

Vermeij, G. J. 1987. *Evolution and Escalation: An Ecological History of Life.* Princeton University Press: Princeton, NJ.

Wenz, W. 1938. Gastropoda. Pp. 1–1639. In: O. Schindewolf (ed.), *Handbuch der Palazoologie.* Vol. 6. Berlin.

Climate Change and Biology: A Proposal for Scientific Impact Assessment and Response

Robert W. Buddemeier

Abstract. Anthropogenically-induced global climate change is widely perceived as combining with local and regional environmental alterations to produce the potential for massive extinctions—a major threat to global biodiversity for which the scientific community has as yet no credible and potentially effective response. I propose a strategy centered around the use of triage, a disaster-oriented technique for optimizing resource allocation, as a prompt and effective iterative approach to both evaluating the problem and mitigating it to the maximum extent consistent with available resources. The paper provides background and examples as well as suggested guidelines for implementation.

INTRODUCTION

This paper is an outgrowth of a talk given in the symposium on Global Warming at the 4th International Congress of Systematic and Evolutionary Biology. The original talk focused on the nature and probable importance of threats to organisms and to biodiversity posed by global environmental change. The talks of the other participants and the spirited discussion following the session made it evident that there is both the perception of an urgent but by no means uniform problem, and widespread frustration over the inability of scientists to devise and apply problem-solving strategies that are at once scientifically-based and effective.

What follows is an outline of a conceptual approach and rationale for applying scientific knowledge and principles to the perceived problem of preserving as much biodiversity as possible in the face of large-scale but not rigorously predictable environmental change. I have chosen to write it in an essay format rather than as a scientific paper, both because there are no new data and much opinion, and because the task of seeking out citations for all of the supporting information was not manageable on the timescale imposed by the publication schedule.

I acknowledge the stimulation and ideas provided by my fellow speakers and the discussants at the Symposium, and by other colleagues and my indefatigable biological advisor, Daphne Fautin. I hope they will find this useful, and I make no representations whatsoever as to any responsibility or agreement on their part.

Dr. Buddemeier is with the Kansas Geological Survey, 1930 Constant Avenue, Lawrence, KS. 66047, USA.

162

STATEMENT OF THE PROBLEM

For the purposes of this paper I assume that there is a significant threat to biodiversity on a global scale as a result of anthropogenic environmental change, and that an important aspect of this threat is the loss of diversity in the global gene pool. Further, I assume that this threat is important enough to merit response, and that at least some aspects of it can be addressed or mitigated on scientific and technical levels. Finally, I propose that we recognize that the magnitude of the potential problem is far greater than the resources available to apply to it, so realistic and cost-effective strategies are need to maximize the effectiveness of scientific contributions.

THE NATURE OF THE THREAT

General

Anthropogenic environmental change as a biotic threat has a number of components. Global change is normally considered to mean primarily climate change induced by the so-called "greenhouse" gases, but is typically expanded to include the issue of stratospheric ozone depletion—a source of environmental stress related to but not directly part of the "greenhouse effect." Truly global processes interact strongly with local and regional changes caused by land use, resource exploitation, population density, etc. One regional environmental problem much discussed as a major threat to biodiversity is the destruction of tropical forests. This destruction contributes to the Greenhouse effect by liberating carbon dioxide, but the biological losses are not the result of uncertain global processes; they are the entirely unmysterious results of setting fire to the landscape and driving large machines across what is left.

It is important to distinguish among different classes of somewhat related problems, since both the problems and their solutions differ in very fundamental ways and require different approaches. In the case of deforestation it is almost trivial to identify the areas and communities at risk, the nature of the threat (in many cases definable at the level of specific individuals or agencies and their operating methods), and at least the broad outlines of the possible solutions (i.e., the use of moral suasion, bribery, or force to dissuade the threat from its preferred course of action). Such approaches are more properly in the domain of the social sciences; therefore the bulk of the subsequent discussion will be devoted to issues of global climate change and the cumulative effects of local and regional environmental alteration, where the physical and biological sciences still have significant contributions to make. Similarly, I make no pretense at specifying approaches to population control or the environmental problems associated with economic development. This does not mean disapproval of political action, recycling programs, or any of the other laudable actions that can be readily justified on the basis of present knowledge; they are worthwhile, but they require little if any new knowledge or changes in the way scientists do business. I focus instead on the ways in which scientists may make a major difference as scientists rather than as activists, educators, or examples. The suggestions that I make deal with adaptation to, rather than prevention of, problems; this reflects my analysis that prevention and mitigation are largely political and engineering problems, while adaptation can still be dramatically furthered by the injection of new scientific ideas and information.

Potential for Global Change

The Greenhouse effect is the result of energy trapping by the radiatively active gases in the Earth's atmosphere. These gases—carbon dioxide, methane, nitrous oxide, chlorofluorocarbons, and others—are transparent to incoming sunlight, but absorb the longer wavelength radiation that the Earth radiates back into space. Thus, with constant solar input, increasing concentrations of these gases in the atmosphere will result in increased heat retention.

The radiative characteristics of these gases are known beyond doubt; so too is the fact that their atmospheric concentrations are steadily increasing. Both direct measurements on time scales of years to decades and retrospective measurements of gases trapped in dated ice cores over time scales of centuries to millennia show that atmospheric concentrations of carbon dioxide and methane have been increasing since the beginning of the Industrial Revolution, with much more rapid increases during the last half-century. The increases correlate well with growth in fossil fuel utilization and with global population increase. Atmospheric carbon dioxide concentrations have risen by approximately 26% since preindustrial times; because both global population and per capita energy use are inexorably rising, reasonable estimates suggest that we will have reached the greenhouse effect equivalent to a doubling of atmospheric carbon dioxide some time prior to the middle of the 21st century.

These documented changes in the concentrations of radiatively active gases mean that a shift in the Earth's energy budget has already occurred; an even greater one is virtually certain to occur as mankind plays out the tragedy of unfettered population growth and economic development. This will unquestionably affect climate; the nature of that effect and its rate of onset are less certain than the fact of its existence. Changes in cloud cover and oceanic heat uptake could essentially cancel the warming effects of the shift in energy balance; however, it is important to recognize that the resulting changes in the hydrologic cycle and in atmospheric and oceanic circulation patterns would constitute a dramatic change in climate even if unaccompanied by any net change in atmospheric mean temperature.

Computer climate models known as general circulation models (GCMs) are used to simulate the climate effects of doubled atmospheric carbon dioxide. They suggest a global warming of a few degrees Celsius, with greater changes (up to 10o C) at higher latitudes. Concomitant changes, according to the rather crude consensus of the different models, may include drier conditions in continental interiors, and rising sea level due to thermal expansion of sea water and some increased melting of glaciers. Such changes, however, represent best scientific guesses rather than confident predictions. The various models do not agree in detail, are relatively crude in terms of their handling of the important issues of air-sea interaction and cloud dynamics, and are very coarse in their spatial resolution. Further, although paleoenvironmental studies suggest that past warm periods share some characteristics with the computer simulations of doubled carbon dioxide (e.g., latitudinal temperature gradients), other aspects such as the distribution of dry and wet conditions do not necessarily agree.

It is reasonable to assume that climate change is in progress, whether currently demonstrable or not. It is further reasonable to assume that global warming is a probable (but not yet certain) outcome of that change. It has been argued, based on some of the climate change scenarios, that the rates of change and the maximum temperatures and sea levels expected are unprecedented in recent geologic and evolutionary history, and that this poses a profound threat to existing organisms and ecosystems, and particularly to biodiversity. Although uncertain in detail, it is this threat—rapid change to unprecedented levels of temperature and sea level, which interact with other local and regional anthropogenic environmental changes—that represents a worst case credible enough to take seriously as a major threat to biodiversity.

TRADITIONAL CONSERVATION APPROACHES

The issue of preserving threatened species is not a new one, but it has been transformed by the much larger scale of the present problem. Approaches available range along a continuum from primarily passive to very active. Examples would include, on the passive end, the cessation of whaling (the whales will do just fine once protected from excessive predation), and on the active end the indefinite maintenance and breeding of captive wild animals in an environment comparable to their natural one. The more "active" the process, the more resource-intensive it will be and the fewer the species that can be accorded such treatment on a fixed budget.

Existing strategies include preservation of living species in controlled environments (botanical or zoological gardens, etc.), preservation of genetic material (e.g., seed banks), and preservation of more or less natural habitats with their included organisms and ecosystems. This last alternative has become the preferred method of protecting communities and specific organisms from local environmental alteration, as it is apparently passive (relying on preventing action rather than taking action) and has the advantage of protecting all of the components of the ecosystem, whether or not they have been identified. Habitat preservation actually has no unique position on the passive-active continuum, since it may range from a simple agreement to avoid a few specific activities (e.g., logging) to very high-cost efforts to safeguard an ecological "island" from the inexorably rising tides of introduced species, resource diversion, and environmental contamination.

Unfortunately, the habitat-preservation approach is also much more vulnerable to climate change than are the other techniques. Large preserves and parks may have dimensions of a few hundred km, but a temperature change of only a few degrees can shift climate zones by far greater distances, essentially decoupling the preserved real estate from the habitat it once supported. Locally protected habitats, such as those associated with endangered species, are generally far too small to provide resilient and sustainable community systems. Preserves with latitudinal dimensions on the same order of magnitude as continents would be required to permit the climatic zones to migrate through "preserved" environments, and even then some discontinuous habitat zones could be squeezed out of existence by global warming (e.g., the alpine regions now found at the tops of some tropical mountains).

An important component of preservation strategy is knowledge, which in turn is based on systematic and ecological research. Research itself does not preserve, but it provides the basis for developing preservation priorities and strategies. I argue that we are unlikely to develop any radically new types of preservation activities and that resources available for preservation will not increase in proportion to the need, so the key to increased effectiveness must be changes in research directions, research output, and the use of research results. This may be seen as a Faustian bargain by those scientists who would prefer to save the world while continuing to pursue their present hobbies, but the generation and use of information is one of the few activities that has leverage (or a multiplier effect, in economic terms) great enough to have much impact on the problem.

ALLOCATION OF LIMITED RESOURCES

Triage is the term applied to the process of sorting casualties and establishing treatment priorities in situations (e.g., warfare, disasters) where the demand for medical treatment exceeds available resources. As the word implies, three categories of casualties are traditionally identified: 1) those whose injuries are minor and who are likely to recover with minimal care and treatment; 2) those whose injuries potentially threaten life or full function, but who can expect complete recovery if treated promptly and effectively; and 3) those unlikely to live or to experience full recovery regardless of treatment. Emergency medical personnel are trained to identify and focus their efforts on group II, since it is there that the greatest payoff per unit effort is expected.

One of the themes of this paper is that a system analogous to triage is needed to allocate research and conservation resources in an era of high ecosystem stress caused by global warming and other anthropogenic environmental change. The biotic effects of global change, especially with respect to biodiversity, are treated with alarm in the biological literature and with something approaching panic in the popular press. That limited resources are available to define, much less solve, the biological problems has been extensively discussed. This general perception of incipient disaster with limited potential for response sets the stage for the application of triage.

I emphasize, however, that the triage I am discussing is a much more intellectual activity than casualty sorting. In most injury-oriented medical emergencies there are well established protocols for diagnosis and sorting; there are also clearly defined and well-tested treatments for the various conditions expected (whether or not these treatments are operationally available to the practitioners). In this respect the parallels with the biotic impacts of environmental change are less than perfect—we may more appropriately use a triage-related public health analogy by asking how to allocate limited supplies of vaccines and medicines in anticipation of a developing epidemic. To answer that question requires some knowledge of the nature of the threat, the vulnerability of various subsets of the population, and the efficacy of the available protective measures and treatments in these same populations; it may also raise the traditional moral dilemma of assigning relative value to different individuals or populations. I will examine these issues individually with respect to climate change, and then return to possible applications of the triage concept.

VULNERABILITY

In the face of change, an organism or a community in reasonable equilibrium with existing conditions can respond in one of three ways: it can adapt *in situ*, it can die (or become extinct if the problem is widespread), or it can migrate to a more favorable location. Because human observations are limited to a relatively brief period of generally stable climate, we have little empirical data from which to predict potential responses to climate change. The challenge, then, is to construct analogies or theoretical arguments that may serve us well in our efforts to identify species or systems that are both vulnerable and salvageable. A major tool that can be used to meet this challenge is the examination of past climate cycles and their effects on both organisms and larger biological systems.

The Quaternary period (the past two million years, consisting of the Pleistocene and Holocene epochs) has been characterized by repeated glacial-interglacial cycles. At the peak of the most recent glaciation (18,000 yrs BP), most high-latitude and high-altitude locations were covered by glaciers, sea level stood about 120 m lower than present with all of the continental shelves exposed, and mean global temperature was significantly lower than at present. Rapid deglaciation and warming occurred between about 12,000 and 6,000 yrs BP; the altithermal (or climatic optimum) period at the end of that interval had temperatures slightly warmer than at present, sea levels close to today's, and was synchronous—probably not coincidentally—with the initial development of extensive human agriculture and urban civilization.

At the previous pronounced interglacial maximum (120,000 yrs BP) temperatures were somewhat warmer than at present; lesser glacial cycles are known to have occurred between then and now, and pronounced cycling also occurred back to the Pliocene-Pleistocene boundary approximately two million years ago. Organisms and communities as we know them today have survived passage through the repeated filters of Quaternary climate change. We can estimate the extent of these changes, but we are unable to specify short-term rates. The methods of paleoenvironmental analysis and geochronology generally do not provide the combined spatial and temporal resolution required to compare past rates of change with actual or anticipated changes during the period of human observation. This is in part the basis for the claim that greenhouse-induced warming will be more rapid than past experiences—we simply lack the data necessary for accurate estimates of past short-term rates of change, so are forced to compare rates based on data with uncertainties on the order of centuries with present-day annual observations. The results of such smoothing are not surprising—present (and near-term future) variability seems more dramatic. The rate of coming changes may be without precedent in recent evolutionary history, but that must be regarded as a possibility rather than a certainty.

The Holocene transgression provides some dramatic comparisons with anticipated future climate changes. Within 6,000 years glaciers had retreated thousands of kilometers in lateral extent and thousands of meters in vertical elevation in the higher mountain ranges, leaving exposed vast expanses of land that had been denuded of vegetation and soil. Sea level had risen over 100 m, flooding vast expanses of continental shelf that had been subaerially exposed for tens of thousands of years. In absolute terms, the inundation of land area that would accompany the melting of all remaining ice would be small compared to that which occurred at the end of the Pleistocene. This massive return of fresh water to the encroaching oceans must have been accompanied by dramatic changes in the salinity and turbidity of coastal and shallow ocean water masses. The time scale involved—a few thousand years—is small compared to that which conventional wisdom assigns to evolutionary processes. Yet the fact and the extent of the change mean that even the most specialized or well-adapted present-day species or associations must have evolved, adapted, or somehow survived in refugia during the Holocene transition and the multiple climate cycles that preceded it.

Viewed superficially, this presents a picture of the modern biota as adroit and hardened survivors, well-conditioned to deal with the trifling perturbation humans have managed to inject into the vast oscillation of the natural environment. This fits well with the increasingly common recognition that modern communities are by no means stable end-products in equilibrium with their environments, but are often the latest of a long series of successional stages, or even simply random assemblages of organisms with similar tolerances. However, the possible real threats to biodiversity of climate change are synergistic with the local and regional habitat and community alterations caused by human land use and species introductions. The combination of a basic assumption of survivability with a complex and novel threat provides the necessary basis from which to develop the concept of triage.

A STRATEGY FOR DECISION-MAKING

I view the present paralysis of organized science in the face of great challenge and opportunity to be in significant measure the result of failure to examine and agree upon fundamental assumptions. For the ensuing strategic exercise, I propose to use the following assumptions and definitions. Others may be employed, but should be specifically identified so that procedures can be identified and evaluated relative to the desired outcomes.

1. Most species can survive and many will survive, with or without human intervention.
2. Some extinctions are inevitable, with or without human intervention.
3. Humans have had, and will continue to have, major effects on global ecosystems

and biodiversity, whether or not those effects are recognized and/or intentional.

4. Any large-scale strategy for preservation of biodiversity will have individual failures and unintended consequences; it must be planned and assessed on the basis of overall objectives.
5. There is a useful correlation between genetic differences and taxonomic separations (e.g., a fish and a mammal will represent more genetic diversity than two species of the same genus of fish).
6. Triage is not a strategy in itself, but rather a central technique for developing and implementing strategy.
7. The time constants of climate change and human activity are such that triage should be an iterative process for targeting research as well as preservation—new knowledge may refine both the strategy and the actions based on its application.

As implied above, the basic strategy is the iterative application to extant taxa of the principle of triage with respect to survivability under (a) present conditions (extrapolations of present trends in land use, pollution, etc.) and (b) the most probable global change scenarios (greenhouse effect, ozone depletion). This review should not start at the species level—the list to be considered is unmanageably large and incomplete, and species within the same genus arguably represent the smallest unit of genetic distance, which is the level at which most of the diversity losses will have to be accepted. Depending on the taxon and the evaluators, the family or genus level is probably the lowest that should be surveyed on an initial pass, in which survivability of any species within the higher grouping would assure survivability of that taxon. Attention could and should be extended to lower levels on subsequent iterations, but as a first approximation the concept would be to survey the major branches of the evolutionary tree before attending to the twigs and leaves. Continued emphasis on species-level investigations without some overall strategy puts much present systematic research into the category of prospective paleontology—describing for future generations the soon-to-be-extinct taxa.

The goals of the initial categorization should be to develop the approach, identify a manageable list of research and information needs, and identify any higher level taxa that may be in triage level II. If priority-setting is to be useful the urgent items should be sufficiently limited in scope and number to be achievable. If everything is threatened, then in all probability nothing will be saved; this implies a major shift in philosophy from treating everything as a worst-case scenario to finding valid reasons for assigning taxa to triage categories I (if reasonably possible) or III (if clearly indicated or if necessary to keep category II manageable). I also want to stress that this proposal is not intended to suggest that all available resources be put into short-term classification, research, and applied preservation efforts. Continued long-term research in many discipline areas will be vital. Ecology is one such area; another is the vital but not presently fashionable subject of mechanisms and controls for the actual and potential short-term rates of adaptation and evolution.

Classification criteria will vary by taxa and over time, as the process is refined, and as climate predictions and data develop. Some practical approaches to an initial, broad-brush classification might be as follows.

In category I (no serious or immediate threat) we might place:

1. All taxa represented by species/genera that are widely distributed, opportunistic, generalists, etc. (i.e., weeds and pests).
2. Those marine taxa that have large ranges and life cycles that are entirely pelagic or abyssal (with the exception of possible ultraviolet light stress in the surface layer, the open/deep ocean is one of the locations best buffered from rapid change).
3. Species that have made it through the Pleistocene climate cycles quite handily and

168

for which those changes are good analogs for present and future environmental changes.

4. Any taxa represented by domesticated species or of significant agricultural, forestry, or fisheries interest (from a practical rather than a scientific standpoint, these taxa will have constituencies and research funds devoted to them far in excess of what the less-applied biological research community can muster).

A sociopolitical comment related to item 4 is that mammals and birds in general have strong preservationist lobbies, so that initial assessment emphasis should be placed elsewhere.

Reasons for including taxa in category III (no intervention is practical) on either an immediate or a provisional basis could include:

1. There is convincing evidence that it is headed for extinction because of irreversible habitat changes, either natural or anthropogenic, or has fallen below threshold population sizes (a good example on both counts is probably the California Condor).

2. It is dependent on an endangered habitat that is not practically manipulatable or reproducible (e.g., permafrost or Arctic sea-ice systems under conditions of extreme high-latitude warming).

3. It is dependent on an association or resource (e.g., obligate pollinator or food source) that is in Category III for independent reasons.

Category II at the first level of consideration consists of those taxa possible to save with some feasible cost and effort. Obvious examples would include taxa—and communities—to which the threat is direct exploitation pressures, habitat fragmentation, or climatic zone movement at rates exceeding apparent migration potentials.

A consideration that will arise rather promptly is that of interactions, which leads to relative valuation. Many species will be found to be dependent on the preservation of their symbionts, obligate food sources, or some particular community structure. For this reason, category assignments of such dependent species will be conditional, with primary attention devoted to the cornerstone species, primary producers, and host species or substrate producers. This in turn provides the basis for a valuation system that considers the "bonus species" potentially saved by investing in individual species.

A final note before turning to some practical considerations: the foregoing discussion has been taxonomically oriented because the presumed audience consists primarily of systematists and because it relates directly to the concept of biodiversity. An entirely analogous and possibly parallel approach would be to apply the triage concept to habitats, communities or ecosystems directly, with valuation based on uniqueness and diversity of the components. Although this may lead more directly to recipes for action, the process will be more involved because the taxonomy and evolutionary history of systems are less precise and less well understood than is the case for organisms.

IMPLEMENTATION

There are many possible ways to envision application of the concepts; the comments I offer are advisory rather than prescriptive. One of the most basic points is that the proposed approach is fundamentally at odds with the institutional culture of academic biology. It requires focus on systematics above the species level, it requires acknowledgment and acceptance of much higher levels of uncertainty than many scientists are comfortable with, it requires prompt, output-oriented review and communication, and it requires a willingness to assign relative values and make decisions. These characteristics tend to be those of management, an activity that many consider antithetical to scientific research.

For the foregoing reasons I suggest that it will be unproductive to attempt to institutionalize the application of this strategy through the existing bioscience establishment at the beginning. If the initial steps can be taken and communicated quickly and competently the idea will generate its own momentum among both scientists and bureaucrats who see the need for some sort of organizing operational principles, and the attendant review and criticism will serve to refine the process. Since the issues are ultimately highly political I also consider it desirable to keep the initial development out of the hands of mission-oriented agencies until it has a conceptual structure and internal logic that will be reasonably robust before the onslaughts of lobbyists.

Although the intellectual and short-term time demands would be high, the process at the first level of organization need not be overly long or expensive. An initial assessment could be done in one to two years by an ad hoc working group using institutional funds or private foundation support, working under the aegis of an institution or small consortium, or possibly a relevant professional society. Self-selection of a philosophically compatible team will be much more effective than inclusive representation in getting the process started; democratization can occur in the second generation. Procedural issues that need to be considered include the following:

1. An informal network for prompt review and information input must be established to provide necessary climatological, ecological, and detailed systematic expertise to the synthesizers of the overview.
2. One of the products of the first-level categorization must be the goals and methods of subsequent iterations.
3. Assumptions, values, and objectives must be clearly stated from the start, both to monitor progress and to provide for constructive review and criticism.
4. The output must be a 'living document,' so traditional publication should probably be a secondary means of communicating the results; electronic information networks can make independent sections available for use and review without waiting for completion of the entire effort or for traditional publication procedures. It may be regarded as an atlas, the various chapters of which are undergoing revision at quite different rates and which only occasionally may be produced as a specific "edition."

SUMMARY

The threat to biodiversity represented by global climate change and large-scale environmental alteration can be understood, and to some extent mitigated, by applying systematic procedures for evaluating vulnerabilities and for allocation of resources to the most productive areas of research and preservation activities. The triage concept, applied at taxonomic levels high enough to permit reasonably rapid iterations of assessments, is a technique central to development of a strategy that will bring the intellectual and informational resources of basic bioscience to bear on one of the most pressing applied issues of our time. It is an approach to which all can ultimately contribute, yet one sufficiently compact and compartmentalizable that it can be implemented quickly and efficiently.

Evolution of Reef-Building Corals During Periods of Rapid Global Change

D. C. Potts and R. L. Garthwaite

Abstract. The evolutionary responses of reef-building corals to the frequent and rapid climatic and geophysical fluctuations of the Pliocene-Pleistocene glaciations provide precedents for predicting the range of future responses to global change. Genetic analyses suggest that species of *Porites* in the Pacific have experienced little evolutionary differentiation since the Miocene, but that Caribbean *Porites* appear to be evolving rapidly in the late Quaternary. Different probabilities of speciation are explained in terms of divergent life history characteristics (longevity, generation time, overlapping generations) that are likely to influence the extent of future evolutionary responses to global change.

INTRODUCTION

Many criteria may be useful for predicting whether a species is likely to have substantial evolutionary responses to global change. In this paper, we concentrate on a single idea—that the potential for evolutionary responses will be influenced by the temporal scales of environmental change *relative* to the temporal scales of ecological and evolutionary processes in that species. The environmental and biological time scales are linked by the species' life history characteristics that determine whether, and how rapidly, populations can respond to a particular environmental change.

Perhaps the most important life history characteristic is individual longevity which affects, at the population level, the generation time of the population. Generation time is a measure of the time required for one generation to be replaced by its progeny, and it is commonly defined as the age of the average female after birth of half her progeny (Caswell, 1989). Genetically, generation time is important through its effects on the breeding structure of populations. As individual longevities and population generation times increase in iteroparous species, more generations will overlap at the same time and, therefore, will be likely to interbreed, and the frequency of inbreeding via backcrosses between young adults and their ancestors will also increase.

We will illustrate possible relationships among life history characteristics, environmental change, and probabilities of speciation by summarizing information from several studies of reef-building corals (Scleractinia) that provide evidence of evolutionary responses to environmental heterogeneity on several temporal and spatial scales, including those of the Pliocene and Pleistocene glacial sea-level fluctuations. However, the principles should be applicable to all organisms, terrestrial or marine, that possess suites of life history characteristics similar to those of corals, and are exposed to environmental changes of similar frequency and magnitude.

Drs. Potts and Garthwaite are with the Biology Department and Institute of Marine Sciences, University of California, Santa Cruz, CA 95064, USA.

ASSUMPTIONS

The argument contains many implicit assumptions about the spatial distributions of populations and species, and about the nature and extent of dispersal and gene flow. For this paper, we simply assume that local populations are panmictic and relatively isolated from other populations, but that long-distance dispersal and gene flow do occur periodically. However, three assumptions about global change should be made more explicitly.

First, continual physical, chemical, and biological change is normal. This was well illustrated by Davis (1986) when she plotted North Hemisphere air temperature on different temporal scales in which the units ranged from years, through decades, centuries and millenia to tens and hundreds of thousands of years. Most of these plots show upward, downward, or reversing trends with time. It was only when time units exceeded 100,000 years that there were some grounds for beginning to describe average temperatures statistically as fluctuating about a relatively constant mean. It is likely that most species have been exposed repeatedly to ecological and selective stresses that are consequences of temperature regimes changing on many different time scales. Presumably, many other environmental factors also change continually in analogous ways on many time scales.

Second, many environmental changes associated with global warming are not new. The Quaternary has seen repeated fluctuations in global temperatures, atmospheric CO_2, ice volumes, sea levels, etc. It is likely that atmospheric ozone also has fluctuated greatly. While past environmental changes differed in the sense that they probably were not caused by anthropogenic disturbances, the organisms present were responding proximally to many of the same environmental factors that are changing today.

Taken together, these two assumptions suggest that there should be precedents in the Quaternary evolutionary record for predicting responses to the environmental changes anticipated in the near future. At the very least, there should be qualitative precedents for many of these environmental factors, and there certainly are quantitative precedents for the proposed magnitudes of some of the changes (e.g., temperature, sea level). What we don't know yet is whether the rates of change likely to be experienced are unprecedented. In addition, there may be no precedents for many kinds of chemical pollution by novel molecules, and the particular combinations of environmental changes expected over the next century also may be unprecedented. However, since organisms must respond to the whole suite of environmental stresses that they experience, not just to single factors, and since many of the factors changing globally today are very familiar to most organisms (e.g., temperature, CO_2, sea level), previous consequences of environmental perturbation are likely to be useful indicators of future responses.

The third assumption is that each species responds independently to its environment. The genetic variation present in a species, and that species' life history characteristics, ultimately will lead to a unique, species-specific evolutionary response. While species with similar genetic and life history characteristics are likely to respond in similar ways, it is also possible for closely related and even co-existing species that have divergent characteristics to have different evolutionary responses to the same environmental stresses.

POSSIBLE EVOLUTIONARY RESPONSES

We have sought evidence of evolutionary adaptation of reef-building corals to temporal and spatial environmental stresses on at least three different scales.

The first is *local adaptation*, which is essentially the rapid, short-term, genetic fine-tuning of a population to local conditions by natural selection favoring the best-adapted individuals that happen to be present at the time. These are the short-term responses, usually reversible, by which populations genetically "track" prevailing conditions, and which occur in almost every generation. All populations that contain ecologically relevant genetic variation can be expected to respond on this scale.

The second scale is *regional differentiation* within species into ecotypes, geographic races, or sub-species. This implies longer time scales, sufficient for evolutionary responses to the different conditions that may prevail in different parts of the species' range. Regional differentiation may persist for long periods, and may or may not be reversible. It implies some combination of habitat differentiation, geographic isolation, and/or interruption of gene flow. Many populations of one species may be involved in regional differentiation.

The third scale is *speciation*, which may be regarded as the ultimate evolutionary response to heterogeneous environments in the sense that populations have diverged to the point where they are reproductively isolated or genetically incompatible. Speciation is a permanent, irreversible response that can be expected only in occasional populations of a few species.

LOCAL ADAPTATION

In initial experiments on natural selection in the corals *Acropora palifera* and *A. cuneata* at Heron Island on the Great Barrier Reef, we simulated environmental change using reciprocal transplants of the same genotypic arrays of cloned corals among five different habitats. We monitored patterns of survival, growth and mortality for several years (Potts, 1978, 1984b). Growth rates differed significantly among habitats and, within each habitat, there were significant and persistent differences in growth among corals originally collected from different habitats (Potts, 1984b). Mortality patterns also varied substantially among the five habitats: major mortality events occurred at different times in each habitat; different factors were responsible for each event; and at least some of these factors acted non-randomly, favoring genotypes derived from certain habitats (Potts, 1977, 1984b).

These data were consistent with the hypothesis that the reef consists of a mosaic of habitats, each of which exerts a qualitatively unique selective regime that acts rapidly within each generation to select for quite different kinds of individuals (physiologically, morphologically, and ecologically). Two measures of habitat favorability, survival and growth (Table 1), gave identical rankings of the habitats (Potts & Swart, 1984; Potts, 1984b). In the best habitat, the protected outer reef flat, there was little selection; practically every coral transplanted there survived and grew rapidly. In the worst habitat, at the base of the outer reef slope, fewer than 15% survived and growth rates were less than 20% of those in the best habitat: only specialists able to tolerate chronically adverse

Table 1. Relative favorability of five habitats at Heron Island, ranked by the average survival and growth of experimental corals, and expressed as proportions of performance in the most favorable habitat (after Potts, 1984b).

| Habitat | Relative favorability | | | Favored phenotypes |
	Survival	Growth	Mean	
Outer flat	1.00	1.00	1.00	Virtually all phenotypes
Crest	.89	.63	.76	Mechanical strength, competitive vigor
Lagoon	.69	.59	.64	Rapid vertical growth
Inner flat	.34	.54	.44	General physiological vigor
Reef slope	.15	.19	.17	Tolerate chronic bad conditions

physical conditions could live there. Other habitats selected for other kinds of corals with distinctive morphological, physiological, and ecological properties (Potts, 1984b).

We concluded that natural coral populations at Heron Island must contain large amounts of genetic variation; that they are very responsive to complex regimes of disruptive selection acting on short, ecological time scales; and that each generation "tracks" prevailing conditions closely through processes of local adaptation, within the limits imposed by the genetic variation existing in the population's gene pool.

REGIONAL DIFFERENTIATION AND SPECIATION

Trying to evaluate this potential of populations for rapid evolutionary adaptation within the broader context of the widely dispersed reefs of the Pacific Ocean as a whole led us to expect that larger scale, regional differentiation should also be common, including speciation in more isolated reef systems. If great distances and adverse currents, plus land exposed during low glacial sea levels, created substantial barriers to dispersal and gene flow, then the recurrent global disturbances of the Plio-Pleistocene should have provided frequent opportunities for intra- and inter-specific differentiation (Potts, 1983).

Therefore, we scanned the general taxonomic and paleontological literature, looking for three kinds of evidence for such differentiation: 1) marked increases in species richness of scleractinian corals during the Pleistocene; 2) extensive regional differentiation (e.g., geographic races/sub-species) within species; and 3) concentrations of endemic species in more isolated, especially peripheral, reef systems. Surprisingly, the generalized patterns among Pacific corals seem to contradict all three expectations for extensive differentiation.

First, generic and species richnesses today seem to be about the same as they were in the late Miocene or early Pliocene (Vaughan & Wells, 1943; Wells, 1956). The Indo-West Pacific today has a total of about 90 genera and over 500 species of reef-building corals (Veron, 1986), and many of these species appear to have persisted without morphological change from the Pliocene to the Recent (Veron & Kelley, 1988). Paleontologists regard reef-building corals as particularly long-lived, with an average species persisting in the fossil record for over 20 million years (Ma) (Stanley, 1979).

Second, many species are widely distributed throughout the Indo-Pacific. Despite many attempts to describe regional sub-species and geographic variants within species, few are widely accepted by recent authors. Local intra-specific, morphological variation on one reef often encompasses much of the array of variation described for a species throughout its entire geographic range. For example, the scleractinian faunas of Japan and the southern Great Barrier Reef are very similar at both specific and intra-specific levels (Veron, 1988).

Third, there are very few peripherally endemic coral species in the Pacific Ocean. There is only one probable endemic (*Porites panamensis*) in the Eastern Pacific, and no endemic species are recognized among the more than 150 species of reef-building corals in French Polynesia (Chevalier, 1981; Pichon, 1985). The major exception may be Hawaii, which is isolated oceanographically by adverse currents as well as by distance. Hawaii has an impoverished fauna of about 42 coral species that includes several probable endemics, mainly in the genus *Porites* (Maragos, 1977; Grigg, 1983, 1988).

Another general trend was that these patterns of limited intra- and inter-specific differentiation among the corals building the Pacific reefs seemed quite different from those among other organisms inhabiting those reefs. In several major groups, including fishes and molluscs, the total number of species has increased substantially during the Quaternary, geographic races or sub-species are often recognized within species, and peripherally endemic species are common. For example, the number of oceanic cone-shells has doubled, to almost 500 species, since the Pliocene (Kohn, 1985) and up to 30% endemism is common in the fishes of Hawaii, French Polynesia, and other geographically isolated Pacific reefs (Randall, 1976; Springer, 1982). Despite a limited fossil record for reef fishes, their potential powers of dispersal and the rapid radiations documented in freshwater fishes (e.g., Fryer & Iles, 1972) indicate that many endemic reef fish species may be very young (late Pleistocene or Holocene).

A MODEL

To explain more limited Quaternary speciation among corals than among other reef taxa during the global disturbances of Plio-Pleistocene glaciations, we developed a model based on differences in their life history characteristics (Potts, 1983, 1984a, 1985). Especially on the continental shelves where >50% of modern reefs exist (Potts, 1983), coral biomass and ecological activity are concentrated in very shallow (<20 m) water (Wells, 1956; Potts et al., 1985). Since the mid-Pliocene, high-frequency (10^3–10^4y) sea-level fluctuations of about 20 m amplitude have been superimposed on the major glacial cycles (about 125,000 years and >100 m) (Shackleton & Opdyke, 1977). Reconstructions of eustatic sea levels for the last 140,000 years (Chappell, 1981) suggest that any one level on the shelf was available for active coral growth for an average of only 3200 years before the sea rose or fell >20 m (Potts, 1984a). Therefore, frequent and total habitat destruction and extinction of local populations, either through exposure to air or through submergence at depths unsuitable for shallow-water corals, must have been very common. New populations established elsewhere by long-distance dispersal probably were founded by propagules dispersing from many parental populations.

One life history characteristic that distinguishes corals from most fishes or molluscs is the extreme longevity of some individuals. Under certain conditions, individual longevity may be translated into very long generation times for a population, and many generations will then overlap in the breeding population. Because most corals have indeterminate growth and apparently no physiological senescence, old corals tend to be very large and fecund, and tend to contribute disproportionately to the gamete pool. In addition, cloning through fragmentation of colonies often enhances genotypic survival and intensifies the effects of large size, high fecundity, and gametic dominance of older genotypes. In such populations, backcrossing between young and old adults, possibly several generations apart, may be commonplace.

Some immediate genetic consequences for life histories that include a breeding structure with extensive backcrossing are: that rare alleles will tend to persist and to increase in frequency in the population; that rates of fixation and elimination of alleles will be reduced; and that directional selection will proceed very slowly (Falconer, 1981). Consequently, genetic variation may tend to accumulate to very high levels within these populations. We proposed that under these circumstances, 3200 years is too short for much genetic divergence to occur, even in isolated populations, before the gene pools are mixed up again during colonization of new sites, and that as a consequence, probabilities of differentiation and speciation have been reduced or even completely inhibited in many corals since the Pliocene. By contrast, organisms with shorter lives, limited overlap of generations, and little or no backcrossing may be able to diverge genetically to the point where probabilities of speciation are relatively high during the same period of isolation.

TESTS OF MODEL

Pacific

To test these ideas, we examined populations of several species of *Porites* on another Australian reef, Pandora Reef. This reef supports many very large, old colonies that approach the demographic maxima of longevity and size for corals, and it lies close inshore where shallow, turbid conditions probably are more representative of the lower sea levels typical of Plio-Pleistocene shelf habitats than are modern, offshore reefs exposed to more oceanic conditions. *Porites* is also a species-rich genus that has been the dominant frame-builder of the Indo-Pacific since the Eocene, about 55 Ma ago.

Porites colonies at Pandora Reef form massive domes that may reach 10 m in diameter with minimum ages of 700 years (Potts et al., 1985; unpublished data). While good estimates of mean generation times are not yet available, Caswell (1989) has estimated generation times for tropical trees with analogous life histories. Three different measures of generation time gave estimates ranging from centuries to >1000 years for these trees, values that are of the same order of magnitude as that proposed for corals. The potential for interbreeding among generations is very high in *Porites* because young colonies become reproductive when less than 5–10 years old.

The prediction of high intra-specific genetic variation was supported by electrophoretic data from 14 enzyme loci. Observed heterozygosities (H_{obs} = .24—.32; D. C. Potts & R. L. Garthwaite, unpublished data) were higher than those reported for most other invertebrates (mean H_{obs} = .10±.09(SD); Nevo et al., 1984). However, the heterozygosities in *Porites* were similar to those of other long-lived organisms in which backcrossing among overlapping generations is probable, such as clonal cnidarians (Smith & Potts, 1987; Sole-Cava et al., 1985) and terrestrial plants (Hamrick et al., 1979).

In a separate phylogenetic analysis of several species collected on Pandora and nearby reefs (Fig. 1A), the known divergence of *Porites* from its ancestral genus *Goniopora* (about 55 Ma ago) was used to calibrate the electrophoretic data as a molecular clock. Assuming

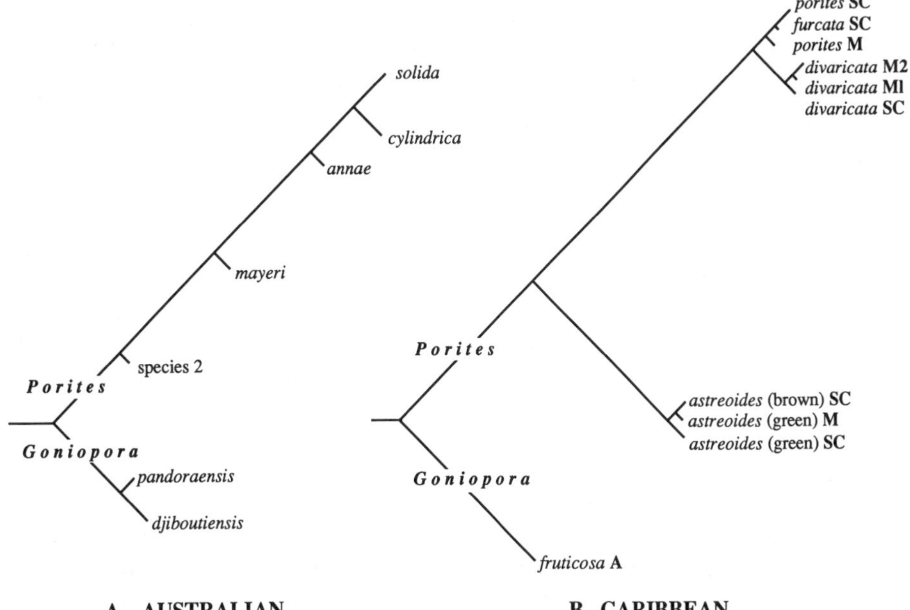

Figure 1. Phylogenetic relationships among species of *Porites* from A.—Australia, and B.—the Caribbean. Dendrograms were constructed from electrophoretic data using FREQPARS (Swofford & Berlocher, 1987). Australian *Goniopora* species were used as outgroups in each analysis. A = Australia; M = Miami; SC = St. Croix.

176

constant average mutation rates, estimates of divergence times between pairs of *Porites* species ranged from about 5 Ma to over 25 Ma, which places their common ancestors in the Miocene. This suggests that at least these lineages of corals did not evolve substantially in the Quaternary. Therefore these data from Australian *Porites* conform, demographically, genetically, and evolutionarily, to assumptions and predictions of the model.

Caribbean

Our most recent study examined whether corals with less extreme life history characteristics also conform to the model. In the Caribbean, between two and five species of *Porites* are generally recognized, all of which tend to form relatively small colonies (Wells & Lang, 1973; Cairns, 1982; Zlatarski & Estalella, 1982). They grow relatively rapidly, but rarely reach 1 m diameter, and most colonies probably survive for only a few years to a few decades. There are two common morphologies, either branching (e.g., *P. porites*, *P. furcata*, *P. divaricata*) or encrusting plates and small domes (*P. astreoides*), both of which are very different from the large (up to 10 m), persistent, massive domes of some Indo-Pacific species.

In 1989, we analyzed four populations of *Porites* from Miami and five from St. Croix at 40 electrophoretic loci. Using one Australian *Goniopora* species as an outgroup to include the divergence of these genera 55 Ma ago, our estimates from preliminary analyses indicate that all the branching forms such as *P. porites*, fall into a single lineage that diverged from the non-branching *P. astreoides* line about 35 Ma ago (Fig. 1B). This divergence time is consistent with the earliest fossil records of both morphologies in the Oligocene (Zlatarski & Estalella, 1982). Subsequent branching in each lineage is too recent for estimating divergence times from existing data.

Within the *P. astreoides* lineage, two color morphs (green and brown) at St. Croix had no fixed allelic differences but did have significantly different allelic frequencies within each population, each of which was in Hardy-Weinberg equilibrium. This suggests that while they are obviously very closely related, there is no gene flow between them, even though they live together in close proximity in the same habitat. Similar electrophoretic patterns link co-existing populations, identified morphologically as *P. porites* and *P. furcata*, from St. Croix; no fixed alleles separated them, but there were significant differences in allelic frequencies with no evidence of gene flow. These pairs of populations may be examples of late Pleistocene or even Holocene divergence and speciation within *Porites* in the Caribbean.

A hypothesis, that Pacific and Caribbean *Porites* have differed in probabilities of speciation since the Pliocene, is consistent with differences in their life history characteristics. Comparisons of the data from both oceans support the assumption that whether, and how fast, an organism responds evolutionarily to major environmental change will be influenced by life history characteristics. The longevities, generation times and, probably, breeding structures of Caribbean *Porites* seem to resemble life histories of many molluscs or fishes more closely than they resemble those of the massive Indo-Pacific *Porites*. Caribbean *Porites* with relatively ephemeral colonies seem to be responding to late Quaternary environments by genetic divergences leading towards speciation; but the Pacific *Porites* with persistent massive colonies show little evidence of Quaternary evolution.

This conclusion, that different probabilities of speciation in Caribbean and Indo-Pacific *Porites* are linked to divergent life history characteristics, is confounded by the fact that these corals are also isolated in different oceans that have experienced increasingly divergent histories since the Oligocene. To test whether life history characteristics alone are sufficient to explain different speciation patterns, it will be necessary to examine corals with both life histories in the same environment. Fortunately, this can be done in the Indo-Pacific, because it also contains many *Porites* species with morphologies and, presumably, life histories similar to those of Caribbean *Porites*, that live on the same reefs as massive

Porites species like those examined at Pandora Reef. In particular, clusters of endemic *Porites* species have been described on both the peripheral, isolated reefs of Hawaii (Vaughan 1907; Maragos 1977; Grigg, 1983) and the central reefs of the Philippines (Nemenzo 1955, 1971, 1976; Veron & Hodgson, 1989). Future genetic studies will test the hypothesis that many of these species have Pleistocene to Holocene origins.

GENERAL CONCLUSIONS

The genetic compositions of populations of modern corals, and the extent of their genetic divergences from common ancestors, are consistent with an interpretation that these animals responded evolutionarily in more than one way to the global and local environmental changes associated with Plio-Pleistocene glaciations. Differences in the extent of regional differentiation and speciation can be explained by life history characteristics likely to be sensitive to the relative scales of ecological processes and environmental changes. For massive Indo-Pacific species of *Porites*, potential longevities and generation times (10^2–10^3 years) may be so close to the scales of habitat destruction (10^3–10^4 years) that the latter acts more as ecological "noise" than as an evolutionary "driving" mechanism. But for the apparently more ephemeral (10^1–10^2 years) Caribbean *Porites*, the potential for evolutionary divergence may be much greater when exposed to similar scales of habitat disturbance.

Probable responses of reef-building corals to future global changes may include similar arrays of outcomes. The successful survival of most corals throughout the Quaternary must be attributed partly to their extensive intra-specific genetic variation that permits rapid local adaptation to conditions encountered at a particular site in any one generation. But in species with life histories and breeding structures that prevent fixation or loss of genetic information, high probabilities of local success may be linked to low probabilities of permanent genetic change, evolutionary divergence, and speciation.

ACKNOWLEDGMENTS

We are grateful for financial support from the Australian Marine Science and Technologies Grants Scheme (MST 83/1342), the NSF Biological Oceanography and US/Australia Cooperative Science Programs (OCE–8502086), the UCSC Faculty Research Committee, and UCSC's Dean of Graduate Studies and Research. Figure 1 is based on unpublished genetic analyses conducted by R. L. Garthwaite. We thank Laurel R. Fox for critical comments on the manuscript.

LITERATURE CITED

Cairns, S. D. 1982. Stony corals (Cnidaria: Hydrozoa, Scleractinia) of Carrie Bow Cay, Belize. Pp. 271–302. *In:* K. Rutzler & I. G. MacIntyre (eds.), *The Atlantic Barrier Reef Ecosystem at Carrie Bow Cay, Belize, I. Structure and Communities.* Smithsonian Contrib. Mar. Sci. No. 12.

Caswell, H. 1989. *Matrix Population Models.* Sinauer: Sunderland, MA. 328 pp.

Chappell, J. 1981. Relative and average sea level changes, and endo-, epi-, and exogenic processes on the earth. Pp. 411–430. *In: Sea Level, Ice, and Climatic Change.* Int. Ass. Hydro. Sci. Publ. No. 131.

Chevalier, J.-P. 1981. Reef scleractinia of French Polynesia. *Proc. 4th Intl. Coral Reef Symp., Manila.* 2:177–182.

Davis, M. B. 1986. Climatic instability, time lags, and community disequilibrium. Pp. 269–284. *In:* J. Diamond & T. J. Case (eds.), *Community Ecology.* Harper & Row: New York.

Falconer, D. S. 1981. *Introduction to Quantitative Genetics.* 2nd edition. Longman: London.

Fryer, G. & T. D. Iles. 1972. *The Cichlid Fishes of the Great Lakes of Africa: Their Biology and Evolution.* Oliver & Boyd: Edinburgh.

Grigg, R. W. 1983. Community structure, succession and development of coral reefs in Hawaii. *Mar. Ecol. Prog. Ser.* 11:1–14.

Grigg, R. W. 1988. Paleoceanography of coral reefs in the Hawaiian-Emperor chain. *Science* 240:1727–1743.

Hamrick, J. L., Linhart, Y. B. & J. B. Mitton. 1979. Relationships between life history characteristics and electrophoretically detectable genetic variation in plants. *Ann. Rev. Ecol. Sys.* 10:173–200.

Kohn, A. J. 1985. Evolutionary ecology of *Conus* on Indo-Pacific coral reefs. *Proc. 5th Intl. Coral Reef Congress, Tahiti* 4:139-144.

Maragos, J. E. 1977. Order Scleractinia: stony corals. Pp. 158-241. *In:* D. H. Devaney & L. G. Eldredge (eds.), *Reef and Shore Fauna of Hawaii. Section 1: Protozoa through Ctenophora.* Bishop Mus. Spec. Publ. No. 64.

Nemenzo, F. 1955. Systematic studies on Philippine shallow water scleractinians: I. Suborder Fungiida. *Natur. Appl. Sci. Bull* 15:3–84.

Nemenzo, F. 1971. Systematic studies on Philippine shallow water scleractinians: VII. Additional forms. *Natur. Appl. Sci. Bull.* 23:141–209.

Nemenzo, F. 1976. Some new Philippine scleractinian corals. *Natur. Appl. Sci. Bull.* 28:229–276.

Nevo, E., Beiles, A. & R. Ben-Shlomo. 1984. The evolutionary significance of genetic diversity: ecological, demographic and life history correlates. *Lecture Notes in Biomathematics,* No. 53. Springer-Verlag: New York.

Pichon, M. 1985. Scleractinia. *Proc. 5th Intl. Coral Reef Congress, Tahiti* 1:399–403.

Potts, D. C. 1977. Suppression of coral populations by filamentous algae within damselfish territories. *J. Exp. Mar. Biol. Ecol.* 28:207–216.

Potts, D. C. 1978. Differentiation in coral populations. *Atoll Research Bull.* 220:55–74.

Potts, D. C. 1983. Evolutionary disequilibrium among Indo-Pacific corals. *Bull. Mar. Sci.* 33:619–632.

Potts, D. C. 1984a. Generation times and the Quaternary evolution of reef-building corals. *Paleobiology* 10:48–58.

Potts, D. C. 1984b. Natural selection in experimental populations of reef-building corals (Scleractinia). *Evolution* 38:1059–1078.

Potts, D. C. 1985. Sea-level fluctuations and speciation in scleractinia. *Proc. 5th Intl. Coral Reef. Congress, Tahiti* 4:127–132.

Potts, D. C., Done, T. J., Isdale, P. J. & D. A. Fisk. 1985. Dominance of a coral community by the genus *Porites* (Scleractinia). *Mar. Ecol. Prog. Ser.* 23:79–84.

Potts, D. C. & P. K. Swart. 1984. Water temperature as an indicator of environmental variability on a coral reef. *Limnol. Oceanogr.* 29:504–516.

Randall, J. E. 1976. The endemic shore fishes of the Hawaiian Islands, Lord Howe Island and Easter Island. *Trav. Doc. ORSTOM* 47:49–73.

Shackleton, N. J. & N. D. Opdyke. 1977. Oxygen isotope and paleomagnetic evidence for early northern hemisphere glaciation. *Nature* 270:216–219.

Smith, B. L. & D. C. Potts. 1987. Clonal and solitary anemones (*Anthopleura*) of western North America: population genetics and systematics. *Mar. Biol.* 94:537–546.

Sole-Cava, A. M., Thorpe, J. P. & J. G. Kaye. 1985. Reproductive isolation with little genetic divergence between *Urticina felina* and *U. eques. Mar. Biol.* 85:279–284.

Springer, V. G. 1982. Pacific plate biogeography, with special reference to shore fishes. *Smithsonian Contrib. Zool.* 367:1–182.

Stanley, S. M. 1979. *Macroevolution: Pattern and Process.* Freeman: San Francisco, CA. 332 pp.

Swofford, D. L. & S. H. Berlocher. 1987. Inferring evolutionary trees from gene frequency data under the principle of maximum parsimony. *Syst. Zool.* 36:293–325.

Vaughan, T. W. 1907. Recent Madreporaria of the Hawaiian Islands and Laysan. *U.S. Nat. Mus. Bull.* 59:1–427.

Vaughan, T. W. & J. W. Wells. 1943. Revision of the suborders, families, and genera of the Scleractinia. *Geol. Soc. Amer. Spec. Pap.* 104:1–363.

Veron, J. E. N. 1986. *Corals of Australia and the Indo-Pacific.* Angus & Robertson: Sydney, Australia. 644 pp.

Veron, J. E. N. 1988. Comparisons between the hermatypic corals of the southern Ryukyu Islands of Japan and the Great Barrier Reef of Australia. *Galaxea* 7:211–231.

Veron, J. E. N. & G. Hodgson. 1989. Annotated checklist of the hermatypic corals of the Philippines. *Pac. Sci.* 43:234–287.

Veron, J. E. N. & R. Kelley. 1988. Species stability in reef corals of Papua New Guinea and the Indo-Pacific. *Mem. Ass. Australas. Palaeontols.* 6:1–69.

Wells, J. W. 1956. Scleractinia. Pp. F328-F440. *In:* R. C. Moore (ed.), *Treatise on Invertebrate Paleontology. Coelenterata.* Geol. Soc. Amer. & University Kansas Press: Lawrence, KS.

Wells, J. W. & J. C. Lang. 1973. Systematic list of Jamaican shallow-water scleractinians. *Bull. Mar. Sci.* 23:55–58.

Zlatarski, V. N. & N. M. Estalella. 1982. *Les Scléractiniaires de Cuba.* Académie Bulgare des Sciences: Sofia, Bulgaria. 472 pp.

Possible Effects of Global Warming on Marine Foodwebs at Low Temperature

EXTENDED ABSTRACT

W. J. Wiebe and L. R. Pomeroy

There has been extraordinary interest this past year in the topic of global warming. In this paper we present results of a study in a low temperature, marine embayment and suggest that even a small change in temperature could produce drastic changes in this and possibly other marine foodwebs.

The site for this work was Conception Bay, Newfoundland. This is the easternmost of a series of large bays on the coast of insular Newfoundland. The main axis is NNE to SSW, with the mouth opening directly into the open North Atlantic Ocean. The maximum depth is 290 m with a sill depth of 174 m. The embayment is 64 km long and 32 km wide at its mouth. There are commercial fisheries for cod, snow crab, lobster, capelin, herring, flounder, mackerel, and squid. Ice is present seasonally, with peak coverage in March and April, although there is great interannual variability. The annual temperature range in surface water is about 16°C, although intermediate and bottom water remains <0°C throughout the year. Tides are semi-diurnal with small amplitude. Circulation is essentially wind-driven.

This bay, while only at 47°N latitude, contains in the spring southerly flowing, Arctic-originating, Labrador current water, with a surface to bottom temperature <−1.5°C. This is as far south as this Arctic-originated water travels undiluted by warmer currents.

When the ice clears and light levels rise in the spring, a massive bloom of microalgae, dominated by diatoms, erupts. This bloom remains until one of two events occur: 1) inorganic nutrients, particularly silicate and nitrogen, disappear causing the algae to sink to the sea floor, or 2) air temperatures rise, the surface water stratifies and herbivorous zooplankton begin to consume the bloom. The length of time for the bloom is also controlled by storms, which act to mix the surface layers, return inorganic nutrients to the surface, and thus extend the length of the bloom.

For the past three years we have been studying microbial activities during, and immediately after, the bloom. In addition, we have examined some physiological aspects of bacterial pure cultures from these and Arctic ocean waters.

First, let us review briefly the food web structure in the ocean and how this changes in a spring bloom. A simplified, classical marine foodweb is presented in Figure 1 in which only the major flows are noted (June through September). The salient points are that: 1) the phytoplankton grow and also release dissolved organic carbon (DOC) during

Drs. Wiebe and Pomeroy are with the Departments of Microbiology and Zoology, and the Institute of Ecology, University of Georgia, Athens, GA 30602, USA.

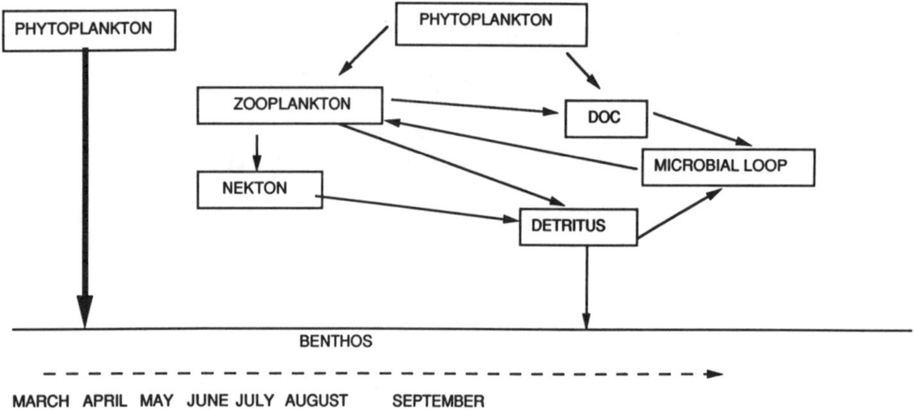

Figure 1. Spring bloom and pelagic foodweb model.

photosynthesis; 2) DOC is used by a variety of microheterotrophs, here lumped together in a box called the "microbial loop"; 3) Zooplankton, mostly small crustacea, graze on the algae and are in turn consumed by nektonic organisms; and 4) these activities release some DOC and, via death and "sloppy feeding," contribute to the detritus in the ocean. Some of this detritus falls to the sea floor, contributing to the growth of the benthic macro and microorganisms.

The microbial loop plays an important part in this story. It consists of bacteria and several different sizes of protozoan species. When organic compounds flow through this loop, most of them are transformed to CO_2, leaving little available energy for microorganisms (Pomeroy & Wiebe, 1988). Thus, this loop results in a loss of photosynthate without biomass accumulation. While there are arguments over how much organic matter flows through the microbial loop, it can account for a significant amount of the primary production; there are numerous reports where more than 50% of the primary production is channeled through this shunt (e.g., Smith et al., 1986). This is over-simplified but serves to point out the complex interactions and potential losses that can occur.

In the spring bloom in Conception Bay zooplankton and the microbial loop are virtually absent (see March, April in Fig. 1). The result is that algae grow rapidly until they run out of some limiting nutrient and sink to the benthos, where they are consumed and buried by the epi- and infauna. Thus, there is a direct link from algae to large animals, with no intervening losses. Benthic feeding fishes, such as cod, thus have a very short path to the primary producers.

In 1986 Pomeroy and Deibel proposed that *because* the microbial loop was inactive during the spring bloom, much more algal production was available to large animals and that this is one major reason that most of the large fisheries of the world are found in cold, generally high latitude waters, even though on the face of it, it is paradoxical that high biomass communities should be concentrated in cold waters where growth rates must be less than in warmer climates.

There were several perplexing aspects to the original idea, and over the past few years we have examined this problem in more detail. The first question was whether the microbial loop was indeed inactive. We examined this problem in three ways: 1) *in situ* bacterial growth rates, 2) measurement of total community respiration, and 3) grazing effects on bacterial growth.

1. *In situ* bacterial growth rates. Tritiated thymidine, used to measure increases of bacteria DNA (Moriarty, 1986) and tritiated leucine, a measure of bacterial protein produc-

tion (Kirchman et al., 1986) and hence growth, were used to assess the question of bacterial growth. Briefly, at bloom temperatures bacterial growth rates, measured by either technique, were generally very low within the bloom, when compared to other published data from other sites, and were virtually undetectable in waters below the euphotic zone.

2. Measurement of total microbial respiration. Pomeroy and Deibel (1986) found virtually undetectable respiration rates for samples incubated at *in situ* temperatures for up to three weeks. We extended this experiment to examine several higher temperatures, asking whether there was something intrinsic about these low temperatures that proscribed growth. When samples were incubated at $-1.5, 0, 3$, and $15°C$ for up to 10 days, we found that temperatures higher than $-1.5°C$ yielded measurable respiration rates, but the $-1.5°C$ incubation showed again almost no O_2 loss over the period.

3. Grazing effects on bacterial growth. A technique that has been used for several years to examine natural growth rates of bacteria in the absence of protozoan grazing involves diluting a water sample with filtered sample water (e.g., Geider, 1989). At high dilutions of a sample, for example 100 ml of sample + 900 ml of filtered water, grazers became uncoupled from prey because of the increased space between them. Under these circumstances, bacteria are free to grow—presumably at their natural rates—and accumulate. One can measure growth rate either by using a radioactive tracer or by measuring the accumulation of cells over time; we used both techniques. The prediction is that with increasing dilution the bacterial growth rate would *increase* up to the dilution that completely uncoupled them from the grazers. In fact, we either achieved a rate commensurate with the dilution factor, or, at lower dilution rates, e.g., 25 and 50% diluted, a reduced growth rate. Thus, there was no evidence that protozoan grazing had any effect on bacterial growth rates. Taken together, these experiments suggested that the microbial loop was not functional.

The question now became, "What ecological factor or factors prevented the onset of the microbial loop, and—given the large amount of pure cultures data that clearly demonstrate microbial growth at $O°C$ and less—why were *these* bacteria so inactive?"

One factor that differed greatly between *in situ* conditions for growth and pure culture conditions is that the latter contain grams of organics per liter, while *in situ* concentrations are only about one milligram per liter, not all of which is readily usable. So we set about examining not only the effect of temperature but also of substrate concentration on bacterial metabolism. Two types of experiments were conducted, one in which natural water samples were enriched with varying concentrations of organic matter—in this case protease peptone and yeast extract, which are common growth materials for bacterial cultures—and the other in which pure cultures isolated from the Conception Bay bloom and the Arctic Ocean were again enriched variably with the same organic material.

For surface waters during the bloom, with an ambient water temperature of $-1.5°C$ and no addition of substrate there was virtually no O_2 loss. In fact, until over 0.75 mg C/l was added, the response was minimal, while at the highest concentration, 75 mg C/l, respiration was rapid, and all O_2 was removed rapidly. At $3°C$ all samples showed some respiration, while at $15°C$ O_2 consumption was rapid at all nutrient concentrations. Thus, there appeared to be an interaction between substrate concentration and temperature with regard to microbial respiration, and this effect was much greater at low than at high temperature.

To examine this further, pure cultures of bacteria were isolated from the site and their growth response to combinations of nutrient and temperature measured. As an example, the results from one marine pseudomonad are described. Between 0 and $10°C$, generation times varied little with respect to differing nutrient concentrations, with Q_{10} of 2 or less. However, at $-1.5°C$ there was a clear nutrient effect. The high nutrient concentration medium (75 mgC/l) supported a growth rate almost equal to the $O°C$ sample. The 0.75

182

mgC/l generation time increased from 20h to 33h, while the 0.075 mg/l culture generation time rose from 20h to about 48h. These results and those from other cultures suggest that there is an especially strong nutrient/temperature interaction confined to temperatures below 0°C.

What are the implications from this work with regard to possible global warming? In over a dozen papers over the past few years, investigators have suggested that a major difference between temperate and high latitude spring blooms is that the latter contribute proportionately more of the primary production directly to the benthos than do the former (e.g., Grebmeier et al, 1988). All investigators see this contribution as an important factor in the maintenance of a large epi- and infauna, and large benthic fisheries. The bloom dynamics shortcircuit the large metabolic losses associated with the pelagic food web.

There are several probable reasons this happens. First, algal productivity appears to be *relatively* insensitive to temperature and thus is maintained at rapid rates at less than 0°C. Second, and this may be the critical feature, copepod development—these are the dominant crustacean zooplanktons that graze the algae—at $-1°C$ is on the order of 70 days. They cannot begin to graze the algae effectively until at least the third or fourth nauplii stage. Third, the microbial loop, as described, appears greatly reduced or inoperative.

Let us now consider how this pattern might change with a 2–4°C rise in surface water temperature. First, a rise in temperature would hasten the onset of water column stratification. The spring bloom dynamics would rapidly shift to the more classical model (see Fig. 1). If temperatures rise, the timing would shift to a progressively shorter spring bloom and thus to less productivity reaching the benthos.

Second, with a rise in temperature, we believe that the microbial loop would be induced rapidly, based on the temperature/nutrient data. This has two potential consequences. One would be to stimulate inorganic nutrient regeneration and thus maintain algae in the water column long enough for copepods to begin grazing. This would stimulate nutrient regeneration further and reduce the standing crop of algae, although not necessarily productivity. Again the result would be the loss of algae to the benthos.

There is another probable consequence of the onset of the microbial loop and that is to stimulate copepod nauplii growth by the development of protozoa and bacteria as prey, and thus hasten their development to actively grazing adults.

All of these effects, some straightforward and some more subtle, would act to maintain the algae in the water column for longer periods and increase the probability of their being grazed. The result is that progressively less primary production would reach the benthos. This would not probably decrease total gross primary production for the system; in fact, it might increase it. But the production available to large benthic organisms would decrease because of the induction of the full pelagic food web. We believe that this would result in a significant decrease in benthic fish and shellfish production—cod, lobster, snow crabs, and flounder, and perhaps an increase of pelagic species such as capelin, mackerel, herring, and squid.

ACKNOWLEDGMENTS

This work was supported by the National Science Foundation Grant No. OCE-8709809.

LITERATURE CITED

Geider, R. J. 1989. Use of radiolabeled tracers in dilution grazing experiments to estimate bacterial growth and loss rates. *Microb. Ecol.* 17:77–87.

Grebmeier, J. M., McRoy, C. P. & H. M. Feder. 1988. Pelagic-benthic coupling on the shelf of the northern Bering and Chukchi Seas. I. Food supply, source and benthic biomass. *Mar. Ecol. Prog. Ser.* 48:57–67.

Kirchman, D. L., Newell, S. Y. & R. E. Hodson. 1986. Incorporation versus biosynthesis of leucine: implications for measuring rates of protein synthesis and biomass production by bacteria in marine systems. *Mar. Ecol. Prog. Ser.* 32:47–59.

Moriarty, D. J. W. 1986. Measurement of bacterial growth rates of nucleic acid synthesis. *Adv. Microbiol. Ecol.* 9:245–292.

Pomeroy, L. R. & D. Deibel. 1986. Temperature regulation of bacterial activity during the spring bloom in Newfoundland coastal waters. *Science* 233:359–361.

Pomeroy, L. R. & W. J. Wiebe. 1988. Energetics of microbial foodwebs. *Hydrobiologia* 159:7–18.

Smith, R. E. H., Harrison, W. G., Irwin, B. & T. Platt. 1986. Metabolism and carbon exchange in microplankton of the Grand Banks (Newfoundland). *Mar. Ecol. Prog. Ser.* 34:171–183.

*EVOLUTION IN A RAPIDLY CHANGING ENVIRONMENT:
GLOBAL WARMING—SYMPOSIUM*

Late Cenozoic Sea Level Fluctuations and the Diversity and Species Composition of Insular Shallow Water Marine Faunas

Gustav Paulay

Abstract. In the Late Pliocene, sea level and climatic fluctuations intensified; how did this affect shallow water faunas? And how did these major fluctuations affect faunas once they had become routine in the Late Pleistocene? First I discuss the effects of fluctuations on coral and mollusk faunas world-wide for the Late Pliocene, then examine how regressions and transgressions affected species composition and distribution in the Central Pacific during the Late Pleistocene. Increased rates of extinction appear to have been largely restricted to times immediately following climatic deterioration in the Late Pliocene, but even then were restricted to certain areas of the world ocean. Although extinction was rare by the mid-Pleistocene, studies in the Central Pacific indicate that many species underwent small or large scale distributional changes as rising and falling sea levels rearranged the complex reef systems surrounding tropical islands.

INTRODUCTION

The frequent climatic and sea level fluctuations of the late Cenozoic drastically altered the nature and distribution of shallow water habitats and their biota. What effect did the onset of major fluctuations have on the persistence and distribution of marine organisms? What subsequent effects did recurring perturbations have? What effects will they have in the future?

I will discuss the effects that late Cenozoic climatic and sea level fluctuations had on marine invertebrates, especially corals and bivalve mollusks, during certain critical times. First I will briefly review how the major intensification of glaciations in the Late Pliocene affected coral and mollusk faunas worldwide. Then I will explore how recurrent glacio-eustatic regressions and transgressions affected the distribution and species composition of marine invertebrates in the Late Pleistocene, by which time large scale sea level cycles were routine. For this second part I will focus on tropical oceanic islands, especially in the Central Pacific Ocean. These islands provide well delineated, simple, replicate microcosms ideally suited for studies on the effects of sea level fluctuations.

The drastic cooling cycles that characterize the Quaternary began rather abruptly about 2.5 Ma, with the initiation of major northern hemisphere glaciations and a concomitant increase in the extent of Antarctic glaciers (Shackleton et al., 1984; Webb, 1990). Cyclic sea level fluctuations intensified at this time, giving rise eventually to the

Dr. Paulay is with the Department of Paleobiology NHB-121, Smithsonian Institution, Washington, DC 20560, USA.

100–150 m amplitude, 100,000 years frequency cycles that dominate the Late Pleistocene (Shackleton & Opdyke, 1976; Chappell & Shackleton, 1986). The extent of climatic cooling was unprecedented, but Late Pliocene-Quaternary sea level fluctuations, though greater in amplitude and perhaps frequency than previous fluctuations, were not novel; similar, if less drastic cyclic changes in sea level occurred throughout the Neogene (e.g., Hoddell et al., 1986; Haq et al., 1987; Greenlee & Moore, 1988; Bloementhal & deMenocal, 1989).

The onset of glaciations at 2.5 Ma was accompanied by increased extinction rates in some, but not all areas of the world ocean (Table 1). In particular, North Atlantic faunas suffered great losses in species diversity, while Indo-Pacific faunas appear to have survived relatively unscathed. Thus, while 32% of Pliocene hermatypic Scleractinia became extinct in the Caribbean by the Early Pleistocene (Frost, 1977), only 11% of Pliocene coral species in Papua New Guinea (Veron & Kelley, 1988) and 10–16% of putatively Pliocene corals on the south Polynesian island of Niue (Paulay, 1988) went extinct. Similarly, 32% of the mollusk genera went extinct in the Caribbean since the Pliocene, compared with 15% in the East Pacific (Vermeij & Petuch, 1986). While Pliocene bivalve species experienced massive extinction in both the Eastern and Western Atlantic, with losses of 46% and 80% respectively, they experienced rates of extinction no greater than background levels (as judged by Lyellian curves) in California and Japan, where 30% and 25% of species disappeared (Stanley & Campbell, 1981; Stanley, 1982; Raffi et al., 1985). Limited data from putative Pliocene deposits on Niue Island indicate similarly low, background levels of extinction among bivalve species (20% extinction) (Paulay, 1988, in prep.). New Zealand mollusks also show no increase in rates of extinction during the Pliocene, although they did experience increased extinction rates in the late Miocene and mid Pleistocene (Beu, 1990). Another area with potentially severe Late Pliocene extinction may be South Africa, where Kensley & Pether (1986) found that about half the mollusk species in Late Pliocene–Early Pleistocene deposits are not known living, an extinction rate comparable to that experienced in the North Atlantic.

Table 1. Pliocene extinction in corals and molluscs. Percent of species from Pliocene faunas not now known to be living. EA: North-east Atlantic including Mediterranean; WA: North-west Atlantic including Caribbean; EP: East Pacific; WP: various West Pacific localities. See text for references.

Region	EA	WA	EP	WP
Coral species		32		10–16; 11
Mollusk genera		32	15	
Bivalve species	46	80	30	20; 25

The restriction of massive extinctions to the North Atlantic and the pattern of extinction there have been interpreted as indications that climatic cooling, resulting from the massive glaciation of the northern hemisphere, was the main cause of extinction (Stanley & Campbell, 1981; Stanley 1984; Raffi et al., 1985); however changes in productivity may also have been important (Vermeij & Petuch, 1986; Vermeij, 1989). Analyses of sea surface temperatures at the time of maximum extent of glaciers, 18 Ka, show that while nearshore areas of the North Atlantic were considerably cooler in glacial times than today, large areas of the Indo-West Pacific were the same temperature as, or only slightly cooler than, today (CLIMAP 1976; Moore et al. 1980). Although these estimates are based on the last glacial cycle, the effects of initial Late Pliocene cooling were likely comparable due to similarities in the locations of continental ice sheets. The fact, then, that this initial cooling was probably accompanied by little or no change in water temperature in the Indo-West Pacific, and that the sea level fluctuations associated with it were, though larger in amplitude than previous ones, not novel to the region, may explain why extinction rates did not increase in the area during the Plio-Pleistocene.

By the mid Pleistocene, extinction had taken its toll in most areas, leaving only those species that could tolerate the conditions of frequently recurring glacial cycles; even in the Atlantic, extinction was now occurring only at background levels (Stanley, 1982). While the continuing glacial cycles did not eliminate these remaining species, they may have nonetheless greatly affected them: many species, especially those with wide distributions, likely experienced local extinction in parts of their ranges and large changes in distribution due to recurrent glacial and interglacial periods.

LATE PLEISTOCENE SEA LEVEL CHANGES

I will now examine how habitat alteration by Late Pleistocene sea level fluctuations altered species composition and distributions on islands in the Central Pacific. I will argue that species were periodically eliminated from these islands during certain sea stands due to poor habitat conditions, but recolonized them during intervening sea stands when conditions were favorable; thus a dynamic fauna of ever-changing species composition existed.

Most Central Pacific Islands are surrounded by well developed reefs of similar physiography, which can be divided into steep outer reef slopes seaward of the reef crest and variously developed inner reefs (defined to include fringing and barrier reefs, lagoons and moats) landward (or centrally on atolls). Inner reefs contain most (often >95%) of the shallow water habitat-area, are dominated by soft sediments of varied composition, vary greatly among islands in both the extent and types of habitats they offer, and are relatively protected from oceanic influence, having more stagnant, calm waters. Outer reef slopes are steep and narrow, dominated by hard substrata, have only limited and rather uniform soft sediments, are relatively similar among islands, and are exposed to the physical force and water of the open ocean (Paulay, 1990).

Changing sea levels would have had only moderate impact on outer reef slopes, whose biota, although displaced by the changes, would generally encounter similar habitats up and down the slope. Inner reefs, however, would be drastically altered. During times of glacial sea level minima inner reef habitats were entirely stranded, as all inner reefs are shallower than the maximum extent of glacial sea level falls. During trangressions inner reefs were inundated and underwent a series of changes in response to sea level rise, reef growth and sedimentation (see below).

REGRESSIONS

Local extinction during glacio-eustatic falls in sea level may be caused by qualitative or quantitative loss of habitat or climatic change. I have shown elsewhere (Paulay, 1988, 1989, 1990) that in the Central Pacific only the first of these processes was important in causing local extinction. Climatic cooling in the area was minimal and appears to have been unimportant in altering species distributions. While the stranding of inner reefs caused the extinction of numerous species restricted to inner reef habitats, it did not eliminate species that were not so restricted, in spite of the great reduction in habitat area and accompanying reduction in population size. Thus tectonically uplifted islands that have effectively lost their inner reef habitats, and serve as analog-models of islands during low sea stands, do not have a more impoverished fauna than expected from the loss of species restricted to inner reef habitats.

Thus, a quantitative decrease in habitat area appears to have been unimportant in causing local extinction. However, studies of the habitat preferences of species on recent reefs reveal that the loss of specific inner reef habitats, especially those dominated by soft substrata, caused numerous species to disappear from Central Pacific islands. The intensity of local extinction due to habitat loss varies among taxa, reflecting their habitat preferences. Corals, which generally exhibit their greatest diversity and abundance on oceanic fore reef habitats, have few (3–10%) species restricted to inner reefs and were little affected by sea level changes (Paulay, 1988). In contrast, 26–33% of bivalves, a group that thrives in lagoons and moats, are restricted to inner reef habitats and presumably eliminated during low sea stands. The effect of habitat preference is also evident when comparing soft sediment with hard bottom faunas. Soft substrata are disproportionately better represented and more diverse in composition in inner reefs than on outer reef slopes, while hard substrata are more evenly distributed. While about 47% of the bivalve species inhabiting soft sediments are restricted to inner reefs, only 28% of those inhabiting hard substrata are so restricted (Paulay, 1990).

Similar results to these based on the zonation of recent reefs are yielded by a comparison of pre- and post-uplift faunas on Niue Island. Uplift on Niue effectively lowered local sea level, transforming an atoll with a spacious lagoon into a limestone island analogous to Pacific atolls during low sea stands. Local extinction through uplift was about three times as high in bivalves as in corals, although soft bottom bivalves did not suffer significantly more extinction than hard bottom forms. The majority of bivalves that underwent local extinction are species restricted to inner reef habitats, indicating that loss of this habitat was the major agent of extinction during uplift (Paulay, 1988, in prep.).

The proportion of species restricted to inner reefs varies among islands according to the size and diversity of inner reef habitats, ranging in bivalves from 10% for islands where narrow, largely intertidal reef flats constitute the inner reef, to 38% for islands with wide, deep barrier reefs and lagoons (Paulay, 1990).

TRANSGRESSION, STILLSTAND, AND REEF GROWTH

Transgressions flooded inner reefs and set in motion a series of changes in these reef systems. I will discuss faunal response to changes in reef form during and following transgressions, using the Holocene sea level rise and recent atolls as models. Considerable changes in the species composition and community structure of an island take place during a transgression, as succession occurs from a stranded, dry inner reef to one widely open to oceanic waters, to one increasingly closed to them. Overall, as species with different habitat requirements alternately colonize and are extirpated from constantly changing inner reefs, the diversity of an island's biota first increases rapidly, then levels off, and in some cases decreases as lagoonar conditions deteriorate.

To facilitate discussion I divide reef development during and following the Holocene trangression into four successive, contiguous stages (see Hopley, 1982, Figs. 8.11 and 9.2 for similar examples from the Great Barrier Reef). Prior to trangression subaerial erosion and subsidence lowers exposed reefs. Since the last interglacial period (125 ka) Central Pacific reefs were lowered on average by 5–20 m (Paulay & McEdward, 1990). Inner reefs are then completely inundated by rapidly rising sea levels that outpace reef growth, creating reefs submerged by several meters at their shallowest (stage I) (cf. Marshall & Jacobson, 1985). As reef growth catches up with sea level around the reef's perimeter, a typical atoll (or barrier/fringing reef) develops (stage II). In most atolls exchange between the lagoon and the surrounding ocean, although substantially decreased since stage I when the reef was deeply submerged, remains through wide, shallow reef tracts encircling much of the atoll. Further exchange with the ocean may also occur through deep passages that penetrate through the reef from lagoon to fore reef on atolls with so called "open lagoons" (Salvat, 1967; Chevalier, 1981). The development of islets from storm-tossed rubble on top of the encircling reef tract, commencing in stage II, may, in atolls lacking deep, navigable passes, largely close communication between the lagoon and surrounding ocean (stage III). A well studied example of an atoll with such a nearly closed lagoon is Taiaro Atoll (Tuamotu Is; Poli & Salvat, 1976; see below). Finally, the lagoon, after losing all connections with the surrounding ocean, may become unsuitable for normal marine life (stage IV); this occurred in a few, mostly equatorial atolls with relatively shallow and often small lagoons. On some atolls, lagoons fill in (e.g., Enderbury in the Phoenix Islands; Tracey, 1980), others become brackish (e.g., Clipperton in the Eastern Pacific; Ehrhardt, 1976), fresh (e.g., Washington in the Line Islands; Wentworth, 1931), or hypersaline (e.g., Christmas in the Line Islands; Wentworth, 1931). In the South Pacific, stages III and IV of the Holocene transgression were reached during the relative stillstand of the sea which has occurred since ca. 6–5 ka, and were further aided by an approximately 1 m drop in sea level in the last 1000 years (Pirazzoli & Montaggioni, 1988). A sea level rise due to global warming would reverse these trends, reopening and deepening inner reef habitats.

Newly flooded inner reefs (stage I) are colonized by species that survived low sea stands on local outer reef slopes as well as by lagoonar species invading from more remote refugia, where they persisted through glacial low stands. The latter group of species which, as discussed above, includes numerous bivalves but few corals, adds substantially to the diversity of the island's fauna. On atolls, but not on high islands, this increase in diversity is partly counteracted by the loss of species restricted to intertidal habitats, which are eliminated as all reef tracts become submerged. The terrestrial faunas of atolls are also extirpated by rising sea levels; thus those seen today are only Holocene in age (cf. Thibault, 1974). Occasional exceptions to the demise of intertidal and terrestrial atoll biotas occur where stage I is bypassed due to the persistence of emergent reefs through the post-glacial transgression. Almost all such exceptions are the result of tectonic uplift (e.g., Anaa and Niau atolls, Tuamotu Islands; Veeh, 1966; Pirazzoli et al. 1988), though in low rainfall areas near the equator, where subaerial erosion rates are low, last interglacial reefs may occasionally remain emergent without evident uplift (e.g., Malden in the Line Islands; Tracey, 1980).

Only a few atolls are in the submerged, stage I condition today, and as very little is known about their faunas, there is little direct information on how lagoonar faunas and communities change as atolls develop from stage I to stage II. However, by considering the obvious habitat changes that must take place, one may speculate on the likely changes in faunal composition. Thus intertidal and supratidal forms can recolonize atolls as reefs reach the surface and new islets form. Species that require quiet, protected habitats are also potential arrivals at this time, as the inner reef evolves from one entirely exposed to the open ocean and thus flushed of fine sediments, to a more closed lagoon supporting habitats dominated by silts and muds (Renaud-Mornant et al., 1971; Adjas et al. 1990).

Some species are likely restricted to stage I reefs, as distinct faunas often characterize submerged banks and seamounts at various depths (B. A. Marshall, pers. comm., 1985). Such species would be locally extirpated as atolls reached stage II. Considerable apparent Holocene extinction in the Enewetak lagoonar molluscan fauna provides potential support for this hypothesis. Approximately 27% of the fossil mollusks (N=115) from Holocene sections of Enewetak lagoon cores are not among the 1116 species known to live on the atoll today (Kay & Johnson, 1987, their Table 1). Enewetak is an enormous atoll that has experienced little reef infilling or lagoon closure and is in a stable stage II state with large, navigable passes. Thus, the major environmental change during the Holocene that most likely accounts for this considerable Holocene extinction was the transition from a several meters deep submerged reef (Szabo et al., 1985) to the typical atoll form evident today.

Faunal change accompanying the closure of atoll lagoons from stage II is well documented, characterized by a sequential loss of species and the addition of few if any colonists. Corals in inner reef habitats are particularly sensitive to lagoon closure, as most species prefer areas exposed to open ocean (Paulay, 1988, 1989). Thus atolls with large passes support considerably more coral species than do those lacking passes, with the highest diversities generally occurring near pass entrances because certain species are limited in lagoons to areas exposed to oceanic waters (Chevalier, 1981). While corals are more sensitive to poor water circulation and disappear more rapidly from closing lagoons than do mollusks, they nonetheless generally persist on islands because most survive on the outer reef slope. Among mollusks, gastropods disappear more rapidly from closing lagoons than do bivalves, such that closed lagoons are often dominated by bivalves—in marked contrast to the usually low bivalve:gastropod ratios of oceanic islands (cf. Kay, 1967). Moreover, certain bivalves, most strikingly large, suspension-feeding and photosymbiotic species (especially *Tridacna maxima* (Röding), *Chama limbula* Lamarck, *Pinctada maculata* (Gould), *Arca ventricosa* Lamarck, and *Fragum fragum* (Linné)), greatly increase in abundance in and typify closed lagoons (Salvat, 1967; Richard, 1985).

The successional loss of species from lagoons undergoing closure is especially well documented for Taiaro Atoll (Tuamotu Island), where water exchange between lagoon and surrounding ocean occurs presently only during major storms (Poli & Salvat, 1976). Although only a single scleractinian (*Porites lobata* Dana) is known to live in the lagoon today, more than 15 species of corals lived there in the recent past (ca. 850 BP), when the lagoon was more open to surrounding waters (Chevalier, 1976). The present day lagoonar malacofauna of Taiaro includes 28 species, dominated (71%) by bivalves; another 17 lagoonar mollusk species (38% of total fauna) are known only from subfossil remains, where bivalves comprised slightly less than half of the species (Richard, 1976).

As the separation of lagoon from surrounding ocean becomes complete and permanent, the lagoon loses its macrofauna if it dries up or becomes fresh, brackish, or hypersaline. Lagoons that are entirely closed but retain waters of normal salinity may, however, hold a limited fauna; the entirely closed lagoon of Puka Puka Atoll (Tuamotu Island) has a large population of the pearl oyster *Pinctada maculata* (Salvat, 1967). Holocene fossils can indicate the last species to survive in dying lagoons. Limited fossil material available from five such islands shows an abundance of photosymbiotic, chemosymbiotic, and deposit-feeding species and a paucity of suspension feeders relative to other bivalve communities in the Central Pacific area (Tables 2 and 3). While on typical islands, about 65–75% of the bivalve fauna consists of suspension-feeding species (calculated from appendix in Paulay, 1990), only about 35% of the species are strictly suspension feeders in dying lagoons (Table 3). This rarity of suspension feeders is reminiscent of the relative decline of suspension-feeding bivalves after the Cretaceous/Tertiary extinction (Sheehan & Hansen, 1986) and may reflect a more rapid collapse of planktonic than benthic food webs.

Table 2. Feeding mode of Holocene bivalves from dry atoll lagoons. Number of species of suspension feeding, chemosymbiotic, photosymbiotic, and deposit feeding bivalves in Holocene deposits from atolls with dry or brackish (Clipperton) lagoons. Starbuck, Enderbury (Phoenix Is.), and Jarvis (Line Is.) based on collections in the Dept. of Paleobiology, U.S. National Museum of Natural History (USNM), Christmas (Line Is.) based on USNM collections and collections kindly made by Kay Kepler; Clipperton from Salvat & Ehrhardt (1970).

Island	Suspension	Chemosymb.	Photosymb.	Deposit
Starbuck	0	1	2	2
Enderbury	1	1	0	0
Jarvis	3	0	0	2
Christmas	2	2	2	5
Clipperton	1	2	0	0
Total	6	3	2	6

Table 3. Suspension feeding bivalves from Pacific islands. Proportion of predominantly suspension feeding species in bivalve faunas of selected Central Pacific islands. N: number of species. Data for Niue (E of Tonga), Rarotonga, Aitutaki (Cook Is.), and Society Is. include all habitats and are based on Paulay (in press). Data for Takapoto and Taiaro (Tuamotu Is.) are based on lagoonar habitats only and are from Richard (1976) and Richard et al. (1979). Dying lagoons are based on the summed fauna of atoll lagoons presented in Table 2.

Island	Suspension feeders	N
Niue	76%	75
Rarotonga	68%	53
Aitutaki	69%	61
Society Is.	75%	92
Takapoto lagoon	79%	34
Taiaro lagoon	67%	27
Dying lagoons	35%	17

GLOBAL WARMING

Greenhouse warming is expected to cause sea level to rise in the immediate future at a rate of several tens of cm per hundred years (e.g., Kerr, 1989; Buddemeier, this volume), a rate comparable to that of average reef growth through the Holocene trangression (6m/ty) (e.g., Davies et al., 1985). At such rates, some reefs may not be able to keep up with rising sea levels at all, while others will keep up in sections of the reef optimal for coral growth but will still become generally more open to water exchange. Increased exchange with the surrounding ocean will cause the reversal of the trends in habitat structure, species composition, and community structure elaborated on above. Islands with the most closed reef-lagoon systems (stages III & IV) are expected to be the most affected, as even a slight rise in sea level will reopen their lagoons to the surrounding ocean, creating stage II reefs. Because the areas of most rapid growth on a reef are the outer reef flat, reef crest, and shallow sections of the outer reef slope (Smith & Harrison, 1977), a shallow tract separating inner reef from fore reef (stage II) is expected to be maintained wherever reef growth keeps pace with rising sea levels. Where reefs cannot keep up with sea level, stage I reefs will result.

BIOGEOGRAPHIC EFFECTS: A DYNAMIC FAUNA

Changing sea levels continually alter the species compositions of shallow water insular faunas as different habitats are created and eliminated. When a change in sea level causes local changes in species composition to occur synchronously on similar islands over a wide area, the result can be major changes in species distribution. For example, during times of major glacial regression all inner reefs, and consequently all species restricted to inner reefs, are eliminated throughout the Central Pacific, where separation of outer and inner reefs is well defined. By the Late Pleistocene, recurring sea level fluctuations rarely cause global extinction; species eliminated from the Central Pacific survive elsewhere, most likely in the complex reef systems surrounding large Western Pacific islands, where inner reef-type habitats persist through low sea stands (Paulay, 1990). Thus the distributions of such species alternate between being widespread throughout the

Pacific during high sea stands, and being restricted to areas west of the Central Pacific islands during low stands. This implies that inner reef specialist species, if present in the Central Pacific today, dispersed there subsequent to the Holocene trangression.

Is there direct evidence for such large scale changes in species distributions? Range constrictions of species previously more widespread have been frequently documented throughout the Indo-West Pacific, as well as in other areas of the world ocean. For example, three species of tridacnid bivalves today restricted to the Western Pacific eastward of Sumatra occurred during previous interglacial high stands on Aldabra (Seychelles) and (two species) in East Africa (Taylor, 1978, Crame, 1986). Ostergaard (1928) and Kosuge (1969) document the local extinction of several mollusks from the Hawaiian islands since the late Pleistocene, while Paulay & Spencer (1988) show that several corals that lived on Henderson Island (Pitcairn Islands) in the Pleistocene are now restricted to island groups to the west. An especially striking example is the mussel *Septifer excisus* (Wiegmann), known from Pleistocene deposits on Henderson, Mangaia (Cook Islands), and Oahu (Hawaiian Islands), but today restricted to localities westward of and including Guam (Paulay, unpublished). Some of these range constrictions have been attributed to sea level changes (Taylor, 1978, Crame, 1986), while others appear to be due to other causes (Paulay & Spencer, 1988). More rigorous studies are needed to investigate the importance of sea level fluctuations in Pleistocene range constrictions.

In contrast to the wide-scale changes in habitat and species distributions caused by regressions, habitat alterations caused by transgression and reef growth are generally island specific (except for the initial flooding of inner reefs); they may result in major changes in the species compositions of island faunas, but extinction and colonization are on an island by island rather than a regional scale.

SUMMARY

There is no evidence to date that shallow water marine faunas in the Indo-Pacific suffered unusually high extinction rates during the intensification of glacial cooling and sea level fluctuations in the late Pliocene, although North Atlantic corals and mollusks suffered a major extinction at that time. Sea level fluctuations have, nonetheless, conspicuously altered the species compositions of insular Pacific marine faunas and caused marked shifts in the distributions of species.

Bivalves are very sensitive to sea level-caused habitat alterations, as many species are restricted to inner reef biotopes on Pacific islands. Such species disappear from the Central Pacific during regressions as inner reefs dry out, and colonize when lagoons reappear. Changes in inner reef development during trangressions precipitate a succession of bivalve communities: species diversity first increases as inner reefs become submerged, then decreases as they lose communication with surrounding waters.

In contrast, the survival of coral species on individual islands is little affected by sea level fluctuations because most species survive on outer reef slopes, habitats that persist through major changes of sea level. Inner reef coral communities change drastically however, as the degree of communication between inner reef and oceanic waters is continually altered by changes in sea level and by reef growth and sediment movement. Coral diversity decreases with increasing reef closure.

ACKNOWLEDGMENTS

I thank B. V. Holthuis for comments on the manuscript and Kay Kepler for collecting Holocene fossils in Christmas Island. This work was supported by the National Science Foundation (BSR-8514056 to G. Paulay and BSR-8700523 to A. J. Kohn), a Smithsonian

Institution Postdoctoral Fellowship, and grants from the Hawaiian Malacological Society and the Lerner-Grey Fund for Marine Research.

LITERATURE CITED

Adjas, A., Masse, J.-P. & L. F. Montaggioni. 1990. Fine-grained carbonates in nearly closed reef environments: Mataiva and Takapoto atolls, Central Pacific Ocean. *Sedimentary Geology* 67:115–132.

Beu, A. G. 1990. Molluscan generic diversity of New Zealand Neogene stages: extinction and biostratigraphic events. *Palaeogeography, Palaeoclimatology, Palaeoecology* 77:279–288.

Bloementhal, J. & P. deMenocal. 1989. Evidence for a change in the periodicity of tropical climate cycles at 2.4 Myr from whole-core magnetic susceptibility measurements. *Nature* 342:897–900.

Chappell, J. & N. J. Shackleton. 1986. Oxygen isotopes and sea level. *Nature* 324:137–140.

Chevalier, J.-P. 1976. Madréporaires actuels et fossiles du lagon de Taiaro. *Cahiers du Pacifique* 21:253–264.

Chevalier, J.-P. 1981. Reef Scleractinia of French Polynesia. *Proceedings of the Fourth International Coral Reef Symposium, Manila* 2:177–182.

CLIMAP Project Members. 1976. The surface of the ice age earth. *Science* 191:1131–1137.

Crame, J. A. 1986. Late Pleistocene molluscan assemblages from the coral reefs of the Kenya coast. *Coral Reefs* 4:183–196.

Davies, P. J., Marshall, J. F. & D. Hopley. 1985. Relationships between reef growth and sea level in the Great Barrier Reef. *Proceedings of the Fifth International Coral Reef Congress, Tahiti* 3:95–103.

Ehrhardt, J. P. 1976. Hydrobiologie du lagon de Clipperton. *Cahiers du Pacifique* 19:89–112.

Frost, S. H. 1977. Miocene to Holocene evolution of Caribbean province reef-building corals. *Proceedings of the Third International Coral Reef Symposium* 2:353–359.

Greenlee, S. M. & T. C. Moore. 1988. Recognition and interpretation of depositional sequences and calculation of sea-level changes from stratigraphic data—offshore New Jersey and Alabama Tertiary. Pp. 329–353. *In:* C. K. Wilgus, H. Posamentier, C. A. Ross & C. G. St. C. Kendall (eds.), *Sea-level Changes: An Integrated Approach.* Society of Economic Paleontologists and Mineralogists Special Publication: Tulsa, OK.

Haq, B. U., Hardenbol, J. & P. R. Vail. 1987. Chronology of fluctuating sea levels since the Triassic. *Science* 235:1156–1167.

Hodell, D. A., Elmstrom, K. M. & J. P. Kennett. 1986. Latest Miocene benthic ^{18}O changes, global ice volume, sea level and the 'Messinian salinity crisis'. *Nature* 320:411–414.

Hopley, D. 1982. *The Geomorphology of the Great Barrier Reef: Quaternary Development of Coral Reefs.* John Wiley & Sons: New York.

Kay, E. A. 1967. The composition and relationships of marine molluscan fauna of the Hawaiian Islands. *Venus* 25:94–104.

Kay, E. A. & S. Johnson. 1987. Mollusca of Enewetak Atoll. Pp. 105–146. In: D. M. Devaney, E. S. Reese, B. L. Burch & P. Helfrich (eds.), *The Natural History of Enewetak Atoll. Volume II Biogeography and Systematics.* U.S. Department of Energy.

Kensley, B. & J. Pether. 1986. Late Tertiary and Early Quaternary fossil Mollusca of the Honderklip area, Cape Province, South Africa. *Annals of the South African Museum* 97:141–225.

Kerr, R. A. 1989. Bringing down the sea-level rise. *Science* 246: 1563.

Kosuge, S. 1969. Fossil mollusks of Oahu, Hawaii Islands. *Bulletin of the National Science Museum, Tokyo* 12:783–794.

Marshall, J. F. & G. Jacobson. 1985. Holocene growth of a mid-Pacific atoll: Tarawa, Kiribati. *Coral Reefs* 4:11–17.

Moore, T. C. Jr., Burckle, L. H., Geitzenauer, K., Luz, B., Molina-Cruz, A., Robertson, J. H., Sachs, H., Sancetta, C., Thiede, J., Thompson, P. & C. Wenkam. 1980. The reconstruction of sea surface temperatures in the Pacific Ocean of 18,000 BP. *Marine Micropaleontology* 5:215–247.

Ostergaard, J. M. 1928. Fossil marine mollusks of Oahu. *Bernice P. Bishop Museum, Bulletin* 51:1–32, pl. 1–2.

Paulay, G. 1988. *Effects of Glacio-eustatic Sea Level Fluctuations and Local Tectonics on the Marine Fauna of Oceanic Islands.* unpublished Ph. D. dissertation. University of Washington: Seattle, WA.

Paulay, G. 1989. Effects of glacio-eustatic sea level fluctuations on insular coral faunas. *American Zoologist* 29:90A.

Paulay, G. 1990. Effects of Late Cenozoic sea-level fluctuations on the bivalve faunas of tropical oceanic islands. *Paleobiology.* 16:415–434.

Paulay, G. & L. R. McEdward. 1990. A simulation model of island reef morphology: the effects of sea level fluctuations, growth, subsidence and erosion. *Coral Reefs* 9:51–62.

Paulay, G. & T. Spencer. 1988. Geomorphology, palaeoenvironments and faunal turnover, Henderson Island, S.E. Polynesia. *Proceedings of the Sixth International Coral Reef Symposium, Australia* 3:461–466.

Pirazzoli, P. A., Koba, M., Montaggioni, L. F. & A. Person. 1988. Anaa (Tuamotu Islands, Central Pacific): an incipient rising atoll? *Marine Geology* 82:261–269.

Pirazzoli, P. A. & L. F. Montaggioni. 1988. Holocene sea-level changes in French Polynesia. *Palaeogeography, Palaeoclimatology, Palaeoecology* 68:153–175.

Poli, G. & B. Salvat. 1976. Etude géomorphologique et biologique de l'atoll fermé de Taiaro (Tuamotu, Polynésie Française). 3. Etude bionomique d'un lagon d'atoll totalement fermé: Taiaro. *Cahiers du Pacifique* 19:227–251.

Raffi, S., Stanley, S. M. & R. Marasti. 1985. Biogeographic patterns and Plio-Pleistocene extinction of Bivalvia in the Mediterranean and southern North Sea. *Paleobiology* 11(4):368–388.

Renaud-Mornant, J. C., Salvat, B. & C. Bossy. 1971. Macrobenthos and meiobenthos from the closed lagoon of a Polynesian atoll. Maturei Vavao (Tuamotu). *Biotropica* 3: 36–55.

Richard, G. 1976. Transport de matériaux et évolution récente de la faune malacologique lagunaire de Taiaro. *Cahiers du Pacifique* 21:265–282.

Richard, G. 1985. Richness of the great sessile bivalves in Takapoto lagoon. *Proceedings of the Fifth International Coral Reef Congress, Tahiti* 1: 368–371.

Salvat, B. 1967. Importance de la faune malacologique dans les atolls polynésiens. *Cahiers du Pacifique* 11:7–49.

Salvat, B. & J. P. Ehrhardt. 1970. Mollusques de l'ile Clipperton. *Bulletin du Museum National d'Histoire Naturelle 2e Serie* 42:223–231.

Shackleton, N. J., Backman, J. Zimmerman, H., Kent, D. V., Hall, M. A., Roberts, D. G., Schnitker, D. Baldauf, J. G., Desprairies, A., Homrighausen, R., Huddlestum, P., Keene, J., Kaltenback, A. J., Smith. 1984. Oxygen isotope calibration of the onset of ice-rafting and history of glaciation in the North Atlantic region. *Nature* 307:620–623.

Shackleton, N. J. et al. 1984. Oxygen isotope calibration of the onset of ice-rafting and history of glaciation in the North Atlantic region. Nature 307:620–623.

Shackleton, N. J. & N. D. Opdyke. 1976. Oxygen isotope and paleomagnetic stratigraphy of Pacific core V28-239, Late Pliocene to latest Pleistocene. *Geological Society of America Memoirs* 145:449–464.

Sheehan, P. M. & T. A. Hansen. 1986. Detritus feeding as a buffer to extinction at the end of the Cretaceous. *Geology* 14:868–870.

Smith, S. V. & J. T. Harrison. 1977. Calcium carbonate production of the *Mare-incognitum*, the upper windward slope, at Enewetak Atoll. *Science* 197:556–559.

Stanley, S. M. 1982. Glacial refrigeration and Neogene regional mass extinction of marine bivalves. Pp. 179–191. *In:* E. M. Gallitelli (ed.), *Paleontology, Essential of Historical Geology.* Mucchi: Modena, Italy.

Stanley, S. M. 1984. Temperature and biotic crises in the marine realm. *Geology* 12:205–208.

Stanley, S. M. & L. D. Campbell. 1981. Neogene mass extinction of Western Atlantic molluscs. *Nature* 293:457–459.

Szabo, B. J., Tracey, J. I., Jr. & E. R. Goter. 1985. Ages of subsurface stratigraphic intervals in the Quaternary of Enewetak Atoll, Marshall Islands. *Quaternary Research* 23:54–61.

Taylor, J. D. 1978. Faunal response to the instability of reef habitats: Pleistocene molluscan assemblages of Aldabra atoll. *Palaeontology* 21:1–30.

Thibault, J.-C. 1974. Les conséquences des variations du niveau de la mer sur l'avifaune terrestre des atolls polynésiens. *Comptes Rendue des Académie des Sciences Sèrie D* 278:2477–2479.

Tracey, J. I., Jr. 1980. Quaternary episodes of insular phosphatization in the Central Pacific. Pp. 247–261. *In:* R. P. Sheldon & W. C. Burnett (eds.), *Fertilizer Potential in Asia and the Pacific. Proceedings of the Fertilizer Raw Material Resources Workshop August 20–24, 1979 Honolulu, Hawaii.* East-West Resource Systems Institute, East-West Center: Honolulu, HI.

Veeh, H. H. 1966. Th^{230}/U^{238} and U^{234}/U^{238} ages of Pleistocene high sea-level stand. *Journal of Geophysical Research* 71:3379–3386.

Vermeij, G. J. 1989. Interoceanic differences in adaptation: effects of history and productivity. *Marine Ecology Progress Series* 57:293–305.

Vermeij, G. J. & E. J. Petuch. 1986. Differential extinction in tropical American molluscs: endemism, architecture, and the Panama land bridge. *Malacologia* 27(1):29–41.

Veron, J. E. N. & R. Kelley. 1988. Species stability in reef corals of Papua New Guinea and the Indo-Pacific. *Memoir of the Australian Palaeontologists* 6:1–69.

Webb, P.-N. 1990. The Cenozoic history of Antarctica and its global impact. *Antarctic Science* 2:3–21.

Wentworth, C. K. 1931. Geology of the Pacific equatorial islands. *Bernice P. Bishop Museum Occasional Papers* 15:1–25.

EVOLUTION IN A RAPIDLY CHANGING ENVIRONMENT: GLOBAL WARMING—SYMPOSIUM

Rapid Change in a Patchy Environment— The "World" from a Plant's-Eye-View

Roy Turkington

Abstract. *Trifolium repens* L. (white clover), a clonal perennial, has a wandering phenotype that can spread through a grassland by up to 25cm annually. The whole plant axis therefore samples a diversity of soil microenvironments and of neighbors, both spatially and temporally. Genetic diversity and phenotypic plasticity within any plant population will allow it to adjust to changing environmental conditions. Their relative importance will depend upon the scale (grain) of the patchiness and rapidity of its change as seen from the plant's-eye-view. Field and glasshouse studies have shown that, when environments are coarse-grained, populations of *T. repens* may adapt to specific soil types and to neighborhoods dominated by a number of different species of grass. In contrast, in fine-grained environments, *T. repens* has a remarkable degree of phenotypic plasticity and will respond to specific genotypes of grass neighbors, to changes in transmitted and reflected light quality (mostly red/far red ratio) imposed by different grasses, and to the strains of *Rhizobium leguminosarum* biovar *trifolii* and *Bacillus polymyxa* in the soil. In addition, the individual plant has the capacity to communicate between its parts when different parts are experiencing microenvironments of varying quality, such that stressed parts of the plant may be supported by more vigorously growing parts of the plant living in better conditions.

INTRODUCTION

Twentieth century man has developed an industry and a technology that allows him to modify the environment in major ways. However, he is not always in control of the modifications he induces. Some of these changes involve the biosphere itself and of current concern are the biological consequences of future climatic change. The major concerns largely focus on the possibility of species extinctions and reduction in biodiversity, but also of interest are both the ecological and evolutionary responses to such climatic change.

An immensely long history of environmental change has shaped the evolution of the species we see today and these continue to evolve in a world that is now changing at a perhaps unprecedented rate. The environment changes on a physiological time scale within the lifetime of the individual organism; on a time scale that covers a few generations during which population densities may fluctuate but genetic change is slight; and on a geological time scale during which enormous genetic change may take place. Here I will consider the first two time scales and ask how a single species, *Trifolium repens* L. (white clover), might respond to changing environments. Coming climatic changes provide scientists with an opportunity for hypothesis development and testing; an opportunity to

Dr. Turkington is with the Botany Department, University of British Columbia, Vancouver, B.C. V6T 1Z4, Canada.

test the biological consequences of large-scale, somewhat predictable and directional climatic changes.

Trifolium repens is a clonal perennial that has a wandering phenotype that can spread through a grassland by up to 25cm annually (Sackville Hamilton & Harper, 1989). The whole plant axis therefore samples a diversity of soil microenvironments (Snaydon, 1962) and of neighbors (Burdon, 1980), both spatially and temporally. MacArthur and Levins (1964) distinguished between environmental variation in which the period of fluctuation is less than the life span of the plant and that in which the period is greater. They called this fine-grained and coarse-grained variation respectively. Phenotypic plasticity and genetic diversity within any plant population will allow it to adjust to changing environmental conditions. Their relative importance will depend upon the scale (grain) of the patchiness and rapidity of its change as seen from the plant's-eye-view. If the individual *T. repens* spends its life wandering through a complex mosaic of patches of different species of grasses, then the *T. repens* will experience a fine-grained environment. In contrast, if the individual *T. repens* spends its entire life, and perhaps a number of generations, in a patch of pasture dominated by a single species of grass, then that *T. repens* will experience its environment in a coarse-grained manner.

A number of studies have been done in a 40-year old pasture in British Columbia and on plants sampled from that pasture. The pasture is composed of grass patches that range up to 1 m^2 but are more typically about 0.25 m^2—a fine-grained environment for *T. repens* (Evans & Turkington, 1988). In contrast, a 100-year old pasture in North Wales (Peters, 1980) has large areas dominated by single grasses; for example, a *Lolium perenne*-dominated site is about 150 m^2 (Turkington & Harper, 1979a)—a coarse-grained environment for *T. repens*. Thus, in the British Columbian pastures we would predict that the *T. repens* would respond physiologically (phenotypic plasticity) to the fine-grained mosaic environment. However, in the Welsh pasture we would predict both a plastic response to local environments and that genetically-based microevolutionary differences would arise among different individuals in the *T. repens* population growing for prolonged periods with different grass neighbors.

PHYSIOLOGICAL RESPONSES

Evans and Turkington (1988) collected 100 ramets (cuttings) of *Trifolium repens* from each of four grass neighborhoods in the British Columbian pasture. These were transplanted to a common garden, without competitors, and later scored for 12 morphological characters. For ten of the characters, a significant proportion of the variation between sampled ramets was accounted for by the identity of the neighboring grass species with which the *T. repens* had been growing in the pasture. These same clovers were grown for two more years in a common garden and were scored for morphological characters as before. In no case was a significant proportion of the variation in characters attributable to the previous grass neighbor. The clear conclusion is that the divergence patterns in *T. repens* morphology have a physiological rather than a genetic basis. A similar study in North Wales (Turkington, 1989b) also showed a physiological response to local grass neighborhoods and demonstrated that the grass-neighbor-specific differences could be retained in common garden conditions for at least one year. In a more detailed study using plants from the British Columbian pasture, Aarssen and Turkington (1985) showed that differences were apparent between subpopulations of *T. repens* that were growing with different genotypes of the grass *Lolium perenne*. These differences were still evident after one year, but similar studies by Evans et al. (1985) in Aberystwyth would suggest that the differences were probably phenotypic.

The question now arises as to the mechanism by which these morphological dif-

196

ferences are generated. Two primary arguments may be made—light quality and soil micro-organisms. Vegetation canopies reduce the quantity of light and change the quality of light *transmitted* through to lower-stratum vegetation. The quality of light *reflected* from different grass species also varies (Marcuvitz, unpublished). Solangaarachchi (1985) showed that the quality (measured as a red/far-red ratio) of light filtered through different grass canopies affects branching patterns in *T. repens*. A decrease in photosynthetically active radiation, and in red/far-red, both independently inhibit the development of axillary buds (Harvey, 1979), and may influence petiole length (Black, 1957), and internode length (Trautner & Gibson, 1966; Solangaarachchi & Harper, 1987; Thompson & Harper, 1989).

Chanway et al. (1989, 1990) attempted to determine the role of micro-organisms in these systems, specifically *Rhizobium leguminosarum* biovar *trifolii*. *Rhizobium* is a nitrogen-fixing bacteria that forms symbiotic associations with *T. repens*. Chanway et al. (1989, 1990) demonstrated that the effects described above only occur when the micro-organisms are present. Strangely, it is the specific association of *Rhizobium* and *L. perenne* genotypes that have the greatest impact on *T. repens* growth even though the *Rhizobium* is symbiotic with the *T. repens*. This significant grass × *Rhizobium* interaction was also shown in earlier studies by Thompson et al. (1989). These studies imply that some of the observed patterns in *T. repens* populations are generated by grasses directly, or indirectly, influencing soil micro-organisms' populations, which in turn influence the growth of *T. repens*. Specifically, *L. perenne* promotes the proliferation of those strains of *Rhizobium* that result in smaller *T. repens*, while the grass *Holcus lanatus* indirectly promotes larger *T. repens*.

Hartnett and Bazzaz (1983) have shown physiological integration between different inter-connected parts of the same clone of *Solidago canadensis*. Parts of a plant exploring areas with low resource levels will be supported by other physiologically connected ramets experiencing higher resource levels, i.e., the spatial environmental variation encountered by different ramets will be integrated by the clone (see Ashmun et al., 1982). It is not clear how far these arguments can be applied to *T. repens* for which studies by Solangaarachchi (1985), Newton (1986), Shivji and Turkington (1989) and Turkington et al. (in press) have shown evidence of independent responses by different parts of the intact clone growing in different environments. In the Turkington et al. (in press) study, a plant of *T. repens* was allowed to grow such that one half of the plant (all of the ramets on one side of the main axis) was growing with one species of grass while the other side of the same *T. repens* plant was growing with a different species of grass. Different *T. repens* ramets on the same plant responded independently to the microenvironment provided by their own neighboring grasses. In contrast, all apical meristems on the plant reacted similarly, showing a unified response and integrating the effects of the different microenvironments experienced by the whole clover plant.

MICROEVOLUTIONARY CHANGE

Turkington and Harper (1979b) collected ramets of *T. repens* from within patches dominated by four different grasses in the Welsh pasture. The *T. repens* populations were grown in all possible combinations with the four species of grass in greenhouse flats containing a standard potting compost. Each *T. repens* tended to grow best with the grass from which it had originally been sampled, i.e., a principal diagonal effect. It is evident that the different grasses impose different constraints on the growth of *T. repens* and that the *T. repens* populations had differentiated into subpopulations defined by the identity of the grass neighbor. In addition to the greenhouse studies the *T. repens* from the four grass patches were also replanted back into patches of each of the four grasses in the pasture. Again each *T. repens* grew best in its natural grass patch. Given other studies on *T. repens* from this pasture (Cahn & Harper, 1976; Burdon, 1980; Gliddon & Trathan, 1985), it seems likely that the observed patterns have a genetic basis and represent true microevolutionary change. However, formal genetic analyses were not done. Later, a more rigorous field study was carried out to test the genetic argument, again without formal genetic analysis. Turkington (1989a), following a procedure proposed by Connell (1980), observed the growth of two presumed competitors, *T. repens* and *Lolium perenne,* in areas where they coexist (called the sympatric site) and in areas where the *L. perenne* was naturally absent but the *T. repens* was present (called the allopatric sites). Various treatments using transplants, replants, species removals, and controls, were carried out (Fig. 1). If we are to conclude that microevolution has occurred rather than merely a plastic or conditioning response, two conditions must be met. First, with its natural competitor, *L. perenne,* removed, the growth of the sympatric individuals of *T. repens* should *not* increase relative to growth with *L. perenne* still present (i.e., Figs. 1 & 2, treatment 5 not different from 4). Second, sympatric individuals of *T. repens* should show no significant difference in growth when transplanted to the allopatric site compared to being replanted within the sympatric site (i.e., Figs. 1 & 2, treatment 6 should not be different from 4); but if there is a difference, then growth in the allopatric site should be less than that of the natural allopatric population (i.e., Fig. 2, treatment 6 < 3). The assumption underlying these predictions is that if a genetic change causing divergence in the *T. repens* population had occurred in sympatry, then, on removal of the competing species, the sympatric population should *not* quickly expand or shift its niche position back to what it presumably had been before evolution. The allopatric *T. repens* population is assumed to have the same niche position that the sympatric population used to have before evolution. In this study each of these conditions were met, suggesting that the observed divergence in the *T. repens* population, in response to different grass neighbors, probably has a genetic basis.

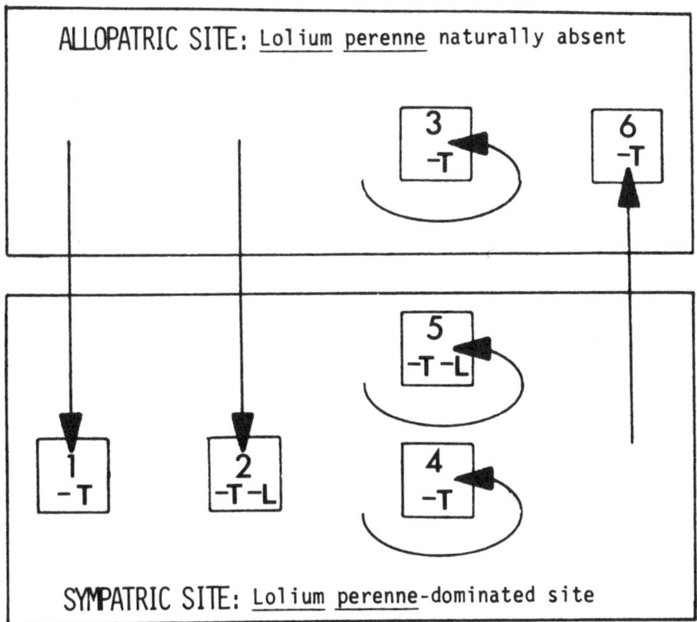

Figure 1. Outline of experimental design to test for a genetic basis to observed patterns of population divergence in *Trifolium repens*. Treatment numbers are in boxes. All treatments in which *T. repens* or *Lolium perenne* was removed (–T, –L, respectively) had the indigenous populations of these removed prior to the introduction of experimental *T. repens*. In all treatments, *T. repens* was collected from the base of the arrow and transplanted, or replanted, at the head of the arrow into the various treatments.

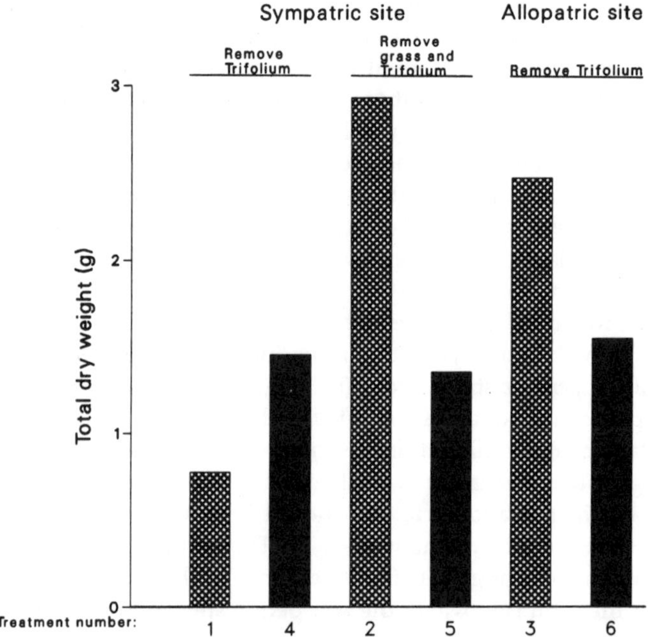

Figure 2. Total plot dry weight of *Trifolium repens* collected from an allopatric site (▩) and a sympatric site (■), and then transplanted or replanted into plots with different treatments (see Fig. 1 for details of treatments).

Also, in the Welsh pasture Gliddon and Trathan (1985) estimated the genetic variation within seven subpopulations of *T. repens* and *L. perenne*. All 15 polymorphic loci studied in *T. repens* and 21 out of 23 in *L. perenne* showed significant heterogeneity of gene frequency, indicating that there was genetic differentiation between the samples.

A FINAL COMMENT

I initially predicted that in a fine-grained environment we would expect *T. repens* to show physiological responses to environmental change, and in a coarse-grained environment the responses would be microevolutionary. All of the studies I have cited are consistent with these predictions. Perhaps the strongest support for the hypotheses is indirect. Mehrhoff (1989) tested for genetic differentiation within populations of *T. repens* in the British Columbian pasture—the fine-grained environment—and could find no evidence of such patterns. He found extensive population variation and any "home-site" advantage that he did find was attributable to short-term phenotypic carry-over effects rather than genetic changes.

CONCLUSION

A number of experimental studies on grassland species suggest that in certain circumstances rapid microevolutionary change is possible. The classical Park Grass Experiment at Rothamsted, England, was designed to study the effect of various fertilizer and lime treatments on hay production. This system was used by Snaydon and various coworkers for ecological genetics studies. The evidence from these experiments suggest that population differentiation in the grass *Anthoxanthum odoratum* had taken place within 120 years and that some change was detectable after only 7 years (see Davies & Snaydon, 1976). Other studies (see Briggs & Walters, 1984, page 314; Evans & Turkington, 1988) show that in many grassland plants, including *Trifolium repens*, detectable microevolutionary change is evident within a decade.

There is an enormous variety of plant adaptations to environmental change, yet there are some environmental variations to which no simple physiological response is adequate (see "Response of pasture plants to temperature" McWilliam, 1978). When the tolerance limits of individuals are exceeded they will be unable to continue growth and reproduction; the population size may fluctuate or may become extinct. This paper has not considered the extinction option because a 3–4°C rise in global temperatures is not likely to have a major impact on such a widespread and versatile species such as *T. repens*. However, such a temperature change may have quite critical effects on habitat distribution and could then result in local extinction of some *T. repens* populations.

LITERATURE CITED

Aarssen, L. W. & R. Turkington. 1985. Biotic specialization between neighbouring genotypes in *Lolium perenne* and *Trifolium repens* from a permanent pasture. *Journal of Ecology* 73:605–614.

Ashmun, J. W., Thomas, R. J. & L. F. Pitelka. 1982. Translocation of photoassimilates between sister ramets in two rhizomatous forest herbs. *Annals of Botany* 49:403–415.

Black, J. N. 1957. The influence of varying light intensity on the growth of herbage plants. *Herbage Abstracts* 27:89–98.

Briggs, D. & S. M. Walters. 1984. *Plant Variation and Evolution.* Cambridge University Press: Cambridge. 412 pp.

Burdon, J. J. 1980. Intra-specific diversity in a natural population of *Trifolium repens*. *Journal of Ecology* 68:717–735.

Cahn, M. G. & J. L. Harper. 1976. The biology of the leaf mark polymorphism in *Trifolium repens* L. 1. Distribution of phenotypes at a local scale. *Heredity* 37:309–325.

Chanway, C. P., Holl, F. B. & R. Turkington. 1989. Effect of *Rhizobium leguminosarum* biovar *trifolii* genotype on specificity between *Trifolium repens* and *Lolium 'perenne*. *Journal of Ecology* 77:1150–1160.

Chanway, C. P., Holl, F. B. & R. Turkington. 1990. Specificity of association between *Bacillus* isolates and genotypes of *Lolium perenne* and *Trifolium repens* from a grass-legume pasture. *Canadian Journal of Botany* 68:1126–1130.

Connell, J. H. 1980. Diversity and the coevolution of competitors, or the ghost of competition past. *Oikos* 35:131–138.

Davies, M. S. & R. W. Snaydon. 1976. Rapid population differentiation in a mosaic environment. III. Measures of selection pressures. *Heredity* 36:59–66.

Evans, D. R., Hill, J., Williams, T. A. & I. Rhodes. 1985. Effects of coexistence in the performance of white clover-perennial ryegrass mixtures. *Oecologia* 66:536–539.

Evans, R. C. & R. Turkington. 1988. Morphological variation in a biotically patchy environment: evidence from a pasture population of *Trifolium repens*. *New Phytologist* 109:369–376.

Gliddon, C. & P. Trathan. 1985. Interactions between white clover and perennial ryegrass in an old permanent pasture. Pp. 161–169. *In:* J. Haeck & J. W. Woldendorp (eds.), *Structure and Functioning of Plant Populations, Vol. 2.* North Holland Publishing Co.: Amsterdam.

Hartnett, D. C. & F. A. Bazzaz. 1983. Physiological integration among ramets in *Solidago canadensis*. *Ecology* 64:779–788.

Harvey, H. J. 1979. *The Regulation of Vegetative Reproduction.* Ph.D. thesis. University of Wales.

MacArthur, R. H. & R. Levins. 1964. Competition, habitat selection and character displacement in a patchy environment. *Proceedings National Academy of Sciences USA* 51:1207–1210.

McWilliam, J. R. 1978. Response of pasture plants to temperature. Pp. 17–34. *In:* J. R. Wilson (ed.), *Plant Relations in Pastures.* CSIRO: East Melbourne.

Mehrhoff, L. A. 1989. *The Evolutionary Consequences of Interactions Between Plants in Permanent Pastures.* Ph.D. thesis. The University of British Columbia, Vancouver.

Newton, P. C. D. 1986. *The Establishment, Growth and Fate of White Clover Plants with Special Reference to the Physiology of Stolon Growth.* Ph.D. thesis. University of Wales.

Peters, B. 1980. *The Demography of Leaves in a Permanent Pasture.* Ph.D. thesis. University of Wales.

Sackville Hamilton, N. R. & J. L. Harper. 1989. The dynamics of *Trifolium repens* in a permanent pasture. 1. The population dynamics and nodes per shoot axis. *Proceedings Royal Society London, Series B* 237:133–173.

Shivji, A. & R. Turkington. 1989. The influence of *Rhizobium trifolii* on growth characteristics of *Trifolium repens:* integration of local environments by clonal genets of *T. repens. Canadian Journal of Botany* 61:1080–1084.

Snaydon, R. W. 1962. Microdistribution of *Trifolium repens* and its relation to soil factors. *Journal of Ecology* 50:439–447.

Solangaarachchi, S. M. 1985. *The Nature and Control of Branching Patterns in White Clover* (*Trifolium repens* L.). Ph.D. thesis. University of Wales.

Solangaarachchi, S. M. & J. L. Harper. 1987. The effect of canopy filtered light on the growth of white clover *Trifolium repens. Oecologia* 75:372–376.

Thompson, L. & J. L. Harper. 1988. The effect of grasses on the quality of transmitted radiation and its influence on the growth of white clover *Trifolium repens. Oecologia* 75:343–347.

Thompson, J. D., Turkington, R. & F. B. Holl. 1989. The influence of *Rhizobium leguminosarum* biovar *trifolii* on the growth and neighbour relationships of *Trifolium repens* and three grasses. *Canadian Journal of Botany* 68:296–303.

Trautner, J. L. & P. B. Gibson 1966. Fate of white clover axillary buds at five intensities of sunlight. *Agronomy Journal* 58:557–559.

Turkington, R. 1989a. The growth, distribution and neighbour relationships of *Trifolium repens* in a permanent pasture. V. The coevolution of competitors. *Journal of Ecology* 77:717–733.

Turkington, R. 1989b. The growth, distribution and neighbour relationships of *Trifolium repens* in a permanent pasture. VI. Conditioning effects by neighbours. *Journal of Ecology* 77:734–746.

Turkington, R. & J. L. Harper. 1979a. The growth, distribution and neighbour relationships of *Trifolium repens* in a permanent pasture. I. Ordination, pattern and contact. *Journal of Ecology* 67:201–218.

Turkington, R. & J. L. Harper. 1979b. The growth, distribution and neighbour relationships of *Trifolium repens* in a permanent pasture. IV. Fine scale biotic differentiation. *Journal of Ecology* 67:245–254.

Turkington, R., Sackville Hamilton, N. R. & C. Gliddon. 1991. Within-population variation in localized and integrated responses of *Trifolium repens* to biotically patchy environments. *Oecologia* 86:183–192.

Biodiversity, Conservation, and Global Change

John R. Ware

Abstract. Discussion Group 16 was organized to provide graduate students a forum for discussing some of the salient environmental issues facing biologists today. In particular, focus was placed on the following questions. Is biodiversity important; if so, why and how can it be preserved? Given the potential impacts of global climate change on biodiversity, what should the role of the scientist be and what can graduate students do to contribute to solving current environmental problems? These are certainly not new issues; nevertheless, the ICSEB IV organizing committee felt it important to provide an opportunity for graduate student discussion and opinion. This paper presents some of the highlights of that discussion.

INTRODUCTION

The Discussion Group on Biodiversity, Conservation, and Global Change was organized particularly to elicit the views and opinions of graduate students attending ICSEB IV on some current environmental issues of controversy and concern to biologists. Approximately 25 graduate students, together with five senior participants, assembled on the evening of July 3rd in an informal atmosphere to discuss issues related to the following questions:

1. Is biodiversity important and, if so, why? How can biodiversity be preserved? How can laymen, who provide the funding for science, be convinced of the importance of biodiversity and its conservation?
2. The traditional role of the scientist is as an observer and reporter of facts. Given the potential effects of global climate change, should scientists act in a new role, that is, in an "unscientific manner?"

These issues are hardly new and have been the subject of considerable discussion in the recent literature (for example, Naess, 1986, McNeely, 1988, Nations, 1988, Norton, 1988, Randall, 1988, Deshmukh, 1989, Lovejoy, 1989, Noss, 1989, Pitelka & Raynal, 1989, Schlesinger, 1989, Crouch, 1990). Nevertheless, the organizers of ICSEB IV felt that it was important to provide a forum for discussion on these themes and to obtain the opinion of the graduate students who provide the basis for future biological research.

The discussion that followed the introduction to these topics was both lively and informative, with agreements on basic issues and disagreements on particulars the rule. It was not possible, nor did I attempt, to connect opinions, recommendations, or suggestions with names. The anonymity of the discussers is therefore preserved. In addition, the

Dr. Ware is with Atlantic Research Corp., 1375 Piccard Dr., Rockville, MD, and the Department of Zoology, University of Maryland, College Park, MD 20742, USA.

ebb and flow of the interchange was not always completely coherent and correlated. Therefore, I have taken the liberty of coalescing and rearranging topics in a more or less logical order rather than presenting a strictly chronological report. I apologize to any of the participants whose opinions I have omitted or not reported with precision and accuracy.

DISCUSSION

There are approximately 1.5 million named species with perhaps as many as 30 million more remaining to be named (May, 1988, Wilson, 1988). The estimate of the number of unnamed species is driven almost totally from estimates of the number of insect species in tropical forests. Because these tropical forests are currently the subject of mass destruction, species are becoming extinct at an unprecedented rate and at a pace perhaps a million times greater than the speciation rate (May, 1988). In addition, global climate changes, driven almost entirely by anthropogenic greenhouse gases, may result in global temperature increases of between 2 and 6° over the next century (Schneider, 1989) (although estimates of temperature increases appear to be steadily decreasing [Aldhous, 1990]), further amplifying survival pressure on the remaining species.

The basic pressure on the environment comes from the sheer weight of numbers of the human population. In the ultimate sense, the global loss of biodiversity will be caused by human pollution and increases in population leading to transformation of undisturbed areas into farm land. To these might be added the trend towards homogeneity of world agriculture, which will lead to a further reduction in supporting species such as pollinators, seed dispersers, nitrogen-fixing bacteria, and so forth (Norgaard 1988).

The question put to the discussion group was, bluntly: "So what? Why should we, as scientists or soon to be scientists, be concerned about the loss of a few species?" The answers given to this question spanned a considerable range and included:

- A destruction or attenuation in the size of the gene pool reduces the potential of species to evolve to meet changing global conditions.

- There may be economic and pragmatic value to species that are yet to be discovered, e.g., medicinal and food plants.

- Because of the interactions between species, the stability of whole ecosystems may be affected by the eradication of even a few keystone species.

- The process of extinction is irreversible and the consequences are unknown, we cannot distinguish the "important" species from the "unimportant" ones.

- Biodiversity is the stuff of our science and all species are of importance to those who study and gain knowledge from them.

While the above arguments have at least the appearance of science about them, it was suggested that they are all essentially anthropocentric; that we are all continuing in our self-centered ways. Additionally, several weaknesses in these arguments may be amplified by opponents of ecological movements. For example, the economic argument for maintaining current levels of biodiversity is difficult to defend. Fewer than two dozen plants and animals provide 80% of the world's food (Nations, 1988) and we derive medicines from only about 120 of the estimated 250,000 to 750,000 plant species (Farnsworth, 1988).

It was proposed that the real reason that we should preserve species is that we have no more right to survive than they do; that each species has an inherent worth, independent of any value we might put on them. It was pointed out that the parallels between the tenets of deep ecology and several Eastern religions (Nations, 1988) which hold that all living things are sacred, may be considered commendable by some and frivolous by others. But whatever view one takes of the ethical argument, it can hardly be considered "scientific."

There continued to be some discussion on these themes, with one side maintaining that all possible justifications for preserving biodiversity should be used while the other side expressed concern that the use of flimsy arguments weakened the case. Despite continued disagreement over some minor issues, the closest to a consensus reached among the students was that the ethical argument was the most appropriate and would be, perhaps, the most effective and compelling. All other arguments are centered on how humans can survive and reproduce even more rapidly when uncontrolled human population growth is the main problem. However, as one of the visiting "grey beards" so aptly observed, it was apparent that not one of the students present really appeared to believe, "cross your heart and hope to die," that preserving the integrity of the global ecosystem is essential or that maintaining biodiversity is critical.

The current of the discussion then changed in the direction of the "whys" biodiversity was being threatened in so many locations. Human population pressure was accepted as the ultimate cause; but much blame was assigned to the more developed countries. Even though the population density is high in the developing countries, the influence per capita of the more developed countries is much higher. Further, the developed countries have exported technology in the form of equipment, chemicals, and methodologies that have provided the means by which the developing countries have been able to destroy their own heritage. It was felt that people in such countries and those with older cultures had learned to live in harmony with their environment and therefore have had less negative impacts on ecosystems. A student from Mexico stated that the technology that Mexico has received from the developed countries is bulldozers—"we have learned to cut down our trees and form a monoculture and the people are losing their traditions."

Judeo-Christian doctrines were also blamed for some aspects of the current problems. The biblical dictum to "go forth and multiply" may be interpreted by a few as "the world is ours to exploit."

With regard to solutions, opinions become even more diverse. Generally, the problem areas can be divided into those associated with the less and those associated with the more, developed countries. Within the developing countries, several approaches were suggested. First, the people must be educated to realize the importance of biodiversity and their local ecosystems to both themselves and the rest of the world. One method of providing this education is to show people the consequences of their intended actions that have occurred in other areas. Second, we must provide concrete solutions to the real problems these people face; show them how to obtain more benefits from their resources with less impact. Third, we must convince them of the long term deleterious impact of environmental destruction and the benefits of preservation and conservation. Debt-for-nature swaps were mentioned but criticized for addressing too small a portion of the problem.

Certainly no one is destroying the environment for the sole purpose of being perverse; economics actuate most of the devastation. Because the situation is driven, in large part, by individual gain with the cost being borne by society as a whole, economic solutions were suggested in the form of "carrot and stick." Not only should there be punishments for destruction, but there should be rewards for preservation.

The irony of a group of western-civilization bred and raised students discussing and proposing solutions to the problems of developing countries was not lost on the assembly. It is difficult not to be self-centered in these discussions and to treat the problems as if they belonged to someone else who must bear the burden of the solution. The possibility of gathering an international group of students was mentioned.

Within the more developed countries, the problem was perceived as being one of obtaining grass-roots support for environmental issues; but to do what is not exactly clear. The sentiment with regard to scientific solutions was more or less that "technology has gotten us into the present predicament and it is unlikely that technology will rectify the situation." Science also has a real credibility problem with the public in general. It was pointed out that scientists have a long history of crying "WOLF!" and that, somehow or another, solutions are always found. Two examples were presented. Scientists in the late 19th century predicted that the extinction of the sperm whale, which appeared imminent, would lead to the collapse of a civilization whose mechanisms were built on the lubrication sperm whale oil provided. Again, scientists in the early 20th century were said to have predicted that, by the year 2000, horse manure would be piled to the rooftops in New York city. Neither of these dire predictions have come true. More recently, it was stated, the fiasco of the Hubble telescope is large in the public eye as an example of the inabilities of scientists to cope with large-scale technical problems.

However, there appears to be a paradox with regard to the attitude of the general public towards scientists as a whole. First there is the "cry wolf" syndrome mentioned above. At the same time, the non-scientific community appears to be of the opinion that science will develop a new technology, as it always has in the past, that will make the current problems resemble the sperm whale oil and horse manure plights of the past century.

Also relevant is the fact that predictions of global climate change continue to be modified in less dramatic directions and the models used for the predictions are "riddled with uncertainty" (Lindzen, 1990). Where we once had predictions of a 2 meter sea level rise by 2050, the predictions are now less than 0.5 meter. Some models even predict a reduction in sea level. Similarly, improvements in general circulation models have resulted in a lowering of predictions of global temperature increase from 4–6 C° by the middle of the 21st century to a rise of 0.3 C°/decade or less (Aldhous, 1990).

What role should scientists play in attempting to resolve environmental issues? It was noted that the British Ecological Society has taken the view that it should have nothing to do with conservation—that conservation and science don't mix. Nevertheless, the consensus among the discussants was that scientists should play a more active role than simply gathering the information that others will use to build the cases for preservation and conservation. However, exactly what that role should be was not evident and the discussion became quite lively at this point. Suggestions included the formation of a lobbying group and that scientists could take a more dynamic and resourceful role in expressing their views. A considerable difficulty related to these views may be that scientists will give up their supposedly unbiased positions, so that by lobbying as a group we may destroy our appearance of objectivity. A repeated theme was to the effect that "legislation is needed"; the contents of such legislation was unspecified.

It was argued that graduate students have little influence. The ability to persuade and convince others can only by attained by having demonstrated the ability to do good science and one must first establish a reputation before one can act effectively. One student explained that he must go through two years of Master's work, four more years for a Ph. D., and then another ten years to establish a reputation before persons in authority will consider him worth listening to. By that time it may be too late to make a meaningful contribution to solving current problems.

One excellent suggestion is that we can act indirectly through education. Many of the graduate students are teaching or assisting now and will go into a career that involves teaching. Teaching introductory biology courses will bring them into contact with students who will ultimately enter many fields: engineering, law, politics, and so on. By demonstrating to these people our enthusiasm and respect for the environment, some will rub off on them.

A suggestion was made by another of the "grey beards" present that each student should examine his own research work to determine whether it is relevant to the problems that face the world now or whether it is simply an exercise one goes through to obtain a degree. An interesting reaction to this suggestion was that, apparently in at least some universities, there is a bias against conservation-related studies and conservation biology is looked down on as not being sufficiently theoretical to merit degree work.

It was also asserted that we don't prove things in science, we disprove them. We can and do accumulate masses of facts; but when we are asked to prove the need for preserving species that task is beyond us. We want to say that we cannot afford to lose species but when we look around us we see systems that appear to be working very well with much reduced diversity. That is why, after making all the usual arguments about why we must preserve biological diversity, we tend to fall back on the ethical argument. It comes down, in the end, to the fact that we have no right to destroy any species.

CONCLUSION

Despite an already long day, many of the participants were prepared to continue all night, or at least as long as the beer held out. But, as all good things must, eventually it came time to draw the discussion to a close. As always, it is the editor's privilege (and duty) to insert a few concluding remarks. In the final analysis, it does not appear that a single convincing argument had been put forth that would persuade a totally objective observer that preservation of biodiversity was a critical issue affecting the long-term survival of the human race. It may be that no such demonstration is possible.

One view of the role of the scientist is that he must be a strictly objective observer and reporter. This view has been well stated by Dr. Robert Buddemeier and I cannot improve on his thoughts: there are good reasons for scientists to study various environments and species and the knowledge gained may be useful to efforts aimed at mitigating human impacts and preserving the natural environment. But science itself neither demands nor justifies such mitigation and preservation. In the long run we do everyone a disservice if we, as scientists, yield to the temptation to portray our advocacy as having an objective basis (Buddemeier, pers. comm.).

So we find ourselves on the horns of a dilemma. If we act as proponents of environmental preservation we yield our scientific objectivity. But if we, who presumably should have a better basis for making decisions with regard to the environment than the general public, do not cry out, then who will? I doubt that there was a single graduate student whose response to that question would not be that we, the scientific community, must act. Preservation of scientific objectivity may be a little like coming home and finding your house on fire. While the objective scientist would make observations and take data, most of us would have a compelling urge to shout "FIRE!"

After the recent fanfare and publicity over the celebration of Earth Day 1990 it is easy to forget that we have been here before. The first Earth Day was the 22nd of April 1970 and Wickstrom (1971), writing one year later, noted that the ecological reaction had reached its peak and was already running downhill to met the common fate of all causes in America. What we have accomplished, when all is said and done, is to have "transported us from an unaware, polluting public to an aware, polluting public" (Wickstrom, 1971).

ACKNOWLEDGMENTS

I would like to thank the organizers of ICSEB IV for providing an opportunity for the graduate students to speak up and have their opinions heard and published. I am particularly grateful to Dr. Marjorie Reaka-Kudla, Chair of the Program Committee, for suggesting the topic and to Dr. James Reveal, Co-President of ICSEB IV, for his interest and contributions. And to all the graduate students who contributed and whose opinions I have done my best to convey, my special thanks and my deep apology for any omissions, oversights, or errors.

LITERATURE CITED

Aldhous, P. 1990. Modest response to climate change threat. *Nature* 345:373.

Crouch, M. L. 1990. Debating the responsibilities of plant scientists in the decade of the environment. *The Plant Cell* 2:275–277.

Deshmukh, I. 1989. On the limited role of biologists in biological conservation. *Cons. Biol.* 3:321.

Farnsworth, N. R. 1988. Screening plants for new medicines. Pp. 83–97. *In:* E. O. Wilson & F. M. Peter (eds.), *Biodiversity*. National Academy Press: Washington, D.C.

Lindzen, R. S. 1990. Some remarks on global warming. *Environ. Sci. Technol.* 24:424–426.

Lovejoy, T. 1989. The obligations of a biologist. *Cons. Biol.* 3:329–330.

May, R. M. 1988. How many species are there on earth? *Science* 241:1441–1449.

McNeely, J. A. 1988. *Economics and Biological Diversity: Developing and Using Economic Incentives to Conserve Biological Resources*. International Union for Conservation of Nature and Natural Resources: Gland, Switzerland. 236 pp.

Naess, A. 1986. Intrinsic value: will the defenders of nature please rise? Pp. 504–515. *In:* M. E. Soulé, (ed.), *Conservation Biology: The Science of Scarcity and Diversity*. Sinauer Associates, Inc.: Sunderland, MA.

Nations, J. D. 1988. Deep ecology meets the developing world. Pp. 79–82. *In:* E. O. Wilson & F. M. Peter (eds.), *Biodiversity*. National Academy Press: Washington, D.C.

Norgaard, R. B. 1988. The rise of the global change economy and the loss of biological diversity. Pp. 206–211. *In:* E. O. Wilson & F. M. Peter (eds.), *Biodiversity*. National Academy Press: Washington, D.C.

Norton, B. 1988. Commodity, amenity, and morality: the limits of quantification in valuing biodiversity. Pp. 200–205. *In:* E. O. Wilson & F. M. Peter (eds.), *Biodiversity*. National Academy Press: Washington, D.C.

Noss, R. F. 1989. Who will speak for biodiversity? *Cons. Biol.* 3:202–203.

Pitelka, L. F. & D. J. Raynal. 1989. Forest decline and acidic deposition. *Ecology* 70:2–10.

Randall, A. 1988. What mainstream economists have to say about the value of biodiversity. Pp. 217–223. *In:* E. O. Wilson & F. M. Peter (eds.), *Biodiversity*. National Academy Press: Washington, D.C.

Schlesinger, W. H. 1989. The role of ecologists in the face of global change. *Ecology* 70:1.

Schneider, S. H. 1989. The greenhouse effect: science and policy. *Science* 243:771–779.

Wickstrom, C. E. 1971. The credibility gap and the ecology reaction. *Limnol. Oceanogr.* 16:998–1000.

Wilson, E. O. 1988. The current state of biological diversity. Pp. 3–18. *In:* E. O. Wilson & F. M. Peter (eds.), *Biodiversity*. National Academy Press: Washington, D.C.

Introduction to the Symposium

Alan J. Kohn, Charles Mitter, and Brian D. Farrell

Perceptions of the importance of diversification vary widely among systematists and evolutionary biologists. At one extreme, it is the most critical macroevolutionary process. By constraining as well as promoting change in lineages, diversification profoundly affects the entire subsequent course of evolution. The significance of this view is evident in the prominence granted to diversification as a theme in important recent books, e.g., Vermeij (1987), Levinton (1988). The view of Hughes (1989: 59) perhaps epitomizes the opposite pole: "The so-called higher taxa, such as divisions, phyla, classes, orders and families, are simply filing systems for human use.... they are manipulated with zest as if their re-arrangement would somehow elucidate biologic evolution. No penetrating results are to be expected from such calcuations." It seems fair to conclude that most systematists and evolutionists agree that diversification is a major component of macroevolution relevant to broad scales of taxonomy, space and time.

PATTERNS AND RATES

The taxonomic scale ranges from "first-order radiations" (Davis, 1981) of species within a genus to the higher taxa mentioned above. First order radiations are sometimes amenable to direct study, as they occur over geologically short time intervals and can be related to the Recent biota. Knowledge of extant taxa provides a strong basis for comparative analysis, and extant taxa that also have a fossil record are especially promising foci for study. At the levels of the highest taxonomic categories, science fiction and the validity of Hughes' statement above increase.

The temporal scale reflects the three "tiers" of Gould (1985), recently expanded to four by Bennett (1990) and extending from "ecological moments" to eras (Table 1). Spatial scales relevant to taxonomic diversification also range widely, over some ten orders of magnitude from the millimeter-sized worlds of the smallest Monera to earth's entire biosphere (Kohn, 1989).

CAUSES AND CONSEQUENCES

We regard the following as the most important hypotheses that have been invoked to explain varying diversification rates and patterns:

1. Historical contingency: who got there first, and who survived mass extinctions, etc.; contingency is on extrinsic factors.

Dr. Kohn is with the Department of Zoology, University of Washington, Seattle, WA 98195, USA. Drs. Mitter and Farrell are with the Department of Entomology, University of Maryland, College Park, MD 20742, USA.

Table 1. Evolutionary time scales relevant to diversification processes.

Scale	Time or Periodicity (years)	Processes and phenomena
Ecological time	10^{-5}–10^{3}	Natural history, genetics, population and community ecology, individual selection
Milankovitch cycles	10^{4}–10^{5}	Variation in climate and sea level; community, habitat and biogeograpic reorganization
Geologic time	10^{6}–10^{7}	Geographic isolation, speciation, species durations; macro-evolutionary trends
Mass extinctions	10^{7}–10^{8}	Catastrophic impacts? Loss of species and higher taxa; faunal and floral replacement

After Gould (1985), Kohn (1989), Bennett (1990)

2. Large-scale environmental change: continental dispersion and dispersal by plate tectonic activity, marine transgressions and regressions, vulcanism.
3. Ecological opportunism: invasion of empty habitats and establishment of new adaptive zones, for example.
4. Key innovations: novel intrinsic attributes associated with diversification early in the evolution of a clade and with greater taxonomic and structural diversity than sister taxa.
5. Developmental and structural constraints: limitations imposed by ontogenetic pathways and mechanical attributes of biological materials on the architecture or design of organisms.
6. Genetic flexibility: changes that affect the rate of generation of innovations and heritable variation.
7. Ecological and geographic amplitudes: generalists are less subject than specialists to range fragmentation, speciation, and extinction.
8. Breeding system and population structure: the effects of these on speciation rates are important only to first-order radiations, but they may affect extinction rates at higher levels.
9. Taxon cycles and other ecological/biogeographic speciation "engines."

In order to examine the causes of diversification patterns and rates and to test the alternative hypotheses listed above, participants in the symposium identified several approaches and several methodological issues. We end this introduction by listing these, as they indicate the themes of the presentations that follow:

Approaches

1. Comparative study of the diversity changes of higher taxa through the stratigraphic record, and their environmental, morphological and ecological correlates.

2. Detailed ecological, anatomical, genetic, and behavioral study of particular groups and the key features hypothesized to promote their diversification.

3. Comparative study across taxa of present-day diversities and their correlates, including replicate comparisons of present-day diversities between sister groups.

Methodological Issues

1. How far and by what methods can we resolve the mechanisms underlying macro-evolutionary diversification patterns?

2. By what methods can we statistically demonstrate heterogeneity of diversification patterns and rates?

3. By what methods can we statistically demonstrate effects on diversification rates? What are the complementary strengths and weaknesses of these methods?

A final and broader issue is that change in biotic diversity, a dominant and pervasive pattern and theme throughout all of organic evolution, is now disproportionately and increasingly strongly influenced by one species, *Homo sapiens*, and it is thus also a major concern for the continued support of life on earth.

LITERATURE CITED

Bennett, K. D. 1990. Milankovitch cycles and their effects on species in ecological and evolutionary time. *Paleobiology*, 16:1–10.

Davis, G. M. 1981. Introduction to the Second International Symosium on evolution and adaptive radiation of Mollusca. *Malacologia* 21:1–4.

Gould, S. J. 1985. The paradox of the first tier: an agenda for paleobiology. *Paleobiology* 11:2–12.

Hughes, N. F. 1989. *Fossils as Information.* Cambridge University Press: Cambridge.

Kohn, A. J. 1989. Natural history and the necessity of the organism. *American Zoologist* 29:1095–1103.

Levinton, J. 1988. *Genetics, Paleontology, and Macroevolution.* Cambridge University Press: Cambridge.

Vermeij, G. 1987. *Evolution and Escalation.* Princeton University Press: Princeton, NJ.

Diversity in the Phanerozoic Oceans:
A Partisan Review

J. John Sepkoski, Jr.

Abstract. This paper summarizes work on documenting and modeling large-scale patterns of animal diversity in the oceans through the nearly 600 myr of the Phanerozoic. These patterns are usually documented with higher taxa, especially genera and families; evidence from modeling, the modern world, and the fossil record all indicate that such taxa can adequately reflect underlying species diversities, despite problems of paraphyly and incomplete sampling. Families as well as lower taxa exhibit a general pattern of low diversity during the Cambrian, higher but not continuously increasing diversity during the later Paleozoic, and then expanding diversity during the Mesozoic and Cenozoic. Each increase is associated with diversification of a quasi-distinct set of unrelated taxonomic classes, termed "evolutionary faunas." Increase in global diversity between faunas is reflected in expansion of utilized ecospace and increases in alpha and beta diversities. The faunas dominated different sectors of the marine environment during the Paleozoic Era and tended to expand toward the offshore as they diversified. A compact model of this diversification can be constructed with the empirically justified assumption that speciation rates decline as diversity increases. If each evolutionary fauna has independent linear functions for speciation and extinction, a coupled logistic model results that can interrelate seemingly disparate patterns of radiation, extinction, and faunal change at widely spaced times through the whole of the Phanerozoic.

INTRODUCTION

Taxonomic diversity refers to the number of taxa in a clade, place, and/or time (Valentine, 1985). The taxa counted are species or their presumed surrogates of higher rank, such as genera and families. Clades may be true holophyletic groups or may be approximated by traditional higher taxa (which, even if paraphyletic, may be useful if one is interested in diversities of surviving plesiomorphic species). Places can range from essentially point localities, such as quadrats, to expansive complexes of ecosystems, such as the world ocean. Finally, times can range from single moments, such as the here-and-now, to units of geologic time, which span millions of years.

Many other definitions of diversity are available. Counts have been made of varieties of morphologies, behaviors, life history traits, etc. within clades, places, and times. More commonly, taxa are not just counted but scaled to relative abundance to provide indices of dominance (see Pielou, 1975; Magurran, 1988). However, over the past decades, the greatest theoretical success has been made with simple taxonomic diversity (e.g., MacArthur & Wilson, 1967; Raup et al., 1973; Maurer, 1989), and this is the quantity that will be of concern here.

Dr. Sepkoski is with the Department of the Geophysical Sciences, University of Chicago, 5734 South Ellis Avenue, Chicago, IL 60637, USA.

Diversity is of interest because it varies among clades, places, and times. It is believed that study of this variation can lead to some understanding of the ecological and evolutionary rules that govern the composition of biotas and the successes of higher taxa. The fundamental question invoked in many studies of diversity is that of Hutchinson (1959): "Why are there so many species?" However, there has been little theoretical progress in answering this question (cf. May, 1988). The more frequently attacked problem is, in fact, "Why are there so many species here rather than there?"; here being the Coleoptera, the Tropics, or the Tertiary.

Approaches to studying diversity are nearly as varied as the phenomena being investigated. The fundamental division is between studies of phylogenetic diversity—the global number of species within a clade—and of ecologic diversity—the number of species within a clade at a locality or local community. Both kinds of studies can be time specific, investigating diversity patterns in one time plane (most often the Recent) or time-integrative, examining patterns of change over time, either by direct evidence from the fossil record or, less commonly, by indirect evidence such as estimated divergence times of sister taxa. There are large differences in methodology and epistomology between workers making time-specific studies (mostly biologists) and those conducting time-integrative studies (mostly paleontologists). Biologists tend to have better samples (although never absolutely complete—witness current questions about tropical diversity) and more information about the biology of individual species, but they tend to focus investigations at smaller taxonomic and spatial scales and to demand more detailed explanations of pattern (i.e., more emphasis on proximate versus ultimate causation). Paleontologists, on the other hand, are plagued by taphonomic problems of incomplete and often biased samples and of ignorance of the biology of extinct species; often they focus on the highest taxa over huge spatial and temporal scales, and they are as excited by pregnant correlations as by apparent causations. Also, paleontological explanations often emphasize extinction as well as speciation, frequently viewing the fossil record as the graveyard of species that didn't make it up to our arbitrary moment in time. Biologists, on the other hand, tend to emphasize mostly speciation or dispersal and their consequences, often ignoring any role played by taxa that have disappeared before the present.

I am a paleontologist, and I bear both the benefits and the shortcomings that training and perspectives from the fossil record endow. Below, I briefly summarize some of my work on the evolution of diversity in marine faunas. The patterns, correlations, and explanations that I present are of the coarsest nature: higher taxa in the world ocean over millions of years of time. What interests me is not so much the proximate causes of phylogenetic and ecologic diversity at any moment as the "boundary conditions," as it were, that constrained global diversification and within which more proximate processes and historical events operated.

DATA ON FOSSIL DIVERSITY: TWO HERESIES

Paleontologists frequently appear guilty of two heresies in their investigations of fossil diversity: they most often analyze higher taxa—especially families—rather than species (an ecologists' heresy), and usually pay little attention to whether or not these taxa are true clades (a taxonomists' heresy).

The Ecologic Heresy

Higher taxa are used because it is difficult to compile good data on species for any large taxon, and even when done the data are notoriously incomplete and biased. For the marine fossil record, it has been estimated variously that only 1 to 10% of animal species that ever lived have been described. These described species are biased toward common

212

animals with mineralized skeletons. Furthermore, they have been differentially collected from various areas (especially around major universities), times (especially more recent geologic systems), habitats (especially shallow marine environments), and taxonomic classes (especially those useful to biostratigraphy). Simpson (1960), Raup (1972, 1976a, 1979a), and Paul (1982, 1985) provide excellent reviews of these problems.

Higher taxa reduce (but do not eliminate) these problems since it takes only one individual of one species to establish the presence of a genus, family, etc. (Raup, 1979a). Since there are many fewer higher taxa than species, the sample of genera or families must be more complete, and probably less biased. Thus, these higher taxa are usually used as surrogates for species.

There is empirical justification for this substitution. Comparative studies of both modern and fossil diversity have found strong correlations between diversities of species and of genera or families (e.g., Stehli et al., 1967; Cook, 1969; Flessa, 1975; Raup, 1975; Jablonski & Flessa, 1986). Figure 1 is an example, showing the relationship between numbers of extant species and families of marine echinoids and bivalves on oceanic islands and segments of continental shelf. Note that the two very different classes appear to follow the same trend. But note also that the trend, though tight, is curvilinear such that families underestimate species at higher diversities (see also Simberloff, 1970, 1974; Raup 1975; Järvinen, 1982). Thus, there may be the problem that diversities of higher taxa are low relative to species during rapid increases, such may be the case in the Cenozoic (Signor, 1982, 1985, 1990). This problem is not evident among higher ranks at the other end of the Phanerozoic, however, and all taxonomic levels from genus to class exhibit congruent diversity patterns through the Cambrian (Sepkoski, 1992).

Nonlinearities in diversities also cause higher taxa to underestimate the magnitudes of species kills during mass extinctions (Raup, 1979b; Carr & Kitchell, 1980; Sepkoski, 1989a): if just one species in a polytypic genus or family survives the event, the whole taxon does. Thus, for example, 50% of families but 80% of genera disappeared during the great end-Permian mass extinction. Raup (1979b) used these numbers with taxonomic rarefaction to arrive at the estimate that 96% of species became extinct at that time.

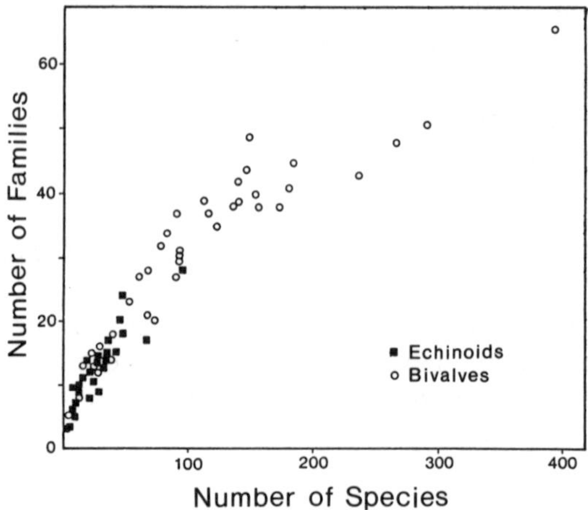

Figure 1. Numbers of families as a function of species for marine bivalves and echinoids on 15 modern islands and 22 shelf areas. There is a strong but nonlinear relationship between families and species, with representatives of the two very different phyla falling along the same trend. After Jablonski & Flessa (1986).

The Taxonomic Heresy

The strong correlation between species and higher taxa would seem to negate the modern systematists' complaint that traditionally defined genera or families are often arbitrary and therefore tell us nothing about evolutionary pattern (e.g., Patterson & Smith, 1987, 1989). But only very preliminary work has been done to investigate the problem of how taxonomic method affects perceived patterns of species diversity. An example is shown in Figure 2. The top of this figure illustrates a portion of a simulated evolutionary tree produced by the Monte Carlo approach introduced by Raup et al. (1973). The tree has been divided into taxa by randomly selecting model lineages to be progenitors of "paraclades" (for more detail, see Sepkoski, 1978, 1989a). The logic here is more or less congruent with traditional taxonomy: evolutionary novelties can occur throughout a phylogenetic tree, and all species which share that novelty, but not any other, are classified together. The algorithm produces taxa that can be monotypic or polytypic and monophyletic or paraphyletic, but not polyphyletic.

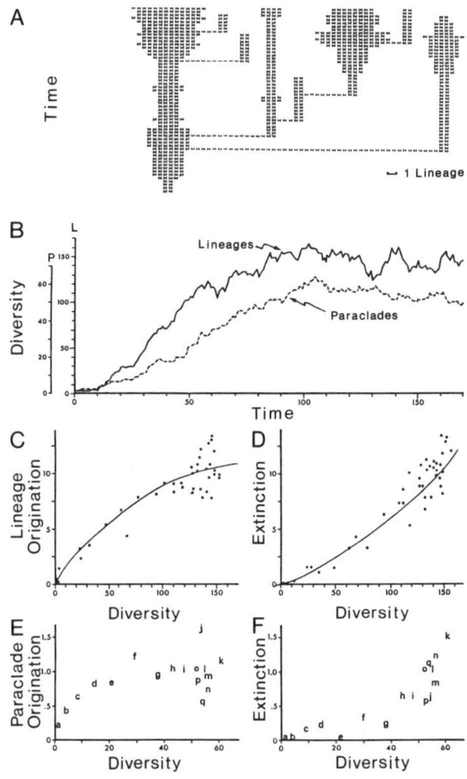

Figure 2. Relationships between species and higher taxa in a Monte Carlo simulation of a phylogenetic system. *A.* A portion of the simulated phylogeny with species lineages grouped together within randomly defined "paraclades" (sensu Raup, 1985); the spindle diagrams show the ancestry of the paraclades and the diversity within each through time. Note that some paraclades are paraphyletic and/or monotypic. *B.* Diversity of simulated lineages and paraclades through time. Lineage diversity was governed by a logistic generating function; diversity of the arbitrary paraclades parallels the lineage pattern although with damped variation. *C* and *D.* Total rates of origination and extinction as functions of diversity for the lineages. The solid curves represent the generating functions, and the points show realized rates calculated over intervals of four time units in the Monte Carlo simulation. *E* and *F.* Parallel plots of total rates of paraclade origination and extinction as functions of diversity, calculated over intervals of 10 time units (denoted in sequence by letters). The rates for the arbitrary paraclades exhibit the same parabolic functions as the underlying lineages. *A.* After Sepkoski (1989a); *B–F* after Sepkoski (1978).

214

Clearly, these simulated taxa break the rules of modern phylogenetic systematics. But they do not break the correlation between diversities of lower and higher taxa. Figure 2B illustrates these diversities. The lineages were generated using a logistic function, giving an initial sigmoidal radiation followed by mild fluctuations about an equilibrium. This logistic pattern is closely followed by the arbitrary paraclades, despite rampant paraphyly. The correspondence also holds for the underlying rates of origination and extinction. Figure 2C and D show the generating functions and patterns of origination and extinction of the simulated lineages; the variance results from the Monte Carlo methods employed. Figure 2E and F illustrate patterns of origination and extinction for the arbitrary paraclades; despite paraphyly and size variation, the paraclades exhibit the same basic relationship between evolutionary rates and diversity as the underlying lineages. Thus, the simulation results suggest that arbitrarily defined taxa need not distort the signal present in species. In fact, simulations of imperfect sampling (Sepkoski, 1978) suggest that the paraclades may be more faithful to underlying diversification patterns than fossil species when sample sizes are small.

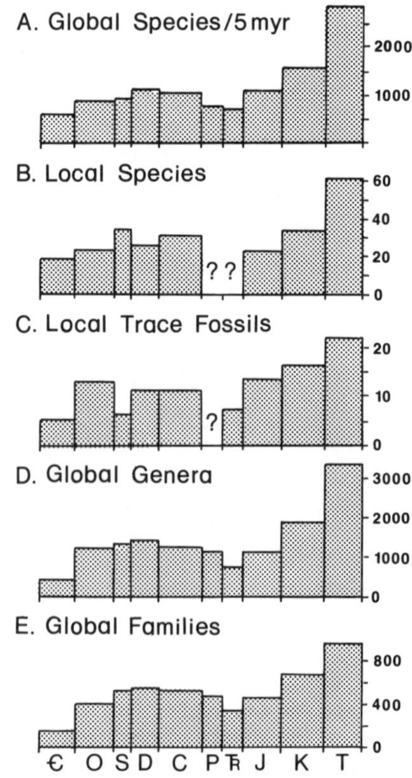

Figure 3. Five estimates of marine animal diversity through the Phanerozoic. *A*. Global numbers of species per five million years (the average duration of a marine species) for each geologic period, from data presented by Raup (1976b). *B*. Average numbers of skeletal species in local assemblages (i.e., alpha diversity), compiled by Bambach (1977); question marks indicate periods for which sufficient data were unavailable. *C*. Average numbers of trace fossils (named burrows, tracks, and trails) in local assemblages, from the smaller data set assembled from personal field observations by Seilacher (1974). *D*. Global numbers of genera averaged per geologic period, based on unpublished data (see Sepkoski, 1989a). *E*. Global numbers of families averaged per geologic period, based on data in Sepkoski (1982a) with updates into 1989. Modified from Sepkoski et al. (1981).

Data from the Fossil Record

The evidence presented above is admittedly circumstantial, although I have never found contrary evidence. But what about large-scale patterns in the fossil record? These exhibit similar correlations between species and higher taxa. Figure 3 illustrates five estimates of Phanerozoic fossil diversity reduced to their least common denominator of number of taxa per geologic period. Three of the estimates are for species diversity: Raup's (1976b) numbers for global diversity of described species; Bambach's (1977) data on alpha diversity, or average numbers of species of body fossils within local communities; and Seilacher's (1974) data on average numbers of trace fossil ichnospecies (named burrows, tracks, and trails) within localities. Bambach's data are important because they circumvent problems in Raup's global data concerning how much area is sampled and with what intensity, even though Raup's data contain far more species. Seilacher's data, which also estimate alpha diversity through time, are particularly important despite the small sample size because the majority of trace fossils were produced by soft-bodied organisms that leave no skeletal fossils. Still, the pattern is qualitatively similar to the other, independent estimates of species diversity with low diversity in the Cambrian, higher but not steadily increasing diversity in the Ordovician to Permian, and then increasing diversity through the Mesozoic to a maximum in the Cenozoic (Sepkoski et al., 1981). Thus, there is no evidence that the diversity of marine soft-bodied animals had a history significantly different from skeletalized animals. Evidence from the Middle Cambrian Burgess Shale is also consistent with this: approximately 15% of the roughly 120 genera occur in normal shelly assemblages (Conway Morris, 1986), which is well within the range of modern marine communities (Johnson, 1964).

The lower two histograms in Figure 3 illustrate estimates of global diversity at the genus and family levels. These data are not independent of one another nor of Raup's data, since they ultimately come from the same monographic sources. But what they show again is that higher taxa do not distort the signal of species; the patterns of low diversity in the Cambrian, higher in the later Paleozoic, low again in the Triassic, and then increasing through to the Cenozoic appear in all data sets. Thus, higher taxa can act as surrogates for species, and hypotheses of diversity tested with data for genera and families need not be explanations tied to those taxonomic levels, as some have mistakenly inferred (e.g., Hoffman 1989).

GLOBAL PATTERNS OF TAXONOMIC DIVERSITY

The strong correlations among the various estimates of Phanerozoic marine diversity in Figure 3 suggest that the most comprehensive and temporally precise data set is the best for detailed analysis. This is the data set for families, the taxonomic level that has been used traditionally to examine global fossil diversity (e.g., Newell, 1967; Valentine, 1969). The history of familial diversity is illustrated in Figure 4 at the highest available temporal resolution: number of families per geologic stage, each averaging about 8 myr in duration.

Total Diversity

The curve for familial diversity can be divided into three main phases of animal diversification: 1) an initial, Vendian to Cambrian phase encompassing the early radiation of metazoans across the Precambrian-Cambrian boundary and a subsequent slowing of diversification in the mid to late Cambrian; 2) a later, Paleozoic phase, beginning with the Ordovician radiations that tripled diversity from Cambrian levels, followed by a 200-myr interval of near steady state, and terminated by the great end-Permian mass extinction that reduced families to less than half previous levels; and 3) a Mesozoic-Cenozoic phase

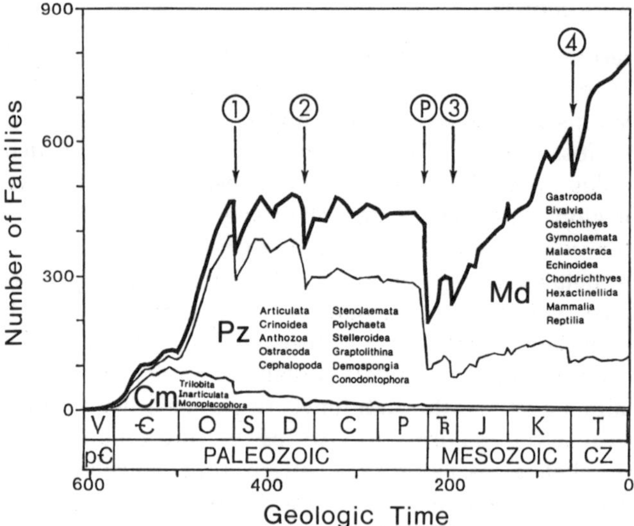

Figure 4. Diversity of well-skeletonized marine animal families over 77 stratigraphic stages of the Phanerozoic. The upper curve depicts the total number of families known from the fossil record. Arrows indicate five major mass extinctions recognized by substantial reductions in diversity: P = end Permian, 1 = end Ordovician, 2 = Late Devonian, 3 = end Triassic, and 4 = end Cretaceous. Fields below the upper curve illustrate the diversities of the three great evolutionary faunas, with the principal constituent classes listed: Cm = Cambrian fauna, Pz = Paleozoic fauna, and Md = Modern fauna. Geologic eras and periods are indicated along the abscissa (Cz = Cenozoic); time is in millions of years. Modified from Sepkoski (1982b) and Sepkoski & Miller (1985).

encompassing the rebound from the Permian event and the continuing increase in marine familial diversity to the present.

The first two phases of diversification appear very similar, differing only by a scale factor: they begin with radiations and end with near steady state. The radiation at the beginning of the Paleozoic phase was longer than that of the Vendian-Cambrian phase, produced more families, and was followed by a longer plateau in diversity. Interestingly, the two big mass extinctions prior to the Permian are followed by immediate rebounds that quickly restored diversity to previous levels; this suggests that the diversity plateau of the Paleozoic reflected some sort of dynamic equilibrium or quasi-equilibrium.

The Mesozoic-Cenozoic phase superficially appears different from the earlier phases, with its long rise in diversity toward the Recent. However, this difference may be more a function of our arbitrary vantage point—the Recent—than of the dynamics. In fact, the overall rate of diversification seems to have slowed through the post-Paleozoic (ignoring rebounds after mass extinctions). It is possible to fit a curve to the declining diversification that would eventually level out, making the Mesozoic-Cenozoic phase appear as a scaled-up version of the preceding phases. (In 1984, I published a calculation which gave the estimate that marine familial diversity would grow to within 5% of a limiting value in approximately 125 myr, barring mass extinction; a distinguished colleage wrote immediately and congratulated me upon making the safest prediction in the history of paleobiology.)

Evolutionary Faunas

It is interesting that each of the three phases of diversification may be similar in form. It is even more interesting that each is contributed by a distinct set of taxonomic groups. A factor analysis of numbers of families within classes through time produced three groups of taxa that had temporally distinct histories (Sepkoski, 1981; see also Flessa & Imbrie,

1973). Later work on environmental occurrences of these groups demonstrated that they also had distinct spatial distributions (Sepkoski & Sheehan, 1983; Sepkoski & Miller, 1985), as is discussed further below. The diversity histories of these three groups, which I have termed "evolutionary faunas," are illustrated in Figure 5 and in the lower curves in Figure 4.

The Cambrian evolutionary fauna was dominated by such "archaic" classes as trilobites, inarticulate brachiopods, hyoliths (an extinct group possibly related to molluscs), monoplacophorans, and others. Collectively, these classes radiated rapidly during the earliest Cambrian, attained maximum diversity in the Middle and early-Late Cambrian, and then began a long decline that extended through the Ordovician radiations and into the late Paleozoic Era. Very few members of the Cambrian fauna persist today.

The Paleozoic evolutionary fauna was dominated by a larger array of classes including articulate brachiopods, crinoids, corals, ostracodes, cephalopods, stenolaemate bryozoans, and others. These began to radiate during the latest Cambrian Period, virtually exploded in the Ordovician, and reached a peak in diversity in the Silurian to Devonian; thereafter they declined, gently at first but radically at the end-Permian mass extinction. During the Mesozoic, these classes had a fascinating collective history of diversifying rapidly during the Triassic, faltering at the end-Triassic mass extinction, slowly radiating again during the Jurassic, but then going into a long decline in the mid-Cretaceous when the Mesozoic-Cenozoic diversification exceeded the Paleozoic diversity maximum. This pattern will be returned to later.

The Modern evolutionary fauna was dominated by bivalve and gastropod molluscs, gymnolaemate bryozoans, malacostracan crustaceans, foraminfera, echinoids, and marine vertebrates. Many of these phylogenetically diverse classes originated during the Cambrian Period, diversified slowly through the later Paleozoic Era (keeping total familial diversity at near steady state as the Paleozoic fauna gently declined), survived the end-Permian event with comparatively mild extinction, and then diversified to unprecedented levels into the Cenozoic. Thus, the final expansion in marine diversity reflects radiations among ancient higher taxa and not appearances of fundamentally new morphological baupläne (cf. Levinton, 1988).

These three "great" evolutionary faunas of the Phanerozoic marine record share a pattern of increasing characteristic diversity but decreasing diversification rate through their sequence (Sepkoski, 1984, 1990). Thus, the rate of large-scale faunal change and diversity increase slows down through time. At the beginning of the Phanerozoic, just after metazoan origins, there is a very rapid sequence of distinct faunas and a diversity increase involving the brief flourishing of the enigmatic Ediacaran fauna (Glaessner, 1984; Seilacher, 1989), the subsequent expansion of the Tommotian small, skeletal taxa (Brasier, 1979, 1989), and finally replacement by the more standard Cambrian groups. It may be useful to subdivide the Cambrian fauna as originally defined and recognize these first two units as "minor," short-lived evolutionary faunas, as depicted in Figure 6. Note that these minor faunas fit neatly into the sequence of increasing level of maximum diversity and decreasing rate of diversification (Sepkoski, 1992).

Ecological Factors in Increasing Diversity

Among the three great evolutionary faunas, differences in diversity can be related qualitatively to differences in basic ecological strategy. Bambach (1983, 1985) has made an extensive study of life modes among the commonly fossilized constituents of the three faunas and has argued that the amount of utilized ecospace has increased with each. He used a simple classificatory system with three dimensions: trophic ecology, life zone, and mode of mobility or attachment (Fig. 7). He found that members of the Cambrian fauna

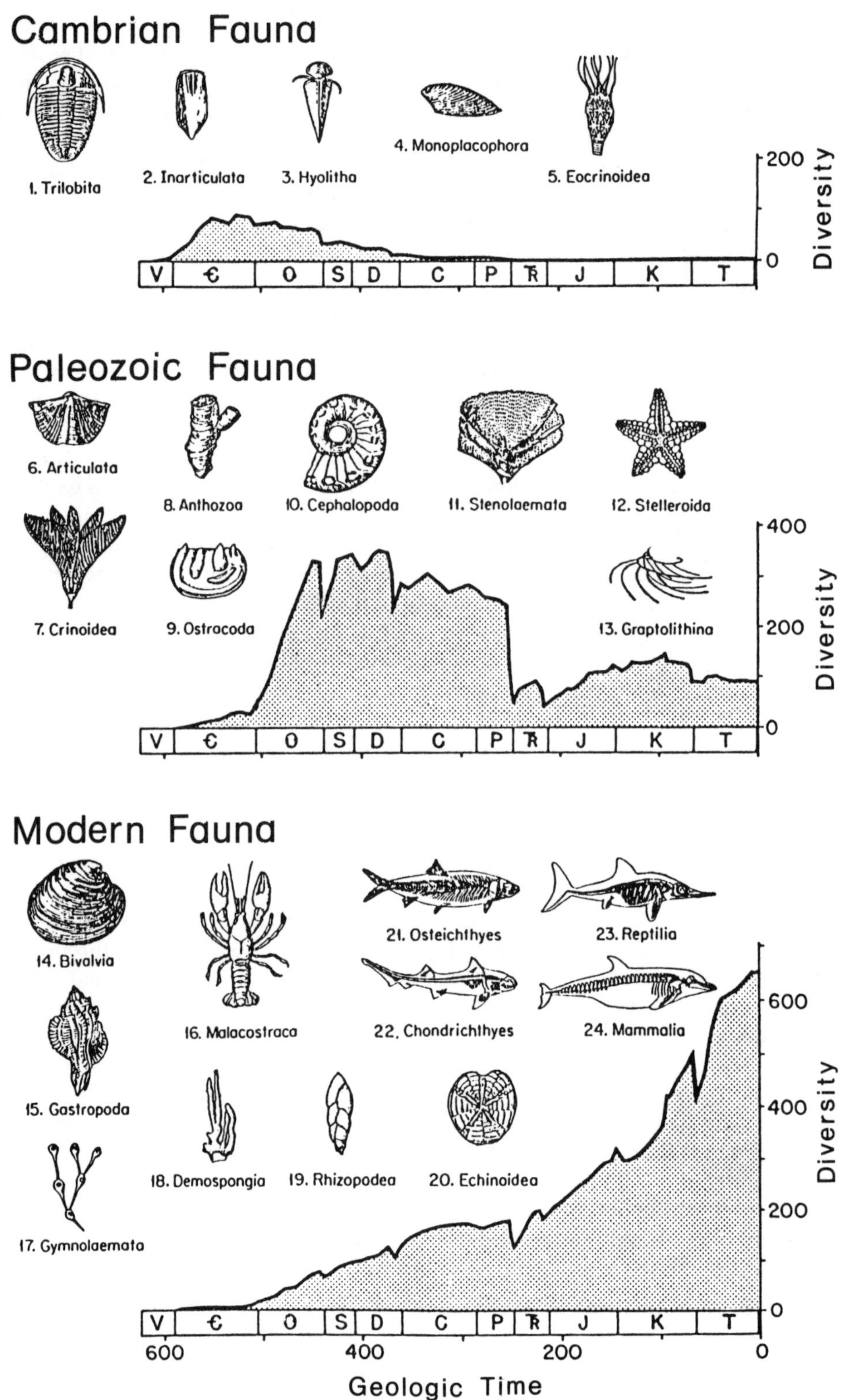

Figure 5. Familial diversity in each of the three great evolutionary faunas through the Phanerozoic. Representatives of important constituents are illustrated for each. (Demospongia should be transferred from the Modern to the Paleozoic fauna if Stromatoporoidea is included.) After Sepkoski (1984).

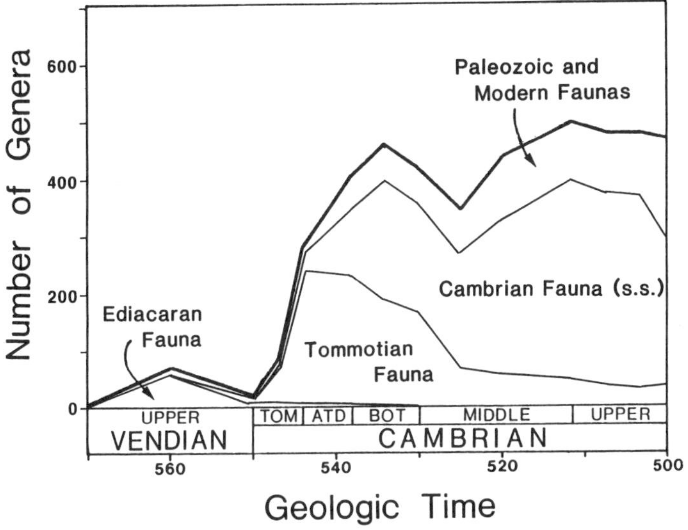

Figure 6. Diversity of animal genera from the Vendian Period (latest Precambrian) through the Cambrian Period. Two additional, minor evolutionary faunas are depicted: the Ediacaran fauna, encompassing most of the fossils of enigmatic soft-bodied animals in the late Vendian, and the Tommotian fauna, including earliest Cambrian problematic shelly fossils as well as monoplacophorans and hyoliths. The Tommotian fauna, as defined, has an asymmetrical history of rapid initial expansion followed by slower decline, much like that seen in the succeeding Cambrian (sensu stricto) and Paleozoic faunas. Note that the curve for total generic diversity is similar in pattern to that for Cambrian familial diversity in Fig. 4. After Sepkoski (1992).

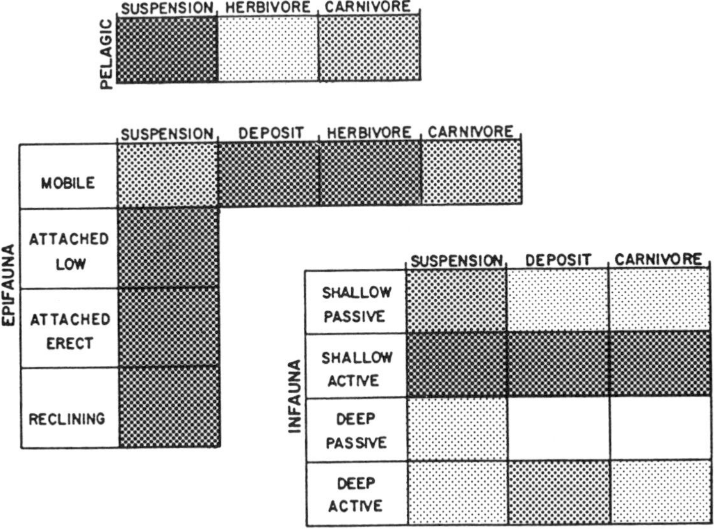

Figure 7. Modes of life of major invertebrate classes, showing an expansion of ecospace through the Phanerozoic. Modes of life are classified by life zone, food, and activity or posture. Members of the Cambrian fauna occupied only cells with the darkest stippling; members of the Paleozoic fauna occupied these cells plus those with intermediate stippling; and members of the Modern fauna occupied all cells with stippling. Modified from Bambach (1983).

220

occupied fewer than half of the categories in this system. Members of the Paleozoic fauna occupied all previously utilized categories plus about 50% more. Paralleling this finding, Ausich and Bottjer (1982, 1985) demonstrated a considerable increase in epifaunal tiering associated with expansion of the Paleozoic fauna (Fig. 8); suspension feeders of the Cambrian fauna usually fed no more than 10 cm above the sediment-water interface whereas the epifauna that dominated the Paleozoic evolutionary fauna fed up to 100 cm above, usually in tiered communities with suspension feeders at several levels. Epifaunal tiering decreased with expansion of the Modern fauna but infaunal tiering increased considerably, with elements burrowing to depths of 100 cm depth or more (see also Thayer, 1983). In Bambach's (1983, 1985) analysis, he also found more ecospace expansion, with members of the Modern fauna utilizing nearly 50% more categories in his classification than did the preceding Paleozoic fauna (Fig. 7).

Increased utilization or finer subdivision of ecospace should be manifested in increased alpha diversity, the number of species in local communities. Indeed, there was an increase on order of 50% in mean numbers of species in fossil assemblages between the Cambrian and the later Paleozoic (Bambach, 1977). However, the increase in global diversity through this interval was about 300% (Figs. 3 and 4). There are two potential sources for this excess diversity: increased ecological differentiation of communities (beta diversity) and increased biogeographic differentiation of faunas (gamma diversity). I analyzed data on beta diversity from the Cambrian to the Permian and found only one major change, a 50% increase in beta diversity during the Ordovician (Sepkoski, 1988). I argued that this reflected increased habitat selection and ecological specialization of the Paleozoic fauna relative to the less diverse Cambrian fauna. Furthermore, I suggested that there were other sources of beta diversity in specialized, species-rich communities not adequately represented in my data, such as reefs, crinoid gardens, bryozoan thickets, and hard-ground assemblages. Thus, I concluded that most of the difference in global diversity between the Cambrian and Paleozoic evolutionary faunas involved ecological expansion and specialization and not biogeographic expansion (see also Bambach, 1985). (For other arguments emphasizing the role of biogeography in global diversification, see Valentine, 1970; Valentine & Moores, 1972; Valentine et al., 1978; Schopf, 1979; Cracraft, 1982, 1985; Signor, 1985).

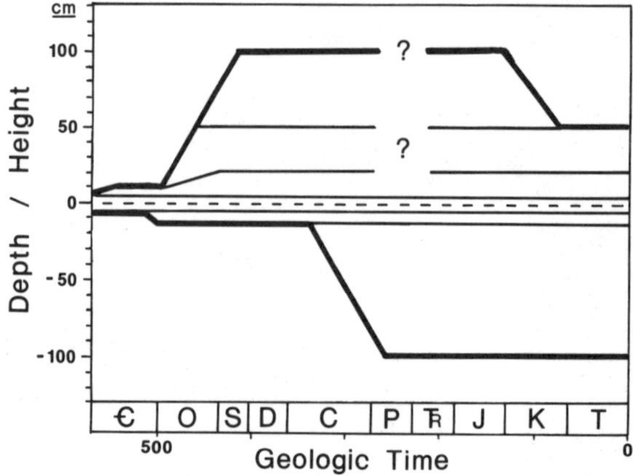

Figure 8. Phanerozoic history of tiering in marine benthic suspension-feeding communities. Tiering is treated here as the vertical distribution of animals above and below the sediment-water interface at 0 cm. The heavy lines indicate maximum average heights above and depths below the interface; lighter lines are lower tiers that occur commonly in the communities. Epifaunal and, to a lesser extent, infaunal tiering increased with the radiation of the Paleozoic fauna during the Ordovician. Infaunal tiering increased later as the Modern fauna expanded, but epifaunal tiering eventually decreased, presumably in response to diversifying predators. Question marks indicate lack of data. After Bottjer and Ausich (1986).

Differences in ecological characteristics can be identified among evolutionary faunas and may have been important in their displacement through time. This is particularly true of differences between the Paleozoic and Modern faunas. The Paleozoic fauna was dominated by immobile suspension feeders such as articulate brachiopods, stenolaemate bryozoans, stalked crinoids, and others. The Modern fauna, in contrast, is characterized by mobile or semi-mobile groups, including most gastropods and many bivalves, malacostracan crustaceans, echinoids, and marine vertebrates. Among these mobile animals are numerous durophagous predators, as emphasized by Vermeij (1977, 1983, 1987). He has argued that diversification of these predators, especially during the "Mesozoic marine revolution," had profound effects on the marine epifauna, forcing a decline in many taxa characteristic of the Paleozoic, the evolution of predator-resistant morphologies among most remaining epifauna, and a switch to infaunal habitats by other groups (see also Aronson, 1989). Expansion of mobile infaunal bivalves, gastropods, echinoids, and malacostracans may have had further detrimental effects on immobile, epifaunal members of the Paleozoic fauna; as Thayer (1979, 1983) has argued, infaunal "bulldozing" and churning of sediment would have made the sediment-water interface a far less stable and therefore more difficult place to live in the absence of adaptations for righting and stabilization.

Analogous differences leading to interference between members of the Cambrian and Paleozoic evolutionary faunas are difficult to identify. Displacement of the Cambrian fauna may have resulted simply from the presumed ecological specialization of typical Paleozoic groups: their species may have been able to insinuate themselves in communities of the more generalized Cambrian animals and usurp resources (Sepkoski, 1979). Such a process might have involved displacive competition, but, given the many million years over which the Cambrian fauna declined, it is more likely to have involved preemptive competition: any time there was a local extinction and a more specialized member of the Paleozoic fauna *happened* to invade and repopulate, it might have been difficult for Cambrian species to regain the pre-empted share of resources.

The inferred sequence of ecological expansion and specialization among the faunas may have also had something to do with the progression of slower diversification rates, especially if low-diversity communities of ecologically generalized species are easier to invade than high-diversity communities of specialists that efficiently utilize resources. If most speciation is a small-population phenomenon (e.g., Mayr, 1963) and if small restricted populations are most vulnerable to extinction (e.g., MacArthur & Wilson, 1967; Diamond, 1984; Stanley, 1986; Simberloff, 1988), then successful speciation is dependent upon the ability of an incipient species to expand its geographic range (cf. Stanley, 1978, 1979). (Here, "successful speciation" means production of a new, widespread species of appreciable geologic longevity and potential for spawning its own daughter species.) In the face of more specialist communities, expansion might become increasingly difficult, with fewer incipient species succeeding. Thus we might expect, to a first approximation, that speciation and diversification rates should decline as ecological specialization and packing increase. This speculation should be testable by comparison of invasions of communities with different alpha and beta diversities and of inferred rates of biogeographic expansion in different diversity situations (cf. Brown & Maurer, 1989).

CRITIQUES OF EVOLUTIONARY FAUNAS

Evolutionary faunas were originally introduced as a descriptive vehicle for summarizing the nature, rate, and course of large-scale faunal change through the marine fossil record. The three "great" evolutionary faunas, as mentioned above, were identified through a Q-mode factor analysis of familial diversity within 91 classes through 63 intervals of Phanerozoic time (Sepkoski, 1981). This work repeated the analysis of an older data set by Flessa and Imbrie (1973), which had been criticized by C. A. F. Smith (1977). He demonstrated that factor analyses of stochastically generated phylogenies of the kind illustrated in Figure 2A can produce factors very similar to Flessa & Imbrie's "diversity associations"; Smith therefore concluded that no special macroevolutionary significance could be attributed to taxa grouped together in multivariate analyses of diversity data.

In my analyses, however, I found patterns that seemed to differ substantially from what Smith generated. In addition to encompassing approximately 91% of the data, the first three factors of the familial data had the following properties: 1) a marked break in the decay of eigenvalues occurred between the third and fourth eigenvectors, suggesting that the data had a nonrandom structure; 2) the composition of the first three rotated factors did not change substantially as more and more factors were added to varimax rotations; however, the composition did change radically when only the first two factors were rotated; and 3) the middle factor, encompassing the Paleozoic fauna, had an asymmetric distribution of loadings through time, differing from the roughly bell-shaped distributions generated by "random" data. I interpreted these properties as indicating that the factors differed "significantly" from expectations for stochastic phylogenies and therefore reflected some underlying organization to the evolution of Phanerozoic marine diversity.

This conclusion has been criticized by Hoffman & Fenster (1986), who essentially repeated Smith's (1977) exercise. They produced an ensemble of stochastically generated clades and performed a Q-mode factor analysis that resulted in three factors encompassing 85% of their data. They concluded, like Smith, that stochastic phylogenies yielded diversity associations (see also Hoffman, 1989). However, they ignored the three points listed above that led me to believe the evolutionary faunas were nonrandom. Kitchell & MacLeod (1988) tested the statistical significance of Hoffman & Fenster's analysis by generating thousands of stochastic phylogenies, grouping and factor analyzing them, and examining frequency distributions of patterns in three-factor solutions. They found that stochastic phylogenies frequently do produce some of the patterns in the empirical analysis but not all. However, they, too, did not address the three points I thought indicated nonrandomness.

An entirely different criticism was made by A. B. Smith (1988). He noted that several of the important classes in the Cambrian fauna—namely, the Inarticulata, Monoplacophora, and Eocrinoidea—are paraphyletic. He therefore suggested that the distinction between the Cambrian and Paleozoic faunas, and the apparently separate radiations of the Early Cambrian and the Ordovician, might be an artifact of taxonomy coupled with a poor fossil record in the Upper Cambrian. This may be true for eocrinoids, which Smith ably demonstrated represent a poorly defined stem group for later pelmatozoans and cystoids. The monoplacophorans may be somewhat similar (Runnegar & Pojeta, 1985), although most of their descendant clades—the Gastropoda, Rostroconchia, and Bivalvia, but not Cephalopoda—seem to have been derived early, and the paraphyletic monoplacophorans continued to diversify. The inarticulate brachiopods gave rise to the articulates early in their history (Rowell, 1981), and good holophyletic clades within the Inarticulata continued to diversify through the later Cambrian. Finally, the hyoliths and diverse trilobites seem to be holophyletic clades (perhaps several clades in the case of hyoliths) that have histories similar to the other groups. Thus, although taxonomic practice may contribute noise to the pattern, the histories of Cambrian classes remain to appear distinct from members of the Paleozoic and Modern faunas.

LOCAL PATTERNS OF TAXONOMIC DIVERSITY

Evolutionary Faunas

There is another interesting manifestation of evolutionary faunas that has further ecologic import: they tended to segregate in space as well as time. Following seminal work by Bretsky (1968, 1969) and Berry (1972, 1974), Sepkoski & Sheehan (1983) and Sepkoski & Miller (1985) investigated the taxonomic compositions of a large number of fossil assemblages described from Paleozoic strata in North America. We again found that the evolutionary faunas separated out in multivariate analyses of the assemblage data. Also, when mapped onto time-environment diagrams that arrayed the communities along a gradient from shoreline to shelf edge, we found regular onshore-offshore shifts in zones of dominance, as illustrated in Figure 9 (see also Jablonski et al., 1983). Members of the Cambrian fauna dominated the diversity of benthic communities across all environments during the Cambrian Period. But with the Ordovician radiations, their zone of dominance became progressively restricted to more offshore environments. In their wake, members of the Paleozoic and Modern faunas diversified. Taxa in the Paleozoic fauna first rose to prominence on the inner shelf and then rapidly expanded their area of high diversity to the outer shelf during the Ordovician; they then dominated mid to outer shelf environments to the end of the Paleozoic Era. Behind them, early members of the Modern fauna expanded in the nearshore. They then slowly expanded their area of dominance offshore, reaching prominence in the mid shelf by the Permian. This slower offshore expansion parallels the slower diversification rate seen in the global history of the fauna. In the

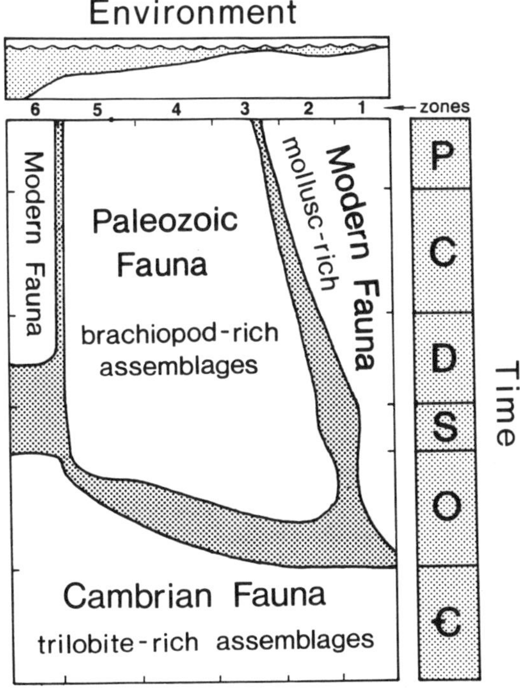

Figure 9. Time-environment diagram illustrating zones of maximum diversity of the three great evolutionary faunas through the Paleozoic Era. The fields indicate where each fauna contributed at least half of the genera in local assemblages; stippling indicates where no fauna dominated. The environmental gradient depicted at top represents a generalized shelf or platform with six zones: 1 = peritidal habitats, 2 = nearshore protected areas, 3 = offshore shoals, 4 = mid shelf, 5 = outer shelf, and 6 = slope and basin. The evolutionary faunas tended to gain dominance in the nearshore and to spread offshore at the expense of previous faunas. After Sepkoski (1991).

middle of the Paleozoic Era, some members of the Modern fauna became established in off-shelf, low-oxygen ("dysaerobic") environments and dominated communities there through the end of the Permian (Kammer et al., 1986).

Similar patterns of onshore-offshore expansion have been documented for Mesozoic higher taxa by Jablonski and Bottjer (1988, 1990a,b,c) and Bottjer and Jablonski (1988). They have demonstrated that an important component of this pattern is a tendency for major evolutionary novelties, as reflected in ordinal and class-level taxonomy, to appear first in shallow-water, nearshore sites; lower-ranked taxa show no such tendency, however. Evolutionary processes that produce this pattern are hardly understood, but it has been suggested that bias toward nearshore origin of novelty might involve high frequencies of genetic transiliences (Templeton, 1980) among small, isolated populations drawn from rapidly dispersing, panmictic species characteristic of shallow-water environments (Jablonski & Bottjer, 1983, 1990a), and/or selection favoring progenesis in frequently disturbed nearshore habitats (McKinney, 1986); Jablonski and Bottjer (1990c) review other possible mechanisms. Offshore expansion of novel taxa could be driven by various processes, although my investigations (Sepkoski, 1989b, 1991) suggest that evolution of resistance to extinction is vital. Higher taxa vary considerably in their average extinction rates (Van Valen, 1973, 1985a; Stanley, 1979; Holman, 1989) for reasons that are not at all clear. But given the observed continuous turnover of species in marine communities, taxa that have lower characteristic extinction rates should be able to hold their ground more successfully and slowly replace species in more extinction-prone groups (Van Valen, 1985b). This will permit extinction-resistant groups to insinuate themselves slowly into more offshore communities and diversify there, producing onshore-offshore patterns of diversification and faunal change (see mathematical model in Sepkoski, 1991).

Individual Taxa

Figure 9 illustrates only the environments of dominance of the evolutionary faunas and thus does not really establish a case for correspondence between local success and global diversity. This case is made in Figure 10. These diagrams illustrate patterns of alpha diversity through the Paleozoic Era for representatives from each of the three evolutionary faunas: ptychopariid trilobites from the Cambrian fauna, orthid brachiopods from the Paleozoic fauna, and pterioid bivalves from the Modern fauna. The time-environment diagrams are contoured and stippled for average alpha diversities of the orders in benthic communities across the shelf. To the left of each diagram is a curve for observed global diversity, measured here at the level of genus.

The diagrams in Figure 10 show the correlation between the distribution and diversity of the orders within local communities and their level of global diversity. For example, ptychopariids are widespread and diverse among benthic communities in the Late Cambrian, at the time of their maximum global diversity; then, in the Ordovician, as they become restricted toward offshore habitats, their global diversity wanes. (The low diversity peak in the Devonian apparently involves a minor radiation in the Old World province of Paleozoic northern Europe and is not reflected in the North American fossil assemblages.) Pterioids, on the other hand, slowly expand the area of maximum alpha diversity from the Late Ordovician through Permian, a pattern closely matched by the slow increase in their global diversity. Orthids fall in between the environmental patterns of ptychopariids and pterioids but again exhibit close correspondence between local and global diversity.

This empirical correspondence suggests that the distinction between ecological and cladal diversity made in the introduction does not demand distinct explanations. Thus, models of global diversity are allowed a basis in local ecological processes. Below, I summarize a model for global diversification that assumes generalized local ecological interactions as controls on global evolutionary rates.

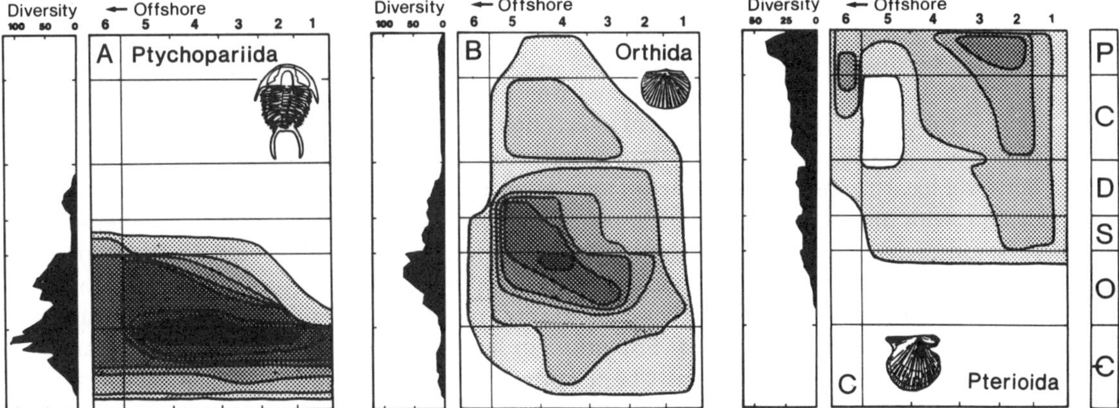

Figure 10. Contoured time-environment diagrams with graphs of global diversity for representative orders from the three evolutionary faunas during the Paleozoic Era. The diagrams are constructed with the same environmental zones as in Fig. 9. Contours indicate the average local (i.e., alpha) diversity for each group in intervals of one genus, with stippling darkening toward higher diversities. The plots to the left of each diagran illustrate the global genus-level diversity for the group. A. Ptychopariid trilobites of the Cambrian fauna. B. Orthid brachiopods of the Paleozoic fauna. C. Pterioid bivalves of the Modern fauna. Each group exhibits a general correlation between local and global diversities as well as some onshore-offshore shift in areas of maximum and minimum alpha diversity. Modified from Sepkoski (1991).

MODELING GLOBAL PATTERNS OF DIVERSITY

Numerous quantitative models of the evolution of global and regional diversity over geologic time have been published (Cailleux, 1950, 1954; Webb, 1969, 1976; Flessa & Imbrie, 1973; Raup et al., 1973; Cisne, 1974; Simberloff, 1974; Rosenzweig, 1975; Stanley 1975, 1979; Dimitriyev, 1978; Sepkoski, 1978, 1979, 1984; Valentine et al., 1978; Carr & Kitchell, 1980; Marshall et al., 1982; Sepkoski & Sheehan, 1983; Walker & Valentine, 1984; Kitchell & Carr, 1985; Raup, 1985; Hoffman & Fenster, 1986; Valentine & Walker, 1986, 1987; Maurer, 1989; Miller & Sepkoski, 1989). Most begin with Stanley's (1975) assumption that diversification is a cladogenetic growth process that can be described by the relationship

$$dD/dt = (o - e)D \tag{1}$$

where D is diversity, t is time, and o and e are the probabilities, or per-taxon rates, of origination and extinction, respectively (for an alternative, see Walker, 1985). If o and e are stochastically constant, the equation predicts exponential diversification, and indeed there are many examples of individual clades that underwent approximately exponential radiation over limited time intervals (Stanley, 1979).

Diversity Dependence

The data in Figures 3 and 4 show that exponential diversification does not dominate the history of total diversity. The pattern of short intervals of rapid increase followed by much slower expansion or virtual steady state suggests that equilibrium processes may be important. This is also suggested by the rebound response of diversity to certain mass extinctions, as noted above.

The simplest way to write an equilibrium into Eq. (1) is to assume that o and e are

226

opposite functions of diversity (MacArthur, 1969; Rosenzweig, 1975; Sepkoski, 1978; Walker, 1985; Maurer, 1989). One might expect extinction to increase with diversity because as more and more species enter communities, average population size must decrease (Levinton, 1979, 1988), and small populations are more vulnerable to extinction, as noted above; increased numbers of interactions, including competition, predation, and disease or parasite transmission, might also increase extinction and/or decrease repopulation after local crashes. Probabilities of speciation, on the other hand, might be expected to decline with diversity, especially if most speciation is a small-population phenomenon: invasion of diverse communities and establishment of peripherally isolated populations may become more difficult, and incipient species in small allopatric populations may have increased difficulty in expanding their geographic ranges and escaping rapid extinction as surrounding diversity increases. For other processes, such as clinal (parapatric) and vicariance speciation, there may also be relationships between rate and diversity: if species in high diversity situations tend to have smaller geographic ranges (Rosenzweig, 1975), then there may be, on average, smaller environmental differences across ranges, lessening chances of clinal speciation (Endler, 1977). There may also be diminished probabilities of random barriers dividing ranges while leaving isolated populations extant, lessening chances of vicariance speciation (Wiley, 1981).

Outcomes of these general expectations are testable with data from the fossil record, and testing has led to some surprises. The logic for a relationship between extinction and diversity seems, at least to me, more plausible than for speciation (although see Walker & Valentine, 1984). However, most investigations have found little correlation between extinction and diversity. For example, Maurer (1989) found virtually no variation in extinction with diversity among species of North American horses through the Miocene. At the family level, Van Valen (1985a) and Gilinsky and Bambach (1987) found that measured probabilities of extinction for a large number of higher taxa remain essentially constant through time (and therefore with respect to diversity). In my own analyses (Sepkoski 1979), I found that total rates of familial extinction (i.e., numbers of extinctions observed over a time interval divided by its estimated duration) varied little with diversity through the Ordovician to Permian (Fig. 11D). I did find a strong correlation for the Cambrian Period, however, but I now suspect this resulted from the frequent biomere extinction events (Palmer, 1982) of the Late Cambrian when the Cambrian fauna was most diverse. Among plants, Knoll et al. (1984) also found only small increases in genus-level extinction with respect to diversity during the Devonian. The reason for this general lack of correlation between extinction and diversity in the fossil record is not clear but it may be that the kinds of species normally sampled are so well established that it takes extraordinary extrinsic, and therefore diversity-independent, perturbations to eliminate them or reduce them to the point where small population vulnerability becomes critical (Raup, pers. comm.)

Although extinction might appear independent of fossil diversity, origination does not. Maurer's (1989) analysis revealed a strong decline in speciation rate among horses as a function of species diversity, with the curve for speciation intersecting the nearly flat curve for extinction at higher diversities. Again at the familial level, Van Valen and Maiorana (1985) and Gilinsky and Bambach (1987) found regular declines in origination rates for various taxa through time, although it is not clear how these relate to diversity. But in my 1979 study, I found a strong relationship between total origination rate and diversity over both the Cambrian Period and the later Paleozoic Era, as illustrated in Figure 11C. Knoll et al. (1984) found a similar relationship for genera of Devonian plant macrofossils.

Hoffman (1985a,b, 1986a,b, 1987, 1989) has not been impressed by these empirical results and has claimed repeatedly that there is no evidence whatsoever for diversity dependence in the fossil record. This is based in part on his rhetorical conflation of argu-

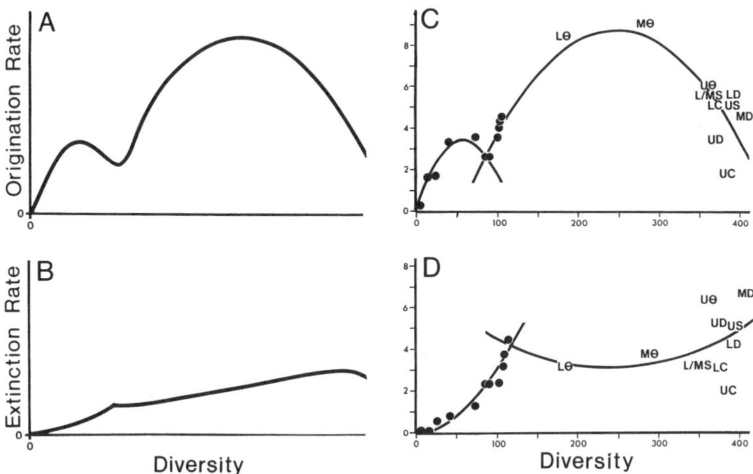

Figure 11. Modeled and observed total rates of origination and extinction as functions of diversity for marine families through the Paleozoic Era. Total rates are defined as the number of originations or extinctions observed during an interval of time divided by the interval's duration. The modeled functions in *A* and *B* were calculated from a pair of coupled logistic equations as in Eq. 3 (see Sepkoski, 1979). The initial maximum in origination rate results from early expansion and then decline of a low-diversity fauna; the second maximum results from replacement by a second fauna with higher equilibrium. Similar relationships hold for extinction rate. Measured rates in *C* and *D* are for all Cambrian families by stages (dots) and all Ordovician to Carboniferous families by series (identifying abbreviations); the curves represent fitted parabolas approximating the modeled functions. These show that Paleozoic families exhibit the patterns expected for a diversity-dependent system in which a rapidly diversifying fauna with a low equilibrium is replaced by a more slowly diversifying fauna with a higher equilibrium. After Sepkoski (1979).

ments for diversity dependence of speciation over evolutionary time with immigration in island biogeographic theory (Hoffman, 1985b). It is also based on his own analysis of "rates" for marine families (Hoffman, 1985a): he divided the Phanerozoic into four intervals, regressed percentages (not rates, as he states) of familial extinction on diversity, and found no significant correlations for either origination or extinction (although, interestingly, there was always a greater negative correlation for origination than for extinction, a pattern expected from the considerations above). His analysis has critical flaws, however. First, he regressed measured per-family percentages, which contain diversity in their denominators, on diversity—a practice that can destroy pattern or induce spurious negative correlations (Sepkoski, 1978). Second, since longer time intervals can accumulate more origination and extinction, he should have divided the percentages by estimated stage durations (even if crude) to produce true rates. Third, and probably most serious, Hoffman used arbitrary divisions of the Phanerozoic to conduct his analysis. As is shown below, diversification with three evolutionary faunas can be complicated, and intervals spanning waxing and waning of more than one fauna can exhibit complex dynamical relationships as low-turnover groups replace high-turnover groups; this is true even if a waning high-turnover fauna has low diversity, since it may still contribute inordinate numbers of extinctions, contrary to Hoffman's (1985a) assertion. The analysis in Figure 11 was designed to accomodate these complexities, and I am surprised Hoffman has never referred to it. (I note that the expectation of diversity dependence for extinction is consistent with Darwin's [1859] metaphor of a wedge for species invading local communities.)

If there is a relationship between speciation and diversity, its formal structure is not immediately obvious, and various functions have been used to describe it (e.g.,

Rosenzweig, 1975; Sepkoski, 1978; Maurer, 1989; see also Walker, 1985). The simplest function, although not necessarily the best, assumes the relationship to be monotonic and linear. If this function is substituted into Eq. (1) and extinction is treated as a constant, a linearly increasing function, or even a linearly decreasing function with smaller slope, the fundamental equation for diversification becomes logistic:

$$dD/dt = rD(1 - D/\hat{D}) \tag{2}$$

where $\hat{D}$ is the equilibrium (i.e., the diversity at which the origination function crosses extinction) and $r = o - e$ for very low diversities.

This model predicts that diversification should have a sigmoidal trajectory through time (Fig. 2B): there should be an initial lag, then a short, approximately exponential radiation, and finally a slowing of diversification and an asymptotic approach to the equilibrium. Indeed, this pattern, or parts of it, are seen in the curve for marine diversity in Figure 4, as argued above.

Three-Phase Kinetic Model

A simple logistic model might work well in homogeneous systems, with groups of clades that share common ecological and evolutionary behavior and thus similar probabilities of speciation and extinction. But the Phanerozoic marine biota consisted of quasi-distinct evolutionary faunas that had different average macroecologies and evolutionary rates. As argued above, the generalist component in such a system might have high rates of speciation but a fairly low limit to its diversity, whereas more specialist faunas might have lower diversification rates but even larger numbers of species (Sepkoski, 1979). Notice that this argument retains diversity dependence but does not demand a set number of niches.

Granting the logic above, it can be argued that each component could diversify logistically but, if interacting in the same environments, will have its diversification rate dampened by the total diversity in the system. This argument leads to a coupled logistic model for diversification:

$$dD_i/dt = r_iD_i(1 - \sum_{j=1}^{n} D_j/\hat{D}_i) \tag{3}$$

where i designates the ith evolutionary fauna and n is the total number of faunas in the system (Sepkoski, 1980, 1984; Sepkoski & Sheehan, 1983; Kitchell & Carr, 1985). (It should be noted that the coupled logistic models of Sepkoski and of Kitchell & Carr are structurally quite similar even though the former used differential calculus to describe the kinetics [for reasons discussed in Sepkoski, 1984] whereas the latter used difference calculus [for reasons discussed in Carr & Kitchell, 1980]. However, the two models have been discussed at times in the literature as if they are quite different because I called mine "equilibrial," emphasizing the contained equilibrium terms, whereas Kitchell & Carr described theirs as "nonequilibrial," emphasizing that solutions to the model with paleontologically realistic parameter values never reach equilibrium over Phanerozoic scales [see Fig. 12]. Their terminology is better.)

The coupled logistic model permits a wonderfully compact description of the pattern of global diversity and the histories of the evolutionary faunas. We do not know how to predict independently the parameter values for Eq. (3), so they must be fitted to the data. But despite the large number of free parameters, there are sufficient data to assess the significance of the fits (Sepkoski, 1984). A fit tuned to the basic pattern for families is illustrated in Figure 12A. (Again let me reiterate that the test of the model assumes that the families constitute an adequate surrogate data base for species; the model is for species,

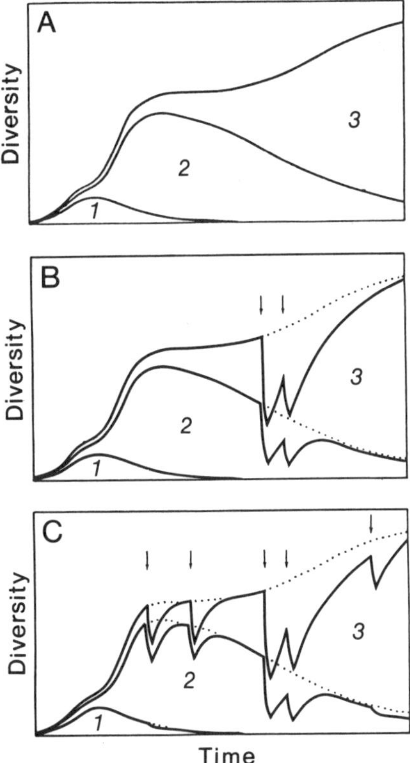

Figure 12. Solutions to three coupled logistic equations (Eq. 3), fitted to diversity patterns of the Phanerozoic evolutionary faunas (Fig. 4). *A.* The simple model, showing the replacement and diversity increase of three successive faunas, each with lower diversification rate and higher equilibrium. *B.* A solution with the same parameters but with two perturbations (arrows) tuned to the timing and total magnitude of the end-Permian and end-Triassic mass extinctions. The perturbations effect a turnover in dominance from phase 2 to phase 3 and are followed by rebounds back to the trajectories of the unperturbed system (dotted lines). *C.* The same solution as above but with five perturbations tuned to the major mass extinctions of the Phanerozoic. After Sepkoski (1984).

not higher taxa.) The fitted model encapsulates many long-term trends in the record, especially over the Paleozoic Era: it produces the first two phases of diversfication; it produces the peak in diversity of the Cambrian fauna slightly before the end of the Cambrian phase; it produces the long, roughly exponential decline of the Cambrian fauna, without any expansion during the Ordovician radiations; it produces the long initial lag of the Paleozoic fauna, followed by an outbreak to effect the Ordovician radiations; and it produces a long interval during which the Paleozoic fauna gradually loses diversity while the Modern fauna gradually expands, leaving total diversity at a nearly constant level. It also produces a secular decline in extinction rate for the entire system (Sepkoski, 1984), similar to the Phanerozoic decline measured by Raup and Sepkoski (1982) and Van Valen (1984).

Perturbations

The simple model does not produce the patterns around the end-Permian mass extinction. But mass extinctions are now generally agreed to reflect extrinsic perturbations of the biosphere, and the logic of the model is strictly intrinsic. Extrinsic perturbations can be easily incorporated into the model (Carr & Kitchell, 1980; Sepkoski, 1980, 1984; Kitchell & Carr, 1985). A perturbation can be treated as a temporary acceleration of extinction, which translates into a transient decline of the equilibrium, $\hat{D}$. If the equilibria of all faunas are reduced by a common proportion, the fauna with the highest value of parameter r suffers most rapidly (Sepkoski, 1984). This closely parallels what we see in the record (Table 1): the Cambrian evolutionary fauna loses proportionally more diversity at the early mass extinctions than do the other two faunas with their lower diversification rates; and the Modern fauna with the lowest rate always loses the least proportional diversity, even during the Mesozoic-Cenozoic interval when it dominates the oceans.

Table 1. Percentage loss of families within evolutionary faunas at each of the five major mass extinctions of the Phanerozoic (Fig. 4). Parentheses indicate percentages calculated from standing diversities of five or fewer families. The Cambrian fauna (Cm), with the highest turnover rate during "normal" times, suffered proportionally more extinction than other faunas when it had substantial diversity. The Modern fauna (Md), with the lowest normal turnover rate, always suffered less than the Paleozoic fauna (Pz). After Sepkoski (1984).

Extinction Event	Evolutionary Fauna		
	Cm	Pz	Md
1) End Ordovician	−44%	−30%	− 4%
2) Late Devonian	−44%	−21%	−10%
P) End Permian	(−40%)	−79%	−27%
3) End Triassic	(0%)	−55%	−10%
4) End Cretaceous	(0%)	−25%	−16%

Figure 12B presents a solution of the coupled model with two perturbations, tuned to depress total diversity to levels after the end-Permian and end-Triassic events. Several things happen that closely resemble empirical patterns. First, there is a switch in dominance from the Paleozoic to the Modern fauna. Again, this is because the Paleozoic fauna has a higher r value; thus, the properties that allowed it to radiate so impressively during the Ordovician may have doomed it in the Permian. But the Paleozoic fauna was not out of the picture at the dawn of the Triassic. In the model solution, it begins a vigorous rebound, diversifying with a per-taxon rate more rapid than the Modern fauna, exactly as seen in the data (Figs. 4 and 5). But it again suffers disproportionately at the second perturbation; although this perturbation in the model is milder than the first, the analog of the Paleozoic fauna actually drops to a resulting diversity lower than following the first perturbation, just as is seen in the data. The model then produces a rebound of the Paleozoic fauna that is slower because the Modern analog is now more diverse. The model Paleozoic fauna continues to diversify slowly until total diversity exceeds the diversity realized in the counterpart of the Paleozoic Era. Thereafter, the diversity of the Paleozoic fauna declines, very close to the trajectory expected from the unperturbed system. In the meantime, the Modern fauna in the model, and the total diversity of the system, rebounds from the two perturbations, at first rapidly and then more slowly, joining the trajectory of the unperturbed model at a point comparable to the mid Cretaceous, where Vermeij (1977) and others have seen major changes in the nature of the marine fauna.

There are several important conclusions to reach here. First, the impressive Mesozoic-Cenozoic rise in diversity is actually a composite of two phenomena: the rebounds from the end-Permian and end-Triassic mass extinctions, and the increase in global diversity that would have resulted anyway from unperturbed expansion of the Modern fauna. Second, the extraordinary end-Permian event, which eliminated up to 96% of marine species, was lost from the memory of the system—at least with regard to diversity—after about 150 myr. The model suggests that the changes in diversity and in evolutionary faunas over the most recent 100 Ma would have occurred anyway, even without the end-

Permian event. Kitchell and Carr (1985) demonstrated with the model that no matter where in time an event of the magnitude of the end-Permian occurred, the prospect was the same: eventual decline of remnants of the Paleozoic fauna and hegemony of the slowly diversifying Modern fauna.

Figure 12C presents one more solution with additional perturbations to simulate the end-Ordovician, late Devonian, and end-Cretaceous mass extinctions. These events perturb the system but do not change the fundamental patterns of diversification. The earlier two events depress diversity but are followed by rebounds back to previous levels, just as seen in the empirical pattern (Fig. 4). The analog of the end-Cretaceous event similarly reduces diversity, as it was constrained to do, but is followed by rebound and then a continuation of the more gradual increase in global diversity.

These last two fits of the model suggest, then, that the largest-scale changes in marine diversity and faunal composition can be rationalized rather easily. Despite the overgeneralized thinking about ecological constraints on evolution, and despite the use of traditional families to test species-level ideas, simple nonlinear mathematical expressions can relate the peculiar behavior of the Paleozoic fauna during the Mesozoic and Cenozoic Eras to patterns of total global diversity. This is far more than any published non-interactive or non-equilibrial model has offered.

Much more needs to be known, however. The link between local and global patterns rests on correlations, and we need a more precise and rigorously tested set of explanations for how local ecological processes relate to biogeographic expanse, and how this is reflected globally; and we need this especially for marine biotas, where speciation, expansion, and extinction are still especially poorly understood. For the paleontological record, theories of diversity desperately need species-level data. Where there are species data for isolated clades, the information is usually inadequate because it is never evident whether the species were radiating into an ecological vaccuum or were interacting with other groups (cf. Miller & Sepkoski, 1989). It is not clear whether we will ever have adequate data for all readily preserved species in a single ecosystem, but in the meantime we need further information on how well arbitrarily defined higher taxa reflect underlying diversification of the fundamental units of evolution and ecology. Polemics about the necessity of monophyletic groupings will not do. Instead, we need careful analyses—both modeling to educate our intuitions and empirical study to provide us with case examples—so that our understanding of what we are studying continues to grow. Finally, we need case studies of individual taxa from both the living world and the fossil record that sort of follow expected patterns but sort of don't, in order to keep us fascinated by the orderliness and orneriness of diversity.

CONCLUSIONS AND SUMMARY

This is an overview of mostly personal work on diversity in the oceans through the ages. In part it is idiosyncratic, in part it is defensive. But I believe the study of evolutionary diversity, with its manifestation in fossil data and modern distributions, can provide important insights into the boundary conditions within which many macroevolutionary processes have operated: the exploration and filling of morphospace, the rate of evolutionary trends, and the effects of species sorting. Some of these processes, of course, will have reciprocal effects on the boundary constraints, as noted by Sepkoski (1984) and Kitchell and Carr (1985).

In this essay, I have attempted to argue several points that need more consideration and analysis:

1. Data from the fossil record, even for higher taxa, contribute valuable information on the macro-scale history of diversity, despite problems of incomplete and biased sampling and of taxonomic inconsistencies.

2. There is a highly corroborated pattern in the history of Phanerozoic marine diversity that involves three phases of diversification induced by differential expansion of three quasi-distinct evolutionary faunas with different evolutionary rates and ecological characteristics: a Cambrian fauna with high diversification rate and generalized ecologies, a Paleozoic fauna with intermediate rate and numerous suspension-feeding epifauna, and a Modern fauna with low diversification rate and diverse ecologies. Each evolutionary fauna has a higher characteristic diversity which seems to relate to a progressively broader utilization of ecospace.

3. Paleontological patterns of global diversity in the oceans correlate strongly with patterns of community-wide diversity, suggesting that the ultimate processes governing diversity are of a local, ecological nature.

4. A very compact mathematical model of global diversity can be constructed on the basis of empirically justified assumptions about diversity dependence of speciation rates. This elementary model, which has the form of coupled logistic functions, successfully interrelates disparate aspects of the histories of the evolutionary faunas as well as seemingly unrelated patterns in global diversification, presuming again that we accept paleontologic families as adequately reflecting the history of biospecies, or at least interacting biospecies.

ACKNOWLEDGMENTS

Research reported here received partial support from NASA grant NAGW-1693.

LITERATURE CITED

Aronson, R. B. 1989. A community level test of the Mesozoic marine revolution theory. *Paleobiology* 15:20–25.

Ausich, W. I. & D. J. Bottjer. 1982. Tiering in suspension-feeding communities in soft substrata throughout the Phanerozoic. *Science* 216:173–174.

Ausich, W. I. & D. J. Bottjer. 1985. Phanerozoic tiering in suspension-feeding communities on soft substrata: implications for diversity. Pp. 255–274. *In*: J. W. Valentine (ed.), *Phanerozoic Diversity Patterns: Profiles in Macroevolution*. Princeton University Press: Princeton, NJ.

Bambach, R. K. 1977. Species richness in marine benthic habitats through the Phanerozoic. *Paleobiology* 3:152–167.

Bambach, R. K. 1983. Ecospace utilization and guilds in marine communities through the Phanerozoic. Pp. 719–746. *In*: M. J. S. Tevesz & P. L. McCall (eds.), *Biotic Interactions in Recent and Fossil Benthic Communities*. Plenum: New York.

Bambach, R. K. 1985. Classes and adaptive variety: the ecology of diversification in marine faunas through the Phanerozoic. Pp. 191–253. *In*: J. W. Valentine (ed.), *Phanerozoic Diversity Patterns: Profiles in Macroevolution*. Princeton University Press: Princeton NJ.

Berry, W. B. N. 1972. Early Ordovician bathyurid province lithofacies, biofacies, and correlations— their relationship to a proto-Atlantic Ocean. *Lethaia* 5:69–84.

Berry, W. B. N. 1974. Types of early Paleozoic faunal replacements in North America: their relationship to environmental change. *Journal of Geology* 82:371–382.

Bottjer, D. J. & W. I. Ausich. 1986. Phanerozoic development of tiering in soft substrata suspension-feeding communities. *Paleobiology* 12:400–420.

Bottjer, D. J. & D. Jablonski. 1988. Paleoenvironmental patterns in the evolution of post-Paleozoic benthic marine invertebrates. *Palaios* 3:540–560.

Brasier, M. D. 1979. The Cambrian radiation event. Pp. 103–159. *In*: M. R. House (ed.), *The Origin of Major Invertebrate Groups*. Academic Press: London.

Brasier, M. D. 1989. Towards a biostratigraphy of the earliest skeletal biotas. Pp. 117–165. *In*: J. W. Cowie & M. D. Brasier (eds.), *The Precambrian-Cambrian Boundary*. Clarendon Press: Oxford.

Bretsky, P. W. 1968. Evolution of Paleozoic marine invertebrate communities. *Science* 159:1231–1233.

Bretsky, P. W. 1969. Evolution of Paleozoic benthic marine invertebrate communities. *Palaeogeography, Palaeoclimatology, Palaeoecology* 6:45–59.

Brown, J. H. & B. A. Maurer. 1989. Macroecology: the division of food and space among species on continents. *Science* 243:1145–1150.

Cailleux, A. 1950. Progression géométrique du nombre des espèces et vie en expansion. *C.R.S. de la Société Géologique de France* 13:222–224.

Cailleux, A. 1954. How many species? *Evolution* 8:83–84.

Carr, T. R. & J. A. Kitchell. 1980. Dynamics of taxonomic diversity. *Paleobiology* 6:427–443.

Cisne, J. L. 1974. Evolution of the world fauna of aquatic free-living arthropods. *Evolution* 28:337–366.

Conway Morris, S. 1986. The community structure of the Middle Cambrian Phyllopod bed. *Palaeontology* 29:423–467.

Cook, R. E. 1969. Variation in species diversity of North American birds. *Systematic Zoology* 18:63–84.

Cracraft, J. 1982. A nonequilibrium theory for the rate-control of speciation and extinction and the origin of macroevolutionary patterns. *Systematic Zoology* 31:348–365.

Cracraft, J. 1985. Biological diversification and its causes. *Annuals of the Missouri Botanical Garden* 72:794–822.

Darwin, C. 1859. *On the Origin of Species by Means of Natural Selection.* John Murray: London. 490 pp.

Diamond, J. M. 1984. "Normal" extinctions in isolated populations. Pp. 191–246. *In*: M. Nitecki (ed.), *Extinctions.* University of Chicago Press; Chicago.

Dimitriyev, V.Yu. 1978. Some aspects of the study of changes in the systematic diversity of fossil organisms. *Paleontological Journal* 12:257–265.

Endler, J. A. 1977. *Geographic Variation, Speciation and Clines.* Princeton University Press: Princeton, NJ. 246 pp.

Flessa, K. W. 1975. Area, continental drift, and mammalian diversity. *Paleobiology* 1:189–194.

Flessa, K. W. & J. Imbrie. 1973. Evolutionary pulsations: evidence from Phanerozoic diversity patterns. Pp. 247–285. *In*: E. H. Tarling & S. K. Runcorn (eds.), *Implications of Continental Drift to the Earth Sciences.* Academic Press: London.

Gilinsky, N. L. & R. K. Bambach. 1987. Asymmetrical patterns of origination and extinction in higher taxa. *Paleobiology* 13: 427–445.

Glaessner, M. F. 1984. *Dawn of Animal Life. A Biohistorical Study.* Cambridge University Press: Cambridge. 244 pp.

Hoffman, A. 1985a. Biotic diversification in the Phanerozoic: diversity independence. *Palaeontology* 28:387–391.

Hoffman, A. 1985b. Island biogeography and paleobiology: in search for evolutionary equilibria. *Biological Reviews* 60: 455–471.

Hoffman, A. 1986a. Neutral model of Phanerozoic diversification: implications for macroevolution. *Neues Jahrbuch für Geologie und Paläontologie Abhandlungen* 172: 219–244.

Hoffman, A. 1986b. Mass extinctions, diversification, and the nature of paleontology. *Revista Española de Paleontología* 1: 101–107.

Hoffman, A. 1987. Neutral model of taxonomic diversification in the Phanerozoic: a methodological discussion. Pp. 133–146. *In*: M. H. Nitecki & A. Hoffman (eds.), *Neutral Models in Biology.* Oxford University Press: New York.

Hoffman, A. 1989. *Arguments on Evolution. A Paleontologist's Perspective.* Oxford University Press: New York. 274 pp.

Hoffman, A. & E. J. Fenster. 1986. Randomness and diversification in the Phanerozoic: a simulation. *Palaeontology* 29:655–663.

Holman, E. W. 1989. Some evolutionary correlates of higher taxa. *Paleobiology* 15:357–363.

Hutchinson, G. E. 1959. Homage to Santa Rosalia or why are there so many kinds of animals. *American Naturalist* 93:145–159.

Jablonski, D. & D. J. Bottjer. 1983. Soft-bottom epifaunal suspension-feeding assemblages in the Late Cretaceous: implications for the evolution of benthic paleocommunities. Pp. 747–812. *In*: M. J. S. Tevesz & P. L. McCall (eds.), *Biotic Interactions in Recent and Fossil Communities.* Plenum: New York.

Jablonski, D. & D. J. Bottjer. 1988. Onshore-offshore evolutionary patterns in post-Paleozoic echinoderms: a preliminary analysis. Pp. 81–90. *In*: R. D. Burke, P. V. Mladenov, P. Lambert & R. L. Parsley (eds.), *Echinoderm Biology.* Balkema: Rotterdam.

Jablonski, D. & D. J. Bottjer. 1990a. The ecology of evolutionary innovation: the fossil record. Pp. 253–288. *In*: M. H. Nitecki (ed.), *Evolutionary Innovations.* University of Chicago Press: Chicago.

Jablonski, D. & D. J. Bottjer. 1990b. The origin and diversification of major groups: environmental patterns and macroevolutionary lags. Pp. 17–57. *In*: P. D. Taylor & G. P. Larwood (eds.), *Major Evolutionary Radiations.* Oxford University Press: Oxford.

Jablonski, D. & D. J. Bottjer. 1990c. Onshore-offshore trends in marine invertebrate evolution. Pp. 21–75. *In*: R. M. Ross & W. D. Allmon (eds.), *Causes of Evolution: A Paleontological Perspective.* University of Chicago Press: Chicago.

Jablonski, D. & K. W. Flessa. 1986. The taxonomic structure of shallow-water marine faunas: implications for Phanerozoic extinctions. *Malacologia* 27:43–66.

Jablonski, D., Sepkoski, J. J., Jr., Bottjer, D. J. & P. M. Sheehan. 1983. Onshore-offshore patterns in the evolution of Phanerozoic shelf communities. *Science* 222:1123–1125.

Järvinen, O. 1982. Species-to-genus ratios in biogeography: a historical note. *Journal of Biogeography* 9:363–370.

Johnson, R. G. 1964. The community approach to paleoecology. Pp. 107–134. *In*: J. Imbrie & N. D. Newell (eds.), *Approaches to Paleoecology.* Wiley: New York.

Kammer, T. W., Brett, C. E., Boardman, II, D. R., & R. H. Mapes. 1986. Ecologic stability of the dysaerobic biofacies during the late Paleozoic. *Lethaia* 19:109–121.

Kitchell, J. A. & T. R. Carr. 1985. Nonequilibrium model of diversification: faunal turnover dynamics. Pp. 277–310. *In*: J. W. Valentine (ed.), *Phanerozoic Diversity Patterns. Profiles in Macroevolution.* Princeton University Press: Princeton, NJ.

Kitchell, J. A. & N. MacLeod. 1988. Macroevolutionary interpretations of symmetry and synchroneity in the fossil record. *Science* 240:1190–1193.

Knoll, A. H., Niklas, K. J., Gensel, P. G. & B. H. Tiffney. 1984. Character diversification and patterns of evolution in early vascular plants. *Paleobiology* 10:34–47.

Levinton, J. S. 1979. A theory of diversity equilibrium and morphological evolution. *Science* 204:335–336.

Levinton, J. S. 1988. *Genetics, Paleontology, and Macroevolution.* Cambridge University Press: Cambridge. 637 pp.

MacArthur, R. H. 1969. Patterns of communities in the tropics. *Biological Journal of the Linnaean Society* 1:19–30.

MacArthur, R. H. & E. O. Wilson. 1967. *The Theory of Island Biogeography.* Princeton University Press: Princeton, NJ. 203 pp.

Magurran, A. E. 1988. *Ecological Diversity and Its Measurement.* Princeton University Press: Princeton, NJ. 179 pp.

Marshall, L. G., Webb, S. D., Sepkoski, Jr., J. J., & D. M. Raup. 1982. Mammalian evolution and the Great American Interchange. *Science* 215:1351–1357.

Maurer, B. A. 1989. Diversity-dependent species dynamics: incorporating effects of population-level processes on species dynamics. *Paleobiology* 15:133–146.

May, R. M. 1988. How many species are there on Earth? *Science* 241:1441–1449.

Mayr, E. 1963. *Animal Species and Evolution.* Belknap Press: Cambridge, MA. 767 pp.

McKinney, M. L. 1986. Ecological causation of heterochrony: a test and implications for evolutionary theory. *Paleobiology* 12:282–289.

Miller, A. I. & J. J. Sepkoski, Jr. 1989. Modeling bivalve diversification: the effect of interaction on a macroevolutionary system. *Paleobiology* 14:364–369.

Newell, N. D. 1967. Revolutions in the history of life. *Geological Society of America Special Paper* 89:63–91.

Palmer, A. R. 1982. Biomere boundaries: a possible test for extraterrestrial perturbation of the biosphere. *Geological Society of America Special Paper* 190:469–476.

Patterson, C. & A. B. Smith. 1987. Is the periodicity of extinctions a taxonomic artefact? *Nature* 330:248–251.

Patterson, C. & A. B. Smith. 1989. Periodicity in extinction: the role of systematics. *Ecology* 70:802–811.

Paul, C. R. C. 1982. The adequacy of the fossil record. Pp. 75–117. *In*: K. A. Joysey & A. E. Friday (eds.), *Problems of Phylogenetic Reconstruction.* Academic Press: London.

Paul, C. R. C. 1985. The adequacy of the fossil record reconsidered. *Special Papers in Palaeontology* 33:7–15.

Pielou, E. C. 1975. *Ecological Diversity.* Wiley: New York. 165 pp.

Raup, D. M. 1972. Taxonomic diversity during the Phanerozoic. *Science* 177:1065–1071.

Raup, D. M. 1975. Taxonomic diversity estimation using rarefaction. *Paleobiology* 1:333–342.

Raup, D. M. 1976a. Species diversity in the Phanerozoic: an interpretation. *Paleobiology* 2:289–297.

Raup, D. M. 1976b. Species diversity in the Phanerozoic: a tabulation. *Paleobiology* 2:279–288.

Raup, D. M. 1979a. Biases in the fossil record of species and genera. *Bulletin of the Carnegie Museum of Natural History* 13:85–91.

Raup, D. M. 1979b. Size of the Permo-Triassic bottleneck and its evolutionary implications. *Science* 206:217–218.

Raup, D. M. 1985. Mathematical models of cladogenesis. *Paleobiology* 11:42–52.

Raup, D. M., Gould, S. J., Schopf, T. J. M. & D. S. Simberloff. 1973. Stochastic models of phylogeny and the evolution of diversity. *Journal of Geology* 81:525–542.

Raup, D. M. & J. J. Sepkoski, Jr. 1982. Mass extinctions in the marine fossil record. *Science* 215:1501–1503.

Rosenzweig, M. L. 1975. On continental steady states of species diversity. Pp. 121–140. *In*: M. L. Cody & J. M. Diamond (eds.), *Ecology and Evolution of Communities*. Belknap Press: Cambridge, MA.

Rowell, A. J. 1981. The Cambrian brachiopod radiation—monophyletic or polyphyletic origins? Pp. 184–187. *In*: M. E. Taylor (ed.), *Short Papers for the Second International Symposium on the Cambrian System*. United States Geological Survey Open-File Report 81–743.

Runnegar, B. & J. Pojeta, Jr. 1985. Origin and diversification of the Mollusca. Pp. 1–57. *In*: E. R. Trueman & M. R. Clarke (eds.), *The Mollusca*, v. 10. Academic Press: Orlando, FL.

Schopf, T. J. M. 1979. The role of biogeographic provinces in regulating marine faunal diversity through geologic time. Pp. 449–457. *In*: J. Gray & A. J. Boucot (eds.), *Historical Biogeography, Plate Tectonics, and the Changing Environment*. Oregon State University Press: Corvallis, OR.

Seilacher, A. 1974. Flysch trace fossils: evolution of behavioral diversity in the deep-sea. *Neues Jahrbuch für Geologie und Paläontologie, Monatshefte* 4:233–245.

Seilacher, A. 1989. Vendozoa: organismic construction in the Proterozoic biosphere. *Lethaia* 33:229–239.

Sepkoski, J. J., Jr. 1978. A kinetic model of Phanerozoic taxonomic diversity. I. Analysis of marine orders. *Paleobiology* 4:223–251.

Sepkoski, J. J., Jr. 1979. A kinetic model of Phanerozoic taxonomic diversity. II. Early Phanerozoic families and multiple equilibria. *Paleobiology* 5:222–251.

Sepkoski, J. J., Jr. 1980. The three great evolutionary faunas of the Phanerozoic marine fossil record. *Geological Society of America Abstracts with Program* 12:520.

Sepkoski, J. J., Jr. 1981. A factor analytic description of the Phanerozoic marine fossil record. *Paleobiology* 7:36–53.

Sepkoski, J. J., Jr. 1982a. A compendium of fossil marine families. *Milwaukee Public Museum Contributions in Biology and Geology*, no. 51. 125 pp.

Sepkoski, J. J., Jr. 1982b. Mass extinctions in the Phanerozoic oceans: a review. *Geological Society of America Special Paper* 180:283–289.

Sepkoski, J. J., Jr. 1984. A kinetic model of Phanerozoic taxonomic diversity. III. Post-Paleozoic families and mass extinctions. *Paleobiology* 10:246–267.

Sepkoski, J. J., Jr. 1988. Alpha, beta, or gamma: where does all the diversity go? *Paleobiology* 14:221–234.

Sepkoski, J. J., Jr. 1989a. Periodicity in extinction and the problem of catastrophism in the history of life. *Journal of the Geological Society, London* 146:7–19.

Sepkoski, J. J., Jr. 1989b. The importance of extinction resistance in onshore-offshore changes in faunal dominance during evolutionary radiations. *Geological Society of America Abstracts with Program* 21(6):A30.

Sepkoski, J. J., Jr. 1990. Evolutionary faunas. Pp. 37–41. *In*: D. E. G. Briggs & P. R. Crowther (eds.), *Palaeobiology: A Synthesis*. Blackwell: Oxford.

Sepkoski, J. J., Jr. 1991. A model of onshore-offshore change in faunal diversity. *Paleobiology* 17:58–77.

Sepkoski, J. J., Jr. 1992. Proterozoic-Early Cambrian diversification of metazoans and metaphytes. *In*: J. W. Schopf & C. Klein (eds.), *The Proterozoic Biosphere: A Multidisciplinary Study*. Cambridge University Press: Cambridge. *In press*.

Sepkoski, J. J., Jr., Bambach, R. K., Raup, D. M. & J. W. Valentine. 1981. Phanerozoic marine diversity and the fossil record. *Nature* 293:435–437.

Sepkoski, J. J., Jr. & A. I. Miller. 1985. Evolutionary faunas and the distribution of Paleozoic marine communities in space and time. Pp. 153–190. *In*: J. W. Valentine (ed.), *Phanerozoic Diversity Patterns: Profiles in Macroevolution*. Princeton University Press: Princeton, NJ.

Sepkoski, J. J., Jr. & P. M. Sheehan. 1983. Diversification, faunal change, and community replacement during the Ordovician radiations. Pp. 673–717. *In*: M. J. S. Tevesz & P. J. McCall (eds.), *Biotic Interactions in Recent and Fossil Benthic Communities*. Plenum: New York.

Signor, P. W., III. 1982. Species richness in the Phanerozoic: compensating for sampling bias. *Geology* 10:625–628.

Signor, P. W., III. 1985. Real and apparent trends in species richness through time. Pp. 129–150. *In*: J. W. Valentine (ed.), *Phanerozoic Diversity Patterns: Profiles in Macroevolution*. Princeton University Press: Princeton, NJ.

Signor, P. W. 1990. Patterns of diversification. Pp. 130–135. *In*: D. E. G. Briggs and P. R. Crowther (eds.), *Palaeobiology. A Synthesis*. Blackwell: Oxford.

Simberloff, D. S. 1970. Taxonomic diversity of island biotas. *Evolution* 24:23–47.

Simberloff, D. S. 1974. Permo-Triassic extinctions: effects of area on biotic equilibrium. *Journal of Geology* 82:267–274.

Simberloff, D. S. 1988. The contribution of population and community biology to conservation science. *Annual Reviews of Ecology and Systematics* 19:473–511.

Simpson, G. G. 1960. The history of life. Pp. 117–180. *In*: S. Tax (ed.), *Evolution after Darwin*, v. 1. University of Chicago Press: Chicago, IL.

Smith, A. B. 1988. Patterns of diversification and extinction in early Palaeozoic echinoderms. *Palaeontology* 31:799–828.

Smith, C. A. F., III. 1977. Diversity associations as stochastic variables. *Paleobiology* 3:41–48.

Stanley, S. M. 1975. A theory of evolution above the species level. *Proceedings of the National Academy of Science USA* 72: 646–650.

Stanley, S. M. 1978. Chronospecies' longevities, the origin of genera, and the punctuational model of evolution. *Paleobiology* 4:26–40.

Stanley, S. M. 1979. *Macroevolution: Pattern and Process*. Freeman: San Francisco, CA. 332 pp.

Stanley, S. M. 1986. Population size, extinction, and speciation: the fission effect in Neogene Bivalvia. *Paleobiology* 12:89–110.

Stehli, F. G., McAlester, A. L. & C. E. Helsley. 1967. Taxonomic diversity of Recent bivalves and some implications for geology. *Geological Society of America Bulletin* 78:455–466.

Templeton, A. R. 1980. Modes of speciation and inferences based on genetic distances. *Evolution* 34:719–729.

Thayer, C. W. 1979. Biological bulldozers and the evolution of marine benthic communities. *Science* 203:458–461.

Thayer, C. W. 1983. Sediment-mediated biological disturbance and the evolution of marine benthos. Pp. 480–625. *In*: M. J. S. Tevesz & P. J. McCall (eds.), *Biotic Interactions in Recent and Fossil Benthic Communities*. Plenum: New York.

Valentine, J. W. 1969. Patterns of taxonomic and ecological structure of the shelf benthos during Phanerozoic time. *Palaeontology* 12:684–709.

Valentine, J. W. 1970. How many marine invertebrate fossil species? A new approximation. *Journal of Paleontology* 44: 410–415.

Valentine, J. W. 1985. Diversity as data. Pp. 3–8. *In*: J. W. Valentine (ed.), *Phanerozoic Diversity Patterns: Profiles in Macroevolution*. Princeton University Press: Princeton, NJ.

Valentine, J. W., Foin, T. C. & D. Peart. 1978. A provincial model of Phanerozoic marine diversity. *Paleobiology* 4:55–66.

Valentine, J. W. & E. M. Moores. 1972. Global tectonics and the fossil record. *Journal of Geology* 80:167–184.

Valentine, J. W. & T. D. Walker. 1986. Diversity trends within a model taxonomic hierarchy. *Physica* 22D:31–42.

Valentine, J. W. & T. D. Walker. 1987. Extinctions in a model taxonomic hierarchy. *Paleobiology* 13:193–207.

Van Valen, L. 1973. A new evolutionary theory. *Evolutionary Theory* 1:1–30.

Van Valen, L. M. 1984. A resetting of Phanerozoic community evolution. *Nature* 307:50–52.

Van Valen, L. M. 1985a. How constant is extinction? *Evolutionary Theory* 7:93–106.

Van Valen, L. M. 1985b. A theory of origination and extinction. *Evolutionary Theory* 7:133–142.

Van Valen, L. M. & V. C. Maiorana. 1985. Patterns of origination. *Evolutionary Theory* 7:107–125.

Vermeij, G. J. 1977. The Mesozoic marine revolution: evidence from snails, predators, and grazers. *Paleobiology* 3:245–258.

Vermeij, G. J. 1978. *Biogeography and Adaptation*. Harvard University Press: Cambridge, MA. 332 pp.

Vermeij, G. J. 1983. Shell-breaking predation through time. Pp. 649–669. *In*: M. J. S. Tevesz & P. J. McCall (eds.), *Biotic Interactions in Recent and Fossil Benthic Communities*. Plenum: New York.

Vermeij, G. J. 1987. *Evolution and Escalation. An Ecological History of Life*. Princeton University Press: Princeton, NJ. 527 pp.

Walker, T. D. 1985. Diversification functions and the rate of taxonomic evolution. Pp. 311–334. *In*: J. W. Valentine (ed.), *Phanerozoic Diversity Patterns. Profiles in Macroevolution*. Princeton University Press: Princeton, NJ.

Walker, T. D. & J. W. Valentine. 1984. Equilibrium models of evolutionary species diversity and the number of empty niches. *American Naturalist* 124:887–899.

Webb, S. D. 1969. Extinction-origination equilibria in late Cenozoic land mammals of North America. *Evolution* 23:688–702.

Webb, S. D. 1976. Mammalian faunal dynamics of the Great American Interchange. *Paleobiology* 2:220–234.

Wiley, E. O. 1981. *Phylogenetics. The Theory and Practice of Phylogenetic Systematics*. John Wiley: New York. 439 pp.

Estimating Probabilities of Origination and Extinction

Norman L. Gilinsky

Abstract. In this paper I estimate the probabilities of familial origination and extinction in twelve phyla and 135 orders of fossil marine invertebrate animals. From among the phyla, the Archaeocyatha had the highest probabilities of familial origination and extinction, and the Annelida had the lowest. More typical "shelly" marine invertebrate groups are arrayed between these extremes in a manner that is largely consistent with the impressions of other workers regarding the pace of evolution in these groups. For the orders, the distribution of origination and extinction probabilities is right skewed, meaning that most orders have relatively low to moderate familial probabilities of origination and extinction. The paucity of orders with high probabilities may reflect a tendency (inferred from simulations) for the diversity to fluctuate substantially when probabilities are high. Such fluctuations might cause higher taxa to become extinct with high frequency and thus leave short, if any, fossil records. Also estimated were the temporal trajectories of the familial origination and extinction probabilities for all of fossil marine invertebrate animal life. Probabilities of extinction increase in association with the five Phanerozoic mass extinction events. Probabilities of origination generally decline for several stages prior to the events of mass extinction. The only mass extinction that is not preceded by a decline of origination probability is the great Cretaceous extinction. It seems to be an extraordinary, singular event.

The probabilities of origination and extinction are estimated using the mathematical theory of discrete branching processes and maximum likelihood. The theory of discrete branching processes is more elementary than the theory of continuous branching processes, yet it allows adequate estimations of the evolutionary parameters when they are estimated by maximum likelihood. Because the topology of life's history is that of a growing, branching tree, the application of the mathematical theory of branching processes to the problem of estimating the probabilities of origination and extinction is both natural and desirable.

INTRODUCTION

Haeckel's evolutionary tree (Haeckel, 1874) is an artful rendering of the history of life as it was understood at the end of the last century (Fig. 1). But the tree of life is more than mere metaphor. Not only does the tree depict the branching nature of the macroevolutionary process, the process of tree growth also has significance for the study of evolution. The terminal shoots of a mature tree are just a small sample of the twigs and shoots that existed during the entire history of the tree. During growth, some grew and prospered, forming sturdy branches that produced new twigs and shoots. Others withered and died. So the form of a mature tree, in a very real sense, is a product of the differential branching, persistence, and death of its constituent shoots, and the vital parts of a mature tree—the

Dr. Gilinsky is with the Department of Geological Sciences, Virginia Polytechnic Institute and State University, Blacksburg, VA 24061, USA.

238

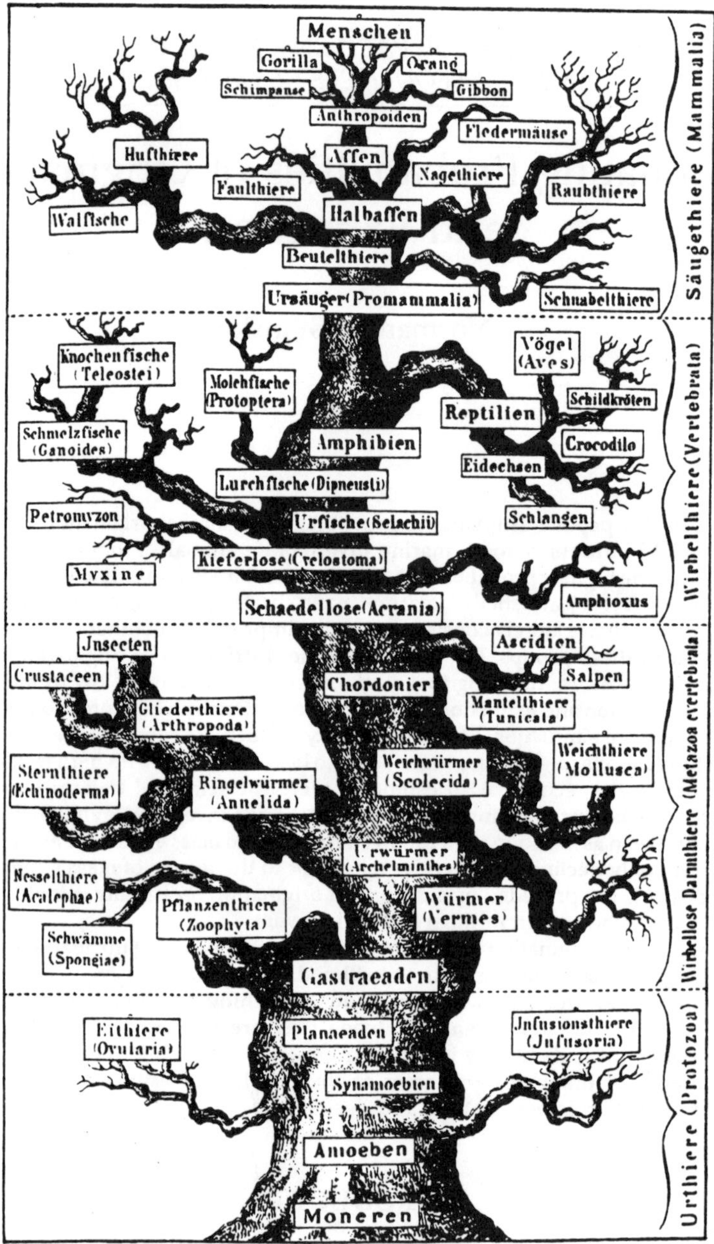

Figure 1. Haeckel's (1874) evolutionary tree of life.

parts that are alive and well today—are the direct descendants of the parts of the tree that were most vigorous in their growth and branching in the past.

And so it is with the diversity of life. The organisms that populate the world today—the mammals, birds, molluscs, arthropods, angiosperms, and others—are those that had a vigorous history of branching, and those that have been most resistant to the forces of extinction. So one of the important aims of paleontology is to establish quantitatively which groups of organisms were vigorous in producing diversity, and which were not, and to develop an understanding of why, biologically, some groups of organisms survived and prospered, while others declined to extinction. In this paper I shall attempt to quantify these propensities in groups of fossil organisms by estimating their probabilities of branching, persistence, and extinction. The estimations will make use of the mathematical theory of discrete branching processes.

Other workers have (implicitly and explicitly) also applied branching theory to paleontological problems. Yule (1924) was probably the first, and Van Valen (1973, 1979), Raup (1978, 1985) and Stanley (1979, 1985) are among those who have applied the theory recently. Their work, which employed the theory of continuous branching processes, was devoted to studying evolutionary rates. Here I attempt to estimate probabilities. A discrete branching model is much more elementary than a continuous one, but it nonetheless allows for adequate estimations of the branching and extinction parameters. Furthermore, a discrete branching model avoids some of the difficulties that researchers have had with continuous branching models, including those of pseudoextinction (Stanley, 1979) and "hidden" speciation events (Raup, 1978). Discrete and continuous branching models are compared in detail in Gilinsky and Good (in press).

THE BIENAYMÉ-GALTON-WATSON BRANCHING PROCESS

Many observers in late 19th century England held that the apparently rapid disappearances of the surnames of aristocrats and notables might signal a decline in the fertility of these groups. The worry, among some, was that this decline in fertility—which was attributed to an increase in the quality of life among the notables—would be mirrored by an increase in the proportional representation of the proletariat in society, and that this would eventually result in a degradation of British culture. But, recognizing that such a conclusion could be hasty because family names can often disappear by chance, Sir Francis Galton and H. W. Watson developed branching theory as a way to quantify the probability of the chance extinction of surnames (Galton, 1873; 1889, Appendix F). In fact, Galton and Watson apparently *re*-developed the theory, the French mathematician Bienaymé having anticipated them (Bienaymé, 1845). Since Bienaymé, Galton, and Watson, the theory of branching processes has been much further developed, and the current literature is now large (Harris, 1963; Kendall, 1966; Mode, 1971).

Branching theory is most readily applied to what might be termed "forward problems." Figure 2 gives an example of a forward problem (that one not from branching theory!). We are given the numbers 2 and 2 as inputs, and we wish to determine their sum as an output. Since there is an easy rule to follow, the answer is obvious: 4. The applications that Raup discussed in his review (Raup, 1985) are all solutions to forward problems, because in each case we *assume* the parameters that govern the operation of the system, the probabilities of origination and extinction, and we predict—using equations from branching theory—some of the characteristics of a hypothetical population of fossil organisms that would result. Such characteristics include the expected number of species at time t, the expected time to extinction of a clade of species, etc.

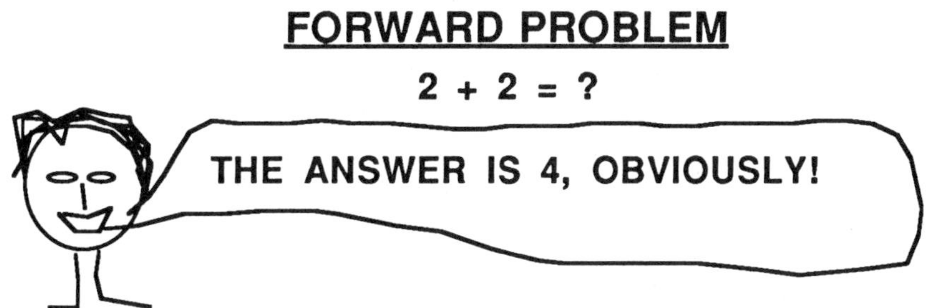

Figure 2. Cartoon depicting the solution of a "forward" problem.

My purpose, however, is to turn this approach on its head by using the diversity histories of real groups of organisms to estimate the probabilities of branching, persistence, and extinction that produced them. Problems of this sort, where the underlying parameters of the system are sought, are often termed "inverse" problems. Inverse problems are in many ways more challenging to solve than forward problems, as Figure 3 suggests, but in a historical science like paleobiology, where we are interested in studying the causal processes that have produced the results we see around us, the solution of inverse problems is more important.

The fundamental basis of the theory of branching processes is the concept of the *probability generating function.* In the present context, a probability generating function is a polynomial in x where the coefficients of the polynomial are the probabilities of occurrence of certain events. Here the probability generating function represents, in a compact form, the probabilities of the diversity going from a specified starting value to any ending value. I shall not attempt to derive the generating functions used here, but they originated with Galton and Watson in the last century, and they are familiar to students of branching theory (see Bailey, 1964, chapter 6, for instance). The generating functions used here correspond to a discrete branching model, which requires that time be divided into "generations" (to be discussed below). A discrete model is more elementary than a continuous model, and it allows for a conceptually straightforward way to estimate the probabilities of branching and extinction.

Since I shall be using a discrete branching model, I will often speak of "generations" of branching, persistence, and extinction. For most organisms, where there is a waiting time between birth and reproduction, the concept of a generation makes sense (even though the generations often overlap). But for species, where there is no waiting time, or where the waiting time may be very short relative to the subsequent lifespan of the species, the concept of a generation is ill-defined. I therefore adopt the following practical approach. I assume that a specific starting lineage can branch *once,* at most, per generation. This entails that the diversity can double, at most, during a generation. I then assume that the number of "generations" that transpired during a time interval is equal to the minimum number of generations that were required to produce the observed change of diversity. If the diversity during a specified time interval went from 3 to 5, for instance, only one generation of branching was required (the diversity less than doubled), but if the diversity went from 3 to 7, two generations were required (the diversity more than doubled but less than quadrupled). This is only an approximation, of course, and its ramifications are discussed more fully in Gilinsky and Good (in press), but it seems to be a reasonable starting point. Further work may produce a better solution.

If we let p_0 be the probability of extinction of a lineage, p_1 be the probability of persistence without branching, and p_2 be the probability of branching into two lineages, and if

Figure 3. Cartoon depicting the solution of an "inverse" problem.

we let v_0 be the diversity (number of coexisting lineages) at generation 0, it follows that the probability generating function for the diversity, or the number of coexisting lineages, at generation 1, given by v_1 is

$$[f(x)]^{v_0} = (p_0 + p_1x + p_2x^2)^{v_0} \tag{1}$$

We assume here that each of the v_0 starting lineages has the same values of the parameters p_0, p_1, and p_2 (these values not necessarily being equal to each other), and that these parameters do not change in time (the model is "time-homogeneous"). The model can be interpreted as assuming that all of the lineages belong to a single clade with clade-specific properties that confer certain intrinsic propensities for branching, persistence, and extinction. The properties, being clade-specific, would mean that the propensities would remain the same throughout the entire history of the clade. Other biological interpretations are possible, and other models, generally more complex, could also be applied. (I shall relax the assumption of time-homogeneity later in the paper.) As mentioned above I assume that a given starting lineage can branch once, at most, during a generation. The model can accommodate multiple branching events during a generation by the same lineage simply by adding more terms to the polynomial but, again, the aim here is to keep the model simple.

To deal with the cases where multiple generations occur during single time intervals (for instance, if the diversity went from 3 to 7 during an interval, as mentioned above), it can be shown that the generating function is to be applied iteratively (Bailey 1964). Specifically, if $[f(x)]^{v_0}$ in Eq. (1) is the generating function for the number v_1 of species present after one generation, then $[f[f(x)]]^{v_0}$ is the generating function for the number v_2 of species present after two generations, $[f[f[f(x)]]]^{v_0}$ is the generating function for the number v_3 of species present after three generations, etc.

Given specified probabilities of origination, persistence, and extinction, the generating functions (Eq. 1, or its iterated derivatives) allow one to calculate exactly the probability of achieving any specified change of the diversity by a process of branching, over any number of generations, as mentioned above. We will eventually solve the inverse problem of obtaining the probabilities of origination, persistence, and extinction for fossil taxa, but let us first become familiar with the generating function of Eq. (1) by illustrating its use in solving a simple forward problem.

Let us assume that the probability of extinction (p_0) is 0.3, that the probability of persistence (p_1) is 0.5, and that the probability of branching (p_2) is 0.2. What is the probability of the diversity going from 3 to 4 during a single time interval? To solve the problem using the generating function, what we do first is to expand $f(x)$ so that we can look at all the terms. In this case, the starting diversity $v_0 = 3$ species, so $f(x)$ must be raised to the 3rd power. The expansion of $f(x)$ is given in Eq. (2):

$$f(x)^3 = (p_0 + p_1x + p_2x^2)^3 = p_0^3 + (3p_0^2p_1)x + (3p_0^2p_2 + 3p_0p_1^2)x^2 \tag{2}$$
$$+ (6p_0p_1p_2 + p_1^3)x^3 + (3p_0p_2^2 + 3p_1^2p_2)x^4 + (3p_1p_2^2)x^5 + p_2^3x^6.$$

To determine the probability of the diversity going from $v_0 = 3$ to $v_1 = 4$ by a discrete branching process, we need only to follow one simple rule. *The probability of the diversity going from v_n to v_{n+1} is exactly equal to the coefficient of x^{v_n+1} in the generating function.* So in this case, the probability of the diversity going from 3 to 4 is the coefficient of x^4, or $(3p_0p_2^2 + 3p_1^2p_2)$, and by plugging the given values of p_0, p_1, and p_2 into this expression, we obtain 0.186 as the probability of the diversity going from 3 to 4.

We can also do the calculation by hand. Figure 4 shows that there are exactly six different tree cross sections that be drawn for a change of the diversity from 3 to 4. (I call these cross sections rather than trees because they could be *parts* of complete trees. See Gilinsky [in press] for a discussion of how the generating function can also be used to identify all of the possible tree cross sections for any given change of the diversity.) Beneath each cross

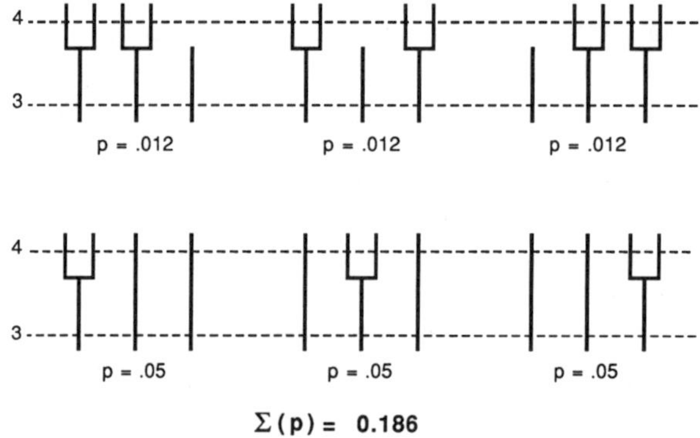

Figure 4. The six possible tree cross sections for a change of the diversity from three to four species. The probability associated with each cross section is given below the cross section. The overall probability of the diversity going from three to four species is the sum of the probabilities for each cross section. The probabilities shown assume that the probabilities of extinction, persistence, and branching are 0.3, 0.5, and 0.2, respectively.

section is the probability of obtaining precisely the pattern in the cross-section, given the values of p_0, p_1, and p_2. For instance, in the cross section at the upper left, two of the original lineages branched and one of them became extinct. Since the fates of the lineages are assumed to be independent in the model, the probability of two lineages branching and one lineage becoming extinct will be the product of the probabilities of the three events, or $0.2 \times 0.2 \times 0.3 = 0.012$. Because there are three different ways for two lineages to branch and for one to become extinct, corresponding to the top row of cross sections, the probability of the diversity going from three to four by two lineages branching and by one becoming extinct is the sum of the three ways of getting there, or $0.012 + 0.012 + 0.012 = 0.036$. By the same reasoning, the bottom row illustrates the other three ways for the diversity to go from three to four: by means of one lineage branching and two lineages persisting. Here the probability for each pattern is 0.05, so that the overall probability of the diversity going from three to four by one lineage branching and two lineages persisting is $0.05 + 0.05 + 0.05 = 0.15$. Therefore, the total probability of the diversity going from three to four when p_0, p_1, and p_2 are given is the sum of all of the possible ways of getting there, or $0.036 + 0.15 = 0.186$, which is the same probability as that obtained using the generating function.

Drawing all of the possible trees also underscores an important theoretical point about the generating function and the analysis that follows: the probability generating function gives the probability of the diversity going from one value to the other *under the theory that evolution is a process of branching.* That is, the model embodies a specific causal view of how evolution proceeds. As the topology of macroevolution is a branching topology, the application of branching theory to the problem is attractive and natural.

Suppose we are interested in the probability of the diversity going from, say, 87 to 93? To solve this problem we need to locate the coefficient of x^{93}. But to do that, we have to raise $f(x)$ to the 87th power (!) and to examine the terms. This seemingly impossible task can be achieved by a method involving the fast Fourier transform. The interested reader can consult Gilinsky and Good (1989, the appendix) for the details. For those who are curious, when the probability of speciation is 0.2, the probability of extinction is 0.3, and the probability of persistence is 0.5, the probability of the diversity from 87 to 93 is 0.055878.

We can now calculate the probability of the diversity going from 3 to 4, or from 87 to 93 (or in fact, from any integer to any other integer). But the geological record is very long indeed, and many taxa have been around for millions and millions of years. How do we determine the probability of obtaining an entire diversity path spun out over a long period of time? If we assume that taxonomic evolution is Markovian, that is, that the probabilities of occurrence of future events depend only upon the present state of the system, it follows that each step in the process can be treated as independent, and that the product rule of probabilities can be applied. Therefore, to determine the probability that a group of organisms will have a particular diversity path, v_n to v_{n+1} to v_{n+2}, etc., we determine the probability for each step and multiply all of the probabilities together.

THE MAXIMUM LIKELIHOOD METHOD OF ESTIMATION

Up to this point, we have been solving forward problems. We can determine the probability of occurrence of any particular diversity path we like, *but only when the parameters are already given.* Since what we want to do is to obtain the extinction, persistence, and branching parameters for fossil taxa when in fact these are not known a priori, the theory that has been developed thus far is not sufficient. How, in fact, do we obtain p_0, p_1, and p_2 when these parameters are not known?

If we can calculate the probability of obtaining the observed diversity path for one set of extinction, persistence, and branching parameters (as we did in the simple example above), we can do it for another set, and another set, and another set, until we obtain a matrix of probabilities (Fig. 5). When using the theory to estimate parameters, as we shall now be doing, these probabilities are called *likelihoods,* hence the Ls in the matrix. Of all the possible sets of evolutionary parameters, the set that is identified by the largest likelihood (the largest probability of occurrence) is the set that best estimates the underlying evolutionary parameters that gave rise to the observed diversity path. So, by exploring large numbers of sets of evolutionary parameters systematically, we can solve the inverse problem and estimate the values of the parameters that produced the observed evolutionary history.

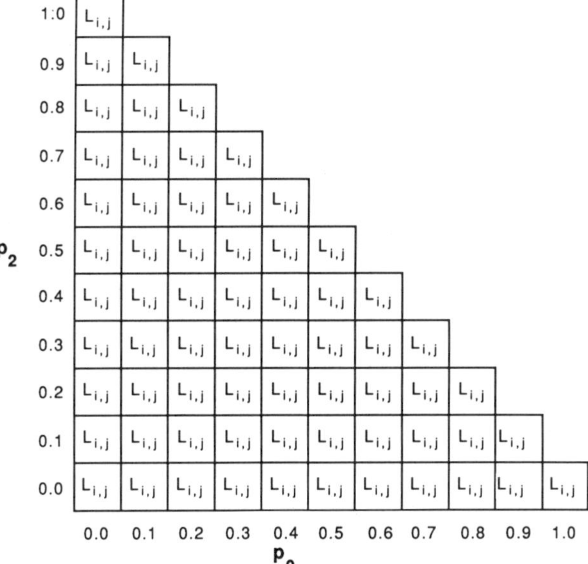

Figure 5. Illustration depicting a likelihood matrix. Each value of $L_{i,j}$ is the probability of achieving a specified diversity path (which could be as little as one step long) given particular "assumed" values of the extinction probability p_0 and origination probability p_2. When the theory (discussed in the text) is used to estimate the parameters p_0 and p_2, the probabilities occupying the cells of the matrix are termed "likelihoods," hence the notation. The maximum likelihood in the matrix identifies the best-estimate values of p_0 and p_2.

AN EXAMPLE: THE ELEPHANTIDAE

Figure 6 shows the stratigraphic ranges of species in the Elephantidae, the elephants. The data are from Maglio (1973). By dividing the history of the elephants into 0.5 million year increments, the diversity path has been obtained. Given this observed diversity history, the fundamental question is: what were the underlying probabilities of extinction, branching, and persistence that produced that history?

The answer is obtained from the likelihood matrix (Table 1). The matrix of numbers contains the probabilities (or likelihoods) of obtaining exactly the observed diversity path for specific probabilities of extinction (the x axis), branching (the y axis), and persistence. We need not show the persistence axis, because the probability of persistence is completely defined by the values of the other 2 parameters. (The three parameters must sum to 1.0.)

The maximum likelihood in the matrix corresponds to a probability of extinction of 0.3, a probability of branching of 0.3, and a probability of persistence of 0.4. These estimates are the best that can be obtained given the resolution of this particular likelihood matrix. But by using a technique that explores the region in the immediate vicinity of the maximum likelihood in more detail, the maximum likelihood parameters have been determined to the thousandths place. So, for the Elephantidae, the maximum likelihood probabilities of origination, extinction, and persistence are: 0.314, 0.296, and 0.390.

The likelihoods can also be represented in the form of a contour plot (Fig. 7). This emphasizes that the maximum likelihood parameters are best *estimates,* and that other combinations of origination and extinction probabilities could have been responsible for the observed diversity history. The 95% confidence contour about the maximum likelihood is shown by the dashed curve. (See Gilinsky and Good [in press] for discussion of how to obtain these contours.) As I shall discuss below, this confidence contour will allow for rigorous comparisons of the parameters from taxon to taxon.

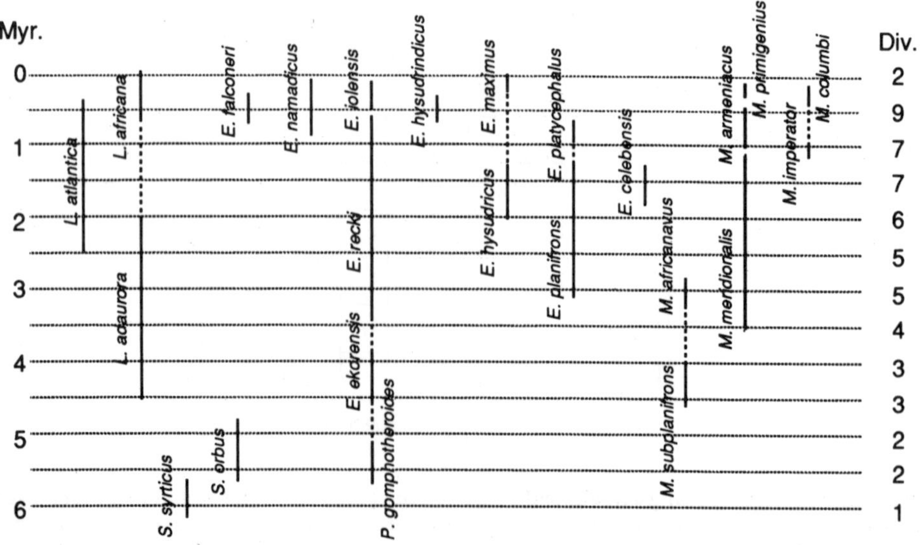

Figure 6. Stratigraphic ranges of lineages in the Elephantidae (From Maglio, 1973). Time is divided into 0.5 million year intervals. The diversity path, shown at the right of the figure, represents the number of lineages that coexisted at each 0.5 million year time line.

Table 1. Each entry is the likelihood of obtaining the observed diversity path for the Elephantidae when the probability of origination is p_2 and the probability of extinction is p_0. The maximum likelihood is 0.329×10^{-10}, and the probabilities of origination and extinction associated with the likelihood are $p_2 = 0.3$ and $p_0 = 0.3$ respectively. By exploring the region surrounding the maximum likelihood in more detail, the more accurate values of $p_2 = 0.314$ and $p_0 = 0.296$ have been obtained.

p_2	0.0	0.1	0.2	0.3	0.4	0.5	0.6	0.7	0.8	0.9	1.0
1.0	0.000										
0.9	0.000	0.000									
0.8	0.000	$0.436{\times}10^{-20}$	0.000								
0.7	0.000	$0.201{\times}10^{-17}$	$0.812{\times}10^{-15}$	0.000							
0.6	0.000	$0.187{\times}10^{-15}$	$0.679{\times}10^{-13}$	$0.186{\times}10^{-12}$	0.000						
0.5	0.000	$0.729{\times}10^{-14}$	$0.104{\times}10^{-11}$	$0.536{\times}10^{-11}$	$0.164{\times}10^{-11}$	0.000					
0.4	0.000	$0.128{\times}10^{-12}$	$0.677{\times}10^{-11}$	$0.228{\times}10^{-10}$	$0.162{\times}10^{-10}$	$0.131{\times}10^{-11}$	0.000				
0.3	0.000	$0.919{\times}10^{-12}$	$0.188{\times}10^{-10}$	$0.329{\times}10^{-10}$	$0.171{\times}10^{-10}$	$0.321{\times}10^{-11}$	$0.932{\times}10^{-13}$	0.000			
0.2	0.000	$0.184{\times}10^{-11}$	$0.153{\times}10^{-10}$	$0.126{\times}10^{-10}$	$0.339{\times}10^{-11}$	$0.415{\times}10^{-12}$	$0.226{\times}10^{-13}$	$0.232{\times}10^{-15}$	0.000		
0.1	0.000	$0.275{\times}10^{-12}$	$0.921{\times}10^{-12}$	$0.306{\times}10^{-12}$	$0.321{\times}10^{-13}$	$0.146{\times}10^{-14}$	$0.312{\times}10^{-16}$	$0.286{\times}10^{-18}$	$0.545{\times}10^{-21}$	0.000	
0.0	0.000	0.000	0.000	0.000	0.000	0.000	0.000	0.000	0.000	0.000	0.000

p_0

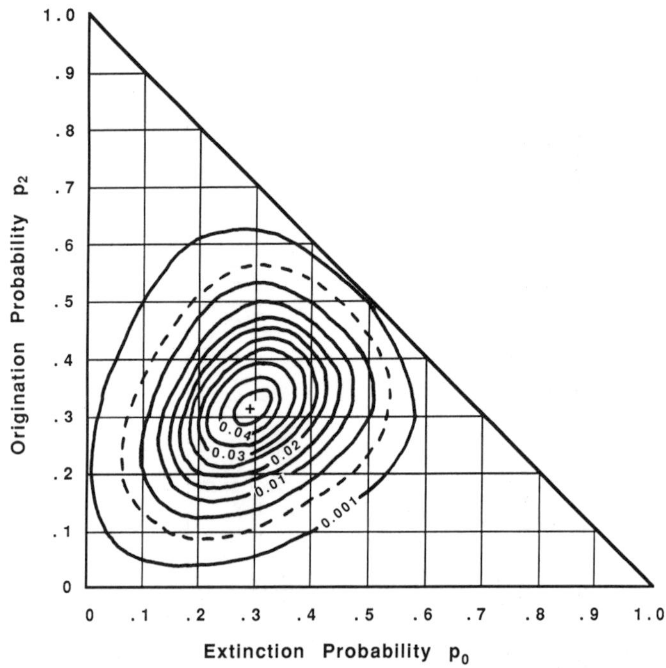

Figure 7. Likelihood matrix for the Elephantidae depicted in the form of a contour plot. The numerical values of the contours ("isolikelihoods") were plotted after normalizing the likelihoods so that they sum to 1.0 (This was done for convenience in plotting. Likelihoods can be multiplied by a constant factor.) The dashed curve is the approximate 95% confidence contour about the maximum likelihood.

A MORE COMPREHENSIVE ANALYSIS: MARINE INVERTEBRATE LIFE

Ideally, all analyses would be applied to species within monophyletic clades, as for the case of the elephants. But species level data are available for only a few fossil taxa; therefore a comprehensive analysis at the species level is not yet possible. But still wishing to attempt some form of comprehensive analysis, I have applied the methods just described to estimating the probabilities of origination, persistence, and extinction of families within 135 orders of marine invertebrate life using the data from Sepkoski's compendium of the stratigraphic ranges of fossil marine families (1982). It must be stressed that the families and orders are in many cases not holophyletic; that is, many of these taxa reflect classical systematic groupings. So the results must be interpreted with care, and it will be important to re-run the analysis again as a true cladistic classification is developed.

A small part of the table of results is reproduced here (Table 2; the entire table of results is given in Gilinsky and Good [in press]). The output includes the estimates of the parameters, and estimates of their standard errors. As can be seen in this portion of the data table, which shows a few selected orders, the Xiphosurida had an estimated familial extinction probability of 0.038, and an estimated familial origination probability of 0.038, values that are at the very low end of the scale of probabilities. The Foraminifera had an estimated familial extinction probability of 0.182 and an estimated origination probability of 0.218, rather low on the scale of probabilities, and the Ammonitida had an estimated familial extinction probability of 0.425 and an estimated familial origination probability of 0.415, values that are at the high end of the probability scale. In fact, the results for these groups confirm what other workers have claimed based on informal, qualitative observa-

Table 2. Estimated familial origination and extinction probabilities, and their standard errors, for several orders of marine invertebrate animals.

Order	Extinction Probability	Origination Probability	S.E. of the Extinction Probability	S.E. of the Origination Probability
Foraminiferida	0.182	0.218	0.033	0.033
Scleractinia	0.067	0.096	0.025	0.025
Tabulata	0.302	0.300	0.049	0.049
Ammonitida	0.425	0.415	0.063	0.063
Arcoida	0.004	0.031	0.004	0.011
Neogastropoda	0.149	0.212	0.098	0.099
Phacopida	0.260	0.250	0.078	0.078
Xiphosurida	0.038	0.038	0.019	0.019

tions. The horseshoe crabs, which are widely touted as "living fossils," turn out to have very low familial probabilities of origination and extinction. The ammonites, which are widely regarded as fast evolvers, turn out to have high familial probabilities of origination and extinction. This confirmation provides support for the methods developed here.

A plot of the estimated probabilities of familial origination and extinction for all 135 orders (Fig. 8) shows that the range of probabilities is very broad; some orders have high probabilities of familial origination and extinction, others have low probabilities, and others fall in between. Furthermore, the probabilities of origination and extinction are highly correlated. This correlation is to be expected for orders that have become extinct (or pseudoextinct, for those that are paraphyletic). For extant orders, probabilities of familial origination generally exceed those of extinction, as they should. Three of the orders of marine invertebrates have been identified on the graph.

A histogram of the results (Fig. 9) shows that most orders of marine invertebrates have relatively low to moderate probabilities of origination and extinction, and many fewer have high probabilities. Although this could not be anticipated a priori, this paucity of orders with high probabilities is reasonable. Gould et al. (1977) showed, using simulations, that artificial clades that have high probabilities of origination and extinction tend to have very unstable diversity histories. As a result, the diversities of these clades frequently wander to the absorbing boundary of 0 (extinction), and they tend to have shorter durations than clades composed of species with lower probabilities of origination and extinction. The results of their simulation suggest that high probabilities of origination and extinction may be rare among fossil taxa because the taxa that have them tend to be short lived and tend to leave little if any fossil record.

Figure 10 shows the results for all 135 orders, as before. But this time living and extinct orders have been labeled separately. While extinct orders may have had either high or low probabilities of familial origination and extinction, living orders have only low probabilities. Put another way, if an order had a high probability of origination or extinction, it invariably did not survive to the present day. Again, this may be because taxa with high probabilities of branching and extinction tend to have unstable diversity paths and readily become extinct.

Also estimated were the probabilities of familial origination and extinction for each of 12 phyla of marine invertebrates, irrespective of the membership of families within orders (Fig. 11). The phyla span a broad range of origination and extinction probabilities. The Archaeocyatha, a group of uncertain taxonomic affinity, had the highest estimated probabilities of origination and extinction. The Hemichordata and Annelida have among the lowest estimated probabilities of familial origination and extinction. The more typical members of fossil marine ecosystems, the molluscs, brachiopods, bryozoa, cnidaria, echinoderms, and arthropods, have relatively high probabilities of origination and extinc-

248

tion, while the protozoans, poriferans, and conodonts have lower probabilities, but not as low as the lowly worms. Again, the probabilities obtained in this study are quite consistent with assessments that have been expressed by other workers, a fact that should be regarded as providing support for the method.

Principally for completeness I also estimated the probabilities of familial origination and extinction for all of marine invertebrate life. The estimate for all of life falls at the high end of the probability scale, and is depicted by a blackened square.

The 95% confidence contours are shown for the Mollusca, the Protozoa, and the Annelida. These confidence regions are much smaller than the one given in Fig. 7 for the elephants. This is because the phyla depicted here have had extremely long histories—the longer the history, the more accurate in general is the estimate of the evolutionary parameters. The lack of overlap of the 95% confidence contours confirms that the Mollusca, the Protozoa, and the Annelida all have statistically significantly different probabilities of origination and extinction.

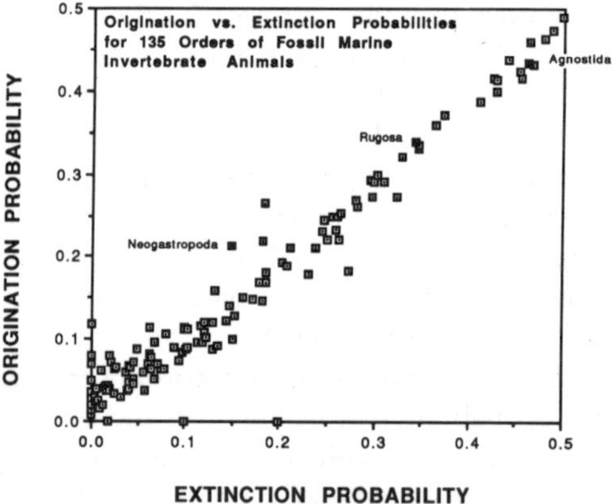

Figure 8. Plot of estimated familial extinction probability vs. estimated familial origination probability for 135 orders of fossil marine invertebrate animals. Three of the orders have been labeled.

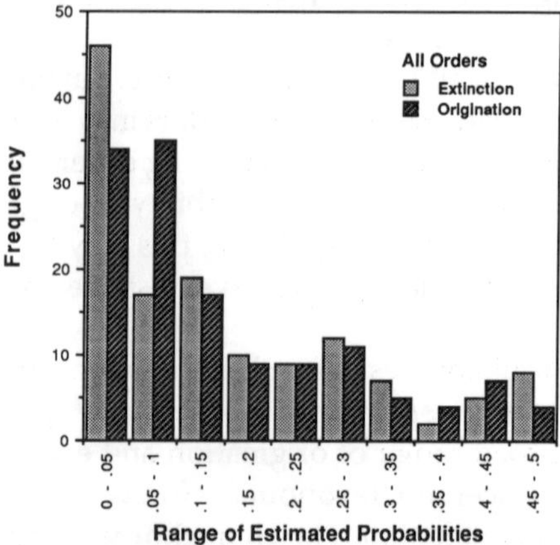

Figure 9. Histogram of estimated familial extinction and origination probabilities for 135 orders of fossil marine invertebrate animals. Most orders have low to moderate values of the probabilities. Relatively few have high probabilities.

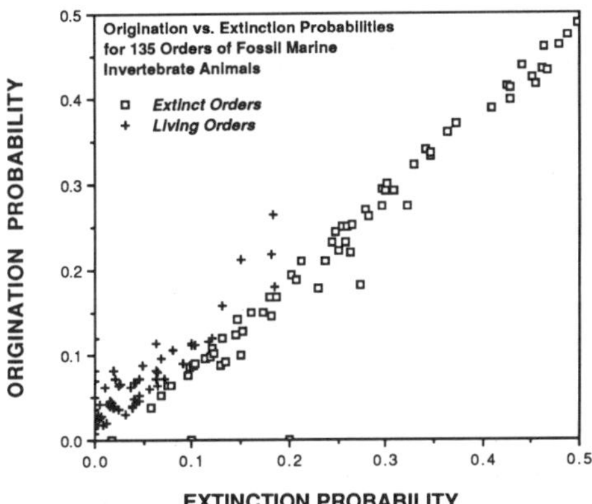

Figure 10. Plot of estimated familial extinction probability vs. estimated familial origination probability for 135 orders of fossil marine invertebrate animals, extinct and living orders labeled separately. Extinct orders cover the entire range of probabilities. All living orders have low probabilities of origination and extinction.

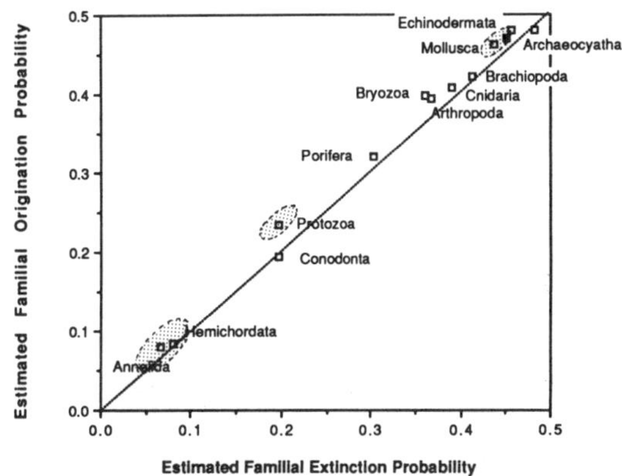

Figure 11. Plot of estimated familial extinction probability vs. estimated familial origination probability for 12 phyla of fossil marine invertebrate animals, irrespective of the membership of families within orders. Approximate 95% confidence contours are plotted for the Annelida, Protozoa, and Mollusca. Filled square gives the estimated familial probabilities of origination and extinction for all of fossil marine invertebrate life.

THE TRAJECTORY OF ORIGINATION AND EXTINCTION PROBABILITIES FOR PHANEROZOIC MARINE INVERTEBRATE LIFE

All of the analyses discussed thus far estimate the origination, persistence, and extinction probabilities by assuming that these parameters are time-homogeneous; that is, that for each group of organisms, the probabilities do not change over their entire geological histories. However, one can also assume that the parameters do change and one can therefore determine a *trajectory* of the parameters through time for any group of organisms for which there are sufficient data on the diversity. Because the trajectory for the diversity of life as a whole can also be obtained, the theory of branching processes can be used to locate times of elevated global extinction probability. The branching theoretic approach can thus serve as an independent test of Raup and Sepkoski's claim (1982; see also Sepkoski, 1986)

250

that the Phanerozoic witnessed 5 major events of mass extinction. One can also ask whether the probability of origination changes in association with the mass extinctions. The relation between extinction and origination during times of mass extinctions has not adequately been explored.

Figure 12B gives the trajectory of the estimated familial extinction probability. Familial extinction probabilities start out fairly low in the Cambrian and Ordovician, and then increase to fluctuating but moderate values for the remainder of the Phanerozoic. The temporal positions of the five Phanerozoic mass extinctions are marked, and indeed, the branching theoretic method does show that the times of major extinctions correspond to substantially elevated probabilities of familial extinction. In addition to the big five, other peaks of high extinction probability are present. Some, but not all, of these correspond to times of heightened extinction rate that have been described elsewhere. So, indeed, mass extinctions seem to be vindicated, but the other times of heightened extinction probability may also have been identified.

Figure 12A gives the trajectory of the estimated familial origination probability. Familial origination probabilities start out relatively high in the Cambrian, drop to low levels for a time, and then rise again during the early part of the Ordovician. These two high periods almost certainly record the expansions of Sepkoski's Cambrian and

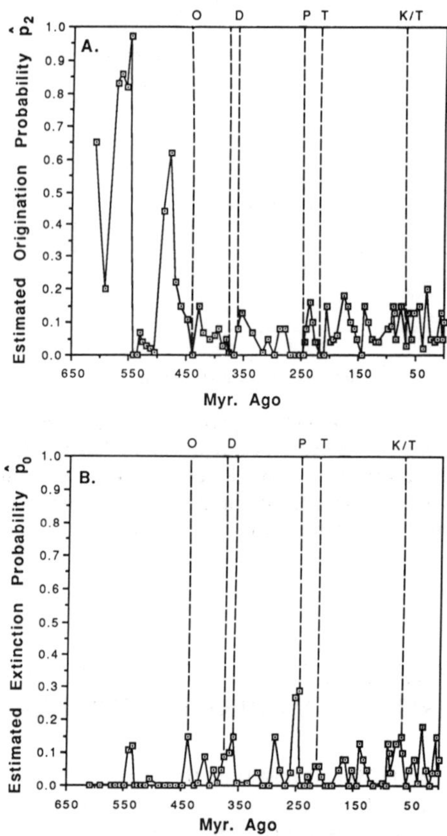

Figure 12. A. Plot of the trajectory of the estimated familial origination probability for all of fossil marine invertebrate life. The times of the five Phanerozoic mass extinction events are shown (O = Ordovician, D = Devonian, P = Permian, T = Triassic, K/T = Cretaceous/Tertiary). B. Plot of the trajectory of the estimated familial extinction probability. The times of the five Phanerozoic mass extinctions are shown; abbreviations are the same as before. The "hats" over p_0 and p_2 denote that the estimates of p_0 and p_2 are based upon data for single time intervals. These estimates are probably less accurate than those based on multiple time intervals.

Paleozoic evolutionary faunas (1984). The temporal positions of the five Phanerozoic mass extinctions events have been marked. Interestingly, in each case but one, the estimated origination probability either declines for several stages prior to the stage of mass extinction, or remains near-zero for several stages prior to the stage of mass extinction. Whether these declines are sampling artifacts or whether in some sense the origination probabilities actually anticipate the mass extinctions is not clear. The only extinction that was not preceded by a decline of origination probability was the Cretaceous/Tertiary. It seems to be an extraordinary, singular event.

SUMMARY AND CONCLUSIONS

Although many researchers have measured rates of taxonomic evolution (Yule, 1924; Simpson, 1944; Kurtén, 1960; Raup, 1978; Stanley, 1979, 1985) the present paper is a first effort to estimate the *probabilities* of origination and extinction that undergird the evolutionary process by making use of the Galton-Watson theory of branching processes in discrete time.

Based upon the analysis attempted here, few orders of marine invertebrates had high familial probabilities of origination or extinction, and all of those that did are extinct. This implies that the composition of the present day biota is fundamentally different than its composition in the distant past, not only in terms of the types of morphologies and physiologies that are present, but also in terms of the capacity for significant evolutionary change. Living taxa, for whatever reason, apparently have a lesser propensity for spinning out significant, new variations than did taxa that have already become extinct.

When looking at the trajectories of familial origination and extinction probabilities through time, the analysis shows that extinction probabilities increase substantially during the known periods of mass extinction. But origination probabilities generally decline for several stages before the stages of mass extinction. Only one extinction is not foreshadowed by a decline of origination probability: the great Cretaceous event.

The model is potentially useful for detailed quantitative comparisons of probabilities of origination and extinction among various groups of organisms. Such comparisons can be carried out because the maximum likelihood approach allows for the construction of 95% confidence contours about the values of the maximum likelihood. Rigorous comparisons of probabilities of origination and extinction may aid in attempts to explore the causes that lie behind differences in the pace of taxonomic evolution among various taxonomic groups.

By taking seriously the metaphor of evolution as a growing, branching tree, the model employed here faithfully reflects the process of macroevolution as it is known to occur. It appears to provide a theoretically defensible way to compute and compare the evolutionary parameters of taxa with sensitivity and rigor.

LITERATURE CITED

Haeckel, E. 1874. Die Gastraea-Theorie, die phylogenetische Klassification des Tierreiches und Homologie der Keimblätter. *Jenaische Zeitschrift fuer Naturwissenschaften* 8:1–55.

Bailey, N. T. J. 1964. *The Elements of Stochastic Processes.* John Wiley and Sons, Inc.: New York. 249 pp.

Bienaymé, I. J. 1845. De la loi de multiplication et de la durée des familles. *Societé Philomathique de Paris, Extraits des Process-Verbaux des Sciences* 5:37–39.

Galton, F. 1873. Problem 4001. *Educational Times (London)* 1 April, 1873, p. 17.

Galton, F. 1889. *Natural Inheritance.* Macmillan: London. 259 pp.

Gilinsky, N. L. 1991. Cross sections through evolutionary trees: theory and applications. *Systematic Zoology.* 40:19–32.

252

Gilinsky, N. L. & I. J. Good. 1989. Analysis of clade shape using queueing theory and the fast Fourier transform. *Paleobiology* 15:321–333.

Gilinsky, N. L. & I. J. Good. In press. Probabilities of origination, persistence, and extinction of families of marine invertebrate life. *Paleobiology.*

Gould, S. J., Raup, D. M., Sepkoski, Jr., J. J., Schopf, T. J. M. & D. S. Simberloff. 1977. The shape of evolution: a comparison of real and random clades. *Paleobiology* 3:23–40.

Harris, T. E. 1963. *The Theory of Branching Processes.* Springer: Berlin. 230 pp.

Kendall, D. G. 1966. Branching processes since 1873. *Journal of the London Mathematical Society* 41:385–406.

Kurtén, B. 1960. Chronology and faunal evolution of the earlier European glaciations. *Societas Scientiarum Fennica Commentationes Biologicae* 21:40–62.

Maglio, V. J. 1973. Origin and evolution of the Elephantidae. *Transactions of the American Philosophical Society, New Series* 63:1–149.

Mode, C. J. 1971. *Multitype Branching Processes.* Elsevier: New York. 330 pp.

Raup, D. M. 1978. Cohort analysis of generic survivorship. *Paleobiology* 4:1–15.

Raup, D. M. 1985. Mathematical models of cladogenesis. *Paleobiology* 11:42–52.

Raup, D. M. & J. J. Sepkoski, Jr. 1982. Mass extinctions in the marine fossil record. *Science* 215:1501–1503.

Sepkoski, J. J., Jr. 1982. A compendium of fossil marine families. *Milwaukee Public Museum Contributions to Biology and Geology* 51.

Sepkoski, J. J., Jr. 1984. A kinetic model of Phanerozoic taxonomic diversity. III. Post-Paleozoic families and mass extinctions. *Paleobiology* 10:246–267.

Sepkoski, J. J., Jr. 1986. Phanerozoic overview of mass extinction. Pp. 277–295. *In:* D. M. Raup & D. Jablonski (eds.), *Patterns and Processes in the History of Life.* Springer-Verlag: Berlin.

Simpson, G. G. 1944. *Tempo and Mode in Evolution.* Columbia University Press: New York. 237 pp.

Stanley, S. M. 1979. *Macroevolution: Pattern and Process.* W. H. Freeman and Company: San Francisco. 332 pp.

Stanley, S. M. 1985. Rates of evolution. *Paleobiology* 11:13–26.

Van Valen, L. 1973. A new evolutionary law. *Evolutionary Theory* 1:1–30.

Van Valen, L. 1979. Taxonomic survivorship curves. *Evolutionary Theory* 4:129–142.

Yule, G. U. 1924. A mathematical theory of evolution, based on the conclusions of Dr. J. C. Willis, F. R. S. *Proceedings of the Royal Society of London* 213(B):21–87.

Diversification Patterns in the Most Diverse Marine Snail Genus

EXTENDED ABSTRACT

Alan J. Kohn

The most diverse marine snail genus is *Conus*. With an estimated 500 extant species, it is probably the largest genus of marine invertebrates, and it thus presents an unusually interesting problem in understanding 'first-order' (Davis, 1981) or genus-level diversification. Through comparative biological studies over the past 35 years, the habits and habitats, feeding, reproduction and development, species diversity, and community ecology of this species-rich yet geologically young genus have become quite well known. However, despite this neontological information and a very respectable fossil record, we remain virtually ignorant of the group's origin and evolutionary history and the historical, environmental, and intrinsic factors that have been important in its diversification.

Here I focus on one part of a broad investigation of evolutionary trends in morphology, distribution, diversity, and ecology in *Conus* following its origin, how these trends might be related, and the various factors that may have influenced them. I discuss the origin of the genus, current understanding of its patterns of radiation in time and space, and the possible roles of key innovations and new adaptive zones in its diversification.

In order to examine the history and patterns of taxonomic diversification, I developed a database from the paleontological literature comprising records indicating geologic age and geographic location of fossil *Conus* species. The entries are currently being examined critically for taxonomic and stratigraphic accuracy; thus far only records from the oldest reports through the Lower Eocene have been so evaluated.

The following conclusions derive from analysis of this database: 1. All 20 Jurassic and Cretaceous species originally described as *Conus* are likely opisthobranchs, or members of the prosobranch family Volutidae, or if *Conus* they are from strata now known to be of, at most, Middle Eocene age. 2. The few literature reports of Paleocene species may also be incorrect. The earliest *bona fide* fossils of *Conus* appear to be Lower Eocene (about 55 mybp) from Britain (*C. concinnus* Sowerby) and France (*C. rouaulti* D'Archiac). Paleobotanical investigations indicate tropical and subtropical climatic conditions in Europe, particularly in the late Lower Eocene (Boulter, 1984). 3. The first substantial radiation of the genus occurred in the Middle and Upper Eocene. Diversity decreased in the Oligocene, one or more major radiations occurred in the Miocene, and diversity again decreased in the Pliocene, followed by the very rapid increase to the present ca. 500 species.

Details of these conclusions, including the biogeographic features of the radiations, are given in Kohn (1990), where the observed patterns are also compared with the predictions of alternative models of taxonomic diversification. The simplest of these is the

Dr. Kohn is with the Department of Zoology, University of Washington, Seattle, WA 98195, USA.

254

exponential or log-linear, modeled as a line plotted between the origin, taken as the time of origin of the taxon as a single species, and the logarithm of the number of species extant at present (Stanley, 1979). This model assumes constant rates of speciation and species survival. The logistic curve models an exponential rate of increase damped as a saturation value is approached. In other models, periods of diversification alternate with periods of stasis or net reduction in diversity when extinction rate exceeds speciation rate. The observed pattern of *Conus* diversification history from the Lower Eocene to the present fits the last model most closely: periods of rapid radiation, especially in the Middle Eocene, Lower Miocene, and Pleistocene, alternate with periods of net extinction and reduced diversity.

"Key innovations" and the occupation of one or more new adaptive zones often characterize new supraspecific taxa. Did the diversification of *Conus* depend on invasion of new adaptive zones involving patterns of resource use, defenses against enemies, etc. different from those of its ancestral and sister taxa? At present this question can only be addressed with a high level of speculation. Likely scenarios include: 1) habitat shifts from fine sediments of continental shelf depths to higher-energy environments of shallow bays and lagoons, and then to hard as well as soft substrata of even shallower coral reef environments; and 2) food shifts from very narrow polychaetes to larger and more robust worms, other molluscs, and fishes.

Novel features that might have been necessary to promote adaptive radiation can be hypothesized by examining the hallmarks of the genus: 1) a detachable, disposable, single-use, hollow, harpoon-like radular tooth used to inject venom into prey via a tubular intraembolic proboscis; 2) a neurotoxic venom, comprising several peptides of low molecular weight, that immobilizes the prey when injected through the radular tooth; 3) a broadly conical or biconical shell with a low spire and aperture with nearly parallel sides; and 4) a heavy, strong, crossed-lamellar last whorl of the shell, with the protected inner walls mainly dissolved.

Not all of these features meet current criteria of key innovations (e.g., Lauder & Liem, 1989; Herrera, 1989). *Conus* shares the toxoglossan radular tooth with numerous taxa in its putatively ancestral family Turridae. Many turrids also have a similar venom apparatus morphology, but information on the chemical nature of their venoms is entirely lacking. Size, shape, thickness, and interior remodeling of the shell may meet the criteria for key innovations, and shell form is thus the most likely candidate for an intrinsic promoter of diversification and the successful occupation of new adaptive zones by *Conus*. The depressed spire and broad, conical form of most modern species allows expansion of the aperture. This may have permitted thickening of the last whorl and concomitant resorption of interior walls without reducing the size of living space for an animal that must engulf whole prey organisms. Attributes of shell form may thus constitute the most important key innovation in *Conus*, but prior possession of the harpoon-like radular tooth and venom may well have been prerequisites for successful diversification.

LITERATURE CITED

Boulter, M. C. 1984. Paleobotanical evidence for land-surface temperature in the European Paleogene. Pp. 35–47. *In:* P. Brenchley (ed.), *Fossils and Climate.* John Wiley & Sons: Chichester.

Davis, G. M. 1981. Introduction to the Second International Symosium on evolution and adaptive radiation of Mollusca. *Malacologia* 21:1–4.

Herrera, C. M. 1989. Seed dispersal by animals: a role in angiosperm diversification? *American Naturalist* 133:309–322.

Kohn, A. J. 1990. Tempo and mode of evolution in Conidae. *Malacologia,* 32:55–67.

Lauder, G. V. & K. F. Liem. 1989. The role of historical factors in the evolution of complex organismal functions. Pp. 63–78. *In:* D. B. Wake & G. Roth (eds.), *Complex Organismal Functions: Integration and Evolution in Vertebrates.* John Wiley & Sons: Chichester.

Stanley, S. M. 1979. *Macroevolution.* W.H. Freeman: San Francisco.

Why Is Life So Diverse?

EXTENDED ABSTRACT

Michael J. Benton

Abstract. The present huge diversity of life—amounting to some 30 million species according to some estimates—has been explained as the result of a variety of processes, many of them incompatible. Specific explanations of diversification fall into four main categories: (a) that the patterns are artefacts of a biased fossil record, (b) that the patterns are real and caused mainly by physical perturbation, (c) that the patterns are real and caused mainly by biotic factors, or (d) that the patterns are real but stochastic. Three out of 14 proposed models for large-scale diversification are tested with data from the fossil record of tetrapods, and the strongest seems to be expansion of adaptive space, followed by subdivision of niches and specialization, and then by provinciality caused by topographic and climatic change.

INTRODUCTION

Life has diversified extensively during the course of geological time, ultimately from a single species to the present total of 30 million, according to recent estimates. It should be possible to establish the broad pattern of the global diversification of life since its origin some 3.7 billion years ago, and to determine the causes of that diversification.

Paleobiologists have plotted the patterns of diversification of relatively well preserved segments of the fossil record, especially benthic marine invertebrates (e.g., Valentine, 1969; Raup, 1972; Sepkoski, 1984). The literal plots of diversification of families and orders generally show low levels in the early Cambrian, a rise during the Cambrian and then a second larger rise during the Ordovician. The level of diversity then remained at a plateau for the rest of the Paleozoic, terminated by a massive decline at the end of the Permian. Since the early Triassic, diversity levels have risen with no sign of a leveling-off. Early debates about the accuracy of the literal plots of diversification have been resolved in favour of the view that they are broadly accurate representations of reality, and that the numerous sources of error do not have an overwhelming effect (Sepkoski et al., 1981).

The aims of this paper are to look at the broad patterns of the diversification of life, to review the numerous available explanations for diversification, and to summarise tests of these postulated causes. This paper is an extended abstract of a fuller account that has been published recently (Benton, 1990). Fuller explanations, and references, may be found there. A critique of adaptive radiation models presented here is not to be found in the longer account.

Dr. Benton is with the Department of Geology, University of Bristol, Bristol, BS8 4TF, UK.

THE CAUSES OF GLOBAL BIOTIC DIVERSIFICATION

Critique of Adaptive Radiation Stories

Typical explanations for diversification are ad hoc models that refer to individual radiation events, such as the radiation of the mammals at the beginning of the Tertiary. Such models may be termed story-like, or narrative, explanations, and they are hard to match with the larger-scale explanations needed for overall biotic increase. The story-like approach has been to describe the pattern and rate of radiation, and to seek adaptive explanations for the success of a group, whether it is supposed to have competitively replaced another group, or to have radiated opportunistically into empty ecospace (Cracraft, 1985; Benton, 1987a).

This kind of approach suffers from numerous difficulties, not the least being problems of simplification, assumptions of adaptation and competition, and untestability. In the case of the mammalian radiation 65 million years ago, the explanations cover most of the mammalian attributes, ranging over endothermy, the possession of hair, lactation, long gestation, parental care, large brains, and adaptable dentitions. The possession of one or more of these 'key adaptations' is said to explain the magnitude, rate, and adaptive variety of the radiation of placental mammals. Five criticisms may be levelled at such a mode of explanation.

1. The ideas are untestable. How is one to determine which one, or more, of the characteristics was important? It is not valid to compare contemporary groups that lack the character or characters in question since these features may not be viewed in isolation in the paleontological record. Can we perform experiments today, by placing mice and lizards in a box in order to test which is the best competitor? Can we test the value of hair by pitting an intact mouse against a half-shaved mouse?
2. Even if a case can be made for the value of a 'key adaptation' in terms of survival, this does not automatically make the link to stating that that was the feature which drove the radiation so successfully.
3. Further, the postulation of a 'key adaptation' does not explain why the radiation of mammals proceeded at such a high rate, why the mammals were able to achieve ten times the diversity of their ecological predecessors, the dinosaurs, nor why the mammals were able to enter so many adaptive zones not exploited by the dinosaurs. Each of these elements of the model involves further assumptions.
4. The 'key adaptation' types of explanations for radiations assume that the mammals were cyphers, or *tabulae rasae*, devoid of biological novelties except for the feature or features of interest. Organisms are complexes of interconnected homeostatic mechanisms, and it is probably impossible to extract one, or a small number of, features to which to ascribe their success.
5. Finally, simplistic 'key adaptation' models refer to individual events, and they allow few generalisations to the higher levels of explanation. Even if mammals radiated in the Paleocene because they had efficient teeth or hair, this does not go further to explain why tetrapods as a whole are ten times as diverse today as they were in the Mesozoic, why life as a whole is so diverse, nor why a particular broad pattern of increase has been followed through time.

Explanations for Global Biotic Diversification

Many explanations have been advanced for the patterns of apparent diversity increase observed in the fossil records of most groups of organisms. These suggestions are listed below under four main headings.

Artifact.   The pattern of increase is an artifact of the patchy quality of the fossil record and the way in which we study it (Raup, 1972):

1. The volume of unmetamorphosed sedimentary rock increases towards the present;
2. the area of exposure of such fossiliferous rock increases towards the present;
3. paleontologists devote more attention to younger faunas and floras and hence name more species, genera, and families;
4. the 'Pull of the Recent', in which our knowledge of modern species enhances our knowledge of close relatives of the recent geological past;

Real and caused by physical factors.

5. increased endemicity (provinciality), as a result of global changes in topography or climate, such as the break-up of Pangaea, changing patterns of ocean circulation, and global climatic gradients (Valentine, 1980);
6. fragmentation of physical habitats, by changes in topography or climatic zones, for example, and associated fragmentation of faunas;
7. continuous change in the physical environment maintains a level of biotic turnover and diversification—'lithospheric complexity' (geographic and climatic barriers to migration, global climatic gradients, topography) is in constant flux and stimulates constant diversification (Cracraft, 1985);

Real and caused by biological factors.

8. increased endemicity, and increased speciation rate, as a result of genetic, physiological, or behavioural changes that reduced gene flow;
9. reduction in the probability of extinction by:
   i. optimization of fitness;
   ii. reduction in levels of diffuse competition;
   iii. increase in the species: family ratio through time;
   iv. increase in mean extinction resistance as the set of families increases in mean age;
10. increase in the overall adaptive space occupied by a clade;
11. subdivision of niches, so that later forms occupy narrower niches and have more specialized adaptations than their forebears;
12. the 'progressive' quality of lineages which leads to increased diversity by:
   i. an internal motor of improvement;
   ii. escalation, or a Red Queen model, in which all lineages specialize in order to remain in evolutionary equilibrium;
   iii. 'stepping up' of the recovery time and scope following mass extinctions—surviving clades of each event are able to radiate faster and further than those surviving earlier events;
   iv. large-scale competition between 'evolutionary faunas' and attainment of ever-higher intrinsic equilibrium levels after each major replacement (Sepkoski, 1984);

Real, but neutral in cause.

13. 'cladistic inevitability'—a clade must have one ancestor by definition and, if the clade survives for long, its diversity is more likely to increase than to remain constant at one (it cannot decrease, or the clade would have disappeared and would not be the subject of a study of diversification); a geometric argument;

14. null, neutral, or stochastic model, that the increase in biotic diversity is the sum of untold numbers of microevolutionary processes acting on all species, and has nothing to do with grand unifying processes (Hoffman, 1989).

Critique

Few paleontologists now subscribe to the idea that diversification is artefactual (nos. 1–4) after the tests by Sepkoski et al. (1981) and others. The abiotic and biotic models (nos. 6–8) are unlikely to apply to the global long-term level, and they are all hard to test. Explanation 9, that increases in diversity are the result of reductions in the extinction rate, is really another way of saying the same thing. In most cases, the observed decline in extinction rates is insufficient to explain the rate of diversification on its own, and for tetrapods at least, extinction rates remain constant, or increase marginally, while diversification proceeds apace. The 'progressive' explanations (no. 12) tend to be teleological, or to involve post hoc assumptions, or to be untestable assumptions (Benton, 1987a). The neutral or null models (no. 13) have been disputed, and none proposed yet is entirely neutral.

This leaves three explanations of diversification for testing, increasing endemicity (no. 5), increase in adaptive space occupied (no. 10), and subdivision of niches (no. 11). These are all testable, and some preliminary analyses (Benton, 1990) suggest that all three factors have played a role in the diversification of the tetrapods, and especially the last two.

Tests of Postulated Causes of Diversification

The data used in this study were based on the fossil record of marine and non-marine tetrapods (Benton, 1987b), with the addition of data on basic ecological divisions, body size, and geographic distribution. A second data base on well-preserved faunas (fossil Lagerstätten) was also constructed.

It was expected that the break-up of Pangaea after the Triassic would have led to increased endemicity of terrestrial tetrapod families at least (explanation no. 5). However, the present study (Fig. 1) did not support this idea: after some major changes in the Devonian, Carboniferous, and Permian, levels of endemicity remained roughly constant throughout the Mesozoic and Cenozoic, which witnessed the breakup of Pangaea. Hence, tetrapod diversification cannot be ascribed to continental drift.

Increases in the breadth of the adaptive space occupied by a clade would seem to be an obvious candidate for an explanation of diversification (no. 10). Indeed, this appears to have been the case (Figs. 2, 3): the relative significance of the mode of life of the earliest tetrapods—freshwater piscivory—declined from 100% in the late Devonian and early Carboniferous to levels of 10–15% in the Tertiary. The addition of diversity since those early times seems to have been mediated largely by the addition of broad new dietary types—insectivory, carnivory, herbivory, omnivory, as well as specialised diets of molluscs, grain, fruit, and so on—and new habitat ranges—terrestrial, marine, burrowing, arboreal, and flying.

The evolution of many tetrapod clades includes as much specialisation, or niche subdivision (no. 11), as it does expansion of adaptive space. An attempt has been made to test the idea of niche subdivision by comparing typical faunas through time. It is valid to compare, for example, an individual, exceptionally well-preserved fauna (Lagerstatte) from the Carboniferous with one from the Eocene, since levels of preservation appear to be equivalent. A measure of the subdivision of niches can be obtained by counting the numbers of species in comparable faunas; that is, groups of tetrapods found in a single deposit in a single locality or small sedimentary basin. A preliminary survey of 100 such faunas, spanning the past 350 million years in Europe and North America (Benton, 1990) shows a marked increase in mean faunal diversities from 18 in the Carboniferous to 35–51 in the Miocene and Pliocene (Fig. 4). The maximum figures rose from 25 in the Carboniferous to 80–100 in the Late Cretaceous and Tertiary. This test requires refinement since a part of the increase in mean and maximum faunal diversities is almost certainly the result of the conquest of new adaptive space, but most of the increase does appear to relate to niche subdivision.

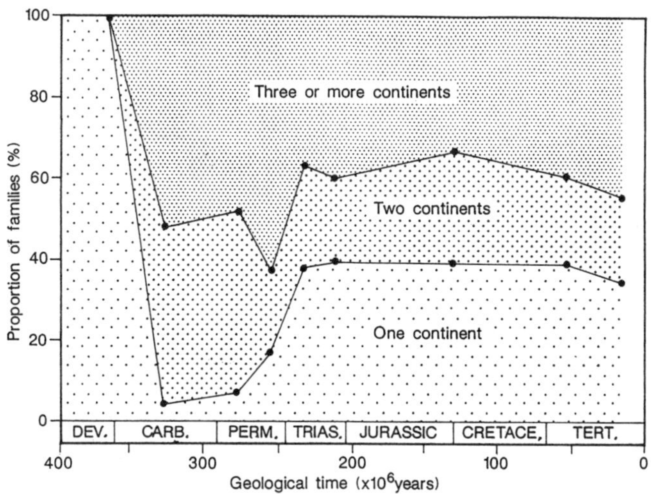

Figure 1. Variations in the ultimate geographic distribution of non-marine tetrapod families that arose during each time interval. Family data were recorded by stratigraphic stage, and were then grouped into broad time intervals in order to provide large enough samples throughout (n = 44–360; mean = 146 families per time interval).

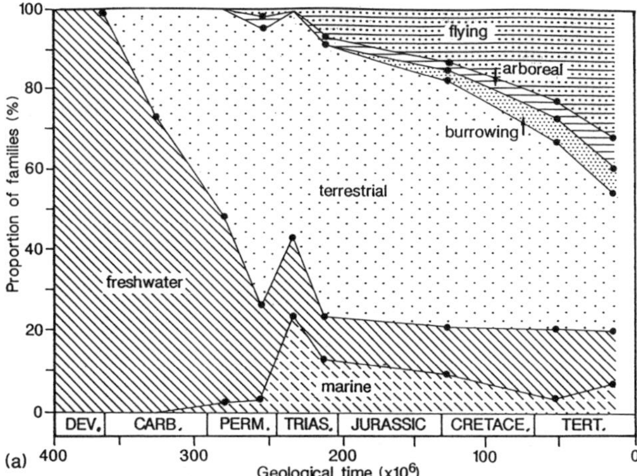

Figure 2. Proportions of broad habitat types of terrestrial and marine tetrapod families through time. The habitats were determined from the primary literature, and they represent the activities of all, or the majority of, species within a family (full details in Benton and Blacker, MS). Families were grouped into broader time units than stages, as in Fig. 1.

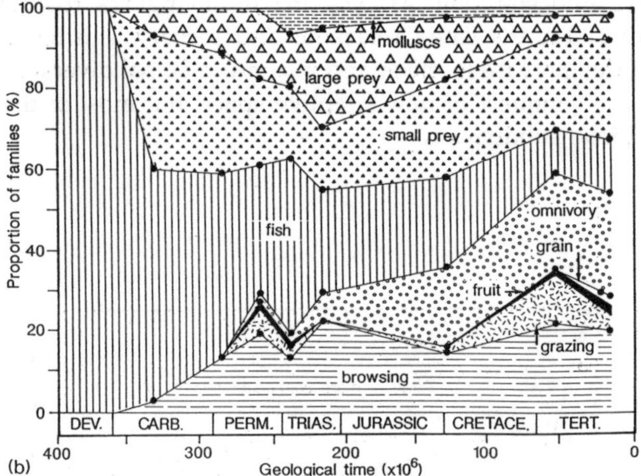

Figure 3. Proportions of broad dietary types of terrestrial and marine tetrapod families through time. The details and implications are as for Fig. 2.

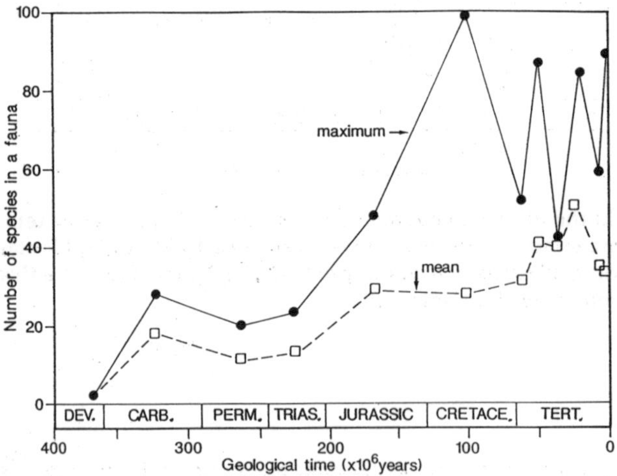

Figure 4. Maximum and minimum mean numbers of tetrapod species in well-preserved terrestrial faunas. This graph is based on a sample of 100 faunas (details in Benton and Blacker, MS), with 1–20 faunas sampled per stratigraphic period (or epoch in the Cenozoic). Each "fauna" is the species list of a single quarry or restricted sedimentary basin. Note the overall rise, with fluctuations in the maximum curve probably largely the result of variations in preservation quality.

CONCLUSION

Most of the diversification of the tetrapods seems to result from the conquest of new adaptive space and the subdivision of niches. Changes in the physical environment seem to have had a smaller effect, in contrast to the view of Cracraft (1985). Long-term physical changes that might have been expected to produce major effects (e.g., continental drift, climatic differentiation) are not adequate agents of diversification, for tetrapods at least. Doubt has also been cast on the role of continental drift and sea level change in the history of marine invertebrates (Jablonski and Flessa, 1986). The global effects of continental drift and marine regression on terrestrial faunas, argued by Bakker (1977), are even harder to confirm.

The relative roles of increased adaptive space and increased specialization in producing diversification are hard to determine, but the former may have had the main influence in tetrapod radiations. This supports the findings of Bambach (1985), who argued that marine invertebrates have increased their diversity mainly by expansions of utilized adaptive space. Sepkoski (1988) also finds that most of the increase in global diversity of marine life was produced by increases in the numbers of broad habitat types occupied.

Certain aspects of diversification patterns may of course be 'on-off' phenomena, dependent on specific historical circumstances, such as the break-up of Pangaea, latitudinal climatic diversification, the diversification of fishes (hence providing new adaptive space for predators), and other environmental changes. There is certainly no evidence for an inevitable motor of change that drives diversity ever upwards.

ACKNOWLEDGMENTS

I thank Anne Blacker for compiling much of the data used in this study, and the Leverhulme Trust for funding this part of the work. Mrs. Pam Baldaro drafted the diagrams.

LITERATURE CITED

Bakker, R. T. 1977. Tetrapod mass extinctions—a model of the regulation of speciation rates and immigration by cycles of topographic diversity. Pp. 439–469. *In:* A. Hallam (ed.), *Patterns of Evolution, As Illustrated by the Fossil Record.* Elsevier: Amsterdam.

Bambach, R. K. 1985. Classes and adaptive variety: the ecology of diversification in marine faunas through the Phanerozoic. Pp. 191–253. *In:* J. W. Valentine (ed.), *Phanerozoic Diversity Patterns; Profiles of Macroevolution.* Princeton University Press: Princeton, NJ.

Benton, M. J. 1987a. Progress and competition in macroevolution. *Biological Reviews* 62:305–338.

Benton, M. J. 1987b. Mass extinctions among families of non-marine tetrapods: the data. *Mémoires de la Société géologique de la France* 150:21–32.

Benton, M. J. 1990. The causes of the diversification of life. Pp. 409–430. *In:* P. D. Taylor & G. P. Larwood (eds.), *Major Evolutionary Radiations. Systematics Association Special Volume* 42. Clarendon Press: Oxford.

Cracraft, J. 1985. Biological diversification and its causes. *Annals of the Missouri Botanical Garden* 72:794–822.

Hoffman, A. 1989. *Arguments on Evolution: a Paleontologist's Perspective.* Oxford University Press: New York. 274 pp.

Jablonski, D. & K. W. Flessa. 1986. The taxonomic structure of shallow-water marine faunas: implications for Phanerozoic extinctions. *Malacologia* 27:43–66.

Raup, D. M. 1972. Taxonomic diversity during the Phanerozoic. *Science* 177:1065–1071.

Sepkoski, J. J., Jr. 1984. A kinetic model of Phanerozoic taxonomic diversity. III. Post-Paleozoic families and mass extinctions. *Paleobiology* 10:246–267.

Sepkoski, J. J., Jr. 1988. Alpha, beta, and gamma: where does all the diversity go? *Paleobiology* 14:221–234.

Sepkoski, J. J., Jr., Bambach, R. K., Raup, D. M. & J. W. Valentine. 1981. Phanerozoic marine diversity and the fossil record. *Nature* 293:435–437.

Valentine, J. W. 1969. Patterns of taxonomic and ecological structure of the shelf benthos during Phanerozoic time. *Palaeontology* 12:684–709.

Valentine, J. W. 1980. Determinants of diversity in higher taxonomic categories. *Paleobiology* 6:444–450.

EVOLUTION
ON
ISLANDS

Introduction to the Symposium

Scott E. Miller

This symposium presents an overview of some of the more interesting current topics in evolutionary research pertaining to island archipelagos. The questions posed to the authors included: What can we learn from island systems? Are they useful models for the study of general processes? How are new techniques or approaches changing old ideas? These questions are investigated by focussing on various taxa in various island groups.

To promote the general interest of the symposium, authors were asked to include a general introduction to their particular island setting, and to place the significance of their study into a broader context when possible. Several papers are reviews of important research programs that have been reported elsewhere in more detail (e.g., Liebherr, Olson, Stuessy et al.), some report on specific research topics (e.g., Holloway, Howarth), and others review current research on particular island systems (e.g., Oromi et al., Paulay, Simon, Thornton). Because of time limitations, we had to focus (except for Paulay) on terrestrial ecosystems. Madagascar and the Philippines were the subjects of independent symposia at the Congress.

As discussed further by Peck and others, islands represent useful models for the study of evolutionary processes. Their special advantages include the possibility of studying simplified ecosystems, assigning relative ages to evolutionary events, and being able to make comparisons of patterns among islands within an archipelago or between different archipelagos. Even so, there is no single modern reference on evolution in island systems. The last general published review (Williamson, 1981) remains useful, but this symposium shows that a massive amount of information has accumulated in the last ten years, and molecular studies now underway promise to further change our understanding. While a comprehensive review is outside the scope of this introduction, I would like to comment briefly on some of the more interesting generalizations that emerge from this symposium.

Despite the popular conception that many island archipelagos (especially Hawaii and the Galapagos) are "well known," there is still a great need for basic data on the distribution, systematics, and ecology of island organisms. This is especially true for invertebrates, but even applies to birds (e.g., Olson)! Evaluation of MacArthur and Wilson's equilibrium theory of island biogeography requires census data *over time*, which are available only in rare cases (Gilbert, 1980). Conservation applications now being attempted on many islands demand quality data on the *current* distribution and status of populations.

Interpretation of evolution and biogeography requires a firm understanding of geography (especially tectonic and eustatic changes over time; e.g., Paulay) and human induced changes (e.g., Olson). Studies of new islands undergoing the early stages of colonization help us to understand these processes (e.g., Thornton).

Dr. Miller is chairman of the Dept. of Entomology, Bishop Museum, Box 19000-A, Honolulu, HI 96817, USA.

266

New methods of data analysis promise better resolution and testability for evolutionary and, especially, biogeographic hypotheses. These methods include cladistic, component, and Brooks parsimony analyses, as well as biogeographic graphs. However, comparative studies warn that no single method should be followed blindly; the most pragmatic method will vary with the history and biology of the taxa under study.

What triggers species-rich groups and adaptive radiations? While such groups may be less common than was once thought, some studies are recognizing presumed polymorphic "species" as new species-rich groups (e.g., Howarth, Olson, Paulay, Peck, Simon, Stuessy et al.). Hopefully, new ecological and molecular studies will illuminate these issues. Studies underway in island cave ecosystems are providing models for rapid evolution in specialized microhabitats (e.g., Howarth, Oromi et al., Simon).

Defining the limits of "species" presents problems in some island organisms. Again, comparative morphological, ecological, and molecular studies now underway (e.g., Simon, Stuessy et al.) should provide new insight. The pattern of high morphological variation but low isozyme and DNA variation within some species-rich island taxa may help illuminate the genetics of formation of such groups (e.g., Simon, Stuessy et al.; Carson, 1987).

Finally, the conservation needs of many of the world's islands challenge biologists to make their studies relevant to saving the biotas they study. Conservation biology requires a better understanding of the population biology of many organisms, as well as ways to minimize the impacts of introduced organisms (e.g., Peck, Simon).

I hope that this symposium generates further interest and research in these and other questions. The world's islands await further evaluation as natural experiments in evolution. I thank the participants for their stimulating contributions. The Congress Program Committee and U.S. National Science Foundation made the symposium possible by supporting travel expenses.

LITERATURE CITED

Carson, H. L. 1987. Colonization and speciation. Pp. 187–206. *In:* A. J. Gray, M. J. Crawley & P. J. Edwards (eds.), *Colonization, Succession and Stability.* Blackwell Scientific Publications: Oxford.

Gilbert, F. S. 1980. The equilibrium theory of island biogeography: fact or fiction? *J. Biogeogr.* 7:209–235.

Williamson, M. H. 1981. *Island Populations.* Oxford University Press: Oxford. 286 pp.

Evolution of Waif Floras:
A Comparison of the
Hawaiian and Marquesan Archipelagoes

Warren L. Wagner

Abstract. Many oceanic islands are volcanic and have originated mid-ocean. All indigenous lineages are the result of long-distance colonization. This paper compares the vascular floras of the Hawaiian and Marquesan archipelagoes. It is difficult to study diversification of island floras because of the unknown losses relating to man's perturbation of insular environments. In Hawai'i about 10% of the historically known angiosperms are extinct, another 40% threatened, and I estimate that another 10-15% went extinct during the Polynesian period. The Hawaiian Islands, and their continuation, the Emperor Seamounts, include a progression of over 100 volcanoes that extend for 6,000 km. The eight main Hawaiian Islands contribute >99% of the land area of 16,885 km², with the island of Hawai'i contributing over half. These islands range from about 500,000 years old to 5.7 million years old, while the northern end of the Emperor Chain is at least 70 million years old. Thus, a group of islands similar to the current main islands has likely existed for at least 70 million years, presenting a continuously available habitat for colonization and speciation for considerably longer than previously assumed. Volcanic activity on a young island like Hawai'i results in a high turnover of habitat caused by lava flows, and lower species diversity (408 taxa) and endemism (26%). The Marquesan Islands are a short archipelago, 355 km long, consisting of 12 islands ranging in age from 1.3 to over 6 million years that equal only 6% of the surface area of the Hawaiian Islands. The Marquesan vascular flora (318 spp., 42% endemic, from 251 colonists) is considerably less diverse than the Hawaiian flora (1140 spp., 86% endemic, from 387-396 original colonists). Both archipelagoes have depressed numbers of monocots and disproportionately species-rich pteridophyte floras, the latter resulting from the high dispersal ability via air of most pteridophytes. High dispersal ability also results in low endemism. Both archipelagoes fit Darlington's rule that a 10-fold increase in area results in a doubling of species, but this does not hold among the Hawaiian Islands themselves, where island age appears to be a significant factor countering size. Further, the island of Hawai'i has not had sufficient time to be fully colonized or for many species to evolve, and if it is removed from consideration, then the islands are especially species-rich relative to other archipelagoes. This relates to the considerable age of the archipelago, which has allowed high speciation and accumulation of taxa. Only a few lineages have diversified in both archipelagoes, these are primarily dicots. In the Hawaiian Islands 542 species (48% of total) have been derived from only 10% of the original colonists, and another 283 species (25%) from over 70% of the presumed colonists. Diversification in the Marquesas Islands has been considerably less because of their smaller size and younger age with 63 species (20% of total) from 10% of the colonists.

INTRODUCTION

Oceanic islands have long been considered to be outstanding natural laboratories for the study of evolutionary biology. Interest in islands for evolutionary studies date to the

Dr. Wagner is with the Botany Department, National Museum of Natural History, Smithsonian Institution, Washington, D. C. 20560, USA.

268

time of Darwin and Wallace, and has continued to the present day (Carlquist, 1974). Many oceanic islands are volcanic in origin and have never been connected to any continent, although sometimes their biota are closely influenced by one. Individual volcanic islands are, on a geologic time-scale, relatively short-lived, and accordingly are inhabited by recent products of evolution or colonization. Thus, their biota usually is differentiated, if at all, to the level of species or sometimes genus, but typically not family. Unique taxa can evolve, and persist, if the islands of an archipelago collectively have considerable size and age. Oceanic islands have an exclusively waif biota; all of the native lineages are the result of long-distance dispersal. Another important feature of oceanic islands is that a tremendous diversity of habitats are compressed, especially those resulting from rainfall gradients.

Our increasing knowledge of plate tectonics, as well as the geology and age of an increasing number of islands, provides a clear framework on which to overlay evolutionary studies. The insular situation provides excellent opportunities to study the relationship among ecological parameters, age, size dispersal, etc. This paper is a preliminary overview of some of the patterns in the Hawaiian and Marquesan archipelagoes, and contrasts some of the differences between them.

Considerable attention has been given in the literature to evolution on oceanic islands, much of it based on the Hawaiian Islands. In particular, dispersal patterns and, especially adaptive radiation have been prevalent topics (Carlquist, 1974, 1980; Crawford et al., 1987; Carr, 1987). The principal motivation for the current study is that new floristic, systematic, and evolutionary studies have been completed or are in progress for both Hawai'i and the Marquesas Islands, providing more substantial data on which to base analyses (e.g., Carr, 1985; Helenurm & Ganders, 1985; Lowrey, 1986; Lammers, in press; Weller et al., 1990; Witter & Carr, 1988). Also, a significant reevaluation of all of the Hawaiian flowering plants has recently been completed (Wagner et al., 1990). New work in the Marquesas Islands has been less substantial, but includes the discovery of new species (Florence, 1986, 1990; Fosberg & Sachet, 1966, 1975, 1981) as well as two remarkable new genera; *Lebronnecia* in the Malvaceae (Fosberg & Sachet, 1966) and the biogeographically anomalous *Plakothira* in the Loasaceae (Florence, 1985). These pivotal studies in both archipelagoes make a new analysis of the patterns of evolutionary diversification timely.

Sources of Data

The data for the angiosperms of Hawai'i essentially follow the taxonomy and hypotheses concerning the number of original colonists of Wagner et al. (1990). The data include several updates since the new flora of Hawai'i went to press. Statistics for the Hawaiian pteridophytes are based on data from W. H. Wagner (1988; pers. comm.). The data for the Marquesas are my own, derived from various sources including Carlquist (1974), Florence (1987), A. C. Smith (1979, 1981, 1985, 1988), the publications by Forest Brown and his wife Elizabeth Brown (F. Brown 1931, 1935; E. Brown & F. Brown, 1931); unpublished information from M.-H. Sachet, R. Oliver, D. Lorence, and J. Florence; and the collections at the US National Museum of Natural Histoy and the Bishop Museum as well as information from numerous taxonomic articles. Because this information is otherwise not available in the form used here a synopsis of the Marquesan vascular flora is presented in the appendix. The hypotheses for the number of lineages represented for each genus are taken from the literature or reflect my own preliminary analyses. The hypotheses for the pteridophytes were compiled with considerable assistance from D. Lorence.

Perturbations

Both the Hawaiian and Marquesan archipelagoes, like most of those in the Pacific, have had major perturbations of the biota, resulting in a tragic loss of species and loss of ecological stability. The lowland, especially leeward, areas of the islands, in both the Hawaiian and Marquesan archipelagoes, have been most significantly altered. On both, indigenous species, especially lowland taxa, are patchy, relict, or extirpated. Today, cattle pastures occupy approximately 50% of the land area of the Hawaiian Islands, plantations and urban areas account for about another 30%, and the remaining 20% harbor native and alien plant communities (Gagné & Cuddihy, 1990). The sheer magnitude of the changes can be appreciated in that approximately half of the species of flowering plants that grow outside of cultivation in Hawai'i and the Marquesas Islands are naturalized. The situation in more mesic upland habitats is somewhat better than in the lowlands, but all of the biota has been greatly impacted by transposed landscapes, introduced ungulates, alien plants, and other introduced organisms. Therefore, the statistics derived from the present floras are probably negatively skewed, especially in the lowlands. The changes and the underlying causes are summarized for Hawai'i in the recent work by Cuddihy and Stone (1990).

The problem relative to the analysis here is that we do not know the extent of the changes in species diversity fostered by the early Polynesians before scientific study and collection of the biota. There is no doubt that the early Hawaiians greatly altered the lowland vegetation of the Hawaiian Islands, particularly during the period of greatest expansion in population and agriculture between 1100 and 1650 A.D. (Kirch, 1985; Cuddihy & Stone, 1990). The situation in the Marquesas Islands is similar (see Adamson, 1936; Decker, 1970), but the changes have not been as extensively studied. The destruction of native ecosystems has accelerated during the past two hundred years in Hawai'i, since the arrival of James Cook, and for nearly twice as long in the Marquesas Islands, largely through continued habitat alteration, and the effects of feral animals, insects, and plants.

The fauna, at least in Hawai'i, has not fared well. Olson and James (1982) found 40 previously unknown extinct species of birds from primarily lowland sites. Radiocarbon dating of the fossils indicates that many of these species became extinct after the Polynesian colonization of the islands. This means that over half of the endemic land birds that were present before the arrival of the Polynesians became extinct prior to the arrival of the Europeans (Olson & James, 1982). On O'ahu, land snail fossil deposits indicate similarly extensive changes in the snail fauna during the Polynesian period, with some species decreasing and others, adapted to disturbance, increasing (Kirch, 1983). Approximately half of the species of Hawaiian land snails are extinct (Gagné & Christensen, 1985). There is no comparable body of fossil studies on the flora, so we can only guess the degree of plant extinction during human occupation of either of these archipelagoes. Among the few paleobotanical studies in Hawai'i are the pollen corings made by Selling (1948). They did not reveal any previously unknown extinct taxa except for one fern, a *Schizaea*. Macrofossils are also rarely encountered in islands. For example, one of the few found in Hawai'i is a mold of *Pteralyxia* (Apocynaceae), an endemic Hawaiian genus, found in ash beds on O'ahu (Caum, 1933) that are between 50,000 and 150,000 years old. Much paleobotanical work needs to be done on oceanic islands.

Thus, we cannot yet reliably estimate the extent to which the alterations on these oceanic islands have caused extinctions. It is clear, however, that distributions were drastically altered. Recent analyses (Wagner et al., 1985; Wagner et al., 1990) suggests that about 10% of the historically known Hawaiian angiosperm flora is extinct, and nearly 40% of the species are threatened. In fact, nearly one-third of the extinctions are in the highly vulnerable group of endemic Campanulaceae. The data indicate that there has been extensive reduction of populations of flowering plants during the past two centuries, and that the reductions vastly exceed extinctions. The persistence of flowering plants appears to be greater than that exhibited by the Hawaiian fauna, and therefore I would estimate that the Hawaiian angiosperm flora present at the time of arrival of the Polynesians was perhaps not much more than about 10–15% greater than now, or about 1200 to 1250 taxa. The loss of plant species during the Polynesian period probably did not exceed the amount observed during the historical period because the pressures and outright elimination of habitat have been far more extensive during the past two hundred years. The data at hand in no way suggest anything like twice as many taxa, as in the case of birds, snails, and other invertebrates.

PHYSICAL AND GEOLOGICAL BACKGROUND

A short overview of the geology and climatic setting of these archipelagoes is needed to understand evolution on islands as well as to assess the differences between the Hawaiian and Marquesan archipelagoes.

Hawaiian Islands

This archipelago is an extensive arc of some 132 islands, reefs, and atolls extending nearly 2,600 km along a southeast to northwest line in the middle of the Pacific Ocean from the youngest island of Hawai'i (154°40' W, 18°54' N) to Kure, the oldest atoll (178°75' W, 28°15' N). The islands are 3,850 km from the California coast, and about 6,400 km from the southern Pacific islands, the source of the ancestors of much of the flora. The closest high islands to Hawai'i are the Marquesas Islands, 3,860 km to the southeast. The amazing diversity of habitats and climatic variation found on the eight principal Hawaiian islands is well known (Price, 1983; Carlquist, 1974, 1980; Wagner et al., 1990). Summaries of the geologic history can be obtained from Stearns (1985) and MacDonald et al. (1983). Geological reviews have just been published by Walker (1990a,b). Considerable progress in understanding Hawaiian geology has been made during the past two decades. Many of these findings significantly alter our concepts of colonization and evolution on oceanic archipelagoes, especially Hawai'i.

The Hawaiian Islands are the emergent upper portions of great volcanoes that rise 5,000–9,000 m from the ocean floor of the Pacific Plate. According to plate tectonic theory most volcanic activity occurs at boundaries between the earth's plates. The Hawaiian Islands are among the best examples of an exception to this rule; they comprise an intra-plate volcanic chain. Wilson (1963) proposed that the Hawaiian Islands were formed as the Pacific Plate moved slowly northwestward over a melting spot, termed a "hot spot." In 1972, Morgan proposed that the Emperor Seamounts were also formed over the Hawaiian hot spot. Collectively, these volcanoes are referred to as the Hawaiian Ridge. The arc of islands and seamounts of the Hawaiian Islands, and of their continuation, the Emperor Seamounts, include a progression of over 100 major volcanoes that extends for 6,000 km across the Pacific Plate. The eight main islands, which make up over 99% of the land area of 16,885 km^2, occupy the southeastern 650 km of the chain from the currently volcanically active Hawai'i Island to the deeply eroded older islands of Kaua'i and Ni'ihau. The island of Hawai'i, one of the world's largest volcanic islands (10,451 km^2), contains over half of the entire surface area of the Hawaiian Islands.

The islands beyond Ni'ihau from Nihoa to Midway and Kure atolls make up the Northwestern Hawaiian Islands. They consist of nine small islands and atolls with only 11.43 km^2 of combined surface area, along with numerous reefs and shoals, covering a distance of 1,650 km. They are presumably erosional remnants of what were formerly high volcanoes. Only Ka'ula, Nihoa, Necker, French Frigate Shoals, and Gardner Pinnacles are still composed of volcanic rock; the others are limestone islands.

Near 32°N, 172°E the Hawaiian Ridge terminates. Beyond this point is a chain of submerged volcanic seamounts known as the Emperor Chain. At their juncture there is a considerable bend. The Emperor Seamounts extend northward for approximately 2,500 km to the Aleutian Trench near the Kamchatka Peninsula, where the Pacific Plate is subducted.

During the past 25 years substantial progress has been made in the dating of the major Hawaiian islands (see MacDonald et al., 1983). The main islands of the chain were all formed during the Pliocene and Pleistocene; they range from about 500,000 to 5.7 million years old. These islands harbor the majority of terrestrial organisms found in the archipelago because of their extensive area and topographic and ecological diversity. The islands over six million years old are substantially eroded and none is presently over 277 m above sea level. All basically have a coastal environment and are inhabited by relatively few species of coastal plants, pelagic birds, a few rare endemic birds, and marine animals. Beyond Midway and Kure (less than 30 million years old) the former islands are now completely eroded to below the current sea level. The "dogleg" between the Hawaiian and Emperor chains occurred about 43 million years ago, and may possibly be related to the separation of Australia from Antarctica (Walker, 1990b). Meiji Guyot is at the northern end of the Emperor Chain with an apparent age of about 70 million years (Scholl & Creager, 1973).

A group of islands with interisland spacing comparable to today's has likely existed for at least 70 million years in approximately the same relative position as the principal Hawaiian Islands of today (Walker, 1990a). As the earlier islands were conveyed off of the hot spot they began to erode, and another new island was formed over the hot spot. This process has created a linear chronological progression of islands unlike any other archipelago in its overall length and consecutive production of new islands. Via interisland colonization there has been habitat continuously available for occupancy for a considerably longer period of time than the few million years previously assumed (e.g., Carlquist, 1974).

During the Pleistocene, climatic fluctuations lowered the sea level several times by about 120 m (Walker, 1990a). The result of the lowered sea level was to unite Moloka'i, Lana'i, Maui, and Kaho'olawe into one larger island known as Maui Nui. The area of this island was enlarged even more by the addition of the land encompassed by portions of the Penguin Bank, an addition nearly the size of Moloka'i. This addition brought land much closer to O'ahu than it currently is to Moloka'i. These Pleistocene modifications have significantly altered distributions and reduced single island endemism on these central islands.

Another significant feature of the islands is subsidence. All of the Hawaiian Islands are slowly subsiding and will eventually become seamounts (Walker, 1990b). For example, there is evidence that a coastal terrace of lava deltas occurs 600 m deep around O'ahu, implying that this subsidence has taken place during the approximately 2 million years since O'ahu became inactive, yielding an average rate of 0.3 mm/yr. (Walker, 1990b). Walker gives an average rate of subsidence for Suiko Seamount (Emperor Chain) of 0.04 mm/yr. The younger islands subside at a more rapid rate. Subsidence at Hilo documented by tide gauge measurements from 1946–1983 averaged 2.4 mm/yr. (Walker, 1990b). We can only speculate that older islands of the main chain were once much higher, like Maui and Hawai'i are today, subsiding to their present levels.

One final feature of young islands should be appreciated. The fresh lava flows of active volcanoes like Kilauea or Mauna Loa on Hawai'i are colonized repeatedly by animal and plants during the active phase. An area will go through a period of development only to be flooded again by new flows. For example, about 40% of the surface of Mauna Loa consists of lavas younger than 1000 years old (Lockwood & Lipman, 1987). This feature of an active island has been underestimated in the past. It almost certainly acts to maintain lower levels of species diversity and endemism.

Marquesas Islands

The geology and geochronology of the Marquesas Islands are not well-studied. Beyond reports by the exploring expeditions, the published geological studies are those of Lacroix (1928), Chubb (1930), and Duncan and McDougall (1974) on geochronology of five of the islands. Other geologic studies are primarily geochemical (Liotard et al., 1986, and papers cited therein). The best description of the physical environment of the Marquesas Chain is still that of Adamson (1936).

The archipelago consists of 12 rugged, relatively small islands at the eastern periphery of French Polynesia. This chain of islands ranges from Eiao (7°53' S, 141°27' W) at the northwestern end to Fatu Hiva (10°35' S, 138°35' W) on the southeastern end. Its axis is nearly parallel to the other archipelagoes of the eastern Pacific (Duncan & McDougall, 1974). The archipelago embodies roughly 17 times less area (6%) than the Hawaiian Archipelago; the larger of the Marquesan islands are slightly smaller than the Hawaiian island of Lana'i. Among the 12 Marquesas Islands three have peaks over 1220 m high. The radiometric ages of the Marquesan Islands range from 1.3 million years for Fatu Hiva to slightly over 6 million years for Eiao. None of these islands are currently volcanically active. The Marquesas Archipelago fits the general pattern in the Pacific where the age progression within an archipelago is toward the northwest, but it is much shorter in length (355 km) than most of the other Pacific archipelagoes.

The Marquesas Islands are further from any continent (4,800 km from Mexico) than any other archipelago, but are less isolated because they are situated at the eastern periphery of a series of archipelagoes. The closest is the Tuamotu Archipelago, which consists entirely of atolls. The flora of the Tuamotu Islands was considered by F. Brown (1931) to have been in large part the source flora for both the Marquesas and Society archipelagoes. However, the Tuamotus are considerably older than either the Marquesas or Society archipelagoes, probably about 37–43 million years old (Clague & Jarrard, 1973). They were weathered to the low island or atoll stage long before the genesis of either the Marquesas or Society Islands and thus could not have served as a primary stepping-stone source area for colonization of the younger archipelagoes. The closest high islands to the Marquesas Islands are the Society Islands, about 1370 km to the southwest. The only Society islands so far dated are at the young end of the archipelago; Tahiti at about 0.5–0.65 millon years old, and close-by Moorea at about 1.65 million (Dymond, 1975). The other principal archipelagoes that are possible source areas for the Marquesas Islands are beyond the Society Islands, including the Austral (ca. 5 to >16 million years old; Clague & Jarrard, 1973), Cook (<1 to >16 million years old; Clague & Jarrard, 1973), Samoan, and Tongan archipelagoes.

The topography of the Marquesas Islands is extremely rugged. There are essentially no developed coastal plains. On most of the islands there is a central ridge, Hiva Oa and Nuku Hiva are exceptions and both have elevated plateau regions. The most striking features of the islands are the towering peaks of Ua Pou, and the long narrow ridge of Fatu Hiva, so narrow and eroded in some places that it has holes through it several hundred meters from the summit. The rugged topography has produced varied habitats on the larger islands, ranging from dry on the leeward sides to mesic valleys and cloud covered summits. Rainfall, based on meager records, ranges roughly from 1,000 to over 2,800 mm per year.

FLORISTIC DIVERSITY

Because the vascular floras of oceanic islands are derived via long-distance colonization they are impoverished relative to comparable subtropical and tropical continental areas. However, substantial variation in floristic components is observed among archipelagoes. A summary of the vascular floras of the Hawaiian and Marquesan Islands is given in Table 1. The numbers given here for Hawai'i are lower than previous estimates (Carlquist, 1974; Fosberg, 1948; St. John, 1973) because of the recent complete reevaluation of the Hawaiian angiosperms (Wagner et al., 1990). The current Marquesan flora, with only 318 vascular species in 177 genera, is much smaller and more typical of oceanic archipelagoes. The reasons for this are twofold. First, the Marquesas Islands have been a relatively short-lived island system. Subsidence and erosion of an island, over about 7–8 million years, reduce it to the low atoll stage leaving little trace of its former flora. In the Marquesas Islands this time-frame represents the entire life of the archipelago. In contrast, in the Hawaiian-Emperor Chain there has been a continuous formation of new high islands for at least 70 million years that has, through interisland colonization and speciation, perpetuated many lineages, resulting in a net accumulation of taxa. Second, the Marquesas Islands comprise considerably less area. This results in a higher turnover of species. There have likely been many other colonizations to the Marquesas Islands that persisted, perhaps speciated, and then went extinct. We have no record of this because of the meager fossil record of plants on oceanic islands.

Do the floras of these archipelagoes have atypical complements of higher order taxa? One way of considering whether these island floras are atypical is to compare them to the worldwide totals of vascular plants. Using the numbers given by Mabberly (1987) for total numbers of genera and species, the Hawaiian angiosperms comprise about 2% of the world's genera, proportionately divided among the dicots and monocots. In contrast, the number of Hawaiian pteridophyte genera comprise 14% of the world's total. The Marquesas Islands exhibit a similar pattern with about 1% of the world's angiosperm and dicot genera, and about 14% of the world's pteridophyte genera. The situation is not similar at the level of species. Hawai'i has 0.4% of the world's angiosperm species, 0.45% of the dicots, and only 0.2% of the monocots. The pattern is basically the same in the Marquesas Islands (0.1% of the angiosperm and dicot species; 0.06% of the monocots). At the species level the pteridophytes are not nearly as abundant as at the generic level, with 2% of the world's species in Hawai'i and 1% in the Marquesas Islands. These figures suggest that both archipelagoes have depressed numbers of monocots and disproportionately rich pteridophyte floras. This much higher representation of pteridophytes relative to angiosperms is expected because, with the exception of *Isoëtes*, *Marsilea*, and *Selaginella*, the original pteridophyte colonists were all dispersed via air flotation. Virtually all ferns have spores that can be readily dispersed in the air over long distances (Tryon, 1970). Therefore a large proportion of pteridophyte taxa are widespread, evident especially at the generic level (Smith, 1972).

Table 1. Summary of the vascular floras of the Hawaiian and Marquesan archipelagoes.

Archipelago	Taxon	Number of families	Number of genera	Number of species (number endemic)	Number of colonizations	Species per colonization	% endemism
Hawai'i	Pteridophytes	27	51	172 (118)	115	1.5	69
	Dicots	73	165	832 (768)	196–205	4.0–4.2	92
	Monocots	14	51	134 (91)	76–77	1.8	68
	Angiosperms	87	216	966 (859)	272–282	3.4–3.6	89
	Vascular plants	114	267	1138 (977)	387–396	2.8–3.0	86
Marquesas Islands	Pteridophytes	26	53	108 (17)	101	1.1	16
	Dicots	53	101	174 (96)	118	1.5	55
	Monocots	6	23	36 (19)	32	1.1	53
	Angiosperms	59	124	210 (115)	150	1.4	55
	Vascular plants	85	177	318 (132)	251	1.3	42

SPECIES AND AREA

The species—area relationship, based on a wide sampling of island data, both for animals and plants, suggests that the rule proposed by Darlington (1957), that a tenfold increase in area leads to a doubling of the number of species, in general holds fairly well (Williamson, 1981). Applying this rule on an archipelago-wide basis to the data here suggests that since there is nearly a 20-fold decrease in area between the Hawaiian and Marquesas archipelagoes the Marquesas vascular flora should consist of about 280 species. Despite the differences in degree of isolation and ecological diversity between these archipelagoes, the actual figure of 318 is close. This pattern appears to be sustained by the expected number of 570 species in the Society Islands (with an area of 1,600 km² — roughly a 10-fold decrease in area to that of Hawai'i) and an actual number of about 620 species. Thus, Hawai'i has more species than the other Polynesian archipelagoes with high islands, but not more than would be expected based on its much larger size.

A quite different pattern emerges when Darlington's rule is applied to the islands within the Hawaiian archipelago (Table 2). The "big island" of Hawai'i has over 60% of the land area of the archipelago, yet has only 408 angiosperm taxa, while Kaua'i (6.4 times less area) has 477 taxa and O'ahu (6.6 times less area) has nearly the same number, 475. There is clearly a different species-area relationship for Hawaiian angiosperms, where island age appears to be a significant factor countering size. The most likely explanation for this anomaly is that the big island has not had sufficient time to be fully colonized or for many species to evolve; therefore, its very large area skews the data for the archipelago. Moreover, because the big island is still volcanically active the surface available for colonization is in an elevated stage of turnover. If the island of Hawai'i is removed from consideration, then the Hawaiian Islands are especially species-rich relative to other oceanic archipelagoes. For example, the island of Lana'i, which is similar in area to the larger islands of the Marquesas Islands, has 251 species of angiosperms, fully 20% more flowering plant species than the entire Marquesan Archipelago.

Table 2. Number of taxa on individual Hawaiian Islands and levels of endemism. Maui Nui consists of the composite of Moloka'i, Lanai, Kaho'olawe, and Maui, which were connected as one island during lowered sea level during the Pleistocene.

Island and area km²	Number of taxa	% endemism
Kaua'i (1,624)	477	43
O'ahu (1,574)	475	33
Maui (1,887)	496	19
Maui Nui (3,040)	587	34
Hawai'i (10,458)	408	26
TOTAL Hawaiian Islands	1110	

These comparisons illustrate that, although the floras of both of these oceanic archipelagoes are impoverished relative to comparable subtropical to tropical continental areas, they fit Darlington's rule. The Hawaiian archipelago, when the large area of Hawai'i Island is removed from consideration, is surprisingly diverse. The explanation for this diversity probably relates to the long, and apparently continuous, succession of large volcanic islands, allowing for the accumulation of species over evolutionary time. These figures also show that a young island that is still in the shield-building stage, such as Mauna Loa or Kilauea, is species-poor for its size. During the alkalic cap and erosional stages volcanic islands accumulate species via colonization and speciation at a faster rate. As the island ages the degree of differentiation of its flora increases as does the level of endemism. Then, as habitat diversity rapidly wanes, numerous taxa go extinct or colonize younger islands of the chain.

COLONIZATION AND DIVERSIFICATION

Several studies of diversification of island biota have expressed diversity using a ratio of one taxonomic level to another higher one (van Balgooy, 1971; Carlquist, 1974; Simberloff, 1970). The ratio typically used is the species per genus (S/G), but van Balgooy (1971) argued that for reasons of taxonomic stability the genera per family (G/F) ratio was better. Carlquist (1974) also used a species per family ratio (S/F). Using the data presented in Table 1 for the vascular plants (angiosperms in parentheses) the following figures are obtained: Hawai'i—S/G = 4.3 (4.6), G/F = 2.3 (2.5), S/F = 10 (11.1); Marquesas Islands— S/G = 1.8 (1.7), G/F = 2.1 (2.1), S/F = 3.8 (3.6). When these ratios are compared to Carlquist's (1974, p. 112), it is clear that van Balgooy was correct concerning stability, because the ratios here are nearly the same as Carlquist's for the Marquesas Islands, but only his G/F ratio is at all close to the ratios given here for Hawai'i. Unfortunately, the genus/family ratio is probably the least directly related to actual diversification in a particular archipelago because it does not differentiate between resident taxa resulting from colonization and from speciation. The differences in the ratios are the result of the massive taxonomic reevaluation of the Hawaiian flora (Wagner et al., 1990).

To more directly examine diversification I prefer to use another ratio: species per colonization (S/C). This ratio corresponds directly with the diversification of lineages, thus making comparisons directly to the products of evolution, eliminating the "noise" of variations in taxonomic opinion and, importantly, the results of multiple introductions of a particular genus. The problem with applying this ratio widely is that one must have fairly detailed knowledge of the flora in order to make reasonable hypotheses of the minimum number of colonizations. It is probably only reasonably applied to relatively small floras, such as those of oceanic islands. The flux in this ratio will result from changing hypotheses of ancestor-descendant relationships as well as from basic taxonomy. It should also be noted that, while the S/C ratio is a more refined way of examining diversification in oceanic archipelagoes, it does not change the general conclusions that have been made in the past using the other approaches (see Carlquist, 1974). These are: a) islands have higher ratios than continental areas; and b) Hawai'i has higher ratios than other islands, indicating that speciation is high. Hawai'i is unique because of its long existence and its great size and related ecological diversity.

The current estimates of the minimum number of colonists (lineages) and S/C ratios are given in Table 1. Greater diversification within the dicots in both archipelagoes is indicated by the higher ratios with 4.0–4.2 for Hawai'i and 1.5 for the Marquesas Islands. The dicot figure for Hawai'i is exceptional. The ratios for all other taxa from both archipelagoes are several times less, and the Marquesan dicot ratio is less divergent from the Hawaiian pteridophyte and monocot ratios.

The same pattern is seen in the percentages of endemism in both archipelagoes (Table 1). The level of endemism is another way to examine diversification, one that accounts for endemic single species lineages. Hawai'i, in general, has higher levels in all taxa than the Marquesas Islands, and again the Hawaiian dicots exhibit the highest percentage, 92% at the species level—the highest of any floristic region in the world. This indicates that, in the Hawaiian flora, speciation far exceeds immigration of new taxa. But, why have the dicots in Hawai'i developed to such an extent?

The answer to this question can reasonably be sought in a more detailed examination of the species-rich lineages. Table 3 lists the most species-rich lineages of vascular plants in the Hawaiian Islands. Slightly over 10% of the vascular genera are listed, comprising 542 species, nearly 48% of the vascular flora. The cut-off for inclusion was a somewhat natural break in the number of species per genus at about 10–12 species. Smaller genera were included in the table if they share a common ancestor with a larger genus, for example the baccate lobeloid genera—*Cyanea, Clermontia, Rollandia,* and *Delissea*—are all presumed

276

Table 3. Most species-rich lineages of vascular plants in the Hawaiian Islands. Data on *Asplenium* and *Dryopteris* from W. H. Wagner, Jr. (pers. comm.). Data for modes of dispersal from Carlquist (1974) with modifications. Modes of dispersal in order of decreasing frequency or ease of dispersal: A = air floatation; DR = Oceanic drift, rare event; BV = birds, seeds or fruits attached externally by viscid substance; BM = Birds, seeds or fruits embedded in mud usually on feet; BB = birds, seeds or fruits attached to feathers mechanically by barbs, bristles, awns or trichomes; BI = birds, eaten and transported internally.

Genus	Number of species	Presumed number of colonizations	Species per lineage	Mode of dispersal
Cyanea	52	1	91	BI or BV
Clermontia	22	—	—	—
Rollandia	8	—	—	—
Delissea	9	—	—	—
Phyllostegia	28	1	54	BI
Stenogyne	21	—	—	—
Haplostachys	5	—	—	—
Melicope	48	1	48	BI
Dubautia	21	1	28	BV or BB
Argyroxiphium	5	—	—	—
Wilkesia	2	—	—	—
Schiedea	23	1	27	BM
Alsinidendron	4	—	—	—
Myrsine	20	1–2	20 or 18, 2	BI
Hedyotis	20	2	19, 1	BI or BM
Bidens	19	1	19	BB
Pritchardia	19	1	19	BI or DR
Cyrtandra	57	4–6	18,11,10,7,4,2 or 25,14,11,2	BM
Chamaesyce	15	1–2	15 or 14,1	BI
Labordia	15	1	15	BI
Lipochaeta sect. Aphanopappus (~ Wollastonia)	14	1	14	BB
Sicyos	14	1	14	BB
Coprosma	13	2	12, 1	BI
Wikstroemia	12	1	12	BI
Peperomia	25	3–4	10, 10, 2, 2	BV
Lobelia	13	2	9, 4	BM
Asplenium (incl. Diellia)	25	16	7, 3, 2, remainder 1	A
Dryopteris	13	5	4, 4, 4, 1, 1	A
Totals	542	47–52	10.4–11.6	

descendants of one ancestor. These smaller genera are listed adjacent to the largest genus. The taxa are arranged in descending order starting with the most species-rich lineage.

The Hawaiian flora long has been known as an outstanding example of diversification and adaptive radiation; however, as shown here, this phenomenon actually involves a very small proportion of the original colonists. The fate of the roughly 387 to 396 original vascular plant colonists is exceptionally variable. Nearly 48% of the vascular species are accounted for by these species-rich lineages, yet they were derived from only 10% of the colonists. These successful genera are distributed in a wide variety of families, with the most conspicuous radiations in the Asteraceae and the baccate Campanulaceae (*Clermontia, Cyanea, Delissea,* and *Rollandia*); the latter group includes 91 species (9% of the native angiosperms), all of which presumably arose from only one colonist. Other notably species-rich groups are in the Gesneriaceae, Lamiaceae, Rubiaceae, and Rutaceae. All of the genera listed in Table 3 are dicots except for one monocot, *Pritchardia,* and two ferns, *Dryopteris* and *Asplenium.* These genera are on the lower end of the list. Moreover, the two

Table 4. Most species-rich lineages of vascular plants in the Marquesas Islands. Data for modes of dispersl from Carlquist (1974) with modifications. Modes of dispersal in order of decreasing frequency or ease of dispersal : A = air floatation; DR = Oceanic drift, rare event; BV = birds, seeds or fruits attached externally by viscid substance; BM = Birds, seeds or fruits embedded in mud usually on feet; BB = birds, seeds or fruits attached to feathers mechanically by barbs, bristles, awns or trichomes; BI = birds, eaten and transported internally.

Genus	Number of species	Presumed number of colonizations	Species per lineage	Mode of dispersal
Psychotria	11	1	11	BI
Bidens	9	1	9	BB
Cyrtandra	8	2	7, 1	BM
Myrsine	6	1	6	BI or BM
Peperomia	7	2	5, 2	BV
Melicope	4	1	4	BI
Cyperus (Mariscus)	6	5	1, 2	BM
Asplenium	12	11	1	A
Totals	63	24	2.6	

fern genera have large numbers of species resulting from multiple colonizations, not from speciation. In fact, there is a trend of increasing number of original colonists per genus from the top of the table to the bottom. Another common feature of the species-rich group is that virtually all of the presumed colonists arrived via bird-dispersal. Those arriving via air-dispersal are members of multiple introduction genera. The dicot genus *Cyrtandra* is at variance with the overall pattern with several hypothesized independent introductions. This anomaly is currently being investigated.

The Hawaiian Islands are spectacular for their radiations of species, presumably relating again to their greater age, diverse array of habitats, and considerable area allowing for the accumulation of unique taxa. In the Marquesas the radiations are less spectacular (Table 4), but virtually the same genera are species-rich. The 12 species of the fern genus *Asplenium* result from multiple introductions, just as they do in Hawai'i. Nearly all of the other genera on the Marquesan list are also on the Hawaiian list. Thus, there is clearly a taxonomic component to the groups that have speciated—specific groups of dicots. This pattern is generally repeated throughout the Pacific. Likewise, the same pattern of bird-dispersal of the colonists in this group is evident.

By contrast, the Hawaiian vascular genera that are not species-rich, which account for about 283 species (98 pteridophytes, 185 angiosperms) and for 71% to 73% of the presumed colonists, have existed in the Hawaiian Islands without speciation or with only a single speciation event, and account for a total of about 25% of the vascular flora. About one-third of the species-poor lineages are presently represented by a single endemic species. These may have arisen virtually concomitant with colonization. With so much emphasis on the examples of spectacular radiation in Hawaiian plants, examples of colonization with little or no subsequent speciation have largely been ignored.

Fully 85% of the pteridophyte lineages are in this group. This is presumably related to their ease of dispersal via air; colonization more closely balances speciation. Less than 67% of the non-differentiated group are angiosperms. These are represented by widely distributed indigenous species and colonization events that have resulted in a single speciation event, with subsequent or concomitant loss of the colonizing species. Many of these single-species lineages are lowland species, while virtually all species-rich lineages occur in upland sites, or at least the colonists is presumed to have colonized an upland site. A large number of these lineages arrived via oceanic drift. Many monocots also fall into this group, which is at least partly skewed by sedges (Cyperaceae). The association of many of the single-species lineages with wetlands suggests dispersal by migrating waterfowl, thus increasing the number of colonizations relative to speciation events. Also, evolutionary canalization may be a factor in specialized plants like monocots. Another factor may be the loss of species never recorded by European scientists, since much of the lowlands in these archipelagoes was severely altered before being studied. A discovery of great impact in the fauna, especially well documented for the birds (Olsen & James, 1982; Olson, this volume), emphasizes the importance of the latter factor.

A subset of the lineages which have not differentiated was considered by Carson (1987). He correctly points out that a large proportion of these lineages are coastal, aquatic or low-elevation species. His suggestion, following Baker (1965), is that these species are preadapted for wide dispersal and their genomes consist of complex, coadapted, heterotic systems that enable them to colonize widely. They are evolutionarily canalized for a specific, widely distributed, patchy environment. These adaptions inhibit speciation, especially that resulting from a founder effect. Some of the included low-diversity lineages may represent lineages that have experienced extinction from perturbations by man.

In summary, there are considerable differences among oceanic island chains, as shown here using the Hawaiian and Marquesan archipelagoes as examples. The Marquesas Islands, because they are relatively short-lived and small in size, are more representative of many of the high island chains of the Pacific. They are unusual in their degree of isolation, high proportions of air dispersed taxa such as ferns, and their impressive, yet relatively small, array of unique taxa: *Plakothira, Oparanthus, Pelagodoxa,* and *Lebronnecia.* Few of the colonists have diversified to any considerable extent. The most diverse lineage is the genus *Psychotria,* a genus that has speciated virtually everywhere it has colonized. The principal limits to diversification appear to be the relatively short existance of the Marquesas Archipelago, its small size, and its proximity to nearby source areas.

The Hawaiian Islands, in great contrast, are unique in their considerable size and great collective age as an archipelago. The chain comprises a great arc of islands and seamounts originating from a hot spot. This hot spot has produced a set of islands comparable to the current set of eight high islands for a period of at least 70 million years. These features have provided both the ecological conditions and area for considerable diversification to take place, and a progression of closely spaced islands that allow, through interisland colonization, the persistence and accumulation of diverse lineages. Relatively few of the original colonists have radiated (about 10%). Virtually all of the diverse lineages are dicots, and members of genera that have radiated elsewhere to a lesser extent. Diversity of certain fern genera as well as some other angiosperm genera is partly the result of autochthonous speciation and, more significantly, partly the result of high dispersal ability and multiple introductions.

ACKNOWLEDGMENTS

I am grateful to Scott E. Miller for inviting me to participate in this symposium, and for the idea to compare the Hawaiian and Marquesan floras. I thank Warren H. Wagner, Jr. for providing me with much information on Hawaiian pteridophytes; David H. Lorence and J. Florence for helping me compile the list of Marquesan pteridophytes; and Royce Oliver and Michael E. Sisson for considerable assistance with the Marquesan specimens at the United States National Museum of Natural History used in part to compile the appendix. I appreciate the assistance with, and comments on, the manuscript by M. Sisson, and comments by F. Raymond Fosberg on the Marquesan appendix.

LITERATURE CITED

Adamson, A. M. 1936. Marquesan insects: environment. *Bernice P. Bishop Mus. Bull.* 139:1–73.

Baker, H. G. 1965. Characteristics and modes of origin of weeds. Pp. 147–172. *In:* H. G. Baker & G. L. Stebbins (eds.), *The Genetics of Colonizing Species.* Academic Press: New York.

Balgooy, van, M. M. J. 1971. Plant-geography of the Pacific as based on a census of phanerogam genera. *Blumea Suppl.* 6:1–222.

Brown, F. B. H. 1931. Flora of southeastern Polynesia. I. Monocotyledons. *Bernice P. Bishop Mus. Bull.* 84:1–194.

Brown, F. B. H. 1935. Flora of southeastern Polynesia. III. Dicotyledons. *Bernice P. Bishop Mus. Bull.* 130:1–386.

Brown, E. D. & F. B. H. Brown. 1931. Flora of southeastern Polynesia. II. Pteridophytes. *Bernice P. Bishop Mus. Bull.* 89: 1–123.

Carlquist, S. 1974. *Island Biology.* Columbia University Press: New York. 660 pp.

Carlquist, S. 1980. *Hawaii, a Natural History. Geology, Climate, Native Flora and Fauna above the Shore Line.* 2nd ed. Pacific Tropical Botanical Garden, Lawai, Hawaii. 468 pp.

Carr, G. D. 1985. Monograph of the Hawaiian *Madiinae* (Asteraceae): *Argyroxiphium, Dubautia,* and *Wilkesia. Allertonia* 4:1–123.

Carr, G. D. 1987. Beggar's ticks and tarweeds: masters of adaptive radiation. *Trends Ecol. Evol.* 2:192–195.

Carson, H. L. 1987. Colonization and speciation. Pp. 187–206. *In:* A. J. Gray, M. J. Crawley & P. J. Edwards (eds.), *Colonization, Succession and Stability,* Blackwell Scientific Publications; Oxford.

Caum, E. L. 1933. Notes on *Pteralyxia. Occas. Pap. Bernice P. Bishop Mus.* 10(8):1–24.

Chubb, L. J. 1930. Geology of the Marquesas Islands. *Bernice P. Bishop Mus. Bull.* 68:1–71.

Clague, D. A. & R. D. Jarrard. 1973. Tertiary Pacific plate motion deduced from the Hawaiian-Emperor Chain. *Geol. Soc. Amer. Bull.* 84:1135–1154.

Crawford, D. J., Witkus, R. & T. F. Stuessy. 1987. Plant evolution and speciation on oceanic islands. Pp. 183–199. *In:* K. M. Urbanska, (ed.), *Differentiation Patterns in Higher Plants.* Academic Press, London.

Cuddihy, L. W. & C. P. Stone. 1990. *Alteration of Native Hawaiian Vegetation: Effects of Humans, Their Activities, and Introductions.* University of Hawaii Cooperative National Park Resources Studies Unit, distributed by University of Hawaii Press. 138 pp.

Darlington, P. J. 1957. *Zoogeography.* Wiley, New York.

Decker, B. G. 1970. *Plants, Man, and Landscape in Marquesan Valleys, French Polynesia.* Ph. D. dissertation, University of California, Berkeley. 324 pp.

Duncan, R. A. & I. McDougall. 1974. Migration of volcanism with time in the Marquesas Islands, French Polynesia. *Earth Plantary Sci. Letters* 21:414–420.

Dymond, J. 1975. K-Ar ages of Tahiti and Moorea, Society Islands, and implications for the hot-spot model. Geology 3:236–240.

Florence, J. 1985. Sertum polynesicum I. *Plakothira* Florence (Loasaceae), genre nouveau des îles Marquises. *Bull. Mus. natn. Hist. nat., Paris* 4ᵉ sér., 7, section B, Adansonia, n° 3:239–245.

Florence, J. 1986. Sertum polynesicum II. Rubiaceae nouvelles des îles Marquises (Polynésie Française). *Bull. Mus. natn. Hist. nat., Paris* 4ᵉ sér., 8, section B, Adansonia, n° 1:3–11.

Florence, J. 1987. Endémisme et évolution de la flora de la Polynésie Française. *Bull. Soc. Zool. France* 112:369–380.

Florence, J. 1990. Sertum polynesicum IV. Deux espèces nouvelles de *Melicope* J. R. & G. Forster (Rutaceae) de l'île de Nukuhiva (Marquises, Polynésie Française). *Bull. Mus. natn. Hist. nat., Paris* 4e sér., 12, section B, Adansonia, n° 1:29–35.

Fosberg, F. R. 1948. Derivation of the flora of the Hawaiian Islands. Pp. 107–119. *In:* E. C. Zimmerman, *Insects of Hawaii.* Vol. 1. University of Hawaii Press, Honolulu.

Fosberg, F. R. & M.-H. Sachet. 1966. Plants of southeastern Polynesia. *Micronesica* 2:157–158.

Fosberg, F. R. & M.-H. Sachet. 1975. Polynesian plants studies 1–5. *Smiths. Contr. Bot.* 21:1–25.

Fosberg, F. R. & M.-H. Sachet. 1981. Polynesian plants studies 6–18. *Smiths. Contr. Bot.* 47:1–38.

Gagné, W. C. & C. C. Christensen. 1985. Conservation status of native terrestrial invertebrates in Hawai'i. Pp. 105–126. *In:* C. P. Stone & J. M. Scott (eds.), *Hawai'i's Terrestrial Ecosystems: Preservation and Management.* Coop. Natl. Park Resources Stud. Unit, University of Hawaii: Honolulu.

Gagné, W. C. & L. W. Cuddihy. 1990. Vegetation. Pp. 45–114. *In:* W. L. Wagner, D. R. Herbst & S. H. Sohmer. *Manual of the Flowering Plants of Hawai'i.* University of Hawaii Press and Bishop Museum Press (Special Publication 83): Honolulu.

Helenurm, K. & F. R. Ganders. 1985. Adaptive radiation and genetic differentiation in Hawaiian *Bidens. Evolution* 39:753–765.

Kirch, P. V. 1983. Man's role in modifying tropical and subtropical Polynesian ecosystems. *Archaeol. Oceania* 18:26–31.

Kirch, P. V. 1985. *Feathered Gods And fishhooks. An Introduction to Hawaiian Archaeology and Prehistory.* University of Hawaii Press. 349 pp.

Lacroix, A. 1928. Nouvelles observations sur les laves des îles Marquises et de l'île Tubuai (Polynésie Australe). *Acad. Sci.* Paris 187:365–369.

Lammers, T. G. A revision of Clermontia (Campanulaceae: Lobelioideae). *Syst. Bot. Monogr.* In press.

Liotard, J. M., Barsczus, H. G., Dupuy, C. & J. Dostal. 1986. Geochemistry and origin of basaltic lavas from Marquesas Archipelago, French Polynesia. *Contr. Mineral Petrol.* 92:260–268.

Lockwood, J. P. & P. W. Lipman. 1987. Volcanism in Hawaii. Holocene eruptive history of Mauna Loa Volcano. *U.S. Geol. Sur. Prof. Pap.* 1350:509–535.

Lowrey, T. K. 1986. A biosystematic revision of Hawaiian *Tetramolopium* (Compositae: *Astereae*). *Allertonia* 4:203–265.

Mabberley, D. J. 1987. *The Plant Book: a Portable Dictionary of the Higher Plants.* Cambridge University Press: London. 706 pp.

MacDonald, G. A., Abbott, A. T. & F. L. Peterson. 1983. Volcanoes in the sea: the geology of Hawaii, 2nd ed. University Hawaii Press: Honolulu. 517 pp.

Morgan, W. J. 1972. Plate motion and deep mantle convection. Geol. Soc. Amer. Mem. 132.

Olson, S. L. & H. F. James. 1982. Prodromus of the fossil avifauna of the Hawaiian Islands. *Smiths. Contr. Zool.* 365:1–59.

Price, S. 1983. Climate. Pp. 59–66. *In:* R. W. Armstrong (ed.), Atlas of Hawaii. 2nd ed. University of Hawaii Press: Honolulu.

St. John, H. 1973. *List and Summary of the Flowering Plants in the Hawaiian Islands.* Pacific Trop. Bot. Gard. Mem. 1. 519 pp.

Scholl, D. W. & J. S. Creager. 1973. Geologic synthesis of Leg 19 (DSDP) results: far north Pacific, and Aleutian Ridge, and Bering Sea.

Selling, O. H. 1948. Studies in Hawaiian pollen statistics, Part III. On the late quaternary history of the Hawaiian vegetation. *Special Publ. Bernice P. Bishop Mus.* 39. 154 pp.

Simberloff, D. S. 1970. Taxonomic diversity of island biotas. *Evolution* 24:23–47.

Smith, A. C. 1979. *Flora Vitiensis Nova. A New Flora of Fiji (Spermatophytes only).* Vol. 1. Pacific Tropical Botanical Garden, Lawai, Hawaii. 495 pp.

Smith, A. C. 1981. *Flora Vitiensis Nova. A New Flora of Fiji (Spermatophytes only).* Vol. 2. Pacific Tropical Botanical Garden, Lawai, Hawaii. 810 pp.

Smith, A. C. 1985. *Flora Vitiensis Nova. A New Flora of Fiji (Spermatophytes only).* Vol. 3. Pacific Tropical Botanical Garden, Lawai, Hawaii. 758 pp.

Smith, A. C. 1988. *Flora Vitiensis Nova. A New Flora of Fiji (Spermatophytes only).* Vol. 4. Pacific Tropical Botanical Garden, Lawai, Hawaii, 377 pp.

Smith, A. R. 1972. Comparison of fern and flowering plant distributions with some evolutionary interpretations for ferns. *Biotropica* 4:4–9.

Stearns, H. T. 1985. *Geology of the State of Hawaii.* Pacific Books, Palo Alto, CA. 335 pp.

Tyron, R. 1970. Development and evolution of fern floras of oceanic islands. *Biotropica* 2:76–84.

Wagner, W. H., Jr. 1988. Status of the Hawaiian fern flora. *Fiddlehead Forum* 15:11–14.

Wagner, W. L., Herbst, D. R. & R. S. N. Yee. 1985. Status of the native flowering plants of the Hawaiian Islands. Pp. 23–74. *In:* C. P. Stone & J. M. Scott (eds.), *Hawai'i's Terrestrial Ecosystems: Preservation and Managemant.* Coop. Natl. Park Resources Stud. Unit, University of Hawaii: Honolulu.

Wagner, W. L., Herbst, D. R. & S. H. Sohmer. 1990. *Manual of the Flowering Plants of Hawai'i.* 2 vols. University of Hawaii Press and Bishop Museum Press (Special Publication 83), Honolulu. 1853 pp.

Walker, G. 1990a. Geology. Pp. 21–35 (vol.1). *In:* Wagner, W. L., D. R. Herbst & S. H. Sohmer, *Manual of the flowering plants of Hawai'i.* University of Hawaii Press and Bishop Museum Press (Special Publication 83), Honolulu.

Walker, G. 1990b. Review article: Geology and volcanology of the Hawaiian Islands. *Pacific Sci.* 44:315–347.

Weller, S. G., Sakai, A. K., Wagner, W. L. & D. R. Herbst. 1990. Evolution of dioecy in *Schiedea* (Caryophyllaceae: Alsinoideae) in the Hawaiian Islands: biogeographical and ecological factors. *Syst. Bot.* 15:266–276.

Wilson, J. T. 1963. A possible origin of the Hawaiian Islands. *Canad. J. Phys.* 41:863–870.

Williamson, M. 1981. *Island Populations.* Oxford University Press. Oxford. 286 pp.

Witter, M. S. & G. D. Carr. 1988. Adaptive radiation and genetic differentiation in the Hawaiian silversword alliance (Compositae: *Madiinae*). *Evolution* 42: 1278–1287.

Appendix. Vascular flora of the Marquesas Islands. (* = endemic genus)

Family	Genus	Total number of indigenous species	Number of endemic species	Number of colonizations
		PTERIDOPHYTES		
Adiantaceae	*Doryopteris*	1	0	1
	Pellaea	1	0	1
Aspleniaceae	*Asplenium*	12	1	11
	Loxoscaphe	1	0	1
Aspidiaceae	*Arachniodes*	1	1	1
	Dryopteris	2	1	1
	Polystichum	1	1	1
	Tectaria	3	2	3
Athyriaceae	*Deparia*	2	1	1
	Diplazium	1	0	1
Blechnaceae	*Blechnum*	5	1	4
	Doodia	1	1	1
Cyatheaceae	*Cyathea*	2	1	2
Dennstaedtiaceae	*Histopteris*	1	0	1
	Hypolepis	2	0	2
	Paesia	1	0	1
Gleicheniaceae	*Dicranopteris*	1	0	1
Grammitidaceae	*Calymmodon*	1	0	1
	Ctenopteris	2	0	2
	Grammitis	4	1	3
	Scleroglossum	1	0	1
	Xiphopteris	1	0	1
Hymenophyllaceae	*Hymenophyllum*	3	0	3
	Trichomanes	4	0	4
Lindsaeaceae	*Lindsaea*	4	0	4
	Sphenomeris	1	0	1
Lomariopsidaceae	*Elaphoglossum*	3	2	2
	Lomogramma	1	0	1
Lycopodiaceae	*Lycopodium*	3	1	3
Marattiaceae	*Angiopteris*	2	1	1
	Marattia	1	0	1
Oleandraceae	*Nephrolepis*	3	0	3
	Oleandra	1	0	1
Ophioglossaceae	*Ophioglossum*	2	0	2
Polypodiaceae	*Belvisia*	1	0	1
	Microsorium	2	0	2
	Phymatosorus	3	0	3
	Selliguea	1	0	1
Psilotaceae	*Psilotum*	1	0	1
Pteridaceae	*Pteris*	4	0	4
Schizaeaceae	*Schizaea*	2	0	2
Selaginellaceae	*Selaginella*	3	0	3
Thelypteridaceae	*Amauropelta*	1	0	1
	Amphineuron	1	0	1
	Chingia	1	0	1
	Christella	1	0	1
	Coryphopteris	1	1	1
	Pneumatopteris	1	0	1
	Thelypteris	2	0	2
	Sphaerostephanos	3	0	3
Tmesipteridaceae	*Tmesipteris*	1	1	1

Appendix. Vascular flora of the Marquesas Islands. (* = endemic genus)

Family	Genus	Total number of indigenous species	Number of endemic species	Number of colonizations
Vittariaceae	*Antrophyum*	2	0	2
	Vittaria	2	0	2
Totals Pteridophyte		108	17	101

DICOTS

Family	Genus	Total number of indigenous species	Number of endemic species	Number of colonizations
Aizoaceae	*Sesuvium*	1	0	1
Amaranthaceae	*Achyranthes*	2	1	2
Apiaceae	*Hydrocotyle*	1	1	1
Apocynaceae	*Alstonia*	2	2	1
	Alyxia	1	0	1
	Cerbera	1	0	1
	Lepinia	1	1	1
	Neisosperma	1	1	1
	Ochrosia	2	2	2
	Rauvolfia	1	1	1
Aquifoliaceae	*Ilex*	1	0	1
Araliaceae	*Cheirodendron*	1	1	1
	Meryta	1	1	1
	Reynoldsia	2	2	1
Asteraceae	*Adenostemma*	1	0	1
	Bidens	9	9	1
	Oparanthus	2	2	1
Boraginaceae	*Cordia*	1	0	1
	Heliotropium	1	1	1
Brassicaceae	*Lepidium*	1	0	1
Campanulaceae	*Apetahia*	2	2	1
Celastraceae	*Maytenus*	1	0	1
Chloranthaceae	*Ascarina*	1	1	1
Convolvulaceae	*Ipomoea*	4	0	4
Cunoniaceae	*Weinmannia*	1	0	1
Epacridaceae	*Styphelia*	1	0	1
Ericaceae	*Vaccinium*	1	0	1
Euphorbiaceae	*Claoxylon*	1	1	1
	Chamaesyce	1	1	1
	Glochidion	2	2	1
	Phyllanthus	1	0	1
Fabaceae	*Caesalpinia*	2	0	2
	Canavalia	1	0	1
	Entada	1	0	1
	Erythrina	1	0	1
	Mucuna	1	0	1
	Serianthes	1	0	1
	Sesbania	1	0	1
	Vigna	2	0	2
Flacourtiaceae	*Homalium*	1	1	1
	Xylosma	1	0	1
Gesneriaceae	*Cyrtandra*	8	8	2
Goodeniaceae	*Scaevola*	4	3	2
Hernandiaceae	*Hernandia*	1	1	1
Loasceae	*Plakothira**	3	3	1
Loganiaceae	*Geniostoma*	3	3	1
	Fagraea	1	0	1

Appendix. Vascular flora of the Marquesas Islands. (* = endemic genus)

Family	Genus	Total number of indigenous species	Number of endemic species	Number of colonizations
Loranthaceae	Decaisnina	1	0	1
Malvaceae	Abutilon	1	1	1
	Gossypium	1	0	1
	Hibiscus	1	0	1
	Lebronnecia*	1	1	1
	Sida	1	0	1
	Thespesia	1	0	1
Menispermceae	Stephania	1	0	1
Moraceae	Ficus	1	0	1
	Streblus	1	0	1
Myrsinaceae	Myrsine	6	6	1
Myrtaceae	Eugenia	1	0	1
	Metrosideros	1	0	1
Nyctaginaceae	Boerhavia	2	0	2
	Pisonia	2	0	2
Onagraceae	Ludwigia	1	0	1
Oxalidaceae	Oxalis	1	1	1
Piperaceae	Macropiper	1	0	1
	Peperomia	7	5	2
Plumbaginaceae	Plumbago	1	0	1
Portulacaceae	Portulaca	1	0	1
Primulaceae	Samolus	1	0	1
Rhamnaceae	Alphitonia	1	1	1
	Colubrina	1	0	1
Rhizophoraceae	Crossostylis	1	0	1
Rubiaceae	Coprosma	3	3	1
	Cyclophyllum	1	0	1
	Guettarda	1	0	1
	Hedyotis	1	1	1
	Ixora	4	4	1
	Morinda	1	0	1
	Psychotria	11	11	1
	Psydrax	1	0	1
Rutaceae	Melicope	4	4	1
Santalaceae	Santalum	1	0	1
Sapindaceae	Allophyllus	1	0	1
	Cardiospermun	1	0	1
	Dodonaea	1	0	1
	Sapindus	1	0	1
Scrophulariaceae	Bacopa	1	0	1
Solanaceae	Nicotiana	1	1	1
	Solanum	4	0	4
Sterculiaceae	Commersonia	1	0	1
	Waltheria	2	1	2
Thymelaeaceae	Wikstroemia	2	1	1
Trimenaceae	Trimenia	2	2	1
Ulmaceae	Celtis	1	0	1
Urticaceae	Boehmeria	1	0	1
	Pipturus	3	2	2
	Procris	1	0	1
Verbenaceae	Premna	1	0	1
	Vitex	1	0	1

Appendix. Vascular flora of the Marquesas Islands. (* = endemic genus)

Family	Genus	Total number of indigenous species	Number of endemic species	Number of colonizations
Viscaceae	*Korthalsella*	1	0	1
Zygophyllaceae	*Tribulus*	1	0	1
Dicot Totals		174	96	118
		MONOCOTS		
Arecaceae	*Pelagodoxa**	1	1	1
Cyperaceae	*Carex*	1	1	1
	Cyperus (Mariscus)	6	4	5
	Fimbristylis	2	1	1
	Gahnia	2	2	2
	Machaerina	2	1	2
	Pycreus	1	0	1
	Rhynchospora	1	1	1
	Torulinium	1	0	1
Juncaceae	*Luzula*	1	1	1
Liliaceae, s. l.	*Astelia*	1	1	1
	Dianella	1	0	1
Orchidaceae	*Calanthe*	1	1	1
	Habenaria	2	1	1
	Liparis	1	0	1
Pandancaeae	*Freycenetia*	2	0	2
	Pandanus	1	0	1
Poaceae	*Centosteca*	1	0	1
	Digitaria	2	0	2
	Leptochloa	2	2	1
	Panicum	1	0	1
	Pennisetum	2	2	2
	Schizostachyum ?	1	0	1
Monocot Totals		36	19	32
Vascular Flora Totals		318	132	251

Hawaiian Cave Faunas:
Macroevolution on Young Islands

Francis G. Howarth

Abstract. The Hawaiian islands are the summits of giant submarine volcanoes that are isolated from the nearest comparable land mass by almost 4000 km of open water. Relatively few terrestrial plants and animals dispersed to and colonized the island chain. The voluminous basaltic lava flows that built the islands contain numerous caves, gas bubbles, cracks, and other voids, which anastomose into interconnected subterranean habitats. These subterranean systems are colonized by representatives of the adaptively radiating fauna seeking to exploit abundant food resources sinking out of reach of surface species. Erosion fills these voids, so that the habitat and fauna are most diverse on the younger islands. Terrestrial cave-adapted species have been found on Kauai (<5.5 Ma), Oahu (<3.7 Ma), Molokai (<1.9 Ma), East Maui (<1.0 Ma), and Hawaii (<0.7 Ma). Nearly 50 cave species are known, and at least 12 native arthropod groups have independently invaded the subterranean biome on two or more islands. Close relatives are often still extant and live in rain forests, lava flows, or seacoasts. Caves are strongly zonal habitats, with five zones (entrance, twilight, transition, deep cave, and stagnant air). Cave-adapted species are largely restricted to the two inner zones and to the medium-sized subterranean voids. The latter may be the primary habitat for most cave animals, a fact with important implications for studies in evolutionary ecology and conservation biology. The macroevolutionary adaptations displayed by cave species include reduction of eyes, wings, pigments, cuticle thickness, spination, fecundity, and attenuation of appendages. Positive morphological, physiological, and behavioral adaptations enable the animals to maintain water balance, breathe stressful gas mixtures, locate food, and reproduce within the unusual environment. Cave adaptation appears to be a general and predictable process, occurring in a relatively brief time (much less than 0.5 Ma), wherever a suitable system of underground voids exists for evolutionary time in both temperate and tropical areas, in both lava and limestone caves, and on both islands and continents.

INTRODUCTION

The Hawaiian islands are the summits of giant submarine volcanoes, which formed sequentially as the Pacific Plate moved northwestward across the Hawaiian Hot Spot (Clague & Dalrymple, 1987). The oldest island, Kure Atoll, is about 35 million years (Ma) old, but the chain is at least 70 Ma in age where it is being subducted into the Aleutian Trench. The eight main southeastern high islands range in age from less than 5.5 Ma for Kauai to less than 0.7 Ma for the still volcanically active Hawaii Island. Each island or island group has always been isolated from the others by deep straits 40 km or more wide. Frequent voluminous basaltic lava flows built each island in turn, in a process which continues today (Clague & Dalrymple, 1987).

The Hawaiian Islands are the most isolated high islands in the world, being nearly 4,000 km from the nearest comparable land: the Marquesas Islands to the south and California to the northeast. Relatively few terrestrial animals and plants managed to disperse

Dr. Howarth is with the J. Linsley Gressitt Center for Research in Entomology, B. P. Bishop Museum, P. O. Box 19000-A, Honolulu, HI 96817 USA.

to and colonize this island chain. An estimated 400 colonizers gave rise to the known arthropod fauna, which currently comprises over 5,000 described species (Howarth, 1990). Most colonists belong to typically vagile groups, while many typical moisture- and darkness-loving groups of the continents are conspicuously poorly represented.

About 180 distinct ecosystems are recognized in Hawaii (Gagné & Cuddihy, 1990; Howarth, 1990). This surprisingly large habitat diversity results from the physiographic development of the high islands and from the parallel but separate development of communities on each island. The higher mountains deflect the trade winds as well as generate surface winds creating steep environmental gradients between windward rain forests and leeward deserts. For example, Mauna Kea and Mauna Loa on Hawaii Island extend from tropical lowlands to cold stony deserts, with frequent snows at their summits over 4000 m above sea level. The gentle slopes with few surface streams seen on the younger volcanoes, especially Kilauea and Mauna Loa, contrast sharply with the deeply eroded river valleys on the older volcanoes.

The sequential ages, relatively simple geology, great physiographic and climatic variation, repetition of environmental regimes and ecosystems on neighboring islands, few successful colonists, and many examples of adaptive radiation make the islands ideal natural laboratories or Rosetta Stones for deciphering evolutionary processes (Simon, 1987; Howarth, 1990). Each island acquired the ancestors of its terrestrial biota from transoceanic dispersal, and each can be viewed as a minicontinent, being a microcosm of the evolutionary and ecological processes occurring on the continents. Native animals and plants have successfully filled most niches through the process of adaptive radiation. Swarms of closely related species exploit different habitats, and some of the best examples of adaptive radiation found anywhere occur in the Hawaiian Islands.

HAWAIIAN CAVE FAUNA

Following the important pioneering studies in caves in Europe and North America, obligate cave animals were thought to occur almost exclusively in limestone caves in temperate regions beyond the limits of maximum Pleistocene glaciation (Barr, 1968; Culver, 1982; Barr & Holsinger, 1985). They were considered relicts that evolved only after their surface populations became extinct due to changing climates. Furthermore, the macroevolutionary processes of cave adaptation appeared to take a long time.

Given the isolation and relative youth of the high Hawaiian islands in general and the extreme youth of Hawaiian lava tubes in particular, it had long been assumed that specialized cave animals did not exist in Hawaii. Lava tubes were thought to be too ephemeral and too often polluted with volcanic gasses to harbor a specialized fauna. Food resources found in continental caves, such as the organic matter brought in by animals that roost in caves and sinking rivers, are absent from Hawaiian caves. Furthermore, the continental animals preadapted for cave life are not likely to cross thousands of kilometers of open ocean to colonize Hawaii. However, one should assume nothing in evolutionary biology. Studies begun in 1971 have revealed a truly remarkable assemblage of cave-adapted animals inhabiting young Hawaiian lava tubes (Howarth, 1987).

The Hawaiian cave animals evolved from native surface species by adaptive shifts. Thus they fit the pattern of adaptive radiation displayed by many other groups in the native biota. Nearly 50 species of terrestrial cave arthropods have been discovered to date (Howarth, 1987; 1988). They are not evenly distributed but are most diverse on the younger islands, with 26 species known from Hawaii, 16 from Maui, two each from Molokai and Kauai, and one from Oahu. They evolved independently from different ancestors on each island, and at least 12 native groups have produced cave species through parallel evolution on separate islands. Close relatives are often still extant and live

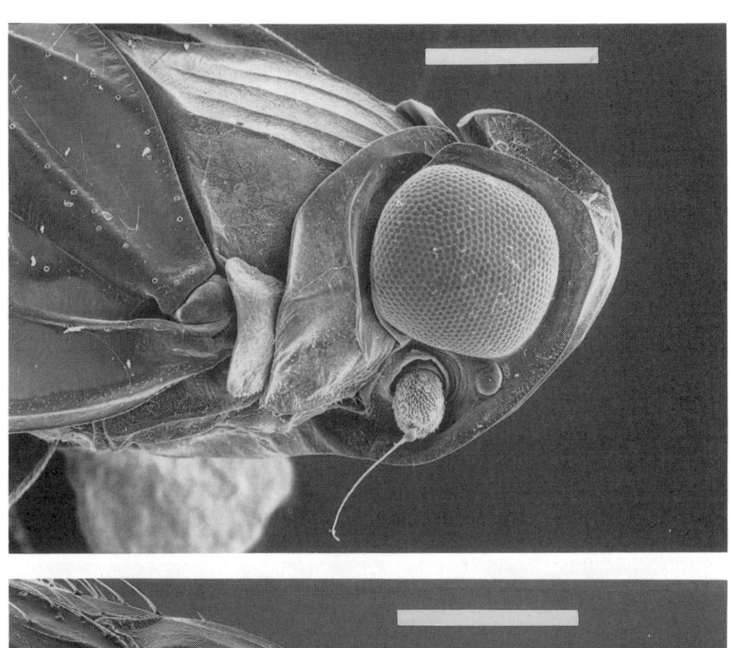

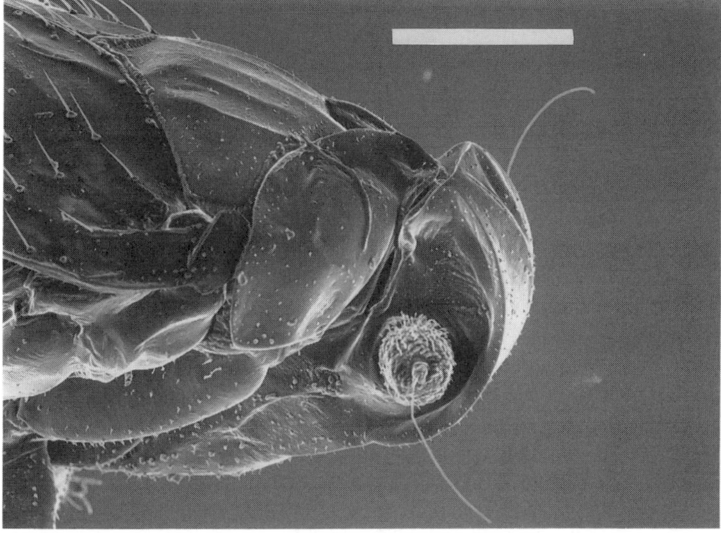

Figure 1. Scanning electron micrographs of the heads of two closely related species of cixiid plant-hoppers endemic to Hawaii Island. Top: *Oliarus filicicola*, a big-eyed rain forest species with dark, cryptic color pattern and functional wings. Bottom: *O. polyphemus*, an eyeless white subterranean species. Note the drastic changes in head and prothorax shape to compensate for the loss of eyes, the enlarged antennal base, and the vestigial tegula, which is fused to both the thorax and fore wing in the cave species. Scale bar equals 0.5 mm. SEM by E. Seling, Museum of Comparative Zoology, SEM Laboratory, Harvard University.

in rain forests, lava flows, or seacoasts (Howarth, 1987; 1988). Often the only morphlogical differences separating these cave-adapted species from their closest presumed surface relatives are those associated with cave adaptation; e.g., the planthoppers, *Oliarus poly-phemus* and *O. inaequalis* (Fennah, 1973) (Fig. 1); the thread-legged bugs, *Nesidiolestes ana* and *N. selium* (Gagné & Howarth, 1975) (Fig. 2); the earwigs, *Anisolabis howarthi* and *A. hawaiiensis* (Brindle, 1980); and the rock crickets, *Caconemobius varius* and *C. fori* (Gurney & Rentz, 1978).

The most surprising adaptive shifts among Hawaiian cave animals have occurred in the family Lycosidae, the big-eyed wolf spiders (Gertsch, 1973). Among the better sighted spiders, the native big-eyed surface species live on young lava flows, on mountain sum-

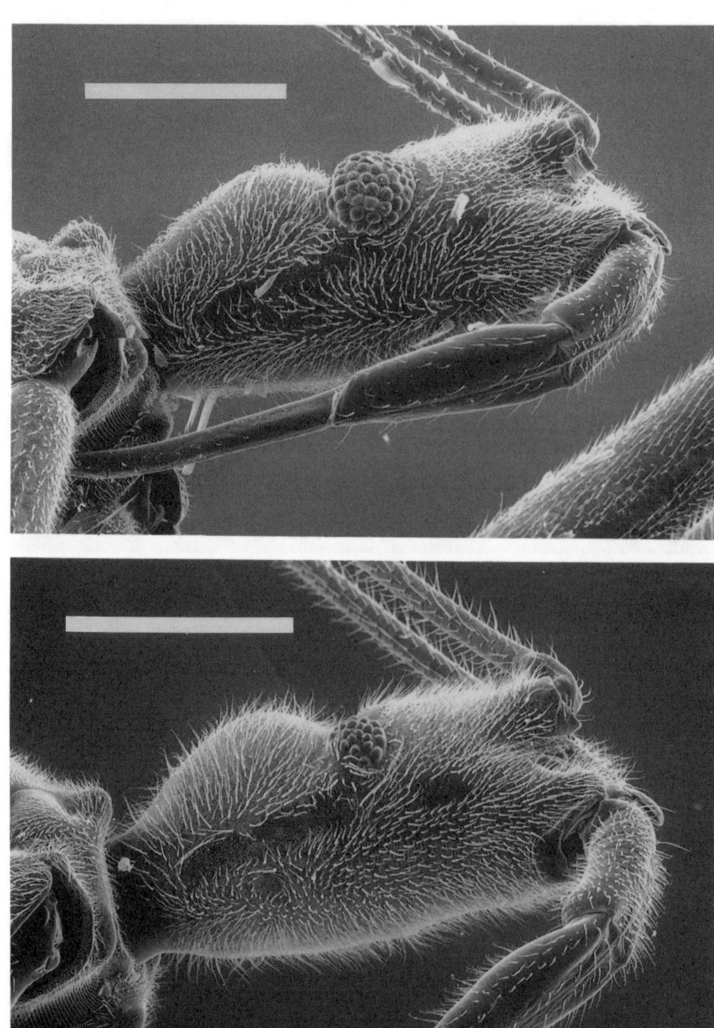

Figure 2. Scanning electron micrographs of the heads of two closely related species of thread-legged bugs endemic to Hawaii Island. Top: *Nesidiolestes selium,* an epigean rain forest species with dark, cryptic color pattern and large, pigmented, functional eyes. Bottom: *N. ana,* a blind subterranean species with pale straw body color and small, weakly pigmented to unpigmented, vestigial eyes. Scale bar equals 0.5 mm. SEM by E. Seling, Museum of Comparative Zoology, SEM Laboratory, Harvard University.

mits, and in forests. The remarkable small-eyed, big-eyed hunting spider, *Lycosa howarthi,* lives only underground within young lava flows on Hawaii Island. Even more remarkable, the no-eyed, big-eyed hunting spider, *Adelocosa anops,* evolved independently and is found only in the youngest lava tubes on Kauai Island, which have been dated at about 600,000 years old.

Cave animals in general are K-strategists compared to their surface relatives (Howarth, 1983), and the Hawaiian species show similar tendencies. For example, lycosid spiders have a complex behavioral repetoire, including maternal care. Epigean lycosid females in Hawaii usually care for over 300 spiderlings per clutch. These remain with the mother for over a month. The small-eyed cave lycosid, however, has about 40 spiderlings per clutch, which stay with her for about two weeks. The no-eyed, big-eyed wolf spider has

Figure 3. Adult female of *Oliarus polyphemus*, a blind, flightless, subterranean cixiid planthopper. Photo by W. P. Mull.

only 25 large, precocious spiderlings per clutch, which leave their mother only a few days after emerging from the egg sac.

Within the cixiid planthopper genus *Oliarus*, there are about 80 described endemic Hawaiian species. Most are cryptically colored, big-eyed, flighted species inhabiting Hawaiian forests. At least five separate lines have invaded caves on three islands, and show varying degrees of cave adaptation (Howarth et al., 1990). *Oliarus polyphemus* (Fig. 3) is eyeless, colorless, micropterous, and flightless, and inhabits young lava tubes on eastern Mauna Loa on Hawaii Island. Its claws are enlarged presumably to walk on wet rocky substrates. It belongs to the *inaequalis* group of species, and its closest relative may be the Hawaii Island rain forest species *O. inaequalis*. *O. priola* has independently evolved on Maui Island from a different surface ancestor, yet it is separable from *O. polyphemus* primarily by genitalic characters (Fennah, 1973). Such a degree of macroevolutionary convergence is surprising. Even though these species share morphological characters far outside the norm of the genus *Oliarus* (Fig. 1) and fit the morphological definition of sibling species (Mayr, 1976), they are independently derived from separate ancestral species within the genus *Oliarus*.

EVOLUTION

The macroevolutionary adaptations displayed by cave species include reduction of eyes, wings, pigmentation, cuticle thickness, spination, fecundity, and attenuation of appendages. Specializations involve modified pilosity, sensilla, and tarsal structure. Positive morphological, physiological, and behavioral adaptations enable the animals to maintain water balance, breathe unusual gas mixtures, reproduce, and locate food within their environment (Ahearn & Howarth, 1982; Howarth, 1983; Howarth & Stone, 1990). The adaptations found in the Hawaiian cave species are remarkably similar to those of cave animals in other regions, including temperate continental caves, indicating that selection pressures must be similar in all such environments, and thus cave adaptation must be a general and predictable process among animals adapting to exploit underground resources. However, the cave environment is so foreign to human experience that it is difficult to conceptualize the system.

Behavior must play an important role in the adaptation of cave species by allowing an animal to cope with the three-dimensional maze in total darkness. Obviously the normal cues, used by surface species to live and find mates and food, are useless underground. Most cave animals move slowly in a random walk, then remain in the vicinity of a suitable food resource (Juberthie-Jupeau, 1983). Many species are sit-and-wait predators.

Geology

The subterranean habitat is rigidly defined by the geological setting; thus, by studying cave geology, one can better understand the ecology and evolution of specialized cave animals. The voluminous basaltic lava flows that built the islands occur in two main forms, a'a and pahoehoe. A'a is cooler with less gas and less kinetic energy than pahoehoe and is therefore more viscous. Molten a'a has a thick fluid interior and flows like a caterpillar tractor tread, with the broken clinkery surface crust tumbling in front of, and buried by, the flow. Cooled a'a thus has a solid core and a layer of porous clinker on both its top and bottom. Pahoehoe initially flows in an amoeba-like manner with many toes covering previous flows layer upon layer. Each new layer welds poorly with the previous one, creating numerous interconnected voids between each layer. As the edges of the flow cool, pahoehoe flows become restricted within channels. Overflows from the channels make levees, which deepen and furthur channel the lava stream. These streams roof over to become lava tubes as the surface freezes. The roof insulates the chamber, allowing the tube to transport the molten lava many kilometers from the vent. Segments of these lava tubes may be preserved as caves after the flow ceases. Thus basaltic lavas contain numerous caves, trapped gas bubbles, cracks, and other voids, which anastomose into interconnected subterranean habitats, and which may be colonized by subterranean animals. The medium-sized voids in these systems are called the mesocaverns (Howarth, 1983).

Zonation of the Cave Environment

A clue to the evolution of the Hawaiian cave animals comes from the observed distribution of specialized animals within caves. Obligate cave species are almost universally restricted to deeper passages where the air remains saturated with water vapor, and many are found only in dead-end passages beyond tortuous crawlways. To test the generality of this observation I compared the cave environment with the distribution of cave animals in Charcoal Cave, on Hawaii Island at 800 m. I established three study sites at 45 m, 110 m, and 185 m respectively from the cave entrance—that is, in the twilight, transition, and deep cave zones—and recorded the temperature, relative humidity, and potential evaporation rate at each site (Howarth, 1982). Even though the temperature remained relatively constant, the relative humidity plummeted whenever the absolute humidity on the surface dropped below that within the cave and meteorological events on the surface pushed surface air into the cave. The drop in humidity occurred most frequently at night when the outside temperature fell below the cave temperature (Howarth, 1982).

The potential evaporation rate correlated well with the hygrothermograph data. The decrease in relative humidity was attenuated deeper into the cave. The potential evaporation rate at the deep zone station was negligible and only 0.15 cm^3 per day, which was 36% of the rate at the transition zone station and 8% of the rate at the twilight zone station. During this period cave adapted species were only found beyond 160 m from the entrance where the potential evaporation rate was less than 0.16 cm^3 per day (Howarth, 1982). These results corroborated the hypothesis that the distribution of cave-adapted species is defined by the water saturated atmosphere. Comparative physiological studies between cave and related surface species of lycosids and crickets demonstrated that their rate of water loss is strongly correlated with the saturation deficit of their natural habitat and with their degree of cave adaptation (Hadley et al., 1981; Ahearn & Howarth, 1982).

These results confirm that caves are strongly zonal habitats. Five zones (Fig. 4) are characterized by the abiotic and biotic environment (Howarth, 1983; Howarth & Stone, 1990). The entrance zone, where the surface and underground meet, is often richer than either habitat because of the better soil and moisture availability. The twilight zone extends from the limit of vascular plants to total darkness. The area of total darkness can be divided into three distinct zones. The transition zone is a broad boundary zone where meteorologic events on the surface are still important in characterizing the microclimate. The deep cave zone maintains a relatively stable air mass that remains nearly constantly saturated with water vapor. The stagnant air zone exchanges air with the surface only slowly. Within the stagnant air zone, the atmosphere also remains constantly saturated, but the concentration of oxygen and carbon dioxide may fluctuate dramatically, usually by the *in situ* decomposition of organic material (Howarth & Stone, 1990). Gas concentrations within this zone approach those of the surrounding voids in the rock.

The extent of each zone is governed by the location, size, and shape of the entrances and passages; however, the boundaries between the zones are dynamic. The deep cave and stagnant air zones usually occur only deep within caves or in dead-end passages beyond a constriction such as an n- or u-shaped passage (Fig. 4). Such n- and u-shaped passages act as traps for water vapor and carbon dioxide. Cave-adapted species are largely restricted to the two innermost zones. Only a few caves extend into the stagnant air zone, although unenterable passages may be nearby, for example, the passage depicted beneath the entrance in Figure 4.

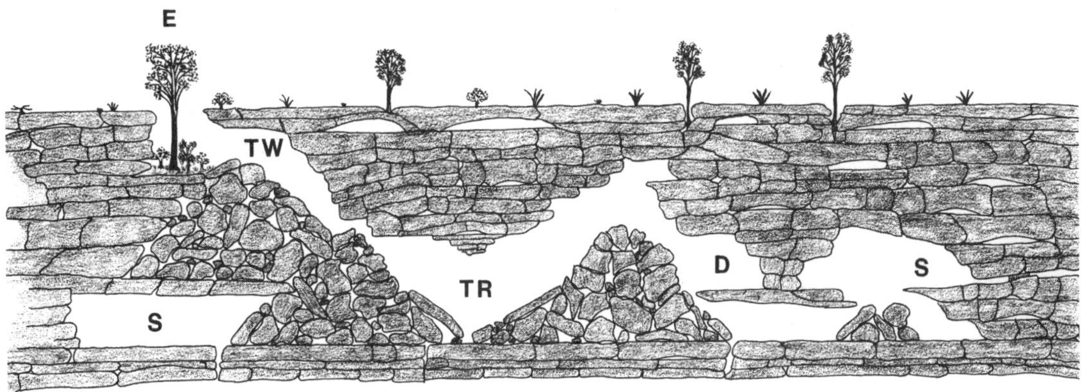

Figure 4. Profile view of a representative lava tube showing the five habitat zones. Length greatly condensed. Key: D = dark zone; E = entrance zone; S = stagnant air zone; TR = transition zone; TW = twilight zone. Drawn by N. C. Howarth.

Colonization of Young Lava Tubes

Another clue to the evolution of cave animals can be found on recent lava flows. For example, the nearly barren rocky surface of the 1881 lava flow at 1800 m elevation on Mauna Loa supports scattered shrubs only 1–2 m tall and is a desert for rain forest species. However, going underground into Emesine Lava Tube in the 1881 lava flow at the same location reveals an abundance of tree roots. The small shrubs growing on the young lava surface require huge root systems to maintain themselves in the xeric environment. Here is the rain forest. Rain forest animals, such as the cixiid planthoppers, whose nymphs feed on tree roots, are preadapted for underground life, but the adults are epigean and restricted to rain forests. The surface of young lava flows are uninhabitable for them, and the underground resources are out of reach for surface species. However, once an adaptive shift occurs and a reproducing and dispersing population becomes established underground, a whole new resource becomes exploitable.

This scenario is corroborated both by the geological evidence that basalt has numerous voids and by the discovery that cave animals colonize young lava tubes surprisingly fast. The most diverse fauna, with 11 obligate cave species, occurs within Kazumura Cave, which is in a 500-year-old flow on Kilauea. The 100-year-old Emesine Lava Tube already supports seven obligate cave species (Gagné & Howarth, 1975). Cave-adapted *Caconemobius* crickets have been collected from six-year-old lava tubes in the Mauna Ulu eruption on Kilauea. These observations support the hypothesis that cave animals live in and disperse through the medium-sized voids in young lava. Also, our physiological studies indicated that the animals were adapted to the stagnant air environment expected to occur within the medium-sized voids or mesocaverns in the lava (Howarth, 1983).

Colonization of Mesocaverns

To confirm that cave species could use these mesocavernous voids, we set pitfall traps in artificial caves dug in road cuts, following the pioneering methods of Juberthie and colleagues at the Laboratoire souterrain du CNRS, Moulis, France (Juberthie et al., 1980; Juberthie, 1983). A number of Hawaiian cave animals have now been collected from roadcuts on both Maui and Hawaii Islands, including: tree crickets, *Thaumatogryllus* spp., which are closely related to native surface tree crickets and not to the continental cave crickets (Gurney & Rentz, 1978; Otte, 1990); several species of rock crickets of the genus *Caconemobius*; a new species of phorid fly in the genus *Megaselia*; and *Oliarus* planthoppers.

No longer can caves be considered a separate biotope from the voids within the surrounding rock. From a biological perspective the subterranean biome can be divided into three habitats based on pore size (Howarth, 1983). The microcaverns include soil pore spaces and generally range less than 1 mm in width. The macrocaverns are cave passages of traditional biospeleological research and range in size from about 20 cm to over 100 m in width. In between these two classes are the mesocaverns. In suitable rock strata these spaces are numerous enough to capture and transport large amounts of food energy but small enough to limit air exchange with the surface. This food resource in an unusual habitat fuels the evolution of cave animals.

Divergence of Hawaiian Cave Species

Initially the biological and geological evidence indicated that no barriers to the dispersal of cave species existed on the younger volcanoes on Hawaii Island, since most cave species were widespread. However, the emerging picture underscores the critical importance of accurate systematic studies in evolutionary biology research. Recent critical re-analysis of the behavior and morphology of two cave groups revealed that what we believed to be a single widespread species may actually be subdivided into separate distinct populations.

Speculating that behaviorial strategies adopted by organisms for locating and recognizing mates might provide clues to the evolution of cave species, H. Hoch, M. Asche, and I have been recording the substrate-borne sounds produced by cave and surface cixiid planthoppers. We discovered that the mating songs of *Oliarus polyphemus* from Kaumana Cave at 290 m elevation on eastern Mauna Loa are distinct from the songs of *O. polyphemus* from Pahoa Cave only 20 km away on eastern Kilauea at 50 m elevation. Since these songs are mate recognition signals, our findings indicate that these two populations are separate cryptic acoustical species (Howarth et al., 1990). These studies are continuing and show that even neighboring cave populations often have different song repertoires. Whether these diverged after a single colonization event or represent multiple invasions of caves by one or more surface ancestors has not yet been determined.

A similar story is emerging with the rock crickets of the endemic Hawaiian genus *Caconemobius*. All known species exploit wet, bare rocky habitats (lava flows, caves, seacoasts, and cliffs). The lava cricket, *Caconemobius fori*, known only from young, unvegetated pahoehoe surfaces at Kilauea, has three outer and three dorsal subapical spines on the hind tibial margin. It moves onto new flows less than a month old and disappears when the first vascular plants appear (Howarth, 1979). *Caconemobius fori* is closely related to an obligate cave species, *C. varius*, which usually also has three outer and three dorsal spines on the hind tibia and which lives in nearby underground habitats, and to the undescribed *Caconemobius* sp. A, with three outer and usually only one dorsal subapical spines on the hind tibia. These two cave species are sympatric and occur over a wide area of Kilauea and Mauna Loa. A third cave *Caconemobius*, with two outer and no dorsal subapical tibial spines, overlaps in range on Kilauea with both *C. varius* and *C.* sp. A, but it is rarely seen and probably prefers to remain in the mesocaverns.

The distribution and morphometrics of these cave cricket populations on Hawaii Island reveal some interesting patterns (F. D. Stone, pers. comm.). The more cave-adapted *C.* species A appears to be monomorphic over its whole range, suggesting that it is able to disperse throughout the young lava. *Caconemobius varius* on the other hand displays significant morphological differences among the cave populations studied. There are no obvious geological barriers separating these populations. The caves are all are less than 600 years old, and the intervening lavas are mostly less than 1000 years old, and all are less than 20,000 years old.

DISCUSSION

These results show a) that cave adaptation is a general process and can be expected to occur wherever a suitable system of underground voids exists for evolutionary time in both temperate and tropical areas, in both lava and limestone caves, and on both islands and continents; b) that cave adaptation can occur in a relatively brief time (much less than 0.5 Ma); c) that significant exploitable resources are often lost to neighboring ecosystems; d) that even apparently barren and inhospitable environments on young oceanic islands may be exploited by native species; and e) that medium-sized voids may be the primary habitat for most cave animals, a fact with important implications for studies in evolutionary ecology and conservation biology.

These conclusions have been corroborated by studies in the Canary Islands (Oromi et al., 1986; this volume; Martin et al., 1989), where about 60 obligate subterranean species have been found in terrestrial habitats to date (Oromi et al., this volume). These show varying degrees of cave adaptation, and, as is true in Hawaii, some groups have independently invaded caves on more than one island. Also, as in Hawaii, the hypogean species inhabit both caves and mesocaverns, with some species preferring either the shallower or deeper voids (Oromi et al., 1986; this volume). In contrast to the situation in Hawaii, the shallow mesocavernous habitat (the *milieu souterrain superficiel* [Juberthie, 1983]) is apparently better represented in older substrates on the Canary Islands (Oromi et al., this volume), perhaps due to the cooler, more temperate and periodically drier climate, which would slow the weathering process.

The Canary Islands, which are up to 22 million years old, are much older than the high Hawaiian Islands; however, sporadic eruptions have periodically rejuvenated the subterranean habitats on most of the islands right up to recent times. The greater age, more complex geological and climatic history, and less isolated position in relation to source areas have imparted to the Canary Islands a greater diversity of taxa colonizing its caves, a greater number of taxon cycles, and a greater proportion of apparent relict species than in Hawaii. At least seven hypogean species have their closest relatives still extant in neighboring surface habitats and appear to have evolved by parapatric speciation, as is envi-

294

sioned for most of the Hawaiian fauna. However, 25 species in 14 genera have no living surface relatives in the Canary Islands and appear to be relicts (Oromi et al., this volume). The relationships of the remaining species remain obscure.

In the Canary Islands, as in Hawaii, the most diverse cave fauna occurs on the island with the largest area of young basaltic lava flows, i.e., Tenerife. Also as in Hawaii, some closely related cave species appear to be allopatric within the young volcanic areas of Tenerife in spite of the lack of any known barriers to dispersal between them. Whether these represent multiple invasions or divergence after cave colonization remains to be determined (Oromi et al., this volume).

Basaltic lava has numerous voids ranging in size from caves to cooling cracks, flow unit layers, buried clinker, and gas vesicles, which create extensive anastomosing systems of voids. Large amounts of food energy are continually being brought into these voids by percolating ground water, sinking streams, penetrating roots, dispersing animals, and gravity. These subterranean systems are rapidly colonized by representatives of the adaptively radiating fauna seeking to exploit abundant food resources sinking out of reach of surface species. Erosion fills these voids, so that the habitat and fauna are most diverse on the younger islands.

Until recently, the few cave animals known from lava tubes were considered to be exceptional (Barr, 1968). It now seems ironic, both on theoretical grounds and from the accumulating empirical evidence, that basaltic volcanic areas are more likely to harbor cave-adapted species than do limestone areas in most climatic regions (Howarth & Stone, 1990). Young basalt is formed with such a well-developed system of interconnected voids that, if volcanism is at least intermittantly continuous enough to create new habitat before erosion destroys these voids, then the vast mesocavernous habitat can support much larger populations of obligate cave species than can comparable sized limestone areas.

The assumption that shrinking resources "force" a population to adapt to a new niche is on weak theoretical ground. New species evolve from a position of strength, since natural selection predicts that organisms are becoming increasingly better adapted to their environment. Adaptive shifts and exploitation of new resources are characteristics of successful species and occur where the resources are abundant (Howarth, 1988).

Rather than being relicts of climatic changes, most obligate cave species appear to have become highly specialized in order to exploit resources within the system of interconnected medium-sized voids and to colonize adjacent caves only where their unique environment is found. Their environment is too inhospitable for most organisms, being perpetually dark and humid, lacking common environmental cues, and with a complex maze-like form, lethal or sublethal gas mixtures, and scattered food resources. Cave-adapted species appear to evolve through adaptive shifts from species that are frequent accidentals in the underground. If significant exploitable resources are available, eventually an accidental may establish a reproducing population, and if successful, may become cave-adapted by parapatric speciation. In conclusion, the emerging picture from our Hawaiian lava tube research reveals that the light at the end of the tunnel is a kaleidoscope.

ACKNOWLEDGMENTS

I thank N. C. Howarth, F. D. Stone, H. Hoch, and M. Asche for helpful discussions; S. E. Miller for the invitation to participate in the symposium; W. P. Mull for the use of his photographic slides of animals; E. Seling, Museum Comparative Zoology, SEM Laboratory, Harvard University, for the scanning electron micrographs; and R. Gillespie, S. E. Miller, and D. Polhemus for reviewing the manuscript. This study was supported by National Science Foundation Grant No. BSR-85-15183 and the Hawaii Bishop Research Institute.

LITERATURE CITED

Ahearn, G. A. & F. G. Howarth. 1982. Physiology of cave arthropods in Hawaii. *J. Experimental Zoology* 222:227–238.

Barr, T. C., Jr. 1968. Cave ecology and the evolution of troglobites. *Evolutionary Biology* 2:35–102.

Barr, T. C., Jr. & J. R. Holsinger. 1985. Speciation in cave faunas. *Annual Review Ecology and Systematics* 16:313–337.

Brindle, A. 1980. The cavernicolous fauna of Hawaiian lava tubes, 12. A new species of blind troglobitic earwig (Dermaptera: Carcinophoridae), with a revision of the related surface-living earwigs of the Hawaiian Islands. *Pacific Insects* 21:261–274.

Clague, D. A. & G. B. Dalrymple. 1987. The Hawaiian-Emperor volcanic chain, part 1, geologic evolution. Pp. 5–54. *In:* R. W. Decker, T. L. Wright & P. H. Stauffer, (eds.), *Volcanism in Hawaii.* U.S. Geol. Surv. Prof. Pap. 1350. Washington.

Culver, D. C. 1982. *Cave Life. Evolution and Ecology.* Harvard University Press: Cambridge, MA. 190 pp.

Fennah, R. G. 1973. The cavernicolous fauna of Hawaiian lava tubes, 4. two new blind *Oliarus* (Fulgoroidea: Cixiidae). *Pacific Insects* 15:181–184.

Gagné, W. C. & L. Cuddihy. 1990. Vegetation. Pp. 42–92. *In:* W. L. Wagner, D. R. Herbst & S. H. Sohmer (eds.), *Manual of Flowering Plants of Hawaii.* University of Hawaii Press & Bishop Museum Press: Honolulu.

Gagné, W. C. & F. G. Howarth. 1975. The cavernicolous fauna of Hawaiian lava tubes, 7. Emesinae or thread-legged bugs (Heteroptera: Reduviidae). *Pacific Insects* 16:415–426.

Gertsch, W. J. 1973. The cavernicolous fauna of Hawaiian lava tubes, 3. Araneae (spiders). *Pacific Insects* 15:163–180.

Gurney, A. B. & D. C. Rentz. 1978. The cavernicolous fauna of Hawaiian lava tubes, 10. Crickets (Orthoptera: Gryllidae). *Pacific Insects* 18:85–103.

Hadley, N. F., Ahearn, G. A. & F. G. Howarth. 1981. Water and metabolic relations of cave-adapted and epigean lycosid spiders in Hawaii. *J. Arachnology* 9:215–222.

Howarth, F. G. 1979. Neogeoaeolian habitats on new lava flows on Hawaii Island: an ecosystem supported by windborne debris. *Pacific Insects* 20:133–144.

Howarth, F. G. 1982. Bioclimatic and geologic factors governing the evolution and distribution of Hawaiian cave insects. *Entomologia Generalis* 8:17–26.

Howarth, F. G. 1983. Ecology of cave arthropods. *Annual Review of Entomology* 28:365–389.

Howarth, F. G. 1987. Evolutionary ecology of aeolean and subterranean habitats in Hawaii. *Trends in Ecology and Evolution* 2:220–223.

Howarth, F. G. 1988. The evolution of non-relictual troglobites. *International Journal of Speleology* 16:1–16.

Howarth, F. G. 1990. Hawaiian terrestrial arthropods: an overview. *Bishop Museum Occasional Papers* 30:4–26.

Howarth, F. G., Hoch, H. & M. Asche. 1990. Duets in darkness: species-specific substrate-borne vibrations produced by cave-adapted cixiid planthoppers in Hawaii (Homoptera: Fulgoroidea). *Mémoires de Biospeologie* 17:77–80.

Howarth, F. G. & F. D. Stone. 1990. Elevated carbon dioxide levels in Bayliss Cave, Australia: implications for the evolution of obligate cave species. *Pacific Science* 44:207–218.

Juberthie, C. 1983. Introduction, le milieu souterrain: etendue et composition. *Mémoires de Biospeologie* 10:17–65.

Juberthie, C., Delay, B. & M. Bouillon. 1980. Extension du milieu souterrain en zone non-calcaire: description d'un nouveau milieu et de son peuplement par les coleoptères troglobies. *Mémoires de Biospeologie* 7:19–52.

Juberthie-Jupeau, L. 1983. Etude experimentale de la presence de coleoptères bathysciinae souterrains autour de substances alimentaires. *Mémoires de Biospeologie* 10:369–376.

Mayr, E. 1976. *Evolution and the Diversity of Life. Selected Essays.* Belknap Press: Cambridge, MA. 721 pp.

Martin, J. L., Izquierdo, I. & Oromi, P. 1989. Sur les relations entre les troglobies et les espèces epigées des Iles Canaries. *Mémoires de Biospeologie* 16:25–34.

Oromi, P., Medina, A. L. & M. L. Tejedor. 1986. On the existence of a superficial underground compartment in the Canary Islands. *Proc. IX Intern. Congr. Speleol., Barcelona* 2:147–151.

Otte, D. 1990. Speciation in Hawaiian Crickets. Pp. 482–526. *In:* D. Otte & J. A. Endler (eds.), *Speciation and its Consequences.* Sinauer Associates, Inc.: Sunderland, MA.

Simon, C. 1987. Hawaiian evolutionary biology: an introduction. *Trends in Ecology and Evolution* 2:175–178.

*EVOLUTION ON ISLAND ARCHIPELAGOES:
THE EMERGING PICTURE—SYMPOSIUM*

Organelle DNA Studies of the Evolution of Recently Derived Species Complexes in the Hawaiian Islands

Chris Simon

Abstract. Evolutionary biologists, taking advantage of recent advances in DNA amplification, restriction mapping, and sequencing techniques, have begun to use Hawaiian species to explore patterns and processes of evolutionary change. One thousand species of flowering plants and two thousand species of lower plants contain many examples of radiation from one or a few colonizing species. The most well-known flowering plant radiations are currently being explored using chloroplast DNA (cpDNA). Hawaiian ferns, an interesting contrast to more species-rich taxa, are also being explored at the molecular level. Among the terrestrial fauna, the most successful colonizers were insects, snails, and birds. In each of these groups, mitochondrial DNA (mtDNA) studies have played, or will increasingly play, an important role in reconstructing evolutionary events and in planning conservation strategies.

INTRODUCTION

The Hawaiian Islands are located some 4000 km from the nearest large land mass. They exist at the southern-most extension of a long archipelago formed as the Pacific plate moved northwestward over a stationary hot spot that now lies under the island of Hawaii (Clague & Dalrymple, 1987). The species that colonized the islands and survived are a small subset of the taxa present on the surroundings islands and mainland. These successful colonists have, in a short period of time evolutionarily speaking, evolved into a wide variety of endemic species, some of which have adopted habits unlike those of their relatives elsewhere (Simon et al., 1984; Simon, 1987; Howarth, 1990). Ninety-five to ninety-nine percent of Hawaiian terrestrial species are endemic. Evolutionary biologists, taking advantage of recent advances in DNA amplification, restriction mapping and sequencing techniques, have begun to use Hawaiian species to explore patterns and processes of evolutionary change. Ultimately, these studies will contribute to a growing body of evolutionary knowledge that will be applicable to practical problems such as the conservation of endangered species and ecosystems. In this paper, I present a review of evolutionary studies that employ organelle DNA to reconstruct evolutionary hypotheses for Hawaiian terrestrial species.

Dr. Simon is with the Department of Ecology and Evolutionary Biology, University of Connecticut, Storrs, CT 06269, USA.

FLOWERING PLANTS

In the Hawaiian Flora there are many examples of diverse radiations from one or a few colonizing ancestors (Wagner et al., 1990). Probably the best and most well studied example is that of the Hawaiian silversword complex (Asteraceae: Madiinae). In his 1985 monograph of the genus, Carr recognized 28 species in three endemic genera: *Wilksea* (2 species), *Argyroxiphium* (3 species), and *Dubautia* (23 species) (Carr, 1985). Within *Dubautia* there is considerable divergence in leaf shape and growth form. Plants range from mat-forming sclerophylls to leafy shrubs, to small trees, and even to vines. *Dubautia* contains two chromosomal groups: n = 13 (9 species), and n = 14 (14 species). *Wilksea* and *Argyroxiphium* species all exhibit n = 14. Carr suggested that the n = 13 forms are most derived. Hybridization is common. Eight intergeneric and 26 infrageneric hybrids have been found in the field and in some instances natural backcross progeny have been found. Carr has experimentally produced 19 different intergeneric hybrids and 38 infrageneric hybrids and suggests that any two members of the alliance could readily hybridize.

Chloroplast DNA studies (Baldwin et al., 1990) for 24 of the 28 species in the silversword complex, using 16 restriction endonucleases, showed that, as suggested by the chromosomal pattern, divergence within *Dubautia* exceeds that of all intergeneric comparisons. Phylogenetic analysis of these data suggests that the n = 13 *Dubautia* group was derived but not monophyletic (i.e., it contained one n = 14 taxon). In general, there was a unidirectional migration from older to younger islands within the n = 13 group. The Kauai species of *Dubautia* and *Wilkesia* formed a monophyletic lineage. The tree was rooted using two North American tarweeds that have been suggested as potential ancestors. As Carr suggested, the cpDNA analysis provided evidence for hybrid speciation and extensive introgression among certain *Argyroxiphium* and *Dubautia* taxa. Nuclear DNA studies are necessary to clarify the extent to which hybridization has influenced the evolution of this group.

The Hawaiian lobeliads comprise seven genera, six of which are endemic. The group is of great evolutionary interest in containing the largest genus derived from a single immigrant in the native Hawaiian flora (i.e., the genus *Cyanea* with 52 endemic species) (Wagner et al., 1990). This genus and the others exhibit a remarkable adaptive radiation in leaf shape and floral morphology, with the majority of species endemic to single islands. Thomas Givnish and Kenneth Sytsma (University of Wisconsin, Madison) are using cpDNA restriction fragment analysis to construct a phylogeny that will allow the exploration of adaptive radiation in leaf form, flower shape, and flower color in the fleshy-fruited Hawaiian lobeliad genera (*Cyanea, Clermontia, Delissea,* and *Rollandia*). In addition, they hope to establish the minimum number of inter-island dispersal events required to account for the observed geographic distribution of these four genera. They will use the estimated ages of islands to study rates of evolution and calculate the time of the initial colonization of fleshy-fruited species rooted with outgroups from: Hawaii (capsular fruited lobeliads), Central America (*Centropogon, Burmeistera*), Borneo (*Pratia*) and Tahiti (*Sclerotheca*). With help from many Hawaii botanists, they have collected leaf tissue from over 130 geographic populations, some of which represented species in which only one individual is currently known. Chloroplast DNA has been extracted from these samples and restriction analysis has begun. Future work will explore relationships in native species of the genus Scaevola (Goodeniaceae).

Bidens (Asteraceae) is a worldwide genus with centers of diversity in Africa, the New World, and Polynesia (Wagner et. al., 1990). The genus has undergone extensive adaptive radiation on the Hawaiian Islands, where it is represented by 19 species and eight subspecies. These occupy a great diversity of habitats from sea level to about 2200 m, including coastal dunes and bluffs, cliffs, dry cinder cones, montane ridges, rainforests, and bogs (Carr, 1987). Species differ in many morphological features (leaf shape, floral

298

characters, achene characters, and growth form) and extensive ecological differentiation has also occurred. The most unusual *Bidens* is *B. cosmoides* from Kauai; it has large flower heads up to 8 cm in diameter with styles that are exerted 2–3 cm beyond the corollas and nectar volumes over 30 times that of other *Bidens* species (Ganders & Nagata 1983). Like the silverswords, *Bidens* species have retained the ability to interbreed (Ganders & Nagata, 1984; Ganders, 1988). Surveys of isozyme loci, flavonoids, and polyacetylenes have revealed very little differentiation among the taxa. Unpublished surveys of restriction site variation in rDNA and cpDNA by Kaius Helenurm (Duke University) and Steve O'Kane in collaboration with Barbara Schaal (Washington University) revealed low sequence divergence. Using 25 restriction endonucleases they found 37 restriction site changes that provided some phylogenetic information indicating that species found on the younger islands are more closely related. Further work is in progress.

A fourth study of Hawaiian plants combines phylogenetic investigations with evolutionary ecological studies of reproductive systems. Steve Weller and Ann Sakai (University of California, Irvine) are collaborating with Warren Wagner (Smithsonian Institution) and Pam and Doug Soltis (Washington State University) to study the evolution of the 26 species in the genera *Schiedea* and *Alsinidendron* (Caryophyllaceae: Alsinoideae) to determine whether various stages in the evolution of dioecy have evolved *in situ* in the Hawaiian Islands. The 26 species in this complex exhibit a diverse array of characteristics not found in the Alsinoideae elsewhere, including woody tissue, specialized nectaries, the liana habit, and fleshy structures surrounding the fruit (Wagner et al., 1990). Their study is unique in that it will combine data from developmental floral morphology, seed and pollen morphology, inflorescence and leaf architecture, chromosomal variation, allozymes, and chloroplast DNA. The allozyme and cpDNA phylogeny will be used to help determine which morphological characters are affected by convergent evolution associated with the evolution of dioecy. In addition, they hope to identify ecological factors such as habit and pollination syndromes that may be associated with the evolution of dioecy.

FERNS

The Hawaiian genus *Adenophorus* (Grammitidaceae) is one of only three endemic fern genera. It contains two subgenera, one with deeply pinnatifid leaves (six species) and one with more simple leaves and vegetative root buds (four species). Tom Ranker (University of Colorado) has begun to use RFLP and sequence variation in cpDNA in these species and their Pacific relatives to explore phylogenetic relationships. This information will be combined with data from nuclear rDNA and morphology. In addition, he is conducting studies of population genetic variability via enzyme electrophoresis to examine levels and distribution of genetic variability within and among populations and islands in order to estimate gene flow and assess mating systems. These studies will be combined with detailed analyses of laboratory-cultured and natural populations of gametophytes to determine the intrinsic and extrinsic factors controlling mating behavior and genetic diversity. The results of these populational studies will mapped onto the results of the phylogenetic analysis to examine patterns of mating system evolution.

BIRDS

The Hawaiian honeycreepers of the subfamily Drepanidinae are often cited as a classic example of adaptive radiation. They presumably originated from a single ancestral species but there has been uncertainty as to the identity of their closest living relatives (Freed et al., 1987). Rob Fleischer and Cheryl Tarr (University of North Dakota) have conducted an analysis of restriction site variation in mitochondrial DNA from 12 honeycreepers and three outgroup taxa (two emberizines [a tanager and a non-Hawaiian honeycreeper] and one cardueline [house finch]). Twelve of the 16 restriction enzymes examined showed variation among the 28 individuals. Using both distance-and character-based phylogenetic analyses, they showed that honeycreepers clustered much closer to the cardueline outgroup than to the emberizine. This agrees with earlier DNA-DNA hybridization and allozyme evidence. The "creepers," *Paroreomyza* and *Oreomystis,* are the oldest and most differentiated lineage, and the "red birds" (apapane, akohekohe, and i'iwi) are the most derived. The amakihis are the sister taxon of the akepa and the anianiau. Additional work employing PCR and sequencing is being done in collaboration with Ned Johnson (University of California, Berkeley).

Other research on the Hawaiian honeycreepers focuses on conservation biology and genetics. Like many Hawaiian organisms, these unique birds are extremely vulnerable (Freed et al., 1987). Sixteen undescribed honeycreepers are known only as fossils. Fossils also indicate that extant species formerly occupied larger ranges (Olson & James, 1984). More than half-a-dozen honeycreepers have gone extinct since European contact; some of these are preserved in museums. Introduced diseases such as avian pox and avian malaria have been linked to the decline. Robert Feldman, Rebecca Cann, Leonard Freed and Jaan Lepson (University of Hawaii) are collaborating in a coordinated population biological/genetic study, the goal of which is to seek clues to the nature of genetic resistance factors which might be employed to save endangered species. They have used mtDNA to explore genetic variability in fragmented populations of six species that are locally abundant at the Hakalau Forest National Wildlife Refuge on the eastern slope of Mauna Kea. These are, the Hawaii akepa, Hawaii creeper, Akiapola'au (all three endangered), and the i'iwi, apapane, and common amakihi. DNA from blood, feather, and skin has been amplified enzymatically using mitochondrial primers specific for ribosomal, cytochrome B, and control region sequences. They plan to use these sequences to construct maternal genealogies and to compare current population variability to that present in past populations represented by museum specimens. They hope to use this information to identify superior genotypes for captive breeding and translocation experiments, especially genotypes that might be resistant to avian diseases.

SNAILS

A diverse assemblage of land snails has evolved in Hawaii (Solem, 1990). The best studied of these are the endemic Hawaiian achatinelline tree snails, which have radiated into four genera and approximately 100 species. On-going morphometric and demographic studies of three species in this subfamily by Michael Hadfield, Stephen Miller and coworkers at the University of Hawaii have demonstrated that conspecific populations in trees only a few meters apart can often be characterized by different shell banding patterns or life-history characters such as survivorship, size, age at first reproduction, and fecundity (Hadfield & Miller, 1989). Their data indicate that achatinelline tree snails occur in small, highly inbred populations which they suggest were founded by one or two individuals. They hypothesize that these founder events have driven the rapid achatinelline radiation. Additionally, they suggest that this segregation of a species' polymorphism into small, strongly monomorphic demes also decreases resistance to environmental perturbations, thus increasing the probability of local extinctions. They are continuing to monitor population parameters and plan to incorporate genetic data from mitochondrial DNA to characterize divergence of populations and species. They hope to use this information to refine laboratory rearing strategies and develop a release protocol for establishing new field populations.

300

SPIDERS

Over the past three years Rosie Gillespie (University of Hawaii) has rediscovered an adaptive radiation of one of the most abundant and conspicuous invertebrate groups in Hawaii: the spider genus *Tetragnatha*. Preliminary studies on the morphology, ecology, and behavior of the group have revealed over 60 distinct species endemic to Hawaii with more morphological and ecological variation than is present in the genus in the rest of the world.

During 1989, Henrietta Croom (University of the South), working in Stephen Palumbi's laboratory (University of Hawaii), used *Drosophila* oligonucleotide primers to amplify and sequence the mtDNA coding for rRNA of the small subunit of a circumtropical, introduced *Tetragnatha (T. mandibulata)*. Tetragnathid specific PCR primers were made from this information and used to sequence mtDNA from 50 individuals representing nine of the endemic species. Comparison of a 204-base-pair region of the third domain of this DNA has shown that all endemics are 25% divergent from the introduced species. Members of the same species vary from one another by less than 2%, while interspecific variation ranges from 4 to 14%. Both the morphological and molecular data point to a radiation from a single introduction with some subgroups exhibiting remarkable similarity in their gross morphology, ecology, and genitalic structure, while others are very diverse.

INSECTS

The Hawaiian Drosophila are probably the best known Hawaiian species, thanks to the pioneering work of Hampton Carson, Elmo Hardy and Kenneth Kaneshiro (University of Hawaii) (Carson et al., 1970; Carson & Kaneshiro, 1976; Carson, 1982). This is the most diverse Hawaiian genus, with more than 346 described species and others waiting to be described. In the family Drosophilidae in Hawaii, there are estimated to be more than 700 species. The largest and most well known of the Hawaiian drosophilids are the picturewings with 108 species, 103 of which are incorporated into the chromosome banding phylogeny devised by Hampton Carson (Carson, 1987). Chromosomal inversion patterns and morphology can be used to group the picture wings into more than eight subgroups. The picturewings are known for their elaborate courtship displays and diverse morphology (Kaneshiro & Boake, 1987). In addition to the picturewings, the Hawaiian members of the genus *Drosophila* can be grouped into a modified mouthparts group (>100 species), a modified tarsus group (>100 species), a fungus feeder or white-tipped scutellum group (>60 species), and several smaller groups, such as antopocerus, ateledrosophila, nudidrosophila, and engiscaptomyza, which have fewer than 20 species each (Kaneshiro & Boake 1987).

Rob DeSalle, Val Giddings, Allan Templeton, and John Hunt have employed DNA data to help elucidate patterns of phylogenetic relationships. Most of these published studies have concentrated on the relationships within the planatibia subgroup of the picturewings and are reviewed in DeSalle and Hunt (1987). Hunt's studies have concentrated on the ADH locus in the nuclear genome and DeSalle et al.'s on the mitochondrial DNA genome. Current work by both DeSalle and Hunt concentrates on elucidating relationships among the higher groups of Hawaiian *Drosophila* and eventually exploring relationships with putative mainland ancestors. In addition, DeSalle is exploring relationships among populations of *D. sylvestris* on the island of Hawaii and the biogeography of species formation in several groups of Hawaiian *Drosophila* that have all radiated from Maui, Lanai, and Molokai.

Crickets are among the most accessible and numerous insects in Hawaii. Thanks to the detailed and careful work of Dan Otte (Otte, 1989), we know that the Hawaiian endemic crickets appear to be derived from four original colonizing species: an American tree cricket (subfamily Oecanthinae, which has radiated into three genera and 54 species), an American swordtail cricket (subfamily Trigonidiinae, which has diversified into three genera and more than 150 species), and two Western Pacific ground crickets (subfamily Nemobiinae, with two independently-colonizing genera, one containing one species, and the other more than six).

The swordtail genus *Laupala* is the subject of a dissertation study by Kerry Shaw under the direction of Alan Templeton (Washington University). The genus *Laupala* (containing more than 30 species), along with its close relative *Anaxipha* (containing more than 100 species), are two of the most common insects in the Hawaiian wet forests; their high pitched, incessant calls are the dominant sound in native and some disturbed habitats. *Laupala* is most diverse on the youngest island, Hawaii (Otte, 1989). Because *Laupala* species are differentiated by few if any morphological features and differ only in the pulse rate of their simple songs, mtDNA data will be particularly important in understanding their evolution. Shaw has used double stranded sequencing of PCR amplified products to examine sequence difference in the large and small ribosomal subunits and to construct a preliminary hypothesis of phylogenetic relationships. Her results suggest that Kauai and Oahu species are indeed the oldest, with the Hawaiian populations being most derived; however evolutionary lineages proposed by Otte on the basis of morphology were scattered about the tree rather than constituting monophyletic groups. The ultimate goal of her study is to examine patterns of song evolution and historical biogeography.

CAVE ORGANISMS

The fact that the Hawaiian Islands are of known age and are arranged in a chronological sequence gives us an a priori scale upon which to superimpose evolutionary events. The age of an island, of course, is not a single point in time (Clague & Dalrymple, 1987). A given island may grow continuously over the course of more than a million years, with older flows being covered by subsequent flows. An island may be composed of more than one volcano. The biotic and physical character of an island constantly changes as it grows, subsides, erodes, and is colonized by a succession of species. The average age of an island, in combination with molecular data, has been used to give a rough idea of the age of species, but always with a qualification; the problem of species moving among islands (island hopping) confounds estimates of species' age (Beverly & Wilson, 1984, 1985).

One ecosystem in Hawaii where island hopping is not a problem is the subterranean ecosystem. Once a species evolves into a cave or deep lava crack, it is unlikely to evolve out. The extreme morphological and physiological adaptations necessary for cave existence (Howarth 1983, 1987) result in high levels of convergence such that molecular data are the only form of characters useful for phylogeny construction. My laboratory, in collaboration with Frank Howarth (Bishop Museum), Fred Stone (University of Hawaii, Hilo), Hannelore Hoch and Manfred Asche (Marburg University), Ken Kaneshiro (University of Hawaii) and Svante Paabo (University of Munich), is gathering ecological, behavioral, and nucleotide sequence information for selected mitochondrial gene regions in two very different cave-adapted insect groups: plant hoppers (*Oliarus*; Cixiidae) and crickets (*Caconemobius*; Nemobiinae). Our goal is to reconstruct evolutionary histories and to examine rates of evolution of molecular, morphological, and behavioral traits. The biological differences between the two groups form an interesting contrast. The endemic planthopper genus *Oliarus* has undergone extensive adaptive radiation in Hawaii after the colonization of a single ancestor (Howarth et al., 1987). This group, including cave- and surface-dwelling species, presently contains about 80 named taxa (species and sub-

302

species). At least six evolutionary lines within *Oliarus* have invaded caves: two on Hawaii Island, three on Maui, and one on Molokai. Epigean relatives of the cave-dwelling Hawaiian *Oliarus* species are still extant on the surface (15 taxa on Hawaii, 11 on Maui, and six on Molokai) (Hoch, pers. comm.). However, no systematic study has attempted to link cave with surface species.

Cave planthoppers display characteristics typical of cave adaptation: reduction of eyes, wings, and pigment. The wing is fused to the supporting sclerite. The prothorax is expanded to take the place of the eye. The three ocelli are missing, the bases of the antennae are enlarged, and the setae are more developed. The five carinae on the thorax identify them as *Oliarus.*

Hannelore Hoch and Manfred Asche, in collaboration with Frank Howarth, have studied song variation among populations of one species of *Oliarus* that is common in Hawaii Island lava caves. They found that although the various cave populations do not show any obvious differences in their external morphology, they do differ dramatically in song type. This song variation among close caves suggests that gene flow is slower than expected, given connections among caves via underground cracks and voids (Howarth et al., 1990).

Juvenile cave-adapted *Oliarus*, like their above-ground relatives, feed on roots. Unlike the surface forms, adult cave plant-hoppers also feed on roots. The abundance of roots in caves could reduce the necessity to migrate in search of food, thus reducing gene flow among caves. In contrast, cave crickets of the genus *Caconemobius*, which are scavengers, may be more mobile and, therefore, provide an interesting comparison to the planthoppers.

The genus *Caconemobius* is wide-spread on Pacific and Indian Ocean seashores. In Hawaii, one species inhabits the shores of all islands. On Maui and Hawaii, species have been found inhabiting lava cracks and lava tube caves. Otte (1989) suggests that speciation in this genus is the result of the shore species repeatedly penetrating into island interiors along lava flows, into cracks and crevices and eventually caves.

Increasing cave adaptation can be seen from the shore species, *Caconemobius* sp. C (none) to the surface-crack species *C. fori*, to the twilight zone species *C. varius*, to the deep cave zone species *C.* sp. B, *C.* sp. A, and *C. howarthi.* In addition to eye reduction and pigment loss, there is a loss of jumping ability (reduction in leg dimensions), and a reduction in ovipositor size (Fred Stone, unpubl. data). DNA data will allow us to construct phylogenetic hypotheses, to examine rates of evolution within and among genes and lineages, and to discover how many cave invasions have occurred. A comparison of among-cave differentiation in plant-hoppers and crickets will provide information on gene flow, population boundaries, and size of habitat necessary to preserve the species.

There are many more Hawaiian terrestrial species of which we know relatively little. The pioneering studies described in this paper will provide a foundation for future studies that seek to compare biogeographic patterns and dates of origin across taxonomic groups in order to describe the commonalities and peculiarities of evolution on islands.

ACKNOWLEDGMENTS

I thank the authors mentioned above who kindly provided unpublished information describing research in progress. Projects described herein are supported by grants from the Hawaii-Bishop Research Foundation, the Hawaii Audubon Society, the MacArthur Foundation, the Matsuda Fund of the University of Hawaii Foundation, the National Geographic Society, the National Science Foundation, the Nature Conservancy of Hawaii, and Sigma Xi. C. Simon was supported by NSF DEB 88–22710 during the writing of this paper.

LITERATURE CITED

Baldwin, B. G., Kyhos, D. W. & J. Dvorak. 1990. Chloroplast DNA evolution and adaptive radiation in the Hawaiian silversword alliance (Asteraceae-Madiinae). *Annals of the Missouri Botanical Garden* 77:96–109.

Beverly, S. M. & A. C. Wilson. 1984. Molecular evolution in *Drosophila* and higher Diptera. I. Microcomplement fixation studies of a larval hemolymph protein. *J. Mol. Evol.* 21:1–13.

Beverly, S. M. & A. C. Wilson. 1985. Ancient origin for Hawaiian Drosophilinae inferred from protein comparisons. *Proc. Natl. Acad. Sci. USA* 82:4753–4757.

Carr, G. 1985. Monograph of the Hawaiian Madiinae (Asteraceae): *Argyroxiphium, Dubautia,* and *Wilkesia. Allertonia* 4:1–123.

Carr, G. 1987. Beggar's ticks and tarweeds: masters of adaptive radiation. *Trends in Ecology & Evolution* 2:192–195.

Carson, H. L., Hardy, D. E., Spieth, H. T. & W. S. Stone. 1970. The evolutionary biology of the Hawaiian Drosophilidae. Pp. 437–543. *In:* M. K. Hecht & W. C. Steere (eds.), *Essays in Honor of Theodosius Dobzhansky.* Appleton-Century-Crofts, New York.

Carson, H. L. & K. Y. Kaneshiro. 1976. *Drosophila* of Hawaii: systematics and ecological genetics. *Annual Review of Ecology & Systematics* 7:311–345.

Carson, H. L. 1982. Evolution of *Drosophila* on the newer Hawaiian volcanoes. *Heredity* 48:3–25.

Carson, H. L. 1987. Tracing ancestry with chromosomal sequences. *Trends in Ecology and Evolution* 2:203–207.

Clague, D. A. & G. B. Dalrymple. 1987. *The Hawaiian-Emperor Volcanic Chain.* Pp. 5–54. *In:* R. W. Decker, T. L. Wright & P. H. Stauffer (eds.), *Volcanism in Hawaii* U.S. Government Printing Office: Washington, DC.

DeSalle, R. & J. A. Hunt. 1987. Molecular evolution in Hawaiian Drosophilids. *Trends in Ecology & Evolution* 2:212–216.

Freed, L. A., Conant, S. & R. C. Fleischer. 1987. Evolutionary ecology and radiation of Hawaiian passerine birds. *Trends in Ecology & Evolution* 2:196–203.

Ganders, F. R. 1988. Adaptive radiation in Hawaiian *Bidens.* Pp. 99–112. *In:* L. V. Giddings, K. Y. Kaneshiro & W. W. Anderson, (eds.), *Genetics, Speciation, and the Founder Principle.* Oxford University Press: Oxford.

Ganders, F. R. & K. M. Nagata. 1983. Relationships and floral biology of *Bidens cosmoides. Lyonia* 2:23–31.

Ganders, F. R. & K. M. Nagata. 1984. The role of hybridization in the evolution of Bidens on the Hawaiian Islands. Pp. 179-194. *In:* W. F. Grant (ed.), *Biosystematics.* Academic Press Canada: Ontario.

Hadfield, M. G. & S. E. Miller. 1989. Demographic studies on Hawaii's endangered tree snails: *Partulina proxima. Pacific Science* 43:1–16.

Howarth, F. G. 1983. Ecology of cave arthropods. *Ann. Rev. Entomol.* 28:365–389.

Howarth, F. G. 1987. Evolutionary ecology of aeolian and subterranean habitats in Hawaii. *Trends in Ecology & Evolution* 2:220–223.

Howarth, F. G. 1990. Hawaiian terrestrial arthropods: an overview. *Bishop Museum Occasional Papers* 30:4–26.

Howarth, F. G., Hoch, H. & M. Asche. 1987. Speciation in *Oliarus* (Homoptera: Fulgoroidea: Cixiidae) from Hawaiian lava tubes. *Proc. 6th Auchenorrhyncha meeting,* Turin (Italy), 1987:255–257.

Howarth, F. G., Hoch, H. & M. Asche. 1990. Duets in darkness: species specific substrate vibrations produced by cave-adapted cixiid planthoppers in Hawaii (Homoptera: Fulgoroidea). *Mémoires de Biospeologie* 17:77–80.

Kaneshiro, K. Y. & C. R. Boake. 1987. Sexual selection and speciation: issues raised by Hawaiian *Drosophila. Trends in Ecology & Evolution* 2:207–212.

Olson, S. L. & H. F. James. 1984. The role of Polynesians in the extinction of the avifauna of the Hawaiian Islands. Pp. 768–780. *In:* P. S. Martin & R. G. Kline (eds.), *Quaternary Extinctions.* University of Arizona Press: Tucson.

Otte, D. 1989. Speciation in Hawaiian crickets. Pp. 482–526. *In:* D. Otte & J. A. Endler (eds.), *Speciation and Its Consequences.* Sinauer Associates, Inc.: Sunderland, MA.

Simon, C., Gagne, W. C., Howarth, F. G. & F. J. Radovsky. 1984. Hawaii: a natural entomological laboratory. *Bulletin of the Entomological Society of America* 30:8–17.

Simon, C. 1987. Hawaiian evolutionary biology: an introduction. *Trends in Ecology & Evolution* 2:175–178.

Solem, A. 1990. How many Hawaiian land snails are left? What can we do for them? *Bishop Museum Occasional Papers* 30:27–40.

Wagner, W., Herbst, D. R. & S. H. Sohmer. 1990. *Manual of the Flowering Plants of Hawaii.* University of Hawaii Press & Bishop Museum Press: Honolulu. 2 vols. 1853 pp.

Henderson Island:
Biogeography and Evolution
at the Edge of the Pacific Plate

Gustav Paulay

Abstract. Henderson is a recently (<1 Ma) uplifted limestone island. Together with two small atolls and volcanic Pitcairn, it forms the Pitcairn Group. These islands lie at the southeast edge of the Indo-West Pacific region, thus posing both dispersal and climatic limitations for potential colonists. As a result, Henderson has a depauperate and unstable fauna, with unusually high short and long term changes and inter-island variation evident in the coral fauna. Despite the island's easterly location, practically the entire indigenous biota is derived from the west. Endemicity is low (2–3%) for the marine fauna and moderate for the land biota: 18% for plants, 3% for flies, 3% for Lepidoptera, 75% for weevils, 100% for land snails, and 67% for land birds. The youth and topographic uniformity of the island appear to preclude intra-island speciation, and even inter-island speciation between Henderson and Pitcairn is uncommon. Consequently no local evolutionary radiations are known. Although briefly inhabited prior to western contact, Henderson lacked human occupation historically and is remarkably undisturbed.

INTRODUCTION

Recent work on the geology and biology of Henderson and neighboring islands in the Pitcairn Group allows for an analysis of the biota of these remarkable islands. Located near the southeast corner of the Indo-West Pacific region, of varied geologic origin, and ranging from almost pristine to moderately disturbed, these islands provide interesting material for the study of how physical factors affect species composition, distribution, and evolution. Below I review the physical characteristics of these islands and then discuss the origin, diversity, stability, evolution, and disturbance of their marine and terrestrial biotas. I focus on Henderson Island, the largest, least disturbed, and best studied island in the group, and discuss the other islands for various comparisons.

GEOLOGY, GEOGRAPHY, AND ENVIRONMENTAL HISTORY

The Pitcairn Group, consisting of Oeno, Pitcairn, Henderson, and Ducie islands (Fig. 1), is the easternmost archipelago on the Pacific tectonic plate, lying south of the Tropic of Capricorn. The Pitcairns are geographically a continuation of the Tuamotu-Gambier Archipelago, the nearest island of which, Temoe Atoll, lies ca. 390 km westward. The nearest land east of the Pitcairns (ca. 1570 km), Easter and Sala y Gomez on the Nazca plate, are the easternmost outposts of the Indo-West Pacific region. The four Pitcairn islands appear to be the result of the activities of two 'hotspots'. One, located near Pitcairn,

Dr. Paulay is with the Dept. of Paleobiology NHB-121, Smithsonian Institution, Washington D.C. 20560, USA.

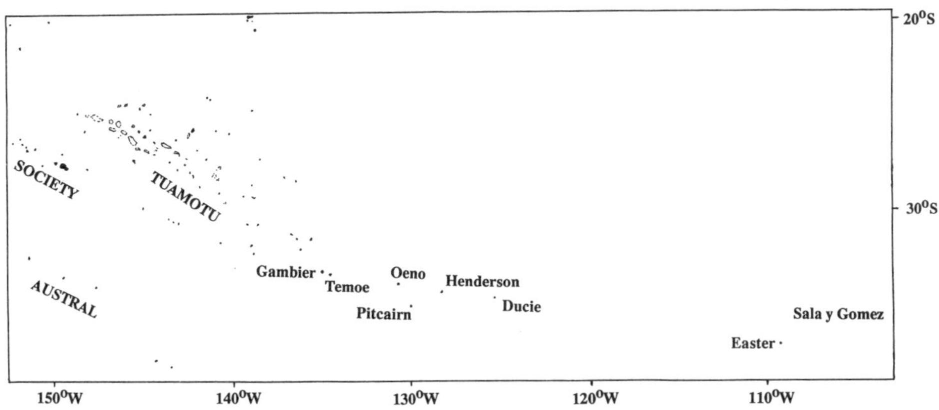

Figure 1. Map of the Pitcairn Group and neighboring islands.

gave rise to that island 0.46-0.93 Ma (Duncan et al., 1975) and appears to have also given rise to Gambier, Temoe, and some of the southern Tuamotu islands. The other appears to be responsible for Oeno, Henderson, and Ducie, estimated to be 16, 13 and 8 Ma, respectively (Spencer, 1989).

Ducie is a small atoll about 2 km in diameter, with one larger and three smaller islets (Rehder & Randall, 1975). It is the easternmost island on the Pacific plate as well as the southernmost atoll in the world. It is also remarkable in having a vascular flora composed of a single species, *Tournefortia argentea* L.f. (Fosberg et al., 1989). Oeno is a small atoll, about 4 km in diameter, located at the western end of the group. It has a reasonably diverse terrestrial biota for an atoll, including some species usually found only on high islands, such as an endemic form of *Bidens* and the widespread rail *Porzana tabuensis* (Gmelin). Pitcairn is a high (330 m), small (4.5 km^2), lush volcanic island, rising precipitously from the ocean. It is the southernmost island in the group, and unlike the other islands it lacks substantial reef development. Pitcairn is the only permanently inhabited island and its terrestrial biota, although potentially the most diverse, is also the most disturbed.

Henderson Island (128°22'W, 24°19'S) is an uplifted atoll ca. 34 m high and 36 km^2 in area, with its original atoll form well preserved. The island interior, which was the site of the former lagoon, is a flat basin ca. 22 m above sea level. A ridge encircling the central depression and rising gradually to a height of 30–34 m is the remnant of the old atoll's reef crest. Seaward of the crest a rugged slope leading to the shore is derived from the fossil fore reef of the original atoll as well as subsequent apron reefs added to the island core during high sea stands in the Late Pleistocene. Henderson has very poor soil development throughout, ground cover being dominated by pebble- to cobble-sized unconsolidated coral rubble, with stretches of solid limestone at the plateau margin, crest, and outer slope. A remarkable feature of the fossil lagoon is the apparent absence of sands or finer sediments. Both topography and plant communities show the greatest variation at the island's periphery, while the interior is relatively uniform (Spencer & Paulay, 1989; Paulay & Spencer, 1989).

The geologic history of Henderson can be reconstructed as follows: volcanic origin perhaps 13 Ma, subsidence and erosion to an atoll form reached ca. 11–6 Ma, and uplift to a limestone island <1 Ma. Uplift of Henderson is attributed to lithospheric flexure around Pitcairn during its origin <1 Ma, a date supported by geomorphological and paleontological evidence (McNutt & Menard, 1978; Spencer, 1989; Spencer & Paulay, 1989).

The availability and extent of terrestrial habitats and the type of marine habitats available on Henderson and neighboring islands changed markedly through the Pleistocene,

306

not only because of uplift, but also due to large (ca. 100 m) and frequent (ca. every 100,000 years) eustatic sea level fluctuations (Shackleton & Opdyke, 1976; Chappell & Shackelton, 1986). Thus atolls, like Ducie and Oeno as well as pre-uplift Henderson, alternated between atoll forms with wide, shallow lagoons and limited, low-lying land areas during interglacial high sea stands, and high limestone island forms, like Henderson today, during times of glacial low stands. As subsidence and erosion rapidly reduce the height of exposed reefs during low sea stands (Paulay & McEdward, 1990), post-glacial rising sea levels usually completely flood atolls, leaving no terrestrial, or even intertidal, habitats. Most present atoll landscapes are the result of Holocene reef growth and subsequent sub-aerial sediment accumulation (e.g., Sackett & Potratz, 1963; Labeyrie et al., 1969; Marshall & Jacobson, 1985). Thus, the present terrestrial biota of most atolls is at most 5–8000 years old.

Despite the age of Henderson (estimated at 13 Ma; Spencer, 1989), permanent land habitats have only been continuously available on the island since uplift commenced. Once uplifted, the interior, fossil lagoon of the island was apparently never re-flooded, although subsequent sea level fluctuations added additional fossil reefs onto the island perimeter (Spencer & Paulay, 1989). Neither Ducie nor Oeno appear to have exposed, last interglacial sediments (Spencer, 1989), and both almost certainly were flooded during the Holocene before developing their current islets.

Below I examine how the physical properties outlined above affected the species composition, biogeography and evolution of the Henderson biota, discussing to some degree the whole Pitcairn Group. I first examine the marine fauna, based on studies of corals, echinoderms, and marine mollusks (Paulay, 1989), and then discuss the terrestrial biota, focusing primarily on the following taxa, for which the most complete information is available: land plants (St. John, 1987; Fosberg et al., 1989), Diptera (Mathis, 1989), Lepidoptera (Holloway, 1990), Curculionidae (Paulay, unpublished), land snails (based on literature survey and Bishop Museum collections identified by C. M. Cook Jr. and Y. Kondo), and birds (Schubel & Steadman, 1989). Introduced species are excluded from all analyses except where otherwise stated, and in the Lepidoptera where they have not been identified.

MARINE FAUNA

The location of the Pitcairn Group at the southeast corner of the Indo-West Pacific region imposes dispersal and climatic limitations on the marine fauna. As predominating currents in the South Central Pacific are from the east, these islands lie upstream and upwind of all Indo-West Pacific source areas with the exception of depauperate Easter Island and Sala y Gomez. Thus both initial colonization and continued recruitment of species with pelagic propagules can be problems. The islands' position south of the Tropic of Capricorn in relatively cool waters imposes further, physical limitations for stenothermal, tropical species. These limitations are reflected in depauperate, temporally unstable, and geographically variable faunas and communities.

Although located at the eastern, upstream end of the Indo-West Pacific region, the marine fauna of Henderson and the entire Pitcairn Group is derived almost exclusively from the west. Thus, all 54 coral species, 198 marine molluscs, and 58 echinoderms recorded from the Pitcairn Group are widespread Indo-West Pacific species, or are related to Indo-West Pacific species if they are not known elsewhere (Paulay, 1989). One possible exception is the asteroid *Allostichaster peleensis* Marsh, known at present only from Pitcairn and Henderson, but with relatives both in Australia–New Zealand and South America (Marsh, 1974).

The western derivation of the marine fauna of the Pitcairn Group is not unexpected, as island stepping stones lead to the group from the west but not from the east, and because the Indo-West Pacific fauna dominates over the East Pacific fauna throughout the insular Pacific, especially in coral reef habitats. Even isolated, oriental islands like Easter (Wells, 1972; Rehder, 1980; Newman & Foster, 1983) and Hawaii (Newman 1986; Kay & Palumbi, 1987) have, with minor exceptions (Barnard, 1970), almost exclusively Indo-West Pacific marine faunas. The dominance of the Indo-West Pacific fauna is felt even off the American coast: 54% of the Clipperton Island marine molluscs are Indo-West Pacific forms (Salvat & Ehrhardt, 1970) and 83% of the Galapagos stylasterid corals have western affinities (Cairns, 1986).

The Pitcairns' marine fauna is impoverished compared to those of islands immediately westward. While 1159 species of molluscs are recorded from French Polynesia (Richard, 1985), only 198 are known from the Pitcairn Group. Similarly the 54 species of corals known from the Pitcairns are but a fraction of the 168 species known in French Polynesia (Pichon, 1985; Paulay, 1989). These trends in diversity reflect in part the fact that there has been less collecting in the Pitcairns than in French Polynesia, and that there are fewer islands in the former group. The trends probably also reflect, however, the dispersal and ecological limitations inherent in the islands' southeasterly location.

These limitations are made further evident by an unusually unstable fauna. Abundant and well preserved fossil coral communities dating from the ?mid-Pleistocene, when Henderson was an atoll, allow for a study of changes in species composition through time. Of the coral species that inhabited old Henderson Atoll, 50% (10 of 20 species) were not found living on the island today, and 30% are not known living in the entire Pitcairn Group. The extent of coral turnover on Henderson is considerably greater than the ca. 15% turnover in coral species on more centrally located Niue, a similar, but older (?Pliocene) uplifted atoll, and the ca. 5% turnover in corals since the late Pleistocene in the Cook Islands. Only one of the ten local extinctions can be attributed to changes in habitat availability accompanying uplift; the other nine species live on outer reef slopes and occur on other uplifted islands (Paulay & Spencer, 1988; Spencer & Paulay, 1989: Table 1).

All ten locally extinct corals are widespread Pacific or Indo-West Pacific species that occur today on neighboring Tuamotu-Gambier Islands; their extinction on Henderson represents only a small contraction of the edge of their range (Paulay & Spencer, 1988). Although such small-scale range contractions in corals indicate that local conditions are responsible for the extinctions, other species known only as fossils from Henderson have undergone extinction throughout the Pacific plate. For example, the mussel *Septifer excisus* (Wiegmann), known as a Pleistocene fossil in the Hawaiian and Cook as well as Henderson Islands, survives today only in the Marianas and westward (Paulay, unpublished).

Large short term changes in coral cover demonstrate a potential mechanism of local extinction in the islands. A major coral kill devastated the reefs on Ducie in the late 1960s, apparently due to incursion of cool water masses; by 1970 coral cover was <1% (Rehder & Randall, 1975). Recovery was rapid however, and in 1987 we found ca. 60% coral cover, indicating that under normal circumstances corals can thrive on these islands (Paulay, 1989).

Substantial differences between the coral faunas and communities of adjacent Ducie and Henderson may reflect their vulnerability to periodic mortality and/or the vagaries of dispersal among these upstream islands. While Ducie and Henderson share only 36% (N = 45 species) of their coral fauna, Rarotonga and Aitutaki (separated by about the same distance) in the more centrally located Cook Islands share 78% (N = 73 species) of their coral fauna (Paulay, 1989).

The only moderate isolation of the islands and the instability of the fauna, together with the great dispersal ability of most marine invertebrates, create a fauna with only slight endemism. Only eight of 310 species of corals, mollusks and echinoderms are unknown from outside the Pitcairn Group, but five of these were dredged in deep water, a habitat that is particularly poorly sampled in Polynesia (Paulay, 1989). At least one of the apparent endemics appears to be real and the result of recent speciation: an octopus species collected on Henderson Island is closely related to, but distinct from, the widely distributed (East Africa to French Polynesia) *Octopus cyanea* Gray (M. Norman, pers. comm., 1990). Randall (1978) estimated that about 2% of the fish species are restricted to the Group.

The low endemicity of marine invertebrates in the Pitcairn Group is comparable to that of other moderately isolated archipelagoes in southeastern Polynesia. Marine endemics are rare in all but the most isolated islands or archipelagoes, such as Hawaii, Marquesas and Easter in the Pacific (e.g,. Rehder, 1968, 1980; Kay & Palumbi, 1987). Further, within-archipelago and intra-island speciation is essentially impossible for marine organisms even in those island groups (Kay & Palumbi, 1987).

TERRESTRIAL BIOTA

The terrestrial biota also shows a predominantly western derivation. All 60 native plants known from Henderson have clearly western affinities, except *Bidens,* which originally colonized Polynesia from the Americas. However, *Bidens* is widespread and diverse in Polynesia and may have colonized Henderson from other islands in the region (Carlquist, 1966). Of 22 species of identified native flies on Henderson, all but one have western or endemic distributions, and the last is cosmopolitan. All identified land snails, weevils, and land birds of Henderson have Polynesian affinities. A largely western derivation is typical for the terrestrial biota of most Pacific islands (e.g., Zimmerman, 1942; Gressitt, 1956; van Balgooy, 1971) although in Hawaii and Easter some groups also have eastern affinities (Zimmerman, 1948; Skottsberg, 1956).

The terrestrial biota of Henderson shows moderate endemism (Table 1). Among the invertebrates, volant insects such as flies and moths have relatively few endemics (5% and 8% respectively), while weevils, all endemic species of which are flightless, and land snails are predominantly endemic (75% and 100%). Bird endemicity is also high (67%), and 18% of the land plants are endemic.

These proportions are biased and probably somewhat inflated due both to extinction and to limited collecting and taxonomic study. Collecting on Henderson has been more thorough for some groups (e.g., plants and birds) than others (e.g., insects except Diptera). More importantly, collecting on neighboring Pitcairn, with which Henderson may share some species now regarded as Henderson endemics, has been generally poor for groups other than birds and plants. Moreover, while Henderson has remained relatively undisturbed, Pitcairn has been greatly ravaged (see below), and some species that appear to be endemic to Henderson today may have occurred on Pitcairn in the past. That pat-

Table 1. Henderson endemics. # spp.: includes only species whose endemic status can be evaluated, and excludes (except for Lepidoptera where data were unavailable) introduced species. # endemic, % endemic: based on # spp.. effort: relative collecting effort and taxonomic knowledge of group on Henderson and adjacent islands.

Taxon:	# spp.	# endemic	% endemic	effort
Diptera	22	1 (3)[1]	5 (14)	avg.
Lepidoptera	13	1	8	poor
Curculionidae	4	3	75	poor
land snails	8	8[2]	100	avg.
land birds	6[3]	4	67	good
land plants	60	11[4]	18	good

[1] 2 species are archipelago endemics shared with Pitcairn
[2] 2 species at subspecific, 1 at varietal level
[3] includes 2 species now extinct on Henderson (Schubel & Steadman 1989)
[4] 1 species at subspecific, 2 at varietal level

terns of endemicity can be greatly altered by human-caused extinction is becoming increasingly clear from studies of fossil insular avifaunas: many species known historically to be limited to single islands or archipelagoes have recently been found to have had much wider ranges before the arrival of humans (e.g., Steadman, 1989).

Only four terrestrial species are known at present to be confined to Pitcairn and Henderson; this low number may reflect, in part, poor collecting on both islands and extinction. The species are *Dacus setinervis* Maloch, *Gaurax* n. sp. (Diptera; Mathis, 1989), *Atylana parmula* Fennah (Hemiptera; Fennah, 1958), and the shrub *Glochidion pitcairnense* (F. Brown) St. John (Fosberg et al., 1983).

Unlike on most Polynesian high islands, where evolutionary radiations are common and lead to often spectacular diversification, especially among insects and land snails (e.g. Pacific Entomological Survey 1932–1939, 1935; Zimmerman, 1948; Solem, 1976; Cooke & Kondo, 1960), essentially no radiations are known in the Pitcairn Group. Species-to-genus ratios, so strikingly elevated in the terrestrial faunas of most Pacific high islands, are near one in the indigenous Henderson biota: 1.1 (60/55) for plants, 1.1 (7/6) for birds, 1.0 (7/7) for weevils, 1.0 (16/16) for Lepidoptera, 1.1 (13/12) for land snails, and 1.2 (22/18) for Diptera. Even groups that usually radiate on other high islands, such as *Rhyncogonus*, *Miocalles*, ampagioid weevils, and endodontid and achatinellid land snails, although represented in the Henderson fauna by endemic species, have not speciated locally. With one doubtful exception (see below), genera in the Henderson biota never have more than one endemic species on the island.

Although intra-island speciation commonly leads to diversification on even relatively small high islands in Polynesia (e.g., Zimmerman, 1948; Solem, 1976; Murray & Clarke, 1980; Paulay, 1985), it is unknown on Henderson, with one questionable exception. Distant (1913) described three very similar species of *"Devagama"* (Hemiptera, Issidae) from Henderson, based on Tait's collections (which total 150 insect specimens; Holloway, 1990), but only one issid species was later identified from Henderson by Fennah (1958—who does not cite Distant's work), based on Zimmerman's much more extensive collections. Issids, represented in Polynesia only by the genus *Atylana*, are not known to have more than two species on even the largest islands in Southeastern Polynesia (Fennah, 1958). Thus it seems more likely that Tait's species are actually variants of a single form.

One possible example of intra-island speciation is known from mountainous Pitcairn: the land snails *Philonesia (Pitcairnia) pitcairnensis* Baker and *P. (P.) filiceti* (Beck). The only other known species of *Pitcairnia*, *P. (P.) mangarevae* Baker, is endemic to Mangareva Island (Baker, 1938). Relationships among these three species have not been resolved. St. John (1987) described two species of *Pandanus* endemic to Pitcairn, but these actually represent the widespread *Pandanus tectorius* Park (F. R. Fosberg pers. comm., 1990).

There are also few apparent cases of inter-island speciation, where possible sister species are endemic to different islands in the group (Table 2). These include the warblers *Acrocephalus vaughani taiti* (Ogilvie-Grant) on Henderson and *A. v. vaughani* (Sharpe) on Pitcairn (Fosberg et al., 1983); the composites *Bidens hendersonensis* var. *hendersonensis* Sherff on Henderson, *B. h.* var. *oenoensis* Sherff on Oeno, and possibly *Bidens mathewsii* Sherff on Pitcairn; and the shrubs *Canthium barbatum* var. *christiani* f. *pitcairnense* Fosb. on Pitcairn and *C. b.* f. *calcicola* Fosb. on Henderson (Carlquist, 1966; St. John, 1987; Fosberg et al., 1989). In addition, the Henderson endemic land snail *Tornatellides oblongus parvulus* Cooke & Kondo, is apparently closest to *Tornatellides oblongus oblongus* (Anton), a subspecies known from numerous islands in southern Polynesia from the Cooks to Pitcairn. Cooke & Kondo (1960) mention, however, that the distinctiveness of the Henderson subspecies may be ecophenotypic.

The striking paucity of local speciation in the Pitcairns is likely due to several factors. Both Henderson and Pitcairn are young (0.5–1 Ma), and thus little time has been available for inter- or intra-island speciation. Further, Pitcairn is a very small island and consequently provides limited opportunity for intra-island speciation; it is also relatively unstudied and considerably disturbed, such that some examples of local speciation may have been missed. Henderson, although of substantial size, exhibits minimal topographic variation, providing little opportunity for intra-island isolation of populations.

Table 2. Potential sister species pairs in the Pitcairn Group.

Henderson	Pitcairn	Oeno
Acrocephalus vaughani taiti	*A. v. vaughani*	
Bidens hendersonensis var. *hend.*	*B. mathewsii*	*B. h.* var. *oenoensis*
Canthium barbatum f. *pitcairnense*	*C. b.* f. *calcicola*	
Tornatellides oblongus parvulus	*T. o. oblongus*[1]	

[1]widespread in southeast Polynesia

HUMAN DISTURBANCE

Human impact in the Pitcairn Group has been varied. Both Pitcairn and Henderson show evidence of Polynesian habitations, but both were found abandoned when first visited by Europeans in 1767 and 1606 respectively (Anonymous, 1963; Fosberg et al., 1983). Prehistoric habitations on Henderson spanned at least from 1160 +/−110 A.D. to 1455 +/−105 A.D. (Sinoto in Fosberg et al., 1983), while Pitcairn appears to have been inhabited by 1100 A.D. (Bellwood, 1987). The prehistoric population on Henderson was limited by a harsh environment: limited water, poor soils, no volcanic rocks, and poor reef development (Kirch, 1984). The archeological record shows a shift from tools made of imported materials such as basalt and pearl shell to locally available materials like coral and *Tridacna* shell (Sinoto in Fosberg et al., 1983). In contrast, lush conditions on Pitcairn allowed the development of an apparently large population that has left ample evidence of its presence, not only in an abundance of stone tools and remnants of dwellings and burial sites, but also in the form of rock engravings, megalithic statues and maraes (religious centers) (Anonymous, 1963; Heyerdahl & Skjölsvold, 1965). Why Pitcairn was later abandoned is unclear. While Henderson has remained uninhabited in historic times, Pitcairn was settled by the Bounty mutineers in 1790 and has been almost continually occupied by a substantial population since.

The limited aboriginal population of Henderson and the absence of humans for the last 400 years have left the island relatively undisturbed: few species have been introduced (Table 3) and presumably few native species have gone extinct, especially when compared to Pitcairn. Undisturbed native vegetation covers the entire island except where small coconut groves occur near the north and northwest landing sites. Of five species of introduced plants recorded from Henderson, two have not been collected since the first half of the century and may now be extinct, and the other three are rare and localized (Paulay & Spencer, 1989). Three of the introductions are aboriginal and two are historic (cf. Imada et al., 1989). Considering that Polynesians inhabited the island for a few hundred years, the number of aboriginally introduced plants is especially small. In comparison, Polynesians introduced at least 25 species of plants even to isolated Hawaii (Imada et al., 1989). Poor edaphic conditions may be partly responsible for the small number of introduced plants on the island, by preventing the establishment of many potential colonizers. The land snail fauna has two possible aboriginal introductions among 15 species: *Lamellidea oblonga* (Pease) and *Tornatellinops variabilis* (Odhner) (Cooke & Kondo, 1960). Interestingly, *Lamellaxis gracilis* (Hutton), widely distributed in Polynesia by prehistoric humans (Christensen & Kirch, 1986), is not known from Henderson or Pitcairn.

In contrast to Henderson, Pitcairn has only limited areas of native vegetation left, and 132 introduced species are recorded by St. John (1987), based mostly on pre-1935 collections. Several additional species have established since (Fosberg et al., 1989). The differential human impact on these islands is also evident in the fly fauna: while 11–21% of the Henderson Diptera are introduced, 28–36% of the Pitcairn flies are (Table 3) (Mathis, 1989, pers. comm. 1990). Interestingly, such a difference is not evident in the ant fauna. As all ants inhabiting Eastern Polynesian Islands are introduced (Wilson & Taylor, 1967), they are useful indicators of human presence. Six ant species have been recorded from both Henderson and Pitcairn (Wilson & Taylor, 1967). Whether this equality is due to lesser collecting on Pitcairn (3 vs. 6 days by the Mangarevan Expedition on whose collections these data are based) or to other factors is unclear at present.

Despite the apparent pristine conditions of Henderson, at least three of the original seven species of resident land birds have been extirpated since the arrival of humans (Steadman & Olson, 1985; Schubel & Steadman, 1989). Nevertheless four land birds, including a flightless rail (*Porzana atra* North), still inhabit Henderson while Pitcairn has only a single extant species.

Table 3. Introduced species on Henderson and Pitcairn. Number of species introduced by humans and the percentage of the total fauna they represent.

	Henderson	Pitcairn
Diptera	3–6 (11–21%)	10–13 (28–36%)
land plants	5 (8%)	132 (66%)

CONCLUSIONS

The varied and diverse biotas of Central Pacific islands often do not conform to the simple rules of the theory of island biogeography (e.g., Solem, 1973; Williamson, 1981; Olson & James, 1982; Wagner, 1991). Colonization, persistence, and in situ diversification all influence the species composition and diversity of island biotas, and are greatly dependent on the physical and geographic characteristics of the island. Island age, size, topographic heterogeneity, location, environmental history (subsidence, uplift, erosion, climatic and sea level fluctuations), and intensity of human disturbance all exert a substantial influence on island biotas.

The biota of the Pitcairn Group demonstrates the importance of several of these factors. The islands' location at the periphery of the insular Central Pacific, combined with their lack of great isolation, appear to be responsible for a depauperate yet relatively unendemic biota. Their position at potentially harsh high latitudes and upstream of potential source areas yields an unstable and locally variable marine fauna. The limited amount of time (<1 Ma) during which these islands have been continuously above sea level has also constrained the development of an endemic land biota; together with the small size of Pitcairn and topographic uniformity of Henderson, it has limited both intra- and inter-island speciation.

ACKNOWLEDGMENTS

This work is based largely on the results obtained during the 1987 Smithsonian expedition to Henderson Island and the Pitcairn Group, with Gary Graves, Wayne Mathis, Sue Schubel, and Tom Spencer. I thank the crew of the R/V Rambler for access to the islands, Wayne Mathis for discussion of the islands' Diptera fauna, and Bern Holthuis for comments on the manuscript. Funding from the Smithsonian Institution is gratefully acknowledged.

LITERATURE CITED

Anonymous. 1963. *A Guide to Pitcairn.* South Pacific Office: Suva, Fiji.

Baker, H. B. 1938. Zonitid snails from Pacific Islands Part 1. Southern genera of Microcystinae. *Bernice Pauahi Bishop Museum Bulletin* 158:1–102.

Barnard, J. L. 1970. Sublittoral Gammaridea (Amphipoda) of the Hawaiian Islands. *Smithsonian Contributions to Zoology* 34:1–286.

Bellwood, P. 1987. *The Polynesians. Prehistory of an Island People.* Revised edition. Thames and Hudson: London.

Cairns, S. D. 1986. Stylasteridae of the Galapagos Islands. *Smithsonian Contributions to Zoology* 426:1–42.

Carlquist, S. 1966. The biota of long-distance dispersal. II. Loss of dispersal ability in Pacific Compositae. *Evolution* 20:30–48.

Chappell, J. & N. J. Shackleton. 1986. Oxygen isotopes and sea level. Nature 324:137–140.

Christensen, C. C. & P. V. Kirch. 1986. Nonmarine mollusks and ecological change at Barbers Point, O'ahu, Hawai'i. *Bishop Museum Occasional Papers* 26:52–80.

Cooke, C. M., Jr. & Y. Kondo. 1960. Revision of Tornatellinidae and Achatinellidae (Gastropoda, Pulmonata). *Bernice Pauahi Bishop Museum Bulletin* 221:1–303.

Distant, W. L. 1913. On a small collection of Rhynchota made by Mr. David R. Tait at Henderson's Island. *Annals and Magazine of Natural History* series 8, 11:554–557.

Duncan, R. A., McDougall, I., Carter, R. M. & D. S. Coombs. 1975. Pitcairn Island—another Pacific hot-spot? *Nature* 251:679–682.

Fennah, R. G. 1958. Fulguroidea of South-eastern Polynesia. *Transactions of the Royal Society of London* 110:117–220.

Fosberg, F. R., Paulay, G., Spencer, T. & R. Oliver. 1989[1990]. New collections and notes on the plants of Henderson, Pitcairn, Oeno, and Ducie Islands. *Atoll Research Bulletin* 329:1–18.

Fosberg, F. R., Sachet, M.-H. & D. R. Stoddart. 1983. Henderson Island (Southeastern Polynesia): summary of current knowledge. *Atoll Research Bulletin* 272:1–47.

Gressitt, J. L. 1956. Some distribution patterns of Pacific Island faunae. *Systematic Zoology* 5:11–32.

Heyerdahl, T. & A. Skjölsvold. 1965. Notes on the archaeology of Pitcairn. Pp. 3–8. *In:* T. Heyerdahl & E. N. Jr. Ferdon (eds.), *Reports of the Norwegian Archaeological Expedition to Easter Island and the Eastern Pacific. Vol. 2 Miscellaneous Papers.* Rand McNally: New York.

Holloway, J. D. 1990. The Lepidoptera of Easter, Pitcairn and Henderson Islands. *Journal of Natural History.* 24:719–729.

Imada, C. T., Wagner, W. L. & D. R. Herbst. 1989. Checklist of native and naturalized flowering plants of Hawai'i. *Bishop Museum Occasional Papers* 29:31–87.

Kay, E. A. & S. R. Palumbi. 1987. Endemism and evolution in Hawaiian marine invertebrates. *Trends in Ecology and Evolution* 2:183–186.

Kirch, P. V. 1984. *The Evolution of Polynesian Chiefdoms.* Cambridge University Press: Cambridge.

Labeyrie, J., Lalou, C. & G. Delibrias. 1969. Etude des transgressions marines sur l'atoll de Mururoa par la datation des différents niveaux de corail. *Cahiers du Pacifique* 13:59–68.

Marsh, L. M. 1974. Shallow water asterozoans of Southeastern Polynesia I. Asteroidea. *Micronesica* 10:65–104.

Marshall, J. F. & G. Jacobson. 1985. Holocene growth of a mid-Pacific atoll: Tarawa, Kiribati. *Coral Reefs* 4:11–17.

Mathis, W. N. 1989[1990]. Diptera (Insecta) or true flies of the Pitcairn Group (Ducie, Henderson, Oeno, and Pitcairn Islands). *Atoll Research Bulletin* 327:1–15.

McNutt, M. & H. W. Menard. 1978. Lithospheric flexure and uplifted atolls. *Journal of Geophysical Research* 83:1206–1212.

Murray, J. & B. Clarke. 1980. The genus *Partula* on Moorea: speciation in progress. *Proceedings of the Royal Society of London* B211:83–117.

Newman, W. A. 1986. Origin of the Hawaiian marine fauna: dispersal and vicariance as indicated by barnacles and other organisms. Pp. 21–49. *In:* R. H. Gore and K. L. Heck. (eds.), *Crustacean Biogeography.* Crustacean Issues 4., Balkema: Boston.

Newman, W. A. & B. A. Foster. 1983. The Rapanuian faunal district (Easter and Sala y Gomez): in search of ancient archipelagoes. *Bulletin Marine Science* 33:633–644.

Olson, S. L. & H. F. James. 1982. Prodromus of the fossil avifauna of the Hawaiian Islands. *Smithsonian Contributions to Zoology* 365:1–59.

Pacific Entomological Survey. 1932, 1935, 1939. Marquesan insects. I.-III. *Bernice Pauahi Bishop Museum Bulletin* 98, 114, 142.

Pacific Entomological Survey. 1935. Society Islands insects. *Bernice Pauahi Bishop Museum Bulletin* 113.

Paulay, G. 1985. Adaptive radiation on an isolated oceanic island: the Cryptorhynchinae of Rapa revisited. *Biological Journal of the Linnean Society* 26:95–187.

Paulay, G. 1989[1990]. Marine invertebrates of the Pitcairn Islands: species composition and biogeography of corals, molluscs, and echinoderms. *Atoll Research Bulletin* 326:1–28.

Paulay, G. & L. R. McEdward. 1990. A simulation model of island reef morphology: the effects of sea level fluctuations, growth, subsidence and erosion. *Coral Reefs* 9:51–62.

Paulay, G. & T. Spencer. 1988[1989]. Geomorphology, palaeoenvironments and faunal turnover, Henderson Island, S. E. Polynesia. *Proceedings of the 6th International Coral Reef Symposium, Australia* 3:461–466.

Paulay, G. & T. Spencer. 1989[1990]. Vegetation of Henderson Island. *Atoll Research Bulletin* 328:1–19.

Pichon, M. 1985. French Polynesian coral reefs, reef knowledge and field guides. Fauna and flora: Scleractinia. *Proceedings of the 5th International Coral Reef Congress, Tahiti* 1:399–403.

Randall, J. E. 1978. Marine biological and archeological expedition to Southeast Oceania. *National Geographic Society Research Reports* 1969 Projects:473–495.

Rehder, H. A. 1968. The marine molluscan fauna of the Marquesas Islands. *Annual Reports for 1968 of the American Malacological Union* :29–32.

Rehder, H. A. 1980. The marine mollusks of Easter Island (Isla de Pascua) and Sala y Gomez. *Smithsonian Contributions to Zoology* 289:1–167.

Rehder, H. & J. E. Randall. 1975. Ducie Atoll: its history, physiography, and biota. *Atoll Research Bulletin* 183:1–40.

Richard, G. 1985. French Polynesian coral reefs, reef knowledge and field guides. Fauna and flora: Mollusca. *Proceedings of the 5th International Coral Reef Congress, Tahiti* 1:412–445.

Sackett, W. M. & H. A. Potratz. 1963. Dating of carbonate rocks by Ionium-Uranium ratios. *In:* S. O. Schlanger, Subsurface geology of Eniwetok Atoll. *Geological Society of America Professional Papers* 260-BB:1053–1065.

St. John, H. 1987. *An Account of the Flora of Pitcairn Island: with New Pandanus species.* Privately published: Honolulu, Hawaii. 67pp.

Salvat, B. & J. P. Ehrhardt. 1970. Mollusques de l'île Clipperton. *Bulletin du Musée National d'Histoire Naturelle* 2e série 42:223–231.

Schubel, S. E. & D. W. Steadman. 1989[1990]. More bird bones from Polynesian archaeological sites on Henderson Island, Pitcairn Group, South Pacific. *Atoll Research Bulletin* 325:1–18.

Shackleton, N. J. & N. D. Opdyke. 1976. Oxygen isotope and paleomagnetic stratigraphy of Pacific core V28–239, Late Pliocene to latest Pleistocene. *Geological Society of America Memoirs* 145:449–464.

Skottsberg, C. 1956. Derivation of the flora and fauna of Juan Fernandez and Easter Island. Pp. 193–439. *In:* C. Skottsberg (ed.), *The Natural History of Juan Fernandez and Easter Island. Vol. I. Geography, Geology, Origin of Island Life.* Almquist & Wiksells Boktryceri Ab: Uppsala, Sweden.

Solem, A. 1973. Island size and species diversity in Pacific island land snails. *Malacologia* 14:397–400.

Solem, A. 1976. *Endodontoid land snails from Pacific Islands (Mollusca: Pulmonata: Sigmurethra). Part I. Family Endodontidae.* Field Museum of Natural History: Chicago.

Spencer, T. 1989[1990]. Tectonic and environmental histories in the Pitcairn Group, Palaeogene to present: reconstructions and speculations. *Atoll Research Bulletin* 322:1–41.

Spencer, T. & G. Paulay. 1989[1990]. Geology and geomorphology of Henderson Island. *Atoll Research Bulletin* 323:1–50.

Steadman, D. W. 1989. Extinction of birds in Eastern Polynesia: a review of the record, and comparisons with other Pacific island groups. *Journal of Archaeological Science* 16:177–205.

Steadman, D. W. & S. L. Olson. 1985. Bird remains from an archeological site on Henderson Island, South Pacific: man-caused extinctions on an "uninhabited" island. *Proceedings of the National Academy of Sciences* 82:6191–6195.

van Balgooy, M. M. 1971. Plant geography of the Pacific. *Blumea Supplement* 6.

Wagner, W. L. 1991. Evolution of waif floras: a comparison of the Hawaiian and Marquesan Archipelagoes. *In:* E. C. Dudley (ed.), *Proceedings of the Fourth International Congress of Systematic and Evolutionary Biology.* Pp. 267–284.

Wells, J. W. 1972. Notes on Indo-Pacific scleractinian corals. Part 8. Scleractinian corals from Easter Island. *Pacific Science* 26:183–190.

Williamson, M. 1981. *Island Populations.* Oxford University Press: Oxford.

Wilson, E. O. & R. W. Taylor. 1967. The ants of Polynesia (Hymenoptera, Formicidae). *Pacific Insects Monographs* 14:1–109.

Zimmerman, E. C. 1942. Distribution and origin of some eastern oceanic insects. *American Naturalist* 76:280–307.

Zimmerman, E. C. 1948. *Insect of Hawaii. Volume 1. Introduction.* University of Hawaii Press: Honolulu.

EVOLUTION ON ISLAND ARCHIPELAGOES:
THE EMERGING PICTURE—SYMPOSIUM

Patterns of Avian Diversity and Radiation in the Pacific as Seen Through the Fossil Record

EXTENDED ABSTRACT

Storrs L. Olson

In the past two decades, increased paleontological exploration of the islands of the Pacific has revealed prehistoric man-caused extinctions on a scale never previously imagined. Habitat destruction, along with predation by man and introduced mammals, removed a minimum of 30% to 80% of the avifauna within the past 2000 years or less on most of the prehistorically inhabited islands for which a fossil record is available (Olson, 1989). Such massive extinctions are assumed to have been pervasive, affecting organisms other than birds, and occurred even on islands not currently populated by man, such as Henderson (Steadman & Olson, 1985; Schubel & Steadman, 1989). Consequently, biogeographical and systematic conclusions based on historically known distributions are bound to be to some degree inaccurate, or entirely erroneous (Olson, in press; Steadman, 1986).

Apart from the story of extinction itself, what does the fossil record, as it presently stands, tell us about patterns of avian diversity and adaptive radiation in the Pacific?

In New Zealand, where the first fossil birds were described from the Pacific, there are many extinct lineages documented in the fossil record. The most famous of these is the moas, which were large, wingless herbivores with characters associating them with the so-called "ratite" birds such as emus and ostriches, although these characters may possibly be convergent (Olson, 1983; 1985). The number of species recognized was once exaggerated but has been continually dwindling through systematic revisions so that now there are some 11 species in six genera admitted (Worthy, 1990). I personally doubt that this diversity is entirely the result of *in situ* radiation from a single ancestral species but is more likely due to multiple invasions by several volant ancestors, which may have been closely related. This remains to be proven, however.

Fossils from New Zealand also document the extinction of the Apterornithidae, a family of flightless gruiform birds possibly related to the kagu, *Rhynochetos*, of New Caledonia; *Cnemiornis*, a flightless "goose" related to the Cape Barren Goose, *Cereopsis*, of Australia; *Euryanas*, a flightless or nearly flightless duck probably congeneric with the Australian Wood Duck, *Chenonetta*; *Harpagornis*, a giant eagle that doubtless preyed on large flightless birds; an owlet-nightjar, *Aegotheles novaezealandiae*, with its presumed

Dr. Olson is with the Department of Vertebrate Zoology, National Museum of Natural History, Smithsonian Institution, Washington, D.C. 20560, USA.

closest relative in New Caledonia; several flightless rails (Rallidae); a swan; several ducks; a hawk; a pelican; a crow; and an extinct genus with two species of the wren-like Acanthisittidae (Millener, 1988).

The overall pattern of avian diversity in the main islands of New Zealand has resulted for the most part from repeated colonizations from the mainland. Thus, each of the 13 species of ducks and geese (Anatidae) and 10 species of rails (Rallidae) represents an independent colonization, mainly from Australia. A single lineage may colonize more than once, as for example, in the rails *Gallirallus, Porphyrio,* and *Fulica.* Presumably because there have been at most three, and often only two, main islands in New Zealand to provide opportunities for genetic isolation, there has been very little adaptive radiation here, and when such does occur, the result is usually no more than three species per radiation, as in the case of kiwis (Apterygidae), the wattlebirds (Callaeidae), and the passerine genus *Mohoua* (sensu lato, Olson, 1990). Discounting moas, the only apparent exception is the New Zealand "wrens" (Acanthisittidae), with four genera and six species. But no close mainland relatives of this peculiar family have ever been identified and it has not been confirmed that the Acanthisittidae resulted from a single invasion rather than two or more.

In New Caledonia the fossil record is still very incomplete but at least 11 extinctions have been recorded (Balouet & Olson, 1989), including a giant, flightless galliform bird, *Sylviornis,* which is probably derived from the Megapodiidae but is best considered a separate endemic family. Despite its deep, somewhat raptorial-looking beak, the abundance of *Sylviornis* in fossil deposits indicates that it could not have been at the top of the food chain and it has been postulated to have been a herbivore (Balouet, 1986). The family Strigidae, represented by a species of *Ninox,* is the only other family found as a fossil that does not occur on the island today. A megapode, two pigeons, and a snipe belong to genera that no longer occur in New Caledonia. All other extinct species belong to genera that still exist in the island. No adaptive radiations are apparent among the birds of New Caledonia, which is essentially a single island. Instead, each of the endemic species arose from a separate colonization, the only apparent exception being the kagu, *Rhynochetos,* of which a second fossil species has been described (Balouet & Olson, 1989).

On the opposite side of the Pacific, in the Galápagos, which were not inhabited prehistorically, the fossil record shows only local extinctions of species of birds, all of which are known or assumed to have taken place in the historic period (Steadman, 1986). The only completely extinct lineage of vertebrate not known historically is the large cricetid rodent *Megaoryzomys* (Steadman & Ray, 1982), although bones of these rodents have been radiocarbon dated to historic times (Steadman et al., in press). There are three *in situ* radiations of birds ranging in size from two species (vermilion flycatchers, *Pyrocephalus*), and four species (mockingbirds, *Mimus*), to some 13 species and many more subspecies of the renowned Galápagos finches, which Steadman (1982) considers to belong to a single genus, *Geospiza.* These radiations appear to be very recent. If Steadman (1982) is correct that the mainland grassquit *Volatinia* is congeneric with *Geospiza* (sensu lato) then none of the birds of the Galápagos differs at the generic level from its mainland ancestors. There are no flightless species among any of the land birds. The only real departures from mainland forms are in the bill shapes of the finches and mockingbirds. Presumably the great number of islands in the archipelago is responsible for the diversity of Galápagos finches. The youthfulness of this radiation and the fact that the subspecies, species, and "genera" form essentially a continuum, with considerable hybridization, make this assemblage unique among insular birds.

The Hawaiian Islands combine aspects of the great divergence from ancestral forms seen in New Zealand, with adaptive radiations even more diverse than in the Galápagos. The fossil record documents the extinction of 35 named species and some 22 possible additional species that await description, pending further study or collection of better specimens (Olson & James, 1982a,b; 1984; in press; James & Olson, in press). These come from five of the main Hawaiian Islands, but the fossil record is reasonably complete from only two of these, Oahu and Maui.

In the Hawaiian Islands, the niche of large herbivores was occupied by geese (at least two extinct species) and the moa-nalos (Olson & James, in press), which were heavy-bodied, flightless derivatives of ducks, with strangely shaped bills. These are divided into three genera and four species. Whether these are products of a single colonization from the mainland or several has not been determined. The niche of kiwis on islands of the Maui group was occupied by flightless ibises of the genus *Apteribis*, which contains at least two and perhaps three species. Flightless rails are represented by at least ten species, probably from at least three separate colonizations. An extinct genus of long-legged bird-eating owl is known from four species on as many islands. Other raptors included an extinct eagle, *Haliaeetus*, and a strikingly modified harrier of the genus *Circus*, with very shortened wing elements. Two species of crows, *Corvus*, were added to the one known historically, and the already impressive radiation of Hawaiian finches, Drepanidini (the "Hawaiian honey-creepers" of old), have been augmented by four new genera, fourteen new named species and up to eight yet unnamed species.

Compared to the Galápagos, the avifauna is much more diverse, with many endemic genera, and with most of the endemic species having diverged far more from their main-land ancestors. Compared to New Zealand, there has been much more *in situ* radiation. There are no really "ancient," relictual elements in the Hawaiian fauna comparable to the Acanthisittidae and Apterornithidae in New Zealand or the Rhynochetidae in New Caledonia, none of which has any close living mainland relatives.

Factors contributing to the character of the Hawaiian avifauna are its great isolation from the continents, the large size and large number of islands, the ecological and climatological diversity of the islands, and their relative age, i.e., older than the Galápagos but younger than New Zealand. The fact that the Hawaiian islands are situated so as to have received colonists from at least three different directions (North America, Asia, Australasia/Oceania), has also doubtless contributed to the makeup of its avifauna.

Fossil avifaunas in the remainder of the Pacific are known almost entirely through the recent work of David W. Steadman who has analyzed material from the Marquesas, Pitcairn, Cook, Society, and Tonga groups (Steadman, 1989). Many of these bones are from archeological middens, hence conferring a decided bias in the sample towards edible, catchable species, and also towards larger species in the case of material collected by archeologists who did not use fine enough screen to recover smaller bones. Time of deposition may also affect species composition in archeological samples representing time periods after significant human-caused extinction had already taken place.

Steadman has documented the extinction of many previously unknown species, as well as many range extensions of extant species. Although considerably greater diversity within genera is indicated, so far most of the new species belong to families or genera already well represented in Polynesia. These include megapodes of the genus *Megapodius*, rails of the genera *Gallirallus*, *Porzana*, and *Porphyrio*; pigeons of the genera *Ducula* and *Gallicolumba*; parrots of the genus *Vini*; and starlings of the genus *Aplonis*. New elements for the Polynesian region include the pigeons, *Caloenas*, previously known only as far east as Micronesia and New Caledonia; and *Macropygia*, not known east of Vanuatu (Steadman, in press).

Although large flightless herbivorous birds evolved in New Zealand, New Caledonia, and the Hawaiian Islands, this niche in the Galápagos being filled by tortoises, no such herbivores are known from central or eastern Polynesia. Nor have flightless, long-billed probing birds, such as kiwis and *Apteribis*, been discovered outside of Hawaii and New Zealand. Except for a species of *Accipiter* from Tonga (D. W. Steadman, pers. comm), no raptorial birds have yet been reported from fossil deposits on islands outside the historically known range of hawks and owls.

Except in Hawaii, there are few recognized adaptive radiations of passerines in Polynesia, apart from such possible examples as the diversity of monarchine flycatchers (Myiagridae) in Fiji (Olson, 1980). The fossil record has so far added nothing that would change this impression.

Why should the patterns of great evolutionary divergence or of adaptive radiations such as seen in the avifaunas of Hawaii, New Zealand, New Caledonia, and the Galápagos, not be evident elsewhere in the Pacific? To some extent this may be attributable to the effects of larger land area and/or greater degree of isolation of these archipelagos. But in part it may be an artifact of inadequate paleontological knowledge. The large, high islands of Fiji and Samoa, for example, are essentially unexplored and it would be difficult to imagine that some great evolutionary novelties do not await the paleontologist who first discovers a significant source of fossils, say, on Viti Levu.

ACKNOWLEDGMENTS

I am most grateful to D. W. Steadman for unpublished information and to him and Helen F. James for valuable comments on the manuscript.

LITERATURE CITED

Balouet, J. C. 1986. *Sylviornis*, des géants débonnaires. Pp. 40–41. *In*: J. C. Balouet. Premiers colons de Nouvelle-Calédonie. *L'Univers du Vivant* 1986(7):35–47.

Balouet, J. C. & S. L. Olson. 1989. Fossil birds from late Quaternary deposits in New Caledonia. *Smithsonian Contributions to Zoology* 469:1–38.

James, H. F. & S. L. Olson. Descriptions of thirty-two new species of birds from the Hawaiian Islands: Part II: Passeriformes. *Ornithological Monographs* 46. In press.

Millener, P. R. 1988. Contributions to New Zealand's late Quaternary avifauna. 1: *Pachyplichas*, a new genus of wren (Aves: Acanthisittidae), with two new species. *Journal of the Royal Society of New Zealand* 18:383–406.

Olson, S. L. 1980. *Lamprolia* as part of a South Pacific radiation of monarchine flycatchers. *Notornis* 27:6–10.

Olson, S. L. 1983. Lessons from a flightless ibis. P. 40. *In*: H. F. James & S. L. Olson. Flightless birds. *Natural History* 92:30–40.

Olson, S. L. 1985. The fossil record of birds. Pp. 79–238. *In*: D. S. Farner, J. R. King & K. C. Parkes (eds.), *Avian Biology*. Vol. 8. Academic Press: New York & London.

Olson, S. L. 1989. Extinction on islands: man as a catastrophe. Pp. 50–53. *In*: D. Western & M. Pearl (eds.), *Conservation Biology for the Next Century*. Oxford University Press: New York & Oxford.

Olson, S. L. 1990. Comments on the osteology and systematics of the New Zealand passerines of the genus *Mohoua*. *Notornis* 37.

Olson, S. L. In press. The prehistoric impact of man on biogeographical patterns of insular birds. *In*: International Symposium on Biogeographical Aspects of Insularity. Rome, May 1987. To be published by Accademia Nazionale dei Lincei.

Olson, S. L. & H. F. James. 1982a. Fossil birds from the Hawaiian Islands: evidence for wholesale extinction by man before Western contact. *Science* 217:633–635.

Olson, S. L. & H. F. James. 1982b. Prodromus of the fossil avifauna of the Hawaiian Islands. *Smithsonian Contributions to Zoology* 365:1–59.

Olson S. L. & H. F. James. 1984. The role of Polynesians in the extinction of the avifauna of the Hawaiian Islands. Pp. 768–780. *In*: P. S. Martin & R. G. Klein (eds.), *Quaternary Extinctions. A Prehistoric Revolution*. University of Arizona Press: Tucson.

Olson, S. L. & H. F. James. Descriptions of thirty-two new species of birds from the Hawaiian Islands: Part I: Non-Passeriformes. *Ornithological Monographs* 45. In press.

Schubel, S. E. & D. W. Steadman. 1989. More bird bones from Polynesian archeological sites on Henderson Island, Pitcairn Group, South Pacific. *Atoll Research Bulletin* 325:1–18.

Steadman, D. W. 1982. The origin of Darwin's finches. *Transactions of the San Diego Society of Natural History* 19:279–296.

318

Steadman, D. W. 1986. Holocene vertebrate fossils from Isla Floreana, Galápagos. *Smithsonian Contributions to Zoology* 413:1–103.

Steadman, D. W. 1989. Extinction of birds in eastern Polynesia: a review of the record, and comparisons with other Pacific island groups. *Journal of Archaeological Science* 16:177–205.

Steadman, D. W. New species of *Gallicolumba* and *Macropygia* (Aves: Columbidae) from archeological sites in Polynesia. *Los Angeles County Museum of Natural History Science Series*. In press.

Steadman, D. W. & S. L. Olson. 1985. Bird remains from an archaeological site on Henderson Island, South Pacific. *Proceedings of the National Academy of Sciences USA* 82:6191–6195.

Steadman D. W. & C. E. Ray. 1982. The relationships of *Megaoryzomys curioi*, an extinct crecetine rodent (Muroidea, Muridae) from the Galápagos Islands, Ecuador. *Smithsonian Contributions to Paleobiology* 51:1–23.

Steadman, D. W., Stafford, Jr., T. W., Donahue, D. J., & A. J. T. Jull. Holocene vertebrate extinction in the Galápagos Islands. *Quaternary Research*. In press.

Worthy, T. H. 1990. An analysis of the distribution and relative abundance of moa species (Aves: Dinornithiformes). *New Zealand Journal of Zoology* 17:213–241.

The Galapagos Archipelago, Ecuador: With an Emphasis on Terrestrial Invertebrates, Especially Insects; and an Outline for Research

Stewart B. Peck

Abstract. Oceanic islands are model systems for estimating the dynamics and observing the results of dispersal, evolutionary differentiation, and ecological structuring of biotas. The Galapagos Islands have been important to the development and testing of much evolutionary and ecological theory. Eighty-seven percent of the area of the islands is now protected by the Republic of Ecuador as the Galapagos National Park. They are undoubtedly the world's best protected tropical volcanic oceanic archipelago, and offer exceptionally well-preserved systems for further short- or long-term studies of evolutionary-ecological processes, including dispersal, colonization, adaptation, competition, and extinction. The Charles Darwin Research Station has conducted and facilitated research there for over 25 years. As a result, the breadth and depth of knowledge of this archipelagic biota is perhaps equalled or exceeded only by that of the Hawaiian Islands.

The evolutionary and ecological dynamics of the plants and vertebrates are fairly well known. In comparison, the arthropods are the poorest known component of the terrestrial biota. Yet, it is the arthropods that are the most diverse and have the greatest research potential for general conclusions on island biology. Long-term studies were started in the 1980s by researchers from Belgium, Austria, and Canada. The numbers of known insect species in the Galapagos increased by over 400 (30%) in that decade. The problems of colonization have resulted in arthropod assemblages that are simpler than those of the continent. But, are their evolutionary and ecological dynamics the same as for the island vertebrates and insects on archipelagos elsewhere? Are there adaptive radiations in the insects? There seems to be an exceptional opportunity to study the process of parapatric speciation in eyeless cave-soil arthropods, carabid beetles, and lycosid spiders. Are communities structured differently from those on other archipelagos or on the South American continent? What are the plant-invertebrate interactions? At present there seem to be unusually few species-specific insect-plant (or insect ectoparasite-vertebrate host) relationships, contrasting with the many on the South American continent.

The Galapagos National Park Service and Charles Darwin Research Station invite international research proposals and scientific collaboration with Ecuadorian personnel, students, and researchers.

INTRODUCTION

"The natural history of these islands is eminently curious, and well deserves attention. Most of the organic productions are aboriginal creations, found nowhere else; there is even a difference between the inhabitants of the different islands; yet all show a marked relationship with those of America, though separated from that continent by an open space of ocean, between 500 and 600 miles in width. The archipelago is a little world within

Dr. Peck is with the Department of Biology, Carleton University, Ottawa, Ontario K1S 5B6, Canada. This article is contribution 456 of the Charles Darwin Research Foundation.

320

itself, or rather a satellite attached to America, whence it has derived a few stray colonists, and has received the general character of its indigenous productions. Considering the small size of these islands, we feel the more astonished at the number of their aboriginal beings, and at their confined range. Seeing every height crowned with its crater, and the boundaries of the lava-streams still distinct, we are led to believe that within a period, geologically recent, the unbroken ocean was here spread out. Hence, both in space and time, we seem to be brought somewhat near to that great fact—that mystery of mysteries—the first appearance of new beings on this earth." (Charles Darwin, 1845. *The Voyage of the Beagle*)

The floras and faunas of islands have played a central role in the development of ecological, evolutionary, and biogeographical thought. The Galapagos Archipelago holds a central position because its biota was instrumental in generating in the mind of Charles Darwin the revolutionary idea that species might be "mutable," that they might change through time, leading eventually to the theory of evolution by natural selection. The Galapagos have continued to attract the attention of natural scientists ever since Darwin. Many of these have gone to the islands only as a pilgrimage of homage to the place or to Darwin and his ideas. Many others have continued the careful and patient biological studies that Darwin could not accomplish or even imagine in the five weeks that he was there in 1835.

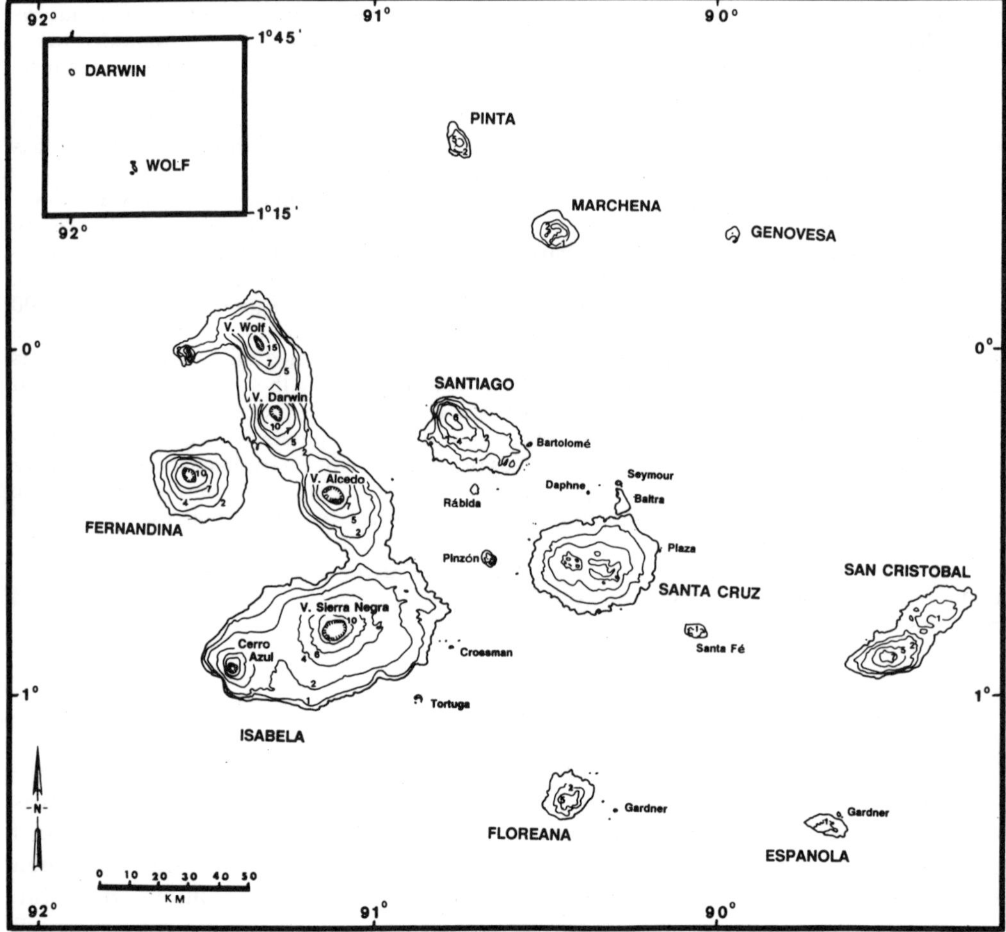

Figure 1. The Galapagos Archipelago (numbers on contour lines indicate elevation in hundreds of meters).

BRIEF DESCRIPTION OF THE ISLANDS

The archipelago consists of 13 large and some 32 smaller named islands (Fig. 1), 7882 km^2 in total area, lying astride the equator, from 900 to 1000 km off the western coast of Ecuador. The islands are strictly volcanic in origin and became available for terrestrial colonization from 3–4 million years ago. They are strictly oceanic in nature, never having had dry-land connections with the South American mainland. All life on the islands has had to cross an oceanic watergap to reach them. They are one of the few Pacific island groups that was not settled by humans before their discovery by the Spanish in 1535.

Climatically the islands are semi-arid at their lower elevations. The islands with higher elevations have a humid forest zone at 300–600 m. During the cool/dry or "garua" season, from July to December, this zone is usually embedded in perpetual drizzle and clouds. January to June is the warm/wet season. Elevations above 700 m are not occupied by native and endemic woody vegetation (i.e., they are above treeline, contrasted with treeline at about 3500 m in the Andes of mainland Ecuador; Peck & Kukalova-Peck 1980).

BRIEF HUMAN HISTORY

For the first 300 years after their discovery the islands were principally a hiding place for pirates or a victualling station for whaling vessels. Human occupation began in the 1830s and faunal and habitat destruction included the harvest of giant tortoises and fur seals, clearing of native vegetation for plantations and farms, and the intentional or accidental introduction of many foreign plants and animals, ranging from pigs, goats, cattle and donkeys to rats, cats, and dogs. Steadman (1986) gives evidence from vertebrate fossils of the impact of the introductions. From the 1830s to 1924 many North American (principally the California Academy of Sciences) and European scientific or surveying expeditions visited the islands and provided the initial taxonomic basis for understanding the biotas. I call this the "Early Natural-History Phase" of the study of the islands. In 1923 the islands were visited by William Beebe of the New York Zoological Society. His resulting book about the voyage (*Galapagos: World's End,* Beebe 1924) initiated what I call the "Romantic Phase" of the island's history. It involved more scientific exploration, but, more importantly, started several small but significant waves of colonization of Europeans, and brought the archipelago to the attention of the world at large (Treherne, 1983). This eventually generated a movement to protect and preserve the islands. A far-sighted Ecuadorian national decree in 1959 set aside 87% of the area of the islands for protection in the Galapagos National Park. This action initiated the "Modern Phase," involving a carefully balanced mix of protection, research in ecology and population biology, and tourism. The Charles Darwin Research Foundation for the Galapagos Islands (constituted in 1959) supported protection, and its internationally operated Charles Darwin Research Station on Isla Santa Cruz has addressed basic research and applied problems of biotic conservation since 1964. The Station has been a base for extensive field studies in population biology, especially on the plants and vertebrates. In 1968 the Galapagos National Park Service was established for protection and management of the Park. The islands now have some 6000 permanent human inhabitants (on six islands) principally involved in administration, the armed forces, fishing, farming, ranching, and servicing of the tourist industry (which now sees about 30,000 visitors a year).

The Galapagos are now the best preserved and protected tropical oceanic volcanic archipelago in the world. They are an ideal site for short and long-term studies on patterns and processes in the ecological and evolutionary dynamics of island biotas. They truly do deserve the often used phrase: "A living laboratory of organic evolution."

The islands, their biotas, and human history have generated an enormous volume of descriptive and analytical literature of both a popular and scientific flavor. This ranges from the scientifically arcane to a real-life murder mystery (Treherne, 1983) to scientifically based futuristic-fantasy (Vonnegut, 1985). It is not possible here to review all Galapagos literature, but to indicate only some of the best of recent examples, and to suggest that these can lead to earlier literature. The human history of the islands is covered by Hickman (1985). The best overall natural history analysis (unfortunately now in need of updating) is Thornton (1971). The best recent natural history field guide is Jackson (1985). Large scientific summary volumes are those of Berry (1984), Bowman et al. (1983), and Perry (1984).

Rather than provide yet another review of what is known, I wish to now emphasize what remains to be learned. Plants and vertebrates have been favorites for study on islands, partly because they are conspicuous and relatively stationary. Insects and other terrestrial arthropods have not received such attention. I contend that this is a grave mistake, because their species diversity, individual abundance, and multiplicity of life styles offer many more replicates of experiments in island living, and statistically much larger and more significant data sets for making generalized conclusions. In the future, I think that insects will set the standards for generalizations about island life, while other organisms may prove to be the exceptions.

TERRESTRIAL ARTHROPODS: RENEWED INITIATIVES

In the 1980s, three independent, long-term, synthetic programs of study were undertaken on the systematics, ecology, and evolution of the terrestrial arthropods of the Galapagos Archipelago. These are: 1) by L. Baert, K. Desender, and J.-P. Maelfait, Royal Institute of Natural Sciences, Entomology (KBIN), Brussels, Belgium, with a focus on spiders and carabid beetles; 2) by Heinrich and Irene Schatz, University of Innsbruck, Innsbruck, Austria, with a focus on oribatoid mites and tenebrionid beetles, and 3) by myself and several colleagues, with a focus on several arthropod groups, especially beetles. These projects have now become coordinated to complement each other. They started with compiling summaries of known data on the composition of the faunas, and some of these are presented in Tables 1, 2, and 3.

The first goal of the studies is to achieve a modern and systematically sound inventory of the species of terrestrial arthropods. This will form the core of a data base which, at least for the insects, will contain: 1) literature citations for all species known from the islands; 2) if it is an introduced, native, or endemic species; 3) its distribution on individual islands and beyond the islands; and 4) bionomic data on adult or immature food materials, habitat, elevational zonation, seasonality, plant or animal hosts or associates, and useful sampling methodologies.

Table 1. Summary of the growth of knowledge on the insect fauna of the Galapagos Archipelago, Ecuador.

All Insects				
Year	Orders	Families	Genera	Species
1835–1966[1]	19	129	395	618
1966–1977[2]	22	164	531	883
1977–1990[3]	24	221	790	1339

[1]data from Linsley & Usinger, 1966.
[2]data from Linsley, 1977; growth due to expeditions of Galapagos International Scientific Program, and of N. and J. Leleup.
[3]Data from above plus literature search in Entomology Abstracts since 1977 (up to March 1990) and unpublished results of 1985 and 1989 collections of S. and J. Peck, B. Landry, and B. Sinclair.

Table 2. Summary of insect fauna known from the Galapagos Archipelago (as of January, 1990). Numbers will change as the fauna becomes better known.

Hexapod Orders	Numbers Known from the Galapagos Archipelago			
	Families	Genera	Species	% endemic species
Collembola	5	11	16	56
Diplura	2	2	3	0
Archeognatha	1	1	1	0
Thysanura	2	3	3	66
Odonata	3	6	7	14
Orthoptera	3	14	26	68
Mantodea	1	1	1	100
Blattaria	3	12	14	20
Isoptera	2	2	6	33
Dermaptera	2	4	5	33
Embidina	1	1	1	100
Zoraptera	1	1	1	100
Psocoptera	15	22	40	50
Phthiraptera	4	25	61	?
Hemiptera	20	61	109	71
Homoptera	12	48	92	59
Thysanoptera	3	6	8	0
Neuroptera	3	5	8	75
Coleoptera	56	219	371	52
Strepsiptera	1	1	1	0
Lepidoptera	22	166	293	45
Diptera	43	138	198	?
Siphonaptera	2	2	2	33
Hymenoptera	14	39	62	?
Totals	221	790	1339	

Table 3. Known terrestrial arthropods (excluding Hexapoda) of the Galapagos (from Baert et al., 1989a,b, Conlon & Peck in prep.; Schatz & Schatz, 1988; Schultz & Peck in prep., Shear & Peck 1987; Shear & Peck in prep.; and personal data).

	families	genera	species	species thought to be		
				endemic	native	introduced
Scorpiones			2	2		
Pedipalpi	1	1	1			
Araneae	28	57	80	46		
Solifugae	1	1	1	1		
Pseudoscorpiones			16	11	5	
Opiliones	1	1	1	1		
Acari						
Gamasina			17			
Uropodina						
Ixodina			13			
Actinedida			6			
Acaridida			1			
Oribatida	49	76	118			
Crustacea						
Amphipoda	1	3	4	1	3	
Isopoda	10	13	17	4	6	7
Chilopoda	4	8	10	5	3	2
Diplopoda	7	9	9	1	0	8

324

The compilation of this data base will be of use to management of the National Park in that it will help identify needed areas of study and identify faunal changes. It will also be available as a basis for analysis and synthesis of ecological, biogeographic, and evolutionary patterns in the composition, structure, dynamics, distribution, and origins of the faunas.

THE COMPARATIVE APPROACH IN SYNTHETIC BIOLOGY

Many first-principles of ecological and evolutionary biology cannot be directly observed or manipulated. We observe and attempt to interpret the results of processes that have operated over long spans of time. An advantage of islands is that the processes have often operated in comparative isolation as independent natural experiments. We compare the results of these experiments. The generality of principles emerges when independent island experiments produce similar results. The fine introductions to such generalities of island biology by Carlquist (1965, 1970, 1974, 1981) should be studied by all beginning students of island biology. I will explore below some aspects of the ecology and evolution of terrestrial invertebrates of the Galapagos by comparing them with plants and vertebrates of the Galapagos and other archipelagos. More and better data sets are needed from other islands in the world to increase the utility and power of the comparative approach.

ISLAND COLONIZATION

The Galapagos are neither especially young nor old islands. Radiometric dating shows that the older southeastern islands of Española and San Cristobal may have been available for terrestrial colonization for about three million years, and that the islands are roughly progressively younger to the west, with Fernandina being the youngest and most volcanically active of the complex.

The islands are distant enough from biotic source areas that the arrival of colonists is a fairly infrequent event. There is now no serious evidence to suggest that the islands ever had direct land-bridge connections with the mainland. They are true oceanic islands and all colonists had to cross an oceanic barrier.

It is a platitude to state that transport modes of colonists are: 1) in or on the ocean, 2) actively or passively in the air by flying or wind transport, 3) hitchhiking on or in other organisms such as birds, and 4) intentional or unintentional introduction by man (Table 4). These transport modes have been considered for some Galapagos organisms. The modes are of differing importance to different higher taxa.

1. It is obvious that the marine fauna has come by sea, with fur seals and penguins from southern South America, and sea lions from California or Mexico. Rafting on the seas has resulted in all reptile colonization, and may be the most important mode for most terrestrial arthropods, which arrived as neuston and on flotsam. Estimates have been made for the beetles and this mode may account for about 60% of the original beetle colonists.

It can no longer be stated (as does Jackson, 1985: p. 203) that a cryptozoic soil fauna is absent. Over 40 eyeless soil arthropods are now known, many of which had eyeless ancestors that must have arrived by rafting (Peck, 1990).

2. Flight or wind transport can account for all bird colonizations except for the penguins. It may seem surprising that it may account for only 9% of angiosperm colonizations (Porter, 1976). Many spiders must have arrived by ballooning of the juveniles. Many small insects have undoubtedly been blown by winds as aerial plankton and about 39% of

Table 4. Modes of dispersal of colonist organisms to the Galapagos Archipelago; as percent of the total number of species in taxon, grouped as either natural or human introductions (data from Jackson, 1985; Porter, 1984; Landry, unpubl.).

	natural introductions				introduced by man		
	No. sp.	ocean drifting, rafting	wind or flight transport	on or in other animals	No. sp.	intentional; "escapes"	unintentional; "weeds"
all plants	413	8%	32%	59%	194	30%	70%
angiosperms	306	11%	9%	79%			
all birds	64	1.5%	98.5%	0%			
reptiles	22	100%	0%	0%	1		100%
beetles	313	60%	39%	1%	28		100%
butterflies & moths	219	0%	100%	0%	73		100%

the beetles and 30% of the lepidopterans may have arrived in this way. Some insects such as dragonflies, larger moths, and butterflies (45% of the Lepidoptera fauna) are strong long-distance fliers and arrived by random (but wind-directed?) flight.

3. Transport on or in other animals. It is estimated that 79% of the angiosperms arrived by hitchhiking on birds as propagules either on or in their bodies (Porter, 1976). Insect ectoparasites such as the 61 species of Phthiraptera chewing bird-lice (=Mallophaga) undoubtedly arrived on bird hosts. Fleas and sucking mammalian lice (=Anoplura) are conspicuously underrepresented with only one and two endemic species respectively. Some insects probably arrived as parasitic immatures on or in their host insects. These are one strepsipteran (in leafhoppers), one meloid beetle (on *Xylocopa* bees), one rhipiphorid beetle (in wood-boring beetle larvae), and several dryinid wasps (in leafhoppers).

4. Man has intentionally brought many domestic animals and cultivated or weedy plants to the Galapagos. I know of no examples of intentional introduction of arthropods, but there are many examples of unintentional introductions. The first may have arrived with the first landings by Bishop Tomas de Berlanga and his party in 1535 in the form of dermestid and *Necrobia* (clerid) beetles and cockroaches, all associated with man and stores in his sailing ships. Pirates or whalers may have brought an alleculid beetle (and other drywood beetles?) in firewood from the mainland. Some 28 beetles (most living in stored products) and 11 species of cockroaches have been introduced to date (but all may not have become permanently established). The number of introductions on plants of thrips, aphids, leafhoppers, and scale insects remains to be estimated. Soil on plant roots has introduced at least millipeds, silverfish, symphylans, earthworms, beetles, and ants. The little red fire ant, *Wasmannia auropunctata*, is the most disastrous arthropod introduction to date and seems to have a most serious impact on native arthropods (Lubin, 1984). Even with strict control, the increased commerce with the continent and tour ships to and between the islands will continue to dilute the native and endemic faunas with introductions (Silberglied, 1978).

There is general agreement that the dominant source of the colonizing ancestors was from Mexico, or Central or South America. This has been long appreciated in plants (Svenson, 1946), and has been well analyzed for the angiosperms (Porter, 1976). To my knowledge, the only attempt to chemically compare in detail the island species with possible continental ancestors has been for the *Opuntia* cacti (Anderson & Walkington, 1968). The western Pacific seems to be an ancestral source area for only a very few species of animals; a land snail (Vagvolgyi, 1974), a stenocephalid (*Dicranocephalus*) true bug (Froeschner, 1985), and some Psocoptera.

Heatwole and Levins (1971) tried to quantify the importance of flotsam transport of terrestrial animals to islands. Schoener and Schoener (1984) have provided a more recent review and more data. Cheng and Birch (1978) used neuston nets to demonstrate the presence of living insects drifting on the sea surface a considerable distance from land. These experimental approaches to measuring flotsam transport should be attempted on the many boats that are continually travelling between islands in the Galapagos. Beaches should be examined immediately after storms to see what insect taxa have arrival potential (Howden, 1977). We suspect that the arrival on an island of live propagules is not very frequent, and that successful colonization after arrival is an even rarer event, but no one has yet even attempted to estimate or to measure the actual immigration rate.

Different taxa, such as different families of insects, undoubtedly have different dispersal, colonization, and differentation potentials. Leston (1957) tried to quantify this and showed that mirid bugs have a higher "spread potential" than other true bug families. Simberloff (1978) reviewed the dynamics of island immigration and extinction.

Electrophoretic allozyme studies have shown that the islands have been colonized by only a single lineage of Darwin's finches and giant tortoises. However, allozymes also show multiple invasions of *Tropidurus* lava lizards and *Phyllodactylus* geckos (Patton, 1984). Such studies on insects would be of great interest.

ESTABLISHMENT OF COLONISTS

After arrival, the propagule must survive and then reproduce. A male and female, or an inseminated female, or a self-fertilizing individual is the minimum requirement for reproduction. Galapagos insects seem to have more parthenogenetic species than continental communities, probably because they have a higher probability to reproduce than do single sexual organisms. Early colonists encountered a harsh climate and young and barren lava. The early plants were most likely hardy, weed-like species with wide ecological tolerances. Early insect colonists probably had similar properties, and were generalist predators, scavengers, or herbivores. What is the sequence of establishment of these trophic groups? Plants and insects with reproductive interdependence are unlikely to arrive together, so those colonists would have perished. There are still few close plant-insect relationships and an unusually large proportion of plants do not require insect pollination; conspicuous flowers are notable by their absence.

Heavy rainfalls accompanying strong "El Niño" events increases the chance of transport of marine rafts of drift which bear propagules from flooded rivers on the continent, and provide more favorable (less arid) circumstances upon arrival in the islands. Overcast and cooler temperatures during the garua season may favor colonist survival, but the drier community conditions in the arid zone at this season may factor out this advantage. It seems an unappreciated fact that high relative humidities, dew, and garúa mists in the arid zone in the night-time may be the most important factor that allows the insect fauna to be as rich as it is. Night-time insect activity and abundance is ever so much greater than in the heat and bright sun of the day.

Colonists on islands may be ecologically released by escape from their continental predators, parasites, and competitors. This is evident in some vertebrates but has not been recognized in insects. At present, intense predation in the littoral zone by birds and lizards on propagule insects arriving by sea must seriously depress establishment success.

DYNAMICS OF DIFFERENTIATION AND SPECIATION

"In July opened first note-Book on 'Transmutation of the Species' ... Had been greatly struck from about month of previous March (1836) on character of S. American fossils—& species on Galápagos Archipelago.—These facts origin (especially later) [i.e., the Galápagos] of all my views."

Charles Darwin, *Evolutionary Notebook*, 1837

There are many genera which have undergone appreciable subspeciation or speciation in the Galapagos (Table 5). But few of these have equaled the dramatic swarms of species of insects, snails, or birds that occur in Hawaii. There is no notable speciation in Galapagos *Drosophila* (Carson et al., 1983). The discovery of a remarkable diversity of singing crickets in Hawaii (at least 1200 times richer per unit area than the continental US; Otte, 1989) is a potential yet to be explored in the Galapagos. Darwin's finches provide the standard examples of speciation through the paradigm of allopatric separation. Accompanying the speciation in the finches has been a remarkable adaptive radiation, which has also become a classic textbook example. Steadman (1982) may have identified the ancestral species of the finches in the blue-black grassquit, *Volatinia jacarinia*, of South America, and suggests that the radiation may have occurred in as little as 100,000 years. But the topic of adaptive radiation has not been carefully addressed or documented for the other components of the biota except *Opuntia* cactus, *Scalesia* composites, and the tortoises.

Table 5. Notable examples of endemic species or subspecies swarms (not all are monophyletic) in the terrestrial biota of the Galapagos (data from Jackson, 1985; Linsley, 1977; Patton, 1984; Peck, unpublished). Single colonizations for insects and beetles remain to be demonstrated by cladistic or other methods.

Example	Number of species or subspecies	Number of ancestral colonization
Vertebrates		
Oryzomine rice rats	8(?)	3(?)
Darwin's Finches	13	1
Elephantopus tortoises	11 subspecies	1
Tropidurus lava lizards	7	2
Phyllodactylus geckos	7	2 or 3
Dromicus snakes	3 sp, 8 ssp	1 ?
Plants		
Scalesia composites	20	1
Camaesyce euphorbs	9	1 ?
Opuntia cactus	14	2
Mollugo (Molluginaceae)	9	?
Invertebrates (other than insects)		
Bulimulus (*Naesiotes*) land snails	65	1
Insects (other than beetles)		
Dagbertia mirid bugs	9	1
Philotis issid plant bugs	16	1
Nesosydne delphacid planthoppers	7	1
Euxesta otitid flies	8	1
Nocticanace canaceid flies	8	1
Beetles		
Pterostichus carabids	11	2 ?
Physorinus elaterids	7	1
Stomion tenebrionids	9	1
Ammophorus tenebrionids	12	1
Pedonoeces tenebrionids	16	1
Pantomorus weevils	6	1
Pseudopentarthrum weevils	6	1

Are these few examples indicative of a generalization, or are they exceptions? In other words, how frequent has adaptive radiation been in the Galapagos? How many of the species or subspecies swarms in the archipelago are monophyletic and have undergone significant ecological-morphological separation? Can birds "hasten" the process more than other organisms through gene-based behavioral niche partitioning coinciding with non-random mating, based on mate selection by morphological or behavioral cues? Grant and Grant (1989a) imply that this may aid or speed allopatric speciation in the finches, but they could not support the likelihood of sympatric speciation. What about other organisms?

Are there insect equivalents to Darwin's finches? Within three genera of tenebrionid beetles (*Ammophorus, Stomion, Pedonoeces*) there are cases of sympatry that are accompanied by different mean body sizes. Possible niche partitioning in these beetles is an open research topic in insect population ecology.

Several cases are now known in Galapagos terrestrial arthropods where sister species have adjacent ranges. This occurs in litter dwelling *Pterostichus* beetles in southern Isabela, Santa Cruz, and Santiago (Desender et al., 1990) and lycosid spiders on Santa Cruz (Maelfait & Baert, 1986). No ecological or geographic barriers lie between these sister species. Several other species pairs have eyed-epigean and eyeless-hypogean sister species that are now separated by habitat but not by geographic distance (Peck, 1990). This pattern of eyed and eyeless sister species also occurs in the Hawaiian and Canary island archipelagos. This may be a system resulting from parapatric speciation, but to prove this over the alternative hypothesis (allopatric speciation and secondary contact of the sisters) may be impossible. Nevertheless, the recurring pattern is intellectually exciting.

Not all groups of organisms present similar amounts of endemism. This is obviously a result of differing vagilities and amount of gene flow between continental and island populations. In insects good dispersers have low levels of endemism, poor dispersers have higher levels (Table 2). Comparison of the Galapagos and Hawaiian archipelagos show much greater mean species formation from a single colonist ancestor in Hawaii (Table 6). This is probably the result of Hawaii's greater age, area, ecological diversity, and

Table 6. Comparisons of total numbers of native and endemic species and mean numbers of species descending from an ancestral colonizing species on two archipelagos (data from Carlquist, 1985; Coppois 1984, Howarth, 1990, Jackson, 1985; Loope et al., 1988; Peck, 1990; Smith, 1966; and Zimmerman, 1948). Comparisons of equivalent groups are unfortunately not always possible. The differences between Hawaii and Galapagos result from the sum of physical properties such as island age and isolation, and biological properties such as habitat diversity, taxon vagility, and genetic characteristics.

Archipelago	Taxon	Number present species	Estimated number colonizing species	Mean number species per ancestor
Hawaii				
	flowering plants	1442	272	5
	reptiles	0	0	0
	land and shore birds	70	15	5
	land molluscs	1064	22	48
	all insects	5005	250	20
	beetles	1388	70	20
Galapagos				
	vascular plants	541	411	1.3
	reptiles	20	10	2
	land birds	57	43	1.3
	land molluscs	80	16?	5?
	all insects	1339	?	?
	beetles	312	239	1.3

isolation (arrival is less frequent, but genetic dilution of island populations by mainland genomes is also less frequent).

DYNAMICS OF ISLAND BIOGEOGRAPHY

Many studies have examined the components determining species diversity on the Galapagos. This was first done for plants and for Darwin's Finches (Hamilton & Rubinoff, 1963, 1964, 1967; Hamilton et al., 1963; Thornton, 1967). The results were equivocal. Area is important for the low islands, but is not the sole determinant in the high islands (discussed in Carlquist, 1965: p. 67). New analyses came after publication of the well-known equilibrium model of island biotas by MacArthur and Wilson, and an improved flora of the islands (Wiggins & Porter, 1971). Extensive data on the plants allowed dissection of the determinants of equilibrium diversity by island and habitat area (at present and at lower sea levels), elevation, and habitat diversity, etc. Johnson and Raven (1973) found that only island area is significant. Numbers of endemic plant species are highest in the arid and transition zone and lowest in the littoral and mesic zone, based on zone-specific immigration and extinction rates, and the youth of the moist upland climates. Simpson (1974) found that plant species diversity in the Galapagos is more significantly correlated with area and distance measures during Pleistocene low-sea-levels, rather than with these measures for the present-day islands. But re-analysis of the data showed that the best predictor of species number on an island is the number of collecting trips to it rather than its area, elevation, or isolation (Connor & Simberloff, 1978).

It is premature to address these topics to the terrestrial arthropods, but they may ultimately provide a better analytical data base. The insects may be the most useful for such study but many taxonomic questions must be settled first. The data base now being compiled will allow the testing of determinants of island species diversity in the archipelago against island age, present and past area, habitat diversity, habitat area, elevation, isolation relative to the continent and within the archipelago, etc. The Galapagos seem to be undersaturated in beetle species numbers when compared to other tropical oceanic islands (Fig. 2).

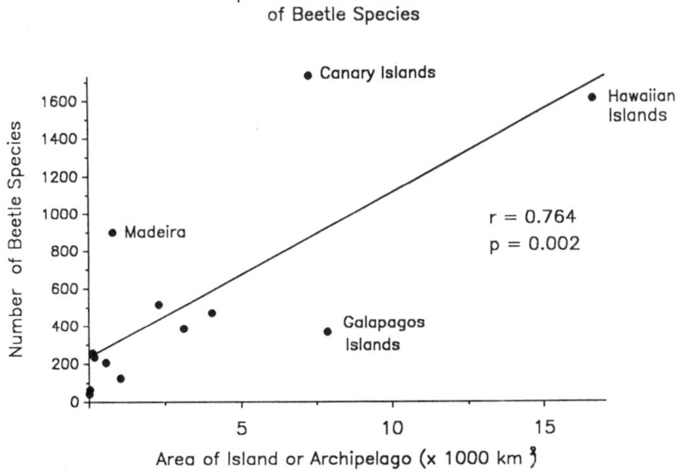

Figure 2. Linear regression of beetle species number against island or archipelago area. Data from Peck (1991) for tropical and subtropical oceanic islands in the Pacific and Atlantic Oceans. Is the "under saturation" of the Galapagos due to the harsh climate or poor collecting, or both?

Are terrestrial arthropods really of greater diversity in the more favorable habitats of the humid zones of the higher islands (as stated by Jackson, 1985: p. 203) even though these habitats have markedly less age and area? The question is interesting.

COMMUNITY AND POPULATION ECOLOGY

Communities

Terrestrial communities in the Galapagos are usually characterized according to the elevation-controlled zonation of the flora (Wiggins & Porter 1971). The islands may possess some of the strongest or most compressed floristic zonation to be found anywhere in the world, passing through a littoral, an arid, a transition, a humid forest, a sub-paramo shrub, and an above treeline fern-sedge (pampa) zone in an elevational rise of less than 700 m (Fig. 3). The best study to date on the zonation of terrestrial invertebrates is Coppois' (1984) study of the remarkably diverse bulimulid land snails on Santa Cruz. Beetles seem to show some zonation (Fig. 2) with fewer species being known from the higher elevation zones (which have less age and area). Spider communities have been characterized in a preliminary analysis by Baert and Maelfait (1986). The number of beetle species known from the different zones is in Table 7.

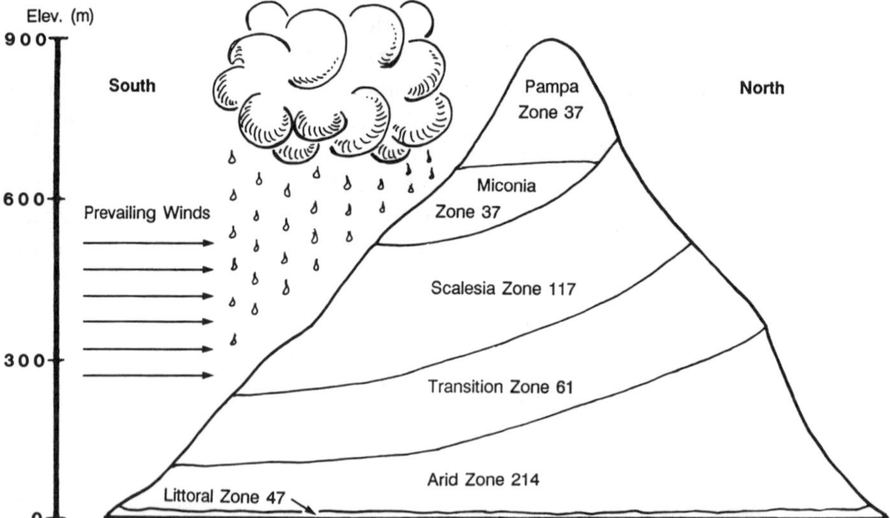

Figure 3. Major vegetational zones of Santa Cruz Island. Zonation is controlled by elevation and moisture, which is related to the prevailing winds. Low islands have only a littoral and arid zone. Other high islands may lack the *Scalesia* and *Miconia*, and have equivalent zones dominated by other woody genera. Numbers indicate the number of species of beetles known to occur in each zone, on all islands combined. The transition zone species number is lower than the zone above it simply because it is less-well studied.

Table 7. Categories of beetle species and numbers known to occur in different elevational vegetation zones of the Galapagos. Most species occur in more than one zone (Peck, 1990).

	littoral	arid	transition	*Scalesia*	*Miconia*	pampa
endemics	27	152	36	72	23	28
native	15	39	17	27	12	9
introduced	5	23	8	18	2	0
totals	47	214	61	117	37	37

It must be realized that these zones have not been stable through time. Colinvaux and Schofield (1976a,b), using pollen and spores recovered from cores of lake sediments and bogs, have shown that the present flora of the humid zones was markedly reduced or absent about 10,000 BP, at the end of the Pleistocene. This means that during Pleistocene glacials the arid-zone was much larger and the moist-zone much smaller than at present. The present moist-zone plant communities may have been assembled in less than 10,000 years and many of the occupants may be recent colonists. This has yet to be addressed for arthropods but it can be predicted that moist-zone habitats should have a lower number of endemic species per unit area of habitat. A preliminary analysis suggests that this is true for the beetles, because their mean body size is smaller in the *Scalesia* forest zone, suggesting an important influence of recent aerial transport from the continent.

How have these communities been assembled? The islands present multiple experiments in community assembly. Are there deterministic rules or is it by chance (Connor & Simberloff, 1979)? Differences in both plant and insect occupants of similar zones on different islands suggest that there is a considerable random component. This may have been first observed by Thornton (1971). Are the insects of the Galapagos "structured" by elevation the same way as in continental South America (Janzen et al., 1976)?

Are the island communities species poor? By how much and in what way? Do the lower numbers of species result in more simple community structures and fewer species interactions? Are the lower species numbers mostly a result of the harsh climate, or the filtering effect of over-water dispersal? Are the species "packed" differently?

Populations

Many populations of Galapagos vertebrates have been studied in detail. These include sea lions, fur seals, land and marine iguanas, and many sea birds. Darwin's finches have received the most detailed attention through prolonged studies, at first by D. Lack, then R. Bowman, and now by P. and B. Grant (1989a,b) and their colleagues. Here the value of isolated small-island populations becomes pre-eminent. The finch populations have no or very low, immigration and emigration. Individuals of known geneology and morphology can be followed through their life time, and their Darwinian fitness (reproduction) measured. The environment can impose harsh periods of intense selection during fluctuations between very wet or very dry years. Because of this, impressive evidence has been assembled to show that bird species are capable of evolutionary change within observable time periods (Grant & Grant, 1989b).

But, our understanding of the population biology of terrestrial invertebrates is now equivalent to the level of understanding of Darwin's finches before the studies of David Lack. We have only general notions of how the populations of invertebrates respond to seasonal climatic fluctuation. Because of their numbers and small body size we may never be able to follow individuals and generations on a daily, seasonal, and yearly basis as is possible in birds.

Competitive interactions are not understood in the arthropods. We know only broad food categories for most. Are the feeding niches of endemic and native species broader than their continental sister groups? Do phytophagous species feed on a wider range of plants on the islands, in spite of there being a narrower range of available plant species? Becker (1975) analyzed the feeding characteristics of beetles on continents and islands, and found that continents have more food specialists (herbivores) than generalists (predators and scavengers), and that this is reversed for islands. This holds true for Galapagos beetles (Table 8). The assumption is that generalist feeders can more readily colonize islands than can specialists. Will it hold for other arthropod groups?

Plant-insect interactions seem to be fewer than on the continent. There are fewer insect-pollinated plants, and only one species of bee (Linsley, 1966; Linsley et al., 1966; McMullen, 1986).

Table 8. Known or assumed predominant feeding categories for beetle species on the Galapagos (Peck, 1990). Generally, feeding specialists such as fungivores, herbivores, and parasites are less well represented on islands than on continents, and feeding generalists like predators and scavengers are better represented (see Becker, 1975 for discussion).

	Number	Percent
predator	99	28.4
scavenger	140	40.1
fungivore	34	9.7
herbivore	73	20.9
parasite	3	0.9
Total	349	100

Co-evolution

There are often co-evolved species-specific relationships between orchids and pollinating insects. Eleven species of orchids are known on the islands, but the pollinators seem to be generalists. Other co-evolved systems may exist between vertebrates and their insect ectoparasites. Two species of endemic lice are known from the two surviving endemic rice rats, but with the extinction of the other six or so species of native rats knowledge of their possible lice was lost. Some 60 species of Mallophaga chewing-lice have been reported from Galapagos birds. This should be a rich data base for host specific studies but the louse systematics may need modern revision. It now seems that widely distributed birds such as migrants and sea birds have widely distributed lice. Endemic birds seem to have endemic lice but there is not much host specificity. For instance, many endemic lice occur both on several species of mockingbirds and on Darwin's finches. Around the world bird lice generally show a host specificity. Not so in the Galapagos. For instance, the little ground finch has had twenty species of lice reported from it; more than any other bird species in the world. Is this a real situation? What does it mean?

Many insect assemblages are attacked by large numbers of hymenopteran parasitoids. The family Ichneumonidae is very diverse in temperate but not tropical latitudes. In the Galapagos only one genus with six species is known. From the literature the total Hymenoptera fauna seems impoverished and dominated by a diversity of ants. In actuality, the parasitoid micro-hymenoptera are present in some abundance but they have not been studied. These Hymenoptera, plus the micro-lepidoptera, that may also show patterns in host association or specificity, are now the least understood components of the Galapagos insect fauna.

SOME IMPORTANT RESEARCH TOPICS FOR FUTURE STUDY OF GALAPAGOS TERRESTRIAL ARTHROPODS

It seems appropriate to indicate here some potential areas of future research. Study of the following can contribute to the growing data base on Galapagos organisms and to broader understanding of biological processes and patterns in island biology in general.

1. Continue to learn what taxa of organisms are present, and analyse and synthesize the evolutionary (sister-group) relationships of the taxa between the islands and the continental sources. The micro-lepidoptera and micro-hymenoptera are

now the poorest known species-rich groups of insects, and the Heteroptera (true bugs) the best known at the alpha systematics level (Froeschner, 1985).

2. Learn the bionomics of the species: seasonality, food items, habitat specificities, etc.

3. In genera with many species conduct a search for morphological, behavioral, ecological, seasonal, etc., indicators of niche partitioning and adaptive radiation, using the many studies on Darwin's finches as comparative models.

4. In genera with many species use electrophoretic and other biochemical techniques to independently determine if the taxa are monophyletic (one ancestral colonization) or descended from two or more ancestral species colonizations, and to determine the biochemical sister-group relationships.

5. Are the arid and other life zone "communities" less diverse in species than on the Ecuadorian mainland, e.g., are they structured differently; are the "communities" assembled in a random or deterministic manner; are niche dimensions "broader"?

6. There are acoustic-communication differences in birds, especially the finches, but what about the songs of the Galapagos katydids and crickets? Are there sibling song-species? Are there ecological or behavioral shifts from their continental congeners, or between separate islands, or within a single island as in Hawaii?

7. Investigate unstudied habitats, such as the insect faunas of very young lava flows, which have a surprisingly special fauna living off of wind-blown debris in the Hawaiian and the Canary Islands, and on Anak Krakatau (New & Thornton, 1988).

8. Investigate host plant-insect, ectoparasite-host, and parasitoid-host interactions and host specificities and contrast these with the continent.

9. Investigate the incidence of flotsam transport to and between the islands. This should be most rewarding during times of flooding on the continents, when debris is fed into the westward trending currents, or between the islands during heavy El Niño floods.

10. Monitor introduction and establishment of new insects in man-altered and undisturbed habitats. Control or eradication of established non-native insects presents many applied problems and questions in insect population biology.

It may now be too late to address many or most evolutionary topics on most of the tropical oceanic islands of the world. But, if Ecuador continues to receive international encouragement and support these and other topics in biology may be addressed in the future in the Galapagos. They will yield additional data and insights into the workings of evolution and communities.

The Galapagos National Park Service and Charles Darwin Research Station invite international research proposals and scientific collaboration with Ecuadorian personnel, students, and researchers. Address inquiries to: Director, Charles Darwin Research Station, Puerto Ayora, Isla Santa Cruz, Galapagos, Ecuador.

THE EMERGING PICTURE OF ISLAND BIOLOGY

Islands and archipelagos are truly separate worlds. Each has a unique geographic position and an individual history. The physical determinants controlling the historical assembly of an island's biota are, at least, its age, isolation, area, elevation, present and past climate, and position relative to wind and water movements from source areas of colonists. Within this framework, predictions can be made about an island's biota. In this we can see a sameness between islands. But beyond this, few predictions can be made because of the

334

stochastic or random nature of dispersal, establishment of colonists, genetic properties of the colonists, varying selection regimes, and the non-deterministic manner in which communities seem to be assembled or structured over time.

Another way of saying this is that the fascination of islands to the ecologist and evolutionary biologist should lie not in a reductionist sameness rooted only in physical and mathematical laws, but in the delightful diversity of results that can be generated by the random biological properties involved in the process of synthesizing the biotas of each individual island world.

ACKNOWLEDGMENTS

This contribution is dedicated to Professor E. Gorton Linsley and the late Professor Robert Usinger who promoted an overview of the insects of the Galapagos by compiling their checklists, which have served as an introductory data base. Lic. Fausto Cepeda, Intendente, Galapagos National Park Services, and the Direccion Forestal, Ministry of Agriculture, Ecuador permitted field studies in the Park. Dr. Gunter Reck and Dr. Daniel Evans (as directors of the Charles Darwin Research Station, Isla Santa Cruz), Sylvia Harcourt, Sandra Abedrabbo, and many Galapaguenos furnished indispensable logistical support for field work. Jarmila Peck, Bernard Landry, and Bradley Sinclair were excellent field companions, and all reviewed the manuscript, as did H. F. Howden, Scott Miller, and I.W. B. Thornton. Many Ecuadorians and scientists of many nationalities (far too many to name individually) have helped me understand the islands, their biotas, and their dynamics. Field work was supported by operating grants from the Natural Sciences and Engineering Research Council of Canada.

LITERATURE CITED

Anderson, E. F. & D. L. Walkington. 1968. A study of some Neotropical Opuntias of caostal Ecuador and the Galapagos Islands. *Noticias de Galapagos* 12:18–22.

Baert, L. & J.-P. Maelfait. 1986. Spider communities of the Galapagos Islands (Ecuador). *Actas X Congr. Arachnol. Jaca, Espana.* 1:183–188.

Baert, L., Maelfait, J.-P. & K. Desender. 1989a. Results of the Belgian 1986-expedition: Araneae, and provisional checklist of the spiders of the Galapagos archipelago. *Bull. Royal Sci. Natur. Belg., Entomol.* 58:29–54.

Baert, L., Maelfait, J.-P. & K. Desender. 1989b. Results of the Belgian 1988-expedition to the Galapagos islands: Araneae. *Bull. Inst. Royal Sci. Natur. Belg., Entomol.* 59:5–22.

Becker, P. 1975. Island colonization by carnivorous and herbivorous Coleoptera. *J. Anim. Ecol.* 44:893–906.

Beebe, W. 1924. *Galapagos, World's End.* G. P. Putnam and Sons, New York. 443 pp.

Berry, R. J. (ed.). 1984. *Evolution in the Galapagos.* Academic Press: London. 260 pp. (Reprinted from *Biol. J. Linnean Soc.* (London) 21: no. 1 and 2, 1984.)

Bowman, R. I., Benson, M. & A. E. Leviton (eds.). 1983. *Patterns of Evolution in Galapagos Organisms.* Pacific Division, AAAS, California Acad. Sci.: San Francisco. 568 pp.

Carlquist, S. 1965. *Island Life.* Natural History Press: Garden City, N.Y. 451 pp.

Carlquist, S. 1970. *Hawaii, a Natural History.* Natural History Press: New York. 463 pp.

Carlquist, S. 1974. *Island Biology.* Columbia University Press, New York. 660 pp.

Carlquist, S. 1981. Chance dispersal. *Amer. Sci.* 69:509–516.

Carson, H. L., Val, F. C. & M. R. Wheeler. 1983. Drosophilidae of the Galapagos Islands, with descriptions of two new species. *Int. J. Entomol.* 25:239–248.

Cheng, L. & M. C. Birch. 1978. Insect flotsam: an unstudied marine resource. *Ecol. Entomol.* 3:87–97.

Colinvaux, P. A. & E. K. Schofield. 1976a. Historical ecology in the Galapagos Islands. I. A Holocene pollen record from El Junco Lake, Isla San Cristobal. *J. Ecol.* 64:989–1012.

Colinvaux, P. A. & E. K. Schofield. 1976b. Historical ecology in the Galapagos Islands. II. A Holocene spore record from El Junco Lake, Isla San Cristobal. *J. Ecol.* 64:1013–1026.

Connor, E. F. & D. Simberloff. 1978. Species number and compositional similarity of the Galapagos flora and avifauna. *Ecological Monog.* 48:219–248.

Connor, E. F. & D. Simberloff. 1979. The assembly of species communities: Chance or competition? *Ecology* 60:1132–1140.

Coppois, G. 1984. Distribution of bulimulid land snails on the northern slope of Santa Cruz Island, Galapagos. *Biol. J. Linn. Soc.* 21:217–227.

Desender, K., Baert, L. & J.-P. Maelfait. 1990. Evolutionary ecology of carabids in the Galapagos Archipelago. *Proc. Europ. Carabidologist Meet.* In press.

Froeschner, R. 1985. Synopsis of the Heteroptera or True Bugs of the Galapagos Islands. *Smith. Cont. Zool.* 407:1–84.

Grant, P. R. 1986. *Ecology and Evolution of Darwin's Finches.* Princeton University Press: Princeton, NJ. 458 pp.

Grant, P. R. & B. R. Grant. 1989a. Sympatric speciation and Darwin's Finches. pp. 433–457. *In:* Otte, D. & J. A. Endler (eds.)., *Speciation and its consequences.* Sinauer Associates, Sunderland, MA.

Grant, B. R. & P. R. Grant. 1989b. Evolutionary dynamics of a natural population: The large cactus finch of the Galapagos. University Chicago Press, Chicago, IL. 368 pp.

Hamilton, T. H. & I. Rubinoff. 1963. Isolation, endemism, and multiplication of species in the Darwin Finches. *Evolution* 17:388–403.

Hamilton, T. H. & I. Rubinoff. 1964. On models predicting abundance of species and endemics for the Darwin finches in the Galapagos Archipelago. *Evolution* 18:339–342.

Hamilton, T. H. & I. Rubinoff. 1967. On predicting insular variation in endemism and sympatry for the Darwin finches in the Galapagos Archipelago. *Am. Nat.* 101:161–171.

Hamilton, T. H., Rubinoff, I. Barth Jr., R. H. & G. L. Bush. 1963. Species abundance: natural regulation of insular variation. *Science* 142:1575–1577.

Heatwole, H. & R. Levins. 1971. Biogeography of the Puerto Rican Bank: Flotsam transport of terrestrial animals. *Ecology* 53:112–117.

Hickman, J. 1985. *The Enchanted Islands: The Galapagos Discovered.* Anthony Nelson Ltd.: Oswestry, England.

Howarth, F. G. 1990. Hawaiian terrestrial arthropods: an overview. Bishop Mus. Occ. Pap. 30. In press.

Howden, H. F. 1977. Beetles, beach drift and island biogeography. *Biotropica* 9:53–57.

Jackson, M. 1985. *Galapagos: A Natural History Guide.* University Calgary Press: Calgary, Canada.

Janzen, D. H., Atarof, M., Farinas, M., Reyes, S., Rincon, N., Soler, A., Soriano, P. & M. Vera. 1976. Changes in the arthropod community along an elevational transect in the Venezuelan Andes. *Biotropica* 8:193–203.

Johnson, M. P. & P. Raven. 1973. Species number and endemism: The Galapagos Archipelago revisited. *Science* 179:893–895.

Leston, D. 1957. Spread potential and the colonization of islands. *Syst. Zool.* 6:41–46.

Linsley, E. G. 1966. Pollinating insects of the Galapagos Islands. Pp. 255–232. *In:* R. I. Bowman, (ed.), *Proceedings of the Symposia of the Galapagos International Scientific Project.* University of California Press: Berkeley-Los Angeles.

Linsley, E. G. 1977. Insects of the Galapagos (Supplement). *Occ. Pap. Cal. Acad. Sci.* 125:1–50.

Linsley, E. G., Rick, C. M. & S. G. Stephens. 1966. Observations on the floral relationships of the Galapagos carpenter bee. *Pan-Pac. Entomol.* 42:1–18.

Linsley, E. G. & R. L. Usinger. 1966. Insects of the Galapagos Islands. *Proc. Cal. Acad. Sci.* (4) 33:113–196.

Loope, L., Hamman, O. & C. P. Stone. 1988. Comparative conservation biology of oceanic archipelagos; Hawaii and the Galapagos. *BioScience* 38:272–282.

Lubin, Y. 1984. Changes in the native fauna of the Galapagos Islands following invasion by the little red fire ant *Wasmannia auropunctata*. *Biol. J. Linn. Soc.* 21:229–242.

Maelfait, J.-P. & L. Baert. 1986. Observations sur les Lycosidae des Iles Galapagos. *Mém. Soc. Royal Belge Entomol.* 33:139–142. McMullen, C. K. 1986. Observations on insect visitors to flowering plants of Isla Santa Cruz. Part II. Butterflies, moths, ants, hoverflies, and stiltbugs. *Noticias de Galapagos (Charles Darwin Foundation).* no. 43:21–23.

New, T. R. & I.W. B. Thorton. 1988. A pre-vegetation population of crickets subsisting on allochthonous aeolian debris on Anak Krakatau. *Phil. Trans. R. Soc. Lond.* B 322:481–485.

Otte, D. 1989. Speciation in Hawaiian crickets. Pp. 482–526. *In:* D. Otte & J. A. Endler (eds.), *Speciation and its Consequences.* Sinauer Associates: Sunderland, MA. 679 pp.

Patton, J. L. 1984. Genetical processes in the Galapagos *Biol. J. Linn. Soc.* 21:97–111.

Peck, S. B. 1990. Eyeless arthropods of the Galapagos Islands, Ecuador: composition and origin of the cryptozoic fauna of a young tropical oceanic archipelago. *Biotropica.* 22:366–381.

Peck, S. B. 1991. Beetle (Coleoptera) faunas of tropical oceanic islands: with emphasis on the

Galapagos Archipelago, Ecuador. *In:* M. Zunino (ed.)., *Advances in Coleopterology.* In press.

Peck, S. B. & J. Kukalova-Peck. 1980. A guide to some natural history field localities in Ecuador. *Stud. Neotrop. Fauna Environ.* 15:35–55.

Peck, S. B. & J. Kukalova-Peck. 1990. Origin and biogeography of the beetles (Coleoptera) of the Galapagos Archipelago, Ecuador. *Can. J. Zool.* 68:1617–1638.

Perry, R. 1984. *Key Environments: Galapagos.* Pergamon Press: New York. 336 pp.

Porter, D. M. 1976. Geography and dispersal of Galapagos vascular plants. *Nature* 264:745–746.

Porter, D. M. 1984. Relationships of the Galapagos flora. *Biol. J. Linn. Soc.* 21:243–252.

Schatz, H. & I. Schatz. 1988. Arachnological research in the Galapagos Islands (Ecuador) with special reference to the Oribatida (Acari). European Assoc. *Acarologists Newsletter* 1(2):4–10.

Schoener, A. & T.W. Schoener. 1984. Experiments on floatation of insular anoles, with a review of similar abilities in other terrestrial animals. *Oecologia* 63:289–294.

Shear, W. A. & S. B. Peck. 1987. Millipeds (Diplopoda) of the Galapagos Islands, Ecuador. *Can. J. Zool.* 65:2640–2645.

Silberglied, R. 1978. Inter-island transport of insects aboard ships in the Galapagos Islands. *Biol. Conserv.* 13:273–278.

Simberloff, D. S. 1978. Colonisation of islands by insects: immigration, extinction, and diversity. Pp. 139–153. *In:* L. A. Mound & N. Waloff (eds.), *Diversity of insect faunas. Symp. Royal Entomol. Soc. London* 9.

Simpson, B. B. 1974. Glacial migrations of plants: island biogeographical evidence. Science 185:698–700.

Smith, S. G. 1966. Land snails of the Galapagos. Pp. 240–251. *In:* R. I. Bowman (ed.), *The Galapagos: Proceedings of the symposia of the Galapagos International Scientific Project.* University of California Press: Berkeley.

Steadman, D.W. 1982. The origin of Darwin's Finches (Fringillidae, Passeriformes). *Trans. San Diego Soc. Natur. Hist.* 19:279–296.

Steadman, D.W. 1986. Holocene vertebrate fossils from Isla Floreana, Galapagos. *Smith. Contr. Zool.* 413:1–103.

Svenson, H. K. 1946. Vegetation of the coast of Ecuador and Peru and its relation to the Galapagos Islands. *Amer. J. Bot.* 33:394–498.

Thornton, I.W. B. 1967. The measurement of isolation on archipelagos, and its relation to insular faunal size and endemism. *Evolution* 21:842–849.

Thornton, I. 1971. *Darwin's Islands, a Natural History of the Galapagos.* Natural History Press: Garden City, NY. 322 pp.

Treherne, J. 1983. *The Galapagos Affair.* Jonathan Cape Ltd.: London. 269 pp.

Vagvolgyi, J. 1974. *Nesopupa galapagoensis*, a new Indo-element in the land snail fauna of the Galapagos Islands (Pulmanata; Vertiginidae). *The Nautilus* 88:86–89.

Vonnegut, K. 1985. *Galápagos, a Novel.* Dell Publishing, New York. 295 pp.

Wiggins, I. L. & D. M. Porter. 1971. *The flora of the Galapagos Islands.* Stanford University Press, Stanford, CA.

Zimmerman, E. C. 1948. *Insects of Hawaii. Introduction.* Vol. 1. University of Hawaii Press: Honolulu. 206 pp.

Evolution of the Endemic Vascular Flora of the Juan Fernandez Islands

EXTENDED ABSTRACT

Tod F. Stuessy, Daniel J. Crawford, and Mario Silva O.

The Juan Fernandez Islands are located 600 km west of continental Chile at 33° S latitude. Lying in an east-west orientation are the two principal islands: Masatierra, closer to the mainland, and Masafuera, located 150 km further west into the Pacific. Both islands are relatively small, 78 and 59 km², respectively. Radiometric dating has determined the geological ages of the islands to be 3.8–4.2 my for Masatierra and 1–2.4 my for Masafuera (Stuessy et al., 1987). Erosion has reduced the original area substantially, and probably also the number of ecological zones of the island. On Masatierra, perhaps as much as 80% of the original land mass has disappeared and broad valleys characterize the landscape. The younger island, Masafuera, has also suffered reduction in size, but less so, with a loss of only about 10% of the original island. The deep amphitheater-headed valleys (or quebradas) of Masafuera are typical for initial stages of young volcanic islands.

Within the relatively small geographic area of the Juan Fernandez archipelago are found 362 vascular plant species including 51 ferns, 66 monocots, and 245 dicots; 77 families and 213 genera occur here. Among these are one endemic family (Lactoridaceae), 12 endemic genera, and 127 endemic species, yielding 11% endemism at the generic level and 60% at the specific level. Of the endemic species 23 are ferns, 14 are monocots, and 90 are dicots. Of the endemic dicots, 29 species are Compositae; they make up 32% of the endemic dicot flora. Ninety percent of the endemic angiosperms are perennial and 64% of the dicots are woody, i.e., rosette-trees or trees. These endemic taxa occur in all of the major ecological zones in the islands, which include fern forests, dry forests, the alpine zone, open ridges and cliffs, dry open slopes, deep ravines (quebradas), and along the shore. The endemic species are particularly abundant in the dry forest and on the open ridges and cliffs.

Cytological surveys of chromosome numbers of endemic Juan Fernandez taxa have been completed to determine the degree of chromosome number change during speciation in the archipelago as well as to learn if polyploidy has accompanied establishment in the islands. Thirty-eight species have been surveyed from 69 populations (Sanders et al., 1983; Spooner et al., 1984; Sun et al., 1990) representing 42% of the endemic dicots (none of the endemic monocots has yet been counted successfully). These data reveal that 31% of the endemic species counted show neither ancient nor recent polyploidy, i.e., they show

Drs. Stuessy, Crawford, and Silva are with the Department of Plant Biology, Ohio State University, Columbus, OH 43210, USA, and Department of Botany, Universidad de Concepcion, Concepcion, Chile.

no obvious polyploidy from continental relatives nor are the relatives at a presumed polyploid level. About two-thirds of the flora, however, are probably pre-colonization polyploids, i.e., they and their ancestors are at chromosomal levels of $n = 12$ or higher. Only two species are perhaps recent polyploids: *Ugni selkïrkii* (Myrtaceae), and *Spergularia confertiflora* (Caryophyllaceae). Three species of *Wahlenbergia* (Campanulaceae) are perhaps of aneuploid origin with $n = 11$; generic relatives are apparently $n = 9$ (Sanders et al., 1983). Important points about these data are that no reticulate evolution (i.e., hybridization or polyploidy) has occurred in the evolution of the endemic species within the archipelago. Chromosomal change from progenitors to derivatives in the islands also seems to have been minimal. Furthermore, there is no evidence of intra- or interisland chromosomal change within particular lineages. These results coincide with those obtained recently from the Bonin Islands by M. Ono (in press) as well as those documented by Carr (1978, 1985) for the Hawaiian flora. Although rapid morphological evolutionary change is the rule in oceanic archipelagos, this is not usually accompanied by change in chromosome number. Why this should be is unclear, but it may relate to the non-adaptive nature of large-scale chromosomal alterations in the face of strong directional selection in rapidly changing environments.

Flavonoid chemical studies from leaves of taxa endemic in the islands have revealed contrasting evolutionary scenarios. In some genera of Juan Fernandez, such as *Robinsonia* and *Dendroseris* (both Compositae), the flavonoid profiles are relatively constant among and within species. In other genera, mainly *Erigeron* (Compositae), *Peperomia* (Piperaceae), and *Gunnera* (Gunneraceae), which are large and broadly distributed Neotropical genera, the variation within species is great. This type of dramatic variation is also found in *Bidens* (Compositae) of the Hawaiian Islands (Ganders et al., 1990), also a widespread genus. There appears to be a correlation with more flavonoid variation occurring in endemic species of widely distributed genera. In Juan Fernandez the endemic genera might have existed there longer; their more species-rich nature also suggests this. Possible explanations for these trends might be loss of populational variation through directional selection into more mature and narrow ecological settings (more recently evolving species would still be undergoing numerous experiments in adaptive radiation and retain much infraspecific variation). One might also expect the maintenance of considerable populational variation through much recombination in newly evolving environments of oceanic islands.

Studies on genetic variation at isozyme loci have been carried out in *Dendroseris* (Compositae; Crawford et al., 1987), *Chenopodium* (Chenopodiaceae; Crawford et al., 1987), and *Wahlenbergia* (Campanulaceae; Crawford et al., 1990). In the first two genera, very little variation exists among congeneric species. In *Wahlenbergia*, however, *W. berteroi* and *W. masafuerae* have high genetic identity (0.947). The third species, *W. fernandeziana*, has lowered identities with the other two taxa (0.68–0.77). Further, the variation found in *W. berteroi* and *W. masafuerae* is a subset of the variation found in *W. fernandeziana*. Based on all evidence, it is hypothesized that the restricted genetic variation in *W. berteroi* and *W. masafuerae* resulted from a founder effect as this evolutionary line diverged from the more primitive *W. fernandeziana*. The most important point from these studies is that within endemic taxa there is very little infraspecific and interspecific variation. Divergence in isozymes has clearly been less than with morphological traits. Changes in structural genes, therefore, apparently have lagged behind modifications in regulatory genes. It may also be that high levels of recombination in open environments help promote the appearance of new morphological forms. These results are similar to those obtained from the Hawaiian islands (for a review, see Crawford et al., 1987).

Our preliminary studies with chloroplast DNA show low levels of variation among species of *Dendroseris* (Compositae), the largest and most morphologically diverse genus of the archipelago. Current studies are focusing on the relationship of this group to mainland relatives such as *Hieracium* and *Hypochaeris*. Dr. Robert Jansen of the University of Connecticut is surveying all of the genera of the tribe *Lactuceae* to which *Dendroseris* belongs to help identify its outgroup. This will help reveal the different changes in chloroplast and nuclear DNA in *Dendroseris* within the archipelago.

The overall perspective on evolution of the vascular flora of the Juan Fernandez Islands is one of limited chromosomal and genetic variation, but marked morphological divergence in response to rapid ecological changes. Although there is no obvious indication of marked ecological zones now on Masatierra (i.e., the older island), studies of *Erigeron* on Masafuera (the younger island) reveal that ecological partitioning does appear to have occurred there (Valdebenito et al., in prep.) One can speculate that this might also have been true for Masatierra some 3 million years ago. Studies done in other island archipelagos, particularly the Hawaiian islands by Carr and associates (e.g., Carr & Kyhos, 1981, 1986), have revealed that the endemic species of Compositae can be crossed easily, even between related genera. All these facts suggest that evolution in oceanic archipelagoes has occurred with a minimum of genetic change, and as such these settings represent good opportunities for examining the genetic and developmental bases of speciation in flowering plants.

LITERATURE CITED

Carr, G. D. 1978. Chromosome numbers of Hawaiian flowering plants and the significance of cytology in selected taxa. *Amer. J. Bot.* 65:236–242.

Carr, G. D. 1985. Additional chromosome numbers of Hawaiian flowering plants. *Pacific Sci.* 39:303–306.

Carr, G. D. & D. W. Kyhos. 1981. Adaptive radiation in the Hawaiian silversword alliance (Compositae: Madiinae). I. Cytogenetics of spontaneous hybrids. *Evolution* 35:543–556.

Carr, G. D. & D. W. Kyhos. 1986. Adaptive radiation in the Hawaiian silversword alliance (Compositae: Madiinae). II. Cytogenetics of artificial and natural hybrids. *Evolution* 40:959–976.

Crawford, D. J., Stuessy, T. F., Lammers, T. G., Silva O., M., & P. Pacheco. 1990. Genetic differentiation and evolution of *Wahlenbergia* (Campanulaceae) in the Juan Fernandez Islands. *Bot. Gaz.* 151:119–124.

Crawford, D. J., Stuessy, T. F. & M. Silva O. 1987. Allozyme divergence and the evolution of *Dendroseris* (Compositae: Lactuceae) on the Juan Fernandez Islands. *Syst. Bot.* 12:435–443.

Crawford, D. J., Stuessy, T. F. & M. Silva O. 1988. Allozyme variation in *Chenopodium sanctae-clarae*, an endemic species of the Juan Fernandez Islands, Chile. *Biochem. Syst. Ecol.* 16:279–284.

Crawford, D. J., Whitkus, R. & T. F. Stuessy. 1987. Plant evolution and speciation on oceanic islands. Pp. 183–199. *In*: K. M. Urbanska (ed.), *Differentiation Patterns in Higher Plants*. Academic Press: London.

Ganders, F. R., Bohm, B. A. & S. P. McCormick. 1990. Flavonoid variation in Hawaiian *Bidens*. *Syst. Bot.* 15:231–239.

Ono, M. Endemism in the flora of the Bonin (Ogasawara) Islands with special reference to the mode of dispersal. *Aliso.* In press.

Sanders, R. W., Stuessy, T. F. & R. Rodriguez. 1983. Chromosome numbers from the flora of the Juan Fernandez Islands. *Amer. J. Bot.* 75:799–810.

Spooner, D. M., Stuessy, T. F., Crawford, D. J. & M. Silva O. 1987. Chromosome numbers from the flora of the Juan Fernandez Islands. II. *Rhodora* 89:351–356.

Stuessy, T. F., Sutter, J. F., Sanders, R. W. & M. Silva O. 1984. Botanical and geological significance of potassium-argon dates from the Juan Fernandez Islands. *Science* 25:49–51.

Sun, B. Y., Stuessy, T. F. & D. J. Crawford. 1990. Chromosome counts from the flora of the Juan Fernandez Islands, Chile. III. *Pacific Sci.* 44:258–264.

Patterns of Moth Speciation in the Indo-Australian Archipelago

J. D. Holloway

Abstract. A small sample of cladistic analyses of Lepidoptera groups is used to investigate biogeographic patterns in the Indo-Australian tropics. Component analysis and Brooks Parsimony analysis are used to assess whether there is a single pattern of area-relationships common to the cladistic analyses. The results are compared with previous cladistic and phenetic biogeographic analyses for various animal groups. It is suggested that, though many of these methodologies yield a result, treating such results as a cladogenic geological hypothesis will fall down in the face of contrary geological evidence.

An alternative, R-mode, approach is supported, directed at producing a classification of cladograms recognising sets of common pattern types. In the Indo-Australian area these are as likely to stimulate hypotheses of earth history. This point is illustrated with reference to Sulawesi.

The representation of Sulawesi in the sample of cladistic analyses for Lepidoptera is complex and various, suggestive of frequent episodes of dispersal through its history as a land area. This is explored by comparing an estimate of dispersability for a sample of Lepidoptera groups with the size of the Sulawesi fauna relative to that of Borneo and with the percentage endemism in the Sulawesi fauna. The results are consistent with a hypothesis of a predominantly dispersive origin for the fauna. Sulawesi generally has a richer fauna of more dispersive groups, but less dispersive ones show compensatory faunal enrichment through speciation within Sulawesi, and generally have a higher proportion of endemics.

Four genera are studied to explore this further. All are diverse in Sulawesi, one through sharing a high number of species with Sundaland, possibly of dispersive origin, and the other three through extensive speciation with Sulawesi. For all genera there is a clear zonation of species with altitude in Borneo. This persists to a large extent in Sulawesi for the dispersive group but it breaks down for the genera where speciation has occurred. All three endemic groups have a close relationship to lowland zone species in Sundaland but in themselves show no clear altitudinal segregation and span a much wider range of altitude than their Sundanian relatives.

Whilst the speciation pattern in the archipelago as a whole is predominantly allopatric, that within islands such as Sulawesi and New Guinea may involve non-allopatric speciation processes.

INTRODUCTION

Consideration of speciation process in archipelagos is not possible without an appreciation of pattern based on an adequate sample of natural groups of organisms for which a modern systematic analysis is available. For the Oriental and Australasian tropical archipelagos our sample is still woefully small. This precludes any wide-ranging assessment of pattern. In this paper I will add to the sample for the Lepidoptera, assess the types of pattern that appear to be emerging, with a consideration of the sorts of methodology that may enable us to progress further, and make some initial observations on differences in speciation process between and within islands. Amongst the Lepidoptera, I will focus on the subfamily Ennominae of the Geometridae.

Dr. Holloway is with the International Institute of Entomology, 56 Queen's Gate, London SW7 5JR, UK.

The Indonesian archipelago, and its extension into eastern Melanesia, owes its present complexity to the convergence of the Indian Ocean and Pacific tectonic plates on each other and on the spur of southeast Asia. Development of microplates has predominated in eastern Melanesia, with archipelagic convergence and shear in the New Guinea region, and accretion of island complexes against the Asian continent in the west. The island of Sulawesi, at the core of the archipelago, epitomises this complexity: it was probably formed in some isolation from surrounding lands through the fusion of Oriental and Australasian geological components, subjected to the strong left lateral shearing process induced by the motion of the Pacific plate against New Guinea and its neighbors. Reviews of the geology of the region may be found in Hamilton (1979, 1989), Holloway (1979, 1986b), Audley-Charles (1987), Ewart (1988) and Cracraft (in press).

The geological evidence indicates that processes of area increase, uplift, convergence, and fusion of archipelagos have been dominant over those of erosion, subsidence, divergence, and fragmentation. However, these latter processes are implicit in the purely biological concept of the sundering of a hypothetical Melanesian Foreland (e.g., Steenis, 1962; Corner 1975), developed to explain patterns of vicariant speciation in supposedly immobile organisms.

The ideal, possibly unattainable, methodology for biogeography should facilitate distinction between patterns originating largely in the mobility of organisms from those originating largely in the mobility of the land areas on which organisms live. The understanding of speciation process is inseparable from this problem. It has been discussed for the Indo-Australian tropics by, amongst others, Holloway (1982, 1984a, 1986a), Schuh & Stonedahl (1986), Duffels (1986) and Cracraft (in press).

The first step is, with the methods of cladistic biogeography (Humphries & Parenti, 1986; Wiley, 1988, Page; 1988), to address the possibility that there is one single underlying pattern of area relationships for the Indo-Australian archipelago, as sought by Schuh & Stonedahl (1986). An attempt was made to apply the methods of component analysis (Page, 1988, 1989; Humphries, 1990) to cladograms generated by myself and others for Lepidoptera groups that are largely allopatric at least in the eastern portion of their range, and thus might be expected to reveal common area relationships if they exist. Three new cladograms for ennomine geometrid groups used in this analysis are discussed in appendices. Brooks Parsimony Analysis was applied to the same data, as advocated by Wiley (1988). The results are compared with those obtained for different groups by Schuh & Stonedahl (1986), Vane-Wright (1990) and Cracraft (in press), and with the often criticised but also often used (e.g., Nelson & Platnick, 1981; Wiley, 1988; Page, 1989) phenetic analysis of Holloway & Jardine (1968).

Absence of a clear underlying pattern promotes the possibility that the raw cladograms (the original data) include much pattern developed through dispersal events. Even if pattern is evident, a dispersive explanation cannot necessarily be excluded, particularly if that pattern is indicative of an earth history hypothesis outweighed by a mass of contrary geological evidence (Cox, 1990). Such a pattern might arise through processes of dispersal and speciation common to the groups concerned interacting with a particular juxtaposition of land areas in time. Duffels (1986) referred to this as dispersive vicariance; it will lead to the sort of geographical snapshot pattern sought by Holloway (1982).

Indications that such pattern types can be categorised for the Indo-Australian tropics will be discussed in conjunction with a review of some evidence that, for at least the Lepidoptera, dispersal process has made a dominant contribution to biogeographic pattern development in the area, particularly with respect to Sulawesi. Particular attention will be paid to patterns that include components of speciation within Sulawesi. Evidence of species relationships in these complexes and from ecological transects hints at speciation processes different from those predominant in the archipelago as a whole.

342

THE CLADOGRAM SAMPLE

The cladograms available for Lepidoptera of the Indo-Australian tropics are few. Many have features that render them uninformative for biogeographic analysis, such as poor cladistic resolution, a large number of uninformative widespread taxa coincident over all or part of the range of the genus, or a high degree of localised sympatry. Many of the cladograms I presented or reproduced in a study of Melanesian distributions (Holloway, 1984a) fall into one or other of these categories.

I have selected for analysis those which display reasonable cladistic structure, a high degree of endemism or restriction to a few areas by the species, and total ranges that are

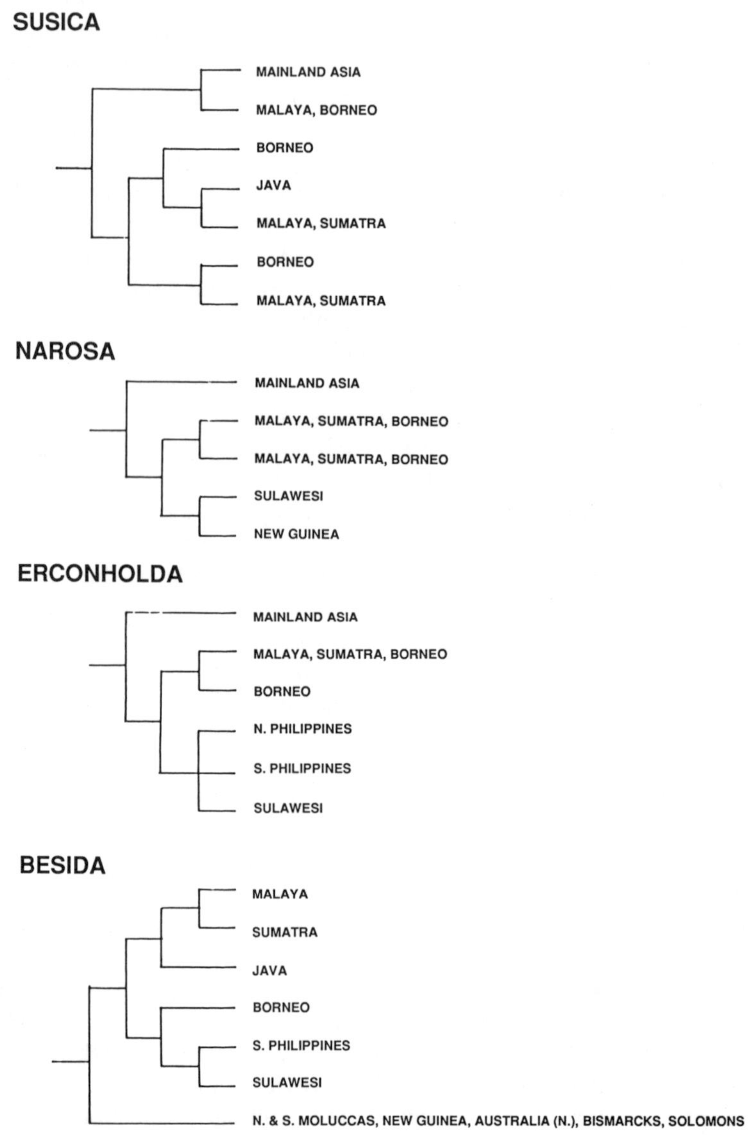

Figure 1. Area cladograms for moth groups used in analyses of general area patterns: *Susica* (Limacodidae) from Holloway (1982) and modified according to Holloway (1986a); *Narosa* (Limacodidae), simplified after Holloway (1987; see also Fig. 15), with additional records for Sundanian species; *Phalera* subgenus *Erconholda* (Notodontidae), from Holloway (1987); *Besida* (Notodontidae) with monotypic Australasian sister-genus *Ortholomia*, from Holloway (1987).

widely Oriental, show discrimination either of Sundanian land areas (Malaya, Sumatra, Java, Borneo) or extensions with endemism east to the Australasian tropics (Figs. 1–3). A number of these (for the limacodid genus *Narosa* and the notodontids *Besida* and *Erconholda*) were presented in my paper at the predecessor of this symposium (Holloway, 1987). Two further limacodid genera are included: *Darna* (Holloway 1986a) and *Susica* (Holloway, 1982). The butterfly genus *Idea*, used by both Vane-Wright (1990) and Humphries (1990) in analyses of general area relationships, also seemed very suitable for inclusion.

Three geometrid genera in the Ennominae were analysed specifically for this paper. *Zeheba* and *Probithia* were chosen because they spanned the region, showing some degree of allopatry with endemism, yet were not highly species rich. Unfortunately they showed a high degree of morphological uniformity such that cladistic analyses on extensive suites of characters were not possible. Justification for the structure in the cladograms for these genera in Figure 4 will be found in Appendix 1.

The third genus, *Bracca*, was marked as potentially interesting by Holloway (1984a) and proved to be morphologically rich and tractable. Therefore a full cladistic analysis was performed using J.S. Farris' Hennig86 programme on thirty-five characters of wing markings and male and female genitalia. The analysis is presented in detail in Appendix 2. The strict consensus tree for the 112 trees resulting from the branch and bound option is presented in Figure 5. It will be discussed in more detail later.

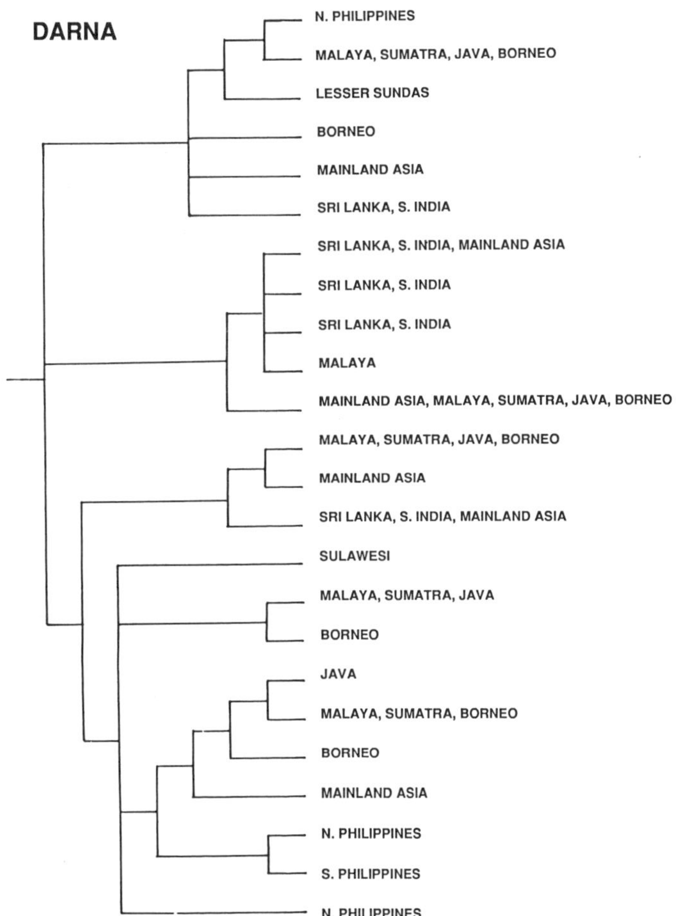

Figure 2. Area cladogram for *Darna* (Limacodidae) from Holloway (1986a) with some additional species and area records.

344

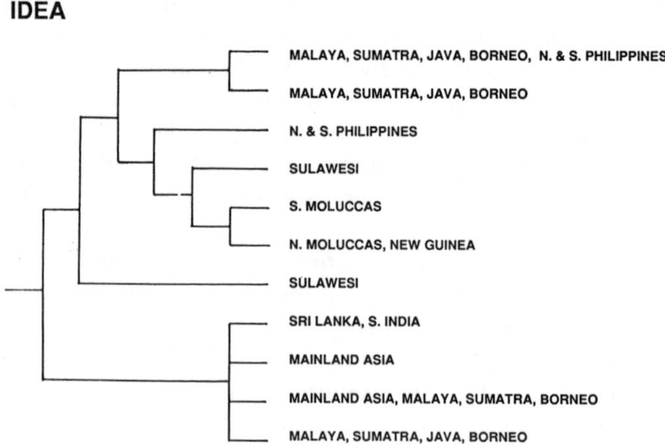

Figure 3. Area cladogram for *Idea* (Nymphalidae: Danainae) from Vane-Wright (1990).

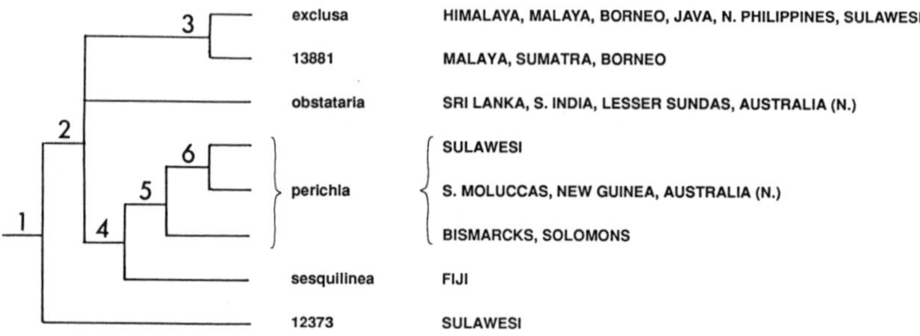

Figure 4. Cladograms for *Zeheba* and *Probithia* (Geometridae: Ennominae). Justification for the clades as numbered is in Appendix 1. Note: where numbers occur instead of species names in the text or figures, these refer to British Museum genitalia preparations for undescribed species.

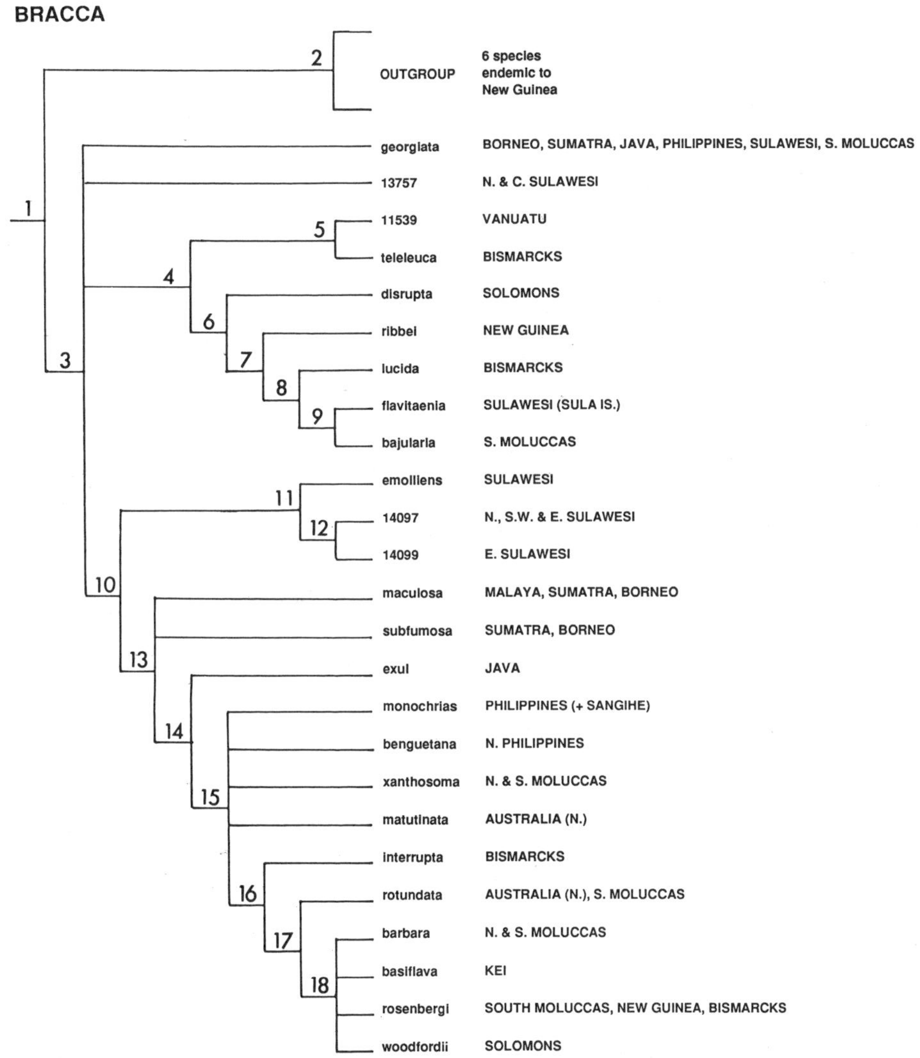

Figure 5. Cladogram for the genus *Bracca* (Geometridae: Ennominae), derived as described in the text and Appendix 2 (in which the numbered clades are discussed).

AREA ANALYSIS

Areas Incorporated

Areas were selected in a manner so as to incorporate as much information from the Lepidoptera cladograms as possible yet permit comparison with area cladograms already in the literature. Sri Lanka and southern India are combined, as all species occur in both, though they were treated separately by Schuh & Stonedahl (1986). Mainland Asia embraces northern India and South-east Asia excluding Malaya and southern Thailand. The Philippines are divided on a north/south basis, again to facilitate comparison with Schuh & Stonedahl and because most records to hand were from either Luzon or Mindanao. Sulawesi includes the Sula Islands but not Sangihe. The division between the north and south Moluccas is south of Obi; the islands of Kei and Tenimber are excluded. The Solomons are treated in a geographical sense including Bougainville Island.

346

Component Analysis

An attempt was made (by Chris Humphries) to run the cladograms through Page's (1988, 1989) Component program but a number of difficulties were encountered. The number of areas involved and the complexity of some of the cladograms require a large computer memory. This is compounded by the incomplete resolution of many of the cladograms. For example, the *Bracca* cladogram could be resolved to give any of 14,175 trees and that for *Darna* 10,125 trees. Though *Probithia* could be fully resolved in only three different ways (the trichotomy of clade 2), the widespread taxa treated under Assumption 2 yield well over a thousand possible trees.

Humphries (1990) performed an analysis of the *Idea* and *Narosa* trees which gave no resolution under strict consensus but much more for majority or Nelson consensus. The result associates the Philippines and Sulawesi in a clade with Australasian tropical areas that have a sister-relationship to a Malaya + Sumatra and Borneo clade.

Humphries has suggested (pers. comm.) that the complexity of the analysis could be reduced by attempting to recognise groups of areas (such as those of Sundaland) that would be treated as a unit in an analysis of broader geographic scope, or by addressing some specific questions such as the relationships of Sulawesi to surrounding areas of endemism.

Attempts to apply component analysis to these cladograms under Assumption 2 should be pursued. This is the methodology considered most likely to extract any underlying pattern of area relationships, so any failure will be of significance. But I would venture one suggestion on Assumption 2. Data on taxonomic relationships of organisms are used to produce hypotheses of biogeographic relationship for areas. Page (1989) argued that these concepts should not be confused when applying Assumption 2. He stated that Assumption 2 makes no claims about the relationships of taxa, countering the argument that widespread taxa, where defined by autapomorphies (probably the majority), should be treated under Assumption 0 (as they are in the coding for Brooks Parsimony Analysis). It would seem reasonable however, prior to application of Assumption 2, to rework the original taxonomic analyses treating as potentially distinct all "widespread taxa" in each area of endemism recognised. This was not done for this paper.

Brooks Parsimony Analysis

The Lepidoptera cladograms were coded for Brooks Parsimony Analysis (Wiley, 1988; Craw, 1989) in two ways. In both, the representation of each clade of a cladogram among the areas represented was scored as 1 (apomorphic); endemics (autapomorphies for areas) were omitted as they would not affect the result, but terminal taxa represented in more than one area were scored as apomorphic for those areas. Thus, for *Probithia* (Fig. 4), the *perichla* clade would yield characters with apomorphies: Bismarcks + Solomons; S. Moluccas; S. Moluccas + New Guinea + Australia + Sulawesi + Bismarcks + Solomons. The areas Fiji and Vanuatu were excluded as they only occur once in any of the cladograms.

The difference in the two codings was in the way areas unrepresented on a complete cladogram were coded. In one analysis they were coded as missing data; in the other they were coded as zero (plesiomorphic). The first method is that used by Craw (1989); Wiley (1988) differed in coding as zero all areas in which terminal taxa were not represented, yet, for the rest of the cladogram, areas not in the total range were coded as missing data. Coding areas outside the total range of a cladogram as missing data acknowledges the possibilities that the taxon concerned (rather than any sister-group or outgroup) could have been represented in these areas and has gone extinct, or is represented but undiscovered. This approach has an underlying assumption of potential cosmopolitanism for all whole cladograms.

Coding areas without representation as zero makes the assumption that the ingroup was never represented in these areas, i.e., the observed current range of the ingroup is the ancestral range; representation in other areas is by a hypothetical sister-group, or a set of paraphyletic outgroups. This might be the more consistent approach given that there may be missing data in the form of extinctions and undiscovered taxa within the ingroup, where zeros are always used in the first approach.

The problem of undersampling and the problem of extinction could be paralleled by those of character loss, reversal, and genetic masking in systematics, but perhaps here is an instance where the analogy between cladistic classifications of organisms and cladistic classification of areas could be overstretched.

Both sets of data were analysed using Hennig86 with both unweighted and successive approximations character weighting options. Weighting is likely to give stress to repeated patterns or generalised tracks and might thus provide a more reliable hypothesis of any underlying vicariant area-relationships. The trees were rooted by use of a hypothetical, fully zero-coded outgroup. The results are illustrated in Figures 6 and 7.

The unweighted branch and bound analysis for the 'missing data' option gave four trees of length 161, a Consistency Index of 62, and a Retention Index of 74. The strict consensus tree is illustrated in Figure 6, the four trees varying only in the relationships of Sumatra, Malaya and Borneo, and the relative positions of the S. Moluccas and New Guinea.

The weighting option gives a single tree of length 742, a Consistency Index of 85, and a Retention Index of 91. There has been resolution of Sumatra, Malaya, and Borneo, and considerable reshuffling of the Australasian areas (Moluccas eastwards), but otherwise the structure is similar. Instability amongst the Australasian areas might be expected, as the sample includes far fewer taxa for these areas, and therefore they are often scored with missing data.

The unweighted branch and bound analysis strict consensus tree (of four trees) for the 'zero coded' option and the weighted analysis unique tree are shown in Figure 7. The first, with a length of 191, Consistency Index of 52, and Retention Index of 71, shows greater

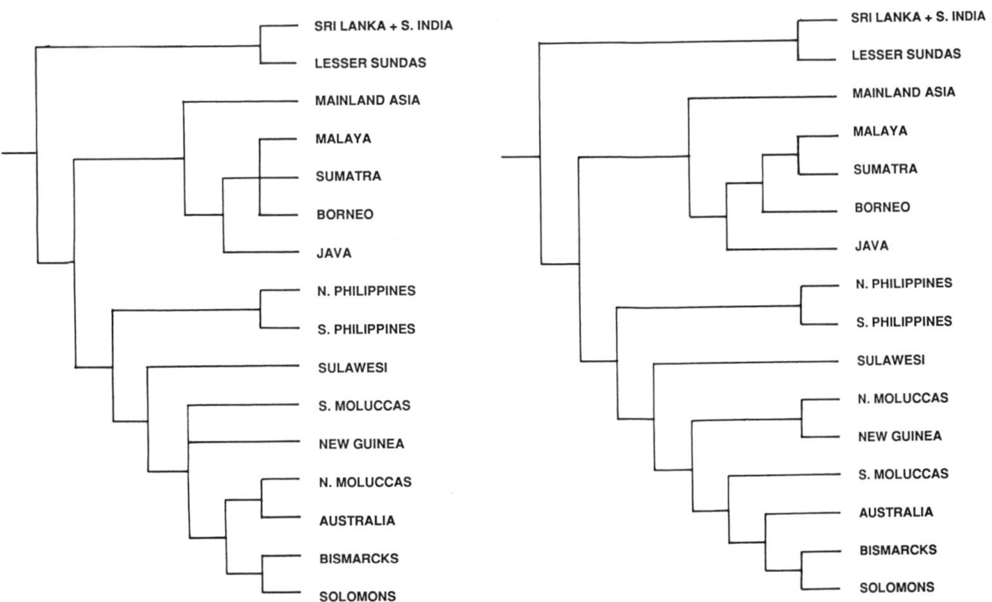

Figure 6. General area cladogram from Brooks Parsimony Analysis coded with the missing data option: unweighted (left) and weighted (right) as discussed in the text.

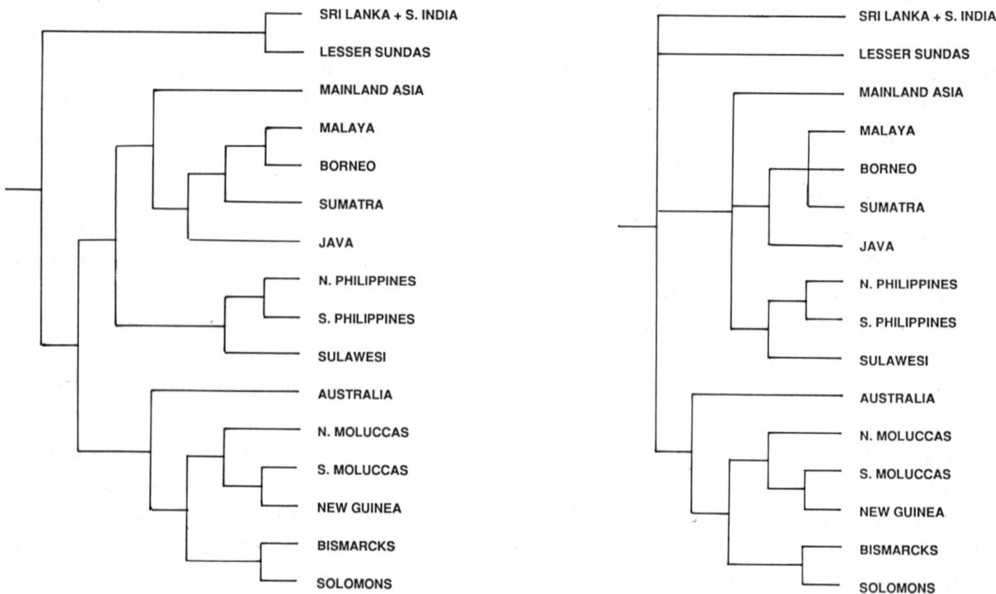

Figure 7. General area cladogram from Brooks Parsimony Analysis coded with the zero option: unweighted (left) and weighted (right) as discussed in the text.

resolution than the second (length 612, Consistency Index 80 and Retention Index 92) but essentially similar structure. Major differences from the missing data option are seen in the Philippines/Sulawesi sister relationship and its association with Oriental rather than Australasian areas, and the segregation of Australia in the latter.

Differences between the results of the two coding options are perhaps predictable if one considers the effects on the Philippines plus Sulawesi. The sample of cladograms includes those where one or both of these areas is included in some sort of sister-relationship with Australasian areas (*Idea, Bracca, Narosa, Probithia*), and those where the sister-relationship is with Oriental areas (*Darna, Besida, Erconholda*). In *Besida* the sister-group (the monotypic genus *Ortholomia*) is actually Australasian, but the other two groups lack relatives in those areas (*Darna*) or have only distant relationships (*Erconholda* is a sub-genus of *Phalera*; the *amboinae* group is the only representation of *Phalera* in Australasia and typifies the 'widespread allopatric' distribution pattern illustrated in Figure 9). Coding for missing data makes implicit the possibility that *Darna* and *Erconholda* of the second three cladograms have missing taxa in Australasia that are sister to those in Sulawesi. The zero coding effectively categorises them as equivalent to *Besida*. Association with Oriental areas is reinforced by data from widespread taxa in *Idea*, the upper section of the *Darna* clado-gram, and in *Zeheba* and *Probithia*.

Wiley's (1988) development of the method goes on to relate cladograms in the raw data to the parsimony analysis cladogram, interpreting differences in each to events of dispersal, extinction, etc. I will not attempt to do this here as it is obvious it would be a lengthy and tedious process. If the method provides a valid approach to deriving under-lying area relationships for an archipelago, then it is evident that, for Lepidoptera of the Indo-Australian tropics, repetitive events of dispersal followed by vicariance have been numerous and complex.

These Brooks Parsimony results can be compared with other analyses of area relation-ships in the Indo-Australian tropics. These are illustrated in Figures 8–10. The analysis by Schuh & Stonedahl (1986) on Hemiptera groups, using cladistic biogeographic methods,

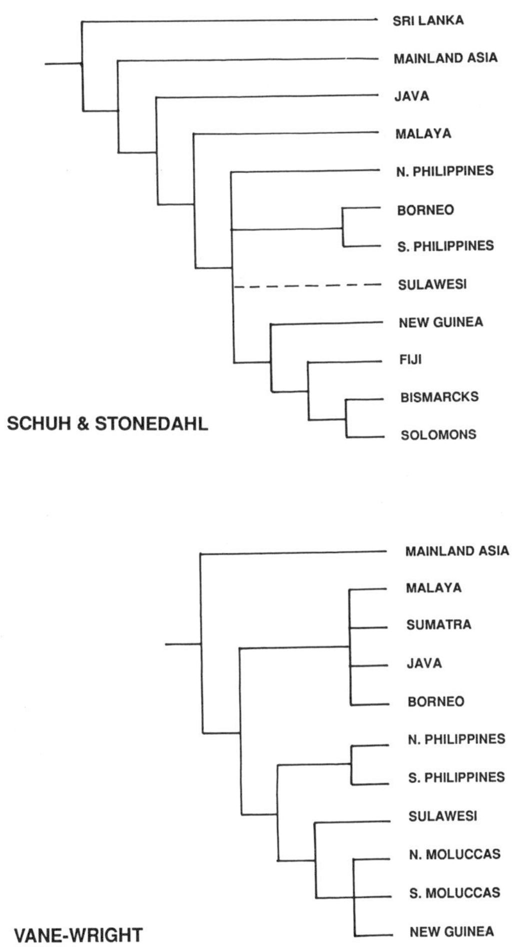

Figure 8. General area cladograms for Hemiptera (Schuh & Stonedahl (1986), somewhat simplified) and butterflies (Vane-Wright, 1990).

and the more intuitive general area cladogram presented by Vane-Wright (1990) for butterfly groups have much structural similarity with the 'missing data' option results described above. But both place mainland Asia in an outlying position relative to Sundaland, Wallacea, and Australasia. Sundaland is not delimited as a unit in the Schuh and Stonedahl analysis, but is by Vane-Wright. The cladogram for birds by Cracraft (in press) has mainland Asia as sister to the Sunda group, with Java as outlying area in the latter, but resolution of the other three Sundanian areas is different. Sulawesi and the Philippines appear in a sister-relationship, though this analysis excludes most of the Australasian areas. The bird result has, as Cracraft noted, similarities with the phenetic results of Holloway & Jardine (1968), that for butterflies, somewhat simplified, being illustrated in Figure 10. The resolution within the terminal triplet of Sundaland areas is different from that of Cracraft, matching that obtained in the weighted analysis, but the rest of the structure is the same (though including the Lesser Sundas).

The unweighted zero-coded result is even more similar to the Holloway & Jardine phenetic analysis, differences being in the position of the Lesser Sundas (with only sparse characterisation in this analysis—three taxa present) in the Malaya/Sumatra/Borneo grouping (the Cracraft bird result is the only other possible arrangement) and in the relationships between New Guinea and the Moluccas.

350

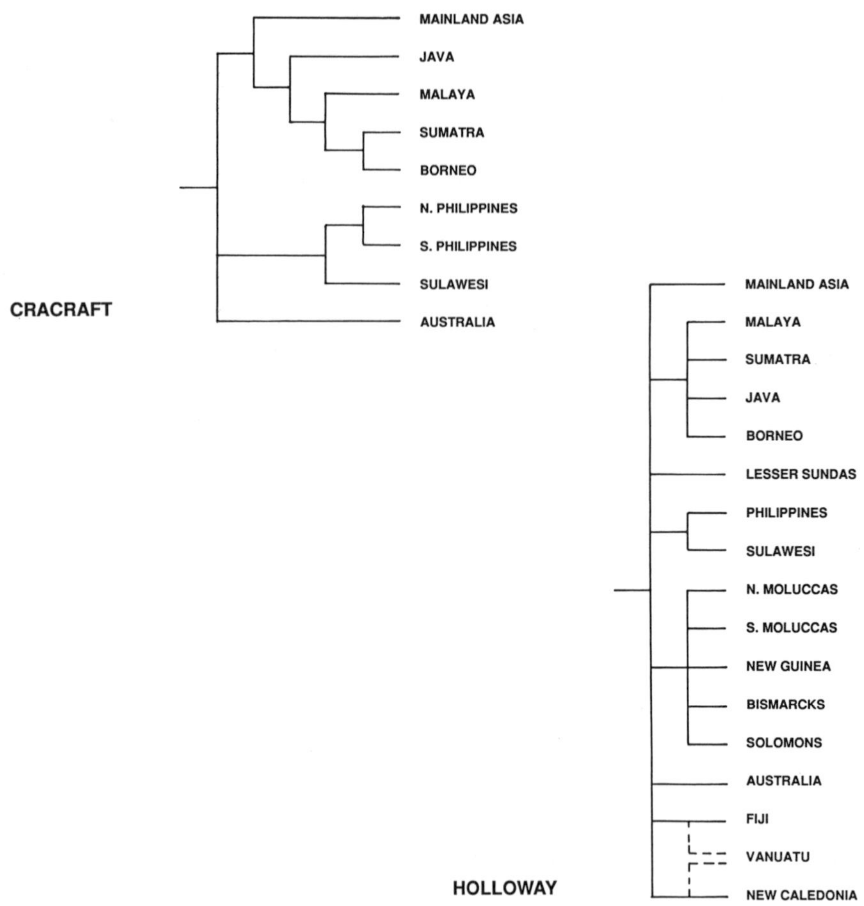

Figure 9. General area cladograms for birds (Cracraft, in press) and widespread allopatric groups of Lepidoptera species (Holloway, 1982, 1987).

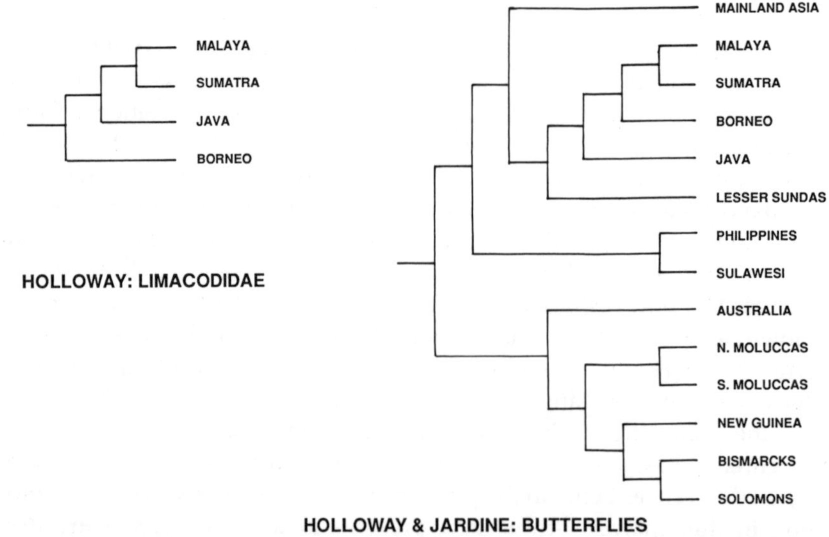

Figure 10. General area cladogram for Sundanian Limacodidae groups from Holloway (1986a, 1987), and phenogram (simplified) for butterflies from Holloway & Jardine (1968).

Yet another of the possible arrangements of the Sundaland quartet was indicated for limacodid moths by Holloway (1986a, 1987). The final example (Fig. 9) is a diagram (Holloway, 1987) that illustrates the common plan in a series of monophyletic, morphologically homogeneous groupings of largely allopatric species of butterflies and moths that span the whole area. It shows low resolution but is compatible with most of the other general area cladograms in its groupings of areas.

Discussion

All these analyses were performed on limited sets of data (except the phenetic one) and using a range of methodology. Yet they have numerous features in common. It is possible, therefore, that increasing the data set in each case will render the results even more convergent, though the application of component analysis needs to be pursued further.

If such an underlying common pattern of area relationships does emerge even from the most rigorous methods such as component analysis, does this represent an hypothesis of earth history as Schuh & Stonedahl maintained, or does it represent a common demoninator response of dispersal and speciation process to the present or past juxtaposition of land areas? Is the lack of resolution in the diagram of Figure 9 a product of taxonomic intractability of morphologically homogeneous groups, or can it be interpreted as resulting from an event of rapid dispersal followed by vicariance—in contrast, say, to a slower island to island process of dispersal from west to east, with concurrent vicariance and some subsequent dispersal as suggested by Clade 13 in *Bracca*?

The position of Sulawesi in both the general and individual cladograms is highly unstable. It can occur within the Sundanian grouping of areas (*Besida*, with S. Philippines; position of latter is similar in Schuh & Stonedahl), as a basal branch on its own (*Probithia*, *Bracca* clade 11, *Darna*, *Idea*), in an independent sister-relationship with the Philippines (*Erconholda*, Cracraft, Holloway, Holloway & Jardine, Brooks Parsimony zero coding) and in a sister-relationship to Australasian areas (*Narosa*, *Idea*, *Probithia*, *Bracca* clade 6, Brooks Parsimony 'missing data', Vane-Wright; also Humphries (1990)). It features in two of these positions in *Idea*, *Probithia*, and *Bracca*, and also features in the distributions of a number of widespread species (in *Bracca*, *Probithia*, and *Zeheba*).

The hypothesis that it is a biologically composite area (e.g., Craw & Weston (1984) citing Croizat) looks insecure when one considers that floral and faunal composition is mostly uniform over the island (Balgooy, 1987; Holloway, 1987), and that groups that have speciated within the island show local endemism patterns that straddle the main geological components and are almost entirely of Oriental affinity (Holloway, 1987; Duffels, 1990; Butlin & Monk, 1990; Polhemus & Polhemus, 1990). Examples from the Lepidoptera are given in the second half of this paper. An alternative explanation is that Sulawesi has never been sundered as a land area from a sister-area (though sea-floor spreading is thought to have opened the Makassar Strait between Sulawesi and Borneo in the mid-Tertiary (Hamilton, 1989)), but has interacted through dispersal events with neighbouring areas (rarely only with Borneo), though the juxtaposition of these has probably changed through time. The likelihood of an origin through dispersal for at least a major part of the Sulawesi fauna will be discussed in the next section.

The pattern of general area relationships emerging, if confirmed by further study, does not appear to be contributing much to our knowledge of earth history in the area concerned. The crescendo nature of that history may be one reason. We may need a more subtle definition of our areas of endemism, such as in the Philippines, Sulawesi, and New Guinea. But the general patterns, certainly for the Lepidoptera cladograms analysed here, represent a significant simplification of the raw data—a lowest common denominator that may have but little to offer over a phenetic analysis. Indeed, a phenetic approach includes the option of applying non-metric, multidimensional scaling techniques that may lead to results relevant to earth history (Holloway & Jardine, 1968).

Increasing the sample of cladograms may not provide a clearer picture of biotic area relationships in relation to earth history, but it may permit analyses of a somewhat different nature directed at classifying cladograms rather than areas, an R-mode rather than a Q-mode classification (Connor, 1988). The methodology and statistics have yet to be developed but the arguments of Simberloff (1987) and Page (1988) may provide a starting point. A first approximation for such a classification may lie in the sort of phenetic R-mode analysis of generic distributions and centers of species richness (or massing centers) proposed by Holloway (1969). But ultimately the sort of classification of trees based on some pairwise distance measure as envisaged by Page must be attempted. Then, as Page stated, distinct clusters of area cladograms may imply there is evidence for different sets of area relationships. These may reflect primarily gross tectonic processes or common pathways of dispersal with speciation on a template of changing geography through time.

A subjective assessment of the Lepidoptera cladograms presented here and previously (e.g., Holloway, 1982, 1984a) suggests a number of patterns may be shown to be prevalent. The first pattern is the simple widespread array of largely allopatric, morphologically very similar species, variations on the theme of the Lepidoptera cladogram in Figure 9. I have interpreted this pattern in the past as resulting from vicariance following rapid dispersal of an ancestral taxon through the archipelago. In many instances there are later dispersal events, usually by taxa on the larger land areas and in a predominantly easterly direction (e.g., the complexes in the noctuid genera *Aplotelia*, *Paectes*, and *Aegilia* mapped by Holloway (1985)). The Indian/Lesser Sunda sister-relationship seen in *Probithia* and *Zeheba* here may represent a parallel pattern, but of seasonal tropical rather than humid tropical habitats. It may also involve Australia.

Largely allopatric Australasian tropical groupings that are more complex with higher endemism are seen in clades within *Avitta*, *Polyura*, *Graphium* and *Parantica*, (illustrated by Holloway (1984a)), but with various sister-relationships. In the case of *Avitta* the sister clade is a largely allopatric Oriental one. Broader relationships often include clades with few (*Avitta*) or many (*Polyura*, *Graphium*) Oriental taxa that are mostly widely distributed and showing a high degree of sympatry. More complex Australasian patterns may reflect past geography in the form of two distinct Melanesian island arcs. These are shown most strongly by cicada groups discussed by Duffels (e.g., 1986) but may also be apparent for some Lepidoptera (Holloway, 1984a). For example, the *Damias* complex of genera in the Arctiidae includes a putative two-arc pattern, an allopatric Indo-Australian array, and an Oriental group.

In the genus *Bracca*, analysed here, there is an endemic New Guinea (inner arc) group, an outer arc group (clade 4) that does not bear a direct sister-relationship to the inner arc one, and a complex clade where the basal branches are Oriental with a triplet of species endemic to Sulawesi, and the more distal ones localised in areas of tropical Australasia. This could be represented as a more elaborate progression of the situation in *Idea* and *Narosa*. Allopatry is high in the terminal clades but sympatry increases on progression down the stem. Yet the two complexes endemic to New Guinea and Sulawesi respectively show a high degree of sympatry.

This would appear to be general to the groups studied here: sympatry is greater in clades endemic to currently continuous areas or those, such as Sundaland with mainland Asia, relatively recently continuous (Pleistocene sea-level fall).

In the case of the group of *Bracca* restricted to New Guinea, there is a high degree of local sympatry with four of the six species recorded from Mt. Goliath and two of these also known from the Mambare River area (where the two remaining species also occur)—localities well separated in the island. A similar situation is seen for the inner arc groups of the cicadas studied by Duffels (1983: Fig. 4) though some of the subgroups contain largely allopatric taxa. Consideration of speciation process in New Guinea is of great importance as biotic diversity in this geologically young island has developed to a degree that rivals or surpasses that of older, more tectonically stable areas. This prompts the question of whether, given allopatric speciation is probably the mode in the development of inter-island patterns, does this also hold in intra-island patterns? The second part of this paper will address this question with respect to Sulawesi, with further comment on the hypothesis that its biota is mostly derived by dispersal across marine barriers.

THE ENRICHMENT OF THE LEPIDOPTERA FAUNA OF SULAWESI

There are a number of pointers to a derivation by dispersal of a high proportion, if not all, of the Lepidoptera of Sulawesi. The first is that, despite the proximity of Sulawesi to Borneo and suggestions of geological union at some time in the past, faunistic and floristic links are much weaker between these areas than between Sulawesi and lands to the north, south and east (Balgooy, 1987; Holloway, 1987). The only exclusive sister-relationship between Sulawesi and Borneo noted so far for Lepidoptera is in the butterfly genus *Eurema* (Yata, 1990), suggested to be a recent dispersal event. Yet the insect fauna is considered to be predominantly Oriental in nature (papers in Knight & Holloway, 1990). The 'filtration' effect of the Makassar Strait is evident in some of the ennomine geometrid groups discussed later.

The second pointer is the somewhat mercurial nature of the island in its appearance on area cladograms. It has no clearly predominant area relationships but associates most frequently with the Philippines or Moluccas, with some association with the Lesser Sundas and Java also evident (Balgooy, 1987; Vane-Wright, 1990). Some endemic members of the fauna represent basal branches of their generic cladograms (e.g., in *Idea*). At the other extreme are more widespread species shared mainly with Oriental areas. A full exploration of the Brooks Parsimony analyses presented earlier along the lines of Wiley (1988) would no doubt involve many episodes of dispersal to Sulawesi.

I have suggested elsewhere (Holloway, 1987, 1990a) that, for various taxonomic groups, the proportions of endemics in Sulawesi and the size of the fauna relative to that of Borneo (as an exemplar of the probable South-east Asian source fauna for Sulawesi) bear a relationship to the dispersability of the groups concerned that supports an hypothesis of origin through dispersal rather than vicariance, or at least that dispersive origins predominate.

Means of assessing the dispersability of a group are much less satisfactory in this situation than they were for Norfolk Island (Holloway 1977, 1990b, and references cited above). The measure used here is the proportion of widespread species, defined as those spanning both of the Lines of Wallace and Weber (identified as major discontinuities in butterfly distribution by Holloway & Jardine, 1968), in the Bornean fauna for each group (or the Malayan fauna for butterflies). An alternative measure might be an assessment of the mean number of areas in the ranges of Bornean members of a group, but the widespread proportion offers a ready-reckoner.

Moth groups were selected where data were readily available in 'The Moths of Borneo' series (e.g., Holloway, 1986a). Malayan statistics were used for butterflies on the assumption that the Bornean and Malayan faunas are approximately comparable, as these are readily available in Corbet & Pendlebury (1978). Figures for Sulawesi are derived mainly from work in hand on material collected during Project Wallace (Holloway, Robinson & Tuck, 1990; R. I. Vane-Wright, pers. comm.).

The data are tabulated in Table 1 and plotted in Figures 11 and 12. The plot of the size of the Sulawesi fauna relative to that of Borneo against the dispersability index (Fig. 11) indicates that groups in Sulawesi generally show greater parity with Borneo if they have a high dispersability index. Exceptions (e.g., Danainae, Pieridae, Arctiinae) are those where there has been some endemic speciation, although the Papilionidae are anomalous in this.

The trend indicates that good dispersers have enriched Sulawesi more rapidly than poor dispersers, but that the latter have often acquired compensatory richness by speciating within Sulawesi.

This is supported by Figure 12, which indicates that the poor dispersers have a much higher level of endemism than the good ones; this endemism tends to be enhanced by speciation within Sulawesi. An hypothesis of vicariant development of the Sulawesi biota, where dispersability would be irrelevant, might lead to the expectation of a more uniform level of comparison between the two biotas and of speciation events within Sulawesi.

The contrast between more and less dispersive groups can be perceived at a generic level. Four genera that are diverse in Sulawesi are illustrated in Figures 13–16, showing the distribution with altitude of species in Borneo, Sulawesi and, where relevant, Seram. The first, *Luxiaria*, represents the dispersive element of the fauna; the other three attained their diversity mainly or entirely through speciation within Sulawesi. The data on altitude are from transect surveys by the author (Holloway, 1970, 1984b, 1989, in press; Holloway, Robinson & Tuck, 1990) augmented by data from material in the British Museum (Natural History).

The transect surveys have indicated a zonation pattern common to all three areas: a lowland forest fauna extending up to, and overlapping, a distinct lower montane fauna at about 900-1200 m, with an upper montane zone above that. Lepidoptera diversity peaks at between 600 m and 1000 m as a whole and for many individual taxonomic groups with a

Table 1. Comparative statistics for various Lepidoptera family groups (including some subfamilies and tribes) in the faunas of Malaya, Borneo, and Sulawesi as discussed in the text. The danaine fauna of Sulawesi is in fact greater than that of Borneo (29; Vane-Wright, 1990).

Family	Species in Borneo or Malaya*	% wides in Borneo or Malaya	Species in Sulawesi	$\dfrac{\text{Sul} \times 100}{\text{Bor/Mal}}$	% endemic species in Sulawesi	Genera with endemic speciation
Papilionidae	46*	9	32	70	47	
Pieridae	47*	26	45	96	56	Delias, Cepora
Nymphalidae	278*	9	155	56	54	Moduza, Lohora, Parantica
Danainae	36*	19	34	92	43	Parantica
Satyrinae + Amathusiinae	85*	6	47	55	70	Lohora
Lycaenidae	403*	11	154	38	39	
Hesperiidae	255*	7	82	32	22	
Limacodidae	100	2	32	32	75	Narosa, Parasa Griseothosea
Notodontidae	124	3	52	42	52	Higena, Ambadra
Saturniidae	22	9	12	55	75	Antheraea
Lasiocampidae	62	0	12	19	100	Trabala
Bombycidae	15	0	8	53	75	
Zeuzerinae	22	41	15	68	13	
Smerinthini	25	4	11	44	55	
Sphingini	9	67	8	89	0	
Macroglossinae	60	42	41	68	12	
Euteliinae	73	33	34	47	9	
Stictopterinae	89	31	42	47	5	
Plusiinae	16	56	16	100	6	
Aganainae	16	44	13	81	31	
Arctiinae	39	13	33	85	52	Utetheisa, Lemyra, Spilosoma, Satara

range of trophic specialisations (Holloway, Robinson & Tuck, 1990). This humped diversity profile is common to the three areas surveyed.

An analysis of species associations for Borneo (Holloway, 1989) revealed a clear distinction of a lowland association, with a number of components specific to forest variants, a lower montane association, and an upper montane one. Similar analyses for Sulawesi and Seram have yet to be performed, but indications are that such associations will be less clearly defined. This is evident to some extent in the genera examined here.

In the ennomine geometrid genus *Luxiaria* (Fig. 13) five out of ten species are shared with Borneo, which has 12 species. Of the rest, one belongs to the subgenus *Euippe* where the species belong to a widespread allopatric array from Sundaland to the Solomons. The *submonstrata* complex is similarly widespread but includes a distinct subspecies in Sulawesi and a distinct species, *L. paganata*. The species 12483 and 12484 in Seram are members of the group, and further taxa occur in New Guinea. Also in Seram is a member of the *Euippe* group and *L. subrasata*, found also in Borneo and Sulawesi.

Zonation is readily apparent in Borneo, though *subrasata* is the only definite lower montane species. Zonation is less clear though still evident in Sulawesi, but appears to have broken down in Seram. Bornean species segregate into comparable zones on Sulawesi.

The only host records located are from the Melastomataceae (*Melastoma, Memecylon*).

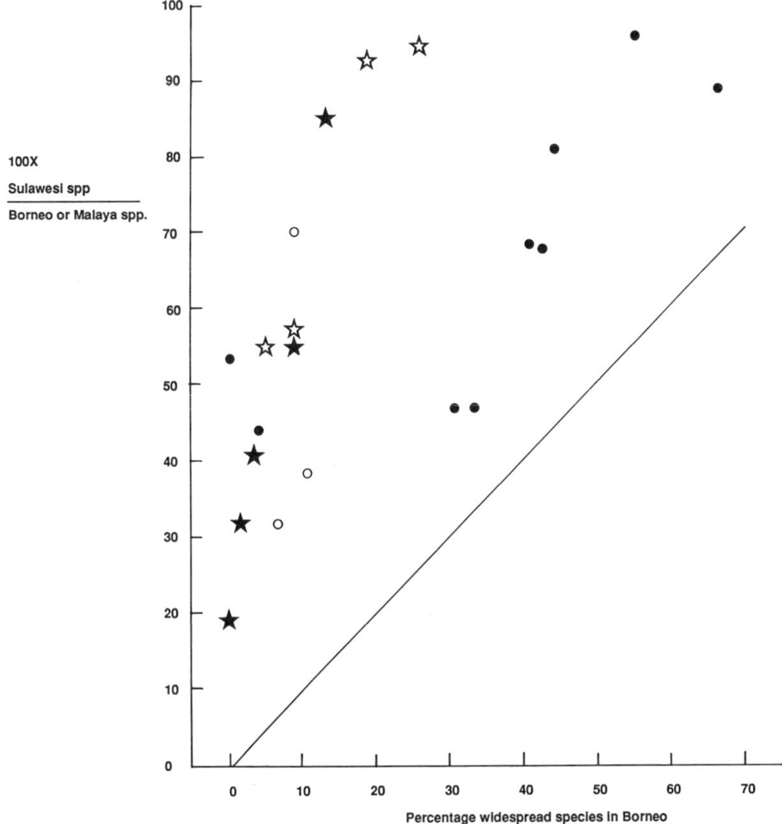

Figure 11. Size of the Sulawesi fauna compared with that of Borneo or Malaya for various Lepidoptera groups plotted against percentage of widespread taxa in those groups in the faunas of Borneo or Malaya. The oblique line indicates the expected fauna if only all the widespread species are represented in Sulawesi. Stars indicate groups where speciation within Sulawesi is known to occur. Open symbols indicate butterfly groups (where figures for Malaya are used). Solid ones indicate moth groups. Data from Table 1.

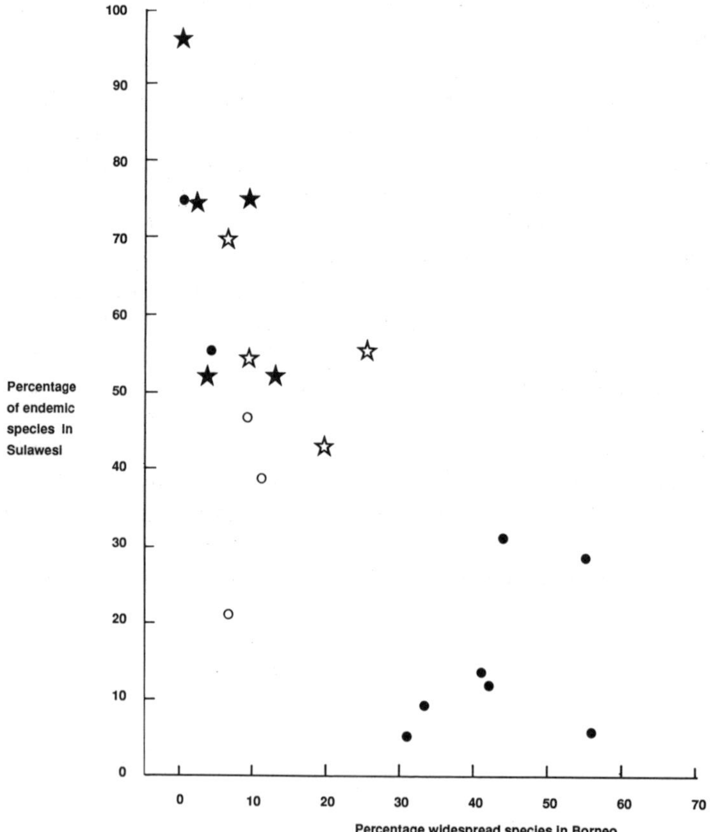

Figure 12. Percentage of endemics in the Sulawesi fauna for various Lepidoptera groups plotted against the percentage of widespread species in the faunas of Borneo or Malaya. Symbols as in Figure 11, and data from Table 1.

Diplurodes (Fig. 14) is another ennomine genus, represented in Borneo by three sub-groups: *Diplurodes, Ectropidia,* and *Necyopa.* The first is very diverse, represented by 14 species. *Ectropidia* has five or six species (the *pustulata* group needs further investigation) and *Necyopa* has four. Only 11 species are known from Sulawesi. Two are in the subgenus *Diplurodes* and appear to be most closely related to individual lowland species in Borneo, though the Sulawesi taxa are montane. The rest of the fauna forms a monophyletic group in *Ectropidia,* most closely related to an undescribed species in Sumatra and to the more widespread Sundanian lowland species *D. exprimata.* This group is represented by two species in Seram, one of which extends to New Guinea. No cladistic analysis of the Sulawesi group has been attempted, as the characters separating the taxa are minor ones of wing markings and setation of the male genitalia. However, there is a subgroup of three where the uncus of the male genitalia is a distinctive 'wing-nut' shape and the more distal coremata of the male abdomen are lost. The total group is defined by the presence of robust setae in groups of one or two along the ventral margin of the valve. The group has phylogenetic parallels with the limacodid *Narosa* (discussed next) in its complexity in Sulawesi and *Idea*-type speciation pattern into New Guinea; *Narosa,* however, has not been recorded from Seram.

Subgenus *Necyopa* is not recorded east of Sundaland. Host-plant records for *Diplurodes* are mostly from Fagaceae but also from Dipterocarpaceae.

The *Diplurodes* group offers circumstantial support for a hypothesis of dispersive

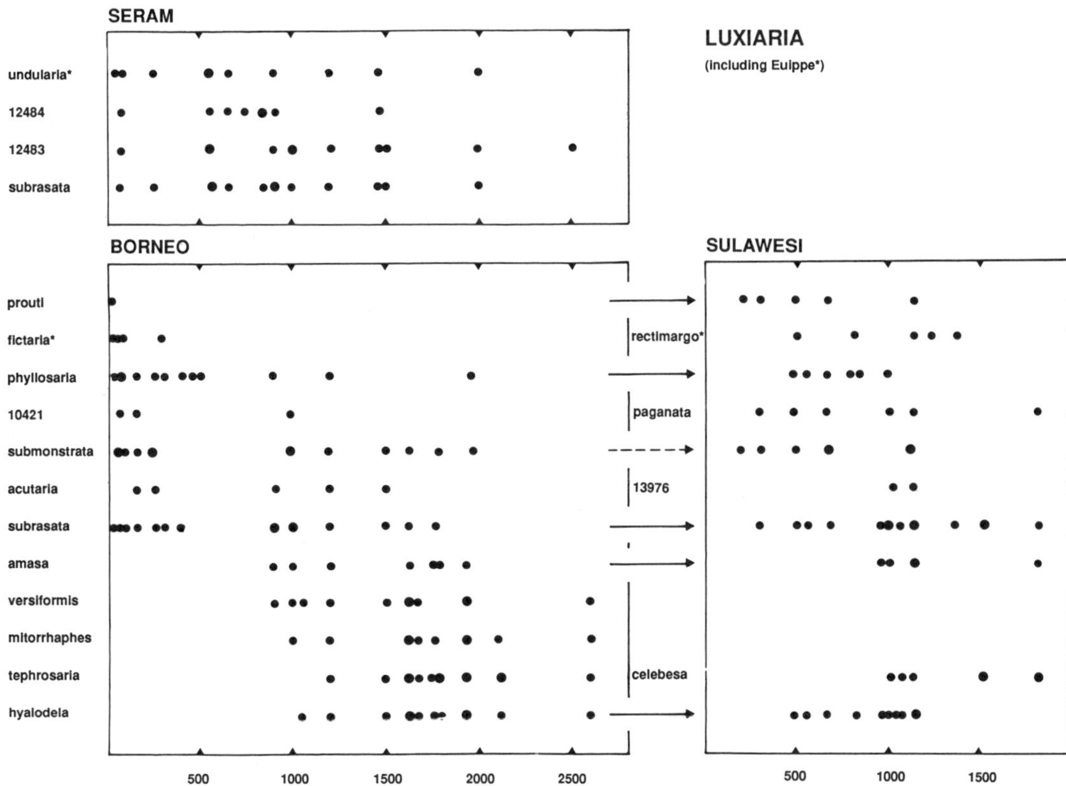

Figure 13. Altitude ranges (in metres) of species of *Luxiaria* (Geometridae: Ennominae) in Borneo, Sulawesi, and Seram. Larger circles indicate the species is common at that altitude.

derivation of the Sulawesi biota. A vicariant argument that the low representation of Bornean groups in Sulawesi is a result of extinction on the latter sits ill with the evidence of extensive speciation of the group on the island and the sympatry and greater range of altitude in the taxa concerned, as discussed later.

The limacodid genus *Narosa* was discussed as an example of speciation within Sulawesi by Holloway (1987); a cladogram of relationships was presented. This is reproduced, with the addition of two further Sulawesi taxa, in Figure 15 (that in Fig. 1 is a simplified version highlighting area relationships). Altitude records for the taxa are illustrated, and localisation within Sulawesi is indicated. As with *Diplurodes*, the closest Bornean (Sundanian) relatives are lowland, but the Sulawesi species together, and in several cases individually, span a much greater range of altitude.

Narosa species are tree-feeders as larvae, with no indications of specialisation or the preference for palms seen in other limacodids (Holloway, 1986a).

The final genus to be discussed is another ennomine geometrid, the genus *Astygisa* (=*Apopetelia*). The larvae of this and allied genera feed on Rhamnaceae. All species are Oriental, and most of the morphological diversity is represented in mainland Asia taxa. Five species occur in Borneo, all lowland except one upper montane one. Sulawesi is richer in species with ten: one is shared with Sundaland (*circularia*) but the others form a monophyletic group that is sister to the Oriental-Sundanian species *A. vexillaria*. Most of these Sulawesi species are undescribed. (See Appendix 3).

The male genitalia present sufficient features for a cladistic analysis to be attempted on the Sulawesi group plus *vexillaria*. Fourteen characters were coded, mostly of male genitalia, and application of Hennig86 yielded 14 trees of length 29, a Consistency Index of

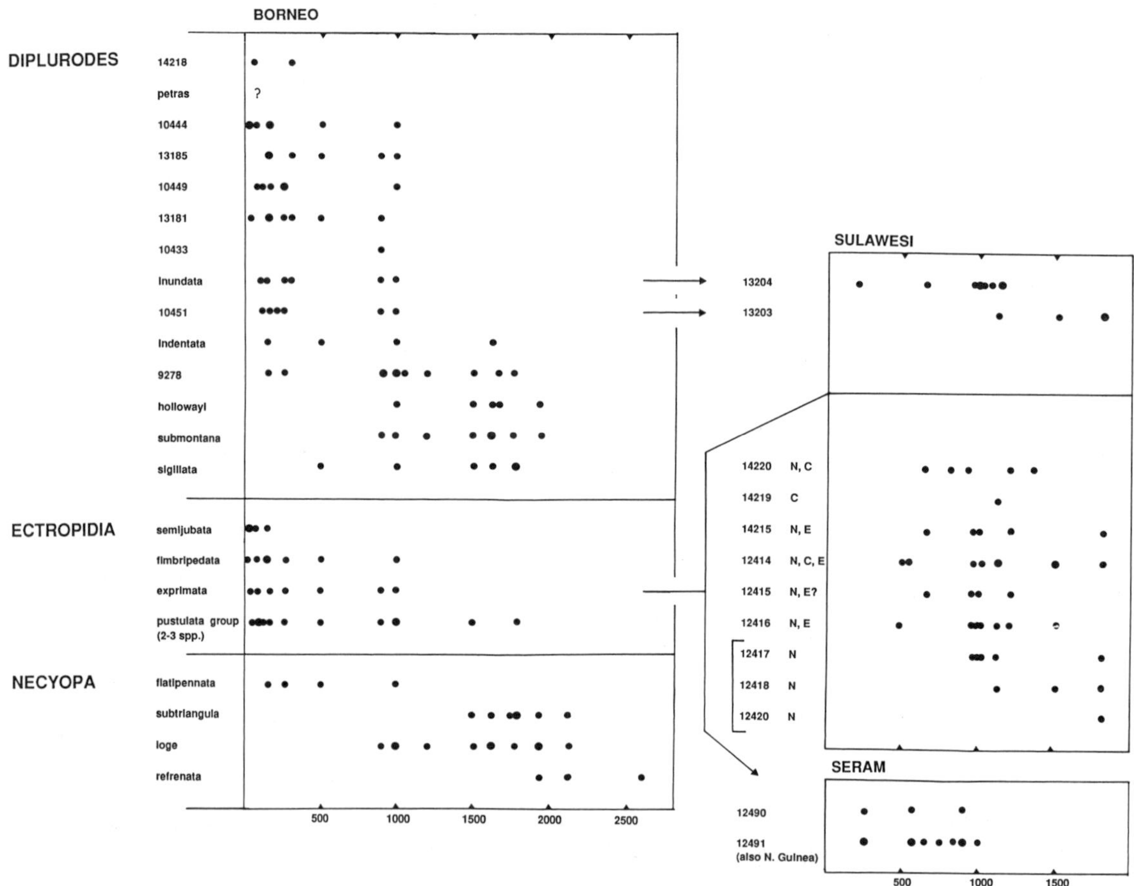

Figure 14. Altitude ranges for species of *Diplurodes* (Geometridae: Ennominae) in Borneo, Sulawesi, and Seram, indicating the subgenera as discussed in the text. Larger circles indicate the species is common at that altitude. The Sulawesi complex of *Ectropidia* has the 'wing-nut uncus' trio bracketed and shows localisation within the island. The name *celebensis* Debauche, described as a subspecies of *semiparata* Walker, is probably referable to species 12417.

65, and Retention Index of 72. The strict consensus tree is shown in Figure 16 with alternative arrangements of the subgroup of five species seen each in half of the trees. A more limited selection of female specimens was available, showing fewer characters (five), and only tentatively associated with males. The results were consistent with those obtained for males, but are not discussed in detail. An account of the analysis and characters used is presented in Appendix 3.

The Sulawesi taxa, as in the other examples of speciation within the island, extend over a much greater range of altitude, often individually, than does the Sundanian sister taxon in Borneo, where it is strictly lowland.

These four genera provide a contrast between instances where the fauna of Sulawesi is enriched primarily by dispersal events from Sundaland (*Luxiaria*) and instances where enrichment is primarily by endemic speciation (the other three genera). In the former case the pattern of altitudinal zonation is more or less maintained (interestingly more weakly in Seram, an island at a greater distance from the putative source and with fewer taxa). In the latter, the closest Sundanian relative is in each case a lowland species, but in Sulawesi the species show no strong zonation but range, often very widely, from the lowlands up into the montane zones. The trio with the modified uncus in *Diplurodes*, however, appears to be restricted to montane zones.

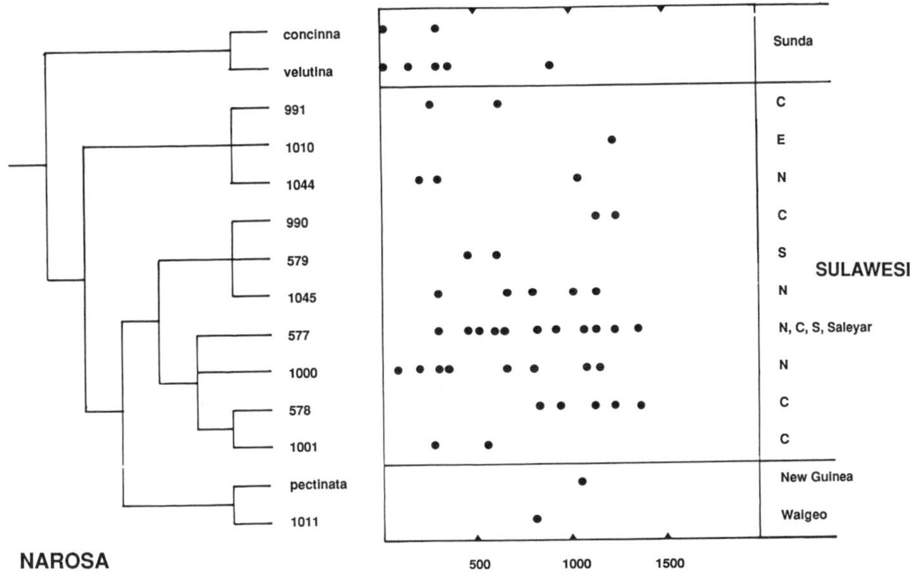

Figure 15. Cladogram for the *velutina* group of *Narosa* (simplified and indicating a mainland Asian sister-relationship in Fig 1), showing the altitude ranges of all species. Larger circles indicate the species is common at that altitude. Sunda records are from Borneo and Sumatra. Localisation within Sulawesi is indicated at the right of the diagram.

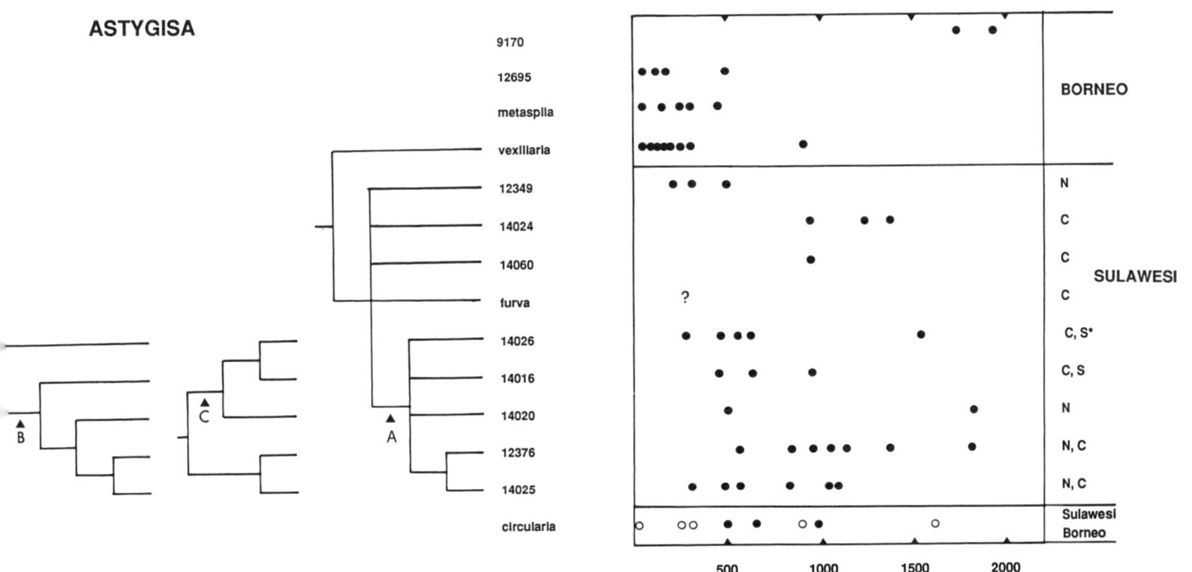

Figure 16. Strict consensus tree for species of *Astygisa* (Geometridae: Ennominae) as discussed in the text and Appendix 3, with alternatives for clade A indicated to the left. The altitude ranges for all Bornean and Sulawesi species are indicated, as is localisation for the latter (the species with an asterisk may occur on Timor). Larger circles indicate the species is common at that altitude. *A. circularia* occurs in both Borneo (open circles) and Sulawesi (solid circles).

360

Thus speciation in Sulawesi appears to have been accompanied by a relaxation of altitudinal constraints. The three groups also show interesting patterns in relation to the localisation of the taxa. They offer no support for taxon cycle or taxon pulse hypotheses (reviewed by Liebherr & Hajek, 1990).

The geological history of Sulawesi offers some indication of a more archipelagic past (papers in Knight & Holloway, 1990). An hypothesis of allopatric evolution of endemic species groups through this archipelago, sympatry arising with increasing coalescence, therefore seems plausible. Comparisons have been made with the current geography of the Philippines and the degree of endemism on the component islands (Vane-Wright, 1990; Holloway, 1990b and comments therein).

The groups with endemic speciation discussed here show a high degree of sympatry. The data are mostly restricted to the Palu, northern arm, Morowali, and Gunung Lompobattang areas, but these localities represent four of the internal areas of endemism identified by Duffels (1990) and Polhemus & Polhemus (1990). Nevertheless, seven out of nine *Astygisa* occur in the Palu area and four in the north. Out of nine *Ectropidia (Diplurodes)*, all occur in the north and four in the Marowali area. In *Narosa* five out of ten species occur in the Palu area and four in the north.

Examination of the terminal groupings in the cladograms reveals a mixture of allopatry and sympatry. In *Narosa* there are two triplets that are totally allopatric and one duplex where both species are found in the Palu area but with weak indication of altitudinal segregation. In *Ectropidia* sympatry is very high, with the trio with a modified uncus all restricted to the northern arm of the island. In *Astygisa* the terminal duplex is wholly sympatric in the north and center, with the other three species of the subgroup of five present either in the north or centre. One of the alternative trees for this terminal group also shows increased sympatry of terminal groups. The danaine butterfly genus *Parantica* has an endemic group in Sulawesi that shows a similar pattern to these moth groups (Ackery & Vane-Wright, 1984; Vane-Wright, pers. comm.).

In the cicada genus *Dilobopyga* (Duffels, 1990) there is also a high degree of sympatry in terminal groupings: in north Sulawesi in the *opercularis* group, and in central Sulawesi for the *minahassae* group and to some extent for the *chlorogaster* terminal group. In contrast, the aquatic heteropteran genus *Ptilomera* (Gerridae) has a group of endemic species distributed almost allopatrically over the island (Polhemus & Polhemus, 1990).

It is evident that speciation processes in these groups endemic to Sulawesi have been complex, with pointers to some form of speciation that is not strictly allopatric, accompanied by an expansion of niche breadth at least with regard to altitudinal zonation. It is interesting also to note the apparent conflict in characters generating the alternative trees for the terminal group of five in *Astygisa* (clade A): excluding the terminal duplex, the alternatives show a complete reversal of the order of branching. This may merely be a property of the rather limited data set, but it might also be indicative of some mode of parapatric or sympatric speciation.

CONCLUSIONS

If vicariant patterns of speciation were predominant in the Indo-Australian tropics, our task would be relatively simple in determining an underlying pattern of area relationships; speciation process would be allopatric. The actual situation appears to be far more complex, with dispersal and speciation processes interacting with an archipelago in a state of formation and growth, the juxtaposition of its component islands rendered fluid by the processes of plate tectonics. This poses a much greater challenge to both pattern and process biogeographers and will require a methodology that can identify in a statistically valid manner sets of overlapping area patterns. These may provide us with snapshots of the geography of the past that could be sequenced to reveal some of the complexities of earth history.

The sample of biogeographic patterns described earlier pinpoint Sulawesi as a key area for understanding the development of Indo-Australian biotas. General data comparing Sulawesi with Borneo, admittedly crude, point to a spectrum of faunal enrichment processes, from relatively rapid saturation of the fauna from multiple episodes of colonisation by more dispersive groups, to less frequent episodes of colonisation by poorly dispersive groups that have often resulted in compensatory faunal enrichment through speciation within Sulawesi.

Examples of speciation within Sulawesi show a number of common features. The sister-relationship in each case is to a lowland species (or species group) in Sundaland. The component species in Sulawesi occupy a much greater altitude range with little indication of segregation or zonation over that range. There is a high degree of sympatry in most groups that is evident even in terminal groups of cladistic analyses, suggesting that speciation has not always been allopatric.

The processes of faunal enrichment in Sulawesi may have parallels with those in the much larger island of New Guinea. Here there are much more definite indications of a biota with composite origins, particularly with regard to ancient Australian elements and biotic patterns reflecting geological hypotheses of two Melanesian arcs. The Melanesian arc patterns themselves offer contrasts in speciation process, with outer arc components reflecting archipelagic, allopatric modes of speciation and inner arc components with a high degree of sympatry similar to that of groups that have speciated with Sulawesi. These processes have led to development of high diversity within a relatively short span of geological time. Studies of speciation process in New Guinea and other biogeographic and ecological aspects should therefore be given priority.

ACKNOWLEDGMENTS

Particular thanks are due to Ian Kitching for running several data sets for me on Farris' Hennig86 program and advising me on interpretation of the results. Also, Chris Humphries kindly explored the complexities of running the raw cladograms on Page's Component program. Dick Vane-Wright provided unpublished data on the butterflies of Sulawesi and neighboring areas. I am grateful to all three, together with Kevin Gaston, Scott Miller, Gaden Robinson, and Gary Stonedahl, for discussion on aspects of this paper and for comments on drafts of the text.

The work was undertaken as core research with the International Institute of Entomology (an Institute of CAB International), with access to the collections of the Natural History Museum, London. The paper is based in part on material collected during Project Wallace, sponsored by the Royal Entomological Society of London and the Indonesian Institute of Sciences (Results of Project Wallace No 124). My participation in Project Wallace was supported partly by a British Ecological Society Travelling Fellowship.

I am indebted to my wife, Phillipa, for keyboarding the text and preparing the lettering for the figures.

LITERATURE CITED

Ackery, P. R. 1987. The danaid genus *Tellervo* (Lepidoptera, Nymphalidae) — a cladistic approach. *Zoological Journal of the Linnean Society* 89:203–274.

Ackery, P. R. & R. I. Vane-Wright. 1984. *Milkweed Butterflies, Their Cladistics and Biology.* British Museum (Natural History): London. 425pp.

Audley-Charles, M. G. 1987. Dispersal of Gondwanaland: relevance to evolution of the angiosperms. Pp. 5–25. *In:* T. C. Whitmore (ed.), *Biogeographical Evolution of the Malay Archipelago.* Clarendon Press: Oxford.

Balgooy, M. M. J. van. 1987. A plant geographical analysis of Sulawesi. Pp. 94–102. *In:* T. C. Whitmore (ed.), *Biogeographical Evolution of the Malay Archipelago,* Clarendon Press: Oxford.

Butlin, R. K. & K. A. Monk. 1990. Catantopine grasshoppers of Sulawesi. Pp. 89–94. *In:* W. J. Knight & J. D. Holloway (eds.) *Insects and the Rain Forests of South East Asia (Wallacea).* Royal Entomological Society: London.

Connor, E. F. 1988. Fossils, phenetics and phylogenetics: inferring the historical dynamics of biogeographic distributions. Pp. 254–269. *In:* J. K. Liebherr (ed.), *Zoogeography of Caribbean Insects.* Cornell University Press: Ithaca, NY.

Corbet, A. S. & H. M. Pendlebury (revision by J. N. Eliot). 1978. *The Butterflies of the Malay Peninsula,* 3rd Edition. Malayan Nature Society: Kuala Lumpur. 578pp.

Corner, E. J. H. 1975. *Ficus* in the New Hebrides. *Philosophical Transactions of the Royal Society of London,* B 272:343–367.

Cox, C. B. 1990. New geological theories and old biogeographical problems. *Journal of Biogeography* 17:117–130.

Cracraft, J. in press. From Malaysia to New Guinea: evolutionary biogeography within a complex continent-island arc contact zone. *Proceedings of the International Ornithological Congress,* 1988.

Craw, R. C. 1989. Quantitative panbiogeography: introduction to methods. *New Zealand Journal of Zoology* 16:485–494

Craw, R. C. & P. Weston, 1984. Panbiogeography: a progressive research program? *Systematic Zoology* 33:1–13.

Duffels, J. P. 1983. Distribution patterns of Oriental Cicadoidea (Homoptera) east of Wallace's Line and plate tectonics. *GeoJournal* 7:491–498.

Duffels, J. P. 1986. Biogeography of Indo-Pacific *Cicadoidea:* a tentative recognition of areas of endemism. Cladistics 2:318–336.

Duffels, J. P. 1990. Biogeography of Sulawesi cicadas (Homoptera: Cicadoidea). Pp. 63–72. *In:* W. J. Knight & J. D. Holloway (eds.) *Insects and the Rain Forests of South East Asia (Wallacea).* Royal Entomological Society: London. 343 pp.

Ewart, A. 1988. Geological history of the Fiji-Tonga-Samoan region of the S. W. Pacific, and some palaeogeographic and biogeographic implications. *Entomonograph* 10:15–23.

Fletcher, D. S. 1979. Geometridae. Volume 3 in: I. W. B. Nye (ed.), *The Generic Names of Moths of the World.* British Museum (Natural History): London. 243pp.

Hamilton, W. 1979. Tectonics of the Indonesian region. *U.S. Geological Survey Professional Paper* 1078:1–345.

Hamilton, W. 1989. Convergent-plate tectonics as viewed from the Indonesian region. Pp. 655–698. *In:* A. M. C. Sengor (ed.), *Tectonic Evolution of the Tethyan Region.* Kluwer: Dordrecht, Boston, London.

Holloway, J. D. 1969. A numerical investigation of the biogeography of the butterfly fauna of India, and its relation to continental drift. *Biological Journal of the Linnean Society* 1:373–385.

Holloway, J. D. 1970. The biogeographical analysis of a transect sample of the moth fauna of Mt Kinabalu, Sabah, using numerical methods. *Biological Journal of the Linnean Society* 2:259–286.

Holloway, J. D. 1976. *The Moths of Borneo with Special Reference to Mount Kinabalu.* Malayan Nature Society: Kuala Lumpur. 263 pp.

Holloway, J. D. 1977. *The Lepidoptera of Norfolk Island, Their Biogeography and Ecology.* Series Entomologica 13. W. Junk: The Hague. 291 pp.

Holloway, J. D. 1979. *A Survey of the Lepidoptera, Biogeography and Ecology of New Caledonia.* Series Entomologica 15. W. Junk: The Hague. 588 pp.

Holloway, J. D. 1982. Mobile organisms in a geologically complex area: Lepidoptera in the Indo-Australian tropics. *Zoological Journal of the Linnean Society* 76:353–373.

Holloway, J. D. 1984a. Lepidoptera and the Melanesian Arcs. Pp. 129–169. *In:* F. J. Radovsky et al. (eds.), *Biogeography of the Tropical Pacific.* Bishop Museum Special Publication 72.

Holloway, J. D. 1984b. The larger moths of Gunung Mulu National Park; a preliminary assessment of their distribution, ecology and potential as environmental indicators. Pp. 149–190. *In:* A. C. Jermy, & K. P. Kavanagh (eds.) *Gunung Mulu National Park, Sarawak, Part II. Sarawak Museum Journal* 30, Special Issue 2.

Holloway, J. D. 1985. The Moths of Borneo: Family Noctuidae: Subfamilies Euteliinae, Stictopterinae, Plusiinae, Pantheinae. *Malayan Nature Journal* 38:57–317.

Holloway, J. D. 1986a. The Moths of Borneo: Key to families; Families Cossidae, Metarbelidae, Ratardidae, Dudgeoneidae, Epipyropidae and Limacodidae. *Malayan Nature Journal* 40:1–166.

Holloway, J. D. 1986b. Origins of Lepidopteran faunas in high mountains of the Indo-Australian tropics. Pp. 533–556. *In:* F. Vuilleumier & M. Monasterio (eds.) *High Altitude Tropical*

Biogeography. Oxford University Press: New York.

Holloway, J. D. 1987. Lepidoptera patterns involving Sulawesi: what do they indicate of past geography? Pp. 103–118. *In:* T. C. Whitmore (ed.) *Biogeographical Evolution of the Malay Archipelago.* Clarendon Press: Oxford.

Holloway, J. D. 1989. Moths. Pp. 437–453. *In:* H. Lieth & M. J. A. Werger (eds.). *Tropical Rain Forest Ecosystems. Ecosystems of the World 14B.* Elsevier: Amsterdam.

Holloway, J. D. 1990a. Norfolk Island and biogeography for the nineties: ideas from a dot on the map. *Journal of Biogeography* 17:113–115.

Holloway, J. D. 1990b. Sulawesi biogeography—discussion and summing up. Pp. 95–102 In. W. J. Knight & J. D. Holloway (eds.) *Insects and the Rain Forests of South East Asia (Wallacea).* Royal Entomological Society: London. 343 pp.

Holloway, J. D. in press. Aspects of the biogeography and ecology of the Seram moth fauna. In. J. Proctor & I. Edwards (eds.), *The Natural History of Seram.*

Holloway, J. D. & N. Jardine 1968. Two approaches to zoogeography: a study based on the distributions of butterflies, birds and bats in the Indo-Australian area. *Proceedings of the Linnean Society of London* 179:153–188.

Holloway, J. D., G. S. Robinson & K. R. Tuck. 1990. Zonation in the Lepidoptera of northern Sulawesi. Pp. 153–166. In. W. J. Knight & J. D. Holloway (eds.) *Insects and the Rain Forests of South East Asia (Wallacea).* Royal Entomological Society: London. 343 pp.

Humphries, C. J. 1990. The importance of Wallacea to biogeographical thinking. Pp. 7–18. *In:* W. J. Knight & J. D. Holloway (eds.) *Insects and the Rain Forests of South East Asia (Wallacea).* Royal Entomological Society: London. 343 pp.

Humphries, C. J. & L. Parenti. 1986. *Cladistic Biogeography.* Clarendon Press: Oxford. 350 pp.

Knight, W. J. & J. D. Holloway (eds.) 1990. *Insects and the Rain Forests of South-East Asia (Wallacea).* Royal Entomological Society: London. 343pp.

Liebherr, J. K. & A. E. Hajek. 1990. A cladistic test of the taxon cycle and taxon pulse hypotheses. *Cladistics* 6:39–59.

Nelson, G. & N. Platnick. 1981. *Systematics and Biogeography, Cladistics and Vicariance.* Columbia University Press: New York. 567 pp.

Page, R. D. M. 1988. Quantitative cladistic biogeography: constructing and comparing area cladograms. *Systematic Zoology* 37:254–270.

Page, R. D. M. 1989. Comments on component-compatibility in historical biogeography. *Cladistics* 5:167–182.

Polhemus, J. T. & D. A. Polhemus. 1990. Zoogeography of the aquatic Heteroptera of Celebes: regional relationships versus insular endemism. Pp. 73–86. *In:* W. J. Knight & J. D. Holloway (eds.) *Insects and the Rain Forests of South East Asia (Wallacea).* Royal Entomological Society: London. 343 pp.

Schuh, R. T. & G. M. Stonedahl. 1986. Historical biogeography in the Indo-Pacific: a cladistic approach. *Cladistics* 2:337–355.

Simberloff, D. 1987. Calculating probabilities that cladograms match: a method of biogeographic inference. *Systematic Zoology* 36:175–195.

Steenis C. G. G. J. van 1962. The land-bridge theory in botany. *Blumea* 11:235–372.

Vane-Wright, R. I. 1990. The Philippines—key to the biogeography of Wallacea? Pp. 19–34 *In:* W. J. Knight & J. D. Holloway (eds.) *Insects and the Rain Forests of South East Asia (Wallacea).* Royal Entomological Society: London. 343 pp.

Warren, W. 1894. New genera and species of Geometridae. *Novitates Zoologicae* 1:366–466.

Wiley, E. O. 1988. Parsimony analysis and vicariance biogeography. *Systematic Zoology* 37:271–290.

Yata, O. 1990. Cladistic biogeography of *Eurema* subgenus *Terias* (Lepidoptera: Pieridae), with particular reference to Wallacea. Pp. 43–48. *In:* W. J. Knight & J. D. Holloway (eds.), *Insects and the Rain Forests of South East Asia (Wallacea).* Royal Entomological Society: London. 343 pp.

APPENDIX 1. THE GENERA *ZEHEBA* MOORE AND *PROBITHIA* WARREN

Zeheba and *Probithia* belong to the *Eutoea* Walker/*Luxiaria* Walker group of genera in the Ennominae. The group is particularly diverse in the Indo-Australian tropics. It will be discussed in more detail in part 11 of *The Moths of Borneo,* currently in preparation.

Both genera have the valves of the male genitalia deeply divided as in the rest of the group of genera, but *Zeheba* has the two arms separated by a lobe-like structure and the ventral arm with a subapical flange as in *Eutoea* and *Luxiaria.* The lobe is diagnostically elongate in *Zeheba.* Other features of generic significance are listed under (1) below.

Cladistic hypotheses for the two genera are presented in Fig. 4, the characters of facies and male genitalia of significance for each clade being listed below.

Zeheba

1. Robust build; wings partially translucent with broad brown borders; sexually dimorphic, males with crenulate, rounded hindwing borders, females with the border strongly angled centrally; forewings bifalcate; central lobe of valve elongate such that dorsal and ventral arms are well separated; aedeagus vesica with opposed, elongate, reversed cornuti, one ligulate, one narrow with a weak subapical barb; bilateral asymmetry of ventral arm of valve prevalent, but possibly homplasious, in clades 4 and 5 and 13949; eighth abdominal sternite with distal margin sclerotised and shallowly concave.

2. Ventral arm of valve sinuous, distally down curved; dorsal arms short, not extending beyond apex of uncus.

3. Ventral lobe of ventral arm of valve narrow, acute (the distal and ventral lobes on the left valve in 13949 form an angle within which there is a small additional spur).

4. Strong bilateral asymmetry of the ventral arms of the valves, the left having the lobe roundedly triangular, basally much constricted, and the right having it broad-based, skewed towards the apex such that that corner is acute (much more so in lucidata than in 12671; the lobes are more delicate in the former, the left less rounded).

5. Ventral arms of valves straight or gently upcurved, of even width, with lobes and processes small; largest lobe a triangular flap at a half or two-thirds, in addition to the sub-apical ventral one that is usually reduced to a small spur.

6. Lobe between valve arms narrow, rather than circular, with a small dorsal angle.

Probithia

1. Medial fasciae of both wings include discal spot rather than being distal to it; sub-basal lobe on sacculus dorsally, on ventral arm of valve (diagnostically developed slightly asymmetrically and setose in 12373); aedeagus terminates in two equal sclerotised strips.

2. Hindwing with angled margin; pectinations of male antennae reduced; lobe with short, coarse setae ventrally at centre of dorsal arm of valve; uncus slightly 'shouldered'; aedeagus vesica with a distal short cornutus and a central long one, both apically blunt and scobinate (a long distal one only in 12373).

3. Two darker markings at anterior end of forewing postmedial united into a chevron rather than well separated.

4. Lobe on costal arm of valve with distal angle more produced than more basal one (as distinct from less produced (clade 3) or rounded (*obstataria*); the lobe has a tendency to be elongate.

5. Fringes of wings dark; markings on forewing costa strong. Bismarcks and Solomons *perichla* are large, those from the latter with particularly elongate lobe on the costal arm of the valve.

6. Reversal of elongation of lobe on costal arm (character of clade 4). Sulawesi specimens are strongly marked in dark brown along the exterior of the hindwing post-medial yet this is weaker (cf. stronger) on the underside.

These characters are not particularly satisfactory. Clades 2, 3, and 5 have the strongest definition.

APPENDIX 2. ANALYSIS OF THE GENUS *BRACCA* HÜBNER

Checklist of taxa and synonyms [with genus of original description]

Generic synonyms: *Cosmethis* Hübner syn. n.; *Duga* Walker syn. n.; *Arycanda* Walker syn. n.; *Panaethia* Guenée syn. n.; *Tigridoptera* Herrich-Schäffer syn. n.; full information on these genus-group names is in Fletcher (1979).

Outgroup:

Bracca xanthogramma Prout comb. n. [*Arycanda*] New Guinea; (Mt. Goliath, Weyland Mts, Snow Mts).

Bracca brunneotacta Warren comb. n. [*Arycanda*]. New Guinea (Mambare R., Arfak Mts.).

Bracca commixta Warren comb. n. [*Arycanda*]. New Guinea (Mt. Goliath, Mambare R., Angabunga R.).

Bracca vinaceostrigata Prout comb. n. [*Arycanda*]. New Guinea (Mt. Goliath). Probably sister-species of the next species.

Bracca fulviradiata Warren comb.n . [*Arycanda*]. New Guinea (Mt. Goliath, Mt. Kunupi, Mambare R., Angabunga R.)

Bracca fritillaria Warren comb. n. [*Arycanda*]. New Guinea (Aroa, Dinawa, Mambare and Angabunga rivers, Mt. Kebea, Wau, Fak Fak).

Ingroup (in order of appearance on cladogram):

Bracca georgiata Guenée comb. n. [*Panaethia*] (synonym: *discata* Warren syn. n. [*Arycanda*]). Borneo, Natuna Is., Sumatra, Sulawesi. *B. georgiata pervasata* Walker stat. & comb. n. [*Panaethia*] (synonyms: *ptochopis* Meyrick [*Tigridoptera*]; *subradiata* Warren syn. n. [*Tigridoptera*]). Java, Philippines, Sangihe, Seram. The typical race has the black punctate fasciae with spots extended longitudinally; longitudinal dull orange streaks are present in *pervasata*.

Bracca sp. n. 13757 (BMNH geometrid genitalia preparation number). North and central Sulawesi.

Bracca sp. n. 11539. Vanuatu.

Bracca teleleuca Prout comb. n. [*Cosmethes*]. Bismarck Is.

Bracca disrupta Warren [*Xanthomima*] (synonyms: *partita* Warren, *isabellina* Warren, *woodfordi* Druce (1888)). Solomons.

Bracca ribbei Pagenstecher. New Guinea.

Bracca lucida Swinhoe. Bismarcks.

Bracca flavitaenia Warren. A single male from Sula Mangoli, here treated as part of the Sulawesi archipelago.

Bracca bajularia (Clerck) (synonym: *scissa* Walker). S. Moluccas.

Bracca emolliens Walker comb. n. [*Eusemia*] (synonym: *atrocoerulea* Felder). Sulawesi, probably throughout.

Bracca sp. n. 14097. South-west, north and east Sulawesi.

Bracca sp. n. 14099. Known from males only, east Sulawesi.

Bracca maculosa Walker comb. n. [*Arycanda*] (synonyms: *absorpta* Warren syn. n. [*Arycanda*]; *omissa* Warren syn. n. [*Arycanda*]; *apicinigra* Bastelberger [*Arycanda*] Sumatra, Malaya, Borneo. *B. maculosa radiolata* Warren stat. & comb. n. [*Tigridoptera*]. Palawan.

Bracca subfumosa Warren comb. n. [*Arycanda*]. Borneo, Sumatra.

Bracca exul Herrich-Schäffer comb. n. [*Tigridoptera*]. Java.

Bracca monochrias Meyrick comb. n. [*Tigridoptera*] (synonym: *cuneiplena* Swinhoe syn. n. [*Tigridoptera*]). Known from males only from Sangihe and north and south Philippines. Material is limited, but there are indications of subspecific variability. Absence of basal black spots on the forewing (character 32) and the configuration of the subbasal black spots suggests the following taxon, represented only by females, may be synonymous; the taxa were treated separately in the analysis but were placed in a position not inconsistent with an hypothesis of synonymy.

Bracca benguetana Schultze comb. n. [*Tigridoptera*]. Females only. North Philippines (Luzon). Males from the Smithsonian Institution examined subsequently confirmed synonymy with *monochrias*, syn. n.

Bracca xanthosoma Warren comb. n. [*Craspedosis*]. North and South Moluccas including Obi; possibly in New Guinea but confirmation needed.

Bracca matutinata Walker comb. n. [*Panaethia*]. Queensland.

Bracca interrupta Butler comb. n. [*Tigridoptera*]. Bismarck Is.

Bracca rotundata Butler comb. n. [*Tigridoptera*]. Queensland. *B. rotundata buruensis* Prout comb. n. [*Cosmethes*]. S. Moluccas.

Bracca barbara Stoll comb. n. [*Phalaena geometra*] (synonyms: *australasiae* Donovan; *pinguis* Walker). North and South Moluccas including Obi, ?Australia (confirmation needed).

Bracca basiflava Warren comb. n. [*Cosmethes*]. A single female from Kei; general facies indicates sister-relationship with *B. rosenbergi*.

Bracca rosenbergi Pagenstecher comb. n. [*Abraxas*] (synonym: *rana* Swinhoe). S. Moluccas, New Guinea (to Gebe Is. in west). Bismarck Is.

Bracca woodfordii Butler (1887) comb. n. [*Eusemia*] (synonyms: *ampliplaga* Warren; *floridata* Warren; *siriella* Druce; *disparilis* Prout). Solomons.

General Comments

The New Guinea complex of species in *Bracca* is used as an outgroup in the absence of any obvious sister-genus. These species exhibit plesiomorphic features of genitalia and facies (as seen in other Ennominae of the Boarmiini complex) where the ingroup is strongly apomorphic. They share with the ingroup the presence of longitudinal dull orange streaks on the wings, which have a dull grey-blue ground colour, and presence of a definitive setose lobe near the anterior end of the transverse sclerotised band in the valve of the male genitalia. Inclusion of these New Guinea species with the ingroup would have rendered coding of the male genitalic characters selected for the ingroup much more difficult as the outgroup taxa show a much more extreme range of variation. They are defined as a natural group by two character-states in the female genitalia: 1(0); 3(0).

The family-group category Braccinae of Warren (1894) is based on the genus. No earlier reference to Braccinae has been located, and the taxa placed in the Braccinae by Warren are mostly not closely related to *Bracca*. A study of tribal categories in the Ennominae is being undertaken for part 11 of *The Moths of Borneo*, but as yet no firm indication of the status of this family-group name has transpired.

The monophyly and extent of *Bracca* as recognised here was originally noted by Holloway (1984a). The male genitalia of Bornean species were illustrated by Holloway (1976: Figs. 662, 663). A number of taxa currently in *Arycanda* and not listed above should be transferred to *Craspedosis* Butler, as a sample dissected had genitalia typical of that genus (Holloway, 1976: Fig. 665; 1984a).

A striking feature of the taxa now included is the variety of the wing pattern. Most of the outgroup and the Vanuatu taxon of the ingroup have a typically cryptic, crenulate fasciation, but pattern in the in-group includes: blue-grey species with linear or punctata transverse fasciae (also seen in *B. fritillaria* of the outgroup) and longitudinal dull grey-orange streaks; species where dull grey orange is increased and white patches occur; species with blackened wings and white patches, most extreme in *B. woodfordii*; a species with wings entirely dark grey (*B. xanthosoma*); a group (clade 6) more or less extensively marked with chrome-yellow. *Bracca woodfordii* and perhaps *B. barbara* may be part of the *Tellervo* Kirby mimicry complex (Ackery, 1987), and the blue-grey patterned taxa have parallels in the geometrine genus *Dysphania* Hübner and the agaristine noctuid genus *Longicella* Jordan, as well as in *Craspedosis*, mentioned above.

Larvae (of *B. rotundata* in BMNH; no host data attached) are also strikingly marked, mainly black, with orange rings and white patches.

Characters Used in the Analysis

Female genitalia

1. Signum: bicornute (0); disc-like, ringed with spines completely (1); cordate, with gap in ring of spines at the distal notch of the 'heart'; (2); weak, squarish, with distal gap in marginal spines (3). State (0) defines the outgroup. State (2) defines clade 13. State (3) is homoplasious in the Vanuatu species and in clade 12. State (1) is therefore indicated to be plesiomorphic and is indeed found commonly in other Ennominae.

2. Signum: without spines centrally (0); with irregular spines centrally (1); with central spines in v-formation (2); with transverse flange (3). State (2) defines clade 16. State (3) is an autapomorphy of *B. emolliens*. State (1) appears to be plesiomorphic (and is seen widely in Ennominae). The 'loss' state (0) is in the outgroup and the Vanuatu species.

3. Bursa copulatrix sclerotisation: a band, tongue-like, extending longitudinally from the neck of the bursa (0); ringing the basal part of the bursa sphere (1); narrow, longitudinal, not in neck of bursa (2); absent (3). Each of the character states is unusual except (3), which defines clade 5 and may represent a secondary loss. State (0) distinguished the outgroup, state (2) the ingroup, with state (1) defining clade 6 and thus possibly a modification of state (3).

4. Transverse ridges of spines on bursa sclerotisation: absent (0); widely spaced (1); spaced apart more than one time and up to two times the width of each ridge (2); more closely spaced (3). Presence of these spines defines the ingroup. The two narrowly spaced states (2,3) together define clade 13 but are not clearly segregated within it. They should probably be recorded as a single state.

5. Ductus bursae with: one zone of sclerotisation (0); two short zones (1); a basal long zone and distal short one separated by a globose membraneous structure (2). State (1) defines clade 14. State (2) defines clade 6.

6. Ductus bursae: not basally broadened (0); distinctly broadened basally (1). State (1) defines clade 14 and may be correlated with state (1) of the previous character.

7. Ostium: sclerotised, not bilobed (0); unsclerotised or weakly so (1); strongly sclerotised and bilobed (2); set in a pouch or pocket, not bilobed (3); set in a quadrate pocket and bilobed (4). State (1) defines clade 11. State (2) defines clade 15. State (3) occurs in *georgiata* and sp. n. 13757; these appear as sister-species in some of the alternative trees. State (4) defines clade 5.

8. Lateral plates of lamella vaginalis: fused anterior to ostium (complexly so as part of 7(4), (0); fused at ostium (1); more or less separate (2); fused posterior to ostium (3). State (2) defines clade 10, though state (3) is homoplasious in *maculosa*, occurring also in *georgiata* and sp.n. 13757. State (1) defines clade 10 and is homoplasious in *exul*. There is a reversal to state (0) in clade 5, but through the complexity of 7(4).

9. Ostial opening: posterior to lamella vaginalis (0); central to lamella vaginalis (1); anterior to lamella vaginalis (2). An anterior position is generally a derived state in Ennominae, and here states (1) and (2) define the ingroup, with (2) homoplasious in clades 5 and 14 and the species *georgiata* and *emolliens*.

10. Apodemes of ovipositor lobes: less than, or equal, in length to three times that of the ovipositor lobes (0); significantly more than three times in length (1). State (1) defines clade 6.

Male abdomen and genitalia

11. Bifurcate corema between sternites 7 and 8: absent (0); present (1). State (1) defines the ingroup.

12. Bifurcate corema between sternites 6 and 7: absent (0); present (1). State (1) defines the ingroup but is reversed to state (0) in the Vanuatu species 11539.

13. Corema on valve sacculus: absent (0); weak (1); strong (2).State (1) defines clade 13, transformed to state (2) at clade 15.

14. Hairlike setae at base of interior of saccus: angled (0); not angled (1). State (1) occurs throughout the ingroup except for clade 6 where state (0) may represent a reversal: both states are seen in other Boarmiini.

15. Small group of setae interior to valve costa and just basal to transverse sclerotised band: present and numerous (0); two adjacent, two on single base or one only (1); absent (2). State (1) defines clade 15 but reverses to (0) in *matutinata*. In *exul* there were three setae on a single base, coded as (0); interestingly this taxon is sister to clade 15. Total loss (2) is an autapomorphy for *teleleuca*.

16. Setose lobe at centre of valve costa: absent (0); weak (1); pronounced, narrow (2); pronounced, broad (3). Presence of this feature defines the ingroup. State (3) defines clade 15. State (1) defines clade 4 with the exception of the Vanuatu species which has state (2).

17. Sclerotised band from apex of sacculus to subcostal processes: sinuous (0); intermediate (1); straight (2). State (2) is homoplasious in clades 11 and 15, with a reversal (*xanthosoma*) and the only occurrence of state (1) (*monochrias*) in clade 15.

18. Uncus: strong (0); vestigial (1). State (1) defines clade 6 but is also seen homoplasiously in *maculosa*.

19. Uncus: bifurcate (0); not bifurcate (1). State (1) defines clade 6.

20. Distal process of valve transverse sclerotised band: strongly extended along costa (0); elongate, narrow, directed towards apex or ventral margin of valve (1); short, narrow (2); short, globose (3). State (1) defines clade 5. State (3) defines clade 6 but also occurs, probably homoplasiously, in *georgiata*. A non-costal position (all states but (0)) defines the ingroup.

21. Spines of distal process of valve: slender, rather needle-like (0); moderate, approximate (1); robust, approximate (2); robust, dispersed (3). State (3) defines clade 6. State (2) appears plesiomorphic for the ingroup, with state (1) general to clade 10 with the exception of *exul* with state (2).

22. Spines of distal process: 5 or more (0); 3 or 4 (1); 1 or 2 (2). State (2) defines clade 5, but is also seen in *exul* and *subfumosa*. State (1), a lesser reduction, is found scattered amongst the ingroup except for clade 6 which has state (0).

23. Interior (basad) extension of transverse sclerotised band: absent (0); weak (e.g.. represented by a slight tooth or angle, (1); strong (2); inwardly scrolled (3). State (3) defines clade 6. State (2) defines clade 15 with the exception of *barbara* where reversal to state (0) has occurred. State (1) defines clade 10 with transformation to (2) in clade 15. State (3) is indicated not to be homologous with (1+2).

24. Strong setae on interior extension: absent (0); multiple (1); single (2). State (2) occurs in clade 6, in most taxa of clade 15 (absence in *barbara* and *woodfordii* could be interpreted as a secondary loss), and probably homoplasiously in subfumosa. Occurrence in the two clades is probably not homologous because the extension of the band in each case may not be homologous (see character 23). State (1) occurs in *xanthosoma* and can be interpreted as an autapomorphic tranformation of state (2).

25. Sclerotised part of sacculus (or heavy setae arising from it) extends: well distal to origin of transverse band (0); only about as far as transverse band (1). State (1) occurs in all the ingroup except for *georgiata*, sp. n. 13757 and *subfumosa*.

26. Heavy setae on sacculus: multiple (more than two) with separate bases (0); single or double with bases more or less fused (1); absent (2). State (1) occurs homoplasiously in clades 5 and 14, though with state (2) in *monochrias* and *xanthosoma* of the latter. State (2) occurs also in *subfumosa* and all of clade 6.

Facies characters

27. Ground colour of wings: blue-grey (0); grey (1); yellow and/or white (2). State (1) is an autapomorphy for *xanthosoma*. State (2) is general to clade 6 but homoplasious in *woodfordii: woodfordii* lacks the bright yellow seen in all members of clade 6.

28. Dull orange (to brown; in taxa of clade 12) longitudinal streaks on wings: present in at least some races (0): absent from all races (1). State (1) is general to clade 4 but is also seen in *emolliens, subfumosa, xanthosoma* and *woodfordii,* and the loss of the feature is therefore somewhat homoplasious. This is not surprising considering the great variety of wing pattern exhibited by the group.

29. Wing fasciae (excluding submarginals): crenulate (0); mostly entire (1); mostly punctate or broken (2); not evident (3). State (1) occurs homoplasiously in clade 4 (with reversal in the Vanuatu taxon to (0)) and clade 17. State (2) is probably the groundplan state for the ingroup but is also seen in *B. fritillaria* of the outgroup. State (3) is seen in *xanthosoma* and *woodfordii* which both have highly divergent facies.

30. Hindwing discal spot: present (0); absent (1): obscured by dark colour (?). Loss occurs in two out of three taxa of clade 8 and in *woodfordii*. The character is obscured in *xanthosoma.*

31. Hindwing submarginal spot in space posterior to vein M3: present (0); absent (1); in linear submarginal band (2); position obscured (?). State (1) defines clade 16 and is homoplasious in *lucida* of clade 6. State (2) is an autapomorphy of *disrupta*. The area is obscured in *xanthosoma* and *woodfordii.*

32. Forewing basal black spots (subbasal and antemedial ones usually present also): present (0); absent (1); obscured (?). State (1) defines clade 14 but with reversal in *barbara* and *matutinata*. The area is obscured in *xanthosoma* and clade 6.

33. Major white patches on forewing: absent (0); marginal (1); subapical (2); postmedial (3). State 1 defines clade 12. State (2) is an autapomorphy of *teleleuca*. State (3) is homoplasious (but probably not homologous) in clades 6 (except for *disrupta* where the spot is yellow and coded (0)) and 18.

34. Major white patches on hindwing: absent (0); marginal (1); submarginal (2); extensive over basal half (3). State (1) defines clade 12. State (2) defines clade 18, but is transformed to state (3) in *woodfordii*. State (3) is homoplasious in *flavitaenia* of clade 6.

35. Abdomen with boundary between ochre and ground colour: irregular (with dorsal dark patches running into ochre zone) (0); circumferential (1). State (1) defines the ingroup but is reversed in the Vanuatu taxon which generally has a plesiomorphic facies (see also character 29). The irregularly defined ochre distal part of the abdomen is also seen in many taxa of *Craspedosis.*

370

Character Matrix

Outgroup	00000	00000	00000	00000	00000	00000	00000
georgiata	11210	03320	11010	20003	20000	00020	00001
sp. 13757	11210	03310	11010	20002	21000	00020	00001
sp. 11539	303?0	04020	10010	20001	22001	10100	00001
teleleuca	113?0	04020	11012	10001	22001	10110	00201
disrupta	11112	00111	11000	10113	30321	22110	2?001
ribbei	11112	00111	11000	10113	30321	22110	0?301
lucida	11112	00111	11000	10113	30321	22110	1?301
flavitaenia	?????	?????	11000	10113	30321	22111	0?331
bajularia	11112	00111	11000	10113	30321	22110	0?301
emolliens	33220	01220	11010	22002	10101	00120	00001
sp. 14097	31210	01210	11010	22002	10101	00020	00111
sp. 14099	?????	?????	11110	22002	11101	00020	00111
maculosa	21230	00310	11110	20102	20101	20020	00001
subfumosa	21230	01210	11110	20002	22120	00120	00001
exul	22231	10120	11110	20002	22101	10020	01001
monochrias	?????	?????	11211	31002	10221	20020	01001
benguetana	21221	12210	?????	?????	?????	?0020	01001
xanthosoma	21220	00220	11211	30002	11211	2113?	??001
matutinata	21221	12220	11210	32002	10221	10020	00001
interrupta	22231	12220	11211	32002	10221	10020	11001
rotundata	22231	12220	11211	32002	10221	10010	11001
barbara	22231	10210	11211	32002	10101	10010	10321
basiflava	22231	12220	?????	?????	?????	?0010	11321
rosenbergi	22231	12220	11211	32002	10221	10010	11321
woodfordii	22221	12220	11211	32002	11201	12131	??331

Results

The results of the analysis are presented partially in the main text, and the strict consensus tree in Figure 5. This was obtained using Hennig86. Option mhennig gave three trees of length 126, a Consistency Index of 57, and a Retention Index of 77. The branch and bound option gave 112 trees of similar statistics to the mhennig tree; the strict consensus tree is derived from this second option. It offers a robust classification with a satisfactory degree of resolution and level of conflict between characters.

The characters themselves have already been discussed above. Particularly well-supported features of the classification are clades 5, 6, 11, 12, 13, 14, and 15.

APPENDIX 3. ANALYSIS OF SULAWESI SPECIES OF *ASTYGISA* WALKER

Astygisa is an ennomine geometrid genus that belongs to a tropical and subtropical complex of genera that feed on Rhamnaceae and that should probably be assigned to the tribe Caberini. *Apopetelia* Wehrli is a synonym. The complex will be discussed in more detail in part 11 of *The Moths of Borneo*, currently in preparation.

An important feature of *Astygisa* is the presence of four intersegmental pairs of coremata in the male abdomen. The genus has its greatest morphological diversity in the Asian mainland, but is also exceptionally rich in Sulawesi. All except one of the Sulawesi species form a monophyletic group, sister to the widespread Oriental species *vexillaria* Guenée, as discussed in the main text. A cladistic analysis of males was performed, using mainly genitalic characters, with *vexillaria* as outgroup. This outgroup taxon shares with the ingroup presence of a dorsal harpe or spine subbasally on the sacculus, but this is small and shows bilateral symmetry. The valves are expanded distally in *vexillaria* and also have prominent setation as in the ingroup.

In addition to the features of males listed below as definitive of the ingroup, the females have a distinctive array of processes surrounding the ostium bursae. There is an anterior labrum-like structure, a pair of serrate lateral processes, one of which is lost in some species, and a central, but asymmetric tongue-like structure, forked in two species.

The Walker species *theclaria* is a member of the Sulawesi group but the type is a female, and further investigation is needed to connect this name to a male of the group as species are best defined on features of the male genitalia. There is a specimen of one of the species labelled as from Timor, but this record needs confirmation.

The characters used in the male analysis, with notes on their contribution to clade definition, are listed below. The clades referred to are those indicated in Figure 16.

1. Uncus of male genitalia at apex: not hooked or coiled (0); weakly hooked or coiled (1); strongly coiled (2). The apomorphic states are found in all of clade A and homoplasiously in 14060. State 2 defines clade C.

2. Dorsobasal process of uncus (extending anteriorly into abdomen): directed in plane of uncus (0); directed to right (1). State 1 defines clade A.

3. Dorsobasal process of uncus: very small (0); large, rounded (1); large, triangular (2); large tricornute. A large process (all apomorphic states) defines the ingroup. State 1 defines clade C. State 2 is an autapomorphy of 14060.

5. Apex of valve: without clublike setae (0); with clublike setae (1). State 1 defines the terminal sister pairs.

6. Left harpe of valve: erect (0); flexed strongly distally to lie parallel to surface of sacculus (1). State 1 defines clade B.

7. Valve harpes with: bilateral symmetry (0); weak bilateral asymmetry (1); strong bilateral asymmetry (2). Asymmetry is seen in all of clade A, with strong asymmetry in clade B, but homoplasiously also in 12349.

8. Left valve with strong dorsolateral process: absent (0); present (1). This character appears to be homoplasious.

9. As 8, but for right valve. This is also homoplasious.

10. Valves with strong setae centrally on dorsal zone (from which may extend the dorsolateral process with setae): absent (0); present (1). State 1 occurs in all taxa except 12349, 14024, and *vexillaria*.

11. Distal parts of valves bilaterally: symmetric or weakly asymmetric (0); strongly asymmetric (1). State (1) defines the terminal triplet of clade B.

12. Pale discal spot of hindwing: weak or absent (0); moderate or strong (1). State 1 is seen in the terminal sister pair but also in 12349 and 14024. It is similarly homoplasious in taxa not found in Sulawesi.

13. Pale patch at centre of forewing margin: weak or absent (0); moderate to strong (1). State 1 defines clade A.

14. Harpe/spines on sacculus: small (0); large (1). State 1 defines the ingroup.

15. Cornuti in aedeagus vesica: present (0); absent (1). State 1 defines the ingroup.

Character Matrix

vexillaria	00000	00000	00000
sp. 12349	00300	02000	01011
sp. 14024	00300	00110	01011
sp. 14060	10210	00001	00011
furva	00300	00111	00011
sp. 14016	21110	12111	00111
sp. 14020	21110	12001	10111
sp. 14026	21110	01111	00111
sp. 12376	11301	12101	11111
sp. 14025	11301	12101	11111

Evolutionary Patterns in the Flora and Vegetation of New Caledonia

EXTENDED ABSTRACT

Porter P. Lowry II

INTRODUCTION

The origin, evolution, and phytogeographic history of the flora of the French Territory of New Caledonia have intrigued botanists for the past 125 years (Balansa, 1873; Brousemiche, 1884; Schlechter, 1905; Sarasin, 1917; Guillaumin, 1921, 1924, 1953a, b, 1954, 1964; Däniker, 1929, 1931, 1939; Good, 1955; Baumann-Bodenheim, 1956, 1988, 1989a, b, c, 1990; Virot, 1956; van Balgooy, 1960, 1971; Thorne, 1965, 1969; Raven & Axelrod, 1972, 1974; Holloway, 1979; Raven, 1980; Morat et al., 1981, 1984, 1986; Jaffré et al., 1987, in press; Schmid, 1987). In particular, New Caledonia's flora is characterized by remarkable levels of specific and generic endemism, and comprises an estimated 3138 native species of angiosperms (flowering plants) in an area of approximately 17,000 sq. km (Morat et al., 1984, 1986). Among the indigenous plants are particularly high numbers of relict representatives of some ancient, Australasian lineages that have either become extinct or are now poorly represented in other areas.

New Caledonia is located in the southwest Pacific about 1200 km east of Queensland, Australia, and ca. 1500 km NNE of New Zealand. The territory comprises: a single large island, Grande Terre, which is approximately 390 km long, averages about 50 km in width, and is oriented NNW to SSE; a number of smaller, geologically associated islands located to the NNW (Iles Bélep, Ile Baaba, Ile Balabio) and SSE (Ile Ouen and Ile des Pins); the Loyalty Islands, which are geologically unrelated (see below) and situated about 200 km to the NE of Grande Terre, covering about 2,000 sq. km; and a number of small outlying islands (the Chesterfields, Walpole, Hunter and Matthew).

This paper summarizes current information on the evolutionary history of New Caledonia's flora, and examines recent evidence on the development of its two most important and widespread vegetation types: the rain forests, and the maquis, a characteristic heathy, scrub-like formation that is almost exclusively restricted to New Caledonia's extensive ultrabasic soils.

Dr. Lowry is with the Missouri Botanical Garden, P.O. Box 299, St. Louis, MO 63166-0299, U.S.A. and is stationed at the Laboratoire de Phanérogamie, Muséum National d'Histoire Naturelle, 16, rue Buffon, 75005 Paris, France.

GEOLOGY, TECTONIC HISTORY, AND SOILS

The geology and tectonic history of New Caledonia have had a profound effect on the evolution of the area's flora and vegetation. With the exception of the Loyalty Islands, New Caledonia comprises an exposed section of a highly complex orogenic belt located along the Norfolk Ridge and flanked on both sides by deep ocean (Guillon, 1974; Lapouille, 1981; Paris, 1981a, b). It is clearly continental in origin, with the most recent tectonic data indicating that it separated from Australia about 65 m.y. ago, and moved to the NE to reach its present position about 50 m.y. ago (Coleman, 1980).

Geological evidence suggests that, during the Tertiary, and especially the Paleocene and middle Eocene, New Caledonia underwent a series of submersions, such that by the late Eocene almost all of the island was covered to a thickness of about 2000 m with peridotites, a type of igneous rock formed from oceanic crust that was thrust up during previous tectonic movement northeastward (Moores, 1973; Guillon, 1975). Today, after considerable weathering, peridotites (and related serpentinites) still cover about 5500 sq. km, or nearly ⅓ of New Caledonia (Guillon & Routhier, 1971; Guillon, 1975). It is not clear, however, whether *all* of New Caledonia's land area was submerged at the same time: botanical evidence, and in particular the tremendous floristic diversity, suggest that at least some land must have remained above water and served as a refugium throughout, although the emergent areas may have been situated to the south or west of present-day New Caledonia and are now submerged. Furthermore, it would be difficult to imagine that all of the numerous archaic lineages represented in New Caledonia's native flora reached the area by long-distance dispersal over the substantial water barrier that existed throughout Paleogene times.

Associated with New Caledonia's extensive peridotites and serpentinites are the ultrabasic soils that are so characteristic (Latham, 1975, 1981, 1986). These soils have exceptionally high levels of Fe and Mg, and of certain heavy metals such as Ni, Cr, Co and Mn (elements that are generally toxic to plants), as well as low levels of N, P, K, Ca and Al (Jaffré, 1976; Jaffré et al., 1987). Because of these chemical imbalances, and especially the ratio of Ca to Mg, ultrabasic soils present extreme edaphic conditions to plants, and have thus almost certainly played an important role in the evolution and diversification of the New Caledonian flora.

The Loyalty Islands, by contrast, are geologically very distinct. Comprising three main islands—Ouvéa, Lifou and Maré—they have a volcanic basalt basement dating from about 30 m.y. ago, which is covered by a layer of coral that was formed during an earlier period of submergence, but is now exposed as a result of a relatively recent general uplifting of the islands, probably beginning in the early Pleistocene, perhaps 1 m.y. ago (Dubois et al., 1973).

Because of the comparative youth of the Loyalties, coupled with their fairly uniform, calcareous soils and low topography, these islands have a rather depauperate flora composed entirely of species that are also found on the Grande Terre (and in many cases in other areas such as Vanuatu). Thus, since the present-day flora of the Loyalties is essentially a small sub-set of that occurring on the Grande Terre that has been derived through relatively recent long-distance dispersal events, and consequently is of rather limited interest, it will not be given further consideration in the present paper.

GEOGRAPHICAL FACTORS

New Caledonia is situated between about 19°30′S and 22°40′S latitude, and consequently has a typically tropical climate. A relatively high mountain chain runs along its entire length, with many massifs exceeding 1000 m in height, and five summits reaching above 1500 m [Mt. Panié (1628 m), Mt. Humboldt (1618 m), Mé Maoya (1508 m), Mt. Colnett (1505 m), and Mt. Kouakoué (1501 m)]. The moist trade winds that reach New Caledonia from the east generate a substantial amount of orographic precipitation along the eastern slopes, especially during the summer months between December and April. Rainfall thus exceeds 2,000 mm per year almost throughout the east coast, and reaches 4,000 mm annually in the Mt. Panié and Mt. Humboldt massifs (Service d'Hydrologie de l'ORSTOM & Service Territorial de la Météorologie, 1981). The western part of New Caledonia, by contrast, lies within the rain shadow of the central chain, where annual precipitation is generally less than ca. 1,000 mm, and in some sites even lower (e.g., Ouaco, ca. 800 mm; Uitoé Peninsula perhaps only 600–700 mm).

Thus, New Caledonia has, on the one hand, a set of physical conditions typical of many moderate sized tropical islands (such as its climate, precipitation patterns, and topography) and on the other a number of special features (including its geological age and continental nature, and the large expanses of ultrabasic soils). This particular combination of factors is largely responsible for the impressive plant diversity and the high level of endemism seen in New Caledonia's flora.

FLORISTIC DIVERSITY AND ENDEMISM

Current estimates indicate that almost 75% of New Caledonia's native flowering plant species occur nowhere else, and that fully 13.5% of the genera are likewise strictly endemic (Morat et al., 1984, 1986). Perhaps even more impressive are the levels of endemism found in the island's two primary vegetation types: rain forests, which occur on both ultrabasic soils and other substrates such as limestone, basalt, schist, etc.; and maquis, a heathy, scrub-like vegetation type that is almost exclusively restricted to ultrabasic soils. Together these two formations currently account for nearly 45% of the total area of the Grande Terre (Morat et al., 1981), and probably covered over half the island before the arrival of humans.

Morat et al. (1984) estimated that almost 89% of the species occurring in rain forests are endemic to New Caledonia, and of the 365 genera that have representatives in this vegetation type, fully 22.5% are also endemic. In a separate study of the maquis, Morat et al. (1986) estimated that nearly 93% of the species present in this specialized vegetation are restricted to the island, while almost 21% of the genera with representatives in the maquis are also endemic. (See also the ongoing series *Flore de la Nouvelle-Calédonie et Dépendences*.)

By comparison, overall endemism among the native plants of Hawaii is only slightly higher than on New Caledonia, with recent estimates indicating that 89% of Hawaii's species and about 15% of its genera are restricted to the island chain (Wagner et al., 1990). However, when one considers that the Hawaiian group is much more isolated geographically, it becomes clear just how remarkably distinctive New Caledonia's flora really is.

ORIGIN AND PHYTOGEOGRAPHY OF NEW CALEDONIA'S FLORA

Another important characteristic of New Caledonia's native flora is the comparatively large number of lineages that appear to be remnants of the late Cretaceous-early Tertiary flora that once covered large parts of Australasia. In the process of separating and drifting away from the Australian continent, New Caledonia is thought to have carried with it a sample of this ancient flora, the derivatives of which have been able to survive in its relatively equable climate, but were in large part eliminated from Australia due to well-documented changes in climatic conditions that took place in Neogene times (Raven & Axelrod, 1972, 1974; Holloway, 1979; Raven, 1980; Lowry, 1986a).

There are many examples of relict endemics that seem to fit this pattern, and point to the ancient origins of an important part of the New Caledonian flora. One of the best known and most spectacular is the gymnosperms, in which fully 43 of the 44 native species, representing nearly 7% of the world's entire gymnosperm flora, are known only from New Caledonia (de Laubenfels, 1972). Furthermore, of the 11 angiosperm genera with vesselless wood, a feature often regarded as primitive within the flowering plants, six occur on New Caledonia, and three of them (*Amborella, Exospermum* and *Zygogynum*) are endemic there (Lowry, 1986b; but see also Vink, 1985).

However, the flora is by no means exclusively derived from groups that were present when New Caledonia separated from Australia or that reached the area shortly thereafter. Many taxa appear to have dispersed to New Caledonia relatively more recently as part of a general influx of Indo-Malesian elements into Australasia that took place during the early and middle Tertiary (Holloway, 1979). It is interesting to note that some of these more recent arrivals have radiated into many of largest genera on New Caledonia, including such examples as *Phyllanthus*, with well over 100 species, *Psychotria* (ca. 80 spp.), *Pittosporum* (ca. 40 spp.), and several genera of Myrtaceae.

The phytogeographic relationships of New Caledonia's rain forest and maquis vegetation have been examined in detail (Morat et al., 1984, 1986). Furthermore, since these two formations include over 80% of the total native species, they were considered as a reasonable and representative sample of the New Caledonian flora for the purpose of quantitatively assessing its affinities with other phytogeographic areas. Morat and his co-workers, using genera as the working taxonomic unit, calculated a correlation coefficient that was proportional to the number of genera common to a given region and New Caledonia, and inversely proportional to the total number of territories in which each of the genera occurred. The relative importance of the floristic relationship between New Caledonia and each of the regions was then expressed as a percentage of the total of the correlation coefficients.

The results of these analyses for rain forest and maquis species are very similar, especially for the three most important regions—Australia, New Guinea, and Malesia, in that order. The conclusions reached by Morat et al. support the hypothesis that the New Caledonian flora comprises two primary elements: an ancient Australasian component that was already present when the island separated from Australia; and an element originating in New Guinea and Malesia that reached New Caledonia relatively more recently by long-distance dispersal.

By contrast, comparatively few groups appear to have phytogeographic affinities with other tropical Pacific islands or with more austral areas such as Norfolk Island, Lord Howe Island, and New Zealand. In the case of the tropical Pacific islands, this is probably a reflection of the fact that they represent rather small source and target areas, coupled in most cases with their relatively young age and generally more uniform edaphic conditions. The low affinities of the more austral islands with New Caledonia's flora likely reflect climatic differences more than anything else.

The role played by New Caledonia's ultrabasic substrates in the evolution and diversification of its flora has been discussed in a general way for many decades. In particular, some authors (e.g., Thorne, 1965) have hypothesized that the ultrabasics were critical in allowing many of the ancient Australasian lineages to survive in a sort of local, ecological refugium following their early adaptation to the special edaphic conditions that these substrates presented. Adaptation to the ultrabasics was also seen as the means by which many of the archaic plant groups were able to survive on New Caledonia, giving a degree of ecological separation and protection against more competitive rain forest taxa that must have continually been arriving from other tropical areas to the northwest.

377

It is clear that the flora now found on New Caledonia's ultrabasic substrates, both in rain forest and maquis vegetation, is particularly rich and has a high degree of endemism. A recent study (Jaffré et al., 1987) shows that almost half of the island's native flowering plant species occur on ultrabasics, despite the fact that these soils only cover about ⅓ of the land mass. Furthermore, almost 92% of these species are endemic, considerably higher than the 75% endemism seen in the native flora as a whole. Thus, most of the species that New Caledonia shares with other areas occur on non-ultrabasic substrates, a fact that is not particularly surprising.

Jaffré et al. (1987) also concluded that the flora found on the ultrabasics was derived in several ways, including: 1) early colonization of a small number of pre-adapted species, usually in open habitats; 2) development of forest vegetation types on highly weathered ultrabasic soils whose chemical characteristics resembled those derived from certain other substrates such as schists; and 3) evolution of tolerance to the extreme edaphic conditions of highly basic soils. The initial vegetation on the ultrabasic substrates was, in any case, highly depauperate, and thus presented ecological conditions favorable for subsequent speciation among those groups that were able to colonize and survive in these areas. This secondary diversification is considered to have given rise to a whole neo-endemic component to the flora, which today accounts for a large portion of the species-level diversity seen on ultrabasics.

There is general agreement that species (and in many cases whole lineages) adapted to ultrabasic substrates have benefited, and indeed still do benefit, from some degree of protection against allochthonous floristic elements reaching New Caledonia from other areas. However, the notion that this process has somehow provided a *special*, differential degree of protection to the oldest, archaic, Australasian elements does not appear to be borne out by the evidence. If such a selective protection actually occurred in areas with ultrabasic substrates, one would expect to see a higher proportion of species belonging to primitive groups in these areas than in zones with other substrates, where competition from more recently-arrived, paleo- or pantropical elements would presumably be higher.

This does not, however, appear to be the case: many groups that must have reached New Caledonia well after it separated from Australasia, but prior to or during the deposition of the ultrabasics in the Eocene, show the same kind of pattern of distribution between ultrabasic and non-ultrabasic substrates as the archaic taxa (Jaffré et al., 1987). Thus, while the presence of ultrabasics may have provided a sort of ecological isolation and protection against allochthonous elements that arrived following the Eocene, it did so indifferently with respect to primitive and more advanced groups that were present on New Caledonia prior to that time.

ACKNOWLEDGMENTS

Work on New Caledonia was funded in part by an NSF Doctoral Dissertation Improvement Grant (BSR 8314691), the Division of Biology and Biomedical Sciences of Washington University, St. Louis, and the Missouri Botanical Garden. Additional support provided by the W. Alton Jones Foundation is also gratefully acknowledged. I would also like to thank Dr. S. Miller for inviting me to participate in the ICSEB-IV Evolution on Island Archipelagoes Symposium and Prof. Ph. Morat for valuable comments on the manuscript.

LITERATURE CITED

Balansa, B. 1873. Sur la géographie botanique de l'Océanie et de la Nouvelle-Calédonie. *Bull. Soc. Hist. Nat. Toulouse* 7:327–332.

Balgooy, M. M. J. van. 1960. Preliminary plant geographical analysis of the Pacific. *Blumea* 10:385–430.

Balgooy, M. M. J. van. 1971. Plant geography of the Pacific. *Blumea Suppl.* 6:1–222.

Baumann-Bodenheim, M. G. 1956. Uber die Beziehungen der neu-caledonischen Flora zu den tropischen und den südhemisphärisch-subtropischen bis-extratropischen Flora und die Gürtemässige Gliederung der Vegetation van Neu-Caledonien. *Ber. Geobotan. Inst. Rübel, Zürich:* 64–74.

Baumann-Bodenheim, M. G. 1988. *Systematik der Flora von Neu-Caledonien (Melanesien-Südpazifik). Band 4. Farbfotos.* A. L. Schenk-Baumann: Merenschwand, Switzerland. 139 pp.

Baumann-Bodenheim, M. G. 1989a. *Systematik der Flora von Neu-Caledonien (Melanesien-Südpazifik). Band 5. Schwarzweissfotos.* A. L. Schenk-Baumann: Merenschwand, Switzerland. 100 pp.

Baumann-Bodenheim, M. G. 1989b. *Systematik der Flora von Neu-Caledonien (Melanesien-Südpazifik). Band 6. Thallophyta.* A. L. Schenk-Baumann: Merenschwand, Switzerland. 109 pp.

Baumann-Bodenheim, M. G. 1989c. *Systematik der Flora von Neu-Caledonien (Melanesien-Südpazifik). Band 7. Stelosporophyta.* A. L. Schenk-Baumann: Merenschwand, Switzerland. 191 pp.

Baumann-Bodenheim, M. G. 1990. *Systematik der Flora von Neu-Caledonien (Melanesien-Südpazifik). Band 14. Vergleiche.* A. L. Schenk-Baumann: Merenschwand, Switzerland. 127 pp.

Brousemiche, A. 1884. Considérations générales sur la végétation de la Nouvelle-Calédonie. *Arch. Méd. Navale* 41:250–260.

Coleman, P. J. 1980. Plate tectonics background to biogeographic development in the southwest Pacific over the last 100 million years. *Palaeogeogr., Palaeoclimat., Palaeoecol.* 31:105–121.

Däniker, A. U. 1929. Neu-Caledonien, Land und Vegetation. *Mitt. Bot. Mus. Univ. Zürich* 131:170–197.

Däniker, A. U. 1931. Ergenisse der Reise von Dr. A. U. Däniker nach Neu-Caledonien und der Loyaltäts Inseln. 3. Die Loyaltäts-Inseln und ihre Vegetation. *Vierteljahrsschr. Naturf. Ges. Zürich* 76:170–213.

Däniker, A. U. 1939. Neu-Caledonien. *Vegetationsbilder* 25, 6:1–9.

Dubois, J., Guillon, J. H., Launay, J., Recy, J. & J. J. Trescases. 1973. Structural and other aspects of the New Caledonia-Norfolk area. Pp. 223–235. *In:* P. J. Colman (ed.), *The Western Pacific: Island Arcs, Marginal Seas, Geochemistry.* University of Western Australia Press: Nedlands. 675 pp.

Good, R. 1955. Madagascar and New Caledonia, a problem in plant geography. *Blumea* 6:470–474.

Guillaumin, A. 1921. Essai de géographie botanique de la Nouvelle-Calédonie. Pp. 256–293. *In:* F. Sarasin & J. Roux, *Nova Caledonia, Vol. 1.* C. W. Kreidel: Berlin & Weisbaden.

Guillaumin, A. 1924. Le peuplement botanique de la Nouvelle-Calédonie. *Compt. Rend. 48e Session Assoc. Franç. Avanc. Sci., Liège:* 953–954.

Guillaumin, A. 1953a. Le développement de nos connaissances sur la flore et la géographie botanique de la Nouvelle-Calédonie et des Nouvelles-Hébrides. *Proc. Seventh Pacific Sci. Congr., Auckland,* 5:118–120.

Guillaumin, A. 1953b. Les caractères floristiques de la Nouvelle-Calédonie. *Proc. Seventh Pacific Sci. Congr., Auckland,* 5:120–122.

Guillaumin, A. 1954. A propos de la répartition de quelques Phanérogames de Nouvelle-Calédonie et des Nouvelles-Hébrides. *Compt. Rend. Somm. Soc. Biogéogr.* 31:38–40.

Guillaumin, A. 1964. L'endemisme en Nouvelle-Calédonie. *Compt. Rend. Somm. Soc. Biogéogr.* 358:67–75.

Guillon, J. H. 1974. The geology of New Caledonia and the Loyalty Islands. Pp. 445–452. *In:* A. M. Spencer (ed.), *Mesozoic-Cenozoic Orogenic Belts.* Geol. Soc. London: London. 809 pp.

Guillon, J. H. 1975. Les massifs péridotitiques de Nouvelle-Calédonie. *Mém. ORSTOM* 76:1–120.

Guillon, J. H. & P. Routhier. 1971. Les stades d'évolution et de mise en place des massifs ultrabasiques de Nouvelle-Calédonie. *Bull. B.R.G.M., Sér. 2, sect. IV:*5–38.

Holloway, J. D. 1979. *A Survey of the Lepidoptera, Biogeography and Ecology of New Caledonia.* Junk Publ.: The Hague. 588 pp.

Jaffré, T. 1976. Composition chimique et conditions d'allimentation minérale des plantes sur roches ultrabasiques (Nouvelle-Calédonie). *Cah. ORSTOM, Sér. Biol.,* 9:53–63.

Jaffré, T., Morat, P., Veillon, J.-M. & H. S. MacKee. 1987. Changements dans la végétation de la Nouvelle-Calédonie au cours du Tertiare: la végétation et la flore des roches ultrabasiques. *Bull. Mus. Natl. Hist. Nat., Paris, Sér. 4, sect. B., Adansonia* 9:365–391.

Jaffré, T. & J.-M. Veillon. (1990) 1991. Etude floristique et structurale de deux forêts denses humides sur roches ultrabasiques en Nouvelle-Calédonie. *Bull. Mus. Natl. Hist. Nat., Paris, Sér. 4, sect. B., Adansonia* 12. In press.

Lapouille, A. 1981. Le sud-ouest du Pacifique: données structurales. Pl. 5. *Atlas de la Nouvelle-Calédonie.* ORSTOM: Paris.

Latham, M. 1975. Les sols d'un massif de roches ultrabasiques en Nouvelle-Calédonie: le Boulinda. Les sols à accumulation ferrugineuse relative. *Cah. ORSTOM, Sér. Pédol.* 13:150–172.

Latham, M. 1981. Pédologie. Pl. 14. *Atlas de la Nouvelle-Calédonie.* ORSTOM: Paris.

Latham, M. 1986. Altération et pédogenèse sur roches ultrabasiques en Nouvelle-Calédonie. *Coll. Etudes & Thèses de l'ORSTOM:* Paris.

Laubenfels, D. J. de. 1972. Gymnospermes. *Flore de la Nouvelle-Calédonie et Dépendences* 4:1–168.

Lowry, P. P., II. 1986a. A systematic study of *Delarbrea* Vieill. (Araliaceae). *Allertonia* 4:169–201.

Lowry, P. P., II. 1986b. A systematic study of three genera of Araliaceae endemic to or centered in New Caledonia: *Delarbrea, Myodocarpus,* and *Pseudosciadium.* Ph.D. Dissertation, Washington University, St. Louis, Missouri. 258 pp.

Moores, E. M. 1973. Plate tectonic significance of alpine peridotite types. Pp. 963–973. *In:* D. M. Tarling & S. K. Runcorn (eds.), *Implications of Continental Drift to Earth Sciences. Vol. 2.* Academic Press: London & New York.

Morat, P., Jaffré, T., Veillon, J.-M. & H. S. MacKee. 1981. Végétation. Pl. 15. *Atlas de la Nouvelle-Calédonie.* ORSTOM: Paris.

Morat, P., Jaffré, T., Veillon, J.-M. & H. S. MacKee. 1986. Affinités floristiques et considérations sur l'origine des maquis miniers de la Nouvelle-Calédonie. *Bull. Mus. Natl. Hist. Nat., Paris, Sér. 4, sect. B., Adansonia* 8:133–182.

Morat, P., Veillon, J.-M. & H. S. MacKee. 1984. Floristic relationships of New Caledonian rain forest phanerogams. Pp. 71–128. *In:* F. J. Radovsky, P. H. Raven & S. H. Sohmer (eds.), *Biogeography of the Tropical Pacific.* Assoc. of Systematic Collections: Lawrence, Kansas & Bernice P. Bishop Museum Press: Honolulu. 221 pp.

Paris, J.-P. 1981a. *Géologie de la Nouvelle-Calédonie. Un essai de synthèse.* B.R.G.M.: Orléans. 278 pp.

Paris, J.-P. 1981b. Géologie. Pl. 9. *Atlas de la Nouvelle-Calédonie.* ORSTOM: Paris.

Raven, P. H. 1980. Plate tectonics and Southern Hemisphere biogeography. Pp. 1–24. *In:* K. Larsen & L. B. Holm-Nielsen (eds.), *Tropical Botany.* Academic Press: London, New York & San Francisco.

Raven, P. H. & D. I. Axelrod. 1972. Plate tectonics and Australasian biogeography. *Science* 176:1379–1386.

Raven, P. H. & D. I. Axelrod. 1974. Angiosperm biogeography and past continental movements. *Ann. Missouri Bot. Gard.* 61:539–673.

Sarasin, F. 1917. Neu-Kaledonien und die Loyalty-Inseln. Reise-Erinnerungen eines Naturforscher.

Schlechter, R. 1905. Pflanzengeographische Gliederung der Insel Neu-Kaledonien. *Bot. Jahrb. Syst.* 39:1–41; 40:20–45.

Schmid, M. 1987. Conditions d'évolution et caractéristiques du peuplement végétal insulaire en Mélanésie occidentale: Nouvelle-Calédonie, Vanuatu. *Bull. Soc. Zool. France* 112:233–254.

Service d'Hydrologie de l'ORSTOM & Service Territorial de la Météorologie. 1981. Eléments généraux du climat. Pl. 11. *Atlas de la Nouvelle-Calédonie.* ORSTOM: Paris.

Thorne, R. F. 1965. Floristic relationships of New Caledonia. *Univ. Iowa Stud. Nat. Hist.* 20:1–14.

Thorne, R. F. 1969. Floristic relationships between New Caledonia and the Solomon Islands. *Philos. Trans. Roy. Soc. Lond.* 255:595–602.

Vink, W. 1985. The Winteraceae of the Old World. V. *Exospermum* links *Bubbia* to *Zygogynum.Blumea* 31:39–55.

Virot, R. 1956. La végétation canaque. *Mém. Mus. Natl. Hist. Nat., Paris, Sér. B, Botanique* 7:1–398.

Wagner, W. L., Herbst, D. L. & S. H. Sohmer. 1990. *Manual of the Flowering Plants of Hawaii. Vol. 1.* University of Hawaii Press & Bishop Museum Press: Honolulu.

The Evolution of the Hypogean Fauna in the Canary Islands

P. Oromí, J. L. Martín, A. L. Medina, and I. Izquierdo

Abstract.   The Canary Islands, lying in the eastern Atlantic Ocean close to North Africa, are entirely volcanic and have a mild, subtropical climate. The fauna is highly endemic as a result of the oceanic situation, and has successfully colonized the subsoil of those zones that are sufficiently humid. Up to the present time some 60 troglobitic species have been found, apart from about 10 stygobiont species in ground water and another 10 species in anchialine caves. All the terrestrial troglobites are single-island endemics, evidently evolved *in situ* from local surface or soil living species. In a few cases the epigean ancestral stock is still extant, but in others the surface living lineage has disappeared completely. The relationships between the subterranean species and their epigean relatives are herein analysed, and the possible parapatric origin for some species is discussed, as well as the relict condition of some others.

THE CANARY ISLANDS: GEOGRAPHIC BACKGROUND

In the northeast sector of the Central Atlantic lie the archipelagos of the Azores, Madeira, Canaries and Cape Verdes, which together form the biogeographic region Macaronesia. The origin of the last three archipelagos is closely connected with the formation of the nearby continental margin (Schmincke, 1976). All three are of volcanic origin and their period of maximum activity was during the Miocene (between 24 and 6 Ma B.P.); eruptions have now ceased in Madeira, in some of the Canaries and in most of the Cape Verdes. The Azores are on the whole much younger (6 Ma maximum); their origin is directly related to the Mid Atlantic Ridge and they are in a geodynamically much more active zone (Feraud et al., 1980).

The Canary Islands form a band about 500 km long and 200 km wide between 27°37′ and 29°25′N and 13°20′ and 18°10′W (Fig.1). The minimum distance between the closest island (Fuerteventura) and the African continent (Cape Juby) is about 110 km, whilst the most distant island (La Palma) is 460 km from the mainland. There are seven main islands in the archipelago (El Hierro, La Palma, La Gomera, Tenerife, Gran Canaria, Fuerteventura and Lanzarote) and several islets, with a total area of 7501 km². All the western and central islands rise to 1400 m above sea level, the highest point being Mt. Teide in Tenerife at 3718 m. Lanzarote and Fuerteventura, on the other hand, are very eroded and have much gentler relief, with a maximum altitude well below 1000 m.

The origin of the Canaries has long been a matter for argument, since the islands are situated close to the continental/oceanic boundary. The existence of a great thickness of sediments between the eastern islands and Africa obscures the true nature of the deep

Drs. Oromí, Martín, Medina, and Izquierdo are with the Department of Animal Biology (Zoology), University of La Laguna, 38206 La Laguna, Tenerife, Canary Islands, Spain.

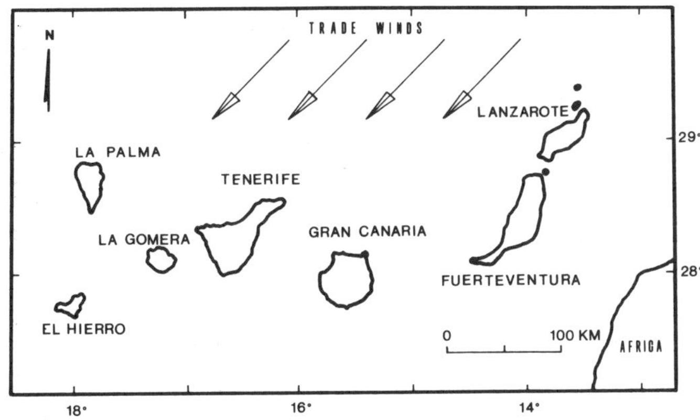

Figure 1. Map of the Canary Islands.

structure of the crust in this area. The unexplained presence of subfossil eggs of ostriches in Lanzarote and Fuerteventura has prompted several authors to consider that these two islands are of continental origin, with later subaerial deposition of volcanic material (Sauer & Rothe, 1972). However, there is no compelling proof for the existence of continental crust beneath the eastern Canaries (Schmincke, 1976). The most recent geological theories (Hernán, 1985) consider that all the islands were built up from the ocean floor, in spite of their closeness to the continent. The hypothesis of a hotspot, which explains the origin of Hawai'i and other archipelagos, has also been applied to the Canaries (Morgan, 1971; Wilson, 1973). However, this idea has been subsequently rejected on the basis of both geochemical considerations (Fuster, 1975) and crustal dynamics (Anguita & Hernán, 1975). The reduction of the age of the islands from east to west (Table 1) (see Cantegrel et al., 1984; Mitchell-Thomé, 1985), combined with the continuation of active vulcanism in the most eastern islands, is explained better by the hypothesis of a propagating fracture (Anguita & Hernán, 1975) than by the classical hotspot model. This fracture would be directly related to the formation of the Atlas Mountains during the Alpine orogeny.

In some heavily eroded places the plutonic basal complex is exposed, but most of the rocks that form the islands are extrusive. Although in earlier periods of activity both acidic and basic rocks were formed, there has been a predominance of the latter in the recent eruptive series. As a result basalts predominate on the surface, forming lava flows that have numerous cracks and are frequently traversed by lava tubes. This complicated network of spaces of various sizes in basaltic terrain (see Howarth, 1983), together with the subsequent formation of the "Milieu Souterrain Superficiel" (MSS) (Juberthie, 1983) or Mesocavernous Shallow Stratum (Ashmole et al., 1990) has produced a subterranean habitat readily colonized by a specialized hypogean fauna. In the Canaries the inhabitable underground occurs both in terrestrial habitats (Martín & Oromí, 1986; Oromí et al., 1986; Hernández et al., 1986) and in freshwater (Stock, 1988a,b) or anchialine habitats (Wilkens et al., 1986).

Table 1. Area, maximum elevation and age of the Canary Islands.

Island	Area (km^2)	Max elev. (m)	Age (Ma)	
Hierro	277	1501	0.8	W
La Palma	728	2426	1.6	
Gomera	378	1487	12.	
Tenerife	2058	3717	15.	
Gran Canaria	1534	1950	16.	
Fuerteventura	1731	807	22.	
Lanzarote	796	670	19.	E

In the older zones where there are outcrops of plutonic rocks or extrusive rocks of salic type, lava tubes are never formed. Although there are also basaltic flows in these areas, massive surface erosion has destroyed the lava tubes (Howarth, 1973; Oromí et al., 1985). Even in the absence of caves, a specialized subterranean fauna occupies the MSS. This is the situation in La Gomera and in the older parts of Tenerife and Gran Canaria (Medina & Oromí, 1990).

The Canarian climate is characterized by its mildness, largely due to the oceanic influence; the monthly mean temperatures vary between 18°C and 21°C depending on altitude. These unusually low temperatures for such latitudes are influenced by cool ocean currents and upwellings and by the cool, humid NE trade winds that blow towards the islands during most of the year. Warmer and drier trade winds coming from the NW are also very constant, but blow at higher levels. Between the upper hot, dry air and the lower humid winds there is an inversion zone, in which extensive cloud banks are formed around 1000 m above sea level on the windward slopes of the higher islands; however, this height varies seasonally and even daily. These clouds do not normally produce heavy rains, but give rise to very humid conditions on the windward slopes. In contrast, the leeward slopes are dry. Winter rains fall from November to March, and are mainly carried by westerly winds formed in oceanic depressions; these mostly affect the western islands. On the other hand, periods with hot and dry winds from the Sahara occur several times a year, carrying abundant dust (as well as flying and flightless invertebrates). These winds mainly affect the two easternmost islands, which are much lower and almost desert.

As a result of this climatic picture, the islands have in general the following vegetation zones: 1) dry to arid subtropical scrub up to about 250 m; 2) humid to semi-arid subtropical scrub and woods from about 250 to 600 m; 3) humid laurel forest in the cloud belt, from about 600 m to 1000 m; 4) humid to dry temperate pine forest from about 1000 to 2000 m; and 5) dry subalpine scrub over 2000 m. On the southern slopes the laurel forest hardly exists, and the transitions are usually at higher altitudes. This altitudinal distribution of climate and vegetation is important in relation to colonization of the subterranean environment because below-ground humidity is essential for life there.

THE SUBTERRANEAN FAUNA OF THE CANARIES

The Canarian fauna is closely related to that of neighbouring Madeira, and to some extent also to those of the other Macaronesian archipelagos (Azores and Cape Verdes). Although elements of the fauna have diverse origins, the main influence is that of the Mediterranean region. The oceanic origin of the Canary Islands, and in most cases the lack of any past connection between them, has led to an evolutionary pattern typical of island groups, with several cases of adaptive radiation. Although there was a greater richness of initial colonizing stocks than in more remote archipelagos such as the Azores or Hawai'i, many animal groups are conspicuous by their absence. There were neither amphibians nor snakes among the original fauna, and apart from bats there were only five mammals: three extinct rodents and two extant shrews (Molina & Hutterer, 1989). Numerous invertebrate families and even whole orders that occur in the mainland are also absent.

Animals adapted to subterranean life encounter special difficulties in reaching oceanic islands, since their morphological and physiological adaptations lessen their capacity for survival outside their habitat (Coiffait, 1958). Although it has been shown that some species considered as subterranean are capable of colonizing islands (Peck, in press), it is clear that these are actually soil organisms. It is difficult to establish valid criteria for classifying subterranean animals by the habitat in which they occur; it is hard to make clear distinctions between edaphobites, troglophiles, troglobites, phreaticoles, stygobionts, etc. (Chapman, 1986a); this is especially true in tropical and subtropical zones, where troglomorphs sometimes occur on the surface (Paoletti, 1980; Peck, 1982; Martín & Oromí, 1987).

In an analysis of the subterranean fauna of the Galapagos, Peck (in press) proposes to unite under the term cryptozoa all the animals that live in spaces at or beneath the earth's surface, and have relevant adaptations. This idea is similar to Chapman's (1986a) concept of stygicoles, which included all the free-living animals that inhabit perpetually dark habitats. However, we still consider that species living in the soil (endogean species or edaphobites) have a different significance from those occurring in the subsoil or in the underground spaces (hypogean species) and that they should not be considered together from the biogeographic point of view. The latter are usually called troglobites or troglomorphs when they are obligate underground inhabitants and show morphological adaptations to this environment. Although endogean and hypogean species show some similar adaptations (especially eye reduction and depigmentation) they often differ in others (for instance body size, physogastry, elongation of appendages and, especially, physiological adaptations). Peck (in press), in his study of the cryptozoa of the Galapagos, considers four groups of animals: 1) species probably introduced by humans; 2) species belonging to eyeless genera or higher taxa that also occur elsewhere; 3) endemic species with close (sister) eyed species living in the same island; and 4) endemic species with no close congeneric eyed or epigean relatives in the island.

The members of group 1 are of no biogeographic interest and need not be considered further. In group 2 are included both non-endemic and endemic species, but the latter have the same biogeographic significance as any endemic epigean species. Palpigrada, Geophilomorpha, Symphyla, Diplura and other arthropods of the Canarian fauna must have arrived in an eyeless condition (this being the most obvious characteristic of surface animals) since they belong to groups entirely without eyes. The same may be said for lower taxa such as families (Cryptopidae, Polyxenidae, Blaniulidae) or genera (the beetles *Entomoculia, Geomitopsis, Langelandia, Paratorneuma,* etc.) lacking eyed representatives. The loss of eyes in each case was well before the colonization of the Canaries. Almost all the cryptozoa in this group are typical soil-dwelling organisms, as mentioned above. One exception is *Cryptops vulcanicus* Zapparoli, a species which, as well as being eyeless shows other clear adaptations to the hypogean environment (Zapparoli, in press).

The species included in groups 3 and 4, on the other hand, are particularly interesting because they have evolved from eyed ancestors that arrived in the archipelago. We can assume, and shall argue below, that their adaptations to subterranean life took place independently on each of the islands where they occur. In the present analysis of the hypogean fauna in the Canaries we will consider only the species belonging to these two groups.

EPIGEAN AND HYPOGEAN SISTER SPECIES

The occurrence of subterranean fauna in temperate zones has been explained classically by allopatric speciation, occurring in general by disjunct distribution caused by Pleistocene glaciation (Vandel, 1964; Barr & Holsinger, 1985). However, for continental zones in the tropics and for oceanic islands (which were not so affected by these climatic changes) an alternative model of parapatric speciation has been proposed (Howarth, 1980). This process is suggested by the existence of pairs (or larger groups) of sister species in the same geographic zone, some of them epigean (with eyes) and others hypogean. Such a situation has been observed in several arthropods in Hawai'i (Howarth, 1987), in Asia and New Guinea (Chapman, 1986b), and in the Galapagos (Peck, in press), as well as in the *Astyanax fasciatus* species complex of Mexican cave fish (Wilkens, 1987). In the Canary Islands we consider that at least seven hypogean species have originated by parapatric speciation, since they live in the same island as closely related epigean species (Table 2).

One of the most spectacular cases is that of *Dysdera*, a genus of spider that has 21 described species (Wunderlich, 1987) and probably more than 50 species as yet to be described (Wunderlich, pers. comm.). In Tenerife three troglobitic species and one troglophile live together in the same complex of lava tubes (Ribera et al., 1985; Ribera & Blasco, 1986), and several other subterranean species live in other parts of the island.

The troglobitic pseudoscorpion, *Paraliochthonius tenebrarum* Mahnert, from Tenerife is very closely related to *P. canariensis* Vachon, an epigean species found in Lanzarote. On the other hand, the other known subterranean species, *P. martini* Mahnert, from El Hierro, seems to be more related to *P. singularis* from the Mediterranean basin (Mahnert, 1989).

The amphipod genus *Palmorchestia* (Talitridae) is endemic to La Palma, and has two species—the eyed *P. epigaea* Stock from the laurel forest, and the microphthalmic, cave-dwelling *P. hypogaea* Stock & Martín.

Two troglobitic centipedes are known in the Canaries, the already mentioned *Cryptops vulcanicus* and *Lithobius speleovulcanus* Serra. Although it has not been decided which are their closest relatives, both genera also have epigean representatives, some of which are endemic (Bröleman, 1900; Eason, 1985).

Table 2. Hypogean species with eyed relatives in the archipelago occurring in the same or in other islands. * = species likely to have originated by parapatric speciation.

| | | sister epigean | |
Hypogean	Island	Same	Other
Pseudoscorpiones			
Paraliochthonius martini	Hierro	no	yes
Paraliochthonius tenebrarum	Tenerife	no	yes
Araneae			
Dysdera esquiveli	Tenerife	yes	yes
Dysdera ambulotenta	Tenerife	yes	yes
Dysdera unguimmanis	Tenerife	yes	yes
Lepthyphantes oromii	Tenerife	yes	yes
Amphipoda			
*Palmorchestia hypogaea**	La Palma	yes	no
Chilopoda			
Lithobius speleovulcanus	Tenerife	yes	yes
Cryptops vulcanicus	Tenerife	?	?
Dermaptera			
*Anataelia troglobia**	La Palma	yes	yes
Blattaria			
Loboptera subterranea	Tenerife	yes	yes
Loboptera anagae	Tenerife	yes	yes
Loboptera cavernicola	Tenerife	yes	yes
Loboptera troglobia	Tenerife	yes	yes
Loboptera fortunata	La Palma	no	yes
Loboptera ombriosa	Hierro	no	yes
Homoptera			
*Tachycixius lavatubus**	Tenerife	yes	no
Coleoptera			
*Eutrichopus martini**	Tenerife	yes	no
Licinopsis obliterata franzi	Hierro	yes	yes
Licinopsis picescens	Hierro	yes	yes
Licinopsis angustula	La Palma	yes	yes
Licinopsis a. schurmanni	Hierro	yes	yes
*Trechus minioculatus**	Hierro	yes	yes
*Trechus benahoaritus**	La Palma	yes	yes
*Thalassophilus subterraneus**	La Palma	yes	yes

The earwig, *Anataelia troglobia* Martín & Oromí, from La Palma, is very closely related to *A. lavicola* Martín & Oromí from La Palma and El Hierro, the two species being distinguished mainly by the markedly troglomorphic characters of the former, which is totally without eyes. The extent to which the two species can coexist in La Palma is shown by the fact that they have been found in the same cave and even in the same trap (Martín & Oromí, 1988). Although *A. lavicola* frequently enters caves, it is a pigmented and eyed insect and is fundamentally lavicolous (Martín et al., 1987). The third species of this endemic genus is *A. canariensis* Bolívar, an eyed and pigmented insect found in La Gomera and Tenerife, usually in recent lava and in dry caves.

Cockroaches of the genus *Loboptera* Brunn., with a wide Mediterranean distribution, are micropterous or virtually wingless but are normally epigean and with appropriate morphological characteristics. In the Canaries seven species are known (Izquierdo & Martín, 1990), of which only one (*L. canariensis* Chopard) is well pigmented and occurs exclusively on the surface. Among the remaining six (Table 2) there is a variable degree of adaptation to subterranean life, especially relating to pigmentation and eye reduction (Martín & Oromí, 1987; Martín & Izquierdo, 1987; Izquierdo & Martín, 1990). In some species, for instance *L. fortunata* Krauss, *L. ombriosa* s.str. Martín & Izquierdo, and *L. anagae* Martín & Oromí, the adaptations are not very advanced; these live below ground but may be found at the surface under certain conditions (Martín et al., 1986). Others, such as *L. ombriosa meridionalis* Martín & Izquierdo and *L. cavernicola* Martín & Oromí, are obligate inhabitants of caves or the MSS: they are more strictly depigmented but still have eyes, although these are reduced. Finally, the two most highly adapted species, *L. subterranea* Martín & Oromí and *L. troglobia* Izquierdo & Martín, show extreme depigmentation and are completely eyeless, as well as showing other clear troglomorphic characters. A recent study by Izquierdo et al. (1990) shows that the stage of morphological adaptation of each species to the subterranean habitat is correlated with reproductive capacity (number of ovarioles and compartments in the ootheca) (Table 3).

Table 3. Number of ovarioles in Canarian *Loboptera* species. EE: exclusively epigean; HE: hypogean and occasionally epigean; EH: exclusively hypogean (after Izquierdo et al., 1990).

Species		
L. canariensis	EE	16–18
L. fortunata	HE	12
L. ombriosa	HE	12
L. anagae	HE	12
L. subterranea (from Icod)	EH	12
L. subterranea (from Teno)	EH	10
L. cavernicola	EH	10
L. troglobia	EH	6

The distributional pattern of the Canarian *Loboptera* deserves a special comment (Fig. 2). *L. fortunata* is found in La Palma, and *L. ombriosa* in El Hierro, with totally allopatric distribution. In El Hierro it appears that geographic subspecies have evolved, with distinct ecology (Martín & Izquierdo, 1987). In Tenerife we find the other five species, with interesting cases of allopatry and sympatry. *L. canariensis* lives above ground throughout much of the island. *L. anagae* and *L. cavernicola* are found only in the northeast (Anaga peninsula), the former in the MSS of the wooded ridges and the latter in a volcanic pit in the low dry zone; in spite of their geographic proximity they have never been found together. Finally *L. subterranea* appears to occupy all the subterranean habitat (MSS and caves) of Tenerife except in the Anaga peninsula (Martín et al., 1986); and it occurs together with *L. troglobia* in Chío Cave.

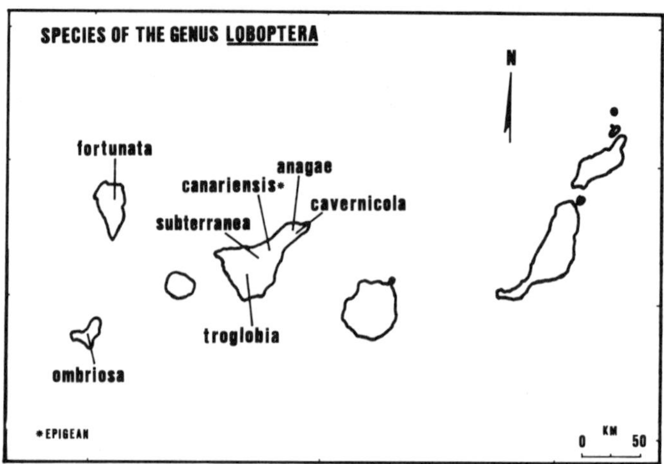

Figure 2. Distribution of *Loboptera* species in the Canary Islands.

In some caves of Tenerife *Tachycixius lavatubus* Remane & Hoch occurs. This troglobitic planthopper (Cixiidae) is closely related to the epigean *T. canariensis* Lindberg, also living on Tenerife and Gran Canaria. As many as 12 cave-dwelling planthopper species (Cixiidae and Meenoplidae) have been found in the Canarian lava tubes and MSS during the last few years (H. Hoch & M. Asche, pers. comm.), most of them not yet described.

The carabid beetles are undoubtedly the arthropod group richest in troglobitic representatives in the Canaries. The endemic Tenerife genus, *Eutrichopus* Tschitschérine, has two epigean and one hypogean species, all three being very closely related (Machado, 1984). *E. fernandezi* Mateu is confined to the humid forests in Anaga; *E. gonzalezi* Mateu is found in similar habitats on the northern slopes except in Anaga (Machado, 1976); and *E. martini* Machado, the troglobitic species, is found only in caves in the north and central part of the island, overlapping with part of the distribution of *E. gonzalezi*.

The genus *Licinopsis* Bedel is also a Canary endemic, and has five species with a complicated subspecific distribution among the four western islands (Machado, 1987). Some of the species, such as *L. alternans* (Dejean) and *L. gaudini* Jeannel, are epigean, although the first has occasionally been found in caves (Martín et al., 1985). The other three species have some morphological adaptations for subterranean life and are primarily hypogean, although they have been found occasionally on the surface (Oromí et al., 1989). They are especially abundant in the MSS, particularly in El Hierro where the three species *L. picescens* (Wollaston), *L. obliterata* (Wollaston) and *L. angustula* Machado (represented by local subspecies) can all be found together. In La Palma *L. angustula* s.str. is abundant in caves and the MSS, while the previously mentioned *L. gaudini* is present on the surface.

Trechus species also provide two cases of sympatry between epigean and subterranean species. In El Hierro *T. flavocinctus gomerae* Jeannel is abundant in the forests, and in the same area *T. minioculatus* Machado is found in the subsoil: it is probably derived from the former (Machado, 1987). In La Palma there are several endemic epigean species of this genus, and also the subterranean species, *T. benahoaritus* Machado. On both these islands hypogean and epigean species have sometimes been collected in the same place.

The trechodine ground beetle, *Thalassophilus whitei* Wollaston, occurs in nearly all the Canary Islands and in Madeira, being typically ripicolous. In La Palma there is *T. subterraneus* Machado, an eyeless species inhabiting the subsoil. Similarly, in Madeira there are two hypogean species, *T. caecus* Jeannel and *Thalassophilus* sp., the latter found in a cave (H. Pieper, pers. comm.).

SPECIES WITH NO EPIGEAN RELATIVES: THE POSSIBLE RELICTS

All the members of group 4 belong to genera that have no surface-living species in the Canaries, or even anywhere else in cases where the genera are endemic. For many groups the epigean fauna in the archipelago is rather well known, and the absence of surface-living species implies extinction of a surface-living ancestor subsequent to the evolution of the hypogean forms, which must be considered as relict species. After the discovery of highly adapted troglobites in regions hardly influenced by glaciations, the relict status of continental subterranean species has sometimes been questioned (Howarth, 1980, 1983; Chapman, 1986a); in the case of oceanic islands, however, the absence of surface relatives implies a relictual situation for troglobites. In the Canaries this situation applies to at least 25 species in 14 genera, collected either in caves or in the MSS (Table 4); several species have now been added to those considered as relicts by Martín et al. (1989).

Pseudoscorpiones. The cavernicolous *Tyrannochthonius superstes* Mahnert is the only representative of the genus in the Western Palearctic. Its closest epigean relatives occur in the Ethiopian Region (Mahnert, 1986), but in Hawai'i and the Galapagos there are also hypogean pseudoscorpions belonging to this genus (Muchmore, 1979; Peck, in press).

Table 4. Hypogean relict species collected in caves or in the MSS with no relatives in the same island; (* = endogean species).

Species	Island	Extant native relatives
Pseudoscorpions		
Tyrannochthonius superstes	Tenerife	unknown
Araneae		
Agroecina n.sp.	Tenerife	troglobite
Isopoda		
Venezillo tenerifensis	Tenerife	unknown
Blattaria		
Loboptera fortunata	La Palma	unknown
Loboptera ombriosa	Hierro	unknown
Hemiptera		
Collartida anophthalma	Hierro	unknown
Homoptera		
Meenoplus cancavus	Hierro	unknown
Meenoplus sp.	La Palma	unknown
Coleoptera		
Wolltinerfia tenerifae	Tenerife	troglobite
Wolltinerfia anagae	Tenerife	troglobite
Pseudoplatyderus amblyops	Gomera	unknown
Canarobius oromii	Tenerife	troglobite
Canarobius chusyae	Tenerife	troglobite
Spelaeovulcania canariensis	Tenerife	unknown
*Lymnastis gaudini**	Tenerife	edaphobite
*Lymnastis gomerensis**	Gomera	unknown
Domene vulcanica	Tenerife	troglobite
Domene alticola	Tenerife	troglobite
Domene sylvatica	Tenerife	troglobite
Domene jonayi	Gomera	unknown
Domene benahoarensis	La Palma	unknown
Apteranopsis canariensis	Tenerife	troglobite
Apteranopsis outereloi	Tenerife	troglobite
Apteranopsis hephaestos	La Palma	troglobite
Apteranopsis tanausui	La Palma	troglobite
Apteranopsis palmensis	La Palma	troglobite
Apteranopsis junoniae	La Palma	troglobite
*Oromia hephaestos**	Tenerife	edaphobite

Araneae. The most abundant and widespread troglobitic spider in Tenerife and probably in Gran Canaria is an undescribed species of *Agroecina*. This clubionid genus also includes one more hypogean species, on La Palma (Wunderlich, pers. comm.).

Isopoda. *Venezillo tenerifensis* Dalens is probably the only troglobitic isopod known so far in the archipelago, occurring in several caves of Tenerife. The eastern islands of Lanzarote and Fuerteventura are inhabited by another close species of this genus, the epigean *V. canariensis* (Dollfuss).

Hemiptera. A remarkable case of geographic disjunction is that of the thread-legged bug, *Collartida anophthalma* Español & Ribes, which inhabits the caves of southern Hierro. All the remaining known species of this genus are in Central Africa except one in Middle East, and only one of them (from Zaire) has some eye reduction (Español & Ribes, 1983). It is notable that this specialized troglomorph is found only in El Hierro, the youngest and westernmost island in the Canaries. However, another species, not yet described, has recently been discovered in caves of La Palma (Hoch & Asche, pers. comm.).

Homoptera. The planthopper, *Meenoplus cancavus* Remane & Hoch, from El Hierro was the first troglobitic meenoplid ever found in the Palearctic (Remane & Hoch, 1988). During recent investigations in the archipelago other new species of troglobitic meenoplids have been discovered in El Hierro and La Palma (H. Hoch & M. Asche, pers. comm.) but not a single epigean species of the family is yet known from the Canaries.

Coleoptera. There are many relict species among beetles, most of them belonging to endemic genera. Among the most remarkable are *Wolltinerfia tenerifae* Machado and *W. anagae* Medina & Oromí, the only representatives of this genus, which is endemic to Tenerife (Machado, 1984, 1985). These species are closely related and inhabit mainly the MSS, though the former has occasionally been found in caves (Hernández et al., 1986). While *W. tenerifae* is eyeless and occupies the northwest quarter of the island, *W. anagae* still has very small eyes and is restricted to the Anaga zone, at the northeast corner (Medina & Oromí, in press).

Pseudoplatyderus amblyops is another eyeless troglobite living in the MSS of La Gomera (Medina & Oromí, 1990). The genus is monotypic and endemic to this island, which lacks lava tubes due to its old age and absence of Quaternary eruptions. *Spelaeovulcania canariensis* Machado, *Canarobius chusyae* Machado and *Canarobius oromii* Machado are troglomorphic ground beetles, the only representatives of these two genera that are endemic to Tenerife (Machado, 1987); in contrast to *Pseudoplatyderus*, these three taxa have been collected in caves but never in the MSS.

Lymnastis gaudini Jeannel and *L. gomerensis* Franz can also be considered as relict ground beetles, since they are the only representatives of the genus in the Canaries. *Lymnastis* species known from elsewhere all have eyes (Jeannel, 1929). *L. gaudini* is restricted to Tenerife, and *L. gomerensis* occurs in La Gomera. These are eyeless, depigmented insects but more polyvalent in the sense that they inhabit caves as well as the MSS or the soil; their morphology, however, is similar to that of typical edaphobites.

The staphylinids are well represented in the hypogean fauna of the Canaries, although some of the species have been collected very rarely. The genus *Domene* Fauvel includes some 16 epigean, Palearctic, continental species and three troglobites in Morocco. There are five troglomorphic species in the Canaries (Fig. 3) and none living on the surface (Oromí & Hernández, 1986). *Domene benahoarensis* Oromí & Martín, endemic to La Palma, is the less troglomorphic and the commoner species (Oromí & Martín, 1990). *Domene jonayi* Hdez. & Medina occurs exclusively in La Gomera, and has so far been collected only in the MSS (Medina & Oromí, 1990). In the lava tubes of the younger parts of Tenerife (north and center) are found *D. vulcanica* Oromí & Hdez. and *D. alticola* Oromí & Hdez., two species that are highly troglomorphic (Oromí & Hernández, 1986). *D. sylvatica* Hdez.& Oromí inhabits the Anaga zone, the oldest part of Tenerife; however, this is a less troglomorphic species still displaying minute eyes, comparable to those of *D. benahoarensis*.

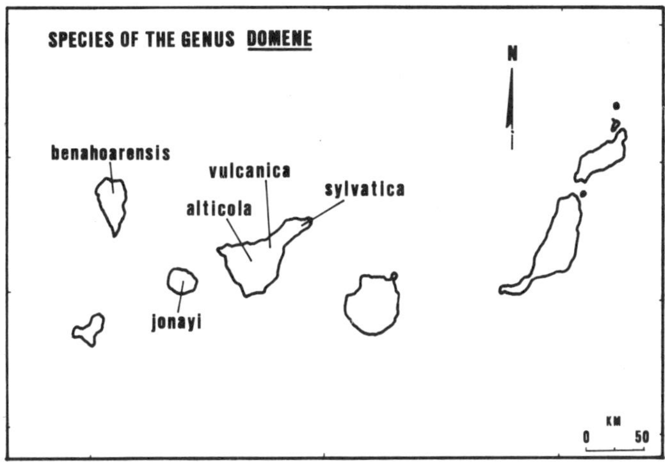

Figure 3. Distribution of *Domene* species in the Canary Islands.

The diversity of troglomorphic species in the genus *Apteranopsis* Jeannel is remarkable: for the moment two are known from Tenerife and four from La Palma, besides those existing in the Maghreb (Hernández & Martín, 1990). No epigean species have ever been found on the islands, and none of the troglobitic species have colonized the Anaga zone in Tenerife. This genus is distributed in Tunisia and Algeria, but in Morocco it is replaced by *Apteranillus,* a closely related genus including several cryptozoic species (Oromí & Martín, 1984).

Finally, it is worth mentioning the weevils *Oromia hephaestos* A. Zarazaga and *O. aguiari* A. Zarazaga, the only members of an endemic genus of Tenerife. This genus belongs to the tribe Cycloterini, with no other species in the Canarian surface fauna (Alonso-Zarazaga, 1987, 1990). These eyeless species have been collected in lava tubes as well as in the deep soil.

ALLOPATRIC SPECIATION

Within the two groups of hypogean taxa discussed above there are a number of genera with several troglomorphic species. In *Dysdera, Loboptera, Canarobius,* and *Apteranopsis* some species are now observed to occur sympatrically, and have been found not only in the same area but even in the same cave. This could mean that they have undergone adaptive radiation leading to a set of different species either simultaneously, or by successive invasions of the underground. The latter hypothesis seems more likely, since in all four genera the species concerned exhibit differing degrees of troglomorphy. The present areas of sympatry could be secondary, perhaps resulting from the disappearance of ancient ecological barriers that prevented gene flow between the two populations.

Other congeneric species, in these genera and some additional ones, show allopatric distributions. It is easy to understand their origin when the species concerned are vicariants in different islands, since the ancestor presumably arrived with eyes and the hypogean species evolved by parallel evolution on the different islands. This could be the case in the respective subterranean species of *Loboptera, Apteranopsis,* and *Domene* occurring on different islands. The microphthalmic condition observed in such species from the westernmost islands, in contrast to the anophthalmic state of their vicariants in Tenerife, supports the hypothesis that former colonizers of each island were eyed.

Vicariance also exists within single islands, as Martín et al. (1986) pointed out for *Loboptera* in El Hierro and Tenerife. In Tenerife there seems to be a clear separation between the northeastern Anaga zone, where *L. anagae* and *L. cavernicola* occur, and the rest of the island, which is inhabited by *L. subterranea* and in part by *L. troglobia*. The other troglobites so far known from Anaga—*Wolltinerfia anagae, Domene sylvatica,* and *Tachycixius* sp.—seem also to be isolated in relation to other species occurring in different zones of Tenerife (Fig. 4). As already mentioned, these species from Anaga are clearly less troglomorphic than their vicariants.

The Anaga peninsula is older and more eroded than the major part of Tenerife, being made up of outcrops of the "Old Series" (Fuster et al., 1968) together with the area of Teno in the northwest (Fig. 5A). Some authors even believed that these two areas were independent islands in the past (Evers, 1964) and were united by later eruptions when the central part of Tenerife was built up. This hypothesis has been abandoned since it is now well known that the Old Series also lies beneath the new rocks in many zones of the island. Anyway, there is evidence that these outcrops of the Old Series have never been covered by the new lavas (Fig. 5B). The junction between the area of Anaga and the rest of Tenerife is a wide zone of clayey alluvial deposits that probably functioned as an effective barrier to subterranean dispersal (Fig. 5C). According to Machado (1976) the powerful volcanic activity that formed the New Series eliminated the fauna of the geographically intermediate zones, leaving the Anaga and Teno massifs as faunistic radiation centers. This hypothesis combined with that of the formation of the underground barrier would explain the existence of sister allopatric troglobites (Fig. 5D). The species from the Teno Massif spread into the new terrains—either before or after colonizing the underground environments—as is demonstrated by the presence of several species in both the new and old areas. On the other hand, the species that invaded the underground environment of Anaga remained in that area because of the barrier.

It is remarkable that in Anaga, one of the geologically oldest parts of Tenerife (Carracedo, 1984), the inhabitants of the subterranean environment are less troglomorph: they usually still display remnants of eyes, pigmentation, etc. This could be due to the absence of a deeper inhabitable underground environment or MSP ("milieu souterrain profond"; Juberthie, 1983) (i.e., the mesocaverns and macrocaverns in the parent rock) because of the filling up of the network of cracks and voids as a result of erosion. The MSS,

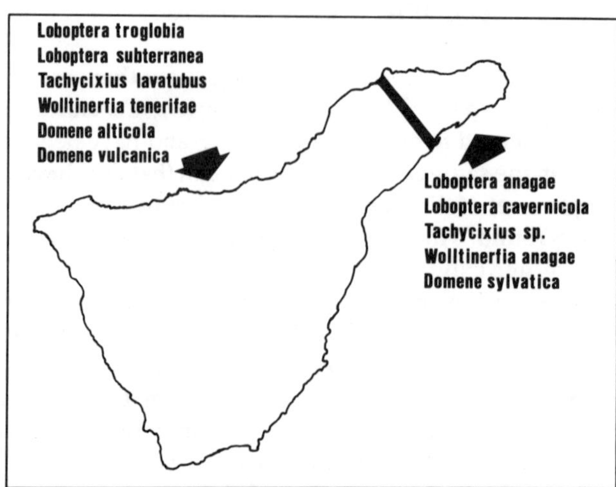

Figure 4. Troglobitic sister species with allopatric distribution in Tenerife.

the only underground environment inhabited in Anaga, seems to be energetically richer than the MSP of the more recent areas. Selective pressures in the MSS might not promote so strongly the energetic economy that leads towards troglomorphy, as seems to happen in the MSP.

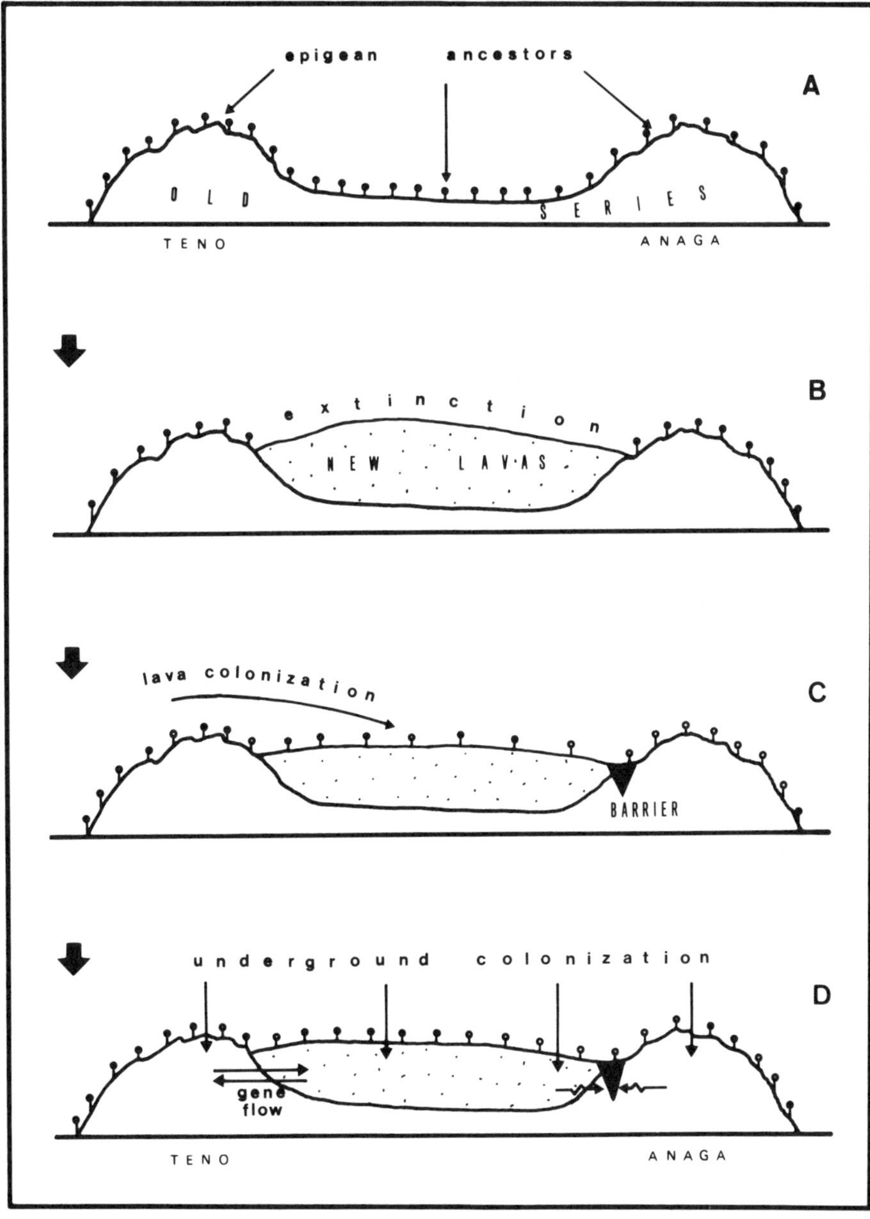

Figure 5. Possible geological changes in the island of Tenerife and their influence on the allopatric evolution of subterranean species (after Machado, 1976, in part). A: The epigean ancestors occupied all the areas. B: Deposition of new lavas and extinction of the epigean ancestors in intermediate zones. C: Formation of a subterranean barrier at the Anaga boundary and biological colonization of new lavas from Teno communities. D: Underground colonization and differentiation of the Anaga troglobitic species.

CONCLUSIONS

The faunistic studies on the underground environment of the Canary Islands during the past decade have led to the discovery of a surprising richness of troglobitic taxa. We now know of more than 60 troglobitic species and subspecies that live within terrestrial caves of the islands, and more than 25 species and subspecies that are found in volcanic MSS.

This diversity has undoubtedly been produced partly by insular speciation and race formation, there being several genera and species with distinct representatives on different islands. But allopatric speciation has also operated within one and the same island, especially in those cases where there may be underground barriers, formed, for instance, by siltation of cracks.

It seems evident, however, that a certain number of species have originated by parapatric speciation. Furthermore, there is another major group of troglobites that could have had a similar origin, although we do not know enough about the systematic relationships between them and their epigean relatives.

It is possible that the regression of the forests during drier periods in the past (Criado, 1984) has caused the isolation of populations in underground environments, resulting in the acquisition of troglomorphic characters. This possibility may explain some particular cases of troglobites that display relict distributions.

On the other hand certain species—sometimes within the same genus—overlap in their distribution, in some places even occurring syntopically. This might indicate that the colonization of underground environments has happened repeatedly over time, leading to the evolution of morphologically close, yet separate, species that differ in their degree of troglomorphy. There are clear examples of this phenomenon in some *Apteranopsis* species in La Palma and some *Loboptera* and *Dysdera* species in Tenerife. In the latter island various modes of speciation gave rise to the richest troglobitic fauna in the Canary Islands.

Finally, also in Tenerife, some closely related species are found to occur in different geographic areas that appear not to be separated by underground barriers. This might indicate that several colonizations of the underground environment took place, perhaps at different times and at different localities. We do not know, however, why the ranges of these subterranean populations do not overlap.

The studies carried out so far do not permit a clear distinction between the MSP and the MSS. We have verified that species found in both environments are usually more abundant in the MSS. On the other hand, there are species common in the MSP that have never been found in the MSS, even when the latter seems to provide suitable conditions.

Concerning the faunal relationships between the hypogean and endogean environments, there are only a few species that are found in both. There is no evidence that any cavernicolous taxa have developed from endogean taxa, with the exception of the centipede, *Cryptops vulcanicus*. This is a rewarding field for further research.

In spite of the richness of the subterranean fauna in the Canary Islands our knowledge of it is far from complete and new taxa are continuously being discovered. A multidisciplinary approach is currently being used to try to elucidate the origin of the taxa and the ways in which they have colonized underground environments. In this we are analysing the possible relationships between subterranean and lavicole (aeolian) habitats, the influence of the particular island situation, and the structure of the ecosystems of each island.

ACKNOWLEDGMENTS

We wish to express our gratitude to Myrtle and Philip Ashmole (Edinburgh), Hannelore Hoch, and Manfred Asche (Marburg) for the comments on the manuscript and the English translation, and to Ramón Oromí (La Laguna) for his practical support.

LITERATURE CITED

Alonso-Zarazaga, M. A. 1987. *Oromia hephaestos* n. gen., n. sp. de edafobio ciego de las Islas Canarias (Col., Curculionidae, Molytinae). *Vieraea* 17 (1–2):105–117.

Alonso-Zarazaga, M. A. 1990. Un nuevo edafobio ciego de Canarias: *Oromia aguiari* n. sp. (Col., Curculionidae, Molytinae). *Vieraea* 18:267–274.

Anguita, F. & F. Hernán. 1975. A propagating fracture model versus a hot spot origin for the Canary Islands. *Earth Planet Sci. Lett.* 27:11–19.

Ashmole, N. P., Ashmole, M. J. & P. Oromí. 1990. Arthropods of recent lava flows on Lanzarote. *Vieraea* 18:171–187.

Barr, T. C. & J. R. Holsinger. 1985. Speciation in cave faunas. *Ann. Rev. Ecol. Syst.* 16:313–337.

BrölemanN, H. W. 1900. Voyage de M. Ch. Alluaud aux Iles Canaries. (Novembre 1889-Juin 1890), Miriapodes. *Mém. Soc. Zool.* France 13:431–452.

Cantagrel, J. M., Cendrero, A., Fuster, J. M., Ibarrola, E. & C. Jamond. 1984. K-Ar chronology of the volcanic eruptions in the Canarian Archipelago: Island of La Gomera. *Bull. Volcanol.* 47 (3):597–609.

Carracedo, J. C. 1984. Etapa en la formación de las Canarias. Pp. 40–54. *In: Geografía de Canarias I.* Ed. Interinsular Canaria. S/C de Tenerife.

Criado, C. 1984. El relieve erosivo. Pp. 105–142. *In: Geografía de Canarias I.* Ed. Interinsular Canarias. S/C de Tenerife.

Chapman, P. 1986a. A proposal to abandon the Shiner-Racovitza classification for animals found in caves. *Proc. 9th Int. Congr. Speleol., Barcelona* 2:179–182.

Chapman, P. 1986b. Non-relictual cavernicolous invertebrates in tropical Asian and Australasian caves. *Proc. 9th Int. Congr. Speleol., Barcelona* 2:161–163.

Coiffait, H. 1958. *Los coléoptères du sol.* Supp. 7, *Vie et Milieu,* Bull. Lab. Arago, Banyuls-sur-Mer. 204 pp.

Eason, E. H. 1985. The Lithobiomorpha (Chilopoda) of the Macaronesian islands. *Ent. Scand.* 15 (1984):388–400.

Evers, A. M. J. 1964. Das Entstehungsproblem der Makaronesischen Inseln und dessen Bedeutung für die Artenstehung. *Entomol. Blätter.* 60 (2):81–87.

Español, F. & J. Ribes. 1983. Una nueva especie troglobia de Emesinae (Heteroptera, Reduviidae) de las Islas Canarias. *Speleon* 26–27:57–60.

Feraud, G., Kaneoka, I. & C. J. Allegre. 1980. K/Ar ages and stress pattern in the Azores: geodynamic implications. *Earth Planet. Scl. Lett.* 46:275–286.

Fuster, J. M., Araña, V., Brandle, J. L., Navarro, M., Alonso U. & A. Aparicio. 1968. *Geology and volcanology of the Canary Islands.* Inst. "Lucas Mallada": Int. Symp. Volcanol., Tenerife. 218 pp.

Fuster, J. M. 1975. Las Islas Canarias: un ejemplo de evolución espacial y temporal del vulcanismo oceánico. *Estudios Geológicos* 31:439–463.

Hernán, F. 1985. Diferentes hipótesis sobre la génesis de las Islas Canarias. Seminario sobre "Vulcanismo, origen, procesos y productos volcánicos". U. I. M. P. S/C de Tenerife. (unpubl. data).

Hernández, J. J., Martín, J. L. & A. L. Medina. 1986. La fauna de las cuevas volcánicas en Tenerife (Islas Canarias). *Act. IX Int. Congr. Speleol., Barcelona* 2:139–142.

Hernández, J. J. & J. L. Martín. 1990. Tres nuevas especies de *Apteranopsis* (Col., Aleocharidae) troglobias de la Isla de La Palma (Canarias). *Annls. Soc. Ent. Fr.* 26(4):585–594.

Howarth, F. G. 1973. The cavernicolous fauna of Hawaiian lava tubes. 1. Introduction. *Pacific Insects* 15 (1):139–151.

Howarth, F. G. 1980. The zoogeography of specialized cave animals: a bioclimatic model. *Evolution* 34:394–406.

Howarth, F. G. 1983. Ecology of cave arthropods. *Ann. Rev. Entomol.* 28:365–389.

Howarth, F. G. 1987. The evolution of non-relictual tropical troglobites. *Int. J. Speleol.* 16:1–16.

Izquierdo, I., Oromí, P. & X. Bellés. 1990. Reduction of reproductive capacity (number of ovarioles) in hypogean species of the genus *Loboptera* Brunn. & W. (Blattaria, Blattellidae) in the Canary Islands. *Mém. Biospéol.* 17.

Izquierdo, I. & J. L. Martín. 1990. Una nueva especie anoftalma de *Loboptera* Brunner W. en la isla de Tenerife (Islas Canarias) (Blattaria, Blattellidae). *Fragmenta Entomologica.* 22(1):19–25.

Jeannel, R. 1929. Un *Limnastis* aveugle de Tenerife (Col. Carabidae). *Mem. r. Soc. esp. Hist. Nat.*, 15:825–828.

Juberthie, C. 1983. Le milieu souterrain: étendue et composition. *Mém. Biospéol.* 10:17–65.

Machado, A. 1976. Introduction to a faunal study of the Canary Islands' Laurisilva, with special reference to the ground-beetles (Coleoptera, Caraboidea). Pp. 347–411. *In:* G. Kunkel (ed.), *Biogeography and Ecology in the Canary Islands.* Junk Publ.: The Hague.

Machado, A. 1984. Pterostíquidos anoftalmos nuevos de las Islas Canarias: descripción de *Wollastonia* n. gen. (Col. Caraboidea). *Nouv. Rev. Ent. (N. S.)* 1(2):129–137.

Machado, A. 1985. *Wolltinerfia* nom. nov. pro *Wollastonia* Machado, 1984 (Col. Carabidae). *Nov. Rev. Entomol.* (NS.) 2 (1):113.

Machado, A. 1987. Nuevos Trechodinae y Trechinae de las Islas Canarias. *Fragmenta Entomologica* 19(2):323–338.

Mahnert, V. 1986. Une nouvelle espèce du genre *Tyrannochthonius* Chamb. des Iles Canaries, avec remarques sur les genres *Apolpolium* Beier et *Calocheirus* Chamberlin (Arachnida, Pseudoscorpiones) *Mém. Soc. R. Belge Ent.* 33:143–153.

Mahnert, V. 1989. Les Pseudoscorpions (Arachnida) des grottes des Iles Canaries, avec description de deux espèces nouvelles du genre *Paraliochthonius* Beier. *Mém. Biospéol.* 16:41–47.

Martín, J. L., Oromí, P. & J. Barquín. 1985. Estudio ecológico del ecosistema cavernícola de una sima de origen volcánico: la Sima Robada (Tenerife, Islas Canarias). *Endins* 10–11:37–46.

Martín, J. L., Izquierdo, I. & P. Oromí. 1986. The genus *Loboptera* Brunn. W. (Blattaria, Blattellidae) in the Canary Islands and its distribution in the underground compartment. *Act. IX Int. Congr. Speleol., Barcelona* 2:142–145.

Martín, J. L. & P. Oromí. 1986. An ecological study of Cueva de los Roques lava tube (Tenerife, Canary Islands). *J. Nat. Hist.* 20 (2):375–388.

Martín, J. L. & P. Oromí. 1987. Tres nuevas especies hipogeas de *Loboptera* Brum. & W. (Blattaria, Blattellidae) y consideraciones sobre el medio subterráneo en Tenerife (Islas Canarias). *Annls. Soc. Ent. Fr. (N. S.)* 23(3):315–326.

Martín, J. L., Oromí, P. & I. Izquierdo. 1987. El ecosistema eólico de la colada volcánica de Lomo Negro en la isla de El Hierro (Islas Canarias). *Vieraea* 17:261–270.

Martín, J. L. & I. Izquierdo. 1987. Dos nuevas formas hipogeas de Loboptera (Blattaria, Blattellidae) en la isla de El Hierro (Islas Canarias). *Fragmenta Entomologica* 19(2):301–310.

Martín, J. L. & P. Oromí. 1988. Dos nuevas especies de *Anataelia* Bol. (Dermaptera, Pygidicranidae) de cuevas y lavas recientes del Hierro y La Palma (Islas Canarias). *Mém. Biospéol.* 15:49–59.

Martín, J. L., Izquierdo, I. & P. Oromí. 1989. Sur les relations entre les troglobies et les espèces épigées des Iles Canaries. *Mém. Biospéol.* 16:25–34.

Medina, A. L. & P. Oromí. 1990. First data on the superficial underground compartment on La Gomera (Canary Islands). *Mém. Biospéol.* 17:87–91.

Medina, A. L. & P. Oromí. 1991. Descripción de *Wolltinerfia anagae* n.sp. (Coleoptera, Carabidae) y consideraciones sobre la fauna del medio subterráneo superficial del Macizo de Anaga (Tenerife). *Mém. Biospéol.* 18:215–218.

Mitchell-Thomé, R. C. 1985. Radiometric studies in Macaronesia. *Bol. Mus. Mun. Funchal.* 37(167):52–85.

Molina, O. & R. Hutterer. 1989. A cryptic new species of *Crocidura* from Gran Canaria and Tenerife, Canary Islands (Mammalia: Soricidae). *Bonn. Zool. Beitr.* 40(2):85–97.

Morgan, W. J. 1971. Convention plumes in the tower mantle. *Nature* 230:42–43.

Muchmore, W. B. 1979. The cavernicolous fauna of Hawaiian lava tubes. 11. A troglobitic pseudo-scorpion (Pseudoscorpionida: Chthoniidae). *Pacific Insects* 20:187–190.

Oromí, P. & J. L. Martín. 1984. *Apteranopsis canariensis,* un nuevo coleóptero cavernícola de Tenerife (Staphylinidae). *Nouv. Rev. Ent. (N. S.)* 1(1):41–48.

Oromí, P., Hernández, J. J., Martín, J. L. & A. Lainez. 1985. Tubos volcánicos en Tenerife (Islas Canarias). Consideraciones sobre su distribución en la isla. *Act. II Simp. Reg. Espeleol. Burgos:* 85–93.

Oromí, P. & J. J. Hernández. 1986. Dos nuevas especies cavernícolas de *Domene* de Tenerife (Islas Canarias) (Col., Staphylinidae). *Fragmenta Entomologica* 19(1):129–144.

Oromí, P., Medina, A. L. & M. L. Tejedor. 1986. On the existence of a superficial underground compartment in the Canary Islands. *Act. IX Congr. Int. Espeleol, Barcelona* 2:147–151.

Oromí, P., Medina, A. L. & J. L. Martín. 1989. The genus *Licinopsis* Bedel (Col., Caraboidea) in the Canary Islands and its distribution in the underground compartment. *Mém. Biospéol.* 16:35–40.

Oromí, P. & J. L. Martín. 1990. Una nueva especie de *Domene* (Col., Staphylinidae) de cavidades volcánicas de La Palma (Islas Canarias). *Vieraea* 18:21–26.

Paoletti, M. G. 1980. La diffusion des troglobies dans les cavernes et le sol des Prealpes Venitiennes (Italie Nord-Orientale). *Mém. Biospéol.* 7:63–75.

Peck, S. B. 1982. Occurrence of *Ptomaphagus cavernicola* in forests in Florida and Georgia (Col., Leiodidae; Cholevinae). *Florida Entomol.* 65(3):378–379.

Peck, S. B. Eyeless arthropods of the Galapagos Islands, Ecuador: composition and origin of the cryptozoic fauna of a young, tropical, oceanic archipelago. *Biotropica.* In press.

Remane, R. & H. Hoch. 1988. Cave-dwelling Fulgoroidea (Homoptera: Auchenorrhyncha) from the Canary Islands. *J. Nat. Hist.* 22:403–412.

Ribera, C., Ferrández, M. A. & A. Blasco. 1985. Araneidos cavernícolas de Canarias. II. *Mém. Biospéol.* 12:51–66.

Ribera, C. & A. Blasco. 1986. Araneidos cavernícolas de Canarias. I. *Vieraea* 16:41–48.

Sauer, E. G. F. & P. Rothe. 1972. Ratite eggshells from Lanzarote. Canary Islands. *Science* 176:43–45.

Schmincke, H. U. 1976. The geology of the Canary Islands. Pp. 67–184. *In:* G. Kunkel (ed.), *Biogeography and Ecology in the Canary Islands.* Junk Publ.: The Hague.

Stock, J. H. 1988a. A new *Rhipidogammarus* (Crustacea, Amphipoda) from Tenerife: first record of the genus outside the Mediterranean region and its biogeographic implications. *Hydrobiologia* 169:279–292.

Stock, J. H. 1988b. Stygofauna of the Canary Islands. 8. Amphipoda (Crustacea) from inland ground-waters of Fuerteventura. *Bull. Zoöl. Museum.* 11(2):105–116.

Vandel, 1964. *La Biologie des Animaux Cavernicoles.* Gauthier-Villars: ed. Paris. 619 pp.

Wilkens, H., Parzefall, J. & T. M. Iliffe. 1986. Origin and age of the marine stygofauna of Lanzarote, Canary Islands. *Mitt. Hamb. Zool. Mus. Inst.* 83:223–230.

Wilkens, H. 1987. Genetic analysis of evolutionary processes. *Int. J. Speol.* 16(1–2):33–57.

Wilson, J. I. 1973. Mantle plumes and plate motions. *Tectonophysics* 19:149–164.

Wunderlich, J. 1987. *The Spiders of the Canary Islands and Madeira.* Triops Verlag: Langen. 435 pp.

Zapparoli, M. 1990. *Cryptops vulcanicus*, a new species from lava tubes in the Canary Islands (Chilopoda, Scolopendromorpha). *Vieraea.* 19:153–160.

Krakatau—
Studies on the Origin
and Development of a Fauna

I. W. B. Thornton

INTRODUCTION

The explosive 1883 eruption of Krakatau (in Sunda Strait, 44 km from Sumatra and Java) resulted in the loss of two-thirds of the island and the biological sterilization of the remnant (Rakata) and the two adjacent islands, Sertung and Panjang (Fig. 1). Biological monitoring of the recolonisation of the archipelago since then was carried out by the Dutch, but zoological surveys did not begin until 1908 (Jacobson, 1909), followed by two

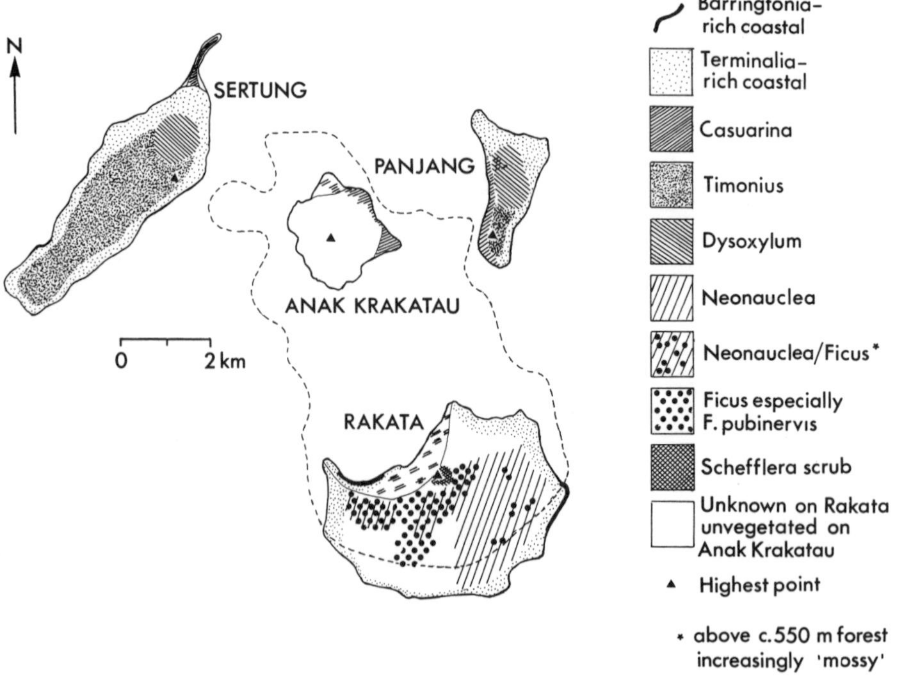

Figure 1. The Krakatau Archipelago, 1980s, showing main vegetation types. Modified from Whittaker et al., 1989.

Dr. Thornton is with the Department of Zoology, La Trobe University, Bundoora, Victoria, Australia 3083.

groups of surveys in 1919–1923 and 1929–1933 (Dammerman, 1948), and a survey of birds only in 1951 (Hoogerwerf, 1953). A fourth island, Anak Krakatau (Krakatau's Child), in the centre of the group, emerged in 1930 and is an active volcano (Fig. 1). All islands but Anak Krakatau are now covered with mixed secondary forest; about 200 species of vascular plants have been recorded in recent years (Whittaker et al., 1989).

Since 1951 only three research groups have worked on animals on the archipelago: Institut Teknologi Bandung (1982), Kagoshima University (1983), La Trobe University-L.I.P.I. (1984–86); one member of the 1983 and 1988 British botanical expeditions worked on butterflies (Table 1). There have so far been 37 participants in the three La Trobe expeditions, and the working up of material has involved about a score of additional specialists.

The aims of the La Trobe-L.I.P.I. expeditions were: 1) to characterize the 1980s fauna and provide a datum point for comparison with past and future surveys and to complement other recent surveys (for example, neither Bandung nor Japanese groups surveyed birds, bats or many invertebrate groups); 2) to assess the extent to which various animal groups are approaching an equilibrium number of species; 3) to characterize the fraction of the source faunas that have become established on the islands after a century; and 4) to assess the effect of Anak Krakatau's emergence on the faunal dynamics of the archipelago.

Table 1. Biological expeditions to the Krakataus, 1980s.

1979	Hull University-LIPI (Botany)	R. J. Flenley
1982	ITB Bandung (Biology)	H. Ibkar-Kramadibrata
1983	Hull University-LIPI (Botany)	K. Richards
1983	Kagoshima University-LIPI (Biology)	H. Tagawa
1984	La Trobe University-LIPI (Zoology)	I. W. B. Thornton
1985	La Trobe University-LIPI (Zoology)	I. W. B. Thornton
1986	La Trobe University-LIPI (Zoology)	I. W. B. Thornton
1988	Oxford University-LIPI (Botany)	R. J. Whittaker

THE ARCHIPELAGO

Brief Review of Physical and Vegetational Changes Since 1883

Physical. There have been three main types of physical change: 1) coastline changes—formation of cuspate headlands and their movement by marine erosion and deposition; an extreme example is the narrow northern Sertung spit (Fig. 1); 2) emergence in 1930 and subsequent growth of the new island Anak Krakatau (Fig. 1); and 3) eruptions of Anak Krakatau and their physical effects on the other islands—only Sertung and Panjang have been significantly affected. All these changes have effects on the development of the biota.

Vegetation. The following simplified summary is from the publications of teams from Kagoshima University, Hull University (and their successors), and the Bogor Herbarium (Tagawa et al., 1985; Whittaker et al., 1989), from 1979 to the present.

Within three years of the explosive eruption (1886) blue-green algae were present on Rakata as a substrate for a fern cover; after 14 years (1897) savannah grassland had replaced the fern cover in the lowlands, and there were patches of *Casuarina* woodland, *Ficus*, and other forest trees, the fern cover persisting only on the high slopes. An *Ipomoea* association had developed on the beaches, often with a *Barringtonia* community behind it. From 1919 to the 1930s the casuarina-savannah grassland was replaced by mixed secondary forest (species-poor) and the canopy began to close. 'Inland' forests of Sertung

398

and Panjang are dominated by *Timonius compressicaulis* or *Dysoxylum gaudichaudianum* (the latter being less extensive on Sertung), and those of Rakata by *Neonauclea calycina* and *Ficus* species, particularly *F. pubinervis*, and by *Dysoxylum,* now increasing in importance in the lowlands. Rakata's wet cloud forest near the summit is dominated by *Sauraria nudiflora, Schefflera polybotrya* and *Ficus ribes.*

The two greatest changes affecting animal colonisation were the change from grassland to woodland, and canopy closure.

There is some difference of opinion between the botanical groups regarding the relative importance, in determining the components and dominants of the forest on different islands, of chance first arrival of dominant species and differential deflection of plant succession by Anak Krakatau's eruptions. The Japanese workers emphasise the former, the British workers the latter. There are also differences in predictions of the future course of succession. The Japanese predict that Rakata's forests will change in time to become *Dysoxylum*-dominated, as are some areas now on Sertung and Panjang (Tagawa et al., 1985). The British group believes these forest areas on Sertung and Panjang are young, disturbance-induced, secondary associations and that Rakata's vegetation will change by becoming more patchy (including patches of *Dysoxylum*) but *Dysoxylum* will not become the forest dominant on Rakata (Whittaker et al., 1989).

Development of the Fauna and Approach to Equilibrium

The resident land bird fauna was used by MacArthur and Wilson (1967) as an example of colonisation of an island resulting in a dynamic equilibrium in species numbers by 1919, i.e., in 36 years. Subsequent surveys by Hoogerwerf (1953) in 1951 and by our group from 1984 to 1986 show that, in fact, species numbers of land birds have increased by 25% since 1919, and only now do they appear to be approaching equilibrium. The change in slope of the curve in 1919 did not herald equilibrium, but rather a change in rate of increase of species numbers (Fig. 2). Similar curves (although with one less datum point) are found for Blattodea, Odonata, and reptiles (Fig. 2), and for nymphalid and hesperiid Lepidoptera (Fig. 3), in which rate of increase of species numbers has slowed markedly over the last 50 years. Other groups, however, for example, lycaenid Lepidoptera (Fig. 3), ants and other aculeates and braconid Hymenoptera (Fig. 4), and several groups of Diptera, Thysanoptera, Neuroptera, and land molluscs (Fig. 5), are building up species numbers at a rate similar to that in the first half century since 1883 (Thornton et al., 1990a).

For most animal groups the second intersurvey period (1908–1919), when forests were forming, was the time of highest immigration rates and the third period (1920s and early 1930s), when the canopy was closing, was the time of highest extinction rates. After canopy closure there is some evidence of a change in resident land birds to more specialised feeders, and clear evidence of an increase in the proportion of true forest bird immigrants and a decline in the proportion of open-country species (Thornton et al., 1990b). During the period of forest formation many important bat or bird-dispersed plant species first appeared (Whittaker et al., 1989).

The colonisation curve of resident land birds for the whole archipelago differs from that for the island of Rakata, rising steadily, but less steeply than previously, after 1921 when forests were becoming established (Fig. 2). The curve for Rakata (Fig. 6) levels from 1921 to 1951, after which it rises again. Disturbances to the archipelago as a result of Anak Krakatau's activity affected Rakata very much less than the other islands (its vegetation being shielded by the high northern cliff), and open habitat persisted on the archipelago as a whole (largely as a result of the emergence of Anak Krakatau and the physical dynamism of Sertung's spit) but not on Rakata itself, providing an 'ecological rescue' for some open-country species (see below). The rise in the curve for Rakata since 1951 coincides with changes in Rakata's lowland forest flora recently detected by Whittaker's group (Whittaker et al., 1989).

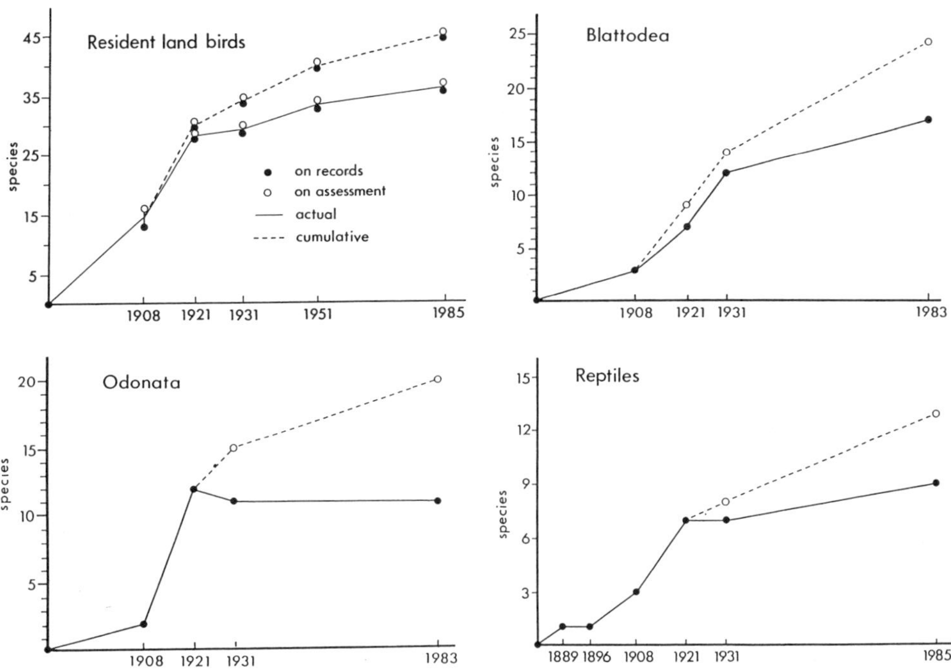

Figure 2. Colonization curves for the Krakatau Islands of resident land birds, Blattodea, Odonata, and reptiles. Continuous line, actual numbers at time of survey; dotted line, cumulative numbers. For all but resident land birds, open circles—cumulative numbers, filled circles—actual numbers.

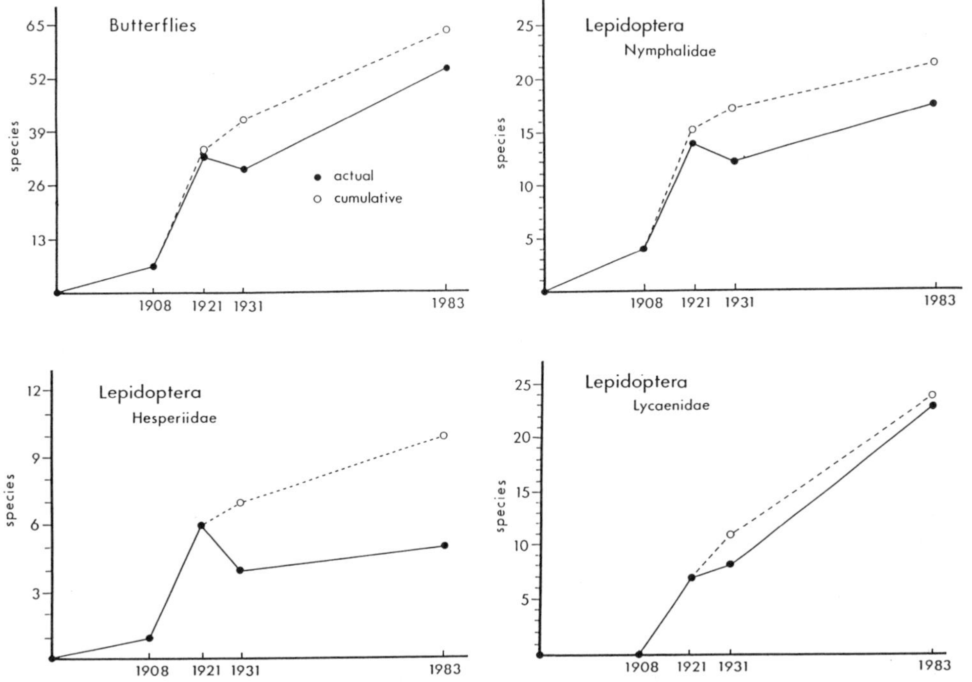

Figure 3. Colonization curves for the Krakatau Islands of butterflies and three families of Lepidoptera. Continuous line and filled circles, actual numbers; dotted line and open circles, cumulative numbers.

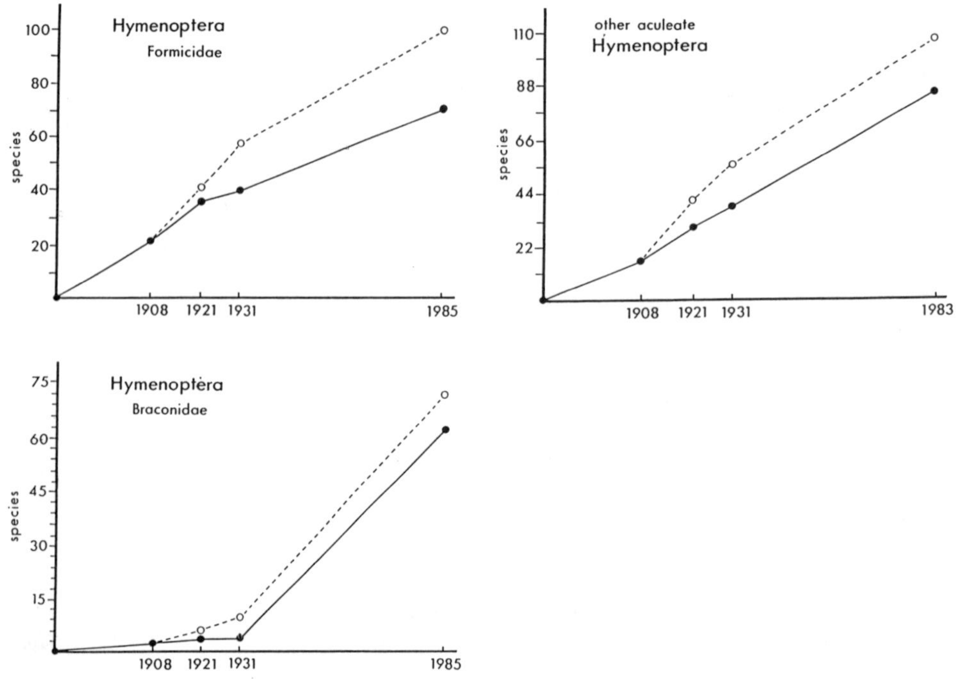

Figure 4. Colonization curves for the Krakatau Islands of three groups of Hymenoptera. Symbols as for Fig. 3.

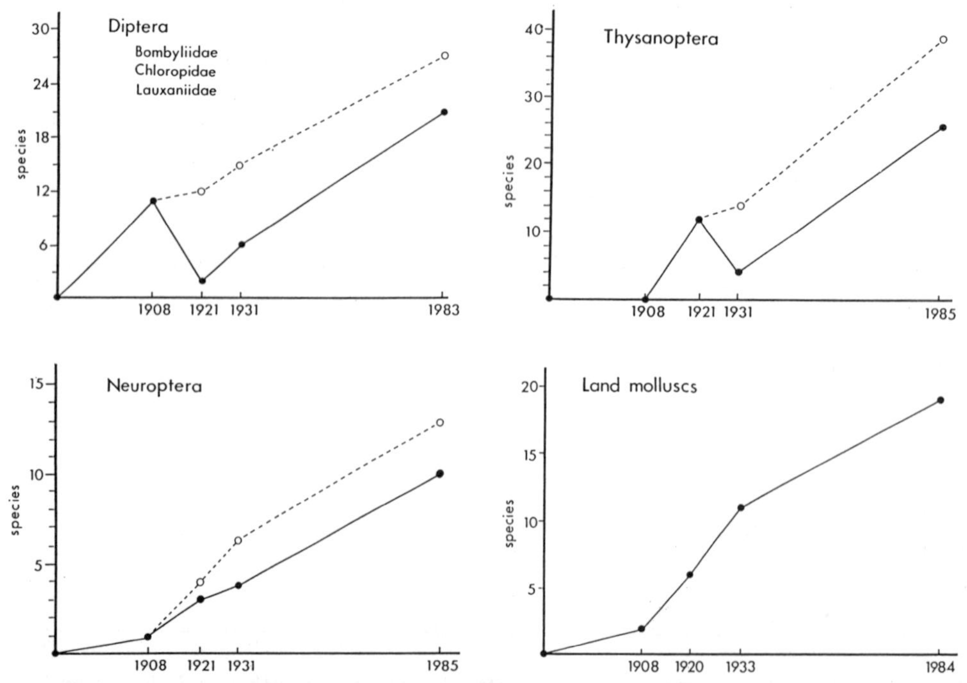

Figure 5. Colonization curves for the Krakatau Islands of three groups of insects, and land molluscs. Symbols as for Fig. 3.

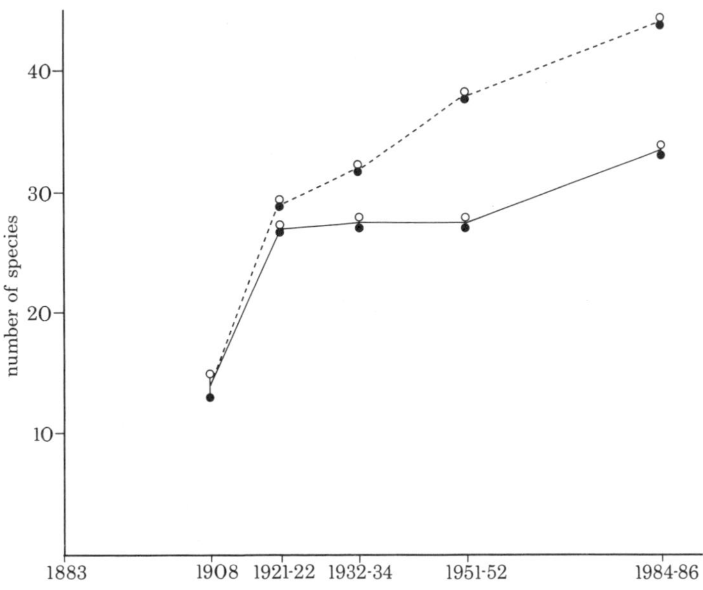

Figure 6. Colonization curve, resident land birds, Rakata. Symbols as for birds, Fig. 2.

The Successful Animal Colonists

In general, successful colonists have been species with wide geographical distributions and broad ecological tolerances; this is particularly well seen in the land birds, bats, reptiles, and some groups of insects (Psocoptera, chloropid Diptera and some butterfly groups) (Thornton et al., 1990a).

Preclusion of Potential Colonisers by Lack of Appropriate Habitat

The fraction of the theoretical mainland species pool that can be regarded as an 'effective' pool (i.e., having potential habitat in the archipelago) obviously changes with time when the target is undergoing successional change. Such change may even be so rapid as to prevent members of the 'effective' pool exploiting the transient opportunity.

There is some evidence that the grassland phase was too short for the establishment of several animals, for example sphecid wasps (Yamane, 1983) and birds of open country such as old world warblers, buttonquails, sparrows, and munias (Thornton et al., 1990a, b).

Many animals characteristic of mature forest on the adjacent mainland have never become established on the Krakataus, which still have no primary forest (indeed, no true primary forest tree species). Many smaller terrestrial mammals, bats, reptiles, nine families of forest birds, and several groups of insects usually associated with deep forest in the Sunda Strait area have never colonised the archipelago (Thornton et al., 1990a). Some stenophagous insects, such as some highly vagile butterflies, have not become established because the islands still lack their food plants (New et al., 1988).

The archipelago's almost total lack of permanent fresh water has precluded the establishment of five families of birds known from the Ujung Kulon peninsula and the archipelago lacks Plecoptera, gyrinid beetles, and psychodid and simuliid Diptera. The aquatic groups Trichoptera and Ephemeroptera were not recorded until 1984 and 1986, respectively. Other aquatic insect groups—dytiscid and hydrophilid beetles, and hydrometrid, corixid, gerrid, and notonectid hemipterans—have been found only rarely, and then usually associated with brackish lagoons that existed on Sertung from about 1915 to 1934. The only aquatic insects that are at all well represented are the Odonata, and most of these are probably non-resident, although we found a libellulid dragonfly larva in a pool on Sertung in 1986 (Thornton & New, 1988).

The Nature of Turnover in Animal Species

Whittaker et al. (1989) noted that turnover of plant species largely involved ephemerals, species lost because of habitat destruction, those introduced by humans, and successional turnover. They found little 'stochastic' turnover in plants. From our studies, the same is generally true for animals (Thornton et al., 1990a). Apart from extinctions that can be ascribed to habitat destruction as a result of Anak Krakatau's activity (a coastal skink on Sertung) and erosion of beach associations (a coastal gecko), most have been the result of succession from grassland to forest, notably hesperiid Lepidoptera ('skippers'), several sphecid wasps, jassids, fulgorids and Diptera, a gecko (*Lepidodactylus lugubris*), and several grassland or open-country birds.

Ecological Refuges and Windows

The mobility of Sertung's spit, which holds its plant succession at an early phase, and the emergence in 1930 of Anak Krakatau, have provided the archipelago with open country that is out of synchrony with the succession on the rest of the archipelago, where secondary forest has superseded grassland and casuarina woodland. Ecological refuges have thus been provided in these areas for animal species that may otherwise have been expected to become extinct as their preferred habitat declined on the archipelago through successional processes (Thornton et al., 1988a). We identify a lycaenid butterfly, four vespid, sphecid and scoliid wasps, an ant, a dove, a crow, a coucal, a waterhen, and possibly the savanna nightjar (*Caprimulgus affinis*), and pied triller (*Lalage nigra*) (both now rare elsewhere on the archipelago), as persisting on the archipelago because of the provision of these refuges.

For mainland open-country species that missed the opportunity to colonize during the relatively short grassland phase, these areas have re-opened an 'ecological window.' We believe an owl has been able to take advantage of this successfully, and know of two other open-habitat bird species (a crow and a shrike) that have arrived but not established themselves. The re-opening of such a window, of course, would have allowed the 'rescue effect' of Brown & Kodric-Brown (1977) to operate on the archipelago for open-country species that otherwise may have declined in numbers to critical levels.

We believe the successional asynchrony resulting from these two physical events (on Anak Krakatau and Sertung) will, on balance, reduce turnover and delay the achievement of equilibrium on the archipelago.

ANAK KRAKATAU

This island emerged in 1930 and is now over 190 m high and 2 km in diameter. Only some 5% of its area is vegetated (about 17 ha) and it is an active volcano. It suffered a self-sterilizing eruption in 1952, and several destructive ones since then. The island is thus only three to four decades old, biologically. The vegetation is confined largely to two forelands (one a grassland, the other a casuarina woodland) with casuarina-dominated vegetation connecting them along the shore. Between 1979 and 1983, 82 vascular plants were recorded on the island (Whittaker et al., 1989).

Anak Krakatau as an Analogue of Early Succession on the Archipelago

Although Whittaker's group believes that Anak Krakatau cannot be regarded as presenting a 're-play' of colonisation processes on the archipelago in the first four decades after 1883 (Whittaker et al., 1989), we find that as far as animals are concerned there are several interesting parallels (Thornton et al., 1990a).

The present (three or four decades since 1952) size of the fauna of Anak Krakatau in several invertebrate groups is similar to that of Rakata three or four decades after 1883, and Anak Krakatau's acquisition rate for resident land birds since 1952 is very close to that for the other islands in the first four decades after 1883 (0.75 spp yr^{-1} and 0.74 spp yr^{-1} respectively). Moreover, over 80% of Anak Krakatau's present resident land bird species (20 of the 24 species present) had become established on the archipelago by 1933, 50 years after 1883 (Thornton et al., 1990b). Only four of the 11 land bird species to arrive on the archipelago in the last 50 years are now present on Anak Krakatau, but nine of the 11 that arrived in the first 25 years after 1883 are present on Anak Krakatau. This pattern is supported by analyses of 11 of the 17 animal groups for which there are adequate data. Among the exceptions are the less vagile land molluscs and oniscidean Crustacea, and groups requiring established host or prey populations for successful colonisation, such as braconid Hymenoptera, predaceous wasps, and lycaenid butterflies (Dr. K. Maeto in litt., Yamane, 1983, New et al., 1988).

The only fruit-eating birds recorded from the archipelago in the first 25 years after 1883 were two obligatory frugivores (green-winged and pink-necked pigeons, *Chalcophaps indica* and *Treron vernans*) and four facultative frugivores (large-billed crow *Corvus macrorhynchos*, black-naped oriole *Oriolus chinensis*, and yellow-vented and sooty-headed bulbuls *Pycnonotus goiavier* and *Pycnonotus aurigaster*) (Table 2). All but the last (which has not been recorded since) have been present in every subsequent survey and constitute five of the eight fruit-eating birds to have been recorded from Anak Krakatau.[1] The green-

Table 2. First record on the Krakatau archipelago (Ks) and Anak Krakatau (AK) of fruit-eating birds (from Dammerman 1948, Hoogerwerf 1953, Zann et al. 1990a, b).

		Ks	AK
FI	*Pycnonotus aurigaster*	1908*	—
IF	*Pycnonotus goiavier*	1908	1982 (H)
FI	*Oriolus chinensis*	1908	1982
IF	*Corvus macrorhynchos*	1908	1983*
F	*Treron vernans*	1908	1986
FS	*Chalcophaps indica*	1908	1986**
FI	*Aplonis panayensis*	1919	1985
FI	*Eudynamys scolopacea*	1919****	—
F	*Ducula bicolor*	1919	—
FI	*Dicaeum trigonostigma*	1919	—
FS	*Macropygia phasianella*	1933	1985
F	*Ptilinopus melanospila*	1951	—
FI	*Pycnonotus plumosus*	1951	—
F	*Ducula aena*	1984	—
FI	*Zoothera interpres*	1984	1986***

*	—no further records
**	—evidently an unsuccessful colonization, only remains found
***	—unsuccessful colonization, only one immature
****	—not seen since 1951
F	—eats fruit
I	—eats insects/arthropods
S	—eats seeds
(H)	—seen on Anak Krakatau by Hoogerwerf in 1951, one year before the island's self-devastating eruption

1. Note added in proof: *Ptilinopus melanospila* was seen on Anak Krakatau in 1991.

Table 3. First record of bats on the Krakatau archipelago (Ks) and Anak Krakatau (AK) (from Tidemann et al., 1990).

		Ks	AK
Megachiroptera			
F	*Cynopterus sphinx angulatus*	1919	1982
F	*Cynopterus horsfieldi lyoni*	1920*	—
F	*Rousettus amplexicaudatus*	1933	1986
F	*Cynopterus tittaecheilus*	1974	—
F	*Macroglossus sobrinus*	1974*	—
F	*Pteropus vampyrus*	1985	1986**
F	*Cynopterus brachyotis brachyotis*	1985	—
Microchiroptera			
I	*Hipposideros diadema*	1928*	—
C	*Megaderma spasma*	1982	—
I	*Hipposideros larvatus*	1984	—
I	*Myotis muricola*	1984	—
I	*Emballoneura monticola*	1984	—
I	*Rhinolophus celebensis javanicus*	1985	—
I	*Hipposideros cineraceus*	1985	—
I	*Scotophilus kuhlii temminckii*	1985	—
I	*Rhinolophus pusillus*	1986	—

* —no further records
** —single vagrant
F —frugivore
C —carnivore
I —insectivore

Table 4. First record of reptiles on the Krakatau group (Ks) and Anak Krakatau (AK) (from Dammermann, 1948, Rawlinson et al., 1990).

	Ks	AK
Varanus salvator	1889	1982
Hemidactylus frenatus	1908	1982
Python reticulatus	1908	—
Emoia atrocostata	1919*	—
Lepidodactylus lugubris	1920*	—
Crocodylus porosus	1924**	—
Mabuya multifasciata	1924	—
Cosymbotus platyurus	1928*	—
Gekko gecko	1982	—
Chrysopelea paradisi	1982	1984
Gekko monarchus	1984	—
Ramphotyphlops braminus	1984	—
Hemiphyllodactylus typus	1985	—

* —not recorded since 1934
**—incidental records only

winged pigeon, however, failed to become established on Anak Krakatau and the survival of the large-billed crow is unlikely.

Similarly (Table 3), the first of 13 species of bats to become established on the archipelago were fruit bats (lesser dog-faced fruit bat *Cynopterus sphinx* and Geoffroy's roussette *Rousettus amplexicaudatus*) in 1919 and 1933. These are the only bat species yet to have colonised Anak Krakatau. Insectivorous and carnivorous microchiropterans, eight species of which successfully colonised the archipelago late, 50–100 years after the eruption, are still absent from Anak Krakatau (Tidemann et al., 1990).[1]

1. Note added in proof: the insectivorous emballonurid *Taphozous longimanous* was netted on Anakk Krakatau in 1990.

Of reptiles (Table 4), the 5-banded monitor (*Varanus salvator*) was seen on the archipelago in 1889, six years after the eruption, and, with the gecko *Hemidactylus frenatus* and a python, *Python reticulatus*, in 1908, made up the archipelago's early reptile fauna (five more species were recorded in the period 1919–1928). The monitor and the gecko are two of the three reptiles now present on Anak Krakatau (along with a late, successful snake colonist, *Chrysopelea paradisi*) (Rawlinson et al., 1990).

Thus the sequence of colonisation of Anak Krakatau by land vertebrates shows remarkable parallels with that for the archipelago since 1883.

The Role of Animals in Forest Diversification

The limited casuarina woodland on Anak Krakatau has been significantly enriched in recent years. Five forest trees absent in 1979 were recorded in 1983, including three that were important components of succession on the other islands and are bird/bat dispersed: *Macaranga tanarius*, *Dysoxylum guadichaudianum* and *Timonius compressicaulis* (Whittaker et al., 1989).

Six of the archipelago's 15 fruit-eating birds were recorded from Anak Krakatau in 1980s surveys; the Philippine glossy starling (*Aplonis payanensis*) (a disperser of *Macaranga*), large-billed crow, black-naped oriole, yellow-vented bulbul, pink-necked pigeon, and Sunda Islands cuckoo-dove (*Macropygia phasianella*). We also found green-winged pigeon remains (the island has three raptors: Oriental hobby *Falco severus*, barn owl *Tyto alba* and white-bellied sea-eagle *Haliaeetus leucogaster*), and an immature ground thrush, *Zoothera interpres*. The lesser dog-faced fruit bat is known to have used the island since 1982 and Geoffroy's rousette since 1986, and these species have probably been of great importance in the colonisation of Anak Krakatau by figs.

Of the archipelago's 18 species of figs, two, *Ficus fulva* and *Ficus septica*, were recorded in 1979 but were not seen in fruit until 1985 (Table 5). *Cynopterus* bats were seen eating figs and defecating and regurgitating their seeds on the island in 1984 when no fruits were on the trees, so that we know they act as fig-dispersers within the archipelago. Also in 1984 we netted four species of agaonid fig-wasps, including the pollinator of *F. septica*, on Anak Krakatau, and in 1985 a fifth. In 1985 we also collected a sycoenine seed-predator fig chalcid in water traps set on bare ash on the windward side of the island far from vegetation, and in 1986 we showed that wind-borne fall out on to the island is considerable (Thornton et al., 1988b). Although in 1985 we could not find agaonids in the fig fruit, they were present the following year, when we also found two torymid fig inquilines. Thus, fig wasp communities as well as fig-eating birds and bats are now established on the island, and diversification of its woodland should accelerate.

An Ash-Lava Aeolian Ecosystem

The aerial fall-out on to Anak Krakatau is exploited in barren areas by a community dominated by a crepuscular flightless cricket (*Speonemobius* sp.) that does not occur in the small vegetated area (Fig. 7). Lycosid spiders, earwigs, and mantids are minor components of this community, and aerial insectivores—three dragonfly species, two swallows, a swift, and a swiftlet—exploit this fall-out before it settles (New & Thornton, 1988).

The existence of this ecosystem is interesting for two reasons. First, it provides a conduit for energy from outside the island itself and short-circuits the usual plants-herbivores-carnivores energy route. Second, it shows parallels with other such systems found on Mt. Etna (Wurmli, 1974) and the Canaries (Ashmole & Ashmole, 1988), and a very close parallel with that on lava flows high on the island of Hawai'i (Howarth, 1979), which is also dominated by a nocturnal, flightless cricket of a related genus. Such ecosystems may be expected to evolve where there is a long history of volcanic activity and where barren areas receiving fall-out have long been available for exploitation.

Table 5. Role of animals in development of mixed forest from *Casuarina* woodland on Anak Krakatau, 1979–86.

			1979	1982	1983	1984	1985	1986
td	*F. fulva* (T)	figs	n.f.	n.f.	n.f.	n.f.	fruit	fruit
td	*F. septica* (T)		n.f.	n.f.	n.f.	n.f.	fruit	fruit
t	*C. papaya*		—	—	✓	✓	✓	✓
td	*M. tanarius* (T)		—	—	✓	✓	✓	✓
td	*D. gaudichaudianum* (T)	other	—	—	✓	✓	✓	✓
d	*Melastoma affine*	plants	✓	✓	✓	✓	✓	✓
td	*Passiflora foetida*		—	✓	✓	✓	✓	✓
d	*T. compressicaulis* (T)		—	✓	✓	✓	✓	✓
d	*Cayratia trifolia*		—	—	✓	✓	✓	✓
	C. sphinx		—	✓	✓	e/dfigs	✓	✓
	R. amplexicaudatus	fruit bats	—	—	—	✓	✓	✓
	P. vampyrus		—	—	—	—	—	straggler
	C. bisulcatus (pollinator of *F. septica*)		—	—	—	s. net	✓	✓
	Blastophaga sp.	agaonid	—	—	—	s. net	✓	✓
	Liporrhopalum sp.	fig	—	—	—	s. net	✓	✓
	Platyscapa sp.	pollinators	—	—	—	s. net	✓	✓
	Waterstoniella spp. (2)		—	—	—	—	s. net	✓
	B. inopinata	fig chalcids	—	—	—	—	—	r. *F. fulva*
	D. macroptera	sycoecine fig seed predator	—	—	—	—	w/p traps	✓
	P. grandii	torymid	—	—	—	—	—	r. *F. fulva*
	Sycoryctes sp.	inquilines	—	—	—	—	—	r. *F. fulva*
	O. chinensis oriole		✓	✓	✓	✓	✓	
	P. goiavier bulbul	facultative	✓	✓	✓	✓	✓	
	C. macrorhynchos crow	frugivores	—	—	1 pair	—	—	
	A. panayensis starling		—	—	✓	✓	?	
	M. phasianella cuckoo dove	obligatory	—	—	—	1 pair	1 pair +	
	T. vernans pigeon	frugivores	—	—	—	—	1 pair +	
	Z. interpres thrush	deep forest bird	—	—	—	—	1 individual	

(T)—important forest tree; d—bird dispersed; t—bat dispersed; n.f.—not fruiting; e/d figs—eating figs and defecating fig seeds; s. net—sweep net capture; r. *F. fulva*—reared from syconia of *F. fulva*; w/p traps—water trap and pit trap captures.

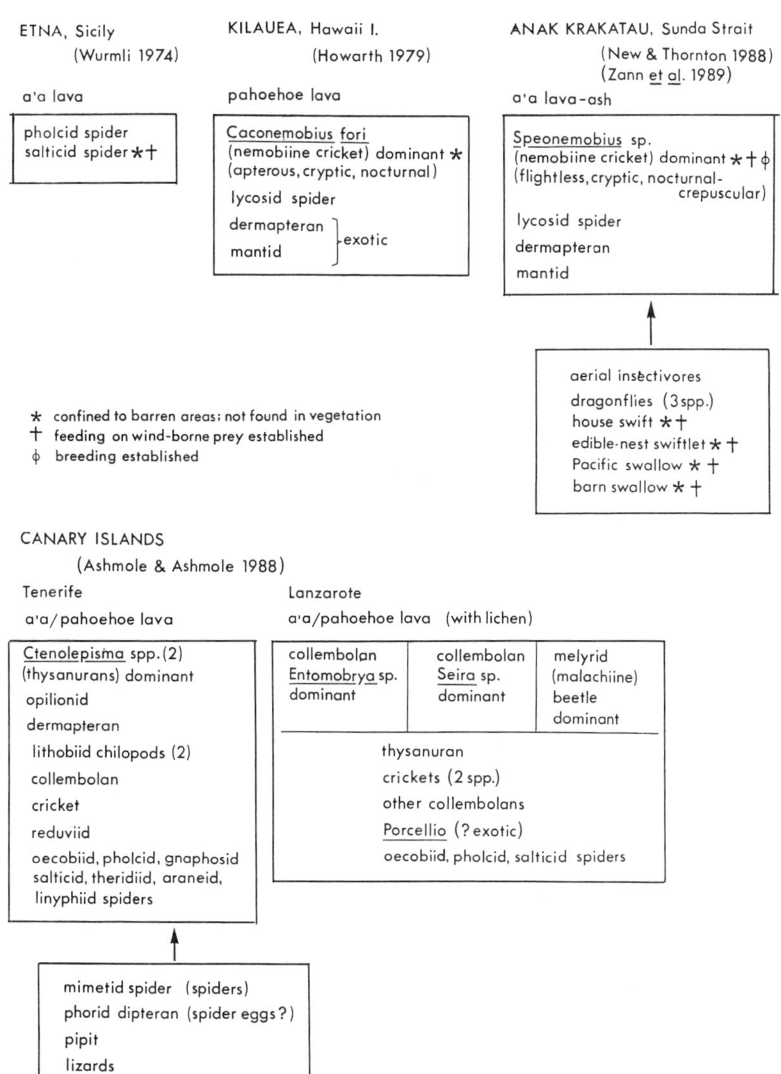

Figure 7. Composition of lava-ash communities exploiting wind-borne fall-out on Etna, Hawaii, Anak Krakatau, and the Canary Islands.

LITERATURE CITED

Ashmole, M. J. & N. P. Ashmole. 1988. Arthropod communities supported by biological fallout on recent lava flows in the Canary Islands. *Ent. Scand. Suppl.* 32:67–88.

Brown, J. H. & A. Kodric-Brown. 1977. Turnover rates in insular biogeography: effect of immigration on extinction. *Ecology* 58:445–449.

Dammerman, K. W. 1948. The fauna of Krakatau 1883–1933. *Verh. K. Ned. Akad. Wet. (Tweede sectie)* 44:1–594.

Hoogerwerf, A. 1953. Notes on the vertebrate fauna of the Krakatau Islands, with special reference to the birds. *Treubia* 22:319–348.

Howarth, F. G. 1979. Neogeoaeolian habitats on new lava flows on Hawaii island: an ecosystem supported by windborne debris. *Pacif. Insects* 20:133–144.

Jacobson, E. R. 1909. Die niewe fauna van Krakatau. *Jversl. Topogr. Dienst. Ned.-Indie* 4:192–206.

MacArthur, R. H. & E. O. Wilson. 1967. *The Theory of Island Biogeography*. Princeton University Press: Princeton, NJ.

New, T. R., Bush, M. B., Thornton, I. W. B. & H. K. Sudarman. 1988. The butterfly fauna of the Krakatau Islands after a century of colonization. *Phil. Trans. R. Soc. Lond.* B 322:445–457.

New, T. R. & I. W. B. Thornton. 1988. A pre-vegetation population of crickets subsisting on allochthonous aeolian debris on Anak Krakatau. *Phil. Trans. R. Soc. Lond.* B 322:481–485.

Rawlinson, P. A., Widjoya, A. H. T., Hutchinson, M. N. & G. W. Brown. 1990. The terrestrial vertebrate fauna of the Krakatau Islands, Sunda Strait, 1883–1986. *Phil. Trans. R. Soc. Lond.* B 328:3–28.

Tagawa, H., Suzuki, E., Partomihardjo, T. & A. Suriadarma. 1985. Vegetation and succession on the Krakatau Islands, Indonesia. *Vegetatio* 60:131–145.

Thornton, I. W. B. & T. R. New. 1988. Freshwater communities on the Krakatau Islands. *Phil. Trans. R. Soc. Lond.* B 322:487–492.

Thornton, I. W. B., Zann, R. A., Rawlinson, P. A., Tidemann, C. R., Adikerana, A. S. & A. H. T. Widjoya. 1988a. Colonization of the Krakatau Islands by vertebrates: equilibrium, succession, and possible delayed extinction. *Proc. Natn. Acad. Sci. U.S.A.* 85:515–518.

Thornton, I. W. B., New, T. R., McLaren, D. A., Sudarman, H. K. & P. J. Vaughan. 1988b. Air-borne arthropod fall-out on Anak Krakatau and a possible pre-vegetation pioneer community. *Phil. Trans. R. Soc. Lond.* B 322:471–479.

Thornton, I. W. B., New, T. R., Zann, R. A. & P. A. Rawlinson. 1990a. Colonization of the Krakatau Islands by animals: a perspective from the 1980s. *Phil. Trans. R. Soc. Lond.* B 328:131–165.

Thornton, I. W. B., Zann, R. A. & D. G. Stephenson. 1990b. Colonization of the Krakatau islands by land birds and the approach to an equilibrium number of species. *Phil. Trans. R. Soc. Lond.* B 328:55–93.

Tidemann, C. R., Kitchener, D. J., Zann, R. A. & I. W. B. Thornton. 1990. Recolonization of the Krakatau Islands and adjacent areas of West Java, Indonesia, by bats (Chiroptera) 1883–1986. *Phil. Trans. R. Soc. Lond.* B 328:123–130.

Whittaker, R. J., Bush, M. B. & K. Richards. 1989. Plant recolonization and vegetation succession on the Krakatau Islands, Indonesia. *Ecol. Monogr.* 59:59–123.

Wurmli, M. 1974. Biocenoses and their successions on the lava and ash of Mt. Etna, Part I. *Image Roche* 59:32–40.

Yamane, Sk. 1983. The aculeate fauna of the Krakatau Islands (Insecta, Hymenoptera.) *Rep. Fac. Sci. Kagoshima Univ. (Earth Sci. & Biol.)* 16:75–107.

Zann, R. A., Walker, M. V., Adhikerana, A. S., Davison, G. W., Male, E. B. & Darjono. 1990a. The birds of the Krakatau Islands (Indonesia) 1984–86. *Phil. Trans. R. Soc. Lond.* B 328:29–54.

Zann, R. A., Male, E. B. & Darjono. 1990b. Bird colonization of Anak Krakatau, an emergent volcanic island. *Phil. Trans. R. Soc. Lond.* B 328:95–120.

Biogeographic Patterns of Antillean Insects: The Emerging Picture of Antillean Geohistory and Faunal Diversification

EXTENDED ABSTRACT

James K. Liebherr

The Caribbean region comprises a ring of islands surrounding the Caribbean Sea. The islands vary greatly in size and distance from the mainland. The geographical complexity of the region is a reflection of its complex, and not completely understood, geological history. Biogeographers have based explanations of the biotic distributional patterns on two very different hypotheses, each entailing very different assumptions about the geographical and geological history of the terrestrial land masses. Emphasizing the dispersive abilities of organisms, some biogeographers have assumed that organisms colonized the island areas over water, and that the islands themselves arose from the sea somewhere near their present positions. This may be called the stabilist (Donnelly, 1988), or dispersalist hypothesis (Darlington, 1957, Fig. 63). Under this hypothesis, organisms are assumed to have colonized the Antilles from mainland source areas, with island size and isolation determining the numbers of taxa present in any one area. Corroboration of this hypothesis is based on phylogenetic patterns that are consistent with assumptions concerning dispersal, e.g., presently neighboring geographic areas should contain closely related taxa. The dispersal hypothesis is considered acceptable by its adherents when the majority of cases are explainable given its assumptions (Perfit & Williams, 1989). A second hypothesis, called the mobilist (Donnelly, 1988) or vicariance hypothesis (Rosen, 1975) assumes that the Caribbean land masses have been transported by eastward movement of the Caribbean plate from their Mesozoic positions in proximity to the Central American mainland (Rosen, 1985). Faunal elements present on these land masses from as early as Eocene time are hypothesized to have been constant residents since then, and their phylogenetic relationships are predicted to conform to area relationships that date back to the Eocene. Under the vicariance hypothesis, only biogeographic patterns observed in taxa existing at the start of movement of the Antillean land masses can corroborate the hypothesis. Corroboration is based on concordance of taxon-area cladograms (Savage, 1982). Taxa that have colonized over water since the islands became isolated from the mainland provide no information regarding the geological history of the region.

Advocates of the dispersal hypothesis argue that geological models only weakly support the possibility that some of the Antilles were constantly subaerial since the Eocene,

Dr. Liebherr is with the Department of Entomology, Comstock Hall, Cornell University, Ithaca, NY 14853, USA.

and thus dispersal from the mainland must explain all current distributions. Moreover, as island faunas are characterized by taxonomically isolated relictual taxa, they must not contain taxa of continental origin. Only the discovery of Eocene fossils indicating that the islands then possessed a continental fauna is deemed supportive of the mobilist hypothesis by advocates of dispersalism (Perfit & Williams, 1989). Under the vicariance hypothesis, biogeographic patterns are compared prior to dispersal hypotheses, with concordance of patterns implying a common cause. Dispersal is considered a process that predates vicariance, and thus cannot provide information about area relationships.

Caribbean insect taxa have becoming increasingly better known over recent years. Extant taxa have been studied, and cladistic hypotheses proposed (Buskirk, 1985; Matile, 1982; Grimaldi, 1988; Hamilton, 1988; Liebherr, 1986, 1988a). Oligocene amber from Hispaniola has produced numerous insect fossils (Baroni-Urbani & Saunders, 1980; Wilson, 1988). Whereas many Antillean insect taxa are most closely related to other North, Central, or South American taxa, others are most closely related to African or Australasian taxa (Liebherr, 1988b), indicating that their progenitors could have been resident on the Antilles since Mesozoic times.

Both mobilist/vicariance and stabilist/dispersal hypotheses for the Caribbean set forth sets of area relationships that can be portrayed as cladograms (Liebherr, 1988b). These sets of relationships differ, permitting their comparison with taxon-area relationships derived from taxonomic cladograms (Platnick & Nelson, 1978). Cladograms are compared using component analysis (Nelson & Platnick, 1981), or biogeographic graphs (Mickevich, 1981; Liebherr, 1988b). The results agree using the two methods, but each has its advantages and limitations. Component analysis preserves the directional information in a cladogram, but becomes cumbersome when taxa are widespread or patterns are partially redundant. Biogeographic graphs provide a visual means to compare the relationships between areas, but they do not contain directional information, and so area connections may represent dispersals or disjunctions among monophyletic or paraphyletic assemblages of areas.

Two examples of ground beetle (Coleoptera: Carabidae) taxa are used to illustrate the method. The *Platynus jaegeri* group comprises 32 species; 14 from Hispaniola, four from the northern Lesser Antilles, four from Jamaica, two from Cuba, and eight from Central America and Mexico. The taxon-area cladogram places the Lesser Antillean areas near the base of the cladogram, Hispaniolan, Jamaican, and Cuban areas in the middle of the cladogram, and mainland, Hispaniolan and Jamaican areas near the apex of the cladogram (Liebherr, 1988b). This taxon-area cladogram shares five of eight possible components with the vicariance-based area cladogram, but only three components with the dispersal-based area cladogram. When the cladograms are converted to biogeographic graphs, seven of eight possible connections are shared between the vicariance and taxonomically based graphs, whereas only three of eight connections are shared between the dispersal and taxonomically based graphs.

The *Platynus ovatulus* group is composed of 38 species, six from Hispaniola, one on Cuba and the Bahamas, and 31 species distributed from Arizona to northern South America (unpubl. data). The taxon-area cladogram places the seven Antillean species as sister group to two unresolved Mexican-northern Central American groups of one and six species. The outgroups of this clade contain North and Central American species. The biogeographic graph supports the Antillean clade as a single derivation of mainland stock that diversified in both Cuba and Hispaniola at approximately the same time. This pattern is more concordant with a dispersal origin than with Eocene vicariance.

Given that *jaegeri* group relationships are more compatible with a vicariant Caribbean origin, whereas *ovatulus* group relationships are more likely due to dispersal, other attributes of the groups may be examined to test these hypotheses of Antillean origin. The *jaegeri* group species are all vestigially winged, and are principally restricted to higher elevation montane habitats. Analysis of the pattern of habitat use in light of the cladistic relationships of the species suggests that the group primitively occurred in high elevation habitats (Liebherr & Hajek, 1990). Conversely, of the 38 *ovatulus* group species, 21 are apparently comprised entirely of macropterous individuals, and 17 exhibit some form of brachyptery. Eight species have been collected more than 20% of the time in flight (percentage of series collected in flight or at lights). Nine species possess geographic ranges over 500 km in maximal expanse. Whereas many species are restricted to montane forests, others inhabit lowland regions, so that no clear trend in habitat use is evident based on phylogeny. It is clear the the biological attributes of these insects are compatible with their hypothesized distributional history in the Caribbean. Further understanding in other taxa, both insect and otherwise, should be based on a primary analysis of distributional patterns in light of phylogeny, followed by investigation of subsidiary process-level hypotheses that might support conclusions reached by the analysis of pattern.

LITERATURE CITED

Baroni-Urbani, C. & J. B. Saunders. 1980. The fauna of the Dominican Republic amber: the present status of knowledge. *Transactions of the 9th Caribbean Geological Conference* (Santo Domingo, Dominican Republic) 1: 213–223.

Buskirk, R. E. 1985. Zoogeographic patterns and tectonic history of Jamaica and the northern Caribbean. *Journal of Biogeography* 12:445–461

Darlington, P. J., Jr. 1957. *Zoogeography: The Geographical Distribution of Animals.* John Wiley & Sons: New York.

Donnelly, T. W. 1988. Geologic constraints on Caribbean biogeography. Pp. 15–37. *In:* J. K. Liebherr (ed.), *Zoogeography of Caribbean Insects.* Cornell University Press: Ithaca, NY.

Grimaldi, D. A. 1988. Relicts in the Drosophilidae (Diptera). Pp. 183–213. *In:* J. K. Liebherr (ed.), *Zoogeography of Caribbean Insects.* Cornell University Press: Ithaca, NY.

Hamilton, S. W. 1988. Historical biogeography of two groups of Caribbean *Polycentropus* (Trichoptera: Polycentropodidae). Pp. 153–182. *In:* J. K. Liebherr (ed.), *Zoogeography of Caribbean Insects.* Cornell University Press: Ithaca, NY.

Liebherr, J. K. 1986. *Barylaus,* new genus (Coleoptera: Carabidae) endemic to the West Indies with Old World affinities. *Journal of the New York Entomological Society* 94:83–97.

Liebherr, J. K. 1988a. Biogeographic patterns of West Indian *Platynus* carabid beetles (Coleoptera). Pp. 121–152. *In:* J. K. Liebherr (ed.), *Zoogeography of Caribbean Insects.* Cornell University Press: Ithaca, NY. 285 pp.

Liebherr, J. K. 1988b. General patterns in West Indian insects, and graphical biogeographic analysis of some circum-Caribbean *Platynus* beetles (Carabidae). *Systematic Zoology* 37:385–409.

Liebherr, J. K. & A. E. Hajek. 1990. A cladistic test of the taxon cycle and taxon pulse hypotheses. *Cladistics* 6:39–59.

Matile, L. 1982. Systématique, phylogénie et biogéographie des Dipteres Keroplatidae des Petites Antilles et de Trinidad. *Bulletin de la Muséum National d'Histoire Naturelle* (4 ser.) 4:189–235.

Mickevich, M. F. 1981. Quantitative phylogenetic biogeography. Pp. 202–222. *In:* V. A. Funk & D. R. Brooks (eds.), *Advances in Cladistics—Proceedings of the First Meeting of the Willi Hennig Society.* New York Botanical Garden: New York.

Nelson, G. & N. Platnick. 1981. *Systematics and Biogeography: Cladistics and Vicariance.* Columbia University Press: New York.

Perfit, M. R. & E. E. Williams. 1989. Geological constraints and biological retrodictions in the evolution of the Caribbean Sea and its islands. Pp. 47–102. *In:* C. A. Woods (ed.), *Biogeography of the West Indies, Past, Present & Future.* Sandhill Crane Press, Inc.: Gainesville, FL.

Platnick, N. I. & G. Nelson. 1978. A method for analysis of historical biogeography. *Systematic Zoology* 27:1–16.

Rosen, D. E. 1975. A vicariance model of Caribbean biogeography. *Systematic Zoology* 24:431–464.

Rosen, D. E. 1985. Geologic hierarchies and biogeographic congruence in the Caribbean. *Annals of the Missouri Botanical Garden* 72:636–659.

Savage, J. M. 1982. The enigma of the Central American herpetofauna: dispersals or vicariance? *Annals of the Missouri Botanical Garden* 69:464–547.

Wilson, E. O. 1988. The biogeography of the West Indian ants (Hymenoptera: Formicidae). Pp. 214–230. *In:* J. K. Liebherr (ed.), *Zoogeography of Caribbean Insects.* Cornell University Press: Ithaca, NY.

Evolution on Islands and Conservation:
Philippines

S. H. Sohmer

The purpose of this discussion group was to bring up to date our understanding of what we do, and do not, know of terrestrial Philippine biodiversity, and to begin thinking of ways in which we can collectively conceptualize new insights and understanding. We wanted also to try to understand the trends the future may hold, so we may be best able to conserve what is left of one of the earth's more significant manifestations of diversity. To these ends eight participants gathered and, after presenting encapsulated summaries of the groups with which they were conversant, engaged in a free-form discussion with members of the audience.

The Philippines probably represents the single worst case scenario in Southeast Asia in terms of the past, present, and future potential loss of biological diversity. Aside from Southeast Asian countries such as Singapore and Hong Kong, which obviously never had much in the way of biological diversity, the Philippines archipelago represents one of the primary areas of the phytogeographic region called Malesia by botanists. This area comprises, along with the Philippines, Malaysia, the sultanate of Brunei, Indonesia, and the island of New Guinea, and collectively represents one of the principal cradles of global biodiversity. Certainly this region represents the richest source of life in the biological history of Southeast Asia that has radiated into other areas. The Pacific area has been one of the major recipients of this largess for it has, over geological time, obtained its principal revenue, in terms of genetic diversity, from this region. The Philippines has played a very important role as part of this region and has, due to its relatively large size, numerous islands, and highly dissected topography, served as a fertile ground for the testing of new evolutionary trends and opportunities and not as a cul de sac into which biological material from the central parts of Malesia migrated. The archipelago also represents a series of stepping stones into part of the Pacific that has no doubt aided long-distance dispersal.

Despite the fact that modern organismal biologists have labored in the archipelago for well over a century, there remains an utter ignorance for many groups of not only what was there but, more importantly, what is left. One would think, for example, that we would know a great deal about the flowering plants. However, even this important group, whose number of species, compared to a group such as the insects, do not represent an insurmountable barrier, is poorly known despite the herculean efforts of E. D. Merrill, A. D. Elmer, and their Philippine collaborators early in this century. We can't even tell you to this day how many species of flowering plants exist or existed in the Philippines. We cer-

Dr. Sohmer is the Assistant Director, Research and Scholarly Studies, and Chairman of the Botany Department, Bishop Museum, P.O. Box 19000-A, Honolulu, HI 96817, USA.

tainly can't tell you how many of these species are left and which are endangered. Again and again we seem to come back to the work of such botanists as Merrill and Elmer from the first two decades of this century. Again and again we see lists that are based on their information simply because more complete and more contemporary information have not been available. It is extremely disheartening and frustrating to view this in the context of the critical pressures to which the natural resource base of the country has been, and will be, subjected. We do not have the luxury of time. We do not have the luxury of getting the last word on this information. And yet, the quick and superficial approaches that have been used for some of these groups, although better than nothing, have led, perhaps, to a false sense of security in some areas and to a false sense of urgency in others. It is really time to assess what it is we wish to do with the few years remaining to those of us interested in knowing and preserving representative samples of the cornucopia of the biological diversity once represented, and perhaps still represented, in the Philippines.

Of the 30 million hectares of land of the Philippine Archipelago, we have reason to believe that about 94% of it was originally forested. It was estimated in August of 1989 by various reports, including the Master Plan for Forestry Development, that there was only a little over two million hectares of original forest left. This would certainly mean, if true, and there is no reason to doubt these figures, that over 90% of the original forests are now gone. We could thereby hypothesize that 50% or more of the original endemic terrestrial life that existed in those forests is no longer here among us, based on modern theories of biogeography. As I have stated elsewhere, my own work with a group of forest plants led me to hypothesize that 30 to 40% of those species were extinct or reduced to such numbers that they were no longer functionally viable in an evolutionary sense. When you come right down to it, these are potentially horrendous figures. It is not horrendous because such things have not occurred in the past. The history of man and the increase of his numbers and technology (for he has been truly fruitful and has multiplied and "conquered" the Earth) is replete with such events: the Middle East, South China, the Indo-Gangetic Plain, the temperate forests of North America and Europe. The difference is the degree of what is being lost in the tropics as compared to some of these other areas. We all know that the figures for the tropics are logarithmically higher in terms of diversity than an equal-sized piece of ground anywhere in the temperate world. In the Lamao Forest Reserve in the Philippines, for example, 120 different species of trees were found in a plot of land slightly over one hectare in size.

Historically speaking, those individuals from the temperate world who are battling to preserve global biodiversity do not stand on very firm ground when pointing fingers at our colleagues in the tropics. Obviously, what has happened in the history of temperate lands can very easily lead to cries of hypocrisy being leveled against those who now want to prevent what is often viewed as development in those countries that are fortunate enough to have significant amounts of tropical forests left. Man's collective problem is that the loss of these forests and their ecosystems is contributing to massive global environmental changes that ultimately affect everyone no matter where they live. It is for this reason that when one speaks of conservation in a place like the Philippines, as an example, we have to assume that the Philippines will receive support that will help them make the decisions to maintain forests and ecosystems that would otherwise be developed. People have spoken before of a "rent" that countries in the developed world will have to pay to maintain natural areas in the tropics. I believe this may well be a good working hypothesis.

To summarize, we need to know more about the status of terrestrial biological diversity in the Philippines, and we need to know it fast in order that we might establish a firmer basis upon which to make predictions for conservation planning and the preservation of the remaining Philippine genetic resources. In the discussion, E. S. Fernando summarized the vegetation of the Philippines and gave special attention to the situation concerning the flowering plants. Benito Tan provided the best basis of knowledge available of the mosses of the Philippines—a group often overlooked but extremely important both in terms of numbers and ecology in the moist and montane forests of the Philippines. Michael G. Price did the same for the fern flora. The terrestrial group with the largest number of species, and the most to lose from the extinctions taking place, and which have taken place, is the insects. Victor Gapud provided us with present data concerning the status of this important group based on work he and his colleagues have carried out at the University of the Philippines–Los Baños over many, many years. Christopher Starr gave us insight into the diversity and the conservation of the social insects in the Philippines. Lawrence Heaney told us about the situation concerning the mammals of the Philippines and Robert Kennedy did the same for the bird life. Finally, Domingo Madulid talked about the status of conservation efforts in the Philippines, particularly as involving the terrestrial plant life.

The discussion, after a number of individuals expressed their view of the potentially dire straits in which remnant natural ecosystems exist in the Philippines today, focused on what can realistically be accomplished. Taken as a given was the importance of the often expressed view that the education of the Philippine public must be a high priority for any solutions to be effective. It was also accepted that the inhabitants of the areas where remnant natural areas are still found have to be part of the solution, and must be brought into the conservation process. Without such indigenous participation there can be no realistic long-term solution.

PHYLOGENETIC
PROCESSES

The Evolution of Bacteria: Does Sex Matter?

J. Maynard Smith

INTRODUCTION

Prokaryotes are known to exchange chromosomal DNA in a variety of ways—by transformation (the active uptake of DNA present in the medium and its incorporation into the chromosome), by transduction (the transfer of DNA by phage particles), and by conjugation (transfer mediated by plasmids). These processes have been widely used in the study of prokaryote genetics, but until recently little was known of their role in the field or in evolution. In the last ten years, however, there has been a dramatic increase in our knowledge of these topics. Two generalizations have emerged. First, electrophoretic studies show that bacterial populations have a clonal structure (Selander et al., 1987), which has been interpreted as indicating that such populations consist of a series of independently evolving clones. Second, individual bacterial genes often display a mosaic structure, suggesting the occurrence of local genetic exchange between homologous genes (DuBose et al., 1988; Dykhuisen & Green, 1986; Milkman and Crawford, 1983; Spratt, 1988; Spratt et al., 1989). At first sight, these two observations appear con-tradictory. One aim of this lecture is to show that they are not. They are, however, the exact opposite of what we are used to in sexual eukaryotes, in which recombination between gene loci is common, but between homologous genes is usually thought to be rare. This makes it hard to think about the evolution of prokaryotes, at least for those, like myself, used to sexual eukaryotes.

I will briefly review the evidence both for clonal and for mosaic structure. I will then discuss what they mean for bacterial evolution. This leads to the question of whether there are such things as bacterial "species," and to the problem of "inverse taxonomy"—that is, the problem of deducing evolutionary mechanisms from the pattern of variation in nature.

THE CLONAL STRUCTURE OF BACTERIAL POPULATIONS

Electrophoretic studies first demonstrated the existence of a clonal population struc-ture in *Escherichia coli.* However, I will present data from another bacterium, *Neisseria meningitidis,* because we have particularly clear evidence of mosaic gene structure in this bacterium, and it is important to have evidence of both phenomena from the same genus. Caugant et al. (1987) examined 15 loci in 650 isolates of *N. meningitidis.* All 15 loci showed

Dr. Smith is with the School of Biological Sciences, The University of Sussex, Falmer, Brighton, BN1 9QG, England.

electrophoretically detectable polymorphism, with an average number of alleles per locus of seven. Compared to sexual eukaryotes, this is a very high level of variability. An immense number of electrophoretic types (ETs) could in principle be distinguished, yet 61 ETs were found more than once among the 650 isolates, and 19 ETs were found in more than one continent and 15 years apart. These figures imply that recombination leading to crossing over between distant gene loci is a rare event. It does not, of course, prove that the members of a single ET are genetically identical at all loci.

THE MOSAIC STRUCTURE OF BACTERIAL GENES

As an illustration of mosaic structure, I take the gene codong for a "penicillin binding" protein, PBP2, in the genus *Neisseria*. Changes in this gene are concerned with the evolution of resistance to penicillin. In most bacteria, this is achieved by the acquisition of a plasmid bearing a gene codong for lactamase, an enzyme that breaks down penicillin. However, in *Neisseria* (and also in *Streptococcus*) the common form of resistance involves changes in chromosomal genes. Penicillin kills bacteria by binding to several high molecular weight proteins, or PBPs, which act in the final stages of cell wall synthesis. Resistance can be acquired by changes in these proteins. Some 1500 base pairs of the gene codong for one of these proteins have been sequenced from five species of *Neisseria*. Two of these, *N. meningitidis* and *N. gonorrhoeae*, are important human pathogens; the other three, *N. flavescens*, *N. lactamica*, and *N. polysaccharae*, are harmless commensals. *N. flavescens* is always resistant to penicillin; in the other species, both resistant and sensitive strains are found, but sensitivity is almost certainly the primitive condition. Resistance has evolved in response to the strong selection pressure arising from the recent widespread use of penicillin, by the acquisition of blocks of DNA from the naturally resistant species, *N. flavescens*, and also from an as yet unidentified sixth species (in fact, this sixth species has now been identified, and partial sequences are available but have not yet been analyzed).

An analysis of these sequences is shown in Figure 1 and Table 1. A paper describing the statistical methods used is in preparation; the reality of the block structure is not in doubt, although the precise boundaries of the blocks are in some cases open to .question. The interpretation is that resistance has evolved by the introduction, probably by transformation, of blocks of DNA, typically of the order of 100–1000 bp. from resistant into sensitive strains. The evolutionary changes have involved at least six "species," differing in sequence by up to 20% of nucleotides.

Figure 2 shows a possible ancestry for most of the strains that have been sequenced; a similar, but simpler, ancestry could readily be devised for the *N. gonorrhoeae* and *N. polysaccharae* strains not included in Figure 2. The basic assumption made in constructing this ancestry was that, if the same apparent crossover point occurs in two strains, to the limit of accuracy met by the existence of 432 polymorphic sites, this is to be explained by common ancestry; that is, specific crossovers have occurred only once.

A very similar picture has emerged from a study of penicillin resistance in *Streptococcus pneumoniae* (Dowson et al., 1989), although in that case the donor species have not yet been identified. The transfer of blocks of DNA has also been important in the evolution of the IgA gene in *Neisseria*, which codes for an enzyme that breaks down human IgA (Halter et al., 1989); of the genes determining capsular type in *Haemophilus* (Kroll & Moxon, 1990), and of a gene codong for the highly antigenic pili in *Neisseria* (Seifert et al., 1988). These are all genes likely to be under strong selection for change. Also, the three genera—*Neisseria, Pneumococcus* and *Haemophilus*—are among the bacteria that have evolved a specific competence for transformation—that is, for the uptake and chromosomal incorporation of DNA present in the surroundings. This raises the ques-

Table 1. Percent nucleotide differences between *Neisseria* species.

	N men.	N flav.	N ?
N. meningitidis	—		
N. flavescens	12.6	—	
N. ?	19.5	16.0	—
N. lactamica central region	15.1	24.4	23.9
No. gonorrhoeae	2.0	—	—

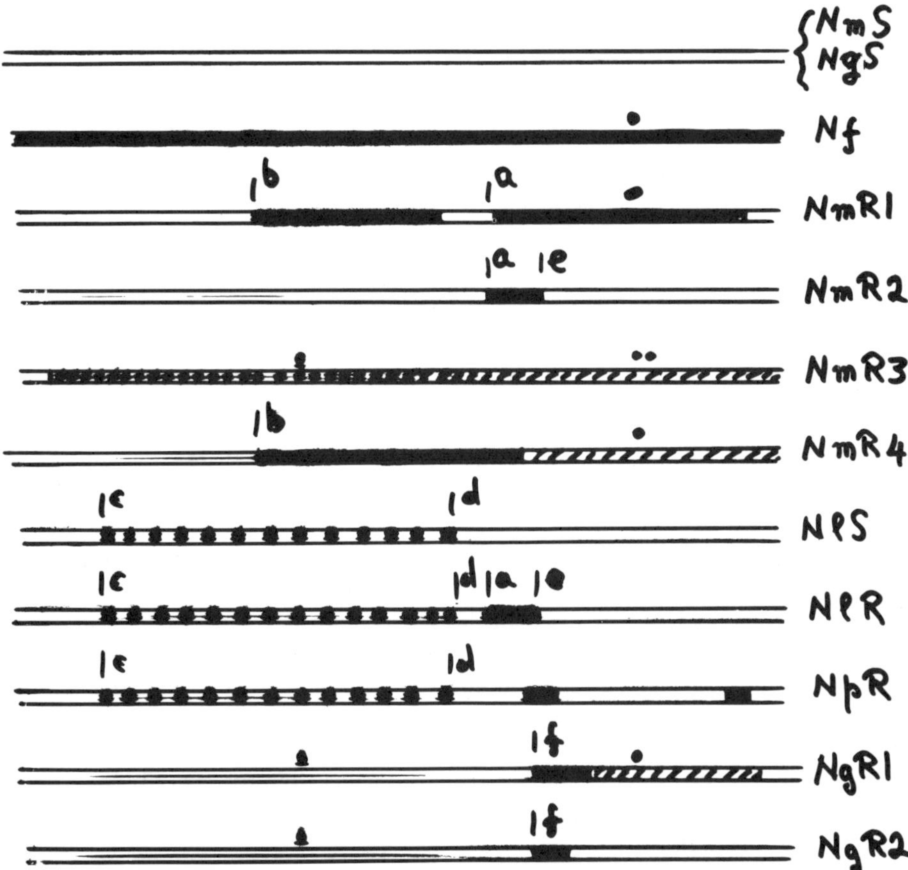

Figure 1. Mosaic structure of 1500 bp of the PBP2B gene, coding for a penicillin-binding protein in *Neisseria* (based on Spratt, 1988, Spratt et al., 1989, and unpublished data). NgS, NmS, penicillin-sensitive *N. gonorrhoeae* and *N. meningitidis;* NF, *N. flavescens* (naturally resistant); NmR1—NmR4, resistant strains of *N. meningitidis;* NlS, NlR, sensitive and resistant strains of *N. lactamica;* NpR, resistant strain of *N. polysaccahrae;* NgR1, NgR2, resistant strains of *N. gonorrhoeae.*

a—a, ... f—f, represent crossover points that occur between the same pair of polymorphic sites (this usually specifies the site to within 10 nucleotides or less).

; DNA similar to sensitive *N. meningitidis.*

; DNA similar to *N. flavescens.*

; DNA differing from *N. meningitidis* by 12% sequence divergence.

; DNA of unknown origin, differing from *N. meningitidis* by 14% sequence divergence.

• ; single additional codon.

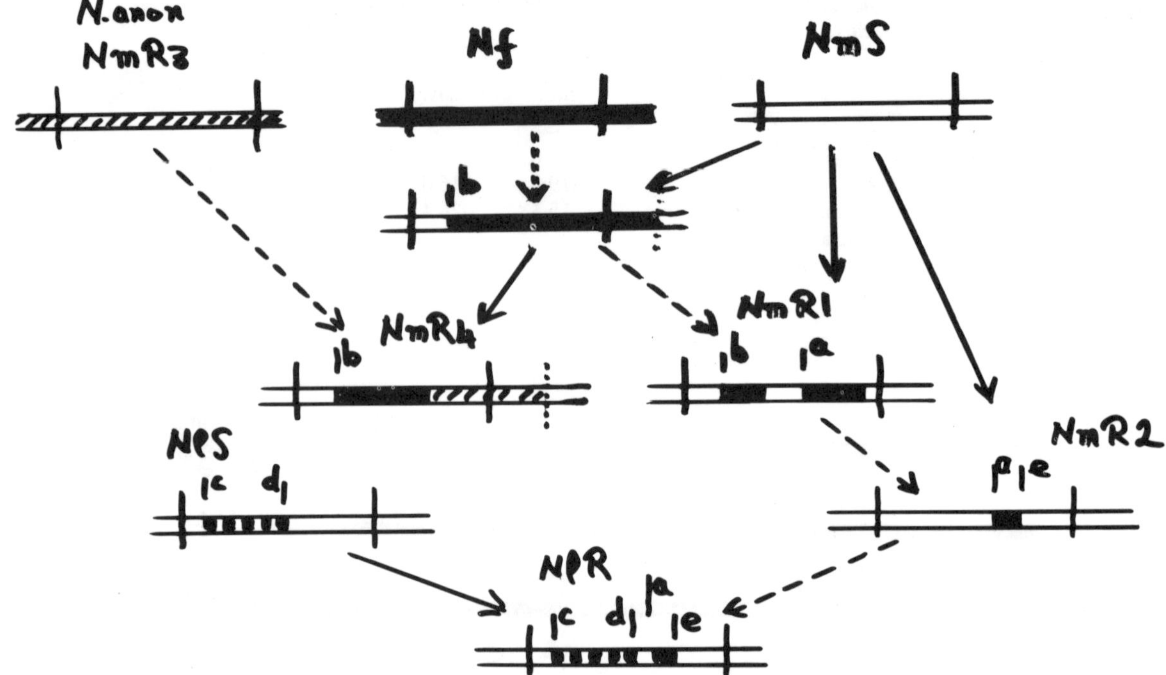

Figure 2. A possible ancestry of some of the strains of *Neisseria* shown in Fig. 1. Full-line arrows indicate the recipient parent, and broken-line arrows the parent contributing the introduced DNA. Notation as in Fig. 1.

tion whether a comparable mosaic structure exists in bacteria, such as *E. coli*, which are not competent for transformation, and for genes that are not known to have been under strong directional selection.

Three enzyme loci in *E. coli*—phoA (DuBose et al., 1988), gnd (Dykhuizen & Green, 1986) and trp (Milkman & Crawford, 1983)—have been sequenced from ten or so strains. In all cases, there is evidence of a mosaic structure, although it is less obvious than that illustrated in Figure 1, and requires statistical analysis to reveal (Sawyer, 1989). In the case of the phoA locus, Dubose et al. have suggested a possible ancestry for the sequenced strains, involving four events in which a short block of DNA from one strain was introduced into another. There is also evidence of recombination at the *pap* gene cluster (Plos et al., 1989), and for a 3500 bp region close to the trp locus (Stoltzfus et al., 1988). The transfer events that gave rise to the mosaic structure were probably mediated by conjugation plasmids, or by phage vectors.

CLONAL POPULATION STRUCTURE AND MOSAIC GENE STRUCTURE— ARE THEY COMPATIBLE?

Conjugation can lead to recombination between large chromosomal regions, carrying many genes; that is, it can have results similar to crossing over in sexual eukaryotes. The electrophoretic data suggest that this type of event must be rare in prokaryotes. However, the gene sequence data show that the transfer of short blocks of DNA from a donor to a related recipient cell has played an important role in the evolution of bacterial genes. There is no incompatibility between the two conclusions. A local transfer event would not alter the electrophoretic type of either strain, except in the rare case that one of the electrophoretic markers was involved. Provided that local transfers occur but are not too frequent, we expect to observe both clonal structure on a large scale, and mosaic structure on a small scale.

In most cases, electrophoretic and sequence data have been collected from different strains. An interesting exception concerns capsular type in *Haemophilus influenzae* (Kroll & Moxon, 1990). The genes for the synthesis of capsular polysaccharide contain a central region that is specific to each of the capsular types, flanked by regions that are common to all types. Capsulated strains of *H. influenzae* have a clonal population structure, and similar ETs usually have the same capsular type. Strains with capsular type b, however, are found in two distantly related ETs. Nucleotide sequencing of parts of the region from a type b isolate from each lineage showed 0/250 differences in the central region and 97/795 (12%) differences in one of the flanking regions. The conclusion is that the central type-specific region of a type b strain has been transferred into a related strain that differs by about 12% of nucleotides, without altering the ET of the recipient.

THE EVOLUTIONARY GENETICS OF PROKARYOTES

In conclusion, I want to raise a number of questions to which the answers are still unclear. The first concerns the significance of clonal structure. Certainly, it implies that large-scale crossing over must be rare. But suppose that such crossing over never takes place. Then, if a unique favourable mutation occurs, it would spread to fixation, and in so doing would render the whole population homogeneous for the alleles that happened to be present in the cell in which the original mutation occurred. Hence we would expect to find bacterial populations to be electrophoretically uniform, and not highly variable, as is in fact the case. Of course, in a sufficiently large population point mutations will not be unique events. But if novelties arise by rare local gene transfer, they could well be effectively unique, and hitch-hiking would lead to homogeneity. The situation is more complex than this simple argument implies. Clearly the model of a unique favourable mutation spreading without recombination through a whole species must be wrong. Milkman & Stoltzfus (1988) have extended the model by including a low frequency of crossing over. I suspect that we should also allow for population structure if we are to explain the data, but it is not clear to me what the appropriate model should be.

A related question is whether there is any unit corresponding to the "species." Among sexual eukaryotes, the species has two defining characteristics. It is a group of organisms that are able to exchange genetic material with one another, but which rarely exchange genes with members of other species. If there is such a unit among prokaryotes, it is wider than the named species such as *Streptococcus pneumoniae* or *Neisseria meningitidis*: it more nearly corresponds to the genus *Neiserria*, or to *Escherichia + Shigella*. But it is not clear that there is any such unit, because it is not clear that there is anything in prokaryotes corresponding to the reproductive isolating mechanisms of sexual eukaryotes. For those types of gene transfer, such as transformation, that depend on homologous recombination, only rather similar organisms can exchange genes—perhaps up to approximately 20% sequence divergence. But this does not imply that there need be genetic discontinuities. Thus A might be too different from E to exchange genetic material, but nevertheless A could exchange with B, B with C, C with D, and D with E: if so, A would not be genetically isolated from E, and no clear species boundaries need exist.

Thus there are two related questions: are there prokaryote species, and, if there are, what is the nature of the isolating mechanisms between them? Most microbiologists, I think, are convinced of the reality of bacterial species, but I remain unpersuaded, if only because, in some cases, the "species" they recognize are exchanging genetic material. In any case it would be premature to try to identify isolating mechanisms until the reality of

422

bacterial species has been established. It may be helpful, however, to describe three conceivable models of prokaryote evolution (Fig. 3):

A. A "fractal tree." If there is no genetic recombination of any kind, then not only does each new population arise by the splitting of a pre-existing population, but each individual arises by the splitting of a parental individual, and each gene by the splitting of a parental gene: the structure is "tree-like" at every scale of magnification.

B. A "large-scale tree." In sexually reproducing organisms, new species usually arise by the splitting of pre-existing species (I ignore the case of allopolyploidy), but each individual has two parents. Thus the structure is tree-like at a large scale, but net-like at a small scale.

C. A "locally-continuous tree." Suppose, as suggested above, that individuals can exchange genes with others not too distant genetically from themselves, but that there are no isolating mechanisms other than those set by genetic distance. The structure will then be that in Figure 3C.

Strictly, only type B need have separate branches: A and C could resemble a liverwort rather than a tree. But there are in practice likely to be gaps between branches also in types A and C, arising by historical accident or ecological necessity.

It is not clear to me which, if any, of these models is appropriate to prokaryote evolution. The answer can only come, I think, from a combined study of natural variation and of molecular mechanisms. We will have to develop a new way of thinking about the world, which might be called "inverse taxonomy" (for examples of this way of thinking, see Eigen et al., 1988). Although taxonomists claim to be studying the pattern of nature, in practice they usually assume that the pattern is that of a large-scale tree: any hierarchical classification in biology in effect assumes that the objects to be classified arose by a branching process. The task of the classifier is then reduced to deciding the precise pattern of branching. In contrast, the task of inverse taxonomy is to deduce from the observed pat-

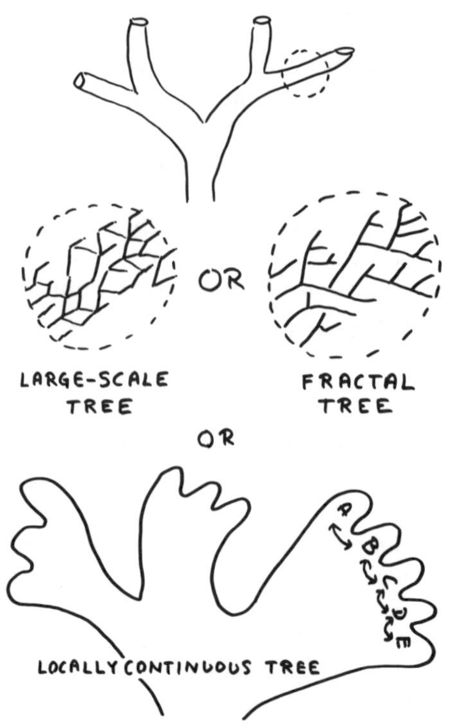

Figure 3. Three kinds of phylogenetic tree.

tern of variation what is the appropriate structure. At a small scale, is a tree or a net the appropriate model? At a large scale, is a large-scale tree or a locally continuous tree appropriate? Must we in addition take into account distant transfer events?

LITERATURE CITED

Caugant, D. A., Mocca, L. F., Frasch, C. E., Froholm, L. O., Zollingen, W. D. & R. K. Selander. 1987. Genetic structure of *Neisseria meningitidis* populations in relation to serogroup, serotype and outer membrane protein pattern. *J. Bact.* 169:2781–92.

Dowson, C. G., Hutchison, A., Brannigan, J. A., George, R. C., Hausman, R. C., Linares, D., Tomacz, A., Maynard Smith, J., & B. G. Spratt. 1989. Horizontal transfer of penicillin-binding protein genes in penicillin-resistant clinical isolates of *Streptococcus pneumoniae. Proc. Natl. Acad. Sci.* USA 86:8842–46.

DuBose, R. F., Dykhuizen, D. E. & D. L. Hartl. 1988. Genetic exchange between natural isolates of bacteria: recombination within the phoA gene of *Escherichia Coli. Proc. Natl. Acad. Sci.* USA 85:7036–40.

Dykhuisen, D. E. & Green, L. 1986. DNA sequence variations, DNA phylogeny and recombination in *E. coli. Genetics* 113:571, and an unpublished manuscript.

Eigen, M., Winkler-Oswatisch, R. & A. Dress. 1988. Statistical geometry in sequence space: a method of quantitative comparative sequence analysis. *Proc. Natl. Acad. Sci.* USA 85:5913–17.

Halter, R., Pohlner, J. & T. F. Meyer. 1989. Mosaic-like organization of IgA protease genes in *Neisseria gonorrhoeae* generated by horizontal genetic exchange in vivo. *EMBO J.* 8:2737–44.

Kroll, J. S., & E. R. Moxon, 1990. Capsulation in distantly related strains of *Haemophilus influenza* type b: genetic drift and gene transfer at the capsulation locus. *J. Bact.* 172:1374–79.

Milkman, R. & I. P. Crawford. 1983. Clustered third-base substitutions among wild strains of *E. coli. Science* 221:378–80.

Milkman, R. & A. Stoltzfus. 1988. Molecular evolution of the *Escherichia coli* chromosome. II Clonal segments. *Genetics* 120:356–66.

Plos, K., Hull, S. I., Hull, R. A., Levin, B. R. & I. Orskov. 1989. Distribution of the P-associated pilus (*pap*) region among *Escherichia coli* from natural sources: evidence for horizontal gene transfer. *Infect. Immun.* 57:1604–11.

Sawyer, S. 1989 Statistical tests for detecting gene conversion. *Mol. Biol. Evol.* 6:562–38.

Seifert, H. S., Ajioka, R. S., Marchal, C., Sparling, P. F. & M. So. 1988. DNA transformation leads to pilin antigenic variation in *Neisseria gonorrhoeae. Nature* 336:392–95.

Selander, K., Caugant, D. A. & T. S. Whittam. 1987. *Genetic structure and variation in natural populations of* Escherichia coli *and* Salmonella typhimurium. *Am. Soc. Microbiol.* Washington, D. C.

Spratt, B. G. 1988. Hybrid penicillin-binding proteins in penicillin-resistant strains of *Neisseria gonorrhoeae. Nature.* 332:173–76.

Spratt, B. G., Zhang, Q. -Y., Jones, D. M., Hutchison, A., Branningan, J. A. & C. G. Dowson. 1989. Recruitment of penicillin-binding protein genes from *Neisseria flavescens* during the emergence of penicillin resistance in *Neisseria meningitidis. Proc. Natl. Acad. Sci.* USA 86:8988–92.

Stoltzfus, A., Leslie, J. E. & R. Milkman. 1988. Molecular evolution of the *Escherichia coli* chromosome. I. Analysis of structure and natural variation in a previously uncharacterised region between trp and tonB. *Genetics* 120:345–58.

Genetic Analysis of Simian Immunodeficiency Viruses and Their Relationship to Human Immunodeficiency Viruses

Yen Li, Robert Steen, Carole Butler, Patricia Fultz, Preston Marx, and Ronald Desrosiers

Abstract. The simian immunodeficiency viruses (SIVs) are a diverse group of nonhuman primate lentiviruses that are the closest known relatives of the human immunodeficiency viruses (HIVs). Genetic analysis of primate lentivirus isolates to date reveals that they may be classified into four groups: SIVagm, SIVsmm/SIVmac/HIV-2, HIV-1/SIVcpz, and SIVmnd (agm, African green monkey; smm, sooty mangabey monkey; mac, macaque monkey; cpz, chimpanzee; mnd, mandrill). We have demonstrated that the genetic diversity among SIVagm isolates is much greater than that observed previously for HIV-1 or SIVmac isolates. Extensive genetic variability among SIVagm isolates and the high prevalence of African green monkey infection without disease suggest that the virus may have been in the African green monkey population for a long time. HIV-2, SIVmac, and SIVsmm form a discrete, separate subgroup of primate lentiviruses. We have now shown that genetic diversity among SIVsmm isolates is also extensive, similar to SIVagm. The apparently recent appearance of HIV-2, the genetic similarity of SIVsmm and HIV-2, the co-location of sooty mangabey natural habitat with HIV-2 endemic area, the lack of disease in sooty mangabeys and its presence in humans, and the nature of the genetic diversity together suggest cross-species transmission from mangabeys to humans for the origin of HIV-2.

INTRODUCTION

Two related but distinct lentiviruses, HIV-1 and HIV-2, have been shown to cause AIDS in humans. Related lentiviruses of nonhuman primates (SIVs) have been isolated from several Old World primate species, including captive macaques (mac) of Asian origin (Daniel et al., 1985; Benveniste et al., 1986), captive and wild caught African green monkeys (agm) (Ohta et al., 1988; Daniel et al., 1988), captive sooty mangabey monkeys (smm) (Fultz et al., 1986; Murphy-Corb et al., 1986), and wild caught mandrills (mnd) (Tsujimoto et al., 1988). However, there is no known example of natural infection of New World primates by SIVs.

The origin and natural history of the human lentiviruses remain uncertain. HIV is apparently not an endogenous virus that is transmitted vertically. Two theories can be proposed to explain the origin of HIV. In the first, the presence of HIV in the human

Drs. Li, Steen, Butler, and Desrosiers are with the New England Regional Primate Research Center, Harvard Medical School, Southborough, MA 01772, Dr. Fultz is with the Yerkes Regional Primate Research Center, Emory University, Atlanta, GA 30322, Dr. Marx is with the California Regional Primate Research Center, University of California, Davis, CA 95618, USA. Please address correspondence to Dr. Li.

population was previously undetected because it was limited to isolated populations with little chance to spread to outside groups. Increased mobilization, the growth of large cities, and some practices of modern society may have introduced virus into larger segments of the population. The second hypothesis holds that HIV was derived from a similar virus in monkeys by cross-species transmission, which suggests a more recent introduction of the virus into the human population.

Molecular analyses have revealed that the genetic diversity among SIVagm isolates is much more extensive than previously reported for HIV-1 or SIVmac (Li et al., 1989a, b; Johnson et al., 1990). The extensive genetic diversity and high prevalence of SIV infection in African green monkeys without disease suggest that the virus may have been in the green monkey population for a long time, perhaps longer than HIV has been in the human population. Here, we describe results with SIVsmm which show that the genetic diversity of SIVsmm is similarly extensive.

MATERIAL AND METHODS

Virus and Cells

Lymphocytes from infected animals were separated from red blood cells and serum by Ficoll-Hypaque centrifugation (Daniel et al., 1985; Fultz et al., 1986; Marx et al., 1990). Viruses were isolated by cocultivation of the infected animal lymphocytes with cultured peripheral blood lymphocytes (PBL) in IL-2 or with CEMX174, a human T and B somatic hybrid cell line.

DNA Isolation

Total cellular DNA was isolated by standard procedures (Li et al., 1984). Briefly, infected culture cells were collected by centrifugation, and washed twice with phosphate buffered saline (PBS). Cells were then resuspended in lysis solution containing 0.15 M NaCl, 0.05M EDTA, 0.5% SDS, and 200 μg/ml proteinase K (Boeriger-Mannheim Biochemicals) and incubated at 50°C for 12 hr. The clear lysate was gently extracted three times with buffer saturated phenol and dialyzed against two liters of dialysis buffer that consisted of 10mM Tris-HCl pH 7.6, 1mM EDTA (TE) with at least 5 changes of the dialysis buffer. DNA concentration was determined by spectrophotometric methods assuming that an absorbance value of 1.0 at 260 nm is equivalent to 50 μg/ml.

Oligonucleotide Primers

Polymerase genes of HIV-1$_{BRU}$ (Wain-Hobson et al., 1985), HIV-2$_{ROD}$ (Guyader et al., 1987), SIVmac142 (Chakrabati et al., 1987) and SIVagmTyo1 (Fukasawa et al., 1988) were compared to choose two oligonucleotides as the primers for subsequent polymerase chain reaction (PCR). Primers were synthesized in a Cyclone DNA synthesizer (Milligen/Biosearch model 8400). Near the 5' end of each primer, a 6 base EcoRI recognition sequence was added to facilitate the subsequent molecular cloning of the PCR product. The sequence of the sense-strand oligonucleotide was 5'-GGGC*GAATTC*GGGAGCAATGGTGGGCGGATTACTGGC-3'; the sequence of the antisense-strand oligonucleotide was 5'-GCGATG*GAATTC*TGCTGCTTCCCCTTTCC-3'.

Polymerase Chain Reaction

A 1.2 kbp *pol* DNA fragment was amplified by PCR (Saiki et al., 1985, 1988). 500 ng to 1 μg of infected cellular DNA (template) was mixed with 0.2 μM of each of the two primers and other necessary reagents from the GeneAmp kit (Perkin-Elmer Cetus). The PCR

426

conditions were as follows: denaturation at 94°C for 1 min, annealing at 42°C for 2 min, and polymerization at 72°C for 1 min with auto extension for 5 sec in each cycle. The PCR was carried out for 30 cycles under a thin layer of mineral oil to prevent evaporation.

Molecular Cloning

PCR amplified *pol* fragments were cloned into a plasmid vector using standard protocols (Sambrook et al., 1989). Briefly, the DNA fragment was digested with the restriction endonuclease EcoRI, ligated to EcoRI-digested and alkaline phosphatase treated plasmid vector pGEM-4Z(f) (Promega). The ligation product was then used to transform *E. coli*. Positive colonies were identified by *in situ* hybridization using a SIVmac *pol* gene probe under reduced stringency of hybridization. Plasmid DNA was prepared by standard procedures (Sambrook et al., 1989).

Nucleotide Sequencing

Nucleotide sequences within the purified recombinant plasmid DNA were determined by the "dideoxy" chain termination method of Sanger et al. (1977) using sequenase (U.S. Biochemical). The nucleotide sequence was analyzed using the Pustell sequence analysis program (IBI-Pustell).

RESULTS AND DISCUSSION

Two SIVsmm isolates from distinct sources were analyzed. SIVsmm-7 was isolated from a captive sooty mangabey monkey housed at Yerkes Regional Primate Research Center (Fultz et al., 1986), whereas SIVsmmLIB-1 was isolated from a pet sooty mangabey in Liberia (Marx et al., 1990). The PCR amplified 1.2 kbp DNA fragment corresponding to the carboxy terminal one-third of the polymerase gene was sequenced. The predicted amino acid sequences were compared with the corresponding sequences of the polymerase polyproteins of SIVsmmH4, HIV-1_{BRU}, HIV-2_{ROD}, HIV-2_{GH-1}, HIV-2_{ST}, HIV-2_{D205}, HIV-2_{BEN}, SIVmac239, SIVmnd, SIVcpz, SIVagm385, and SIVagmTyo1 (Table 1). Within the SIVsmm group, *pol* amino acid sequence identity ranged from 83 to 90%, whereas SIVmac shared 87 to 92% amino acid sequence identity with SIVsmm. HIV-2 isolates shared 81–90% amino acid sequence identity with SIVsmm and 85–94% with each other. Thus, genetic relatedness of SIVmac and HIV-2 to SIVsmm is similar to the genetic relatedness among SIVsmm isolates themselves. HIV-2 isolates are in some cases as close to SIVsmm isolates as to other HIV-2 isolates. Similarly, SIVsmm isolates are in some cases as close to HIV-2 isolates as to other SIVsmm isolates. HIV-2, SIVsmm, and SIVmac are thus not distinct groups at the genetic level. Not only do SIVsmm, SIVmac, and HIV-2 form a single subgroup of primate lentiviruses, but also the genetic diversity of SIVsmm is as extensive as that observed previously for SIVagm.

We have compared the entire *pol* gene of SIVsmm-H4, SIVmac239, HIV-2_{ROD}, HIV-1_{BRU}, and SIVagmTyo1 and found that the percentage of amino acid sequence identity was quite close to that obtained by comparing the 1.2 kbp of *pol* sequences used in the current study (data not shown). Therefore, we conclude that the 1.2 kbp *pol* sequence appears to be representative of the entire *pol* sequence for such comparisons.

The extensive genetic diversity of SIVsmm and the fact that naturally infected sooty mangabeys do not develop apparent disease suggest that sooty mangabeys perhaps are the natural host for the virus and the SIVsmm might have been in the sooty mangabey population for a long time. The apparently recent appearance of HIV-2, the genetic similarity of SIVsmm and HIV-2, the co-location of sooty mangabey natural habitat with HIV-2 endemic areas, the lack of disease in sooty mangabeys and its presence in humans,

Table 1. Amino Acid Identity in *pol* of Primate Lentiviruses.

	SIVsmm-LIB-1	SIVsmm-7	SIVsmm-H4	SIVmac-239	HIV-2ROD	HIV-2GH-1	HIV-2ST	HIV-2D205	HIV-2BEN	SIVmnd	SIVagmTyo1	SIVagm385	HIV-1BRU
SIVsmm-7	83												
SIVsmm-H4	90	85											
SIVmac-239	87	87	92										
HIV-2ROD	83	82	89	86									
HIV-2GH-1	83	81	88	86	92								
HIV-2ST	85	84	90	89	94	91							
HIV-2D205	83	84	87	85	87	86	85						
HIV-2BEN	84	81	88	86	93	95	90	87					
SIVmnd	61	58	63	62	62	61	61	63	61				
SIVagmTyo1	61	55	64	65	63	63	62	63	62	62			
SIVagm385	53	51	56	55	54	53	54	55	54	52	78		
HIV-1BRU	59	56	63	60	60	61	60	61	60	61	63	54	
SIVcpz	59	57	63	61	61	60	60	61	60	61	64	53	86

The sequences used for comparison correspond to SIVsmm-H4 *pol* amino acids 573–969 (HIV-1BRU *pol* amino acids 574–970). All sequences were obtained from the human retrovirus and AIDS database at the Los Alamos National Laboratory. The SIVsmm-LIB-1 sequence described in this report has been submitted to the database (accession number M62651).

Table 2. Open Reading Frames of Primate Lentiviruses.

Gene	HIV/SIVcpz	HIV-2/SIVmac/SIVsmm	SIVagm	SIVmnd
gag	+	+	+	+
pol	+	+	+	+
env	+	+	+	+
vpu	+	−	−	−
vif	+	+	+	+
vpx	−	+	+	−
vpr	+	+	−	+
tat	+	+	+	+
rev	+	+	+	+
nef	+	+	+	+

and the nature of the genetic diversity together suggest cross-species transmission from mangabeys to humans for the origin of HIV-2.

Recently, a lentivirus was isolated from a chimpanzee (designated SIVcpz) in Gabon (Peeters et al., 1989). Genetic analysis of this isolate revealed that it is more closely related to HIV-1 than to any other SIV isolates or HIV-2 (Huet et al., 1990). However, the genetic relatedness of SIVcpz is outside the range of variation of known HIV-1 isolates. Continued investigation, including surveys of wild monkey populations and genetic analysis of more SIVcpz isolates, will be needed to determine whether chimpanzees may have been a source of cross-species lentivirus transmission to humans (Desrosiers, 1990).

Hayami and his colleagues isolated an SIV from a wild caught mandrill (designated SIVmnd), also from Gabon (Tsujimoto et al., 1988). Sequence analysis of this isolate placed it in a fourth group of primate lentiviruses, equally distant from HIV-1/SIVcpz, HIV-2/SIVmac/SIVsmm, and SIVagm (Tsujimoto et al., 1989). However, the extent of genetic diversity among SIVmnd isolates remains to be determined. Known primate lentiviruses may thus be classified into four groups. Each group is approximately equidistant to the others, sharing only about 55 to 60% amino acid sequence identity in *pol*. Within a group, different isolates exhibit 80% or more sequence identity in *pol*. Furthermore, each of the four groups has a distinct genomic organization (Table 2). For example, the HIV-1/SIVcpz group lacks the *vpx* gene, the HIV-2/SIVmac/SIVsmm group lacks the *vpu* gene, SIVagm lacks both *vpu* and *vpr*, and the SIVmnd group lacks *vpu* and *vpx*. The genomic organizations shown in Table 2 lead us to propose that an ancestral lentivirus existed before Old World primate speciation and that lentiviruses have co-evolved with specific natural hosts, e.g., African green monkey, sooty mangabeys, and perhaps other primates yet to be identified. Continued study of SIVs from these and additional simian species will provide important information on the evolutionary history of this exceptionally diverse group of primate lentiviruses.

ACKNOWLEDGMENTS

We wish to thank Diane Schmidt and Cynthia Troup for virus propagation and cell culture; Dean Regier for help in computer analysis; and Nancy Adams and Joanne Newton for preparing the manuscript. This work was supported in part by grants from NIH and AmFAR.

LITERATURE CITED

Benveniste, R. E., Arthur, L. O., Tsai, C. -C., Sowder, R., Copeland, T. D., Henderson, L. E. & S. Oroszlan. 1986. Isolation of lentivirus from a macaque with lymphoma: comparison with HTLV-III/LAV and other lentiviruses. *J. Virol.* 60:483–490.

Chakrabarti, L., Guyader, M., Alizon, M., Daniel, M. D., Desrosiers, R. C., Tiollais, P. & P. Sonigo. 1987. Sequence of simian immunodeficiency virus from macaque and its relationship to other human and simian retroviruses. *Nature* 328:543–547.

Daniel, M. D., Letvin, N. L., King, N. W., Kannagi, M., Sehgal, P. K., Hunt, R. D., Kanki, P. J., Essex, M. & R. C. Desrosiers. 1985. Isolation of T-cell tropic HTLV-III-like retrovirus from macaques. *Science* 228:1201–1204.

Daniel, M. D., Li, Y., Naidu, Y. M., Durda, P. J., Schmidt, D. K., Troup, C. D., Silva, D. P., MacKey, J. J., Kestler, H. W., III, Sehgal, P. K., King, N. W., Ohta, Y., Hayami, M. & R. C. Desrosiers. 1988. Simian immunodeficiency virus from African green monkeys. *J. Virol.* 62:4123–4128.

Desrosiers, R. C. 1990. HIV-1 origins: a finger on the missing link. *Nature* 345:288–289.

Fukasawa, M., Miura, T., Hasegawa, A., Morikawa, S., Tsujimoto, H. Miki, K., Kitamura, T. & M. Hayami. 1988. Sequence of simian immunodeficiency virus from African green monkey, a new member of the HIV/SIV group. *Nature* 333:457–461.

Fultz, P. N., McClure, H. M., Anderson, D. C., Swenson, R. B., Anand, R. & A. Srinivasan. 1986. Isolation of a T-lymphotropic retrovirus from naturally infected sooty mangabey monkeys (*Cecocebus atys*). *Proc. Natl. Acad. Sci. USA* 83:5286–5290.

Guyader, M., Emerman, M., Sonigo, P., Clavel, F., Montagnier, L. & M. Alizon. 1987. Genome organization and transactivation of the human immunodeficiency virus type 2. *Nature* 326:662–669.

Huet, T., Cheynier, R., Meyerhans, A., Roelants, G. & S. Wain-Hobson. 1990. Genetic organization of a chimpanzee lentivirus related to HIV-1. *Nature* 345:356–359.

Johnson, P. R., Fomsgaard, A., Allan, J., Gravell, M., London, W. T., Olmsted, R. A. & V. M. Hirsch. 1990. Simian immunodeficiency viruses from African green monkeys display unusual genetic diversity. *J. Virol.* 64:1086–1092.

Li, Y., Holland, C. A., Hartley, J. W. & N. Hopkins. 1984. Viral integration near C-*myc* in 10–20% of MCF 247-induced AKR lymphomas. *Proc. Natl. Acad. Sci. USA* 81:6808–6811.

Li, Y., Naidu, Y., Fultz, P., Daniel, M. D. & R. C. Desrosiers. 1989a. Genetic diversity of simian immunodeficiency viruses. *J. Med. Primatol.* 18:261–269.

Li, Y., Naidu, Y. M., Daniel, M. D. & R. C. Desrosiers. 1989b. Extensive genetic variability of simian immunodeficiency virus from African green monkeys. *J. Virol.* 63:1800–1802.

Marx, P. A., Li, Y., Lerche, N. W., Sutjipto, S., Gettie, A., Yee, J. A., Brotman, B., Prince, A. M. Hanson, A., Webster, R. G. & R. C. Desrosiers. 1991. Simian immunodeficiency virus from a West African pet sooty mangabey related to human immunodeficiency virus type 2. (submitted).

Murphy-Corb, M., Martin, L. N., Rangan, S. R. S., Baskin, G. B., Gormus, B. J., Wolf, R. H., Andes, W. A., West, M. & R. C. Montelaro. 1986. Isolation of an HTLV-III-related retrovirus from macaques with simian AIDS and its possible origin in asymptomatic mangabeys. *Nature* 321:435–437.

Ohta, Y., Masuda, T., Tsujimoto, H., Ishikawa, K., Kodama, T., Morikawa, S., Nakai, M., Honjo, S. & M. Hayami. 1988. Isolation of simian immunodeficiency virus from African green monkeys and seroepidemiologic survey of the virus in various non-human primates. *Int. J. Cancer* 41:115–122.

Peeters, M., Honore, C., Huet, T., Bedjabaga, L., Ossari, S., Bussi, P., Cooper, R. W. & E. Delaporte. 1989. Isolation and partial characterization of an HIV-related virus occurring naturally in chimpanzees in Gabon. *AIDS* 3:625–630.

Saiki, R. K., Scharf, S., Faloona, F., Mullis, K. B., Horn, G. T., Erlich, H. A. & N. Arnheim. 1985. Enzymatic amplification of β-globin genomic sequences and restriction site analysis for diagnosis of sickle cell anemia. *Science* 230:1350–1354.

Saiki, R. K., Gelfand, D. H., Stoffel, S., Scharf, S. J., Higuchi, R., Horn, G. T., Mullis, K. B., & H. A. Erlich. 1988. Primer-directed enzymatic amplification of DNA with a thermostable DNA polymerase. *Science* 239:487–491.

Sambrook, J., Fritsch, E. F. & T. Maniatis. 1989 *Molecular Cloning. A Laboratory Manual.* Second edition. Cold Spring Laboratory Press: Cold Spring Harbor, NY.

Sanger, F., Nicklen, S. & A. R. Coulson. 1977. DNA sequencing with chain-terminating inhibitors. *Proc. Natl. Acad. Sci. USA* 74:5463–5467.

Tsujimoto, H., Cooper, R. W., Kodama, T., Fukasawa, M., Miura, T., Ohta, Y., Ishikawa, K. -I., Nakai, M., Frost, E., Roelants, G. E., Roffi, J. & M. Hayami. 1988. Isolation and characterization of simian immunodeficiency virus from mandrills in Africa and its relationship to other human and simian immunodeficiency viruses. *J. Virol.* 62:4044–4050.

Tsujimoto, H., Hasegawa, A., Maki, N., Fukasawa, M., Miura, T., Speidel, S., Cooper, R. W., Moriyana, E. N., Gojobori, T. & M. Hayami. 1989. Sequence of a novel simian immunodeficiency virus from a wild-caught African Mandrill. *Nature* 341:539–541.

Wain-Hobson, S., Sonigo, P., Danos, O., Cole, S. & M. Alizon. 1985. Nucleotide sequence of the AIDS virus, LAV. *Cell* 40:9–17.

Protistan Evolution Inferred from Ribosomal RNA Sequences

EXTENDED ABSTRACT

J. H. Gunderson and M. L. Sogin

Morphological evidence drawn from light microscopical and ultrastructural studies has not been very successfully applied to the question of how the major groups of protists are related to each other. This is due in part to the relative lack of visible structure in some groups, such as the amoebae, and to the difficulty of recognizing homologous structural features in other groups. Ribosomal RNA sequences provide a very useful alternative way of infering relationships.

Although trees have been produced from 5S and 5.8S sequences, the larger RNAs can be expected to provide more accurate measures of relationship. Trees based on 18S and 28S rRNA sequences are similar to each other, but frequently quite unlike those produced from the smaller rRNAs. Trees constructed from the longer sequences by using maximum parsimony and distance matrix methods (the two most commonly used algorithms for producing trees from rRNA sequences) are remarkably alike, and frequently identical.

In trees produced from complete small subunit rRNA gene sequences, the deepest branches in the eukaryotic world are those leading to several small groups of colorless anaerobic organisms (the diplomonads, microsporidians, and trichomonads).

The primary divisions among the aerobic eukaryotes are correlated to a great extent with mitochondrial crista type. The first aerobic organisms to break off are those with discoid cristae. This group includes the euglenoids, the kinetoplastids, and the vahlkamp-fiids (a small group of amoeboflagellates including such genera as *Naegleria*). Then a large number of different groups having tubular mitochondrial cristae (ciliates, oomycetes, chrysophytes, apicomplexans, dinoflagellates, and various amoebae) appear in rapid succession. Cellular and acellular slime molds appear as isolated lineages emerging from near the base of this large group. The dinoflagellates form a supraphyletic assemblage with the ciliates and apicomplexans. These three groups share the peculiar characteristic of having a cell surface composed of three unit membranes, which may be a morphological indication of a common ancestry for these forms.

Another assemblage, which emerges slightly higher up in the tree, consists of the heterokont algae and their relatives. The diatoms, chrysophytes, xanthrophytes, and oomycetes belong to this group. Although the oomycetes have frequently been regarded

Dr. Gunderson is with the Department of General Biology, Vanderbilt University, Nashville, TN 37235. Dr. Sogin is with the Marine Biological Laboratory, Woods Hole, MA 02543, USA.

as being more closely related to the true fungi, biochemical and ultrastructural studies had indicated a relationship with the heterokont algae, an idea which is supported by rRNA sequence comparisons.

The major groups branching off most recently are those forms with lamellar mitochondrial cristae, which include the three kingdoms of multicellular eukaryotes. The multicellular forms themselves are very closely related assemblages that are dwarfed by the age and genetic diversity of the protistan world from which they only recently emerged. The correlation between crista type and position in the tree is not perfect. *Acanthamoeba*, which has tubular cristae, emerges from near the base of the green alga–higher plant group. The red alga *Gracilaria*, which has lamellar cristae, emerges as an isolated lineage among the forms having tubular cristae. If these organisms are accurately placed, it would indicate that a given crista type has arisen on more than one occasion. On the other hand, it is noteworthy that both these organisms emerge rather near the boundary separating organisms having these two crista types. Therefore, this may reflect the imprecision of rRNA-based trees and give an indication of the degree to which lineages may be misplaced by this type of study. It is not presently possible to decide which of these two alternatives is the more likely.

The rRNA trees support the endosymbiotic hypothesis of the origin of mitochondria and chloroplasts. Mitochondria and chloroplast sequences group with sequences from purple bacteria and blue-green algae, and the deepest branches yet found among the eukaryotes are those leading to colorless anaerobic forms. The idea that an algal group is ancestral to the other eukaryotes is specifically refuted; none of these lineages are among the deepest branches of the tree. The red algae and dinoflagellates, the two algal groups most frequently considered ancestral or "primitive," have actually arisen comparatively recently.

The general picture of protistan evolution emerging from rRNA comparisons is quite different from the traditional viewpoint, particularly with respect to the relative positions of the anaerobic forms and the algae. rRNA comparisons do not only change the alignment of previously recognized natural groups. Small, morphologically rich groups which were defined by a constellation of characters (e.g., the ciliates) are maintained as phylogenetic units in the RNA trees. However, "the flagellates" and "the amoebae" are completely dismembered, which suggests that traditional methods were not adequate to define these groups. For instance, the amoeba *Naegleria* appears within the deepest branches leading to aerobic organisms, while *Acanthamoeba* doesn't appear until approximately the time of the emergence of the multicellular kingdoms. These quite unrelated organisms have been regarded as much too closely related to each other, on the basis of superficial characteristics.

Making Phylogenetic Sense
of Biochemical and Morphological Diversity
Among the Protists

Mark A. Ragan and Arthur R. Lee, III

Abstract. The increasing availability of nucleotide sequences of genes encoding ribosomal RNAs or biosynthetic enzymes has led to an optimism that phylogenetic relationships among structurally diverse eukaryotes may soon be settled. However, topologically different trees have been deduced from genes for glyceraldehyde-3-phosphate dehydrogenase and for ribosomal RNAs; and among the latter, parsimony and distance-matrix analyses typically yield significantly incompatible trees.

Characters associated with certain biosynthetic pathways and with a few highly conserved ultrastructural features are widely distributed among eukaryotes, functionally conservative, and sufficiently stable to serve as external checks on proposed DNA-sequence trees. Unfortunately the biochemical and ultrastructural data are not highly self-consistent, and their straightforward analysis provides little support for any of the proposed sequence trees, nor for a clear alternative.

Both the internal congruence of the biochemical and ultrastructural data, and their support of some sequence-based trees, can be improved by postulating multiple lateral transfers of character states, for example as putatively associated with endosymbiotic origins of organelles. However, when this is done on an ad hoc basis, we run risks of reinforcing our own biases and of missing optimal solutions. Against this background we introduce to biological literature the use of BLUDGEON, a software package which includes parsimony-based heuristics originally developed to detect and trace lines of contamination (laterally transferred characters) in the analysis of the manuscript traditions of medieval texts.

INTRODUCTION

Among the assemblage known as the protists may be found a remarkable diversity in basic body construction, ontogeny, genetic systems, physiology, and biochemistry. Although useful for low-level taxonomy, this diversity has greatly complicated attempts to reconstruct the deeper phylogenetic structure of protistan lineages; quite often, potential phylogenetic characters are simply absent, or have been modified to the point where homology is doubtful or interpretation becomes subjective. Comprehensive hypotheses of phylogenetic relationships among protists have thus traditionally been cobbled together from a miscellany of macromorphological, ultrastructural, nutritional, and other characters. Not surprisingly, most protistan phylogenetic trees have been of low predictive utility, and far from robust in the face of new data.

Dr Ragan is with the Institute for Marine Biosciences, National Research Council of Canada, 1411 Oxford St, Halifax, Nova Scotia B3H 3Z1, Canada. Rev. Lee is rector of the Episcopal Church of the Holy Spirit, P. O. Box 817, Safety Harbor, FL 34695 USA.

In this context, the biochemical-unity paradigm of Kluyver and Donker (1926) as repopularized by Florkin (1974) has borne a special significance for protistan phylogenetics, raising expectations that among the biochemically definable features of the DNA/RNA/ribosome machinery, central metabolic pathways, and cytostructural proteins might be found stable, universally distributed characters with enough diverse, homologizable character-states to populate the protistan data-matrix in more-informative ways.

Photosynthetic-pigmentation differences were one of the first characters used in algal taxonomy (Sorby, 1873); indeed, for better or worse, pigmentation plays an important role in algal taxonomy even today. But most attempts to use other biochemical characters available in the 1950s and early 1960s (vitamin and mineral requirements, uptake and metabolism of nitrogenous compounds) foundered because these characters are unstable: within natural populations, some clones are auxotrophs, some are not. The undisciplined application of such unstable characters to reconstruction of protistan phylogeny tended to give physiological and biochemical characters a bad name.

The recent successes in molecular phylogenetics have come about because it has been possible to find specific macromolecules (e.g. rRNAs, tubulins, glyceraldehyde-3-phosphate dehydrogenases) that are presumably homologous (derived from common ancestral genes), of wide phyletic distribution, functionally conservative, unlikely to have been transferred laterally among organisms, and unaffected by environmental fluctuation (Ragan, 1989). In the case of rRNAs, moreover, a large database is available.

Coded macromolecules (RNAs and proteins) are not, however, the only characters which can meet these criteria. Morphological features associated with mitosis, for example, and ultrastructure of the flagellar root and associated cytoskeleton, have now been examined in a wide range of protists; in a few cases, immunological or molecular-sequence studies have been used to support the proposition that individual structures are homologous. In addition, adequate data have been established for a limited number of biochemical pathways, either directly through labelling and enzyme studies, or indirectly through characterization of end-products. Some of these characters not only meet most or all of the above phylogenetic requirements, but moreover represent multiple nuclear genes; lysine-biosynthetic pathways, for instance, involve some seven or eight enzymes which, where investigated, are encoded by at least this many genes on different chromosomes (Patte, 1983; Bhattacharjee, 1985).

The most powerful results have come from analysis of nucleotide sequences of nuclear rRNAs, where alignment can be based on secondary structures and on comparison among phyletically diverse organisms. Unfortunately, different methods of phylogenetic-tree construction frequently give different topologies even from identical alignments (Patterson, 1989). This problem is found with 5S and 28S as well as 18S rRNAs; it is not limited to broad-scale comparisons among diverse eukaryotes, but occurs also in more tightly focused problems such as the relative branching order of plants, fungi, and animals.

Nor is there satisfactory concordance among trees based on different macromolecules. A recent phylogeny of glyceraldehyde-3-phosphate dehydrogenases, constructed using the method of operator metrics, disagrees with rRNA trees in separating filamentous ascomycetes from ascomycetous yeasts (Smith, 1989). A parsimony tree of β-tubulin amino acid sequences (MacKay, unpublished) is incongruent with the rRNA and glyceraldehyde-3-phosphate dehydrogenase trees.

These results leave open the possibility that single-gene-product trees do not sample enough of the genome to give an adequate picture of organismal evolution, or individually are otherwise compromised by problems of unequal crossing-over, copy-correction (Dover, 1982), substitutional saturation (Meyer et al., 1986), or paralogy (Ragan 1988). There appears to be a widespread consensus that molecular-sequence trees should

434

(at the very least) be carefully evaluated vis-à-vis known morphological and other characters.

Preliminary steps in this direction have been taken for rRNA phylogenies of protists (Ragan, 1989). When phyletic distributions of pathways leading to δ-aminolevulinic acid, lysine, sterols, or various carotenoids are plotted onto different rRNA gene-sequence trees, it becomes obvious that distributions of the former characters cannot be reconciled in any straightforward manner with the rRNA trees without postulating multiple origins, wholescale losses, or lateral transfers from other organisms of biochemical characters (and presumably the associated genes). Multiple origins of these pathways seem unlikely in view of the gene-sequence complexity presumably required, while multiple losses seem both physiologically inexplicable in some cases, and unparsimonious as a working assumption (although these alternatives cannot be ruled out until appropriate primary-sequence analyses have been conducted).

Lateral transfers among organisms may better explain some of the observed incongruence. In *Euglena gracilis*, for example, biosynthesis of δ-aminolevulinic acid for chlorophylls occurs via the glutamate pathway, whereas δ-ALA committed to mitochondrial hemes is made via the synthase reaction; only the latter route appears to operate in some trypanosomes. If these pathways are too complex to have arisen independently more than once, and if, in fact, euglenoids and trypanosomes share a short common history as indicated by several rRNA trees (Wolters & Erdmann, 1988; Sogin et al., 1989), then *Euglena* must have picked up its glutamate-ALA pathway by lateral transfer, presumably during the endosymbiotic event giving rise to its chloroplast. This is consistent with the presence of glutamate-ALA biosynthesis in cyanobacteria and green algae. It is interesting that the synthase pathway, which requires forward operation of the oxidative TCA cycle, has been well-characterized among prokaryotes only among α-purple bacteria (Jordan & Shemin, 1972; Chen et al., 1981; Sato et al., 1985), a potential source of endosymbiotic mitochondria. *Giardia* and microsporidia, organisms whose rRNAs are among the most deeply branching known among eukaryotes, have no mitochondria and are believed to be devoid of hemes or other ALA-derived compounds. Thus it is possible that the ancestral eukaryote had no ALA biosynthesis, and that both ALA-biosynthetic pathways are organellar markers whose specific distribution in eukaryotes is the result of multiple lateral transfers (Ragan, 1989).

Many biochemical and morphological data not only fail to support the current eukaryotic rRNA-gene trees in any straightforward manner, but moreover are inconsistent among themselves. Tabulating the distributions of 59 stable biochemical and ultrastructural characters (Table 1) in 30 protistan lineages gives a data-matrix which is about 85% populated (Table 2). Straightforward cladistic analysis of this matrix, using HENNIG86 and PAUP, yields several trees (e.g., Fig. 1) which, while generally

Table 1. Morphological and biochemical characters used to construct data matrix (Table 2).

01	ALA synthase	0= absent	1= present
02	C-5 pathway to ALA	0= absent	1= present
03	AAA lysine biosynthesis	0= absent	1= present
04	DAP lysine biosynthesis	0= absent	1= present
05	*meso*-DAP lysine biosynthesis	0= absent	1= present
06	epoxysqualene cyclase product	0= no synthesis	1= lanosterol 2= cycloartenol
07	hopanoids	0= absent	1= present
08	sterol C-24α stereochemistry	0= absent	1= present
09	sterol C-24β stereochemistry	0= absent	1= present
10	DHFR-TS protein	0= absent	1= bifunctional or high-MW 2= separate polypeptides

Table 1. Continued.

11	chitin / chitin synthase	0= absent	1= present
12	Fe superoxide dismutase	0= absent	1= present
13	Mn superoxide dismutase	0= absent	1= present
14	CuZn superoxide dismutase	0= absent	1= present
15	cellulose	0= absent	1= present
16	α–1,4 glucans (starches)	0= absent	1= present
17	nuclear membrane	0= absent	1= present
18	nuclear-membrane pores	0= absent	1= present
19	mitochondria	0= absent	1= present
20	mitochondrial cristae	0= absent	1= tubular
			2= flattened or discoidal
			3= vesicular
			4= discoidal with constricted base
21	9+2 structures	0= absent	1= present
22	mitotic spindle location	0= absent	1= extranuclear
			2= intranuclear
23	nuclear membrane in mitosis	0= absent	1= close 2= fenestrate
			3= open
24	chloroplasts	0= absent	1= present
25	chloroplast outer membranes	0= absent	1= two 2= three 3= four
26	chloroplast nucleoid	0= absent	1= point 2= ring 3= skein
27	flagellar length	0= absent	1= different 2= same
			3= single flagellum
28	flagellar surface	0= absent	1= smooth
			2= tubular mastigonemes
			3= scales 4= hairs
29	Golgi type	0= absent	1= not true Golgi
			2= unstacked true Golgi
			3= true dictyosome Golgi
30	histones	0= absent	1= present
31	5-OH-methyluracil in DNA	0= absent or not reported 1= present	
32	80S nuclear ribosomes	0= absent	1= present
33	separate 5.8S rRNA	0= absent	1= present
34	microbodies (peroxisomes, glyoxysomes)	0= absent	1= present
35	glycosomes	0= absent	1= present
36	paraflagellar/paraxial rod	0= absent	1= present
37	macronucleus + assoc. chars.	0= absent	1= present
38	kinetoplast	0= absent	1= present
39	actin	0= absent	1= present
40	calmodulin	0= absent	1= present
41	phycobiliproteins	0= absent	1= present
42	bacteriochlorophylls	0= absent	1= present
43	chlorophyll a	0= absent	1= present
44	chlorophyll b	0= absent	1= present
45	chlorophyllide c_1	0= absent	1= present
46	chlorophyllide c_2	0= absent	1= present
47	β-carotene	0= absent	1= present
48	monocyclic glucosidic carotenoids	0= absent	1= present
49	dicyclic xanthophylls	0= absent	1= present
50	4-keto carotenoids	0= absent	1= present
51	epsilon-carotenoids	0= absent	1= present
52	5,6-epoxycarotenoids	0= absent	1= present
53	allenic carotenoids	0= absent	1= present
54	carotenoid acetates	0= absent	1= present
55	acetylenic carotenoids	0= absent	1= present
56	8-keto carotenoids	0= absent	1= present
57	C-37 carotenoids	0= absent	1= present
58	19-acyloxycarotenoids	0= absent	1= present
59	2-hydroxycarotenoids	0= absent	1= present

436

Table 2. Data-matrix of character states (Table 1) in protists.

	00000000011111111112222222222333333333344444444445555555555
	12345678901234567890123456789012345678901234567890123456789
Giardia	————————1———11001—000111–0–10000011000000000000000000
Microsporidia	————————1———1100111000001–00000000—000000000000000000
Trypanosomes	10——10—101——01114121000143101101101–1000000000000000000
Euglenophyceae	1110020—1011–001114121123143101110100–10011001011011011010
Naegleria	————2000————1–11141–100021—0——000011000000–000000000000
Entamoeba	————————01——11100021000—0–01–000001–000000–000000000000
Physarum	———1001–0–1–0–11111??0001131011–000011000000–000000000000
Dictyostelium	———10——1—111111101200000–1011–000011000000–000000000000
Plasmodium	—00000001000000111111000—3–011–0000—00000000000000000000
Dinoflagellates	———001–0——1111111111121143011110100—0010011010011110100
Ciliates	100000100111——1111112100021310111001011000000–000000000000
Oomycota	—01–1001–1—11011111220001231011–0000–1000000–000000000000
Phaeophyceae	———2001–0——1–1111111132123–011100001–0010111010011101000
Chrysophyceae	01——2011–1——10111113132123101–10000–10010111010111111000
Bacillariophyceae	—01—010–1——01111113132323–011–0000—0010111010111111000
Rhodophyceae	01——2011–0–1–11111202211003101–10000—1010001010100000000
Chytridiomycetes	—1001001–1——01111212100031310 1–10000–10000000000000000000
Zygomycetes	—1001001–1—10–1112021000021011–00001–00000010000000000000
Asco-yeasts	10100100121–11011112021000002101110000110000 0–000000000000
Asco-filamentous	—1001001——11111120210000021011–0000–10000010000000000000
Basidiomycetes	—100100121–110111120230000021011–000011000001011000000000
Acanthamoeba	———20011——1111121–3000—3–011100001–000000–000000000000
Chlorophyceae	0101—0111011–11111211311121310111000011001100101111000011
Charophyceae	———011–0—1111112113111233–01–1000011001100101111000011
Ulvophyceae	———011–0——111112112111213–01–10000—00110010111110000 11
plants	?101120111011111111211311121310111000011001100101111000011
animals	1?000100021–11111112113000–131011100001100000000010000 10000
eubacteria	11011010020111100000000000000000000000010100011110000000 01
cyanobacteria	010100–00–0——010000000000000000000000000 1101000111001000001
archaebacteria	–111–0–00–011–010000000000000000000000–1000001000000000000

biologically reasonable, have consistency indices only in the range 0.47 to 0.49 (retention indices (Farris, 1989), where calculated, are 0.73 to 0.74). If a significant part of this inconsistency is indeed due to putative lateral transfers, perhaps associated with endosymbiotic events as illustrated above for δ-ALA biosynthesis, then it presumably becomes necessary to remove or filter out this contamination in order to reveal the underlying phylogenetic pattern. Identifying these contaminants "by eye" is unsatisfactory, however, owing to the likelihood that we would only reinforce our own prejudices while still not necessarily discovering optimal solutions.

We are unaware of prior analysis of this problem in evolutionary or comparative-biological contexts. However, the problem of lateral transfer of these biological characters is formally equivalent to the problem of contamination in the transmission of hand-copied medieval manuscripts (Lee, 1989, 1990). Before the invention of the printing press, manuscripts were necessarily copied by hand. In so-called closed traditions, scribes copied only from single exemplars without ever incorporating alternative readings from other manuscripts that might be available. It seems, however, that few textual traditions were closed in this sense; scribes usually copied from manuscripts in other branches of the tradition, contaminating their texts with "laterally transferred" readings. The software package BLUDGEON has been designed to identify and analyze such contamination heuristically (Lee, 1990).

In brief, BLUDGEON consists of two packages ancillary to conventional parsimony programs, such as those in PHYLIP or PAUP. We have used only the first stage, BLUDG1,

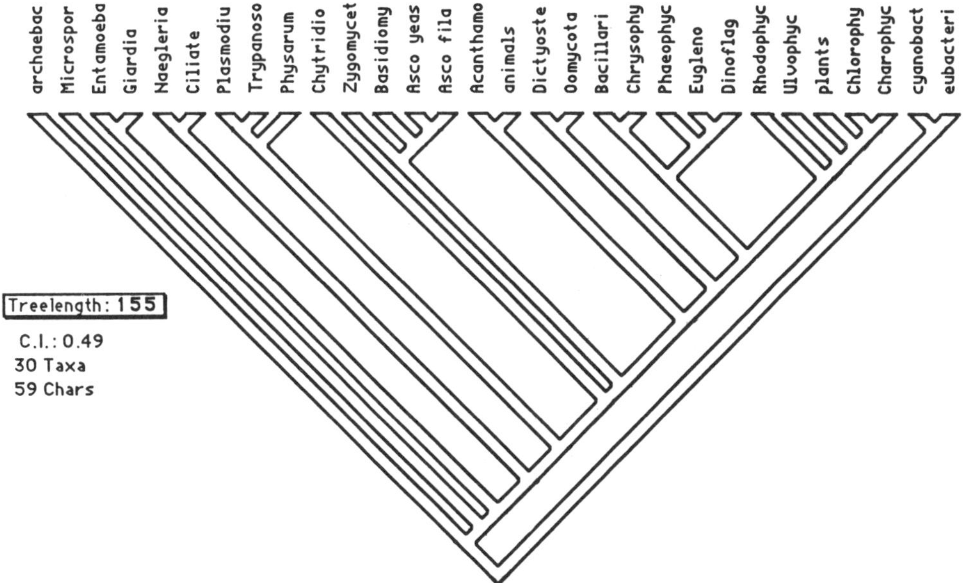

Figure 1. The most-parsimonious tree relating data in the penultimate version of Table 2, which was found using PAUP. Other trees showing many of the same topological features were found using HENNIG86; however, tree lengths and consistency indices are difficult to compare owing to slightly different character-coding requirements.

the variant filter. This package identifies lineages most likely to be affected by contamination; finds which characters are most likely to have arisen laterally, based on separation of shared from idiosyncratic variants; identifies the other taxa or nodes sharing these contaminate variants; and describes the pattern of these occurrences so as to allow the contaminate influences to be mapped onto the tree of primary descent. It also produces a modified matrix which can be recycled through the parsimony program. The second stage, BLUDG2, finds closest matches between nodes of the initial and decontaminated trees, allowing more-accurate mapping of the contamination (Lee, 1990).

METHODS AND MATERIALS

A character-state matrix (Table 2) was compiled from the primary literature; leading references may be found in Ragan & Chapman (1978), Ragan (1989), and Margulis et al. (1990). Data were recoded into binary form using FACTOR in PHYLIP version 2.8. All biochemical and morphological characters are treated as unordered; serial weighting was not attempted. BLUDGEON version 1.3 was used in conjunction with the BOOTM (100 replicates) and MIX programs of PHYLIP version 2.8, or with version 2.4.1 of PAUP. BLUDGEON is being further developed for use with PHYLIP version 3.4, PAUP, HENNIG86, and MacClade; for more information contact Arthur Lee.

RESULTS AND DISCUSSION

Applying BLUDGEON to our inconsistent 59 x 30 matrix of biochemical and ultrastructural characters, we find that *Euglena* is identified as the lineage likely to have the highest degree of contamination (Fig. 2). This result would surprise no protistologist:

```
BLUDG1:
Ordered list of degrees of difference using shared variants only

Node          Parent  Difference
______________________________________________________________

Euglenophy      16    21.69  *********************
cyanobact        3    16.87  *****************
Rhodophyc       13    15.66  ***************
animals         23    15.66  ***************
Trypanosom      29    15.66  ***************
Basidiomyc      10    14.46  **************
ciliate         26    14.46  **************
eubacteria       3    13.25  *************
archaebact       2    12.05  ************
Asco-yeast      11    12.05  ************
dinoflagel      16    12.05  ************
    19          14    12.05  ************
Bacillario      18    10.84  **********
plants          21    10.84  **********
Chlorophyc      21    10.84  **********
Zygomycet       11     9.64  *********
    17          15     9.64  *********
Charophyc       20     9.64  *********
Acanthamoe      23     9.64  *********
Chytridiom      25     9.64  *********
Plasmodium      28     9.64  *********
Oomycetes       29     9.64  *********
    .            .      .   .
    .            .      .   .
    .            .      .   .
```

Figure 2. Output from BLUDG1 analysis of the penultimate version of Table 2, showing investigated lineages most contaminated by lateral transfer (based on consensus parsimony tree produced using PHYLIP). The penultimate version of Table 2 does not differ from the final version (Table 2) in character-states of representatives of Euglenophyceae, trypanosomes, Chlorophyceae, or cyanobacteria.

Euglena has long been suspected to be a protozoan that has incorporated a chlorophyll *b*-containing organism. "Suspect" characters in *Euglena* were identified (Fig. 3) as ALA synthase, the C-5 pathway to ALA, mitochondrial cristae, mitotic spindle location, behaviour of the nuclear membrane during mitosis, 9+2 structures, Golgi type, 80S ribosomes, presence of a separate 5.8S rRNA, the chloroplast nucleoid, chlorophyll *b*, and presence of 4-keto, 8-keto, and 19-acetyloxycarotenoids, as well as absence of cellulose and starch. Most of these characters can reasonably be associated with organelles; some that cannot, such as Golgi type or 80S ribosomes, may be artefacts directly or indirectly related to absence of data.

The identification of cyanobacteria as the second most-contaminated lineage (Fig. 2) underscores the lack of polarity in the analysis: BLUDGEON does not distinguish direction of putative transfer.

BLUDGEON further identifies the most-likely source of this contamination (Fig. 3). In the tree produced by PHYLIP, *Euglena* and *Trypanosoma* are not returned as sister groups (see also Figure 1); thus BLUDGEON's identification of the trypanosome lineage as the source of greatest contamination may be taken as corroboration of the 18S rRNA distance and parsimony trees, which specifically relate small-subunit rRNAs of *Euglena* and trypanosomes (Wolters & Erdmann, 1988; Sogin, 1989). Nine of the above 16 "suspect" characters in *Euglena* found sources in *Trypanosoma*.

The second most-important source of contamination in *Euglena* is identified with the green algal lineage, followed closely by plants and Ulvophyceae; although a matrix decontaminated of putative trypanosome character-states was not recycled through the parsimony, consensus-tree and variant-filter programs, this secondary source appears responsible at least for the presence of chlorophyll *b* and 4-keto- and 19-

```
BLUDG1
NODE: Euglenophy     PARENT: 16

               TOTAL DIFFERENCE =         27.71 %
               IDIOSYNCRATIC VARIANTS =    6.02 %
               SHARED VARIANTS =          21.69 % >> ABOVE THRESHOLD
```

```
BASIS FOR SEARCH ON NODE = Euglenophy

Node         Diffs   %Diff  Search string
Euglenophy     23    27.71  11... ...?? ..... ?00.. ..111 11..1 ..... .....
                            .11.. ..... 1.... 11... ..?.. ..1.. ...1. ....1
                            .1.
-------------------------------
ORDERED LIST OF MATCHES

Node       Match Left %Matched Search string
Trypanosom   15    8   65.22  1.... ...?? ..... ?.0.. ...11 11..1 ..... .....
                              ..... ..... 1.... 11... ..?.. ..... ..... .....
                              ...
Chlorophy    13   10   56.52  .1... ..... ..... ?..... ..1.1 1...1 ..... .....
                              ..... ..... 1.... 11... ..?.. ..1.. ...1. .....
                              .1.
plants       11   12   47.83  .1... ..... ..... ..... ..1.1 1...1 ..... .....
                              ..... ..... 1.... 11... ..... ..1.. ...1. .....
                              .1.
Chrysophy    10   13   43.48  .1... ..... ..... ?.0.. ...1 1...1 ..... .....
                              .1... ..... 1.... ..?.. ..... ..... ....1
                              ...
Bacillario   10   13   43.48  ..... ..... ..... ?.0.. ...1 1...1 ..... .....
                              .1... ..... 1.... .1... ..?.. ..... ..... ....1
                              ...
Ulvophyc     10   13   43.48  ..... ..... ..... ?..... ..1.. 1...1 ..... .....
                              ..... ..... 1.... ..... ..?.. ..1.. ...1. .....
                              .1.

      .        .    .    .     .
      .        .    .    .     .
      .        .    .    .     .
```

Figure 3. Output from BLUDG1 analysis, showing most-parsimonious interpretation of sources of laterally transferred character states (based on consensus parsimony tree produced using PHYLIP).

acetyloxycarotenoids in *Euglena*. Cyanobacteria are not identified as a likely source of contamination, although available in the matrix. We interpret this result as preliminary support for the argument, based originally on ultrastructural evidence, that the *Euglena* plastid had a green algal origin (Gibbs, 1978; Lefort-Tran, 1981).

A software switch in the variant filter BLUDG1 allows analysis not only of shared but also of idiosyncratic variants. As such an analysis departs from a parsimony approach, the results may not be applicable to phylogenetic reconstruction per se. To demonstrate its use, we analyzed the same 30-taxa matrix (Table 2) from which pigmentation characters (columns 42–59) had first been deleted. BLUDGEONing the resulting consensus tree produces an ordered list of most-variant taxa and nodes (Fig. 4) with a number of heterotrophic protists at the top. Reanalysis based only on shared variants (results not shown) demonstrates that idiosyncratic variants contribute overwhelmingly in this analysis. It is interesting that, based on a completely independent analysis of rRNA sequences and operon structure, Wolters (1991) has identified *Naegleria*, *Acanthamoeba*, *Physarum*, *Dictyostelium*, and *Plasmodium* as "high-evolutionary-rate" organisms likely to be misplaced on rRNA gene trees. Although it is reasonable to identify high variance with high evolutionary rate, other explanations are possible, including one from textual analysis: that there have been great losses (extinctions) in the branch(es) of the tradition leading to the taxa in question.

We caution that specific results, whether or not interpretable biologically, depend very critically on the topology of the consensus tree returned by the parsimony program

```
Nodes and leaves, sorted in descending order of variance:

All variants included, both shared and idiosyncratic...

                            1         2         3         4         5         6
Name        Variance 12345678901234567890123456789012345678901234567890123456789(
========    ======== ---------+---------+---------+---------+---------+---------+

Entamoeba     50.00  ***********************************************************
Giardia       48.28  *********************************************************
Naegleria     46.55  *******************************************************
Acanthamo     44.83  *****************************************************
Microspor     41.38  *************************************************
Bacillari     36.21  *******************************∗******************
Physarum      36.21  *************************************************
Dictyoste     34.48  ******∗************************************
Trypanoso     34.48  ******************************************
Ulvophyc      32.76  ********************************
Charophyc     32.76  ********************************
Dinoflag      32.76  ********************************
Plasmodiu     32.76  ********************************
Rhodophyc     31.03  ******************************∗
Phaeophyc     29.31  ****************************
Chytridio     29.31  ****************************
Eugleno       24.14  ***********************
Chrysophy     24.14  ***********************
Oomycota      24.14  ***********************


    .            .    .
    .            .    .
    .            .    .
```

Figure 4. Output from BLUDG1 analysis, showing ordered list of most-variant taxa (based on consensus parsimony tree produced using PHYLIP) when both shared and idiosyncratic variants are scored. Owing to an error in the version of BLUDGEON used, the variance values, and hence the order of the listing, should be considered approximate.

selected for use with BLUDGEON. Bootstrapping reveals that many of the deeper branches in our provisional nonmolecular trees (including Figure 1) are very weakly supported. Reanalysis using PAUP (instead of PHYLIP) with BLUDGEON yielded essentially the same result; however, many uncertainties remain concerning interpretation, homology, and coding of character states, and about phyletic distribution of the data underlying our analysis (Table 2). Use of BLUDGEON in any biological context must be considered experimental, and we did not pass partly decontaminated data-matrices through iterative filterings. For these reasons we do not put forward a putatively decontaminated protistan phylogenetic tree at this juncture.

Instead we wish only to propose that lateral transfer of characters, brought about, for instance, by endosymbiosis, need not hopelessly confuse phylogenetic reconstruction. At least some of the resulting inconsistencies can be recognized and untangled impartially, whether by a parsimony-based heuristic approach such as embodied in BLUDGEON, or perhaps by other methods. Lack of agreement with presumably uncontaminated trees, such as those based on rRNAs, could then be turned to our advantage in tracing putative pathways of endosymbiotic gene-transfer.

ACKNOWLEDGMENTS

We thank R. MacKay for analysis of β-tubulin sequences, T. Cavalier-Smith and D. Lipscomb for assistance with Table 2, J. Felsenstein for supplying the source code for PHYLIP, J. Wolters for communicating results prior to publication, and D. F. Spencer for critically reviewing the manuscript. P. H. A. Sneath and W. H. Day independently recognized the logical equivalence of lateral transfer in the biological and textual contexts. We especially thank W. H. Day for having suggested this most enjoyable collaboration. Issued as NRCC No. 31934.

LITERATURE CITED

Bhattacharjee, J. K. 1985. α-Aminoadipate pathway for the biosynthesis of lysine in lower eukaryotes. *CRC Crit. Rev. Microbiol.* 12: 131–151.

Chen, J., Miller, G. W. & J. Y. Takemoto. 1981. Biosynthesis of δ-aminolevulinic acid in *Rhodopseudomonas sphaeroides. Arch. Biochem. Biophys.* 208:221–228.

Dover, G. 1982. Molecular drive: a cohesive mode of species evolution. *Nature* 299:111–117.

Farris, J. S. 1989. The retention index and the rescaled consistency index. *Cladistics* 5:417–419.

Florkin, M. 1974. Concepts of molecular biosemiotics and of molecular evolution. Pp. 1–124. *In:* M. Florkin & E. H. Stotz (eds.), *Comparative Biochemistry* 29A. Elsevier: Amsterdam.

Gibbs, S. 1978. The chloroplasts of *Euglena* may have evolved from symbiotic green algae. *Can. J. Bot.* 56:2883–2889.

Jordan, P. M. & D. Shemin. 1972. δ-Aminolevulinic acid synthase. Pp. 339–356. *In:* P. D. Boyer (ed.), *The Enzymes,* 3rd edition, vol. 7. Academic Press: New York.

Kluyver, A. J. & H. J. L. Donker. 1926. Die Einheit in der Biochemie. *Chemie Zelle Gewebe* 13:134–190.

Lee, A. R. 1989. Numerical taxonomy revisited: John Griffith, cladistic analysis and St. Augustine's *Quaestiones in Heptateuchum. Studia Patristica* 20:24–32.

Lee, A. R. 1990. BLUDGEON: a blunt instrument for the analysis of contamination in textual traditions. Pp. 261–292. *In:* Y. Choueka (ed.), *Computers in Literary and Linguistic Research,* vol. 3. Champion-Slatkine: Paris.

Lefort-Tran, M. 1981. The triple layered organization of the *Euglena* chloroplast envelope (significance and function). *Ber. Dt. Bot. Ges.* 94:463–476.

Margulis, L., Corliss, J. O., Melkonian, M. & D. J. Chapman (eds). 1990. *Handbook of Protoctista.* Jones & Bartlett: Boston. 914 pp.

Meyer, T. E., Cusanovich, M. A. & M. D. Kamen. 1986. Evidence against use of bacterial amino acid sequence data for construction of all-inclusive phylogenetic trees. *Proc. Natl Acad. Sci. USA* 83:317–220.

Patte, J.-C. 1983. Diaminopimelate and lysine. Pp. 213–228. *In:* K. M. Hermann & R. L. Somerville (eds), *Amino Acids: Biosynthesis and Genetic Regulation.* Addison Wesley: Reading, MA.

Patterson, C. 1989. Phylogenetic relations of major groups: conclusions and prospects. Pp. 471–488. *In:* B. Fernholm, K. Bremer & H. Jörnvall (eds), *The Hierarchy of Life.* Elsevier: Amsterdam.

Ragan, M. A. 1988. Ribosomal RNA and the major lines of evolution: a perspective. *BioSystems* 21:177–188.

Ragan, M. A. 1989. Biochemical pathways and the phylogeny of the eukaryotes. Pp. 145–160. *In:* B. Fernholm, K. Bremer & H. Jörnvall (eds), *The Hierarchy of Life.* Elsevier: Amsterdam.

Ragan, M. A. & D. J. Chapman. 1978. *A Biochemical Phylogeny of the Protists.* Academic Press: New York. 317 pp.

Sato, K., Ishida, K., Mutsushika, O. & S. Shimizu. 1985. Purification and some properties of δ-aminolevulinic acid synthases from *Protaminobacter ruber* and *Rhodopseudomonas spheroides. Agric. Biol. Chem.* 49:3415–3421.

Smith, T. L. 1989. Disparate evolution of yeasts and filamentous fungi indicated by phylogenetic analysis of glyceraldehyde-3-phosphate dehydrogenase genes. *Proc. Natl Acad. Sci. USA* 86:7063–7066.

Sogin, M. L., Edman, U. & H. Elwood. 1989. A single kingdom of eukaryotes. Pp. 133–143. *In:* B. Fernholm, K. Bremer & H. Jörnvall (eds), *The Hierarchy of Life.* Elsevier: Amsterdam.

Sorby, H. C. 1873. On comparative vegetable chromatology. *Proc. R. Soc. London* 21:442–483.

Wolters, J. 1991. The troublesome parasites: molecular evidence that apicomplexa belong to the dinoflagellate-ciliate clade. *BioSystems* 24, in press.

Wolters, J. & V. A. Erdmann. 1988. Cladistic analysis of ribosomal RNAs—the phylogeny of eukaryotes with respect to the endosymbiotic theory. *BioSystems* 21:209–214.

Molecular Evolution of Archaebacteria

Frank T. Robb, Harold J. Schreier, and Allen R. Place

From both evolutionary and genetic perspectives, research on the archaebacteria is presently at an exciting stage. The group as a whole has been elevated to the status of a newly named domain, the Archaea (Woese et al., 1990), which is considered to represent the deepest branching prokaryotes known. This domain has been proposed as a new highest level taxon; its acceptance indicates that the archaebacteria are separated widely from other groups that are considered traditionally to be Kingdoms, viz, the animals and green plants.

Since the formal program at ICSEB did not include any specific meetings addressing archaebacterial evolution, it was decided that a workshop-style meeting with a mix of short review talks and round table discussions would fill a need. In deference to the many examples of isolation of Archaea from extreme environments, the slogan "Life at the Edge" was adopted for the workshop. This report is a narrative of the presentations and discussions that occurred during this meeting, which was held during the holiday day of the ICSEB, July 4. The meeting was organized by Frank Robb, Allen Place, and Harold Schreier at the Center of Marine Biotechnology, University of Maryland. As Frank Robb remarked in the introductory address, this type of gathering was very different from the traditional Independence Day celebrations. Despite this, the business of the day would be both interesting and entertaining.

The first presentation, by John Baross from the University of Washington in Seattle, addressed the natural history of thermophilic Archaea, the newly named Crenarcheota. Baross' speciality is the study of high temperature environments, in terms of organisms present, the geochemistry that determines the microbial flora and conditions most appropriate for recovery of isolates. He described two new isolates, ES4, a hyperthermophile with similar properties to the *Thermococcus/Pyrococcus* group, and ES1, a moderate thermophile that grows well without sulfur. The latter property is unusual in this group. A video tape taken from an expedition to Juan de Fuca Ridge hydrothermal vents and the discovery of the new formation, named a flange, from which ES4 was isolated, was presented.

A major source of controversy has been the evolutionary relationship of the Eukaryotes (now Eucarya) to the Archaea, and James Lake addressed the essence of this controversy. His presentation compared available numerical methods for consideration of sequence data. Almost all of the major findings concerning the early evolution of cells are

Drs. Robb, Schreier, and Place are with the Center of Marine Biotechnology, 600 East Lombard Street, Baltimore, MD 21202, USA. Dr. Robb is also with the Department of Microbiology, University of Maryland, College Park, MD 20742, USA. Dr. Schreier is also with the Department of Biological Sciences, University of Maryland, Baltimore County, 5401 Wilkens Ave., Baltimore, MD 21228, USA.

based on sequence data from the 16S ribosomal RNA database, which came into being as a result of the pioneering studies of Carl Woese and his colleagues. Clearly, the assumption that evolution has proceeded at equal rates in widely diverse groups over long periods of time is difficult to defend, and this has led to two hypotheses concerning the primordial life form. Lake maintains that the Eocyte was similar to, and gave rise to the Archaea and the Bacteria. Woese and colleagues support an "archaebacterial tree" in which the Bacteria and the Archaea branched prior to the evolution of the Archaea. Lake detailed the differences between evolutionary parsimony analysis (which supports the "Eocyte tree") and the more generally used distance matrix and maximum parsimony methods. It appears that the controversy will persist until more sequence data from highly conserved genetic loci, for example genes encoding translation factors, are added to the database. Ribosomal RNA has stretches of sequence whose divergence is relatively unconstrained, resulting in what Woese refers to as noise.

The study of molecular genetics of Archaea is also entering a new era heralded by the addition of methods for genetic manipulation of two major groups of Archaea, namely halobacteria and methanogens. The presentation by Ford Doolittle was an elegant exposition of the state of the art of molecular genetics in *Haloferax volcanii,* a halobacterium. As a result of the efforts of Doolittle's group this system is the most advanced in any of the Archaea, and has enabled the 3.8 mbp genome of *Haloferax* to be ordered into superimposed physical and genetic maps. The ability to generate physical maps of prokaryote genomes will shortly allow a very rapid increase in our understanding of the relationships between different groups of Archaea. Interestingly, the genome of *Haloferax* is 25% smaller than an average Eubacterium such as *Escherichia coli.* Doolittle and his coworkers have also developed procedures for transforming *Haloferax* by chromosomal as well as plasmid DNA. This has also enabled them to create a shuttle vector that replicates in both *E. coli* and *Haloferax,* that may be the key element in understanding gene regulation in Archaea. Recently, the sequence of three genes encoding enzymes (TrpC, TrpB, and TrpA) for tryptophan biosynthesis (Lam et al., 1990) has revealed that these proteins have diverged at least as far from bacteria as they have from the homologous Eukaryote tryptophan pathway.

Jorn Wolters presented the Berlin 5S ribosomal database, a valuable resource in defining shorter range phylogenetic distances. Interestingly, most of the Archaea have an rRNA operon structure that is similar to Bacteria, and differences between the domains are not as clear when these functional gene groupings are considered. It has become a widely held belief that the earliest organisms were extreme thermophiles. Jorn Wolters pointed out that, from the Archaean perspective, organisms such as Eukarya inhabiting environments at "normal" temperatures, pressures, and ionic strength may in fact have evolved from extremophiles, and it is the mesophiles that should be considered to be "at the Edge."

During the lunch hour, two discussion groups were held focusing on the isolation of extreme thermophiles (led by John Baross), and the use of a new software package for sequence comparison. Laurie Achenbach, who led the latter discussion, arranged a hands-on display for the software package entitled the "GDE", presently available free on enquiry from its originators, Steve Smith, Ross Overbeck, Carl Woese, and Gary Olsen. It is extremely powerful, with on-line access to Genbank and is ideal for predicting secondary structure and construction of RNA sequence alignments.

David Stahl presented an overview of the new technology arising out of the extensive 16S rRNA database. The use of oligonucleotide probes designed to hybridize to 16S rRNA sequences with controlled specificity is revolutionizing the experimental approach to microbial population structure. As a tool for identifying and enumerating Archaea in natural populations, and detecting the minor components (many of which cannot presently be cultured), these probes are going to be invaluable.

In the final session, chaired by Robert Kelly, Eddie Chang described the chemistry and

444

applications of the ether linked lipids that are exclusive to Archaea. Cytoplasmic membranes in Archaea are not bilayers such as those universally found in the other domains. Rather the lipids are dumbbell shaped molecules with polar heads. Lipid vesicles derived from *Sulfolobus* membranes have extraordinary properties, including extremely low permeability. This may account for the total domination of the hyperthermophiles by archaebacteria, since low membrane permeability and structural integrity may be critical to survival at temperatures approaching 100°C. Michael Adams described the biochemistry of novel ferredoxin and hydrogenase from the "model" hyperthermophile *Pyrococcus furiosus*. Both of these enzymes show extraordinary thermostability and novel catalytic properties, and give important clues about the novel metabolic pathways that may operate in hyperthermophiles. For example, the isolation of a pyruvate oxidoreductase from *Pyrococcus* that is coupled to hydrogenase may be an example of redox chemistry without cofactors. Since NAD and NADP are unstable at 100°C *in vitro* (the growth optimum of *Pyrococcus*) it is not clear how *Pyrococcus* maintains pools of these cofactors, or in fact whether the organism has a different metabolic emphasis in which dehydrogenases are less important. A major metalloenzyme currently named the "Red Protein," with tungsten as a coordinated metal group, has also been described recently.

Robert Kelly described the growth of extreme thermophiles in continuous culture, and the identification of polysulfides as the probable sulfur reduction substrates in the hyperthermophiles. Isolation of glucosidases and proteases having exceptional thermostability is easily achieved, and these proteins have great potential in biotechnology. The proteases can maintain activity (and presumably structural integrity) indefinitely at near 100°C in the presence of ionic detergents.

Rita Colwell concluded the session with an overview of the meeting and the speculation that the meeting had barely scratched the surface of the incongruous and noncongruent life forms that probably exist at the extremes of pressure and temperature.

ACKNOWLEDGMENTS

We gratefully acknowledge financial support from the Office of Naval Research, Grant N00014–86-K-0696, and from Beckman Instruments. Research in the laboratories of F. Robb and H. Schreier is supported by grant N00014–90-J–1823 from the Office of Naval Research.

LITERATURE CITED

Lam, W. L., Cohen, A., D. Tsouluhas, D. & W. F. Doolittle. 1990. Genes for tryptophan biosynthesis in the archaebacterium *Haloferax volcanii. Proc. Natl. Acad. Sci.* USA 87:6614–6618.

Woese, C. R., Kandler, O. & M. L. Wheelis. 1990. Towards a natural system of organisms: proposal for the domains Archaea, Bacteria and Eucarya. *Proc. Natl. Acad. Sci. USA* 87:4576–4579.

The Origin of the Metazoa

Claus Nielsen

The round table discussion had attracted an audience of about 30. The four speakers each gave an introduction to their views and lively discussions took place both during the presentations and afterwards.

Claus Nielsen presented his view of the early phylogeny of metazoans. Ciliary structures and the overall similarity of choanoflagellate cells and choanocytes of sponges were considered to support a sister-group relationship between the two groups. Multicellularity characterizes the metazoans, but a number of synapomorphies on the ultrastructural/biochemical levels were identified, viz., spermatazoa, septate junctions (present only in certain cell types in sponges), collagen, and ciliary necklaces with 3 strings. *Trichoplax* and Eumetazoa were considered sister groups and together to be the sister group of the sponges. *Trichoplax* and eumetazoans exhibit the following synapomorphies: cross-striated ciliary rootlets and epithelia sealed with septate junctions. Synapomorphies of the Eumatazoa (Cnidaria + "Bilateria") comprise basal membranes, special sensory cells, nerves, and chemical synapses.

Diana L. Lipscomb (Department of Biological Sciences, George Washington University, Washington, DC, USA) took a starting point in a cladistic analysis of 84 unicellular and multicellular eukaryotes (Lipscomb, 1989—In B. Fernholm, K. Bremer & H. Jörnvall (eds.): *The Hierarchy of Life*, pp. 161–178. Elsevier Biomedical) which indicates that there is a distinct metazoan (= animal) clade that includes *Pelomyxa* and the Chytridiomycota. The taxon is united by mitosis in which the nuclear membrane breaks down by fragmentation, but this is also seen, for example, in higher plants. The characters usually said to exclusively unite the multicellular animals are the presence of choanocyte-like cells, the presence of diplosomal bodies in choanoflagellates and metazoan monociliate cells, and a UU odd base pair at position 81:95 of the 5S rNA. The two first characters are, in her opinion, weak. The resemblance between the choanocyte-like cells in Eumetazoa and the choanocytes of sponges and choanoflagellates is superficial because the flagellar roots and the appendages are different at the ultrastructural level. Diplosomal bodies are found in many unicellular organisms indicating that their presence is a plesiomorphy. The UU odd base pair does appear to be a character found in all eumetazoans, sponges, and dicyemid mesozoans (*Pelomyxa* has not been investigated). Choanoflagellates, *Pelomyxa*, and chytrids are not known to possess this character. These three groups group with the sponges because they all have microtubules radiating from the basal body; these microtubules are united by concentric rings of electron-dense material. It can only be concluded that either the similarities in the flagellar root systems are convergent features and the UU odd base pair indicates close relatedness among multicellular animals, or the

Dr. Nielsen is with the Zoologisk Museum, University of Copenhagen, Universitetsparken 15, DK-2100 Copenhagen, Denmark.

UU odd base pair has been secondarily lost in the chytrids and the flagellar root system is the synapomorphy that unites the taxa. A choice between these two hypotheses cannot be made at this time.

Dennis Goode (Department of Zoology, University of Maryland at College Park, MD, USA) pointed out that at least 34 separate origins of multicellularity can be identified. Protists use three alternative routes to multicellularity: aggregation of cells, nuclear division to form a coenocytic organism followed by cytoplasmic division, and cell division to form a colony. The Metazoa and Metaphyta appear to have originated from colonial protists that become multicellular by dividing to form colonies, but the origins of the fungi are unclear. An analysis of characters shared between animals and protists suggests that the choanoflagellates are closest to the metazoans, with the chytrids as the next closest protist group. Characters shared by choanoflagellates and metazoans include flat mitochondrial cristae, phagocytosis, cytokinesis by constriction around a dense midbody of microtubules, centifugally beating, uniflagellated vegetative cells or sperm, and clonal colony or embryo development by uninucleate cell division. Choanoflagellates and sponges also have similar flagellar rootlets composed of arrays of radial microtubules, whereas other metazoans have striated flagellar rootlets fibers with a periodicity of 50–78 nm rarely, if ever, seen in the protists. The Porifera appear to have originated from a near ancestor of the extant choanoflagellates, and the Eumetazoa and Placozoa from related and as yet unidentified choanoflagellate-like colonial flagellates.

Christian Bardele (Zoologisches Institut, Eberhard-Karls-Universität, Tübingen, Germany) pointed out that comparative freeze-fracture investigations of the flagellar/ciliary membrane have revealed very conservative, genetically fixed arrays of intermembranous particles. This character, which is of interest at phylum or even sub-kingdom levels, mirrors the high diversity of protists and multicellular organisms seen in ribosomal RNA sequence data. In addition to freeze-fracture data published earlier (Bardele, 1983—*Journal of Submicroscopical Cytology* 15:263–267) it was found that dicyemid mesozoa have a double-stranded ciliary necklace as do all ciliates and opalinids. This links the dicyemids with the protists and not with the lower invertebrates, which have a triple-stranded necklace. The triple-stranded necklace of sponges, anthozoans, scyphozoans, hydrozoans, ctenophores, and numerous bilaterians support the monophyletic origin of the Metazoa. On the other hand, the triple-stranded necklace may be a plesiomorphic character since this type is also observed in unicellular and colonial chlorophytes now known to have originated close to the Metazoa. The choanoflagellate *Sphaeroeca* is very different from the sponges but resembles the chytrid *Rhizoplyctis*. Thus, at least as far as the freeze-fracture data are concerned, the "choanoflagellate to sponge line" is still open to debate. On the other hand, the Steiböck-Hadzi "ciliate to turbellarian line" has turned out to be wrong.

The discussions showed surprisingly high degrees of agreement among the participants. It was pointed out that a number of characters need to be investigated in more detail, for example mitosis in choanoflagellates and sponges, and the distribution of different types of collagen. Special studies of life cycles of choanoflagellates are much needed, especially to reveal if sexual reproduction occurs after all.

It was finally agreed by all that the cellularization theories for the origin of the Metazoa, i.e., the origin of multicellular animals from ciliates or ciliate-like protists, are contradicted by all the available evidence, and that such theories can be left out of contemporary discussions and referred to the history of zoology.

Homeosis and the Evolution of Plants

U. Posluszny, R. Sattler, J. P. Hill, M. K. Komaki, J. M. Gerrath
N. Lehmann, and B. K. Kirchoff

POSLUSZNY

This discussion group was organized in order to explore what homeosis means in plants and the possible role homeosis plays in the evolution of plants. Although the term homeosis has been around since the beginning of the twentieth century (First coined by Bateson in 1894) its specific implication and relation to plant development has not been fully investigated until relatively recently, spurred on primarily by work with homeotic mutants such as *Pisum* and *Arabidopsis*. What is still unclear is what "homeosis" really means in plants where meristematic regions can be subtly changed from one organ to another with varying degrees of genetic input.

J. P. HILL—COMPARATIVE DEVELOPMENT OF WILD TYPE AND HOMEOTIC PISTILLATA MUTANT FLOWERS OF *ARABIDOPSIS*

Homeosis, broadly defined, is the translocation of the likeness of one part of an organism to a position where it does not normally occur (1). In higher plants, the polar nature of vegetative and floral meristematic shoot growth over time leads to a commonly observed class of homeotic phenotypes that have phenotypic features shifted basipetally or acropetally along the shoot axis (2,3). For instance, examples of the apparent replacement of one or more whorls of floral organs by a whorl normally located in a different position are well known in the angiosperms (references in 1,4). Genetic mutations that elicit these abnormalities provide opportunities to examine the ontogenetic basis for homeotic forms and deepen our understanding of the genetic controls of normal development.

The pistillata (*pi*) floral mutant of *Arabidopsis thaliana* (Brassicaceae) has sepal-like organs ("petals") in the position of normal wild type (WT) petals (3–6). The mutant is also male sterile and has a teratological gynoecium. Comparative developmental study of WT

Dr. Posluszny is with the Department of Botany, University of Guelph, Guelph, Ontario, N1G 2W1, Canada. Dr. Sattler is with the Biology Department, McGill University, Montreal, Quebec, Canada. Dr. Hill is with the Department of Botany and Plant Sciences, University of California, Riverside, CA 92521, USA: his present address is at the Department of Botany, University of Georgia, Athens, GA 30602, USA. Dr. Komaki is with the Division of Gene Expression and Regulation, National Institute for Basic Biology, Okazaki 444, Japan. Dr. Gerrath is with the Department of Horticulture, University of Guelph, Guelph, Ontario; and the Department of Biology, University of Waterloo, Waterloo, Ontario, Canada. Dr. Lehmann is with the Biology Department, McGill University, Montreal, Quebec, Canada. Dr. Kirchoff is with the Department of Biology, University of North Carolina, Greensboro, NC 27412, USA.

448

and *pi* flowers revealed that the initiation and early ontogeny of WT petals and homeotic *pi* "petals" are the same, and the first evidence of developmental divergence is at stamen initiation in *pi* flowers (3). Abnormal patterns of cell division in the *pi* floral meristem distal to *pi* "petal" primordia, and the subsequent differentiation of gynoecial cell fates by these cells, underlie the abnormal structure of the gynoecium in mature *pi* flowers. Ontogenetic divergence of homeotic *pi* "petals" relative to WT petals is evident when organs reach 90 μm in length. The post-initation onset of sepal differentiative events in the position of *pi* "petals" produces organs with a final form that is anatomically and morphologically intermediate between WT sepals and petals; a one-for-one sepal-for-petal replacement does not occur. One simple explanation is that stamen initiation and some major aspects of petal determination occur simultaneously, and the WT gene product is required at this time. Alternatively, normal gene expression may be first for stamen initiation, and again at some later time for normal petal differentiation to occur.

Pistillata "petals" exemplify the direct origin of a novel form without passing through a series of smaller, intermediate ontogenetic steps. This principle forms the basis for Bateson's (7) proposal that phenotypic discontinuities between species and limited evidence of transitional morphological forms between related taxa might be the result of homeotic transformations during evolution. The existence of regulatory genes which cause homeotic phenotypes suggests that evolution at these loci may be an important mechanism of morphological change in plants (8–10). Homeotic transformations among the limited number of organ systems in plants is one possible origin for convergent morphologies of non-homologous organs that are specialized for very similar functions. However, conclusions about the scope of homeosis in plant evolution remain premature; greater information is needed about the nature of plant homeotic genes and the specific developmental effects of these genes.

Literature Cited

1. Sattler, R. 1988. Homeosis in plants. *Amer. J. Bot.* 75:1606–1617.
2. Poethig, R. S. 1988. Heterochronic mutations affecting shoot development in maize. *Genetics* 119:959–973.
3. Hill, J. P. & E. M. Lord. 1989. Floral development in *Arabidopsis thaliana:* a comparison of the wild type and the pistillata mutant. *Can. J. Bot.* 67:2922–2936.
4. Meyerowitz, E. M., Smyth, D. R., & J. L. Bowman. 1989. Abnormal flowers and pattern formation in floral development. *Development* 106:209–217.
5. Pruitt, R. E., Chang, C., Pang, P. P.-Y., & E. M. Meyerowitz. 1987. Molecular genetics and development in *Arabidopsis. In:* W. Loomis (ed.), *Genetic Regulation of Development.* Alan R. Liss: New York.
6. Bowman, J. L., Smyth, D. R., & E. M. Meyerowitz. 1989. Genes directing flower development in *Arabidopsis. The Plant Cell* 1:37–52.
7. Bateson, W. 1894. *Materials for the Study of Evolution.* MacMillan: London.
8. Raff, R. A. & T. C. Kaufman. 1983. *Embryos, Genes, and Evolution.* MacMillan: New York.
9. Hilu, K. W. 1983. The role of single-gene mutations in the evolution of flowering plants. *Evol. Biol.* 16:97–128.
10. Gottlieb, L. D. 1984. Genetics and morphological evolution in plants. *Amer. Nat.* 123:681–709.

M. K. KOMAKI—GENES CONTROLLING HOMEOTIC CHANGES OF *ARABIDOPSIS* FLOWERS

Works on the genes controlling flower morphogenesis in *Arabidopsis thaliana* provide interesting data on homeotic issues. *A. thaliana* is a small plant in the Brassicaceae and is easy to grow as its generation time is as short as 6–8 weeks. It is also advantageous that analysis at the molecular level is easy for *Arabidopsis* as this plant has an extremely small genome for a higher plant.

During the last several years, many mutant plants have been isolated and characterized (1–4). Among the mutations, at least four genes, *ap2, ap3, pi,* and *ag,* showed homeotic alterations among floral organs. That is, if a function of a gene is damaged and lost by mutation, a certain abnormality in morphology occurs.

We have constructed all combinations of double mutant plants among eight representative genes at seven loci (5). From the results, it has been proved that the number and arrangement of floral organs on a whorl are determined by a system independently from another system that determines the identity of floral organs by homeotic genes.

Spatial and temporal distribution of the expression of homeotic genes are critical to set four kinds of floral organs (sepals, petals, stamens and carpels) in the development of flowers. Here I present a model which explains that three factors are necessary to determine four kinds of identity of floral organs in *Arabidopsis* and that three factors are produced by *AP2, PI* (and coworking *AP3*) and *AG* genes, respectively; their spatial distribution is regulated by each other. This model gives us a perspective for further research, including that at the molecular level. It also gives expectations about which homeotic changes may easily occur among floral organs and which may not.

Literature Cited

1. Haughn, G. W. & C. R. Somerville. 1988. *Dev. Genet.* 9:73–89.
2. Komaki, M. K., Okada, K., Nishino, E. & Y. Shimura. 1988. *Development* 104:195–203.
3. Bowman, J. L., Smyth, D. R. & E. M. Meyerowitz. 1989. *Plant Cell* 1:37–52.
4. Kunst, L., Klenz, J. E., Martinez-Zapater, J. & G. W. Haughn. 1989. *Plant Cell* 1:1195–1208.
5. Komaki, Okada and Shimura (manuscript in preparation).

J. M. GERRATH—ARE TENDRILS HOMEOTIC ORGANS?

My work on ontogeny of tendrils in members of the Vitaceae (grapes), *Pisum sativum* L. (garden pea), and *Passiflora quadrangularis* (passion vine) has led me to attempt to unify some of these observations within the framework of homeosis. This paper explores the question of whether or not one can use homeosis to explain how tendrils may have arisen.

Conventional pea leaves are constructed on a pinnately compound plan, except that tendrils are found in the position where one would normally find a terminal leaflet and the one or two most distal pairs of leaflets. Single gene mutants exist in which there are only leaflets (acacia), or only tendrils (afilia) (Gottschalk, 1970). Tendrils and leaflets develop from primordia that are apparently similar at inception. However, tendrils remain cylindrical, whereas leaflets become dorsiventrally flattened. Biochemical studies (Côté & Grodzinski, 1990) indicate that the suite of compounds present in mature tendrils may be more representative of those usually found in juvenile leaf tissues. This fact, when coupled with the ontogenetic evidence, suggests that pea tendrils may reflect a process of juvenilization of the leaflets. Semi-leafless pea tendrils have been explained in terms of homeosis (Gould et al., 1986) and certainly in this instance one can clearly pinpoint a genetic mutation which causes a substitution of one organ type in place of another. These pea mutants appear to conform to the definition of homeosis.

In the Vitaceae, the leaf-opposed uncommitted primordium (Gerrath & Posluszny, 1988) may develop into either a tendril or an inflorescence. Intermediates between the two are not uncommon. Work by Srinivasan and Mullins (1981) has shown that exogenously applied gibberellic acid results in initiation of the uncommitted primordium and its subsequent development into a tendril, whereas cytokinin application stimulates branching of the uncommitted primordium, leading to the production of inflorescences. Their work suggests the immediate mechanism of control of the two different developmental pathways. However, there do not seem to be any reports of genetic control of uncommitted primordium development in the Vitaceae. The observation that in some taxa, such as *Ampelopsis brevipedunculata*, the tendrils appear consistently to terminate in a poorly-formed flower (Gerrath & Posluszny, 1989) suggests the possibility that tendrils might reflect a juvenilization of the inflorescence.

In passion vine, tendrils and flowers develop from a common axillary primordium, which subsequently bifurcates unequally to produce two different organs, each of which undergoes a separate developmental pathway. There are several published reports of tendril ontogeny (see Shah & Dave, 1971 for a review), but none establish the homology of the tendril itself. I do not know of any studies on the genetics or physiology of the tendrils of these plants. Thus at this time it is not possible even to speculate about the controlling mechanisms in this system, and I only present the data to indicate the wide range of variation in modes of tendril formation.

Attempting to use homeosis as a unifying concept in understanding how tendrils may have evolved requires certain assumptions. First, one must recognize that tendrils are defined functionally, not positionally. Although one could state that in general they are cylindrical organs with a determinate growth pattern, any plant organ that possesses the capacity to coil at some stage during its development can be considered a tendril. In some instances the organ in question may function otherwise at certain stages in its development. For instance, Tucker and Hoefert (1968) report that grape tendrils function as hydathodes before they are able to coil, and Côté et al. (1990) report on the significant photosynthetic contribution made by pea tendrils after the tendril itself has lost the capacity to coil. In some species of Vitaceae, it is the peduncle of the inflorescence that may coil and function as a tendril (Gerrath & Posluszny, 1989). These observations result in a muddying of our concepts of a tendril.

Second, one must decide to what extent tendrils are to be viewed as organs in their own right. Unlike floral parts such as stamens and petals, tendrils have usually been viewed as modified shoot homologues (grape, Bugnon, 1953; passion vine, Shah & Dave, 1971) or modified leaflets (Meicenheimer et al., 1983). This is not surprising, given (as this overview shows) that tendrils can arise in a number of different fashions. If one takes the view that tendrils are modified from some other structure, one probably cannot think of tendrils as homeotic organs. However, simply because it is often possible to ascertain the likely organ of origin of a tendril does not mecessarily mean that a tendril should not be thought of as a distinctive organ. If tendrils are distinct, then they can be viewed as organs which on some occasions, to a greater or lesser degree, may replace other, more normally-occurring structures. This approach should allow for a more unifying conceptualization of the modes by which tendrils develop.

Given that tendrils form from primordia that usually develop into leaf or stem structures, coupled with the fact that they are functionally specialized, it seems reasonable to assume that tendrils are more recently evolved plant organs than, for instance, leaves. If this is so, then tendrils could give us the insight into possible mechanisms by which other plant organs may have formed.

Although tendrils are easily recognized, unlike many other organs such as leaves, their functional definition overrides other considerations. This means that their homology and ontogeny are of secondary importance when we conceptualize about them. This seems to me to be an important fact to remember when one is dealing with questions of homology, ontogeny, and evolution in other organs. It may very well be that the essentialism that tends to form the basis of much morphological thinking only appears to be valid for other organs because they are more clearly and well-established as independent plant organs.

To bring us back to my examples, what all three of these systems have in common is the possibility of divergent pathways of development from a primordium, the fate of which is not easily discernible. In this sense, one may view tendrils as homeotic organs; ones in which the developmental pathway an organ primordium will follow is not apparent at inception. In addition, in the broad sense, since it is also often apparent from what structure the tendril is derived, this new tendril organ could be seen as an example of homeotic replacement. However, given our scanty knowledge of the genetic, biochemical, and physiological mechanisms that control the switch from one organ to another, we are dealing here with speculation. I hope that these examples will form a useful framework for further work and discussion.

Literature Cited

Bugnon, F. 1953. *Recherches sur la ramification des Ampélidacées.* Publications de l'Université de Dijon 11, Presses Universitaires de France: Paris.

Côté, R. & B. Grodzinski. 1990. Photosynthesis, photorespiration and partitioning in leaflets, stipules and tendrils of *Pisum sativum* L. *Current Research in Photosynthesis.* 4:79–82.

Côté, R., Gerrath, J. M., Posluszny, U. & B. Grodzinski. 1990. Leaf development in conventional and semi-leafless peas. (*Pisum sativum* L.). Abstract #P-1, Canadian Botanical Association Meetings, Windsor, Ontario.

Gerrath, J. M. & U. Posluszny. 1988. Morphological and anatomical development in the Vitaceae I. Vegetative development in *Vitis riparia. Cana. J. Bot.* 66:209–224.

Gerrath, J. M. & U. Posluszny. 1989. Morphological and anatomical development in the Vitaceae. V. Vegetative and floral development in *Ampelopsis brevipedunculata. Can. J. Bot.* 67:2371–2386.

Gottschalk, W. 1970. Möglichkeiten der Blattevolution durch Mutation and Rekombination. Ein Modell für die Entwicklung and Weiterentwicklung des Leguminosenblattes. *Z. Pflanzenphysiol.* 63:44–54.

Gould, K., Cutter, E. G. & J. P. W. Young. 1986. Morphogenesis of the compound leaf in three genotypes of the pea, *Pisum sativum. Can. J. Bot.* 64:1268–1276.

Meicenheimer, R. D., Muehlbauer, F. J., Hindman, J. L. & E. T. Gritton. 1983. Meristem characteristics of genetically modified pea (*Pisum sativum*) leaf primordia. *Can. J. Bot.* 61:3430–3437.

Shah, J. J. & Y. S. Dave. 1971. Morpho-histogenic studies on tendrils of *Passiflora. Ann. Bot.* 35:627–635.

Srinivasan, C. & M. G. Mullins. 1981. Physiology of flowering in the grapevine—a review. *Am. J. Enol. and Vitic.* 32:47–63.

Tucker, S. C. & L. L. Hoefert. 1968. Ontogeny of the tendril in *Vitis vinifera. Am. J. Bot.* 55:1110–1119.

N. LEHMANN—HOMEOSIS IN THE PERIANTH AND ANDROECIUM

I am currently investigating the role of homeosis in floral evolution in four genera of plants. 1) I have compared a single-flowering species of *Begonia* with a double-flowering cultivar in which the stamens are replaced with petaloid structures. I found that early development of single and double flowers was similar with respect to the primordial pattern of morphogenesis and primordial shape. Later stages diverged, the primordia of the double-flowering cultivar forming structures that ranged from somewhat stamen-like to petal- and even sepal-like. 2) *Sanguinaria canadensis*, in the Papaveraceae, has eight petals rather than four as is usual in the family and may be an example of petals replacing stamens during evolution. The four additional petals are initiated in the relative positions where the first four stamens form in *Chelidonium majus* (a related genus). Early development of the additional petals share shape features with early stages of stamen development, later becoming more petal-shaped. 3) *Calla palustris* (Araceae) lacks a perianth and has from 10–12 stamens per flower while related genera have 4–6 tepals forming a perianth and 4–6 stamens per flower. In *Calla*, 4–6 stamens occupy inner floral positions and 4–6 occupy outer positions. In related genera tepals are found in the outer positions. Here the homeotic change is the replacement of tepals with stamens. 4) *Actaea rubra* and *A. pachypoda*

(Ranunculaceae) have variable numbers of petals and stamens and I am currently studying development to see if petals and stamens replace each other positionally.

These examples suggest that homeosis has played a role in floral evolution, the last three genera being examples of naturally occurring homeosis. Early developmental studies are crucial in obtaining evidence of this, as are assumptions of the ancestral condition. Information about ontogeny can be used to test existing phylogenetic hypotheses as well as to suggest possible reconstructions. As homeosis becomes more well-known to botanists, its role in evolution will become clearer.

BRUCE K. KIRCHOFF—HOMEOSIS IN THE FLOWERS OF THE ZINGIBERALES

Homeosis has played an important role in the macroevolution of the flowers of the plant order Zingiberales. The Zingiberales are a monophyletic order of monocotyledons of uncertain affinity. Based on overall similarity, the order may be divided into two informal groups consisting of four families each. The Banana Group consists of the Musaceae (bananas), Strelitziaceae (bird-of-paradise), and the two monogeneric families Heliconiaceae (*Heliconia*) and Lowiaceae (*Orchidantha*). The Ginger Group consists of the remaining four families: Zingiberaceae (gingers), Costaceae, Marantaceae (prayer plant family), and Cannaceae.

The flowers of the Banana Group resemble those of a "typical" monocotyledon. Their floral parts are arranged in whorls of three, and, with one exception, all of the members of a single whorl are similar in structure. Homeosis has played a role in floral evolution in only one family of this group. In the Heliconiaceae, a stamen of the outer whorl has been replaced by a staminode.

The flowers of the Ginger Group are greatly modified compared to a "typical" monocotyledon flower. The main source of difference is in the highly modified androecium. The number of stamens that produce pollen is reduced to one (Zingiberaceae, Costaceae) or one half (Cannaceae, Marantaceae), and the other androecial members are transformed into petaloid structures, which have various forms and degrees of fusion. Homeosis has been an important mechanism of floral evolution in this group. In the Zingiberaceae, two members of the outer androecial whorl are replaced by a lip, and the inner androecial whorl is represented by two staminodes. Most of the androecium of the Costaceae has also been replaced by petaloid structures. In this family, the single fertile stamen is often attached to an enlarged petaloid "filament." The androecia of the Cannaceae and Marantaceae are constructed along very similar lines. Both consist of one half of one fertile anther and three to four staminodes. The other half of the fertile anther is replaced by a petaloid appendage that is fused to the fertile locules. The staminodes proper are relatively unmodified in the Cannaceae and resemble petals more so than in the Marantaceae. The staminodes of the Marantaceae are variously modified to function in pollination.

The above examples serve to show the relevance of a study of homeosis to the evolution of at least one major group of plants. In both the Banana and Ginger Groups, members of the androecial whorls have been replaced by structures that resemble petals. However, these organs are not petals. The staminodes are often much larger, thinner or thicker, and more brightly colored than the petals. What relevance do these observations have to the definition of homeosis?

Since the staminodes do not assume the total "form or character" of any perianth members, a strict definition of homeosis must be ruled out. The staminodes merely resemble the perianth in some respects, being considerably different in others. Many additional examples of replacement by an organ similar, but not identical to, a serially homologous organ could be cited in plants. With this in mind, it is reasonable to extend the definition of homeosis to include replacement by an organ like, but not identical to, some other organ. Holmes (1979; as discussed in Sattler (1988)), proposed a suitably broad definition: homeosis is "the assumption by one part of likeness to another part." I support this broad definition of homeosis to include replacement by an organ that is not part of the serial homology of the plant.

Literature Cited

Holmes, S. 1979. *Henderson's Dictionary of Biological Terms.* Van Nostrand Reinhold: New York.
Sattler, R. 1988. Homeosis in plants. *American Journal of Botany* 75: 1606–1617.

SYNOPSIS OF DISCUSSION

Discussion during presentation by Rolf Sattler

Rolf Sattler: Definitions of homeosis are not very clear. For example, if, as Bateson states, homeosis only occurs among homologous parts, then the question of what is homologous dominates. I prefer the broad definition of "likeness" because it allows for partial homeosis. There are assumptions that a) homeosis can occur during phylogeny (the usual view) and b) that it can also occur during ontogeny. It is likely an epigenetic phenomenon. The definition I prefer is that it is the partial or total replacement of a structure by another one of the same organism. Certain terms need to be clarified. For example, "morphic translocation" is more or less equal to homeosis, and is no longer used. In the same way, "transference of function" is very close to homeosis. "Heterotopy" and "heterochrony" both overlap with homeosis, but are broader terms. Homeosis is heterotopy or heterochrony that leads to total or partial replacement of one part by another one of the same organism. Scott Poethig: It is impossible to separate space and time. For instance, the change from sepal to petal may be an example of heterchrony. I would prefer mechanistic neutrality. The mechanism of change is what is important. By using the term "homeosis", are we creating an artifice that is mechanistically not useful? Sattler: But a comparative morphologist needs a broader view. Poethig: Are "homeotic" changes interesting with regard to evolutionary plant development or mechanisms of plant development? Sattler: One cannot generalize. Victor Albert: Perhaps two different terms are useful, because there are two different ways of looking at the same thing. Maybe we should restrict the definition to a certain level of organization. Is homeosis a morphological and heterochrony a developmental term? What does this do to changes of levels of homology?

Discussion during presentation by Scott Poethig

Sattler: You (Poethig) think of a one to one organ replacement. Is it not a partial replacement? Poethig: Right. No term is better. Shoots are both reproductive and vegetative.

Discussion during presentation by Jeffrey Hill

Posluszny: At what point does the partial substitution become a total substitution? Hill: I don't think there can be a resolution. All we can try to do is to describe what we see with as few value-laden terms as possible. Sattler: Maybe we should use the term "developmental hybrid?" Hill: Are we transferring organ types to the same function, or is it a replacement? Albert: It is very important that we determine exactly what is homologous. Tracy McLellan: I object to the use of a term such as "hybridization" in morphology. We

454

should use "intermediate." Poethig: What we are seeing is primordia that start as the same thing, but have a different fate. They may differ either in position, fate, processes of development, or order of decision. The homeotic genes function at the end.

Discussion during presentation by Jean Gerrath

Albert: How can you call replacement with a new organ homeosis? Gerrath: I'm not sure you can, but it may be one way to explain how new organs are regenerated.

Discussion during presentation by Naida Lehmann

Posluszny: In your example of *Calla* could space play a role? Lehmann: No, because you can have crowding of primordia without homeosis.

Discussion during presentation by Bruce Kirchoff

Posluszny: If a structure looks and functions like a petal, why is it an androecium? Kirchoff: It's not. It's only an androecium when compared positionally with floral structures in other genera.

Fossil Evidence for the Evolution of Modern Humans in Africa

EXTENDED ABSTRACT

G. Philip Rightmire

Studies of variation at the molecular level provide important information bearing on the evolution of *Homo sapiens*. Work on mtDNA sequences, RFLPs of nuclear DNA, and classical genetic markers is reported in this symposium, and much of this evidence can be interpreted to support a "single location of origin" for modern human groups. At the same time, there are questions about the geography of such an origin, and when it occurred. Here the prehistoric record yields more clues. Fossil remains uncovered at several sites in Africa suggest that this region must be considered a likely source. Particularly significant are traces of occupation in caves at Klasies River Mouth in South Africa. Human bones from Klasies appear modern anatomically, and some of the specimens must be more than 100,000 years in age.

More archaic representatives of *Homo sapiens* are also known from Africa and from other parts of the Old World. Many of these discoveries consist of crania and jaws, and the archaic skulls differ from recent ones in being more robust, with thick brows, low frontal bones, more angled or projecting occiputs, and little chin development. It is important to note that there are significant geographic differences. The Neanderthals of Europe are distinctive morphologically, both in facial form and in proportions of the braincase. These people seem to persist in the record through much of the Late Pleistocene. In western Asia, the Neanderthals are similar in most respects to those of Europe but may be relatively late arrivals. The lower Moustrian levels at Kebara Cave are dated to about 60,000 years, and sites such as Shanidar are probably about the same age. Other people from this region, found at Qafzeh Cave and Skhul in Israel, are not Neanderthals, although they are quite ancient. On the basis of thermoluminesence applied to burned flints and electron spin resonance measurements from animal teeth, these remains are 90,000 to 100,000 years old. The skeletons are robust but modern-looking, and not likely to be part of any lineage including Kebara or Shanidar. Still further to the east, there is less material, but certainly the cranium from Dali is one representative of archaic *Homo sapiens* from China. This individual has a short, broad face quite different from that of the Neanderthals.

In Africa, archaic populations are again distinctive anatomically. Broken Hill (or Kabwe) from Zambia is one of the most famous fossils, remarkably well preserved. The brows are very thick, and the vault is long and low. The face is especially broad in its upper parts, and the nasal region does not project in the same way as in Neanderthals. Broken

Dr. Rightmire is with the Department of Anthropology, SUNY at Binghamton, Binghamton, NY 13902, USA.

Hill is just one of several fossils from Africa that date to the later part of the Middle Pleistocene. Other specimens are known from South Africa and Tanzania.

The key question, of course, is which, if not all, of these archaic populations evolved toward later people. The issue is whether evolutionary continuity can be established from the prehistoric record, for one or more of the geographic areas considered. Anthropologists hold rather diverse opinions on this point. I myself think that the western Neanderthals have a number of unique features, and it is hard to grant that they contributed much to the make-up of Upper Paleolithic populations. In the Far East the fossils are sparse and poorly dated. In any case, individuals like Dali show no special resemblance to recent Chinese people.

For sub-Saharan Africa, the record is again far from satisfactory. There is nothing approaching the density of material (fossils, long archaeological sequences) present in Europe during the Late Pleistocene. However, there is an interesting collection of hominids from sites including Florisbad in South Africa, Laetoli in Tanzania and the Omo in Ethiopia. Some of these crania are less archaic than Broken Hill in their anatomy, and exhibit features that foreshadow the modern condition. Still, differences are substantial, and we seem to be left without very convincing evidence for continuity, in Africa as for other regions.

At this point, I will hasten to add that there are more discoveries, which prove quite interesting. One well-known site is Border Cave in South Africa, where there are Middle Stone Age deposits probably dating to the beginning of the Late Pleistocene. Parts of several human skeletons have been recovered from the cave, and there is one damaged adult cranium. This cranium is modern in its anatomy, as are the (two) adult mandibles. This would be an important observation, if the bones were firmly in association with the Middle Stone Age tools. A problem is that the cranium and one of the jaws were collected by guano diggers, and there is continuing doubt about the exact provenience of these bones in the deposits. If the cranium is not so old as has been claimed, then it cannot inform us about the origins of modern people.

Another locality is Klasies River Mouth, situated on the south coast of Africa, some 40 km to the west of Cape St. Francis. The main site at Klasies consists of several caves and shelters, which lie in a spectacular setting. These caves were excavated by Ronald Singer and John Wymer in 1967–1968, and much material was collected at that time, including human bones. More recently, the main site has been worked by Hilary Deacon and a crew from the University of Stellenbosch. These 1984–1989 excavations have been aimed at clarifying stratigraphy and dating. It is now well established that the caves were inhabited nearly 120,000 years ago, during regression from the maximum high sea level stand of the Last Interglacial. Two main stages of subsequent deposition can be recognized, and multiple short episodes of human occupation are associated with hearths, shell middens and other traces of activity. Apparently the site was abandoned about 60,000 years ago, perhaps because of climatic and related ecological changes leading to a less productive human habitat.

Human skeletal remains recovered in a secure Middle Stone Age context include several (damaged) lower jaws, along with other skull parts and bones from the shoulder, arm, and foot. Measurements, coupled with anatomical comparisons, confirm that these Klasies specimens differ from the fossils of archaic people. Although the mandibles are not much smaller than those of some Neanderthals, the corpus tends to be deeper at the front than posteriorly, and a chin is relatively prominent. None of the jaws exhibits a superior transverse torus, and only one shows traces of an inferior torus. These internal buttresses are common in Neanderthals. The specimens that are complete enough to allow reconstruction of the posterior tooth row in relation to the ramus show no retromolar space. This impression that the morphology is modern extends also to the face and vault. No heavy browridge is developed. Postcranial elements from the 1967–1968 excavations are indistinguishable from the bones of living people, and an ulna discovered in 1985 has none of the distinctive features of Neanderthals.

This material from Klasies River Mouth constitutes about the strongest direct evidence we have that points to an African origin for *Homo sapiens sapiens.* At the same time, it is apparent that the Klasies folk had not acquired all of the behavioral attributes of recent humans. They made tools similar to those produced by the Neanderthals, and there are indications that they were inexpert (or cautious) hunters. An important question is when and how such people developed the skills that would allow them to expand in numbers and replace other, more archaic groups. Another problem is how the Klasies population relates to those at Skhul and Qafzeh. Some differences are present in the anatomy of the African fossils as compared to the skeletons from the Near East. This must be checked further, to ascertain whether signs of such (regional) differentiation correspond to what is observed today between San or African blacks and the people of western Asia. Alternatively, both the Klasies and the Skhul/Qafzeh assemblages may have to be regarded as part of a robust but "modern" stock, already dispersed geographically but not yet molded by selection to produce the patterns of variation that we see in recent times. If this is the case, then it will be difficult to recognize either the Klasies hominids or the Qafzeh people as directly ancestral to living populations.

I am pleased to acknowledge support for this research from the L. S. B. Leakey Foundation.

Relationships Among Living Human Populations Determined from Classical and DNA Polymorphisms

J. L. Mountain and L. L. Cavalli-Sforza

Abstract. Frequencies of 120 classical genetic markers (44 loci, including blood groups, and protein and enzyme polymorphisms) in 42 human populations were analyzed previously; the resulting genetic tree correlated well with archaeological and linguistic data. More recently, 99 DNA polymorphisms, primarily restriction fragment polymorphisms (RFLPs) and one polymerase chain reaction polymorphism (PCRP), have been studied for five populations. While the number of populations included to date is limited, the number of loci is large enough to ensure that the behavior of single genes does not adversely influence the results. The maximum likelihood genetic tree generated from these data supports the hypothesis of an initial African/non African split. This maximum likelihood model has been extended to incorporate admixture, and the hypothesis of a European admixture evaluated.

The puzzle of the origin of modern humans is particularly complex; with tools of genetics we may solve one small piece, and with tools of archaeology another, but without communication and collaboration among paleoanthropologists, geneticists and others we may never begin to put those pieces together. As population geneticists we analyze various types of genetic data in order to investigate human evolution. It is a challenging project because even though the history is present, in the sequence of the DNA, the picture has been altered to some extent by mutation and recombination at the DNA level and by migration, natural selection, and random genetic drift at the population level.

There are several ways of circumventing these difficulties. One can avoid the scrambling due to recombination by working with mtDNA, and thereby look at one part of the human evolution picture. MtDNA data have been used successfully by Stoneking and Cann (1989) to provide an estimate for the date when the common ancestor of modern humans lived of between 50 ky (kiloyears = thousand years) and 500 ky ago (a more conservative estimate than the 142 ky—285 ky range suggested earlier (Cann et al., 1987)). This relatively recent date is evidence against the multiregional hypothesis of the origin of modern humans; one would expect more variability among present day mtDNA types under the multiregional hypothesis (Stoneking & Cann, 1989). While one can use mtDNA sequence or restriction site data in order to investigate the history of mutations of a molecule there are two main problems with using mtDNA to study population events. First, mtDNA in many respects behaves as a single site, or locus, and is therefore subject to sub-

J. L. Mountain and L. L. Cavalli-Sforza are with the Department of Genetics, Stanford University, Stanford, CA 94305, USA.

stantial stochastic error that depends on the number of mutations involved. Minimization of such stochastic error as well as other types of error through the use of many independent loci is essential. An additional problem with building trees from individual data is that the tree gives a mutational rather than a population history. Mutational events necessarily precede population events (Nei, 1987). For these reasons we prefer to use gene frequency data in order to investigate the evolution of modern humans. With a large number of genetic markers we hope to ensure that even if natural selection affects a few, the overall picture remains undistorted.

CLASSICAL MARKERS

Along with collaborators we have looked at two types of genetic data so far. Until a few years ago data consisted of frequencies of blood group types and of protein and enzyme markers—the "classical" genetic markers. These markers give a hazy view of differences at the DNA level; even if two individuals share a similar protein, the DNA sequences that code for the protein in each individual might differ. The body of such data collected so far is enormous, and certainly worth analyzing fully. Cavalli-Sforza, Piazza, and Menozzi are currently publishing a detailed analysis of gene frequency data from about 3000 world populations. In 1988 we summarized that data set, constructing a tree based on the degree of genetic relationship between 42 aboriginal populations, using the average linkage algorithm (Cavalli-Sforza et al., 1988). Fig. 1 shows a condensed version of that tree, constructed from the same data but using a different distance measure. The tree supports the hypothesis that the oldest genetically recognizable split of modern humans was into two groups, one which eventually evolved into the group of all African populations and another which evolved into the collection of all non-African populations. The most simple interpretation is that modern humans evolved in, or near, Africa. The second major fission in the tree is also interesting; it separates Europeans, Amerindians, and northern Asians from southeastern Asians and populations of Oceania. We hope to investigate this fission further using other data.

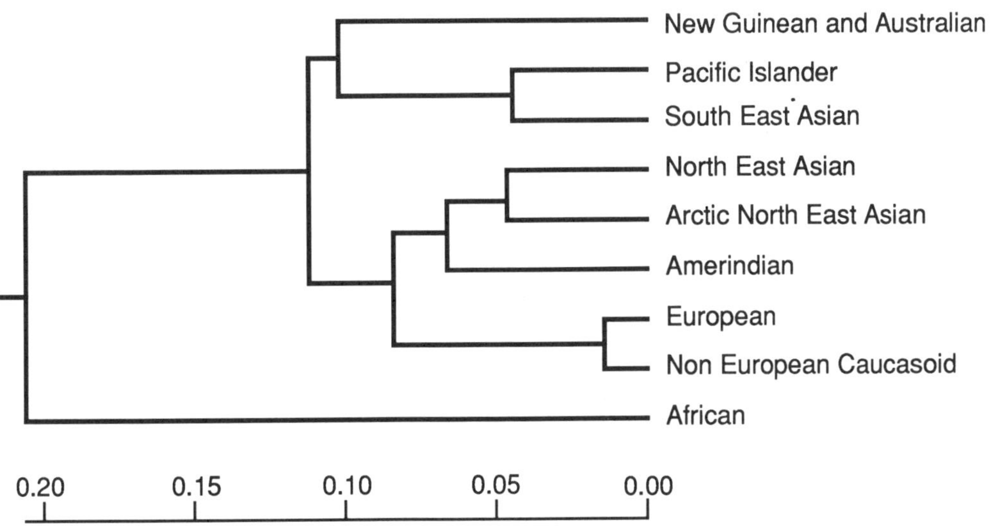

Figure 1. Average linkage tree for nine world populations, based on data for up to 120 classical polymorphisms (see Cavalli-Sforza et al., 1988, Fig. 1, for details on genetic systems). Genetic distance is based on F_{ST}, as suggested by Reynolds et al., 1983. This tree is a condensed version of that presented in Cavalli-Sforza et al., 1988.

460

Comparison with Archaeological Data

While the tree gives no information regarding the actual dates of population fissions, it does give an estimate of the nature and relative order of these fissions. We compared these relative dates with those estimated from archaeological data. The ratio of genetic distance ($F_{ST} \times 1000$) to time (ky) is roughly constant for the three available comparisons (2.15: Cavalli-Sforza et al., in preparation). Using this ratio and the genetic distance between Amerinds and northeast Asia (about 0.066) we can estimate the date of the human expansion to North America. The genetic estimate is about 31 ky ago, whereas archaeological estimates range from 15 ky ago to 35 ky ago.

Comparison with Linguistic Data

The genetic tree was compared to a set of 17 linguistic phyla and several linguistic superphyla, proposed by Joseph Greenberg and others (Ruhlen, 1987). There, too, we found remarkable agreement; the linguistic families correspond roughly with the low level genetic clusters, and the union of the two proposed linguistic superphyla (Nostratic and Eurasiatic), which may actually turn out to be one, corresponds with rare exception to the north Eurasian genetic cluster. The similarity may seem odd, given the temporal distance between genetic evolution and linguistic evolution, but genes and languages often travel together in populations. Genetic isolation generally implies linguistic isolation, and conversely a language barrier is often a barrier to genetic exchange, so that the history of major fissions is similar.

DNA POLYMORPHISMS

More recently, in collaboration with members of the Kidd laboratory at Yale, we have begun to study genetic differences closer to the DNA sequence by looking at restriction fragment length polymorphisms (RFLP's) (Bowcock et al., 1987, Bowcock et al., 1991a). While this technique does not give the precise sequence of a region, it does show how the sequences of two individuals differ. There are several advantages to using DNA polymorphisms rather than the "classical" markers described earlier. In particular, there is an essentially unlimited number of DNA polymorphisms. To date in our two labs 15 aboriginal populations have been tested for up to 130 polymorphisms, or markers. Chimpanzees have also been tested for some of these markers; on rare occasions they have alleles that are common in humans, indicating that these polymorphisms are ancient. The most complete analysis of these data so far has been for five populations which have each been tested for 100 markers at 73 genetic sites, or loci. These include two African Pygmy populations, one of which is probably mixed with non-Pygmy Africans (Bantu, or closely related peoples), a Melanesian population from the Bougainville islands, a Chinese population primarily from southern China, and a Caucasoid population of European descent. We look forward to extending our analysis to the other ten populations currently being tested in our two labs. By using 100 markers we hope to ensure that behavior of a single gene does not distort the results (i.e., due to natural selection). All but two of the 22 non-sex chromosomes are represented by at least one polymorphism among these 100. Each polymorphism tells a story; the consensus of all such stories tells us something about the evolution of our species.

METHODS

The techniques we use to analyze gene frequency data differ greatly from those used to study differences between species, or differences between individuals. The methods for analyzing gene frequencies are most appropriate for studying the evolution of populations within a species. One of the first steps is to calculate from the population gene frequencies the variance between the populations, as measured by F_{ST} (which standardizes the variance to a range between 0 and 1). The average F_{ST} for these five populations (0.139) is very similar to that of the classical markers described above (0.118), indicating that the apparent divergence among the five is similar to that among the 42 populations (Bowcock et al., 1991b). A second step is to determine how genetically different populations are, one from another. In order to do so we estimate a genetic distance, based on the F_{ST} variance between the populations, for each pair of populations (Reynolds et al., 1983). For short time spans, i.e., for populations within a species, other distance measures are highly correlated (Chakraborty & Tateno, 1976). Table 1 gives the F_{ST} distance. Note that the distance between the two Pygmy populations is the smallest (0.043), and that between Europeans and Chinese is next smallest (0.093).

Table 1. Estimates of genetic distances between five populations, based on 99 DNA polymorphisms. Corrections for sampling error due to the small number of individuals tested were made. PygC = Pygmies from the Central African Republic; PygZ = Pygmies from Zaire; Eur = Caucasoids of European origin; Chi = Chinese; Mel = Melanesians. Modified from Bowcock et al., 1991b.

	PygC	PygZ	Eur	Chi	Mel
PygC	—				
PygZ	0.043	—			
Eur4	0.141	0.142	—		
Chi4	0.235	0.235	0.093	—	
Mel4	0.242	0.265	0.148	0.171	—

Genetic Tree

While the genetic distance matrix summarizes the information on genetic relationships among these populations, relationships can be seen more easily if we use the matrix to construct a genetic tree. Fig. 2 shows one such tree, constructed using a maximum likelihood algorithm (Felsenstein, 1973, Thompson, 1975). The maximum likelihood method gives a tree very similar to that of the average linkage algorithm that was used to build the classical marker tree described above. The model assumes that evolutionary rates in different branches are roughly constant and that after divergence two populations do not interact genetically. This tree splits the small but representative group of world populations into African populations on the one hand and non African populations on the other. In this respect the tree is consistent with the classical marker tree. Lower splits are not yet possible to compare, because this Chinese sample includes both northern and southern Chinese. Once again we can compare the relative times of fissions to the dates estimated from archaeological data. If we assume that the African - non-African split took place about 100,000 years ago, an approximate estimate based on archaeological data, then the split of Melanesians from Europeans and Chinese occurred about 73 ky ago, the split between Europeans and Chinese took place about 45 ky ago, and the two Pygmy populations diverged roughly 22 ky ago.

While the tree is the most likely under the model assumed, it does not appear to fit the distance matrix very well. We tested the fit by determining the expected distance matrix, given the tree structure, and comparing that to the observed matrix. Both Pygmy popula-

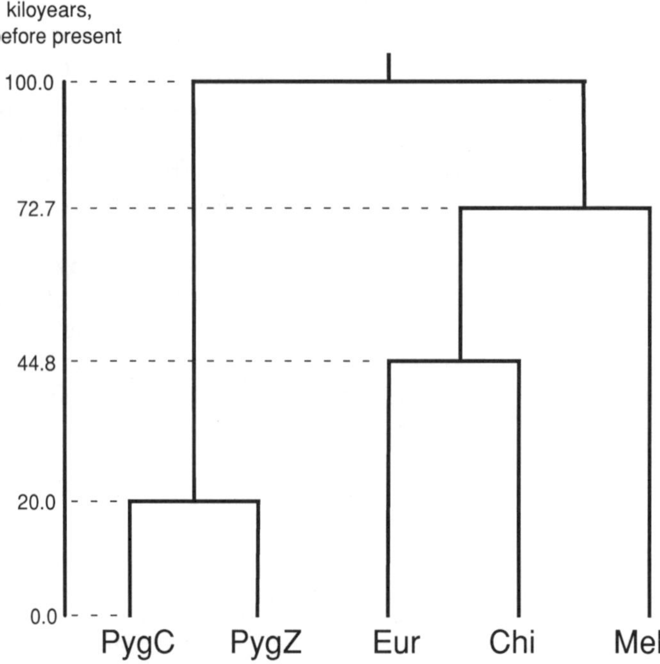

Figure 2. Maximum likelihood tree for five populations constructed from distance matrix of Table 1. This model assumes constant evolutionary rates. Standard errors were estimated using the bootstrap. All times were calibrated against the segment from the divergence of the two Pygmy populations to the present, considered as fixed and taken to be 20 ky. This leads to a time of 100 ky ago for the first major split. Modified from Bowcock et al., 1991b.

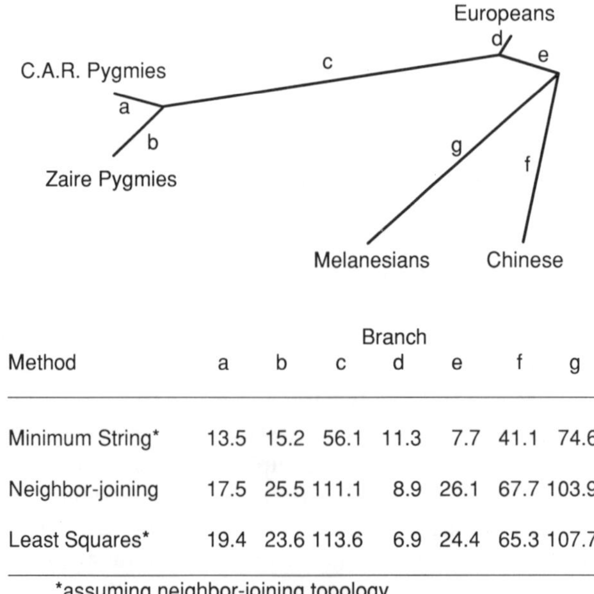

Method		a	b	c	d	e	f	g
	Branch							
Minimum String*		13.5	15.2	56.1	11.3	7.7	41.1	74.6
Neighbor-joining		17.5	25.5	111.1	8.9	26.1	67.7	103.9
Least Squares*		19.4	23.6	113.6	6.9	24.4	65.3	107.7

*assuming neighbor-joining topology

Figure 3. Tree constructed using Nei's neighbor-joining method, from genetic distances of Table 1. Branch lengths were estimated using three different methods, including "neighbor-joining" (Saitou & Nei ,1987), "minimum string", and "least squares" (Cavalli-Sforza & Edwards, 1964, Cavalli-Sforza & Edwards, 1967, Kidd & Sgaramella-Zonta, 1971, Fitch, 1971). From Bowcock et al., 1991b.

tions, for example, are expected to be equidistant from the three populations in the other cluster. In the observed matrix the European population is much closer to the Pygmy groups than are the Melanesians or the Chinese. Thus, the tree does not fit the observed data well. This is an indication that either the evolutionary rate during the time period under consideration varied substantially, or the populations did not remain isolated after they diverged from one another.

In order to examine this point further we constructed a tree using methods which do not require the assumption that evolutionary rates are constant. One such method is Nei's neighbor joining method; the tree is shown in Fig. 3. It differs from the maximum likelihood tree primarily in the length of the European branch. Such a short branch could be due either to a slower evolutionary rate for the European population or to an admixture event creating the European population. The first hypothesis is difficult to test without further evidence. While we cannot directly test the second hypothesis either, we can check to see whether the maximum likelihood method incorporating this possibility provides a tree which fits the distance matrix more closely.

European Admixture

The maximum likelihood tree was constructed for a second time, allowing for admixture (Fig. 4). Based on an African - non-African split occurring about 100 ky ago, the European admixture was estimated to be 35% ancestral African and 65% ancestral Chinese, and to have taken place approximately 30 ky ago. The estimated times for other fissions have also been revised; in particular, the Chinese appear to have diverged from the Melanesians about 68 ky ago. The revised tree fits the data much more closely than does the tree of Fig. 2.

While the admixture hypothesis is only one of several plausible explanations for the

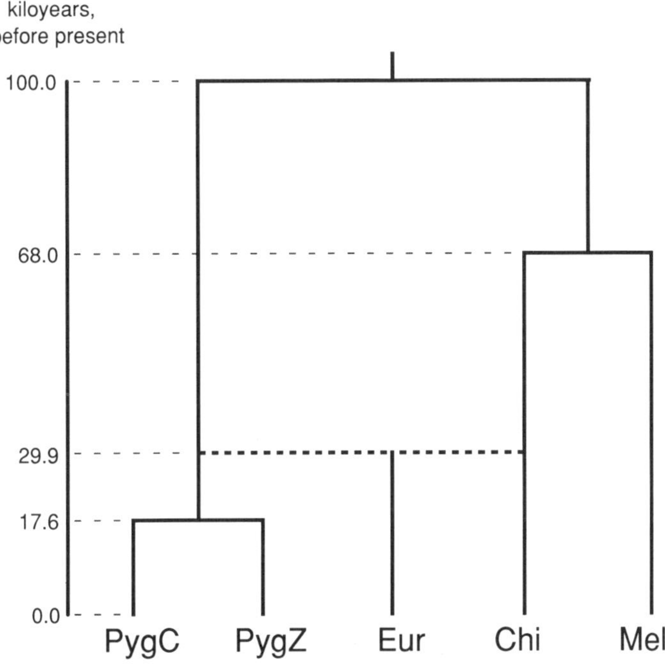

Figure 4. Maximum likelihood tree, assuming a model of admixture between ancestral Africans and ancestral Asians, fitting the distance matrix of Table 1. Standard errors were calculated by the bootstrap. All times were calibrated against the segment from the divergence of the two Pygmy groups to the present, considered as fixed and taken to be 17.6 ky. Modified from Bowcock et al., 1991b.

genetic data, there is external geographic and historical evidence that favors this hypothesis. Assuming that anatomically modern humans from west Asia replaced the European Neanderthals, Europe was first populated by modern humans roughly 40 ky to 30 ky ago. Other major migrations into Europe took place during the Neolithic transition, from the Middle East, and more recently, from the steppes of Eurasia. The regions which contributed most heavily to the colonization of Europe are geographically located between Africa and Asia. These regions are likely to have been mixed, genetically, at the time of the first major expansion. Our tree suggests that such admixture, whether it was due to one or more independent events, whether it occurred over a short or long period of time, was a major genetic phenomenon.

Population Sizes

From the F_{ST} variance described earlier we can estimate the effective population sizes in the various branches of the tree. For an F_{ST} of 0.139 to arise over a period of 100,000 years, the effective population sizes in all the branches would have been about 13,000, corresponding to a total census of roughly 40,000 (Bowcock et al., 1991b). Further evidence is necessary to confirm this estimate.

DISCUSSION

The African - non-African divergence suggested by the trees based on classical and DNA polymorphisms is fairly robust; both trees have been tested for sampling error using the bootstrap (Efron, 1982, Felsenstein, 1985), and other studies using some of the same data on a more limited number of populations (three) have produced similar results (Nei & Livshits, 1989). But the conclusion is based on certain assumptions, including that of constant evolutionary rates. While the overall demographic trend throughout history has been one of steady growth, population bottlenecks have occurred, and may distort the evolutionary picture. Our present goal is to clarify that picture further by increasing the number of populations in our sample, by adding Australians, South American Indians, Japanese, and other populations, and by increasing the complexity of our models. The incorporation of admixture into the maximum likelihood model is just one step in that direction. Finally, we must also keep up with the advances made in archaeology and in other fields in order to further clarify the picture of the evolution of modern humans.

ACKNOWLEDGMENTS

This work was supported by National Institutes of Health Grant GM 20467.

LITERATURE CITED

Bowcock, A. M., Bucci, C., Hebert, J. M., Kidd, J. R., Kidd, K. K., Friedlaender, J. S. & L. L. Cavalli-Sforza. 1987. Study of 47 DNA markers in five populations from four continents. *Gene Geography* 1:47–64.

Bowcock, A. M., Hebert, J. M., Mountain, J. L., Kidd, J. R., Rogers, J., Kidd, K. K. & L. L. Cavalli-Sforza. 1991a. Study of an additional 58 DNA markers in five populations from four continents (in preparation).

Bowcock, A. M., Kidd, J. R., Mountain, J. L., Hebert, J. M., Carotenuto, L., Kidd, K. K. & L. L. Cavalli-Sforza. 1991b. Nuclear DNA polymorphisms and human evolution. *Proc. Natl. Acad. Sci. USA* 88:839–843.

Cann, R. L. Stoneking, M. & A. C. Wilson. 1987. Mitochondrial DNA and human evolution. *Nature* 325:31–36.

Cavalli-Sforza, L. L. & A. W. F. Edwards. 1964. Analysis of human evolution. Pp. 923–933. *In*: S. J. Geerts (ed.), *Genetics Today, Proceedings of the 11th International Congress of Genetics*, vol. 2. Pergamon Press: New York.

Cavalli-Sforza, L. L. & A. W. F. Edwards. 1967. Phylogenetic analysis: models and estimation procedures. *Am. J. Hum. Genet.* 19:223–257.

Cavalli-Sforza, L. L., Piazza, A., Menozzi, P., & J. Mountain. 1988. Reconstruction of human evolution: bringing together genetic, archaeological, and linguistic data. *Proc. Natl. Acad. Sci. USA* 85:6002–6006.

Cavalli-Sforza, L. L., Piazza, A. & P. Menozzi. *History and geography of human genes* (in preparation).

Chakraborty, R. & Y. Tateno. 1976. Correlation between some measures of genetic distance. *Evolution* 30:851–853.

Efron, B. 1982. *The Jacknife, Bootstrap, and Other Resampling Plans*. Society for Industrial and Applied Mathematics: Philadelphia, Pennsylvania. 92 pp.

Felsenstein, J. 1973. Maximum-likelihood estimation of evolutionary trees from continuous characters. *Am. J. Hum. Genet.* 25:471–492.

Felsenstein, J. 1985. Confidence limits on phylogenies: an approach using the bootstrap. *Evolution* 39:783–791.

Fitch, W. M. 1971. Toward defining the course of evolution: minimum change for a specified tree topology. *Syst. Zool.* 20:406–416.

Kidd, K. K. & L. A. Sgaramella-Zonta. 1971. Phylogenetic analysis: concepts and methods. *Am. J. Hum. Genet.* 23:235–252.

Nei, M. 1987. *Molecular Evolutionary Genetics*. Columbia University Press: New York. 512 pp.

Nei, M. & G. Livshits. 1989. Genetic relationships of Europeans, Asians and Africans and the origin of modern *Homo sapiens*. *Hum. Hered.* 39:276–281.

Reynolds, J., Weir, B. S. & C. C. Cockerham. 1983. Estimation of the coancestry coefficient: basis for a short-term genetic distance. *Genetics* 105:767–779.

Ruhlen, M. 1987. *A Guide to the World's Languages*. Stanford University Press: Stanford, California. 433 pp.

Saitou, N. & M. Nei. 1987. The neighbor-joining method: a new method for reconstructing phylogenetic trees. *Mol. Biol. Evol.* 4:406–425.

Stoneking, M. & R. L. Cann. 1989. African origin of human mitochondrial DNA. Pp. 17–30. *In*: P. Mellars & C. Stringer (eds.), *The Human Revolution: Behavioural and Biological Perspectives on the Origins of Modern Humans*. Princeton University Press: Princeton, New Jersey.

Thompson, E. A. 1975. *Human Evolutionary Trees*. Cambridge University Press: Cambridge. 158 pp.

The Evolution of Regionality in Modern Humans

C. B. Stringer

Abstract. Fossil evidence suggests that the skeletal morphology shared by modern humans originated in Africa (or perhaps the adjacent area of the Levant) prior to 100,000 years ago. However, early, anatomically modern fossils do not usually show clear "regionality" in the sense of being clearly identified with their present day regional successors. When combined with the results of genetic analyses of existing populations, this suggests that the development of regionality was a much more recent phenomenon than the origin of the modern human anatomical pattern.

INTRODUCTION

It is becoming clear that modern humans did not evolve simultaneously in many areas or spread rapidly through the Old World following their first known appearances in the fossil record. Skeletal remains of hominids closely related to or actually representing modern humans are known from Israel, South Africa, and perhaps also Ethiopia and Kenya, dating from at least 80,000 years ago. Yet the earliest modern humans known from Europe, China, Malaysia and Australia may be only half that age (Stringer, 1989, in press).

What is especially interesting is the apparent persistence of "archaic" (non-modern) hominids in Europe and the Middle East (Neanderthals), and Indonesia (the Ngandong or Solo hominids), after the origin of modern *Homo sapiens,* with the presumption that similar populations may also have been present in areas with a poorer fossil record for this period, such as China. Potentially, these late archaic populations could have hybridised with dispersing early modern humans, as is argued in gene flow/hybridisation models, providing the possibility of a two-stage origin for modern human variation. The establishment of the basic modern human anatomy by about 100,000 years ago could have been followed by a "hybridisation phase," lasting until about 30,000 years ago, that added a number of regional morphological features to the early modern human populations. However, this would mean that regionalisation was clearly established by 30,000 years ago. There is possible, but disputed, evidence of such a hybridisation phase in fossils from eastern Europe and Australia (Smith et al., 1989), but late archaic and early modern fossils from western Europe and the Middle East show little or no evidence of gene flow from one population to the other.

The present fossil record suggests that the *origin* of modern humans was an African phenomenon (with the possible inclusion of the adjacent Levant in the zone of origin), and that archaic hominids from elsewhere were generally replaced, with perhaps a minor

Dr. Stringer is with the Department of Palaeontology, Natural History Museum, London SW7 5BD, England.

input into the early modern gene pool. The objection that groups of anatomically modern hunter-gatherers could not have completely replaced groups of archaic hunter-gatherers can be challenged from various lines of evidence, including many examples of replacements of one mammalian species by another that is closely related.

THE DEVELOPMENT OF REGIONALITY

It is significant, when considering models which posit a long period of development for regional differentiation in modern humans, that no reasonably objective means of recognising the modern pattern of regionality in crania has yet succeeded in detecting its unequivocal presence in samples older than about 15,000 years (i.e., ancient, anatomically modern crania do not readily group with recent populations from the same region). My own recent investigations have employed the IDCRAN package developed by Richard Wright (Anthropology, University of Sydney). Using Principal Components Analyses performed on Howells' database of measurements of some 2500 recent crania (Howells, 1989), additional individual crania can be compared by computation of Euclidean distances to provide the 50 nearest neighbour crania in shape from Howells' sample, as well as overall affinities to whole population samples. When European Upper Palaeolithic and Mesolithic crania were tested, European "regionality" was much more strongly expressed in the Mesolithic than in the Upper Palaeolithic crania (as is also found in postcranial data such as brachial and crural indices: Trinkaus, 1981). The Zhoukoudian Upper Cave crania 101 and 103 (measured from casts) were predominantly identified with recent African, Australasian, Ainu or Polynesian crania, rather than those of "classical" Mongoloids. Cerro Sota 2 from the late Pleistocene/early Holocene of Chile was, by contrast, clearly related to Mongoloid crania, especially those of Amerindians. Late Pleistocene/early Holocene Australian crania tested as casts or originals were divided into those with clear Australasian affinities (Cohuna, Talgai and a Coobol Creek specimen) and those without (Keilor, primarily affined with African and European crania).

While lack of affiliation to modern successors might be explicable in cases where the archaic ("grade") differences override the regional ("clade") similarities (e.g., in Neanderthals), this hardly seems to be an appropriate explanation for material which is only 10–35,000 years old in clades that supposedly have an antiquity of at least an order of magnitude greater. Thus, if the Zhoukoudian Upper Cave crania really represent members of a Chinese "clade" with an antiquity of up to a million years, it is surprising that they show so few signs of "local" affinities.

Moreover, multiregionalists have consistently ignored data on modern human variation that suggest caution is required in interpreting regional characters. First, there seems little reason why traits such as a supratoral sulcus (claimed as a stable Chinese regional character spanning some 700,000 years) or a zygomaxillary ridge (claimed as an Australasian trait spanning a similar time interval) should have been maintained over these periods while the morphology of the face and supraorbital torus was being significantly remodelled. Second, these traits vary within and among modern populations, and there seems little reason to select mainland Chinese and Australian aborigines as the representative "clade" members for character selection, rather than, say, Ainu and Amerindians or New Guineans and Tasmanians, respectively. Presumably, these populations would also trace descent from the same ancient Chinese and Japanese ancestors under the multiregional model, and their distinct morphologies must be taken into account in examining supposed regional "clade" characters (Stringer, in press).

468

CONCLUDING REMARKS

The apparent lack of "regionality" in early modern human crania suggests that such regionality was either undeveloped or only incipient before 20,000 years ago. When combined with the incompatibility between the geographic pattern of present day genetic relationships and that which can be reconstructed from the middle Pleistocene fossil record, this suggests that modern "regionality" is a recent development. Attempts at projecting the genetic patterns of modern humanity onto the middle Pleistocene fossil record lead to several serious problems. One is that analyses of nuclear DNA or its products consistently show a close relationship between European and East Asian populations (e.g., Cavalli-Sforza et al., 1988; Long et al., 1990; Nei & Livshits, in press). Such a close relationship is not apparent in the fossil hominid records of these areas until the advent of modern humans. Another is that several mtDNA analyses of Khoisan samples show their distinctiveness from other modern populations (e.g., Johnson et al., 1983; Vigilant et al., 1989; Harihara & Saitou, 1989). Given their overall morphology and similarities to the rest of humanity, is it really credible that this could be a reflection of 500,000 years or more of separate evolution? It seems much more likely that the genetic and morphological differences between recent populations have arisen within the last 200,000 years, and that many have evolved very much more recently than that, following the morphological origin of our species (in Africa?) some time prior to 100,000 years ago. The more recent the development of regional characters, the less likely it is that they represent evidence of a heritage from local archaic ancestors through either direct descent (multiregional model) or gene flow from relict populations (gene flow/hybridisation model).

LITERATURE CITED

Cavalli Sforza, L. L., Piazza, A., Menozzi, P. & J. Mountain. 1988. Reconstruction of human evolution: bringing together genetic, archaeological and linguistic data. *Proceedings National Academy Sciences USA* 85:6002–6006.

Harihara, S. & N. Saitou. 1989. A phylogenetic analysis of human mitochondrial DNA data. *Journal of the Anthropological Society of Nippon* 97:483–492.

Howells, W. W. 1989. Skull shapes and the map. *Papers of the Peabody Museum, Harvard* 79 (whole volume).

Johnson, M. J., Wallace, D. C., Ferris, S. D., Rattazzi, M. C. & L. L. Cavalli-Sforza. 1983. Radiation of human mitochondrial DNA types analyzed by restriction endonuclease cleavage patterns. *Journal of Molecular Evolution* 19:255–271.

Long, J. C., Chakravarti, A., Boehm C. D., Antonarakis S. & H. H. Kazazian. 1990. Phylogeny of human Beta-Globin haplotypes and its implications for recent human evolution. *American Journal of Physical Anthropology* 81:113–130.

Nei, M. and G. Livshits. In press. Evolutionary relationships of Europeans, Asians and Africans at the molecular level. *In:* N. Takahata & J. Crow (eds.), *Population Biology of Genes and Molecules.* Baifukan: Tokyo; Springer: Heidelberg.

Smith, F. H., Falsetti, A. B. & Donnelly, S. M. 1989. Modern human origins. *Yearbook of Physical Anthropology* 32:35–68.

Stringer, C. B. 1989. Documenting the origin of modern humans. Pp. 67–96. *In:* E. Trinkaus (ed.), *The Emergence of Modern Humans.* Cambridge University Press: Cambridge.

Stringer, C. B. In press. Replacement, continuity and the origin of *Homo sapiens. In:* G. Bräuer & F. H. Smith (eds.), *Continuity or Replacement? Controversies in the Evolution of Homo Sapiens.* Balkema: Rotterdam.

Trinkaus, E. 1981. Neanderthal limb proportions and cold adaptation. Pp. 187–224. *In:* C. B. Stringer (ed.), *Aspects of Human Evolution.* Taylor & Francis: London.

Vigilant, L., Pennington, R., Harpending H., Kocher T. D. & A. C. Wilson. 1989. Mitochondrial DNA sequences in single hairs from a southern African population. *Proceedings National Academy Sciences USA* 86:9350–9354.

MORPHOLOGY AND DEVELOPMENT IN EVOLUTION

What Functional Morphology Cannot Explain: A Model of Sea Urchin Growth and a Discussion of the Role of Morphogenetic Explanations in Evolutionary Biology

Clifford Baron

Abstract. Embedded in the perspective of evolutionary functional morphology is a teleological viewpoint similar in kind to the final causation invoked by Georges Cuvier. The difference is that modern functional morphology has replaced final causation with natural selection, the assumption being that the explanatory territory of natural selection is similar to that covered by final causation. Unpacking the logic of natural selection reveals that this is not the case: while final causation explains the *generation* of form (though in a metaphysical rather than a material sense), natural selection only explains why particular organisms, *once present in a population,* predominate in a population over time. There is, therefore, no causal explanation for the structures of organisms in a theory of evolution whose entire explanatory power is based on natural selection.

The *morphogenetic approach* to the study of the evolution of organismal form fills the explanatory void vacated by final causation. To illustrate the application and explanatory content of the approach, I present here the results of a morphogenetic model of the skeletal morphology of sea urchins. The model shows how specific features of morphology may be explained outside the context of functional morphology and its implicit populational approach. By extension, I argue that, *in principle,* the theory of natural selection is without explanatory value in causal explanations of organismal form. Thus, there are two distinct provinces within the bounds of evolutionary morphology, based on the fact that *natural selection alters the organismal composition of populations and thereby acts in the realization of particular lineages, while morphogenesis acts in the realization of organisms and their structures.* The implications of this conclusion have had little impact on evolutionary explanations of form, due largely to the ability of the "populational perspective" to set the terms of the debate. I conclude that any meaningful reconciliation of populational (functional) and organismal (morphogenetic) perspectives must respect the unique explanatory territory of each.

INTRODUCTION

A principal goal of this symposium is to define the limits of functional morphology in evolutionary explanations of form. This statement of purpose, however, is deceptively concise. The phrase "evolutionary explanations of form" is loaded with ambiguity and controversial assumptions. When we say "evolutionary," what, exactly, is evolving? Are we referring to the series of transformations from ancestors to descendants in particular lineages, or are we speaking of the processes by which particular lineages and not others have been realized (i.e., have survived)? I will argue that *form* has different implications in

Dr. Baron is with the Department of Integrative Biology, University of California, Berkeley, CA 94720, USA.

these two senses of evolution, on the one hand referring to the structural features of individual organisms, and on the other referring to the statistical composition of populations. I contend that this distinction accounts in large part for a long and persistent division in evolutionary biology that is rooted in the failure of evolutionists to recognize the limits of "functional explanations," broadly construed. As these limits proscribe two different realms of evolutionary inquiry, what I call the populational and the organismal, these approaches must specify clearly their aims and emphases. I present here results of a study that in broad outline characterizes the organismal approach to the study of the evolution of morphological features, an approach I call "morphogenetic."

TELEOLOGY, FINAL CAUSES, AND POPULATIONAL THINKING

While Aristotle codified the teleological premises of causation in western thought, it was Georges Cuvier, following Kant's lead, who fully integrated *final causation* into the study of organismal morphology (Russell, 1916). Cuvier firmly believed that function *causes* structure, in the sense that the functional efficiency and harmony of organic structures are the manifestations of the intent of the Creator (Appel, 1987). Cuvier, in fact, went beyond Aristotle in insisting on the exclusivity of final causes. As Cuvier's critiques of studies of teratology reveal, the formulation of theories for the efficient cause of organismal structure were not a part of the French anatomist's thinking, for in his view there were no causes besides the teleological (Appel, 1987). Indeed, Cuvier avoided the issue of the genesis of structures at some cost to his professional reputation; he clung to the doctrine of preformation (which denied that structures form anew with each generation) for some time after the evidence for epigenetic development was irrefutable (Appel, 1987). Final causation, as an explanator of the *generation of organismal structure*, was the ground on which Cuvier's "principles of design" were founded, and it was, in this view, the morphologist's task to discover these principles in the organisms studied.

Though its demise was slow to reach biology, final causation was one of the major casualties of the Enlightenment, having been replaced by Newtonian materialism (Smart, 1963). Following this philosophical reformation of the sciences, Darwinism argued that the design-like quality of organisms is *not inconsistent* with a materialist philosophy of nature, i.e., that final causes are *not necessary* to explain the adaptedness of structures. Thus, natural selection replaced (or attempted to replace) final causes as the ground on which the teleological (or "teleonomic" *sensu* Mayr, 1974) principles of design could still be used to interpret organic structure. Therefore, far from burying Cuvier, Darwinism resuscitated Cuverian functional analysis, replacing Cuvier's problematic belief in final causes with materialistic natural selection; in this sense, natural selection reformed teleology (Saunders, 1989; O'Grady, 1984). At this point the question naturally arises, does the theory of natural selection really provide an adequate foundation for a *causal* analysis of form or are our teleonomic explanations lacking in some fundamental ways that Cuvier's analyses, in their appeal to final causes, were not?

Final causation, which explains originations as the result of purposeful design, is surely not consciously subscribed to by many biologists. Biologists do, however, frequently make statements like the following, which is difficult to defend as a causal explanation of form without appealing to final causes: "Birds have wings *because* in populations ancestral to birds those individuals having wings were favored by natural selection. Subsequent generations retained wings through inheritance and the advantages these structures conferred (whether by virtue of their capacity for flight, thermoregulation, prey capture, whatever) assured that individuals having them would be favored by natural selection." This is a fairly careful wording of what most functional morphologists would say is a statement explaining why birds have wings. Two parts of this statement, however, render it inadequate (indeed, wrong) as a causal explanation for birds having wings. The first is its reference to specific structures, i.e., "wings." If we take the term "because" to mean "accounting for the particular structural attributes that, in association, are termed 'bird wings,'" then the statement is clearly false. Unless we adopt final causation or a somewhat radical Lamarckism, natural selection cannot account for these particular entities we call "bird wings" at all, because it plays no causal role in the appearance of structures (wings are already assumed to be present in the ancestral population). Nor, for that matter, does natural selection explain the repeated generation of forms with each generation (i.e., heritability, *sensu* Wagner, 1989; see also Endler, 1986). As the rest of the functional explanation makes clear, the best we can expect of natural selection is a bias in the preservation of individuals that *have* wings.

The second flaw in this functional explanation is its reference to a specific lineage of organisms, i.e., birds. By referring to a monophyletic lineage of organisms we are referring to an individual entity, unique by virtue of a singular sequence of ancestor-descendant relationships. There is, however, an implicit and problematic assumption in the question, "Why do birds have wings?," namely that organisms composing this same bird lineage could be wingless. Given the criteria of individuality (particularly, that individuals have unique histories) (Ghiselin, 1974), this cannot be the case. Had different organisms (e.g., those without wings) been favored in populations ancestral to birds, the lineage that we call "birds" would not have been realized, regardless of whatever birdlike features might have been realized in this alternate lineage. This is not to say that wings *define* the bird lineage; it is to say that an alternate lineage, *without wings*, is not the same lineage as that other possible lineage with wings (Aves). Thus, the explanatory statements quoted above implicitly assume a false premise, i.e., that another lineage (in which, *incidentally*, the composing organisms do not have wings) would still be the lineage we call Aves. Therefore, had a lineage of wingless organisms arisen in the populations that were immediately ancestral to birds, it would not, by definition, be the bird lineage. Thus, it is oxymoronic to refer to a lineage of organisms without wings as birds, unless the lineage arose within the established bird lineage.

The question, "Why do birds have wings?," can be revised in two ways to avoid this oxymoronic implication. First, the question may be restated as a question about the "choice" by natural selection of the bird lineage among those lineages possible in populations ancestral to birds. Specifically, this revision inquires, "Why did those individuals having wings in populations ancestral to birds predominate over other individuals without wings?" Significantly, this is no longer a question requiring an explanation for a structure; this is a question about the *effects* of a structure (wings) on the organismal composition of a population and by extension the subsequent realization of a lineage composed of organisms having this structure. Thus, the object to be explained has shifted from organismal *structures* to the *composition of populations*.

The second reinterpretation of the question "Why do birds have wings?," may be restated as a question about the *retention* of wings in descendent populations of birds. To even consider a functional explanation for this question, one must first assume that it is morphogenetically possible for birds to "lose" wings. Assuming this *is* possible, the original functional explanation outlined above must appeal to differential survival and reproduction of individuals with and without wings. Once again, this is obviously an explanation for the realization of particular *lineages* and the consequent absence of others. This explanation does not, nor can it be expected to, address the issue of why the organisms in particular lineages *have* specific structures. Therefore, the proper sequence of causal connections is the following: "Morphogenesis generates structures which are instrumental in the performance of particular functions. Differences in the performance of these functions among individuals *cause* biases in survival and reproduction which determine the organismal composition of populations and changes in the composition of these populations over time." The two revisions of the original question outlined here make clear that *it is morphogenesis that causes structures, structures that cause differences in performance, and differences in performance that cause biases in survival and reproduction.*

In light of these considerations, I argue that evolutionary explanations of form should address two distinct questions. The first question, and that which functional explanations at least partly address (but only in the best of cases), is "Of the possible lineages emanating from populations ancestral to a given lineage, why was a particular lineage (e.g., Aves) realized?" The second question is the domain of morphogenetic explanations: "Why do the organisms composing a given lineage (e.g., Aves) have specific structures (e.g., wings)?" The separation of these two questions is to make clear *that natural selection acts in the realization of particular lineages; morphogenesis acts in the realization of organisms and their structures.*

Philosophers of science have critiqued the role of functional or teleological explanations of biological structures, yet these conclusions have made little impact on how biologists construct their evolutionary explanations of form. As Nagel (1979) comments, though there are no formal criteria for the recognition of teleological statements, there is little disagreement as to whether a given statement is teleological. Loosely delimited, teleological statements directly or implicitly attribute the presence of an object to the functional effects of the object. Hempel (1965) objects strongly to the implied *intentionality* of such statements as well as to the circularity of the relationship between the explanation and the thing explained. He demands rigid adherence to the "nomological-deductive" (i.e., covering-law) model of explanation and finds no room for teleological explanations of biological structures. Nagel's analysis (1979), while less sweeping in its condemnation of functional explanations, also objects to their use in *causal* explanations of form (i.e., causal explanations for the presence of particular structures in particular organisms), but allows a positive role for teleological explanations for the examination of the *effects* of structures on organisms bearing them. Ruse (1973) and Sober (1984), on the other hand, apply a causal analysis of functional (i.e., teleological) statements only to explanations of the composition of *populations* of organisms. Sober, in particular, specifically emphasizes that functional explanations interpreted through the action of natural selection are implicitly statements of populational, not organismal or intergenerational (in his words, "developmental") change. Sober notes that "the theory of natural selection created a new object of explanation by placing the population fact in a new contrastive context" (Sober, 1984, p. 150), thereby echoing Mayr's (1982) views about the breakthrough of the "populational way of thinking." But while Mayr's view emphasizes the importance to Darwinian thought of subjugating processes that account for the transformation of organisms to the "ultimate causation" of populational transformation, Sober maintains that populational and organismal transformations have distinct explanations. It is important to note here that underlying Mayr's formulation of "ultimate causation" is an implicit shift from a statement of *cause* to a statement of *ultimate ends.* As O'Grady (1984) has pointed out, this is manifest in explanatory statements in which "because" is replaced by phrases such as "so that" or "in order to", which replacement clearly exposes the teleological intentions of the author. Such confusion of cause with ultimate ends is clearly abetted by the conflation of organismal and populational levels of explanation. Therefore, although there are strong philosophical objections to the use of functional explanations in *causal* explanations for specific morphological features, there are compelling justifications for their use in explanations of the organismal composition of populations.

It is the thesis of this paper that the explanatory limits of evolutionary functional morphology must be observed and respected by morphologists, particularly the assertion that functional statements cannot explain (in a causal sense) the particularities of organismal form. The explanation of particularities of form demands a return to a *causal science of morphology,* or morphogenesis, which is similar in many respects to some versions of biological structuralism (e.g., Goodwin 1982, 1989, but not to the strictly formal approach of Saunders, 1989). The approach is rooted in Goethe's description of morphology as "the science which should discover the basis of organic *bildung* (structure)" (Russell, 1916, p. 50), and is a descendant of Roux's *Entwicklungs-mechanik* (developmental mechanics). Its present practitioners are mostly developmental biologists whose concerns, unfortunately, do not generally extend to post-embryonic morphology or to evolutionary biology (e.g., Mittenthal & Jacobson (1990) and references therein). Increased computational and experimental sophistication has allowed these workers to study morphogenetic processes that were formerly intractable. Evolutionary morphology, so long dominated by the paradigm of natural selection, is in an historically unique position to exploit the morphogenetic approach and to benefit from morphogenetics' unique insights into the transformation of organic structures.

THE MORPHOGENETIC APPROACH DESCRIBED

The morphogenetic approach attempts to articulate in explicitly mechanistic terms those processes that determine organic form. Though superficially similar to the "theoretical morphology" of Raup and Michelson (1965) in its focus and jargon (e.g., both are concerned with the parameterization of morphology), morphogenetics is not principally concerned with formal descriptions of structure. Thus, where theoretical morphology describes "morphospace" in terms of those parameters that are necessary for a *mathematical* description of form, morphogenetics defines morphospace in terms of parameters that play a role in morphogenetic processes. In fact, there is no necessary (or even likely) reason that the two approaches should at all overlap in their description of a particular system. For this reason, the distinction between the two approaches must be kept in mind. In particular, it is "morphogenetic space" that defines those forms that are *possible* for a given developmental system; the mathematical morphospace described by Raup and Michelson (1965), has an undefined relation to mechanism and cannot, therefore, be said to represent a range of *possible* forms, as some have suggested (e.g., Gould, 1984). In the common reliance on arbitrary parameters in the description of form, Raup's approach is more similar to statistical descriptions of form that compose the field of morphometrics (e.g., Bookstein et al., 1985).

In practice, however, there is some ambiguity in the designation of particular theories as either "morphogenetic" or "morphometric." For instance, one may regard D'Arcy Thompson's use of transformation grids purely as a geometrical tool for the description of differences in form (e.g., Bookstein et al., 1985), but Thompson himself clearly believed that these deformed grids were indicative of biomechanical (albeit insufficiently articulated) causes of changes in shape (Thompson, 1942). This ambiguity between formal description and causal mechanism is common among mathematical theories of pattern formation (Turing, 1952; Thom, 1975; French et al., 1976; Murray, 1977, 1981; Meinhardt, 1982; Farmer et al., 1984; Frankel, 1989) in which model parameters may have either dubious or multifarious interpretations as biological entities. When this is the case, it is usually because the theory is insufficiently articulated in terms of *efficient causes* (in the Aristotelian sense) of structure, and thus depends heavily on formal description. A sufficiently developed morphogenetic theory must articulate efficient causes. That is to say, that to be useful as an explanation for specific organic structures, a morphogenetic theory must specify those *biological* entities that interact, and the manner in which they do so to generate structure (e.g., Briére & Goodwin, 1988; Oster et al., 1983; Oster et al., 1988). To the extent that pattern formation models do this successfully, they are morphogenetic theories.

A MORPHOGENETIC MODEL OF REGULAR ECHINOIDS

I present here results of a morphogenetic model to account for the pattern of growth and shape change in adult sea urchins (Echinodermata: Echinoidea). Perhaps owing to their radial symmetry and the geometric intricacies of their plated construction, the familiar "regular" sea urchins (comprising the Echinacea, the Diadematacea and the Cidaroidea) have long been subjects of morphogenetic speculation. The globular form of regular sea urchins suggested to D'Arcy Thompson (1942) that these animals assume the form of sessile drops of fluid under the influence of gravity. Unfortunately, Thompson devoted more discussion to the similarity of appearance between drops of fluid and sea urchins than he did to specifying the *cause* of the similarity. He did, however, suggest that interspecific differences in form may be due to properties of the skeleton that are analogous to surface tensions that characterize different fluids. Subsequently, several workers have developed the analogy of the echinoid "test" as a "mineralized pneu" (that is, a mineralized vessel under internal pressure) (Seilacher, 1979) whose component plates grow in response to biomechanically-induced expansion. Moss and Meahan (1968) hypothesized that the echinoid skeleton grows as do the bones of the mammalian cranial vault, i.e., under the influence of positive internal pressure from an expanding internal volume. Extending this hypothesis, Seilacher (1979) used the echinoid test to illustrate his concepts of "constructional morphology," hypothesizing that many anatomical features of the echinoid skeleton are simple "epiphenomena" of the "pneu" growth mechanism. Most recently, Dafni (1986) offered some experimental evidence supporting the "pneu hypothesis" and has speculated on an array of forces to which the skeleton may be exposed and their possible manifestations in morphology.

In spite of the attention the pneu hypothesis has attracted, there has been no predictive model for the translation of forces over the test (skeleton) into a pattern of growth. To

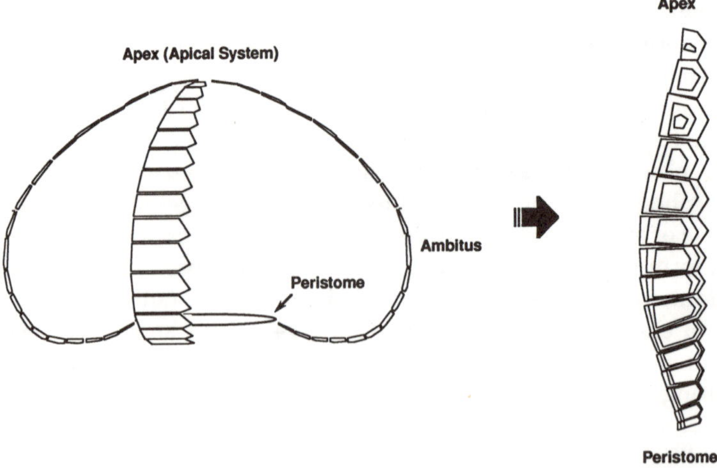

Figure 1. Structure and Growth of the Echinoid Skeleton: This figure shows the general structure and pattern of growth of the skeletons of regular echinoids. The general form in profile (left) is of an oblate spheroid, open at the bottom (i.e., at the peristome) where the test curves slightly upward. The skeleton is a mosaic of plates arranged in 20 meridional columns (only one column pictured at left) radially positioned around the central axis. Plates are ± planar and are added sequentially with growth at the apical system. Though plates nominally thicken with age, they principally grow around their margins (i.e., at their sutures with other plates). Using any of several techniques, "growth rings" can be resolved from individual plates, thus allowing reconstruction of the animal's ontogenetic growth pattern (right). Note that plates closest to the apex grow faster than those closer to the peristome.

explicitly evaluate the implications of the pneu theory, I developed a model for the growth of the echinoid skeleton that depends upon the distribution of mechanical stresses over the skeleton (Baron, in prep). The model explicitly assumes that the deposition of calcite by the skeletogenic cells of echinoids is stimulated either by stresses or, more likely, the strains that stresses induce. (Presumably, such sensitivity to mechanical stimuli may be similar in kind to that of vertebrate osteoblasts. This premise, presently untested, should be amenable to testing by culture of primary mesenchyme *in vitro*.) The structure of the model is diagrammed and explained further in Figure 2.

The model's predictions of the pattern of growth over the skeleton depend on the geometric shape of the skeleton, the mechanical properties of the materials composing the skeleton, and the forces applied to the skeleton. The model also incorporates factors that are both intrinsic and extrinsic to the organism. In particular, the forces modelled include a positive internal pressure, the forces exerted by the locomotory and adhesive activity of the tube feet, and forces exerted by the jaw apparatus. All three of these are intrinsic to the normal activities of sea urchins, but each is likely to be strongly influenced by conditions external to the organism. For instance, the effect on an urchin of living in an area where there are high water velocities is that the animal must cling more tenaciously to the substratum to avoid dislodgement than does an animal living in calmer waters. Thus the forces translated by the tube feet to the skeleton are higher for an animal living in an area of high water velocity (see Results), a condition over which the animal has no direct control. One cannot construct a biologically realistic mechanical model of this system without ignoring what are, in this case, artificial distinctions between the animal and environmental factors that impinge upon it.

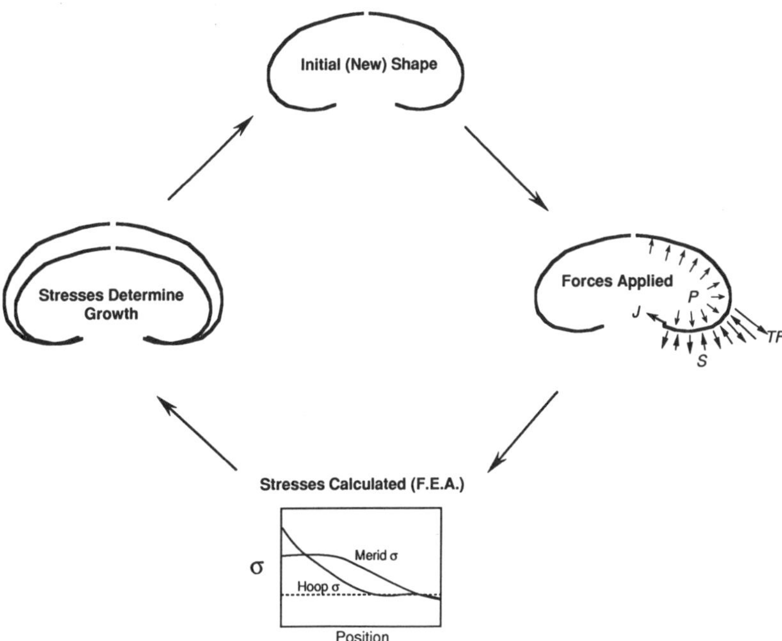

Figure 2. Iterative Structure of the Growth Model: The growth model is based on repeated cycles of loading a geometrical model of the skeleton with a specified array of forces, resolving the stresses over the skeleton using finite element analysis, and then applying these stresses to a growth algorithm to generate a new shape. Forces applied in the model include a positive internal pressure, forces exerted by tube feet and spines, and forces exerted on the peristome by the jaw musculature. The growth algorithm assumes expansion of the skeleton (along the margins of the plates composing the skeleton) where stresses are positive (i.e., tensile). Details of the model in prep.

478

RESULTS OF THE MODEL

The model, if confirmed by experiment, explains in mechanistic terms the generation and transformation of shape in echinoids and suggests how various heritable genetic characteristics and environmental influences may mediate growth. The principal object of explanation in this case is the pattern of skeletal growth through ontogeny, which may be general for a wide range of regular echinoid taxa. In general, the growth of plates composing the skeletons of regular echinoids is highest for those plates on the apical (upper) surface of the test and declines towards the peristome (see Figs. 1 & 3) (Pearse & Pearse, 1975; Märkel, 1976, 1981; Deutler, 1926; Dafni & Erez, 1982). Raup (1968) took a formal approach to the analysis of the echinoid growth gradient and showed that a logistic curve could describe meridional growth of plates (his analysis did not include hoop growth). Presented in Figure 3 are Raup's logistic curve together with measurements of growth from *Strongylocentrotus purpuratus* and the pattern of growth predicted by the present model. With the exception of hoop growth on the oral (bottom) side of the test, there is close correspondence between simulated and actual patterns.

As one would expect from a growth model that depends on the distribution of stresses around the test, manipulating the magnitudes of forces applied in the model has a strong effect on the predicted growth response. In living urchins, the tube feet adhere to the substratum and thus apply tensile pulling forces to the oral surface of the test. By increasing the magnitude of such forces in the model, the effects of increased exposure to fluid drag are mimicked by the simulations. The results of such simulations (Fig. 4) are consistent with observations of intraspecific variation in at least four different families: individuals in

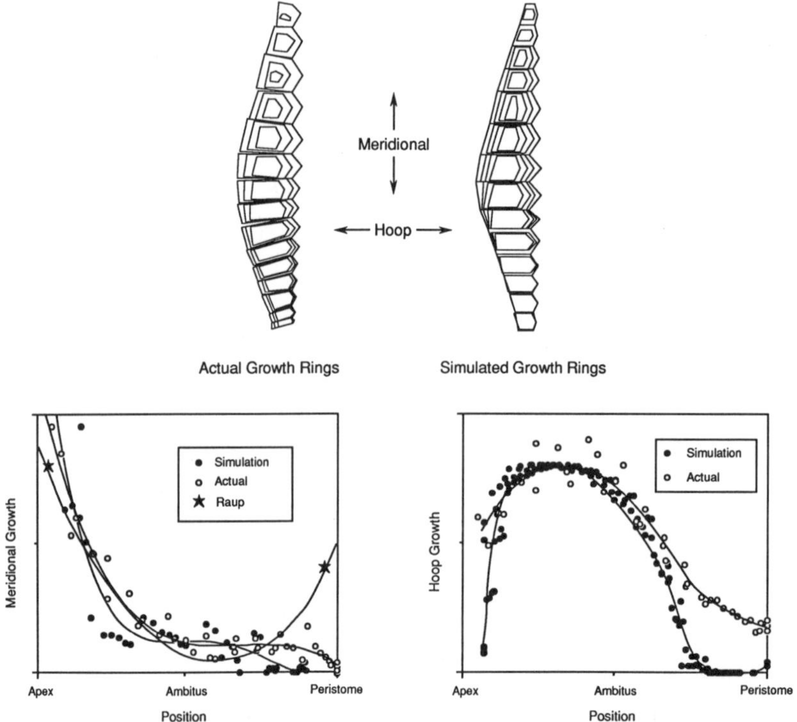

Figure 3. Simulated vs. Actual Growth Results: Actual growth rings in a single meridional column of plates from *Strongylocentrotus purpuratus* are compared with growth rings simulated by the mechanical model (top). At bottom, meridional (left) and hoop (right) components of growth are graphed by position. Included in the graph of meridional growth is a reproduction of Raup's logistic curve for comparison (see text).

more exposed environments tend to have lower height/diameter ratios (i.e., are "flatter") than those in less exposed environments (Lewis & Storey, 1984; McPherson, 1965; Moore, 1935; Nichols, 1982; personal observation of *S. purpuratus*).

The model also addresses a question of particular concern to analyses of interspecific differences in form: what heritable structural features play a role in the morphogenetic process and what is their effect on form? Among echinoid taxa there are heritable differences in the number of plates. The model predicts that with each iteration, the amount of growth along an entire meridional column will depend upon the number of plates in that column. If there are more plates in a meridional column, then there are more sutures, which are where growth occurs (Fig. 5a). Since an increase in height of the skeleton is due primarily to meridional growth, increasing the number of "plates" in the model increases the height (relative to diameter) of the model form that results (Fig. 5b). The positive allometries of height/diameter (i.e., urchins assume a more globose shape (higher H/D) as they grow and add plates) in most species of regular echinoids (Telford, 1985; Ebert, 1988) support this prediction, as does the observation that taxa with many meridional plates tend to have high height/diameter shapes, regardless of size (Baron, in prep.). A parallel effect should result from alteration of the number of meridional columns composing the skeleton, thus altering the number of hoop growth sutures. Normally, there is no variation in this number among extant echinoids (it is always 20), but teratological specimens having either fewer or more meridional columns do occur naturally (Chadwick, 1924; Dafni, 1980; Koehler, 1924) and may appear in culture (Hinegardner, 1975), possibly allowing evaluation of this prediction. Specifically, the model predicts those individuals having fewer than 20 meridional columns of plates to be more globose than those having 20 or more columns.

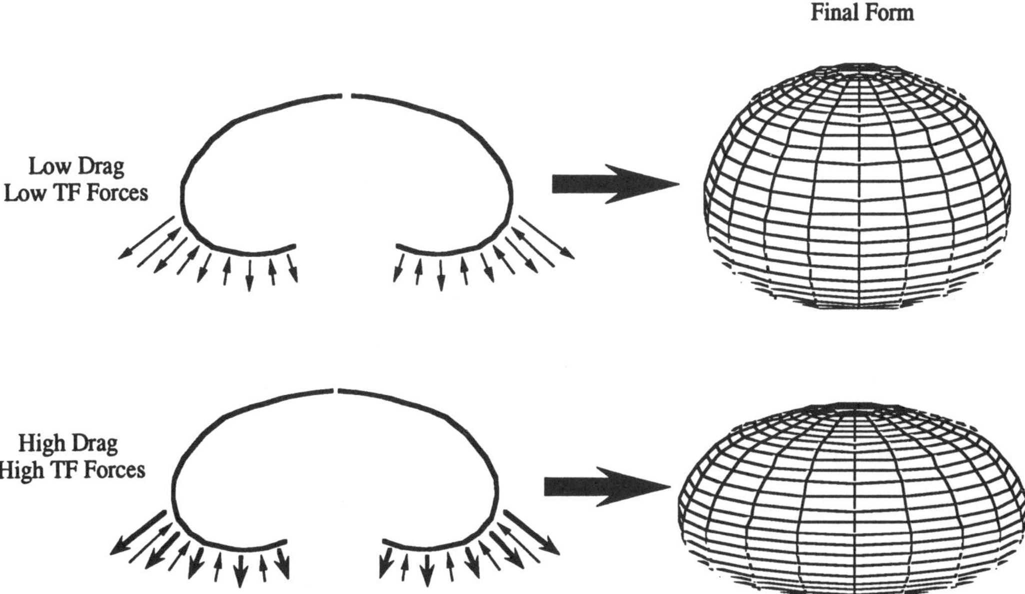

Figure 4. Effects of Exposure on Form: Diagrammed above are the effects on skeletal form of variation in the magnitude of tube foot loads in the simulations. Simulations run with high tube foot loads (bottom) mimic conditions experienced by urchins in more exposed habitats (i.e., high drag conditions; see text). The final shapes of such models have lower height/diameter ratios than those simulated with low tube foot loads (low drag conditions). High and low drag final forms have comparable volumes.

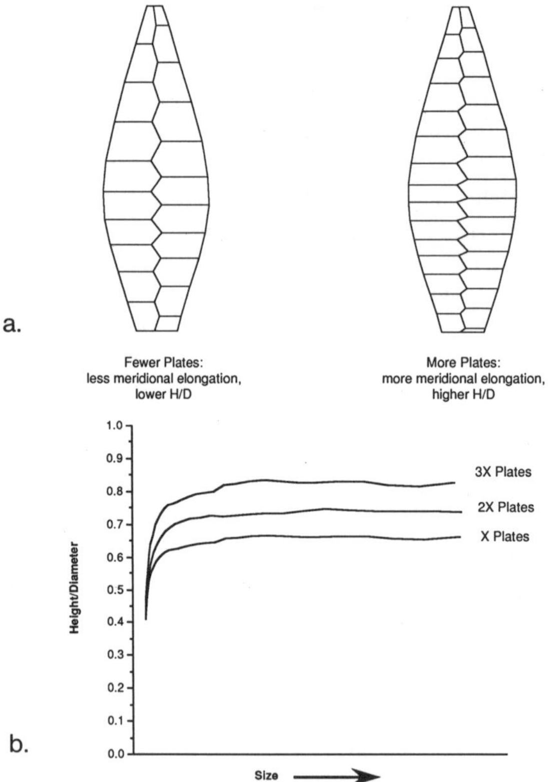

Figure 5. Effect on Shape of Number of Plates: a. Illustrates two double meridional columns of plates of equivalent length: the double column on the right has 1.5× the number of plates as the one on the left. Owing to its greater number of meridional sutures, a model having more plates will grow more in the meridional direction with each iteration of the simulation and will therefore achieve a higher H/D ratio. b. Shows graphically the effects on simulated shape change of varying numbers of plates. Pictured are three simulations in which the number of plates varies threefold. Size is a measure of volume. Each of the three simulations appears to converge on a particular H/D value with growth, with this "terminal H/D" value greater for simulations with greater numbers of plates.

THE MORPHOGENETIC APPROACH TO THE EVOLUTION OF FORM IN SEA URCHINS

This morphogenetic model of sea urchin growth demonstrates how such models may explain particular morphological features and their transformation. Assuming the proposed sensitivity of skeletogenic cells to stress (or strain), the model may explain the characteristic growth gradient of regular echinoids, its alteration in response to particular environmental influences, and the effects on growth of specific heritable morphological features represented by specific controlling parameters in the model. Many other influences on growth (e.g., nonlinearities in the response of cells to stresses, resorption of skeletal material, physiological or behavioral fluctuations of internal pressure) are not addressed in the examples presented here, but are readily accommodated by the model (Baron, in prep.). Other aspects of echinoid morphogenesis, such as the initial condensation of plates in the larval rudiment and processes that control the rate of plate addition at the top of meridional columns, are not addressed by the model. A complete morphogenetic explanation would include accounts of these and many other phenomena as well. Still, in its account of post-embryonic growth and shape change, the model is internally consistent and predictive. Most important, the model is testable since it predicts specific changes in growth pattern for varying arrays of applied forces. Though without experimental evaluation of its predictions the model cannot be taken as a true representation of how growth occurs in echinoids, it is consistent with our present knowledge of growth and shape change and it can be evaluated by specific experiments.

In light of this model, the question arises, "In what sense is such a morphogenetic model a part of evolutionary explanation as opposed to a purely developmental explanation of form?" As mentioned earlier, the morphogenetic approach to evolutionary explanations of structure seeks to explain not the persistence of preservation of organisms having particular morphological features, but the *existence* and *transformation* of these structures through time (i.e., through the history of a lineage). Thus, for this particular model to be successful as an evolutionary explanation, it must succeed in accounting for the diversity of forms within the Echinoidea and their transformation from ancestors to descendants. To the extent the model is consistent with not only experimental evaluation on extant taxa but also with observations of the morphology of extinct taxa, it can be used to explain the evolution of the forms of echinoids. (It should be noted here that there has been no attempt as yet to apply the model to the irregular echinoids, but with sufficient elaboration of the model to include the many apomorphies of this group, application is not unfeasible.) Even at the "population level of explanation," such morphogenetic explanations for the particularities of form in a group may be used in analyses of selection: the morphogenetic account explains the variation on which natural selection operates, and potentially allows "reconstruction" of those structural variants among which natural selection has "chosen." It must be stressed, however, that such linkage between populational and organismal levels is not *necessary* to an internally consistent evaluation of the effects of natural selection on populations. The populational approach can, as it has historically, simply "take as given" those variants that are present in a population.

THE DOMAINS OF FUNCTIONAL (POPULATIONAL)
AND MORPHOGENETIC (ORGANISMAL)
APPROACHES IN EVOLUTIONARY MORPHOLOGY

While the view that natural selection is not an explanator of morphology may not be shared by the participants in this symposium (see Discussion Group Comments, this volume), unpacking the logic of natural selection reveals the limits of its application to evolutionary explanations of form (e.g., Endler, 1986; Sober, 1984; Ruse, 1973). These limits are directly attributable to the fact that natural selection *assumes* the appearance of variant morphologies in populations, without offering any *explanation* for either these originations or for structures as biological entities. A tacit belief in final causation (or Lamarckism) is the precarious position of the functional morphologist who denies the importance of morphogenetic processes to his own evolutionary explanations of organismal structure. Final causation rushes in to fill the void of accounting for structure that is unfilled by natural selection, just as it did for Cuvier and his functional explanations. (Those who argue that variation is explained by mutation and recombination implicitly assume that we understand development far better than we do. Genes are no more an explanation for morphological features than are pulses of heat that may elicit phenocopies of particular mutations (see also Horder (1989)). Most emphatically, morphogenesis is *epigenetic*.)

Natural selection undoubtedly acts to alter the frequency of various morphologies in populations, but it should never be overlooked that natural selection explains changes in populations, not morphologies. Differential survival and reproduction of individuals within populations affects the lineal trajectory of a population through time, i.e., that population's *particular* sequence of ancestor-descendant relationships. Had different individuals survived and reproduced at some previous point in the history of a population, then the resulting lineage *would have a different identity from that lineage that actually exists.* As noted earlier, this conclusion stems from the criteria of individuality, among which is the assertion that individuals have unique histories. The point here is that with each generation there are as many potential lineal pathways for a population as there are possible breeding combinations of individuals, but few, if any, of these are realized. Thus when we speak of natural selection "transforming" a lineage (i.e., a sequence of ancestor-descendant relationships), we are using a sort of shorthand for what we know to be the actual effects of natural selection. A lineage is not "transformed" by natural selection, it is "chosen" among many possible alternate lineages by probabilistic natural selection. Thus, the agenda of the Darwinian program is to explain why, of the multitudes of possible (as distinguished from conceivable) lineages in the history of life, only particular lineages among these have survived.

This, however, cannot be said to constitute the agenda of all of evolutionary biology. In particular, in the history of any particular lineage (whether or not the component organisms have been favored by natural selection) there is demonstrable change in form from ancestors to descendants. Natural selection may be used to explain the relative "success" or persistence of organisms having particular characteristics, but it is incapable of explaining either these characteristics or their transformation. This latter task, when focused on the structure of organisms, has been the historical province of morphology from the time of German *Naturphilosophie.* Specifically, an organismal-based study of evolutionary morphology has as its goal the explanation of the *particularities* of organismal structure, rather than the *effects* of these structures on the composition of populations, which is the province of evolutionary functional morphology.

Many functional morphologists will certainly object to the inclusion of their discipline within population biology. As an "organism-based" discipline, functional morphology is the study of how organismal structures work, i.e., how structures are implemented in the performance of particular activities, and thus how structures affect performance. The units of study are typically individual organisms and although conclusions are often based on analyses of numerous individuals, the aim, usually, is to characterize those factors that affect the performance of individuals, not populations.

Notwithstanding this strong organismal basis in the study of animal function, once the conclusions of a functional study are used to make inferences about evolution through natural selection, the functional morphologist is in the realm of population biology. There are two principal reasons for this leap, both of which result from the causal connection between performance and selection. The first and most obvious reason is that the functional morphologist must assume that his study subjects are *representative* of the range of variation within the population (or species) of interest. Without this assumption there is no reason to believe that the functional morphologist's measures of performance can be related to natural selection.

The second and more subtle point is that natural selection, unlike performance, is a statistical phenomenon of populations. For example, while increasing the thickness of, say, the tibia confers greater stiffness and strength on that bone in *any* individual, the potentially advantageous effects of such a change on survival and reproduction do not accrue to every individual having thick tibia. That is to say that sometimes relatively *unfit* individuals survive and reproduce, due to the probabilistic nature of natural selection (*Fitness* here refers to some composite measure of performance; fitness as a measure of survival and reproduction often leads to tautologous formulations of cause and effect in natural selection). Thus, when the functional morphologist invokes the superior performance conferred by a particular morphological feature to explain the evolutionary "success" of organisms having of this structure, he is implicitly making a probabilistic argument about those factors that contribute to the composition of populations.

It follows, for instance, that if we could replay the history of a particular population, different individuals would survive with each generation than had survived the first time around, although the survivors of the two "histories" would *likely* have similar features. Simply because different individuals survive and reproduce when history is replayed, however, does not invalidate the conclusions of the functional morphologist. This is because his evolutionary inferences are about the probable composition of *populations;* the survival and reproduction of *particular* individuals is irrelevant. Thus the evolutionary functional morphologist in his appeal to natural selection is aligned firmly in philosophy, if not method, with the populational approach rather than the organismal. The first seeks to explain the composition of populations and the second seeks to explain the structures of organisms. Any assertion that one approach explains the domain of the other should be regarded skeptically. Unfortunately, the assertion that natural selection explains the particularities of organismal form has not enjoyed such circumspect evaluation, at least from biologists.

In a causal account of the evolution of organismal structures, the direction of cause is critical; as many have pointed out, if we are to avoid tautologous or causally inert formulations of natural selection, it must be that morphologies *cause* differences in survival and reproduction (i.e., fitness) (Endler, 1986; Sober, 1984). Thus, it is morphology that is a *cause* of natural selection, not the converse, which is how Darwinian selection is commonly invoked to "explain" structures (e.g., the common locution that natural selection "builds" structures (Dobzhansky, 1955; Mayr, 1963; Stebbins & Ayala, 1981; Gould & Vrba, 1982)). Those who would argue that selection does indeed cause changes in morphology are either 1) misconstruing organismal structures as characteristics of populations, or are, 2) as noted earlier, Lamarckians. This argument does not lead to the exclusion of natural selection from one's world view as some strict structuralists would have it (e.g., Webster, 1989). Instead, it asserts that the domain of natural selection cannot be extended to account for what is properly in the domain of morphogenetic explanation, nor can the morphogenetic approach explain the organismal composition of populations over time.

Viewed in this way, the touted division of evolutionary morphology into "internalist" and "externalist" contingents (e.g., Maynard-Smith et al., 1985; Wake & Roth, 1989) is ill-conceived and subtly deceptive. The real division is between those who wish to explain the features of organisms and those whose focus is the composition of populations. The internalist/externalist dichotomy, whose origin is really from the populational perspective (see below), cannot be maintained.

THE PERVASIVENESS OF THE POPULATIONAL PERSPECTIVE: ITS INFLUENCE ON OUR PERCEPTIONS OF "CAUSATION" AND ON THE CONTRIBUTIONS OF THE ORGANISMAL APPROACH

Given the division in evolutionary biology into populational and organismal approaches, how is it that the populational view has successfully annexed the domain of explanation of the features of organisms? The answer appears to lie in the vaunted hierarchy of causation that populationists invoke. Most versions of the populational perspective imply a version of "downward causation", i.e., that the processes occurring within populations (namely natural selection) are efficient causes of organismal structures. The argument for downward causation reads something like this: "Natural selection, in contributing to the realization of particular lineages, helps explain why some

484

lineages, and not others, have survived to the present. Had these lineages not survived, neither would there exist the organisms composing these lineages. Therefore, natural selection helps explain the realization of particular organisms with their attendant structures." (This is consistent with the view expressed by Vrba and Gould (1986) that selection at a given hierarchical level always exerts "downward causation" on the levels below.)

Though this line of reasoning appears sound, this appearance is aided by the omission of a careful consideration of the logical form of natural selection. Natural selection is a syllogism: its necessary propositions include the propositions that variation exists and that this variation has a *causal influence* on organisms' probabilities of survival and reproduction. That is, the direction of cause is "upward", from structure to performance to survival and reproduction to the composition of the population. Furthermore, simply because organisms occupy a "lower" hierarchical level than populations, it does not follow that selection accounts for the structures *of the organisms that exist at a given time.* This is because the only sense in which causation is "downward" is in the context of populations: natural selection *causes* biases in the *representation* of organisms, structures and genes *in the populations* so affected, but *has no causal effect on the properties of constituent organisms, structures and genes of these populations.*

The direction of cause may be illustrated by analogy to a hypothetical observation in the inorganic world, the study of which is less encumbered by teleological baggage. Suppose that geologists discover that a particular isotope of an element always predominates over an alternative isotope of the same parent element (a "populational problem" since it is concerned with frequency of representation). These scientists find that the probability of decay of the parent element into either of the two daughter isotopes is equivalent (a "morphogenetic problem"), but that the more prevalent isotope is more stable than the less prevalent form. Now, suppose these scientists wanted to explain the structure of the more stable isotope. It is very unlikely that the community of geologists would accept as explanation for this structure the observation that the isotope is stable. To the contrary, it is the structure of the isotope that accounts for its stability and thus its predominance over the less stable form (i.e., causation is "upward"). Furthermore, the predominance of one isotope over another, *regardless of the cause of this predominance,* in no way explains the structure of either one. The explanation of the structure of the isotope must be in terms that account for the transformation of this isotope from its parent (i.e., in terms of radioactivity, geochemistry and physics). I can think of no compelling reason why explanations for organic structures should not be of this type. While the "agent of selection" is clearly different for organisms and for isotopes (e.g., differential mortality or reproduction for organisms vs. variation in structural stability for isotopes) this does not justify a redirection of causes.

As this redirection of Darwinian causation illustrates, the "hardening of the synthesis" (Gould, 1982) went much further than simply developing the Darwinian perspective to its fullest logical extent. A strong case can be made that the populational view has deeply colored biologists' views of causal processes in evolution, to the extent that evolutionists freely invoke types of "causal scenarios" that would be accepted in no other science. Undoubtedly, this state is abetted by the populationist's free and uncritical use of terms such as *adaptation* and *selection,* which in some contexts may refer to *processes,* in others to *outcomes,* and in still others, to *states of being* (cf. Lambert & Hughes, 1989). Wittgenstein saw clearly how sciences, as they become entrenched in particular "ways of seeing," alter the meaning of observations they set out to explain:

Nothing is commoner than for the meaning of an expression to oscillate, for a phenomenon to be regarded sometimes as a symptom, sometimes as a criterion, of a state of affairs. And mostly in such a case the shift in meaning is not noted. In science it is usual to make phenomena that allow of exact measurement into defining criteria for an expression; and then one is inclined to think that now the proper meaning has been *found*. Innumerable confusions have arisen in this way. (Wittgenstein, 1970, Article 438)

There are many examples of subtle perversions of meaning and blurring of distinctions that have expanded the "explanatory territory" of the populational perspective (e.g., genetic determinism in "accounting" for organismal structure; the *phenomenological* and *correlational* methods and assumptions of quantitative genetics interpreted as true representations of heredity, performance and selection; ever-expanding units of selection, and their interpretation in terms of "upward" and "downward" causation). Perhaps the most insidious manifestation of the influence of the populational perspective, though, is its determination of the "relevance" of the various contributions of the organismal approach.

The population perspective has largely dictated which facets of the organismal perspective are to be taken seriously. The most informative example is the concept of "developmental constraint." Generally associated with an internalist or organismal perspective, "developmental constraint" is probably the most widely invoked organismal concept in present evolutionary discourse. Remarkably, this concept has been refashioned to fit the populational perspective: a "developmental constraint" is construed as the lack, within a population, of additive genetic variance for a particular trait (e.g., Maynard-Smith et al., 1985). This populational view of developmental constraint can be generalized as the "limits to variation" thesis, i.e., that development imposes limits on form such that particular functional optima cannot be attained. Through such remolding into the framework of the populational approach, "developmental constraint" has gained the imprimatur of populationists as an explanation of last resort for suboptimal structures, and therefore enjoys widespread attention.

Arguably, though, the *prohibition* of particular structures is but a peripheral conclusion of a thorough morphogenetic analysis, a conclusion of principal interest mainly to those concerned with natural selection and its limits (i.e., populationists, including functional morphologists who apply their inferences to evolution through natural selection). Most emphatically, *the contribution of the organismal approach is not in explaining what has been prohibited in the course of evolution, but in explaining the particular forms realized over the course of evolution.* Unfortunately, the willingness of those advocating a morphogenetic approach to allow the populational view to define their contributions has led to a trivialization of the fundamental differences, emphasized throughout this paper, that characterize these two approaches. The timid view that prevails among many morphogeneticists (and those who support their approach) is that natural selection is a creative force, but that it has limits that are proscribed by developmental constraints, among other things (e.g., Gould, 1982; Gould & Vrba, 1982; Gould & Lewontin, 1979; Wake & Larsen, 1987; Maynard-Smith et al., 1985; Alberch, 1980, 1982; Thomson, 1988). The inevitable result of such thinking is the characterization of the differences between the explanations of the populational and organismal approaches as being simply differences in emphasis: organismalists allegedly emphasize and populationists de-emphasize the importance of limits to variation, a perspective that is manifest in the internalist/externalist dichotomy (e.g., Wake & Roth, 1989). It is true that this difference in emphasis exists, *but only in the realm of populational questions.* By allowing populationists to supply the context in which the organismal approach is evaluated, many who support the study of morphogenetics have undermined the approach by not emphasizing that morphogenetics has a unique explanatory content, namely the explanation of the particularities of organismal structure.

486

CONCLUSION

In contrast to the "limits to variation" view of the relationship between selection and morphogenesis, I assert that natural selection *in principle* cannot be a creative force in the context of organisms and their structures, and that this fact accounts for the irreducible differences in the explanatory realms of the organismal and populational approaches. At issue here is not the extent to which development plays a role in morphological evolution. In a large and embracing sense, changes in development are morphological evolution; in Lambert's rephrasing of Dobzhansky's famous dictum, "Nothing in evolution makes sense except in the light of biology." (Lambert from Goodwin, 1989, p. 51). To borrow and alter Jacob's famous metaphor (Jacob, 1977), *morphogenesis* is the tinkerer in the evolutionary process and *natural selection* is the tinkerer's marketplace where goods are either sold or die on the shelf. The organismal and populational approaches do indeed explain completely different aspects of evolution. The advocates of the two approaches may coexist, not because at bottom they address similar problems which eventually may be reconciled, but because their realms of inquiry are distinct in subject, method and explanatory content. Any *rapprochement* must recognize these fundamental divisions.

ACKNOWLEDGMENTS

I am deeply indebted to the thoughts of all those who, either in discussion or in their writing, have helped clarify my thinking about the issues addressed in this paper. Thanks to those who have offered comments and suggestions on this manuscript: D. Carrier, K. DeQuieroz, H. Greene, M. A. R. Koehl, E. Lessa, and M. Wake. There remain many assertions to which these people object; those flaws that remain in logic or locution are, of course, my own. Special thanks to M. Wake for the invitation to participate in this symposium on the role of functional morphology in evolutionary biology, to D. Carrier and R. Emlet for many fruitful discussions of these issues, to K. DeQuieroz for the quotation from Wittgenstein, and to all those, who, by their reaction (both positive and negative) to the thesis of this paper, convinced me that it was worth developing.

LITERATURE CITED

Alberch, P. 1980. Ontogenesis and morphological diversification. *American Zoologist* 20:653–667.

Alberch, P. 1982. Developmental constraints in evolutionary process. Pp. 313–332. *In:* J. T. Bonner (ed.), *Evolution and Development.* Springer-Verlag: New York.

Appel, T. A. 1987. *The Cuvier-Geoffroy Debate: French Biology in the Decades Before Darwin.* Oxford University Press: New York. 305 pp.

Bookstein, F., Chernoff, B., Elder, R., Humphries, J., Smith, G. & R. Strauss. 1985. *Morphometrics in Evolutionary Biology.* The Academy of Natural Sciences of Philadelphia: Philadelphia. 277 pp.

Briére, C. & B. C. Goodwin. 1988. Geometry and dynamics of tip morphogenesis in *Acetabularia. Journal of theoretical Biology* 131:461–475.

Chadwick, H. C. 1924. On some abnormal and imperfectly developed specimens of the sea urchin *Echinus esculentus. Proceedings of the Zoological Society of London* 94:163–172.

Dafni, J. 1980. Abnormal growth patterns in the sea urchin *Tripneustes* cf. *gratilla* (L.) under pollution (Echinodermata: Echinoidea). *Journal of Experimental Marine Biology and Ecology* 47:259–279.

Dafni, J. 1986. A biomechanical model for the morphogenesis of echinoid tests. *Paleobiology* 12:143–160.

Dafni, J. & J. Erez. 1982. Differential growth in *Tripneustes gratilla* (Echinoidea). Pp. 71–75. *In:* J. M. Lawrence (ed.), *International Echinoderm Conference, Tampa Bay.* A. A. Balkema: Rotterdam.

Deutler, F. 1926. Über das Wachstum der Seeigelskeletts. *Zool. Jb. Anat.* 48:119–200.

Dobzhansky, T. 1955. *Evolution, Genetics and Man.* John Wiley and Sons: New York.

Ebert, T. A. 1988. Allometry, design and constraint of body components and of shape in sea urchins.

Journal of Natural History 22:1407–1425.

Endler, J. A. 1986. *Natural Selection in the Wild.* Princeton University Press: Princeton, NJ. 336 pp.

Farmer, D., Toffoli, T. & S. Wolfram. 1984. *Cellular Automata.* North-Holland Physics Publishing: Amsterdam.

Frankel, J. 1989. *Pattern Formation: Ciliate Studies and Models.* Oxford University Press: New York. 314 pp.

French, V., Bryant, P. J. & S. V. Bryant. 1976. Pattern regulation in epimorphic fields. *Science* 193:969–981.

Ghiselin, M. T. 1974. A radical solution to the species problem. *Systematic Zoology* 23:536–544.

Goodwin, B. C. 1982. Development and evolution. *Journal of theoretical Biology.* 97:43–55.

Goodwin, B. 1989. A structuralist research programme in developmental biology. Pp. 49–61. *In:* B. Goodwin, A. Sabatani & G. Webster (eds.), *Dynamic Structures in Biology.* Edinburgh University Press: Edinburgh, Scotland.

Gould, S. J. 1982. Darwinism and the expansion of evolutionary theory. *Science* 216:380–387.

Gould, S. J. 1984. Morphological channelling by structural constraint: convergence in styles of dwarfing and gigantism in *Cerion,* with a description of two new fossil species and a report on the discovery of the largest *Cerion. Paleobiology* 10:172–194.

Gould, S. J. & R. Lewontin. 1979. The spandrels of San Marco and the Panglossian paradigm: a critique of the adaptationist programme. *Proceedings of the Royal Society, London B* 205:581–598.

Gould, S. J. & E. S. Vrba. 1982. Exaptation: a missing term in the science of form. *Paleobiology* 8:4–15.

Hempel, C. G. 1965. *Aspects of Scientific Explanation.* The Free Press: New York.

Hinegardner, R. T. 1975. Morphology and genetics of sea urchin development. *American Zoologist* 15:679–689.

Horder, T. J. 1989. Syllabus for an embryological synthesis. Pp. 315–348. *In:* D. B. Wake & G. Roth (eds.), *Complex Organismal Functions: Integration and Evolution in Vertebrates.* John Wiley & Sons: New York.

Jacob, F. 1977. Evolution and tinkering. *Science* 196:1161–1166.

Koehler, R. 1924. Anomalies, irregularitiés et déformations au tests chez les echinids. *Annales de l'Institute Oceanographique, Monaco* Nouvelle série 1:159–480.

Lambert, D. M. & A. H. Hughes. 1989. Key words and concepts in structuralist and functionalist biology. Pp. 62–76. *In:* B. Goodwin, A. Sabatani & G. Webster (eds.). *Dynamic Structures in Biology.* Edinburgh University Press: Edinburgh, Scotland.

Lewis, J. B. & G. S. Storey. 1984. Differences in morphology and life history traits of the echinoid *Echinometra lacunter* from different habitats. *Marine Ecology Progress Series* 15:207–211.

Märkel, K. 1976. Structure and growth of the coronal skeleton in *Arbacia lixula* Linné (Echinodermata, Echinoidea). *Zoomorphologie* 84:279–299.

Märkel, K. 1981. Experimental morphology of coronal growth in regular echinoids. *Zoomorphology* 97:31–52.

Maynard-Smith, J., Burian, R., Kauffman, S., Alberch, P., Campbell, J., Goodwin, B., Lande, R., Raup, D. & L. Wolpert. 1985. Developmental constraints and evolution. *Quarterly Review of Biology* 60:265–287.

Mayr, E. 1963. *Populations, Species, and Evolution.* Harvard University Press: Cambridge, MA.

Mayr, E. 1974. Teleological and teleonomic, a new analysis. *In:* R. S. Cohen & M. W. Wartofsky (eds.), *Methodological and Historical Essays in the Natural and Social Sciences.* D. Reidel: Dordrecht-Holland/Boston.

Mayr, E. 1982. *The Growth of Biological Thought: Diversity, Evolution and Inheritance.* Belknap Press: Cambridge, MA.

McPherson, B. F. 1965. Contributions to the biology of the sea urchin *Tripneustes ventricosus. Bulletin of Marine Science* 15:228–244.

Meinhardt, H. 1982. *Models of Biological Pattern Formation.* Academic Press: London.

Mittenthal, J. E. & A. G. Jacobson. 1990. The mechanics of morphogenesis in multicellular embryos. Pp. 295–401. *In:* N. Akkas (ed.), *Biomechanics of Active Movement and Deformation of Cells.* Springer-Verlag: Berlin.

Moore, H. B. 1935. A comparison of the biology of *Echinus esculentus* in different habitats. *Journal of the Marine Biological Association, United Kingdom* 20:109–128.

Moss, M. L. & M. Meehan. 1968. Growth of the echinoid test. *Acta Anatomica* 69:409–444.

Murray, J. D. 1977. *Nonlinear Differential Equation Models in Biology.* Clarendon Press: Oxford.

Murray, J. D. 1981. On pattern formation mechanisms for lepidopteran wing patterns and mammalian coat markings. *Philosophical Transactions of the Royal Society, London B* 295:473–496.

Nagel, E. 1979. *Teleology Revisited and Other Essays in the History of Science.* Columbia University Press: New York. 352 pp.

Nichols, D. 1982. A biometrical study of populations of the European sea-urchin, *Echinus esculentus*

488

(Echinodermata: Echinoidea) from four areas of the British Isles. *Australian Museum Memoirs* 16:147–163.

O'Grady, R. T. 1984. Evolutionary theory and teleology. *Journal of theoretical Biology* 107:563–578.

Oster, G. F., Murray, J. D. & A. Harris. 1983. Mechanical aspects of mesenchymal morphogenesis. *Journal of Embryology and Experimental Morphology* 78:83–125.

Oster, G. F., Shubin, N., Murray, J. D. & P. Alberch. 1988. Evolution and morphogenetic rules: the shape of the vertebrate limb in ontogeny and phylogeny. *Evolution* 42:862–884.

Pearse, J. S. & V. B. Pearse. 1975. Growth zones in the echinoid skeleton. *American Zoologist* 15:731–753.

Raup, D. M. 1968. Theoretical morphology of echinoid growth. *Journal of Paleontology* 42:50–63.

Raup, D. M. & A. Michelson. 1965. Theoretical morphology of the coiled shell. *Science* 147:1294–1295.

Ruse, M. 1973. *The Philosophy of Biology.* Hutchinson University Library: London. 231 pp.

Russell, E. S. 1916. *Form and Function* (1982 edition). University of Chicago Press: Chicago. 383 pp.

Saunders, P. T. 1989. Mathematics, structuralism and the formal cause in biology. Pp. 107–120. *In:* B. Goodwin, A. Sabatani & G. Webster (eds.), *Dynamic Structures in Biology.* Edinburgh University Press: Edinburgh, Scotland.

Seilacher, A. 1979. Constructional morphology of sand dollars. *Paleobiology* 5:191–221.

Smart, J. J. C. 1963. *Philosophy and Scientific Realism.* Routledge & Kegan Paul: London.

Sober, E. 1984. *The Nature of Selection.* MIT Press: Cambridge, MA. 383 pp.

Stebbins, G. L. & F. J. Ayala. 1981. Is a new evolutionary synthesis necessary? *Science* 213:967–971.

Telford, M. 1985. Domes, arches and urchins: the skeletal architecture of echinoids (Echinodermata). *Zoomorphology* 105:114–124.

Thom, R. 1975. *Structural Stability and Morphogenesis.* The Benjamin/Cummings Publishing Co.: Reading, MA.

Thompson, D. W. 1942. *On Growth and Form.* Cambridge University Press: Cambridge, UK.

Thomson, K. S. 1988. *Morphogenesis and Evolution.* Oxford University Press: New York. 154 pp.

Turing, A. M. 1952. The chemical basis of morphogenesis. *Philosophical Transactions of the Royal Society, B* 237:37–72.

Vrba, E. S. & S. J. Gould. 1986. The hierarchical expansion of sorting and selection: sorting and selection cannot be equated. *Paleobiology* 12:217–228.

Wagner, G. P. 1989. The origin of morphological characters and the biological basis of heredity. *Evolution* 43:1157–1171.

Wake, D. B. & A. Larson. 1987. Multidimensional analysis of an evolving lineage. *Science* 238:42–48.

Wake, D. B. & G. Roth. 1989. Introduction. Pp. 1–5. *In:* D. B. Wake & G. Roth (eds.), *Complex Organismal Functions: Integration and Evolution in Vertebrates.* John Wiley & Sons: New York.

Webster, G. 1989. Structuralism and Darwinism: concepts for the study of form. Pp. 1–15. *In:* B. Goodwin, A. Sabatani & G. Webster (eds.), *Dynamic Structures in Biology.* Edinburgh University Press: Edinburgh, Scotland.

Wittgenstein, L. 1970. *Zettel.* G. E. M. Anscombe & G. H. von Wright (eds.), G. E. M. Anscombe, transl. University of California Press: Berkeley, CA.

Biomechanics and the Adaptive Significance of Multicellularity in Plants

Karl J. Niklas and Donald R. Kaplan

Abstract. A review of the mechanical attributes of plant cells and tissues indicates that the partitioning of the protoplast by cell walls confers structural, as well as physiologic and reproductive advantages. These attributes suggest that the compartmentalization of the protoplast by the production of cell walls or their analogs was elaborated upon during the course of plant evolution, resulting in stiffer plant bodies. This elaboration had as one of its consequences the evolutionary appearance of the multicellular plant body, which is viewed to be the result of a highly specialized mode of development in which cytokinesis and nuclear division are precisely correlated. Contra the tenets of the Cell Theory, the protoplast of the plant body is viewed as a single indivisible feature. Accordingly, notions predicated on the developmental and evolutionary corollaries of the Cell Theory are rejected. Among these are the notions that the plant body is fundamentally divided into cells of initially equal developmental and morphological rank, that morphogenesis and differentiation involve 'competitive interactions among cell lineages', and that the individuality of the organism resulted from similar competitive interactions.

INTRODUCTION

This paper is concerned with the mechanical significance of multicellularity in plants and explores selectionist arguments based on the biomechanical consequences of different patterns of development on plant evolution. These arguments are predicated on the fact that most plant cells possess a cell wall (deposited by the protoplast external to the plasma membrane) whose geometry and chemical composition dictate the mechanical attributes of cells and the texture of tissues (Esau, 1977). The rigid infrastructure of cell walls within the plant body (called the apoplast) is mechanically analogous to the skeletons found in many animals (Niklas, 1989). However, unlike metazoan evolution in which the acquisition of a skeleton followed or paralleled the specialization of cells, the evolution of the apoplast predates the occurrence of multicellularity and cellular specialization in plant lineages. While the mature cell wall is typically rigid and stiff, it is flexible and malleable when produced by actively growing cells (Taiz, 1984). Thus, mechanical differences in cell walls can be the result of developmental differences in the age of cells. Additionally, the patterns in which the protoplast orients cellulosic microfibrils within successively deposited wall layers significantly affects preferred directions of cell expansion; hence they influence the geometry of the apoplast within a tissue and the morphology of an individual cell (see Neville, 1986).

Dr. Niklas is with the Section of Plant Biology, Cornell University, Ithaca, NY, 14853, USA. Dr. Kaplan is with the Department of Plant Biology, University of California, Berkeley, CA, 94720, USA.

490

Although the geometry, location, and chemical composition of cell walls represent the most obvious physical manifestations of growth and development, they cannot be interpreted as the principal agency through which morphogenesis and histogenesis are effected. This agency must reside within the protoplast as evidenced by (1) the production of the apoplast by the living protoplast, (2) the lack of correlation sometimes seen between the geometry of the apoplast within a plant body and the morphology of the plant body, (3) the fact that correlative growth can be sustained despite the interruption of normal cell division, and (4) the expression of morphological complexity by nonmulticellular organisms. Thus, for example, coenocytic, semicoenocytic, and multicellular plant species can show morphological convergence, while anatomical convergence among multicellular species can belie morphological dissimilarity. Logically, therefore, form (morphology) and structure (anatomy) potentially provide two independent biological attributes upon which Darwinian selection can operate. An emergent conclusion is that the concepts of the organism and of the cell are logically independent of one another (Kaplan, 1987a, b). Thus, multicellularity must be seen as a consequence of a highly specialized mode of development in which cytokinesis and nuclear division are highly correlated (Kaplan & Hagemann, submitted). The thesis developed here is that, in addition to conferring physiologic and reproductive advantages, the partitioning of the plant protoplast into cells provided mechanical advantages upon which Darwinian selection operated. The assertions made in this paper are summarized in Table 1.

Table 1. Philosophical assertions concerning plant "cellularity." Assertions f–i provide the principal foci for biomechanical analysis of plants.

a. The concepts of the organism and the cell are logically independent of one another.
b. Multicellularity is not a requisite condition for the expression of morphological complexity.
c. Multicellularity is the result of a highly specialized developmental scheme in which cytokinesis and nuclear divisions are precisely correlated.
d. The protoplasm is the fundamental organismic unit.
e. In multicellular plants, the protoplasm is incompletely partitioned by an apoplastic (cell wall) infrastructure.
f. In addition to providing an opportunity for the physiologic and reproductive specialization of the protoplasm, the apoplastic infrastructure provides mechanical support.
g. The apoplast is a shared-primitive condition in all plants, i.e., unicellular, colonial and multicellular organisms have an apoplastic-symplastic structure.
h. The "internalization" of the apoplast has occurred in at least two different ways (siphonous and multicellular plants).
i. Only one of these ways (multicellularity) permits the acquisition of large size in a terrestrial habitat.
j. Multicellularity was an exaptation for the colonization of the land by plants.

PLANT BIOMECHANICS

Plants are constructed out of two basic types of materials: the living protoplasm (symplast) and the semirigid or rigid cell wall (which is the most significant contributor to the apoplast). These two components differ substantially in their individual mechanical behavior and exhibit remarkable properties when they operate together. The symplast and cell walls are mechanically distinguishable in terms of their capacities for recoverable deformation. This is illustrated by contrasting the mechanical behavior of a linearly elastic solid and a Newtonian fluid. When an external force is applied to the former, molecules resist displacement and, provided bonds are not broken, the solid resumes its original shape and dimensions when the force is removed. Thus, within a fairly broad range of stress σ (force per area) and strain ε (deformation), the relationship between stress and strain is linear and can be defined by an unique proportionality constant E called the elastic or Young's modulus: $E = \sigma/\varepsilon$. (At some stress level, all solids either fail or undergo plastic deformations, and the relationship between stress and strain becomes nonlinear. Under normal circumstances, however, plants mechanically function within the proportional limits of their structural—mechanically supportive—materials.) By contrast, when a Newtonian fluid is stressed, the force is dissipated by molecular displacements, and the relationship between stress and strain depends upon the rate of deformation, expressed by the dynamic viscosity μ. Thus, a Newtonian fluid has no capacity for recoverable deformation.

The time required for a stress to deform or "relax" a material's structure such that strains are permanent is reflected by the relaxation time T_R. The relaxation time for an ideal fluid is zero, that of an ideal solid is infinity. Most materials for which T_R is measurable are called viscoelastic materials. Their behavior lies somewhere between that of a fluid and a solid. The symplast and the cell walls are viscoelastic in their mechanical behavior; both have molecular structures that undergo relaxation (nonrecoverable deformation). However, their capacities for relaxation are not equivalent. During the active growth phase, cell walls mechanically behave as a solid when subjected to high rates of stress and, when fully mature, cell walls have a substantial linear elastic range of behavior. By contrast the symplast behaves as a fluid, albeit with a variable and wide range in viscosity.

The cell wall and the symplast operate together as a hydrostatic mechanism (Niklas, 1989). This is most evident when plants are deprived of water and wilt. When fully turgid, the symplast exerts a pressure on cell walls. This internal pressure reduces the extent to which cell walls can buckle under the weight they normally sustain (Fig. 1A). When deprived of water, the symplast is reduced in volume and the pressure it exerts on cell walls decreases (Fig. 1B). As a consequence, cell walls can deform under the compressive loading from the weight above them. Intuitively, the relative thickness of cell walls must play a role in the extent to which any tissue deforms (wilts) under a given loading condition (self-weight). The extent to which a tissue or entire organ deforms as turgor pressure declines would be expected to decrease as the average cell wall thickness increases. This intuition can be reinforced by a mathematical treatment of the hydrostatic behavior of tissues and cells.

Referring to Figure 2, if r_i and r_o represent the inner and outer radii of a cylindrical cell wall and if P_i and P_o represent inner (hydrostatic) and outer (ambient) pressures exerted on the cell wall, then the radial and circumferential stress (σ_r and σ_ϑ) within the cell wall at any radius r are given by the equations

$$\sigma_r = \{[r_i^2\, r_o^2\, (P_o - P_i)]\, r^{-2} + P_i r_i^2 - P_o r_o^2\}\, (r_o^2 - r_i^2)^{-1}$$

$$\sigma_\vartheta = \{[r_i^2\, r_o^2\, (P_o - P_i)]\, r^{-2} + P_i r_i^2 - P_o r_o^2\}\, (r_o^2 - r_i^2)^{-1}$$

The circumferential stress is the "hoop" stress developed within the cell wall. This stress is resisted by cellulosic microfibrils that in the newest cell wall layers are preferentially oriented to girdle the cell wall.

Since the elastic modulus E is stress divided by strain,

$$ E = \frac{(\sigma_\vartheta - \nu\,\sigma_r)}{\varepsilon_\vartheta} $$

where ν is Poisson's ratio and ε_ϑ is the circumferential strain. From these relationships we can see that, dependent upon the internal pressure, the stresses developed within the cell wall produce a uniform extension or contraction in the direction of the longitudinal axis of the cell wall and, when the turgor pressure P_i is higher than the ambient pressure P_o, the radial stress is compressive and the circumferential stress is tensile. Also, the tensile stress is always greatest at the inner surface of the cell wall (at r_o), it is always greater than P_i, and it approaches P_i as the outer radius r_o increases.

Referring once again to Figure 2, the point at which the hydrostatic influence of the symplast becomes negligible as the average wall thickness t of the symplast increases can be estimated based on three dimensionless ratios: one that relates the wall thickness to cell radius (t/r_o), one that relates the inner to the outer pressure (P_i/P_o), and one that relates the circumferential stress to the ambient (outer) pressure (σ_ϑ/P_o). The last of these three ratios reflects the effective elastic modulus which is a function of the circumferential stresses developed within the cell wall. When σ_ϑ/P_o is plotted as a function of t/r_o for a variety of P_i/P_o, it is seen that the hydrostatic influence on cell wall rigidity decreases as the cell wall thickness increases. Figure 2 indicates that when the cell wall thickness is roughly 20% that of the cell radius ($t/r_o = 0.2$), the values of σ_ϑ/P_o for a range of internal pressures converge.

Thus, from the first principles of mechanical engineering, the extent to which the cell wall-symplast operates as a hydrostatic mechanism depends upon the volume fraction of each component within the organism. When the volume fraction of the symplast is relatively high, the hydrostatic mechanism can predominate. Conversely, when the volume fraction of the apoplast is high, cell walls play the leading role as the support mechanism. In the former case, the stiffness of the cell or tissue is dependent upon the availability of water and is controlled in part by ecological factors, while in the latter case the cell or tissue is mechanically secured against variability in the availability of water to the plant.

Lester Sharp, the leading cytologist of his day, clearly understood this when he wrote: "This suggests that mechanical support is one of the chief functions in connection with which cell walls were developed in plants" (Sharp, 1926, p. 78). Implicit in this suggestion is that the ratio of the apoplast to symplast differs among tissues and that tissue systems with higher ratios can lead to stiffer body plans.

Figure 3 illustrates both of these points. In this figure, the critical aspect ratios of cylindrical beams of tissues are plotted as a function of the elastic moduli of various tissues for which ratios of apoplast to symplast have been computed. The critical aspect ratio is the theoretical limit to which a cylindrical piece of tissue can be extended in length before it bends significantly from the vertical under its own weight (Niklas, 1990). It is a crude measure of the relative proportions that a plant body can grow based on each of the tissues for which data are available. Parenchyma and collenchyma have the lowest critical aspect ratios and the lowest ratios of apoplast to symplast. These two tissues are hydrostatic in their mechanical behavior. By contrast, sclerenchyma and secondary xylem (wood) have some of the highest aspect ratios. They also have the highest ratios of apoplast to symplast ratios.

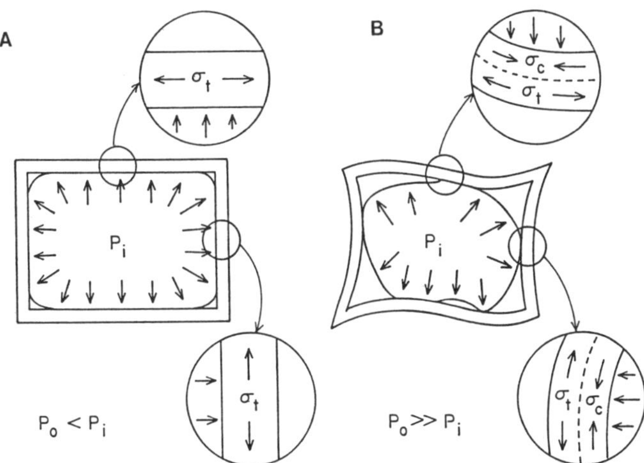

Figure 1. Hydrostatic pressure and cell wall mechanical stresses. A. The fully inflated protoplast exerts a hydrostatic pressure (P_i) upon its cell wall, placing the cell wall in tensile stress (σ_t) and limiting the extent to which the cell wall can deform when a compressive load is applied externally (not shown). B. When the protoplast is deflated (highly exaggerated in this figure), the cell wall is free to deform when a compressive stress is applied (see arrows in the upper and side inserts) and undergoes tensile (σ_t) and compressive (σ_c) stresses during buckling.

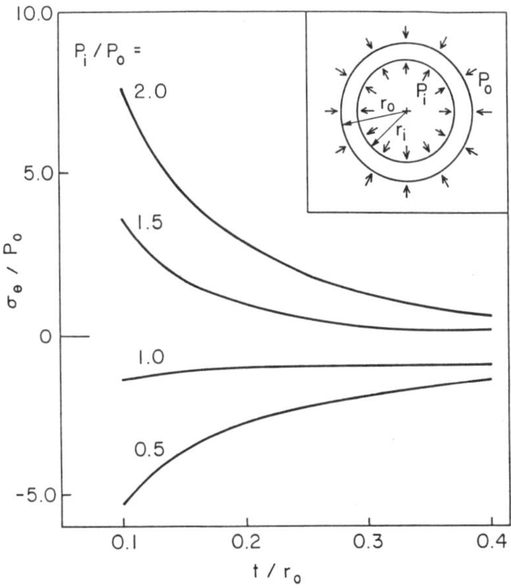

Figure 2. Relationships among the "hoop" or circumferential stresses (σ_ϑ developed within a cylindrical cell wall, the ratio of the internal, hydrostatic pressure (P_i) and external, ambient pressure (P_o), and the ratio of the cell wall thickness (t) to external cell wall radius (r_o). (The geometry of pressure application and cell wall dimensions is shown in the upper-right insert, which is a transection through the cell.) The plots of dimensionless ratio of "hoop" stress to external pressure (ordinate) versus the dimensionless ratio of wall thickness to external radius (abscissa) for various pressure ratios indicate that "hoop" stresses are influenced by the pressure ratio when the cell wall thickness is small ($<$ 20% of the external cell wall radius). As the cell wall thickness increases in comparison to the external radius of the cell wall, the differences among the stresses developing within the cell wall as a result of differences between the internal and external pressures on the cell converge (to roughly zero). These plots indicate that thin-walled cells mechanically operate as hydrostatic "devices." Thick-walled cells do not operate as hydrostats.

494

By way of comparison, the elastic modulus and the aspect ratio of a cylinder of pure dehydrated cellulose are plotted in Figure 3. With respect to its density, cellulose is the strongest material known. As can be seen, all the plant tissues represented in Figure 3 have a lower elastic modulus than that of cellulose. This is due to the fact that wet cellulose has a lower elastic modulus than dry cellulose and the cell walls of tissues within a living plant are hydrated to some degree. At any given moisture content, however, lignified plant fibers are weaker than their delignified counterparts. Thus, superficially, lignification would appear to be a counterintuitive "mechanical strategy." However, lignin is hydrophobic and its presence in cell walls reduces the extent to which walls can be hydrated. Thus, lignified cell walls are somewhat weakened but are less likely to vary in moisture content, hence less likely to vary in strength, i.e., lignification can be viewed as a design factor conferring a margin of safety against the reduction in tissue stiffness as a result of cell wall hydration.

To a very limited extent, Figure 3 contains an evolutionary "vector" in that the earliest multicellular plants, whether aquatic or terrestrial, consisted essentially of parenchyma. During the course of terrestrial plant evolution, collenchyma, the primary vascular tissue system, and finally the secondary vascular system made their appearance. Thus, in general, the diagonal line in Figure 3 (going from the bottom left to the upper right) reflects the evolutionary sequence in the appearance of different tissues. The general trend in time is one in which the plant body could be made stiffer and larger by the incorporation of newly innovated tissues, as well as a transition from a hydrostatic support mechanism involving thin walled tissues to one dominated by the apoplast.

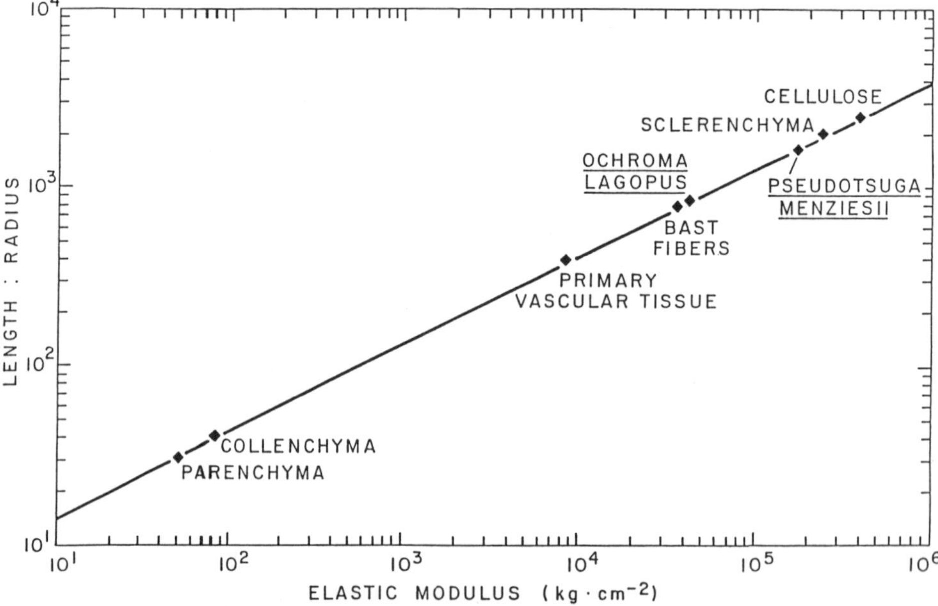

Figure 3. Critical aspect ratios of cylinders composed of various tissues and pure cellulose plotted as a function of the elastic modulus of the materials from which the cylinders are constructed. The critical aspect ratio is the maximum ratio of the cylinder's length to radius that can be achieved before the cylinder deflects from a vertical orientation under its own weight. The elastic modulus is the ratio of stress to strain within the linear elastic region of a material's mechanical behavior. The critical aspect ratios span three orders of magnitude; the elastic modulus spans five orders of magnitude. Parenchyma and collenchyma are hydrostatic tissues with low elastic moduli; primary vascular tissues (including bast fibers), wood (*Ochroma lagopus*, Balsa; *Pseudotsuga menziesii*, Douglas fir), and sclerenchyma are non-hydrostatic tissues that mechanically operate as cellular solids. Pure cellulose (dry) has the highest elastic modulus plotted among all these "materials." The extent to which a cylinder can "grow" vertically (maximize the critical aspect ratio) cannot be distinguished on the basis of whether the tissues used in its construction are from primary or secondary growth.

THE MECHANICS OF CELLULAR SOLIDS

There are plant tissues in which the symplast is essentially lacking. Wood (secondary xylem), cork (phellum), and some sclerenchyma are examples. The mechanical behavior of these tissues is remarkably like the behavior of artificially constructed materials called cellular solids which are constructed of walled partitions or struts that separate gas-filled voids (Ashby, 1983; Gibson & Ashby, 1982). Cellular solids have an overall density that is significantly less than the density of the materials from which they are constructed. Thus they are very light in weight, yet they are very strong. Examples of commercially fabricated cellular solids are polystyrene, polyurethane, and rubber latex foams. Because of their thermal and mechanical properties, these materials are used for insulation, impact-resistance, and wherever strength and weight need to be optimized—functions that parallel those of cork, wood, and sclerenchyma.

For the purposes of this discussion, the most significant features of cellular solids are summarized in Figure 4, which shows the stress-strain relationship for a typical cellular solid subjected to compressive loadings. Figure 4 shows that the cellular solid has three principal regions of mechanical behavior. The first region is one in which the stress-strain curve is linear, indicating that the solid exhibits elastic behavior defined by a single, unique elastic modulus E. When stresses exceed a certain limit, a second region of the stress-strain curve becomes evident in which the solid behaves in a non-linear fashion and strain increases despite little or no increase in stress. The second region is followed by a third in which the cellular solid "densifies." That is, a disproportionate increase in stress is required to produce an equivalent strain. What is most relevent here is that the stress-strain relationship for a cellular solid depends upon the magnitude of the applied load and

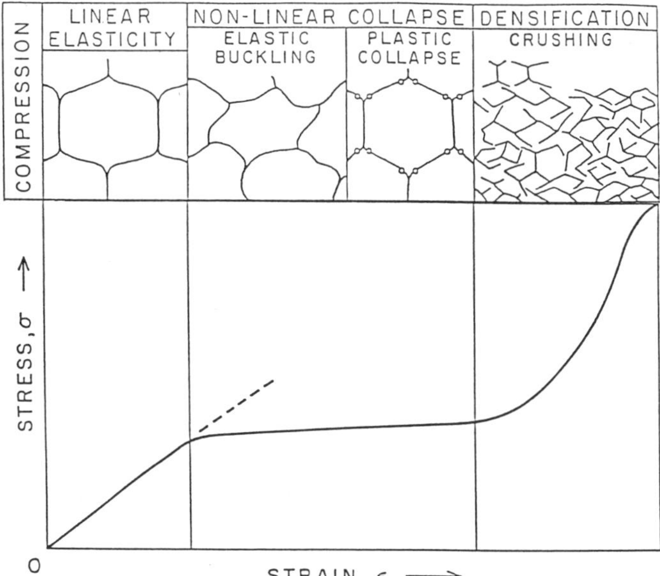

Figure 4. Stress-strain diagram for a "honeycomb" cellular solid subjected to compressive stress. (Geometric "responses" of the cellular solid are diagrammed above each region of the stress-strain diagram.) Under compressive loadings, the stress-strain diagram of a cellular solid typically exhibits three regions: (1) an initial, linear eleastic response, where strain is proportioinal to stress and an unique tangent modulus (the elastic modulus) may be calculated, (2) a non-linear "collapse" response, where elastic buckling or plastic collapse may occur (the location of plastic "hinges" developing in "walls" is shown in the upper insert), and (3) a region of densification when the cellular units within the cellular solid are crushed together and the density of the cellular solid increases dramatically.

496

that at very high stresses the tangent modules (the instantaneous elastic modulus, which is a measure of stiffness) of the stress-strain curve increases, often dramatically. Accordingly, the mechanical behavior of a cellular solid, like wood or cork, is versatile and is responsive to the loading conditions experienced. However, this versatility is not dependent upon the availability of water to living cells. This is advantageous when the mechanical environment in which a cellular solid operates is variable and unpredictable. Plants exist in just such an environment. They are subjected to a variety of dynamic loadings due to wind pressure and falling debris. Similarly, aquatic plants are frequently subjected to tensile and torsional shear stresses due to the movement of water. The magnitude of dynamic loadings in both habitats often varies dramatically, as can the duration and direction of application.

SIZE, CELLULARITY, AND HYDROSTATICS

The thesis to be developed in this section is that plant bodies that mechanically operate as hydrostats are limited in size, dependent upon the availability of water. Beyond a given size, an apoplastic infrastructure would be mechanically desirable.

Figure 2 illustrates the consequences on cell wall stresses of thickening the wall of an idealized, internally pressurized cylindrical shell (=cell). However, what is not addressed by this consideration are the mechanical consequences on the ability of a cylindrical structure to stand vertically as the thickness of the structure's wall is decreased. The structure in question could be a simple cylindrical cell, or a hollow cylindrical stem, or a solid cylindrical stem consisting of an outer more rigid "rind" and an inner less rigid "core". For the latter two structures, "cell wall thickness" would refer to the thickness of the hollow stem and the thickness of the "rind", respectively.

Brazier (1927) mathematically treated the mechanics of a thin walled shell and the influence of wall thickness on the maximum bending stress that the shell could sustain before it underwent irreversible mechanical failure. When a cylindrical thin walled shell bends under loading its terete transectional geometry undergoes deformation and becomes ovalized. (The semi-minor axis of the elliptic transection is oriented normal to the plane of bending.) As a result of ovalization, the maximum bending moment required to buckle the cylinder decreases, often dramatically. This phenomenon is now called Brazier buckling in recognition of Brazier's mathematical treatment of the subject. Figure 5 plots the Brazier bending moment as a function of the shell-wall thickness for three tubes constructed from materials differing in their elastic moduli. For our purposes the most significant aspect of the relationship is that the Brazier bending moment M_{cr} dramatically decreases as the shell-wall thickness decreases, regardless of the "stiffness" of the material used to construct the tube's wall. When placed in apposition, Figures 3 and 5 indicate that for internally pressurized cylinders there are theoretical limits to the extent that the wall thickness can be reduced for any given internal (turgor) pressure. A thin walled cylindrical cell bends easily when its internal pressure is decreased, a hollow stem bends easily when the wall thickness of the stem is decreased, and a solid stem bends easily when the thickness of its "rind" of tissue and the turgor pressure of its "core" decrease.

Qualitatively, biomechanical theory indicates that a hollow tubular cell or a stem becomes mechanically unstable when hydrostatic pressures are reduced unless these structures are internally reinforced. Multicellularity is one way in which this can be achieved. By concentrating cell wall material toward the perimeter of a cylindrical plant structure's transection the susceptibility of the structure to bending failure is reduced. Another way to achieve the same result is to place struts spanning lateral, external cell walls within the cylindrical plant organ, e.g., nodal septa in grass stems and *Equisetum* shoots; trabeculae in *Caulerpa*.

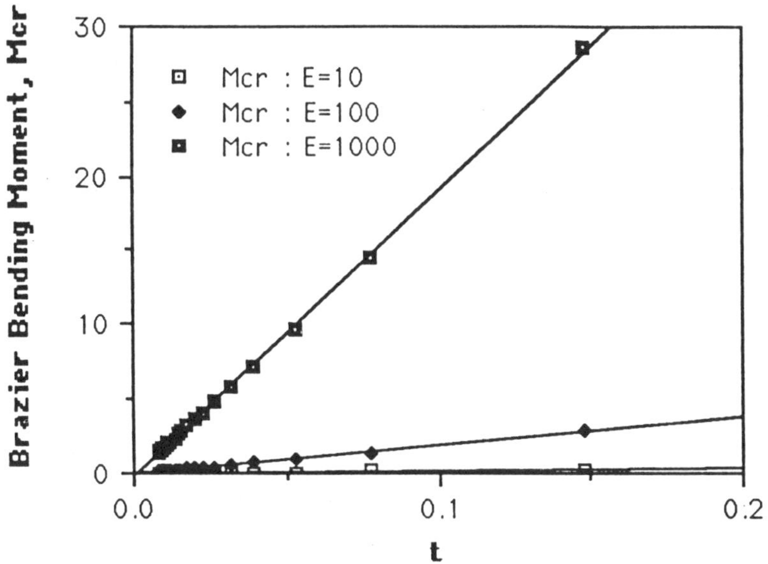

Figure 5. Brazier bending moment (M_{cr}) plotted as a function of the shell-wall thickness (t) of a hollow tube for three tubes made from materials differing in their elastic moduli (E). Brazier bending results from the ovalization of the tube as it bends under an applied bending stress. The Brazier bending moment is the maximum bending moment that a tube can sustain before it buckles. (A moment is the product of the load and the distance over which it operates.) Regardless of the elastic modulus of the tube's material, the Brazier bending moment decreases as the shell-wall thickness decreases. Thus, "thin" tubes are susceptible to buckling.

THE CELL AND ORGANISMIC THEORIES

Much of current theory regarding development and evolution is dependent upon the intellectual legacy of the Cell Theory by its implicit tendency to describe everything of a biological nature in terms of individual cells that collectively define the "republic" of the organism. Since one of the objectives of this symposium is to interrelate the areas of morphology, biomechanics and evolution, and since one of the objectives of this paper is to argue that multicellularity conferred mechanical advantages to plants but is not a requisite condition for morphological complexity, a review of the Cell Theory and its counterpart, the Organismal Theory, is a propos. The salient features of both theories are summarized in Table 2.

The Cell Theory was one of the great biological generalizations of the nineteenth century, although it was not without immediate detractors. It is commonly attributed to the botanist Schleiden (1838) and the zoologist Schwann (1839), but became widely adopted when Nägeli (1844) argued that growth occurs through cellular division. Further support for Nageli's observations quickly followed, and led to Virchow's now famous proclamation ". . . *omnis cellula e cellula*" (Virchow, 1855).

The principal propositions of the Cell Theory are (1) that the cell is morphologically and physiologically the elementary biological unit, (2) that all cells are initially of equivalent morphological rank, and (3) that each organism is an aggregate of cells (for substantive reviews see Heidenhain, 1907; Sharp, 1926; Sinnot, 1960). The developmental corollaries of the Cell Theory are (1) the organism results from the collaborative efforts of cells, and (2) morphogenesis the result of the collective actions of many special cells. The

Table 2. Developmental and phylogenetic corollaries of the Cell Theory and the Organismic theory.

The Cell Theory
    Developmental Corollaries
        a. All living things are made up of cells.
        b. Each cell is an individual of equal morphological rank.
        c. The multicellular organism is an aggregate of cells.
        d. The properties of the organism are the sum of many cells.
        e. Ontogeny is the cooperative effort of many cells.
    Phylogenetic Corollaries
        a. Unicellular organisms are "elementary."
        b. Elementary units formed colonial organisms through an acquired failure to separate after multiplication.
        c. Cells within colonial organisms became increasingly interdependent and eventually produced the multicellular organism.
The Organismal Theory
    Developmental Corollaries
        a. Ontogenesis is the property of the organism as a whole.
        b. Growth and differentiation are properties of the protoplasm.
        c. Development may or may not involve septation of the protoplasm.
        d. When septation occurs, the cells are subordinate parts of the whole.
        e. Ontogenesis is the resolution of the whole into parts.
    Phylogenetic Corollaries
        a. Unicellular and multicellular organisms are non-septate and septate individuals, respectively.
        b. Unicellular and multicellular organisms are homologous.
        c. Colonial organisms are derived, not primitive, organisms.
        d. Division of labor is effected by cellularization.

phylogenetic corollary of the Cell Theory is that the organism evolved from uninucleate cells that through time formed loose aggregates of cells (by a failure to disaggregate after a period of multiplication), followed by true multicellularity in which cells became functionally interdependent. One of the greatest difficulties of the Cell Theory was appreciated by its early advocates—plasmodial and coenocytic organisms existed with a capacity to construct bodies of definite and often complex form. This difficulty is evident in Sachs' Energid Theory, advocated in the 1890s, which viewed each nucleus with its immediate portion of cytoplasm as a cell unit operating within the collective coenocyte.

As the Cell Theory developed, increasing attention was paid to the composition and structure of cells. Payen (1846) and Cohn (1850) independently concluded that the living substance or "sarcode" of the animal (Dujardin, 1835) and the "protoplasm" of the plant were the same, leading Schultze (1861) to assert that the protoplasm was the fundamental unit of biological organization. Schultze's Protoplasm Doctrine was the precursor to the Organismal Theory, which essentially views the organism as a continuous mass of protoplasm (the symplast) that may or may not be incompletely partitioned into cells during its ontogeny. This provided developmental and phylogenetic corollaries countering those of the Cell Theory. Thus, plant and animal cells are viewed as having secondary importance; cells are the result, not the cause of organization; and the unicellular and multicellular organism are placed in parity with one another, since the latter is the septated equivalent of the former. Perhaps the most articulate expression of the developmental content of the Organismic Theory was de Bary's (1862) now famous phrase ". . . *die Pflanze bildet Zellen, nicht die Zelle bildet Pflanzen.*"

The difference between the Cell Theory and the Organismal Theory is fundamental to our view of development and morphogenesis. The cell as a functional unit is clearly important, and the analysis of cell growth and differentiation is essential to the study of many developmental processes. However, the solution to the problem of organization, which is the central problem for all of biology, requires the study of organized systems as wholes rather than the study of the units of which they are composed.

THE COENOCYTE

Development provides an understanding of how a phenotype is achieved and, since Darwinian selection offers an explanation for the persistence and relative prevalence of phenotypes, it is argued that an understanding of development will allow an evaluation of the relative importance of natural selection to evolution. However, this argument is valid only if developmental patterns and the phenotypes they engender are uniquely correlated, and provided natural selection operates principally at the level of morphogenesis and its structural consequences. There is ample evidence to doubt that both of these stipulations are true. As discussed in this section, analogous phenotypes can result from disparate developmental processes, and Darwinian selection must operate on the organism *in toto* (encompassing physiologic, reproductive, and other attributes in addition to those of form and structure). Given this, an alternative strategy for evaluating the importance of natural selection to evolution is to document the frequency of convergence. The significance of development to evolution can be evaluated by the investigation of developmental schemes that produce these convergences. In this context, comparative morphology should not be seen to seek "ultimate causations," as it is sometimes said to do, but rather as a discipline that draws attention to convergence as evidence for adaptation, without relying on an inextricable link between development and form as one of its principal assumptions.

A review of the structural attributes of cells of multicellular organisms indicates that an apoplastic infrastructure confers mechanical benefits. However, these benefits shed little or no light on the developmental mechanism giving rise to this infrastructure nor do they provide an unambiguous basis upon which evolutionary modifications of the plant body can be interpreted as adaptive. Fortunately, coenocytes provide another group of organisms in which an apoplastic infrastructure has evolved and the convergences seen between coenocytes and multicellular organisms can shed light on developmental processes of plants. The coenocyte has evolved in at least two separate plant lineages, the Chlorophyta and the Chrysophyta. Within the former, the Dasycladales and Caulerpales manifest morphological convergence with multicellular plants.

The Caulerpeaceae is a monotypic family consisting of species that have a general plan of construction with a prostrate, more or less cylindrical "rhizome" that bears numerous well-branched "rhizoids" below and a number of upright assimilatory "shoots." *Caulerpa fastigiata* is one of the more morphologically simple species, with vertical shoots that are irregularly branched (Fig. 6A). By contrast, *C. taxifolia* and *C. cupressoides* have assimilatory shoots that are leaflike in appearance (Fig. 6B-C). Morphological variability in habit parallels the morphological range observed among species. When exposed to wave-action, some rock-dwelling species produce reduced shoots and densely felted rhizomes and rhizoids (Fig. 6D). Morphological transitions from radial to bilateral symmetry occur in some species as a result of changing light intensity.

The apoplast of *Caulerpa* is distinguished by the abundant development of strands of cell wall material (trabeculae) that traverse the central lumen in all parts of the thallus (Fig. 6E). The trabeculae develop either connected *ab initio* to the cell wall near the growing points of the thallus or at first free at one or both ends within the symplast. They gradually thicken by the apposition of cell wall layers (Fig. 6F). At the cellular level, the formation of trabeculae and the apoplastic infrastructure within multicellular species are essentially identical, while from a functional perspective trabeculae serve analogous roles. Noll (1888) established that the trabeculae provide an apoplastic route for the transport of mineral salts, while Janse (1890) speculated that they mechanically function to resist high turgor pressures. Experiments by Niklas indicate that trabeculae significantly increase the stresses required to bend portions of the thallus. Thus, the apoplastic infrastructure of

500

Caulerpa mechanically functions in much the same way as it functions in multicellular organisms. The genus also illustrates that multicellularity is not a requisite condition for morphological complexity.

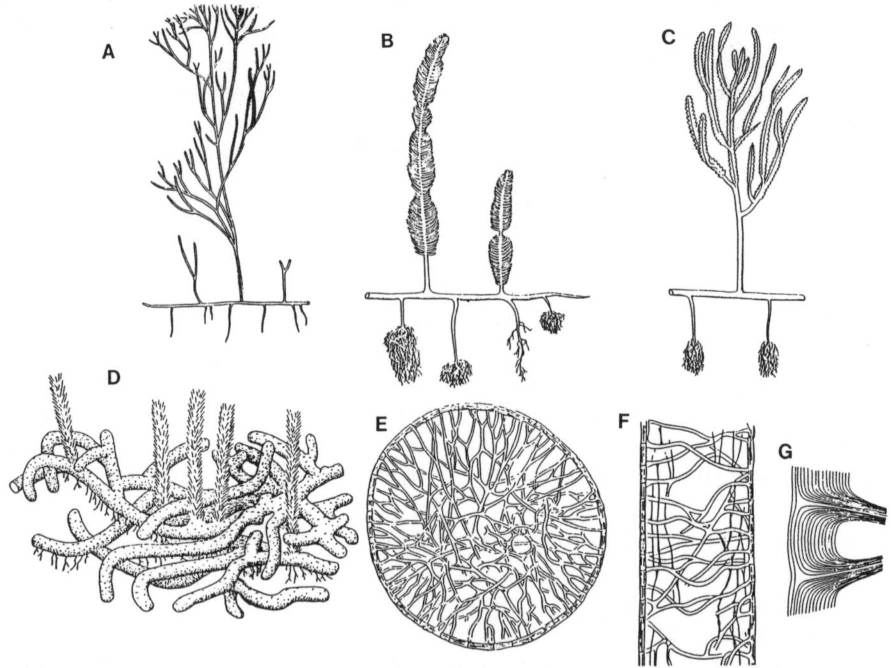

Figure 6. Morphological (A–D) and anatomical (E–G) complexity in the siphonaceous algal genus *Caulerpa*. A. *C. fastigiata*. B. *C. taxifolia*. C. *C. cupressoides*. D. Horizontal "rhizomatous-like" elements of the thallus covered with "rhizoid-like" cellular extensions. E. Transection through vertical component of a cell showing internal "strut-like" extensions of the primary cell wall (called trabeculae). F. Longitudinal section through a cell showing trabeculae. G. Details of the primary cell wall layers interconnecting trabecular extensions within the cell (to right) and the external cell wall (to left).

CELLS OR ORGANISMS?

The formation of cellular partitions and the division of nuclei within the plant body show various degrees of correlation (Kaplan, 1987 a, b; Kaplan & Hagemann, submitted). They may be totally independent, as in coenocytic plants like *Caulerpa* and *Vaucheria*; they may be correlated, as in semi-coenocytic plants like *Cladophora*; or they may be intimately related such that they appear to be the same process, as in most multicellular plants. In the latter case, growth appears to be a matter of cell replication and enlargement because the cell wall provides the most obvious markers by which growth can be recognized. However, when coenocytic and semicoenocytic plants are considered, it becomes clear that the characteristics of cell walls are the physical manifestations of differentiation within the protoplasm. When cell wall patterns are used as "surrogates" by which growth and development are measured, the underlying mechanisms responsible for growth and development are obscured.

Multicellularity is a highly specialized consequence of development. There are at least three lines of evidence that support this view. (1) The protoplasm of unicellular organisms can and often does show remarkable differentiation. Among many protists, the distribution of ultrastructural features is highly structured, indicating that the presence or absence of cellular partitions is subordinate to regional differentiation of the protoplasm. (2) Coenocytic organisms can express levels of morphological complexity convergent with those seen among multicellular organisms. The morphology of algae such as *Acetabularia* and *Caulerpa* is reminiscent of that of vascular plants. (3) Correlative growth in multicellular organisms is unrelated to mitotic activity, suggesting that variability in mitotic activity is not, in general, the immediate cause of differential growth. The classic experimetns of Haber and Foard and their colleagues (Haber et al., 1961; Foard & Haber, 1961; Haber, 1962; Haber & Foard, 1963, 1964; Foard et al., 1965), as well as Lillie's (1906) work on *Chaetopterus* larvae, led to the conclusion that growth and development must be understood at the level of the organism and that the mechanisms underlying morphogenesis and development are embedded within the protoplasm. Accordingly, attempts to explain these mechanisms on the basis of patterns of cellular partitioning are necessarily incomplete at best and misleading at worse.

The implicit acceptance of the developmental and phylogenetic corollaries of the Cell Theory characteristically colors the interpretation of developmental and evolutionary data. Three examples illustrate this point: (1) the canonical view of evolution within the Volvocaceae expressed in many textbooks, that implicitly treats *Gonium, Pandorina, Eudorina* and *Volvox* as a linear evolutionary sequence and, therefore, as evidence for the evolution of a multicellular organism (such as *Volvox*) through the failure of cells to disaggregate after cell division, (2) the biophysical interpretation that shoot morphogenesis (phyllotactic patterns) in angiosperms is dictated by stress fields generated within the cell walls of superficial cells (see Green, 1980), and (3) the notion that the "relative competitive mechanism of cell lineages" within an organism will influence the extent to which genetic variation arising during ontogeny will be inherited (see Buss, 1987). The linear view of volvocine evolution is clearly refuted by the available traditional (Starr, 1984, p. 265) and molecular data (D. L. Kirk, personal communication) that indicate much more complicated intergeneric relationships, while the application of the evolutionary corollaries of the Cell Theory to these algae is also refuted likewise by the appearance of intercellular cytoplasmic bridges during the early ontogeny of the genera typically considered to be "colonial" (Viamontes et al., 1979 and references cited therein). Similarly, the data relevant to the research agenda of (2) and (3) are as easily interpreted to indicate that shoot morphogenesis generates stress fields within the walls of epidermal cells (i.e., that stress fields are the consequence, not the cause of morphogenesis) and that "cell lineages" reflect the manner in which the symplast can ontogenetically partition genetic variations among nuclei (i.e., "cell lineages" are the physical manifestation of ontogenetic patterns rather than the reverse). The adoption of an organismic view of shoot morphogenesis and the manner in which germ-lines are sequestered in an individual does not alter the fact that stress-fields occur or that "cell lineages" are manifest in multicellular organisms. Significantly, however, what is altered is the fundamental level at which we infer proximate causalities for the observed phenomena. Cellular patterns provide tangible and relatively easily measured features of growth and differentiation whose generative mechanisms lie within the symplast.

LITERATURE CITED

Ashby, M. F. 1983. The mechanical proeprties of cellular solids. *Metallurgical Trans.* 14A:1755–1769. (The 1983 Institute of Metals Lecture; Metallurgical Society of the American Institute of Mechanical Engineering.)

Buss, L. W. 1987. *The Evolution of Individuality.* Princeton University Press: Princeton, NJ.

Brazier, L. G. 1927. On the flexure of thin cylindrical shells and other "thin" sections. *Proc. Royal Soc. London, Series A* 116:104–114.

Cohn, F. 1847. Nachträge zur Naturgeschichte des *Protococcus nivalis,* Kutzing (1850). *Acad. Caes. Leop. Nova Acta* 22:605–764.

de Bary, A. 1862. Die neuesten Arbeiten über Entstehung und vegetation der niederen Pilze, insbesondere Pasteur's Untersuchungen. *Flora* 45:355–365.

Dujardin, F. 1835. Recherches sur les organismes inférieurs. *Ann. Sci. Nat. (Zool.)* 4:343–376.

Esau, K. 1977. *Anatomy of Seed Plants.* (2nd edition) John Wiley & Sons: New York.

Foard, D. E. & A. H. Haber. 1961. Anatomic studies of gamma-irradiated wheat growing without cell division. *Amer. J. Bot.* 48:438–446.

Foard, D. E., Haber, A. H. & T. N. Fishman. 1965. Initiation of lateral root primordia without completion of mitosis and without cytokinesis in uniseriate pericycle. *Amer. J. Bot.* 52:580–590.

Gibson, L. J. & M. F. Ashby. 1982. The mechanics of three-dimensional cellular solids. *Proc. Roy. Soc. London A* 383:43–59.

Green, P. B. 1980. Organogenesis—a biophysical view. *Ann. Rev. Plant Physiol.* 31:51–82.

Haber, A. H. 1962. Nonessentiality of concurrent cell divisions for degree of polarization of leaf growth. I. Studies with radiation-induced mitotic inhibition. *Amer. J. Bot.* 49:583–589.

Haber, A. H. & D. E. Foard. 1963. Nonessentiality of concurrent cell divisions for degree of polarizatioin of leaf growth. II. Evidence from untreated plants and from chemically induced changes of the degree of polarization. *Amer. J. Bot.* 50:937–944.

Haber, A. H. & D. E. Foard. 1964. Further studies of gamma-irradiated wheat and their relevance to use of mitotic inhibition for developmental studies. *Amer. J. Bot.* 51:151–159.

Haber, A. H., Carrier, W. L. & D. E. Foard. 1961. Metabolic studies of gamma-irradiated wheat growing without cell division. *Amer. J. Bot.* 48:431–438.

Heidenhain, M. 1907. *Plasma and Zelle.* Fischer Verlag: Jena.

Janse, J. M. 1890. Die Bewegungen des Protoplasma von *Caulerpa prolifera. Pringsheim, Jahrbüch. Wiss. Bot.* 21:163–284.

Kaplan, D. R. 1987a. The significance of multicellularity in higher plants. I. The relationship of cellularity to organismal form. *Amer. J. Bot. (Botanical Society of America Ohio State Meeting)* 74:617.

Kaplan, D. R. 1987b. The significance of multicellularity in higher plants. II. Functional significance. *Amer. J. Bot. (Botanical Society of America Ohio State Meeting)* 74:618.

Kaplan, D. R. & W. Hagemann. Submitted. The relationship of cell and organism in higher plants: are higher plants really multicellular? *BioScience.*

Lillie, F. R. 1906. Differentiation without cleavage in the egg of the annelid *Chaetopterus. J. Exp. Zool.* 3:153–268.

Nägeli, C., von. 1844. Vegetable cells (translation from the German). *Ray Society* 1(1).

Neville, A. C. 1986. The physics of helicoids: multidirectional "plywood" structures in biological systems. *Phys. Bull.* 37:74–76.

Niklas, K. J. 1989. Mechanical behavior of plant tissues as inferred from the theory of pressurized cellular solids. *Amer. J. Bot.* 76:929–937.

Niklas, K. J. 1990. Biomechanics of *Psilotum nudum* and some early Paleozoic vascular sporophytes. *Amer. J. Bot.* 77:590–606.

Noll, F. 1888. Über die Funktion der Zellstoffasern der *Caulerpa prolifera. Arb. Bot. Inst. Würzburg* 3:459–465.

Payen, A. 1846. Mémoires sur les développments des végétaux. *Paris mém. Savans Étrang.* 9:1–254.

Sharp, L. W. 1926. *An Introduction to Cytology.* (2nd edition) McGraw-Hill: New York, 581 pp.

Schleiden, M. J. 1838. Beitrage zur Phytogenesis. *Muller, Archiv: 137–177.*

Schultze, M. 1861. Über Muskelkorperchen und das was man eine Zelle zu nennen habe. *Reichert, Archiv: 1–27.*

Schwann, T. 1839. *Mikroskopische Untersuchungen über die Übereinstimmung in der Struktur und dem Wachstume der Tiere und Pflanzen.* Ostwalds' Klassiker der Exakten Wissenschaften Nr. 76. Wilhelm Engelmann: Leipzig.

Sinnott, E. W. 1960. *Plant Morphogenesis.* McGraw-Hill: New York.

Starr, R. C. 1984. Colony formation in algae. Pp. 261–290. *In:* H.-F. Linskens & J. Heslop-Harrison (eds.), *Cellular Interactions.* Encyclopedia of Plant Physiology, New Series 17. Springer-Verlag: Berlin.

Taiz, L. 1984. Plant cell expansion: regulation of cell wall mechanical properties. *Ann. Rev. Plant Physiol.* 35:585–657.

Viamontes, G. I., Fochtmann, L. J. & D. L. Kirk. 1979. Morphogenesis in *Volvox:* analysis of critical variables. *Cell* 17:537–550.

Virchow, R. 1855. Cellular-Pathologie. *Virchow, Archiv.* 8:3–39.

*FUNCTIONAL MORPHOLOGY, BIOMECHANICS,
AND EVOLUTIONARY PROCESS—SYMPOSIUM*

Comparative Biomechanics and the Evolutionary Diversification of Flying Insect Morphology

Robert Dudley

Abstract. While the extraordinary morphological diversity of flying insects is widely recognized, functional consequences of this diversity are poorly understood. A biomechanical analysis evaluates the kinematics and aerodynamics of insect flight. In turn, the consequences and constraints of wing and body morphology can be interpreted for performance parameters of more general significance, such as maneuverability and the energetic costs of flight. To illustrate the utility of the comparative biomechanical approach, I explore the aerodynamic consequences of the evolution of asynchronous flight muscle, possession of which characterizes three of the four largest insect orders (Coleoptera, Diptera, and Hymenoptera) and over 75% of all insect species. The higher wingbeat frequencies associated with asynchronous muscle have permitted a reduction in wing area relative to body mass (increased wing loading) and an increase in flight speed. These trends promote independence of flight trajectories from ambient air motion. Freed of their aerodynamic duties, the forewings of beetles have undergone elaboration as protective devices (the elytra), while in flies the hindwings have become specialized as gyroscopic halteres. For virtually all species of flying insects flight behavior, wing kinematics, and even flight speeds under natural conditions are unknown. Broad comparative surveys of flight biomechanics will be necessary if we are to understand the functional significance of patterns of morphological evolution in the winged Insecta.

INTRODUCTION

The ability to fly appears to have conveyed a decisive advantage upon those animals that possess it, and the ensuing capabilities for dispersal have long been cited as a major factor promoting diversification of the Insecta. Considering species numbers alone, winged insects dominate the biosphere, and their high degree of morphological diversity has long been a source of discomfiture for entomologists. Given the importance of flight for various aspects of insect life history, it is then somewhat surprising that, for most insect species, we know virtually nothing about flight behavior and performance under natural conditions. Knowledge of wing kinematics, aerodynamics of flight, and functional morphology of the flight apparatus is also severely limited, considering the number of extant species. A biomechanical analysis can be used to evaluate the functional correlates of morphological diversity in the pterygote Insecta. Such an approach can assess both the effectiveness of existing design and the likelihood of past scenarios of morphological evolution.

Because the essential morphological and kinematic data for most flying insects are

Dr. Dudley is with the Smithsonian Tropical Research Institute, P.O. Box 2072, Balboa, Republic of Panama.

504

simply unavailable, it would be premature to attempt a synthesis of insect flight biomechanics similar to the elegant analyses now available for bats and birds (e.g., Norberg & Rayner, 1987; Rayner, 1987). Instead, I want to focus on just one particular feature of the insect flight apparatus, and to explore its biomechanical consequences and evolutionary implications. One of the most significant developments in the evolution of flying insects has been the repeated appearance of asynchronous flight muscle. This muscle type, known also as myogenic or fibrillar muscle, is characterized by repeated contractions of the muscle in response to only one nervous impulse. By contrast, in the more primitive synchronous muscle type there is a one-to-one correspondence between nervous impulses and muscle contraction (see Usherwood, 1975; Tregear, 1977; and Pringle, 1981 for general reviews of insect muscle).

Constraints on the rate of nervous stimulation to synchronous flight muscle impose an upper limit to the muscle operating frequency of about 100 Hz, although there are some important exceptions. For example, the synchronous stridulatory muscle of some homopterans can reach contraction rates as high as 550 Hz (Josephson & Young, 1985). Also, the flight muscle of some aleyrodids (Homoptera) is classified as synchronous using morphological criteria, but exhibits contraction frequencies substantially higher than 100 Hz when the insect is in free flight (Wootton & Newman, 1979). In general, however, the stretch-induced contractions of asynchronous flight muscle permit much higher muscle contraction rates to be attained than would otherwise be possible. Wingbeat frequencies are correspondingly elevated, with typical values for such asynchronous fliers as bees and flies in the range of 100 to 300 Hz (Greenewalt, 1962). Rates of contraction as high as 1000 Hz have been recorded from some asynchronous flight muscle (Sotavalta, 1953). By contrast, wingbeat frequencies of such synchronous fliers as odonates and orthopterans are typically less than 50 Hz (Greenewalt, 1962).

Whereas the ability to fly is typically cited as a fundamental factor promoting insect diversification (e.g., Daly et al., 1978; Borror et al., 1981), I will instead argue here that high wingbeat frequencies facilitated by asynchronous flight muscle are the principal feature conveying evolutionary advantage to flying insects. I first will survey the taxonomic distribution of this muscle, and will show that there is a strong correlation between insect species richness and the presence of asynchronous flight muscle. This approach presupposes that asynchronous flight muscle is a morphological innovation that facilitates diversification of insects into new "adaptive zones" (see Mitter et al., 1988 and references therein for discussion of the "adaptive zone" concept). Aerodynamic and biomechanical consequences of the high wingbeat frequencies that follow acquisition of asynchronous flight muscle will then be investigated. It will be shown that, among other consequences, possession of asynchronous flight muscle has indirectly facilitated adaptive use of wings in such non-aerodynamic roles as the protective elytra of beetles and the gyroscopic halteres of flies. Finally, I will discuss some ecological and evolutionary implications of high wingbeat frequencies and the attendant phenomena of reduced wing area and increased flight speeds.

TAXONOMIC DISTRIBUTION OF ASYNCHRONOUS FLIGHT MUSCLE

Asynchronous muscle has evolved independently at least nine or ten times in the Insecta (Cullen, 1974). Of an approximate total of 770,000 described insect species, more than 75% possess asynchronous flight muscle (Fig. 1; these and the following data on insect diversity are obtained from Richards & Davies, 1977; Daly et al., 1978; and Parker, 1982). The endopterygote orders Coleoptera, Diptera, and Hymenoptera (with the possible exception of some Symphyta (Daly, 1963)) are all characterized by asynchronous flight muscle (Pringle, 1949; Roeder, 1951; Boettiger, 1957). These three orders represent

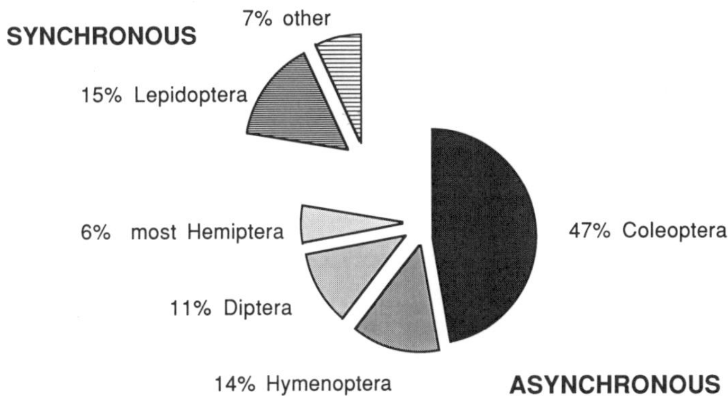

Figure 1. Taxonomic distribution of asynchronous and synchronous flight muscle in pterygote insects. See text for further details.

an approximate total of about 515,000 species, more than two-thirds of all pterygote insects. Among the exopterygotes, heteropterans, thysanopterans, and some psocopterans and homopterans possess asynchronous flight muscle (Cullen, 1974). Endopterygote insect orders also include the Lepidoptera, a group with synchronous flight muscle comprising more than 110,000 species. The Lepidoptera are an important counterexample to the general association of asynchronous flight muscle and species richness, and will be discussed in this context later. The remaining species of winged insects, approximately 75,000 in number, either possess synchronous flight muscle or have not been investigated.

The preceding designations of flight muscle type must be accepted cautiously, as such assessments have been made principally on morphological grounds, and only a miniscule fraction of the world's insect fauna has been investigated physiologically for the presence of asynchronous flight muscle (Josephson & Young, 1985). Statements on insect diversity must of course be similarly regarded. However, the numbers presented here are conservative estimates for the relative fraction of asynchronous insect fliers, as current knowledge of tropical beetles (Erwin, 1982) and of chalcidoid Hymenoptera (Noyes, 1978) probably underestimates their actual numbers substantially (see also Stork, 1988).

One of the obviously confounding factors in such an analysis is that the four largest insect orders (Coleoptera, Diptera, Hymenoptera, and Lepidoptera) are also endopterygote orders, displaying distinct larval, pupal, and adult life stages. The holometabolous condition, with its separation of larval and adult lifestyles, is usually cited as an important factor promoting adaptive radiation in the Insecta (e.g., Hinton, 1948; Hinton, 1963; Hennig, 1981). There are, however, major exceptions to this trend. Four endopterygote orders of flying insects (Neuroptera, Mecoptera, Megaloptera, and Raphidioptera) include from one to two orders of magnitude fewer species than do the four largest endopterygote orders, and also have far fewer species than do large exopterygote orders such as Hemiptera and Orthoptera. Association between complete metamorphosis and species richness is thus not unambiguous.

In this regard, the diverse exopterygote order Hemiptera is of particular interest, in that both synchronous and asynchronous muscle types are represented. Hemiptera is the largest exopterygote order (approximately 80,000 species), and is composed of two sub-

506

orders, the Heteroptera (35,000 species) and the Homoptera (45,000 species). Cullen (1974) presented a detailed analysis of the ultrastructure and taxonomic distribution of asynchronous flight muscle in the Hemiptera. All of the heteropterans examined possessed asynchronous muscle, while of the homopterans, representatives of the Aphididae, Cicadellidae (Jassidae), and Psyllidae possessed asynchronous muscle. These three families combined represent about 55% of all homopteran species; the Cicadellidae alone, with approximately 20,000 species, includes 44% of all homopteran species. Approximately 75% of all hemipteran species are thus characerized by asynchronous flight muscle.

BIOMECHANICAL CONSEQUENCES OF HIGH WINGBEAT FREQUENCIES

It is clear that there is a strong (but far from perfect) association between possession of asynchronous flight muscle and species richness for a variety of insect taxa. What selective advantages might asynchronous flight muscle confer to the flying insect? Relative to synchronous muscle, one direct advantage may be energetic (Josephson & Young, 1985). By virtue of their neurogenic activation, synchronous muscles incur substantial costs of calcium cycling, and of construction and maintenance of the associated network of sarcoplasmic reticulum. Stretch-activated contraction of asynchronous muscle substantially reduces these costs. Relative myofibril volume can also be increased if the sarcoplasmic reticulum is reduced, thereby increasing mechanical power output (Josephson & Young, 1985).

Because the stretch-activation of asynchronous flight muscle can dramatically increase rates of contraction, it permits a substantial increase in wingbeat frequency. What are the aerodynamic and biomechanical consequences of such an increase? To answer this question, I will rely upon the quasi-steady analysis of insect flight aerodynamics. This approach assumes that the dynamic motions of flapping wings can be reduced to a series of static conditions of steady-state flow, and that conventional aerodynamic theory can accordingly be used to estimate the forces that ensue (Ellington, 1984a). Whereas unsteady aerodynamic effects are a prominent feature of flight in many insects (e.g., Ellington, 1984c; Ennos, 1989a; Dudley & Ellington, 1990b), use of the quasi-steady analysis permits an initial assessment of the consequences of an increase in wingbeat frequency. I will also invoke two widely used aerodynamic parameters: wing loading, and the advance ratio of the wings. Wing loading is the ratio of body weight to total sustaining wing area, and represents the average pressure exerted by the wings on the air. The advance ratio in forward flight can be defined as the ratio of the forward airspeed of the animal to the flapping velocity of the wing tip (Ellington, 1984b), and is equal to $V/2\Phi nR$, where V is the forward airspeed, Φ is the amplitude of the wingbeat, n the wingbeat frequency, and R the wing length.

In quasi-steady theory, aerodynamic forces are proportional to the product of the area of the airfoil in question and the square of its velocity. Because the velocity of moving wings is directly proportional to their flapping frequency (and also to the wingbeat amplitude), an increase in wingbeat frequency results in a disproportionate increase in aerodynamic forces (Fig. 2a). The magnitude of this increase depends, however, upon the forward airspeed and thus upon the advance ratio. At higher forward airspeeds, aerodynamic forces on the wings are less dependent upon the wing flapping velocity, and the relative effect of an increase in wingbeat frequency, although still significant, is less pronounced (Fig. 2b). The evolution of asynchronous flight muscle will thus have greatest aerodynamic consequences for slow-flying insects.

In even the fast forward flight of insects, forces required from the beating wings are primarily required to offset the body weight (see Dudley & Ellington, 1990b). If equiva-

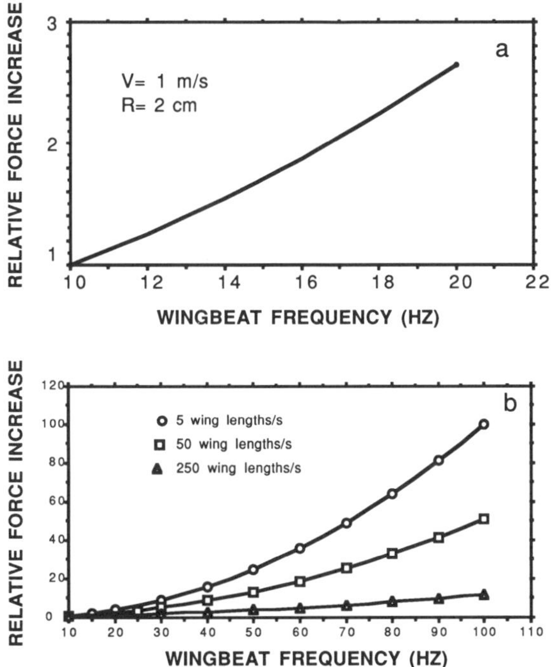

Figure 2. a) The increase in quasi-steady aerodynamic force arising from increased wingbeat frequency, relative to the force produced at 10 Hz. Forward airspeed is 1 m/s, and the assumed wing length of the insect is 2 cm. b) Aerodynamic force increase relative to that at a wingbeat frequency of 10 Hz, for three values of relative forward speed: 5, 50, and 250 wing lengths/s. These values correspond to advance ratios of 0.12, 1.2, and 6, respectively. An advance ratio of 0.12 is close to hovering flight, while the value of 6 represents fast forward flight. A stroke plane angle of 45° and a stroke amplitude of 120° were assumed in a) and b). Effects of the induced velocity are ignored in a) and b), but the results are qualitatively unchanged with its inclusion. Similarly, different stroke amplitudes and stroke plane angles do not affect the qualitative results.

lent forces are to be produced by the wings, an increase in wingbeat frequency permits a reduction in wingbeat amplitude, a decrease in muscle strain, and perhaps most importantly, a decrease in wing area. The relevant data to evaluate an evolutionary reduction of wingbeat amplitude in asynchronous fliers are not available. Strain rates and the energy expended per wing stroke of asynchronous flight muscle are generally lower than those of synchronous muscle, but mechanical power output of asynchronous muscle is typically greater because of its much higher contraction frequencies (see Ellington, 1985; Casey, 1989). It is the potential reduction in wing area, however, that has the greatest consequences for the evolution of flying insects.

By definition, reduction in wing area increases the insect's wing loading. Wing loading is the ratio of a mass to an area, and for isometrically designed organisms increases in proportion to mass$^{1/3}$. Although there are some important exceptions, flying animals generally conform to this prediction (Greenewalt, 1975). By equating the weight of a flying animal with the average lift force produced by the beating wings, it can be shown that the characteristic flight speed increases with the square root of the wing loading (Lighthill, 1977), or with mass$^{1/6}$ for isometric design. This result has also been confirmed empirically for flying animals (Greenewalt, 1975). Mediated by a reduction in wing area, higher wingbeat frequencies thus permit an increase in wing loading and characteristic flight speeds. The latter result in particular has important ecological consequences that will be discussed later.

Because less wing area is needed to fly, one evolutionary possibility afforded by a higher wingbeat frequency is that an entire wing pair may be gradually modified for purposes other than aerodynamic force generation. This key option permitted by the possession of asynchronous flight muscle is characteristic of two major insect orders, the Diptera and Coleoptera. In flies, the hindwings have evolved into gyroscopic halteres which function in stabilization of flight (Pringle, 1948). Halteres do not have a direct aerodynamic role because of their dramatically reduced size relative to the forewings. In most beetles, the forewings have become hardened cuticular sheaths, the elytra, that act to protect the body and folded wings. The elytra may make small contributions to aerodynamic force production, and in some species undergo low amplitude flapping motions, but typically produce much less lift than do the beating hindwings (e.g., Nachtigall, 1964; Schneider & Hermes, 1976). The conventional view of elytra in beetles is that they act as protective devices to resist crushing forces imposed by predators (Crowson, 1981). Indeed, this is one of the few hypotheses routinely invoked to explain the extraordinary diversity of beetles. Somewhat surprisingly then, no analysis of the structural characteristics of elytra appears to have been performed. This would be a productive area for biomechanical analyses of antipredatory morphology similar to those conducted on shelled marine invertebrates (see Vermeij, 1978; Vermeij, 1987).

The thickened hemielytra of Heteroptera probably also serve a protective function (Wootton & Betts, 1986), but here again empirical results are lacking. It is of particular interest that the cicadellids (leaf hoppers), in contrast to other homopterans, generally have thickened forewings with a distinct clavus and corium, similar to the forewings of heteropterans. As mentioned previously, cicadellids have asynchronous flight muscle and are one of the largest homopteran families. In the Orthoptera, an exopterygote order possessing synchronous flight muscle, the forewings (or tegmina) are somewhat reduced in size relative to the hindwings, and are generally thickened. After the Heteroptera, Orthoptera is the most species-rich exopterygote order, and it is tempting to speculate that this abundance of species may be in part due to the mechanical protection afforded by the tegmina. Of course, forewings may also function in camouflage and crypsis.

Lepidoptera is the third largest order of pterygote insects, but is characterized by synchronous flight muscle. In part, lepidopteran diversity may be a consequence of their larval feeding ecology. Lepidoptera are unique in being an almost exclusively phytophagous insect order (Powell, 1980), and phytophagy and lineage diversification are strongly associated in many insect groups (Mitter et al., 1988). Lepidoptera may, however, represent a trend exactly opposite to the theme of this paper, namely an evolutionary reduction in wingbeat frequency. In the Rhopalocera (butterflies), this tendency has been carried to extremes, and wingbeat frequencies are on the order of 10 Hz or lower for large butterflies (Betts & Wootton, 1988; Dudley, 1990). Relative wing area must, however, then be increased to produce equivalent aerodynamic forces, and wing loadings of lepidopterans are generally low compared to other insects (see Greenewalt, 1962; Bartholomew & Heinrich, 1973; Heinrich, 1986; Dudley, 1990). Once enlarged, wings may potentially assume other roles, although substantial costs may ensue. The large, and at times brightly colored, wings of butterflies often function secondarily as devices of sexual advertisement (Silberglied, 1984). Heightened apparency to conspecifics also increases visibility to predators, however, and avian predation appears to impose significant selective pressures on butterflies. Correspondingly, butterflies (and diurnal Lepidoptera in general) tend to have either erratic flight paths (a presumed protean adaptation; Humphries & Driver, 1970) or to be unpalatable (see Chai, 1986 for examples from a Neotropical butterfly assemblage). Moths, by contrast, have occupied primarily the nocturnal habitat, which remained essentially predator-free until the relatively late appearance of the Chiroptera in the Eocene (Jepsen, 1970). As primitive Lepidoptera are known from the fossil record as early as the Lower Cretaceous (Whalley, 1977), adult moths may have enjoyed as many as 50 million years of evolution prior to the appearance of their principal aerial predators.

As mentioned previously, some Psocoptera and also Thysanoptera possess asynchronous flight muscle, as suggested by morphological criteria (Cullen, 1974). Both of these exopterygote orders are fairly species-poor, with about 1700 and 4000 species, respectively. As such, these orders are also exceptions to the general association of species richness and the possession of asynchronous muscle. One possibility is that morphological criteria alone are inadequate to establish asynchronous muscle performance. This limitation potentially extends to all discussion of the taxonomic distribution of asynchronous muscle, and can only be evaluated when the appropriate physiological experiments are performed.

What are the consequences of an increase in wingbeat frequency and a reduction in wing area for maneuverability? We have little detailing information on the comparative flight performance of different insects. Odonates, a group with synchronous flight muscle, use phase shifts and unsteady interactions between fore- and hindwings to effect at times remarkable maneuvers (Somps & Luttges, 1985; Alexander, 1986; Ruⁿppell, 1989). The Odonata are also the only order of pterygote insects to have retained the ancestral condition of homonomous (equal-sized) wings. Among asynchronous fliers, dipterans and hymenopterans would appear to be the most maneuverable insects (e.g., Wagner, 1986; Zeil & Wittmann, 1989). At higher wingbeat frequencies, aerodynamic forces and the associated torques acting upon the insect's body are more dependent upon the flapping velocity of the wing than upon the forward airspeed. This may result in a faster response of such forces and torques to alterations in wing kinematics, and may thereby result in greater capacity for rapid changes in flight speed and direction.

There may be substantial costs associated with increased wingbeat frequencies, particularly those stemming from inertial forces acting on the wing. Wings must be strengthened to resist deformation arising from these inertial forces (Wootton, 1981; Brodskii & Ivanov, 1983; see also Ennos, 1989b). Additionally, fore- and hindwings must be coupled together with greater complexity to ensure that they move in synchrony (Schneider & Schill, 1978). This problem obviously does not arise for beetles or flies, but in the Hymenoptera the fore- and hindwings are tightly coupled by the hamuli, small hooks on the costal margin of the hindwing. The inertial power required to accelerate the wing mass and wing virtual mass also will increase, in proportion to the square of the wingbeat frequency (Elington, 1984c). Inertial power requirements may thus potentially impose prohibitive energetic costs upon asynchronous fliers. However, the existence of a variety of morphological elements in the flight apparatus that act to store elastic energy, particularly the flight muscle itself (Alexander & Bennet-Clark, 1977; Ellington, 1984c), suggests that the entire kinetic energy of the oscillating wings may be stored at the conclusion of a half-stroke and then used to reaccelerate the wings. Inertial power requirements for asynchronous insect fliers would then be negligible (e.g., Dudley & Ellington, 1990b), although we do not know the actual extent of elastic energy storage for any flying insect.

ECOLOGICAL AND EVOLUTIONARY IMPLICATIONS

Asynchronous flight muscle may be a morphological innovation that, by indirectly increasing insect flight speeds, promotes entrance into new adaptive zones. An enhanced capacity for movement between habitats probably promotes ecological diversification, and may thereby foster speciation. Although greater vagility clearly increases gene flow between populations and reduces their genetic divergence, it may also permit greater specialization on temporary and spatially distant resources, which would promote population differentiation. An increase in flight speed will also increase the effectiveness of such

510

varied flight behaviors as mate location, courtship, searching for host plants, nectaring at flowers, and prey location. Increased aerodynamic force production will also result in heightened capacity for acceleration, which may convey decisive advantage in predator-prey interactions. More generally, the adaptive character of an evolutionary increase in locomotor speed has been discussed by Manton (1977) for terrestrial arthropods and Vermeij (1987) for terrestrial vertebrates (see also Gans, 1989).

For flying animals, a major consequence of an increase in locomotor speed is that the direction and trajectory of the flight path are less affected by ambient air motions. While this may seem to be of minor significance to large terrestrial vertebrates such as ourselves, most insects are very small and are readily subject to the vagaries of the winds. May (1978) estimated that the average insect body length is on the order of 4–5 mm, and this is almost certainly an overestimate, given our poor knowledge of very small flying insects such as chalcidoid Hymenoptera. Insects less than 10 mm in length typically fly at airspeeds less than 1 m/s (see Lewis & Taylor, 1967; Johnson, 1969), speeds which are low compared to the ambient winds likely to be encountered. Consider, for example, the wind speeds typical of tropical rain forests, the haunt of most insect species. At mid-level heights of these forests, ambient airspeeds are on the order of 1 m/s, while at the canopy level and above they are frequently much higher (see Haddow & Corbet, 1961; Allen et al., 1972; Thompson & Pinker, 1975; Aoki et al., 1978). For small insects flying in this environment, wind motions will dominate the flight trajectory. Under such circumstances, even small increases in flight speed will convey substantial advantage whenever flight involves a specific destination, as in mate location or searching for nectar sources.

The existence of small insects is in part promoted by evolution of asynchronous flight muscle, as wingbeat frequency and wing length are, in general, inversely correlated (Lighthill, 1977). Small insects typically require wingbeat frequencies well in excess of 100 Hz to sustain flight (see Greenwalt, 1962; Byrne et al., 1988). The quasi-steady aerodynamic analysis is, however, less appropriate for such insects, as unsteady aerodynamic mechanisms are prominent in their flight (e.g., Weis-Fogh, 1973; Wootton & Newman, 1979; Zanker & Götz, 1990). Also, the use of wings as airfoils is progressively impeded at Reynolds numbers below 100, because drag forces increase dispropor-tionately relative to lift (Horridge, 1956). Pringle (1957) noted that even for very small insects the Reynolds number of the beating wings probably exceeds 100. Again, detailed data relevant to this point are not available, although the Reynolds number for the bearing wings of the small chalcid hymenopteran *Encarsia formosa* is between 10 and 20 (Weis-Fogh, 1973).

One of the most significant events in the evolution of the terrestrial biota was the rise of the angiosperms in the Cretaceous and their associated coevolution with insect pollinators (see Crepet, 1979, Crepet, 1983). Asynchronous insect fliers may have had an important role in this coevolutionary process. We know little about early modes of pollina-tion, but asynchronous fliers such as beetles, flies, and hymenopterans (particularly bees) are among the dominant contemporary invertebrate pollinators (Faegri & Van der Pijl, 1971; Kevan & Baker, 1983). By increasing the foraging range and rate of flower visitation, faster flight may promote evolution of trap-lining by pollinators and outcrossing of plants. In tropical forests, for example, tree species typically occur at low adult densities (Ashton, 1969; but see Hubbell, 1979), and bees capable of flying long distances are major pollinators (Janzen, 1971; Frankie, 1975; Frankie, 1976; Roubik, 1989). Among pollinators, the ability to hover is associated with an increased rate of flower visitation (Heinrich, 1983). Hovering is very demanding aerodynamically (Ellington, 1984a), and is facilitated by high wingbeat frequencies. Among the Lepidoptera, hawkmoths (Sphingidae) are important pollinators, and exhibit both high wingbeat frequencies and hovering flight (Bartholomew & Casey, 1978), albeit within the constraints imposed by synchronous flight muscle.

CONCLUSIONS

Key aerodynamic consequences of the evolution of asynchronous flight muscle and higher wingbeat frequencies are increases in insect flight speed and in wing loading. An evolutionary increase in flight speed heightens effectiveness of various flight behaviors, permits the evolution of small insects, and may have promoted the coevolution of angiosperms and their many insect pollinators. The reduction of wing area permitted by high wingbeat frequencies may have facilitated the evolution of halteres in Diptera and of elytra in Coleoptera. Cuticular armor, in the form of elytra, hemielytra, and tegmina, appears to convey mechanical protection to some insect taxa, but structural analyses of these forms of bioarmor are lacking. For most insect groups, we need physiological evaluation of the morphological criteria used to differentiate synchronous and asynchronous flight muscle. Finally, the very limited data on insect flight characteristics and functional morphology need to be vastly increased. Fortunately, detailed kinematic and biomechanical data are becoming available for a variety of insect taxa (e.g., Ellington, 1984b; Betts, 1986; Ennos, 1989a, Dudley & Ellington, 1990a; Dudley, 1990). However, such studies generally examine flight in wind tunnels or enclosures, and we have but limited knowledge of insect flight patterns and behavior in natural contexts (see, however, DeVries & Dudley, 1990; Dudley & DeVries, 1990). Our knowledge of the flight activity of tropical insects is particularly limited. Most flying insects in tropical rain forests live in the forest canopy (Sutton, 1983), a milieu about which we have little knowledge, even of its faunistic composition. As is well known, this "last biotic frontier" (Erwin, 1983) is now being destroyed at unprecedented rates. Sophisticated biomechanical analyses will do little for biologists if our organisms of interest have permanently disappeared from the surface of the earth.

ACKNOWLEDGMENTS

I am indebted to the Smithsonian Tropical Research Institute for the privilege of extended residence on Barro Colorado Island, and for continued support of biomechanical research on tropical insects. Numerous members of the S.T.R.I. scientific community contributed helpful comments on the manuscript. This work was supported by a Smithsonian Institution Postdoctoral Fellowship.

LITERATURE CITED

Alexander, D. E. 1986. Wind tunnel studies of turns by flying dragonflies. *Journal of Experimental Biology* 122:81–98.

Alexander, R. M. & H. C. Bennet-Clark. 1977. Storage of elastic strain energy in muscle and other tissues. *Nature* 265:114–117.

Allen, L. H., Lemon, E. & L. Muller. 1972. Environment of a Costa Rican forest. *Ecology* 53:102–111.

Aoki, M., Yabuki, K. & H. Koyama. 1978. Micrometeorology of Pasoh forest. *Malayan Nature Journal* 30:149–159.

Ashton, P. S. 1969. Speciation among tropical forest trees: some deductions in the light of recent evidence. *Biological Journal of the Linnean Society of London* 1:155–196.

Bartholomew, G. A. & T. M. Casey. 1978. Oxygen consumption of moths during rest, pre-flight warmup and flight in relation to body size and wing morphology. *Journal of Experimental Biology* 76:11–25.

Bartholomew, G. A. & B. Heinrich. 1973. A field study of flight temperature in moths in relation to body weight and wing loading. *Journal of Experimental Biology* 58:123–135.

Betts, C. R. 1986. The kinematics of Heteroptera in free flight. *Journal of Zoology, London* (B) 1:303–315.

512

Betts, C. R. & R. J. Wootton. 1988. Wing shape and flight behaviour in butterflies (Lepidoptera: Papilionoidea and Hesperioidea): a preliminary analysis. *Journal of Experimental Biology* 138:271–288.

Boettiger, E. G. 1957. The machinery of insect flight. Pp. 117–142. *In:* B. T. Scheer (ed.), *Recent Advances in Invertebrate Physiology.* University of Oregon Press: Portland.

Borror, D. J., Delong, D. M. & C. A. Triplehorn. 1981. *An Introduction to the Study of Insects.* Saunders College Publishing: Philadelphia, PA. 852 pp.

Brodskii, A. K. & V. D. Ivanov. 1983. Functional assessment of wing structure in insects. *Entomological Review* 62:35–52.

Byrne, D. N., Buchmann, S. L. & H. G. Spangler. 1988. Relationship between wing loading, wing-beat frequency and body mass in homopterous insects. *Journal of Experimental Biology* 135:9–23.

Casey, T. M. 1989. Oxygen consumption during flight. Pp. 257–272. *In:* G. J. Goldsworthy & C. H. Wheeler (eds.), *Insect Flight.* CRC Press: Boca Raton, FL.

Chai, P. 1986. Field observations and feeding experiments on the responses of rufous-tailed jacamars (*Galbula ruficauda*) to free-flying butterflies in a tropical rainforest. *Biological Journal of the Linnean Society of London* 29:161–189.

Crepet, W. L. 1979. Insect pollination: a paleontological perspective. *BioScience* 29:102–108.

Crepet, W. L. 1983. The role of insect pollination in the evolution of the angiosperms. Pp. 29–50. *In:* L. Real (ed.), *Pollination Biology.* Academic Press: Orlando.

Crowson, R. A. 1981. *The Biology of the Coleoptera.* Academic Press: London. 802 pp.

Cullen, M. J. 1974. The distribution of asynchronous muscle in insects with special references to the Hemiptera: an electron microscope study. *Journal of Entomology* 49A:17–41.

Daly, H. V. 1963. Close-packed and fibrillar muscles of the Hymenoptera. *Annals of the Entomological Society of America* 56:295–306.

Daly, H. V., Doyen, J. T. & P. R. Ehrlich. 1978. *Introduction to Insect Biology and Diversity.* McGraw-Hill: New York. 564 pp.

DeVries, P. J. & R. Dudley. 1990. Morphometrics, airspeed, thermoregulation and lipid reserves of migrating *Urania fulgens* (Uraniidae) moths in natural free flight. *Physiological Zoology* 63:235–251.

Dudley, R. 1990. Flight biomechanics of Neotropical butterflies: morphometrics and kinematics. *Journal of Experimental Biology* 150:37–53.

Dudley, R. & P. J. DeVries. 1990. Flight physiology of migrating *Urania fulgens* (Uraniidae) moths: kinematics and aerodynamics of natural free flight. *Journal of Comparative Physiology* A 167:145–154.

Dudley, R. & C. P. Ellington. 1990a. Mechanics of forward flight in bumblebees. I. Kinematics and morphology. *Journal of Experimental Biology* 148:19–52.

Dudley, R. & C. P. Ellington. 1990b. Mechanics of forward flight in bumblebees. II. Quasi-steady lift and power requirements. *Journal of Experimental Biology* 148:53–88.

Ellington, C. P. 1984a. The aerodynamics of hovering insect flight. I. The quasi-steady analysis. *Philosophical Transactions of the Royal Society of London* B 305:1–15.

Ellington, C. P. 1984b. The aerodynamics of hovering insect flight. III. Kinematics. *Philosophical Transactions of the Royal Society of London* B 305:41–78.

Ellington, C. P. 1984c. The aerodynamics of hovering insect flight. VI. Lift and power requirements. *Philosophical Transactions of the Royal Society of London* B 305:145–181.

Ellington, C. P. 1985. Power and efficiency of insect flight muscle. *Journal of Experimental Biology* 115:293–304.

Ennos, A. R. 1989a. The kinematics and aerodynamics of the free flight of some Diptera. *Journal of Experimental Biology* 142:49–85.

Ennos, A. R. 1989b. Inertial and aerodynamic torques on the wings of Diptera in flight. *Journal of Experimental Biology* 142:87–95.

Erwin, T. L. 1982. Tropical forests: their richness in Coleoptera and other Arthropod species. *The Coleopterists Bulletin* 36:74–75.

Erwin, T. L. 1983. Tropical forest canopies, the last biotic frontier. *Bulletin of the Entomological Society of America* 29:14–19.

Faegri, K. & L. van der Pijl. 1971. *The Principles of Pollination Ecology.* Pergamon Press: Oxford. 248 pp.

Frankie, G. W. 1975. Tropical forest phenology and pollinator plant coevolution, Pp. 192–209. *In:* L. E. Gilbert & P. H. Raven (eds.), *Coevolution of Plants and Animals.* University of Texas Press: Austin.

Frankie, G. W. 1976. Pollination of widely dispersed trees by animals in Central America, with an emphasis on bee pollination systems. Pp. 151–159. *In:* J. Burley & B. T. Stiles, *Variation, Breeding, and Conservation of Tropical Forest Trees.* Academic Press: London.

Gans, C. 1989. Stages in the origin of vertebrates: analysis by means of scenarios. *Biological Reviews of* the *Cambridge Philosophical Society* 64:221–268.

Greenwalt, C. H. 1962. Dimensional relationships for flying animals. *Smithsonian Miscellaneous Collections* 144:1–46.

Greenewalt, C. H. 1975. The flight of birds. *Transactions of the American Philosophical Society* 65, part 4.

Haddow, A. J. & P. S. Corbet. 1961. Entomological studies from a high tower in Mpanga Forest, Uganda. II. Observations on certain environmental factors at different levels. *Transactions of the Royal Entomological Society of London* 113:257–269.

Heinrich, B. 1983. Insect foraging energetics. Pp. 187–214. *In:* C. E. Jones & R. J. Little (eds.), *Handbook of Experimental Pollination Biology.* Scientific and Academic Editions: New York.

Heinrich, B. 1986. Comparative thermoregulation of four montane butterflies of different mass. *Physiological Zoology* 59:616–626.

Hennig, W. 1981. *Insect Phylogeny.* John Wiley & Sons: Chichester. 514 pp.

Hinton, H. E. 1948. On the origin and function of the pupal stage. *Transactions of the Royal Entomological Society of London* 99:395–409.

Hinton, H. E. 1963. The origin and function of the pupal stage. *Proceedings of the Royal Entomological Society of London* A 38:77–85.

Horridge, G. A. 1956. The flight of very small insects. *Nature* 178:1334–1335.

Hubbell, S. P. 1979. Tree dispersion, abundance, and diversity in a tropical dry forest. *Science* 203:1299–1309.

Humphries, D. A. & P. M. Driver. 1970. Protean defence by prey animals. *Oecologia* 5:285–302.

Janzen, D. H. 1971. Euglossine bees as long-distance pollinators of tropical plants. *Science* 171:203–205.

Jepsen, G. L. 1970. Bat origins and evolution. Pp. 1–64. *In:* W. A. Wimsatt (ed.), *Biology of Bats.* Vol. I. Academic Press: New York.

Johnson, C. G. 1969. *Migration and Dispersal of Insects by Flight.* Methuen & Co Ltd: London. 763 pp.

Josephson, R. K. & D. Young. 1985. A synchronous muscle with an operating frequency greater than 500 Hz. *Journal of Experimental Biology* 118:185–208.

Kevan, P. G. & H. G. Baker. 1983. Insects as flower visitors and pollinators. *Annual Review of Entomology* 28:407–453.

Lewis, T. & L. R. Taylor. 1967. *Introduction to Experimental Ecology.* Academic Press: London. 401 pp.

Lighthill, J. 1977. Introduction to the scaling of aerial locomotion. Pp. 365–404. *In:* T. J. Pedley (ed.), *Scale Effects in Animal Locomotion.* Academic Press: London.

Manton, S. M. 1977. *The Arthropoda: Habits, Functional Morphology, and Evolution.* Clarendon Press: Oxford. 527 pp.

May, R. M. 1978. The dynamics and diversity of insect faunas. Pp. 188–204. *In:* L. A. Mound & N. Waloff (eds.), *Diversity of Insect Faunas.* Blackwell Scientific Publications: Oxford.

Mitter, C., Farrell, B. & B. Wiegmann. 1988. The phylogenetic study of adaptive zones: has phytophagy promoted insect diversification? *The American Naturalist* 132:107–128.

Nachtigall, W. 1964. Zur Aerodynamik des Coleopterenflugs: wirken die Elytren als Tragflügel? *Verhandlungen der Deutschen Zoologischen Gesellschaft (Kiel)* 58:319–326.

Norberg, U. M. & J. M. V. Rayner. 1987. Ecological morphology and flight in bats (Mammalia: Chiroptera): wing adaptations, flight performance, foraging strategy and echolocation. *Philosophical Transactions of the Royal Society of London* B 316:335–427.

Noyes, J. S. 1978. On the numbers of genera and species of Chalcidoidea (Hymenoptera) in the world. *The Entomologist's Gazette* 29:163–164.

Parker, S. P. 1982. (ed.) *Synopsis and Classification of Living Organisms.* McGraw-Hill: New York. 2 volumes.

Powell, J. A. 1980. Evolution of larval food preferences in Microlepidoptera. *Annual Review of Entomology* 25:133–160.

Pringle, J. W. S. 1948. The gyroscopic mechanisms of the halteres of Diptera. *Philosophical Transactions of the Royal Society of London* B 233:347–384.

Pringle, J. W. S. 1949. The excitation and contraction of the flight *Locomotion.* Academic Press: London.

Pringle, J. W. S. 1957. *Insect Flight.* Cambridge University Press: Cambridge. 132 pp.

Pringle, J. W. S. 1981. The evolution of fibrillar muscle in insects. *Journal of Experimental Biology* 94:1–14.

Rayner, J. M. V. 1987. Form and function in avian flight. Pp. 1–66. *In:* R. F. Johnston (ed.), *Current Ornithology,* Vol. 5. Plenum Press: New York.

Richards, O. W. & R. G. Davies. 1977. *Imm's General Textbook of Entomology, Tenth Edition.* Chapman and Hall: London. 1354 pp.

Roeder, K. D. 1951. Movements of the thorax and potential changes in the thoracic muscles of insects during flight. *The Biological Bulletin* 100:95–106.

Roubik, D. W. 1989. *Ecology and Natural History of Tropical Bees.* Cambridge University Press: Cambridge. 514 pp.

Rüppell, G. 1989. Kinematic analysis of symmetrical flight manoeuvres of Odonata. *Journal of Experimental Biology* 144:13–42.

Schneider, P. & M. Hermes. 1976. Die Bedeutung der Elytren bei Vertretern des Melolontha-Flugtyps. *Journal of Comparative Physiology.* 106:39–49.

Schneider, P. & R. Schill. 1978. Der Gleitdoppelmechanismus bei vierflügeligen Insekten mit asynchronen Flugmotor. *Zoologische Jahrbücher Abteilung für allgemeine Zoologie und Physiologie der Tiere* 82:365–382.

Silberglied, R. E. 1984. Visual communication and sexual selection among butterflies. Pp. 207–223. *In:* R. I. Vane-Wright & P. R. Ackery. *The Biology of Butterflies.* Academic Press: London.

Somps, C. & M. W. Luttges. 1985. Dragonfly flight: novel uses of unsteady separated flows. *Science* 228:1326–1329.

Sotavalta, O. 1953. Recordings of the high wing-stroke and thoracic vibration frequency in some midges. *The Biological Bulletin* 104:439–444.

Stork, N. E. 1988. Insect diversity: facts, fiction and speculation. *Biological Journal of the Linnean Society of London* 35:321–337.

Sutton, S. L. 1983. The spatial distribution of flying insects in tropical rain forests. Pp. 77–91. *In:* S. L. Sutton, T. C. Whitmore & A. C. Chadwick (eds.). *Tropical Rain Forest: Ecology and Management.* Blackwell Scientific Publications: Oxford.

Thompson, O. E. & R. T. Pinker. 1975. Wind and temperature profile characteristics in a tropical evergreen forest in Thailand. *Tellus* 27:562–573.

Tregear, R. T. 1977. (ed.) *Insect Flight Muscle.* North Holland: Amsterdam. 367 pp.

Usherwood, P. N. R. 1975. (ed.) *Insect Muscle.* Academic Press: London. 621 pp.

Vermeij, G. J. 1978. *Biogeography and Adaptation.* Harvard University Press: Cambridge, MA. 332 pp.

Vermeij, G. J. 1987. *Evolution and Escalation.* Princeton University Press: Princeton, NJ. 527 pp.

Wagner, H. 1986. Flight performance and visual control of flight of the free-flying housefly (*Musca domestica* L.) II. Pursuit of targets. *Philosophical Transactions of the Royal Society of London* B 312:553–579.

Weis-Fogh, T. 1973. Quick estimates of flight fitness in hovering animals, including novel mechanisms for lift production. *Journal of Experimental Biology* 59:169–230.

Whalley, P. 1977. Lower Cretaceous Lepidoptera. *Nature* 266:526.

Wootton, R. J. 1981. Support and deformability in insect wings. *Journal of Zoology, London* 193:447–468.

Wootton, R. J. & C. R. Betts. 1986. Homology and function in the wings of Heteroptera. *Systematic Entomology* 11:389–400.

Wootton, R. J. & D. J. S. Newman. 1979. Whitefly have the highest contraction frequencies yet recorded in non-fibrillar flight muscles. *Nature* 280:402–403.

Zanker, J. M. & K. G. Götz. 1990. The wing beat of *Drosophila melanogaster.* II. Dynamics. *Philosophical Transactions of the Royal Society of London* B 327:19–44.

Zeil, J. & D. Wittmann. 1989. Visually controlled station-keeping by hovering guard bees of *Trigona (Tetragonisca) angustula* (Apidae, Meliponinae). *Journal of Comparative Physiology* A 165:711–718.

FUNCTIONAL MORPHOLOGY, BIOMECHANICS,
AND EVOLUTIONARY PROCESS—SYMPOSIUM

The Functional Basis of Intraspecific Trophic Diversification in Sunfishes

Peter C. Wainwright, George V. Lauder, Craig W. Osenberg,
and Gary G. Mittelbach

Abstract. Research on the patterns of transformation in the vertebrate feeding mechanism has historically focused on interspecific comparisons. A major generalization that has emerged from this work is that morphological features of the feeding mechanism tend to change frequently during evolution while muscle activity patterns that operate the feeding mechanism during prey capture and prey handling tend to be conserved. As a contrast to these interspecific comparisons, we present a case study of intraspecific trophic divergence in the molluscivorous pumpkinseed sunfish, *Lepomis gibbosus*. We have explored the effect an environmentally determined diet shift has on the morphology and muscle activity patterns associated with snail predation by comparing pumpkinseeds from two lakes. In one lake snails are abundant and comprise the bulk of pumpkinseed diets, while in the other lake snails are rare and pumpkinseed diets are comprised almost entirely of soft-bodied invertebrates. Most pharyngeal jaw muscles and bones were significantly larger in pumpkinseeds inhabiting the snail-rich lake. In contrast, only one of five muscles examined exhibited functional (activity pattern) variation between lakes. This muscle, the pharyngocleithralis externus muscle, showed a major difference in its pattern of use, yet it varied least morphologically between lakes. Thus, patterns of change in muscle morphology are not congruent with changes in the snail crushing motor pattern. The morphological and motor pattern changes underlie differences between populations in the ability to feed on hard-shelled gastropods. These results are consistent with previous interspecific studies of the evolution of the feeding mechanism of fishes, and support the conclusion that, in spite of the highly integrated and complex nature of the feeding mechanism in fishes, morphology and muscle activity patterns frequently change independently of each other.

INTRODUCTION

The evolution of complex biomechanical systems has been an enduring problem in evolutionary biology. Functional units within organisms are often highly complex, tightly integrated networks, and many authors have noted the potential difficulty in modifying these systems during evolution (Darwin, 1859; Dawkins, 1986; Wake & Roth, 1989). Classic examples within vertebrates include the visual system (Darwin, 1887; Levine, 1985), the locomotor system (Goslow et al., 1989; Alexander, 1982), the hearing system (Turner, 1980; Fay & Popper, 1985), and the focus of this paper, the feeding mechanism (Frazetta, 1975; Lauder et al., 1989). These functional systems are all recognized as being

Drs. Wainwright and Lauder are with the Department of Ecology and Evolutionary Biology, University of California, Irvine, CA 92717, USA. Dr. Wainwright's present address is Department of Biological Sciences, Florida State University, Tallahassee, FL 32306, USA. Dr. Osenberg is with the Marine Science Institute, Department of Biological Sciences, University of California, Santa Barbara, CA 93106, USA. Dr. Mittelbach is with the Kellogg Biological Station, Department of Zoology, Michigan State University, Hickory Corners, MI 49060, USA.

516

tightly integrated and complex, yet vertebrates, in general, are characterized by broad diversity within each system. Thus, any attempt to understand the nature of vertebrate diversity must include an understanding of how complex systems transform during evolution.

One approach to this problem that has proven to be especially fruitful is to separately consider the evolution of different levels of design of the system under evaluation (Liem, 1989; Lauder, 1990; Wainwright & Lauder, in press). For example, the design of the vertebrate feeding mechanism can be rendered into at least three components (Lauder, 1990; Wainwright & Lauder, in press): peripheral morphology (e.g., muscle and bone organization), physiological properties of the peripheral morphology (e.g., contractile properties of feeding muscles), and motor patterns (the patterns of muscle activity that drive feeding behaviors). By analyzing the components of the feeding mechanism separately it is possible to ask how evolutionary transformations at each level of design have contributed to evolutionary alterations in overall feeding performance: is the diversification of feeding habits, so often observed among even fairly closely related taxa, due primarily to alterations in trophic morphology, the physiological capacity of the morphology, or changes in motor pattern, or, are some levels of design more conservative during evolution than others?

Recent comparative research on the evolution of the feeding mechanism in fishes and aquatic salamanders has resulted in the key generalization that trophic diversification is generally associated with morphological transformations and only rarely with changes in the motor pattern used during prey capture and handling. Quantitative comparisons of the motor pattern among members of fairly closely related groups have usually found that fewer than 10% of the motor pattern parameters measured differ significantly among species (Shaffer & Lauder, 1985; Wainwright & Lauder, 1986; Sanderson, 1988; Wainwright, 1989a; Westneat & Wainwright, 1989), though this pattern may break down in very broad phylogenetic comparisons (Wainwright et al., 1989). In contrast, differences in diet and feeding performance are usually associated with morphological changes (Sanderson, 1988; Motta, 1988; Wainwright & Lauder, in press). In some cases even major morphological transformations occur in the face of motor pattern conservatism (e.g., Westneat & Wainwright, 1989).

The data base on the evolution of the feeding mechanism in lower vertebrates is built primarily on interspecific comparative analyses. Few data exist on the process of trophic diversification within species (Liem & Kaufman, 1984; Meyer, 1990). Comparative analyses characterize the diversity present in extant species and, coupled with a well corroborated phylogenetic hypothesis, can give an estimate of the sequence of changes in the feeding system and its components. What these studies do not generally permit, however, is a view of the process of change in the feeding system and the environmental conditions that bring it about.

In this paper we present a summary of a series of investigations that we have conducted on different *populations* of a single species, the pumpkinseed sunfish, *Lepomis gibbosus*. Adult pumpkinseeds are typically trophic specialists on gastropod molluscs that they crush in their pharyngeal jaws. We have taken advantage of natural variation among lakes in the abundance of snails, and documented the influence of an environmentally imposed diet shift on the morphology and the snail crushing motor pattern of the pharyngeal jaw apparatus. The combined morphological and motor pattern analyses permit us to contrast the patterns of change at these two design levels in the same muscles. Do muscles that show the greatest morphological response also tend to show the greatest motor pattern response, or are morphological changes uncorrelated with changes in muscle activity?

Snail crushing in pumpkinseeds offers a particularly promising system to study intraspecific trophic diversification because of extensive previous research on both the functional morphology of snail crushing (Lauder, 1983a, b; Wainwright & Lauder, in press) and the general feeding ecology of this species (Keast, 1978; Mittelbach, 1984, 1988; Osenberg & Mittelbach, 1989). Only rarely are studies of polymorphic species made in the light of such detailed knowledge of the functional morphology and ecology of a complex functional system (i.e., Liem & Kaufman, 1984; Meyer, 1989, 1990). The existing data strengthen our ability to interpret the functional and ecological consequences of morphological differences and, because the functional morphology of this system is well understood, we are able to infer the mechanistic processes that produced the morphological and motor pattern variation.

The System

The pumpkinseed sunfish, *Lepomis gibbosus*, and its sister species, the redear sunfish, *L. microlophus*, are the only molluscivorous members of the endemic North American freshwater fish family Centrarchidae. Mollusc crushing in these species is one of the few well-documented cases in which the novel trophic habit has involved the evolution of a novel pattern of muscle activity (Lauder, 1983a, b, 1986). Snail crushing is associated with a unique motor pattern not usually found in non-snail crushing sunfishes. In addition to neuromuscular specializations the pumpkinseed and redear also exhibit morphological modifications, including hypertrophy of the pharyngeal jaw muscles, bones, and teeth between which snails are crushed.

Gastropods commonly make up a large fraction of adult pumpkinseed diets in nature (often greater than 70% by volume or dry weight; Sadzikowski & Wallace, 1976; Keast, 1978; Mittelbach, 1984; Osenberg & Mittelbach, 1989). Small pumpkinseeds, which are unable to effectively crush snails (Mittelbach, 1984), feed on insect larvae and other soft-bodied invertebrates (Sadzikowski & Wallace, 1976; Keast, 1978; Mittelbach, 1988). Although pumpkinseeds show considerable specialization for feeding on snails, their prey selection is flexible and responds to changes in resource levels (Werner & Hall, 1979; Osenberg, Mittelbach & Wainwright, in press). In our studies we have used pronounced differences in pumpkinseed diets between two natural lakes in southern Michigan (USA) to examine the impact of consuming gastropods on the development of the snail crushing motor pattern, pharyngeal jaw morphology, and snail crushing ability. Wintergreen lake has a very dense population of pumpkinseeds and a depauperate snail fauna (Osenberg et al., in press). In nearby Three Lakes II (hereafter simply called Three Lakes), snail abundances are more typical of the region (over an order of magnitude higher than in Wintergreen lake).

METHODS

In this paper we will summarize our comparisons of several aspects of the feeding biology of pumpkinseeds from Wintergreen and Three Lakes: dietary habits, feeding ability, trophic morphology, and the muscle activity patterns used during snail crushing. Below we briefly describe the methodology used in these analyses. More detailed descriptions are presented elsewhere (Wainwright et al., 1991; Osenberg et al., in press).

Fish were collected from the two lakes for dietary analysis on four dates throughout 1988 and 1989. On each sampling date between 30 and 40 pumpkinseeds, ranging in size from 28–130 mm standard length (SL) were collected and the contents of their stomachs were later identified, counted, and measured under a dissecting microscope. Previously determined length-mass regressions were used to convert prey dimensions to dry mass.

The ability of pumpkinseeds from the two lakes to handle and consume snails was compared in laboratory experiments (Osenberg et al., in press). Here we present data for pumpkinseeds feeding on two snail species, *Amnicola limosa*, a relatively thick-shelled

518

gastropod, and *Physa*, a thinner-shelled snail (laboratory estimates of the force required to crush the snails averaged 10.1 N vs. 2.7 N for *Amnicola* and *Physa* respectively, Osenberg et al., in press). Fish ranging in size from 63 to 125 mm SL were collected from Wintergreen and Three Lakes and housed separately in laboratory aquaria for one week prior to the feeding experiments. Snails were presented to fish individually and the time required to crush each snail was measured. Data were analyzed with ANCOVA using lake as the grouping variable and standard length as the covariate.

A morphological analysis was carried out on 33 fish (40–132 mm SL) collected from Three Lakes and 20 individuals (45–109 mm SL) from Wintergreen lake (Wainwright et al., 1991). Each fish was preserved, ten muscles and five bones were dissected from the pharyngeal jaw region, and each was weighed to the nearest 0.01 mg. Several additional shape variables were measured from two of the bones (the shape analysis is discussed in Wainwright et al., 1991). Nine of the ten muscles function during snail crushing, with the levator posterior being the primary force producing muscle of the system. One muscle, the sternohyoideus, functions during prey capture but not during snail crushing and we used this muscle as a control to ensure that differences between lakes were not simply present in all muscles but were instead specific to muscles that function during snail crushing. Of the five bones, three are directly involved in snail crushing, while the other two were selected as controls. Muscle and bone masses were $\log_{10}$ transformed and regressed against body mass, and comparisons were made between the lake populations with ANCOVA.

Patterns of muscle activity during snail crushing were recorded in electromyographic experiments following methods outlined in detail elsewhere (Wainwright & Lauder, 1986; Wainwright, 1989a). Seven fish collected from Three Lakes (103–120 mm SL) and six fish from Wintergreen lake (99–117 mm SL) were first anesthetized and then fine-wire bipolar electrodes were implanted directly into the belly of five pharyngeal jaw muscles. The five muscles were a subset of the ten muscles examined in the morphological analysis. The five paired electrode wires were then bundled together into a cable that was sutured to the back of the fish. Following recovery from anesthesia, fish were offered snails (*Physa* sp.) that were eagerly accepted. Voltages produced during contractions of the pharyngeal muscles at prey capture and snail crushing were amplified and recorded on a multi-track FM tape recorder. Several feedings were analyzed from each individual for a total of 59 feedings from Three Lakes fish and 51 from Wintergreen fish. The analogue myogram recordings were transformed to digital computer files and 16 variables were measured from each snail crushing event that characterized the overall timing and intensity of muscle activity. A nested ANOVA was used to analyze each EMG variable, with individuals nested within lake.

Diet and Feeding Performance

Pumpkinseeds in Three Lakes show a pronounced diet shift during ontogeny (Mittelbach, 1984; Osenberg et al., in press). Fish smaller than about 45 mm SL have few snails in their stomachs (mean ± SE percent of diet based on prey biomass: 3.3% ± 1.6) and instead feed on soft-bodied invertebrates (primarily insect larvae such as Chironomidae). Between 45 and 75 mm SL there is a drastic increase in the amount of snails eaten, with fish above 75 mm feeding almost entirely on gastropods (87.2% ± 2.6). Other studies of pumpkinseeds in midwestern U.S.A. lakes have found patterns similar to those seen in the Three Lakes population (Seaburg & Moyle, 1964; Sadzikowski & Wallace, 1976).

Wintergreen pumpkinseeds show a strikingly different dietary pattern. The classic dietary switch that is seen in Three Lakes fish is absent in the Wintergreen population, with snails comprising only 1.5% ± 0.8 of the diet of pumpkinseeds ≥ mm SL (Osenberg et al., in press). Thus, throughout ontogeny Wintergreen fish feed mostly on insect larvae and other soft-bodied invertebrates.

The difference in dietary patterns between the two populations is reflected in their ability to handle gastropod prey (Osenberg et al., in press). When fed the thick-shelled gastropod *Amnicola* Wintergreen fish required significantly longer to crush the snail shell than fish from the Three Lakes population (Fig 1; ANCOVA lake effect test, $P<0.01$). In contrast, there was no difference between populations in handling time when fed the thinner-shelled gastropod *Physa* (Fig. 1; ANCOVA lake effect test, $P > 0.05$).

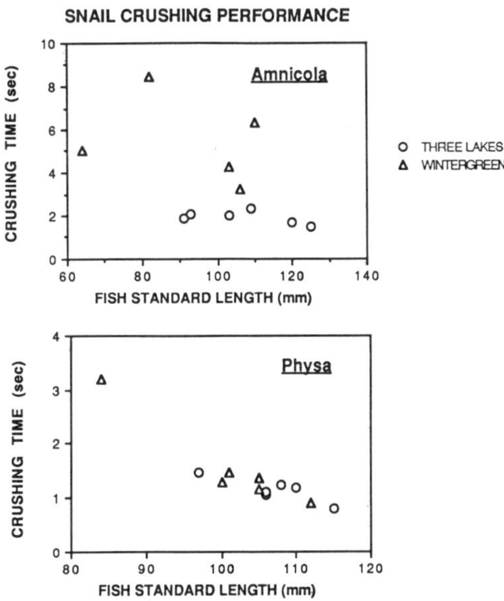

Figure 1. Results of feeding performance experiments conducted with pumpkinseeds from Wintergreen lake and Three Lakes. Crushing time when feeding on the hard-shelled gastropod *Amnicola* is significantly longer in Wintergreen than in Three Lakes fish. There is no difference between populations in the time required to crush the shell of the softer-shelled gastropod *Physa*. See text for statistics.

Mechanism of Snail Crushing

The functional morphology of snail crushing in pumpkinseed sunfish has been discussed in detail elsewhere (Lauder, 1983a; Wainwright, 1989b; Wainwright & Lauder, in press) but a brief account is presented here to provide a context in which to interpret the morphological and motor pattern differences between fish of the two lakes. Snails are first captured by the oral jaws, using suction to draw the prey into the buccal cavity, and are then passed to the pharyngeal jaw apparatus for processing. The key movement in snail crushing behavior is the depressive action of the upper pharyngeal jaw (PB3) against the relatively stationary lower pharyngeal jaw (CB5; see Fig. 2). This action compresses the snail shell between the jaws, ultimately crushing the shell. Upper jaw depression is accomplished through rotation of the fourth epibranchial about the insertion of the obliquus posterior muscle on its midventral aspect (Wainwright, 1989b). The rotating fourth epibranchial articulates with the dorsal surface of the upper pharyngeal jaw and presses it ventrally. Several muscles, principally the levator posterior, fourth levator externus, and third obliquus dorsalis function to depress the upper jaws. At the same time that the upper jaw is depressed it is also retracted posteriorly by the retractor dorsalis muscle. During crushing the lower jaws are situated such that the posterior region is more dorsal than the

520

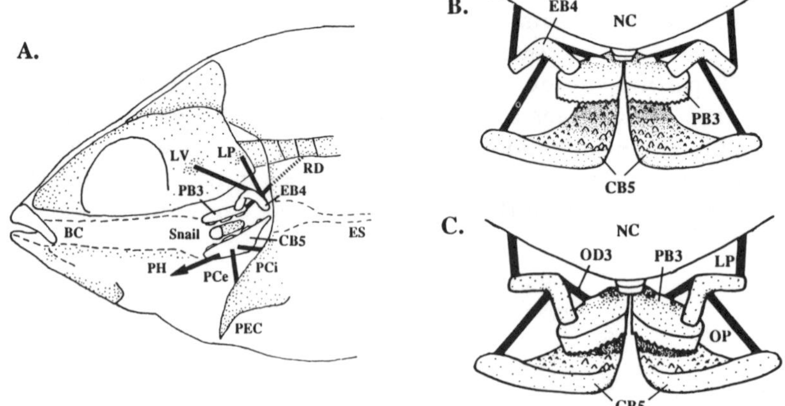

Figure 2. Illustrations of the snail crushing mechanism in the pumpkinseed sunfish. A) Schematic lateral view of the head illustrating the position of the pharyngeal jaw apparatus and the muscles that function during its use. Muscles are shown as thick black lines indicating their attachments. B) Posterior view diagram of the pharyngeal jaws illustrating the mechanism of upper jaw depression that is crucial to snail crushing. During snail crushing the lower jaw is held relatively stationary and the upper jaw exerts the primary crushing force as it is pressed firmly against the snail shell. Upper jaw depression is caused by rotation of epibranchial 4 about the insertion site of the obliquus posterior muscle. This rotation causes the epibranchial to press against the dorsal surface of the upper jaw, forcing it downward. This crushing action is produced by several muscles, principally the levator posterior, levator externus 3/4, and the obliquus dorsalis. Abbreviations: BC, buccal cavity; CB5, fifth ceratobranchial or lower pharyngeal jaw; EB4, fourth epibranchial; ES, esophagus; LP, levator posterior; LV, fourth levator externus; NC, neurocranium; OD3, third obliquus dorsalis; OP, obliquus posterior; PB3, third pharyngobranchial or upper pharyngeal jaw; PCe, pharyngocleithralis externus; PCi, pharyngocleithralis internus; PEC, pectoral girdle; PH, pharyngohyoideus; RD, retractor dorsalis. Figure is from Wainwright et al. (1991).

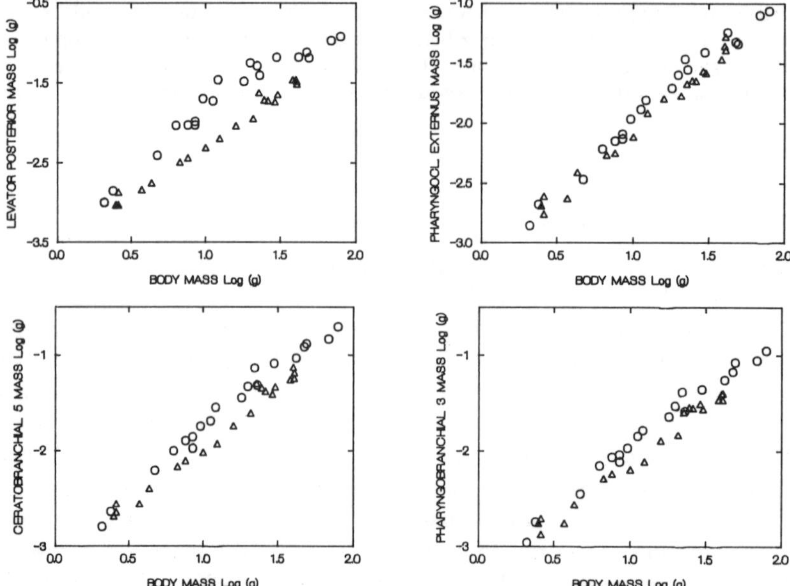

Figure 3. Plots of muscle and bone mass against body mass for pumpkinseeds from Wintergreen lake (triangles) and Three Lakes (circles). The levator posterior and pharyngocleithralis are muscles of the pharyngeal jaws, and the 5th ceratobranchial and 3rd pharyngobranchial are the toothed lower and upper pharyngeal jaw bones respectively. Among the four structures plotted, only the pharyngocleithralis externus muscle is not significantly larger in the Three Lakes population. Data from Wainwright et al. (1991).

Table 1. Results of ANCOVAs for the $\log_{10}$ masses of muscles and bones from the pharyngeal jaw apparatus (PJA) and head of pumpkinseeds from Three Lakes and Wintergreen lake. Body mass was the covariate in each case. Muscles and bones are divided into separate categories depending on whether they are involved in snail crushing function. Data from Wainwright et al. (1991).

	ANCOVA		
Structure	Slopes	Intercept	Ratio of ** adjusted means
muscles:			
PJA muscles:			
levator posterior	NS	109.3*	2.33
levator internus 3	NS	35.1*	1.48
pharyngohyoideus	NS	17.0*	1.39
obliquus dorsalis 3	NS	23.4*	1.37
retractor dorsalis	NS	13.6*	1.31
pharyngocl. internus	NS	16.5*	1.29
levator externus 3/4	NS	6.7	1.19
levator internus 2	NS	5.1	1.14
pharyngocl. externus	NS	4.4	1.10
not PJA muscle:			
sternohyoideus	NS	1.5	1.10
bones:			
elements of the PJA:			
ceratobranchial 5	NS	47.5*	1.52
pharyngobranchial 3	NS	44.9*	1.47
epibranchial 4	NS	41.6*	1.65
not elements of the PJA:			
certobranchial 1	NS	2.4	0.80
epibranchial 1	NS	0.9	0.81

* = for the muscles this is $P < 0.005$, and for the bones this is $P < 0.0083$ (these are Bonferroni corrections of $P < 0.05$ for each data set).
** = using a pooled slope for the two populations, this column lists the ratio of predicted muscle and bone masses for fish of an average body mass from each population. Ratios are predicted values for Three Lakes fish divided by predicted values for Wintergreen fish.

anterior region. The snail is thus held against this surface while the upper tooth plates press ventrally and posteriorly on the snail.

Electromyographic studies of the patterns of muscle activity exhibited during snail crushing (Lauder, 1983a, b) have revealed a novel motor pattern in the pumpkinseed and redear sunfish, *Lepomis microlophus*. During crushing all pharyngeal jaw muscles are active simultaneously in intense bursts. Thus, in addition to the upper jaw depressors many antagonistic muscles are active during crushing, presumably to stabilize the jaws during the forceful exertion of the crushing action.

Pharyngeal Jaw Morphology

Pumpkinseeds in Wintergreen lake eat many fewer snails than fish in Three Lakes and are less proficient at handling gastropod prey. In association with these observations we found extensive differences between the two populations in the morphology of the pharyngeal jaw apparatus. Of the nine pharyngeal jaw muscles that were examined, six were significantly larger in Three Lakes fish, with the other three showing a trend in this direction (Table 1; Fig. 3). The primary crushing muscle, the levator posterior, showed the biggest difference, being over twice as large in Three Lakes fish as in Wintergreen fish. The sternohyoideus, which is not activated during snail crushing (Lauder, 1983a), showed no difference between lakes.

522

Significant differences were also observed between lakes in the masses of pharyngeal jaw bones. All three pharyngeal jaw bones were significantly heavier in Three Lakes fish than they were in Wintergreen pumpkinseeds (Table 1; Fig. 3). Two branchial arch bones that are not part of the crushing apparatus showed no difference between populations (Table 1). A more detailed analysis of pharyngeal jaw bone morphology, including considerations of tooth and bone shape is presented elsewhere (Wainwright et al., 1991).

Motor Pattern

Electromyographic data were collected from five of the muscles included in the morphological analysis; the levator posterior, levator externus 3/4, the retractor dorsalis, the pharyngocleithralis internus, and the pharyngocleithralis externus (Wainwright et al., ms.) During snail crushing, pumpkinseeds from Three Lakes exhibited muscle activity patterns characterized by long, simultaneous bursts of activity in four of the five muscles (Fig. 4). The pharyngocleithralis externus muscle (PCe) showed a very different pattern of activation from the other four muscles. This muscle exhibited repeated, short bursts of activity between the time the snail was captured and the beginning of crushing. During the crushing event the PCe muscle showed less activity than the other muscles.

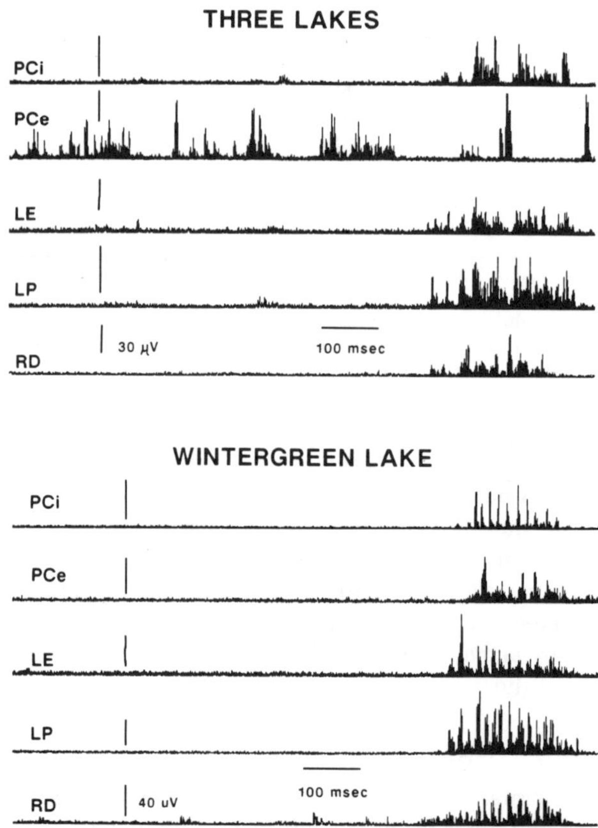

Figure 4. Sample electromyogram from a Three Lakes pumpkinseed and a Wintergreen individual. The myograms have been rectified and the area under each curve is shaded in. The simultaneous bursts of activity that occur in most muscles in each panel correspond to the snail crushing event. Of 16 variables that were measured on myograms during snail crushing only those describing the activity of the PCe muscle were significantly different between populations. Abbreviations: PCi, pharyngocleithralis internus muscle; PCe, pharyngocleithralis externus muscle; LE, levator externus 3/4 muscles; LP, levator posterior muscle; RD, retractor dorsalis muscle. Figure modified from Wainwright, et al. (ms).

Wintergreen pumpkinseeds exhibited a motor pattern that appears very similar to the Three Lakes pattern in four of the five muscles. However, activity of the PCe muscle in Wintergreen fish showed marked differences from that seen in Three Lakes fish (Fig. 4). Activity of this muscle prior to the crushing event was very rare in Wintergreen individuals, and during crushing the PCe was activated in a pattern similar to that seen in the other four muscles.

These differences are born out by statistical analyses. Out of a total of 16 variables that were measured from the electromyographic recordings of each snail crushing event, only 4 varied significantly between lakes (Table 2). Of these four variables, one was the time between the capture of the snail and the onset of the crushing event (significantly longer in Three Lakes fish), while the other three were measures of activity in the PCe muscle. The total integrated area of activity in the PCe prior to crushing was greater in Three Lakes fish. The integrated area and duration of activity of the PCe during crushing were significantly less in Three Lakes pumpkinseeds. Among the other four muscles no integrated area, activity duration, or relative timing variables showed significant lake differences (Table 2).

Table 2. Results of nested ANOVAs (individuals nested within lakes) contrasting 16 electromyographic variables measured from recordings from five pharyngeal jaw muscles made during snail crushing by pumpkinseed sunfish from two lakes. Data analyzed for seven fish from Three Lakes and six fish from Wintergreen lake. Table entries are F-ratios from each test. Data are from Wainwright et al. (ms).

Variable	Lake Effect	Individual
Activity Duration:		
PCe1 DUR	28.8**	4.5**
PCi DUR	2.8	2.6**
LE DUR	0.2	4.6**
LP DUR	0.7	1.7
RD DUR	0.7	1.5
Timing:		
STRIKE-LP	7.5*	3.0**
LP-PCe1	2.4	3.0**
LP-PCi	0.8	2.9**
LP-LE	1.0	3.3**
LP-RD	0.7	3.5**
Rectified Integrated Area:		
PCe pre AREA	18.1**	4.9**
PCe AREA	5.4*	7.7**
PCi AREA	0.3	22.5**
LE AREA	1.3	15.4**
LP AREA	1.3	25.9**
RD AREA	2.5	11.6**

Degrees of freedom for most lake effect tests $= (1, 11)$; degrees of freedom for most individual effect tests $= (11, 95)$. For RD DUR, LP-RD, and RD AREA the lake effect degrees of freedom $= (1, 9)$; the individual effect degrees of freedom $= (9, 80)$.
$* = P < 0.05; ** = P < 0.01$.

DISCUSSION

Our studies on the pumpkinseeds from these two lake populations have revealed a distinct polymorphism in the functional morphology of the snail crushing pharyngeal jaw apparatus. Both the morphology of the pharyngeal jaw muscles and bones and the motor

524

pattern used during snail crushing show important differences between populations. Six of the nine pharyngeal jaw muscles were significantly larger in Three Lakes pumpkinseeds, with the other three muscles showing a trend in this direction. Similarly, pharyngeal jaw bones are larger in Three Lakes fish. In the motor pattern analysis, four muscles exhibited the same activity pattern in the two lakes, while the PCe muscle showed a major, qualitative shift in its activity during snail crushing. In pumpkinseeds from Three Lakes this muscle is repeatedly activated for short, consecutive bursts in the time prior to crushing. In contrast, activity prior to crushing is rare for this muscle in Wintergreen fish, and the PCe is instead activated during crushing much like the other four muscles. In association with these morphological and neuromuscular differences is a reduction in the ability of pumpkinseeds from Wintergreen lake to handle hard-shelled snail species.

Below we discuss the consequences of the polymorphism for pharyngeal jaw function. This is followed by discussions of the possible causes of the polymorphism, and finally by a comparison of this intraspecific data set with previous interspecific studies of the evolution of the feeding mechanism in aquatic-feeding lower vertebrates.

Consequences of Polymorphism for Pharyngeal Jaw Function

Previous studies of the functional morphology of snail crushing in pumpkinseeds and other perciform fishes (Lauder, 1983a; Wainwright, 1989a, b) have identified the key anatomical and neuromuscular elements of snail crushing and make it possible to infer the consequences of the particular morphological and motor pattern differences that were found.

The muscle mass differences between populations probably result in differences in the force producing capacity of those muscles. Within ontogenetic series muscle mass has been demonstrated to be tightly associated with the force producing capacity of individual muscles (Marsh, 1988; Bennett et al., 1989; Thomason et al., 1990). Similarly, when comparing homologous muscles among closely related species, muscle mass has been found to provide an accurate estimate of relative strength (Powell et al., 1984; Wainwright, 1988). This relationship between muscle size and force producing capability is not unexpected. The maximum tension that a muscle can develop is a combined function of its force producing capacity per unit of cross sectional area of muscle tissue (stress) times the physiological cross sectional area (e.g., Calow & Alexander, 1973). If muscle shape, the degree of fiber pinnation, and stress properties do not change during the growth of a muscle, then an increase in muscle mass will result in an increase in the physiological cross sectional area of the muscle, and an increase in the total force that the muscle can develop. Available data for ectothermic vertebrates show that stress is not influenced by body size (Marsh, 1988; Bennett et al., 1989).

Previous studies identify the levator posterior muscle of generalized perciform fishes, like sunfishes, as one of the primary muscles involved in generating the forceful actions used during prey crushing by the pharyngeal jaws (Lauder, 1983a; Wainwright, 1989a, b). Snail crushing performance in fishes has been shown to be limited directly by the amount of force that an individual can exert against the gastropod shell (Wainwright, 1987; Wainwright, 1988; Osenberg & Mittelbach, 1989). Hence, the force capabilities of the levator posterior muscle can be expected to provide a reasonable indicator of the potential crushing force of an individual fish. We estimated the cross sectional area of the levator posterior muscles examined in our morphological analysis and found that the expected force producing capability of this muscle was over twice as large in Three Lakes fish as in Wintergreen fish (Wainwright et al., 1991).

Similarly, the heavier and more robust (Wainwright et al., 1991) pharyngeal jaw bones in Three Lakes fish may be better able to withstand the stress that will be associated with forceful pressing of these elements against gastropod shells. Though we have not measured the mechanical properties of the jaw bones, a correlation between bone robustness and strength has been observed in other systems (e.g., the two Woo et al., 1981 papers). In general then, the broad-scale differences between lakes in the masses of all pharyngeal muscles indicate that Three Lakes fish have substantially stronger pharyngeal jaws than fish from Wintergreen lake. The difference between populations in the ability to feed on hard-shelled snails (Fig. 1) is probably largely due to these differences in crushing strength.

Though the levator posterior can be identified as a key muscle in generating crushing forces, there were no differences between lakes in the motor pattern of this muscle. Hence, the capacity for generating crushing force is greater in Three Lakes fish, yet fish from the two lakes employ the levator posterior muscle in a similar fashion. The only muscle that showed motor pattern differences, the PCe muscle, functions to retract the lower pharyngeal jaw ventrally, away from the shell of a gastropod that is in position to be crushed (Fig. 2). The repeated bursts of activity that occur in this muscle prior to the crushing event thus reflect repeated cycles of lower pharyngeal jaw depression. We hypothesize that this action of the lower jaw and the water motion that it may create within the buccal cavity assist in manipulating the gastropod shell as the fish prepares for a crushing attempt. Thus, the motor pattern polymorphism seen between lakes does not appear to relate directly to the act of snail crushing, but rather to the manipulation of gastropod prey that occurs prior to crushing.

Pharyngeal Jaw Polymorphism: Plasticity or Evolution?

A key issue in this case study concerns the causal basis of the observed polymorphism. Clearly, the difference between lakes in the abundance of snails ultimately underlies the trophic divergence that we have documented. However, whether this divergence results entirely from developmental plasticity, in response to the abundance of snail prey, or has a genetic basis is an issue that can not be resolved with certainty at this time. The answer to this question is significant because it determines the extent to which the observed transformation in pharyngeal jaw function can be viewed as a model of evolutionary change in functional morphology of the feeding mechanism.

Though the issue remains unresolved until we can address it directly in common environment experiments, two observations suggest that the polymorphism is primarily due to phenotypic plasticity and not due to genetic differences between the populations (a more detailed discussion is presented in Wainwright et al., 1991). The first concerns the history of Wintergreen lake. Available evidence suggests the rarity of snails is a recent phenomena (post 1977) brought about by a change in the Wintergreen fish community (Osenberg et al., in press). The second observation is that all of the changes, morphological and neuromuscular, can be seen as resulting from ontogenetic interactions of the feeding mechanism with prey use. By frequently crushing snails during ontogeny, fish in Three Lakes exert a training effect on their pharyngeal jaw muscles (Chapman & Troup, 1970; Ashton & Singh, 1974). The salient feature of pharyngeal jaw muscle activity in the pumpkinseed is that nearly all muscles are active for long, intense bursts during crushing bouts (Fig. 4; Lauder, 1983a, b). Hence, even muscles that would function alone to abduct the jaws are active during the strong adduction that occurs during snail crushing, presumably to stabilize the jaws during forceful exertion. We have estimated that adult pumpkinseeds in Three Lakes eat over 100 snails per day during the summer months (Osenberg, unpublished data). Forceful exertions of most of the pharyngeal jaw muscles at such a frequency could readily induce the relative hypertrophy of muscles seen in Three Lakes fish compared to their Wintergreen counterparts.

The motor pattern differences may also have a basis in the relative experience of fish in the two lakes. Pumpkinseeds have previously been shown to be able to adjust the motor pattern used during prey capture following continued exposure to a novel prey over a period of several weeks (Wainwright, 1986). This ability to "fine tune" the motor pattern in

526

response to repeated exposure to specific prey types may occur during the ontogeny of Three Lakes pumpkinseeds. We suggest that Wintergreen pumpkinseeds exhibit the basic snail crushing motor pattern, characteristic of this species (Lauder, 1983a, 1986), and that Three Lakes fish "fine tune" this distinctive motor pattern, specifically by altering the use of the PCe muscle to enhance their ability to manipulate and position snails prior to crushing them. This hypothesis could be tested in the laboratory by maintaining groups of fish from each population on diets of snails or no-snails, and examining the effect that these diets have on the snail crushing motor pattern.

Inter- and Intraspecific Trophic Divergence

One of the most striking results from our combined analysis of muscle morphology and motor pattern is that the pattern of change seen at the morphological level is incongruent with the pattern of change seen in the motor pattern (Fig. 5). Specifically, the muscle that shows the biggest morphological difference between lakes, the levator posterior muscle, shows one of the least indications of a lake effect in the motor pattern. Also, of the five muscles studied electromyographically, only the PCe muscle showed a significant motor pattern lake effect, though this muscle showed the least indication of a morphological lake effect (Fig. 5).

This result has two important implications. First, it suggests that predicting the function of structures entirely from anatomical observations can be a very hazardous undertaking. In the present case one would clearly have been wrong to expect the patterns of divergence in motor pattern to reflect those found in the morphology of the muscles. Directly measuring the function (use) of structures is a crucial component of comprehensive analyses of complex systems that has no replacement.

Secondly, this case study demonstrates the independent nature of these two levels of biological design. In spite of the highly complex, integrated quality of the pharyngeal jaw apparatus, morphology and motor pattern can transform independently. Change at one level of design does not depend on, nor does it necessarily affect, change at the other level.

Our study of the pumpkinseed polymorphism is only the second that we are aware of in which the analysis of an intraspecific trophic polymorphism has included an analysis of muscle activity patterns. Liem and Kaufman (1984) reported that the trophic poly-

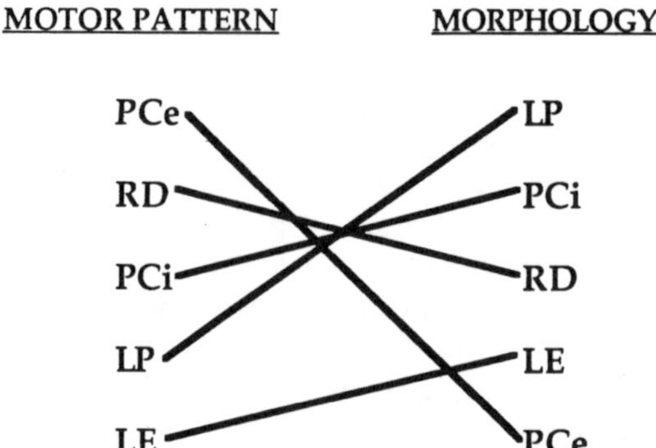

Figure 5. A comparison of rankings of population differences for five pharyngeal jaw muscles in morphological and motor pattern analyses. Within each column the ranking of muscles was determined by averaging the F-ratios from ANOVA significance tests for the lake effect. Thus, the top muscle in each column is the most distinctly different between lakes, based on morphological or motor pattern criteria. The bottom muscle in each column is the least different between lakes. Lines connect the muscles in each list and illustrate the incongruence of morphological and motor pattern divergence in the pharyngeal jaws of the pumpkinseed. Abbreviations are as in Fig. 4.

morphism seen in the Central American cichlid *Cichlasoma minkleyi* involves several morphological modifications of the pharyngeal jaw apparatus, but they found no motor pattern differences during prey capture and manipulating behaviors. With the exception of the PCe muscle motor pattern, which we found to be activated differently during snail crushing, our results are generally very similar to those of Liem and Kaufman. In addition, our results are largely in agreement with what has been found in interspecific comparative studies of the feeding mechanism in other lower vertebrates. Typically, morphological aspects of the feeding mechanism will vary extensively among closely related species, while the motor pattern exhibited during suction feeding or pharyngeal jaw function tends to be strongly conserved (Shaffer & Lauder, 1985; Wainwright & Lauder, 1986; Sanderson, 1988; Wainwright, 1989a; Westneat & Wainwright, 1989). Hence, we see in comparative studies that: (1) predicting patterns of change in motor pattern from anatomy is difficult, and (2) morphology and muscle activity can evolve independently. One of the concerns over the evolution of complex systems focuses on the potential difficulty in modifying them because of the constraining effects created by their highly integrated, interconnected nature. At least in the case of the feeding mechanism of aquatic feeding lower vertebrates it seems that some levels of design can transform independently of others and they are not constrained by the complex nature of the system.

ACKNOWLEDGMENTS

We thank Marvalee Wake for organizing this symposium and the National Science Foundation for providing traveling expenses. Support for the research described in this paper was provided by NSF grants BSR 87-96261 to G. G. M., BSR 89-05867 to G. G. M. and C. W. O., and BSR 85-20305 and DCB 87-10210 to G. V. L.

LITERATURE CITED

Alexander, R. M. 1982. *Locomotion of Animals*. Blackie: Glasgow.

Ashton, T. E. J. & M. Singh. 1974. The effect of training on maximal isometric back-lift strength and mean peak voltage of the erector spinae. Pp. 448–452. *In*: R. C. Nelson & C. A. Morehouse (eds.), *Biomechanics IV*. University Park Press: Baltimore.

Bennett, A. F., Garland, T. J. & P. L. Else. 1989. Individual correlation of morphology, muscle mechanics, and locomotion in a salamander. *Am. J. Physiol.* 256:R1200–R1208.

Calow, L. J. & R. M. Alexander. 1973. A mechanical analysis of a hind leg of a frog (*Rana temporaria*). *J. Zool., Lond.* 171:293–321.

Chapman, A. E. & J. D. G. Troup. 1970. Prolonged activity of lumbar erector spinae. An electromyographic and dynamometric study of the effect of training. *Ann. Phys. Med.* 10:262–269.

Darwin, C. 1859. *On the Origin of Species by Means of Natural Selection or the Preservation of Favoured Races in the Struggle for Life*. Murray: London.

Darwin, F. 1887. *The Life and Letters of Charles Darwin*. Murray: London.

Dawkins, R. 1986. *The Blind Watchmaker*. W. W. Norton & Co.: London.

Fay, R. R. & A. N. Popper. 1985. The octavolateralis system. Pp. 291–316. *In*: M. Hildebrand, D. M. Bramble, K. F. Liem & D. B. Wake (eds.), *Functional Vertebrate Morphology*. Belknap Press: Cambridge, MA.

Frazetta, T. H. 1975. *Complex Adaptations in Evolving Populations*. Cambridge University Press: Cambridge.

Goslow, G. E., Jr., Bennett, A. F., Blickman, D. M., Bramble, D. M., Duncker, H. -R., Fischer, M. S., Hinchliffe, J. R., Jenkins, F. A., Jr., Szekely, G., van Mier, P. & J. J. Videler. 1989. Group report: How are locomotor systems integrated and how have evolutionary innovations been introduced? Pp. 205–218. *In*: D. B. Wake & G. Roth (eds.), *Complex Organismal Functions: Integration and Evolution in Vertebrates*. John Wiley & Sons: New York.

528

Keast, A. 1978. Trophic and spatial interrelationships in the fish species of an Ontario temperate lake. *Env. Biol. Fish.* 3:7–31.

Lauder, G. V. 1983a. Functional and morphological bases of trophic specialization in sunfishes (Teleostei, Centrarchidae). *J. Morphol.* 178:1–21.

Lauder, G. V. 1983b. Neuromuscular patterns and the origin of trophic specialization in fishes. *Science* 219:1235–1237.

Lauder, G. V. 1986. Homology, analogy and the evolution of behavior. Pp. 9–40. *In:* M. H. Nitecki & J. A. Kitchell (eds.), *Evolution of Animal Behavior.* Oxford University Press: New York.

Lauder, G. V. 1990. Functional morphology and systematics: studying functional patterns in an historical context. *Ann. Rev. Ecol. Syst.* 21:317–340.

Lauder, G. V., Crompton, A. W., Gans, C., Hanken, J., Liem, K. F., Maier, W. O., Meyer, A. Preseley, Rieppel, O. C., Roth, G., Schluter, D. & G. A. Zweers. 1989. Group report: how are feeding systems integrated and how have evolutionary innovations been introduced? Pp. 97–115. *In:* D. B. Wake & G. Roth (eds.), *Complex Organismal Functions: Integration and Evolution in Vertebrates.* John Wiley & Sons: New York.

Levine, J. S. 1985. The vertebrate eye. Pp. 317–337. *In:* M. Hildebrand, D. M. Bramble, K. F. Liem & D. B. Wake (eds.), *Functional Vertebrate Morphology.* Belknap Press: Cambridge, MA.

Liem, K. F. 1989. Respiratory gas bladders in teleosts: functional conservatism and morphological diversity. *Amer. Zool.* 29:333–352.

Liem, K. F. & L. S. Kaufman. 1984. Intraspecific macroevolution: functional biology of the polymorphic cichlid species *Cichlasoma minckleyi.* Pp. 203–215. *In:* A. A. Echelle & I. Kornfield (eds.), *Evolution of Species Flocks.* University of Maine Press: Orono.

Marsh, R. L. 1988. Ontogenesis of contractile properties of skeletal muscle and sprint performance in the lizard *Dipsosaurus dorsalis. J. Exp. Biol.* 137:119–139.

Meyer, A. 1989. Cost of morphological specialization: feeding performance of two morphs in the trophically polymorphic cichlid fish, *Cichlasoma citrinellum. Oecologia* 80:431–436.

Meyer, A. 1990. Ecological and evolutionary aspects of the trophic polymorphism in *Cichlasoma citrinellum* (Pisces, Cichlidae). *Biol. J. Linn. Soc.* 39:279–299.

Mittelbach, G. G. 1984. Predation and resource partitioning in two sunfishes (Centrarchidae). *Ecology* 65:499–513.

Mittelbach, G. G. 1988. Competition among refuging sunfishes and effects of fish density on littoral zone invertebrates. *Ecology* 69:614–623.

Motta, P. J. 1988. Functional morphology of the feeding apparatus of ten species of Pacific butterfly fishes (Perciformes: Chaetodontidae): an eco-morphological approach. *Env. Biol. Fish.* 22:39–67.

Osenberg, C. W. & G. G. Mittelbach. 1989. Effects of body size on the predator-prey interaction between pumpkinseed sunfish and gastropods. *Ecol. Monogr.* 59:405–432.

Osenberg, C. W. G. G. Mittelbach & P. C. Wainwright. In press. Two-stage life histories in fish: the interaction between juvenile competition and adult performance. *Ecology.*

Powell, P. L., Roy, R. R., Kanim, P., Bello, M. A. & V. Edgerton. 1984. Predictability of skeletal muscle tension from architectural determination in guinea pig hindlimbs. *J. Appl. Physiol.* 57:1715–1721.

Sadzikowski, M. R. & D. C. Wallace. 1976. A comparison of the food habits of size classes of three sunfishes (*Lepomis macrochirus, L. gibbosus* and *L. cyanellus*). *Amer. Midl. Natur.* 95:220–225.

Sanderson, S. L. 1988. Variation in neuromuscular activity during prey capture by trophic specialists and generalists (Pisces: Labridae). *Brain, Behav. Evol.* 32:257–268.

Seaburg, K. C. & J. B. Moyle. 1964. Feeding habits, digestion rates, and growth of some Minnesota warm water fishes. *Trans. Amer. Fish. Soc.* 93:269–285.

Shaffer, H. B. & G. V. Lauder. 1985. Aquatic prey capture in ambystomatid salamanders: patterns of variation in muscle activity. *J. Morphol.* 183:273–284.

Thomason, J. J., Russell, A. P. & M. Morgeli. 1990. Forces of biting, body size, and masticatory muscle tension in the opossum *Didelphis virginiana. Can. J. Zool.* 68:318–324.

Turner, R. G. 1980. Physiology and bioacoustics in reptiles. Pp. 117–134. *In:* A. N. Popper and R. R. Fay (eds.), *Comparative Studies of Hearing in Vertebrates.* Springer-Verlag: New York.

Wainwright, P. C. 1986. Motor correlates of learning behavior: feeding on novel prey by the pumpkinseed sunfish. *J. Exp. Biol.* 126:237–247.

Wainwright, P. C. 1987. Biomechanical limits to ecological performance: mollusc-crushing in the Caribbean hogfish, *Lachnolaimus maximus* (Labridae). *J. Zool., Lond.* 213:283–297.

Wainwright, P. C. 1988. Morphology and ecology: functional basis of feeding constraints in Caribbean labrid fishes. *Ecology* 69:635–645.

Wainwright, P. C. 1989a. Prey processing in haemulid fishes: patterns of variation in pharyngeal jaw muscle activity. *J. Exp. Biol.* 141:359–375.

INTRODUCTION

This paper is about diversification of feeding design in birds. If zoomorphology must explain diversification of organismic systems, then explanation may be done by deductive transformation. Such deductive explanation describes by which transformation a known mechanism is to be changed to turn it in a mechanism that is to be explained. For example, deductive transformation may be a matter of changing a model of the known mechanism according to some kind of scaling, maximizing a specific functional requirement, or changing the balance of different functional requirements. The mechanism of the system from which the deduction is started is called the initial mechanism.

In short the present deductive approach is: physical laws and internal organismic constraints specify the optimal shape into which an initial mechanism should be modified if it must meet changed functional requirements in a maximally economical manner. This means, for deduction of transformation, that a model of an initial mechanism is maximized for a specific functional capacity and that initial assumptions and predicted modifications are tested against real systems which have that capacity. The degree of coincidence determines the explanatory value. This reasoning follows the approach of Hempel (1965) and Dullemeijer (1974, 1985).

Adopting a deductive procedure means that explanation is restricted to biomechanics. The fact that maximization is used as the transformation function introduces polarity (this may be, but is not, per se, a historical polarity). Thus, a simulated transformation describes a feasible mechanical domain of modifications of a mechanism from an initial state into a specifically maximized state. The domain is a mechano-space that may have the shape of a pathway, branching pattern, or space; it therefore informs about order, ranges of compromises, and alternatives of modifications.

A major problem in zoomorphology is to integrate these domains in evolutionary explanation. One reason for this problem is that neither causation nor timing are inherent in the domain. Therefore, when the domain is used to develop evolutionary hypotheses, presence of adequate morphogenetic mechanisms of organismic innovation must be assumed. Thus, the domain does not inform whether epigenetically induced macromutations (Lovtrup, 1987), environmentally induced changes with subsequent genetic assimilation (Matsuda, 1987), internal energetic mechanisms (e.g., Gutmann, 1989), or external selection mechanisms (e.g., Mayr, 1989) operate. The domain, however, becomes a framework for an historical hypothesis if it meets four conditions: 1) the initial mechanism for the deduction is the same as the ancestral one, 2) the boundary conditions include internal constraints as well as proper ontogenetical developmental mechanisms (e.g., heterochronical mechanisms such as dissociation, Alberch & Alberch, 1981, and ontogenetic repatterning, Wake & Roth, 1989); 3) the transformation function must not violate physical laws and must be consistent with realistic evolutionary themes, so that the polarity in the deduction becomes a feasible historical polarity; and 4) the organismic concept of the optimal design—that necessarily must be applied in formal mechanical analyses (e.g., Alexander, 1982; Stephens & Krebs, 1986)—must be extended to that of sufficient design, accepting that it is open, dynamic, and stochastic. Application of these conditions allows the use of the deduced domain for narrative explanation (cf. Bock, 1985) based upon an inductive/comparative method.

This paper intends to evaluate some transformation hypotheses and assess what they contribute to the understanding of pathways and space for the evolution of the avian feeding design. The aim is restricted to review here some recent hypotheses about development of feeding designs in Neornithes. Four hypotheses will be considered: 1) a pecking mechanism generalized and despecialized from *Gallus* and *Columba* represents the ancestral feeding mechanism in Neornithes; 2) probing mechanisms are modified pecking mechanisms in which probing requirements are maximized; 3) filter feeding

532

mechanisms are modified pecking mechanisms in which filter feeding requirements are maximized; and 4) probing and filter feeding mechanisms have evolved from pecking mechanisms (Gerritsen and Meiboom, 1986; Zweers et al., 1991).

Viewed from the organismic level of organization alone, there are major shortcomings in the development of these hypotheses. They start from isolated adult systems and therefore ignore systemic constraints from the complexity of integrated systems, phylogeny, and ontogeny. That means that potentials for phenotypic plasticity must be considered since they are a prerequisite for any change. Therefore the following hypothesis is reviewed: different appearances of the mallard's mouth design are modifications in a continuum of the integrated peck, filter feed, and drink design in which different proportions of performances of these mechanisms are maximized (Kooloos et al., 1989; Kooloos and Zweers, 1991).

From the analysis of mallards I consider if the ancestral duck's feeding system has a wide mechano-space allowing phenotypic plasticity. Then, an examination of how trophic diversification in ducks is released from a strong phylogenetic constraint, due to a necessary change in the drinking mechanism (that results from the fact that the duck's pumping mechanism obstructs the ancestral way of drinking) is presented. Finally, there is a discussion as to whether or not ontogenetic development is able to keep up with the transformations proposed in the adult systems. A final hypothesis is examined: ontogenetic repatterning of behavioral elements in the initial drinking system facilitates development of adequate pumping and drinking mechanisms (Heidweiller et al., 1991).

METHODOLOGY

The deductions in the projects reviewed follow a sequence of steps (Fig. 1). First, a system is selected that will serve as the initial system for the deduction of a mechanistic transformation. Later the same system must also serve as the ancestral system to allow the introduction of historical polarity and to develop an evolutionary narrative, so it must be selected carefully from generally accepted opinion about relative ancestry. Second, extensive anatomical descriptions of the system must be present, so that the isomorphy between original anatomy and the models developed can be tested. Then, the initial system is modeled by applying optimal design in a formal analysis of the mechanism(s) involved. The parameters that represent the initial conditions of anatomy, operation and performance are now defined. The boundary conditions for the simulation of the initial model are also determined. They must include the internal anatomical constraints due to characters of elements as well as of the total system.

The next step is maximization of the initial model for changed functional requirements. A (branching) pathway of subsequent modifications or a continuum of modifications develops from the simulation. Predicted pathway and space are tested by comparison with: 1) real systems which have the functional capacity for which the initial system is maximized; 2) empirical evidence from technical design analogues; and 3) trends in morphoclines of systems that possess the specified functional capacity increasingly better. Finally, an historical narrative is developed by way of a comparative method and by applying sufficient design, considering the domain as a framework for possible trophic diversification.

The present nomological deduction (cf. Hempel, 1965; Kahane, 1986) requires formal analysis of the systems studied (e.g., Stephens & Krebs, 1985). Extensive description of the methodology and terminology is published elsewhere (Zweers 1985b, 1991). See Bock (1985) and Dullemeijer (1985) for a discussion of deductive and inductive/comparative methodologies.

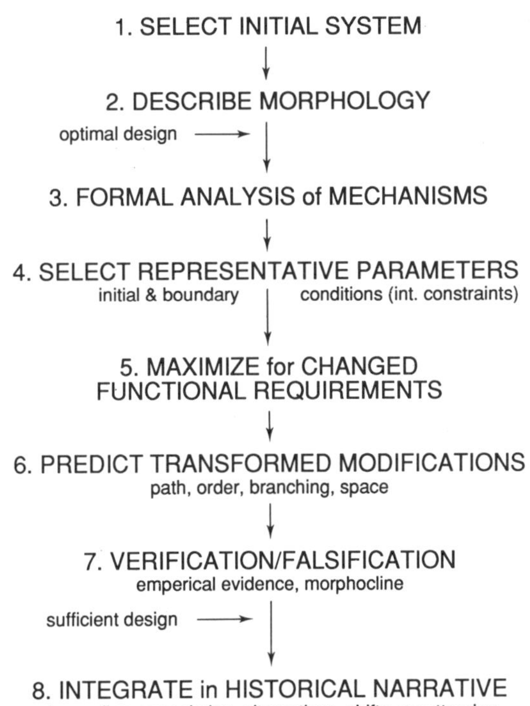

Figure 1. Diagram illustrating the subsequent steps in the methodology of deductive analysis of transformation. See text for explanation.

PATHWAYS OF TRANSFORMATION FROM PECKING TO PROBING MECHANISMS

Pecking as Ancestral Mechanism in Neornithes

The pecking mechanism is, for two reasons, assumed to be the ancestral mechanism for feeding in Neornithes by Zweers et al. (1991). They show that the pecking mechanism is the general feeding mechanism since it is found in all Neornithes independent of whether they have another mechanism of feeding. Further the general feeding cycle of the pecking mechanism resembles closely the main elements of the general feeding cycle in lower tetrapods (Bramble & Wake, 1985).

A model of the initial pecking mechanism was developed from taxa that were close to the neornith ancestor. Cladistic systematists suggest that the Galliformes and Anseriformes are one lineage of the earliest dichotomy in the monophyletic Neognathae (Cracraft, 1988; but see Olson, 1985, about Anseriformes). Evolutionary systematists do not deny that the Galliformes are an ancient order; further, there is general agreement that the Columbiformes are also (pers. comm. Bock). Most species in these orders display the pecking mechanism as the major feeding mechanism, therefore *Gallus* and *Columba* were selected to develop the description and model of the initial/ancestral pecking mechanism.

A short remark on terminology follows. The term 'initial' is used with reference to the model of the system that is selected as the starting point for the deductive transformation (e.g., the model of the pecking mechanism in this section; the model of the complex of pecking, drinking, and filter feeding mechanisms in the next section). The initial condi-

534

tions of the model for deduction may be the same as the anatomical and behavioral elements of the (assumed) ancestral mechanism (e.g., the avian pecking mechanism). If the initial model represents the ancestor the term 'ancestral' is used when discussing evolution.

Initial and Boundary Conditions for Transformation

Models were developed in earlier formal analyses of the pecking mechanism in chicken and pigeon (Berkhoudt, 1985, Zweers, 1982, 1985b, respectively). Zweers et al. (1991) listed the initial conditions for anatomy and behavior of the pecking mechanism by generalization and de-specialization of these models. Generalization was a process of formulating what the models had in common and de-specialization was a matter of leaving out elements that most particularly serve the improvement of the pecking mechanism.

The initial osteology and arthrology were defined largely consistent with the *Archeopteryx* characters (cf. Bühler, 1985), as well as in accord with the synapomorphies of braincase and jaws to the level of the Neognathae (cf. Cracraft, 1988). Further, the integument and its touch, taste, and glandular organs are primarily defined as they appear in chicken and pigeon, except that pecking specializations such as curved beak tips and erectability of the pharyngeal flaps were omitted. The behavioral elements were integrated from chicken and pigeon in a long list of specified motions of jaws, tongue, larynx, mouth floor and pharynx floor, including, for each, the specific function they have in the transport of food. Most prominent are the catch and throw by head jerks, the lingual slide and glue, and the lingual wing transport (i.e., cranioinertial, wet adhesion and lingual inertial transport, respectively, cf. Bramble and Wake, 1985). These mechanisms may occur in mixtures of the patterns and they are carefully tuned by a stepwise control via vision, touch, and taste perception (Deich et al., 1985; Zeigler et al., 1980; Zweers, 1985b).

Also, the boundary conditions for change in the simulation were defined. They were developed from general comparative knowledge. The following may serve as examples; cross sectional profiles of all bones are considered modifiable, but the overall connection scheme of the bones is non-modifiable. This means that the initial palaeognath, prokinetic skull may be modified in any rhynchokinetic condition as described by Zusi (1985). Further, the clustering and position of touch and taste organs as well as their specific modality may be modifiable. However, taste organs must be close to glandular orifices and may not pierce through keratin covers (Berkhoudt, 1985). Also, amplitudes and frequencies of motions are modifiable, and phase shifts may occur, while behavioral elements may be omitted or interpolated from other patterns, but the main order of the behavioral transport elements is non-modifiable.

Functional Requirements for Probing

The criteria for the transformation deduction were selected from analyses of probing sandpipers (*Calidris* sp.). Probing has two functional capacities that set different requirements upon the beak and mouth: 1) inspection capacity of water and air filled holes to find, and grasp prey, and 2) penetration capacity to enter substrates, find, and grasp prey. Functional analyses (Gerritsen, 1988) have shown that the organization of penetration differs from pecking behavior in five respects: 1) A prolonged downward head approach occurs by which the beak tips penetrate a substrate. Downward penetration may change into horizontal motions through the substrate during pursuit of prey. 2) Beak tips have a "direct touch" capacity, since penetration may be interrupted when beak tips hit upon a prey. 3) Gerritsen and Meiboom (1986) have calculated, from ecological selection experiments in large aviaries, that the probability of direct hits upon prey hidden in a substrate was far too small to have directed the selection of the best foraging location. They concluded that sandpipers have an additional "remote touch" capacity that allow them to monitor prey at a distance of at least 2 cm from the beak tips. 4) Burrowed prey is grasped. 5) Grasped prey is retracted from the substrate by head elevation.

Maximizing the Initial Pecking Mechanism for Probing Requirements

Many aspects are involved when the pecking mechanism is maximized for penetration, inspection, handling burrowed prey, and remote touch. Zweers et al. (1991) have focused on maximizing the mechanical organization of the beak, the jaw apparatus and the beak tips, and on the sensorial organization of the beak tips. To illustrate how conclusions were reached some isolated features are mentioned from their more extensive analysis.

Penetration. Maximizing for penetration requires a shape of beak and tips that minimizes penetration costs, in addition to a jaw apparatus that safely conducts penetration forces onto the braincase. The slenderness of the beak turned out to be a balance of keeping costs for increasing penetration depth low and jaw strength sufficient. The initial beak shape was concluded to be modified into a lengthened, slenderized, flattened beak with a broad base. Given the initial conditions of minimum weight and bilaterality it was also concluded that a bifurcation occurs in the deduction: beaks can be flattened either laterally (vertical) or dorsoventrally (horizontal). This means, as a consequence, that vertically compressed, in contrast to horizontally compressed, beaks cannot elevate the symphyseal area independent from the jaw rami, due to a large vertical diameter of the cross sectional profile of the rami.

The organization of the jaw apparatus must also change to conduct penetration forces economically and safely onto the braincase. Simple mechanical considerations lead to a list of seven modifications required in the jaw apparatus. Among these are a shift rostrad of the flexible joints in the maxilla and a slanting backward of the quadrate and bringing it in line with the mandibular ramus, to ensure economic force conduction and prevent disjunction; and an extended and ventrocaudally directed angular region is required to ensure proper space, attachment area, and orientation of the strongly increased depressor muscle. These modifications were verified as a trend developing in a morphocline of increasingly better penetrating taxons.

Maximization of the beak tip shape was examined empirically. Using a simple apparatus, releasing pairs of beak models into a tray filled with wet sand, and complex statistics, allowed the determination of the best shape for deepest penetration. It was concluded that a sharp, slim, straight, wedge-shaped and closed beak with an extended lightweight symphysis is the best modification. These conditions are best met in oyster-catchers (Haematopidae) which have vertically flattened beaks. However, other deep penetrating taxa, such as many Scolopacidae, only partly meet these conditions for economical penetration. For example, sandpipers have dorsoventrally flattened beak tips, spoon-shaped and somewhat bulging. The bifurcation in the deduction may also have consequences for grasping prey and remote touch capacities.

Inspection capacity. Like the requirement of the least costly deep penetration, inspection also requires a needle-like beak, but unlike the requirement for strength in penetration, maximizing inspection area requires a curved rather than a straight beak. This is shown by calculating the volume of reach behind a small opening of curved and straight beaks. Curved beaks have a much larger volume of reach. A compromised modification may be expected depending upon the ratio of required use of penetration and inspection capacities. A compromise may also be seen when straightening of a curved beak occurs prior to penetration. This is observed in high speed films of *Calidris alba*.

Handling burrowed prey. Two aspects are involved: opening the beak while penetrating and holding prey while retracting. Beak opening requires minimizing bending forces on the beak, thus minimizing the displaced substrate. This is a matter of minimizing the size of the movable jaw portion. Again, a bifurcation occurs in the deduction, since minimizing size is achieved by either minimizing the width of beaks, resulting in a wedge shape as observed in oystercatchers, or minimizing the length of the movable beak portion. The latter modification results in beak tips that are elevated independent from the rest of the beaks. Calculations show that, given a fixed gape, the displaced volume of substrate decreases when the joints shift distad along the jaw rami.

This modification is a matter of decoupling the symphyseal region from the jaw rami and developing a motion driving mechanism for that region. Decoupling the symphyseal area requires the development of flexible zones caudal to that area. They must have a slenderness ratio of intermediate columns, about 20–60, since they must be neither so stocky that they do not bend nor so flexible that they buckle easily. Values of this order are observed in upper and lower jaws of sanderling and purple sandpiper. The lower jaws have a rather long, flexible, somewhat more stocky zone, which is explained by the protection function the mandibular bone has for the large ramus mandibularis of the trigeminal nerve.

The driving mechanism is, in the maxilla, the same as in the initial mechanism and is present by coincidence because of the rostral shift of the flexible zones in the maxillary bars. This shift transforms the kinematically one bar unit of the upper beak in the initial mechanism into a kinematically four bar unit in the modification. These modifications are observed in all Scolopacidae as one of the five kinds of rhynchokinesis that Zusi (1985) describes.

The decoupling in the mandible is examined empirically by using cardboard models in which the cross sectional profiles of the jaws are changed from a U-shape via a V-shape to a horizontal shape. They show that widening and outward rotation of the caudal ends of jaws result in depression of the very tip without affecting the rami. The required V-profiles were verified in microsections of several sandpipers.

Remote touch.   The initial touch mechanism records prey by direct touch alone. Within the boundary conditions Zweers et al. (1991) designed three modifications for remote touch mechanisms monitoring polychaete worms like *Nerine* of 2–5 cm long and 1–2 mm diameter at 2 cm from the beak tips. Here, a real engineering approach was used since first the end-product, a remote-touch monitoring beak tip organ, was designed and then a feasible pathway from the initial simple direct-touch organ to the deduced complex remote-touch organ was developed.

A first modification is a touch organ of compound facets, combining direction sensitive sensors in facets that record only when a source is present in the sector they can "view." Calculation shows that a minimum of 16 sectors is needed on both beak tips for accurate location of prey. The maximally available aperture at a sandpiper-sized beak tip is about 0.4 mm. This value must be larger than the wave length to be effective, however, wave lengths range from 0.2 to 1.5 m. Thus this modification must be abandoned.

A second possible modification is a fields touch organ measuring phase shift or difference in arrival time. Touch corpuscles on either side of the beak monitor a maximum difference in arrival time of 5.10^{-6}s. Corpuscles of Herbst respond maximally at 1000 Hz; spikes travel about 4 cm to the brain in about 1 ms. Thus an extremely elaborate evaluation system is required to decode source direction by monitoring difference in arrival time. This modification may have developed, but the next one is less elaborate and therefore more probable.

A third modification is a clusters touch organ measuring differences in intensity. The ratio of intensity of a wave signal as it passes two touch corpuscles on either side of a beak tip is 0.76. A difference of 25% in signal intensity requires a level of accuracy in the direction evaluating system which is much more likely than in the previous modification. Minimum numbers of clusters and of corpuscles of Herbst required per cluster to allow a

360° view were calculated. Generally speaking, some hundreds of corpuscles are required, organized in at least 15 clusters at both beak tips, each comprising a minimum of 10 corpuscles per cluster. Counts in microsections of *Calidris* species and data from Bolze (1968) clearly show that these conditions are present in many Scolopacidae.

PATHWAYS OF TRANSFORMATION FROM PECKING TO FILTER FEEDING MECHANISMS

Functional Requirements for Filter Feeding

The criteria for transformation deduction were selected from earlier analyses of filter feeding in ducks (*Anas*) and flamingos (*Phoenicopterus*) (e.g., Allen, 1956; Jenkin, 1957; Zweers et al., 1977; Kooloos et al, 1989). Filter feeding has three major capacities which make it different from pecking. The system has transport capacity for water in and out the mouth, capacity to filter particles of preferred size, and discrimination capacity for size and hardness of floating particles. For example, ducks pump water in and out the mouth with a frequency of 10–20 Hz and flamingos pump at about 5 Hz; simultaneously they select food particles of preferred size from mixtures of particles of different sizes. Mallards are able to select by the same behavior, in 60 seconds, 50 peas from 50 equally sized clay balls randomly submerged in wet sand. These discrimination capacities are clearly managed by direct touch organs at the inflow openings. Further, since catch and throw motions are not present and a slide and glue mechanism does not work under wet conditions, the system must have an alternative mechanism to transport food into the pharynx.

Maximizing Initial Pecking for Filter Feeding Requirements

Only some main lines are reviewed, so that the branching pattern of the deduction becomes clear. Filter feeding is not possible without water flow, therefore the requirement that suspensions must move in and out of the mouth by fast repetitive cycles of motions is considered. The initial pecking mechanism could not directly be transformed into a water transporting system. However, quite a different situation is present when some lengthening and slenderizing of the beak has occurred first due to maximizing for probing requirements. Then an epiphenomenon, called the "tweezer effect," occurs, which allows as a coincidence transport of water along the beak rami. A drop between the rami of a tweezer runs up through the mouth when the rami are delicately spread and it runs back to the tips when they are closed. This epiphenomenal feature is observed in high speed films of sandpipers (*Calidris*) ("fetch and carry transport" in *C. canutus* cf. Gerritsen, 1988). Any particle present in the drop will run with it. This is observed in *Phalaropes*, where a delicate tuning of the drop position allows the transport of shrimp ("surface tension transport" cf. Rubega and Obst, personal communication).

Possibilities for water transport are illustrated in Figure 2. A bifurcation in the deduction occurs. One modification is the fetch and carry mechanism; continued maximization of the water transport capacity, however, requires a different mechanism than surface tension alone. A pumping mechanism is the primary possibility. The reason is that a simple repatterning of jaw and lingual motions causes a major functional shift to a potentially strong pumping mechanism. This is the feature that, if continued lingual retraction occurs when a delicately tuned beak opens to transport a drop by surface tension, allows a more powerful mechanism, suction, to add to the surface tension transport (see 3 in Fig. 2).

By different repatterning of tongue and jaw motions again a bifurcation occurs in the deduction, now for a kind of pumping mechanism (Fig. 2). The choice depends upon how the lingual motion continues: 1) either the tongue protracts in elevated position, pushing the water back distad and expelling it, making the system a back and forth pump, which is

538

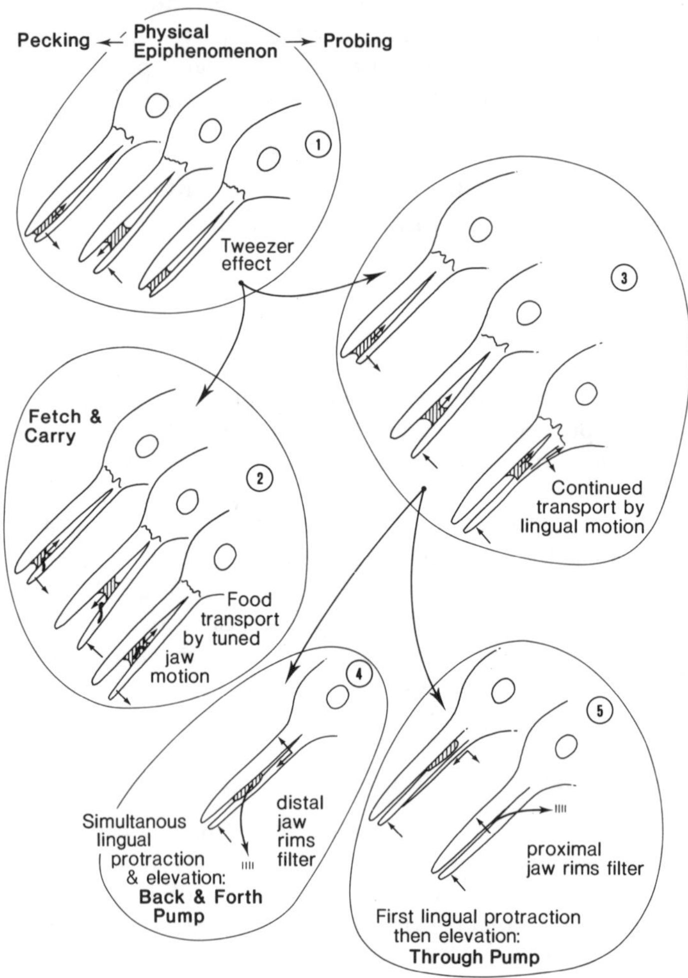

Figure 2. Diagram illustrating the branching pattern of feeding mechanisms from a pecking mechanism, which is slightly maximized for probing. Major bifurcations are illustrated that may occur during further maximization for either "fetch and carry" transport, "back and forth pumping" or "through pumping" requirements. See text for explanation (changed after Zweers et al., 1991).

basically verified in flamingos, or 2) the tongue protracts in depressed position, running underneath the water mass and then elevating during retraction, pushing the water caudad and expelling it, making the system a through pump, which is basically verified in ducks.

It is beyond the scope of this paper to review further deductions—they are worked out by Zweers et al. (1991)—but a few major points may be mentioned. Development of keratin lamellae in the in- and outflow openings will certainly specialize the system. Maximizing the pump capacity requires a strong increase of equally powerful adductor and depressor muscle complexes. They in turn require increased areas for attachment, and, moreover, economically fast repetitive jaw motions require parallel running quadrate and major working lines of the adductor and depressor muscle complexes. This requires a modification in which the quadrate is positioned vertically and where both the processus postorbitalis and angularis posterior run horizontally and are largely extended. This condition is found in ducks and to a lesser extent in flamingos.

EVOLUTION FROM PECKING TO PROBING
AND FILTER FEEDING MECHANISMS

The domain of deduced modifications of feeding mechanisms serves as a framework for the development of an hypothesis for the evolution of the probing and filter feeding mechanisms from the initial/ancestral pecking mechanism. This domain is called a mechano-space; a morpho-space represents only the form features of a mechano-space. The mechano-space is assumed to represent feasible pathways for order and diversity in evolution of probing and filter feeding, given a strictly defined set of initial and boundary conditions that operate as phylogenetic constraints. The optimal design and deterministic maximization are now extended to the sufficient design and chance-based mechanisms for change. These include the presence of behavioral capacities to anticipate new epiphenomenal features and to invade into new areas due to these new mechanisms. The branching points in the pattern are regarded as evolutionary steps. Some major lines are reviewed here to elucidate what kind of further progress is necessary to evaluate systemic constraints from complexity, history, and ontogeny.

A hypothetical initial pecking mechanism is defined. Six major steps are recognized in the mechano-space (Fig. 3).

1. The initial pecking mechanism allows pecking specializations in several directions: e.g., speeding up swallowing as a result of predator pressure in open fields (e.g., pigeons

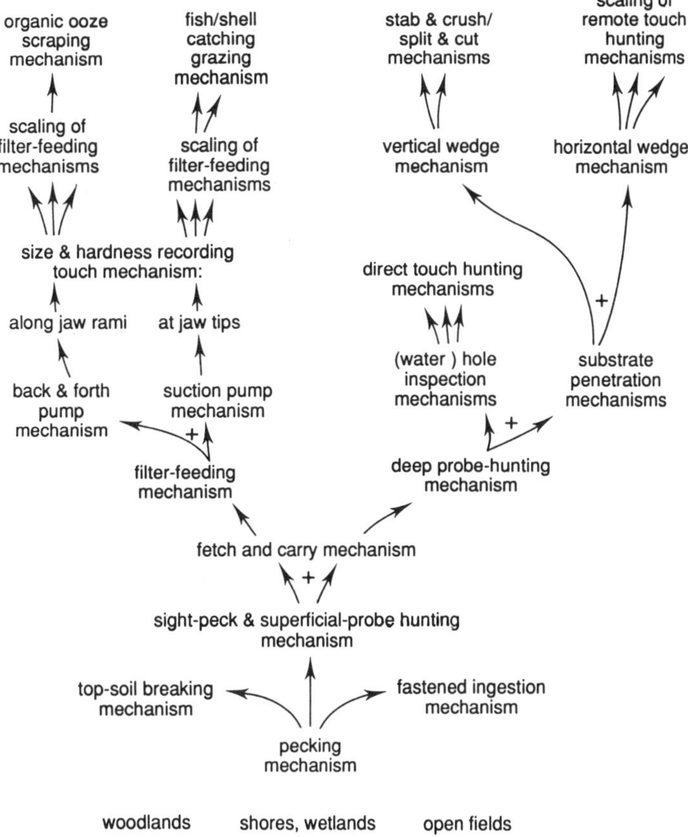

Figure 3. Diagram illustrating a mechano-space of feasible pathways deduced from the ancestral/initial avian pecking mechanism by maximizing it for the respective functional requirements of probing and filter feeding mechanisms. See text for explanation (changed after Zweers et al., 1991).

540

and other Columbiformes); breaking top soil to exploit the rich food supply in the very top soil of moist woodlands (e.g., like fowl and other Galliformes).

2. Birds hunting by a sight-peck technique, which developed as the birds crossed shores and discovered the rich food supply on top of the flats during low tide. They developed run-and-peck plus superficial probe hunting mechanisms, that initiated lengthening of legs and as a compensation, lengthened and slenderized beaks. The flexible zones at the base of the beak shifted rostrad to any rhynchokinetic condition to absorb shocks better. Additional decurving and recurving of the beaks opened a wide array of potential niches, so that radiation of this feeding kind may have split off. (e.g., like plovers and other Charadriidae).

3. By accidental deeper penetrations the enormous food supplies in the wet top soil were discovered. Penetration became rewarding and continued lengthening of beaks occurred. Accidentally, the "tweezer effect" may have occurred as an epiphenomenon, transporting a drop along slenderized and lengthened beaks. This was the initial step required for surface tension transport in fetch and carry mechanisms as well as for suction transport in pumping mechanisms. Both directions may have split off here, since both represent a functional shift that occurs by repatterning of the jaw and tongue motion patterns, despite absence of morphological changes. For example, sight peck hunting with the fetch and carry mechanism may have developed (e.g., like in Phalaropidae).

4. Improved direct-touch perception and continued lengthening of the beak for improved penetration led to another bifurcation. Direct touch hunting with lengthened slender beaks led to inspection of water masses, so that head sweeping, in which prey is caught by a direct hit, became rewarding. Radiation may have occurred here by decurving, recurving, straightening or widening the beaks (e.g., like in ibisbill, *Ibidorhyncha* sp.; avocet, *Recurvirostra* sp.; stilt, *Himantopus* sp.; spoonbill, *Platalea* sp.). A second radiation may have split off here by compromising this behavior with all kinds of possible combinations of moderate inspection probing by direct touch and a curved beak, with moderate penetration probing by a reinforced curved beak (e.g., like ibises, Threskiornithidae).

5. Another bifurcation developed via continued improvement of deep penetration. Either vertical or horizontal wedge-shaped beaks decrease penetration costs. Vertically flattened beaks possess, as an epiphenomenon, the potential to enter the siphon of mussels and to cut the adductor muscles by the sharp beak tips (e.g., like oystercatchers, Haematopidae).

Horizontally flattened beaks have as an epiphenomenon the potential to decouple the symphyseal region from the rest of the jaws, since flattened jaw rami allow independent elevation of the symphyseal region. Rhynchokinetic conditions provide the upper jaw with a decoupled symphyseal region, while a cross sectional V-shape and flexible zones in the mandibles provide the lower jaw with a decoupled tip area. Symphyseal regions are necessarily reinforced to withstand compression at penetration and provide therefore, as an epiphenomenon, space and support for development of a remote-touch organ. Remote touch decreases penetration costs dramatically, so that all kinds of compromises may have developed in which the beak tip shape of a strict wedge is compromised by the spherical shape required for the remote-touch organ at the tips. A radiation may have developed in which curvature and scaling features also influence this beak shape (e.g., like in Scolopacidae).

6. Beyond the fourth branching point, where ibis- and phalarope-like mechanisms branch off, developed another bifurcation point due to two kinds of functional shifts that may result from repatterning jaw and lingual motions. Either a back and forth pump (e.g., like in flamingos, Phoenicopteridae), or a through pump develoed (e.g., like in ducks, Anatidae).

The proposed pattern is compared to independently derived patterns such as proposed by Feduccia (1978, 1980) and Olson & Feduccia (1980a,b) for relationships of waterfowl, flamingos and waders; and patterns in waders proposed by Strauch (1978). Feduccia and Olson regard Presbyornithidae as connecting waders and ducks. Their conclusions are based on untested assumptions about filter feeding. It was shown from functional analyses that, if interpretation is restricted to skull materials, the Presbyornithidae are best characterized as an early offshoot in radiating filter feeding ducks, possessing a duck-like pumping mechanism. Strauch (1978) proposes a scheme that is largely comparable to the present one, except that the branch to the Scolopacidae (except Phalaropidae) in his diagram beyond the fourth node should be placed at the fifth node of the Charadriidae, the branch beyond the Haematopidae, to make them coincident. The difference is due to our giving primacy of connecting less rather than more complexity in the case of superficial probe and direct touch hunting compared to remote touch hunting.

MECHANO-SPACE CONSTRAINED BY INTERNAL COMPLEXITY

Previous considerations have not yet included the integrated complexity of feeding systems. Constraints may result from other systems that must operate simultanously. This section evaluates whether internal constraints leave space for change in the complex mouth design of the duck. By a deductive approach, the hypothesis is reviewed that different appearances of the mouth design are modifications in a continuum in which different proportions of performances of pecking, drinking, and filter feeding are maximized. This section is a review of from work of Kooloos and of Zweers and his coworkers.

Methodology

The methodological steps are similar to those described in the previous projects. The difference is that the system is regarded as a multi-role system. Consequently, a shift in proportion of several functional performances is maximized, rather than one functional capacity being maximized. The mallard's pecking, filter feeding and drinking mouth is the initial system (not the ancestral one). Four functional components are selected which make up the initial conditions in each of the three mechanisms. The major variable parameter in each of them is defined. They are: 1) storage capacity of the rostral mouth cavity (RMC), 2) transport capacity of the rostral mouth tube (RMT), 3) storage capacity of the caudal mouth cavity (CMC), and 4) transport capacity of the caudal mouth tube (CMT). The anatomical elements of these four functional components connect the mouth tips and the pharynx. A maximization procedure is developed that formulates a feasible space for the three mechanisms in which the capacities of the components may change and consequently the proportions of their performances, but in which they do not exclude each other from operation. The meaning of such a mechano-space for the development of evolutionary hypotheses is evaluated.

Initial Models

Earlier analyses have shown that each of these mechanisms operates by a two step procedure. After uptake at RMC transport occurs first through RMT to the CMC, where food or water is stored, and then further transport occurs through the CMT to the pharynx.

Pecking. Mallards elicit, after grasping a food mass, a series of standard cycles of jaw and lingual motions. A cycle is not adjusted to food size and serves two transport mechanisms: a catch and throw mechanism through the RMT, and a conveyer shaker mechanism via the CMT. A simulation procedure was developed that calculates how many cycles are required to transport a certain amount of seeds of specific size into the pharynx. The simulated and measured performances for seeds of 2–7 mm are similar, while deviations for larger/smaller particles were explained by observed changes in behavior (Kooloos & Zweers, 1991).

542

Drinking. Mallards release a tip-up drinking behavior. Water is transported to the pharynx during the phase that beak tips are immersed in the water. A series of jaw and lingual motion cycles occurs. Beak closure gathers a water dose in the RMC, protraction of the elevated tongue makes the RMT a tube into which water flows by capillary action; continued closure and protraction push water through the RMT into the CMC. This is repeated until the CMC is full, then a tube is shaped along the CMT in which water runs by capillarity and is pushed through the CMT into the pharynx by the decreasing volume of the CMC. A simulation model was developed that tests the assumptions of the capillarity. It was shown that only at the moment when water is observed to be transported are conditions fulfilled for capillarity (Kooloos & Zweers, 1989).

Filter feeding. Filter feeding is mechanically made up of three major components: a water pumping mechanism, a filtering mechanism, and a mechanism that transports filtered particles into the pharynx. The pumping mechanism has the nature of a through pump in which water enters at the beak tips and is expelled along the jaw rims by a suction pressure mechanism. The lingual bulges serve during an elevated retraction as a piston with a closed valve to suck water in at the front, and to expel the previously sucked water mass at the rear of the mouth. Lamellae standing in the outflow serve as a filter mechanism by either direct impact or by causing vortices which centrifuge small particles out the flow against the tube walls formed by the lamellae. Subsequently, elevated lingual hairs brush the particles away from the lamellae and lingual scrapers transport them caudally into the pharynx in a way similar to that in pecking. A simulation model was developed to calculate the change in volume of the mouth cavities, and hence the pump capacity. Amplitude, initial position and phase shifts of jaw rotation, elevation, and depression of all lingual elements could be simulated. Experimental data verify that when measured amplitudes, etc., were simulated, the predicted water expulsion was the same as the measured one for millet, poppy seeds, and milo in mallards (Kooloos et al., 1989).

Simulated Changes

First, Kooloos and Zweers (1991) explain that whenever the pumping mechanism works the drinking mechanism is also operable. Thus, for analysis of internal constraints we are left with pecking and filter feeding mechanisms. For that analysis the performances that result from alterations in the RMC, RMT, CMC and CMT are calculated for a variety of food sizes. For example, the volume of the RMC was changed from 0.1 to 1.6 cc for three seed sizes. Calculations show that the pecking performances increase rapidly and stay then constant at the level of measured data, despite further increase of the storage capacity. Apparently, a further device in the pecking mechanism acts as a bottleneck for further increase of the overall performance. A stepwise inspection of the simulation procedure shows that CMC acts as that bottleneck.

Feasible Space for Change in a Complex System

The procedure to develop a feasible space for change, a mechano-space, is as follows (Fig. 4). Performances of pecking and filter feeding were set along two axes. Four lines were drawn, connecting the points of calculated performances of pecking and straining for increasing values of RMC, RMT, CMC and CMT, respectively. The real measured value is indicated (asterisk in Fig. 4). Now the maximal alterations, determined in the internal constraint analysis, are drawn at the end of each line (arrow heads in Fig. 4). For example, increasing of RMC beyond 0.65 cc (at performances of 0.6 for pecking and 1.6 gr/s for filter feeding in Fig. 4) does elicit the feature of pipe separation, hampering the pump mechanism (upper arrow head in Fig. 4). ("Pipe-separation" is the feature that occurs if the gape increases so far that air rather than water is sucked at the distal jaw rims). The area within the four arrow heads not passing the outer lines is proposed as a feasible region in

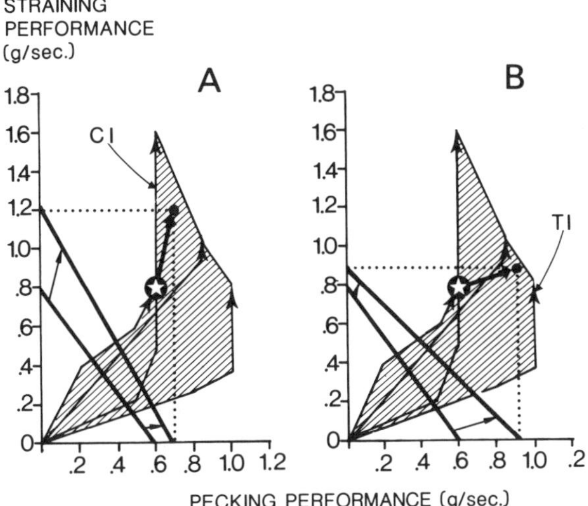

Figure 4. Diagram illustrating a mechano-space of a feasible continuous area deduced from the mallard's feeding system. A feasible space for change is deduced in which the filter feeding, drinking and pecking mechanisms may alter without hampering each other. See text for explanation (changed after Kooloos and Zweers, 1991).

which alterations in the designs of pecking and filter feeding do not hamper the mechanical organization.

Change in elements required for a change in performance can now be predicted from the feasible region. For example, increase of filter feeding capacity requires primarily an increase in RMC, since the point of intersection shifts almost parallel to the increase in RMC (Fig. 4A). This means that the storage capacity of the rostral mouth cavity is increased primarily by widening and lengthening this area. Similarly, increase in pecking performance requires primarily an increase in RMT, since the point of intersection shifts almost parallel to increase in RMT. This means that the transport capacity of the rostral mouth tube is increased by shortening the tube, causing a shortened midportion of the beak.

Restricting a test to the mentioned examples, Kooloos and Zweers (1991) show that shovelers and tufted ducks have pecking and straining behaviors that are similar to those of the mallard. Shovelers perform better in filter feeding small particles than mallards, they spend 40% of their foraging time straining, while mallards spend 7% (Szijj, 1965). Shovelers have a relatively widened and lengthened RMC, which verifies the prediction from the feasible region.

Tufted ducks perform better in pecking larger particles. They feed almost exclusively on mussels (Draulans, 1982). They dive and hoe through the mud, performing straining patterns, but mussel intake is a matter of pecking transport mechanisms. Tufted ducks have, relative to mallards, a shortened mid portion of the mouth, which is predicted from the feasible region.

POTENTIALS FACILITATING DIVERSIFICATION

Previous sections did not deal with all requirements that must be met to release trophic diversification. Some specific aspects are evaluated here by discussing the question: what makes the duck's ancestor system have a high diversification potential? Occurrence of phenotypic plasticity, and examples of release from systemic phylogenetic and ontogenetic constraints are considered.

Phenotypic Plasticity and Ancestral Mechano-space

Pecking, drinking, and filter feeding systems and, also, the four functional components constituting the mouth constrain each other. Nevertheless, a mechano-space was deduced to be left in which the mechanisms may alter without hampering each other. This is an important conclusion since, if the mechanics do not allow any shift in compromises neither phenotypic plasticity, nor diversification by either establishing shifts in compromises, or by the integration of epiphenomenal new features is possible. For diversification, a versatile feeding system with a wide mechano-space is required. Conversely, occurrence of phenotypic plasticity reflects presence of a mechano-space for change. Presence of phenotypic plasticity has been observed in mallards. For example, Pehrsson (pers. comm.) studied habitat discrimination in mallards in Sweden. He observed two "morphotypes." One type was, by its habitat, forced to feed by filtering. The beak of this type had a relatively lengthened and widened appearance. The other type was, by habitat, forced to feed by pecking. This type had a somewhat shortened, stout, and broad-based beak. These observations are expected from the feasible space for change of mechanisms in the mallard developed earlier.

Mallards have a large mechano-space, in this case a large feasible space in which filter feeding, pecking and drinking mechanisms may alter and do not hamper each other. The duck's ancestor had a mouth design much simpler than that of the mallard, since no secondary adaptations for filter feeding were present. The mutual tolerance of the ancestral pecking, drinking, and filter feeding mechanisms must therefore have been even larger.

A second aspect is important here. As soon as a filter feeding mechanism has been established, the feeding system has, as an epiphenomenon, a sufficient grazing mechanism also. The reason is that the grasping jaw elements of pecking, the enlarged tongue mass of pumping and the lamellae of filtering provide the system with proper grass-holding elements, while the transporting elements of filter feeding (the lingual combs and spines) provide the system with an adequate grass transport mechanism. Making the mechanisms operable requires no more than a repatterning of the behavioral elements to establish this functional shift (personal observations). This feature enlarges the ancestral mechano-space even more.

Phylogenetic Systemic Constraints

Systemic constraints may occur if a change is required from one drinking mechanism into another. This must occur during phylogeny, since the ancestral drinking mechanism in Neornithes is assumed largely similar to that in *Gallus* (Heidweiller & Zweers, 1990) and duck-like drinking is, despite being tip up drinking also, of quite a different nature. The special kind of tip up drinking in mallards is a necessity that results from the specific requirements of the pumping mechanism in filter feeding (Kooloos & Zweers, 1989). For example, development of a large lingual cushion, which is required for improvement of filter feeding, obstructs the ancestral chicken-like water transport dorsally along the tongue into the pharynx. Thus, improvement of filter feeding must exist along with a takeover of the ancestral chicken-like drinking mechanism by the duck-like one, and at the moment that filter feeding is incorporated, the duck-like drinking mechanism must be set into action. It was shown that whenever the filter feeding mechanism is effective also the duck-like drink mechanism is operable also (Kooloos & Zweers, 1991). Hence, development of filter feeding mechanisms is released from a systemic constraint during phylogeny due to drinking, since the requirements for the anatomy from duck-like drinking are met as an epiphenomenon. Setting the mechanism into action is reduced to repatterning some behavioral elements.

Ontogenetic Systemic Constraints

Systemic constraints may also result from the possibility that the sequence of ontogenetical stages can not keep up with the evolutionary development of filter feeding and the required change of the drinking mechanism. If the chicken-like ontogenetical development of drinking is assumed the ancestral mechanism, the following may be said. Development of drinking was shown to be a matter of morphological scaling which, for physical reasons, forces the system to introduce extra transport and storage mechanisms to ensure proper intake (Fig. 5; Heidweiller et al., 1991). These authors explain that at first capillary action, bringing in the water until the larynx, and a constricting mouth floor squeezing the water caudad, suffice as a drinking mechanism. Then, at a later stage, increasing length and widening of the oropharynx force the system to integrate extra mechanisms for transport and storage. One aspect is that in order to ensure proper capillary action between the tongue and mouth roof, the tongue must be elevated. Due to this elevation an extra possibility of a water transport mechanism lateroventral to the tongue (Fig. 5) develops, again as an epiphenomenon. This extra transport potential meets the requirements of development of mechanisms for filter feeding as well as for duck-like drinking.

The extra transport potential is a bilateral tube that lies lateroventral and caudally along the tongue. The tubes act by capillarity for water inflow and can be squeezed for caudad transport of drinking water; also they can be kept closed easily to squeeze water out from the caudal mouth cavity during pumping, rather than keep it in, and they further serve as a transport mechanism for filtered food (Kooloos & Zweers, 1989; Kooloos et al., 1989). Hence, the ancestral chicken-like ontogeny of drinking provides a moment in early development that releases trophic diversification from a strong systemic ontogenetical constraint. The release potential is present at the moment that bifurcation may occur: either ancestral chicken-like development of drinking, or simultaneous further development of duck-like drinking and filter feeding mechanisms is continued.

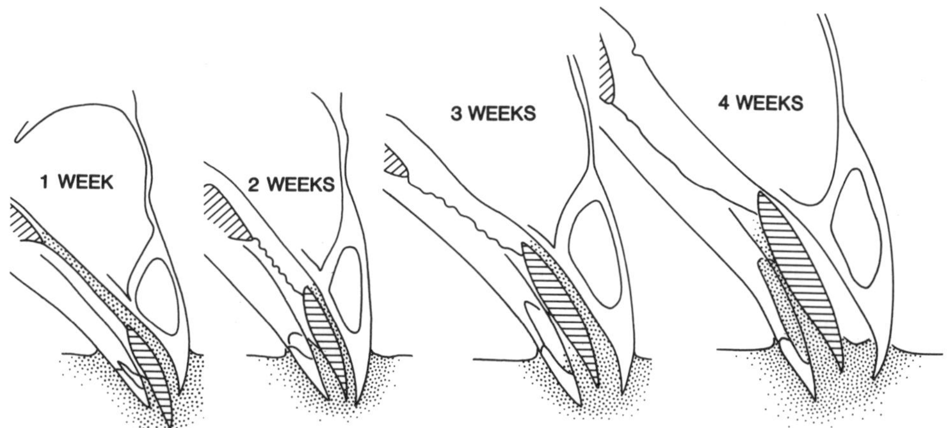

Figure 5. Outline diagrams of radiograms made of four stages of development of drinking in chicken. The radiograms represent a stage at or just prior to (4 weeks) the water intake of the drinking cycle during immersion of the beak tips. The sequence illustrates that due to widening and lengthening of the oropharynx, in addition to capillarity, extra intake mechanisms are required. The formation of the capillary tube above the tongue, required for water intake, has as a consequence that at the developmental stages of 2–4 weeks an extra bilateral capillary tube ventrolateral to the tongue develops. This feature occurs as an epiphenomenon. It allows the development of duck-like drinking and filter feeding mechanisms. See text for explanation (changed after Heidweiller et al., 1991).

546

CONCLUSIONS

Deductive analysis of the avian feeding system's mechano-space shows feasible pathways for the evolution of probing and filter feeding from pecking mechanisms, elucidating trophic diversification by predicting that decoupled mechanisms, functional shifts, changed compromises, dichotomies, or epiphenomenal features may occur. Also, a feeding system's mechano-space, shaped like a continuous space and elucidating phenotypic plasticity by predicting feasible change in complex multi-role systems, can be analyzed in this way. A large capacity for diversification in ducks is shown to be due to a versatile, potentially multi-role, ancestral feeding system. Three facilitating factors were found: 1) The system has a large mechano-space allowing a wide phenotypic plasticity since pecking, filter feeding and drinking are mutually compatible. 2) A grazing potential is present as an epiphenomenon. 3) Phylogenetically, a required take-over potential of the assumed ancestral chicken-like drinking mechanism by the duck-like one is present as an epiphenomenon. 4) Ontogenetically, a required potential to change the ancestral chicken-like drinking development into a simultaneous duck-like drinking and filter feeding development is present as an epiphenomenon. The latter three aspects are a matter of repatterning behavioral elements.

ACKNOWLEDGMENTS

The author has benefited very much from discussions with Drs. H. Berkhoudt, W. J. Bock, A. F. C. Gerritsen, J. Heidweiller, and J. Kooloos. Drs. J. C. Vanden Berge, M. H. Wake, and H. P. Zeigler are gratefully acknowledged for their comments upon an earlier version of this paper. M. Rubega, and Drs. B. Obst, and O. Pehrsson are thanked very much for allowing reference to their ideas about surface tension transport in phalaropes and to habitat discrimination in mallards, respectively. Mr. M. Brittijn is acknowledged for making the illustrations.

LITERATURE CITED

Alberch, P. & J. Alberch. 1981. Heterochronic mechanisms of morphological diversification and evolutionary change in the neotropic salamander, *Bolitoglossa occidentalis* (Amphibia: Plethodontidae). *J. Morphol.* 167:249–264.

Alexander, R.McN. 1982. *Optima for Animals*. Edward Arnold: London. 112 pp.

Allen, R. P. 1956. The flamingos: their life history and survival. *Research Report 5*. National Audubon Society.

Berkhoudt, H. 1985. Structure and function of avian taste receptors. *In:* A. S. King & J. McLelland (eds.), *Form and Function in Birds*. Academic Press: London.

Bock, W. J. 1985. The nature of explanations in morphology. *Amer. Zool.* 28:205–215.

Bolze, G. 1968. Anordnung und Bau der Herbstchen Körperschen in Limicolen Schnabeln in Zusammenhang mit der Nahrungsfindung. *Zool. Anz.* 181: 21–355.

Bramble, D. M. & D. B. Wake. 1985. Feeding mechanisms in lower Vertebrates. Pp. 230–261. *In:* M. Hildebrand, D. M. Bramble, K. F. Liem & D. B. Wake (eds.), *Functional Vertebrate Morphology*. Belknap Press of Harvard University Press: Cambridge, MA.

Bühler, P. 1985. On the morphology of the skull of *Archeopteryx*. Pp. 135–140. *In:* M. K. Hecht, J. H. Ostrom, G. Viohl & P. Wellenhofer (eds.), *The Beginnings of Birds*. Proc. Int. *Archeopteryx* Conference, Eichstatt, 1984.

Cracraft, J. 1988. The major clades in birds. *In:* M. J. Benton (ed.), *The Phylogeny and Classification of Tetrapods*. Vol. 1: *Amphibians, Reptiles, Birds*. Clarendon Press: Oxford.

Deich, J. D., Klein, B. & H. P. Zeigler. 1985. Grasping in the pigeon: mechanisms of motor control. *Brain, Behav. Evol.* 25:85–98.

Draulans, D. 1982. Foraging and size selection of mussels by tufted duck, *Aythya fuligula*. *J. Anim. Ecol.* 51:943–956.

Dullemeijer, P. 1974. *Concepts and Appraoches in Animal Morphology*. Van Gorcum: Assen, The Netherlands.

Dullemeijer, P. 1985. Diversity of morphological explanation. *Acta Biotheor.* 34:111–123.

Feduccia, A. 1978. *Presbyornis* and the evolution of ducks and flamingos. *Amer. Sci.* 66:289–304.

Feduccia, A. 1980. *The Age of Birds*. Harvard University Press: Cambridge, MA. 196 pp.

Gerritsen, A. F. C. 1988. Feeding techniques and the anatomy of the bill in sandpipers (*Calidris*). Thesis: Leiden.

Gerritsen, A. F. C. & A. Meyboom. 1986. The role of touch in prey density estimation by *Calidris alba*. *Neth. J. Zool.* 36:530–562.

Gutmann, W. F. 1989. *Die Evolution hydraulischer Konstruktionen: Organische Wandlung statt altdarwinistischer Anpassung*. Verlag Waldemar Kramer: Franfurt am Main. 201 pp.

Heidweiller, J. & G. A. Zweers. 1991. Development of drinking mechanisms in the chicken (*Gallus gallus*), (submitted).

Heidweiller, J. & G. A. Zweers. 1990. Drinking mechanisms in the Zebra finch and Bengalese finch. *Condor* 92:1–28.

Hempel, G. C. 1965. *Aspects of Scientific Explanation*. Free Press: New York.

Jenkin, P. M. 1957. The filter feeding and food of flamingos (Phoenicopteri). *Phil. Trans. Roy. Soc. Lond. B,* 240:401–493.

Kahane, H. 1986. *Logic and Philosophy*. Wadsworth: Belmont, CA.

Kooloos, J., Kraaijeveld, A. R., Langeveld, G. E. J. & G. A. Zweers. 1989. Comparative mechanics of filter-feeding in *Anas platyrhynchos, Anas clypeata* and *Aythya fuligula* (Aves, Anseriformes). *Zoomorphol.* 108:269–290.

Kooloos, J. G. M. & G. A. Zweers. 1989. Mechanics of drinking in the mallard (*Anas platyrhynchos,* Anatidae). *J. Morphol.* 199:327–347.

Kooloos, J. G. M. & G. A. Zweers. 1991. Integration of pecking, filter feeding and drinking mechanisms in waterfowl. *Acta Biotheor.* 39(2) In press.

Lovtrup, S. 1987. The theoretical basis of evolutionary thought. *Ann. Sci. Nat. Zool.* 8:219–236.

Matsuda R. 1987. *Animal Evolution in Changing Environments, with Special Reference to Abnormal Metamorphosis*. Wiley: New York.

Mayr, E. 1988. *Toward a New Philosophy of Biology*. Belknap Press of Harvard University Press: Cambridge, MA. 564 pp.

McLelland, J. 1979. Digestive system. Pp. 69–181. *In:* A. S. King & J. McLelland (eds.), *Form and Function in Birds,* Vol. 3. Academic Press: London.

Olson, S. L. 1985. The fossil record of birds. *Avian Biol.* 8:79–238.

Olson, S. L. & A. Feduccia. 1980a. *Presbyornis* and the origin of the Anseriformes (Aves: Charadriomorphae). *Smiths. Contrib. Zool.* 323:1–24.

Olson, S. L. & A. Feduccia. 1980b. Relationships and evolution of flamingos (Aves: Phoenicopteridae). *Smiths. Contrib. Zool.* 316:1–73.

Stephens, D. W. & J. R. Krebs. 1986. *Foraging Theory*. Princeton University Press: Princeton, NJ. 247 pp.

Strauch, J. G. 1978. The phylogeny of the Charadriformes (Aves): a new estimate using the method of character compatibility analysis. *Trans. Zool. Soc. Lond.* 34:263–345.

Szijj, J. 1965. Okologie der Anatiden im Ermatinger Becken. *Vogelwarte* 23: 40–72.

Wake, D. B. & G. Roth. 1989. The linkage between ontogeny and phylogeny in the evolution of complex systems. Pp. 361–377. *In:* D. B. Wake & G. Roth (eds.), *Complex Organismal Functions: Integration and Evolution in Vertebrates*. Wiley: Chichester.

Zeigler, H. P., Levitt, P. W. & R. Levine. 1980. Eating in the pigeon (*Columba livia*). *J. Comp. Physiol. Psychol.* 94:783–794.

Zusi, R. L. 1985. A functional and evolutionary analysis of rhynchokinesis in birds. *Smiths. Contrib. Zool.* 395:1–40.

Zweers, G. A. 1982. Pecking in the pigeon (*Columba livia*). *Behaviour* 81:1173–230.

Zweers, G. A. 1985a. Greek Classicism in living structures? Some deductive pathways in animal morphology. *Acta Biotheor.* 34:249–276.

Zweers, G. A. 1985b. Generalism and specialism in the avian mouth and pharynx. *Fortschr. Zool.* 30: 189–201.

Zweers, G. A. 1988. Holism and neutralism for open systems. *Amer. Zool.* 28:277–288.

Zweers, G. A. 1991. Transformation of avian feeding mechanisms: a deductive approach. *Acta Biotheor.* 39:15–36.

Zweers, G. A., Gerritsen, A. F. C. & P. van Kranenburg-Voogd 1977. Mechanics of feeding in the mallard (*Anas platyrhynchos* L.; Aves, Anseriformes). *Contrib. Vertebr. Evol.* 3:1–109.

Zweers, G. A., Gerritsen, A. F. C. & W. J. Bock. 1991. Morphological modifications for probing and filter feeding mechanisms in the avian pecking mechanism. (submitted).

Functional Analysis
and the Power of the Fourth Dimension
in Comparative Evolutionary Studies

Carole S. Hickman

Abstract. The static form of organisms is a less rich source of comparative data for systematics than is function, which includes the added dimension of time. The use of separate vocabularies for characterizing form and function has created and enforced an artificial distinction between "what is" and "how it works" and retarded the translation of information about function into characters and character states, especially at the level of mechanical properties and alternative states of structure. Lack of unanimity as to what characters are and how they should be defined has further retarded the incorporation of certain kinds of data into systematic analysis. Distinctions are drawn between functional and adaptive characters and between functional and adaptive analysis in order to identify problems and misconceptions in the past usages of these terms. Examples of functional characters under a narrowly restricted definition of function are developed for trochacean gastropods, where alternative states of configuration of the foot and neck lobes in functioning animals extend the amount of information available from the static morphology of preserved specimens.

INTRODUCTION

Form, function, and evolution can be connected in a variety of ways. This paper briefly outlines a simple program for connecting them in a purely systematic context: for analyzing relationships and hypothesizing phylogeny. Because the use of functional characters in systematics is highly controversial and unacceptable to many systematists, the program is introduced through an historical perspective on the form-function relationship, discussion of the questions involved in morphological, functional, and adaptational analysis, limitation of the definition of function, and an argument for the equivalency of functional and morphological characters in phylogenetic analysis. Figure 1 provides a summary of the relationships among the concepts discussed below and a guide to the discussion.

THE SCOPE OF FUNCTIONAL MORPHOLOGY

The relationship between form and function has an evolutionary context, but it can be analyzed without reference to historical explanation. There are three questions involved in defining the relationship that bear no reference to historical explanation. These three questions constitute the traditional scope of the discipline of functional morphology.

Dr. Hickman is with the Department of Integrative Biology and Museum of Paleontology, University of California, Berkeley, CA 94720, USA.

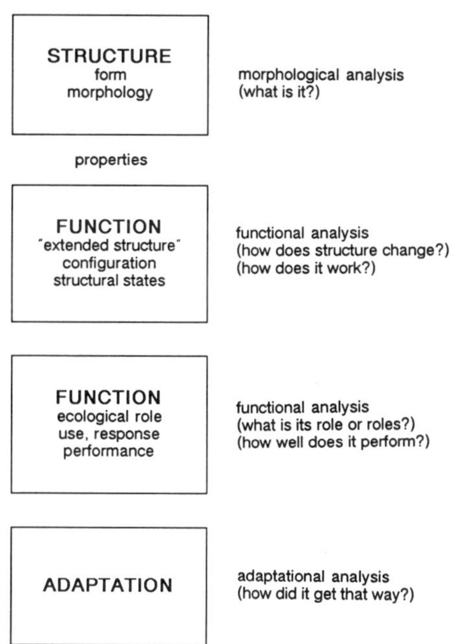

Figure 1. Summary of the distinctions between structure, function, and adaptation and the corresponding questions investigated under morphological, functional, and adaptational analysis. Function is defined at two levels. The argument for functional characters in this paper applies only to the first level, where function is linked to structure via properties and is defined as change in structure without reference to its ecological role or performance. Characters can be defined from the second level of functional analysis as long as functional and adaptational analysis are clearly distinguished.

The initial functional question posed by a structure is *"What is it for?"* This question is a simple extension of the descriptive morphological question *"What is it?"* The initial functional question frequently has more than one answer in the case of structures that perform multiple functions. The rhipodoglossan radula is a complex structure for feeding. On a deeper level, however, it is a structure with the dual function of preparing (rasping) and gathering (sweeping up) food. The wing of a bird is a locomotor structure identified primarily with the function of powered flight, but it may be used as well in soaring, gliding, hovering, and courtship display.

The remaining questions fall into two distinct categories, and it is important to my later arguments to draw a distinction between functional analysis and adaptive analysis.

Functional Analysis

Beyond the simple initial question of function or alternative functions, lie the more intriguing questions *"How does it work?"* and *"How well does it work?"* Questions of how structure performs are explored through diverse techniques of direct observation, comparison, and quantification. For mechanical structures such as the rhipodoglossan radula of a snail or the wing of a bird, function can be modeled on mechanical engineering principles. Performance of mechanical structures can be measured and compared in structures of similar form to provide comparative estimates of goodness (efficiency) relative to an optimally efficient design or paradigm (sensu Rudwick, 1964). Evaluation of the goodness-of-fit of a structure to its function does not require any knowledge or estimate of past goodnesses-of-fit.

Network analysis (Liem, 1980; Lauder, 1981) provides a more complex method for

550

examining patterns of functional interactions among structural elements to provide a broader picture of the organization of structure and function. It is, again, independent of historical analysis or inference of past conditions. It is an equilibrium analysis in the sense of Lauder (1982).

Traditionally most vertebrate and invertebrate functional morphologists have assumed that the relationship between form and function is a result of evolution (*adaptation as a process* driven by natural selection) and that the major contribution of their discipline is to provide a rich, detailed, and increasingly quantitative characterization of the result (*adaptation as a product*) (e.g., Wainwright et al., 1976). In other words, it tends to be an "exclusively interpretive" discipline (Cowen, 1979) that may be mapped onto phylogeny as an adaptive explanation of structure (a history of goodnesses-of-fit), but is not used to generate phylogeny or to study the mechanisms of evolution. Research programs directed at elucidating the functional properties of structure are unencumbered by the philosophical and procedural difficulties of historical adaptive analysis.

Adaptive Analysis

The question *"How did it get that way?"* is an historical question that falls within the purview of evolutionary morphology, evolutionary paleontology, and systematics. It is at once the most exciting and the most difficult question. The form of the question unfortunately has tempted a few evolutionary biologists to adaptive storytelling and the extreme style of selectionist argumentation ridiculed and characterized by Gould & Lewontin (1979). There is more to adaptive analysis. Phylogenetic reconstructions based on a hypothetico-deductive system of analysis of morphological similarity provide hypotheses of relationship that are preliminary to full historical reconstruction. Again, there is more to adaptive analysis. The question is ultimately one of identifying the selective forces underlying structural and functional transformations in populations and involves considerably more than hypotheses of genealogy. There is a large and controversy-laden literature concerned with the philosophy and practice of adaptational analysis (e.g., Bock, 1977, 1981; Bock & von Wahlert, 1965; Gans, 1974, 1988; Cracraft, 1981; Fisher, 1981, 1985; Lauder, 1981, 1982). This is a thoughtful but difficult body of literature, and it is likely that many practicing systematists have been deterred from using functional data because they are confused or intimidated by the formidable semantic and epistemological arguments and complex logic. It is not surprising that functional morphologists working outside an evolutionary context have not taken a greater interest in entering the fray.

THE PROBLEM OF FUNCTIONAL CHARACTERS IN SYSTEMATICS

I suspect that there are a number of reasons why functional data are seldom used in phylogenetic reconstruction. The first reason has to do with the language that functional morphologists use. The second reason (or set of reasons) has to do with a peculiar misconception among systematists about the nature of "functional characters" that has perpetuated a strong prohibition against using functional information of any kind in systematics, and especially in systematics using cladistic methods.

The Nature of Functional Data in Functional Morphology

Functional morphologists use a vocabulary to describe function and to characterize the relationship between form and function that is not isomorphic with the vocabulary used to describe morphology and define traditional morphological characters. This is especially true of relationships that are depicted diagrammatically, such as force reaction patterns and successive positions of structures. It is true of the functional component of properties, involving forces and deformations whose measurement is expressed in rates, ratios, moduli, and algebraic formulations. It is also true of sequences of events that are variously filtered, recorded, amplified, and charted as in the case of pressure changes, electrical patterns of muscle firing, gas concentrations, vocalization patterns, movement patterns, etc. These kinds of functional data are not generated for phylogenetic analysis, and functional morphologists have not been concerned with translating their data into the appropriate form for "doing systematics." It is also the case that functional data frequently are not generated in a comparative framework and cannot be mined from the literature and translated into characters by the systematist.

The Nature of Characters in Systematics

Systematists, on the other hand, have assiduously avoided developing translation procedures even where good comparative data are available. In fact, systematists have not even been able to reach consensus as to what characters are, let alone how they should be defined. There is a massive literature on methods of character weighting, methods of determining character polarity, and methods of character analysis. *Character designation*, however, is largely an art and frequently is referred to as "character choice," as if characters are inherent attributes of organisms that have some objective reality. For example, Hennig (1966), Eldredge and Cracraft (1980), and Wiley (1981), three treatises on cladistic methodology, provide no guidelines for defining characters. Wiley (1981, p. 115–117) briefly reviews the discussion of characters in earlier treatments of systematic methodology (e.g., Davis and Heywood, 1965; Mayr, 1969; Bock, 1977) and concludes that "the earlier discussions of characters and character states are largely semantic."

What is a Functional Character?

If there is confusion as to what characters are and how they should be defined, it should come as no surprise that it is even more difficult to determine what a "functional character" is and why it is objectionable or unsafe. The least confusing definition is that of Wiley (1981, p. 118), who states that *"these are characters that are similar in basic function in such a way that the parts may be compared in terms of that function."* In other words, they are morphological features that are named or recognized on the basis of similar function. The character *"wing"* is given as an example of a functional character. A wing is recognized on the basis of what it does rather than how it is structured. It is a prime target for the criticism that functional characters are likely to be the result of convergence. I would not choose to call this a functional character. It is a morphological character that has been inadequately designated by a term that can be applied to many features that do not bear close comparison at deeper levels of structure (Hickman, 1988).

There is another sense in which the term "functional character" is understood by evolutionary biologists and paleontologists who are concerned with the analysis of adaptation (see Lauder, 1981; Fisher, 1985; Greene, 1986). Functional characters can be equated with adaptive characters in the sense of characters conferring a performance advantage. Such characters are attributes of organisms, but at a different level. When analyzing adaptation as a process, it is heuristically important to separate hypotheses of relationship based on patterns of similarity in morphology from hypotheses of historical origin based on adaptive analysis. I want to make a clear distinction between adaptive characters and functional characters.

The third sense in which functional characters may be defined is in terms of the functional configuration or conformation of a feature, a sequence of structural states that define a function, or the basic properties that connect structure and function. In this sense, function can be expressed in basically structural terms that are readily translated into characters and character states to form character sets that can be analyzed either in combination with morphological characters or separately. The vocabulary for defining functional characters is a vocabulary for describing what structure does (a structural repertoire), not

552

why it does it (a role, purpose, or performance advantage). Functional characters of this third kind cannot be understood without some additional definition of what we mean by function.

LIMITING THE DEFINITION OF FUNCTION

There are several ways to define function. The definition that is adopted here is a simple definition that excludes reference to ecological roles and fitness, in the interest of keeping adaptation out of the picture. Most simply stated, function is changing structure (see Picken, 1960; Wainwright, 1988). Within the time dimension, function is a sequence of structural states. Functional potential may be viewed as the full range of structural states permitted by a given structure. It is, in fact, difficult to draw a line between structure and function. What I am defining here as function is an intermediate-level concept, and some may prefer to consider alternative conformational states of a structure to be structure itself rather than function. If this is the case, what I am calling for is a broader definition of structure and encouraging systematists who normally work with preserved, inanimate specimens to look into and use aspects of structure that are more familiar to functional morphologists and biomechanicians who work with animate specimens or experiment with the potentials of inanimate ones. I might, alternatively, have called this *extended structure.* (see Fig. 1).

Function as an extension of form into the time dimension provides an increase in the amount of information available for character analysis. The dynamic capabilities of structure frequently are not obvious in the study of static morphology. Patterns of shared similarity in function undoubtedly have underpinnings at some finer level of structure, but it is not always feasible to probe all these levels.

The only legitimate criticism of the use of this vast untapped quantity of functional information is the potential for repetitive information. However, there is no difference in the way that morphological characters and functional characters are defined and scored. There is no difference in the procedures for determining polarity, and no difference in the way they are analyzed. The danger of redundancy is no greater than it is for traditional morphological characters. Characters are not inherent attributes of organisms waiting to be discovered and sorted by the systematist. They are defined attributes, and it is the quality of definition that determines the outcome of the analysis, not the source of the information.

EXAMPLES OF FUNCTIONAL CHARACTERS IN GASTROPODS

Two brief examples are offered from the external soft part morphology of trochacean gastropods. The trochacean gastropod foot is a remarkably conservative organ in its appearance in preserved specimens (Fig. 2a) and has not been used in trochacean systematics prior to Hickman and McLean (1990). However, when the foot is observed in living snails, it is clear that it is capable of assuming some distinctive configurations. Two distinctive innovations that help define one clade of highly derived trochaceans are the ability to enroll the foot, coupled with the ability to produce the anterior end of the foot into a pair of lateral horns. These derived states of foot configuration (Fig. 2b) are not visible in preserved feet but are part of the repertoire of shape configurations observed in living animals. Note that I have described changes in form (=function) without saying anything about what they are for. This clade of gastropods lives on seagrasses and marine algae; the horns of the foot are used in probing and reaching by an animal that moves on a discontinuous surface from one blade to another, and the foot enrolls around blades or stipes and clings to portions of its discontinuous substratum. The performance of the foot (e.g., grasping ability) might be measured and compared in a separate adaptive analysis, but it is not part of the character analysis.

A second example is provided by the alternative configurations of the trochacean left neck lobe (a flap of epipodial tissue) and several associated structures. In most trochaceans the neck lobe is a flap with an entire margin, but in several clades it is complexly subdivided. In preserved animals the flap is in a conformation that provides no clue as to possible function or conformation in the living animal (Fig. 3a). When living animals are observed, some remarkable conformations and changes in conformation are observed. Figure 3b, c illustrates two alternative configurations of a neck lobe that can be enrolled into a semitubular structure with a finely-divided meshwork covering the top of the tube. In one case (Fig. 3b) the left cephalic tentacle is incorporated into the mesh. In the other case (Fig. 3c) both the left cephalic tentacle and the left eye stalk and eye are incorporated into the tube, and the cephalic tentacle periodically reaches down and sweeps back and forth across the top of the mesh. Pigment spots are present on the surface of the mesh (morphological character), and the spacing in the mesh changes (functional character) with changes in flow rate and suspended particle load. Shared configurational states and capabilities are useful in distinguishing clades. Again, I have said nothing about the role of these changes in configuration. These are snails that burrow and suspension feed, and the

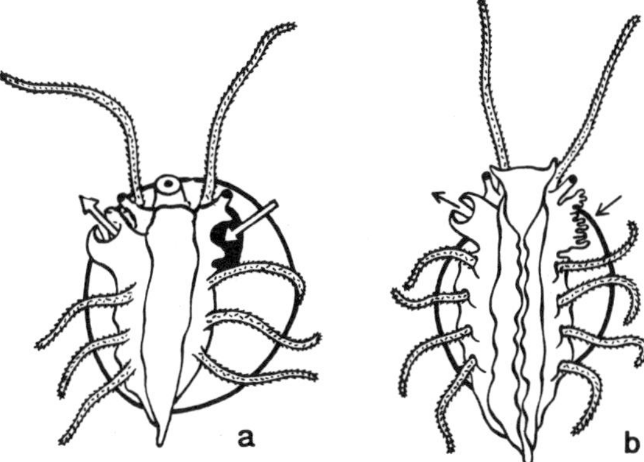

Figure 2. Alternative states in the functional configuration of the trochacean gastropod foot. a. configuration in relaxed or preserved specimens. b. two configurations shared by members of a clade that are able to produce temporary anterior horns and are able to enroll the foot. The two configurations are illustrated together, but are achieved independently during different activities (probing and grasping).

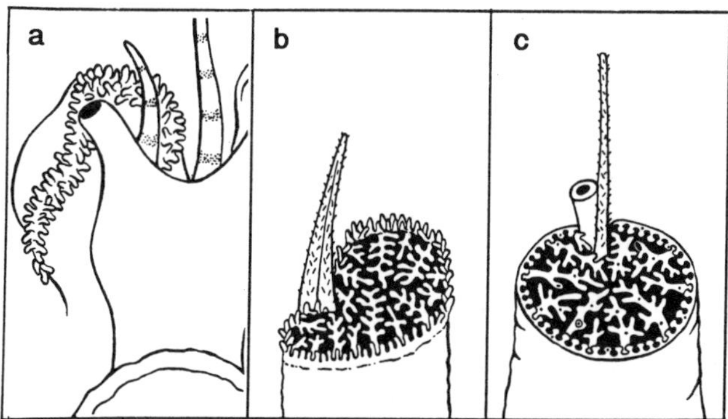

Figure 3. Alternative configurational states of the left neck lobe in trochacean gastropods. a. configuration in a relaxed or preserved specimen. b. active configuration in a clade that enrolls the neck lobe and incorporates the left cephalic tentacle into the semitubular structure. c. active configuration in a clade that enrolls the neck lobe and incorporates the left eye and left cephalic tentacle into the semitubular structure. In this clade the tentacle shares the additional ability to sweep across the tentacular mesh. In both clades the meshwork acts as a sieve and the tubular structure as an inhalant siphon.

neck lobes channel the flow of water into the mantle cavity where the extraction of food particles occurs on the surface of the gills. The mesh prevents the entry of large particles, and the animal that has the ability to sweep the cephalic tentacle across the mesh removes sieved particles that accumulate and begin to clog the mesh.

In choosing to define functional analysis and functional characters at a level of function that excludes references to ecological roles and adaptation, I do not intend to express a lack of interest in adaptive characters, adaptive analysis, and historical explanation. Understanding of evolutionary history is the ultimate goal of evolutionary biology and paleontology, and analysis of the role of adaptation (as a process) in history does not preclude an important role for other evolutionary processes. My intent is to encourage practicing systematists to make greater use of "safe" functional data in a manner that will generate better-supported phylogenies that can be used to test more complex historical hypotheses.

LITERATURE CITED

Bock, W. J. 1977. Foundations and methods of evolutionary classification. Pp. 851–895. *In*: M. K. Heckt et al. (eds.), *Major Patterns in Vertebrate Evolution*. Plenum Press: New York and London.

Bock, W. J. 1981. Functional-adaptive analysis in evolutionary classification. *American Zoologist* 21:5–20.

Bock, W. J. & G. von Wahlert. 1965. Adaptation and the form-function complex. *Evolution* 19:269–299.

Cowen, R. 1979. Morphology, Functional. Pp. 487–491. *In*: R. W. Fairbridge & D. Jablonski (eds.), *The Encyclopedia of Paleontology*. Dowden, Hutchinson & Ross, Inc.: Stroudsburg, PA.

Cracraft, J. 1981. The use of functional and adaptive criteria in phylogenetic systematics. *American Zoologist* 21:21–36.

Davis, P. H. & V. H. Heywood. 1965. *Principles of Angiosperm Taxonomy*. P. Van Nostrand Co., Inc.,: Princeton and New York.

Eldredge, N. & J. Cracraft. *Phylogenetic Patterns and the Evolutionary Process*. Columbia University Press: New York. 349 pp.

Fisher, D. C. 1981. The role of functional analysis in phylogenetic inference: examples from the history of the Xiphosura. *American Zoologist* 21:47–62.

Fisher, D. C. 1985. Evolutionary morphology: beyond the analogous, the anecdotal, and the ad hoc. *Paleobiology* 11:120–138.

Gans, K. 1974. *Biomechanics, an Approach to Vertebrate Biology*. J. B. Lippincott: Philadelphia, PA. 261 pp.

Gans, C. 1988. Adaptation and the form-function relation. *American Zoologist* 28:681–697.

Gould, S. J. & R. C. Lewontin. 1979. The spandrels of San Marco and the Panglossian paradigm: a critique of the adaptationist programme. *Proceedings of the Royal Society of London* 205B:581–589.

Greene, H. W. 1986. Diet and arboreality in the emerald monitor, *Varanus prasinus*, with comments on the study of adaptation. *Fieldiana Zoology*, New Series 31:1–12.

Hennig, W. 1966. *Phylogenetic Systematics*. University of Illinois Press: Urbana, Chicago, London. 263 pp.

Hickman, C. S. 1988. Analysis of form and function in fossils. *American Zoologist* 28:775–793.

Hickman, C. S. & J. H. McLean. 1990. Systematic revision and suprageneric classification of trochacean gastropods. *Natural History Museum of Los Angeles County Science Series* 35:1–169.

Lauder, G. V. 1981. Form and function: structural analysis in evolutionary morphology. *Paleobiology* 7:430–442.

Lauder, G. V. 1982. Historical biology and the problem of design. *Journal of Theoretical Biology* 97:57–67.

Liem, K. F. 1980. Adaptive significance of intra- and interspecific differences in the feeding repertoires of cichlid fishes. *American Zoologist* 20:295–314.

Mayr, E. 1969. *Principles of Systematic Zoology*. McGraw-Hill: New York.

Pickin, L. E. R. 1960. *The Organization of Cells and Other Organisms*. Oxford University Press: London.

Rudwick, M. J. S. 1964. The inference of function from structure in fossils. *British Journal for the Philosophy of Science* 25:27–40.

Wainwright, S. A. 1988. Form and function in organisms. *American Zoologist* 28:671–680.

Wainwright, S. A., Biggs, W. D., Currey, J. D. & J. M. Gosline. 1976. *Mechanical Design in Organisms*. John Wiley & Sons: New York. 423 pp.

Wiley, E. O. 1981. *Phylogenetics*. John Wiley and Sons: New York. 439 pp.

The Impact of Functional Morphology and Biomechanics on Studies of Evolutionary Biology

Marvalee H. Wake

Discussion of the impact of research in functional morphology and biomechanics on studies in evolutionary biology centered around six questions:

1. What is the contribution to biology in general of functional morphology and biomechanics, and what are their specific contributions to studies in evolutionary biology, ecology, development, physiology, and systematics?
2. How do we generate theory about structure and its variation?
3. How do we formulate problems in evolutionary morphology and how do we identify appropriate methodologies?
4. Is function an extension of form, and therefore translatable into characters of systematic utility?
5. When is a phylogenetic hypothesis useful/not useful in functional morphological and biomechanical studies?
6. What is evolutionary morphology?

Discussion initially considered the identification and definition of functional morphology, biomechanics, and evolutionary morphology. It was generally agreed that functional morphology treats mechanisms, how organisms work, based on an understanding of their structure. Some consider biomechanics a subcategory of functional morphology, others view it as a self-standing discipline in which certain tools are used to evaluate the mechanical properties of organisms—functional description or analysis are not necessarily included (Kipp Baron). Gart Zweers thought that biocybernetics, or the logic of the "steering" of systems, is a part of functional morphology; others suggested that control mechanisms, such as neurendocrinological, be included. It was generally felt that the latter two components are appropriate to functional morphology *only when they are placed in a context of understanding organismal function based on morphological data.* If such ideas are included in a general definition of functional morphology, others should be admitted as well, and the definition becomes all-inclusive and loses its meaning, in the opinion of most discussants. It was generally agreed that "evolutionary morphology" is a term meant to be inclusive of data on morphology of development, ecology, physiology, etc., analyzed in an historical context. "Evolutionary" is not meant to be a restrictive descriptor, as, for example, "particle" is of particle physics.

Dr. Wake is with the Department of Integrative Biology, and Museum of Vertebrate Zoology, University of California, Berkeley, CA 94720, USA.

556

Considerable attention was paid to the context in which studies in functional morphology are done. It was generally agreed that the context should be made explicit. Functional morphology can be the study of how organisms work in an environmental and historical context, for example. Several discussants emphasized that functional morphology is a way of defining, describing, modeling, and analyzing adaptation, in its historical (evolutionary) context (Janet Sherman, Christine Janis). Some participants believe that morphologists have concentrated on between-species differences, whereas evolutionists have focused on intra-species variation (Sharon Emerson). Several discussants mentioned that interspecific (comparative) studies should be joined with intraspecific (microevolutionary) studies, and that the value of both approaches has been the integration across these levels, as appropriate to the problem in question. Workers use the comparative method to elucidate patterns of all sorts.

Several discussants thought it important to include the fossil record in such considerations; others pointed out that studies can be uncoupled from the historical context—biomechanics can be done without reference to the fossil record, or to adaptation in a comparative sense (Robert Dudley). A major contribution of paleontology has been the development of constructional morphology as a logical framework for examining the interaction of functional and historical constraints with constructional constraints (the properties of constructional materials and the rules for their assembly). Paleontologists can perform "experiments" with fossils in order to test alternative hypotheses of function. By comparing the performance of fossil structure with the performance of idealized models they can evaluate the functional efficiency of ancient structural designs (Carole Hickman). Interest exists not just in how an organism works, but the functional diversity of organisms. Analysis of fossils in functional morphology has allowed dogma to be revised, and provided major new insights into the function of extant organisms.

Much concern was expressed about the need to introduce functional morphology into other disciplines in biology. Stephen Wainwright commented that functional morphology is useful according to the number of connections it makes, and that functional morphology depends on time (first, time during function [e.g., wing beat]; second, time—lifetime, over which a function develops; and third, time—generations, over which a function evolves). Therefore it must join other branches of biology to maintain its life. Functional morphologists need to branch out, to be more aware of other disciplines and the contributions they can make to them, and, in fact, to make the connections. Knowledge of structure is basic to understanding of phylogenetics, systematics, physiology, and to aspects of ecology and behavior. Connections between structure and function (broadly defined) can be made by identifying properties and qualities, such as strength $=$ stress at break (function)/cross-sectional area (structure), and power $=$ force $\times$ distance/time. We need a vocabulary for categorizing these, and other, qualities, in order to draw attention to them and to discuss them communally. The link of functional morphology to ecology was emphasized as an example—there is much work in testing performance by examining and manipulating animals with morphological/physiological variation (Peter Wainwright).

Carl Gans presented the position that morphology is the handmaiden of all biological science, and therefore fundamental to all analysis. Other discussants responded with the concern that by considering morphology a handmaiden, it loses its own identity as a synthetic discipline within biology. A unique contribution of functional morphology is the elucidation of analogues between the physical and, especially, the technological worlds and the biological.

The thesis presented by Carole Hickman that function is an extension of form, and therefore translatable into characters useful in phylogenetic/systematic analysis, was hotly debated. Points raised in support of the idea were several. Among them are: 1) one should use all characters possible in systematic analysis; 2) by using functional characters in cladistic analysis, one can incorporate hypotheses of functional relationship and evolution in traditional systematics; and 3) one can consider several different levels of function when one uses functional characters. Opponents to the thesis pointed out that there are many problems in interpreting functional characters; 1) often they are complexes of characters, both functional and specifically morphological; 2) it may be difficult to determine whether functional characters are independent or not (true of structural characters as well, in certain situations); 3) many functional states can be most effectively interpreted when they are broken down into the structural features that facilitate function; and 4) determination of homology of functional characters may prove difficult; examination of the underlying morphology may be necessary in order to determine homologies of characters. Participants, though concerned about identification and interpretation of functional characters, expressed interest in finding good examples and rigorous methods of analysis of such characters.

The question of ways to generate theory about structure and its variation met with several suggestions, and little resolution. It was generally agreed that functional and evolutionary morphologists deal with structural and functional complexes. Part of the research effort is identifying complexity and diversity, and then understanding them at several levels of analysis (morphological, physiological, then organismal, and above). Gart Zweers presented a multi-step model for the analysis of a complex function, using feeding and drinking in mallard ducks as the example. This carefully constructed model was of interest to the group, but not all thought it presented a method that could be adapted to a diversity of problems in functional morphology. Other suggestions for approaches to the generation of theory about structure and function included assessing mechanisms of development that explain form and its changes; finding ways to sort out genetically based from environmentally based variation; and looking more carefully at "optimal" versus "adequate" adaptation, the concern being that structure/function need not be "perfect," but only "good enough" to function adequately and to allow organisms to persist.

Despite much diversity of opinion about simple answers to important questions (largely because forward-looking, synthetic fields do not easily permit simplicity, uniformity, and agreement), participants seemed to believe that functional morphology and biomechanics have much to offer evolutionary biology. The very diversity of contribution is a strength. However, more attention must be paid to elucidating an appropriate vocabulary for functional morphology (broadly defined), and to the explicit inclusion of functional morphology with other branches of biology, so that real connections that enhance studies of morphology, ecology, behavior, systematics, evolution, etc. are made. More focused discussions, such as this session at ICSEB IV, of the relevance, perspective, and future of research in evolutionary morphology that bring together scientists who work on different organisms, with different tools, and with different foci, are essential to the development and life of the discipline.

Developmental Processes and Phenotypic Change in Archosaur Limb Evolution

EXTENDED ABSTRACT

Gerd B. Müller

Archosaurs are a monophyletic taxon (Benton & Clark, 1988) including thecodontia, dinosaurs, pterosaurs, and, with living representatives, crocodilians and birds. The evolution of archosaur limbs is characterized by a progressive numerical reduction of skeletal elements, namely in the carpus, the tarsus, and the digits. For example, primitive reptiles (e.g., *Paleothyris*, Carroll & Baird, 1972) possess nine ossified elements in the tarsus while protosuchians have seven, crocodiles four, and birds none. In the carpus the reduction goes from eleven elements in primitive amniotes to two in birds. At the ontogenetic level these trends have been explained on the basis of recapitulatory concepts (Holmgren, 1933; Steiner, 1934), assuming that essentially the same primary pattern of skeletal elements is laid down in all tetrapod limb buds and that only during the subsequent phases of development are the derived patterns realized through large numbers of fusions and deletions. In recent years, however, it was shown that the primary chondrogenic patterns of birds (Hinchliffe, 1977) and of crocodilians (Müller & Alberch, 1990) already have a derived characteristic at the stage of their initial appearance and only very few secondary modifications occur in later ontogeny. The present study analyzes through which processes the primary and the secondary alterations of limb patterns are realized in archosaur evolution.

A comprehensive investigation of the patterns and sequences of skeletogenesis in crocodilian limbs (Müller & Alberch, 1990) shows that the chondrogenic condensations of individual carpalia and tarsalia appear in a well-defined temporal arrangement. In general, the proximal elements appear first, the elements of the distal rows follow in a posterior to anterior sequence, and the centralia—elements between the proximal and distal rows— usually appear last, again in a posterior to anterior sequence. Thus, proximal elements are usually ahead of distal and central ones, and postaxial elements are ahead of preaxial ones.

A comparison of the crocodilian sequences with those observed in the more primitive limbs of chelonians (Burke & Alberch, 1985) shows that essentially the same temporal patterns are followed. However, both the central and the distal rows in the carpus and tarsus of chelonians exhibit the formation of additional elements at the ends of these sequences. In contrast, the comparison of crocodilian limbs with the chondrogenic patterns in the most derived archosaurs, the birds, shows that, in general, the most terminal

Dr. Müller is with the Department of Anatomy, University of Vienna, Währingerstrasse 13, A-1090 Wien, Austria.

elements in these same sequences are missing. No more centralia form at all in birds and the distal carpal and tarsal sequences are drastically abbreviated. Thus, the comparison of chondrogenic sequences suggests that the phylogenetic modification of the primary chondrogenic patterns in archosaur limbs is the consequence of a progressively earlier termination of the developmental processes that give rise to the primary chondrogenic elements. This principle of terminal deletion also applies to digital reduction in the archosaur lineage. It is there supported by limb reduction experiments involving treatment of crocodilian embryos with mitotic inhibitors (Müller, unpublished).

Secondary modifications of the primary chondrogenic patterns do occur at different times and phases of skeletogenesis and entail different types of processes. Separate chondrogenic foci, that are not yet demarcated cartilage elements, can fuse during their increase in matrix deposition and size, eventually forming a single cartilage element. This, for example, is the case in the crocodilian carpus, where a preaxial and a postaxial condensation fuse to form the radiale-intermedium. In the tarsus of birds another type of fusion takes place. Here, distinct cartilage elements, demarcated by perichondrial layers, fuse with the tibia only after they exist as individual entities for a prolonged period of time. Finally, representing a third type of fusion, separate cartilage elements can become combined at an even later stage, when embryonic cartilage is replaced by bone through the ossification process, such as in the metatarsus of birds. In addition to fusion processes, primary patterns can be modified through the complete failure of cartilage elements to ossify. Elements that are bone in primitive limbs can remain cartilaginous in derived forms, as is the case with the centrale of the crocodilian carpus. Thus, secondary modification of chondrogenic patterns is not a uniform mechanism. Rather, it can be noted that each time new processes replace those governing earlier phases of skeletogenesis alterations of the prior patterns are possible.

It appears that in the archosaur lineage the evolutionary modifications of limb patterns are realized at two distinct levels involving fundamentally different mechanisms. One is the level of primary pattern formation. Here the basic mechanism of transformation seems to be heterochrony, in terms of a paedomorphic truncation of condensation-forming processes. Elements that appear late in the sequences of condensation formation are usually the ones lost first in phylogeny (Müller & Alberch, 1990). At a second level, further transformations of limb patterns are realized through the expansion of skeletogenic processes across formerly separate areas. Particular opportunities for such fusions arise when transitions of process take place in skeletogenesis (Müller, 1990). Through the mechanisms of fusion and non-ossification a second reduction in number of limb elements is realized in archosaur evolution.

In a preliminary overview it seems that the second level changes are more closely related to changing functional demands in limb evolution than are the heterochronic processes underlying the first level changes. It remains to be shown whether this cursory observation can be substantiated through more comprehensive empirical studies. If so, this would indicate that the reductive patterns of primary chondrogenic condensations are shaped to a large extent by the internal dynamics of early skeletogenic processes, while the reductive patterns reached through second level processes are more under the influence of adaptational factors. Archosaur limb development, therefore, may represent a useful system for an empirical approach to the distinction between internal and external factors in morphological evolution.

ACKNOWLEDGMENT

This work has been supported by the Austrian Fonds zur Förderung der wissenschaftlichen Forschung.

560

LITERATURE CITED

Benton, M. J. & J. M. Clark. 1988. Archosaur phylogeny and the relationship of the crocodylia. Pp. 295–338. *In*: M. J. Benton ed.), *The Phylogeny and Classification of the Tetrapods, Volume 1: Amphibians, Reptiles, Birds*. Clarendon Press: Oxford.

Burke, A. C. & P. Alberch. 1985. The development and homology of the chelonian carpus and tarsus. *Journal of Morphology* 186:119–131.

Carroll, R. L. & D. Baird. 1972. Carboniferous stem-reptiles of the family Romeriidae. *Bulletin of the Museum of Comparative Zoology* 143:321–363.

Hinchliffe, J. R. 1977. The chondrogenic pattern in chick limb morphogenesis: a problem of development and evolution. Pp. 293–309. *In*: D. A. Ede, J. R. Hinchliffe & M. Balls (eds.), *Vertebrate Limb and Somite Morphogenesis*. Cambridge University Press: Cambridge.

Holmgren, N. 1933. On the origin of the tetrapod limb. *Acta Zoologica* 14:185–295.

Müller, G. B. 1990. Developmental mechanisms at the origin of morphological novelty: a side-effect hypothesis. Pp. 99–130. *In*: M. H. Nitecki (ed.), *Evolutionary Innovations*. The University of Chicago Press: Chicago.

Müller, G. B. & P. Alberch. 1990. Ontogeny of the limb skeleton in *Alligator mississippiensis*: developmental invariance and change in the evolution of archosaur limbs. *Journal of Morphology* 203:151–164.

Steiner, H. 1934. Über die embryonale Hand- und Fuss-Skelett-Anlage bei den Crocodiliern, sowie über ihre Beziehungen zur Vogel-Flügelanlage und zur ursprünglichen Tetrapoden-Extremität. *Revue Suisse de Zoologie*. 41(23):383–396.

The Evolution of Segment Patterning Mechanisms in Insects

Michael Akam, Guy Tear, and Robert Kelsh

Abstract. Molecular probes reveal similar patterns of segmentation and homeotic gene expression in the mature germ bands of *Schistocerca* and *Drosophila* embryos—representatives of insects with short and long germ type embryos respectively. This raises a paradox, for the analysis of segmentation in *Drosophila* has revealed a mechanism that depends primarily on the intracellular diffusion of transcription factors within the syncytial environment of the blastoderm. Such a mechanism cannot operate to generate segments within the cellular environment of the growing germ band in short germ insects. Different, and probably non-homologous, mechanisms must have evolved to generate homologous structures in these different insect groups.

INTRODUCTION

To understand how animal forms evolve, we must determine, not how morphologies have changed, but how mechanisms of development have changed, and how alterations in those mechanisms have been brought about by the alteration of specific genes. Only then will we be able to relate the population genetics of specific loci to the evolutionary history of whole animals. At present, few processes of development are understood in sufficient detail for this approach to be possible. One of the best such cases is the development of segments in the insect embryo.

This paper focuses on developmental and evolutionary questions relating to the different modes of generating segments in short and long germ insects. We argue that the conserved stage of the 'segmented germ band' reflects homologous patterns of gene activity in both short and long germ insects. However, the molecular mechanisms that establish these patterns may be very different in the two groups.

MORE THAN ONE WAY TO BUILD A BODY

Among the arthropods, insects are conservative in basic body plan. Segmental origins of the head are not clear, but in the more posterior part of the body, the group is characterized by a pattern of three mouthpart segments, three thoracic segments, and an abdomen of up to 11 segments (Snodgrass, 1935). This body plan is not always obvious in adults, where overt segmentation of the mouthparts and the posterior abdomen is frequently obscured. But almost all insect groups pass through a stage of development

---

Drs. Akam, Tear, and Kelsh are with the Department of Genetics, Downing Street, Cambridge, CB2 3EH, UK.

562

where this basic pattern is clearly displayed (Fig. 1). This "germ band" stage can therefore be regarded as a "phylotypic" stage in insect development (Seidel, 1960; Sander, 1983).

Different insects generate this germ band in different ways. In most insect groups, repeated cycles of nuclear cleavage without cell division generate an array of syncytial nuclei (∿5000 in *Drosophila*) covering the surface of the egg. (Anderson, 1972; Foe & Alberts, 1983). Cells form by the invagination of membranes from the egg surface around each nucleus, thus converting the syncytial blastoderm to a cellular blastoderm.

In *Drosophila* and most other endopterygote insects, the blastoderm is partitioned into groups of determined cells before gastrulation starts. These map out virtually the complete body plan onto the egg. Mesoderm, neurogenic, and epidermal territories are defined around the dorso-ventral axis; individual segment primordia are arrayed along the antero-posterior axis (Campos-Ortega & Hartenstein, 1985). The ablation of small groups of cells at this stage generates specific segmental defects in locations that correspond to the site of the injury (Lohs Schardin et al., 1979). Thus, by the time that cellularization of the blastoderm is complete, the ability of the embryo to regenerate lost pattern elements is already limited.

We can characterize this mode of development as formation of segments by subdivision in space, the generation and subsequent development of all segments being simultaneous, or very nearly so (Fig. 2). By contrast, many lower insect orders generate segments by sequential growth—a temporal process. These insects also form a syncytial blastoderm, but the distribution of nuclei is not uniform throughout the egg. Most blastoderm nuclei aggregate to form a small germ disc, covering only a small part of the egg, frequently near the posterior pole. The remaining nuclei generate the serosa—an extra-embryonic membrane that envelops the remainder of the egg (Anderson, 1972).

Gastrulation begins soon after the germ disc is formed, and before any overt signs of segmentation. At this stage, ablation experiments suggest that the developing embryo contains primordia for the head, but not for individual segments in the posterior part of the body (Sander, 1976). The cells that will generate these posterior segments are born later, as the germ band extends. In this regard, segment generation in the short germ insects is analogous to that in annelids (e.g., growth from leech teloblasts (Weisblat et al., 1988), and some crustaceans (Dohle & Scholtz, 1988). The long germ pattern—generation of segments by subdivision of the blastoderm—is presumably the derived state, and budding of

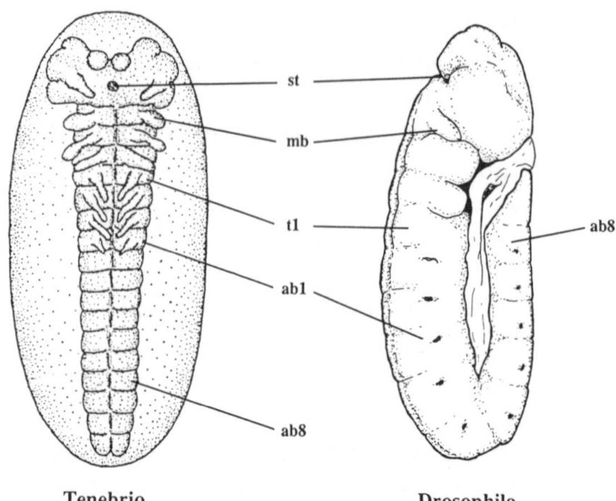

Figure 1. Comparison of the germ band stage of a beetle (*Tenebrio molitor*, after Ullman (1967) and a fly (*Drosophila*). The segmentation in the head of the fly will later be obscured, but at this stage the array of gnathal segments is still clearly visible. (st—stomodaeum; mb—mandibular (first mouthpart) segment; t1—first thoracic segment; ab1—first abdominal segment; ab8—eighth abdominal segment.)

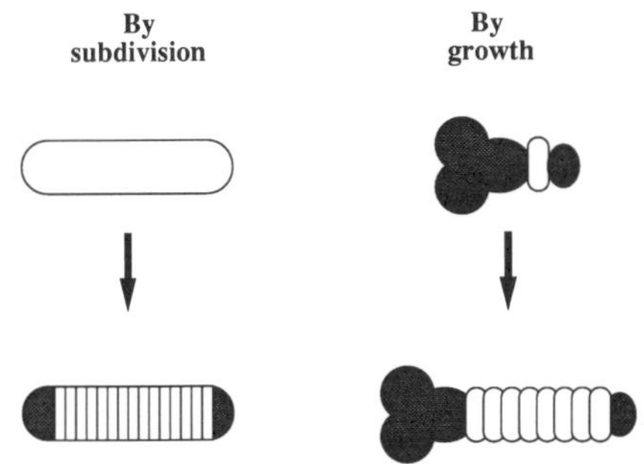

Figure 2. Contrasting modes of segmentation in short and long germ insects. Such long germ insects as *Drosophila* generate segments virtually simultaneously, by subdivision in space. Short germ insects generate segments sequentially, by growth. (It is not clear whether an extreme short germ insect generates all of its segments sequentially—the first few segments to appear do so almost simultaneously (Patel et al., 1989). Conversely it may be argued that *Drosophila*, an extreme "long germ" insect, shows vestiges of sequential segment formation, in that molecular markers for the most posterior segments of the abdomen are not resolved until after gastrulation; the pair rule pattern does not directly generate *engrailed* stripe 15 (Kornberg et al., 1985)).

segments the more primitive. However, at present the "long germ band" mode of development is much the better understood, because of rapid progress in the molecular analysis of *Drosophila* development.

SEGMENT GENERATION IN *DROSOPHILA*

In *Drosophila*, generation of the segment pattern is initiated by localized maternal determinants at the anterior and posterior of the egg (Nüsslein-Volhard et al., 1987; Nüsslein-Volhard & Roth, 1989). These determinants include localized maternal RNA's within the egg and specific signals generated by the somatic cells that surround it. Patterning occurs fast, during nuclear cleavage and formation of the syncytial blastoderm. It is essentially complete by the time cells are fully individualised, at the onset of gastrulation (Akam, 1987; Ingham, 1988).

We can make two generalizations about this patterning process. First, that many of the gene products involved are nuclear proteins that act, directly or indirectly, to regulate transcription of target loci (Carroll, 1990). Secondly, that the spatial co-ordination of the pattern is dependent on the diffusion of these nuclear proteins from their site of synthesis within the syncytium directly to their site of action (Driever & Nüsslein-Volhard, 1988; Stanjovec et al., 1989). I use the term diffusion loosely, to indicate the spreading of these proteins through the cytoplasm. In the structured environment of the egg (Foe & Alberts, 1983), the process is unlikely to be free diffusion. Short half-lives for most of the products, together with specific mechanisms to immobilize transcripts, ensure that the activities of these gene products are localized to specific regions within the egg (Edgar et al., 1987).

At each stage in the cascade of interactions the range of the gene products is reduced. The anterior determinant *bicoid* generates a gradient of protein concentration that extends through more than half of the egg, establishing positional cues that activate the first transcription of segmentation genes in the embryo's own nuclei (Driever et al., 1989; Struhl et al., 1989). The products of these zygotic genes in turn generate secondary gradients extending from their sites of synthesis over a few segments (Gaul & Jäckle, 1989; Stanjovec et al., 1989). The overlapping distributions of these zygotic gene products generate more precise positional cues, until ultimately a repeating pattern of cell states is established that defines the periodicity and limits of individual segments (Ingham, 1988).

564

The same initial cues are used to define the identities of different segments, by activating members of the homeotic gene family at particular positions along the axis of the egg (Irish et al., 1989; Jack & McGinnis, 1990; Reinitz & Levine, 1990). Indeed, in *Drosophila* the generation of the repeating segment pattern, and the specification of unique identities for the segments are essentially similar processes—for periodicity appears to be an epiphenomenon, arising from the independent specification of a series of unique stripes (Akam, 1989).

THE EVOLUTIONARY PARADOX

This understanding of the mechanism used to make segments in *Drosophila* immediately raises a paradox in our attempts to understand the evolution of segment patterning in other insects. The mechanisms that *Drosophila* uses to build segments are essentially intra-cellular, and can only function in a syncytial environment. Once nuclei are separated from one another by cell membranes, transcription factors can only convey information to other cells by altering the activity of genes that effect cell communication. They cannot, directly, generate pattern.

In short germ insects, both the process of building segments and the specification of precise segment identities occur in a cellular environment. At first sight, then, segmentation would seem to proceed by fundamentally different mechanisms in short and long germ insects.

We can consider three possibilities to resolve this paradox:

1. That segments are not homologous in short and long germ insects, in the restricted sense that they do not reflect the establishment of the same underlying patterns of gene activity.
2. That the segments are homologous—that is, in the mature germ band the same or very similar patterns of activity of homologous genes define the array of segments—but that the generative mechanisms are not homologous and involve fundamentally different genetic interactions.
3. That the differences are more apparent than real—for example, that *Drosophila* has eliminated intermediate pathways of cell signalling, but has conserved the role of key regulatory genes in generating and defining segments.

In the past, few have cast doubts on the homology of segments in different insects, but possibility 2 above has been raised explicitly (Sander, 1983) to account for the different behaviour of different insect embryos in response to experimental manipulation.

SEGMENT GENERATION IN A SHORT GERM INSECT

Two molecular markers now allow us to argue that the mature germ band in the short germ insect *Schistocerca* (*Orthoptera;* a grasshopper) is homologous, in the sense defined above, to that of *Drosophila*. One of these is the protein *engrailed*. In *Drosophila,* the expression of the *engrailed* gene marks the posterior stripe of cells in every segment, and plays a major role in defining and maintaining the boundaries between segments throughout development (Kornberg, 1981; Kornberg et al., 1985). *Engrailed* is activated just as cells form, in response to specific combinations of regulatory molecules established in individual nuclei during syncytial blastoderm stages (Ingham, 1988).

Using an antibody that reacts with *engrailed*-related proteins in a wide range of species, Patel et al., (1989) have shown that the distribution of *engrailed* protein in the mature germ band of *Schistocerca* corresponds very closely to that in *Drosophila*—marking the posterior part of each segment. These stripes of *engrailed* expression are generated as the germ band extends, shortly before segments become visible. If we can take *engrailed* to be an early marker for the formation of segments, this result confirms that segment patterning is occurring sequentially, and long after the syncytial blastoderm stage.

The *abdominal-A (abd-A)* gene provides a second molecular marker (Tear et al., 1990). In flies, this homeotic gene defines the identity of most abdominal segments. It is expressed from a well-defined boundary in the middle of the first abdominal segment, posteriorly to the end of the segmental region of the abdomen (Karch et al., 1990). The *Schistocerca* homologue of *abdominal-A* is expressed in the same region from an identical anterior boundary (Fig. 3).

In both insects, the anterior boundaries in the mature germ band are coincident with the anterior limit of the *engrailed*-expressing cells in A1 (Karch et al., 1990; Tear et al. 1990). In flies, transcription of the *abdominal-A* gene is initiated in the syncytial blastoderm, throughout the abdominal region of the embryo. Like *engrailed,* it is regulated by the distribution of transcription factors encoded by the segmentation genes.

In *Schistocerca,* expression of *abd-A* initiates as the germ band grows (Tear et al., 1990). The precise anterior boundary of expression is established between cells that were almost certainly born after cellularization was complete. The positional signals that define the uniqueness of this boundary must therefore be generated by different mechanisms in *Schistocerca* and *Drosophila.*

If we can extrapolate from these two cases, it seems that the germ band of *Schistocerca* is, in molecular terms, homologous to that of *Drosophila.* We do not yet know whether genes that act earlier in the segmentation hierarchy of *Drosophila* will have homologous roles in the development of *Schistocerca.* It seems likely that some will not. There is no evidence to suggest that the eggs of Orthoptera contain an anteriorly located maternal determinant—the germ band can develop normally in the absence of the entire anterior half of the egg, and the removal of anterior cytoplasm from the egg has no effects on the pattern of the embryo (Sander, 1976). In other short germ insects, a normal germ band can be formed after all cells in the primary blastoderm germ disc have been killed by irradiation (Sander, 1976). Nuclei migrate to the same location within the egg, and regenerate a germ anlage. Whatever maternal signals are provided must be capable of initiating a further process of pattern generation after the normal time of gastrulation, acting upon nuclei originally present at other locations within the egg.

Schistocerca   Drosophila

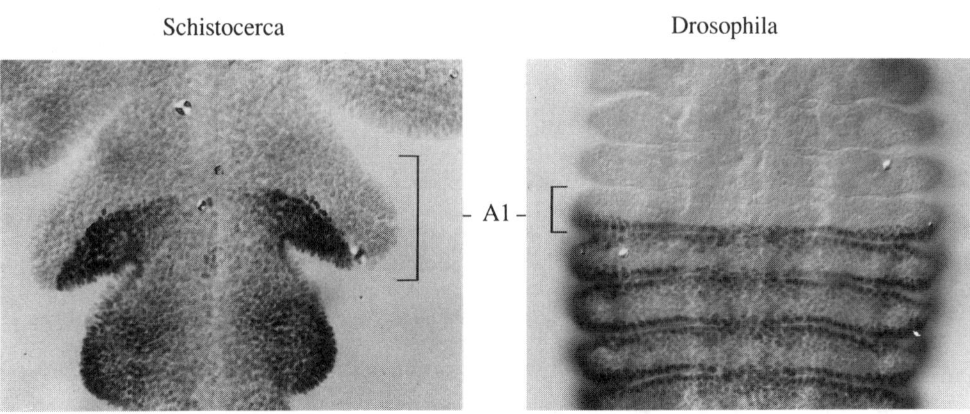

Figure 3. *abd-A* expression in *Drosophila* and *Schistocerca.* Embryos stained with anti-abdominal-A antibodies. An equivalent anterior limit of expression is established in the two species. This lies at the anterior margin of the *engrailed* stripe of A1 (Karch et al., 1990; Tear et al., 1990). The anti-*Drosophila* anti-body was kindly provided by J. Casanova. The anti-*Schistocerca* antibody is described in Tear et al. (1990).

566

PERSPECTIVE

Although in *Drosophila* the pattern of segments is defined by the onset of gastrulation there are later acting mechanisms that regulate the expression of segmentation genes. Two independent regulatory mechanisms control the *engrailed* gene—one acting in the blastoderm, the other acting in the completed germ band (DiNardo et al., 1988). This later acting mechanism mediates the regulation of abnormal segment patterns observed in some mutant *Drosophila* embryos, and may in the wild type serve as an error correcting mechanism to compensate for cell death or inadvertent rearrangement (Martinez-Arias et al., 1988). The set of interacting gene products that mediate this later regulatory pathway include cell surface and extracellular molecules that probably mediate cell communication (Martinez-Arias, 1989). It may be that in *Schistocerca* homologous regulatory interactions serve the primary role of generating the segment pattern during growth, and that in *Drosophila* the primacy of this pathway has been displaced by the invention of a novel, earlier acting regulatory hierarchy that takes advantage of the syncytial environment in the blastoderm.

The same argument cannot help us to understand how the mechanism that patterns homeotic gene expression in *Drosophila* may have evolved from lower forms. In *Drosophila* *engrailed* and other segment polarity genes active in the later regulatory network display a repetitive pattern of activity throughout the length of the germ band. They do not provide unique signals to convey segment identity. Once the products of the maternal and early acting zygotic segmentation genes have decayed, the only known molecular signals that distinguish one segment from another are the patterns of activity of the homeotic genes. The cues that establish these patterns in the *Schistocerca* embryo remain unknown. However, as pointed out above, the generation of each stripe in the repeating segment pattern is essentially equivalent to the activation of a homeotic gene at a unique position along the axis of the embryo. As the mechanism that activates homeotic genes evolved in higher insects to exploit the syncytial environment of the insect egg, so perhaps the segment generating machinery hitched along for the ride.

ACKNOWLEDGMENTS

We thank the Medical Research Council and the Wellcome Trust for the support of research in the authors' laboratory.

LITERATURE CITED

Akam, M. 1987. The molecular basis for metameric pattern in the *Drosophila* embryo. *Development* 101:1–22.

Akam, M. 1989. Making stripes inelegantly. *Nature* 341:282–283.

Anderson, D. T. 1972. The development of hemimetabolous insects. Pp. 96–165. *In:* S. J. Counce & C. H. Waddington (eds.), *Developmental Systems: Insects.* Academic Press: London.

Campos-Ortega, J. A. & V. Hartenstein. 1985. *The Embryonic Development of* Drosophila melanogaster. Springer-Verlag: Berlin.

Carroll, S. B. 1990. Zebra patterns in fly embryos: activation of stripes or repression of interstripes? *Cell* 60:9–16.

DiNardo, S., Sher, E., Heemskerk-Jongens, J., Kassis, J. & P. O'Farrell. 1988. Two tiered regulation of spatially patterned *engrailed* gene expression during *Drosophila* embryogenesis. *Nature* 332:604–609.

Dohle, W. & G. Scholtz. 1988. Clonal analysis of the crustacean segment: the discordance between genealogical and segmental borders. *Development* 104 supplement:147–160.

Driever, W. & C. Nüsslein-Volhard. 1988. A gradient of *bicoid* protein in Drosophila embryos. *Cell* 54:83–93.

Driever, W., Thoma, G. & C. Nüsslein-Volhard. 1989. Determination of spatial domains of zygotic gene expression in the *Drosophila* embryo by the affinity of binding sites for the *bicoid* morphogen. *Nature* 340:363–367.

Edgar, B. A., O'Dell, G. M. & G. Schubiger. 1987. Cytoarchitecture and the patterning of *fushi-tarazu* expression in the *Drosophila* blastoderm. *Genes Dev.* 1:1226–1237.

Foe, V. E. & B. M. Alberts. 1983. Studies of nuclear and cytoplasmic behaviour during the five mitotic cycles that precede gastrulation in *Drosophila* embryogenesis. *J. Cell. Sci.* 61:31–70.

Gaul, U. & H. Jäckle. 1989. Analysis of maternal effect mutant combinations elucidates regulation and function of the overlap of *hunchback* and *Krüppel* gene expression in the *Drosophila* blastoderm embryo. *Development* 107:651–663.

Ingham, P. W. 1988. The molecular genetics of embryonic pattern formation in *Drosophila*. *Nature* 335:25–34.

Irish, V. F., Martinez-Arias, A. & M. Akam. 1989. Spatial regulation of the *Antennapedia* and *Ultrabithorax* homeotic genes during *Drosophila* early development. *EMBO J.* 8:1527–1539.

Jack, T. & W. McGinnis. 1990. Establishment of the *Deformed* expression stripe requires the combinatorial action of coordinate, gap and pair-rule proteins. *EMBO* 9:1187–1198.

Karch, F., Bender, W. & B. Weiffenbach. 1990. *abdominal-A* expression in *Drosophila* embryos. *Genes Dev.* 4:1573–1588.

Kornberg, T. 1981. *engrailed*: a gene controlling compartment and segment formation in *Drosophila*. *Proc. Natl. Acad. Sci. U.S. A.* 78:1095–1099.

Kornberg, T., Siden, I., O'Farrell, P. & M. Simon. 1985. The *engrailed* locus of *Drosophila*: In situ localization of transcripts reveals compartment specific expression. *Cell* 40:45–53.

Lohs Schardin, M., Cremer, C. & C. Nüsslein-Volhard. 1979. A fate map for the larval epidermis of *Drosophila melanogaster*: localized cuticle defects following irradiation of the blastoderm with an UV laser microbeam. *Devl. Biol.* 73:239–255.

Martinez-Arias, A. 1989. A cellular basis for pattern formation in the insect epidermis. *Trends Genet.* 5:262–267.

Martinez-Arias, A., Baker, N. E. P. W. Ingham. 1988. Role of segment polarity genes in the definition and maintenance of cell states in the *Drosophila* embryo. *Development* 103:157–170.

Nüsslein-Volhard, C., Fröhnhofer, H. G. & R. Lehmann. 1987. Determination of antero-posterior polarity in *Drosophila*. *Science* 238:1675–1681.

Nüsslein-Volhard, C. & S. Roth. 1989. Axis determination in insect embryos. Pp. 37–55. *In: Cellular basis of Morphogenesis*. Ciba Foundation Symp. 144. John Wiley and Sons: New York.

Patel, N. H., Kornberg, T. B. & C. S. Goodman. 1989. Expression of *engrailed* during segmentation in grasshopper and crayfish. *Development* 107:201–213.

Reinitz, J. & M. Levine. 1990. Control of the initiation of homeotic gene expression by the gap genes *giant* and *tailless* in *Drosophila*. *Dev. Biol* 140:57–72.

Sander, K. 1976. Specification of the basic body pattern in insect embryogenesis. *Adv. Insect. Physiol* 12:125–238.

Sander, K. 1983. The evolution of patterning mechanisms: gleanings from insect embryogenesis and spermatogenesis. Pp. 137–161. *In: Development and Evolution*. B. Goodwin, N. Holder & C. Wylie (eds.), Cambridge University Press: Cambridge.

Seidel, F. 1960. Körpergrundgestalt und keimstruktur. Eine Erörterung über die Grundlagen der vergleichenden und experimentellen embryologie und deren Gültigkeit bei phylogenetischen Uberlegungen. *Zool. Anz.* 164:245–305.

Snodgrass, R. E. 1935. *Principles of Insect Morphology*. McGraw-Hill: London and New York.

Stanjovec, D., Hoey, T., Warrior, R. & M. Levine. 1989. Sequence-specific DNA binding activities of the gap genes encoded by *hunchback* and *Kruppel* in *Drosophila*. *Nature* 341:331–335.

Struhl, G. Struhl, K. & P. M. Macdonald. 1989. The gradient morphogen *bicoid* is a concentration dependent transcriptional activator. *Cell* 57:1259–1273.

Tear, G. Akam, M. & A. Martinez-Arias. 1990. Isolation of an *abdominal-A* homologue from the locust *Schistocerca gregaria* and its expression during early embryogenesis. *Development.* 110:915–925.

Ullman, S. L. 1967. The development of the nervous system and other ectodermal derivatives in *Tenebrio molitor* L. (Insecta, Coleoptera). *Phil Trans Roy Soc* 252:1–25.

Weisblat, D. A. Price, D. J. & C. J. Wedeen. 1988. Segmentation in leech development. *Development* 104 Supplement: 161–168.

The Genetic Basis for the Evolution of Multicellularity and Cellular Differentiation in the Volvocine Green Algae

David L. Kirk, Marilyn M. Kirk, Kandace A. Stamer, and Allan Larson

Dedicated to our friend and colleague Viktor Hamburger on the occasion of his ninetieth birthday. In addition to his more widely known role as one of the founders and leaders of the field of neuroembryology, Professor Hamburger was one of the first embryologists to recognize and exploit the analytical power of developmental genetics.

Abstract. In *Volvox carteri*, a complete division of labor occurs in which certain vegetative functions are executed only by terminally differentiated somatic cells and reproductive functions are executed by immortal germ cells. This is in marked contrast to the situation in most other volvocine algae, wherein an individual cell is able to execute these two sets of functions sequentially. Differentiation of germ and soma in *V. carteri* begins in the embryo in midcleavage, when a programmed set of visibly asymmetric divisions sets apart large and small sister cells; the smaller cells then give rise to terminally differentiated somatic cells, while the larger cells bypass the vegetative phase of the life cycle and differentiate directly as asexual reproductive cells, or gonidia. Mutational analysis is being used to define the genes controlling the various steps of this developmental program. Asymmetric division requires the function of the "gonidialess" (*gls*) locus; in a *gls⁻* mutant, symmetric cell divisions are unaffected, but no asymmetric divisions occur, all cells are equally small at the end of cleavage, and no gonidia develop. Terminal differentiation of somatic initials requires a functional "somatic regenerator" (*regA*) locus; in a *regA* mutant, somatic cells differentiate normally, but then they redifferentiate as gonidia. (Therefore, a *regA/gls* double mutant resembles a colonial volvocean, in which all cells first differentiate as vegetative cells and then redifferentiate as reproductive cells.) The ability of wild-type gonidial initials to bypass the vegetative phase of the life cycle and differentiate directly as reproductive cells requires the function of a set of "late gonidia" (*lag*) loci; in strains mutant at any one of a series of *lag* loci, presumptive gonidia first differentiate as "large somatic cells" and only later redifferentiate as gonidia. Second site suppressers of the loci just described are being used to identify other genes involved in cytodifferentiation, and attempts to clone and analyze the function of key genes in the program are underway.

In preparation for future attempts to determine the evolutionary origins of these genes that control cytodifferentiation in *Volvox carteri*, we are using comparative studies of aligned rRNA sequences to reconstruct the phylogeny of *Volvox* and its colonial relatives. By this approach, it has been possible to falsify the hypothesis that some or all members of the genus *Volvox* are more closely related to the family Haematococcaceae than they are to *Chlamydomonas* and colonial members of the family Volvocaceae. Although preliminary results indicate that the phylogenetic relationships among the unicellular, colonial, and multicellular volvocine algae may be somewhat more complex than has sometimes been assumed, they also indicate clearly that *Chlamydomonas reinhardtii* and genera in the family Volvocaceae constitute a coherent group of closely related organisms. Thus, this group appears to have the potential to provide novel insights into the molecular genetic mechanisms by which one form of multicellularity and division of labor between fully differentiated cell types may have evolved.

Drs. Kirk, Kirk, Stamer, and Larson are with the Department of Biology, Washington University, St. Louis, MO 63130, USA.

INTRODUCTION

Together with its presumed relatives in the "volvocine lineage" of green algae, *Volvox carteri* provides a potentially powerful model for elucidating how a genetic program for cellular differentiation may have evolved during the derivation of a multicellular organism from a unicellular or colonial ancestor (Kirk & Harper, 1986; Kirk, 1988). In hopes of exploiting this potential, we are currently attempting to define and analyze the genes that play important roles in *V. carteri* cytodifferentiation, and to clarify the phylogenic relationship of this organism to other volvocine algae.

Volvox carteri exhibits a complete division of labor between two fully differentiated cell types: mortal somatic cells and immortal germ cells (Starr, 1970). Each wild-type asexual *V. carteri* spheroid contains about 2000 small *Chlamydomonas*-like somatic cells, and about 16 large, asexual reproductive cells, or "gonidia" (Fig. 1). Whereas the somatic cells are post-mitotic and specialized for motility, the gonidia are non-motile and specialized for cell division. Each gonidium cleaves 11–12 times to generate all of the cells that will be present in an adult spheroid of the next asexual generation; included in this cleavage program are a predictable set of visibly asymmetric divisions that set apart large and small sister cells destined to become gonidia and somatic cells, respectively (Fig. 2). These asymmetric divisions pose graphically the question of how one cell may be genetically programmed to give rise to two extremely different cell types. The fact that many mutants with specific defects in the development of one or another of these cell types can be readily isolated and subjected to Mendelian analysis (Starr, 1970; Huskey et al., 1979; Harper et al., 1987; Kirk, 1990) supports our hope that it may be possible to define the manner in which germ-soma differentiation is programmed into the *V. carteri* genome. If and when any portion of this goal is achieved, it will be of compelling interest to attempt to discern how the components of this program have arisen during diversification of the volvocine algae.

The volvocine algae include a number of unicellular and colonial forms that are frequently aligned in a conceptual series suggestive of an evolutionary progression in which larger size was accompanied by increasing degrees of cellular differentiation (Kochert, 1973). *Chlamydomonas*, *Gonium* and *Pandorina*, for example, possess a single cell type that executes sequentially the vegetative and reproductive functions that are assigned to different cell types in *V. carteri*. In these genera, all cells enlarge 2^n-fold while they are in a motile, vegetative phase, and then they enter an asexual-reproductive phase in which they cleave rapidly n times to produce 2^n daughter cells that repeat the cycle. The principal difference between the unicellular and colonial forms is that in unicells, such as *Chlamydomonas* (Fig. 3A), the sister cells produced by cleavage divisions separate from one another as they begin to differentiate, whereas in colonial forms, such as *Gonium* (Fig. 3B) and *Pandorina* (Fig. 3C), sister cells remain associated to form coherent motile colonies. In *Eudorina* (Fig. 3D), the life cycle resembles that of *Pandorina*, except that under some circumstances the most anterior cells of a *Eudorina* colony remain small, biflagellate, and non-dividing while the rest of the cells enlarge, redifferentiate as gonidia, and cleave. Such a partial division of labor between one set of vegetative cells that redifferentiate as gonidia and another set that remain terminally differentiated somatic cells is a characteristic feature of the life cycle of both *Pleodorina* (Fig. 3E) and most species of *Volvox*. It is only in a few species of *Volvox*, including *V. carteri*, that there is an early and complete dichotomous differentiation of somatic and reproductive cells. The fact that this small, presumably closely related group of algae contains members that span the full range from unicellularity to true multicellularity—with fully differentiated cell types—suggests that the group is capable of providing novel insights into the molecular genetic basis for the evolution of multicellularity and cellular differentiation.

As appealing as it is to align the volvocine algae in a conceptual series suggesting that they constitute a simple, linear evolutionary progression in size and extent of cellular dif-

570

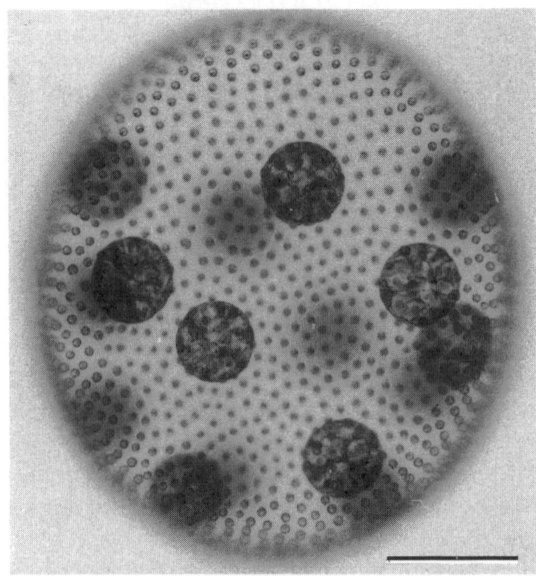

Figure 1. An asexual adult of *Volvox carteri,* consisting of ∿2000 biflagellate *Chlamydomonas*-like somatic cells near the surface, and 16 asexual reproductive cells, or gonidia, in the interior of the spheroid. Gonidia behave as stem cells: each divides to generate an embryo containing all of the gonidia and somatic cells that will be present in an adult of the next generation. The somatic cells, however, are terminally differentiated and eventually undergo programmed senescence and death. Bar: 100 μm.

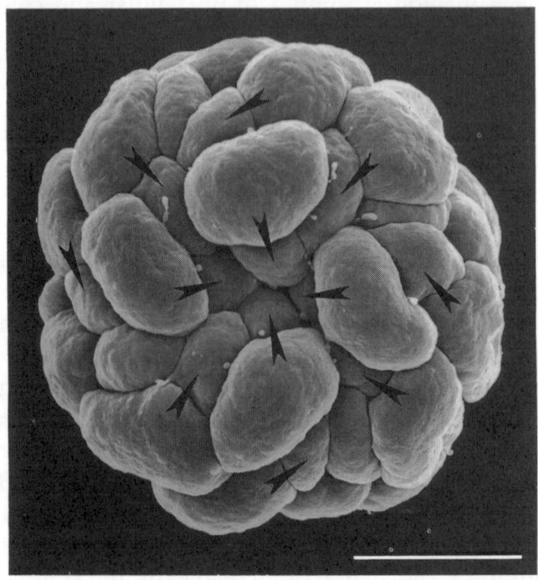

Figure 2. A scanning election micrograph of a 64-cell embryo of *Volvox carteri* that has just executed the differentiative cleavage division that is the first step in the differentiation of somatic cells and gonidia. Arrowheads connect 12 of the 16 large/small sister-cell pairs that have just been set apart by asymmetric division. (The other 4 pairs are below the equator, out of sight in this view.) Each of the smaller cells will divide symmetrically five times and give rise to somatic cells only. Each of the larger cells will divide asymmetrically two more times, generating one additional somatic-cell initial at each division, will then cease dividing several cycles before the somatic-cell initials do, and will generate one gonidium. Bar: 25 μm.

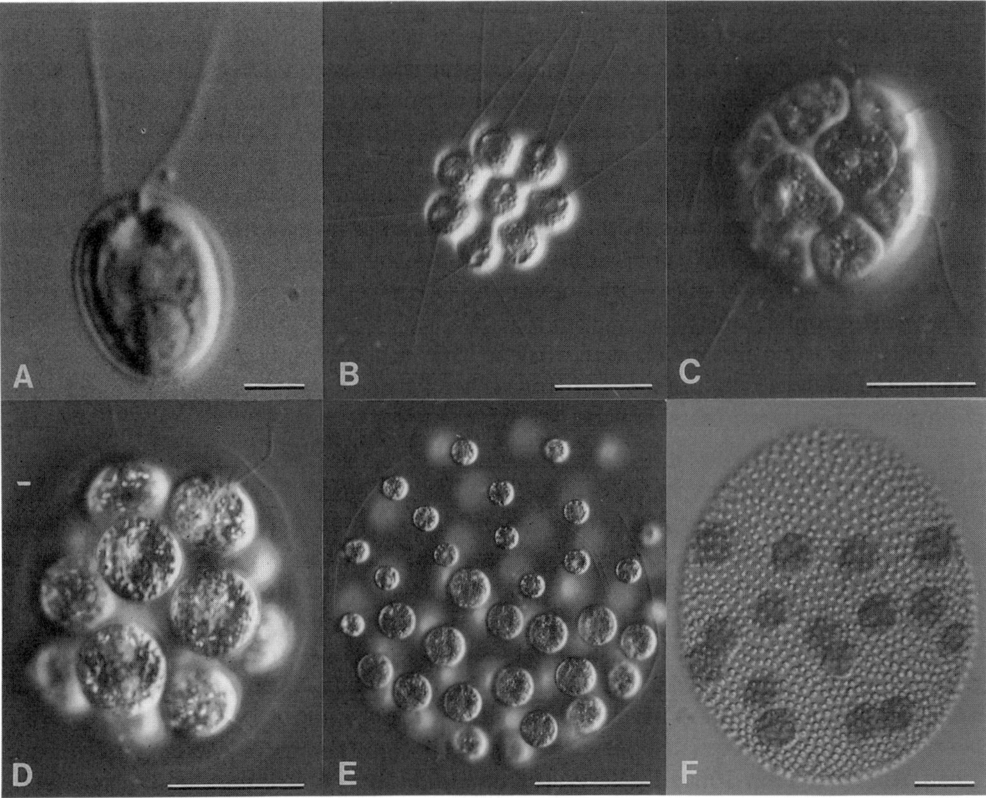

Figure 3. Representative volvocine algae, photographed in life by differential interference contrast microscopy. A. The biflagellate unicell *Chlamydomonas reinhardtii*. B. *Gonium pectorale:* eight coherent *Chlamydomonas*-like cells in a convex disc. C. *Pandorina morum:* 16 cells tightly appressed to form a sphere. D. *Eudorina elegans:* 16 cells loosely packed in a sphere of extracellular matrix. E. *Pleodorina californica:* 64 cells in a sphere of extracellular matrix; initially all cells within a spheroid are similar in size and morphology, but the posterior cells dedifferentiate, enlarge and redifferentiate as gonidia, while the anterior cells remain small, biflagellate somatic cells. F. *Volvox carteri*. Bars: A, 5 μm; B–D, 25 μm; E, F, 50 μm.

ferentiation, there have been no data with which to decide whether this really reflects the phylogenetic history of the group. Therefore, we are investigating the phylogenetic relationships among these volvocine genera by use of ribosomal RNA sequence comparisons. By shedding light on questions such as how many times, and by what routes, the colonial or multicellular way of life may have arisen during diversification of the Volvocaceae, this study should be of interest in its own right. But it should also provide valuable guidelines for future attempts to seek the evolutionary origins of the developmentally important genes of *V. carteri* that we are now attempting to analyze at the Mendelian- and molecular-genetic level.

MATERIALS AND METHODS

Organisms

The taxa for which ribosomal RNA sequences have been analyzed (with strain identifications in parentheses), and the sources from which they were obtained, were as follows: *Chlamydomonas reinhardtii* Dangeard (R3$^+$) and *C. eugametos* Moewus (CC-1419) were obtained from Ursula Goodenough, Department of Biology, Washington Univer-

572

sity; *Chlorella pyrenoidosa* Chick (UTEX 1230), *Eudorina elegans* Ehrenberg (UTEX 12), *Gonium pectorale* Müller (UTEX 826), *Haematococcus lacustris* (Gir.) Rostaf. (UTEX 16), *Pandorina morum* Bory (UTEX 877), *Platydorina caudata* Kofoid (UTEX 850), *Pleodorina californica* Shaw (UTEX 198), *Volvox capensis* Rich and Pocock (UTEX K37), *V. carteri* f. *nagariensis* Iyengar (UTEX 1885), and *V. obversus* (Shaw) Printz (UTEX 1866) were obtained from the University of Texas Culture Collection of Algae, Austin Texas (Richard Starr, Director).

The *Volvox* mutants described here were all derived from strain HK 10 of *V. carteri* f. *nagariensis* Iyengar (UTEX 1885). Some mutants were isolated following mutagenic treatment with N-methyl-N'-nitro-nitrosoguanidine (Huskey et al., 1979), but a number of them arose spontaneously.

With minor exceptions (Larson, Kirk & Kirk, ms. in preparation) all cultures were maintained axenically in the medium and under the conditions previously described for cultivation of *V. carteri* (Kirk & Kirk, 1983).

Sequence Analysis

Details of the methods used to obtain and analyze rRNA sequences will be published elsewhere (Larson, Kirk & Kirk, ms. in preparation), but briefly the methods were as follows: RNA samples were isolated by minor variants of a published method (Kirk & Kirk, 1986). Nuclear-encoded rRNA was sequenced directly, using reverse transcriptase, synthetic DNA oligonucleotides complementary to highly-conserved segments of the rRNA genes to prime the reaction, and dideoxynucleotides for chain termination; confirmation of sequence data was accomplished by use of primers that permitted analysis of overlapping segments of rRNA, as previously reported (Larson & Wilson, 1989; Larson, 1990). Approximately 1600 bp of large and small subunit rRNA sequences obtained by the above methods were aligned as closely as possible to the homologous sequences of representative angiosperms, and Felsenstein's (1985) parsimony-based confidence interval tests were used to test the statistical significance of differences among alternative phylogenetic trees that were constructed using these aligned rRNA sequences.

RESULTS AND DISCUSSION

Phylogenetic Relationships Among the Volvocine Algae

Comparative studies of partial large- and small-subunit ribosomal RNA sequences were used to test a number of hypotheses from the literature regarding the phylogenetic relationships of various volvocacean genera to one another and to certain other green flagellates.

The hypothesis that the genus *Volvox* was derived from the Haematococcaceae, rather than from *Chlamydomonas* and the colonial Volvocaceae (Crow, 1918), was tested using *Chlorella* as an outgroup. The tree that placed *V. carteri* closer to *C. reinhardtii* than to *Haematococcus* was favored by 54 out of 57 phylogenetically informative sites (Fig. 4A), and the tree that placed *V. carteri* closer to *Eudorina* (a colonial volvocacean) than to *Haematococcus* was favored by 44 out of 46 informative sites (Fig. 4B). By the criteria of Felsenstein (1985) these are highly significant results ($p < 0.05$), and thus Crow's hypothesis can be considered falsified. In a similar manner, rRNA sequence data were used to test the hypothesis that the genus *Volvox* is diphyletic, with *V. carteri* and its nearest relatives having been derived from the Volvocaceae, while most of the rest of the genus *Volvox* (including *V. capensis*, which was our test species) was derived from the Haematococcaceae (Fritsch, 1935). Trees linking *V. capensis* to *C. reinhardtii* and *Eudorina*,

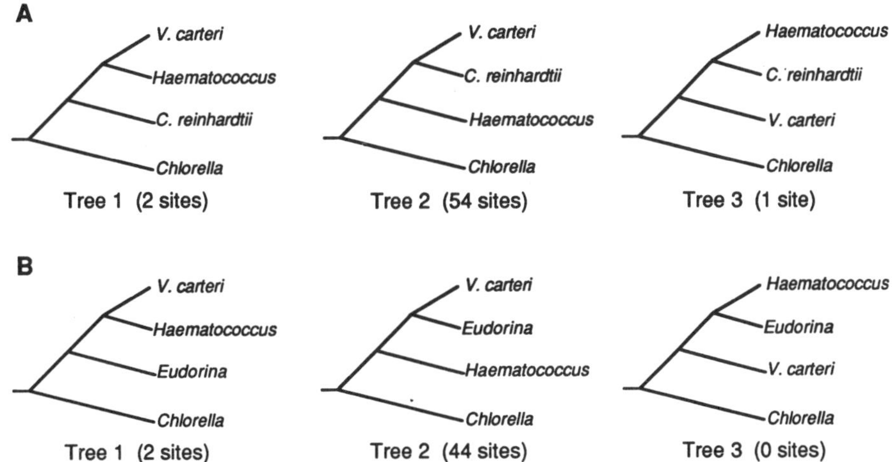

Figure 4. Trees representing alternative hypotheses regarding the phylogenetic affinity of *Volvox carteri* to *Chlamydomonas reinhardtii* (A), or to the colonial Volvocaceae, represented by *Eudorina elegans* (B), versus the Haematococcaceae (represented by *Haematococcus lacustris*) with *Chlorella pyrenoidosa* serving as the outgroup. The numbers of phylogenetically informative sites supporting each of the alternative hypotheses are shown.

rather than *Haematococcus,* were also strongly favored (p<0.05; data not shown); thus the hypothesis of Fritsch (1935) was also falsified.

The hypothesis that *Chlamydomonas reinhardtii* is more closely related to *Volvox carteri* than it is to *C. eugametos* (Adair et al., 1987) was tested using *Chlorella* as the outgroup. The rRNA-based tree that grouped *C. reinhardtii* and *V. carteri* as sister taxa was favored by 49 informative sites, while the tree that grouped *C. reinhardtii* and *C. eugametos* as sister taxa was favored by only 8 sites (data not shown). Thus the hypothesis of Adair et al. (1987) is supported (p<0.05) by these data, indicating that the genus *Chlamydomonas* is paraphyletic with respect to the Volvocaceae.

A preliminary test of the phylogenetic significance of Smith's (1944) subdivision of the genus *Volvox* into four sections was performed by examining the relationships of two members of his section Merillosphaera (*V. carteri* and *V. obversus*) to a member of his section Euvolvox (*V. capensis*), using *Haematococcus* as the outgroup. The tree grouping *V. carteri* and *V. obversus* as sister species was favored by 12 out of 13 informative sites over trees linking one of them to *V. capensis,* in accord with Smith's taxonomic judgment (p<0.05).

Tests of other specific hypotheses regarding relationships within the Volvocaceae are in progress. Meanwhile, the aligned rRNA sequence data have been used to derive the most parsimonious tree linking 13 algal species that we have studied to date, including 9 representative volvocaceans (Fig. 5). The amount of sequence information available so far (∼2000 bp/taxon, for the large and small rRNA subunits combined) is not adequate to attach statistical significance to all details of the branching pattern for the family Volvocaceae that is derived by maximum parsimony and shown here. However, if patterns similar to those in this provisional dendrogram are supported by additional data and statistical analysis, it will indicate that the relationships between the unicellular, colonial, and multicellular volvocine algae are considerably more complex than has generally been envisioned in the past. Nevertheless, the existing sequence data do confirm the view that the volvocine algae (the Volvocaceae and *C. reinhardtii*) constitute a relatively coherent, natural group that is potentially capable of providing insights into the molecular genetic basis for the evolution of multicellularity.

574

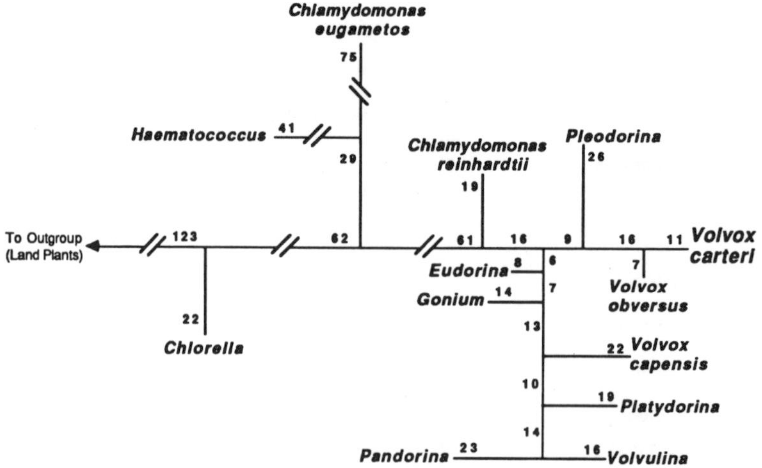

Figure 5. A provisional dendrogram based on comparisons of ∿2000 bp of aligned rRNA sequence data (large and small subunits combined) for 13 species of algae, including 9 representatives of the family Volvocaceae. Except where broken by hash marks, lengths of line segments are proportional to the inferred number of mutational changes between nodes, which are also indicated by the numbers beside the lines.

Genes Controlling Cell Differentiation in *V. carteri*

In *V. carteri*, vegetative and reproductive functions are executed separately by two distinct cell lineages, whereas in unicellular and colonial volvocine algae these functions are executed sequentially by individual cells. Therefore, one strategy for identifying genes that may have been added to the ancestral genome during *Volvox* evolution to account for dichotomous cytodifferentiation is to search for loci at which mutation causes a breakdown of the normal pattern of germ-soma differentiation, and causes one or both cell lineages to retrace the ancestral pathway of "first vegetative and then reproductive."

regA: **a gene repressing reproductive development in somatic cells.** The first mutant phenotype meeting the above criteria to be discovered was the "somatic regenerator" (or "Reg") phenotype initially described by Starr (1970) and then studied in more detail by Huskey & Griffin (1979). The more than 50 Reg mutants that have been studied genetically to date all have lesions that map by recombination to the *regA* locus (Huskey & Griffen, 1979; Kirk et al., 1987; Stamer & Kirk, unpublished).[1]

In Reg mutants, somatic cells and gonidia both differentiate normally initially, but then the somatic cells dedifferentiate, enlarge, and redifferentiate as gonidia that cleave to produce small spheroids containing gonidia and somatic cells; the latter then repeat the regeneration cycle (Fig. 6A–D). In the presence of the glycoprotein pheromone that acts as a potent, species-specific inducer of sexual development (Starr, 1969; Starr & Jaenicke, 1974; Kirk & Kirk, 1986; Mages et al., 1988) Reg somatic cells can either divide to produce sexual spheroids containing gametes, or differentiate directly into gametes (Starr, 1970). Thus, whereas wild-type somatic cells lose the ability to divide ever again once they begin to differentiate, *regA*[−] somatic cells retain all of the reproductive potentials of the species. We conclude that the function of *regA*[+] is to repress all aspects of reproductive development in somatic cells.

---

[1]One strain was used by Huskey and Griffen (1979) to define a *regB* locus; but this strain, which subsequently was lost (Huskey, personal communication), had a quite different, more complex phenotype.

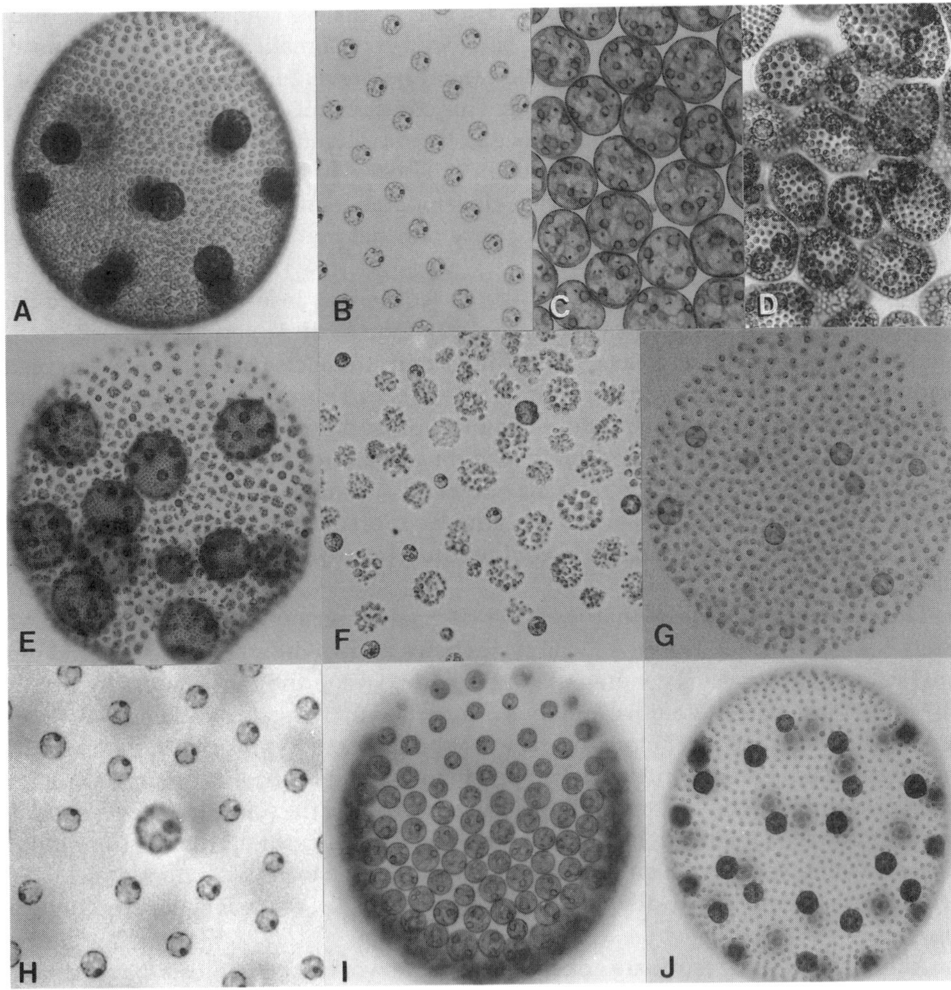

Figure 6. Selected mutants of *V. carteri* that exhibit a disruption of the normal pattern of germ-soma differentiation. A. A "somatic regenerator" (*regA⁻*) mutant at the stage at which all somatic cells except those at the anterior pole (top) have begun to enlarge and redifferentiate as gonidia. B–D. Stages in somatic cell regeneration: B, initially cells develop characteristic somatic-cell features, including the prominent eyespots visible here; C, subsequently, cells resorb their eyespots and flagella, enlarge, and develop the characteristic features of gonidia; D, eventually, these cells cleave to generate miniature spheroids containing both somatic cells and gonidia. E. A "modified regenerator" (*regA⁻/ram⁻*) mutant at the stage at which regeneration is essentially complete. F. A higher magnification view of the surface cells of the spheroid in the preceding micrograph. Some of the somatic cells have remained terminally differentiated. Most of the rest have enlarged only slightly before cleaving, have retained certain somatic cell features, such as eyespots and flagella, and have produced very small spheroids, often containing only somatic cells (compare with Fig. 6D). G. A "late gonidia" (*lagB⁻*) mutant. At a comparable age, the gonidia of a wild-type spheroid would be cleaving to produce juvenile spheroids, but in this mutant the gonidia are just beginning to enlarge and develop gonidial features; they will not divide for about 24 hours. H. A higher power view of the cells of a *lagB⁻* mutant at an earlier stage of development than that shown in the preceding micrograph. The cell in the center is a presumptive gonidium that has developed initially as a "large somatic cell" with a very large eyespot and long flagella (not seen in this focal plane). I. A "gonidialess regenerator" (*regA⁻/gls⁻*) mutant at the same developmental stage as the spheroid in Fig. 6A. Note the absence of "true" gonidia within the spheroid. J. A "multiple gonidia" (*mulB⁻*) mutant. In this strain, differentiative division is delayed one cycle, and occurs in ∿²/₃–³/₄ of the blastomeres; as a consequence, 40–48 gonidia are produced per spheroid. This pattern of cleavage and germ cell production resembles that normally seen in sexually-induced wild-type females, except that in the mutant the germ cells develop as asexual gonidia, not eggs.

If the preceding interpretation is correct, it follows that *regA* must be fully expressed in somatic cells and fully inactive in gonidia, since wild-type somatic cells cannot be caused to divide by any known means, but gonidia cannot be prevented from dividing except by withholding energy or essential nutrients. This conclusion is reinforced by the observation that there is no discernible difference in the developmental behavior of *regA*$^+$ and *regA*$^-$ gonidia. How might such differential, "fully-on"/"fully-off", regulation of the *regA* locus be achieved in the course of normal development of the two cell types? A possible clue came from observations concerning very unusual mutational behavior of the *regA* locus.

At two brief periods of the asexual life cycle, the *regA* locus exhibits extraordinary, locus-specific hypermutability (Kirk et al., 1987). One of these periods is shortly before gonidia are about to initiate cleavage, and the second is after cleavage has been completed and differentiation of the juvenile cells is about to begin. During these intervals, treatment with agents that induce error-prone recombination and repair—but not treatment with conventional point mutagens—result in *regA* mutation rates as high as 10%, but with no detectable elevation of mutation rates for other loci. Although the gene is expressed only in somatic cells, it is hypermutable only in gonidia; moreover, the post-cleavage period in which *regA* becomes hypermutable in gonidia is just before it is to be expressed in somatic cells (Kirk et al., 1987).

The working hypothesis that we have derived to account for these extraordinary results is as follows: (a) The *regA* locus cycles between two different structural and functional states in each asexual generation. (b) Prior to cleavage, the gene is switched from the "off" to the "on" state in gonidia, and thus is inherited in this state by all embryonic cells. (c) After cleavage, and just before the gene is to be activated in presumptive somatic cells (thereby precluding reproductive development), it is switched back to the "off" configuration in presumptive gonidia, permitting these cells to enter the reproductive pathway. (d) If error-prone recombination has been induced in the cells prior to a period of configurational switching, there is a high probability that part of the locus will undergo deletion— thereby permanently inactivating it—rather than the usual, reversible inactivation event (Fig. 7).

Although present results are consistent with the working hypothesis diagrammed in Figure 7, the only definitive test of it will come when *regA* has been cloned, and it is possible to determine whether the locus has a different sequence in gonidia and somatic cells. We are attempting to clone the locus by a chromosome walk from a linked DNA polymorphism (Harper et al., 1987). In addition, we are collaborating with colleagues in the laboratories of Rüdiger Schmitt and Donald Straus, who are exploring alternative approaches to cloning this and other developmentally important loci of *Volvox*.

Recently, we have discovered that a mutation at a locus we have named *ram* (for "*regA* modifier") acts as a second-site suppresser of *regA*. In *regA*$^-$/*ram*$^-$ double mutants, the regenerative potential of somatic cells is suppressed: some remain terminally differentiated like wild-type cells, while others enlarge only slightly and cleave a few times while still possessing differentiated features of somatic cells (Fig. 6E, F). In the absence of a *regA*$^-$ mutation, the *ram*$^-$ mutation is lethal (Stamer & Kirk, unpublished). One working hypothesis consistent with these results is that both of these genes encode transcriptional regulators for the genes that are required for reproduction: *regA* encoding a transcriptional repressor that is active in somatic cells, and *ram* encoding a transcriptional activator that is active in gonidia. Definitive tests of this hypothesis should also become possible once the relevant genes have been cloned.

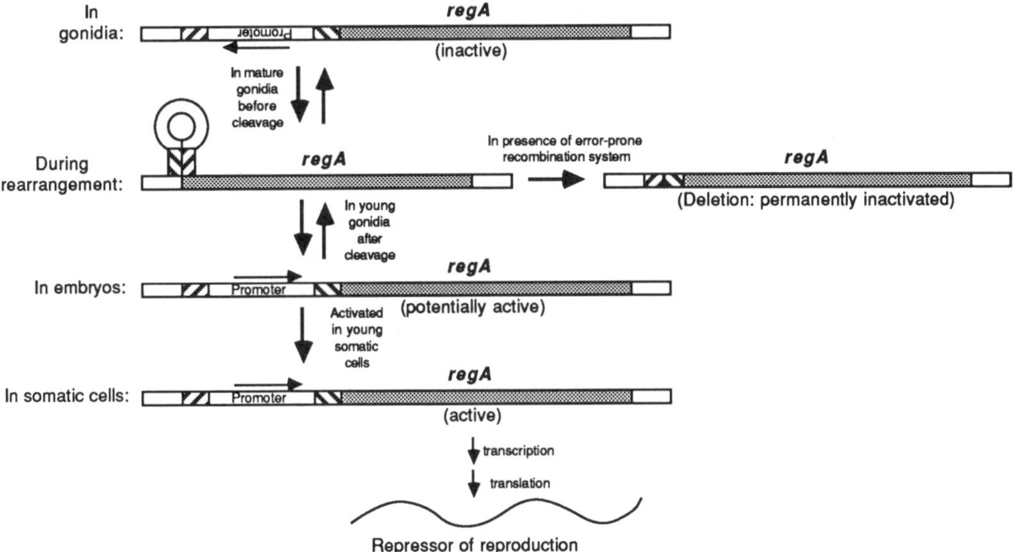

Figure 7. One of several possible variants of a model to account for the unusual mutational behavior of the *regA* locus of *V. carteri*, and its differential expression in somatic cells and gonidia. This model proposes that the promoter can exist in two opposite orientations, one of which permits, and one of which precludes, transcription of the gene. In developing gonidia, the locus is in the inactive configuration, but just prior to cleavage it rearranges to the active configuration, and thus is inherited by all embryonic cells in that form. Following cleavage, the locus is restored to its inactive configuration in gonidia, but remains in the active configuration in somatic cells and is expressed, thus preventing these cells from expressing genes required for reproduction. If error-prone recombination enzymes have been induced by appropriate mutagenic treatment prior to the time that one of the rearrangement events is to take place, the rearranging segment may be deleted, permanently inactivating the gene.

lag: **a set of loci suppressing vegetative development in gonidia.** Mutations at any one of a number of "late gonidia," or *lag*, loci have an effect on gonidial development that is symmetrical with the effects of *regA* mutations on somatic cells. That is, these mutations cause presumptive gonidia to pass through a "somatic cell" phase before they redifferentiate as gonidia. In Lag strains, asymmetric divisions appear to occur fairly normally, but the larger cells produced by these divisions do not withdraw from the division cycle prematurely (3–4 cycles before the somatic initials) as wild-type gonidial initials do, and as a consequence they end up intermediate in size between somatic cells and normal gonidia. Following cleavage, they differentiate as what appear to be large somatic cells, with large eyespots and long flagella (Fig. 6H). But, a day later it becomes obvious that they have undergone a different pathway of determination than their somatic cell neighbors when they enlarge and redifferentiate as gonidia (Fig. 6G). Because similar phenotypes are observed in all strains bearing mutations at any one or two of the (three or more) *lag* loci that are defined by existing Lag strains, we conclude that the *lag* loci control sequential steps in a pathway that must be traversed in order for gonidial initials to stop cleaving before the somatic initials do, and to bypass the vegetative portion of the ancestral volvocacean life cycle. However, the fact that larger cells that eventually undergo gonidial development are produced in all mutants with *lag* lesions eliminates the possibility that the *lag* functions are required for gonidial specification *per se*.

578

gls: **a locus required for asymmetric division.**   Mutation at the "gonidialess," or *gls* locus has no effect on the basic, symmetric cell division process of embryos, but it abolishes the asymmetric divisions by which gonidial and somatic cell lineages are normally set apart. We conclude that gls^+ must encode a component required to shift the plane of division away from the midline of a cleaving cell; studies to establish how this shift in division planes is accomplished at the cytological level are in progress. Second-site suppressers of *gls* are also being identified (Kirk, 1990); it is thought that these may encode other components of the asymmetric division apparatus.

As its name implies, a Gls mutant fails to produce any gonidia. Thus, it can only be maintained as a Reg/Gls double mutant, in which regenerating somatic cells carry the full burden of reproduction (Fig. 6I). In such a double mutant, cells divide about 8 times to produce about 256 cells, all of which resemble the cells of a colonial volvocacean: they differentiate first as vegetative cells and then redifferentiate as gonidia.

The manner in which the *gls* locus probably affects the gonidial specification process may be illuminated by consideration of a newly isolated set of mutants with a somatic-cell-less," or Sls, phenotype. Embryos of Sls mutants cease dividing after only 4 or 5 cycles; thus they never reach the stage at which asymmetric division normally occurs, and they generate only gonidia (data not shown). [These may be variants of the "premature cessation of division," or *pcd* mutants described earlier by Pall (1975).] Thus, failure to execute asymmetric divisions can result in either a gonidialess or a somatic-cell-less phenotype, depending upon the stage—and consequently the size of cells at the time—at which division is terminated. These are but two of many lines of genetic and experimental evidence suggesting that cell size at late stages of cleavage somehow plays a central role in gonidial specification (Pall, 1975; Kirk & Kirk, ms. in preparation). Thus we believe that gls^+ is involved in the gonidial specification process only indirectly—via its effect on the generation of cells large enough to activate the gonidial specification program. We postulate that there is one or more gene that is activated in midcleavage, in cells beyond some threshold size, that is the direct cause of gonidial specification, and that has not yet been identified through our mutational analysis. Undoubtedly lesions in such a function would be lethal on a wild-type background and difficult to detect on a Reg background.

Although the gls^+ function appears to be essential for achieving asymmetric division, the timing and pattern of asymmetric divisions, when they occur, is known to be under control of a number of "multiple gonidia," or *mul* loci (Huskey et al., 1979). For example, in certain strains with lesions at the *mulB* locus, asymmetric divisions are delayed until the 7th cleavage cycle (rather than the 6th), and occur in blastomeres of the anterior ⅔–¾ of the embryo (rather than just the anterior ½); this results in more and smaller gonidia than are seen in wild type (Fig. 6J). This is the pattern normally seen only in sexually-induced, egg-producing females (Starr, 1970). Other Mul mutants exhibit gonidial patterns similar to the pattern normally seen in sperm-producing sexual males, and yet others produce a gonidial pattern that is essentially random in terms of number, position and size of gonidia (Kirk, 1990). It is presumed, but not yet proven, that all of the *mul* loci will be found to be under the epistatic control of the *gls* locus.

A provisional genetic program for germ-soma specification.   The genetic functions described above can be used to build a provisional model that accounts for many of the differences in asexual development between *Volvox carteri* and its unicellular and colonial relatives in the volvocine lineage (Fig. 8). The first step in *V. carteri* germ-soma differentiation appears to be the asymmetric divisions that are permitted by *gls* and patterned by the *mul* functions. Then in the larger cells created by asymmetric division, some as-yet-undefined locus (or loci, indicated in the diagram by "?") is activated late in the cleavage period to commit these cells to gonidial development. This in turn, we propose, activates the *lag* loci, which causes gonidial initials to withdraw from the cell division program

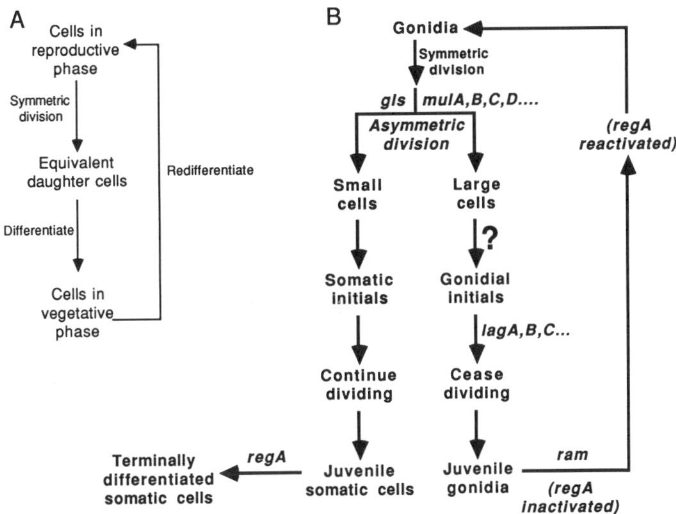

Figure 8. A model indicating proposed roles for certain defined genes in the dichotomous differentiation program of *Volvox carteri*. A. A diagrammatic representation of the asexual life cycle of a representative unicellular or colonial volvocine alga is provided for reference purposes: all cells differentiate first as vegetative cells and then redifferentiate as reproductive cells. B. A diagrammatic representation of the asexual life cycle of *V. carteri*. Mature gonidia first undergo a series of symmetric divisions. Then, under the influence of the *gls* locus and a number of *mul* loci, selected cells undergo asymmetric division, generating large and small sister cells. In mid-cleavage, large cells become committed to gonidial development, presumably by activation of some as-yet-undefined gene, or set of genes, indicated by "?". This leads to activation of the *lag* loci, which causes these cells to withdraw from the cleavage cycle prematurely. Following the conclusion of embryogenesis, *regA* is inactivated (by rearrangement) in the juvenile gonidia, and *ram* is expressed. Because they have expressed the *lag* locus, the juvenile gonidia bypass the vegetative phase of the life cycle, and because they have not expressed the *regA* locus, but have expressed the *ram* locus, they differentiate as gonidia. Just before the mature gonidia are ready to initiate a new cleavage cycle, they restore the *regA* locus to the state in which it is capable of being expressed (see Fig. 7). Meanwhile, since the *lag* loci have not been not activated in the somatic initials, they divide several times more than the gonidial initials do; then they express the *regA* locus and undergo terminal differentiation as somatic cells.

prematurely and differentiate directly as gonidia, bypassing the vegetative phase of the ancestral cycle. A second important consequence of the activation of "?" is postulated to be a rearrangement of the *regA* locus (so as to reversibly inactivate it) and a third is postulated to be activation of the *ram* locus that is thought to be required for expression of the rest of the genes in the gonidial program. Finally, a very late event in the gonidial program (just preceding the next round of embryogenesis) is postulated to be a reversal of the *regA* rearrangement that occurred earlier.

Meanwhile, in the smaller cells produced by asymmetric division, the *lag* loci are not activated, but the *regA* locus is. Failure to activate the *lag* loci permits these cells to enter the vegetative phase of the life cycle, and activation of the *regA* locus locks them in that phase, preventing their return to the reproductive phase, and leading to their eventual senescence and death.

Obviously, such a scheme is likely to turn out to be grossly oversimplified, but it does appear to have heuristic value in identifying as-yet-undefined genetic functions that are likely to exist and be essential for normal development, in making testable predictions about epistatic interactions among the various loci, and in guiding future attempts to explore the differentiation process at deeper cytological and molecular levels.

580

CONCLUSION

By virtue of its developmental simplicity and its accessible genetic system, *Volvox carteri* is beginning to yield insights into the sort of genetic program that may be involved in the dichotomous development of two entirely different cell lineages from sister cells. The availability of powerful modern methods of analysis support the hope that eventually it may be possible to elucidate the mechanisms by which this developmental program is expressed at the cellular and molecular levels. Meanwhile, rRNA-sequence-based phylogenetic analysis reinforces the belief that *V. carteri* is closely related to extant unicellular and colonial volvocine algae, and that, therefore, it may eventually be possible to trace the pathway by which the *V. carteri* genetic program for cellular differentiation arose in the course of evolution.

ACKNOWLEDGMENTS

We wish to express our deep appreciation for the excellent technical assistance of Christine R. Adams and Norma L. Gieg-Sinclair. The research of D. L. K., M. M. K. and K. A. S. was supported by grant GM 27215 from the National Institutes of Health, grant DCB-8615691 from the National Science Foundation, and Collaborative Research Grant 0065/87 from the North Atlantic Treaty Organization; the research of A. L. was supported by grant BSR-8708393 from the National Science Foundation, and BRSG grants from Washington University.

LITERATURE CITED

Adair, W. S., Steinmetz, S. A., Mattson, D. M., Goodenough,U. W., & J. E. Heuser. 1987. Nucleated assembly of *Chlamydomonas* and *Volvox* cell walls. *J. Cell Biol.* 105:2373–2382.

Crow, W. B. 1918. The classification of some colonial chlamydomonads. *New Phytol.* 17:151–158.

Fritsch, F. E. 1935. *The Structure and Reproduction of Algae. Vol. 1.* Cambridge University Press: Cambridge. Pp. 103–104.

Felsenstein, J. 1985. Confidence limits on phylogenies with a molecular clock. *Syst. Zool.* 34:152–161.

Harper, J. F., Huson, K. S., & D. L. Kirk. 1987. Use of repetitive sequences to identify DNA polymorphisms linked to *regA*, a developmentally important locus in *Volvox. Genes & Develop.* 1:573–584.

Huskey, R. J. & B. E. Griffin. 1979. Genetic control of somatic cell differentiation in *Volvox.* Analysis of somatic regenerator mutants. *Dev. Biol.* 72:226–235.

Huskey, R. J., Griffin, B. E., Cecil, P. O. & A. M. Callahan. 1979. A preliminary genetic investigation of *Volvox carteri. Genetics* 91:229–244.

Kirk, D. L. 1988. The ontogeny and phylogeny of cellular differentiation in *Volvox. Trends Genet.* 4:32–36.

Kirk, D. L. 1990. Genetic control of reproductive cell differentiation in *Volvox. In:* W. Wiessner, D. G. Robinson & R. C. Starr (eds.), *Experimental Phycology 1: Cell Walls and Surfaces, Reproduction, Photosynthesis.* Springer-Verlag: Berlin. Pp. 81–94.

Kirk, D. L., Baran, G. J., Harper, J. F., Huskey, R. J., Huson, K. S., & N. Zagris. 1987. Stage-specific hypermutability of the *regA* locus of *Volvox,* a gene regulating the germ-soma dichotomy. *Cell* 48:11–24.

Kirk, D. L. & J. F. Harper. 1986. Genetic, biochemical and molecular approaches to *Volvox* development and evolution. *Int. Rev. Cytol.* 99:217–293.

Kirk, D. L. & M. M. Kirk. 1983. Protein synthetic patterns during the asexual life cycle of *Volvox carteri. Dev. Biol.* 96:493–506.

Kirk, D. L. & M. M. Kirk. 1986. Heat shock elicits production of sexual inducer in *Volvox. Science* 231:51–54.

Kirk, M. M. & D. L. Kirk. 1985. Translational regulation of protein synthesis, in response to light, at a critical stage of *Volvox* development. *Cell* 41:419–428.

Kochert, G. 1973. Colony differentiation in green algae. *In:* S. J. Coward (ed.), *Developmental Regulation: Aspects of Cell Differentiation.* Academic Press: New York. Pp. 155–167.

Larson, A. 1990. A molecular perspective on the evolutionary relationships of the salamander families. *Evol. Biol.* 25:211–277.

Larson, A. & A. C. Wilson. 1989. Patterns of ribosomal RNA evolution in salamanders. *Molec. Biol. Evol.* 6:131–154.

Mages, H.-W., Tschochner, H., & M. Sumper. 1988. The sexual inducer of *Volvox carteri:* primary structure deduced from cDNA sequence. *FEBS Lett.* 234:407–410.

Pall, M. L. 1975. Mutants of Volvox showing premature cessation of division: evidence for a relationship between cell size and reproductive cell differentiation. *In:* D. McMahon & C. F. Fox (eds.), *Developmental Biology: Pattern Formation, Gene Regulation.* W. A. Benjamin: Menlo Park. Pp. 148–156.

Smith, G. M. 1944. A comparative study of the species of *Volvox. Trans. Amer. Micros. Soc.* 63:265–310.

Starr, R. C. 1969. Structure, reproduction, and differentiation in *Volvox carteri* f. *nagariensis* Iyengar, strains HK9 and 10. *Arch. Protistenkd.* 111:204–222.

Starr, R. C. 1970. Control of differentiation in *Volvox. Dev. Biol. Suppl.* 4:59–100.

Starr, R. C. & L. Jaenicke. 1974. Purification and characterization of the hormone initiating sexual morphogenesis in *Volvox carteri* f. *nagariensis* Iyengar. *Proc. Nat. Acad. Sci. U.S.A.* 71:1050–1054.

Development and Evolution— The Emergence of a New Field

David B. Wake, Paula Mabee, James Hanken, and Gunter Wagner

The occasion of the Fourth International Congress of Systematic and Evolutionary Biology, in College Park, Maryland, provided the opportunity for a "Round-table Discussion Group" to explore ideas and opportunities for research on the relationship of development to morphological evolution. What was planned as a tightly organized discussion by a small working group quickly changed when over 200 individuals filled a large, formal lecture hall. A lively discussion ensued, despite the size of the group. The present report is not a summary of that discussion, but an attempt to distill from it the central issues.

OVERVIEW

The most general impression gained by the authors as a result of the four-hour discussion is the widespread acknowledgment within the evolutionary biology community that findings in the area of developmental biology have a central relevance for evolutionary studies. The heralded synthesis that occurred following the integration of genetic principles into evolutionary studies by theorists such as Haldane, Fisher and Wright ultimately failed to incorporate developmental biology (although attempts were made by some, including de Beer and Schmalhausen), almost certainly because development remained too much of a "black box." In recent years, however, the spectacular gains in developmental biology have highlighted: 1) the opportunity to test hypotheses in evolutionary and developmental biology concerning morphological evolution by direct experiments, 2) the opportunity to incorporate a new body of facts and new theories of development into the framework of evolutionary theory, and 3) the need for a phylogenetic and evolutionary perspective in developmental biology. It is the reciprocity between disciplines that interests both evolutionists and developmentalists, and raises the possibility of new kinds of interaction that might substantively benefit both fields. The question is no longer simply "What do facts of development offer to evolutionary biology?",

Dr. Wake is with the Museum of Vertebrate Zoology and Department of Integrative Biology, University of California, Berkeley, CA 94720, USA. Dr. Mabee is with the Department of Biology, Dalhousie University, Halifax, Nova Scotia, Canada B3H 4J1 (present address: Department of Biology, San Diego State University, San Diego, CA 92182, USA). Dr. Hanken is with the Department of Environmental, Population, and Organismic Biology, University of Colorado, Boulder, CO 80309, USA. Dr. Wagner is with the Institute of Zoology, University of Vienna, Althanstrasse 14, A-1090 Vienna, Austria.

but also "What can evolutionary biology offer to students of development?". It is the possibilities for new synergisms, of the heuristic value of interdisciplinary interaction, and of the opportunities of an expanded evolutionary and developmental synthesis that excite the imagination.

There was a general sense in the discussion that the current organization of science, with its sharp subdivisions (often reinforced by the organization of funding agencies), acts as a deterrent to the expansion of what was seen by discussants as a new discipline. This new discipline is emerging at the confluence of three traditional areas of study: development, evolution, and phylogenetics. Within the area of development at least two major foci of major significance for evolution are evident, the first being the genetics of development (with reference to direct and indirect effects on trait differences), and the second being mechanisms of morphogenesis and pattern formation (including the study of cell and tissue interactions and the mechanics of development, among a wide array of investigations). Within the area of evolution the principal focus is quantitative genetics, which is seen as a link between genetic (and developmental) phenomena and evolutionary theory. Emphasis also has been given to the nature and generation of phenotypic variation, in particular discrete alternative morphological states, on which natural selective processes are based. Within the area of phylogenetics the principal foci of investigation are the role of historically acquired functional and developmental constraints, and the evaluation of the frequency and causes of homoplasy (and the attendant message that morphological outcomes are limited), as well as the value of development in understanding character state transformation.

There is as yet no consensus as to whether the intellectual effort of this new field should proceed along strict functionalist (neoDarwinian) lines, with an emphasis on population-level phenomena, along strict structuralist lines, with an emphasis on form-generation, or in an explicitly synthetic framework, in which both perspectives are pursued equally and simultaneously, with priority given to neither. Yet, without doubt, a new field has arisen.

This new area of investigation is synthetic, incorporative and integrative. It may need a label, but not because, as some have said, "when concepts fail, a name will suffice," but because the concepts are beginning to fall into place and the work before us is beginning to take form. One participant has suggested "evolutionary embryology" as the most appropriate identifier for the new field. We need better communication, and we need ways to identify colleagues with similar interests and approaches, perhaps through a dedicated journal. Finally, we need to communicate effectively with administrators and bureaucracies in order to have influence on planning and to assure appropriate levels of funding. Our work is diverse, and broadly distributed across different levels of biological organization, involving diverse taxa. Young people, attempting to find positions in traditional university departmental structures, are most keenly aware of the need for identity.

Despite lively discussion on these matters and many positive suggestions, there was no consensus as to what should be done, but little disagreement that something should be done. Perhaps the success of the general endeavor speaks for itself, for many papers are already published, many students are attracted to the area, and obviously much research is being conducted.

The question before us is whether the area of study should remain interdisciplinary, with all of the diffuse and unorganized features that have characterized it to date, or if a new field should be identified, with some central principles and goals. At least one speaker believes that a central problem impeding a successful synthesis and integration of developmental and evolutionary biology is the absence of a comprehensive theory of development. It may be that those who wish to work in the new field must receive training in the principles and methodologies of cell and molecular biology, the area in which the largely unheralded revolution in development is occurring.

ORGANIZATION OF THE DISCUSSION

The discussion was divided into four segments. Gunter Wagner led a general discussion of the hierarchy of developmental processes and patterns. Three general questions were pursued: 1) What developmental mechanisms are involved in the generation of both morphological order and novelty? 2) Can hierarchical approaches (ontogenetic, phylogenetic, and levels of organization) contribute to our understanding of the relationship between development and morphology in evolution? 3) What is the relationship among genetic, developmental, structural and functional constraints with respect to the evolution of morphology?

A discussion of the developmental origin of evolutionarily significant morphological change was led by James Hanken. Topics considered included: 1) How do developmental systems become decoupled, and how does this decoupling provide opportunities for evolutionary change? 2) How do developmental constraints restrict evolutionary change, and to what degree do they channel (or direct) any morphological change that occurs? 3) What developmental mechanisms underlie heterochrony, and to what degree is heterochrony indicative of the genetic basis of evolutionary change?

Paula Mabee led a general discussion of ontogeny and systematics. Among the questions pursued were: 1) What developmental components of homology can be identified, and can homologous structures have very different ontogenies? 2) How important is heterochrony, and how does it relate to taxic and transformational aspects of morphological evolution? 3) How can ontogenetic information be used in systematics?

The final discussion segment, on prospects for the general area of the relation of development to morphological evolution, was led by David Wake. The questions pursued included: 1) What can evolutionary morphology and systematics offer students of development? 2) What can we expect from new molecular and quantitative genetic approaches applied to development, other than more data? 3) What are the components of a discipline of evolutionary developmental biology?

Discussion was free and wide-ranging, but at the same time both meaningful and provocative. The overview presented here focuses not on the details or formal sequence of the discussion, but on the general principles that emerged.

HETEROCHRONY

Heterochrony immediately arose as a central focus in several of the discussions. Heterochrony (a change in developmental timing, relative to an ancestor) is an evolutionary term that describes the result (a phylogenetic pattern) of a process or combination of processes; it can be known only in a phylogenetic context. Because specific developmental events can be perceived as resulting from shifts in timing of genetic or morphogenetic processes, many workers have discussed heterochrony as if it were a process or even a mechanism, and this has led to both ambiguity and confusion in the literature (evident in such terms as "heterochronic genes"). The term heterochrony is best viewed as a pattern, and a phenomenon that is discernible through comparisons conducted in a phylogenetic context.

Assessments of the importance of heterochrony range from maximal, often ascribing major patterns of lineage evolution to the phenomenon, to minimal or trivial. Some discussants argued that there have been surprisingly few general novelties since the Cambrian (a point commonly made in the literature), in terms of new body plans, tissue types, or morphogenetic rules. For this reason alone, many phylogeneticists think that most major evolutionary events are the result of a phylogenetic reshuffling (in which heterochrony is dominant) of developmental events. Extending this argument, if major events arise from heterochrony, it also is likely that much morphological diversification at lower taxonomic levels is the result of heterochrony. Skeptics counter by claiming that in such arguments heterochrony becomes a universal redescriptor, and that virtually all evolutionary modifications can be couched in this context with no useful outcome. Some see heterochrony as an epiphenomenon at an organizational level above a heterogeneous assemblage of processes at another level. Viewed in a phylogenetic perspective, this is not a problem, but in a more narrow evolutionary mechanism framework it is. However, even when viewed as a result, there is widespread disagreement concerning the importance of heterochrony.

As a general pattern, heterochrony is evident at the level of character evolution, at the level of evolutionary changes in the integration (coupling and decoupling, and the degree to which these occur) of groups of characters, and at the level of the whole organism. The degree to which heterochrony might be useful depends on the research strategy and goals of different investigators.

Classifications are intended to convey information, and the classification schemes that have been used with respect to heterochrony have reached some stability; there appears to be wide agreement on terminology and this has improved communication among biologists investigating the relationship between developmental and evolutionary change. However, workers are beginning to question the usefulness of this terminology. A common accusation is that classification of a pattern often does not contribute to an explanation of its cause. Not all patterns that appear to be, or are construed as, representative of changes in developmental timing, are such. But, because classification systems are necessary for progress (they often motivate and direct research), how can processes of relevance to developmental biologists be most appropriately classified in an evolutionary context? If data are simply amassed without the context of an organizing framework, which a classification can provide, we have no way of determining progress toward some goal, and simply await the emergence of a new paradigm, which may never appear.

Some suggestions were made concerning future classifications of development in relation to evolutionary change. One scheme might classify according to types of developmental mechanisms or known processes. Another might classify according to establishment of systems of coupling and decoupling of developmental processes (such as subsystems within an organism—bone and cartilage in skull formation in a vertebrate, segments within an insect, sequential leaves in an angiosperm). Perhaps we need different classification schemes appropriate to different levels of organization. There is a need for a holistic classification, at the level of whole organisms, that at the same time permits experimental exploration. Some enthusiasm was expressed for the analysis of coupling/decoupling for this reason, because coupling is an evolutionary statement that is simultaneously a statement concerning developmental mechanism. Empirical investigation requires a hypothesis of phylogenetic relationships among investigated taxa, a set of characters that vary in their correlations, characters whose sequence of appearance in ontogeny varies, and an experimental developmental approach for investigating the system. Direct experimental interventions and quantitative genetics might well be incorporated in such a framework of study. In order to interpret the evolution of developmental changes, a phylogenetic context is required.

DEVELOPMENTAL CONSTRAINTS

The notion of constraint is widespread, and has been the subject of much investigation and controversy. The most general argument in its favor is the fact that lineages evolve within evident limits, as perceived by the occupancy of morphospace, but there seem to be few profound constraints (an example is bilaterality in many groups). Developmental con-

586

straints are evident principally in biases in the production of phenotypes. That structures can be identified as homologues is in itself an argument for the reality of developmental constraints. A primary reason for the ubiquity of homoplasy in evolution may be constraints on the production of form. The effects of constraints can be general. An example is the upward causation of genome size on cell size, and its apparent large effects on morphogenesis (manifest directly in changes in developmental rate and indirectly in adult morphology). Another general constraint is the ability of embryos to regulate, for example, by the specific mechanism of intercalary mitotic growth in epimorphic systems in metazoans (e.g., vertebrate limbs, dipteran imaginal discs). This resilient mechanism is responsible for establishing pattern continuity at the cellular level during embryogenesis (positional information, "Bateson's Rule," etc.) and for re-establishing this continuity after perturbation (as during regeneration). Interactions of processes seem to be responsible for constraints. Those workers who invoke developmental constraint at the population and specific levels need to avoid ad hoc arguments and attempt more explicit and experimentally based explanations for limits on evolutionary change.

Developmental constraints can provide evolutionary opportunity. The phenomenon of regulation during development simplifies the kinds of evolutionary transitions necessary for the evolution of complex characters. For example, developmental regulation of the peripheral circulation in vertebrates makes changes in the locomotor system possible without concomitant change in the genetic information for the circulatory system.

A major unsolved problem is the absence of operational criteria for the recognition of developmental constraints. If this concept is to have general utility, and not simply invoked in an ad hoc manner, such criteria are necessary.

An area of important disagreement is the biological basis of developmental constraints. Some discussants were willing to concede that constraints might exist if they are ultimately genetic in nature, while others argued that evolutionarily significant constraints are most likely imposed by developmental processes (e.g., epigenetic interactions) that are far removed from the level of the genes.

Strong arguments were made in favor of the organization of a research program in development and evolution that is explicitly genetic, with an emphasis on specific gene substitutions and on mutations of relatively large effect. Others strongly disagreed, arguing that development is not strictly hierarchical, e.g., the same genes are used over and over again. Such workers view the roles of genes as being more generic, with differences in the regulation of developmental timing and interactions, rather than new genes, being responsible for new morphologies. Although there is strong disagreement over whether new molecules and new genes, or reorganizations of existing molecules and genes, are of primary importance in new morphologies, there is considerable agreement that the time has come to emphasize processes over patterns in understanding the interaction of development, phylogeny and evolution.

SYSTEMATICS AND HOMOLOGY

Ontogeny and systematics have been conceptually related ever since the "three-fold parallelism" of the last century: ontogeny, paleontology (time), and comparative anatomy. The perception of this relationship continues to flourish and evolve. There are, however, some fundamental differences in the approach to ontogeny by developmentalists and systematists, and these need to be understood if a commonality of purpose—to link development and evolution—is to be attained.

In order to examine trait or character evolution, developmental biologists rely on systematics for hypotheses of phylogenetic relationships. Only within a phylogenetic context can questions be asked concerning the evolution of ontogenies, the importance of heterochrony, the role of developmental constraints on evolution, and the extent to which developmental processes are coupled or decoupled in evolution. Developmental processes or patterns, or both, must be "mapped" onto a phylogenetic hypothesis and examined by parsimony methods in order to infer ancestral conditions and pathways of change.

Systematic biologists ask how organisms are interrelated. In order to answer this question, characters must be polarized. Ontogenetic information is used by some systematists to distinguish ancestral from derived states, for ordering character state transformations, for determining homologies among adults, and as a source of new character data (i.e., ontogenies used as characters). All of these uses are controversial and under continued study, but there is general agreement that the empirical information that is generated by developmental biologists is of value to systematists.

Homology is the basis of systematic biology. When one asks whether homologous structures can have different ontogenies, an implicit assumption is that the structures in question are those found in adults, i.e., the endpoints of developmental processes. Because no one doubts that nonterminal changes can take place in developmental sequences, it would seem that ontogeny cannot be used as the sole criterion in recognizing homologues, but there is not agreement on this point and some consider common development as the central determinant of homology. Increased emphasis is being placed on whole ontogenies as constituting the relevant characters of systematists. Determining homology among parts of developmental processes for use in phylogenetic reconstruction, however, remains a major and largely unappreciated problem. The issue of homology may be strictly a theoretical issue to many, but it is a matter of central, crucial, pragmatic concern to systematists. Developmental biologists can help by suggesting criteria for individuating or delimiting parts of developmental processes or of ontogenetic trajectories as homologues.

PROSPECTS

The discussion of prospects in the emerging field featured brief statements by many participants on opportunities and difficulties. Some applauded the audaciousness of modern work in developmental genetics, pointing out that we do learn even from sweeping comparisons of yeast, flies, frogs, and humans, as in the case of homeobox genes. Sometimes we need to be bold. But others argued that we are not after universals, but rather differing levels of generality. The issue becomes how to systematize these levels. Variation occurs in development, and the level at which it is investigated is a function of the question pursued and the system under study. Broad comparisons may be appropriate for evolutionarily conservative phenomena. More fine-scaled comparisons are necessary if one is studying the nature of evolutionary transitions at low phylogenetic levels. The incorporation of systematic biological procedures (fundamentally those of phylogenetic systematics) into all parts of the investigation of developmental processes should have a generally salutary effect.

There was lively debate concerning the recent incorporation of quantitative genetic approaches to development and evolution by various workers. Some argued that most quantitative genetic approaches incorporated simplistic assumptions about gene action. Others defended the quantitative genetic approach on the grounds of its empiricism and utility, pointed out the existence of recent theoretical developments, and urged that quantitative genetic models be tested in the context of new genetic discoveries.

One speaker argued that "truth is the intersection of independent lines," and thus argued for diversity in approach. This was echoed by many speakers, who advocated different approaches including: quantitative, developmental and physiological genetics, that

588

would focus on whole phenotypes and connect to phenomena such as pleiotropy and epistasis; molecular developmental genetics, that would focus on molecular genes of large effect; population genetics of development, which may require new formulations; and phylogenetic analyses of ontogenies, which should give deeper and more meaningful insight into character evolution than at present. Modern developmental biology is "full of details," as one speaker put it. Will the evolutionary generalities be found in the details?

Although there was not unanimity of view, a widespread notion was that a new field of evolutionary developmental biology should not be reduced to the functioning of genes. Rather, central questions should be: how development itself evolves, how multicellularity has arisen and what has led to the origin of individualized parts of the phenotype, how developmental processes influence morphological evolution, how appropriate experiments in development can be conducted in a comparative framework, and how an experimental framework can be formulated in an evolutionary and phylogenetic setting. The synthesis of development, evolution and systematics will require knowledge from all fields in areas ranging from theory to experimental design, methods for testing, and analytical procedures. Pursuit of "pure" developmental or evolutionary questions will continue to produce valuable data, but the accumulation of information from such approaches is not necessarily applicable to the questions generated from a synthetic approach.

The enthusiasm generated by the workshop and the spirited debates that took place are clear signs of the vitality of what is a new and exciting area of biology. We are entering, once again, a period of synthesis with respect to evolution. This is not a unification of information from diverse areas of inquiry, but a synthesis of methods, procedures, and analysis combined with new knowledge. It is a time of opportunity and challenge.

ACKNOWLEDGMENTS

This report was distilled from the oral comments of many speakers, and we have deemed it inappropriate to identify ideas or statements with specific persons, since many of the same ideas were expressed in different ways by different speakers. We have avoided literature citations in order to emphasize the discussion. Individuals whose comments have been incorporated in part into this report include: M. Akam, P. Alberch, K. Bachmann, C. Baron, R. Burian, A. Collazo, W. Dickinson, G. Dover, W. Fink, D. Futuyma, B. Hall, C. Hickman, B. Kirchoff, G. de Jongh, D. Kirk, D. Lipscomb, T. McClellan, T. Miyake, G. Moreno, G. Muller, R. O'Grady, P. Phillips, S. Rachootin, R. Raff, B. Riska, J. Regier, R. Sattler, S. Sessions, L. Van Valen, R. Wassersug, J. Webb, and G. Wray. We apologize to those who made comments but whose names were unrecorded, and to those whose point was misinterpreted or overlooked. We especially thank the following for written and editorial comments: A. Collazo, B. K. Hall, T. McClellan, S. Minsuk, S. Sessions, N. Shubin, and M. Wake.